11th European Conference on Applied Superconductivity

(EUCAS 2013)

Journal of Physics: Conference Series Volume 507

Genoa, Italy
15-19 September 2013

Editors:

Stefania Farinon
Andrea Malagoli
Ilaria Pallecchi
Gianrico Lamura

ISBN: 978-1-63439-287-7
ISSN: 1742-6588

Printed from e-media with permission by:

Curran Associates, Inc.
57 Morehouse Lane
Red Hook, NY 12571

Some format issues inherent in the e-media version may also appear in this print version.

Printed by Curran Associates, Inc. (2014)

Institute of Physics
Dirac House, Temple Back
Bristol BS1 6BE UK

Phone: 44 1 17 929 7481
Fax: 44 1 17 920 0979

techtracking@iop.org

Additional copies of this publication are available from:

Curran Associates, Inc.
57 Morehouse Lane
Red Hook, NY 12571 USA
Phone: 845-758-0400
Fax: 845-758-2634
Email: curran@proceedings.com
Web: www.proceedings.com

TABLE OF CONTENTS

MATERIALS

WIRES, TAPES AND CONDUCTORS

LARGE SCALE

ELECTRONICS

Preface

During the 11th edition of the European Conference on Applied Superconductivity, successfully held in Genoa from 15–19 September 2013, more than one thousand participants from over 40 countries were registered and contributions of 7 plenary lectures, 23 invited talks, 203 oral talks and 550 posters were presented.

The present issue of *Journal of Physics: Conference Series* (JPCS) collects the 218 submitted papers that were peer reviewed and accepted in the Conference Proceedings.

Similarly to the *Superconductor Science and Technology* Special issue: "EUCAS 11th European Conference on Applied Superconductivity" which contains some plenary and invited contributions, as well as some selected contributions, in this issue the papers are sorted according to the four traditional topics of interest of EUCAS, namely Materials (56 papers), Wires and Tapes (47 papers), Large Scale Applications (64 papers) and Electronics (51 papers). While the *Superconductors Science and Technology* special issue focuses on the scientific and technological highlights of the conference, this collection provides an overall view of the worldwide research activity on applied superconductivity, mirroring the main guidelines and the hottest issues, which range from basic studies on newly discovered superconducting compounds to the state-of-the-art advances in large scale applications, wires and tapes fabrication and electronics.

We would like to point out that, among the JPCS contributions, six papers present works financed by ongoing EU–Japan projects, three papers belong to the session on junctions and SQUIDs dedicated to the memory of Antonio Barone and one paper belongs to the session on pinning and flux dynamics dedicated to the memory of John Clem.

Finally, we would like to thank all the people whose careful work contributed to the preparation of this JPCS issue, in particular the session chairs as well as the peer reviewers.

The Editors

Stefania Farinon (Editor in Chief, Large Scale), Ilaria Pallecchi (Materials), Andrea Malagoli (Wires and Tapes), and Gianrico Lamura (Electronics)

Peer review statement

All papers published in this volume of *Journal of Physics: Conference Series* have been peer reviewed through processes administered by the proceedings Editors. Reviews were conducted by expert referees to the professional and scientific standards expected of a proceedings journal published by IOP Publishing.

11th European Conference on Applied Superconductivity (EUCAS2013) IOP Publishing
Journal of Physics: Conference Series **507** (2014) 012001 doi:10.1088/1742-6596/507/1/012001

Vortex pinning by intrinsic correlated defects in $Fe_{1-y}Se$

M L Amigó[1], V Ale Crivillero[1], D G Franco[1], A Badía-Majós[3,4], J Guimpel[1,2] and G Nieva[1,2]

[1] Comisión Nacional de Energía Atómica - Centro Atómico Bariloche and Instituto Balseiro, Universidad Nacional de Cuyo, S.C. de Bariloche, Argentina
[2] CICyT - Consejo Nacional de Investigaciones Científicas y Técnicas, Argentina
[3] Universidad de Zaragoza, Spain
[4] Instituto de Ciencia de Materiales de Aragón, CSIC (Spain)

E-mail: louramigo@gmail.com

Abstract. We present a study on the transport and magnetic properties of superconducting $Fe_{1-y}Se$ single crystals. In the superconducting state, the in-plane electrical resistivity of the crystal is measured for fields up to 16 T and as a function of field direction, in order to understand how the vortex dynamics is affected by the presence of defects. A strong deviation from the slightly anisotropic crystal (electronic anisotropy constant $\gamma \sim 1.08$) is observed as a steep angular dependence, which is interpreted as a signature of the presence of correlated defects. The influence of the correlated defects on the critical current is studied through the angular dependence of the magnetization, and compared to numerical simulations.

1. Introduction

Recently, a new iron-based superconductor family was found, of which $FeSe$, the compound focused on here, is a member [1]. In all cases, the samples present a tetragonal layered structure and a slight anisotropy, $\gamma = m_c/m_{ab}$, with m_c and m_{ab} the effective electronic masses in the c axis and the ab plane, respectively [2].

It is common to find spurious phases in the crystals mixed with the superconducting tetragonal $FeSe$ since the latter only spans a very narrow region of the thermodynamic phase diagram [3]. The spurious phases could include metallic Fe, in excess of this element, or Fe_7Se_8, in its defect. In this work we identify how the presence of Fe_7Se_8 affects the superconducting vortex dynamics in $Fe_{1-y}Se$ single crystals.

2. Crystal characterization

$Fe_{1-y}Se$ single crystals were grown from 1/4 KCl:3/4 $NaCl$ flux in evacuated double quartz ampoules. Structural characterization was made by X Ray diffraction (XRD). Chemical composition was quantified by Energy Dispersive Spectroscopy (EDS). The crystals have a platelet shape and present Fe deficiency. XRD shows the presence of two phases, the superconducting tetragonal $FeSe$, and the ferrimagnetic hexagonal Fe_7Se_8. Tetragonal $FeSe$ shows two preferential orientations of the c axis, one parallel and the other at 57° from the platelet normal. Hexagonal Fe_7Se_8 grows only with the c axis parallel to the normal [4] and presents an ordered array of Fe vacancies [5].

3. Results and discussion

Figure 1 shows in plane electrical resistivity, ρ, measurements for an $Fe_{0.92}Se$ crystal with magnetic field, H, perpendicular to the crystal plane. The onset of the transition temperature for $H = 0$ is $T_c = (11.9 \pm 0.1)$ K. The transition width is broadened by about 35% when H varies from 0 to 16 T. This is a signature for the presence of vortex liquid.

In order to study the influence of correlated defects on the dynamics of the vortex lattice we measured the angular dependence of ρ in terms of the orientation of H. The angle of H is measured with respect to the normal to the platelet, see θ in Fig. 2*a*). We used a constant Lorentz force configuration, i.e. the external field is always at a right angle to the applied current. This avoids spurious angular dependencies of the pinning force related to the non-perpendicularity between current density and magnetic field. Figures 2*b*) and 2*c*) show the

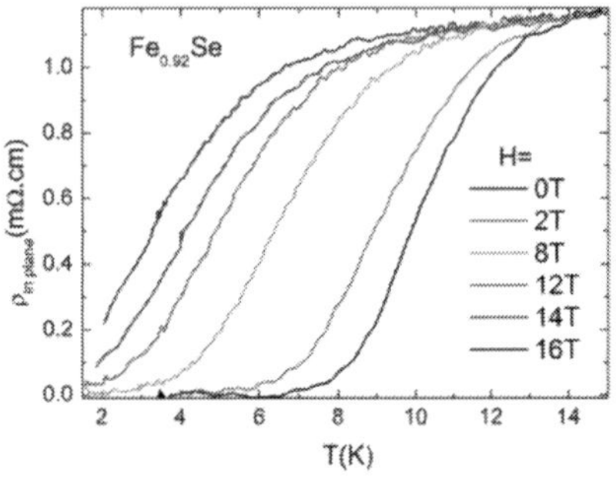

Figure 1. Temperature dependence of the in plane resistivity for a crystal of $Fe_{0.92}Se$ with magnetic field perpendicular to the crystal plane.

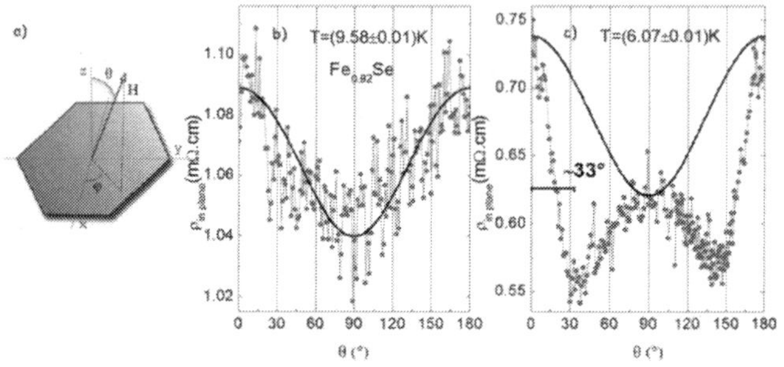

Figure 2. *a*) Definition of the rotation angle, θ (here $\varphi = 0$ and transport current is along y). *b*) and *c*) Angular dependence of the in plane resistivity for an applied field of H =12 T, and temperatures $T = (9.58 \pm 0.01)$ K and $T = (6.07 \pm 0.01)$ K respectively. The solid lines represent the expected behavior for an anisotropic material with point defects for $\gamma = (1.08 \pm 0.05)$.

results for a $Fe_{0.92}Se$ crystal at H =12 T and for $T = (9.58 \pm 0.01)$ K and $T = (6.07 \pm 0.01)$ K. The continuous line shows the expected behavior for a slightly anisotropic material with point defects. This dependence was evaluated in base to a thermally activated ρ [6] and to the scaling for the angular dependence proposed by Blatter and co-workers [7]. The final equation is:

$$\rho(T,H,\theta) = \frac{\rho_0}{T} exp\left[-U(T,H,\theta)T^{-1}\right] \qquad \text{with} \qquad U(T,H,\theta) = U_o(H\varepsilon(\theta))^{\frac{1}{2}-q}\left(1-\frac{T}{T_c}\right)^q \tag{1}$$

where $\varepsilon(\theta) = \sqrt{cos^2\theta + \gamma^{-2}sen^2\theta}$, U is the activation energy for the movement of the vortices, ρ_0 is a constant and q is a parameter that takes values between 0.5 and 2. In figure 2, we use H = 12 T, T_c = (11.9 ± 0.1) K and the corresponding temperature. The parameters $\rho_0 = (13 \pm 2)$ mΩ.cm.K, $q = (1.1 \pm 0.2)$, $U(9.58\,\text{K}, 12\,\text{T}, 0°) = (51 \pm 4)$ K and $U(6.07\,\text{K}, 12\,\text{T}, 0°) = (141 \pm 6)$ K were obtained adjusting $\rho(T, H, \theta = 0°)$ in Fig. 1. Finally, for both temperatures, we get $\gamma = (1.08 \pm 0.05)$ fitting Eq. 1 to the data in Fig. 2, and using the former values for the other parameters.

Notice that at temperatures near T_c the angular dependence is very well described by the equation above. However, for lower temperature, the resistivity presents strong deviations from this behavior. This is observed in all samples, and can be understood in terms of vortex dynamics. At low temperature, the extra minimum at ±33° (Fig. 2*c*)) could signal vortex pinning by correlated defects. The absence of this minimum near the critical temperature (Fig.

2*b*)) is consistent with the idea that vortices move in the flux flow regime and the pinning is less effective.

It is also expected for the correlated defect pinning to be observed in the critical current which should in turn, be observable with magnetization measurements. In order to assess these effects, we measured the longitudinal ($M_L \parallel H$) and transversal ($M_T \perp H$) magnetization of the crystals as a function of the rotation angle θ around an horizontal axis, while keeping the modulus of H constant, for $H = 0.15$ T at 2 K, well below T_c, and at 20 K, around 10 K above T_c.

In these crystals, there are two contributions to the magnetization, one coming from the superconducting $FeSe$, and the other coming from the ferrimagnetic Fe_7Se_8. In order to separate both signals, we assume that they are independent, and that the magnetization of Fe_7Se_8 is approximately constant in our measurements, given that the Neel temperature is much higher ($T_N \sim 455$ K) [5]. Consequently, we obtain the superconducting $FeSe$ contribution to the magnetization by simply subtracting the 20 K curve from the 2 K one.

Figure 3 depicts the results in a plot of $M_i = \pm\sqrt{M_L^2 + M_T^2}$ as a function of the field component normal to the platelet, $H_z = Hcos\theta$, showing the superconducting magnetization loop. The curve has a rich structure due to the contribution of anisotropy, correlated defect pinning and crystal shape. In particular, at the angles where extra minima are observed in ρ, i.e. 33°, 147°, 213° and 327°, the magnetization loop also presents distinctive structure. Theoretically, this structure can be understood in terms of a critical current density with anisotropic behavior. Figure 4 summarizes the main facts in a number of simulations that were performed by a critical state modelling [8]. A θ-rotation experiment for an isotropic platelet-like superconductor would produce a *bumped* cycle as compared to the conventional *Bean-like* behavior for a standard H_z experiment. An extra magnetic moment occurs at $\theta \approx 90°$. This is a purely geometric effect related to the *force free* configuration (due to the zones with induced current parallel to the magnetic field). On the other hand, when an anisotropic

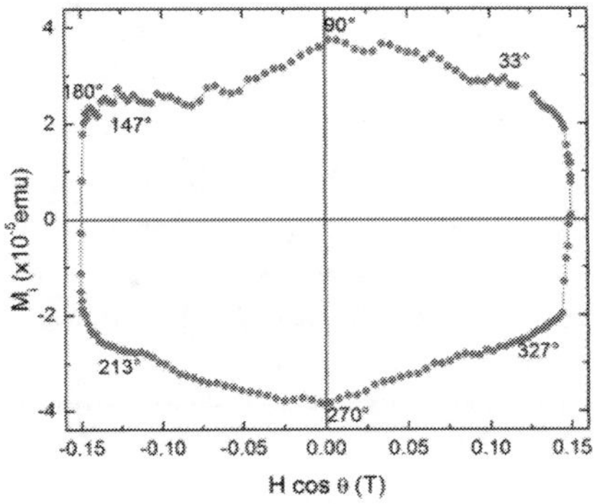

Figure 3. Superconducting contribution to the magnetization, $M_i = \pm\sqrt{M_L^2 + M_T^2}$, as a function of $Hcos\theta$ for $H = 0.15$ T and $T = 2$ K for a crystal of $Fe_{1-y}Se$.

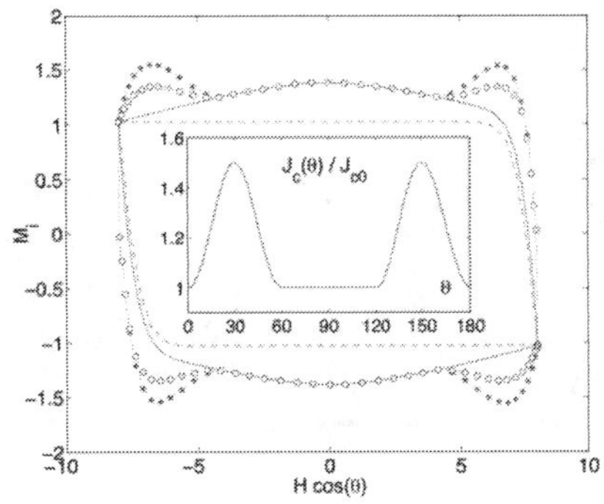

Figure 4. Same as Figure 3, but obtained from critical state simulations. The dashed line corresponds to a conventional Bean model cycle, the continuous line is for the platelet rotating in a magnetic field assuming isotropic critical current. Symbols are obtained by using anisotropic J_c as shown in the inset. A maximum $J_c(\theta)$ of 1.5 and 2.0 was used respectively. In all cases, units for $Hcos\theta$ are $J_c/4\pi$ and units for M_i, $J_c(a/d)$ with a the radius and d the thickness of the platelet.

critical current $J_c(\theta)$ enhanced at $\pm 30°$ is assumed, a superimposed structure appears, with new features at $30°, 150°, 210°, 330°$ as observed in experiments. The more anisotropic $J_c(\theta)$, the more pronounced the effects.

In relation to the crystal microstructure, the correlated defects observed in transport and magnetic measurements may be linked to the phase mixture and the growth directions of the tetragonal phase. Figure 5 presents the crystal structure scheme of our samples, panel *a*) shows the possible directions of the *c* axis of the tetragonal phase compatible with our XRD measurements. The direction of the *ab* plane at $\sim 34°$ from the normal of the crystal is coincident with a dense plane of vacancies in Fe_7Se_8 as can be inferred from the structural study of ref. [5] (see Fig. 5*b*)). This plane is, possibly, the phase boundary between the tetragonal and hexagonal phases and it could behave as a correlated defect for the vortex lattice.

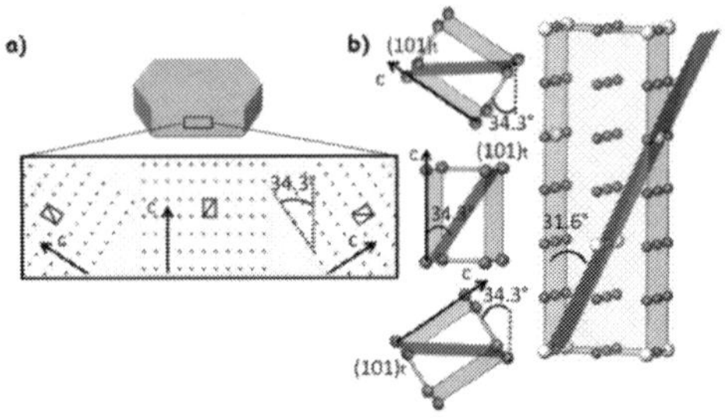

Figure 5. Crystal structure scheme of our samples. *a*) Possible directions of the *c* axis of the tetragonal phase. *b*) Unit cell of the tetragonal phase (*Left*) with the different orientations of the *c* axis and the unit cell of the hexagonal phase (*Right*). We only show the *Fe* atoms in red and the vacancies in white.

4. Conclusion

Electrical resistivity and magnetization results in $Fe_{1-y}Se$ crystals show clear evidence for the existence of superconducting pinning by correlated defects. At the same time, the structural characterization of the samples shows the existence of different phases and orientations. One of the crystalline orientations of the $FeSe$ phase coincides with the direction in which the correlated defects increase the pinning, evidencing that the origin of the defects could be related to phase matching in the samples.

Acknowledgments

AB acknowledges funding by Spanish MINECO and the European FEDER Program (Projects MAT2011-22719, ENE2011-29741) and by Gobierno de Aragon (Research group T12, grant FMI02/11). JG acknowledges funding from ANPCyT PICT2007-824, Universidad Nacional de Cuyo 06/C375. GN acknowledges funding from ANPCyT PICT2007-819, Universidad Nacional de Cuyo 06/C381 and CONICET PIP11220090100448.

References

[1] Hsu F C *et al.* 2008 *Proceedings of the National Academy of Sciences* **105** 14262–14264
[2] Johnston D C 2010 *Advances in Physics* **59** 803–1061
[3] Okamoto H 1991 *Journal of Phase Equilibria* **12** 383–389 ISSN 1054-9714
[4] Amigó M L *et al. to be published*
[5] Kawaminami M and Okazaki A 1970 *Journal of the Physical Society of Japan* **29** 649–655
[6] Palstra T T M, Batlogg B, van Dover R B, Schneemeyer L F and Waszczak J V 1990 *Phys. Rev. B* **41**(10) 6621–6632
[7] Blatter G, Feigel'man M V, Geshkenbein V B, Larkin A I and Vinokur V M 1994 *Rev. Mod. Phys.* **66**(4) 1125–1388
[8] Badía-Majós A, López C and Ruiz H S 2009 *Phys. Rev. B* **80**(14) 144509

Temperature-dependent electric noise level in different iron-based superconductors

C Barone[1], S Pagano[1], E Bellingeri[2], C Ferdeghini[2], M Adamo[3], E Sarnelli[3], F Kurth[4], B Holzapfel[4] and K Iida[4]

[1] Dipartimento di Fisica "E.R. Caianiello" and CNR-SPIN Salerno, Università di Salerno, I-84084 Fisciano, Salerno, Italy
[2] CNR-SPIN Genova, corso Perrone 24, I-16152 Genova, Italy
[3] CNR-ICIB, Via Campi Flegrei 34, I-80078 Pozzuoli, Napoli, Italy
[4] Leibniz-Institut für Festkörper- und Werkstoffforschung (IFW) Dresden, PO Box 270116, 01171 Dresden, Germany

E-mail: cbarone@unisa.it

Abstract. A detailed characterization of the voltage-noise properties has been performed in $FeTe_{0.5}Se_{0.5}$ epitaxial thin films and Co-doped $BaFe_2As_2$ bilayers, deposited by pulsed laser deposition. In all the samples analyzed, the experimental voltage-spectral density has a 1/f noise component. Different behaviors are observed for the bias current and temperature dependencies of this 1/f noise, and are related to specific structural and electric transport properties of the two materials.

1. Introduction

Among the family of new Fe-based superconductors, iron chalcogenides and Co-doped $BaFe_2As_2$ (Ba-122) compounds are of interest for both basic physics and high-field applications. Since their discovery, great efforts have been made to determine their fundamental properties. In particular, the crystalline quality, the surface morphology and the electric transport of these superconducting thin films are generally influenced by several factors such as strain, defect formation and change in growth mode [1, 2].

Due to the high sensitivity at ambient air, experimental results obtained with different methodologies on various Fe-based samples are often inconsistent. For instance, the role played by layer thickness on the increase of the superconducting transition temperature T_c, and the pairing symmetry in pnictides and chalcogenides are still disputed [3, 4]. A possible reason for such discrepancies between experiments could be related to the type of analysis technique used. In this respect, electric noise spectroscopy has already revealed its potentiality for a sensitive and non-destructive analysis of transport processes in advanced materials, such as double perovskites [5], manganites [6], and electron-doped cuprate superconductors [7].

Triggered by this scenario, a detailed voltage-noise characterization of $FeTe_{0.5}Se_{0.5}$ and Co-doped Ba-122 thin films has been performed. The comparison of the experimental data, collected by analyzing several configurations and contacts geometries on patterned samples, allows to identify different fluctuation mechanisms. This may help to shed some light on the peculiar structural properties and electrical conduction processes in the two systems investigated.

2. Experiments

Pulsed laser deposition technique was used to prepare all the thin films here studied. $FeTe_{0.5}Se_{0.5}$ was deposited on single crystal substrates of lanthanum aluminate (LAO) with (001) orientation. The optimized fabrication parameters were: deposition temperature 550 °C, pressure 5×10^{-9} mbar, laser repetition rate 3 Hz, laser wavelength 248 nm, laser fluency 2 J cm^{-2}, spot size 2 mm^2, target-substrate distance 5 cm. Co-doped Ba-122/Fe bilayers were prepared by a two-step process, characterized by a room temperature deposition of Fe on MgO (001) single-crystalline substrates and by a subsequent deposition of Co-doped Ba-122 layers at 750 °C with a laser repetition rate of 10 Hz.

Atomic force microscope (AFM) measurements indicated that the surface was flat with a root mean square roughness of $\simeq$ 1 nm in the case of $FeTe_{0.5}Se_{0.5}$ (typical thickness $\simeq$ 150 nm) [3], and of $\simeq$ 0.8 nm in the case of Co-doped Ba-122 (typical thickness $\simeq$ 100 nm) [8]. Terraces with monoatomic steps of $\simeq$ 0.6 nm in size were resolved in scanning tunnel microscope (STM) images, confirming the good quality of the growth for both the compounds analyzed.

All the samples were patterned by standard UV photolithographic processes and Ar ion milling etching. During the etching process, the samples were cooled at -40 °C to prevent local heating and damage, due to losses of volatile species. The patterned geometry consists in a number of strips having different width (between 2 and 16 μm) and in plane orientation.

The experimental investigations were carried in a closed-cycle refrigerator, at temperatures between 8 and 325 K. A specific procedure was used to minimize the electrical noise generated by the contacts. All the details of the measurement setup and of the outlined procedure are illustrated in [9].

3. Results and discussion

Different transition temperatures have been measured for the two types of superconductors. As shown in Figure 1, for $FeTe_{0.5}Se_{0.5}$ thin films T_c, defined at 90% of normal state resistance, is 16.2 K, that is the same critical temperature as in bulk samples [3]. For Co-doped Ba-122, instead, the T_c is 24.0 K and improves with increasing thickness, as described in [8]. Linear *I-V* characteristics have been observed in the whole temperature range, and for all sample configurations. Possible Joule heating effects have been avoided by the pulsed measurement technique used.

The detailed investigation of the low-frequency voltage noise, whose spectral density traces are shown in Figure 2, has revealed the presence of an evident $1/f^{\gamma}$ dependence (with γ in the range between 1.02 and 1.14), for all the materials. A high-frequency background spectrum

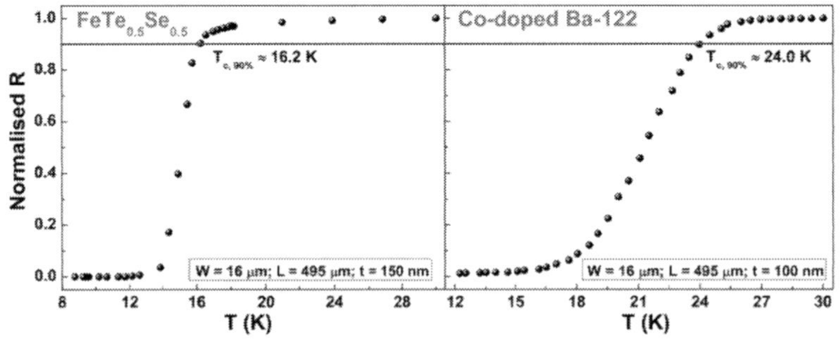

Figure 1. Resistance versus temperature dependence in the superconducting transition regions. The normalized resistance R is computed as $R(T)/R(30\ K)$. The critical temperature T_c is defined at 90% of normal state resistance.

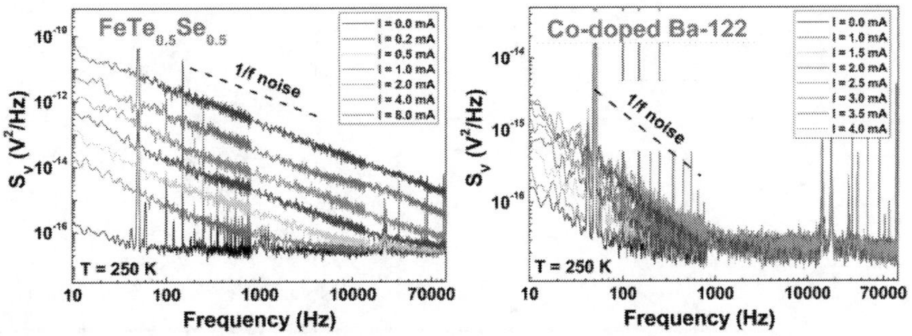

Figure 2. Low-frequency voltage-noise spectra of a 16-μm-wide strip taken at a temperature of 250 K and for different dc bias currents.

given by the electronic chain noise ($\approx 1.4\times10^{-17}$ V^2 Hz^{-1}) added to the sample Johnson noise is also visible in the experimental data of Figure 2.

Despite the similar frequency dependence of the measured spectral densities, $FeTe_{0.5}Se_{0.5}$ and Co-doped Ba-122 thin films show a different behavior in the bias current dependence of the 1/f noise amplitude. In Figure 3 a bias-activated noise region, at high applied electric fields and temperatures, is found for $FeTe_{0.5}Se_{0.5}$ and is characterized by a more than quadratic current scaling of the 1/f component. This is usually associated to an excess of fluctuations, ascribed to different physical mechanisms. One is sliding-charge density waves, which, however, in this material occurs at significantly lower temperatures [10]. A quadratic bias current scaling of 1/f noise is, instead, observed for Co-doped Ba-122 samples (see right panel in Figure 3 for details). This is interpreted in terms of a standard resistance fluctuations model, and can be expressed by the well-known empirical Hooge relation as [11]: $S_V(f)=\alpha_H V^2/n\Omega f^\gamma$. Here, α_H/n is the normalized Hooge parameter, Ω the sample volume, γ the dimensionless frequency exponent, and $V=RI$ the dc voltage bias for Ohmic compounds. The parameter α_H/n, also identified as the noise level, can be extracted by a fitting procedure of the experimental data, as in Figure 3, in the whole investigated temperature range but only for low applied electric fields, in the case of $FeTe_{0.5}Se_{0.5}$ samples. A clear step-increase of the noise level is found in iron chalcogenides by increasing the temperature (see red points linked to left y-axis in Figure 4), probably related to the presence of a structural transition occurring in $FeTe_{0.5}Se_{0.5}$ below 100 K [10]. However,

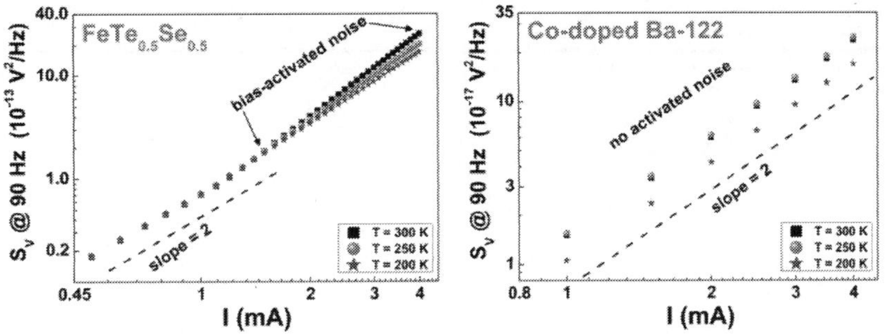

Figure 3. Bias current dependence of the 1/f noise amplitude at 90 Hz. Three different temperatures (300, 250, and 200 K) are shown.

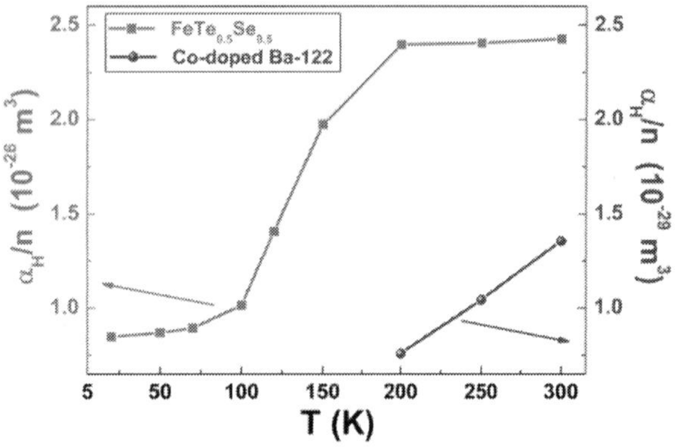

Figure 4. Temperature dependence of the normalized Hooge parameter. The left y-axis refers to $FeTe_{0.5}Se_{0.5}$ samples (red points); the right y-axis refers to Co-doped Ba-122 samples (blue points). The lines are only guides to the eyes. Below 200 K, in 122 samples the noise signal is lower than the experimental background level.

possible effects due to interband coupling cannot be excluded and require the development of a specific theoretical model for these type of superconductors. Lower values of the 1/f noise amplitude, instead, have been found in Co-doped Ba-122 superconductors (see blue points linked to right y-axis in Figure 4), indicating this material as more appropriate for future electronic applications.

4. Conclusions

The voltage-spectral density data, for all the investigated samples, show a 1/f noise component. Different behaviors are observed for the bias current and temperature dependencies in the two materials analyzed. In $FeTe_{0.5}Se_{0.5}$ superconductors, a noise level step-increase and a bias-activated nonlinear noise are found in the high temperature limit. This may be related to the presence of a structural transition occurring at $\sim$ 100 K. In Co-doped Ba-122 superconductors, lower values of the 1/f noise amplitude and of the electrical resistivity are observed. The temperature dependence of the noise level is very similar to that already found in metals. The experimental indications suggest that, the latter material is a better candidate for electronic applications.

Acknowledgments

Financial support from EU project "IRON-SEA" NMP.2011.2.2-6 nr. 283141 is gratefully acknowledged. The authors thank Francesco Lippi for his support during the experiments.

References

[1] Ishida K, Nakai Y and Hosono H 2009 *J. Phys. Soc. Japan* **78** 062001
[2] Mizuguchi Y and Takano Y 2010 *J. Phys. Soc. Japan* **79** 102001
[3] Bellingeri E, Buzio R, Gerbi A, Marrè D, Congiu S, Cimberle M R, Tropeano M, Siri A S, Palenzona A and Ferdeghini C 2009 *Supercond. Sci. Technol.* **22** 105007
[4] Hirschfeld P J, Korshunov M M and Mazin I I 2011 *Rep. Prog. Phys.* **74** 124508
[5] Savo B, Barone C, Galdi A and Di Trolio A 2006 *Phys. Rev. B* **73** 094447
[6] Wu X D, Dolgin B, Jung G, Markovich V, Yuzhelevski Y, Belogolovskii M and Mukovskii Y M 2007 *Appl. Phys. Lett.* **90** 242110
[7] Barone C, Guarino A, Nigro A, Romano A and Pagano S 2009 *Phys. Rev. B* **80** 224405
[8] Iida K, Hänisch J, Trommler S, Haindl S, Kurth F, Hühne R, Schultz L and Holzapfel B 2011 *Supercond. Sci. Technol.* **24** 125009
[9] Barone C, Galdi A, Pagano S, Quaranta O, Méchin L, Routoure J-M and Perna P 2007 *Rev. Sci. Instrum.* **78** 093905
[10] Barone C, Bellingeri E, Adamo M, Sarnelli E, Ferdeghini C and Pagano S 2013 *Supercond. Sci. Technol.* **26** 075006
[11] Kogan S 1996 *Electronic Noise and Fluctuations in Solids* (Cambridge: Cambridge University Press)

Composite superconducting bulks for efficient heat dissipation during pulse magnetization

A Baskys[1], A Patel[1], S Hopkins[1], D Kenfaui[2], X Chaud[2], M Zhang[3] and B A Glowacki[1]

[1] Applied Superconductivity and Cryoscience Group, Department of Materials Science and Metallurgy, University of Cambridge, Cambridge CB3 0FS, UK
[2] CNRS LNCMI CRETA, F-38042 Grenoble 09, France
[3] Department of Engineering, University of Cambridge, 9 J J Thomson Avenue, Cambridge CB3 0FA, UK

E-mail: ab857@cam.ac.uk

Abstract. Pulsed field magnetization is the most practical method of magnetizing a (RE)BCO bulk, however large heat generation limits the trapped field to significantly less than possible using field cooling. Modelling has been used to show that effective heat removal from the bulk interior, using embedded metallic structures, can enhance trapped field by increasing thermal stability. The reported results are for experimental pulsed magnetization of a thin walled YBCO sample with 55 vertical holes embedded with high thermal conductivity wires. A specially designed copper coldhead was used to increase the trapped field and flux of the perforated YBCO by about 12% at 35 K using a multi-pulse magnetization. Moreover, by filling the perforations with copper, the central trapped field was enhanced by 15% after a single-pulse at 35 K. 3D FEM computer model of a perforated YBCO bulk was also developed showing localised heating effects around the perforations during pulse magnetisation.

1. Introduction

Superconducting bulks can act as very compact sources of magnetic field, which makes them attractive for a variety of different applications, such as motors [1] and contactless bearings [2]. However, pulse magnetisation causes rapid heating of the bulk superconductor. Modelling has been used to show that effective heat removal from the bulk using embedded metallic components can enhance the trapped field by increasing the effective thermal conductivity [3]. To implement such structure a composite bulk, comprising of a perforated bulk with embedded copper wires, was assembled and tested. To efficiently sink the heat, extracted from the bulk, a specially designed copper thermal mass was used, which minimises eddy current heating. Furthermore, a 3D critical state magneto-thermal model was developed to study the thermal and magnetic flux dynamics in the composite during pulse magnetization.

2. Experimental pulse magnetization

The sample used in this study had a diameter of 16 mm and a height of 19.5 mm. The sample had a total of 55 perforations in a hexagonal lattice with seven largest holes around the perimeter and the centre having an average diameter of 0.58 mm, the rest 0.42 mm. To improve thermal contact between the copper wires and YBCO, the YBCO bulk was impregnated with Apiezon N thermal grease at 50°C and any trapped air was removed under vacuum. Copper (RRR=30) wires with diameters 0.4 and 0.5 mm were inserted into the bulk. Then, the vacuum treatment was repeated. The assembled composite bulk is shown in figure 1c. To effectively sink the heat extracted from YBCO, a copper thermal mass (referred to as coldhead) was made, the design of which inhibits eddy current formation. The coldhead comprises of 1 mm sheets of copper glued with Stycast resin, compatible with vacuum and low temperatures. Finally, thermal grease was applied on the interface between coldhead and the YBCO sample.

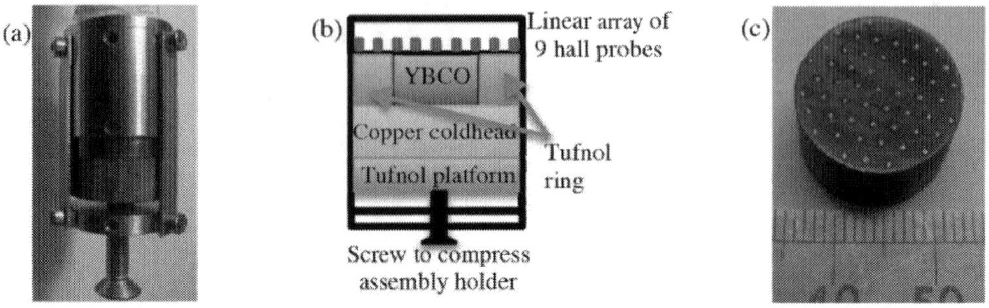

Figure 1. a) Picture of the sample holder with the sample inside; b) schematic diagram of the sample arrangement; c) YBCO sample with copper wires inserted into the perforations.

The sample was magnetized using the MPSC and IMRA techniques [4]. The pulsed field was reduced in steps of 0.2 T until there was no significant improvement in the trapped field or flux observed. Then, the temperature was reduced and the process repeated. The starting and ending fields of each sequence at temperatures tested are indicated in table 1. The pulse magnetisation was performed using the experimental setup developed at the Applied Superconductivity and Cryoscience Group, University of Cambridge [2]. The Hall probes (figure 1b) measured the magnetic flux density 0.5 mm above the sample surface.

Table 1. Sequence of pulses applied to the samples starting at "Start field" and then pulsing with incrementally smaller (by 0.2 T) field until "End field" is reached.

Temp. [K]	Start field [T]	End field [T]
77.4	3.6	1.6
55	5.0	2.6
35	5.8	3.4
15	4.8	1.8
10	4.6	1.8

3. Results and discussion

Figure 2a shows the central trapped field and figure 2b shows the trapped flux above the YBCO sample. It is worth noting that the trapped flux continued to increase even after the central field has saturated, since the trapped field profile has widened. If we take a "bare bulk" (supported only by tufnol) as a baseline we see that the most significant increase in the trapped field is seen after adding a coldhead to the sample assembly; at 35 K the trapped field and flux was increased by 12.4% and 11.9% respectively. By further adding copper into the YBCO perforations there was only a marginal 1.6% improvement in the trapped flux at 35 K. This is likely because the multi-pulse magnetisation techniques already address the heating problem. However, there is marked improvement when comparing the trapped flux and field after the *first* pulse at intermediate temperatures (see figure 3). By inserting wires into the perforations, the central trapped field at 35 K increases by 15%. Hence, the amount of pulses and time needed to fully magnetise the sample can be reduced. Furthermore, filling the perforations does not increase the volume of the sample and the perforations themselves are made for a completely different purpose – to reduce oxygenation time, porosity and cracking [5,6] during bulk YBCO manufacture.

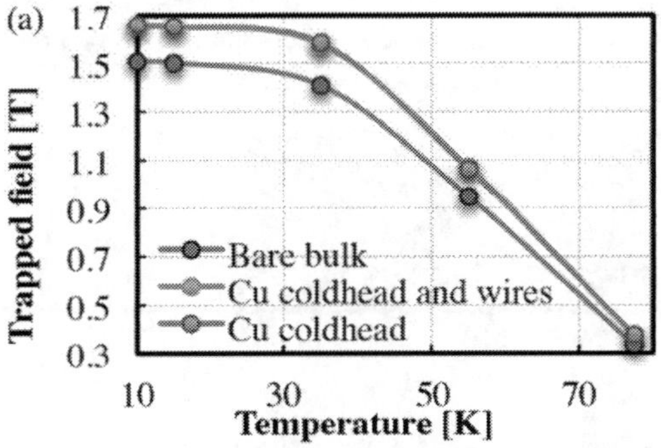

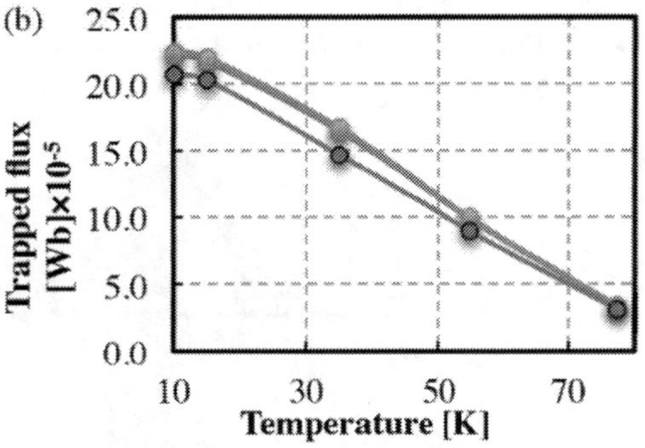

Figure 2. Trapped central field a) and flux b) after multi-pulse magnetization.

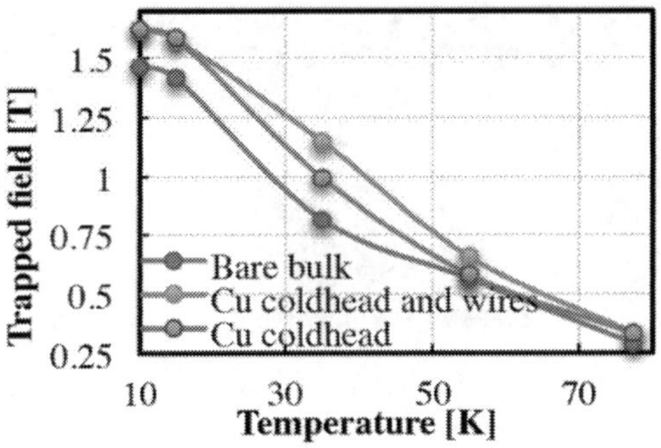

Figure 3. Trapped field after the first pulse at each temperature.

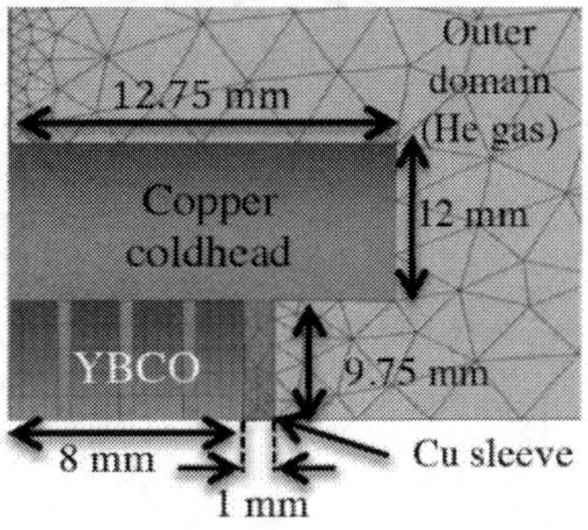

Figure 4. Dimensions of the modelled geometry. Cu sleeve was used in some of the tests.

4. FEM model of the composite bulk

A magneto-thermal FEM model of the perforated YBCO geometry was developed using COMSOL Multiphysics 4.3a. The basis of the model was a 3D critical state model (H - formulation, no thermal effects) developed by Zhang *et al* [7]. Complexity of the geometry modelled, combined with added thermal effects, makes the model very computationally demanding. To minimize element count in the model, symmetries were exploited by modelling only 1/6 of the sample. Furthermore, another symmetry plane along the middle of the sample in the x-y plane (see figure 4) was introduced to reduce the computation time to a manageable 55 h. A 5 T and 31 ms duration pulse was applied to model single pulse magnetization. The equilibrium temperature was set to 40 K and the cooling by He gas was simulated by a cooling source around the periphery of the sample assembly. The source has power, which is proportional to the difference between current and equilibrium temperature and saturates at 30 W when this difference reaches 15 K.

The dynamics of pulse magnetization of the bulk with holes (filled with copper) is illustrated in figure 5. The main difference, compared to the dynamics expected for a normal bulk, is that the current is forced to meander around the perforations where local current density increases, resulting in 'hotspots'. This shows that the copper in the perforations provides cooling where it is needed most. The simulation results also show that by filling perforations with copper, the maximum temperature reached during magnetisation is lower and the equilibrium temperature is reached faster. The final trapped field inside the bulk has no indications of holes present in the sample. However, little difference in the trapped field was seen when modelling the bulk with and without copper filling the holes. This may be due to the cold head geometry not matching the experiments and a relatively low exponent (n = 9) in the *E-J* power law, used in order to stabilize the simulation and reduce the computation time. As a result, the main purpose of the model developed is to give an insight into flux dynamics for a bulk with 3D anisotropy, and only a qualitative link to the experiments. A general point arising from both the model and experiment is that certain forms of symmetrical J_c anisotropy such as an array of holes of zero J_c, does not necessarily prevent large trapped fields.

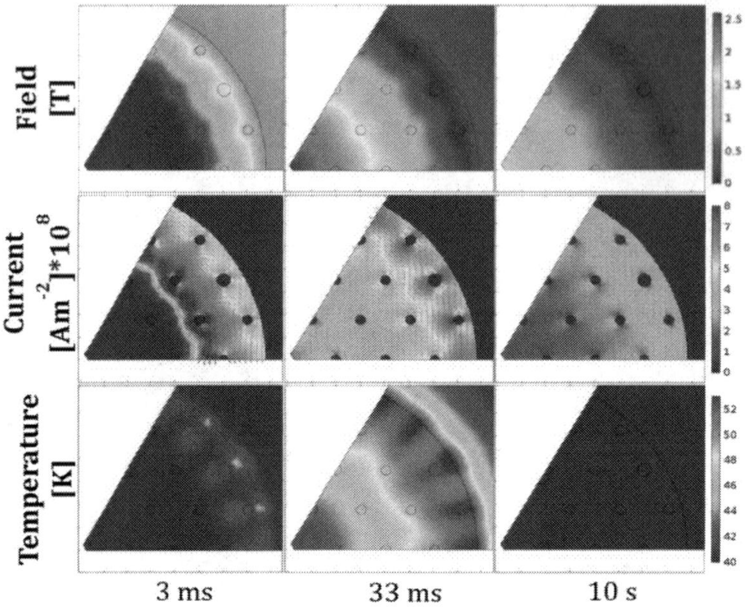

Figure 5. Flux density, current density and temperature in plane through the centre of sample at 3 ms, 33 ms and 10 s after the start of a pulse in YBCO with Cu coldhead and wires in the perforations.

5. Conclusions

A composite YBCO bulk with copper embedded in the perforations was magnetized using pulsed magnetization and the performance of the sample was evaluated. Using the MPSC magnetization technique and filling the perforations with copper, resulted in a modest 1.6% increase in trapped flux (compared to a bulk and coldhead alone). Nevertheless, the performance is substantially increased when considering the trapped field after the first pulse. A 15% increase in trapped field was observed after the first pulse at 35 K, which suggests that such composite samples can be fully magnetized faster and with a lower number of pulses. A 3D critical state FEM model that considers thermal effects was constructed. The model showed localized heating effects around perforations and confirmed that the heat removal was significantly improved by using composite structures.

References

[1] Zhou D, Izumi M, Miki M, Felder B, Ida T and Kitano M 2012 Supercond. Sci. Technol. **25** 103001
[2] Patel A, Giunchi G, Albisetti a F, Shi Y, Hopkins S C, Palka R, Cardwell D A and Glowacki B A 2012 *Phys. Procedia* **36** 937–42
[3] Patel A and Glowacki B A 2012 *Supercond. Sci. Technol.* **25** 125015
[4] Fujishiro H, Hiyama T, Naito T, Yanagi Y and Itoh Y 2009 *Supercond. Sci. Technol.* **22** 095006
[5] Chaud X *et al* 2009 *Phys. C Supercond.* **469** 1200–6
[6] Diko *et al* W 2005 *Supercond. Sci. Technol.* **18** 1400–4
[7] Zhang M and Coombs T A 2012 *Supercond. Sci. Technol.* **25** 015009

Femtosecond spectroscopy in a nearly optimally doped Fe-based superconductors $FeSe_{0.5}Te_{0.5}$ and $Ba(Fe_{1-x}Co_x)_2As_2$/Fe thin film

C Bonavolontà[1,2], L Parlato[1,2], C de Lisio[1,2], M Valentino[1,2], GP Pepe[1,2], I Kazumasa[3], F Kurth[3], E Bellingeri[4] , I Pallecchi[4] , M Putti[4], C Ferdeghini[4], G A Ummarino[5], F Laviano[5]

[1]Department of Physics, University of Naples "Federico II", Italy
[2]CNR-SPIN, Piazzale Tecchio 80, 80125 Naples, Italy
[3] IFW Dresden, P. O. Box 270116, 01171 Dresden, Germany
[4]CNR-SPIN, Corso Perrone 24, 16152 Genova, Italy
[5]Dipartimento di Fisica and CNISM, Politecnico di Torino, Corso Duca degli Abruzzi 24, 10129 Torino, Italy

E-mail: bonavolo@na.infn.it

Abstract: Femtosecond spectroscopy has been used to investigate the quasi-particle relaxation times in nearly optimally doped Fe-based superconductors $FeSe_{0.5}Te_{0.5}$ and optimally doped Ba-122 thin films growth on a Fe buffer layer. Experimental results concerning the temperature dependence of the relaxation time of such pnictides both in the superconducting state are now presented and discussed. Modelling the T-dependence of relaxation times an estimation of both electron-phonon constant and superconducting energy gap in the excitation spectrum of both Fe(Se,Te) and Ba-122 compounds is obtained.

1.Introduction

The recent discovery of superconductivity with transition temperature T_c up to 55 K in iron–pnictide systems [1,2]] has sparked enormous interest in this class of materials. Investigations on this new class of superconductors have revealed a lot of similarities to cuprates, for instance, layered structures, a multiband nature, high upper critical fields and a short coherence length. A common phase diagram for Fe–pnictides suggests the existence of an antiferromagnetic spin density-wave (SDW) ground state that is stabilized by Fermi-surface nesting [3] as well as by strong antiferromagnetic (AFM) spin interactions along the Fe-square-diagonal [4,5].

The coexistence of superconductivity and magnetism in the phase diagram suggests that magnetism plays an important role in determining the superconducting properties and in forming electron Cooper pairs. Then, the importance of the spin degrees of freedom for the superconducting pairing interaction becomes immediately apparent, and investigating the Fe-pnictide compounds from the point of electron-phonon and spin-charge interactions is therefore beneficial for understanding the nature of the superconducting coupling in the doped compounds.

Time-resolved spectroscopy is considered an efficient and useful technique to study the nature of the electronic excitations in superconductors [2] and more recently also in iron pnictides [3,4]. This technique gives information on the low-lying electronic structures, and potentially on the pseudo-gap, due to its capability to distinguish between scattering processes characterized by different relaxation dynamics [5]. In particular, the transient reflectivity change as a function of the time

delay (in the ps range) between the pump and the probe beams hitting the same region of the sample is clearly dependent on the electron relaxation processes which restore the equilibrium. Pump-probe fast spectroscopy has been already proposed for estimating the electron-boson constant λ in novel pnictides materials as in the case of $BaFe_{0.92}Co_{0.08}As_2$ ($\lambda = 0.12$) [6], $BaFe_2As_2$ [7] ($\lambda = 0.15$), and SmFeAsO [8] ($\lambda = 0.18$). This technique can be considering complementary to other conventional techniques (i.e. point contact spectroscopy) to evaluate the superconducting energy gap.
In this work we use a time-resolved spectroscopy on $FeSe_{0.5}Te_{0.5}$ (Fe-11) thin films and optimally Co-doped $BaFe_2As_2$ (Ba-122) epitaxial thin films grown on Fe-buffered MgO(001) substrates to infer about the value of the energy gap in the quasi-particle density of states and the strength of the electron-boson interaction which characterize the superconducting state.

2. Materials and experimental set up

The Fe(Se,Te) films were deposited on single crystal lanthanum aluminates (001) substrates in a ultra high vacuum pulsed laser deposition (PLD) system using as a target a bulk polycrystal with a sharp resistive transition with an onset at T=16.2 K. further experimental details about the sample fabrication are found in [9]. The thin $Fe(Se_{0.5}Te_{0.5})$ films has a thickness of about ≈ 150 nm and the critical temperature of the film was about 19K.
The Co-doped Ba-122/Fe bilayers on MgO substrate were prepared by pulsed laser deposition under ultra high vacuum condition (base pressure of 10^{-9} mbar). The nominal Co content was x=0.08 in $Ba(Fe_{1-x}Co_x)_2As_2$. The respective layer thickness of Co-doped Ba-122 and Fe are 100 nm and 20 nm. In-situ reflection high energy electron diffraction confirmed that both, Fe and Co-doped Ba-122, were grown epitaxially with smooth surface. The detailed films preparation can be found in Ref. [10]. The critical temperature, T_c, defined as 90% of the normal state resistance was T_c=26.3K,
Time resolved spectroscopic measurements were performed using a standard pump-probe technique, based on mode-locked Ti:sapphire laser, delivering pulses at 82 MHz repetition rate, with 810 nm center wavelength, 100 fs duration. The sample was mounted on a cold finger in a temperature controlled liquid helium continuous flow optical cryostat, operating down to 5K. Further details on experimental details can be found in [11]. The photoinduced normalized reflectivity change ΔR/R was measured using a weaker probe pulse (1:20 the ratio between probe and pump beam) at a variable delay time, and as a function of the temperature. The pump fluences were about 5 $\mu J/cm^2$ and 1.6$\mu J/cm^2$ for the Fe-11 and Ba-122 sample, respectively. The different fluences are chosen because of the different signal to noise ratio of the signals due to the different samples surface reflectivity.

3. Results and discussion

In figure 1 the normalized reflectivity change ΔR/R as a function of the delay time between the pump and probe pulse at different temperatures is reported, in the case of Fe(Se,Te) and Ba-122/Fe thin films, respectively.

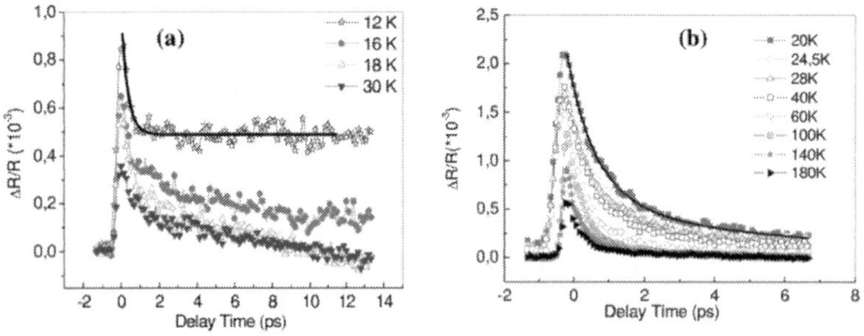

Figure 1. (a) The photoinduced reflectivity ΔR/R vs delay time at different temperatures related to the FeSeTe compound; (b) The ΔR/R vs the delay time at temperatures ranging from 20K to 180K, measured in Ba-122/Fe. The black solid lines in figure 1(a)-(b) represent the single and double-exponential fit curves used to estimate the relaxation time for Fe(Se,Te) and Ba-122/Fe, respectively.

Journal of Physics: Conference Series **507** (2014) 012004 doi:10.1088/1742-6596/507/1/012004

To estimate the relaxation time τ the experimental curves have been fitted by a single and double-exponential function (see the solid line in figures 1(a) and (b)) for the Fe(Se,Te) and Ba-122/Fe thin films, respectively. Since for both Fe-pnictides samples the relaxation time (in particular the slowest time for the Ba-122/Fe thin film) shows a peak close to the critical superconducting temperature T_c, as reported in figures 2(a) and (b), the electron relaxation time data were analyzed according to a temperature dependence of quasiparticle recombination time in a superconductor with a temperature-dependent energy gap $\Delta(T)$ below T_c as described by the equation (1) [12]:

$$\tau_R = \frac{A}{\Delta(T)^2}\ln\left\{\left(\frac{B}{\Delta(T)^2}+e^{-\frac{\Delta(T)}{k_BT}}\right)^{-1}\right\} \qquad (1)$$

where A and B are physical constant related both to phonon characteristic energies and the normal density of states $N(0)$ of the Fe-11 and Ba-122 compound, and $\Delta(T)=\Delta_0[1-(T/Tc)]^{0.5}$ is the temperature dependence of the superconducting energy gap.

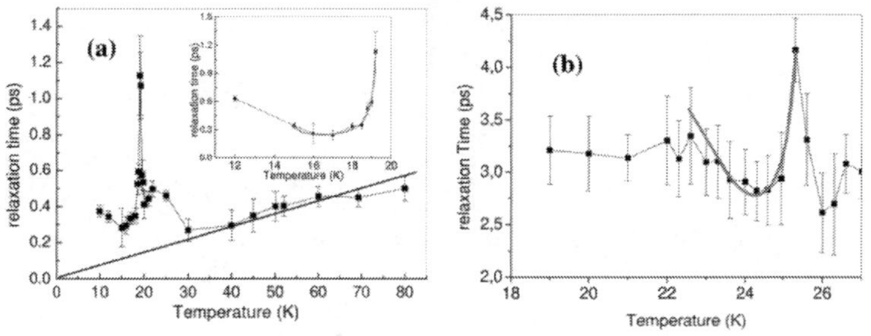

Figure 2. The relaxation time τ as a function of temperature for the sample with T_C=19.1 K and thickness of 150nm. The inset in figure 2(a) shows the fitting procedure applied to the decay time for $T<T_C$. The solid line in figure 2(a) represents the linear dependence between the relaxation time and the temperature starting from T=30K. (b) relaxation time τ_{slow} extracted from double-exponential decay fits of $\Delta R/R$ data in Ba-122/Fe.

In the case of FeSeTe sample, the fitting procedure, showed in the inset of figure 2(a), applied to experimental points recorded at $T<T_c$, gives Δ_0=4.1 ± 0.4meV for the superconducting energy gap extrapolated at T=0K, a value which corresponds to $2\Delta_0/k_BT_c$ = 4.9 ±0.5. This is in agreement with similar determinations already reported by specific heat [13] and Scanning Tunneling spectroscopy measurements [13]. The pre-factor $A\approx 2.9 \pm 0.1\ 10^{-14}$ eV s cm^{-1} in eq.(1) can be written also as follows [14], $A=\hbar\,\omega_{ph}/12\Gamma_{ph}$ where ω_{ph} is a typical phonon frequency, Γ_{ph} is its characteristic phonon line-width. According to Subedi *et al* [14] we can assume Γ_{ph} =2cm^{-1}, and then estimate the characteristic phonon frequency ω_{ph}=32.4±0.4 cm^{-1}= 4.0±0.1 meV.
Moreover, using a linear approximation for τ vs T behavior also at T>80K, according to[15]:

$$\tau = \frac{\pi k_B T}{3\hbar\lambda<\omega^2>} \qquad (2)$$

where $\lambda<\omega^2>$ is the second moment of the Eliashberg function, $\alpha^2F(\omega)$. The equation is expected to be valid for energy higher than $\hbar\omega_0$, where ω_0 is the characteristic Eliashberg –function phonon frequency. Looking at figure 2(a), in the range T>50K, and assuming a similar slope also at larger temperature, we obtain $\lambda<\omega^2>$= 10±2 meV^2, which corresponds to an electron-phonon coupling constant λ~0.6±0.1 (calculated using ω_{ph} previously obtained from the parameter A). We consider this feature only a remarkable element of consistency even if we are not completely within the range of validity of the proposed theory. Moreover, the same results can be obtained using a two-temperature model, as reported in a previous work [16].
For the Ba-122/Fe sample, from the fitting analysis using eq.(1), the value of superconducting gap Δ_0=3.67 ± 0.13meV that corresponds to $2\Delta_0/k_BT_c$ = 3.4 ± 0.1 has been evaluated. In previous time-resolved studies on SmFeAs(O,F) [17,18] and Ba-122 [7,18] iron pnictides superconductors, $2\Delta_0/k_BT_c$ ranging from 3 to 3.7 have been estimated. Our result is in agreement with the superconducting gap value estimated with other techniques [6,7,19]. In addition to evaluate the

electron-phonon coupling constant for the Ba-122/Fe, we use again the value of the fit parameter in eq.(1), namely A =22.2e-15 eV s cm^{-1} to calculate the phonon frequency ω_{ph}, where Γ= 4cm^{-1} [6]. In this way ω_{ph}=127.2 cm^{-1}=15.77 meV and considering $\lambda\langle\omega^2\rangle\sim$50meV^2 from inelastic neutron data[20], the electron-phonon coupling constant is about λ=0.18. This results is in agreement with that obtained for Co optimally doped Ba-122 and SmFeAsO compounds [20,21].

4. Conclusions

In conclusion, we report femtosecond spectroscopy measurements in a nearly optimally doped Fe-based superconductor $FeSe_{0.5}Te_{0.5}$ with $T_c\approx$19K and Ba(FeCo)As with $T_c\approx$26K. Our results indicate the existence of temperature dependence of quasiparticle recombination time in the superconducting state and a temperature-dependent energy gap $\Delta(T)$ at T<T_c. The fitting procedure gives for the Fe(Se,Te) sample Δ_0=4.1±0.4meV for the superconducting energy gap, and an electron-phonon constant $\lambda\approx$0.6, while for the Ba-122/Fe Δ_0=3.67±0.13meV and $\lambda\approx$0.18. These results are in very good agreement with values of the superconducting energy gap and electron-phonon coupling constant obtained with other techniques (i.e. Angle Resolved Photoemission Spectroscopy (ARPES)) and predicted by theoretical models [6,19,21] applied to the pnictides compounds.

Acknowledgments
This work was supported by European Project "IRONSEA", Grant Agreement nr. 283141.

References

[1] Kamihara Y, Watanabe T, Hirano M and Hosono H 2008 *J. Am. Chem. Soc.* **130** 3296
[2] Ren Z A 2008 *Mater. Res. Innovat.* **12** 105
[3] Mazin I I, Singh DJ, Johannes M D and Du MH 2008 *Phys. Rev. Lett.* **101** 057003
[4] Yildirim T 2008 *Phys. Rev. Lett.* **101** 057010.
[5]Gadermaier C, Alexandrov A S, Kabanov V V, Kusar P, Mertelj T, Yao X, Manzoni C, Brida D, Cerullo G and Mihailovic D 2010 *Phys.Rev.Lett.* **105** 257001
[6] Mansart B, Boschetto D, Savoia A, Rullier-Albenque F, Bouquet F, Papalazarou E, Forget A, Colson D, Rousse A, and Marsi M, 2010 *Phys Rev. B* **82**, 024513
[7] Chia E E, Talbayev D, Jian-Xin Zhu, Yuan H Q, Park T, Thompson J D, Panagopoulos C, Chen G F, Luo J L, Wang N L, and Taylor A J, 2010 *Phys. Rev. Lett.* **104**, 027003
[8] Mertelj T, Kusar P, Kabanov V V, Stojchevska L, Zhigadlo N D, Katrych S, Bukowski Z, Karpinski J, Weyeneth S, and Mihailovic D, 2010 *Phys. Rev. B* **81**, 224504
[9] Bellingeri E, Pallecchi I, Buzio R, Gerbi A, Marrè D, Cimberle M R, Tropeano M, Putti M, Palenzona A, and Ferdeghini C, 2010 *Appl. Phys. Lett.* **96**, 102512
[10] Kurth F, Iida K, Trommler S,Hanisch J, Nenkov K, Engelmann J, Oswald S, Werner J,Schultz L, Holzapfel B and Haindl S 2013 *Supercond. Sci. Technol.* **26** 025014
[11] Pepe G P, Parlato L, Marrocco N, Pagliarulo V, Peluso G, Barone A, Tafuri F, Scotti di Uccio U, Miletto F, Radovic M, Pan D, Sobolewski R, 2009 *Cryogenics* **49**, 660
[12] Demsar J, Podobnik B, Kabanov V V, Wolf T and Mihailovic D, 1999 *Phys. Rev. Lett.* **82**, 4918
[13] Sales B C, Sefat A S, McGuire M A, Jin R Y, and Mandrus D, 2009 *Phys. Rev. B* **79**, 094521
[14] Subedi A, Zhang L, Singh D J and Du M H, 2008 *Phys. Rev. B* **78**, 134514
[15] Allen P B, 1987 *Phys.Rev. Lett* **59**, 1460
[16] Bonavolontà C, Parlato L, Pepe G P, de Lisio C, Valentino M, Bellingeri E, Pallecchi I, Putti M and Ferdeghini C, 2013 *Supercond. Sci. Technol.* **26**, 075018.
[17] Mertelj T, Kabanov VV, Gadermaier C, Zhigadlo N D, Katrych S, Karpinski J and Mihailovic D, 2009 *Phys. Rev. Lett.* **102** 117002
[18] Torchinsky D H, Chen G F, Luo J L, Wang N L and Gedik N 2010 *Phys.Rev.Lett.* **105** 027005
[19] Stojchevska L, Mertelj T, Jiun-Haw Chu, Fisher I R, Mihailovic D, 2012 *Phys. Rev. B* **86**, 024519
[20] Kumar S,Harnagea L,Wurmehl S, Buchner B and Sood A K, 2013 **arXiv:1209.3551v2**
[21] Ummarino G A, Tortello M, Daghero D, Gonnelli R S, 2009 *Phys. Rev. B* **80**, 172503

Epitaxial growth of $YBa_2Cu_3O_{7-\delta}$ films on $SrTiO_3$ (100) by direct solution precursor deposition

A. Bustamante[1], Ana M. Osorio[2], Luis De Los Santos Valladares[3,4,5], Jorge Garcia[1], J.C. González[6], C.H.W. Barnes[3], Y. Azuma[4], Y. Majima[4], J. Albino Aguiar[5].

[1] *Laboratorio de Cerámicos y Nanomateriales, Facultad de Ciencias Físicas, Universidad Nacional Mayor de San Marcos, Apartado Postal 14-0149, Lima 14, Perú.*

[2] *Facultad de Química e Ingeniería Química, Universidad Nacional Mayor de San Marcos, Perú.*

[3] *Cavendish Laboratory, Department of Physics, University of Cambridge, J.J. Thomson Av., Cambridge CB3 OHE, United Kingdom.*

[4] *Materials and Structures Laboratory, Tokyo Institute of Technology, 4259 R3-5, Nagatsuta cho, Midori- ku, Yokohama 226-8503, Japan.*

[5] *Laboratório de Supercondutividade e Materiais Avançados, Departamento de Física Universidade Federal de Pernambuco 50670-901, Recife, Brasil.*

[6] *Grupo de Investigación de Superficies, Intercaras y Láminas Delgadas, Instituto de Ciencia de Materiales de Sevilla – CSIC – Univ. Sevilla. Calle Américo Vespucio 49, Isla de la Cartuja.41092 Seville - Spain.*

Abstract

We study the optimal temperature to obtain $YBa_2Cu_3O_{7-\delta}$ epitaxial films grown onto $SrTiO_3$ substrates by direct solution deposition. The samples received heat treatment at 820, 840 and 860 °C, then characterized by XRD, observing the (00l) profiles; and magnetic susceptibility measurements. The $T_{C\text{-onset}}$ for all the samples was 90 K. In addition, the current - voltage (I-V) measurements shows typical tunneling signals corresponding to normal metal-superconducting junctions indicating the films are promising for potential electrical applications.

Keywords: YBCO films; epitaxial growth films; thermal treatment of superconductors.

1. Introduction

During the past two decades, there has been great interest in developing superconducting $YBa_2Cu_3O_7$ (YBCO) films at low sintering temperatures to be used in the second generation superconducting tapes. YBCO epitaxial layers can carry high current densities in the order of 10^6A-cm^{-2} [1]. Chemical solution deposition (ChSD) is a cheap technique to growth YBCO epitaxial films. In a previous work we reported YBCO films grown onto $LaAlO_3$ substrates by ChSD [2]. In this work we study YBCO epitaxial films by XRD, magnetic measurements and I-V curves.

Journal of Physics: Conference Series **507** (2014) 012005 doi:10.1088/1742-6596/507/1/012005

2. Experimental

Acetates of yttrium, barium and copper were mixed and dissolved in an ethanol solution (C_2H_5OH) and distilled water. It was decanted for 12 hours in an oxalic acid solution with ethanol under magnetic stirring at 250 rpm. Thereafter it was dripped carefully on $SrTiO_3$ (100) substrates. Three samples were heated and sintered at 820, 840 and 860 °C for 36 hours without and with flow of oxygen respectively. The heating and cooling rate were 1 °C/min and1.4 °C/min, respectively. Surface morphology images were obtained by Nicon optical microscope. X-ray diffractions (XRD) were performed in a Bruker D8 diffractometer using Cu Kα1 radiation. The magnetic moment curves were obtained by DC Magnetic Property Measurement System (DC-MPMS-SQUID) from Quantum Design. The data were collected in Zero Field Cooling (ZFC) and Field Cooling (FC) modes. For the electrical characterization, two identical gold strips were evaporated on the surfaces of the samples. The strips were rectangular, 50 nm thickness, 100 µm width and they were separated by 10 µm. The current–voltage (*I-V*) measurements were performed in a prober station Grail10-205-LV Nagase Techno-Engineering Co., LTD at room temperature (RT) and in high vacuum 8.6×10^{-3} Pa. Two Be-Cu electrodes (source (S) and drain (D)) touched the center of each gold strips. The current response was sensed in the drain electrode after applying source-drain bias in the range $0 < V_D < +4$ V.

3. Results and discussions

Figure 1 shows the surface morphology. For the sample annealed at 820°C, there is a coating faulty consisting of grain agglomeration and group of bright spots which are probably produced by the reflections of light clusters grains scattered on the surface. Also at 840° C, grain agglomerates are formed on the surface and the substrate surface is better filled. However, at 860 °C the agglomerated grains cover homogeneously the surface of the substrate than the other cases and no bright spots were observed.

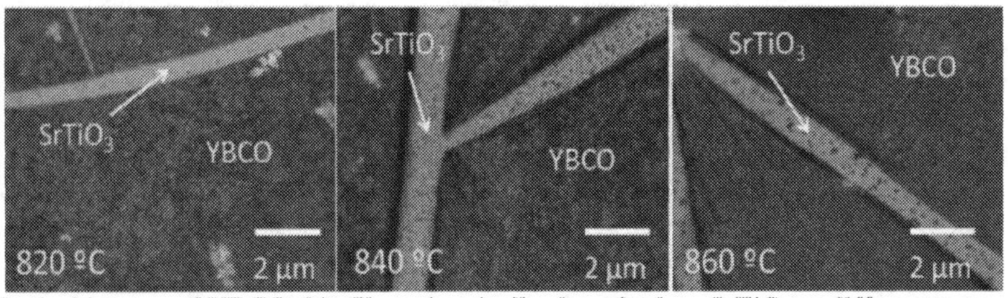

Fig. 1. Optical images of YBCO thin films chemically deposited on $SrTiO_3$ at different temperatures

Figure 2 shows the X-ray diffractograms at semi-logarithmic scale. At 820 °C, we observed mainly the profiles (100) (200) and (300) corresponding to substrate, also the peaks (002), (003) (004) (005) (006) and (007), indicating mainly the YBCO epitaxial growth. The degree of epitaxy (*fc*), was calculated according to: $I^{005}/I^{200} = fc.\, I^{005}_{standard} / (1\text{-}fc).\, I^{200}_{standard}$. At 820 °C was $fc = 0.02$, indicating poor epitaxial growth. In addition, peaks belonging to the Y_2BaCuO_5 are also detected, as indicated in Table 1. There are also profiles corresponding to Y_2O_3 (see legend of Fig. 2) indicating some unreacted precursors. At 840 °C, $fc = 0.61$, we detected the presence of Y_2BaCuO_5, $BaCuO_2$ and Y_2O_3 phases. Whereas at 860 °C, $fc = 0.64$ correlated with a high intensity of YBCO (*00l*) profiles, revealing a good epitaxy

Table 1. Secondary phases and *fc* values.

Sintering temperature	Secondary phases	*fc*
820 °C	Y_2BaCuO_5	0.02
840 °C	Y_2BaCuO_5, $BaCuO_2$	0.61
860 °C	Y_2BaCuO_5, Cu_2BaO_2	0.64

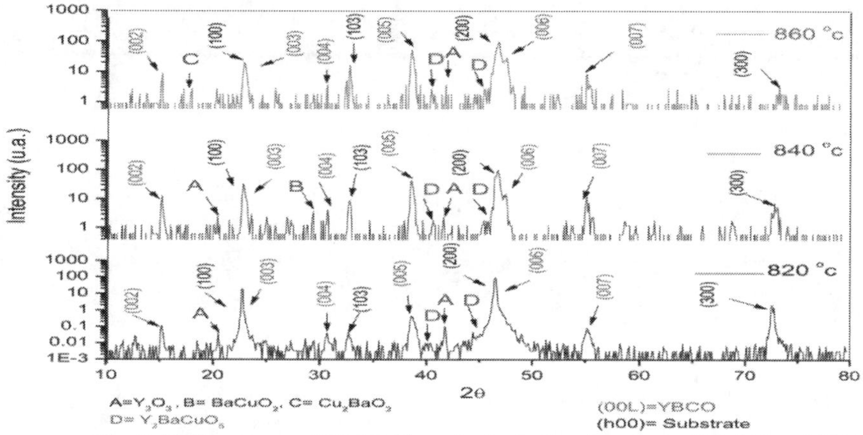

Fig. 2. XRD patterns of YBCO films obtained after sintering at 820, 840 and 860 °C.

Figure 3 shows the magnetic susceptibility curves. Despite the fact that secondary phases have been detected in all the samples (see Table 1), the Meissner effect is clearly observed indicating the superconducting state. At 820 °C, weak signals of the ZFC and FC diamagnetic branches are observed because a poor epitaxy. The ZFC and FC signals are improved when the film is treated at 840 °C. Nevertheless, at 860 °C, there is a great improvement of the ZFC and FC signals because the grains are mainly oriented according to the *c*-axis than the other cases.

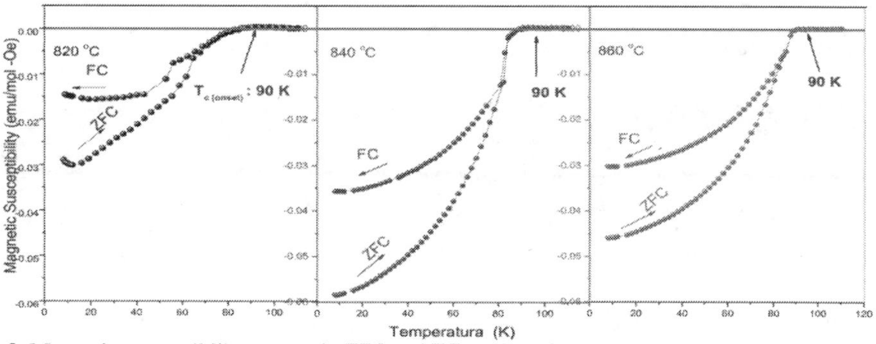

Fig. 3. Magnetic susceptibility curves in ZFC and FC modes with external magnetic field of 300 Oe.

The current - voltage measurements performed at different temperatures for the film obtained at 840 °C are shown in Figure 4. Above the T_C the typical semiconducting tunnelling signal is observed. However, below T_C the signal shows small current flow for $0 < V_D < 2.5$ V because the low energy is not high enough to make the electrons to tunnel to empty target levels, whereas, for $V_D > 3$V the electrons can tunnel from the conduction band of the normal metal (gold) to the empty states above the gap of the superconductor [3].

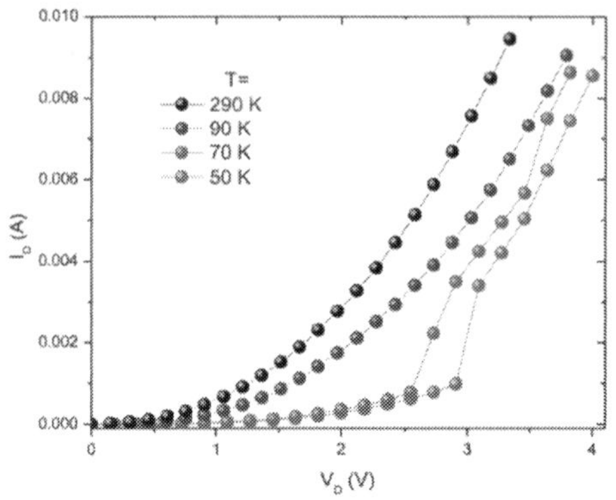

Fig. 4. I-V characteristics of YBCO films obtained at 840 °C

4. Conclusions

We achieved epitaxial growth of YBCO superconductor layers at 820, 840 and 860 °C temperatures. We found the $BaCuO_2$, Cu_2BaO_2 and Y_2BaCuO_5 secondary phases coexisting with the superconducting film. The morphology of the film and susceptibility curves are improved with the sintering temperature, which is intimate related to the increasing of the epitaxy with the temperature. From these three studied temperatures, the most optimal for growing epitaxial YBCO films on $SrTiO_3$ was 860 °C. The current - voltage measurements shows typical tunneling signals corresponding to normal metal-superconducting junctions indicating the films are promising for potential electrical applications.

Acknowledgements

This work was supported in part by the UNMSM through the grant program for research Con-Con (Projects 111301071 and 090701011) for the year 2011. The work in Cambridge has been supported by the Engineering and Physical Science Research Council (EPSRC). The work in Japan was supported by the Japan Society for the Promotion of Science (JSPS). The work in Brazil is funded by CAPES (PNPD/2011) and FACEPE (APQ-0589/1.05-08). J.C. Gonzalez thanks the CSIC (Spain) for his contract JAE-Doc (2009-2012) at the Institute of Materials Science of Seville - CSIC-Univ. of Seville. ABD is grateful to CONCYTEC for funding in the publication.

References

[1] High performance YBCO-coated Superconductors wires. 2004 MRS Bulletin Vol 29 8
[2] Bustamante A. et al 2013. Physics Procedia **36** 256 – 531
[3] Poole C et al 2007 Superconductivity, second edition, Academic Press, ISBN: 978-0-12-088761-3

Preparation of Nb_3Al by high-energy ball milling and superconductivity

Yongliang Chen[1]*, Zhao Liu[1], Pingyuan Li[1], Xiaolan Zhang[1], Suhua Yang[1], Dawei Yang[1], Lupeng Du[1], Yajing Cui[1], Xifeng Pan[2], Guo Yan[2] and Yong Zhao[1,3]

[1] Key Laboratory of Magnetic Levitation Technologies and Maglev Trains, Ministry of Education of China, and Superconductivity and New Energy R&D Center, Southwest Jiaotong University, Chengdu 610031, China

[2] Western Superconducting Technoligies Co., Ltd., Xi'an 710018, China

[3] School of Materials Science and Engineering, University of New South Wales, Sydney 2052 NSW, Australia

E-mail: *llchen@home.swjtu.edu.cn

Abstract. The A15 phase superconductor Nb_3Al has been considered as an alternative to Nb_3Sn for high field and large scale applications. However, to prepare a stoichiometric Nb_3Al with fine grain structures is very difficult. High-energy ball milling is a solid state powder processing technique and is a very useful for preparing Nb-Al alloys (Nb_3Al). The effects of ball milling time and annealing temperature on the formation of Nb_3Al superconducting phase have been studied. Pure Nb and Al powders with stoichiometric ratio of Nb_3Al were mixed and milled, and the charging and milling were performed in an inert atmosphere. Phase formation and structural evolution during high-energy ball milling have been examined by X-ray diffraction. Al disappeared and Nb peaks broadened after about one hour of milling. With increasing milling time, the peaks of Nb became considerably broader and intensities decreased, the Nb-Al solid solution phase was extensive when milled about 3 hours. In order to obtain Nb_3Al superconducting phase, a subsequent anneal was required. We have annealed the as-milled powders at 800-900°C for different times to prepared Nb_3Al superconducting alloy. The results indicated that Nb_3Al with small amount of impurity phase can be obtained on annealing the Nb-Al solid solution phase and the superconducting transition temperature was about 15K, but it is difficult to obtain a homogeneous Nb_3Al phase by annealing the amorphous powder.

1. Introduction

Because of high critical current density in high field and excellent strain/stress tolerance, Nb_3Al superconducting alloy has attracted much interest in experiment studies and device manufacture. Enormous efforts have intensively focused on the preparations and properties of different form (or structure) Nb_3Al conductor and great progress has been made. Up to now, Nb_3Al superconductor can be achieved by many methods, such as powder metallurgy [1, 2], powder-in-tube [3], jelly-roll [4, 5], clad chip extrusion [6], rod-in-tube [7, 8], laser or electron beam irradiation [9, 10] and the rapid-heating, quenching and transformation method (RHQT) [11-13]. According to the phase diagram of the Nb-Al system, the A15 phase is formed by a peritectic reaction at about 2060 °C and shifted to lower Al concentrations with decreasing temperature. So the quenching from high temperature to retain Al in composite is a good technology, and a bcc supersaturated solid solution forms in this process. Post-annealing is necessary to obtain stoichiometric Nb_3Al [14]. An excellent Nb_3Al conductor with multifilamentary structure can be obtained by the RHQT method. However, the RHQT techniques require specialized instruments to provide high reaction temperature and there exist some problems need to be solved.

High-energy ball milling or mechanical alloying is a very effective route in preparing nanosized solid solution or amorphous materials, including metallic alloys, intermetallics, composites and ceramics. It occurs by successive flattening, welding, fracturing and rewelding of powder particles in a high-energy ball mill [15]. High-energy ball milling has been proved to be a useful technique for preparing Nb-Al alloys [16-21]. $NbAl_3$ compound can be obtained directly by mechanical alloying, however, post-annealing is necessary to obtain Nb_3Al superconducting phase [19, 20]. In this work we investigate the effects of ball milling time and annealing temperature on the formation of Nb_3Al superconducting phase.

2. Experimental details

Pure elemental Nb (purity 99.99%, 325 mesh) and Al (purity 99.5% 325 mesh) powders were used as starting materials. Two grams of powder mixture was loaded in the vial with a weight ratio of powder to ball =1:10. The charging of powder was performed in an argon-filled glove box. A SPEX Mixer-Mill (Model 8000) with a stainless steel vial and balls were used. Mixtures with the nominal composition of Nb_3Al were high-energy ball milled in an argon environment. Samples were not exposed to the air until milling has finished. After milling, the mixtures were pressed into round pellet with a diameter of 10 mm and a thickness of 1.5 mm using hardened steel die with uniaxial 12 MPa pressure. Then, the pellet was sealed into a vacuum quartz tube. Finally, the samples were heat treated in a tube furnace and argon was vented in case a permeability or leakage occurred.

The crystal structure was studied by powder X-ray diffraction (XRD) using an X'Pert MRD diffractometer with Cu Kα radiation. All observed reflections were indexed. Lattice constants were determined from LeBail refinements. Microstructure and composition of the sample were analyzed using a field emission scanning electron microscope (FESEM) equipped with an energy dispersive X-ray analysis (EDX). DC magnetization was measured with a SQUID magnetometer (MPMS, Quantum Design) and resistivity measurements were performed with a physical property measurement system (PPMS, Quantum Design).

3. Results and discussion

Figure 1 shows the X-ray diffraction patterns of powder mixture milled for different times. After 1 hours of milling, the peaks of aluminum were barely detectable and only the Nb was present. When the milling time is 3 hours, disappearance of Al peaks and broadening of Nb peaks are apparent. With increasing milling time, the peaks became broader and their intensities decreased and peak shift towards to greater diffraction angles after 5 hours of milling. The disappearance of Al peaks can be considered as dissolution of Al in Nb. The SEM images of sample milled for different hours (not shown here) indicated that the particle size decreased gradually with the increasing of milling time. Decrease of the particle size leads to the increase of surface area.

Figure 2 shows the XRD patterns of the samples milled and annealed for different times. The XRD pattern of sample milled for 3h and annealed for 10h at 800°C was shown in Figure 2(a), the Nb is the dominant phase which can be clearly observed. When annealing temperature increases to 900°C, as shown in Figure 2(b), Nb_3Al phase formed and Nb was observed also. Figure 2(c) shows the XRD patterns of the sample milled for 5h and annealed for 10h at 900°C. the sample is detected to be Nb_3Al phase with small amount of impurity phase and no Nb phase can be observed. Figure 2(d) and 2(e) show the XRD patterns of the sample milled for 3h and 5h respectively and annealed for 20h at 900°C. The better Nb_3Al phase can be obtained as show in Figure 2(d).

The morphology of sample milled for 3h and annealed at 900°C is show in Figure 3. The sample is brittle and the particle size is about 100nm~400nm. Figure 4(a)-4(d) show the temperature dependence

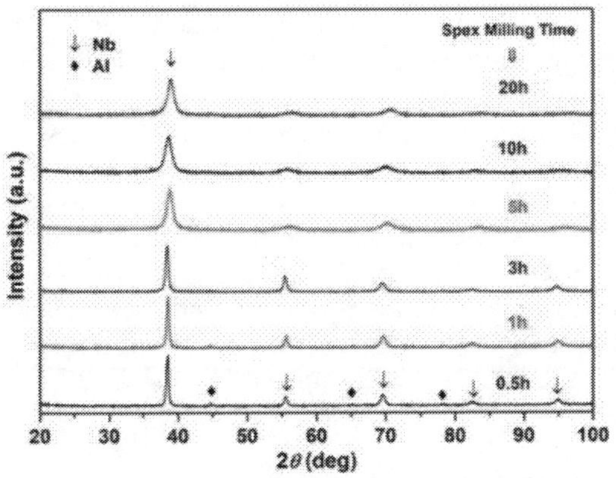

Figure 1. The XRD patterns of the Nb-Al powder mixture milled for different hours.

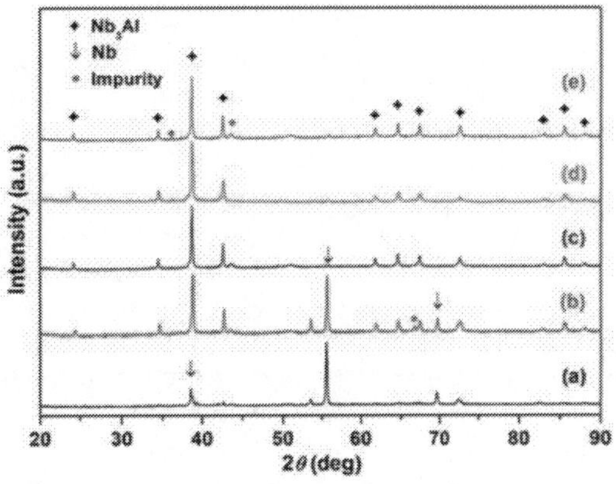

Figure 2. The XRD patterns of the samples milled and annealed for different time.

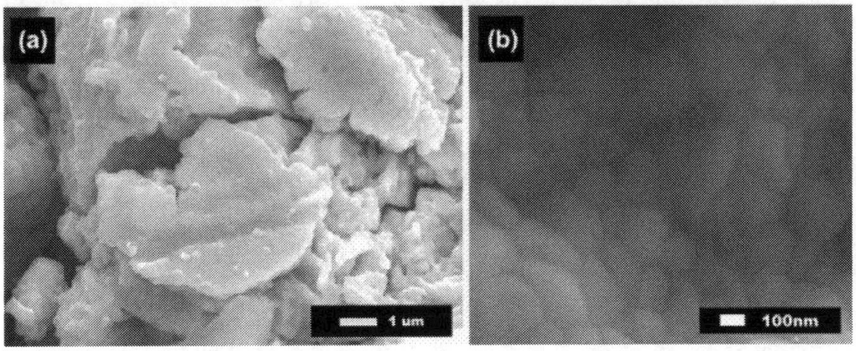

Figure 3. (a) Typical SEM images of the samples milled for 3h, annealed at 900°C, (b) shows enlarged view.

of magnetization for the sample milled for 3 h and 5h, annealed at 900°C for 20h and 40h respectively, and the highest T_c is 14.3 K for sample milled for 3 h and annealed at 900°C for 20h. The increase of milling time and annealing time will decrease the T_c value. EDX analysis indicates that sample compositions deviate from the stoichiometry of Nb_3Al. The inset of Figure 3 shows the temperature dependence of magnetization for two Nb-26at.%Al samples milled for 3h and annealed for 5h at 900°C. The values of T_c are 15K, which is about 3K lower than the optimal T_c value (18K) of stoichiometric Nb_3Al. Lower T_c value arises from the deviation of Nb_3Al phase compositions from

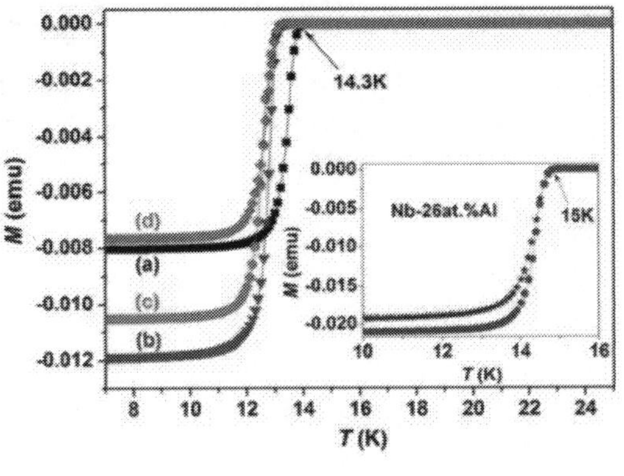

Figure 4. The temperature dependence of magnetization for the Nb_3Al sample annealed at 900°C, (a) milled for 3h and annealed for 20h, (b) milled for 3h and annealed for 40h, (c) milled for 5h and annealed for 20h, (d) milled for 5h and annealed for 40h, inset shows the temperature dependence of magnetization for the Nb-26at.%Al sample milled for 3h and annealed for 5h at 900°C.

A15 stoichiometry. Another reason is the impurity phases. Although little impurity phase was indexed from the XRD patterns, the impurities (Fe or other magnetic impurities) introduced by milling would strongly destroy the superconducting properties. It should be emphasized that, no superconducting transition of Nb was detected which further confirms that the Nb has sufficiently reacted.

4. Summary

The effects of ball milling time and annealing temperature on the formation of Nb_3Al superconducting phase have been investigated. High-energy ball milling is effective in increasing activity of the powders by improving the mixing of the powders, simultaneously reducing the diffusion distance. The impurities are probably responsible for the poor-quality of superconducting properties. The achievement of stoichiometry is essential for producing high T_c for Nb_3Al superconductor.

Acknowledgments

The authors are grateful for the financial support of the National Magnetic Confinement Fusion Science Program (Grant No. 2011GB112001), Program of International S&T Cooperation (Grant No. 2013DFA51050), the National Natural Science Foundation of China (Grant No. 11104224, 11004162, 51271155), the Research Fund for the Doctoral Program of Higher Education of China (Grant No. 20110184120029), the Fundamental Research Funds for the Central Universities (Grant No. 2682013ZT16, SWJTU11ZT31, 2682013CX004, SWJTU11BR063), the Science Foundation of Sichuan Province (Grant No. 2011JY0031, 2011JY0130). Y.L. Chen is also grateful for the financial support of Li Jinsheng scholarship of Superconductivity and New Energy R&D Center.

References

[1] Lohberg R, Eager T W, Puffer I M and Rose R M 1973 *Appl. Phys. Lett.* **22** 69
[2] Watanabe K, Noto K, Morita H, Fujimori H and Muto Y 1989 *IEEE Trans. Magn.* **25** 1984
[3] Akihama R, Murphy R J and Foner S 1980 *Appl. Phys. Lett.* **37** 1107
[4] Bruzzese R, Ricci M, Spadoni M, Sacchetti N and Ceresara S 1994 *Curr. Topics Magn. Res.* **1** 191
[5] Yamada Y, Ayai N, Takahashi K, Sato K, Sugimoto M, Ando T, Takahashi Y and Nishi M 1994 *Adv. Cryog. Eng.* **40** 907
[6] Saito S, Ikeda K, Ikeda S, Nagata A and Noto K 1989 *Proc. 11th Int. Conf. Magnet Technology (Tsukuba)* (Amsterdam: Elsevier) p 974
[7] Inoue K, Iijima Y and Takeuchi T 1988 *Appl. Phys. Lett.* **52** 1724
[8] Takeuchi T, Iijima Y, Kosuge M, Kuroda T, Yuyama M and Inoue K 1989 *IEEE Trans. Magn.* **25** 2068
[9] Kumakura H, Togano K, Tachikawa K, Tsukamoto S and Irie H 1986 *Appl. Phys. Lett.* **49** 46
[10] Schaper W, Kehlenbeck M, Zimmermann I and Freyhardt H C 1989 *IEEE Trans. Magn.* **25** 1988
[11] Iijima Y, Kosuge M, Takeuchi T and Inoue K 1994 *Adv. Cryog. Eng.* **40** 899
[12] Takeuchi T, Iijima Y, Inoue K, Wada H, ten Haken B, ten Kate H J, Fukuda K, Iwaki G, Sakai S and Moriai H 1997 *Appl. Phys. Lett.* **71** 122
[13] Takeuchi T, Tagawa K, Kiyoshi T, Itoh K, Kosuge M, Yuyama M, Wada H, Iijima Y, Inoue K, Nakagawa K, Iwaki G and Moriai H 1999 *IEEE Trans. Appl. Supercond.* **9** 2682
[14] Thieme C L H, Pourrahimi S and Foner S 1989 *IEEE Trans. Magn.* **25** 1992
[15] Suryanarayana C, 2001 *Prog. Mater Sci.* **46** 1
[16] Hellstern E, Schultz L, Bormann R and Lee D 1988 *Appl. Phys. Lett.* **53** 1399
[17] Tracy M J and Groza J R 1992 *Nanostruct. Mater.* **1** 369
[18] Oleszak D, Burzynska-Szyszko M and Matyja H 1993 *J. Mater. Sci. Lett.* **12** 3
[19] Peng Z, Suryanarayana C and Froes F H 1996 *Metall. Mater. Trans. A* **27A** 41
[20] Yoo D J, Hwang S M and Lee S M 2000 *J. Mater. Sci. Lett.* **19** 1327
[21] Mostaan H, Karimzadeh F and Abbasi M H 2012 *Powder Metall.* **55** 142

Journal of Physics: Conference Series **507** (2014) 012007 doi:10.1088/1742-6596/507/1/012007

DC to AC converter on Abrikosov vortices in a washboard pinning potential

Valerij A Shklovskij[1,2] and Oleksandr V Dobrovolskiy[2,3]

[1]Institute of Theoretical Physics, NSC-KIPT, 61108 Kharkiv, Ukraine
[2]Physical Department, Kharkiv National University, 61077 Kharkiv, Ukraine
[3]Physikalisches Institut, Goethe-University, 60438 Frankfurt am Main, Germany

E-mail: `Shklovskij@gmail.com`, `Dobrovolskiy@Physik.uni-frankfurt.de`

Abstract. The nonlinear dynamics of Abrikosov vortices in a cosine dc-biased washboard pinning potential at *nonzero* temperature is theoretically investigated. The problem is treated relying upon the *exact* solution of the Langevin equation for non-interacting vortices by using the Fokker-Planck method combined with the scalar continued fractions technique. The time variation of the local mean vortex velocity $v(t)$ is calculated. The time voltage $E(t) \propto v(t)$ is predicted to oscillate with a dc current-dependent frequency and a tunable pulse shape. Formulas for $v(t)$ are discussed as functions of dc transport current and temperature, in a wide range of the corresponding dimensionless parameters. The derived expressions can be adapted for a number of physical applications utilizing the overdamped motion of a Brownian particle in a tilted cosine potential, e.g., the resistively shunted Josephson junction model.

1. Introduction

Described by the Langevin equation or by the equivalent Fokker-Planck equation, the problem of Brownian motion in a periodic potential arises in different fields of science and engineering, such as solid-state physics, chemical physics, and communication theory. Exemplary systems include Josephson junctions (JJs) [1], superionic conductors [2], ring laser gyroscopes [3], charge density waves [4], phase-locking loops [7] in radioengineering, the magnetization dynamics [6], the diffusion of colloidal particles in periodic structures [5], and the dynamics of Abrikosov vortices in superconductors [8, 9, 10]. The ubiquitous appearance of *pulsed* signal in response to *constant* drive in the presence of a periodic potential is generic to all these domains. For instance, the appearance of the ac voltage in the resistively shunted JJ problem at *zero temperature* was theoretically analyzed as early as forty years ago [1]. However, at *non-zero temperature* this problem has not been solved in either domain so far.

The objective of this work is to report the results of a theoretical treatment of the *time-dependent* nonlinear stochastic dynamics of the Abrikosov vortex in a tilted cosine washboard pinning potential (WPP). We will show that for the cosine WPP, in the approximation of non-interacting vortices, it is possible to derive *exact* expressions for the ac voltage arising in response to a dc transport current. The predicted results can be experimentally examined, e.g., on superconductor thin films with a WPP landscape [11], e.g., similar to that shown in figure 1(a). Besides, the derived expressions can be adapted for many of the aforementioned research domains.

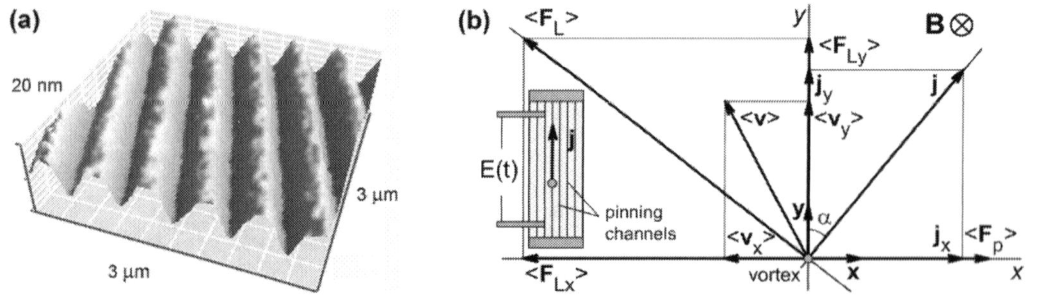

Figure 1. (a) Atomic force microscope image of an exemplary WPP landscape fabricated by focused ion beam milling [11] on the surface of a Nb thin film. (b) The system of coordinates xy with the unit vectors $\mathbf{x}$ and $\mathbf{y}$ is associated with the WPP channels which are parallel to the vector $\mathbf{y}$. In the general case, the transport current density $\mathbf{j}$ is tilted at an angle α with respect to $\mathbf{y}$. $\mathbf{F}_p$ is the average pinning force provided by the WPP, $\mathbf{F}_L$ is the Lorentz force for a vortex, and $\mathbf{B}$ is the magnetic field vector directed perpendicular to the film plane. A typical layout of contacts in the thin-film geometry is shown in the inset.

2. Main results

The geometry of the considered system is sketched in figure 1(*b*). Our objective is to calculate the time voltage arising due to vortices on move under the action of the dc transport current $\mathbf{j}$. We treat that system on the basis of the Langevin equation for a vortex moving with an actual stochastic velocity $\mathbf{v}$ in a magnetic field $\mathbf{B} = \mathbf{n}B$. Here $B \equiv |\mathbf{B}|$, $\mathbf{n} = n\mathbf{z}$, $n = \pm 1$, and $\mathbf{z}$ is the unit vector in the z direction. Neglecting the Hall effect the Langevin equation reads [8, 9, 10]

$$\eta \mathbf{v} = \mathbf{F}_L + \mathbf{F}_p + \mathbf{F}_{th}, \tag{1}$$

where $\mathbf{F}_L = n(\Phi_0/c)\mathbf{j} \times \mathbf{z}$ is the Lorentz force, Φ_0 is the magnetic flux quantum, c is the speed of light, and $\mathbf{j}$ is the dc transport current density. $\mathbf{F}_p = -\nabla U_p(x)$ is the anisotropic pinning force, where $U_p(x)$ is a WPP. As usually [9, 10], the WPP is modeled by $U_p(x) = (U_p/2)(1 - \cos kx)/2$, where $k = 2\pi/a$, a is the period, and U_p is its depth. $\mathbf{F}_{th}$ is the thermal fluctuation force represented by a Gaussian white noise and η is the vortex viscosity constant.

In the general case, equation (1) can be reduced to the system of Fokker-Planck equations [7] for the distribution function $P = P(\mathbf{r}, t)$ associated with the probability density of finding the vortex at the point $\mathbf{r} = \mathbf{r}(x, y)$ at the time t. This stationary distribution function allows one to calculate the probability flux density of the vortex $\mathbf{S}(\mathbf{r}, t) \equiv P(\mathbf{r}, t)\mathbf{v}(\mathbf{r}, t) = const$ and to derive expressions for the experimentally observable ac voltage in terms of a continued fraction. For the sake of simplicity, in the following we consider the one-dimensional vortex motion along the x axis, i.e., when $\mathbf{j}$ flows along the WPP channels as in the inset of figure 1(*b*). Then, the stationary dimensionless Fokker-Planck equation reads [12]

$$S = P(\xi - \sin x) - (1/g)(dP/dx), \tag{2}$$

where $S = P(x)v(x)$, $\xi = F_{Lx}/F_p$ is the dimensionless dc driving force, $kx \to x$ is the new dimensionless coordinate, and $g = U_p/2T$ is the dimensionless inverse temperature. In the right part of equation (2) the first term stands for the deterministic vortex drift in the direction of the driving force, whereas the second term is the contribution due to the thermal fluctuations.

The solution of equation (2) can be found in terms of an ordinary continued fraction as follows. First, we decompose the probability function $P(x)$ into a series of Fourier amplitudes c_m as $P(x) = \sum_{m=-\infty}^{\infty} c_m e^{imx}$, where $c_m = c^*_{-m}$ and the asterisk denotes the complex conjugate. Then, equation (2) can be reduced to the following tridiagonal recurrence equation

$$(\xi - im/g)c_m + (i/2)c_{m-1} - (i/2)c_{m+1} = S\delta_{m0}, \tag{3}$$

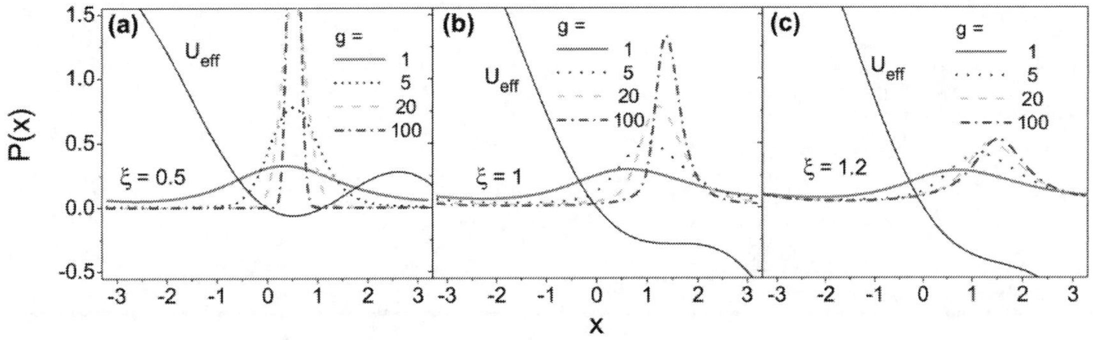

Figure 2. The coordinate dependence of the probability distribution function $P(x)$ calculated by equation (5) at (a) subcritical, (b) critical and (c) overcritical transport currents for a set of inverse temperatures, as indicated. At subcritical tilts, the maximum in $P(x)$ corresponds to the minimum in the effective pinning potential $U_{\text{eff}}(x) \equiv U_p(x) - F_{Lx}x$.

where δ_{m0} is Kronecker's delta and we have used the relation [9] $\langle \sin x \rangle = i(c_{m+1} - c_{m-1})/2$ with $\langle \ldots \rangle$ meaning the statistical averaging. For $m \geq 1$ equation (3) gives an infinite continued fraction for $c_1/c_0 = c^*_{-1}/c_0$. If one then uses this equation for $m = 0$ and the normalization condition for $P(x)$ in the form $\int_0^{2\pi} P(x)dx = 1$ corresponding to $c_0 = 1/2\pi$, it is possible to express $S_m \equiv 2\pi c_m$ as the following continued fraction

$$S_m = 2\frac{1/4}{i\xi + m/g + \cfrac{1/4}{i\xi + (m+1)/g + \cfrac{1/4}{i\xi + (m+2)/g + \ldots}}}, \tag{4}$$

From equations (3) and (4) follows the next expression for the stationary distribution function

$$P(x, g, \xi) = 1/2\pi + (1/\pi)\sum_{m=1}^{\infty} \text{Re}[e^{imx}(S_m \ldots S_1)], \tag{5}$$

whose graphs for various values of the parameters g and ξ are shown in figure 2. Using the *stationary* $P(x)$ one can treat the *time-dependent* periodic dynamics for a vortex moving in the cosine WPP under the influence of a dc drive. Due to the stationarity of the probability current $v(x) = S/P(x)$ we conclude that $t = \int_0^x dx'/v(x')$ or

$$t = \frac{1}{S}\int_0^x P(x')dx', \tag{6}$$

where the time t is scaled in units of the relaxation time $\hat{\tau} \equiv \eta/kF_p$.

From equation (6) one finds $t = t(x)$ and its inversion results in $x = x(t)$. Finally, we arrive at the main result of our study, $v(t) = v[x(t)]$, which determines the time variation of the voltage $E(t) \propto v(t)$. In equation (6), $0 < x < 2\pi$ and $0 < t < \theta$, where 2π and $\theta = 1/S$ are the dimensionless spatial and temporal periods of the vortex motion. The dependences $v(t)$ are plotted in figure 3 in the different dc current regimes for a wide range of g. Consider these curves in more detail. Depending on the bias value ξ the behavior of the curves differs substantially. In contrast to the well-known zero-temperature results in the resistively shunted JJ problem [1], accounting for the thermal fluctuations allows the vortex to overcome the WPP barrier at subcritical dc biases. Overcoming of each WPP barrier is accompanied by a voltage pulse, as depicted in figure 3(a). With increasing bias the time between two neighboring pulses becomes

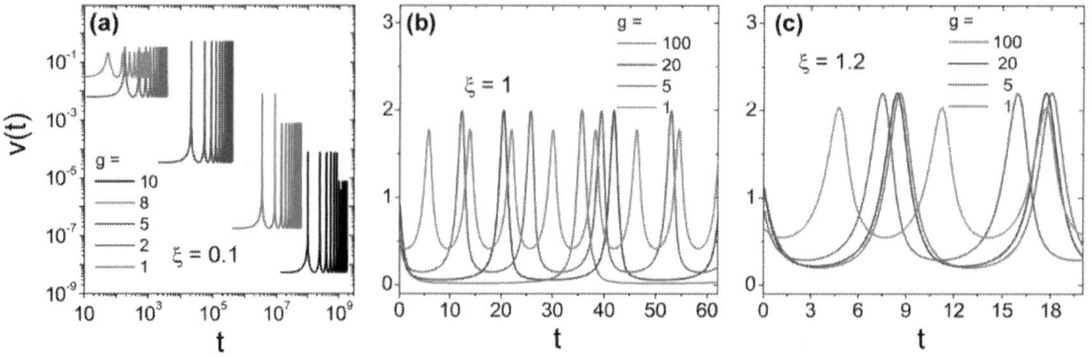

Figure 3. The time dependence of the mean local vortex velocity $v(t)$ at (a) small, (b) critical, and (c) overcritical dc currents for a set of inverse temperatures, as indicated. In (a) the curves are cut off to simplify reading the data.

shorter and a notable "voltage floor" appears in figure 3(b) with decreasing g. Averaging the time voltage over the oscillation period and plotting it versus ξ results in the conventional current-voltage characteristics [1]. With further increasing ξ the "voltage floor" is rising [figure 3(c)] and the pulses become harmonic in the limit of very strong dc currents.

3. Conclusion

In this work we have theoretically treated the nonlinear dynamics of Abrikosov vortices in a cosine dc-biased washboard pinning potential at *nonzero* temperature. The corresponding Langevin equation for non-interacting vortices has been solved by using the Fokker-Planck method combined with the scalar continued fractions technique. The main result of this work is the derived time variation of the local mean vortex velocity $v(t)$ responsible for the observable ac voltage $E(t)$. The latter is predicted to oscillate with a dc current-dependent frequency and a tunable pulse shape ranging from very short, delta function-like peaks in the limit of small dc current ξ to the well-known harmonic form in the limit of strong ξ. The reported model structure can be adapted for a number of physical applications (see Introduction) utilizing the overdamped motion of a Brownian particle in a tilted cosine potential.

Acknowledgments

OVD gratefully acknowledges financial support from the DFG through grant no. DO 1511/2-1.

References

[1] Barone A and Patterno G 1982 *Physics and Applications of the Josephson Effect* (John Wiley & Sons, New York)
[2] Fulde P, Pietronero L, Schneider W R and Strässler S 1975 *Phys. Rev. Lett.* **35**(26) 1776–1779
[3] Chow W W, Gea-Banacloche J, Pedrotti L M, Sanders V E, Schleich W and Scully M O 1985 *Rev. Mod. Phys.* **57**(1) 61–104
[4] Grüner G 1988 *Rev. Mod. Phys.* **60**(4) 1129–1181
[5] Evstigneev M, Zvyagolskaya O, Bleil S, Eichhorn R, Bechinger C and Reimann P 2008 *Phys. Rev. E* **77**(4) 041107
[6] Titov S V, Kachkachi H, Kalmykov Y P and Coffey W T 2005 *Phys. Rev. B* **72**(13) 134425
[7] Risken H 1989 *The Fokker-Planck Equation* (Springer, Berlin)
[8] Shklovskij V A and Dobrovolskiy O V 2006 *Phys. Rev. B* **74**(10) 104511–1–14
[9] Shklovskij V A and Dobrovolskiy O V 2008 *Phys. Rev. B* **78**(10) 104526–1–12
[10] Shklovskij V A and Dobrovolskiy O V 2011 *Phys. Rev. B* **84**(5) 054515–1–12
[11] Dobrovolskiy O V, Begun E, Huth M and Shklovskij V A 2012 *New J. Phys.* **14** 113027–1–27
[12] Shklovskij V A and Dobrovolskiy O V *to be published*

Influence of the spreading resistance on the conductance spectrum of planar hybrid thin film SNS' junctions based on iron pnictides

S Döring[1], S Schmidt[1], S Gottwals[1], F Schmidl[1], V Tympel[1], I Mönch[2], F Kurth[2], K Iida[2], B Holzapfel[2] and P Seidel[1]

[1] Friedrich-Schiller-University Jena, Helmholtzweg 5, 07743 Jena, Germany
[2] IFW Dresden, Helmholtzstr. 20, 01069 Dresden, Germany

E-mail: sebastian.doering.1@uni-jena.de

Abstract. To investigate the superconducting properties of iron pnictides we prepared planar hybrid SNS' junctions in thin film technology with a pnictide base electrode, a gold barrier layer and a lead counter electrode. Our design allows characterization of the electrodes and the junction independently in a 4-probe method. We show how both electrodes influence the measured spectra due to their spreading resistance. While the Pb electrode has a constant resistance above its T_c, the contribution of the pnictide electrode is clearly current-dependent and thus it needs a more advanced method to be corrected. We present an empirical method, which is simple to apply and allows to deal with the spreading resistance in our junctions to recalculate the actual conductance and voltage of one junction at given temperature.

1. Introduction

Within Andreev reflection studies the conductance of a prepared junction or a point contact is measured in dependence on the voltage. Usually the results are modelled within the BTK-theory [1] and its possible extensions like scattering [2, 3], anisotropic order parameter [4] or non-isotropic Fermi surfaces [5, 6]. To apply the BTK-model one has to normalize each conductance spectrum at a given temperature by a spectrum measured at $T \gtrsim T_c$. A problem, which occurs especially when thin films are used is the spreading resistance of the electrodes. This leads to a downshift of the conductance for higher temperatures, which avoids simple normalization and results in an overestimation of gap values from the spectra [7, 8]. Additionally, the exceeding of the critical current of one electrode causes dips in the spectra [9, 10]. In our former work [11, 12] we clearly obtained an influence of the electrode resistances. Thus, it is necessary to characterize both electrodes to subsequently identify the extent of influence on the junction spectra and correct the measured data within an appropriate model.

2. Measurements setup and results

For our junctions we use a photolithographic mask design which allows to measure and characterize each electrode and the junction, respectively, in a 4-probe technique independently from each other. The used contacts for each of these measurements are shown in table 1. The details of junction preparation can be found in [11].

Figure 1. Scanning electron microscope image of a planar SNS' cross type junction. The numbers mark the bonding pads. Their assignment to the electrodes are explained in table 1.

Table 1. Device parts corresponding to bonding pads numbers of figure 1.

setup	device part	current	voltage
I	122 electrode	1-5	2-6
II	Pb electrode	8-4	7-3
III	junction	1-4	2-3

To measure the junction we used setup III from table 1 to obtain the differential resistance over the bias current. The results are shown in figure 2. One can distinguish three different regimes depending on the corresponding temperature. For temperatures lower than the T_c of lead the spectrum shows a central dip with multiple shoulders for low currents. The resistance for high currents saturates at $1\,\Omega$. For increasing temperatures jumps in the resistance occur, which correspond to the critical current of the counter electrode. At 7.5 K the measured spectrum shows no jumps any more and it is shifted parallel by $\approx 1\,\Omega$ upwards. The general shape stays nearly unchanged except for features caused by the gap of lead in the low temperature measurements. This additional $1\,\Omega$ is due to the resistance of the lead counter electrode.

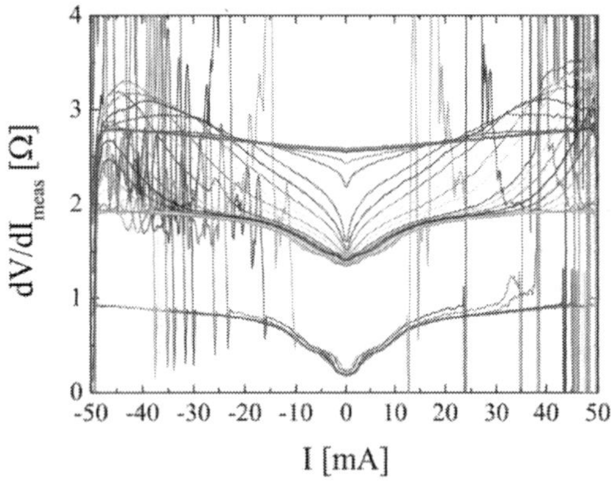

Figure 2. Differential resistance versus bias current of a junction at temperatures between 4.2 K and 23.0 K.

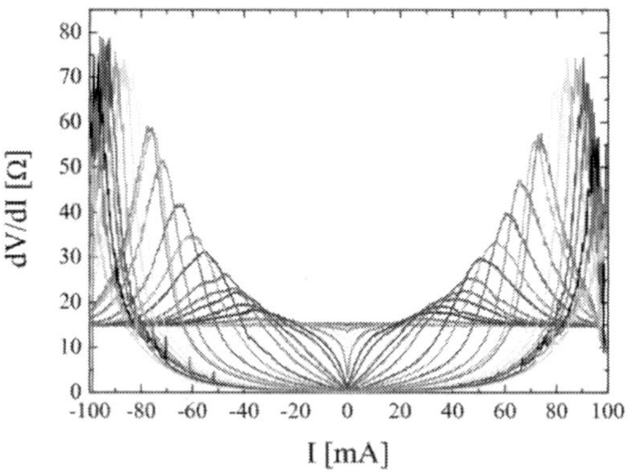

Figure 3. Differential resistance versus bias current of the Ba-122 base electrode of the same junction at temperatures between 4.2 K and 23.0 K.

By increasing the temperature further one can see an additional rise of the resistance for high currents. As can bee seen, the resistance in the normal state is about $2.8\,\Omega$. This means, the Ba-122 base electrode causes another $0.8\,\Omega$ resistance in the junctions spectrum. The rise in resistance increases, leading to peaks at special current values and for even higher currents a decrease of resistance, which closes asymptotically to the normal state value. Additionally, by

increasing temperature the shape around zero current changes from a "u-shaped" to more "v-shaped" or mathematically from a convex function to a concave one.
By switching to setup II of table 1 one can characterize the lead counter electrode. The differential resistance for $T \leq T_c$ shows zero resistance up to a critical current. At this critical current jumps occur at the same currents as the jumps in the junction's resistance when the measurement is done at the same temperature and the same bias direction. This is necessary because of the hysterical behaviour of the electrodes curve. For $I \geq I_c$ the differential resistance has a constant value, for a fixed $T > T_c$ it is current independent, too, but slightly increases with temperature.
On the same junction we can switch the electrical contacts to setup I of table 1 to measure the behaviour of the Ba-122 base electrode at each temperature used for the junction's characterization. A typical behaviour visible in figure 3 is the occurrence of peaks in the differential resistance, which are mainly caused by flux flow and pinning but also possibly influenced by iron buffer layers [13]. The height and the current position decrease with increasing temperature. One can identify these peaks in figure 3 with those in figure 2, because at given temperature the current position of one peak is the same in both figures. This fact together with the change from "u-shaped" to "v-shaped" current dependence of the differential resistance gives strong evidence of an influence of the Ba-122 electrode to the measured junction spectra. The resistance of the electrode increases from zero to $16\,\Omega$ when increasing the temperature from 4.2 K to 23 K.

3. Model

The correction algorithm we use is based on the one described in [10]. For the constant resistance of lead, this is just subtracted from the the measured junction resistance. In the same way the voltage of the junction is corrected by subtracting the Pb electrode's resistance times the bias current from the measured voltage. The value of this resistance can easily be obtained from figure 2 as the first up shift of $1\,\Omega$.
For the Ba-122 electrode in contrast to [10] we do not assume a constant spreading resistance but use its measured (current dependent) resistance characteristics. The extent of the influence $r_E := \Delta R_{junction}/\Delta R_{122}$ can be obtained by comparing the raise of the normal resistance with temperature in figure 2 and figure 3. As aforementioned, in the shown example $\Delta R_{junction} = 0.8\,\Omega$ while $\Delta R_{122} = 16\,\Omega$, thus giving $r_E = 5\,\%$. Now the actual voltage and resistance of the junction can be calculated by subtracting the measured resistance dependence of the Ba-122 electrode times r_E from the measured voltage and resistance, respectively for each temperature:

$$V_{corr} = V_{meas} - I \cdot R_{Pb} - r_E \cdot R_{122}(I) \cdot I \tag{1}$$

$$\left(\frac{dV}{dI}\right)_{corr} = \left(\frac{dV}{dI}\right)_{meas} - R_{Pb} - r_E \cdot R_{122}(I) \tag{2}$$

4. Corrected data

The analysis of a superconductor order parameter is performed in the presentation of differential conductance versus voltage as shown in figure 4a. One can see, that the conductance for high voltages at $T = 4.2\,\mathrm{K}$ goes to $1\,\Omega^{-1}$ corresponding to the resistance of $1\,\Omega$ in figure 2. For higher temperatures one can see the influence of the electrode's resistance by a downshift of the high voltage value lowering as well as broadening of the central conductance peak. While the lowering of the peak and the downshift are connected to a loss of conductance the broadening is due to additional voltage at the electrodes. By using equation (2) to correct the conductance and (1) to correct the voltage one gets the actual values of the junction at each temperature without the influence of the electrode. The obtained spectra are shown in figure 4b. By comparing both spectra before and after the correction one can see, that there is of course no difference

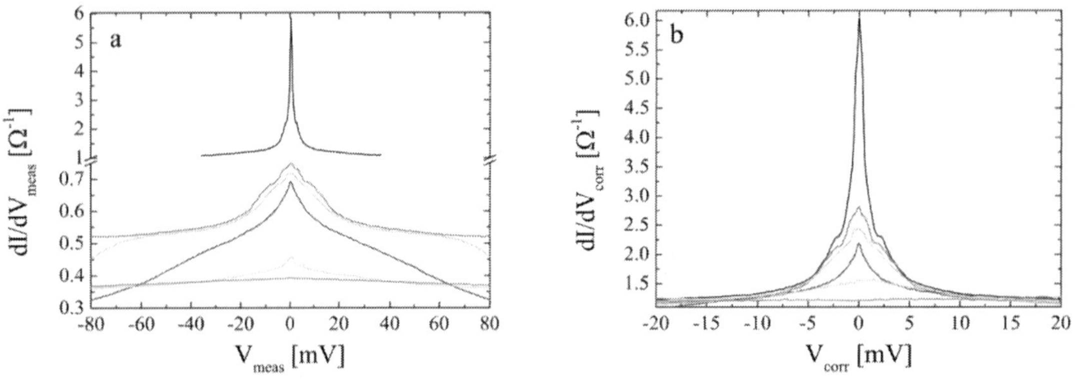

Figure 4. Differential conductance of the same junction as figure 2 as measured (a) and corrected with equations (1) and (2) as described in the text (b) for 4.2 K, 8.0 K, 12.2 K, 16.6 K, 20.9 K and 23.0 K.

at $T = 4.2$ K but a noticeable difference for higher temperatures. Now all conductance curves merge at the same value of $G \approx 1.0\,\Omega^{-1}$ at large voltage and the central peak broadens no more but stays nearly constant in width by changing temperature from 4.2 K to 8.0 K. There is still a loss in its height but this is natural due to the transformation from a SNS' junction to a SNN' junction by exceeding the critical temperature of lead.

Acknowledgments

This work was funded by the DFG within priority program 1458 (project no. SE 664/15-2), by the EC within project IRON-SEA (project no FP7-283141) and by the Landesgraduiertenförderung Thüringen. The authors thank Renato Gonnelli and Dario Daghero from Politecnico di Torino for fruitful discussion.

References

[1] Blonder G E, Tinkham M and Klapwijk T M 1982 *Phys. Rev. B* **25** 4515–4532
[2] Dynes R C, Narayanamurti V and Garno J P 1978 *Phys. Rev. Lett.* **41** 1509–1512
[3] Plecenik A, Grajcar M, Benacka S, Seidel P and Pfuch A 1994 *Phys. Rev. B* **49** 10016–10019
[4] Kashiwaya S, Tanaka Y, Koyanagi M and Kajimura K 1996 *Phys. Rev. B* **53** 2667–2676
[5] Mazin I I 1999 *Phys. Rev. Lett.* **83** 1427–1430
[6] Brinkman A, Golubov A A, Rogalla H, Dolgov O V, Kortus J, Kong Y, Jepsen O and Andersen O K 2002 *Phys. Rev. B* **65** 180517
[7] Woods G, Soulen R, Mazin I, Nadgorny B, Osofsky M, Sanders J, Srikanth H, Egelhoff W and Datla R 2004 *Phys. Rev. B* **70** 054416
[8] Chen T Y, Huang S X and Chien C L 2010 *Phys. Rev. B* **81** 214444
[9] Sheet G, Mukhopadhyay S and Raychaudhuri P 2004 *Phys. Rev. B* **69** 134507
[10] Baltz V, Naylor A D, Seemann K M, Elder W, Sheen S, Westerholt K, Zabel H, Burnell G, Marrows C H and Hickey B J 2009 *J. Phys.: Condens. Mat.* **21** 095701
[11] Döring S, Schmidt S, Schmidl F, Tympel V, Haindl S, Kurth F, Iida K, Mönch I, Holzapfel B and Seidel P 2012 *Physica C* **478** 15–18
[12] Schmidt S, Döring S, Tympel V, Schmidl F, Haindl S, Iida K, Holzapfel B and Seidel P 2012 *Phys. Proc.* **36** 82–87
[13] Iida K, Haindl S, Thersleff T, Hanisch J, Kurth F, Kidszun M, Huhne R, Monch I, Schultz L, Holzapfel B and Heller R 2010 *Appl. Phys. Lett.* **97** 172507

Pulse magnetization and voltage-current characteristic of $YBa_2Cu_3O_{7-x}$ film

G Dorofeev[1], V Drobin[2], H Malinowski[3]

[1]Kurchatov Institute, Moscow, Russian Federation,
[2]JINR, Dubna, Russian Federation,
[3]JINR, Dubna, Poland
E-mail: dorofeevgl@mail.ru

Abstract. The measurements of $YBa_2Cu_3O_{7-X}$ films magnetization are made at ramp rate of the external magnetic field up to 10^5 T/sec at temperature of liquid nitrogen in fields up to 4 T. The linear dependence of the magnetic moment of films on rate of an external magnetic confirms, that in the large electrical fields down to 1 V/cm the voltage-current characteristic can be described by the flux flow resistance ***Rflow*** and the critical current ***Jflow***, similar to traditional low temperature superconductors: $E = Rflow\cdot(J - Jflow)\cdot\Theta(J - Jflow)$, where Θ is the Heaviside function. The experimental data of pulse magnetization of films of thickness 0,2 microns on substrates Al_2O_3 of size 10 X 10 mm with critical currents higher 10^6 A/cm^2 have allowed to determine field dependence of the critical current and the flux flow resistance. In particular the resistance ***Rflow*** is increased approximately by 3 times for increase of an external field from 0,5 up to 4 T, and the ***Jflow*** values approach the critical currents determined from quasi-stationary magnetization.

Introduction

Volt - ampere characteristic (VAC) of the superconducting material contains information about magnetic flux pinning and character of currents flow in a superconductor. VAC is necessary for analysis of the superconducting material workability, for an estimation of thermal losses of energy, for stability superconducting state in different operating conditions, etc. For example, the form of VAC determines the thermal instability of a superconductor [1]. VAC have formed the basis to find out mechanisms of energy dissipation due to the movement of magnetic flux lines in low temperature superconductors at relative high electrical fields (about 1 mV/cm and above) [2,3]. For these reasons, the experimental researches on VAC of HTS ceramics represent certain interest. In particular, previous researches on VAC and magnetization of porous HTS ceramics of 1-st generation have shown that the electrical field is connected to movement of a magnetic flow on a current carrying network (cluster), which was formed by contacts of superconducting grains. The flux flow resistance (***Rflow***) grows with a field **B** linearly (***Rflow ~ B***) or logarithmically (***Rflow ~ lnB***) in low or high magnetic fields respectively [4, 5]. This behavior ***Rflow*** may be due to the suppression of bulk superconductivity, i.e. a result of the transition from three-dimensional to two-dimensional flow of magnetic flux. The characteristics of Bi-Sr-Ca-Cu-O_Pb ceramics without pores produced at high pressure support this picture too [6]. Modern 2-G HTS ceramics have high current carrying capacity. And the high bulk density of heat dissipation makes it difficult to carry out measurements of VAC in relatively high electric fields, where a movement of the magnetic flux is developed. Solution of the problem can be pulsed measurements. Indeed, superconductor magnetic moment **M** in the conditions of full magnetization is proportional to the density of flowing currents in the superconductor. The values of these currents **J** are determined by the local values of the electric field **E**, induced by the change of magnetic field **B** in the superconductor. Thus, measuring the dependence of the magnetic moment of a superconductor on the ramp rate of a magnetic field, we can define a relationship between the electric field and electric current in a superconductor that are the parameters of its volt-ampere characteristics.

The purpose of this work is to demonstrate the use of a pulse magnetization for the search parameters of volt-ampere characteristics of a superconductor on the example of the 2-G HTS Y123 ceramics on a dielectric substrate.

Samples and measurement technique

We measured $YBa_2Cu_3O_{7-X}$ coating ceramics of thickness of about 0,2 microns, grown by laser ablation on Al_2O_3 substrates of sizes of 10 x 10 mm and thickness of 0,3 mm. These epitaxial samples had the critical temperature in the range from 85 to 88K and the width of inductive S - N transition about 1K. The samples were placed in the center of solenoid for pulsed magnetic field (Helmholtz coils) located inside of another solenoid of larger size, that was used to create the supporting magnetic field. The pulsed magnetic field ***Bpul*** and the supporting magnetic field ***Bsup*** were oriented perpendicular to the sample surface. Measurements were made at the temperature of liquid nitrogen.
Measurement procedure was as follows.

1. Big solenoid creates long (about 0,1 sec) pulse relatively slowly changing supporting magnetic field ***Bsup***
2. At the time when the maximum value ***Bsup*** was achieved, the discharge of a capacitor battery at the Helmholtz coils was carried out. Several oscillations of pulsed magnetic field ***Bpul*** are monitored during the time up to 30 μsec.
3. EMF signals of two measuring coils are registered. One coil by a diameter of 1 mm is placed at the center of a sample; the second coil by a diameter of 19 mm is placed in a plane of symmetry of Helmholtz coils, so that the difference between the signals of the two measuring coils is directly related to the currents circulating in the sample.
4. After this the maximum value (amplitude) of the supporting magnetic field ***Bsup***, and/or amplitude and/or direction of pulsed magnetic field ***Bpul***, and/or sample are changed and the measurements are repeated again from step 1.

The rate of change of magnetic field ***dBpul/dt*** achieves a value of about 10^5 T/sec for the voltage of the condenser battery ***Uc*** = 350 V and the diameter of the Helmholtz coils 16 mm. Such magnetic pulse induces electric field above 1 V/cm in a sample size of 10 mm.
As a result we can make measurements of the magnetization of samples in magnetic fields up to 4 T at induction of electric fields up to 1V/cm by changing the parameters of pulse magnetic fields ***Bpul*** and ***Bsup***. Initial results are obtained in the form of EMF signals of two measuring coils as a function of time ***t*** at the various values of the supporting magnetic field ***Bsup***, capacitance of the battery ***C*** and its primary voltage ***Uc***.

Results and discussion

The subsequent integration of EMF signals of measuring coils gives us the rapidly changing in time component of the external magnetic field ***Bpul(t)*** and the difference between ***Bpul(t)*** and field ***Bo(t)*** in the center of the sample ***Bpul(t) - Bo(t) = M(t)*** , i.e. magnetic field of currents induced in the sample. The signals ***Bpul(t)*** and ***M(t)*** for the sample **S4** in a magnetic field ***Bsup*** = 0,3 T and for pulse parameters ***C*** = 10 mF and ***Uc*** = 350 V are presented in Fig. 1A.

Quick and full change of magnetizing of the sample accompanies each change of direction of change of the magnetic field. The same data are presented in the form of dependence ***M*** from the value ***Bpul*** in figure 1B and give us the hysteresis loop of full magnetizing of the sample for different amplitudes ***Bpul*** (various rates of the magnetic field ***dB/dt***), i.e. for different levels of electric field ***E*** in the

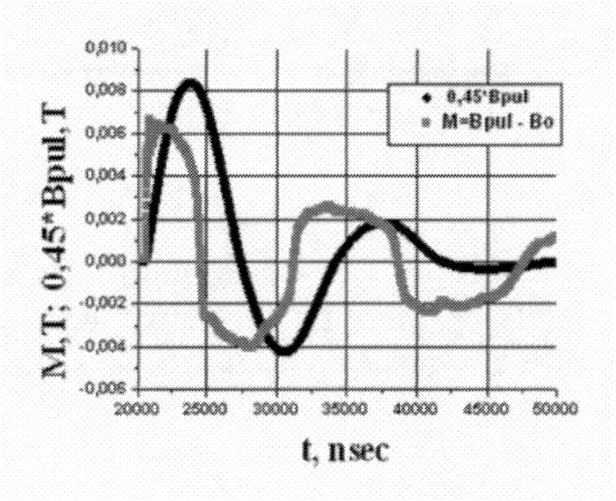

Fig. 1A. Time dependence ***Bsup*** and ***M*** for the sample **S4** in supporting field ***Bsup***=0,3 T.

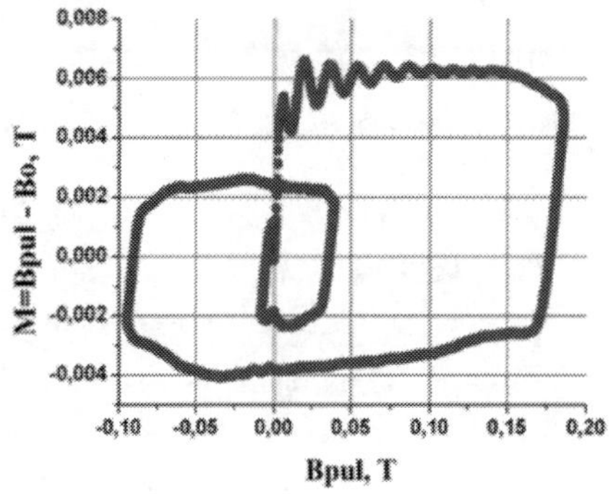

Fig. 1B. The dependence ***M*** on ***Bpul*** for the sample **S4** in supporting field ***Bsup***=0,3 T.

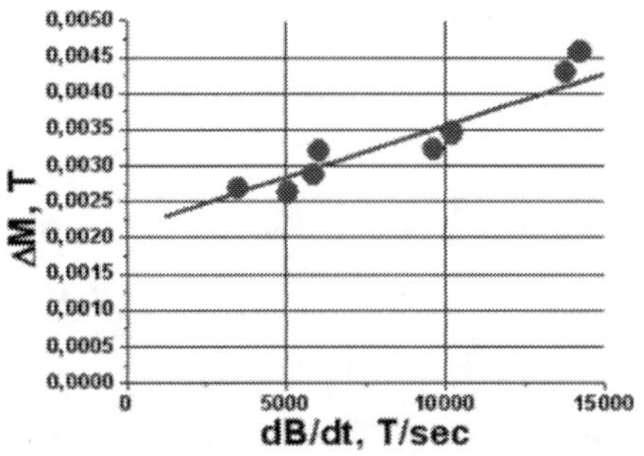

Fig. 2. Dependence ***ΔM*** from the ***dB/dt*** for the sample **S4** in the field ***Bsup*** = 0,75 T

sample. The ripples in the beginning all curves **M(B)** are caused, probably, noise pulse connecting a capacitor battery. The width of the loop of a magnetic hysteresis ***ΔM*** has the lineal increase (with a precision of 10 - 20%) with the growth ***dB/dt***, that is with increasing electric field ***E*** in the sample (see Fig. 2.). This relationship between ***ΔM*** and ***dB/dt*** gives us an opportunity to estimate the slope of the linear plot ***σm= dΔM/d(dB/dt)*** and ***ΔMflow*** as extrapolation of linear part of the curve to the value ***dB/dt=0***. It is obvious that the values ***σm*** and ***ΔMflow*** differ from the parameters of volt - ampere characteristics namely the flux flow conductivity ***σflow*** and the critical current ***Jflow*** only by constant coefficients, which are determined by the details of the experiment (the sample geometry, the placement of sensors, the nature of the current flow, and others). Thus, the dependence ***σm*** and ***ΔMflow*** on the value of the magnetic field ***Bsup*** gives us the field dependence of the flux flow conductivity ***σflow*** and critical current ***Jflow*** in the measured sample. Figure 3 shows ***ΔMflow*** and width of the curve of magnetization in quasi-stationary magnetic field (***ΔMqs)*** for the sample **S4** as a function of the external field.

Journal of Physics: Conference Series **507** (2014) 012009 doi:10.1088/1742-6596/507/1/012009

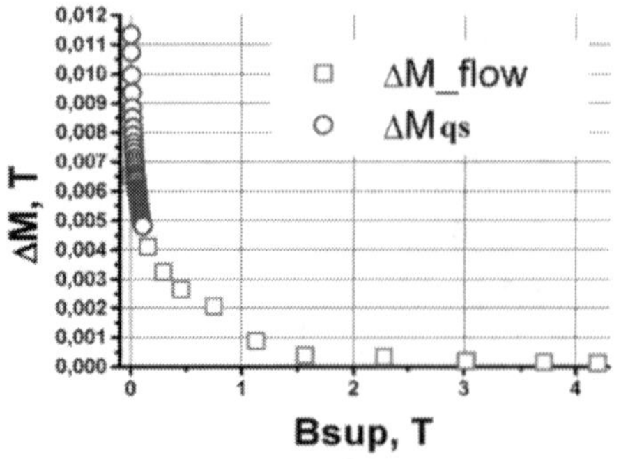

Fig. 3. ***ΔMflow*** and ***ΔMqs*** vs. ***Bsup*** for the sample **S4**

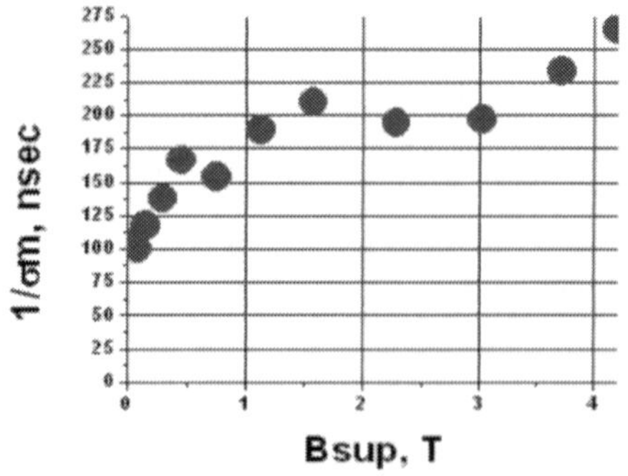

Fig. 4. ***1/σm*** vs. ***Bsup*** for the sample **S4**

We can see that the result of quasi-stationary magnetization (***ΔMqs)*** coincides with those for pulsed magnetization ***ΔMflow***. Width of the magnetization loop ***ΔMqs*** more than 0,01 T corresponds to the value of the current, circulating in the sample over 25 A, i.e. the critical current density is over 10^6 A/cm^2. In figure 4 the field dependence of the inverse ***σm*** (***1/σm*** is proportional to flux flow resistivity ***Rflow***) for the same sample is presented. A weakly increasing resistivity with increasing field is observed.

Conclusions

The experimental results show that pulsed magnetization allows estimating the parameters of volt-ampere characteristics of 2-G HTS ceramics on a dielectric substrate. In particular, it was shown on YBCO samples of thickness of about 0,2 microns, obtained by laser ablation on Al_2O_3 substrates of 10X10 mm^2 size and 0,3 mm thickness mm the following.

1. The form of volt-ampere characteristics of the 2-G HTS $YBa2Cu3O7_{-X}$ ceramics in the range of 10 - 500 mV/cm is close to linear, that is has the form similar to the current-voltage characteristic of low-temperature superconductors.
2. The results of the pulse magnetization (***ΔMflow***) correspond to the results for quasi-stationary magnetization (***ΔMqs***) of 2-G HTS ceramics.
3. The effective flux flow resistance ***Rflow*** weakly depends on the value of the magnetic field similarly to the case of low-temperature superconductors. This behavior of ***Rflow*** may be related with the motion of the vortices of the magnetic flux in granular films along grain boundaries or other reasons. The investigation of mechanisms of dissipation in the HTS ceramics needs a more detailed study of the flux flow.

Acknowledgments

The authors would like to thank Kulikov E. and Nikonov A. for assistance in measurements, and Chernikh I., Shavkin S. and Zanaveskin M. for presented samples and support.

References

[1] Polak M., Hlasnik I., Krempasky L. 1973 *Cryogenics* **13**, pp. 702-711
[2] Bardeen J., Stephen M.J. 1965 *Phys.Rev.* A **140** pp.1197-1207
[3] Clem J.R. 1968 *Phys.Rev.Lett.* **20** pp. 735-738
[4] Dorofeev G.L. et al 1991 *Superconductivity: physics, chemistry, engineering* **4**, pp. 723-725 (in Russian)
[5] Dorofeev G.L. et al 1991 *Superconductivity: physics, chemistry, engineering* **4**, pp. 738-740 (in Russian)
[6] Dorofeev G.L. et al 1991 *Superconductivity: physics, chemistry, engineering* **4**, pp. 1800-1803 (in Russian)

11th European Conference on Applied Superconductivity (EUCAS2013) IOP Publishing
Journal of Physics: Conference Series **507** (2014) 012010 doi:10.1088/1742-6596/507/1/012010

Tunneling STM/STS and break-junction spectroscopy of the layered nitro-chloride superconductors *M*NCl (*M* = Ti, Hf, Zr)

Toshikazu Ekino[1], Akira Sugimoto[1], Alexander M. Gabovich[2], Zhanfeng Zheng[3], Shuai Zhang[3], Shoji Yamanaka[3]

[1]Hiroshima University, Graduate School of Integrated Arts and Sciences, Higashi-Hiroshima, 739-8521, Japan

[2]Institute of Physics, National Academy of Sciences, Nauka Ave. 46, Kyiv 03680, Ukraine

[3]Hiroshima University, Graduate School of Engineering, Higashi-Hiroshima, 739-8527, Japan

E-mail: ekino@hiroshima-u.ac.jp

Abstract. The layered superconductors β-*M*NCl with the critical temperatures T_c = 14 K (*M* = Zr) - 25 K (*M* = Hf) were investigated by means of scanning-tunneling microscopy/spectroscopy and break-junction tunneling spectroscopy. The STM/STS was used to investigate the surface electronic structures in nanometer length scale, while the BJTS was employed to precisely determine the gap characteristics. Both techniques consistently clarified the unusually large size of the superconducting gap. Wide gap distributions with large-scale maximum gap values were also revealed in α-K_yTiNCl with a different crystal structure.

1. Introduction

The layered compounds β-*M*NCl are well known to exhibit superconductivity with high critical temperatures T_c = 14 K (Zr) - 25 K (*M* = Hf) [1,2]. Superconductivity emerges after electron doping of the pristine semiconducting β-*M*NCl [1,2]. Similar doping effect has been reported in another polymorph α-*M*NCl(*M* = Ti) with T_c = 16 K [3]. The *M*N bilayers inserted between Cl layers constitute the main frame of the crystal structure, in which nitrogen *p* and metal *d* states dominate the valence and conduction bands, respectively [4]. Because of the relatively high T_c, a question arose about the relevant superconductivity mechanism. Among experimental methods of investigating such mechanisms, electron-tunneling spectroscopy is an essential technique since it directly probes the energy gap 2Δ. It seems that the huge gap previously found by us to be far above the weak-coupling BCS cannot be explained by the conventional approach [5-7]. In particular, the strong-coupling Migdal-Eliashberg theory fails to explain superconductivity of *M*NCl [4]. In this paper, the superconducting gaps of α- and β-types of *M*NCl having different crystal structures are examined at the nanometer length scale by scanning-tunneling spectroscopy (STS) and are compared with their counterparts found from the break-junction tunneling spectroscopy (BJTS) measurements [5-7].

2. Experimental procedures

Polycrystals of *M*NCl were synthesized by a method reported elsewhere [1-3]. Since the samples were very sensitive to air humidity, they were treated in Ar-filled glove box. The STM (scanning-tunneling microscopy)/STS measurements were done with a modified Omicron low-temperature ultra-high vacuum (UHV) STM apparatus. The sample contained in a UHV sealed box was brought into an UHV chamber with the residual pressure of ~10^{-8} Pa just before the measurements to avoid contamination of the crystal surface. Then, the sample was cleaved along the layers at 3 - 77 K. The *c*-axis-oriented single-crystal domain facet of ~ 100 μm² was used to probe the surface. Before the scanning measurements, a Pt/Ir tip was cleaned by a high-voltage field emission process with the Au single-crystal target, resulting in the local work function of ~ 2 - 4 eV. The BJTS was employed to obtain surface-contamination-free spectra appropriate to fresh and clean superconductor – insulator – superconductor (SIS) junction interfaces at 4.2 K. This is in contrast to SIN (I symbolizes the vacuum and N a normal metal) junctions inherent to STS. The BJTS is a very sensitive probe to detect the gap structure.

3. Results and discussion

Figure 1 shows STM images of the cleaved *ab*-plane at 4.9 K for (a) β-$HfNCl_{0.7}$, (b) β-$ZrNCl_{0.7}$, and (c) α-$K_{0.5}TiNCl$. In the figure, the triangular lattices of bright spots are easily visible against the weakly inhomogeneous background with the vertical deviation of ~ 0.2 nm (a) and ~ 0.07 nm (b) [6,7]. On the other hand, the rectangular lattice spots are seen in Fig. 1 (c). Therefore, we can readily distinguish between the lattice networks. This means that the cleaved surface is clean and stable in contrast to that exposed to the ambient atmosphere where the chemical decomposition easily occurs.

Figure 1. STM images at 4.9 K and the atomic arrangements of (a) β-$HfNCl_{0.7}$ (sample bias $V = 0.1$ V, tunnel current $I = 0.1$ nA), (b) β-$ZrNCl_{0.7}$ (0.1 V, 0.3 nA), and (c) α-$K_{0.5}TiNCl$ (0.1 V, 0.1 nA). Each sampling area is 5 nm x 5 nm.

The nearest-neighbour spot intervals of ~ 0.36 nm for (a) and (b), and the nearest and the next nearest spot intervals of ~ 0.33 nm and ~ 0.41 mn, respectively, for (c) are determined from the two-dimensional fast Fourier-transform analysis. They are consistent with the X-ray diffraction measurements [1-3]. Since the observed interval between the bright spots is consistent with the unit cell size, these spots can be attributed to metallic ions in the conducting *MN* network on the *ab*-plane. This feature is the same for all studied compounds except for the degree of atomic corrugations.

Figure 2 shows the local STS conductance ($dI/dV = G(V)$). The line profiles of $G(V)$ are shown for the line cuts with the intervals of (a) 10 nm and ((b) and (c)) 5 nm. The spatial distributions of the gap values (the gap maps) are also presented for (a) β-$HfNCl_{0.7}$ and (c) α-$K_{0.5}TiNCl$. The $G(V)$ exhibits variations within the *nanometer* length scale, and the majority of $G(V)$ dependences are asymmetric. The curves $G(V)$ for β-$HfNCl_{0.7}$ exhibit broadened gap structures but are rather homogeneous over the sampling area of ~100 nm², while the inhomogeneity occurs on the shorter length scale for β-$ZrNCl_{0.7}$.

Journal of Physics: Conference Series **507** (2014) 012010 doi:10.1088/1742-6596/507/1/012010

One sees that the conductance *G*(*V*) for α-$K_{0.5}TiNCl$ also exhibits substantial nanometer-scale gap distribution. The spatial range of this distribution for (b) and (c) seems to be comparable or shorter than the superconducting coherence length [3]. The origin of the inhomogeneous distribution due to the cleaved layered crystal structure is unlikely because a fairly homogeneous mapping pattern is obtained in β-$HfNCl_{0.7}$. The asymmetric *G*(*V*) may be provoked by modified surface electronic states with the obscured electron spectrum due to the dispersion of incident electrons.

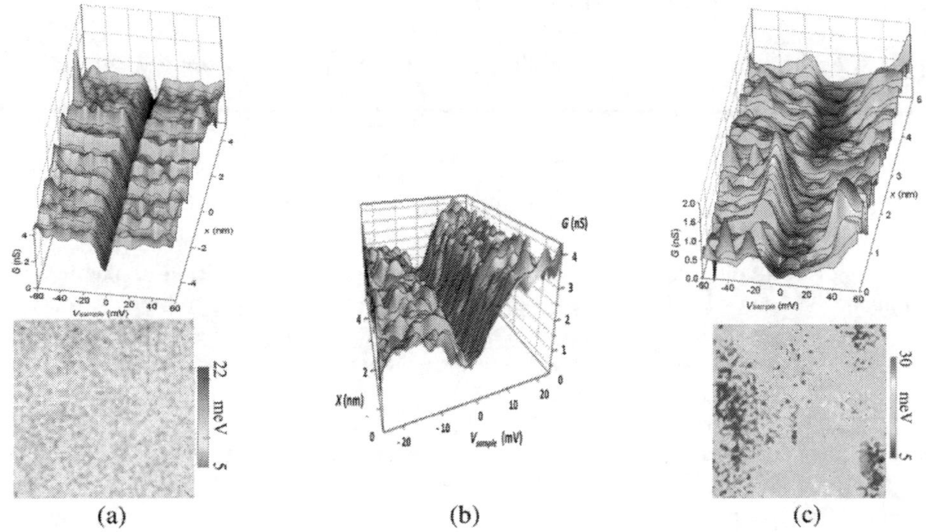

Figure 2. Line profiles of the conductance and gap maps for (a) β-$HfNCl_{0.7}$ (10 nm), (b) β-$ZrNCl_{0.7}$ (5 nm, without the gap map), and (c) α-$K_{0.5}TiNCl$ (5 nm).

The STS gap distribution inferred from the conductance *G*(*V*) was measured at 4.9 K. The gap maps are presented in figure 2 (a) and (c). In figure 3, we show the gap histograms obtained from the STS gap maps displayed in figure 2. The wide spread of the gap values in α-$K_{0.5}TiNCl$ is obviously seen as compared with β-$HfNCl_{0.7}$, which is in accordance with the spatial mapping features in figure 2. For β-$HfNCl_{0.7}$, the spatially averaged gap value defined by the *peak-to-peak* interval of *G*(*V*) is $2\Delta_{avg} \sim 24$ meV over the area of ~ 100 nm^2. The gap distribution spreads up to $2\Delta \sim 40$ meV with a sharp cut-off at $2\Delta \sim 10$ - 12 meV. The gap-to-T_c ratio estimated from these data is an order of magnitude larger than the Bardeen-Cooper-Schrieffer (BCS) value 3.53 appropriate to *s*-wave superconductivity. This conclusion remains valid even if we somewhat overestimated the peak-to-peak value of the gap [6].

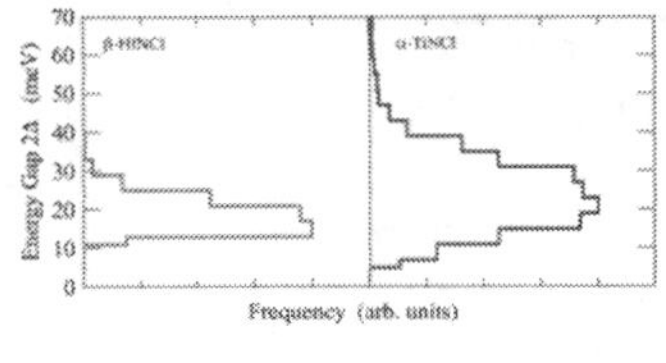

Figure 3. Histogram of the superconducting gaps for α-$K_{0.5}TiNCl$ and β-$HfNCl_{0.7}$.

Similar and widely spread gap values were also found for β-$ZrNCl_{0.7}$ whatever the bulk T_c values. The gap distribution for α-$K_{0.5}TiNCl$ is even much broader. It starts at a few meV and ends above ~ 40 meV with $2\Delta_{avg} \sim 28$ meV. We emphasize that T_c of α-$K_{0.5}TiNCl$ is ~ 10 K lower than that of β-$HfNCl_{0.7}$. Since we observed such a scatter notwithstanding the regular atomic arrangements on the crystal surface, it is most probably due to the electronic inhomogeneity. The fact that electronic inhomogeneity in the α-type crystals is much stronger than in the β-type ones could be due to the difference in the lattice symmetry in the *ab* plane.

In Table 1, the values of T_c, the average gap, and the gap spread are listed for three nitro-chloride superconductors. One can recognize that the averaged gap values $2\Delta_{avg}$ have similar magnitudes regardless of the T_c difference. At the same time, the gap spread is much wider for the lower T_c compound α-$K_{0.5}TiNCl$ than that for β-$HfNCl_{0.7}$.

Table 1. T_c, the average gap $2\Delta_{avg}$, and the gap spread σ for the nitro-chloride superconductors.

	T_c (K)	$2\Delta_{avg}$ (meV)	σ (meV)
α-TiNCl	16	~ 28	10 ~ 20
β-ZrNCl	14	20 - 26	---
β-HfNCl	25	~ 24	4 ~ 10

The STS gap distribution at high *T* is also unusual. The local Cooper pairing in nanometer regions is almost unchanged upon warming, while superconductivity in other surface regions is suppressed much stronger than what would have been expected from the mean-field temperature behaviour. This circumstance may be the origin of the unusual *T*-dependence of the gap values found by BJTS [5]. In figure 4, the STS conductance mapping and the averaged conductance *G*(*V*) are shown at 77 K. In the *G*(10mV) mapping, the conductivity is enhanced on the cation sites and this provides a good indication of the surface cleanliness even at this high temperature. The unusually large gap observed below T_c has already disappeared at 77 K leaving a quite shallow and broad dip in the *V*-shaped background. This feature is somewhat similar to that found in high-*T* superconductors where pseudogaps are observed.

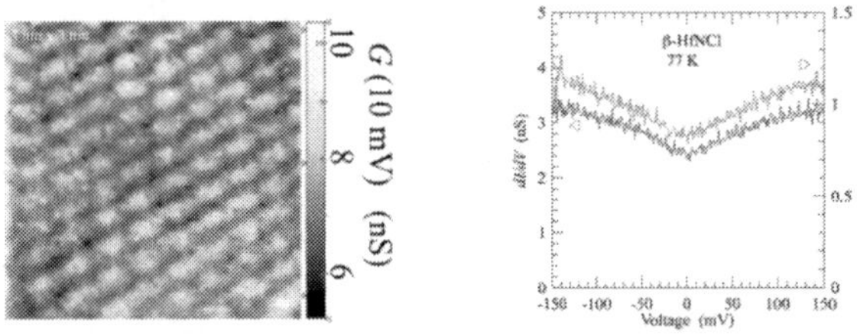

(a) (b)

Figure 4. (a) Conductance *G*(10 mV) mapping (3 nm x 3 nm), and (b) the averaged *G*(*V*) for β-HfNCl at 77 K.

In figure 5, we compare the conductances *G*(*V*) obtained by STS and BJTS. For β-$HfNCl_{0.7}$ (a), the peak positions in the averaged *G*(*V*) for STS is just twice smaller than the peak position of the single measurements of *G*(*V*) by BJTS, which is consistent with the SIN and SIS geometries across the junctions, respectively. On the other hand, the STS gap-edge bends occur at the same position as the BJTS gap edge in β-$ZrNCl_{0.7}$. This is probably due to the fact that the BJTS probes the most stable bulky portion of the sample, while the STS *G*(*V*) is an average over the wide spatial gap distribution.

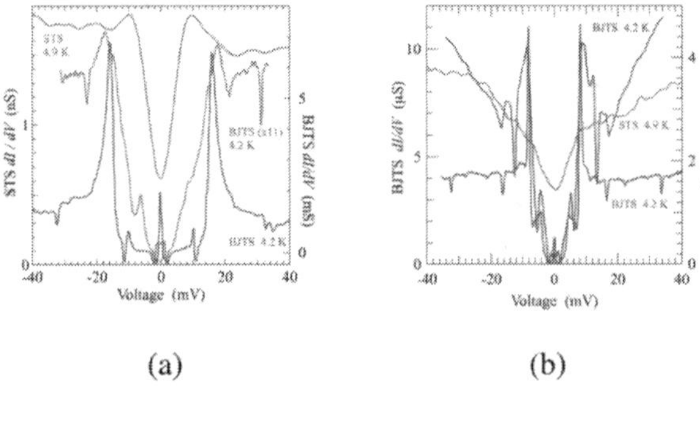

(a) (b)

Figure 5. Comparison of STS and BJTS conductances for (a) β-$HfNCl_{0.7}$ and (b) β-$ZrNCl_{0.7}$. The STS curves are the averaged data, while the BJTS ones are obtained from the different junctions.

We attribute the observed difference in the sharpness of the gap structures as well as in the sub-gap leakage of $G(V)$ between STS and BJTS to the difference between the junction actual interfaces, because for each compound the examined samples were taken from the same batch. Therefore, results of the BJTS measurements support STS results and both methods taken together overcome tough problems concerning the interface quality. The results also mean that there still exists much room for improvement of the STS technology to study the sample interface which is extremely sensitive to an unavoidable change of the surface during necessary STS operations.

The inhomogeneous distribution of superconducting gaps in the nitro-chloride compounds could be connected to the surface electronic instability probed by both the STS and BJTS tunneling techniques. This is because charge carrier densities in such layered compounds can be easily reduced under an external influence or an intrinsic instability (for instance, phase separation at the interface region [8]) resulting in the appearance of inhomogeneous spatial patterns. This situation inevitably leads to the spatial spread of superconducting gaps and, probably, to the suppression of superconductivity in the probed surface areas. In contrast, the deliberate charge carrier injection may enhance superconductivity. Actually, such an experiment resulted in the electric-field-induced superconductivity in the initially semiconducting β-$ZrNCl_{0.7}$ [9]. Therefore, one should bear in mind that in any conventional tunneling technique we not only probe the surface but also disturb it locally possibly changing the parent electronic states.

4. Conclusion

STM/STS and break-junction tunneling measurements of α-$K_{0.5}TiNCl$ and β-$M NCl_{0.7}$ (M = Hf, Zr) were carried out to clarify the nature of superconductivity. We found from STM/STS measurements that substantial nanometer-scale electronic inhomogeneity, including the broad distribution of gap values, occurs in all studied compounds showing, nevertheless, quite regular atomic lattices of the STM images which are well consistent with the X-ray diffraction data. The STS averaged superconducting gap in each compound is extremely large strongly exceeding the *s*-wave weak-coupling BCS prediction as well as conventional strong-coupling values. STS-inferred huge gaps are consistent with those revealed by the break-junction measurements, thereby giving evidence for an unusual nature of superconductivity in the doped layered nitro-chlorides.

Acknowledgement
This work was supported by a Grand-in-Aid for Scientific Research (245403770) and the FIRST Program of the Japan Society for the Promotion of Science (JSPS). The work was partially supported by the Project N 8 of the 2012-2014 Scientific Cooperation Agreement between Poland and Ukraine.

References

[1] S. Yamanaka, K. Kotehama, H. Kawaji, Nature 392 (1998) 580.
[2] S. Yamanaka, J. Mater. Chem. 20 (2010) 2922.
[3] S. Zhang, M. Tanaka, S. Yamanaka, Phys. Rev. B 86 (2012) 024516.
[4] R. Akashi, K. Nakamura, R. Arita, M. Imada, Phys. Rev. B 86 (2012) 054513.
[5] T. Takasaki, T. Ekino, A. Sugimoto, K. Shohara, S. Yamanaka, A.M. Gabovich, Eur. Phys. J. B 73 (2010) 471.
[6] A. Sugimoto, K. Shohara, T. Ekino, Z. Zheng, S. Yamanaka, Phys. Rev. B 85 (2012) 144517.
[7] T. Ekino, A. Sugimoto, A.M. Gabovich, Z. Zheng, S. Yamanaka, Physica C 494 (2013) 89.
[8] Ariando, X. Wang, G. Baskaran, Z.Q. Liu, J. Huijben, J.B. Yi, A. Annadi, A. R. Barman, A. Rusydi, S. Dhar, Y.P. Feng, J. Ding, H. Hilgenkamp, T. Venkatesan, Nat. Commun 2 (2011) 188.
[9] J.T. Ye, S. Inoue, K. Kobayashi, Y. Kasahara, H.T. Yuan, H. Shimotani, Y. Iwasa, Nat. Mater. 9 (2011) 125.

Growth of (001) or (115) Bi-2201 Thin Films by Spin Coating and MOCVD Targeting Future Electronics Applications

K Endo[1], **T Kaneko**[1], **T Moriguchi**[1], **H Takemata**[1], **T Takada**[1], **N Ikenaga**[1], **P Badica**[2]
[1] Research Laboratory for Integrated Technological Systems, Kanazawa Institute of Technology (KIT), Yatsukaho, Ishikawa 924-0838, Japan
[2] National Institute of Materials Physics, Magurele, Atomistilor 105bis, 077125, Romania

E-mail: kendo@nepyune.kanazawa-it.ac.jp

Abstract. Thin films of $Bi_2Sr_2CuO_6$ (Bi-2201) with (001) or (115) orientation were grown on $SrTiO_3$ substrates. These films are expected to be useful as component films of heterostructures and devices (e.g. as insulators). We used two chemical routes, namely spin coating followed by thermal annealing and MOCVD. For both routes we demonstrated growth of c-axis and non-c axis thin films depending on the selected orientation of the $SrTiO_3$ substrate. To our knowledge, films of (115) Bi2201 obtained by spin coating or MOCVD are reported for the first time.

1. Introduction

In the last years, the interest in chemical routes for the growth of high temperature superconducting (HTS) films and the auxiliary films (used together with HTS ones in different applications) significantly increased. This is because films obtained by chemical routes are cheaper than those obtained by other methods. Much of the effort was devoted to development of films produced by chemical routes for coated conductors and only a few works are looking to potential of these films for electronics. In addition, $YBa_2Cu_3O_7$ (Y123) system was much investigated, while for other HTS systems a significantly less level of information is available.

2. Experimental

(001) and (110) $SrTiO_3$ substrates were used for the growth of (001) and (115) $Bi_2Sr_2CuO_6$ (Bi-2201) films, respectively. Substrates are from Furuuchi Chem. Co., LTD.
Coating solution is supplied by Kojundo Chem. Lab. Co., LTD. Spin coating was performed at 3000 rpm for 30 min., and it was followed by heat treatment at 820°C for 1 h for the (001)Bi2201 and at 710°C for 1 h for the (115)Bi2201 .
Metal-organic-chemical-vapor-deposition (MOCVD) was carried out with unique laboratory-designed machine of horizontal cold-wall type [1]. Raw materials were Metal- DPM (DPM is abbreviation for di-pivaloyl-methanate and Metal = Sr, Ca, Cu, Y, Ti) and $Bi(C_6H_5)_3$ (triphenyl-Bi). Growth temperature was 750°C for 1 h for both (001)Bi-2201 and (115)Bi-2201 films.

3. Result and Discussion

We present growth of (00l) and (115) (Bi-2201) thin films by spin coating. Orientation control is obtained by using substrates of (001) and (110) $SrTiO_3$, respectively. The principle based on lattice matching relationship between the substrate and the film is illustrated in Fig. 1. We also compare these films obtained by spin coating with (001) and (115) Bi-2201 films deposited on the same substrates by MOCVD.

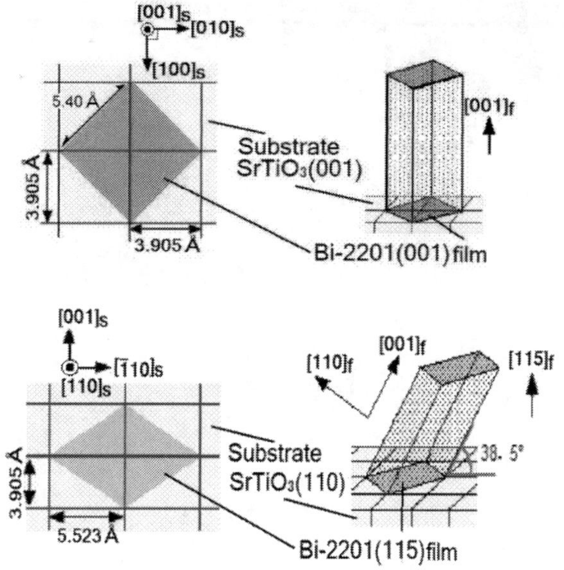

Fig. 1 Lattice matching relationship for the growth of (00l) and (115) Bi-2201 thin films.

X-ray diffraction patterns and SEM images for (00l) and (115) Bi-2201 thin films obtained by MOCVD are shown in Figs. 2, 3, while for films grown by spin coating are in Figs. 4, 5.

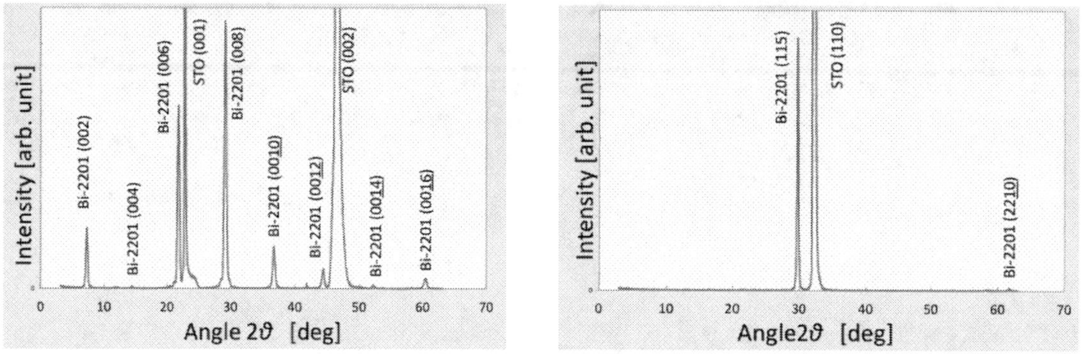

Fig. 2 XRD patterns of left - (001) and right - (115) Bi-2201grown by MOCVD.

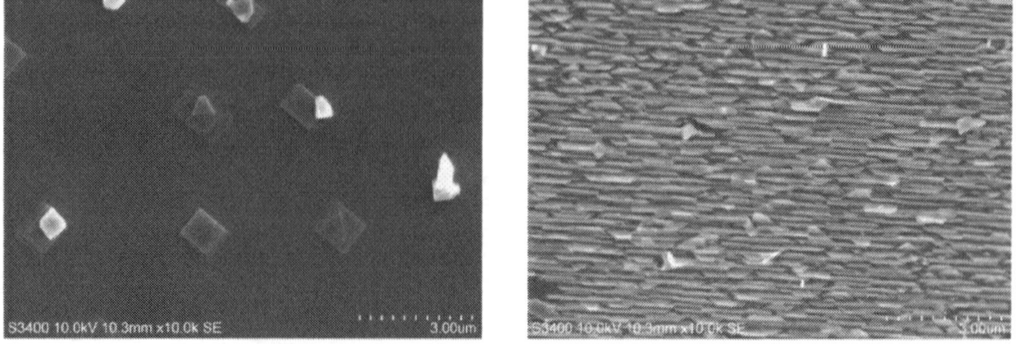

Fig. 3 SEM images of left - (001) and right - (115) Bi-2201grown by MOCVD.

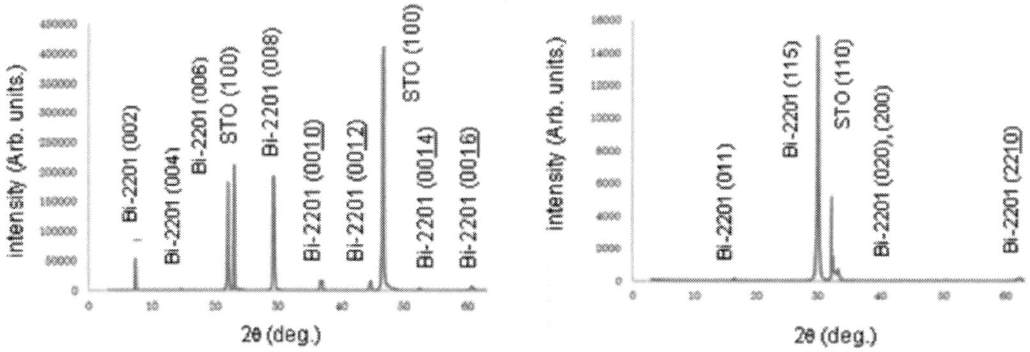

Fig. 4 XRD patterns of left - (001) and right - (115) Bi-2201grown by spin coating.

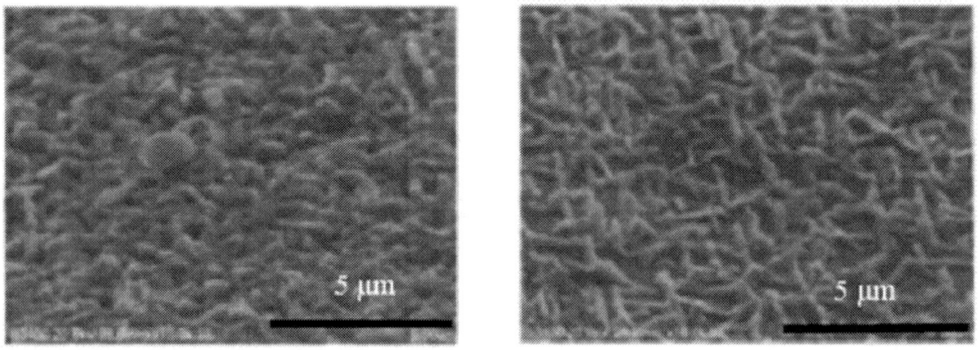

Fig. 5 SEM images of left - (001) and right - (115) Bi-2201grown by spin coating.

The first important observation is that our results demonstrate that orientation control is independent of growth method: for both growth techniques, i.e. MOCVD and spin coating, (001) and (115) orientations and single phase Bi-2201 films are obtained. However, there are also significant differences as follows:

(i) Films obtained by MOCVD have a higher morphological uniformity and a lower roughness (compare Figs. 3 and 5).

(ii) The (115) Bi-2201 grown by MOCVD shows a higher degree of *in-plane* alignment than for the film obtained by spin coating (compare Figs. 3 and 5).

(iii) The (115) Bi-2201 grown by MOCVD shows less residual impurity phases or orientations than for the film obtained by spin coating (compare Figs. 2 and 4).

All Bi-2201 films in this work were not superconducting and are thought as possible candidate insulating layers for design of certain stacked composite heterostructures.

4. Conclusion

Orientation control of Bi-2201 thin films through lattice matching between the film and the substrate was realized when growth of the films is done by spin coating and MOCVD. The quality of spin coated films is lower than for advanced deposition techniques such as MOCVD, but spin coating is a simple and cheap alternative.

Acknowledgement

Work at KIT was performed with financial support from Grant-in-Aid for Scientific Research No. 24560386, Japan. PB acknowledges PCCE 138/2012, Romania.

References

[1] Endo K, Yamasaki H, Misawa S, Yoshida S and Kajimura K 1992 *Nature* **355** 327

Preparation and characterization of PEG/Bi2212 nanocomposites

H Enomoto [1], T Deguchi [2], Y Takano [3], and N Mori [4]

[1] Department of Environmental Science, Osaka Electro-Communication University, Hatsu-cho 18-8, Neyagawa, Osaka 572-8530, Japan

[2] Department of Chemistry, Graduate School of Science, Kyoto University, Kitashirakawa Oiwake-cho, Sakyo-ku, Kyoto 606-8502, Japan

[3] Department of Physics, Nihon University, Kanda-surugadai 1-8, Chiyoda-ku, Tokyo 101-8308, Japan

[4] Department of Electrical and Computer Engineering, Oyama National College of Technology, Nakakuki 771, Oyama 323-0806, Japan

E-mail: h-enomot@isc.osakac.ac.jp

Abstract. Single-phase nanocomposites of $Bi_2Sr_2CaCu_2O_{8+d}$ (Bi2212) with organic polymer have been successfully synthesized for the first time by the foliation-adsorption technique. Colloidal suspension containing exfoliated Bi2212 nanosheets was prepared with a stepwise intercalation method. Poly(ethylene glycol) (PEG) was slowly added into colloids containing Bi2212 nanosheets. X-ray diffraction patterns show that the gallery height of Bi2212 host is expanded from 3.08 nm to 4.91 nm in PEG/Bi2212 nanocomposites. The basal plane expansion of 0.9 nm is in good agreement with the values reported for PEG-containing nanocomposites, suggesting that a PEG bilayer is incorporated within the Bi2212 nanosheets.

1. Introduction

The layered bulk materials are exfoliated into individual sheets, so-called nanosheets, in solution to form colloidal suspensions, being expected to open a new route to synthesize nanoparticles and thin films. One of the most important and attractive aspects of the exfoliated nanosheets is that various nanostructures can be fabricated using them as two-dimensional building blocks. It is even possible to tailor superlattice-like assemblies, incorporating a wide range of materials, such as polymers, into the nanosheets. From a viewpoint of application, the intercalation of large organic molecules provides a new synthetic route to high-T_C superconducting thin-films and nanocomposites by separating superconducting blocks into isolated single sheets.

$Bi_2Sr_2CaCu_2O_{8+d}$ (Bi2212) is the noble high-T_C superconductor and has a layered structure, which consists of perovskite blocks separated by weekly-coupled Bi-O double layers. Bi2212 colloidal suspension consisting of exfoliated nanosheets was first reported by J-H. Choy *et al.* [1-2]. In our group, Bi2212 colloids dispersed in acetone were also obtained from intercalated Bi2212 via the $HgBr_2$/Bi2212 precursors [3-4].

In this report, we describe the first synthesis of single-phase nanocomposites of Bi2212 with organic polymer by the exfoliation-adsorption technique. This technique [5-7] is one of the most attractive ways if the host materials can be separated into the nanosheets and the polymer is soluble in a solvent. Poly(ethylene glycol) (PEG) is adopted as the guest polymer, because PEG is soluble to organic solvent. The exfoliation-adsorption method, however, may require additional process controls, such as ultrasonic exposure and mechanical stirring, to optimize the synthesis of nanocomposites. This is because the exfoliation-adsorption method is highly dependent on the chemical and physical properties of host inorganic materials and guest organic polymers. These conditions therefore need to be optimized for each synthesis.

2. Experimental

The inorganic starting materials used for all preparations described in this work were analytical reagent grade and used without further purification. Tetrabutylammonium bromide (TBA:Br, Aldrich), dimethylsulfoxide (DMSO, Wako Chemicals), and Poly(ethylene glycol) (PEG, average molecular weight 1000, Wako Chemicals) was also used as received.

Bi2212 hosts were prepared by the solid-state reaction of Bi_2O_3, $SrCO_3$, $CaCO_3$, Y_2O_3, and CuO powders (1-2 μm, Soekawa Chemicals) [8-12]. They were mixed in the ratio of Bi:Sr:Ca:Y:Cu = 2:2:0.8:0.2:2. The well-mixed powders were sintered at 800°C for 20 h in air. They were pulverized, mixed, palletized and sintered at 870°C for 24 h with air-flow. This process was repeated twice in order to improve the crystallinity and the superconducting transition temperature T_C.

Colloidal suspension containing exfoliated Bi2212 nanosheets was prepared with a stepwise intercalation method [3-4]. A large molecule, such as alkylammonium ion, cannot be intercalated directly into the narrow Bi-O layers of Bi2212 host. To increase the gallery height between the Bi-O layers, we prepared the $HgBr_2$/Bi2212 precursors prior to the alkylammonium ion intercalation. The intercalated precursor was prepared by the chemical vapour transport technique with Bi2212 and $HgBr_2$ in a vacuum-sealed quartz ampoule. TBA:Br was then intercalated into $HgBr_2$/Bi2212 via the ion exchange method. The molar ratio of $HgBr_2$/Bi2212 precursor and TBA:Br was 1:3. They were mixed with 50 mL of DMSO in a Schlenk-type flask. The mixture was vigorously stirred at 70°C for 6 h. Exfoliation of Bi2212 was carried out by sonicating TBA/Bi2212 in DMSO solution. In typical reactions, 0.3 g of TBA/Bi2212 was exfoliated by ultrasonication at ambient temperature in 300 mL of DMSO. Excess PEG polymer (0.9 g) was slowly added into colloids containing Bi2212 nanosheets. The mixtures were stirred, and solid product was precipitated. The nanocomposites were then collected by centrifugation, washed to remove unintercalated polymer, and dried *in vacuo* overnight.

X-ray powder diffraction (XRD) data on samples were collected on a Rigaku RINT Ultima$^+$ diffractometer, using Ni-filtered Cu Kα radiation, by 0.02° 2θ step/0.6 sec between 2 and 120°.

3. Results and discussion

It is well known that the superconducting properties of the oxide high-T_C superconductors are very sensitive to the hole concentration. In the Bi2212 system, the hole concentration can be easily varied by substituting the divalent Ca ion by trivalent rare-earth ions, *e. g.* Y [13]. In this study, host Bi2212 single crystals were prepared at Y content x = 0.2, where the highest T_C can be obtained at the optimum hole concentration in the $Bi_2Sr_2Ca_{1-x}Y_xCu_2O_{8+d}$ system [8-9].

Various layered compounds have been delaminated into their colloidal single-sheets via the so-called soft chemical techniques, which require neither high temperature nor ultrahigh vacuum for preparations. The resulting nanosheets exhibit novel physical and chemical properties due to an extremely small thickness of around 1 nm. Treatment of 0.3 g of TBA/Bi2212 intercalation compound with DMSO of 300 mL after ultrasonication for 200 min in the ice bath yields uniform and transparent colloidal solution. A clear Tyndall light scattering can be observed, indicating the presence of exfoliated Bi2212 nanosheets dispersed in DMSO. The resulting colloidal suspension is stable for a few days. Then black solid is precipitated.

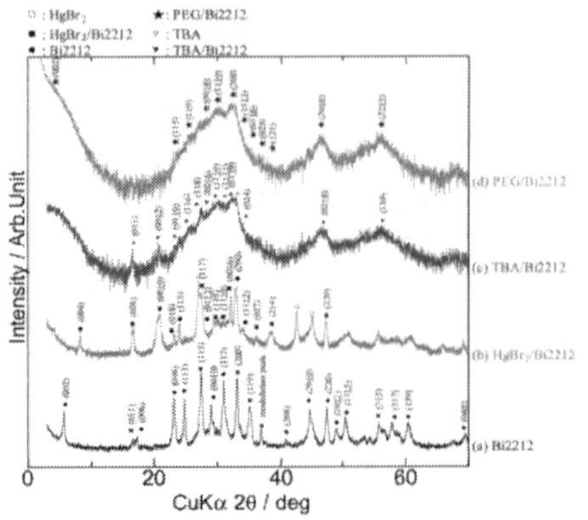

Figure 1. XRD patterns of the host and its nanocomposites; (a) Bi2212 host, (b) $HgBr_2$/Bi2212, (c) TBA/Bi2212, and (d) PEG/Bi2212 nanocomposite prepared at room temperature.

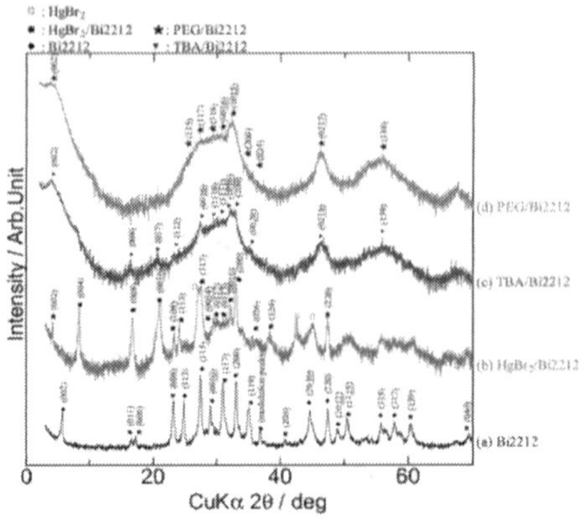

Figure 2. XRD patterns of the host and its nanocomposites; (a) Bi2212 host, (b) $HgBr_2$/Bi2212, (c) TBA/Bi2212, and (d) PEG/Bi2212 nanocomposite prepared at 70°C.

Table 1. XRD data for Bi2212 nanocomposites.

Sample	*a* (nm)	*b* (nm)	*c* (nm)	Δ*c* (nm)
Bi2212	0.541	0.541	3.076	—
$HgBr_2$/Bi2212	0.541	0.541	4.384	1.308
TBA/Bi2212	0.541	0.540	5.101	2.025
PEG/Bi2212[a]	0.543	0.538	4.908	1.832
Bi2212	0.541	0.541	3.076	—
$HgBr_2$/Bi2212	0.541	0.541	4.384	1.308
TBA/Bi2212	0.543	0.539	5.117	2.041
PEG/Bi2212[b]	0.541	0.540	4.876	1.800

[a] Prepared at room temperature.
[b] Prepared at 70°C.

Figure 1 shows the XRD pattern of PEG/Bi2212 nanocomposite together with Bi2212 host and its derivatives. The XRD data obtained are summarized in Table 1. In this study, all XRD peaks are assigned by the basic double-block model, although the $(a + b)/2$ shift of the cuprate block may not occur upon Bi2212 intercalation compounds. The observed lattice expansion along the *c*-axis for $HgBr_2$/Bi2212 precursor, Δc = 1.308 nm, is very near to 1.26 nm of the reported value [3-4]. This suggests that the $HgBr_2$ monolayer is intercalated between the Bi-O double layers of Bi2212 host crystal. The difference between interlayer spacing of TBA/Bi2212 intercalation compounds and unintercalated Bi2212, Δc = 2.025 nm, is in good agreement with twice of 1.05 nm for the TBA-intercalated α-zirconium phosphate [14]. This means that the TBA^+ ions reside the single-layer arrangement in the interlayer gallery of Bi2212. $HgBr_2$/Bi2212 precursor and TBA/Bi2212 intercalation compounds are used to increase the gallery height between the Bi-O layers.

Figure 1(d) shows the XRD pattern of PEG/Bi2212 nanocomposite prepared at room temperature for 7 days. There is no trace of the pristine phase in XRD patterns for PEG/Bi2212 nanocomposite, indicating that PEG polymer is incorporated homogeneously into each Bi2212 layer. It is found that the gallery height of Bi2212 host is expanded from 3.08 nm to 4.91 nm in PEG/Bi2212 nanocomposites. The basal plane expansion of 0.9 nm is in good agreement with the values reported for PEG-containing nanocomposites [6-7]. This suggests that a PEG bilayer is incorporated within the Bi2212 nanosheets.

The experimental conditions of the foliation-adsorption technique need to be optimized for each synthesis. The synthesis time and temperature of PEG/Bi2212 nanocomposites and the molar concentration of Bi2212 colloids are carefully checked. Figure 2(d) shows the XRD pattern of PEG/Bi2212 nanocomposite prepared at 70°C for 7 days. Although the synthesis temperature is elevated from room temperature to 70°C, the other conditions for PEG/Bi2212 nanocomposites are unchanged. A more obvious (002) peak can be found around 4.1° in Fig. 2(d), indicating that the crystallinity of PEG/Bi2212 nanocomposite prepared at 70°C is higher than that prepared at room temperature, although the lattice constants calculated from the assigned XRD peaks are almost same.

4. Summary

The single-phase stage-1 intercalation nanocomposites of $Bi_2Sr_2CaCu_2O_{8+d}$ (Bi2212) with organic polymer poly(ethylene glycol) (PEG) have been synthesized for the first time via $HgBr_2$/Bi2212 precursors and TBA/Bi2212 intercalation compounds. It is found that the gallery height of Bi2212 host is expanded by 1.8 nm in PEG/Bi2212 nanocomposites. TBA/Bi2212 can be exfoliated into single layers in DMSO, which is stable for a few days after ultrasonication. Further study will be needed to prepare the more stable colloidal suspension of Bi2212 nanosheets, which are useful as a building block for the fabrication of a wide variety of functional nanomaterials.

Acknowledgements

The authors gratefully acknowledge Mao Ito, Takashi Uda, and Takanari Kikumoto for their kind assistance in sample preparations. And the authors would like also to thank Prof. Michael M. Lerner of Oregon State University (USA) and Prof. Hajime Ozaki of Waseda University (Japan) for their helpful discussions. A part of this study is supported by a Grant-in-Aid from Osaka Electro-Communication University.

References

[1] Choy J-H, Hwang S-J and Park N-G 1997 *J. Am. Chem. Soc.* **119** 1624
[2] Choy J-H, Kwon S-J, Hwang S-J, Kim Y-I and Lee W 1999 *J. Mater. Chem.* **9** 129
[3] Enomoto H, Iwasaki M, Gyoten T, Takano Y, Sekizawa K, Mori N, Ozaki H, and Lerner M M 2009 *J. Phys.: Conf. Series* **150** 052049
[4] Iwasaki M, Enomoto H, Ozaki H, and Lerner M M 2009 *J. Phys.: Conf. Series* **150** 052081
[5] Enomoto H and Lerner M M 2002 *Mater. Res. Bull.* **37** 1499
[6] Enomoto H, Takai H, Ozaki H, and Lerner M M 2004 *J. Phys.: Condens. Matter* **16** 6375
[7] Enomoto H, Lerner M M, and Ozaki H 2006 *J. Phys. Chem. Solids* **67** 965
[8] Takauchi H, Enomoto H, Shin J S, Yamanaka H, Furukawa H, Takano Y, Mori N and Ozaki H 1990 *Physica B* **165&166** 1227
[9] Yamanaka H, Enomoto H, Shin J S, Kishimoto T, Takano Y, Mori N and Ozaki H 1991 *Jpn. J. Appl. Phys.* **30** 645
[10] Ikuta H, Kishimoto T, Enomoto H, Shin J S and Ozaki H 1993 *Phase Transition* **42** 1
[11] Wakamatsu K, Oikawa Y, Enomoto H, Shin J S, Uchida E and Ozaki H 1994 *Physica C* **235-240** 687
[12] Furubayashi H, Enomoto H, Takano Y, Sekizawa K, Oikawa Y and Ozaki H 1996 *Czechoslovak J. Phys.* **46** Suppl. S3 1419
[13] Tamegai T, Koga K, Suzuki K, Ichihara M, Sakaki F and Iye Y 1989 *Jpn. J. Appl. Phys.* **28** L112
[14] Kim H-N, Keller S W, and Mallouk T E, 1997 *Chem. Mater.* **9** 1414

Vortex structure and magnetic properties of $BaFe_2(As_{1-x}P_x)_2$ single crystals

L Vinnikov[1], **A Friedman**[2], **I. Felner**[3], **A Troshina**[1,4], **Y Yeshurun**[2]
[1] Institute of Solid State Physics, RAS, Chernogolovka, Russia
[2] Institute of Superconductivity, Bar-Ilan University, Ramat-Gan, Israel
[3] Racah Institute of Physics, The Hebrew University, Jerusalem, Israel
[4] Moscow State University, Moscow, Russia
E-mail: freidma@gmail.com

Abstract. Magnetic decoration of $BaFe_2(As_{1-x}P_x)_2$ revealed an hexagonal vortex lattice in overdoped single crystals (x>0.4, T_c=14-18 K). A disordered vortex lattice was observed in a crystals with optimal content of phosphorus (x ≈ 0.4, T_c = 25-30 K). Magneto-optical studies and global magnetization measurement of the optimally-doped crystal have led to an estimated critical current density of ~ 10^5A/cm^2 at 5-12 K in fields up to 0.7 T. This high critical current density suggests strong pinning, probably due to defects, a result consistent with the observed disordered vortex lattice. The critical current density of the overdoped crystal is much smaller, i.e. ~ 10^4 A/cm^2 at 5 K in fields up to 0.1 T, decreasing to virtually zero at 0.3 T.

1. Introduction

The pnictides, discovered in 2008 [1], provoked a great deal of interest owing to a series of unusual properties [2]. These materials share, as a common structural motif, superconducting FeAs or FeSe/Te layers. In most cases, superconductivity was induced by doping charge carriers into a stoichiometric parent compound which shows a collinear antiferromagnetic spin-density-wave ground state. The chemical doping was initially carried out only outside the FeAs layers, [3,4] with the electronic phase diagram resembling that of cuprate superconductors [5,6].

Later, however, it was found that even doping at the Fe site (with Co [7] or Ni [8]) could induce superconductivity. Numerous state-of-the-art experimental techniques developed for the cuprates have been applied to the study of these new families (for review articles see [9,10]). These materials with T_C values up to 55 K, are of great interest for applications due to their lower anisotropies and ultrahigh upper critical fields. Moreover, in the last years a significant progress has been made in the fabrication of iron-based superconducting wires and tapes using the powder-in-tube processing method [11]. It is thus clear that study of pinning and vortex structure in pnictides is of utmost importance for further development of wire technology and their exploitation in future devices.

The iron-based pnictide $BaFe_2(As_{1-x}P_x)_2$ system has recently attracted special interest for its potential large critical current. Its structure strongly depends on phosphorus content. For low doping (x<0.2) and low temperatures, the material exhibits a spin-density wave state – which is equivalent to anti-ferromagnetic order in a metal. For higher doping the materials turn to be superconducting at lower temperatures [12]. There is an optimal doping level which exhibits maximal critical temperature T_C of about 30 K. Recently, it was suggested [13] that the charge state of the doped compounds has an important role in the mechanism of collective pinning in this materials [14]. This suggestion was based on the observation [15] of a sharp decrease (by several orders of magnitude) in J_C in phosphorus underdoped and overdoped single crystals compared to the optimally-doped compound. This decrease in J_C signifies a strong decrease in the vortex pinning energy.

The goal of the work described here was to correlate the vortex structure in optimally-doped and overdoped samples with their critical current. The working assumption of the project was that the vortex lattice expected for a low-J_C material will change to a disordered structure for the high-J_C materials.

2. Experimental

Single crystals were received from Stanford University. The (nominally) optimally-doped crystals have critical temperatures T_c of 25-30 K. The overdoped and underdoped samples show T_c of 14-18 K. The weight of the crystals was estimated from their dimensions. For a reliable estimation of the dimensions we used scanning electron microscope.

The critical current, J_C, was measured exploiting both SQUID susceptometry for global magnetic measurements and magneto-optical technique for local magnetic imaging. The vortex structure in single crystals of $BaFe_2(As_{1-x}P_x)_2$ with different doping levels was studied by high resolution Bitter decoration technique, using magnetic nanoparticles [16]. Helium gas pressure was ~$2x10^{-2}$ Torr. High resolution of about 100 nm was achieved owing to the small average size (< 10 nm) of the magnetic particles. Decorations have been performed at temperatures $T_d \leq 6$ K in applied magnetic fields of 80-200 Oe perpendicular to the 'as grown' surface namely, parallel to the *c* axis. The vortex structures were studied and analyzed by application of Delaunay triangulation and Fourier transformation, allowing estimation of the extent of ordering of each vortex structure.

3. Results and discussion

The left panel of figure 1A shows vortex structure in applied magnetic field of 80 Oe in an overdoped sample of $BaFe_2(As_{1-x}P_x)_2$ with $T_c \approx 18$ K. The central panel shows the same image following Delaunay triangulation computation. This figure is composed of bright and grey and black areas. The bright areas correspond to domains with 6 close neighbors generating regular vortex lattice whereas the grey and black areas correspond to defective domains with number of close neighbors other than 6. In the grey areas the coordination number is 5 or 7 and in black areas corresponds to other coordination numbers. Defects of the vortex lattice are mostly located on domains boundaries allowing estimation of the size of the domains. Ratio of the area of bright zones to the area of grey and black zones can be regarded as an extent of vortex lattice perfection. This ratio was found to be 0.45 for the data of figure 1. Figure 1B shows the isothermal magnetic moment measurements of the same sample at 5 K, indicating low critical current density of 10^3-10^4 A/cm^2 in zero field.

A B

Figure 1. A. Left: Vortex structure at applied magnetic field of 80 Oe in the overdoped sample $BaFe_2(As_{1-x}P_x)_2$ ($T_c \approx 18$ K). **Right:** Delaunay triangulation of the same portion of structure. **B.** Magnetization loop of the same sample at T=5 K.

Figures 2A and 2B exhibit similar data for another overdoped sample, with $T_c \approx 14$ K. In this case, the ratio of the area of bright zones to the area of grey and black zones is 0.75 and J_C at 5 K in the range of 10^3-10^4 A/cm^2.
For an optimally-doped crystals (T_c =25 K) the ratio of the areas of bright to grey and black zones was less than 0.25 (Fig. 3A) and estimation of J_C based on the measurements of magnetic moment of the same sample (Fig. 3B) yields ~ 10^5 A/cm^2 at 5 K and zero field. It is important to note that the ratios of the area of bright zones to the area of grey and black zones are different in different parts of the same crystal; this can be associated with heterogeneities of the phosphorus content (see below).

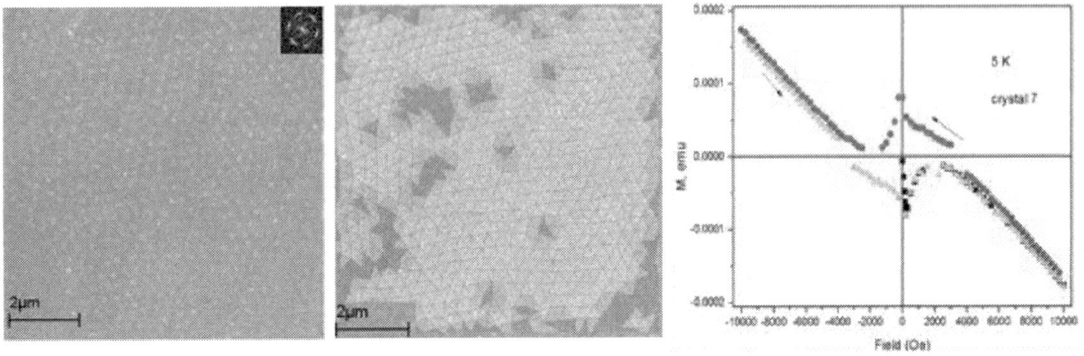

A B

Figure 2. A. Left: Vortex structure at applied magnetic field of 80 Oe in the overdoped sample $BaFe_2(As_{1-x}P_x)_2$ ($T_c \approx 14$ K). **Right:** Delaunay triangulation of the same portion of structure. **B.** Magnetization loop of the same sample at T=5 K.

A B

Figure 3. A. Vortex structure (**left**) and Delaunay triangulation (**right**) of a single crystal with optimal doping ($T_C \approx 25$ K) at applied magnetic fields 20 Oe. **B.** Magnetization loop of the same sample at T=5 K and T=12 K.

Figure 4A shows a magneto-optical image of optimally-doped crystal (T_c = 25 K) and profiles of magnetic field penetration (4B). Estimation of the critical currents density for this crystal, based on the magnetic field profiles, also yields a value of about 10^5 A/cm^2. The critical current density degrades slowly with increasing magnetic field and temperature [17].

The Fourier transforms of the vortex structures of two optimally-doped (T_c = 25 and 29 K) and an overdoped (T_c = 14 K) samples are shown in figure 5A, 5B, and 5C, respectively. While the Fourier transform of the overdoped crystal exhibits 6 well pronounced spots indicative of a regular triangular lattice, the optimally-doped crystals show a continuous circle that corresponds to arbitrary orientation of triangular elements.

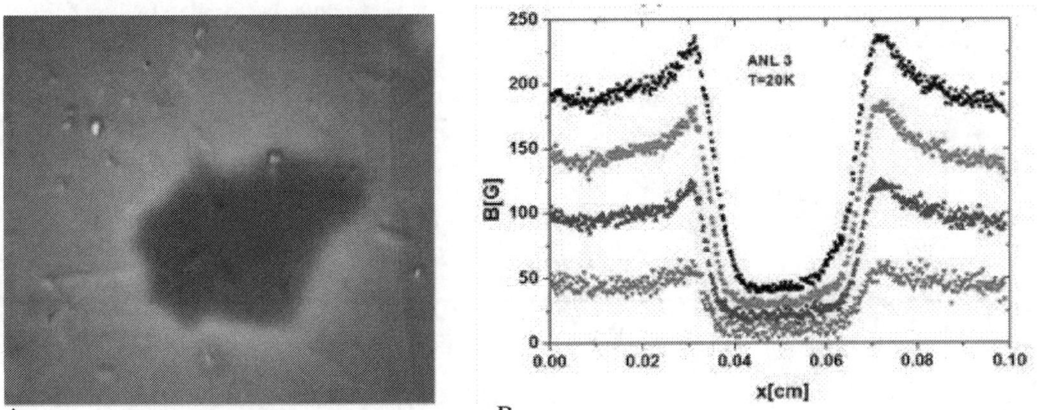

A B

Figure 4. Magneto-optical image of optimally-doped crystal **(A)** and profiles of the magnetic field penetration **(B).**

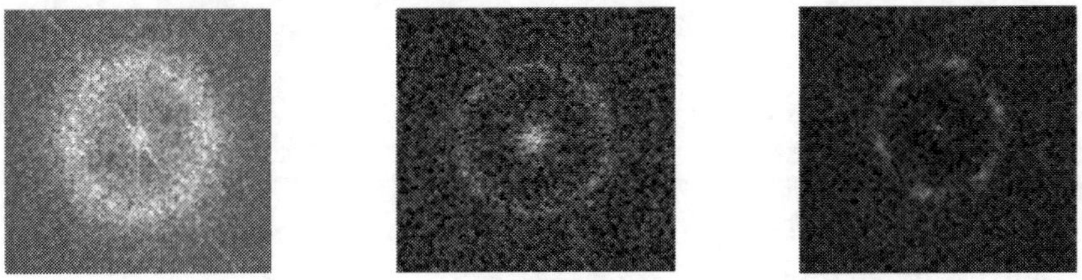
A B C

Figure 5. Fourier transforms of vortex structures of optimally-doped (T_c = 25, and 29 K) (**A** and **B**, respectively) and overdoped (T_c = 14 K) **(C)** samples.

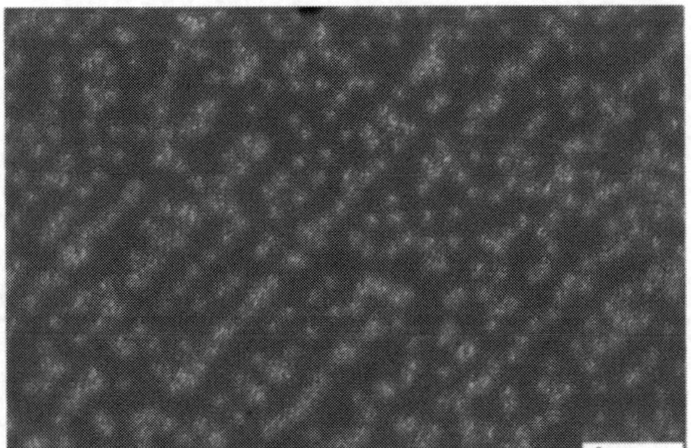

Figure 6. Vortex structure in optimally-doped $BaFe_2(As_{1-x}P_x)_2$ crystal in presumably underdoped area.

In crystals with optimal doping (nominally, x = 0.4) we observed certain areas with linear arrangement of vortices that can be related to twinning typical for underdoped crystals. Twinning in these crystals appears as a result of antiferromagnetic transition. Presence of underdoped areas can be explained by heterogeneities of a phosphorus distribution that were frequently revealed by local X-ray electron probe analysis. As shown in figure 6, the vortices in these areas are situated on the twin boundaries that probably serve as pinning centers.

To summarize, the main finding of this work is the observation of large, well-ordered domains of regular triangular vortex lattice over tens of vortex lattice periods in overdoped $BaFe_2(As_{1-x}P_x)_2$ crystals. Such structure correlates with reduced critical currents density of these crystals. In optimally-doped crystals the size of ordered domains is only a few vortex lattice periods as in other pnictides [18,19] and in iron-containing chalcogenides [20] all characterized by high currents density.

Acknowledgment
This work at Bar-Ilan University and Institute of Solid State Physics, Russian Academy of Science, was supported jointly by the Ministry of Science and Technology of Israel and the Russian Foundation for Basic Research (project no. 110292480). The work at the Hebrew University was partially supported by the Klachky Foundation for Superconductivity. The authors express their thanks to J. Analytis and I.Fisher, Stanford, for providing $BaFe_2(As_{1-x}P_x)_2$ single crystals. An anonymous referee contributed an important comment concerning the comparison of J_C in samples with different T_c.

References

[1] Kamihara Y, Watanabe T, Hirano M, Hosono H 2008 *J. Am. Chem. Soc.* **130** 3296
[2] Paglione J and Greene R L *Nature Physics* 2010 **6** 645
[3] Chen X H, Wu T, Wu G, Liu R H, Chen H and Fang D F 2008 Nature **453** 761
[4] Rotter M, Tegel M and Johrendt D 2008 *Phys. Rev. Lett.* **101** 107006
[5] Luetkens H, Klauss H H, Kraken M, Litterst F J, Dellmann T, Klingeler R, Hess C, Khasanov R, Amato A *Nat. Mater.* 2009 **8,** 305
[6] Yoichi Ando, Seiki Komiya, Kouji Segawa, Ono S, and Kurita Y *Phys. Rev. Lett.* 2005 **93** 267001
[7] Takeshita S, Kadono R, Hiraishi M, Miyazaki M, Koda A, Matsuishi S, Hosono H *Phys. Rev. Lett* 2009 **103**, 027002
[8] Butch N P, Saha S R, Zhang X H *et al. Phys. Rev. B* 2010 **81** 024518
[9] Izyumov Yu A and Kurmaev E Z *Phys. Usp.* 2008 **51**, 1307
[10] Kenji Ishida, Yusuke Nakai, Hideo Hosono *J. Phys. Soc. Jpn.* 2009 **78**, 062001
[11] Ma Y, Gao Z, Qi Y, Zhang X, Wang L, Zhang Z, Wang D, *Physica C* 2009 **469**, 651; Ma Y, *Supercond. Sci. Technol.* 2012 **25** 113001
[12] Shishido H *et al Phys. Rev. Lett.* 2010 **104**, 057008
[13] van der Beek C J, Konchukowski M, Kasahara S *et al. Phys. Rev. Lett.* 2010 **105** 267002; van der Beek C J, Rizza G, Konchukowski M *et al., Phys. Rev. B* 2010 **81**, 174517
[14] Larkin A I, Ovchinnikov Yu N *J. Low Temp. Phys.*, 1979 **34**, 409
[15] Fang *et al. Phys. Rev. B* 2011 **84** 140504(R)
[16] Essmann U, Trauble H, *Phys. Lett.* 1967 **47A** 526 ; Vinnikov L Ya, Grigor'eva I V, and Gurevich L A 1993 *The Real Structure of high-T_c Superconductors* (*Springer Series in Materials Science* vol. 23) ed V Sh Shekhtman (Springer-Verlag, Berlin Heidelberg) 89
[17] $J_C(H = 0)$ is proportional to the condensation energy, and hence – at low temperatures – to T_c^2. This may explain part of the difference in $J_C(H = 0)$ in the different samples. However, these samples also show different field dependence and this is due to pinning.
[18] Eskildsen M R, Vinnikov L Ya, Blasius T D, Veshchunov I S, Artemova T M, Densmore J M, Dewhurst C D, Ni N, Kreyssig A, Bud'ko S L, Canfield P C, and Goldman A I, *Phys. Rev. B* 2009 **79**, 100501(R)
[19] Eskildsen M R, Vinnikov L Ya, Blasius T D *et al. Physica C* 2009 **469**, 529
[20] Vinnikov L Ya, Radaev A V, Veshchunov I S, Troshina A G, Liu Y, Lin C T, and Boris A V *JETP Letters,* 2011 **93** 287–289.

Journal of Physics: Conference Series **507** (2014) 012014 doi:10.1088/1742-6596/507/1/012014

Split coil made of (RE)BCO pancake coils for $I_c(B)$ anisotropy measurements of superconductors

L Frolek[1], **E Pardo, F Gömöry, J Šouc and J Pitel**
Institute of Electrical Engineering, Slovak Academy of Sciences, Bratislava, Slovakia

E-mail: lubomir.frolek@savba.sk

Abstract. Measurement of the $I_c(B)$ anisotropy is standard characterization of superconducting tapes, wires or cables. This contribution presents a split coil consisting on two superconducting pancake coils in order to generate the magnetic field necessary for this kind of measurement. Both coils were made using (RE)BCO - based second generation (2G) coated conductor tape with cross section 0.1 mm x 12 mm. The individual turns of the tape were insulated by a fiberglass tape without impregnation. These coils have identical inner and outer diameter and number of turns. Their inner and outer diameters are 50 mm and 80 mm, respectively, and they have 62 turns. The length of conductor in each coil is approximately 13 m. The distance between both pancake coils is 22 mm. Individual coils and the complete split coil were characterized in liquid nitrogen bath. Their parameters, like the critical currents, *E(I)* characteristics and magnetic field of complete split coil, were measured and interpreted. The split coil can be used up to magnetic fields of 210 mT. The length between the potential taps on the sample can be up to 20 mm, while the magnetic field decrease is lower than 1% on this length.

1. Introduction

Copper split coils commonly used for $I_c(B)$ anisotropy measurements have oftentimes big dimensions and their cooling requires great volume of liquid nitrogen. These coils are starting to overheat and are electrically unstable at higher currents. For these reasons, we constructed a superconducting split coil. This coil was made using (RE)BCO – based second generation (2G) coated conductor tape. The tapes of this type are commercialized and some (RE)BCO coils have been developed [1-2]. This is in line with an effort to use high-temperature superconductors (HTS) in many applications. Their advantage is that these superconductors can operate at much higher temperatures than low-temperature superconductors (LTS). The split coil consisting of two pancake coils was made without impregnation, because epoxy impregnation can substantially degrade its performance [3-4]. Important parameters as dependences of electrical field on current (*E(I)* characteristics) and critical currents (I_c) were measured and compared for each pancake coil separately and also together (in final configuration and in serial connection) in liquid nitrogen bath. For the complete split coil, we measured the axial component of the magnetic field along the radial and axial axes and compared with calculated values. The hysteresis of magnetic field in the centre of split coil was measured and discussed too.

[1] lubomir.frolek@savba.sk

2. Description of the split coil

The split coil consists on two superconducting pancake coils. Both of these coils are connected in series and the distance between them is 22 mm. Each coil have inner diameter 50 mm, outer diameter 80 mm and 62 turns of superconducting tape. The length of conductor in each coil is approximately 13 m. They have been made from Super Power's (RE)BCO coated conductor SCS12050-AP which has a copper stabilizing layer. Cross section of the coated conductor is 0.1 mm x 12 mm and the self-field critical current (I_c) is 480 A at the temperature of liquid nitrogen (77K). Photography of the complete split coil is shown in figure 1 and the dimensioned sketch is shown in figure 2.

Figure 1. Photography of the complete split coil.

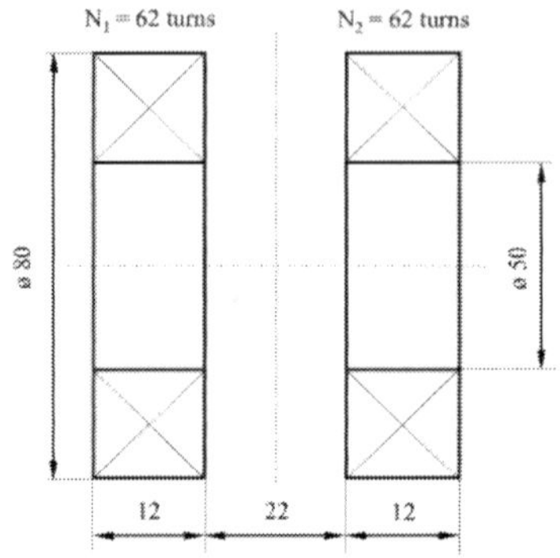

Figure 2. Dimensional sketch of the split coil.

The individual turns in these coils were insulated with a fibreglass tape 0.12 mm thick and 12 mm wide and were wound simultaneously with the superconducting tape. The bobbin of this coil is made from fibreglass and in the centre there is a hole with diameter 20 mm for sample placement. Current leads were located on the inner part of individual coils and the coils were interconnected on their outer part. The complete split coil was made without impregnation and forms a compact unit.

3. Achieved parameters and discussion

3.1. Comparison of the important parameters of individual coils

First were measured *E(I)* characteristics for each pancake coil separately and then for each one pancake coil together in complete split coil (with common supply). The potential taps were soldered on both ends of superconducting tapes. The coil terminal voltage was measured to achieve corresponding electric field in range 1 μV per 1 cm of tape. In order to prevent the damage of coils during measurement, for assessment of safe terminal voltage was taken the length of innermost (first) turn into account, because the innermost turn limits the maximum DC current the coil could carry [5-6]. Measured data were used for determination of critical currents according to power law dependence given by equation (1)

$$E(I) = E_0\left(\frac{I}{I_c}\right)^n \qquad (1)$$

where

E = electric field [μV/cm], E_0 = 1 [μV/cm], I = current [A], I_c = critical current [A], n = parameter.

Critical currents of both coils (I_c) (assuming that the voltage is uniformly distributed on the whole coil) and safe critical currents (I_{sc}) (assuming that the entire voltage is on the first turn) were established.

Using the safe critical currents, the coils are protected, because in reality the entire voltage is not only on the first turn.

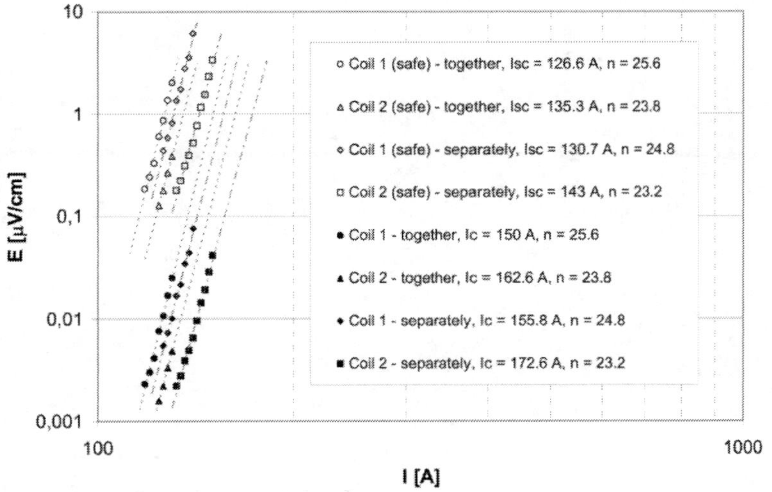

Figure 3. Graph of the $E(I)$ characteristics of both coils measured both separately and together forming the complete split coil using safe critical current (I_{sc}) and usual (I_c) interpretation.

Measured $E(I)$ characteristics with critical currents and n parameters are shown in figure 3. Comparison of the critical currents of the first and second coil in individual cases reveals a difference between values between 6 and 11%, although the coils have identical geometry and were made by identical technology. This is probably due to inhomogeneities or local I_c degradation of the tape [6]. Therefore the maximum supply current of complete split coil was established on 120 A, which is certainly safe current (considering the first coil).

3.2. Important parameters of the split coil

The distribution of magnetic field inside the split coil is one important parameter significant for $I_c(B)$ anisotropy measurements. The system axis and method of measurement are sketched in figure 4. As method of measurement, the Hall probe mapping method was used.

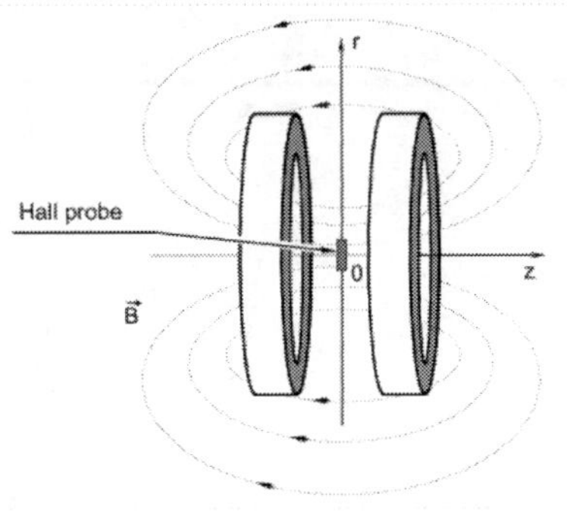

Figure 4. Methodological sketch for measurement of the split coil magnetic parameters. Zero point (0) is the centre of the split coil and the axis system together.

The axial part of magnetic field in axial (z) and radial (r) direction were measured and compared with calculated one assuming uniform current density in the coil. These distributions are shown in figure 5 and figure 6.

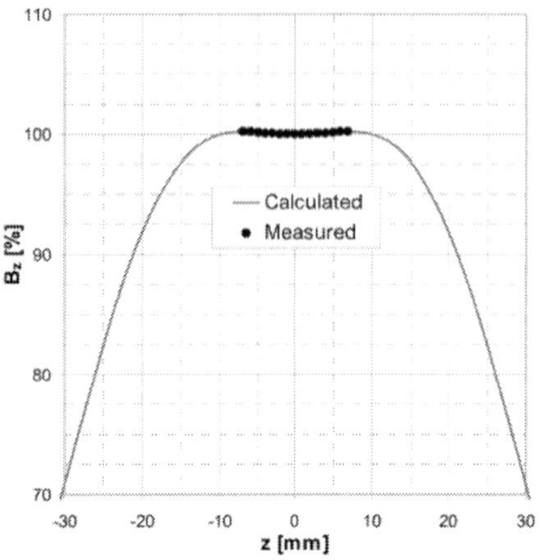

Figure 5. The distribution of magnetic field along the axial direction (on axis z).

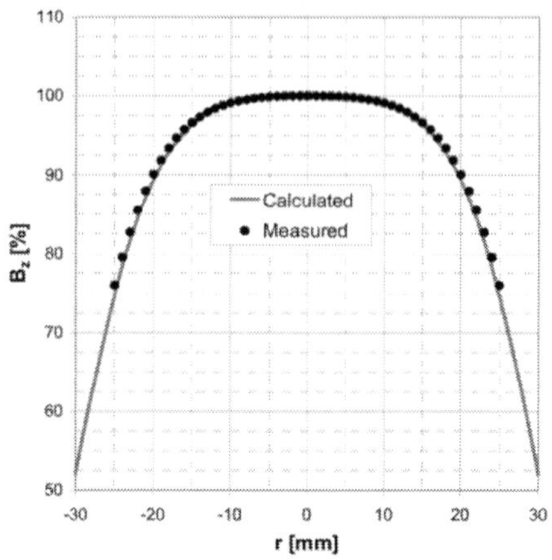

Figure 6. The distribution of magnetic field along the radial direction (on axis r).

The calculations and measurements are in good agreement in both cases. On axis z is homogeneity of magnetic field up to distance ±10 mm better than ±0.3% and on axis r at the same distance is drop of magnetic field lower than 1% (in the centre of the split coil is value of magnetic field 100%). This is important information for $I_c(B)$ anisotropy measurement. If the length between the potential taps on the sample will be 20 mm, decrease of magnetic field on this length will be lower than 1%. For higher precision the length between the potential taps must be reduced.

Another important parameter is hysteresis of magnetic field. This hysteresis of the split coil was measured in Zero point (in the centre of the split coil) and results of measurement are shown in figure 7.

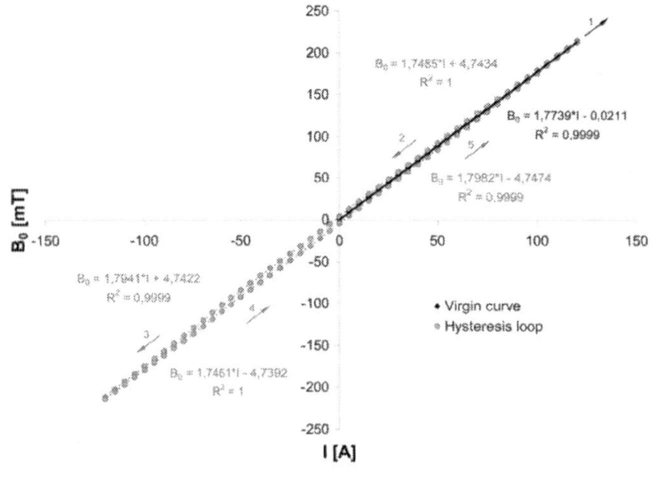

Figure 7. The hysteresis of magnetic field measured in the centre of the split coil (in Zero point). The arrows indicate the sequence of measurements.

From figure 7 is noticeable that the split coil has hysteresis, which can cause inaccuracies at $I_c(B)$ anisotropy measurements. The inaccuracies are the highest at the lowest magnetic fields. Either this fact is necessary to take into consideration during all measurements or all experiments have to start from zero-field cool and the current rising monotonically following the virgin curve (if afterwards a lower magnetic field is desired, the coil has to be previously warmed up). The constant of split coil determined from virgin curve is $k = 1.774$ mT/A. Using supply current of 120A, will be magnetic field of 212.88 mT in the centre of the split coil and so 210 mT is certainly safely achievable magnetic field.

4. Summary and conclusions

A split coil appointed for measurements of $I_c(B)$ anisotropy of superconductors was made using two (RE)BCO pancake coils without impregnation. Based on the parameters of the coils, the safe supply current was established up to 120A. Guaranteed magnetic fields in the centre of the split coil are up to 210 mT and constant is 1.774 mT/A. The length between the potential taps on the sample can be up to 20 mm, while the magnetic field decrease is lower than 1% on this length. The coil has magnetic field hysteresis, therefore, for accurate measurements all experiments have to start from zero-field cool and the current rising monotonically following the virgin curve. The main advantages of this coil in comparison with a copper split coil are: no overheating at high supply currents, small size and low liquid nitrogen consumption at measurement. Proposed future work is to study relaxation effects.

Acknowledgments

The research leading to these results has received funding from the European Union Seventh Framework Programme under grant agreement number NMP3-LA-2012-280432 "EUROTAPES". We also acknowledge the support of EURATOM FU-CT-2007-00051 project co-funded by the Slovak Research and Development Agency under contract number DO7RP-0018-12.

References

[1] Friedman A, Wolfus Y, Kopansky F and Yeshurun Y 2012 *Physics Procedia* **36** 1169
[2] Daibo M, Fujita S, Haraguchi M, Kikuchi K, Iijima Y and Saitoh T 2011 *Physica C* **471** 1436
[3] Takematsu T, Hu R, Takao T, Yanagisawa Y, Nakagome H, Uglietti D, Kiyoshi T, Takahashi M and Maeda H 2010 *Physica C* **470** 674
[4] Yanagisawa Y, Nakagome H, Takematsu T, Takao T, Sato N, Takahashi M and Maeda H 2011 *Physica C* **471** 480
[5] Pardo E, Vojenčiak M, Gömöry F and Šouc J, 2011 *Supercond. Sci. Technol.* **24** 065007
[6] Gömöry F, Šouc J, Pardo E, Seiler E, Solovyov M, Frolek L, Skarba M, Konopka P, Pekarčíková M and Janovec J 2013 *IEEE Transactions on Applied Superconductivity* **23** 5900406

Journal of Physics: Conference Series **507** (2014) 012015 doi:10.1088/1742-6596/507/1/012015

Amorphous FeAs-free $SmFeAsO_{1-x}F_x$ using low temperature sintering with slow cooling

Masaya FUJIOKA[1], Saleem J. DENHOLME[1], Hiroyuki OKAZAKI[1], Keita DEGUCHI[1,2], Satoshi DEMURA[1,2], Hiroshi HARA[1,2], Hiroyuki TAKEYA[1], Takahide YAMAGUCHI[1], Hiroaki KUMAKURA[1,2], and Yoshihiko TAKANO[1,2]

[1]National Institute for Materials Science, Tsukuba, Ibaraki 305-0047, Japan

[2]University of Tsukuba, Tsukuba, Ibaraki 305-0001, Japan

FUJIOKA.Masaya@nims.go.jp

Abstract. We obtained the highest superconducting transition temperature of $SmFeAsO_{1-x}F_x$ at 58.1 K by using low temperature synthesis with slow cooling. We found that the low temperature sintering suppresses the formation of amorphous FeAs, and the slow cooling introduces a high level of fluorine into oxygen site. By using this method, impurity phases are not formed up to x = 0.16, and they gradually increase above x = 0.16. However, the resistivity of the normal state continues to decrease and T_c onset gradually increases with increasing fluorine concentration even when x > 0.16. Therefore, it is suggested that the fluorine is gradually introduced into oxygen site above x = 0.16.

1. Introduction

The first report of $SmFeAsO_{1-x}F_x$ was published in 2008 by Chen *et al* [1]. The superconducting transition temperature (T_c) was reported at only 43 K. Afterward, T_c was immediately improved to around 55 K [2]. Through the various researches, it is found that introduction of a high level of fluorine into oxygen site is effective to obtain the superconductivity with a high T_c. However, up to the present time, overdoped $SmFeAsO_{1-x}F_x$ has not still been reported. Therefore, how to effectively introduce fluorine is important in this system. Table 1 shows the various

Table 1. Synthesis conditions for SmFeAsO1-xFx

$SmFeAsO_{1-x}F_x$	Sintering temperature (°C)	T_c (K)	Reference
x = 0.15	1160	43	[1]
x = 0.35	1160-1180	52	[3]
x = 0.20	1160	54	[4]
x = 0.2 + δ (δ = 0.4)	1100	55	[5]
x = 0.20	1000	56.1	[6]
x = 0.25	900	57.8	[7]

reports for synthesis of $SmFeAsO_{1-x}F_x$. In this table, fluorine concentrations, sintering temperatures and T_c are presented. Interestingly, T_c seems to relate with the sintering temperature. It gradually increases with decreasing sintering temperature. To obtain $SmFeAsO_{1-x}F_x$ with a high superconducting transition temperature, it is necessary to reveal the effect on a low temperature sintering.

2. Low temperature sintering with slow cooling

Figure 1 (a) and (b) show the polished surface of $SmFeAsO_{0.9}F_{0.1}$ sintered at 1200 °C and 980 °C respectively. Although a high temperature synthesis of 1200 °C produces an amorphous FeAs impurity phase, a low temperature synthesis 980 °C does not form this impurity phase. Amorphous FeAs prevents the current from flowing, because it is formed between superconducting grains. Therefore, to enhance the superconducting properties, how to remove this impurity from between superconducting grains should be considered. We found that a low temperature sintering is one of the effective methods to avoid the formation of this impurity phase. Actually, any impurities phases cannot be observed in the sample sintered at 980 °C as shown in figure 1 (b).

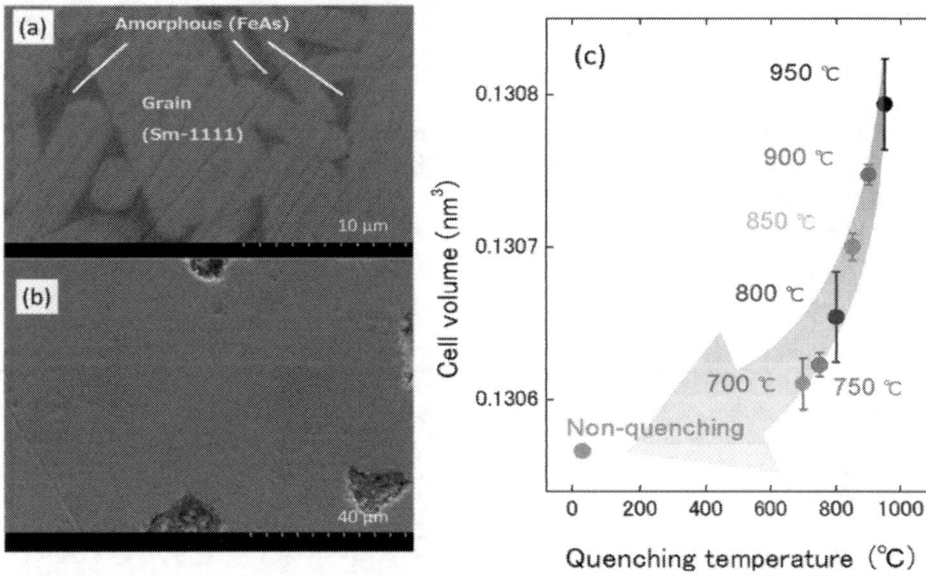

Figure1. (a): SEM image of the polished surface of $SmFeAsO_{0.9}F_{0.1}$ sintered at 980 °C. (b): SEM image of the polished surface of $SmFeAsO_{0.9}F_{0.1}$ sintered at 1200 °C. (c): Cell volume versus quenching temperature.

In addition, slow cooling is also effective to obtain the homogeneous $SmFeAsO_{1-x}F_x$ with a high fluorine concentration. To investigate the change of fluorine concentration during slow cooling, the following experiment was performed. For the preparation of $SmFeAsO_{1-x}F_x$, stoichiometric Sm_2O_3, SmF_3 and two precursors, composed of compounds made of Sm, Fe, and As, were ground in a mortar and compressed into pellets. The obtained pellets were heated at 980 °C for 40 h in an evacuated quartz tube. After that, from a maximum heating temperature, samples were slowly cooled at the rate of 5 °C / h down to 600 °C. During slow cooling , samples were taken out one by one from an electronic furnace and quenched by water at the following temperatures (950, 900, 850, 800, 750, and 700 °C).A subsection Figure 1 (c) shows the cell volume versus the temperature at which the samples were quenched during the slow cooling process. The gradual decrease in cell volume can be observed at a lower quenching temperature. When the fluorine is substituted for the oxygen site, the cell volume is supposed to decrease, because the ion size of fluorine is smaller than that of the oxygen. It is found that fluorine is introduced into the oxygen site not only at the maximum temperature but also during the cooling process. Even if the same nominal amount of fluorine is mixed in synthesizing, it is demonstrated that using low temperature sintering with slow cooling makes it possible to introduce a large amount of fluorine.

Therefore, the samples with various fluorine concentration from x = 0 to 0.30 were prepared by a low temperature sintering with slow cooling. Figure 2 shows the XRD pattern for these samples. Impurity phases start to increase from x = 0.16. The intensities of the peaks for SmOF and SmAs gradually increase when x > 0.16. On the other hand, in all samples where x < 0.16, their intensities are quite small and almost comparable. This means the solid solubility limit is near x = 0.16. However, the resistivity of the normal state continues to decrease and T_c onset also continues to increase with increasing x as shown in figure 3. As a result, maximum T_c is observed at 58.1 K [8]. This suggests that fluorine continues to be gradually introduced into $SmFeAsO_{1-x}F_x$ above x = 0.16. Moreover, the *a* lattice parameter estimated from XRD pattern also shows the gradual decrease. This result also suggests the gradual introduction of fluorine above x = 0.16 [9].

In this study, we found that a low temperature sintering suppresses the formation of amorphous FeAs and a slow cooling introduces a high level of fluorine into oxygen site. By using this method, the record T_c of 58.1 K is obtained in the sample where x = 0.26.

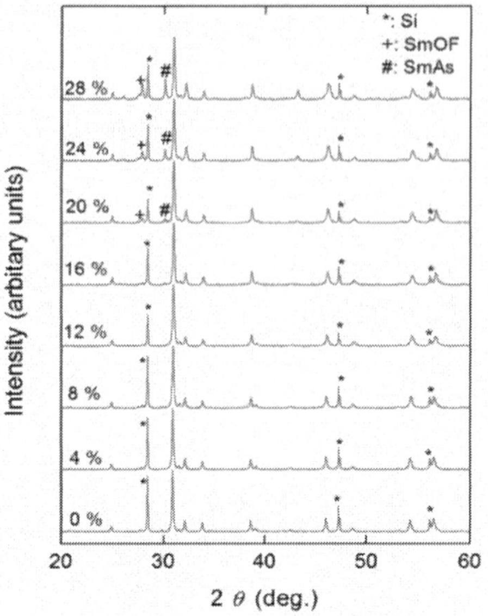

Figure 2. XRD patterns of $SmFeAsO_{1-x}F_x$ (x = 0, 0.04, 0.08, 0.12, 0.16, 0.2, 0.24, 0.28).

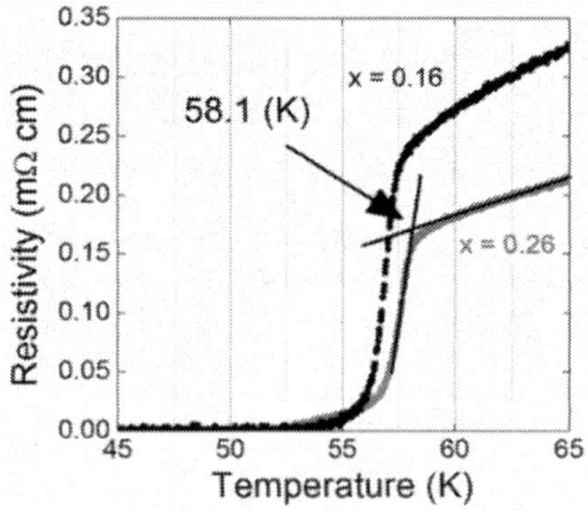

Figure 3. Temperature dependence of resistivity for SmFeAsO1-xFx (x = 0.16, 0.26).

Acknowledgment
This work was supported in part by the Japan Society for the Promotion of Science through 'Funding program for World-Leading Innovative R&D on Science Technology Program (FIRST)' and Japan

Science and Technology Agency through Strategic International Collaborative Research Program (SICORP-EU-Japan).

References

[1] X. H. Chen, T. Wu, G. Wu, R. H. Liu, H. Chen and D. F. Fang: Nature **453** (2008) 761.
[2] Y. Kamihara, T. Nomura, M. Hirano, J. E. Kim, K. Kato, M. Takata, Y. Kobayashi, S. Kitao, S. Higashitaniguchi, Y. Yoda, M. Seto, and H. Hosono; New J. Phys. **12** (2010) 033005.
[3] Z. Gao, L. Wang, Y. Qi, D. Wang, X. Zhang, Y. Ma, H. Yang and H. Wen, Supercond. Sci. Technol. **21**, 112001 (2008).
[4] R. H. Liu, G.Wu, T.Wu, D. F. Fang, H. Chen, S.Y. Li, K. Liu, Y. L. Xie, X. F.Wang, R. L. Yang, L. Ding, C. He, D. L. Feng, and X. H. Chen, Phys. Rev. Lett. **101**, 087001 (2008).
[5] Y. L. Chen, Y. J. Cui, C. H. Cheng, Y. Yang, L, Wanga, Y. C. Li, Y. Zhang and Y. Zhao, Physica C **470**, 989 (2010).
[6] C. Wang, Z. Gao, L. Wang, Y. Qi, D. Wang, C. Yao, Z. Zhang and Y. Ma, Supercond. Sci. Technol. **23**, 055002 (2010).
[7] S. J. Singh, J. Shimoyama, A. Yamamoto, H. Ogino and K. Kishio, IEEE Trans. Appl. Supercond. 23, 7300605 (2013)
[8] M. Fujioka, S. J. Demholme, T. Ozaki, H. Okazaki, K. Deguchi, S. Demura, H. Hara, T. Watanabe, H. Takeya, T. Yamaguchi, H. Kumakura and Y. Takano, Supercond. Sci. Technol. **26**, 085023 (2013).
[9] M. Fujioka, S. J. Demholme, T. Ozaki, H. Okazaki, K. Deguchi, S. Demura, H. Hara, T. Watanabe, H. Takeya, T. Yamaguchi, H. Kumakura and Y. Takano, J. Phys. Soc. Jpn, **82**, 094707 (2013).

Crystallinity and in-field characterization of high-T_c $YBa_2Cu_4O_8$ and $Y_{0.9}Ca_{0.1}Ba_2Cu_4O_8$ epitaxial films fabricated by low temperature flux method

S Funaki, Y Yamada and F Nakayama

Department of Material Science, Shimane University, 1060 Nishikawatsu, Matsue, 690-8504, JAPAN

E-mail: s-funaki@riko.shimane-u.ac.jp

Abstract. $REBa_2Cu_4O_8$ superconductor has an important advantage that the critical temperature (T_c) is improved by Ca substitution for RE site. We fabricated the $YBa_2Cu_4O_8$ (Y124) and $Y_{0.9}Ca_{0.1}Ba_2Cu_4O_8$ (YCa124) films by molten hydroxide method in ambient atmosphere and investigated the crystallinity and in-field characterization. Obtained samples show that the film alignment is biaxial orientation and the grain size is larger than 50 μm, and the irreversibility field of YCa124 film is superior to that of Y124 in all the measured magnetic fields

1. Introduction

For applications of the RE-Ba-Cu-O coated conductor operated at boiling temperature of liquid nitrogen (77.3 K) to various fields such as superconducting cable, transformer, the magnet and superconducting magnetic energy storage (SMES), an improvement of the coated conductors in critical current density (J_c) in magnetic fields is required. In the case of $REBa_2Cu_3O_y$ (RE123; RE: rare earth elements) coated conductors, a biaxial alignment of growth crystals, an introduction of pinning centers and the improvement of a critical temperature (T_c) have been greatly conducted. Although many groups have reported the improvement of T_c in RE123 film by using liquid phase such as Ba-Cu-O flux [1, 2], these special techniques need a high growth temperature during film fabrication and cause degradation of superconducting properties due to impurity diffusion from metallic tape substrate.

The $REBa_2Cu_4O_8$ (RE124) has a rigorous stability of oxygen stoichiometry and no structural phase transition as opposed to RE123 phase. Moreover, RE124 keeps stable at far lower temperature than RE123 in the ambient atmosphere [3]. Although the T_c of RE124 is inferior to RE123, the Y124 has an important advantage that the T_c was improved by Ca substitution for RE site [4]. Single phase and single crystal of bulk RE124 phase are fabricated under high oxygen pressures about 400 atms and temperatures over 1000°C [5]. 1-atm-pressure preparation is possible not only with addition of reaction enhancement reagents [6-8], but also by simple solid state reaction with extremely long annealing time. On the other hand, the preparation methods of single crystalline RE124 compound are reported by using molten alkali hydroxide in ambient atmosphere. Sandford et al. described synthesis of Eu124 single crystal from molten potassium hydroxide [9], Song et al. reported a more feasible

Corresponding author. Tel.: +81-852-32-6406 ; fax: +81-852-32-6406.
E-mail address: s-funaki@riko.shimane-u.ac.jp

process for Y124 growth [10]. Recently, we achieved the synthesis of *c*-axis oriented RE124 epitaxial films fabricated by molten hydroxide method with low heating temperature of ~650°C, and we reported that the T_c of Ca-doped Y124 (YCa124) film achieved 90 K, which is 10 K higher than that of Y124 film [11].

In this investigation, we fabricated the biaxial oriented YCa124 films on $NdGaO_3$ (NGO) substrate by molten hydroxide method and also discussed the crystallinity and in-field characterizations.

2. Experimental

Yttrium oxide, calcium carbonate, barium carbonate and copper oxide powders were used as starting materials, and potassium hydroxide (KOH) was used as a solvent. These starting materials were weighed with a total weight of 10 g with a molar ratio of Y : Ba : Cu = 1 : 2 : 4 and Y : Ca : Ba : O = 0.9 : 0.1 : 2 : 4 (amount of Y / Ca substitution *x*: 0.1) for Y124 and YCa124, respectively, and then mixed and put into alumina crucible with KOH of 100wt% to the raw materials. The NGO (001) single crystalline substrate was also put into the mixture. Then the mixture was heat-treated using a muffle furnace at 641°C for 12 hours in air. The obtained samples were washed in distilled water and ethanol using ultrasonic cleaning to eliminate the KOH and K_2CO_3.

Phase identification and orientation were measured by X-ray diffraction (XRD) pattern with a CuKα source. Grain size and surface morphology were obtained by scanning electron microscope (SEM). In-field characterization was measured by a standard four-probe method.

3. Results and discussion

3.1. Crystallinity

It is well known that the RE-Ba-Cu-O superconducting material has anisotropic physical properties due to layered crystal structure with alternating two-dimensional CuO_2 planes, and the J_c along the CuO_2 planes (// *ab*-plane) is superior to that of along the *c*-axis direction. Furthermore, in the case of RE123, the increase of misorientation angle between two grains causes serious degradation of J_c, even for the *c*-axis oriented film [12].

The *c*-axis orientation of Y124 and YCa124 films are shown in Fig. 1, moreover, Fig. 2 shows in-plane orientation of (a) Y124 and (b) YCa124 films. It is revealed that the Y124 and YCa124 films showed sharp and intense diffraction patterns of the RE124 phase, and Y124 crystals grew epitaxially on NGO (001) substrate with 45° rotations in *ab*-plane direction. From these results, we can conclude that the Y124 and YCa124 film has biaxial alignment.

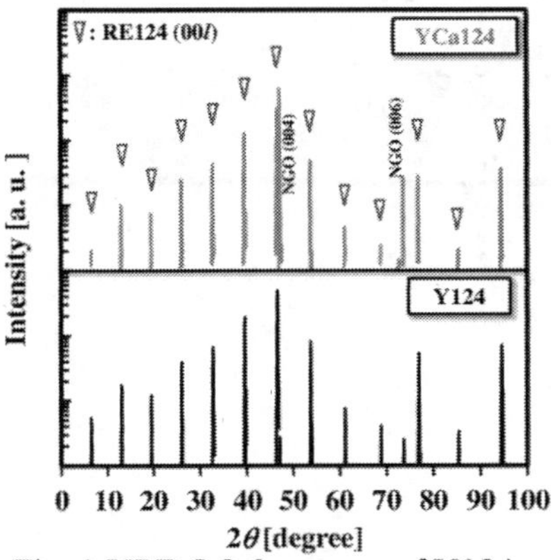

Fig. 1 XRD 2θ-θ patterns of Y124 and YCa124 film

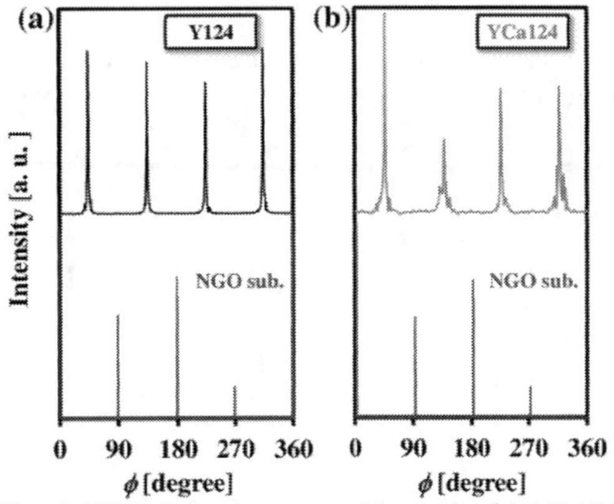

Fig. 2 XRD $2\theta\chi$-ϕ patterns (ϕ-scan) of (a) Y124 and (b) YCa124 film. Diffraction peaks of both film and NGO substrate were observed by the Y124 (108) and NGO (204) reflection, respectively.

Fig. 3 Surface morphology of (a) Y124 and (b) YCa124 films

In order to discuss the crystallinity, we observed surface morphologies of Y124 and YCa124 films by SEM. Fig. 3 (a) and (b) shows SEM image of Y124 and YCa124 films, respectively. Both films have large grain size (> 50 μm). Moreover, it is found that the grain size was widened by Ca doping, and average grain size of YCa124 film achieve 100 μm. These results indicate that molten hydroxide method is novel technique for obtaining single crystalline film.

3.2. In-field characterization

For application of the RE-Ba-Cu-O, the improvement of the irreversibility field (B_{irr}) is important, and B_{irr} was determined for understanding flux pinning states.

R-T curves under *B // c* and *B // ab*-plane for the Y124 film are shown in Fig. 4 (a) and (b), respectively. Broadening of the superconducting transition occurred by an applied magnetic fields. However, the degree of transition broadening largely depended on the *B* direction, and the degree of transition broadening for the Y123 film under *B // c* is obviously larger than that under *B // ab*, which is a typical behavior in Y123 films [13]. In this report, the transition broadening in *B // c* for the Y124 film is larger than that in *B // ab*, as in the case of Y123 films. This indicates that the Y124 film has an intense pinning force due to intrinsic pinning by alternating stacked structure with blocking layer.

Fig. 5 shows the irreversibility line of Y124 and YCa124 films determined from *R-T* curves under various magnetic fields. Interestingly, B_{irr} of YCa124 film is superior to that of Y124 in all the measured magnetic fields. In order to discuss the difference of B_{irr}, we observed high-resolution SEM image of Y124 and YCa124 films. It was found that plenty of etch pits are present only in the YCa124 film (as shown in Fig. 6), and we can speculate that the B_{irr} of YCa124 was enhanced by plenty of dislocation.

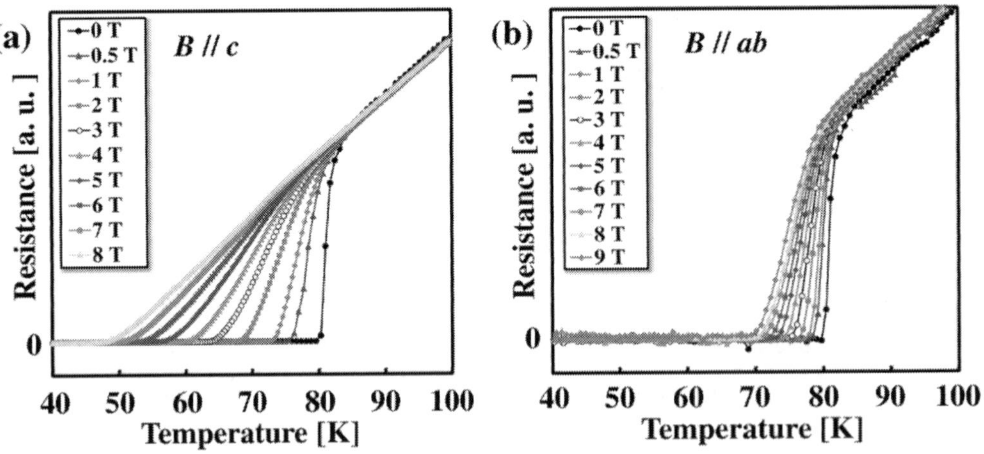

Fig. 4 *R-T* curves for Y124 film of (a) *B // c* and (b) *B // ab* in various magnetic field intensity

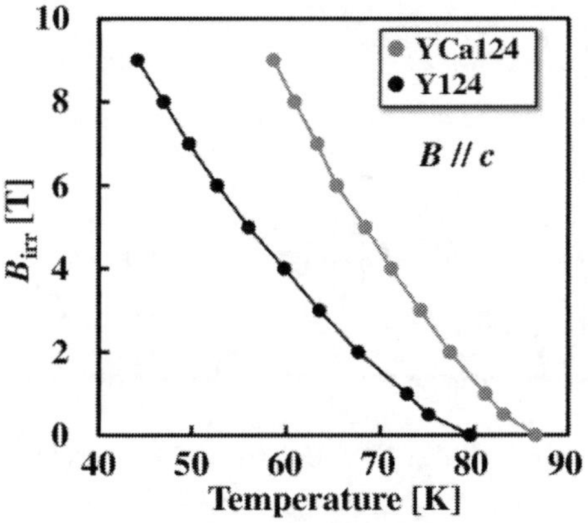

Fig. 5 Irreversibility line of Y124 and YCa124 films under various magnetic fields

Fig. 6 High-resolution SEM image of YCa124 film

4. Conclusion

We fabricated the Y124 and YCa124 films by molten hydroxide method in ambient atmosphere and investigated the crystallinity and in-field characterization. Y124 and YCa124 films show a biaxial orientation on $NdGaO_3$ (001) substrate with 45° rotations in *ab*-plane direction, and both films have large grain size (> 50 μm). Moreover, the grain size of film surface is wider up to 100 μm in the Ca substituted film. Broadening of the superconducting transition in *R-T* curves is occurred by an applied magnetic fields, and transition broadening under *B // c* is larger than that under *B // ab*. This indicates that the Y124 film has an intense intrinsic pinning such as RE123. Accordingly, the molten hydroxide method is suitable for the production of RE124 film with high crystalline perfection and in-field properties, which is an indispensable feature for reliable RE-Ba-Cu-O applications.

Acknowledgment

The authors appreciate of The Center for Integrated Research in Science, Shimane Univ. for in-field measurements.

References

[1] Yun K S, Choi B D, Song J H, Matsumoto Y, Ahmet P, Chikyow T, Kawasaki M and Koinuma H 2002 *Physica C* **378-381** 1202
[2] Ichino Y, Matsuo H, Sudoh K, Yoshida Y and Takai Y 2003 *Jpn. J. Appl. Phys.* **42** L243
[3] Wada T, Suzuki N, Ichinose A, Yaegashi Y, Yamauchi H and Tanaka S 1990 *Jpn. J. Appl. Phys.* **29** L915
[4] Miyatake T, Gotoh S, Koshizuka N and Tanaka S 1989 *Nature* **341** 41
[5] Karpinski J, Kaldis E, Julek E, Rusiecki S and Bucher B 1988 *Nature* **336** 660
[6] Cava R J, Krajewksi J J, Peck jr W F, Batlog B, Rupp Jr L W , Fleming R M, James A C W P and Marsh P 1989 *Nature* **338** 328
[7] Adachi S, Adachi H, Setsune K and Wada K 1991 *Physica C* **175** 523
[8] Tatsumi M, Kawazoe M and Yamamoto S 1996 *Physica C* **262** 261
[9] Sandford D, Marquez L N and Stacy A M 1995 *Appl. Phys. Lett.* **67** 422
[10] Song Y T, Peng J B, Wang X, Sun G L and Lin C T 2007 *J. Cryst. Growth* **300** 263
[11] Funaki S, Nakayama and F, Yamada Y 2012 *Physics Procedia* **27** 284
[12] Dimos D, Chaudhari P, Mannhart J and LeGoues F K 1988 *Phys. Rev. Lett.* **61** 219
[13] Iye Y, Tamegai T, Takeya H and Takei H 1987 *Jpn. J. Appl. Phys.* **26** L1057

Manufacture of Bi-cuprate thin films on MgO single crystal substrates by chemical solution deposition

J-C Grivel[1], C V Bertelsen[1], N H Andersen[2], K Kępa[1,3], T Hlásek[4], K Rubešová[4], H Huhtinen[5] and P Paturi[5]

[1]Department of Energy Conversion and Storage, Technical University of Denmark, Frederiksborgvej 399, 4000 Roskilde, Denmark
[2]Department of Physics, Technical University of Denmark, Fysikvej, 2800 Kgs Lyngby, Denmark
[3]Chemical and Process Engineering Division, Poznan University of Technology, pl. M. Skłodowskiej-Curie 2, 60-965 Poznań, Poland
[4]Institute of Chemical Technology, Department of Inorganic Chemistry, Technická 5, Prague 6, 166 28, Czech Republic
[5]Wihuri Physical Laboratory, Department of Physics and Astronomy, University of Turku, 20014 Turku, Finland

E-mail: jean@dtu.dk

Abstract. $Bi_2Sr_2CaCu_2O_8$ thin films have been deposited on MgO single crystal substrates by spin-coating a solution based on 2-ethylhexanoate precursors dissolved in xylene. Pyrolysis takes place between 200°C and 450°C and is accompanied by the release of 2-ethylhexanoic acid, CO_2 and H_2O vapour. Highly c-axis oriented $Bi_2Sr_2CaCu_2O_8$ as well as Er- or Ho-doped $Bi_2Sr_2(Ca,Ln)Cu_2O_8$ (Ln = Er, Ho) films were obtained after heat treatment at 840°C in air.

1. Introduction

Coated conductor tapes based on $REBa_2Cu_3O_7$ (RE = rare earth) thin films deposited on cube textured substrates can nowadays be produced in long lengths with remarkably high critical current densities (j_c) in excess of 1 MA/cm^2 at 77 K [1,2], which is more than three orders of magnitude higher than in $YBa_2Cu_3O_7$ wires made by the Powder-in-Tube (PIT) technology [3-5]. This large improvement is due to the decrease of the intergrain j_c when the grain misorientation angle exceeds 10° [6]. Nevertheless, the j_c of coated conductors at 77K under external magnetic fields is not high enough for some applications and cooling to 20K – 30K may be needed. In this temperature range, the $Bi_2Sr_2CaCu_2O_8$ (Bi2212) superconductor can also reach very high j_c values under magnetic fields in the form of powder-in-tube wires or ribbons [7-9], with c-axis texture but no in-plane texture. The intergranular j_c of Bi2212 is also lower when the grain boundary misorientation angle exceeds 10° [10], so that the application of a coated-conductor architecture to Bi2212 tapes could result in large performance improvements. Thin films deposited by vacuum techniques show J_c values in excess of 10 MA/cm^2 at 5K [10,11]. Only a few reports dealing with Bi2212 thin films prepared by solution deposition have been published to date [12-19], in spite of the potential of this technique for producing long textured

Journal of Physics: Conference Series **507** (2014) 012017 doi:10.1088/1742-6596/507/1/012017

coated conductors. The films are deposited on MgO single crystals and their T_c is generally between 70K and 84K [13-15,17-19]. In this contribution, we summarize our studies on the pyrolysis of a precursor solution based on 2-ethylhexanoates and on attempts at doping the films with Ho and Er.

2. Experimental details

The coating solutions were prepared by dissolving 2-ethylhexanoate salts of Bi (Alfa Aesar, 70% solution in xylene), Sr (Alfa Aesar, 40% solution in 2-ethylhexanoic acid), Ca (Alfa Aesar, powder), Ho (home made), Er (home made) and Cu (Aldrich, powder) in xylene (Aldrich). For the preparation of Ho- and Er-2-ethylhexanoates, Ho acetate and Er acetate were dissolved in 2-ethylhexanoic acid and excess solvent was evaporated by heating to 90°C. The cation ratio of the coating solutions was Bi : Sr : Ca : (Ho,Er) : Cu = 2 : 2 : 1-x : x : 2 and the solutions had a metal molar concentration of 0.3M.

MgO (100) single crystals from CrysTec GmbH were used as substrates. Details on substrate pre-treatment and film deposition can be found in a previous publication [19]. 10 layers were deposited by spin coating. Annealing of the films was performed in air at 840°C.

Simultaneous thermogravimetry (TG) and differential thermal analysis (DTA) measurements were performed in a STA 449C from Netzsch with a heating rate of 10K/min in air (gas flow 40 ml/min) using Al_2O_3 crucibles. Fourier transform infra-red (FTIR) spectra were recorded in a Bruker Tensor 27 spectrometer coupled to the exhaust line of the TG/DTA device. X-ray diffraction (XRD) patterns were collected in a MD-10 bench-top X-ray diffractometer (for powder samples) or in a Bruker D8 diffractometer (for thin films), both using CuKα radiation. Scanning electron microscopy (SEM) was performed in a TM3000 Tabletop Microscope from HITACHI equipped with a QUANTAX 70 EDS analysis system. T_c was determined from AC-susceptibility measurements.

3. Results and discussion

The TG and DTA traces recorded on the precursor solution dried at 130°C (Fig. 1-left) show that the mass loss takes place in several overlapping steps from 200°C to 450°C. Beyond 450°C and up to 680°C, the sample mass is nearly constant. The mass loss at 500°C (67.3 %) is close to the theoretical value for the decomposition of 2-ethylhexanoates into oxide phases [19].

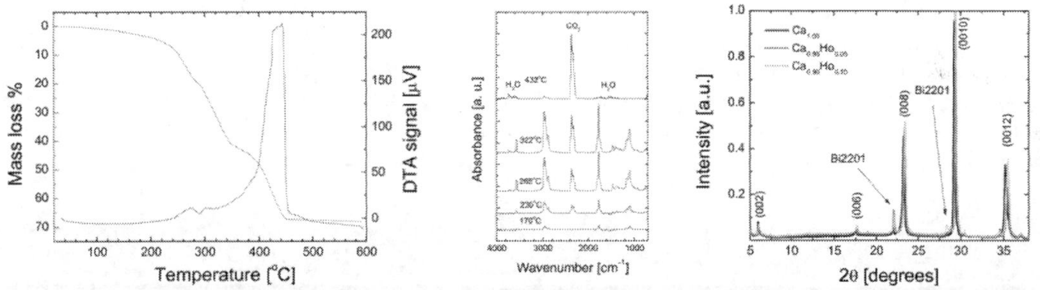

Figure 1: left panel: TG and DTA traces registered on a precursor mixture with $Bi_2Sr_2CaCu_2O_8$ nominal composition after drying at 130°C for 1h. Middle panel: FTIR spectra of the gas evolved during the thermal decomposition of the dried precursor powder. Right panel: XRD patterns of films with $Bi_2Sr_2Ca_{1-x}Ho_xCu_2O_8$ (x = 0.00, 0.05 and 0.10) nominal compositions.

The FTIR spectra of the gas educts (Fig. 1-centre) show that 2-ethylhexanoic acid is released from about 170°C. From 200°C, CO_2 is also detected. The intensity ratio of the CO_2 and 2-ethylhexanoic acid absorption lines increases with temperature. H_2O vapor is also visible from 250°C, suggesting that combustion of 2-ethylhexanoic acid takes place at an increasing rate as the temperature increases.

Although films with a high degree of c-axis orientation were obtained as shown in Fig. 1-right, the T_c of the undoped films is limited to 73 K – 81 K depending on the samples. The T_c of Bi2212 can be varied by annealing under specific conditions [20-27]. We tried to increase the T_c of the film by post annealing in Ar at 600°C (30 min) and in O_2 at 840°C. According to Triscone et al. [25], these conditions should result in optimum superconducting properties in bulk samples. However, as shown in a previous paper [19], both treatments result in a partial decomposition of the Bi2212 phase and no T_c improvements.

According to Vinu et al. [28], low level Ho doping in Bi2212 can result in an increase of T_c. We therefore tried to dope our films with Ho. Er-doping was also attempted due to the similar ionic size and chemistry of Er and Ho. Fig. 2 shows that the XRD lines of Bi2212 are shifted to higher 2θ angles with doping, as expected from a reduction of the c-axis parameter of the structure [28]. However, both Ho- and Er-doping also resulted in the formation of an increased amount of Bi2201 phase.

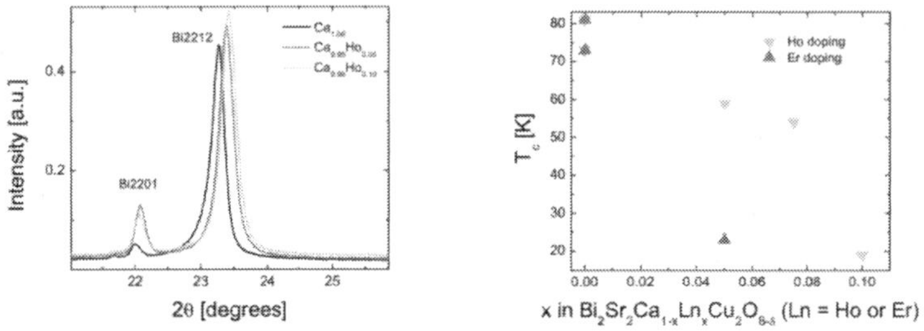

Figure 2: Left panel: detail of the XRD patterns of films with $Bi_2Sr_2Ca_{1-x}Ho_xCu_2O_8$ (x = 0.05, 0.10) nominal compositions. Right panel: T_c values measured on films with various $Bi_2Sr_2Ca_{1-x}Ln_xCu_2O_8$ (Ln = Ho, Er) nominal compositions.

The T_c of the Bi2212 phase was not increased upon doping with either Ho or Er. In contrast, as shown in Figure 2, T_c decreases with doping and this occurs apparently much faster for Er doping compared to Ho doping. The microstructure of the surface of a few selected films is shown in Figure 3. The undoped film is covered mostly with platelet-like grains that have grown parallel to the substrate plane, as reflected by the strong c-axis texture observed in the XRD patterns. Ho doping does not seem to induce significant changes in the surface microstructure. In contrast, the Er doped sample consists of much smaller grains, with evidence for many platelets with ab-axis oriented growth as suggested by the appearance of plate edges in the right-hand picture of Fig. 3.

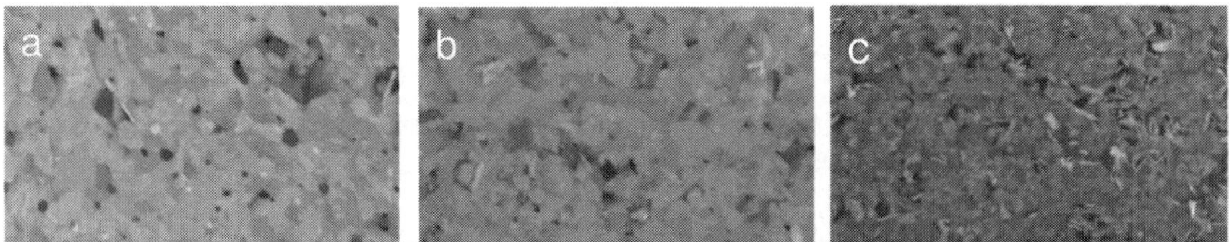

Figure 3: SEM pictures (secondary electron mode) of the surface of films with different stoichiometries. a: $Bi_2Sr_2Ca_1Cu_2O_{8+\delta}$, b: $Bi_2Sr_2Ca_{0.95}Ho_{0.05}Cu_2O_{8+\delta}$ and c: $Bi_2Sr_2Ca_{0.95}Er_{0.05}Cu_2O_{8+\delta}$.

4. Conclusions

$Bi_2Sr_2Ca_{1-x}Ho_xCu_2O_8$ ($0.00 \leq x \leq 0.10$) films with good c-axis orientation were grown on MgO single crystal substrates by means of chemical solution deposition. The T_c of the films was however not larger than 81 K in the best un-doped sample and decreases upon Ho doping. Attempts at doping with Er instead for Ho resulted in a degradation of the microstructure and a sharper decrease of T_c.

References

[1] A.P. Malozemoff et al., Supercond. Sci. Technol. 21, 034005 (2008)
[2] M. Igarashi et al., J. Phys. Conf. Ser. 234, 022016 (2010)
[3] M. Okada et al., Jpn J. Appl. Phys. 27, L185 – L187 (1988)
[4] P. Paturi et al., Physica C 372-376, 779 – 781 (2002)
[5] H. Ignatov et al., J. Supercond. Nov. Magn. 21, 69 – 73 (2008)
[6] D. Dimos et al., Phys. Rev. Lett 61, 219 - 222 (1988)
[7] J. Jiang et al., Supercond. Sci. Technol. 24, 082001 (2011)
[8] S. Jin et al., Physica C 177, 189 – 194 (1991)
[9] H. Miao et al., Cryogenics 38, 257 – 259 (1998)
[10] T. Amrein et al., Phys. Rev. B 51, 6792 – 6795 (1995)
[11] V.V. Metlushko et al., Appl. Phys. Lett. 63, 2821 – 2823 (1993)
[12] A. Gupta and E.I. Cooper, Physica C 158, 225 – 229 (1989)
[13] S.J. Golden et al., J. Am. Ceram. Soc. 74, 123 – 129 (1991)
[14] Y.Y. Tomashpolsky et al., Intergrated Ferroelectrics 29, 179 – 188 (2000)
[15] Y.Y. Tomashpol'skii et al., Inorg. Mater. 40, 960 – 966 (2004)
[16] S. Watanabe et al., Physica C 469, 818 – 821 (2009)
[17] K. Hamanaka et al., Physica C 470, 1457 – 1460 (2010)
[18] K. Murakami et al., Physica C 426 – 431, 1155 – 1158 (2005)
[19] J.-C. Grivel et al., Physica C 486, 32 – 36 (2013)
[20] R. Buckley et al., Physica C 156, 629 – 634 (1988)
[21] J. Zhao and M.S. Seehra, Physica C 159, 639 – 642 (1989)
[22] W.A. Groen and D.M. de Leeuw, Physica C 159, 417 – 421 (1989)
[23] Y. Idemoto and K. Fueki, Physica C 168, 167 – 172 (1990)
[24] C. Allgeier and J.S. Schilling, Physica C 168, 499 – 505 (1990)
[25] G. Triscone et al., Physica C 176, 247 – 256 (1991)
[26] A. Jeremie et al., Physica C 235 – 240, 943 – 944 (1994)
[27] Q.B. Hao et al, Physica C, 1100 – 1102 (2011)
[28] S. Vinu et al., J. Appl. Phys. 063920 (2009)

Transport and optical properties of epitaxial $Nd_{1.83}Ce_{0.17}CuO_{4-\delta}$ thin films

Anita Guarino[1,*], Loredana Parlato[2], Carmela Bonavolontá[2], Massimo Valentino[2], Corrado de Lisio[2], Antonio Leo[1], Gaia Grimaldi[1], Sandro Pace[1], Giampiero Pepe[2], Antonio Vecchione[1], Angela Nigro[1]

[1] CNR-SPIN Salerno and Dipartimento di Fisica "E. R. Caianiello", Universitá di Salerno, Via Ponte Don Melillo, 84084, Fisciano (SA), Italy

[2] CNR-SPIN Napoli and Dipartimento di Fisica Universitá "Federico II" di Napoli, Dipartimento Scienze Fisiche, via Cinthia, 80126 Napoli, Italy

E-mail: * guarino@sa.infn.it

Abstract. Electrical transport measurements and transient optical pump-probe experiments have been performed on epitaxial films of the electron doped $Nd_{1.83}Ce_{0.17}CuO_{4-\delta}$ compound for studying the non-equilibrium carrier dynamics in this material. Samples have been grown on (001)-oriented $SrTiO_3$ substrates by dc sputtering in a mixed atmosphere of both Ar and O_2. X-ray diffraction analysis and scanning electron microscope equipped with a wavelength dispersive spectroscopy detector have been used to characterize the structure and the composition of the thin films. Time-resolved femtosecond pump-and-probe spectroscopy has been also carried out on our samples in the temperature range 4.2K-300 K.

1. Introduction

New superconductors have been discovered over the past century, nevertheless, electronic states and mechanism of superconductivity have not been clarified yet [1]. One of the most discussed issue regards the pseudogap phenomena, that are less explored in the case of the electron-doped cuprates in comparison to the hole-doped systems[2]. Recently, a crossed experimental and theoretical study on optimally doped $Pr_{1.85}Ce_{0.15}CuO_4$ single crystal has shown the evidence of a pseudogap state with antiferromagnetic-like nature [3]. The results obtained by both methods supported competing order parameter fluctuations as the origin of the pseudogap instead of the superconducting scenario [4].

In hole-doped cuprates, time resolved optical spectroscopy has proven to be a powerful method for tracking the onset of both pseudogap and superconducting order with a single technique [5]. During the last year, Hilton and cooworkers have reported time-resolved reflection experiments on the single crystal electron-doped superconductor $Nd_{1.85}Ce_{0.15}CuO_4$, measuring the change in reflectivity induced by an ultrashort optical pulse as a function of time [6]. They associate the positive signal that they observe at critical temperature T_c with superconductivity and the negative signal with the pseudogap. In fact, its onset at 75 K lies on the pseudogap temperature-cerium content line $T^*(x)$ of the doping phase diagram, as determined from the appearance of gaps in optical conductivity [7] and photoemission spectra [8]. In this work, the characterization from structural and compositional point of view of superconducting

$Nd_{1.83}Ce_{0.17}CuO_{4-\delta}$ (NCCO) thin films will be presented. Moreover, the preliminary analysis of pump and probe measurements performed on them will be discussed. Nevertheless, further experiments should be performed in order to better understand the nature of the pseudogap state by correlating it to doping levels in thin films.

2. Experimental details

Dc sputtering technique was optimized to grow NCCO films on (100) $SrTiO_3$ (STO) substrates by using a single stoichiometric target as a sputtering source in an on-axis configuration with the substrate [9]. The sputtering depositions have been carried out in pure Argon (Ar) atmosphere at pressure of 1.7 mbar and substrate temperature of 900 °C. The as-grown samples do not show the superconductive transition. In order to observe the superconductivity, the films have to be thermal treated for several hours (it depends on their thickness) in flowing argon. Compositional and structural characterizations were performed to investigate the quality of the samples by wavelength dispersive spectroscopy (WDS) and x-ray diffraction (XRD), respectively. Standard four probe technique in a Cryogenic variable temperature cryogen-free system has been used to investigate the electrical transport properties of the films. Finally, the femtosecond pump-probe optical reflectivity experiments were performed using a mode-locked Ti:sapphire laser which produced 100-fs-wide, 790-nm-wavelength pulses at a repetition rate of 82 MHz. The pump and probe beams were cross-polarized to eliminate coherent artifacts arising from the interference between two beams. The samples were mounted on a cold finger in a temperature-controlled liquid helium continuous-flow optical cryostat, operating from 5 K to 300 K. Both pump and probe beams have spots close to 70 μm diameter onto the sample. The pump-probe average power ratio was set at 3:1, while the pump fluence ranged from 2.5 $\mu J/cm^2$ to 12.5 $\mu J/cm^2$.

3. Results and discussion

The compositional analysis confirms an over-doped cerium content equal to 0.17 in our samples. Neodymium and copper result in the right stoichiometry with respect this value, while it is not possible the oxygen content evaluation with our equipment. In fact, the use of the STO as substrate prevents the possibility to distinguish the contribution coming from the two oxides [10]. The left panel of Figure 1 shows the behaviour of the resistance versus temperature (r-T) of a film ex-situ thermal treated for 2 hours in flowing argon at 900 °C, with a $T_c = 9K$. The

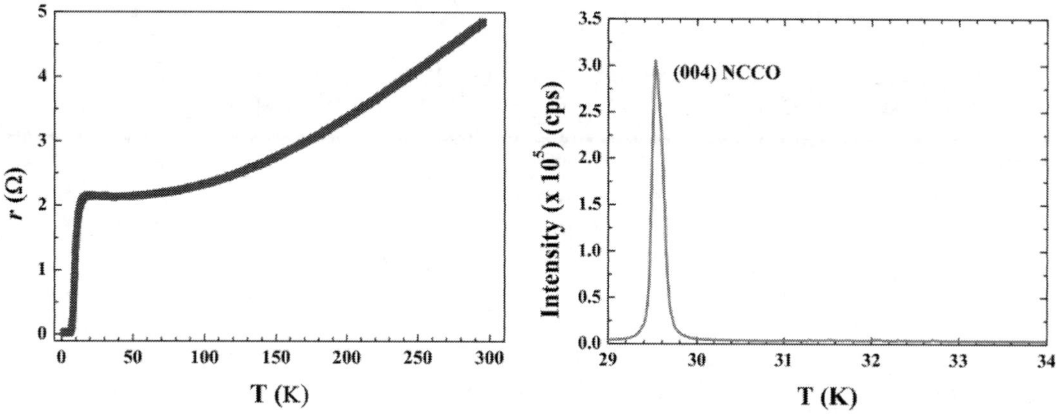

Figure 1. On the left, the r-T behavior of a superconducting NCCO sample. On the right, the corresponding $\theta - 2\theta$ scan around the (004) reflection of the NCCO phase are reported. No spurious phases are detected in the range investigated.

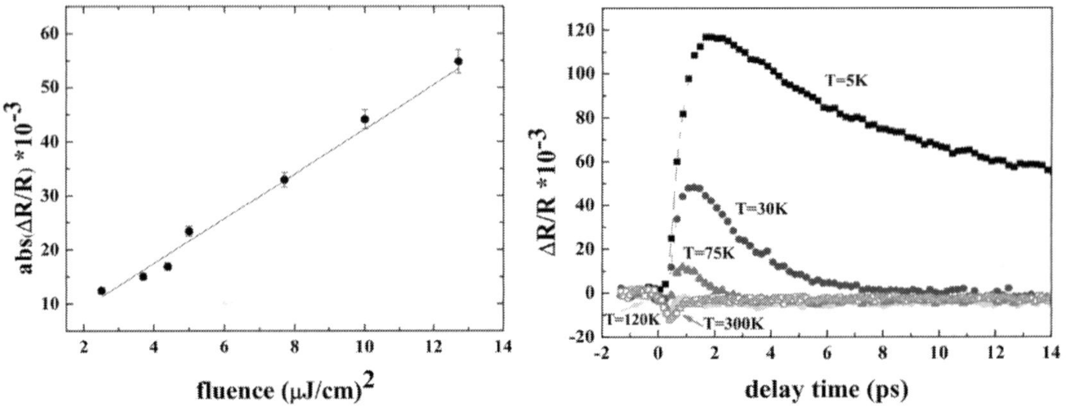

Figure 2. Left panel shows the ΔR/R as a function of pump fluence measured at 300K; the line is a guide for eyes. Right panel reports the ΔR/R as a function of the delay time between the pump and probe pulse at different temperatures.

right panel of Figure 1 reports the corresponding $\theta - 2\theta$ scan around the (004) reflection of the NCCO phase, from 29° to 34 °. This particular range allows to detect the possible presence of spurious phases in the samples. The data show that our well oriented films are free from undesired features due to policrystalline residuals or to the Cu-free phase, $(Nd,Ce)_2O_4$ [9]. In our pump-probe experiments, the optical reflectivity change, ΔR, induced by the pump beam was measured by the probe beam, time delayed with respect to the pump. Thus, by studying the ΔR signal, normalized to the DC value, R, as a function of the time delay between the pump and the probe beams, time-resolved dynamics of the excited carrier relaxation processes can be measured directly. In the left panel of Figure 2, the ΔR/R as a function of pump fluence measured at 300K is plotted. In the range from 2.5 $\mu J/cm^2$ to 12 $\mu J/cm^2$, the behaviour results linear, then samples heating can be consider negligible. The minimum value we have choosen for the fluence in our experiment is comparable with that reported in ref. [6]. The right panel of Figure 2 reports the ΔR/R as a function of the delay time between the pump and probe pulse at different temperatures. At temperatures T<30K, only a positive peak with a long tail for T=5 K is observed, probably due to a slow relaxation time related to the recombination process of excitations in Cooper pairs. At T>30 K a small negative peak appears and by increasing the temperature the positive peak decreases and it disappears at a temperature T>75 K.

In spite of a sign inversion observed at temperatures close to 75K in the transient reflectivity data, our preliminary experiments do not allow us to clearly correlate such a feature to the occurring of a pseudogap component. Further experiments at different doping levels, and hence different strenghts of the superconducting state, and different laser pump fluences are currently in progress aiming to better understand the nature of such physical features.

4. Conclusions

In conclusion, highly oriented over-doped NCCO films with $x_{Ce} = 0.17$ are grown on (100) STO substrates by dc sputtering deposition technique. No spurious phases have been detected after the high temperature ex-situ thermal treatment to obtain the superconductivity. Preliminary analysis of pump and probe measurements point out the decrease of the reflectivity amplitude by increasing the temperature up to its disappearance for T>75 K in overdoped superconducting NCCO films. Further analysis are in progress in order to evaluate the time responses and to interpret the different behavior of the films with respect to the data on the single crystals of the literature.

Acknowledgements

We acknowledge financial support from the European Union Seventh Frame- work Programme (FP7/2007-2013) under grant agreement no. 264098 MAMA and financial support from PON Ricerca e Competitivitá 2007-2013 under grant agreement PON NAFASSY, PONa3-00007.

References

[1] Tohyama T 2012 *Jpn. J. Appl. Phys.* **51** 010004
[2] Armitage N P, Fournier P and Greene R L 2010 *Rev. Mod. Phys.* **82** 2421.
[3] Nekrasov I A, Pavlov N S, Kuchinskii E Z, Sadovskii M V, Pchelkina Z V, Zabolotnyy V B, Geck J, Bchner B, Borisenko S V, Inosov D S, Kordyuk A A, Lambacher M and Erb A 2009 *Phys. Rev. B* **80** 140510.
[4] Kokorina E E, Kuchinskii E Z, Nekrasov I A, Pchelkina Z V, Sadovskii M V, Sekiyama A, Suga S and Tsunekawa M 2008 *JEPT* **107** 828.
[5] Coslovich G, Giannetti C, Cilento F, Dal Conte S, Abebaw T, Bossini D, Ferrini G, Eisaki H, Greven M, Damascelli A and Parmigiani F 2013 *Phys. Rev. Lett.* **110** 107003.
[6] Hinton J P, Koralek J D, Yu G, Motoyama E M, Lu Y M, Vishwanath A, Greven M and Orenstein J 2013 *Phys. Rev. Lett.* **110** 217002.
[7] Onose Y, Taguchi Y, Ishizaka K and Tokura Y 2004 *Phys. Rev. B* **69** 024504.
[8] Matsui H, Takahashi T, Sato T, Terashima K, Ding H, Uefuji T and Yamada K 2007 *Phys. Rev. B* **75** 224514.
[9] Guarino A, Patimo G, Vecchione A, Di Luccio T and Nigro A *Physica C* to be published.
[10] Guarino A, Fittipaldi F, Romano A, Vecchione A and Nigro A 2012 *Thin Solid Films* **524** 282.

Preparation of c-axis textured Bi-2212 thin films on silver substrates by using chemical solution deposition

Shutong Deng[1], Timing Qu[2], Guan Lin[2], Zhenghe Han[1*]

1. Applied Superconductivity Research Center, Department of Physics, Tsinghua University, Beijing 100084, China

2. Department of Mechanical Engineering, Key Laboratory for Advanced Materials Processing Technology, Tsinghua University, Beijing 100084, China

E-mail: z.han@tsinghua.edu.cn

Abstract. The silver sheath material plays an important role in the preparation of Bi-2212 and Bi-2223 tape conductors. In this paper, a chemical solution deposition method was proposed to fabricate the Bi-2212 thin film on Ag substrates after polishing and annealing. The influences of the sintering temperature were investigated. After sintered at 780 °C or 790 °C for 5 h in 8% O_2 (N_2 balanced), samples with almost pure and highly c-axis textured Bi-2212 phase were prepared. And the full width at the half maximum (FWHM) value of the rocking curves for the Bi-2212 (00$\underline{12}$) plane reached below 5.8°.

1. Introduction

High temperature superconductor $Bi_2Sr_2CaCu_2Ox$ (Bi-2212) still merits a great attraction for its high critial current density in a high magnetic field at 4.2 K [1]. And because of its intrinsic Josephson effects, Bi-2212 is one important candidate to realize solid-state sources of tera hertz (THz) radiation [2]. Bi-2212 thin films were successfully fabricated by several methods, such as chemical vapor deposition (CVD) [3], pulsed laser decomposition (PLD) [4], molecular beam epitaxy (MBE) [5], sputtering [6] and metal-organic deposition [7, 8]. It is found that Ag substrate is effective to produce highly textured Bi-2212 microstructures [9, 10], and it can lower the solidification temperature of Bi-2212 [9]. In this study, the Bi-2212 thin film was prepared on Ag substrate by using the chemical solution method. The influences of the sintering temperature on phase assemblage, surface morphology, and the c-axis texture of Bi-2212 thin films were investigated.

2. Experimental

Silver substrates with the size of 5 mm × 5 mm were cut from a silver tape of 0.4 mm in thickness. After pressed at 90 MPa to obtain a flat surface, the silver substrates were annealed at 300 °C in pure N_2 for 30 min to release stress. Then all samples were mechanical polished by 1 μm diamond suspension. Finally, the silver substrates were ultrasonic cleaned in acetone for 8 min and alcohol for 5 min.

The solution preparation followed the same procedure proposed by Tanaka [11]. The starting compositions $Bi(NO_3)\cdot 5H_2O$, $Sr(NO_3)_2$, $Ca(CH_3COO)_2\cdot H_2O$ and $Cu(CH_3COO)_2\cdot H_2O$ were weighed to ensure that the molar ratio of Bi, Sr, Ca and Cu was 2:2:1:2. The pH value was adjusted to 5.5 by adding ammonia water (28%). The solution was stirred at 60 °C for 2 hours and stirred at 90 °C for 1 hour. Finally, 9 mL of the solution was refined under decompression at 45 °C for half an hour. The molar concentration of total metal ions was 0.59 mol/L.

The obtainded solution was spin-coated on as prepared Ag substrate with 6000 rpm for 80 s. After dried at 70 °C for 2 min, the obtained sample was pyrolyzed in air at 400 °C for 5 h. The spin-coating and pyrolysis processes were repeated 6 times in total. Finally, the samples were sintered at different temperatures (T_s) from 730 °C to 850 °C under 8% O_2 (N_2 balanced) for 5 h. The temperature ramping rate of all heating and cooling processes is 200 °C/h.

The thermal stability of the chemical solution was investigated by thermogravimetric and differential scanning calorimeter (TG-DSC) with a heating rate of 10 °C/min up to 900 °C in air. An optical microscope (Nikon DXM1200) was used to observe the surface morphologies of pyrolyzed films. The X-ray diffraction (XRD) patterns and rocking curves for the Bi-2212 (0012) plane were measured by X-ray diffractometer (Rigaku, D/max 2500 V, Cu Kα). The surface morphologies of thin films were observed by a scanning electron microscope (SEM) of LEO 1530 SEM, equipped with an energy-dispersive X-ray spectroscopy (EDX) detector.

3. Results

Fig. 1 shows the TG-DSC curves of the chemical solution. Two abrupt weight losses appeared at 260-360 °C and 370-460 °C in TG curve, accompanied by two exothermic peaks at about 340 °C and 425 °C in DSC curve. Another abrupt weight loss occurred at 530-600 °C in TG curve, with an endothermic peak at about 575 °C in DSC curve.

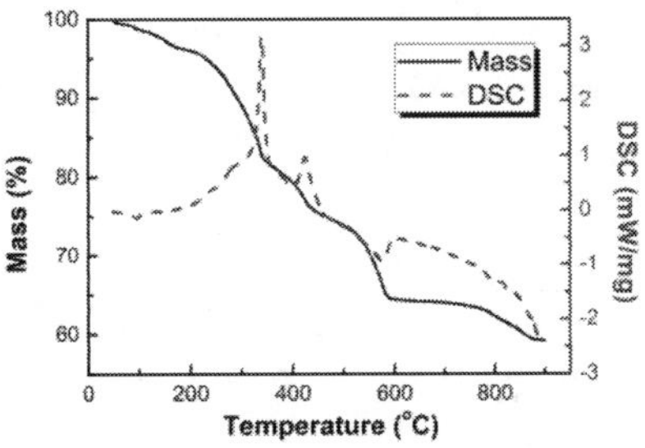

Figure 1. TG-DSC curves of the solution. Solid line for the mass curve and dashed line for the DSC curve.

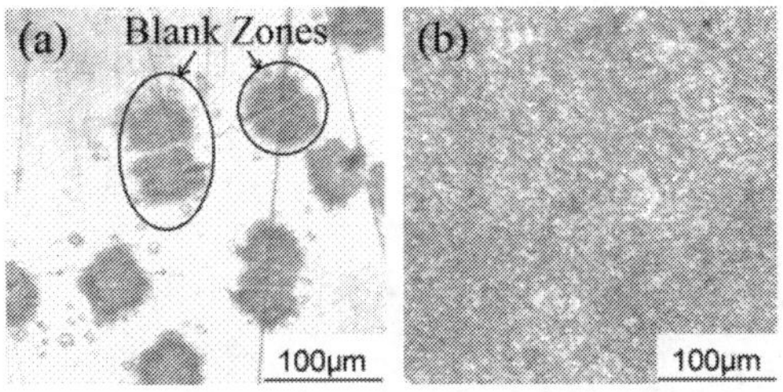

Figure 2. Optical microscope images of pyrolyzed films, (a) after one time coating, (b) after 6 times coating.

The optical images of selected pyrolyzed samples are shown in Fig. 2. Blank zones appeared in the sample after only one time coating, as shown in Fig. 2 (a). The blank zones are quite large, with an average size of about 50 μm. After 6 times repeated spin-coating and pyrolysis processes, the surface became much more homogeneous, as shown in Fig. 2 (b).

Fig. 3 shows the XRD patterns of the samples sintered at different temperatures (730 °C to 850 °C) in 8% O_2 (N_2 balanced) for 5 h. The diffraction peaks of the Bi-2201 phase can be observed in almost all samples, especially in case of high sintering temperature. Almost pure Bi-2212 phase is formed at 780 °C and 790 °C. According to XRD patterns, the volume fractions of Bi-2201 and Bi-2212 phase can be calculated by equation (1) and (2),

$$f_{2201} = \frac{I_{2201}^{(006)}}{I_{2201}^{(006)} + I_{2212}^{(008)}} \quad (1) \qquad f_{2212} = \frac{I_{2212}^{(008)}}{I_{2201}^{(006)} + I_{2212}^{(008)}} \quad (2)$$

where f_{2201} and f_{2212} represent the volume fractions of Bi-2201 and Bi-2212 phase respectively, $I_{2201}^{(006)}$ and $I_{2212}^{(008)}$ represent the intensities of Bi-2201 (006) peak and Bi-2212 (008) peak respectively.

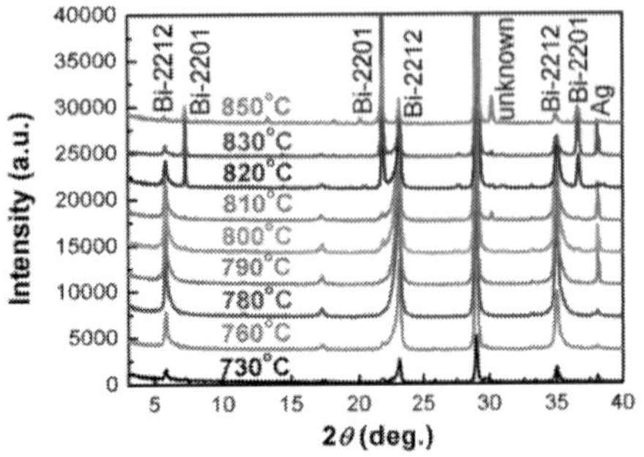

Figure 3. X-ray diffraction patterns of samples at different sintering temperature.

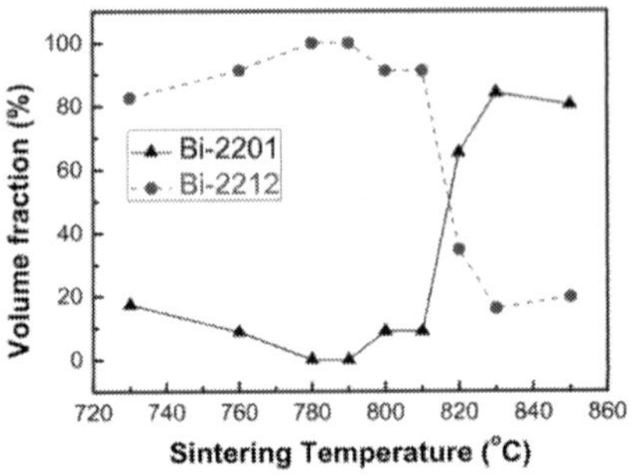

Figure 4. The volume fraction of Bi-2201 and Bi-2212 phases as a function of sintering temperature.

The calculated volume fractions of the Bi-2212 and Bi-2201 phases with different sintering temperature are shown in Fig. 4. f_{2212} reached maximum at 780 °C and 790 °C. It decreased rapidly at above 810 °C. The Bi-2201 phase became the main phase at this temperature region.

Fig. 5 shows the surface morphologies of selected samples sintered at 730 °C, 780 °C, 820 °C and 850 °C, respectively. The grain size increased with increasing sintering temperature. A needle sharp phase appeared at 850 °C, as shown in Fig. 5 (d). This needle phase seems to be Cu free according to the EDX measurement. The molar ratio of Bi, Sr and Ca is close to 9.0 : 12.4 : 6.6.

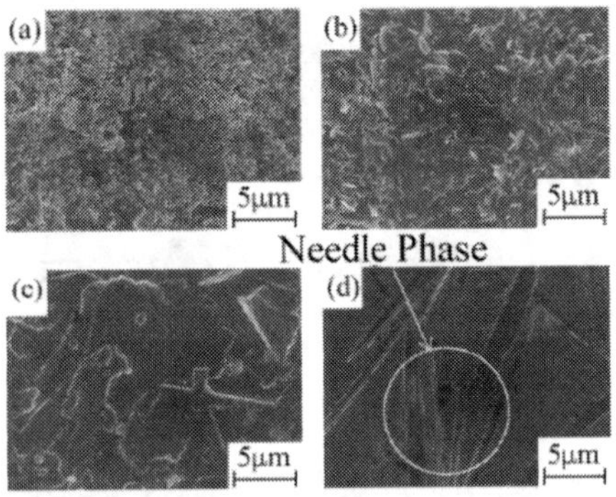

Figure 5. SEM images for BSCCO thin films sintered at (a) 730 °C, (b) 780 °C, (c) 820 °C and (d) 850 °C, respectively.

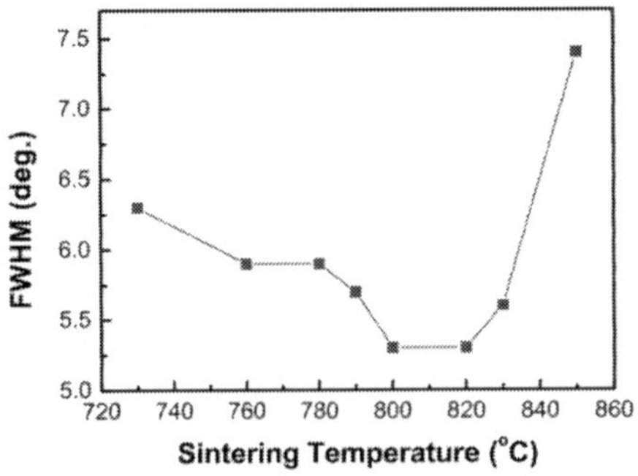

Figure 6. Full width at the half maximum (FWHM) of the rocking curves for the Bi-2212 (0012) plane as a function of the sintering temperature.

The FWHM values of the rocking curves for samples with different sintering temperature (T_s) are shown in Fig. 6. It decreased with increasing T_s in general. The minimum value reached

5.3° at 800 °C to 820 °C, which showed a nice c-axis texuture. But the FWHM value increased rapidly when $T_s > 830$ °C, corresponding to the formation of large amount of the Bi-2201 phase.

4. Discussion

Blank zones appeared on the film which was spin-coated and pyrolyzed only once (Fig. 2 (a)). The low wettability of the solution for Ag substrate and a relatively fast heating rate (200 °C/h) might be the reasons. For the low wettability led to the shrink of the film during the pyrolysis process and a fast heating rate caused bubbles of big size. Both the shrink of the film and the bubbles gave rise to the blank zones. After repeated coating and pyrolyzing, the blank area was covered by new layers, leading to a comparatively smooth surface (Fig. 2 (b)).

Almost pure Bi-2212 phase was formed when sintering temperature was 780 °C or 790 °C, which was lower than the common solidification temperature for Bi-2212 thin film [7, 8]. It is because that silver can lower the solidification temperature of Bi-2212 [9] and low oxygen pressure decreases the melting point of Bi-2212. Below 780 °C, Bi-2201 phase still existed after the sintering process of 5 hours. For Bi-2201 phase is stable at low temperature [12]. Above 790 °C, Bi-2201 phase and a new phase appeared as second phases. This new phase showed a relatively strong intensity in the XRD patterns when sintering temperature was 850 °C (See the marked unknown phase in Fig. 3). Baker [13] has reported that the decomposition of Bi-2212 leads to the appearances of some new phases including $Bi_9Sr_{11}Ca_5O_y$. Its major diffraction peeks are near the angle 30° 2θ. And the peek of the unknown phase we observed here located at the angle 30.2° 2θ. So, the unknown phase might be $Bi_9Sr_{11}Ca_5O_y$ and it was due to the decomposition of Bi-2212 phase. Further study is needed to identify this interpretation.

5. Conclusion

We have successfully fabricated almost pure Bi-2212 thin film on Ag substrate by a chemical solution deposition method. Blank zones on the pyrolyzed film will disappear after repeating the spin-coating and pyrolysis processes for 6 times. The sintering temperature for pure Bi-2212 phase is 780 °C or 790 °C. It is lower than the common solidification temperature for Bi-2212 because of Ag substrate and low oxygen pressure. For the film sintered at 780 °C, rocking curve shows that the FWHM of Bi-2212 ($00\underline{12}$) reflection is 5.8°.

Acknowledgement

This work is supported by the Beijing Natural Science Foundation (2122026) and the International S & T Cooperation Program of China (2010DFA64510).

References

[1] Miao H, Marken K R, Meinesz M, Czabaj B and Hong S 2005 *IEEE Trans. Appl. Supercond.* **15** 2554
[2] Ozyuzer L, et al. 2007 *Science* **318** 1291
[3] Hårsta A and Lundquist S 1996 *Chem. Vapor. Depos.* **2** 109.
[4] Kim S S, Moon J H, Lee B T and Hishita S 2002 *Thin Solid Films* **415** 224
[5] Alami H E and Cavellin C D 2005 *J. Crystal Growth* **277** 170
[6] Yildirim G, Bal S and Varilci A, 2012 *J. Supercond. Nov. Magn.* **25** 1655
[7] Hamanaka K, Tachiki T and Uchida T 2010 *Physica C* **470** 1457
[8] Yu X M, Yang H Z and Qi Y 2011 *Appl. Surf. Sci.* **258** 38
[9] Hasegawa T, Kobayashi H, Kumakura H, Kitaguchi H and Togano K 1994 *Physica C* **222** 111
[10] Kumakura H, Kitaguchi H, Togano K, Muroga T, Sato J and Okada M 1999 *IEEE Trans. Appl. Supercond.* **9** 1804
[11] Tanaka K, Nozue A and Kamiya K 1990 *J. Mater. Sci.* **25** 3551
[12] Hasegawa T, Kitamura T, Kobayashi H, Kumakura H, Kitaguchi H and Togano K 1992 *Appl. Phys. Lett.* **60** 2692
[13] Baker A P and Glowacki B A 1994 *Physica C* **223** 383

The effect of $BaCeO_3$ dopant concentration on magnetically defined B_{irr} and B_{c2} in $YBa_2Cu_3O_{6+x}$ thin films deposited on $SrTiO_3$ substrates

H. Huhtinen[1], H. Palonen[1,2], M. Malmivirta[1], R. Jha[3], V. P. S. Awana[3] and P. Paturi[1]

[1] Wihuri Physical Laboratory, Department of Physics and Astronomy, FI-20014 University of Turku, Finland

[2] The National Doctoral Programme in Nanoscience (NGS-NANO), Turku, Finland

[3] The Superconductivity Division, National Physical Laboratory (CSIR), New Delhi 110012, India

E-mail: hannu.huhtinen@utu.fi

Abstract. The effect of $BaCeO_3$ (BCO) dopant concentration on magnetically obtained irreversibility field B_{irr} and upper critical field B_{c2} is investigated in $YBa_2Cu_3O_{6+x}$ (YBCO) thin films prepared by pulsed laser deposition on $SrTiO_3$ (001) substrates. The unchanged values of B_{irr} and B_{c2} are connected to the structural properties where BCO has been observed to form non-elongated and randomly distributed particles within the whole concentration range of 0 – 10 wt.%. This indicates a different type of flux pinning mechanism for BCO-doped YBCO than for the widely studied $BaZrO_3$-doped YBCO where increased B_{irr} and B_{c2} values were observed.

1. Introduction

In order to optimize the high-T_c superconductors for large-scale applications, the requirement for optimized flux pinning in high magnetic fields is realized by organizing a network of self-assembled isotropic nanodots or anisotropic nanorods in the $YBa_2Cu_3O_{6+x}$ (YBCO) matrix [1–5]. In addition to the improved in-field critical current j_c of YBCO thin films, its intrinsic high anisotropy has been radically modified by adding $BaZrO_3$ (BZO), as it creates film penetrating non-superconducting columnar defects with diameter equal of the vortex core [4, 6]. The situation changes dramatically, when the non-elongated and randomly distributed nanoparticles of *e.g.* $BaCeO_3$ (BCO) are used as a pinning centres [7].

For the undoped YBCO, the upper critical field B_{c2} is roughly six times larger in the field in-plane than in the field out-of-plane direction [8]. In spite of the intrinsic property of B_{c2}, it was also demonstrated that BZO doping clearly increases B_{c2} in all field directions, and the angular dependence of B_{c2} follows the Blatter scaling [9]. When compared to the BZO nanorods as a dopant, the enhancement of B_{c2} is more pronounced in the case of $BaHfO_3$ (BHO) which has almost similar in-plane lattice parameters than BZO, $a = b = 4.17$ Å, but forms slightly thinner but higher density of nanorods in YBCO matrix [10]. This leads to an assumption that BCO, which also has longer lattice parameters ($a = b = 4.40$ Å) than in BZO and BHO as well

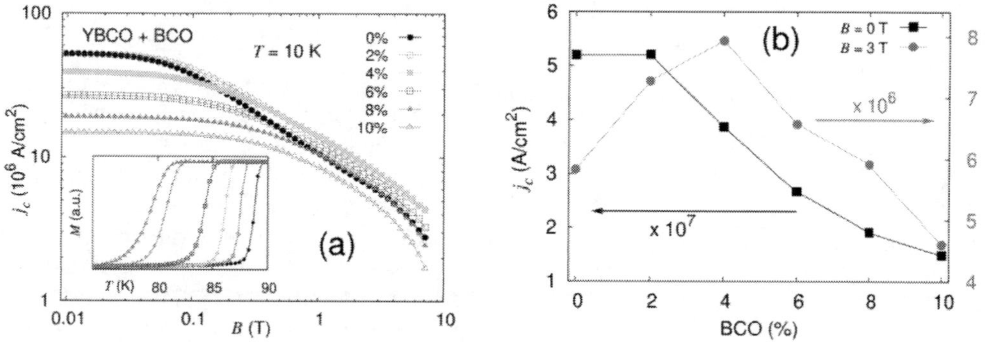

Figure 1. (a) Magnetic field dependence of j_c determined from the hysteresis loops at 10 K and the normalized ac magnetization as function of temperature (inset) for undoped and BCO doped YBCO films. (b) BCO content dependence of j_c in external fields of 0 and 3 T.

as reduced amount of strain, could produce completely different critical current properties in high magnetic fields, as well as different irreversibility field B_{irr} and B_{c2}.

Based on magnetic measurements, we have determined the temperature dependence of B_{irr} and B_{c2} for YBCO films doped with isolated BCO particles. The results are discussed with the conclusions obtained for BZO doped YBCO films having columnar defect network.

2. Experimental and basic magnetic properties

The films were prepared on (100) $SrTiO_3$ (STO) substrates by pulsed laser deposition (PLD) from YBCO targets with varied dopant concentrations containing 0, 2, 4, 6, 8 and 10 wt.% of BCO. Details of the preparation of targets and films as well their structural characterizations are described elsewhere [7]. Because the XRD (004) peak of BCO starting at $2\theta \approx 41°$ is extremely wide, it was concluded that BCO is distributed into YBCO lattice as very small particles in $(00l)$ orientation. According to Scherrer formula, the width of the peak produces an average particle size of ≈ 2 nm. In addition, the $(2\theta, \phi)$ scans of YBCO (212)/(122) peaks indicate biaxial twins up to 10 wt.% of BCO, *i.e.* the strain is notably smaller than in the BZO-doped YBCO.

Magnetic measurements were made with a Quantum Design PPMS system, and the onset critical temperatures $T_{c,onset}$'s were determined with ac magnetization measurements in the range of 10 - 100 K (ac field of 0.1 mT). As can be seen from the inset of figure 1(a), $T_{c,onset}$ decreases systematically with increasing BCO concentration, having the maximum onset value of 89.5 K for 0 % and minimum of 82 K for 10 % BCO doped YBCO. In contrast with results for BZO doped YBCO where the width of the transition ΔT_c stays within 1.3 – 2.0 K [11], ΔT_c increases with increasing BCO content from 1.2 K (0 %) to 3.7 K (10 %), indicating increased amount of lattice mismatch induced strain and possible oxygen deficiency at the interface between dopants and superconducting matrix [7, 10]. The critical current densities j_c at 10 K were determined from the hysteresis loops with the Bean model for rectangular shape film: $j_c = 2\Delta m/[a(1 - a/3b)V]$, where a and b ($b \geq a$) are the width and the length of the sample, V is the sample volume and Δm is the opening of the hysteresis loop [12]. The shape of the $j_c(B)$ curve varies with BCO concentration, showing clearly different magnetic field dependence, as can be seen from figure 1(a). Figure 1(b) shows the BCO concentration dependent j_c data, shown at two different magnetic fields. In zero field, the maximum j_c can be achieved with low BCO content below 2 %, while the optimal concentration is increased up to ≈ 4 % in the applied field range of 3 T. These results are in line with our earlier observations for BCO doped YBCO films [7].

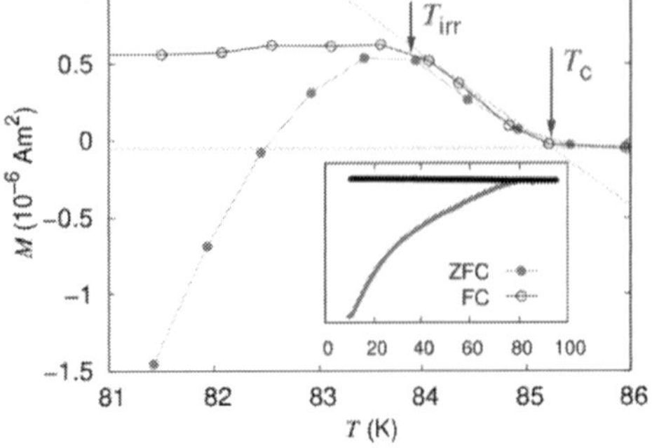

Figure 2. Determination of T_{irr} and T_c from the close-up of ZFC and FC curves as illustrated for the 4 wt.% BCO doped YBCO sample measured in 0.25 T field. Due to subtle diamagnetic signal of the substrate and the sample holder, the horizontal dashed line indicates the zero magnetization level of the sample and the descending dashed line the low temperature region as expressed in ref. [10].

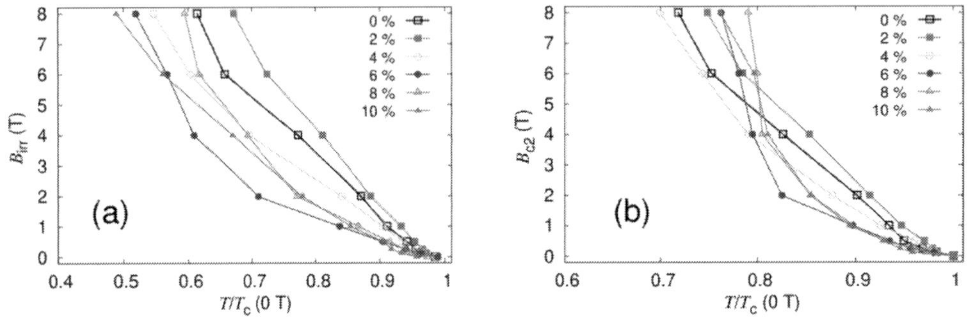

Figure 3. The reduced temperature dependence of B_{irr} (a) and B_{c2} (b) for YBCO thin films doped with various BCO concentration.

PPMS dc magnetization option is used to measure temperature dependence of zero-field-cooled (ZFC) and field-cooled (FC) magnetizations in fields of 0, 0.125, 0.25, 0.5, 1, 2, 4, 6 and 8 T. The irreversibility temperature T_{irr} for each measuring field is determined as a temperature where the ZFC and FC curves start to deviate from each other. The upper critical field B_{c2} is determined by using a method described in [10] where B_{c2} is equivalent to the magnetic field at which linear extrapolation of the FC magnetization from the low and high temperature regions meet (intersection of dashed lines), indicating T_c, as shown in figure 2.

3. Irreversibility field and upper critical field

When looking at the reduced temperature dependence of B_{irr} in figure 3(a), only the smallest amount of BCO (2 %) has even some kind of potential to increase the B_{irr}, while all the other BCO contents give smaller B_{irr} in the whole temperature range. The same situation occurs when B_{irr} is plotted as a function of absolute temperature values. However, when taking into account the sensitivity of the determination of T_{irr}, we can conclude that no clear difference between the curves can be observed and therefore the BCO doping does not have notable influence on B_{irr}. The behaviour is different what we have seen for BZO doped YBCO films, where clearly lower magnetically determined B_{irr} for undoped YBCO is explained by the strong pinning of BZO nanorods [11].

From the temperature dependence of B_{c2} in figure 3(b), we can observe that the curves with different BCO concentrations are situated close to each other without BCO dependent tendency, and therefore we conclude that BCO doping does not have effect on B_{c2} either. It is good to notice that both B_{irr} and B_{c2} are observed from the high temperature range, close to onset T_c, where extremely small BCO pinning centers are not effective. This gives an explanation for the difference where increased $j_c(B)$ was seen in the data measured at 10 K. The behaviour is also in contradiction to the results, where BZO doping has been observed to cause a clear change in B_{irr} and B_{c2}, indicating a change in the coherence length and the band structure of YBCO [9, 11]. This is in agreement with our structural results where strain has been observed to be clearly smaller in BCO doped than in BZO doped YBCO films [7], and it is also in line with recent results where strained regions could prevent the Cooper pair formation thus becoming effective for pinning vortices [13]. However, an interesting detail can be seen that with high BCO content, between 6 and 10 % of BCO, at around $T = 0.85T_c$ increase of B_{c2} is much steeper with decreasing temperature when compared with the curves of low BCO concentration. In order to throw light on this phenomenon and assess the possible mechanism for it, the pulsed magnetic field measurements up to clearly higher fields are needed. In addition, it has been suggested earlier that the area of interface between dopant and the superconducting matrix plays an important role in the enhancement of B_{c2} [10], but according to our results, the shape and the distribution of the dopants seem to be the most critical and therefore the selection of suitable pinning material for different type of applications is the most important factor.

4. Conclusions

In this work, the irreversibility field and the upper critical field were determined from the magnetic measurements for YBCO thin films doped with various concentration of single BCO nanoparticles. Our results show that the temperature dependence of both B_{irr} and B_{c2} remained unchanged when BCO dopant was introduced into the YBCO lattice. The observation differs clearly from the results for YBCO with BZO nanorods, and the phenomenon is connected to the formation of isotropic BCO nanoparticles as well as to the lower strain caused by the BCO nanoparticles in the YBCO matrix.

Acknowledgments

The Wihuri Foundation, Finnish Foundation for Technology Promotion and Finnish Cultural Foundation are acknowledged for financial support.

References

[1] Civale L, Marwick A D, Worthington T K, Kirk M A, Thompson J R, Krusin-Elbaum L, Sun Y, Clem J R and Holtzberg F 1991 *Phys. Rev. Lett.* **67** 648
[2] Blatter G, Feigel'man M V, Geshkenbein V B, Larkin A I and Vinokur V M 1994 *Reviews of Modern Physics* **66** 1125
[3] Polkovnikov A, Kafri Y and Nelson D R 2005 *Phys. Rev. B* **71** 014511
[4] Maiorov B, Baily S A, Zhou H, Ugurlu O, Kennison J A, Dowden P C, Holesinger T G, Foltyn S R and Civale L 2009 *Nat. Mater.* **8** 398–404
[5] Baca F J, Barnes P N, Emergo R L S, Haugan T J, Reichart J N and Wu J Z 2009 *Appl. Phys. Lett.* **94** 102512
[6] MacManus-Driscoll J L, Foltyn S R, Jia Q X, Wang H, Serquis A, Civale L, Maiorov B, Hawley M E, Maley M P and Peterson D E 2004 *Nat. Mater.* **3** 439
[7] Irjala M, Huhtinen H, Jha R, Awana V P S and Paturi P 2011 *IEEE T. Appl. Supercond.* **21** 2762
[8] Wijngaarden R J, Heeck K, Griessen R, Menovsky A A and Menken M J V 1988 *Physica C* **152** 523
[9] Palonen H, Huhtinen H, Shakhov M A and Paturi P 2013 *Supercond. Sci. Technol.* **26** 045003
[10] Matsushita T, Nagamizu H, Tanabe K, Kiuchi M, Otabe E S, Tobita H, Yoshizumi M, Izumi T, Shiohara Y, Yokoe D, Kato T and Hirayama T 2012 *Supercond. Sci. Technol.* **25** 125003
[11] Paturi P, Malmivirta M, Palonen H and Huhtinen H 2013 *Journal of Physics: Conference Series* submitted
[12] Wiesinger H P, Sauerzopf F M and Weber H W 1992 *Physica C* **203** 121
[13] Llordés A, Palau A, Gázquez J, Coll M, Vlad R, Pomar A, Arbiol J, Guzmán R, Ye S, Rouco V, Sandiumenge F, Ricart S, Puig T, Varela M, Chateigner D, Vanacken J, Gutiérrez J, Moshchalkov V, Deutscher G, Magen C and Obradors X 2012 *Nat. Mater.* **11** 329

Flux Trapping and Field Magnet Stability of Bulk Superconductors

YF Zhang[1], W Zhou[1], Y Xu[1], B Li[2], D Zhou[2], K Tsuzuki[2,3], M Watasaki[2], M Miki[2] and M Izumi[2]
[1]Shanghai University of Electric Power, New Pudong, Shanghai 201300, China
[2]Tokyo University of Marine Science and Technology, Tokyo 135-8533, Japan
[3]Toba National College of Maritime Technology, Mie 517-8501, Japan

Abstract. The stability and durability of trapped flux is a focus in investigations of melt-growth (RE)$Ba_2Cu_3O_{7-\delta}$ (RE=rare earth, e.g., Y, Gd, Sm) bulk high-temperature superconductors. In magnetic flux engineering applications, two important issues affect the expected performance of bulk superconductors: flux trapping and flux stability or durability. The introduction of effective pinning with homogeneous spatial distribution is inevitable. The materials process and durability studies have been pursued in the last decade with either a stationary or a momentary applied external field. In machine applications, the magnetic flux stability is influenced by external perturbations. We topically summarize our progress and current status in relation to past and ongoing work.

1. Introduction

Melt-growth (RE)$Ba_2Cu_3O_{7-\delta}$ (RE=rare earth Y, Gd, Sm, etc.) bulk high-temperature superconductors have attracted interest for their potential applications, and trapped flux densities of 17 T at 29 K [1] and 3 T at 77 K [2] have been shown. The enhanced critical current density (J_c) based on flux pinning is a key issue for industrial applications [3]. An improvement in J_c under a magnetic field has been reported by introducing the possible interface between the non-superconducting and superconducting phases. A variety of strategies and techniques have been developed to improve the flux pinning and the homogeneity of materials. During the melt-growth process, intrinsic pinning could be introduced through the effects of stacking faults, low angle grain boundaries, twin boundaries, precipitates, impurity phases, secondary phases, point defects, oxygen vacancies, growth dislocation, structural inhomogeneity, and other objects with different dimensionality. In addition, so-called chemical doping with ionic substitution has been widely investigated, in which Fe, Co, and/or Ga is preferentially substituted to at Cu-O chain sites [4]. On the other hand, for example, Ni or Ti may occupy the Cu-O plane and may affect the superconductivity of the mother phase bulks by way of local lattice distortion [4].

Another strategy to achieve high flux pinning is the addition of fine particles such as nanometre-sized second phase RE_2BaCuO_5 (RE211), $RE_2Ba_4CuMO_y$ (RE2411, Mo, Zr), or other metal oxides. This meets the requirements for a short coherence length close to about 2-4 nm below 77 K [5]. The *in situ* formation of nanosized $BaZrO_3$ or others have been reported. The J_c of the portion taken below the seeding site in the melt-growth bulks is considerably high, and the trapped flux density exhibits a peak value larger than that of conventional permanent magnets. A couple of viewpoints are considered in industrial applications: the maximization of the trapped flux intensity by enhanced J_c, or the optimization of the trapped flux and its stability and durability under a variety of external field and temperature perturbations. In this paper, we summarize and review several key issues and describe our activities.

2. Critical current density

Xu *et al.* [6] extensively studied c-axis-oriented, single-domain $GdBa_2Cu_3O_{7-d}$ (Gd123) melt-textured bulk superconductors. First, they clarified the solubility of the solid solution with $Gd_{1+x}Ba_{2-x}Cu_3O_{7-\delta}$ (x < 0.1), but found a remarkable inhomogeneous distribution and ripening of Gd211. With this limitation in the refinement of the Gd211 addition due to the pushing/trapping effect and ripening, nano-Gd211 inclusion was difficult to induce. Thus, another kind of material was synthesized, the Gd2411 single phase. The J_c value was dependent on the positions because there was an inhomogeneous distribution of these second phases. However, the highest J_c was 84,500 A/cm^2 at a liquid nitrogen temperature under the self-field. Gd2411 has been observed to exist in an RE123 matrix with a size of several nanometres and it improves J_c. Nanosized particles were doped into Gd123 bulks, which displayed enhanced superconducting properties with the optimum additional amount. SEM results showed an inhomogeneous distribution of Gd211 appearing around the seeding site in the bulk, even when the doping amount was quite low. ZrO_2 or SnO_2 nanosized particles reacted with the Gd123 matrix before the melt process to form $BaZrO_3$ or $BaSnO_3$ particles *in situ*, with an average diameter of approximately 50 nm. These played a role in the effective flux pinning. $BaZrO_3$ nanosized particles were embedded in the Gd211 particles and contributed to the size reduction of the latter. Enhanced superconductivity in Gd123 has been achieved for the J_c value in an air process of 100,000 A/cm^2 at 77 K and a self-field [7]. The trapped field was up to 0.7 T for bulk ϕ = 24 mm. Using atomic force microscopy (AFM), nanostripes with a wavelength of 15 nm were observed in the ZnO-doped Gd123 sample. Nanosized particles in the grown Gd123 single domain, together with micro-defects induced by particle doping, can account for the enhanced superconducting properties.

The magnetic particle of Fe-B alloys (e.g., $Fe_{0.69}B_{0.12}Si_{0.14}Nb_{0.03}Cr_{0.01}Cu_{0.01}$) was added by Xu [8] for the first time in Gd123 bulk superconductors. Suitable amounts of magnetic particle inclusions resulted in the enhancement of J_c in the bulk. Xu *et al.* [9] found that the maximum trapped field of the sample with 0.4 mol% Fe-B alloy particles reached 0.16 T, which was 31% higher than that without Fe-B additives. Tsuzuki *et al.* [10] studied the effect of addition for Fe-Si, Fe-Si-Al, and Fe-Si-B-Cr-C alloys in the Gd123+Gd211 matrix. The trapped flux of the Fe-B doped and Fe-Si doped samples considerably exceeded those of the undoped samples. A value of B = 0.16 T was obtained for the local maximum in the centre of the top surface in the Fe-Si and Fe-Si-B-Nb-Cr-Cu doped samples. For the result of Fe-B in the 0.4 mol% doped samples, the value of the trapped magnetic flux was 8.80 μWb, which is 14% higher than that for the undoped samples. The Fe-Si addition marked 8.17 μWb. The bulk Gd123 with doping by several magnetic particles (MP) provided an intensified trapped magnetic flux compared with those without the addition. The reason is possibly that the Fe-containing particles contributed to a reduction in the interfacial energy of Gd211 and increased the nucleation sites with fine dimensions, thus preventing the Gd211 coarsening.

Li *et al.* [11] studied the pinning effect of Gd123 bulk, with different doping amounts of $BaTiO_3$ up to 0.4 mol%. The J_c showed the existence of two optimized doping levels. One was at 0.1 mol% addition, in which the J_c was enhanced around the zero field and the middle field of 1.5 T. The other was 0.3 mol% with a J_c value close to that for the 0.1 mol% doped bulk around the zero field; however, it exhibited a lower peak. The highest J_c of 3.8×10^4 A /cm^2 was achieved in 0.1 BTO [11]. A 0.3 mol% BTO addition mainly contributed to the enhancing flux pinning under a low field. Using a similar method, Li *et al.* [12] studied Gd123 with a TiO_2 addition. The TiO_2 doping resulted in improved T_c and enhanced δl pinning. The $BaTiO_3$ doped sample exhibited a maximum around the zero field (Fig 1). From this it was concluded that the nanometre-sized TiO_2 dilute addition has the ability to enhance flux pinning under a low field.

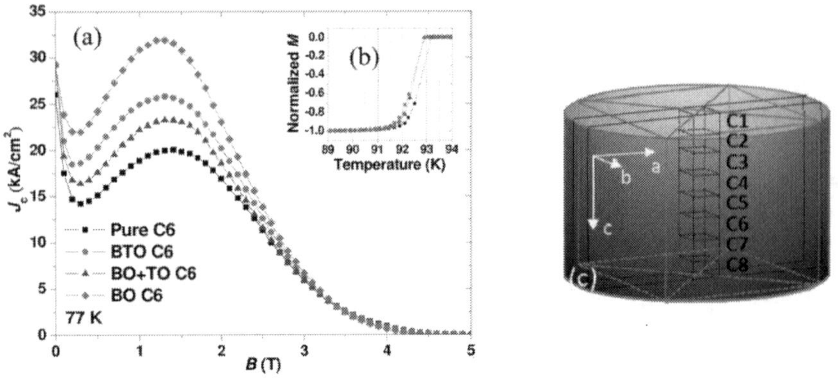

Fig. 1 Critical current density as a function of (a) magnetic field and (b) the critical transition temperature for Gd123 (pure), Gd123 with $BaTiO_3$ (BTO), BaO_2+TiO_2 (BO+TO), and BaO_2 (BO) additions. The samples were taken from C6 as indicated in (c).

Based on the nature of liquid dispersion and the anisotropic growth of the matrix, Li *et al.* [12] addressed the recrystallization of fine inclusions related to the employment of $BaTiO_3$. The recrystallization may overcome the disadvantage of the pushing effect and encourages the *in situ* formation of fine particles other than the starting powder. These results demonstrated that the flux pinning performance was greatly improved, and the self-field J_c value reached 52.5 kA/cm^2. The onset transition temperature was improved as a result of the lower starting growth temperature. These results showed that the BTO doping-related precipitations are independent of the matrix, and the positive effect of recrystallization of the finer inclusions is reasonably verified by growing the bulk at the lower initial growth temperature. The contribution of Gd211 on the enhancement of flux pinning is finite because the main Gd211 particles are still in the micrometre range. Therefore, the significantly increased pinning performance along the *c*-axis in a low field was additionally attributed to the recrystallization of the finer BTO doping-related precipitation, due to the anisotropic crystal growth. As shown in Fig. 1, the aligned peaks should come from the anisotropic crystal growth since it unavoidably causes the rearrangement of fine precipitations.

3. Trapped flux and stability

Zhou *et al.* [13] successfully solved the liquid source leakage problem, even at a T_{max} of 1100°C, as a result of making the seeded infiltration and growth of the Gd123 single grain using a $YBa_2Cu_3O_y$ pressed pellet as the liquid source supply. To develop this process, a simple but essential modification to insert a liquid source pellet at the bottom of the pre-form was made in the top-seed melt growth process. The trapped flux density of the sample without a liquid source substrate increases with the reduced thickness and reaches a peak value of 0.26 T after 2 mm of its thickness has been ground away, and then it decreases monotonously. The trapped flux density of 0.17 T was measured in the sample with a Y123 pellet substrate only 1.5 mm thick, indicating full growth until the very bottom. This is a very effective way to improve the trapped flux ability.

Concerning the trapped flux stability, Fig. 2 exhibits the trapped flux measured at different positions along the normal direction, from the centre of the surface of the Gd123 bulk with 0.4 mol% Fe-B addition. The sample was magnetized with field cooling down to 30 K under a stationary applied field of 5 T. The trapped field was successfully stabilized as a function of time. The results clearly show that the trapped flux would become stable after removing the external static field within 55 min. The results show a specific flux distribution along the c-axis and its stability.

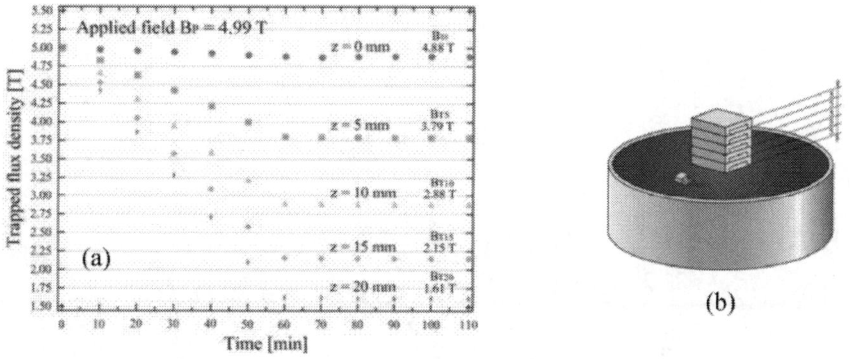

Fig. 2 Trapped magnetic flux density measured by stacked Hall sensors at different positions along the normal to the surface for Gd123 with 0.4 mol% Fe-B addition. The trapped field is successfully stabilized as a function of time at 30 K.

4. Conclusion

By exploring the processing procedure, heat treatment profile, seeding technology, and annealing conditions, the optimal growth conditions for processing high-textured RE-Ba-Cu-O single grains/domains have been achieved. The cold seeding melt-growth technology employing Nd-Ba-Cu-O/MgO thin films as seed crystals allows the batch processing of high performance Gd-Ba-Cu-O bulks. Furthermore, it was found that supplying greater sufficient liquid source during the melt growth benefits the homogenous distribution of Gd_2BaCuO_5 in a large-scale matrix, leading to a significant enhancement of the trapped magnetic field.

Acknowledgments

This work was partly performed using facilities of the ISSP, the University of Tokyo. The authors are grateful to the National Natural Science Foundation of China (NSFC, Grant No.11004129).

References

[1] Tomita M and Murakami M 2003 *Nature* **421** 517.
[2] Nariki S, Sakai N and Marakami M 2005 *Supercond Sci Technol* **18** 126.
[3] Xu K, Zhou D, Li B, Hara S, Tsuzuki K, Zhang Y, Xu Y and Izumi M 2012 *Superconductivity: Recent Developments and New Production Technologies* Chapter 5 ed. Mularidhar Miryara (Nova Science Pub Inc).
[4] Ishii Y, Shimoyama J, Tazaki Y, Nakashima T, Horii S and Kishio K 2006 *Appl. Phys. Lett.* **89** 202514.
[5] *Handbook of High Temperature Superconductivity:Theory and Experiment* 2007 eds. Schrieffer J R and Brooks J S (Springer).
[6] Xu CX, Hu A, Sakai N, Izumi M and Hirabayashi I 2005 *Supercond. Sci. Technol.***18** 229.
[7] Hu A, Xu CX, Izumi M, Hrabayashi I and Ichihara M 2006 *Appl. Phys. Lett.* **89** 192508.
[8] Xu Y, Izumi M, Tsuzuki K, Zhang Y, Xu CX, Murakami M, Sakai N and Hirabayashi I 2009 *Supercond. Sci. Technol.***22** 095009.
[9] Xu K, Tsuzuki K, Hara S, Zhou D, Zhang Y, Murakami M, Nishio-Hamane D and Izumi M 2011 *Supercond. Sci. Technol.* **24** 085001.
[10] Tsuzuki K, Hara S, Xu Y, Morita M, Teshim H, Yanagisawa O, Noudem J, Harnois Ch and Izumi M 2011 *IEEE Trans. Appl. Supercond.* **21** 2714.
[11] Li B, Zhou D, Hara S, Xu K and Izumi M 2013 *IEEE Trans. Appl. Supercond.* **23** 8001804.
[12] Li B, Zhou D, Xu K, Tsuzuki K, Zhang JC and Izumi M 2013 *Physica C* http://dx.doi.org/10.1016/j.physc.2013.06.017
[13] Zhou D, Hara S, Li B, Xu K, Noudem J and Izumi M 2013 *Supercond. Sci. Technol.* **26** 015003

Journal of Physics: Conference Series **507** (2014) 012022 doi:10.1088/1742-6596/507/1/012022

Spark plasma sintering consolidation of $Bi_{1.65}Pb_{0.35}Sr_2Ca_2Cu_3O_{10+\delta}$ ceramic samples

E. Govea-Alcaide[1], **I. F. Machado**[2], **M. S. Torikachvili**[3], **N. R. Dilley**[4], **and R. F. Jardim**[5]

[1]Departamento de Ciencias Básicas, Facultad de Ciencias Técnicas, Universidad de Granma, Apdo. 21, P. O. Box 85100, Bayamo, Cuba
[2]Departamento de Engenharia Mecatrônica e Sistemas Mecânicos, Escola Politécnica, Universidade de São Paulo, 05508-900, São Paulo, SP, Brazil
[3]Department of Physics, San Diego State University, San Diego, CA 92182, USA
[4]Quantum Design, 6325 Lusk Blvd., San Diego, CA 92121, USA
[5]Instituto de Física, Universidade de São Paulo, CP 66318, 05315-970, São Paulo, SP, Brazil

E-mail: `rjardim@if.usp.br`

Abstract. Pre-reacted powders of $(Bi\text{-}Pb)_2Sr_2Ca_2Cu_3O_{10+\delta}$ (Bi-2223) were consolidated by using the spark plasma sintering (SPS) technique under vacuum and at different consolidated temperatures T_D. We have found that the transport properties of SPS samples depend on their oxygen content because the SPS process is performed under vacuum. From these results we have inferred that SPS samples are oxygen deficient and that a post-annealing treatment, performed in air for a brief time interval, is needed to restore such a deficiency. Measurements of temperature dependence of the thermal conductivity, $\kappa(T)$, were performed in all SPS samples including a reference sample sintered at 845 °C in air. The results indicate that the electronic contribution to the thermal conductivity in all samples is very low. Values of κ were found to be higher in SPS samples mostly because of their high volume density of $\sim$ 5.7 g/cm^3. We have also performed measurements of the critical current density as a function of the applied magnetic field, $J_c(B_a)$. All samples display a clear Josephson-like behavior and low values of the superconducting critical current density at zero applied magnetic field. The experimental results indicate that the oxygen deficiency in SPS samples is marked near the grain boundaries. This suggests the occurrence of grains with core-shell morphology, where the width of the shell is consolidation temperature dependent.

1. Introduction

The spark plasma sintering (SPS) is an effective, unconventional method that promotes densification by the simultaneous action of a high direct electric current through graphite dies and uniaxial pressure [1]. In high-T$_c$ cuprates superconductors, the use of SPS can be a difficult task because the process occurs under vacuum, and these materials are very sensitive to the oxygen content. Such an obstacle is less noticeable in compounds of the Bi-Sr-Ca-Cu-O (BSCCO) system [2], making them excellent candidates for the application of the SPS technique. However, up to now there are few studies regarding the use of SPS in the BSCCO system. In the last decade, the influence of the SPS conditions on the formation of $(Bi,Pb)_2Sr_2CaCu_2O_{8+\delta}$ (Bi-2212) and $(Bi,Pb)_2Sr_2Ca_2Cu_3O_{10+\delta}$ (Bi-2223) phases was conducted [3]. In this study, the materials were SPS under different holding temperatures and times, and the uniaxial compacting pressure was only 17.5 MPa. Under theses conditions, the consolidated samples exhibited extra, non-superconducting phases. In addition to this, the SPS samples showed semiconducting-like behavior in the normal state region, broad superconducting transitions, and very low values of the superconducting transition temperature. The authors attributed the results to the presence of extra phases and their distribution within the samples.

Recently, we have reported a systematic study on the influence of the consolidation temperature T_D on the normal and superconducting transport properties of Bi-2223 ceramics samples obtained by using the SPS technique [4]. Similar results as those mentioned in Ref. [3] were obtained. However, after the SPS, the dominant phase in all samples was the Bi-2223. In addition to this, the relative density of the pellets reached $\sim$ 85 % of the theoretical value by using an uniaxial compacting pressure of 50 MPa. We have found that the normal and the superconducting transport properties of the SPS samples were strongly dependent of their oxygen content because the process occurs under vacuum. Under these circumstances, a post-annealing treatment was needed to improve the transport properties of the samples [4].

In this paper, we report on the thermal and electrical properties of $Bi_{1.65}Pb_{0.35}Sr_2Ca_2Cu_3O_{10+\delta}$ ceramic samples consolidated by the SPS technique at different temperatures. We have also studied the role of the post-annealing treatment for the improvement of the superconducting properties of the SPS samples.

2. Experimental procedure

Polycrystalline samples of $Bi_{1.65}Pb_{0.35}Sr_2Ca_2$ $Cu_3O_{10+\delta}$ (Bi-2223) were prepared by the traditional solid state reaction method as described elsewhere [5] The final consolidation of the samples was performed in a SPS 1050 Dr Sinter apparatus [4]. In order to study the influence of the consolidation temperature, T_D, the samples were subjected to two different temperatures T_D = 750 and 830 °C. The maximum uniaxial compacting pressure used in these experiments was 50 MPa. Also, the SPS samples were subjected to an additional post-annealing heat treatment in air at 750 °C during 5 minutes [4]. Additionally, for comparison reasons, $\sim$ 4 g of the starting powder was *cold pressed* inside the SPS apparatus and the resulting pellet was sintered at 845 ^{o}C in air for 2400 min. This sample (**84PA**) will thereafter be referred as the *reference sample.*

The temperature dependence of the electrical resistivity, $\rho(T)$, and transport critical current density as a function of the applied magnetic field, $J_c(B_a)$, were performed by using the procedures reported elsewhere [4, 5]. The thermal conductivity measurements were carried out on parallelepiped-shaped samples with approximate dimensions of 5 $\times$ 1.5 $\times$ 1 mm^3 in the temperature range of 2 - 300 K using the thermal transport option (TTO) of the Quantum Design PPMS system [6].

3. Results and discussion

Our previous results indicated that the SPS method significantly promotes the densification of the pre-reacted powders and produces nearly single phase $(Bi,Pb)_2Sr_2Ca_2Cu_3O_{10+\delta}$ ceramic samples [4]. The density of the *reference sample* (**84PA**), uniaxially pressed at 50 MPa, was only $\sim$ 3.2 g/cm^3, but increased to $\sim$ 5.7 g/cm^3 in SPS samples consolidated at the same low compacting pressure. However, there are important points to be addressed here: the SPS samples are sintered under vacuum, at high temperatures, and in a short time interval [4]. These features affect the properties of the SPS samples, as discussed below.

The $\rho(T)$ data of the samples **83PA** and **83PA75** are displayed in Fig. 1. Both samples were consolidated by SPS at T_D = 830 °C and, in addition, the latter was post-annealed at 750 °C for 5 min in air. The sample **83PA** exhibits a clear semiconductor-like behavior in which $\rho(T)$ increases with decreasing temperature, reaches a maximum at $T \sim$ 118 K, which is followed by a very broad superconducting transition. In contrast, the $\rho(T)$ behavior of sample **83PA75** displays a typical metallic behavior and an abrupt resistive superconducting transition. These different features between samples were interpreted by considering that the

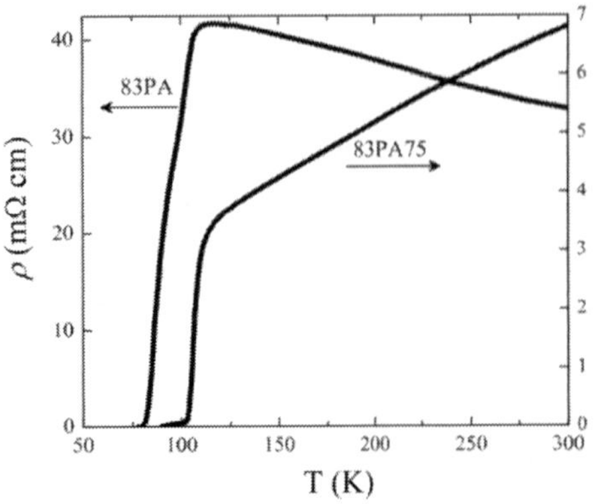

Figure 1. Temperature dependence of the electrical resistivity of the SPS samples **83PA** and **83PA75**.

deoxygenation process in SPS samples preferably takes place near the grain-boundaries, leading to the occurrence of grains with core-shell morphology where the shell is oxygen deficient and that the width of shell increases with increasing T_D [4].

Figure 2 displays the temperature dependence of the thermal conductivity, $\kappa(T)$, of the samples **84PA**, **75PA**, **75PA75**, **83PA**, and **83PA75**. The total thermal conductivity consists of the lattice and the electronic (carrier) contribution, which is connected to the electrical conductivity, in first approximation, by the Wiedemann-Franz law. In all samples, $\kappa(T)$ increases with increasing temperature between 2 to 300 K and SPS samples were found to displays higher ($\sim$ 2 times) values of $\kappa(T)$ when compared with the *reference sample*. Such a feature is more pronounced in the normal-state region ($T \geq T_c^{\kappa}$) and is certainly related to the higher volume density of SPS samples which displays a very low concentration of voids. We have also found that $\kappa(T)$ exhibits a linear behavior in the temperature range $220 \leq T \leq 300$ K (see Fig. 2). We attribute the deviation of such a linear behavior below $\sim$ 220 K to the pseudogap formation, a feature corroborated by changes observed in the same T window in $\rho(T)$ data, probably related to a decrease of the free carriers due to the pseudogap formation [7].

As a first attempt to estimate the electronic contribution to $\kappa(T)$, we have used the the Wiedemann-Franz law, $\kappa_e(T) = (L_0T)/\rho(T)$, where $L_0 = 2.45\times10^{-8}$ WΩ/K^2 is the Lorentz number. Table 1 displays the estimated values of κ_e contribution to the thermal conductivity for all samples at T = 300 K. We first point out that all values are at least one order of magnitude smaller than the measured κ value at this temperature. This indicates that $\kappa(T)$ is mainly governed by the phonon contribution, as described elsewhere [8]. The results also suggest that the large phonon contribution to $\kappa(T)$ is present as well in post-annealed samples in which the electrical resistivity values are lower. Under these conditions, grain-boundaries seems to be a more plausible source of electron scattering for the decrease in $\kappa(T)$ and such a decrease is

Table 1. Some parameters of the samples studied in this work (see the text for details).

Sample	T_{off} (K)	ρ(300K) (mΩ cm)	κ_e(300K) (K)	T_c^{κ} (K)	$J_c(0)$ (A/cm^2)	B_0 (mT)
84PA	101.0	9.3	0.08	106	1.4	1.02
75PA	83.0	10.0	0.07	101	2.4	0.64
75PA75	102.6	4.3	0.17	108	6.5	1.44
83PA	81.0	32.6	0.02	99	1.3	0.42
83PA75	102.6	6.8	0.11	108	3.5	0.82

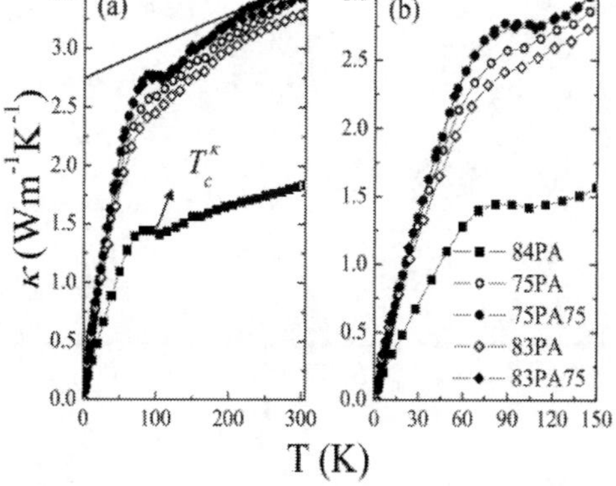

Figure 2. (a) Temperature dependence of the thermal conductivity of the *reference sample* **84PA** and the SPS samples **75PA**, **75PA75**, **83PA**, and **83PA75**. Here, the solid line is a visual guide to show the range where κ data have a linear behavior. (b) An expanded view of the $\kappa(T)$ data in the 2 - 150 K range. T_c^κ is defined as the superconducting critical temperature. The lines are a guide for the eyes (see the text for details).

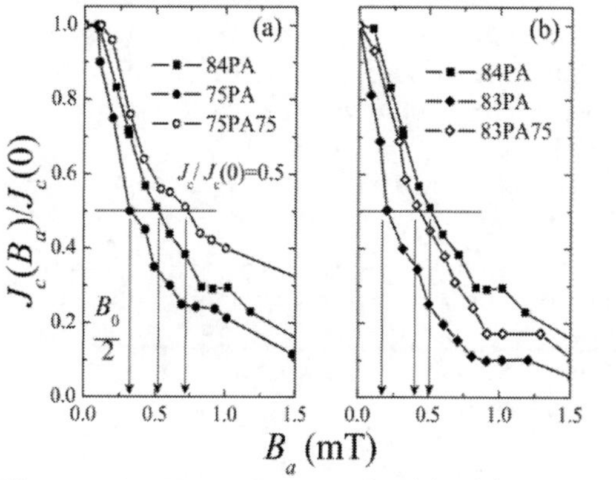

Figure 3. Normalized $J_c(B_a)/J_c(0)$ curves measured in samples **75PA** and **75PA75** (a), and **83PA** and **83PA75** (b). For comparison reasons the curve measured in the sample **84PA** was included in both panels. Measurements were performed at the same reduced temperature $t = T/T_{off} = 0.94$. Dashed lines are guide for the eyes and indicate the position of the critical field of the Josephson junctions (see text for further details).

much less market is all SPS samples with very high density.

The features occurring at the grain-boundaries have enormous influence on the behavior of $J_c(B_a)$. Figure 3 displays the normalized $J_c(B_a)/J_c(0)$ virgin curves measured in samples **84PA**, **75PA**, **75PA75**, **83PA**, and **83PA75**. For comparison reasons, the above measurements were performed at the same reduced temperature, $t = T/T_{off} = 0.94$, very close to the temperature in which the zero resistance state is observed (see Table 1) and the flux penetration is robust. Values of the critical current density at zero applied magnetic field, $J_c(0)$, are reported in the Table 1). Due to the proximity to T_{off}, values of $J_c(0)$ are very low in all samples, providing extra information regarding the relative change of the transport properties after the post-annealing treatment. The critical current density increases $\sim$ 2.7 times in SPS samples obtained at the same consolidation temperature. On the other hand, as compared to the *reference sample*, $J_c(0)$ was found to raise $\sim$ 4.5 times (sample **75PA75**) and only $\sim$ 2.7 (sample **83PA75**). All curves exhibits a clear Josephson-like behavior, in which the ratio $J_c(B_a)/J_c(0)$ abruptly decreases with increasing B_a. We first mention here that values of $J_c(B_a)/J_c(0)$ in the SPS samples **75PA** and **83PA** are smaller than that measured in the *reference sample*, further indicating the major role played by the oxygen content in the transport properties of the materials (see Table 1). We also mention that the decrease in $J_c(B_a)/J_c(0)$, for the same B_a range, is much more pronounced in samples consolidated at 830 °C, indicating that the flux penetration through grain boundaries is significant in these samples.

In order to further address the flux penetration, a particular value of the applied magnetic field B_0, where $J_c(B_a)/J_c(0)$ decreases $\sim$ 50 %, is of interest [9]. Such a magnetic field is referred to as the critical field of the Josephson junctions, $B_0 = \Phi_0/(2\lambda_L L_g)$. Here, Φ_0 is the flux quantum, λ_L is the London penetration depth, and L_g is the grain size. The obtained values of B_0 are listed in Table 1. Despite of the fact that values of B_0 are very small in all samples, a 2-fold increase after the post-annealing treatment is clearly observed. Previous results in Bi-2223 sample yielded a value of 1.5 mT at $T = 77$ K in a sample pelletized at 50 MPa and with volume density of ~ 4.0 g/cm^3 [5]. We mention here that a similar value was found in the **75PA75** sample but at $T = 96.4$ K.

By assuming that the SPS process unaffects significantly the size of the grains, values of B_0 between samples can be related to changes in the λ_L of the materials. In this regard, increasing B_0 results in a decrease of λ_L. The observed behavior of B_0, and consequently of λ_L, can be implicitly related to changes in the width of the oxygen deficiency shell near the grain boundaries, as suggested previously [4]. A preliminary analysis suggests that λ_L is larger in samples consolidated at higher temperatures, further indicating a thicker oxygen-deficient shell of these SPS samples.

4. Conclusions

We have used the spark plasma sintering (SPS) method to consolidated pre-reacted powders of $(Bi,Pb)_2Sr_2Ca_2Cu_3O_{10+y}$ (Bi-2223) compounds. All samples were uniaxially pressed at 50 MPa and the influence of the consolidation temperature on the physical properties was investigated and compared with a sample synthesized by the solid-state reaction method (*reference sample*). The SPS samples are oxygen deficient and are believed to be comprised of grains with a shell-core morphology. The electronic contribution to the thermal conductivity in all samples is very low. Values of κ were found to be higher in SPS samples because of their high volume density (~ 5.7 g/cm^3) as compared with the *reference sample* (~ 3.2 g/cm^3). We have also observed a Josephson-like behavior of $J_c(B_a)$ and have obtained the critical field of the Josephson junctions, $B_0 \propto \lambda_L^{-1}$. A preliminary analysis indicate that increasing B_0 decreases λ_L in the post-annealed samples and that result is related to changes in the width of the oxygen-deficient shell of the superconducting grains.

Acknowledgments

The authors acknowledge financial support from the Brazil's agencies CAPES, FAPESP, and CNPq, NSF Grant No. DMR-0805335, U.S., and Petrobras company.

References

[1] Orrù R, Licheri R, Locci A M, Cincotti A and Cao G, 2009 *Mater. Sci. Eng. B* **63** 127.
[2] Fujii T, Watanabe T and Matsuda A, 2001 *Physica C* **357** 173.
[3] Badica P, Aldica G, Groza J R, Bunescu M C and Mandache S, 2002 *Supercond. Sci. Technol.* **15** 32.
[4] Govea-Alcaide E, Machado I F, Bertolete-Carneiro M, Muné P and Jardim R F, 2012 *J. Appl. Phys.* **112** 113906.
[5] Muné P, Govea-Alcaide E and Jardim R F, 2003 *Physica C* **384** 491
[6] Mun E, Bud'ko S L, Torikachvili M S and Canfield P C, 2010 *Meas. Sci. Technol.* **21** 055104.
[7] Govea-Alcaide E, Muné P and Jardim R F, 2005 *Braz. J. Phys.* **35** 680.
[8] Castellazi S, Cimberle M R, Ferdeghini C, Giannini E, Grasso G, Marre D, Putti M and Siri A S, 1997 *Physica C* **273** 314.
[9] Müller K H and Matthews D N, 1993 *Physica C* **206** 275.

Journal of Physics: Conference Series **507** (2014) 012023 doi:10.1088/1742-6596/507/1/012023

Carbon doping of precursor boron powder for control of normal resistance of MgB_2 bulks for specific use in fault current limiter applications

J Archer and A L L Jarvis*

School of Engineering, University of KwaZulu-Natal, King George V Ave, Durban, KwaZulu-Natal, 4014, South Africa

Jarvis@ukzn.ac.za

Abstract. Magnesium Diboride (MgB_2) bulk superconductor has been manufactured in order to study the enhancement of the normal resistivity for application to superconducting fault current limiters (SFCLs). SFCLs have proven to be a viable means for limiting surge currents by dissipating fault energy as the superconductor quenches. As the current limiting behaviour is determined by the normal resistivity (ρ_n), research has been conducted to evaluate an effective means to increase ρ_n for bulk superconducting MgB_2, which is intrinsically much lower than for high temperature superconductors. Intragranular carbon doping has been confirmed as a means to increase ρ_n, and was implemented by chemical vapour deposition (CVD) on the boron precursor powder by ethylene gas at 600 °C for durations up to 6 hours in a tubular furnace apparatus. In situ manufacturing of MgB_2 bulk was performed using the reactive liquid magnesium infiltration technique. Overall, carbon doping provided a factor of 11.00 increase in the residual resistivity, ρ_0, which provides the initial limiting action to fault currents, for an accompanying decrease in the critical temperature, T_c, of 2 K.

1. Introduction

Globally, as power demand grows and the use of renewable energy sources increase, it is typically met with an increase in grid complexity and distributed generation, which increases fault current levels and makes them harder to predict [1]. The installed switchgear capacity may no longer be sufficient to support this growth, presenting a need for fault current limiter to reduce the magnitude of fault currents before switchgear interruption [2].

In a resistive SFCL under fault conditions, the superconductor quenches and presents to the grid an impedance determined by the normal-state resistivity, ρ_n, dissipating fault energy through resistive losses (joule heating), thus limiting the fault current. Once the fault has been cleared, the FCL can be 'reset' by cooling the superconductor to superconducting state again.

Since it is ρ_n that determines the energy dissipated during fault currents, a high ρ_n is desired for enhanced current limiting properties. A practical problem with the use of MgB_2 in resistive SFCLs is the intrinsically low normal resistivity of MgB_2, resulting in longer lengths of superconductor required, increasing both size and cost. Work has been done to investigate the effect of chemical doping on the superconducting properties of MgB_2, primarily in the form of J_c and upper critical field, H_{c2}, and has revealed increases in these parameters up to a certain limit with an increase in disorder due to doping [3]. The approach taken in this research has been to further dope the specimens to substantially increase ρ_n, making it more suitable for use in FCLs whilst still maintaining reasonable operational conditions.

2. Experimental procedure

2.1. Carbon Doping
MgB_2 specimens were prepared by the reactive liquid infiltration technique, with the precursor boron powder doped by carbon chemical vapour deposition using ethylene gas. For the chemical vapour deposition (CVD) of carbon to the boron precursor powder (Sigma-Aldrich, 95-97%, amorphous, submicron particle size), a purpose-built stainless steel tubular furnace assembly was used in conjunction with the gaseous hydrocarbon ethylene (C_2H_4) as the carbon source. The ethylene gas was introduced to a relative pressure of 80 kPa. The ethylene in the tube was flushed hourly at the 600 °C operating temperature for a total doping period of 6 hours, to obtain high doping concentrations up to 16%. Estimation of the carbon content was achieved by measurement of the boron precursor powder mass before and after CVD with a Mettler Toledo PB3002-S.

2.2. Reactive Liquid Infiltration
Specimens were manufactured by the reactive liquid infiltration (RLI) technique [4] in appropriately designed stainless steel containers. A number of iterations on the designs were made, to improve various aspects; straightforward removal of product, simplified welding of crucibles, increasing internal reaction pressure, and minimising stainless steel volume and machining time.

Five bulk specimens of MgB_2 were produced representing carbon doped boron concentrations from 0-16%. Optical and scanning electron microscopy analysis was performed on the specimens, revealing good density. The samples were sectioned by diamond wafering blade into thin rectangular slices for resistive testing and into rectangular bars for AC susceptibility investigations.

2.3. DC resistivity measurements
DC resistivity measurements were performed on all of the samples with a 4-wire apparatus constructed with gold-plated spring-loaded test pins. This facilitated rapid changing of the samples, and the internal springs ensured electrical contact with the sample despite differing in thermal expansion coefficients in apparatus, and achieved reproducibility after unload-reload cycles. The sample stage was cooled by a two stage GM cryocooler with a Lakeshore 336 temperature controller.

2.4. AC susceptibility
The bar sections were inserted into an AC susceptibility apparatus, to provide contactless measurement of the critical temperature. This was required because the apparatus used for DC resistivity measurements with the limited equipment specifications introduced unavoidable thermal heat leakage, and particularly, joule heating in the sample contacts. This introduced uncertainty into the critical temperatures obtained, so a contactless approach was taken.

3. Results and Discussion

3.1. Sample morphology
Scanning electron microscopy (SEM) was used to investigate the topology of the surface presented after abrasion cutting, enabling some intragranular views as well as some of the intergranular interfaces, the grain boundaries. A scanning electron micrograph is shown in figure 1, and a low magnification optical micrograph is shown in figure 2.

Figure 1. SEM micrograph of an undoped specimen (~100 000 ×).

Figure 2. Optical micrograph showing clear bimodal grain distribution

Smaller grains appeared to be about 100 nm in diameter throughout all the micrographs, thus doping did not appear to affect the grain size. Optical microscopy showed a well-formed MgB_2 phase with few voids, and clear bimodal grain distribution, likely as a result of bimodality in the precursor boron powder [5].

3.2. DC resistivity results

The resistivity results are shown in figure 3, showing a distinct increase in resistivity with an increase in carbon content, with further analysis of the data shown in figure 4, where the overall resistivity has been broken down into the residual and phonon-contributed resistivities. The separation is in accordance with the Rowell criterion [6], with the phonon contribution, $\Delta\rho$, defined as the difference between room temperature resistivity, ρ_{300K}, and the resistivity just above T_c.

Carbon doping appeared to increase both the residual resistivity as well as the phonon-contributed resistivity, implying that the doping had also decreased the intergranular connectivity. Data from each specimen was thus divided by the geometrical factor, A, which was defined as the specimen's phonon contributed resistivity, $\Delta\rho$, divided by the single crystal phonon contribution, $\Delta\rho_{SC}$, of 10 μΩ.cm. After this treatment, a clear increase in the residual resistivity as a function of doping could be observed.

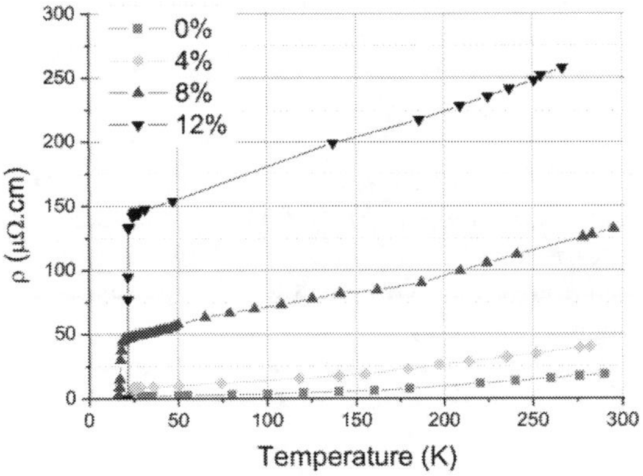

Figure 3. Resistivities of doped specimens as a function of temperature, showing superconducting transitions.

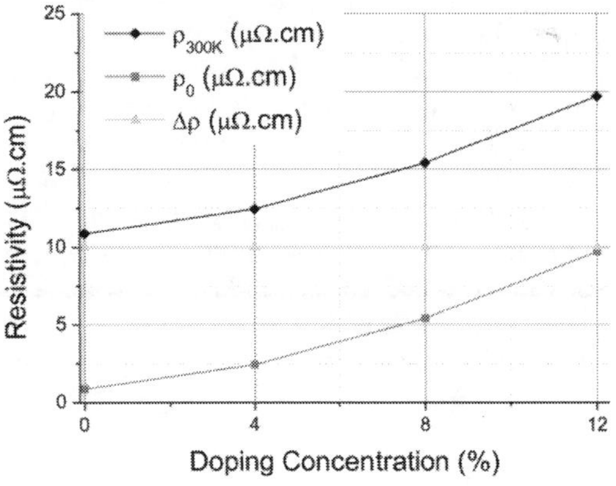

Figure 4. Resistivity contributions from residual and phonon components for the doped specimens, after division by the respective geometrical factors.

3.3. AC susceptibility results
The critical temperatures characterised by AC susceptibility are shown in figure 5, showing only a small overall change in T_c, and are presented in figure 6 with the corresponding normalised resistivity. Without any secondary peaks in the AC susceptibility data, the inter-granular doping that was responsible for the decrease in effective cross sectional area in the resistivity measurements was not observed in the specimens, with the exception of the 8% doped specimen.

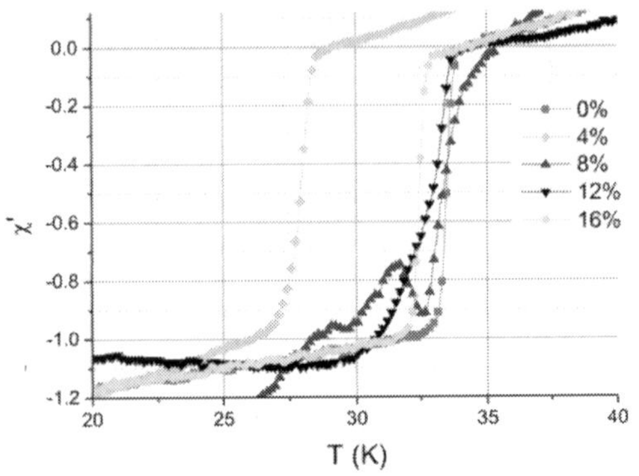

Figure 5. Real susceptibility of specimens doped in the range of 0% to 16%.

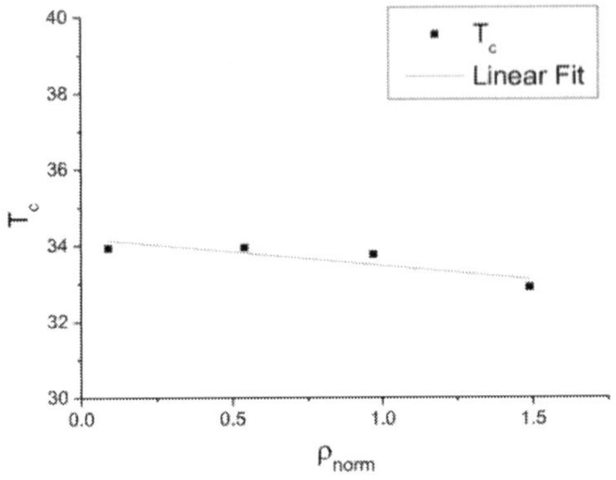

Figure 6. Critical temperature dependence on the normalised resistivity. The solid line represents a linear fit applied to the data.

4. Conclusions
A study of the effect of carbon doping on MgB_2 specimens fabricated via the RLI technique was performed, with the research goal being to increase the normal state resistivity of MgB_2 by the use of carbon doping, particularly to establish if the resistivity could be increased significantly whilst still maintaining reasonable operational requirements.

Boron powder was doped by chemical vapour deposition of carbon, before use in the reactive liquid infiltration technique. Morphology analysis performed by SEM and optical microscopy revealed that all specimens were dense, featuring few voids, and no signs of impurity phases. There was a clear bimodal grain distribution, and doping did not appear to affect the grain size.

The DC resistivity was measured, and AC susceptibility was used to determine critical temperature. Analysis of resistivity measurements established that doping increased the normal state resistivity, via both increases in the residual resistivity, and phonon-contributed resistivity. Even when the geometrical factor was taken into account, the resistivities increased as a function of doping. From the results of the AC susceptibility measurements, clear transitions were evident, and the critical temperature was defined from these low-field measurements.

Overall, carbon doping provided a factor of 11.00 increase in ρ_0, which provides the initial limiting action to fault currents, for an accompanying decrease in the critical temperature, T_c, of 2 K.

References
[1] Papathanassiou S A 2007 *Electric Power Systems Research* **77** 24-34
[2] Ye L, Lin LZ, and Juengst K-P 2002 *IEEE Trans. Appl. Supercond.* **12** 900-3
[3] Young EA and Yang Y 2007 *IEEE Trans. Appl. Supercond.* **17** 2794-97
[4] Giunchi G, Ripamonti G, Cavallin T, and Bassani E 2006 *Cryogenics* **46** 237-42
[5] Chen SK, Yates KA, Blamire MG and MacManus-Driscoll JL 2005 *Supercond. Sci. Technol.* **18** 1473-77
[6] Rowell JM 2003 *Supercond. Sci. Technol.* **16** R17-27

Journal of Physics: Conference Series **507** (2014) 012024 doi:10.1088/1742-6596/507/1/012024

Increasing the sensor performance using Au modified high temperature superconducting $YBa_2Cu_3O_{7-\delta}$ thin films

C Katzer[1], C Stahl[2], P Michalowski[1], S Treiber[2], M Westerhausen[1], F Schmidl[1], P Seidel[1], G Schütz[2] and J Albrecht[3]

[1] Institute of Solid State Physics, Friedrich-Schiller University Jena, Helmholtzweg 5, D-07743 Jena (Germany)
[2] Max-Planck-Institute of Intelligent Systems, Heisenbergstraße 3, D-70569 Stuttgart (Germany)
[3] Aalen University, Beethovenstraße 1, D-73430 Aalen (Germany)

E-mail: christian.katzer@uni-jena.de

Abstract. We prepared planar, galvanically coupled gradiometers whereby the antenna structures of some of them were modified by incorporating Au nanoparticles. Gradiometers with gold modified antennas were compared with conventional ones to investigate the influence of gold induced pinning on the performance of superconducting sensor devices. We found that a local inclusion of gold nanoparticles offers the possibility of increasing the pinning of flux lines in the antenna regions, thus significantly reducing flux noise, especially in the low-frequency range. Since also the properties of grain boundary Josephson Junctions are strongly affected by Au particles, SQUID and antenna regions in gradiometric sensor devices can be separately optimized.

1. Introduction

Many superconducting thin film devices require a spatially resolved current carrying capability due to different boundary conditions. On the one hand, the critical current density and the pinning of flux lines, respectively, should be high to reduce flux noise in the antenna regions of gradiometers. On the other hand, the critical current density of the Josephson Junctions itself must not be too high to ensure a proper functionality. We show that adding thin gold layers during the preparation process of epitaxial $YBa_2Cu_3O_{7-\delta}$ (YBCO) thin films offers the possibility of creating such spatially varying flux pinning properties, allowing to separately optimizing SQUIDs and antennas.

2. Sample Preparation

Unfortunately, the grain boundary itself strongly influences the properties of grain boundary Josephson junctions and thus the performance of gradiometric sensor devices. To achieve comparable conditions we prepared gradiometers on the same $SrTiO_3$ (STO) bicrystal substrate, thus using the same grain boundary. A patterned gold seed layer was deposited on the STO substrates to locally introduce modifications of the pinning only at the antenna areas of some of the gradiometers as can be seen in figure 1. The patterning of this Au seed layer was done using photolithography combined with dc-sputtering. After the gold seed layer had been deposited, the samples were heated to 780 °C to induce a dewetting of the Au layer, leading to a self-assembly of crystalline Au nanoparticles. As no lateral diffusion of nanoparticles was observed, the usage of a prepatterned Au seed layer appears to be suitable for preparing nanoparticles at well-defined positions on the substrate - in our case only at the antenna regions. Using pulsed laser deposition in an oxygen atmosphere a YBCO thin film of 150 nm thickness was deposited on top of the nanoparticles. The gradiometer layouts subsequently were patterned utilizing ion beam etching and photolithography.

Figure 1. Layout of the four gradiometers on one and the same SrTiO3 substrate. Black parts refer to pristine parts of the YBCO thin film, gold ones to YBCO with embedded Au nanoparticles.

3. Results

In contrast to what was reported by Mikheenko et al. [1] the nanoparticles cannot only be found at the substrate/YBCO thin film interface. As we confirmed by transmission electron microscopy (TEM) and scanning electron microscopy (SEM – see figure 2), some of the particles are overgrown by the YBCO matrix but some of them also can be found on top of the YBCO layer or within the superconducting layer itself. Interestingly, in TEM images we do not find any additional defects or disturbances in the epitaxial growth of the YBCO thin film in vicinity of Au nanoparticles. Instead, there only appears to be a local bending of lattice planes around the nanoparticles [2].

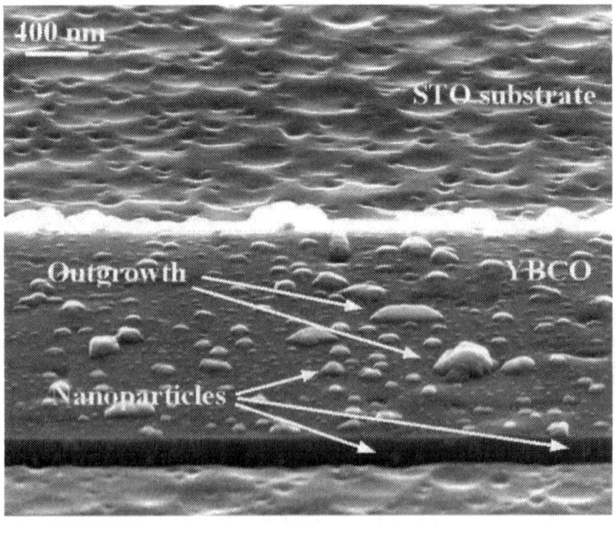

Figure 2. SEM image of a patterned YBCO layer. Note that the sample surface does not only show nanoparticles but also some additional outgrowths which are typical for YBCO thin films. Some nanoparticles can be found within the YBCO layer as it is visible in the cross section.

To characterize the critical current density in our samples we employed magneto-optical Faraday microscopy. Thus, in contrast to transport measurements the local critical current density could be calculated out of the obtained flux density distribution via an inversion of the Biot-Savart law [3]. Using this method, in principle it is possible to achieve a very high spatial resolution in the local j_c measurements. In our case this resolution was restricted to about 5μm. As can be seen in figure 3, the critical current density j_c was significantly increased over the whole temperature range in the Au modified areas (at a temperature of 77 K the enhancement is approximately 70 %). Although the absolute value of $5\cdot10^6$ A/cm^2 is not a record in the field of artificial pinning centres, the reported method still appears to be advantageous compared to conventional ones. The most intriguing argument is that the incorporation of self-assembled Au nanoparticles allows modifying the properties of the grain boundary Josephson junctions at the same time [4]. More specifically, in contrast to the increasing current density of thin films, the critical current density of grain boundary Josephson junctions is decreased, allowing to reduce the noise due to fluctuations of I_c. At the same time the normal state resistance of the junction is increased, compensating the decreasing critical current [4]. Thus, the characteristic I_cR_N product of the junctions remains the same, enabling high transfer functions of the SQUID [5].

It should be noted that the curve progression of the temperature dependency of the critical current density remains the same, indicating that the same pinning mechanism as in pristine YBCO films is responsible for the increasing current density. I.e. individual pinning of flux lines at the gold nanoparticles can be neglected. These findings are in excellent agreement with transport measurements performed on different samples (see figure 4).

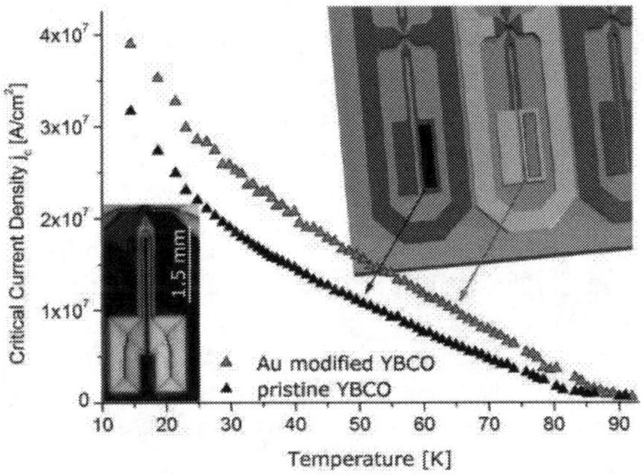

Figure 3. Temperature dependence of the critical current density obtained using magneto-optical imaging.

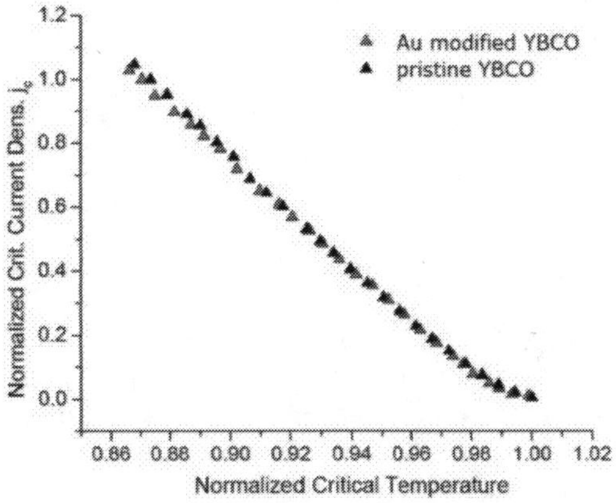

Figure 4. Normalized temperature dependence of the critical current density obtained using transport measurements on different samples.

Any additional pinning of flux vortices should not only increase j_c but also influence the flux noise of the gradiometers. Because the optimization of high temperature superconducting SQUID gradiometers is dependent on various parameters, e.g. the ambient magnetic field, we carried out noise measurements in a locally shielded environment (μ-metal shield) at T = 77K. The measurements were performed in flux locked loop mode with an ac-bias reversal technique. Comparing a gradiometer which had an Au modified antenna structure with a gradiometer already used in biomagnetic measurements, one finds the white noise level to be drastically decreased (see figure 5). We achieved white noise levels between 5 $\mu\Phi_0/\sqrt{Hz}$ and 6 $\mu\Phi_0/\sqrt{Hz}$ with gold modified gradiometers. Compared to the conventional SQUID gradiometer this value is very low, as it is even comparable to the best values achieved for single SQUIDs without an antenna structure [6]. Furthermore, the corner frequency was shifted to lower values, significantly enhancing the performance in the low-frequency range. This is far more interesting, as it was possible to increase the field gradient resolution at 1 Hz from 700 $fT/(cm\sqrt{Hz})$ to 600 $fT/(cm\sqrt{Hz})$ for the gradiometer with a gold modified antenna structure.

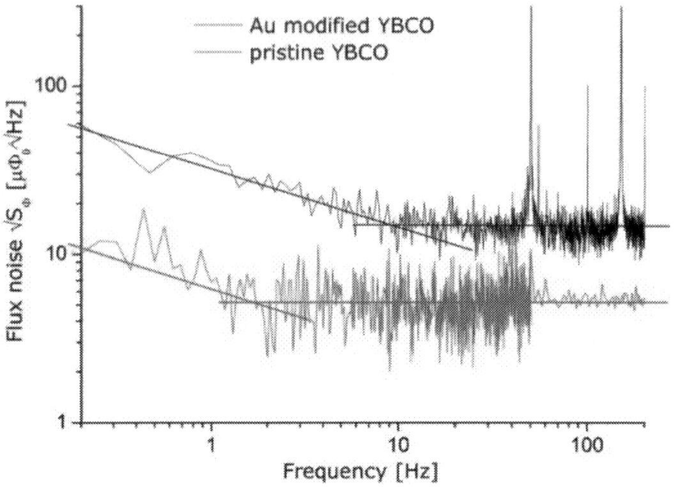

Figure 5. Noise measurements of gradiometers with and without embedded gold nanoparticles in the antenna regions. The corner frequencies are about 2 Hz with Au particles and 10 Hz without Au, respectively.

4. Conclusion

A local inclusion of gold nanoparticles offers the possibility of increasing the pinning of flux lines in the antenna regions, thus significantly reducing flux noise. Since also the properties of grain boundary Josephson junctions are strongly affected by Au particles (lowering the critical current while increasing the normal state resistance), SQUID and antenna regions in gradiometric sensor devices can be separately influenced. Thus, a patterning of the Au seed layer using lithographic techniques seems to be highly suitable for preparing ideal templates for the production of superconducting sensor devices with optimized performance.

Acknowledgement

C. Katzer would like to thank the Landesgraduiertenförderung Thüringen for financial support.

References

[1] Mikheenko P, Abell K S, Sarkar A, Dang V S, Awang Kechik M M, Tanner J L, Paturi P, Huhtinen H, Hari Babu N, Cardwell D A and Crisan A 2010 *J. Phys. Conf. Ser.* **234** 022022

[2] Grosse V, Engmann S, Schmidl F, Undisz A, Rettenmayr M and Seidel P 2010 *Phys. Status Solidi RRL* **4** 97

[3] Jooss C, Albrecht J, Kuhn H, Leonhart S and Kronmüller H 2002 *Rep. Prog. Phys.* **65** 651

[4] Michalowski P, Schmidt M, Schmidl F, Grosse V, Kuhwald D, Katzer C, Hübner U and Seidel P 2011 *Phys. Status Solidi RRL* **5** 268

[5] Wu C H, Sou U C, Chen J C, Chen K L, Yang H C, Hsu M H, Lai T S, Jeng J T, Tsai Y S and Horng H E 2006 *Appl. Phys. Lett.* **88** 102504

[6] Wunderlich S, Schmidl F, Dörrer L, Schneidewind H and Seidel P 1999 *IEEE Trans. Appl. Supercond.* **9** 71

11th European Conference on Applied Superconductivity (EUCAS2013) IOP Publishing
Journal of Physics: Conference Series **507** (2014) 012025 doi:10.1088/1742-6596/507/1/012025

Growth and superconducting properties of $Pb_{1-y}Bi_ySr_2Y_{1-x}Ca_xCu_2O_{7+\delta}$ epitaxial films

Sachio Komori, Ryotaro Inaba, Itsuhiro Kakeya and Minoru Suzuki

Department of Electronic Science and Engineering, Kyoto University, Nishikyo-ku, Kyoto 615-8510, Japan

E-mail: komori@sk.kuee.kyoto-u.ac.jp, inaba@sk.kuee.kyoto-u.ac.jp, kakeya@kuee.kyoto-u.ac.jp, suzuki@kuee.kyoto-u.ac.jp

Abstract. $Pb_{1-y}Bi_ySr_2Y_{1-x}Ca_xCu_2O_{7+\delta}$ (PbBi1212) single-crystal thin films have been grown on $SrTiO_3$ (100) substrates by the ex-situ growth method in which an amorphous film is annealed at 970 °C in a closed ceramic container made of the same material as the film. It is found that PbBi1212 exhibits superconductivity when the Ca concentration x is greater than 0.3. It is also found that a quench into liquid nitrogen from 815 °C is required for the films to exhibit the superconductivity. The highest onset temperature of the superconducting transition attained in the present study is 65 K.

1. Introduction

It has been reported that PbBi1212 shows superconductivity at T_c higher than 90 K.[1, 2, 3, 4] PbBi1212 is of industrial importance because it mainly consists of low-cost elements such as Pb and Ca rather than Bi and Y, which are currently used for high-T_c superconducting wires. However, the superconductivity in PbBi1212 has been investigated only a little because no single-crystal sample has been obtained. As a matter of fact, since the previous reports on PbBi1212 is based on multi-phased bulk sample including Bi2212, some papers[6, 7] claim that PbBi1212 is even non-superconducting. Therefore, the central concern of this study is to grow single-crystal PbBi1212 epitaxial films and to verify its superconductivity and substitution effects. Thin film growth of cuprates including Pb and Bi is highly challenging because their vapor pressure is too high to keep the chemical composition of the film at the growth temperature $\sim$ 700 °C. Although we had tried to grow PbBi1212 by a conventional in-situ sputtering technique, we had never obtained its single-phase films due to re-evaporation of Pb and Bi from the substrate.

In the present paper, we describe an ex-situ growth technique which allows to keep the chemical composition of the film. [8] In this technique, an amorphous film of PbBi1212 deposited by a low temperature sputtering method is annealed at a high temperature of 970 °C in a closed ceramic container made of the same material as the film. We have found that this technique leads to the epitaxial growth of single-phase PbBi1212 thin films. We have also found that the PbBi1212 system exhibits superconductivity with the maximum T_c^{onset} of 65 K with the composition of $x = 0.37$ and $y = 0.00$. The ex-situ growth technique may provide a method to fabricate new generation superconducting wires consisting of ubiquitous elements.

2. Methods

Crystal structures and the compositions of the prepared films were analyzed by using x-ray diffraction (XRD) and energy dispersive x-ray spectroscopy (EDS). Temperature dependence of the electrical resistivity was measured by a conventional DC four probe method. Temperature dependence of the Hall coefficient was measured by the AC four probe method with physical properties measurement system (PPMS), Quantum Design Co. Ltd. Hall coefficient was determined by the linear fitting to the transverse resistivity as a function of external magnetic field ($\parallel c$) between +– 5 T at various temperatures.

Ingredients of sputtering targets were synthesized by the solid state reaction method using high purity (> 99.9%) powders of PbO, Bi_2O_3, $SrCO_3$, Y_2O_3, $CaCO_3$, and CuO. These powders were mixed into cationic compositions of $(Pb_{0.75}Bi_{0.25})_{1.5}Sr_2Y_{1-x}Ca_xCu_{2.6}O_z$ ($x = 0.4 - 0.7$) and calcined twice, firstly at 860 °C for 10 h, and secondly at 880 °C for 10 h in air with pulverization in between. After the calcination, the powders were pressed into cylindrical pellets being 100 mm in diameter and 7 mm in height, and sintered at temperatures as a function of x as $1020 - 200x$ °C for 24 h in air. The sputtering conditions used for depositing PbBi1212 amorphous films on the $SrTiO_3$ (STO) (100) substrates were as follows. The sputtering gas pressure was 100 mTorr (Ar 60 sccm O_2 15 sccm), the anode voltage was 1.4 kV, and the substrate temperature was supposed to be around 200 °C because substrates were not heated in this case. A typical deposition time was set to 2 h which led to a film thickness of 3500 Å.

PbBi1212 polycrystalline containers utilized for the ex-situ growth were prepared in the same way as the sputtering targets at cationic compositions of $Pb_{1-y}Bi_ySr_2Y_{0.3}Ca_{0.7}Cu_2O_z$ ($y = 0 - 0.5$). The mixed powders were calcined three times in this case at 880 °C for 10 h in air, pressed into two cylindrical pellets being 26 mm in diameter and 5 mm in height, and sintered at 1007 °C for 3 h in air. After the sintering we dug a pit in the center of one pellet about $8 \times 8 \times 2$ mm^3 in size, where an amorphous film on a STO substrate was placed for the growth. The other pellet was used as a lid of the growth container. These pellets are polished to be flat with sandpaper and flushed with acetone. If the surface of these pellets is not sufficiently smooth, samples are not heated equally during the growth. In the ex-situ growth, the container with the film was put on the zirconia substrate ($30 \times 30 \times 3$ mm^3) and heated in a muffle furnace at 970 °C for 6 h under O_2 atmosphere and cooled down to room temperature at a rate of 200 °C/h. The EDS analyses on films before and after the growth reveal that the grown film concentrations depend strongly on the compositions of both the sputtering target and the growth container. Pb and Bi concentrations of the films drastically change during the growth and depend strongly on the compositions of the growth container. The concentrations of the other elements of the films depend strongly on those of the sputtering target.

3. Results and discussion

Figure 1 shows the out-of-plane θ–2θ scan XRD results for the single-phase PbBi1212 thin film. Only the c-axis oriented $(00n)$ peaks of PbBi1212 are seen. It was found that both the Ca concentration $x < 0.36$ and the Bi concentration $y < 0.20$ are required to obtain single-phase PbBi1212 films. For a single-phase PbBi1212 thin film, the peaks from other phases were completely absent or less than 0.5% of the PbBi1212 (005) peak magnitude. Impurity phases such as Bi2212 were found in samples whose concentration of Ca or Bi exceeds these limits. In order to prevent the growth of the impurity phase, the Ca concentration of the sputtering target has to be less than 0.5 and the Bi concentration of the growth container has to be less than 0.25.

Figure 2 shows SEM images for the single-phase PbBi1212 film. It is evident that the surface morphologies of the films are very smooth except for some steps. None of cracks, grain boundaries, and impurity phase precipitations were observed on the entire surface of the film.

Figure 3 (a) shows the result of the in-plane $2\theta_\chi$–ϕ scan for the single-phase PbBi1212 film. The peaks of STO (200) and PbBi1212 (200) are identified. This implies that the lattice

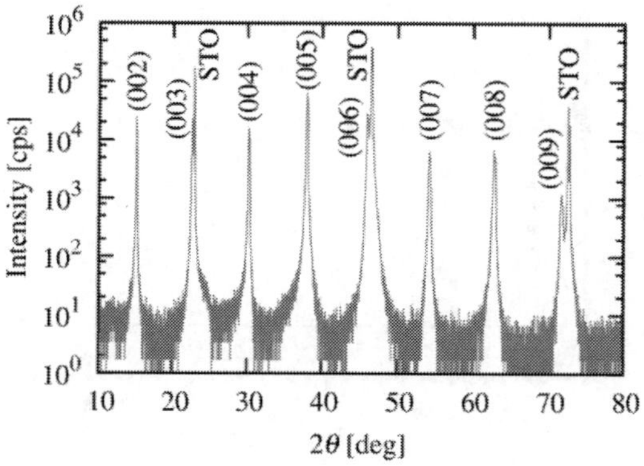

Figure 1. Out-of-plane XRD θ–2θ scans of the single-phase PbBi1212 sample.

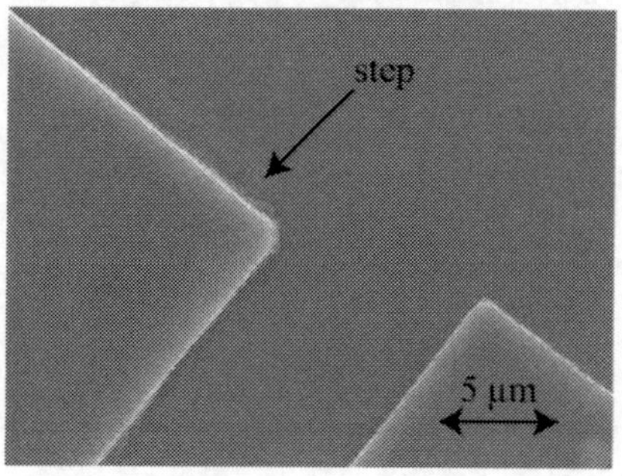

Figure 2. SEM images for the single-phase PbBi1212 epitaxial film.

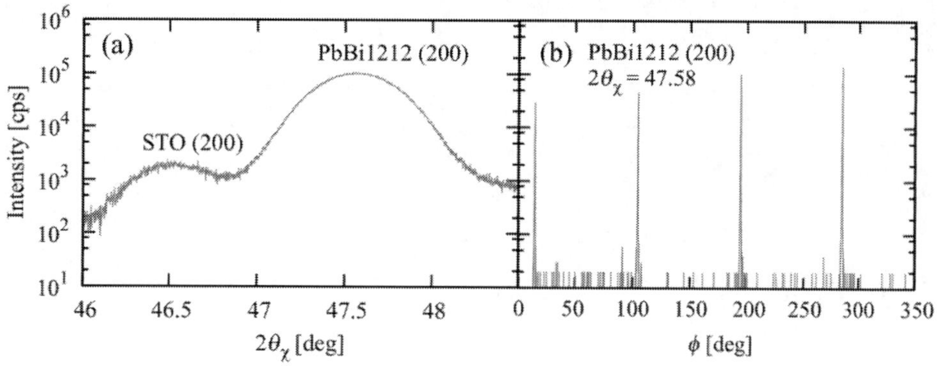

Figure 3. (a) In-plane XRD $2\theta_\chi$–ϕ scan pattern around the (200) peaks of PbBi1212 and STO, and (b) ϕ scan pattern at PbBi1212 (200) ($2\theta_\chi = 47.58°$).

relaxation occurs from the 2% in-plane lattice mismatch between STO and PbBi1212. Figure 3 (b) shows the in-plane ϕ scan result at PbBi1212 (200) ($2\theta_\chi = 47.58°$). Four peaks appearing at every 90° suggest that PbBi1212 has a fourfold in-plane rotational symmetry and tetragonal crystal structure. The difference in peak angle between STO and PbBi1212 was at most 0.7°. This indicates that the PbBi1212 thin films were epitaxially grown on STO (100) substrates.

The resistivity measurements show that as-grown PbBi1212 epitaxial films exhibit semiconductor-like temperature dependence and no trace of superconductivity. However, the films change to show superconductivity after the quenching treatment. In this treatment, the sample is put into the quartz tube, and annealed at 815 °C in air in a muffle furnace. After 1 h annealing, the quartz tube is taken out from the furnace and quenched into liquid nitrogen within 2 seconds. Figures 4 (a) and (b) show temperature dependence of the resistivity $\rho(T)$ and the Hall coefficients $R_{\mathrm{H}}(T)$ for samples with different Ca concentrations. It is suggested that substitution of a Ca^{2+} for a Y^{3+} has an effect to dope a hole. Figure 5 represents the distribution of the Ca and Bi concentrations of the superconducting (non-superconducting) single-phase PbBi1212 films. It was found that the Ca concentration $x > 0.3$ is essential, and lower Bi concentration is preferable for emergence of the superconductivity in PbBi1212.

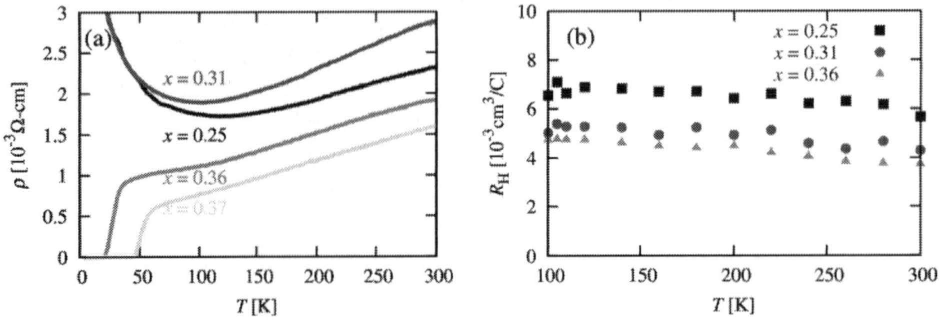

Figure 4. (a) $\rho(T)$ and (b) $R_{\mathrm{H}}(T)$ in the samples with different Ca concentrations x. The black line and squares show data in $x = 0.25$, $y = 0.15$. The blue line and circles show data in $x = 0.31$, $y = 0.13$. The red line and triangles show data in $x = 0.36$, $y = 0.17$. The green line represents data in sample showing the highest T_c^{onset} of 65 K and $T_c^{\rho=0}$ of 48 K with the composition of $x = 0.37$, $y = 0.00$.

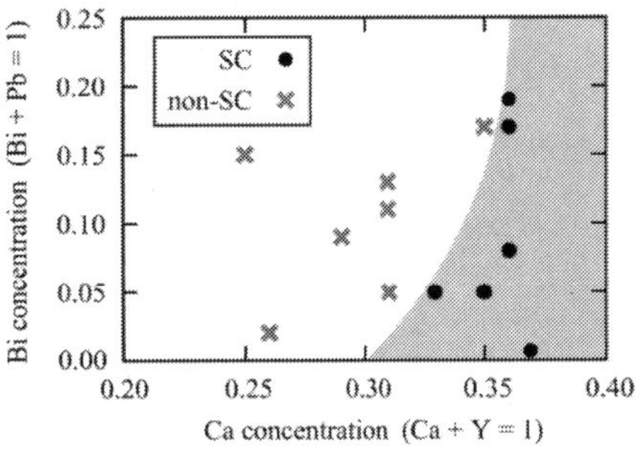

Figure 5. Distribution of Ca and Bi concentrations of the single-phase samples and their superconductivity. The black circles show the superconducting (SC) samples and the red crosses show the non-superconducting (non-SC) samples. The shaded area shows the concentrations where the superconducting films considered to be obtained.

4. Summary

In summary, we confirmed the superconductivity in PbBi1212 by measuring transport properties of epitaxial films prepared by the ex-situ growth method. The single-phase PbBi1212 films were obtained with a Ca concentration of less than 0.36 and Bi concentration of less than 0.20. Ca concentration of greater than 0.3 and the quench treatment are essential to make the epitaxial film superconducting. The increase in Ca and the decrease in Bi concentration lead to the increase in the hole density. The highest T_c^{onset} attained in the present study is 65 K.

Acknowledgements

We thank professor Shizuo Fujita and Dr. Kentaro Kaneko for assistance and providing useful knowledge and information.

References

[1] Bauer A, Zoller P, Glaser J, Ehmann A, Wischert W and Kemmler S 1996 *Physica* C **256** 177
[2] Frank H, Stollmann R, Lethen J, Müller R, Gasparov L V, Zakharov N D, Hesse D and Güntherodt G 1996 *Physica* C **268** 100
[3] Zoller P, Ehmann A, Glaser J, Wischert W and Kemmler-Sack S 1998 *Physica* C **306** 119
[4] Schneider D, Köbel S and Gauckler L J 2000 *Supercond. Sci. Technol.* **13** 1476
[5] Hughes R A, Lu Y, Timusk T and Prestön J S 1991 *Appl. Phys. Lett.* **59** 2597
[6] Kaesche S, Majewski P and Aldinger F 1999 *J. Am. Ceram. Soc.* **82** 197
[7] Wang J, Takano S, Wakata M, Usami R, Hamada K, Fukuoka A and Yamauchi H 1994 *Physica* C **219** 33
[8] Yan S L, Cao H L, Yang X M and Zhou X D 1991 *J. Appl. Phys.* **70** 526

Superconductivity of fullerides with composition $A_nIn_xGa_yC_{60}$ (A=K,Rb,Cs; n=2, 3) synthesized from gallams

V A Kulbachinskii[1], B M Bulychev and R A Lunin, A.V. Sycheva
M.V. Lomonosov Moscow State University, 119991, GSP-1, Leninskie Gory, Moscow, Russia

E-mail: kulb@mig.phys.msu.ru

Abstract. The fullerides with composition $A_nIn_xGa_yC_{60}$ (A=K,Rb,Cs; n=2,3; x,y<1) have been synthesized by a new method using liquid alloys of metals with indium and gallium (gallams) at room temperature. It was found that the fulleride $K_2In_xGa_yC_{60}$ is a superconductor with transition temperature T_c = 24.5 K that exceed T_c for K_3C_{60} (19 K). The fulleride $Rb_2In_xGa_yC_{60}$ is a superconductor with T_c = 26 K. The fullerides with composition $Rb_3In_xGa_yC_{60}$ and $Cs_3In_xGa_yC_{60}$ are not superconductors and crystallizes in orthorhombic lattice.

1. Introduction

Today a significant amount of the experimental data about properties of fullerides of alkali and some other metals are accumulated [1-5]. The first fullerides were synthesized by a method of reactions of the metal vapor with fullerite [1,2,5]. Such method is suitable for alkali and other fusible metals. Also a fullerides synthesis method through an interaction of a metal solution in ammonia with fullerite or melt of alkali metals with fullerite are known [3]. Other method is a fullerides synthesis by reactions of alkali metals with a solution of fullerene in the organic solvent [6]. More complicated fullerides were obtained by an exchange reactions of fullerides of alkali metals with solutions of salts of other metals in organic solvents [7,8]. In all listed methods one of the reagent is in a solid state that reduces a speed of the reaction and demands the execution of the synthesis at the increased temperature and can negatively affect the homogeneity of obtained fullerides. In this work we investigated heterofullerides obtained by a new method of reactions of liquid gallams with a solution of fullerene. In this method both of reagents are liquid that allows executing a reaction at room temperature. Homogeneity of the composition of obtained compounds should be close to the ideal that is especially important for the complicated compounds. In this case, however, indium and gallium can partially remain in the obtained fulleride.

Fullerides synthesized by different methods were studied in Ref. [9]. Superconductive transition temperatures for heterofullerides with similar composition but prepared from amalgams or by exchange reactions do not differ significantly (fig. 1a). It was shown that mercury get into fulleride and some charge transfer from mercury to the fulleride molecule occurs [10,11]. It was found that in homo-, as in heterofullerides transition temperature T_c increases with *fcc* lattice parameter a, as shown in fig. 1b (we added data for 3 samples synthesized from gallams and investigated in this work).

The aim of this work is to study the superconducting properties of a new heterofullerides $A_nIn_xGa_yC_{60}$ (A=K,Rb,Cs; n=2,3; x,y<1), synthesized from gallams. We also tried to find the relations of the composition and the structure of superconducting heterofullerides with the superconducting transition temperature T_c.

[1] To whom any correspondence should be addressed.

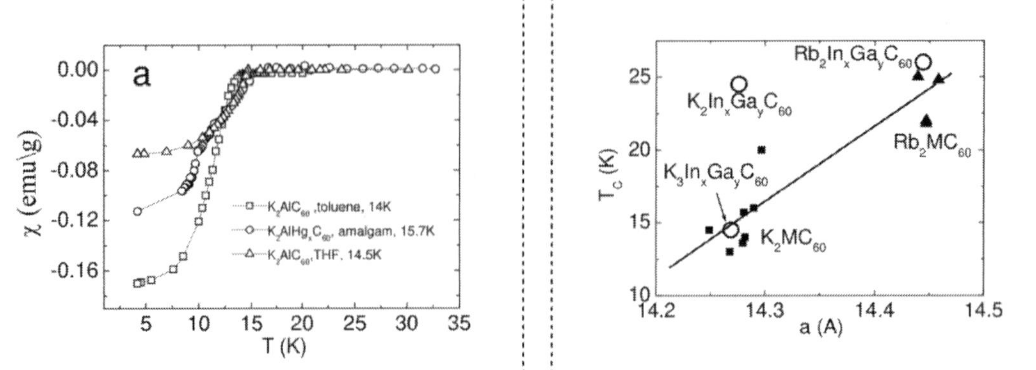

Figure 1. (a) Temperature dependence of the magnetic susceptibility χ of the fulleride K_2AlC_{60} (synthesized in toluene, amalgam and THF) and (b) dependence of the superconducting transition temperature T_c on the lattice papameter a for heterofullerides K_2MC_{60} and Rb_2MC_{60} (M = Be, Mg, Al, Ga). T_c for samples synthesized from gallams are shown by open cycles.

2. Experimental

In this paper we study the samples, synthesized by a new method using liquid gallams (alloys of metals with eutectic alloy Ga:In=70:30). The interaction between components was carried out in the organic solvent (toluene), in which fullerene is soluble, in order to better deliver a metal reagent to the fullerene molecule and to facilitate the subsequent separation of gallams from the target product. The synthesis of the alkali metal fullerides, and subsequent removal of toluene and gallams, drying of products of reaction, and their packaging in glass ampoules were carried out in vacuum in the full glass facilities equipment. In more detail the last technique is described in Ref's [6-8]. In the initial gallams the content of indium and gallium were in 30-70 times higher than alkali metal in order to gallams were liquid, but in the final product $A_nIn_xGa_yC_{60}$ (A=K,Rb,Cs; n=2,3) the content of indium x and gallium y were less then 1.

The measurements of X-ray diffraction were carried out with Guinier G670 HUBER. Temperatures of superconducting transitions of fullerides were defined by low-frequency induction method by measuring the temperature dependence of a magnetic susceptibility in the temperature range 4.2<T<297 K [10,11].

3. Results and discussion

Temperature dependence of the magnetic susceptibility for synthesized samples $K_2In_xGa_yC_{60}$ and $K_3In_xGa_yC_{60}$ is shown in **fig. 2**. The transitions in a superconducting state were observed at temperatures 24.5 K and 14.5 K correspondingly. Until now the fulleride with highest T_c on the base of potassium (which does not include atoms of other alkali metals Rb and Cs) was the $K_2Hg_xC_{60}$ (T_c = 22 K) [10,11]. Thus the used method of synthesis made a superconductor $K_2In_xGa_yC_{60}$ with highest T_c from not a superconductor K_2C_{60}. It indicates that components of gallams aren't inert solvents in reaction of intercalation, but intercalates itself into a fullerides lattice and change their superconducting properties. According to the X-ray data the heterofulleride with the composition $K_2In_xGa_yC_{60}$ was crystallized in the *fcc* lattice with the lattice parameter a=1.427 nm (**fig. 3a**), which is close to the value obtained for K_3C_{60} (a=1.424 nm [4]). The parameters of investigated samples are listed in table 1. The dependence of T_c on the lattice parameter a for superconducting samples synthesized from gallams is shown in fig. 1b by open cycles. The difference in T_c for samples $K_2In_xGa_yC_{60}$ and $K_3In_xGa_yC_{60}$ is the most probably due to the more optimal charge transfer from intercalated atoms to fullerene molecule for the sample $K_2In_xGa_yC_{60}$.

The fulleride $Rb_3In_xGa_yC_{60}$ is not a superconductor. The X-ray diffraction data of the sample indicates the orthorhombic crystal lattice with the lattice parameters shown in table 1.

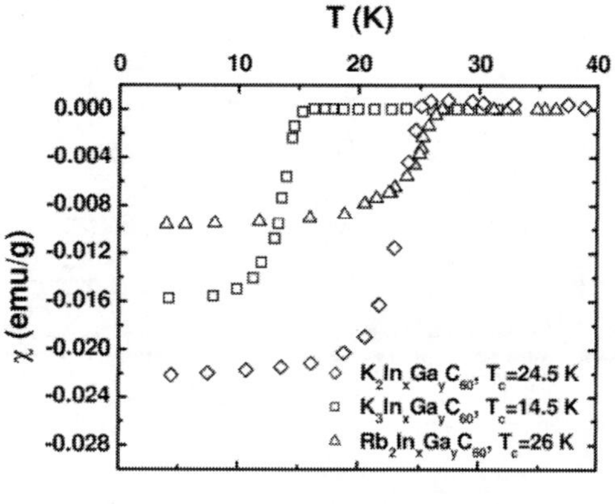

Figure 2. Temperature dependence of magnetic susceptibility of samples $K_2In_xGa_yC_{60}$, $K_3In_xGa_yC_{60}$ and $Rb_2In_xGa_yC_{60}$.

Table 1. Temperature of superconducting transition, type of crystal lattice and lattice parameters of investigated samples.

Composition	T_C (K)	Lattice type	Lattice parameters (Å)
$K_3In_xGa_yC_{60}$	14,5	*fcc*	14.269(2)
$K_2In_xGa_yC_{60}$	24,5	*fcc*	14.276(4)
$Rb_3In_xGa_yC_{60}$	-	orthorhombic	*a*= 9.110(13), *b*=10.105(11), *c*=14.188(15)
$Rb_2In_xGa_yC_{60}$	26	fcc (30%) + orthorhombic (70%)	14.444(1) + *a* = 9.091(4), *b*=10.075(5), *c*=14.173(6)
$Cs_3In_xGa_yC_{60}$	-	orthorhombic	*a*=9.097(5), *b*=10.227(5), *c*=14.159(9)

These data shows that components of gallams were intercalated in fulleride lattice, changed the lattice type and made from superconductor Rb_3C_{60} not superconducting composition. At the same time the fulleride $Rb_2\mathbf{In_xGa_yC_{60}}$ remain the superconductor with T_c = 26K. Temperature dependence of magnetic susceptibility of sample $Rb_2\mathbf{In_xGa_yC_{60}}$ is plotted in fig. 2.

According to the X-ray data, this sample consists of two phases: the first phase is similar to $Rb_{2.92}C_{60}$ (*fcc* lattice) and the second phase is close to $Rb_{0.91}C_{60}$ (orthorhombic lattice). The X-ray diffraction data for $Rb_2Hg_xC_{60}$ is plotted in fig. 3b. The peaks related to the *fcc* phase are indicated by solid lines, and to the orthorhombic phase – by dashed lines. The lattice parameter of the *fcc* phase equals to 1.444 nm (see table 1). This value is close to the value measured for Rb_3C_{60} (a=1.4384 nm [5]). The phase with *fcc* crystal lattice is probably superconducting, because measured T_c is relatively close to the value of superconducting transition temperature for Rb_3C_{60} (28 K [5]). Thus, in the conditions of synthesis method chosen by us the composition «$Rb_2In_xGa_yC_{60}$» is not formed, or even if it is formed then it decay on two known phases – mono- and triple-rubidium fullerides.

The heterofulleride with the composition $Cs_3In_xGa_yC_{60}$ is not a superconductor. According to the X-ray data, this samples crystallize in the orthorhombic lattice (see table 1). The absence of a superconductivity in this fulleride is probably because its crystal lattice is not *fcc*. The absence of superconductivity in cesium fullerides, synthesized from amalgams, was also registered by us earlier [10,11]. In previous work [10] we investigated the fullerides with composition $A_nHg_xC_{60}$ (A=K,Rb; n=2,3; x<1) synthesized by method using liquid alloys of alkali metals with mercury (amalgams). It was found that the fulleride $K_2Hg_xC_{60}$ was superconductor with T_c = 22 K.

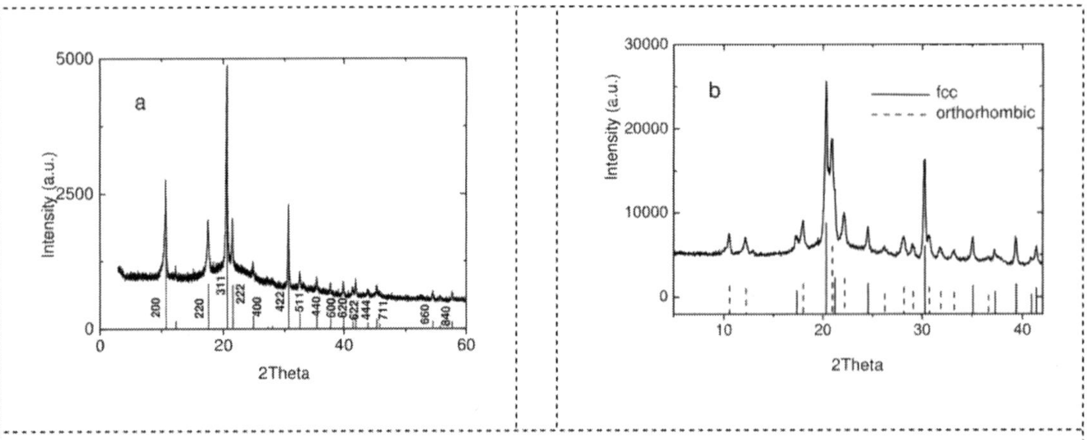

Figure 3. X-ray diffraction of the sample (a) $K_2In_xGa_yC_{60}$ and (b) $Rb_2In_xGa_yC_{60}$. Indicated phases: *fcc* $Rb_{2.92}C_{60}$ (30%) + orthorhombic $Rb_{0.91}C_{60}$ (70%).

The fulleride $K_3Hg_xC_{60}$ was not a superconductor. The fulleride $Rb_2Hg_xC_{60}$ was a superconductor with T_c = 25 K. Thus the alkali metals fullerides synthesized from gallams have higher T_c than analogous fullerides synthesized from amalgams. We also synthesized from amalgams the fullerides with composition $K_2GaHg_xC_{60}$, $Rb_2GaHg_xC_{60}$ and $K_2InHg_xC_{60}$, $Rb_2InHg_xC_{60}$. The fullerides $K_2GaHg_xC_{60}$ and $Rb_2GaHg_xC_{60}$ are superconductors with T_c equal to 20 K and 25 K, correspondingly. The fullerides $K_2InHg_xC_{60}$ and $Rb_2InHg_xC_{60}$ are not superconductors though crystallize in *fcc* lattice. It means that rather gallium than indium is responsible for high T_c for the sample $K_2In_xGa_yC_{60}$.

4. Conclusions

The new heterofullerides with the composition $A_nIn_xGa_yC_{60}$ (A=K,Rb,Cs; n=2,3; x,y<1) were synthesized. Temperature dependence of the magnetic susceptibility was measured in the temperature range from 4.2 K to 297 K and transitions in the superconducting state were detected at temperatures T_c ranged from 14.5 K to 26 K. For the fulleride $K_2In_xGa_yC_{60}$ the T_c is equal to 24.5 K, that is the highest T_c among fullerides on the base of potassium (which does not include atoms of other alkali metals Rb and Cs). Thus, the components of gallams intercalate into a fulleride lattice and improve their superconducting properties.

References

[1] A.F. Hebard, M.J. Rosseinsky, R.C. Haddon, et. al., *Nature*, **350**, 600 (1991).
[2] K. Tanigaki, I. Hirosawa, T.W. Ebessen, et. al., *Nature* **356,** 419 (1991).
[3] O. Gunnarsson, *Reviews of Modern Physics* **69,** 575 (1997).
[4] T. Tanigaki and K. Prassides, *J. Mater. Chem.* **5,** 1515 (1995).
[5] M.J. Rosseinsky, A.P. Ramirez, S.H. Glarum, et. al., *Phys. Rev. Lett.* **66,** 2830 (1991).
[6] Bulychev BM, Privalov VI and Dityat'ev AA (2000) *Russian Journal of Inorganic Chemistry* **45** 931 (2000).
[7] B.M. Bulychev, R.A. Lunin, A.V. Krechetov, V.A. Kulbachinskii, V.G. Kytin, K.V. Poholok, K. Lips and J. Rappich, *J. of Phys. and Chem. of Solids* **65,** 337 (2004).
[8] V.G. Kytin, B.M. Bulychev, A.V. Krechetov, V.A. Kulbachinski, R.A. Lunin, E.A. Konstantinova and Yu.A. Velikodny, *Russian Journal of Inorganic Chemistry* **53,** 30 (2008).
[9] V.A. Kulbachinskii, R.A. Lunin, I.P. Kachan, et.al., *Low Temp. Phys.*, **39**, 76 (2013).
[10] B.M. Bulychev, V.A. Kulbachinskii, R.A. Lunin, V.G. Kytin and Yu.A. Velikodny, Fullerenes, *Nanotubes and Carbon Nanostructures* **18**, 381 (2010).
[11] V.A. Kulbachinskii, B.M. Bulychev, V.G. Kytin and R.A. Lunin, *Low Temperature Physics* **37**, 245 (2011).

Analysis of peak effect in the critical current density and flux pinning properties in iron based superconducting $Ca_{10}(Pt_4As_8)(Fe_{2-x}Pt_xAs_2)_5$ single crystal

D. Ahmad[1,2], B. H. Min[2,3], J. H. Lee[4], G. C. Kim[4], Y. C. Kim[4] and Yong Seung Kwon[1,2]

[1]Department of Emerging Materials Science, Daegu Gyeongbuk Institute of Science and Technology (DGIST)711-873 Daegu, Republic of Korea
[2]DGIST-LBNL Joint Research Center, Daegu Gyeongbuk Institute of Science and Technology (DGIST)711-873 Daegu, Republic of Korea
[3]Center for X-ray Optics, Lawrence Berkeley National Laboratory (LBNL), Berkeley California 94720, USA
[4]Department of Physics, Pusan National University, 609-735 Busan, Republic of Korea

Email: yskwon@dgist.ac.kr

We present the magnetic properties of the iron based superconducting $Ca_{10}(Pt_4As_8)(Fe_{2-x}Pt_xAs_2)_5$ (x =0.01) single crystal as a function of field and time. The single crystals were grown by Bridgman method and showed T_c of 32 K. The results show that the well-known secondary peak effect becomes prominent in certain temperature range in the field dependent magnetic hysteresis. Moreover, the maximum of pinning force density F_{max} versus H_{max} follows a scaling law $F_{p,max} \approx 49.5H_{max}^{3/2}$ indicating the single type pinning mechanism is involved in the sample within our measured temperature range. The time dependence of the magnetization indicates thermally activated flux motion and the relaxation rate reveals a plateau in the temperature region where secondary peak effect is more prominent after that a sudden increase is observed. We found that the magnetization dependence of activation energy U obtained by Maley's method follows a power law in low temperature region with exponent $\mu = 0.15$ which is in good agreement with collective pinning theory.

1. Introduction:

The enigmatic characteristics of superconductivity have a special attraction for the condensed matter physicists. After the discovery of first iron based superconductor [1], a variety of superconducting materials have been reported. Moreover, all of the iron based superconducting materials consist of edge-sharing $FeAs_4$ and $FeSe_4$ tetrahedra along with spacer layers [1,2]. Recently, two new superconducting compounds with composition $Ca_{10}(Pt_3As_8)(Fe_{2-x}Pt_xAs_2)_5$ (10-3-8) and $Ca_{10}(Pt_4As_8)(Fe_{2-x}Pt_xAs_2)_5$ (10-4-8) were discovered [3]. The crystal structure of 10-3-8 compound is a triclinic crystal with space group P-1, whereas 10-4-8 is of higher symmetry tetragonal crystal structure, respectively [3]. The doping behavior of the CaPtFeAs system is similar to that of FeAs systems in which the electron doping outside the FeAs layer significantly enhances the T_c [3].

The critical current density extracted from field dependence magnetic hysteresis gives important information about the vortex matter in a superconductor. The secondary peak effect is a unique feature in the field dependence of magnetic hysteresis in high T_c superconductor. Several reasons have been presented to explain the secondary peak effect at specific temperature and applied magnetic field.

In this work, we analyze the flux pinning properties of $Ca_{10}(Pt_4As_8)(Fe_{2-x}Pt_xAs_2)_5$ with x = 0.01 as a function of applied magnetic field and time. The variation of the critical current density J_c with respect to applied field and temperature is discussed. Moreover, the time dependence of the magnetization is analyzed in the temperature region where prominent secondary peak effect is observed. The differences of the observed behavior with that of other high T_c superconductors are discussed.

2. Experimental Details:

The $Ca_{10}(Pt_4As_8)(Fe_{2-x}Pt_xAs_2)_5$ with $x = 0.01$ was grown by Bridgman method [4]. FeAs precursor, Pt wire and Ca pieces were mixed together according to stoichiometric ratio and put into alumina crucible. The alumina crucible was then put into molybdenum crucible and sealed using arc welding system. Mixing and sealing steps were performed in under argon atmosphere. Single crystal was grown at 1550°C for around 170 hours with slow moving down. Single crystals of maximum cm size were found upon crucible breaking. The structural characterization of the sample was performed using X-ray diffractrometry (XRD) with Cu Kα radiation source. The compositional analysis was performed using an energy dispersive X-ray analysis (EDAX) with scanning electron microscopy (SEM). The field dependence of magnetic hysteresis and magnetic relaxation measurements were carried out with superconducting quantum interference device (SQUID) magnetometer in the case when $H \| c$.

3. Results and Discussion:

The temperature dependence of magnetic susceptibility was measured at 10 Oe. The onset transition temperature $T_{c,onset}$ is about 32 K. A sharp transition indicates the existence of single superconducting phase. At low temperatures, the zero field cooled susceptibility is about $4\pi\chi = -0.9$, which corresponds to almost complete screening of single crystal interior. The field cooled susceptibility is $4\pi\chi = -0.05$ indicating the strong flux pinning at low fields in our single crystal. Fig. 1 presents the field dependence of the magnetic hysteresis in $Ca_{10}(Pt_4As_8)(Fe_{2-x}Pt_xAs_2)_5$ ($x = 0.01$) at several temperatures. Moreover, the magnetic hysteresis curves in our measured temperatures are symmetric. A well-formed secondary peak can also be seen in the specific temperature range which is a characteristic of some of the high T_c superconductors. Recently, the secondary peak was also observed in some iron based superconductors, e.g., $(Ba,K)Fe_2As_2$, $Ba(Fe,Co)_2As_2$, and $Ba(Fe,Ni)_2As_2$ superconductors [5,6]. For instance, in Co-doped Ba-122 system, the secondary peak was interpreted as a crossover from one kind of vortex dynamics to other.

The critical current density J_c in our sample was calculated using Bean Model [7] as $20\,\Delta M[a(1 - a/3b)]^{-1}$, where ΔM is the width of magnetic hysteresis, a and b is the sample dimensions in cm with $a < b$ and J_c is in A/cm^2. The critical current density was estimated about 5×10^4A/cm^2 at 5 K and remains field independent up to 5 T. Fig. 2 displays the pinning force density ($F_p = J_c B$) as a function of the applied field for our $Ca_{10}(Pt_4As_8)(Fe_{2-x}Pt_xAs_2)_5$ sample. With the increase of temperature, the maximum of F_p is shifted to lower applied field as shown in Fig. 2. Moreover, the inset of Fig. 2 exhibits the maximum of pinning force density $F_{p,max}$ as a function of corresponding field H_{max}. The solid line in the inset is a fit to $F_{p,max} \approx 49.5 H_{max}^{3/2}$ which is in good agreement with the experimental data. Therefore, this scaling of $F_{p,max}$ indicates that a single type of pinning mechanism is involved in our sample within measured temperature range in agreement with the previous reports [5,6].

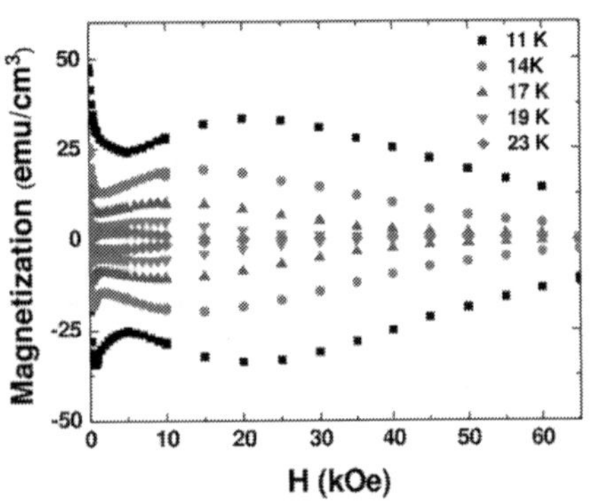

Fig. 1. Field dependence of the magnetization in increasing and decreasing fields in temperature range from 11 to 23 K for $H \| c$.

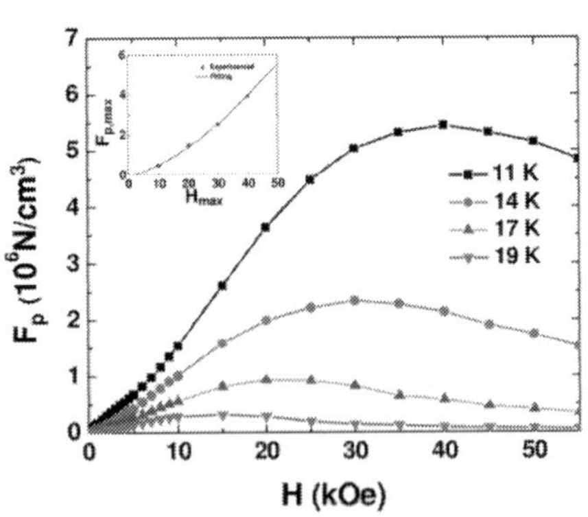

Fig. 2. Field dependence of F_p for temperatures from 11 to 19 K. Inset: $F_{p,max}$ as a function of H_{max}. The solid line is a fit of $F_{p,max} \approx 49.5 H_{max}^{3/2}$.

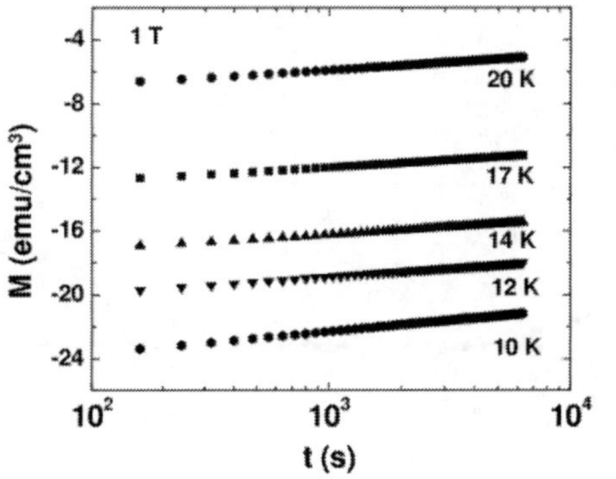

Fig.3. Log-Log plot of magnetization *M* vs time at various temperatures at H = 1T in the case when $H \| c$.

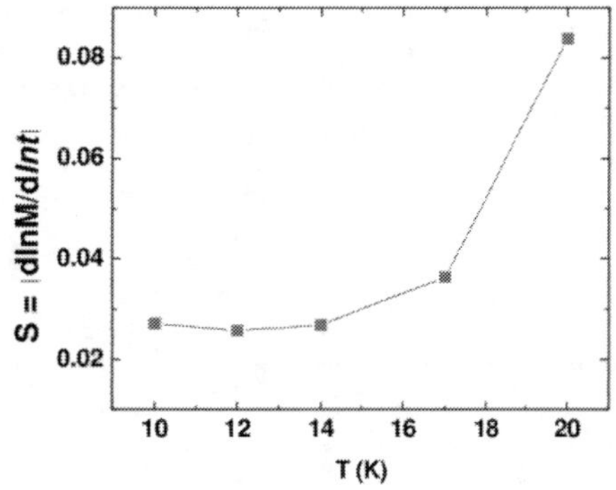

Fig.4. Temperature dependence of the normalized relaxation rate S calculated at H=1T in the case when $H \| c$.

Figure 3 shows a typical time dependence of magnetization at various temperatures in the applied field of 1 T in $Ca_{10}(Pt_4As_8)(Fe_{2-x}Pt_xAs_2)_5$ (x= 0.01) single crystal. Moreover, the logarithmic dependence of M vs t in Fig. 3 is the same as expected in the model of thermally activated flux motion. Fig. 4 shows the relaxation rate S = $|d \ln M/d \ln t|$ obtained at 1 T as a function of temperature for our sample. Moreover, the high relaxation rate in our sample is in similar range reported previously for YBCO [8]. In addition, we can see a plateau in the intermediate temperature range. This plateau in the particular temperature range cannot be explained within the single vortex creep theory with the rigid hopping length predicted by Anderson-Kim model. Recently, similar behavior was observed in $Ba(Fe_{0.925}Co_{0.075})_2As_2$ and $FeTe_{0.6}Se_{0.4}$ superconductors and was attributed to the vortex collective pinning [9,10]. However, our result for vortex creep rate is in contrast to the case for MgB_2 superconductor in which the relaxation rate increases linearly with temperature which indicates a thermally activated hopping of vortices with rigid length [11].

The critical current J_c dependence of the activation energy U is an important parameter for investigating the vortex dynamics in the sample. For instance, a well-known equation for determining the activation energy U was proposed by Maley et.al using the relaxation rate S as $dM/\mathrm{d}t \propto \exp\{-U/kT\}$ calculated in the irreversible region [12]. The activation energy is then finally calculated as $U = k_B\, T(A - \ln|dM/dt|)$, where A is the time independent constant corresponded to the average hopping velocity. Fig. 5 shows the dependence of the activation energy on the magnetization at different temperatures from 10 to 20 K. Moreover, up to 14 K, all the curves are scaled together but we were unable in scaling the activation energy above 14 K. The Maley's method is only valid at low temperature ($T<T_c/3$) where the temperature dependence of the activation energy $U(T)$ is relatively weak. We selected the temperature range in which the secondary peak is prominent which is slightly above $T<T_c/3$ and therefore found an excellent agreement with the temperature range where Maley's method is applicable. Furthermore, a sudden increase of the relaxation rate above 17 K is due to a crossover between different vortex creep regions as shown in Fig. 4 supporting our results in Fig. 5. For analyzing the experimental data points of the activation energy at various temperature, we used the power law from the collective creep theory [13] which is given as $U \propto \{(J_c/J)^{\mu} - 1\}$, where J_c is the critical current density and μ is a critical exponent describing the nature of the pinning barrier.From the fitting we get μ = 0.15 for our temperature range which is in agreement with collective pinning theory and thus confirms the flux creep as a result of the motion of individual vortices [16]. Furthermore, according to collective pinning theory the value of μ will be significantly increased at higher temperatures suggesting a crossover from the single vortex to vortex-bundle pinning state. Therefore, from our results we can conclude that a crossover between different pinning regimes occurs at around 17 K.

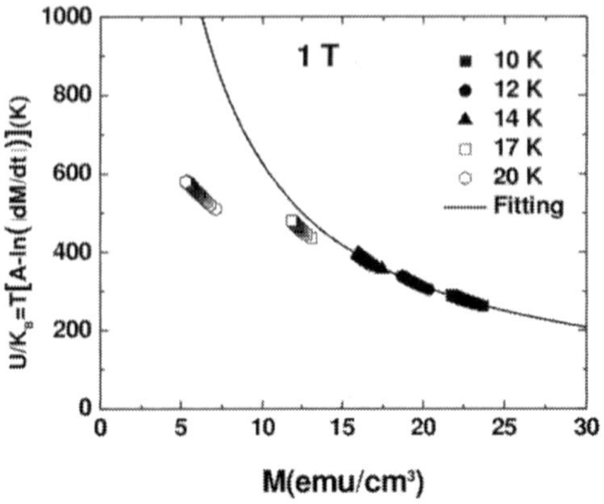

Fig. 5. Magnetization dependence of the activation energy U obtained by Maley's method with parameter A= 19. The solid line is the fitting result by $U \propto \{(J_c/J)^{\mu}-1\}$ with $\mu = 0.15$.

Conclusion:

The superconducting $Ca_{10}(Pt_4As_8)(Fe_{2-x}Pt_xAs_2)_5$ (x=0.01) single crystals were grown by Bridgman method. The transition temperature T_c was found at 32 K measured by temperature dependence of magnetization. The field and time dependence of the magnetization was measured in various fields and temperatures. We found a prominent secondary secondary peak effect in the magnetic hysteresis in particular temperature range in agreement with other high T_c superconductors. A scaling law was followed by F_{max} vs H_{max} indicative of a single type of pinning activated in the sample at least in our measured temperature range. The activation energy $U(J)$ calculated by Maley's method have shown a trend in the favor of the collective pinning model by following a power law.

Acknowledgements:

This work was supported by the Basic Science Research Program (NRF-2013R1A1A2009778) and the Leading Foreign Research Institute Recruitment Program (Grant No. 2012K1A4A3053565) through the National Research Foundation of Korea (NRF) funded by the Ministry of Education, Science and Technology (MEST).

References:

[1] Kamihara Y, Watanabe, Hirano T M, and Hosono H, (2008) *J Am Chem Soc.* 130 3296.
[2] Hsu F C et. al, (2007) *Proc Natl Acad Sci.* 105, 14262.
[3] Kakiya S, Kudo K, Nishikubo Y, Oku K, Nishibri E, Sawa H, Yamamoto T, Nozaka T, and Nohara M, (2011) J. Phys. Soc. Jpn. 80, 093704.
[4] Song Y J, Ghim J S, Min B H, Kwon Y S, Jung M H and Rhyee J S, (2010) *Appl Phys Lett.* 96 212508.
[5] Sun D L, Liu Y, and Lin C T, (2009) *Phys Rev B.* 80 144515.
[6] Yamamoto A et al, (2009) *Appl Phys Lett.* 94 092504.
[7] Bean C P, (1964) *Rev. Mod. Phys.*, **36** 31.
[8] Blatter G et al, (1994) *Rev Mod Phys* 66 1125.
[9] Haberkorn N, Maiorov B, Usov I O, Weigand M, Hirata W, Miyasaka S, Tajima S, Chikumoto N,Tanabe K, and Leonardo Civale, (2012) *Phys Rev B* **85** 014522.
[10] Sun Y, Taen T, Tsuchiya Y, Pyon Y S, Shi Z, and Tamegai T, (2013) *EPL*, **103** (2013) 57013.
[11] Malozemoff A P and Fisher M P A, (1990) *Phys Rev B.* 42 6784 (R).
[12] Maley M P, Willis J O, H. Lessure H, and McHenry M E, (1990) *Phys Rev B.* 42 2639.
[13] Anderson P W, and Kim Y B, (1964) *Rev Mod Phys* **36**, 39.

Journal of Physics: Conference Series **507** (2014) 012028 doi:10.1088/1742-6596/507/1/012028

Thermal expansion of the high-*Tc* superconductors $REBa_2Cu_3O_7$ (RE = Y, Dy) revisited at low temperatures

M Lahoubi[1]

[1] Badji Mokhtar-Annaba University, Faculty of Sciences, Department of Physics, PO Box 12, 23000, Annaba, Algeria

E-mail: mlahoubi@hotmail.fr

Abstract. The linear thermal expansion coefficient $\alpha(T)$ of two high-*Tc* superconductors, $YBa_2Cu_3O_7$ and $DyBa_2Cu_3O_7$ has been revisited in the 4.5-80K temperature range. The X-ray diffraction analysis and microstructural characterization using scanning electron microscopy confirm an orthorhombic single phase with an average of the grain size of about 20μm for both polycrystalline samples. In the case of the Dy compound, a small magnetic excess superimposed to the high phonon contribution is observed in the T^3 term, whereas no anomaly is detected below 40K. The deduced magnetic contribution $\alpha_m(T)$ which increases monotonically with rising temperature is attributed to the CF interactions induced by the 4f electric quadrupole moment of the Dy Kramers ions. A discussion and a comparison with respect to the previous experimental and theoretical works are presented.

1. Introduction

The high-*Tc* superconductors (HTS) such as those of the series $REBa_2Cu_3O_7$ traditionally abbreviated as REBCO or RE-123, with RE a rare earth ion, play until now an outstanding role for the present understanding in the relationship between the magnetic interactions and superconductivity [1 and references herein]. The observed anisotropic magnetic properties of these cuprates have been successfully described in terms of a model which consists of dipolar interactions with crystal field (CF) and exchange interactions [2, 3]. The contribution of the CF interactions induced by the 4f electric quadrupole moment, i.e., another manifestation of the magnetoelastic coupling of CF states with lattice strains to the coefficient of the linear thermal expansion $\alpha(T) = L^{-1}(dL/dT)$ in these orthorhombic REBCO compounds is known to lead theoretically anomalies in the $\alpha(T)$ curves of some RE compounds such DyBCO and PrBCO [2, 3] where the antiferrimagnetic ordering temperature (T_N) of the Dy and Pr sublattices appear at ≈ 1K and in the 14-20K range respectively [1]. $\alpha(T)$-measurements have been previously performed at low temperatures on the DyBCO compound [4, 5] and the question arising now about the manifestation of the magnetic contribution $\alpha_m(T)$ due to the 4f-electrons of the Dy Kramers ions is not completely resolved both experimentally and theoretically.

2. Experimental

The specimens of both compounds have been prepared using the standard ceramic method and characterized by X-ray diffraction (XRD) at room temperature. The initial samples labelled as YBCO (ybc700) and DyBCO (dybc725) have a geometry of a cylinder with the length (L) equal to 8.86 and 8.92mm and with approximately the same diameter (ϕ), 4.45 and 4.47mm respectively. The microstructural analysis and thermal expansion experiments have been performed at the Néel Institute (Grenoble, France) using a scanning electron microscopy (SEM) and a very sensitive capacitive dilatometer of oxygen free high conductivity, (OFHC) 5 copper [6]. Due to the cell dimension requirement, both samples have a final length (L) of order of 4.89mm. The other parts of the samples have been characterized by means of magnetic measurements, i.e., dc Meissner effect at different low fields, taking into account the correction of the demagnetizing field effects. The results of our reinvestigations on some other REBCO compounds with RE = Nd, Sm, Ho, Er and Tm will be published in a forthcoming paper in connection with their thermal expansion properties.

3. Results and discussion

The XRD patterns of both samples are reported in Figure 1. They confirm that the samples are single phase having the typical crystallographic structure of the orthorhombic space group $Pmmm$-(D_{2h}^{1}) No. 47 of these REBCO compounds. A few superconducting parameters are given at the true low magnetic fields (H_{true}), 4.41±0.2Oe and 5.2±0.2Oe respectively for YBCO and DyBCO: the critical temperature, T_c = 90.0±0.2K (resp. 89.5K) with the corresponding transition width, ΔT_c = 2.5±0.2K (resp. 3.1K) and a flux expulsion of 57.5% (resp. 85%). The large amount of flux expulsion obtained for DyBCO can be explained by the SEM micrograph of the DyBCO sample reported in Figure 2 where an average of the grain size within the 10-20μm range with a low level of porosity are revealed. Due to some requirements of the experimental procedure, the $\alpha(T)$ data for the DyBCO sample have been measured with T increasing continuously in the 5-30K temperature range using an increment of about 0.66K for a time of taking the measurement for stabilizing T, varying from 200s at 5K to 300s above 20K. An increment of the order of 1.1-1.6K is chosen in the 30-70K temperature range. The measured $\alpha(T)$ data where their total number is equal to 61 are compared to some smoothed $\alpha(T)$-values from our previous study [5]. The measurements of the $\alpha(T)$ data for the YBCO sample have been performed in the 4.5-80K temperature range with a waiting period of about 200s and using approximately similar increment for T of about 0.55K below 40K. An increment of about 1K is used above 40K and the total number of the $\alpha(T)$ data is equal to 105. The $\alpha(T)$ data have been measured within an experimental error of $\pm 2.5\times10^{-8}K^{-1}$ which is comparable to the sensitivity of other capacitive dilatometers [7, 8]. The temperature dependences of the $\alpha(T)$ data obtained for both samples are reported in Figure 3.

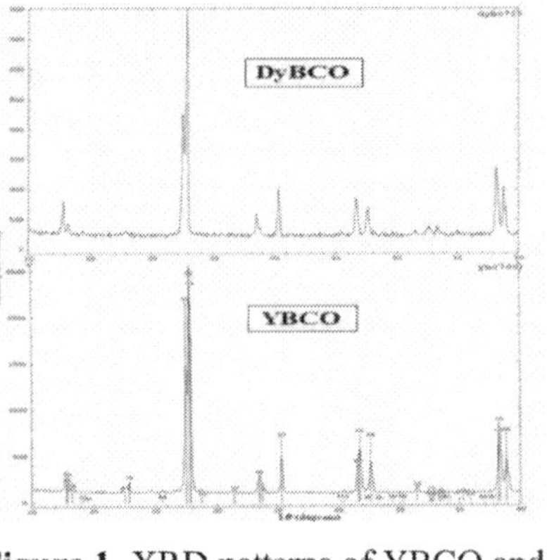

Figure 1. XRD patterns of YBCO and DyBCO polycrystalline samples.

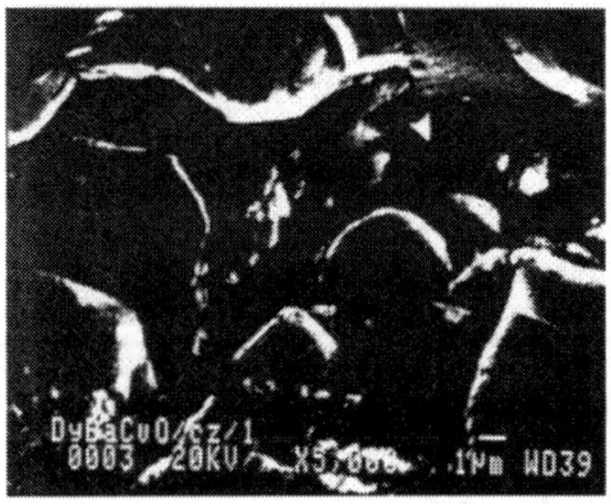

Figure 2. SEM micrograph of DyBCO sample at X5,000 showing grain size in the 10-20 μm range.

The main observation which can be reported here is related to an excess of $\alpha(T)$ for the DyBCO with respect to YBCO where no noticeable anomaly is detected below 40K. However, these results seem considerably different to those found earlier by Schnelle *et al.*, [4]. In fact, their $\alpha(T)$ curve for the DyBCO sample is dominated by a pronounced positive anomaly at 11K. In addition, their $\alpha(T)$ curves

Journal of Physics: Conference Series **507** (2014) 012028 doi:10.1088/1742-6596/507/1/012028

for both DyBCO and YBCO samples cross one another at about 40K and the absolute value of $\alpha(T)$ for the DyBCO sample appears much smaller than that for the YBCO sample above 40K. This sample-dependent anomaly which has been also observed near 12-13K in the specific heat $C_p(T)$ [9] and magnetization $M_H(T)$ curves [10] could be associated probably to the ordering of an impurity phase, namely of Dy_2BaCuO_5 (green phase) according to the literature data [4, 5, 9, 10, 11]. From our various analyses (XRD and SEM characterizations, magnetic and $\alpha(T)$ measurements), one can only assume that our DyBCO sample does not contain any secondary phase that may be produced during the preparation of such ceramic materials. Both measured $\alpha(T)$ data are accurately fitted as mentioned in the literature data for YBCO [12-15], using the same representation for the $C_p(T)$ data which is usually displayed by plotting $C_p(T)$ versus T^3. The plots of $\alpha(T)$ versus T^3 for the DyBCO and YBCO samples are practically linear up to 32 and 33.5K respectively, as shown in Figure 4 where the fitted values of the coefficient (B) of the T^3 terms are indicated: 4.24617×10^{-11} and $3.57593 \times 10^{-11} K^{-4}$.

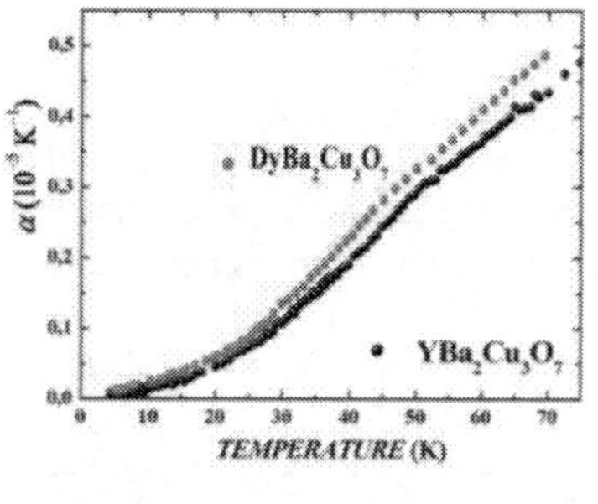

Figure 3. $\alpha(T)$ versus T: upper point for DyBCO (red circle); lower point for YBCO (black circle).

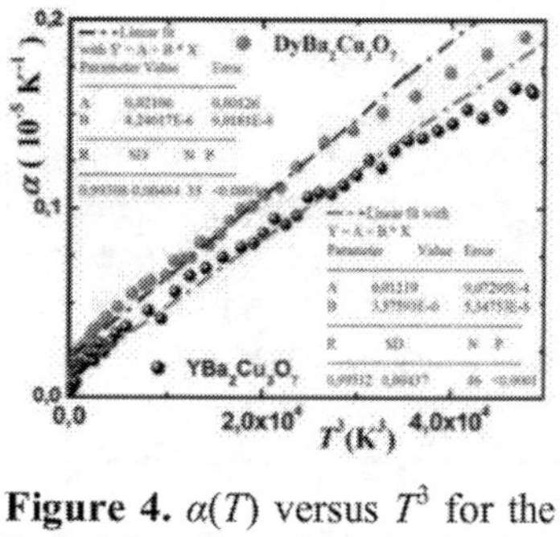

Figure 4. $\alpha(T)$ versus T^3 for the DyBCO and YBCO samples.

The results show that a small magnetic contribution $\alpha_m(T)$ attributed to the 4f electrons of the Dy ions is presents below 40K in addition to the high phonon contribution. These $\alpha_m(T)$ values are defined by the difference α(DyBCO) − α(YBCO), the data of α(YBCO) are serving as a non-magnetic reference by assuming identical lattice contribution. Both $\alpha_m(T)$ and fitted curves are plotted in Figure 5. They increase monotonically with increasing T and follow much the same shape (not show) as the previous calculated Schottky $\Delta C_p(T)$ curve due to the magnetic contribution to $C_p(T)$ measured with $H = 0$ in the low temperature range 4-15K, with doublets at 0, 40 and 100K [16], while the pronounced anomaly [4] reaching a value of about $0.070 \times 10^{-5} K^{-1}$ below 40K has been interpreted theoretically [3].

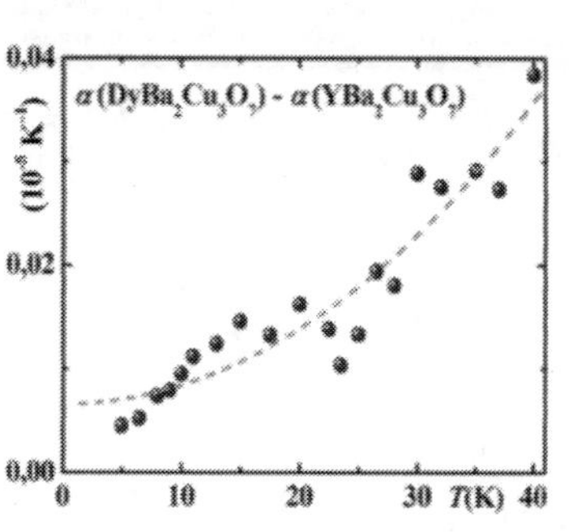

Figure 5. $\alpha_m(T)$ versus T up to 40K. The orange short dash line is the fitted curve.

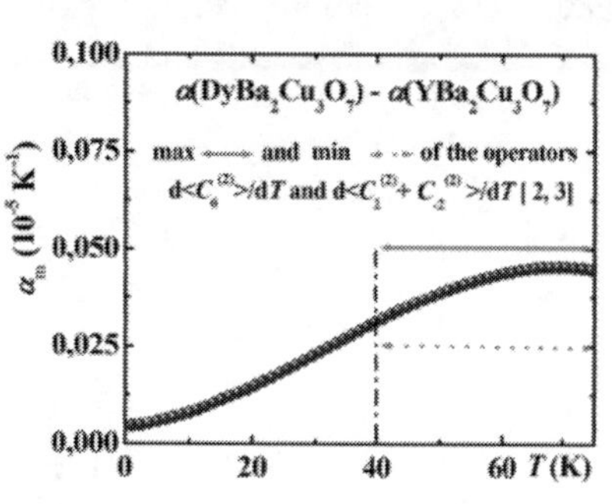

Figure 6. Thermal variation of the calculated curve $\alpha_m^{cal}(T)$ up to 75K.

However, as shown in Figure 6, the thermal variation of the calculated curve $\alpha_m^{cal}(T)$ up to 75K using the subtraction α(DyBCO) − α(YBCO) of the respective fitted curves, appears above 40K inside the maximum and minimum values of the calculated terms of irreductible tensor operators, $\mathrm{d}<C_0^{(2)}>/\mathrm{dT}(T)$ and $\mathrm{d}<C_2^{(2)}+C_{-2}^{(2)}>/\mathrm{dT}(T)$ [2, 3], with sets of the phenomenological CF parameters available for the Dy ion in DyBCO at this time [17]. One can noted that the subtraction between the data at the same temperature T is in itself, a critical procedure, which yields uncertainty but, within our satisfactory resolution of the measurements, a consistency with these limit values is observed in this temperature region. Finally, for a better interpretation of the experimental results in the whole temperature range, the theoretical calculations of the magnetic contribution $\alpha_m^{cal}(T)$ to $\alpha(T)$ using the related formula which is a function of these operators [2, 3] could be revisited by considering new reliable sets of CF parameters from other authors [18].

4. Conclusion

Linear thermal expansion coefficients $\alpha(T)$ for DyBCO and YBCO compounds have been revisited at low temperatures. The small magnetic excess to $\alpha(T)$ data for the DyBCO sample observed in the T^3 term below 40K is superimposed to the high phonon contribution, whereas no anomaly is detected. Theses $\alpha(T)$ data allow determination of the magnetic contribution, $\alpha_m(T)$ which is attributed to the CF interactions induced by the 4f electric quadrupole moment of the Dy Kramers ions. The $\alpha_m(T)$ curve increases monotonically with rising temperature while the calculated curve $\alpha_m^{cal}(T)$ up to 75K based on both fitted $\alpha(T)$ curves, shows that an agreement can be achieved with previous theoretical approaches above 40K. New calculations based on various available sets of the CF parameters of the Dy Kramers ion in DyBCO are needed to provide a full description of $\alpha_m(T)$ in this compound.

Acknowledgements

This work has been partially supported in the frame of a previous collaboration between the Badji Mokhtar-Annaba University and the Néel Institute-UJF, Grenoble. The author thanks J. Hejtmánek (Institute of Physics of ASCR, Praha, Czech Republic) for the preparation and the XRD analysis of both specimens. The author wishes to express gratitude to M. Diviš (Department of Electronic Structures, Charles University, Praha, Czech Republic) for his encouragements during the last icfe-8 Conference to continue the research on these compounds and for useful discussions and suggestions.

References

[1] Lynn J W and Skanthakumar S 2001 in *Hand. Phys. Chem. of Rare Earths*, ed. by Gschneidner K A, Jr Eyring L and Maple M B, vol. **31** Chap.199, Elsevier, Amsterdam, pp. 313–348
[2] Nekvasil V *et al.*, 1995 *J. of Alloys and Compounds* **225** 578
[3] Diviš M *et al.*, 1996 *Physica* B **223–224** 571
[4] Schnelle W *et al.*, 1990 *Physica* C **168** 465
[5] Lahoubi M *et al.*, 1998 *J. of Alloys and Compounds* **275–277** 591
[6] Lacerda A *et al.*, 1989 *Phys. Rev.* B **40** 8759
[7] Schmiedeshoff G M *et al.*, 2006 *Rev. Sci. Instrum.* **77** 123907
[8] Meingast C 2006 *Netsu Sokutei* **33** 167
[9] Lee B W *et al.*, 1988 *Phys. Rev.* B **37** 2368
[10] Ning C *et al.*, 1988 *Solid State Communications* **65** 333
[11] Revaz B *et al.*, 1998 *Phys. Rev.* B **58** 11153
[12] Vieira S *et al.*, 1988 *Solid State Communications* **65**, 1555
[13] Vieira S *et al.*, 1988 *Physica* C **153–155** 1006
[14] Sparn G *et al.*, 1988 *Physica* C **153–155** 1010
[15] Roa-Rojas J and Landínez Téllez D 2002 *Revista Colombiana de Fisica* **34** 9
[16] Van der Meulen H P *et al.*, 1988 *Physica* C **153–155** 1018
[17] Allenspach P *et al.*, 1989 *Phys. Rev.* B **39** 2226
[18] Likodimos V *et al.*, 2001 *Eur. Phys. J.* B **39** 143

Journal of Physics: Conference Series **507** (2014) 012029 doi:10.1088/1742-6596/507/1/012029

Comparison of the pinning energy in $\mathbf{Fe(Se_{1-x}Te_x)}$ compound between single crystals and thin films

A Leo[1], A Guarino[1], G Grimaldi[1], A Nigro[1], S Pace[1], E Bellingeri[2], S Kawale[2], C Ferdeghini[2] and E Giannini[3]

[1] CNR-SPIN Salerno and Dipartimento di Fisica 'E R Caianiello', Università di Salerno, Via Giovanni Paolo II, 132, Stecca 9, I-84084, Fisciano (SA), Italy
[2] CNR-SPIN Genova, corso Perrone 24, I-16152 Genova, Italy
[3] Départment de Physique de la Matière Condensée, Université de Genève, Quai Ernest-Ansermet 24, 1211 Genève, Switzerland

E-mail: `antoleo@sa.infn.it`

Abstract. Among the families of iron-based superconductors, we investigate flux pinning mechanisms in the $Fe(Se_{1-x}Te_x)$ compound. We perform magneto-resistance and current-voltage measurements on single-crystals, as well as on several epitaxial thin films grown on different substrates (CaF_2, $LaAlO_3$). The activation energy is derived as a function of magnetic field, $U(H)$. The influence of magnetic field orientation on the pinning energy activation mechanism is also studied, leading to the anisotropy analysis which reveals low anisotropy in thin films grown on CaF_2 substrate with respect to single crystals and films grown on $LaAlO_3$. Concerning the dominant pinning regime, the exponents of the power law dependence $U_0(H) \propto H^{-\alpha}$ have been evaluated, confirm that weak pinning is a general characteristic of this compound. The single exponent feature, generally noticed on thin films grown on $SrTiO_3$ substrate and associated to a strong single vortex regime, has been observed in thin films grown on $LaAlO_3$, only in the parallel configuration. At the end, this overall comparison can be useful to develop a technological material able to compete with high temperature superconductors.

1. Introduction

About seven years have gone by the discovery of superconductivity in iron-based compounds [1] and this new class of high-T_c superconductors is already looking promising for the applications. Indeed, iron-based superconductors show very high upper critical fields (H_{c2}) even close to the critical temperature (T_c), similarly to MgB_2. Likewise MgB_2, they are also characterized by small anisotropy factors (γ). Moreover, they do not present the technical hurdles (such as the metal-insulator transition or the d-wave symmetry of the order parameter) presented by copper oxide superconductors and show an high robustness to impurities. Up to now, the main technical limit of iron-based superconductors is the low critical temperature (the highest T_c for this class has been reported to be 56 K [2]), while the critical current density (J_c) values are comparable to those of the other high-T_c superconductors (J_c at 4.2 K and 30 T up to 10^5 A cm^{-2} have been reported [3]).

Almost all known iron-based superconductors belong to one of the following four families: a) the 1111 ReOFeAs (Re = Rare earth), b) the 122 AFe_2As_2 (A = alkaline earth), c) the 111 XFeAs (X = Li, Na), and d) the 11 $Fe(Se, Ch)$ (Ch = S, Te). As it is well known [4, 5], the

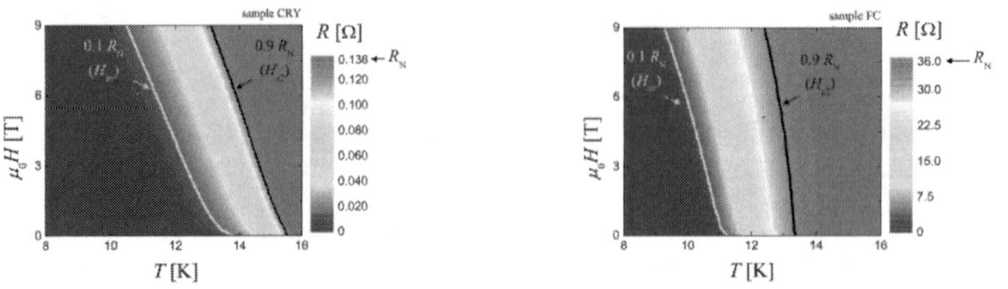

Figure 1. [color online] Field versus temperature (HT) phase diagrams for the single crystal sample [left panel] and for the thin film on CaF_2 substrate [right panel] with enlightened lines at $0.1\,R_N$ (white lines) and $0.9\,R_N$ (black lines).

highest T_c's are shown by the 1111 compounds, while the 122 family presents the highest upper critical fields. Nonetheless, the interest in 11 compounds is increasing [3, 6], mainly due to their low magnetic field anisotropy, the high $H_{c2}(T)$ slope near T_c, the absence of poisonous elements and the possibility to increase the relative low T_c by growing thin films on proper substrate inducing the Volmer-Weber growth mode, which mimic the application of an external pressure (see Ref. [7] and references therein).

In this paper we study the pinning properties of the $Fe(Se_{1-x}Te_x)$ (FST) compound grown as single crystals and thin films on different substrates to show that such thin films can be promising for high-field applications.

2. Experimental details

Single crystals of $Fe_{1+y}(Se_{1-x}Te_x)$ ($y < 0.1$, $0 < x < 0.45$) were grown in self-flux by using a modified Bridgman method from high purity precursors (Fe, Se and Te granules, all with purity $\geq$ 4N) melted inside a vacuum-sealed quartz tube vertically placed in a gradient furnace, and slowly cooled down (more details in Ref. [8]). Single crystals are easily cleaved in the ab-plane and characterized by X-Ray Diffraction (XRD) and Energy Dispersive X-Ray Spectroscopy (EDX) in order to check the crystal quality, composition and homogeneity. The actual composition of the crystals used in this work was found to be $Fe_{1.02}Te_{0.68}Se_{0.32}$ (nominal $x = 0.3$). Thin films were prepared in an ultra-high vacuum Plasma Laser Deposition (PLD) system from a starting target prepared by a two steps procedure from high purity materials (Fe 99.9+%, Se 99.9% and Te 99.999%), with a nominal composition $FeSe_{0.5}Te_{0.5}$, on single crystals substrates with different cell parameters and different chemical composition (more details in Ref. [9]).

Magneto-resistance and current-voltage measurements have been carried out in a Cryogen Free Magnet (CFM) system by Cryogenic Ltd. equipped with a cryogen-free variable temperature insert and a 9 T superconducting magnet. Magneto-resistance measurements have been performed by a standard 4-probe technique. Fig. 1 shows two typical field versus temperature (HT) phase diagrams realized acquiring more than 35 resistance-temperature ($R(T)$) curves, with the applied field direction parallel to the c-axis of the samples ($H\,/\!/\,c$). Current-voltage measurements have been performed in a pulsed current 4-probe technique [7] biasing the sample with rectangular current pulses 2.5 ms wide and a delay of 200 ms.

3. Results and discussion

In the following, we will present results on three different FST samples: a single crystal, which we will identify as sample CRY, and two thin films grown on a $LaAlO_3$ and on a CaF_2 substrate, namely sample FL and sample FC respectively.

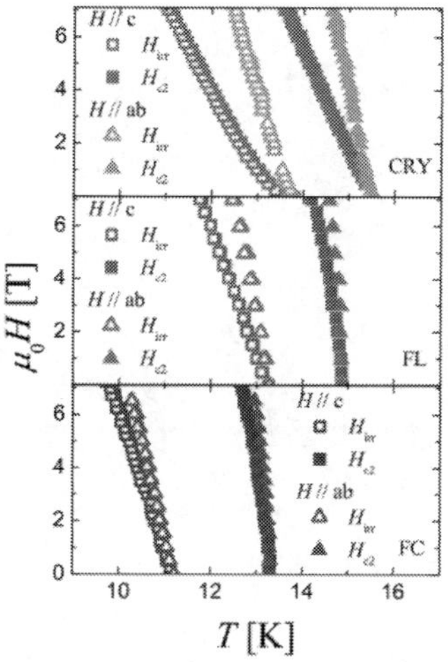

Figure 2. [color online] $H_{c2}(T)$ (full symbols) and $H_{irr}(T)$ (open symbols) curves of the three samples with the applied field parallel to sample c-axis (squares) and to sample ab-plane (triangles).

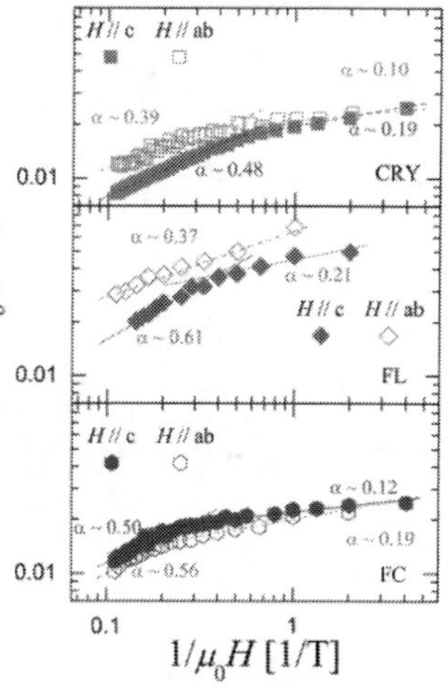

Figure 3. [color online] $U_0(H)$ curves of the three samples with the estimate of the α exponent in the power law $U_0 \propto H^{-\alpha}$.

In Fig. 1 the HT phase diagrams of sample CRY and FC, with $H /\!/ c$, are shown. In both phase diagrams the irreversibility line ($H_{irr}(T)$) and the $H_{c2}(T)$ curve are reported. The $H_{irr}(T)$ line is determined by the 10% criterion of the normal state resistance (R_N) in the $R(T)$ curve, while the $H_{c2}(T)$ line is defined by 90% of R_N. In Fig. 2 we show $H_{irr}(T)$ and $H_{c2}(T)$ data of all three samples both in the case of applied field parallel to sample c-axis and to ab-plane ($H /\!/ ab$). It results that the $H_{c2}(T)$ slope values evaluated near T_c are equal to 3 for sample CRY, to 14 for sample FL and to 6 for sample FC, in good agreement with the literature [5]. Moreover, it is evident the larger field anisotropy of sample CRY with respect to sample FL, while sample FC shows a substantial isotropy. Indeed, the anisotropy factors $\gamma = H_{c2}^{/\!/}/H_{c2}^{\perp}$ evaluated at $0.98\,T_c(0)$ are 1.9 for sample CRY, 1.7 for sample FL and 1.3 for sample FC. We found the same anisotropic behavior in the activation energy for vortex motion as a function of the applied field $U_0(H)$ displayed in Fig. 3. Here, U_0 has been evaluated via the relation $U(H,T) = U_0(H)\,g(t)$ [10], with $g(t) = (1-t)\,(1-t^4)^{1/2}$ and $t = T/T_c$, where $U(H,T)$ values have been deduced from the Arrhenius plots of the $R(T)$ at different applied magnetic fields. The $U_0(H)$ curve for sample FL shows a crossover from a weak field dependence to a power law dependence in the case $H/\!/c$, which can be associated to the crossover from a single vortex pinning regime [11] to a collective-pinning one [12], according to the literature [13]. On the contrary, in the case $H /\!/ ab$ no crossover is present thus pointing out the higher anisotropic behavior of the pinning properties in the films grown on $LaAlO_3$ substrates. For sample CRY, the $U_0(H)$ curves for $H /\!/ c$ and $H /\!/ ab$ exhibit anisotropy, although in both cases a field dependence crossover is observed. Instead, for sample FC no evident anisotropy can be observed in the $U_0(H)$ curves, in the full field range. This feature is confirmed by the pinning force $f_p = F_P/F_{P,max}$ curve as a function of the reduced field $h = H/H_{irr}$ shown in Fig. 4. Here, the H_{irr} values have been evaluated as the extrapolated zero value in the Kramer plots, where $J_c^{1/2}\,H^{1/4}$ is plotted as a function of H [14], with the critical current density J_c defined by the standard 1 μV/cm criterion. In Fig. 4 the $f_p(h)$ curves are shown at three different temperatures (8.0, 8.6, 9.3 K) in the case $H /\!/ c$ and also at 9.3 K for the case $H /\!/ ab$. All three curves can be fitted by a single

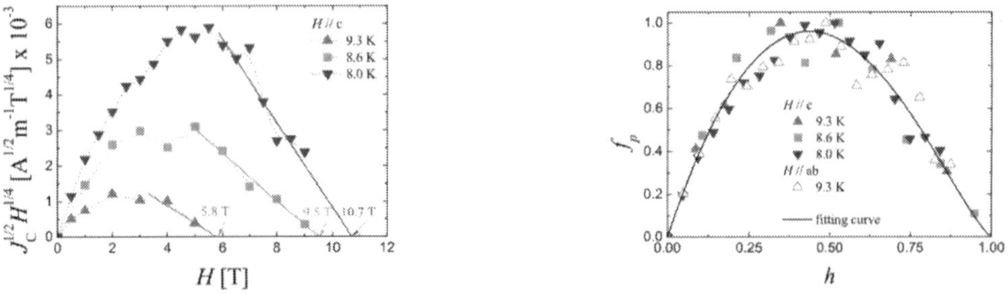

Figure 4. [color online] Left panel: Kramer plots for sample FC in the case $H // c$ at 8.0, 8.6 and 9.3 K, with the indication of the extrapolated zero value of H_{irr}. Right panel: $f_p(h)$ curves for sample FC at the same three temperature in the case $H // c$ (full symbols) and at 9.3 K in the case $H // ab$ (open triangles) with the fitting curve $h\,(1-h)^{1.3}$ (black solid line).

function $C\,h^p\,(1-h)^q$ with $p = 1$ and $q = 1.3$. According to the Dew-Hughes model [15], this behavior can be ascribed to the presence of δT_c volume pins, regardless of the temperature.

4. Conclusions

We have carried out a comparison of the pinning properties of the 11 iron-based compound $Fe(Se_{1-x}Te_x)$ realized in the form of single crystal and thin films on different substrates. We have shown that the thin films deposited on CaF_2 substrate are characterized by an high isotropy, whereas films deposited on $LaAlO_3$ and single crystals have a similar anisotropy. Moreover, the pinning sites in the CaF_2 samples have been basically identified having 3D δT_c nature. Considering that for high field applications high $H_{c2}(T)$ slope near T_c, good in-field isotropy and strong pinning are required, we conclude that $Fe(Se_{1-x}Te_x)$ films grown on proper substrates are good candidates for further developments in coated conductors.

Acknowledgments

The research leading to these results has been partially founded by the PON Ricerca e Competitività 2007-2013 under grant agreement PON NAFASSY, PONa3_00007.

References

[1] Kamihara Y, Hiramatsu H, Hirano M, Kawamura R, Yanagi H, Kamiya T and Hosono H 2006 *J. Am. Chem. Soc.* **128** 1001210013.
[2] Ren Z *et al* 2008 *Chin. Phys. Lett.* **25** 2215.
[3] Si W, Han S J, Shi X, Ehrlich S N, Jaroszynski J, Goyal A and Li Q 2013 *Nature Communication* **4** 1347.
[4] Yang H, Luo H Q, Wang Z S and Wen H H 2008 *Appl. Phys. Lett.* **93** 142506.
[5] Tanabe K and Hosono H 2012 *Japan. J. Appl. Phys.* **51** 010005.
[6] Mele P 2012 *Sci. Technol. Adv. Mater.* **13** 054301.
[7] Kawale S, Bellingeri E, Braccini V, Pallecchi I, Putti M, Grimaldi G, Leo A, Guarino A, Nigro A and Ferdeghini C 2013 *IEEE Trans. Appl. Supercond.* **23** 7500704.
[8] Viennois R, Giannini E, van der Marel E and Černý R 2010 *J. Solid State Chem.* **183** 769-775.
[9] Bellingeri E *et al* 2009 *Supercond. Sci. Technol.* **22** 105007.
[10] Tinkham M 1988 *Phys. Rev. Lett.* **61** 1658-1661.
[11] Blatter G, Feigel'man M V, Geshkenbein V B, Larkin A I and Vinokur V M 1994 *Rev. Mod. Phys* **66** 1125.
[12] Yeshurun Y and Malozemoff A P 1988 *Phys. Rev. Lett.* **60** 2202.
[13] Bellingeri E *et al* 2012 *Supercond. Sci. Technol.* **25** 084022.
[14] Bonura M, Giannini E, Viennois R and Senatore C 2012 *Phys. Rev. B* **85** 134532.
[15] Dew-Hughes D 1974 *Phil. Mag.* **30** 293.

The dependence of resistively measured B_{c2} and B_{irr} on $BaZrO_3$ concentration in $YBa_2Cu_3O_{6+x}$ thin films

M Malmivirta[1], **H Palonen**[1,2], **H Huhtinen**[1] **and P Paturi**[1]

[1] Wihuri Physical Laboratory, Department of Physics and Astronomy, FI-20014 University of Turku, Finland

[2] The National Doctoral Programme in Nanoscience (NGS-NANO), Turku, Finland

E-mail: `mika.malmivirta@utu.fi`

Abstract. The effect of $BaZrO_3$ (BZO) concentration on B_{irr} and B_{c2} in $YBa_2Cu_3O_{6+x}$ (YBCO) thin films was investigated with resistive measurements. It was found that in field out-of-plane configuration both B_{irr} and B_{c2} decrease with the highest BZO concentration of 9 wt% whereas the other samples show no significant change. The angular dependence of B_{irr} of the highest BZO concentration sample has a pronounced *c*-axis peak that is typical for BZO-doped YBCO but the others do not exhibit that clear peak. On the other hand, B_{c2} obeys the Blatter scaling law and the Blatter scaling parameter γ levels off to 3 – 3.5 above few wt% of BZO.

1. Introduction

In the applications of high-temperature superconductors isotropic high critical current is a preferred property. $YBa_2Cu_3O_{6+x}$ (YBCO), however, has intrinsically anisotropic properties. Doping YBCO with non-superconducting second phase, *e.g.* $BaZrO_3$ (BZO), improves flux pinning and reduces its anisotropic properties [1]. The pinning properties affect extrinsic properties like critical current densities and the irreversibility field B_{irr}. The upper critical field B_{c2}, on the contrary, is an intrinsic property. It should not be affected by the second phase unless doping changes some intrinsic properties. However, it has been seen that BZO increases B_{c2} [2,3]. Another intrinsic property, anisotropy, is quantified with the Blatter scaling parameter γ [4] and it can be determined from the measurements of B_{c2} with a straightforward process. The larger γ is, the larger is the anisotropy. Due to the high values of B_{c2}, usually a pulsed magnetic field system is needed to determine the anisotropy parameter.

In this paper B_{irr} and B_{c2} have been determined by resistive measurements of YBCO thin films doped with different concentrations of BZO. The angular dependence of B_{c2} is interpreted in terms of Blatter scaling.

2. Experimental details

Thin films doped with BZO between 0 – 9 wt% were grown with a pulsed laser deposition (PLD) XeCl excimer laser ($\lambda = 308$ nm) system on 5×5 mm^2 $SrTiO_3$ (100) substrates. The process is the same as for the BZO-doped films in [2].

The films were first characterized with Quantum Designs Physical Property Measurement System (PPMS) by determining the critical temperature from the AC magnetization. The critical temperature was defined to be the temperature at which magnetization is 10 % of its

Journal of Physics: Conference Series **507** (2014) 012030 doi:10.1088/1742-6596/507/1/012030

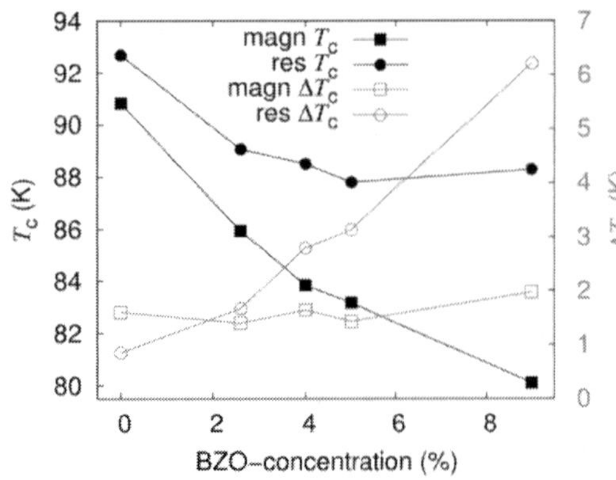

Figure 1. The critical temperatures and the widths of the transitions determined by magnetic and resistive measurements.

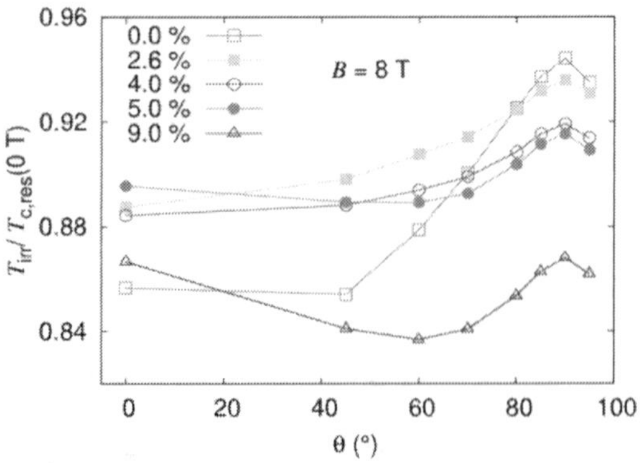

Figure 2. T_{irr} scaled with the resistive critical temperature as a function of the angle between sample and magnetic field, measured at 8 T.

saturation value. The width of the transition was defined to be the temperature difference between 10 % and 90 % of the superconductive state. On resistive measurements, however, the width was between 80 % and 10 % due to the definition used for B_{c2}. Also the DC magnetization was measured at 10 K and magnetic field was varied between −8 and 8 T to determine the critical current of the films using the Bean model. For the resistive measurements of B_{c2} and B_{irr}, the films were etched with photolitography. A standard four probe configuration with a 50 µm wide current stripe was used.

The resistances of the films were measured as a function of temperature between 120 and 40 K with 10 µA probing current. Several scans were taken for one sample. Magnetic field was varied between 8 and 0 T with smaller step size at small fields. Also, angle θ was changed from 0° to 95° focusing on larger θ values. The value $\theta = 0°$ corresponds to the case $B||c$-axis of YBCO.

The temperature T_{irr} corresponding to the irreversibility field at a certain scan was defined to be the the temperature at which the electric field was 1 µV/cm. B_{c2} was defined from scaled data: At first the data was divided with a line which was fitted on the normal state resistivity. Then the temperature corresponding to B_{c2} at certain field was taken to be the point where resistance was 80 % of its normal state value. This has been noted to be approximately the same as finding the intersection of two line fits to steepest transition and normal state [5]. Also, other ways to determine B_{c2} were used, but this was found most reliable for all the samples, fields and angles. It is very close to the method used in [5].

3. Results and discussion

The critical temperature decreases as BZO concentration increases (figure 1), which is typical for BZO-doped films [6]. Magnetic T_c decreases much more (from 92 to 84 K) than resistive T_c (from 93 K to 88 K). On the other hand, the width of the magnetic transition stays at 1 – 2 K whereas resistive widths increase from 1 K up to 6 K. A two-step transition was observed in samples with higher BZO concentration. The second phase got more pronounced with increasing BZO concentration. The same phase has also been noted before as the peak the in magnetic absorption in AC magnetization during the transition had split into two [6]. In resistive measurements a small current is able to choose a route using only the best path, whereas in magnetic measurements the signal comes from the whole sample. However, as the

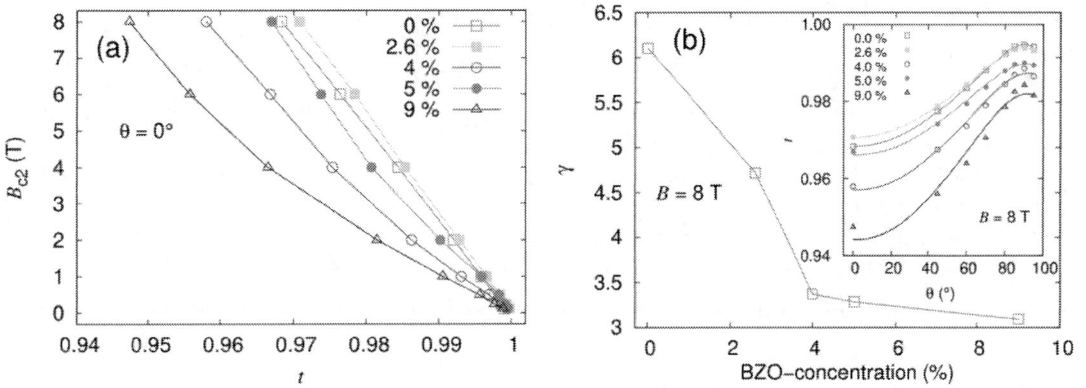

Figure 3. (a) The dependence of B_{c2} on reduced temperature in samples with different BZO concentrations. (b) Inset: The dependence of the reduced temperature t corresponding to B_{c2} at 8 T on the angle between sample and magnetic field. Solid lines are fits to equation (1). The main figure shows that the Blatter scaling γ decreases and levels off as the BZO concentration is increased.

BZO concentration increases, the amount of good quality areas in the film decreases and the transition broadens. The high-T_c phase is so small that it can barely be seen in magnetic measurements. The magnetically determined critical current densities are in the range 25 – 35 MA/cm^2 at 0 T, only the 9 % sample has smaller J_c. This is similar to what has been found before [6]. More details on J_c can be found in [7].

The irreversibility field as a function of the angle between the magnetic field and the sample for different concentrations shows rather different shapes for the various concentrations (figure 2). The change is best seen at the highest measured field 8 T. BZO doping increases the irreversibility temperature in middle concentrations. With the highest concentration of 9 wt% measured here, the irreversibility temperature decreases. Probably the most important reason for this is that the width of the transition is largest. The *ab*-peak, on the other hand, drops monotonically as the BZO concentration increases. The *c*-axis peak seen in higher concentrations is due to correlated columnar nanorods typical for BZO [1].

The order of the curves at $\theta = 0°$ is not monotonic with respect to concentration. However, at fields of a few teslas, the undoped film has the smallest T_{irr} and the others have larger T_{irr} with the exception of the 9 % sample. The undoped sample has been noted to give the smallest B_{irr} in measurements in pulsed magnetic field [8]. However, at fields smaller than 2 T the undoped film has the largest T_{irr} and T_{irr} decreases monotonically with increasing BZO concentration. The same kind of difference between doped and undoped films has been found in magnetic measurements [7]. The optimal BZO concentration increases with increasing field, as has been noted before [8].

The dependence of B_{c2} on reduced temperature t ($t = T/T_{c,res}(0\ \mathrm{T})$) varies with BZO concentration (figure 3 (a)). The sample with the largest BZO concentration has so many defects that it decreases the B_{c2} significantly compared to the other samples. The other small differences originate possibly from the small distinctions in the high-T_c phases and uncertainties in the determination of B_{c2}. Recently it has been found that, in contrast to this result, doping improves B_{c2} in samples measured both with pulsed magnetic field setup [2] and with magnetic measurements [3, 7]. The films used in this work are the same as in [7] and quite similar to samples in [2], so the reason for different behaviour is probably the small probing current used here. For the same reason, the measured properties of B_{c2} are quite strongly affected by the small amount of high-T_c phase where the current percolates.

The angular dependence of B_{c2} (figure 3 (b) inset) exhibits no c-peak in contrast to $B_{irr}(\theta)$-data. The 9 % sample has on the whole angle range the smallest values of the reduced temperature corresponding to B_{c2}. The strain caused by BZO results in degradation of the crystal structure and thus the drop of T_c. In addition, the distorted crystal structure also affects the coherence length ξ which in turn affects the values of the upper critical field.

Using the dependence of B_{c2} near T_c from the Ginzburg-Landau equations [9, p. 53], solving it for $\frac{T}{T_c}$ and applying Blatter scaling [4] for B_{c2} at 0 K, we get

$$\frac{T}{T_c}(\theta) = 1 - \frac{B_{c2}}{2B_0}\left(\frac{\sin^2(\theta)}{\gamma^2} + \cos^2(\theta)\right)^{1/2}, \quad (1)$$

where θ is the angle between sample normal and magnetic field, B_0 is B_{c2} at 0 K with $\theta = 0°$ and γ is the Blatter anisotropy parameter. Fits to equation (1) are solid lines in figure 3 (b) inset and they fit best to the data of low BZO concentration, probably due to uncertainties caused by the small high-T_c phases in higher BZO concentrations. The values of γ obtained from these fits (figure 3 (b)) is 6 for undoped, which is near to what has been found before [10]. For doped YBCO the value levels off to 3 – 3.5 which has also been seen before [2]. However, no change in γ was seen in the case of isotropic BZO nanodots [11].

4. Conclusions

B_{irr} and B_{c2} were determined by resistive measurements of PLD YBCO thin films doped with different concentrations of BZO. It was found that both B_{irr} and B_{c2} decreased when the doping concentration was high enough. However, the results are inconsistent with magnetic measurements and pulsed magnetic field data. This probably results from the small probing current. From the angular dependence of B_{c2} it was found that the value for Blatter scaling γ is 3 – 3.5 for doped samples.

Acknowledgments

Jenny and Antti Wihuri foundation, Finnish Foundation for Technology Promotion and Finnish Cultural Foundation are acknowledged for financial support.

References

[1] MacManus-Driscoll J L, Foltyn S R, Jia Q X, Wang H, Serquis A, Civale L, Maiorov B, Hawley M E, Maley M P and Peterson D E 2004 *Nat. Mater.* **3** 439
[2] Palonen H, Huhtinen H, Shakhov M A and Paturi P 2013 *Supercond. Sci. Technol.* **26** 045003
[3] Matsushita T, Nagamizu H, Tanabe K, Kiuchi M, Otabe E S, Tobita H, Yoshizumi M, Izumi T, Shiohara Y, Yokoe D, Kato T and Hirayama T 2012 *Supercond. Sci. Technol.* **25** 125003
[4] Blatter G, Feigel'man M V, Geshkenbein V B, Larkin A I and Vinokur V M 1994 *Reviews of Modern Physics* **66** 1125
[5] Ando Y, Boebinger G S, Passner A, Schneemeyer L F, Kimura T, Okuya M, Watauchi S, Shimoyama J, Kishio K, Tamasaku K, Ichikawa N and Uchida S 1999 *Phys. Rev. B* **60** 12475
[6] Peurla M, Paturi P, Stepanov Y P, Huhtinen H, Tse Y Y, Bódi A C, Raittila J and Laiho R 2006 *Supercond. Sci. Technol.* **19** 767
[7] Paturi P, Malmivirta M, Palonen H and Huhtinen H 2013 *Journal of Physics: Conference Series* **submitted**
[8] Huhtinen H, Irjala M, Paturi P, Shakhov M A and Laiho R 2010 *J. Appl. Phys.* **107** 053906
[9] Poole Jr C P, Farach H A, Creswick R J and Prozorov R 2007 *Superconductivity, Second Edition* (Academic Press)
[10] Civale L, Maiorov B, Serquis A, Willis J O, Coulter J Y, Wang H, Jia Q X, Arendt P N, MacManus-Driscoll J L, Maley M P and Foltyn S R 2004 *Appl. Phys. Lett.* **84** 2121
[11] Llordés A, Palau A, Gázquez J, Coll M, Vlad R, Pomar A, Arbiol J, Guzmán R, Ye S, Rouco V, Sandiumenge F, Ricart S, Puig T, Varela M, Chateigner D, Vanacken J, Gutiérrez J, Moshchalkov V, Deutscher G, Magen C and Obradors X 2012 *Nat. Mater.* **11** 329

High-T_c Superconducting Cuprates, $(Ce,Y)_sO_{2s-2}Sr_2(Cu_{2.75}Mo_{0.25})O_{6+\delta}$: T_c-increase with apical Cu-O decrease at constant Cu-O planar distance

M Marezio[1], O Chmaissem[2,3], C. Bougerol[4], M Karppinen[5], H Yamauchi[5] and T H Geballe[6]
[1]CRETA/CNRS, Grenoble 38042 cedex 9, France
[2]Northern Illinois University, DeKalb, IL 60115, USA Argonne
[3]National Laboratory, Argonne, IL 60439, USA CEA-CNRS,
[4]Institut Néel, Grenoble 38042, cedex 9, France Aalto University,
[5]FI-00076 Aalto, Finland
[6]Stanford University, Stanford, California 94305, USA

E-mail: massimo.marezio@grenoble.cnrs.fr

Abstract. Evidence for high-T_c cuprate superconductivity is found in a region of the phase diagram where non-superconducting Fermi liquid metals are expected. Cu valences estimated independently from both XANES measurements and bond valence sum calculations are greater than 2.25 and are in close agreement with each other for structures of the homologous series given in the title with s = 1, 2, 3, 4 and 5. Two questions arise from the present perspective: 1) Is all the action in the CuO_2 layers? 2) Is there superconductivity beyond the usual dome? The record short apical oxygen distance found in the homologous series especially in the s = 1 member, at odds with the current theory, suggests the possibility of a new pairing mechanism. The apical Cu-O distance in the s = 1 member decreases upon oxygenation from 2.29 to 2.15 Å while the Cu valence increases to 2.45 Å.

Between 2004 and 2010 the researchers at Tokyo Institute of Technology (Japan) and Aalto University in Helsinki (Finland) synthesized the first 5 members of the cuprate homologous series[1-6]: $(Ce,Y)_sO_{2s-2}Sr_2(Cu_{2.75}Mo_{0.25})O_{6+\delta}$, s = 1, 2, 3, 4, and 5. None of the as-prepared samples (AS) were found to become superconducting, but when the samples were treated under high oxygen pressure (5 GPa, 500° C in the presence of $KClO_3$ or Ag_2O_2) the s = 1 member became superconducting at 88 K, while the s = 2, 3, 4 and 5 members exhibited a constant T_c of about 56 K. The chemical formula and the structural arrangement of the five compounds were determined by conventional x-ray powder diffraction analysis, electron diffraction and high-resolution transmission electron microscopy. The superconducting transition temperatures were determined by FC and ZFC magnetization measurements. Later

Table 1 :Cu valences for the AS and HPO samples estimated from the Cu $L_{2,3}$ absorption edge and the hole concentration ρ on the CuO_2 layers estimated from the O K absorption edge

	v(Cu)* AS	v(Cu)* HPO	$\rho(CuO_2)$**	T_c
s = 1	2.16***	2.46	0.53	88
s = 2	2.13	2.24	0.28	56
s = 3	2.14	2.25	0.26	56
s = 4	2.14	2.26	0.27	56

*Values estimated from the Cu L2,3 absorption edge. ** Values estimated from the O K absorption edge. ***Value obtained also by iodometric titration.

the same group was able to produce very pure samples for the first 4 members, which allowed them to carry out XANES measurements leading to an estimate of the Cu valence and the holes' concentration for these unusual cuprates. These results are summarized in Table 1.

It is interesting to note that the Cu valence of the cations belonging to the CuO_2 layers is definitely high when compared to the values of 2.15/2.2+ obtained for the all other superconducting cuprates.

Fig. 1 represents a three-dimensional structural model of the s = 2 member.

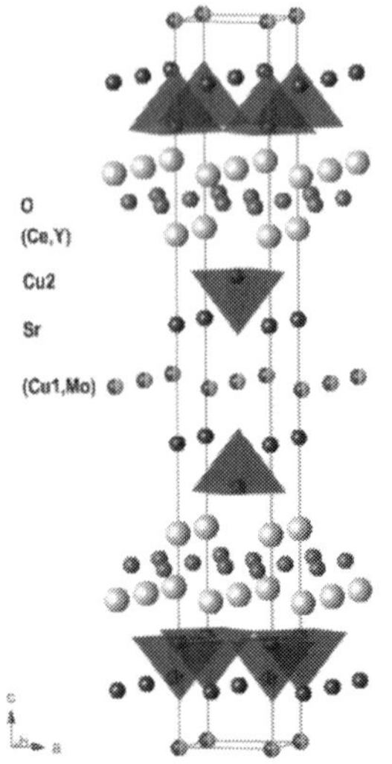

Fig.1: A three-dimensional structural model of the s = 2 member: $(Ce,Y)_2O_2Sr_2(Cu_{2.75}Mo_{0.25})O_{6+\delta}$.The oxygen on the basal plane have been omitted. The distance between 2 CuO_2 layers is 5.998 Å. It increases to 8.746 and to 11.441 Å for the members with s = 3 and 4 respectively, while T_c remains constant at 56 K for both members. The higher-member structures are obtained by inserting additional 2-layer blocks (O_2)(Ce,Y) in between (Ce,Y) and (CuO_2). For example, for the member with s = 3 the sequence between two successive CuO_2 layers is: $(CuO_2)(Ce,Y)(O_2)(Ce,Y)(O_2)(Ce,Y)(CuO_2)$.

The new samples of the Mo-cuprates allowed reliable structural refinements at room temperature. These refinements were performed at the Argonne National Laboratory[7]. The results confirmed the structural arrangement previously determined by x-ray, and electron diffraction techniques and HRED imaging. Moreover, they revealed important structural features concerning the basal plane occupied by 75% of Cu cations, 25% of Mo cations, and δ oxygen atoms. The results are summarized in Table 2.

Table 2: Oxygen stoichiometry of the basal plane δ and the total oxygen content for the first four members of the homologous series $(Ce,Y)_sO_{2s-2}Sr_2(Cu_{2.75}Mo_{0.25})O_{6+\delta}$ from neutron diffraction.

	s = 1 AS	s = 1 HPO	s = 2 AS	s = 2 HPO	s = 3 AS	s = 3 HPO	s = 4 AS
δ	1.36	1.56	1.01	1.41	1.16	1.72	0.97
Total oxygen	7.36(16)	7.56(4)	9.01(8)	9.41(17)	11.16(9)	11.72(15)	12.97(7)

The observation that after the HPO treatment the first member shows a T_c of 88 K while the members with s = 2, 3, 4 and 5 exhibit a constant T_c of ~56 K is a very strong indication that the charge reservoir layers must play an essential role for the superconducting properties. The first member has actually the $YBa_2Cu_3O_7$ structure in which the Ba cations are substituted by Sr and 25% of the planar squarely-coordinated Cu cations are substituted by Mo^{6+} cations. These substitutions make a substantial difference in the structure and consequently in the properties because the Mo^{6+} cations require an octahedral coordination and therefore the basal z = 0 plane can incorporate a larger amount of oxygen than either $YSr_2Cu_3O_7$ or $YBa_2Cu_3O_7$.[8]

In Table 3 the valence of the Cu cations belonging to the CuO_2 layers, determined by the bond valence sum (BVS) from the interatomic distances obtained by neutron diffraction data are reported. They are compared to the corresponding values obtained by XANES experiments. The same trend was confirmed. Note that the total oxygen intake is reported in Table 2 whereas in Table 3 only that of the CuO_2 planes is given. The high-valence value of

2.45 obtained for the s = 1 HPO sample and values greater than 2.25 for the other members indicate that superconductivity can exist in the overdoped region of the generalized cuprate phase diagram.

Table 3: The Cu2 site valence in the $(Ce,Y)_sO_{2s-2}Sr_2(Cu_{2.75}Mo_{0.25})O_{6+\delta}$ samples calculated from neutron diffraction data and XANES experiments.

	s = 1 AS	s = 1 HPO	s = 2 AS	s = 2 HPO	s = 3 AS	s = 3 HPO	s = 4 AS	s = 4 HPO
v(Cu2) BVS	2.21	2.45	2.22	2.31	2.22	2.34	2.20	
v(Cu2) XANES	2.16	2.46	2.13	2.24	2.14	2.25	2.14	2.26

Fig. 2 shows possible structural arrangements of all the basal-planes of AS phases (a) and HPO phases (b). The supercells are only a guide for the eyes. The oxygen atoms in the HPO phases with z = 0 are in red.

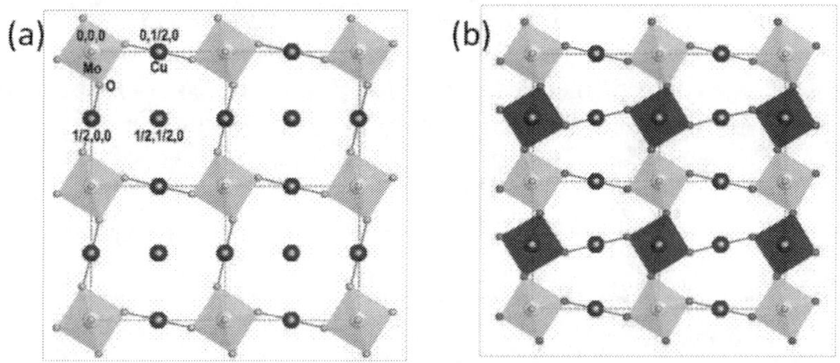

Fig. 2: (a) A possible structural arrangement of the basal plane $(Cu_{0.75}Mo_{0.75})O_\delta$, $\delta = 1$ for all the AS samples. The supercell is indicated. The oxygen atoms are disordered over the xy0 and the more symmetrical ½y0 positions. However, for simplicity only the more symmetrical positions are used and correspond to the small green balls. The large spheres are: Mo (yellow) and Cu (blue). (b) A possible structural arrangement of the basal plane $(Cu_{0.75}Mo_{0.75})O_\delta$, $\delta = 1.5$ for all the HPO samples. Only for the s = 1 member all the oxygen are in the more symmetrical ½y0 positions. See reference 7 for details.

The increase in T_c found upon high-pressure oxygen annealing which causes a marked decrease in the apical Cu-O distance while the in-plane Cu-O distance remains almost constant cannot be explained by any current theory. We suggest that T_c can be enhanced because the longer range repulsive Coulomb interactions in the CuO_2 layers are screened by the highly polarizable charge of the reservoir layers. However, since no measurements of the superfluid density (from muon spectroscopy experiments) have been made there still remains the possibility that an undetected inhomogeneity, such as stripe formation, might explain the results without requiring new physics.[8] These experiments are currently underway in Villingen, Switzerland.

It is important to point out that the apical Cu-O distance in the s = 1 AS member is 2.295(4) Å, very close to that found in superconducting $YBa_2Cu_3O_{6.95}$, 2.301 Å for which $T_c \approx 93$ K.[8] After the HPO treatment resulting in the superconducting $T_c = 88$ K, the apical distance in the first member decreases to 2.15(2) Å, contrary to the two-orbital model of Sakakibara et al[9] or to any other theory that finds T_c to increase with apical Cu-O distance.

As for the metallization of the reservoir block, according to Jorgensen et al.[10] an apical oxygen distance of 2.15 Å and an O-Cu-O corrugation angle of 166° in the CuO_2-square layers suggest that the reservoir block of the s = 1 member, would have a metallic character. The idea of metallic reservoir blocks is not new. As early as 1996 Tallon and coworkers[11] inferred from muon spin relaxation studies that a 50-fold increase in irreversibility field

Journal of Physics: Conference Series **507** (2014) 012031 doi:10.1088/1742-6596/507/1/012031

observed when 25% of the Hg cations are substituted by Re and all the Ba cations are replaced by Sr in HgBa2CaCu2O6+δ could be explained by the metallization of the (Hg0.75Re0.25)Oδ reservoir block.

To explain why T_c decreases on going from the s = 1 to the s = 2 member (88 to 56 K) and then remains practically constant for the s = 3 and 4 we report in Table 4 the thickness of the two blocks existing in the structure of these compounds. They can be viewed as built of two types of blocks: one being the CuO_2(fluorite-block of different thickness)CuO_2, and one of about constant thickness: $CuO_2(SrO)(Cu_{0.75}Mo_{0.25})(SrO)CuO_2$. The fluorite block does not exist in the s = 1 member or in $YBa_2Cu_3O_{6.95}$. In both cases it consists of only one yttrium layer. Two observations can be made from Table 4: 1) the oxygen insertion increases the thickness of the CuO_2(fluorite-block)CuO_2 block even though the extra oxygen is not incorporated into this block. 2) The thickness of the reservoir block, which is the same for all members, decreases when the extra-oxygen is incorporated by the HPO treatment. These observations indicate that these compounds have a strong ionic character because by inserting negative oxygen ions the Coulomb interactions definitely prevail against the size effect. The same observation can be made for $YBa_2Cu_3O_{6+x}$, which is included in Table 4 for comparison. Moreover, the same structure of the "reservoir block" for all members with inserted-fluorite blocks can explain why T_c remains constant for s = 2, 3, and 4 members.

Table 4: Thickness (Å) of the 2 blocks separated by the CuO_2 layer, which make up the structure of the homologous series $(Ce,Y)_sO_{2s-2}Sr_2(Cu_{2.75}Mo_{0.25})O_{6+\delta}$ or $(Cu_{0.75}Mo_{0.25})$-12s2

Sample	CuO_2(fluorite-block)CuO_2	$(CuO_2)(SrO)(Cu_{.75}Mo_{.25}O_x)(SrO)(CuO_2)$
s=1 AS	3.308*	8.242
s=1 HPO	3.434*	8.037
s=2 AS	5.998	8.283
s=2; HPO	6.064	8.161
s=3; AS	8.746	8.237
s=3; HPO	8.801	8.104
s = 4; AS	11.441	8.231
$YBa_2Cu_3O_6$	3.304*	8.513*
$YBa_2Cu_3O_{6.95}$	3.391*	8.269*

* In $YBa_2Cu_3O_6$ and $YBa_2Cu_3O_{6.95}$ and in the s = 1 member the fluorite blocks do not exist, the two CuO_2 layers are separated by a single Y layer. Furthermore, the basal layers of the first two contain only Cu cations.

References

[1]Morita Y *et al* 2004 *Phys. Rev.* B **70** 174515

[2]Karppinen M *et al* 2005 *J. Solid State Chem.* **178** 3464

[3]Grigoraviciute I *et al* 2006 *Solid State Comm.* **137** 601

[4]Grigoraviciute I *et al* 2007 *J. Am. Chem. Soc.* **129** 2593

[5]Grigoraviciute I *et al* 2010 *J. Am. Chem. Soc.* **132** 838

[6]Gao W *et al* 2011 *Solid State Comm.* **151** 1400

[7]Chmaissem O *et al* 2010 *Phys. Rev.* **B 82** 104507

[8]Marezio M *et al* 2013 *APL Mater* **1**, 021103

[9]Sakakibara H *et al* 2010 *Phys. Rev. Lett.* **105**, 057003.

[10]Jorgensen J D *et al* 1996 *"Lecture Notes in Physics"* Published by Springer-Verlag

[11]Tallon JL *et al* 1996 *J. Low Temp. Phys.* **103** 1379.

Low ac field response of Bi-based superconductors with addition of Antimony oxide

I.R. Metskhvarishvili[1], G.N. Dgebuadze[1], B.G. Bendeliani[1], T.E. Lobzhanidze[2], M.R. Metskhvarishvili[3], G.N. Mumladze[1]

[1]Department of Cryogenic Technique and Technologies, Ilia Vekua Sukhumi Institute of Physics and Technology, Mindeli str. 7, 0186 Tbilisi, Georgia
[2]Department of Chemistry, Faculty of Exact and Natural Sciences, Ivane Javakhishvili Tbilisi State University, Chavchavadze Ave. 3, 0179 Tbilisi, Georgia
[3]Department of Physics, Georgian Technical University, Kostava str. 77, 0175 Tbilisi, Georgia
E-mail: metskhv@yahoo.com

Abstract. We report the effects of the addition of Sb_2O_3 on the superconducting properties of Bi-2223 system. Samples with composition $Bi_{1.7}Pb_{0.3}Ca_2Sr_2Cu_3Sb_xO_y$ (x=0,0-0.8 wt.%) are prepared by the solid-state reaction method. Ac susceptibility and high harmonic response of polycrystalline superconductors are measured in weak *ac* and *dc* magnetic fields. Critical state models are used to explain the nonlinear magnetic response and it is found that the low field odd harmonic response is qualitatively explained reasonably well within the Kim-Anderson regime. We observed that, the low level Antimony doping enhances the transport critical current densities J_c.

1. Introduction

The Bismuth based superconductors with substitution or addition of various dopants is an interesting topic for study. Antimony belongs to the fifth group elements and is metalloid. Substitution by antimony or by other fifth group elements improves the superconducting parameters of the high-temperature superconductors.

The effect of the partial substitution of Bi simultaneously by Pb and Sb on the superconductive properties of the Bi-2223 system has been investigated in many works. It has been shown that the optimal sintering temperature is about of 845°C, which yields the highest density value and the highest volume fraction of Bi-2223 phase [1]. It was observed that antimony makes the system more reactive, enhances the kinetics of the reaction and promotes the high-T_c phase [2]. Siddiqi and Akhtar [3] observed that low-concentration of Sb (2%) promotes the formation of high temperature phase (2223). It was found out that Sb has no effect on increasing the transition temperature of this system. The maximum superconducting transition temperature observed in the $Bi_{2-x}Pb_xSr_2Ca_2Cu_3O_y$ system was T_c = 108 K, which was completely independent from the doping of Sb. Iqbal and Mehmood [5] made substitution of Sr by Sb and

[1] metskhv@yahoo.com

concluded that the maximal value of the volume percentage of high temperature (2223) phase was 69.44% for the sample having Sb content of x=0.06. Hongbao [6] studied the partial substitution of Bi by Sb $Bi_{2-x}Sb_xSr_2Ca_2Cu_3O_y$ (x=0.05, 0.1, 0.15, 0.2). All of these samples have superconducting transition temperature of about at 120 K. Unlike the above-mentioned works, the aims of our studies are investigations of the effect of Sb_2O_3 addition on superconductive properties of Bi-2223 system.

2. Experimental

The Sb-doped samples $Bi_{1.7}Pb_{0.3}Sr_2Ca_2Cu_3Sb_xO_y$ were prepared by the conventional solid-state reaction method. The concentrations of Sb were in the range of x=0.000, x=0.002, x=0.004, x=0.006 and x=0.008 identified as 0, 0.2, 0.4, 0.6 and 0.8 wt.%. The powders of Bi_2O_3 (99.9% Aaldrich), PbO (99.9% Sigma-Aldrich), $SrCO_3$ (99.9% Aaldrich), $CaCO_3$ (99.9% Oxford ChemServe), CuO (98% Sigma-Aldrich) and Sb_2O_3 (99% Aldrich) were weighed in necessary atomic ratios. These oxides were thoroughly mixed and ground by using a mortar and pestle and charged in alumina crucibles. The mixtures were first calcined in air by using of the temperature programmed muffle furnace (KSL-1100X-S, MTI-corporation, temperature accuracy ±1°C).The first heat treatment schedule was: 120°C for 4 h, 200°C for 2 h, 400°C for 2 h, 600°C for 2 h, 800°C for 2 h and then was quenched to room temperature. The powder was reground and calcined in air at 800°C for 30 h. The resulting powders were ground and pressed into the pellets of 10 mm diameters and 2 mm thickness by a hydraulic press (Holzmann-maschinen, Type: WP10H) with about 900 MPa. These pellets were sintered in air at 850 for 100 h. The samples were allowed to cool at natural furnace cooling. The prepared patterns were characterized by X-ray diffraction (XRD, Dron-3M) with Cu$K\alpha$ radiation.

We have studied the real part of *ac* susceptibility χ' and the magnitude of the amplitudes of the high harmonics. In our experiment we discuss the case when collinear constant and variable fields $H(t) = H + hcos\omega t$ are applied to the sample. In this case, the induction $B(t)$ could be expanded into Fourier series as $B(t) = \frac{a_0}{2} + \sum_n (a_n cosn\omega t + b_n sinn\omega t)$. In case $h \ll H$, the expression for a_n and b_n coefficients [7], which are determined in our experiment are:

$$a_0 = 2\mu_{eff}H, \quad a_1 = \frac{\mu_{eff}h^2}{4\pi j_c(H)d}, a_{2k+1} = 0, \quad k \geq 1; \qquad (1)$$

$$b_{2k+1} = -\frac{\mu_{eff}h^2}{8\pi^2 j_c(H)d}\frac{1}{(k^2-1/4)(k+3/2)}; \ a_2 = \frac{\mu_{eff}h^3}{32\pi d}\frac{d}{dH}\left\{\frac{1}{j_c(H)}\right\}, \quad a_{2k} = 0, \quad k > 2;$$

$$b_{2k} = -\frac{\mu_{eff}h^3}{16\pi^2 d}\frac{d}{dH}\left\{\frac{1}{j_c(H)}\right\}\frac{k}{(k^2-1/4)(k^2-3/2)}$$

Here, μ_{eff} is the effective permeability of the ceramic material, which takes into account the fact that the field does not penetrate into the granules, h is a variable field, d is a thickness of the sample, H is a constant field and j_c is the critical current density. Errors in high harmonics measurement could be as high as ~ 2% when the measuring signal is lower than 0.2 μV. However, the error does not exceed 0.5% for higher amplitudes [8].

The phase method was used to study the real parts of the linear susceptibility [9].The errors in the determination of χ'at higher frequencies than 1 kHz does not exceed 1%.

3. Result and discussion

The X-ray diffraction patterns for all samples are plotted in Figure 1. As one can see, the samples consist of a mixture of Bi-2223 and Bi-2212 phases as the major constituents. From the XRD results one will notice that the low Sb concentration samples (0.2 wt.% and 0.4 wt.%) was dominated by high-T_c 2223 phase. However, in higher doping (0.6 wt.% and 0.8 wt.%) concentration, the low T_c-phase 2212 is dominant.

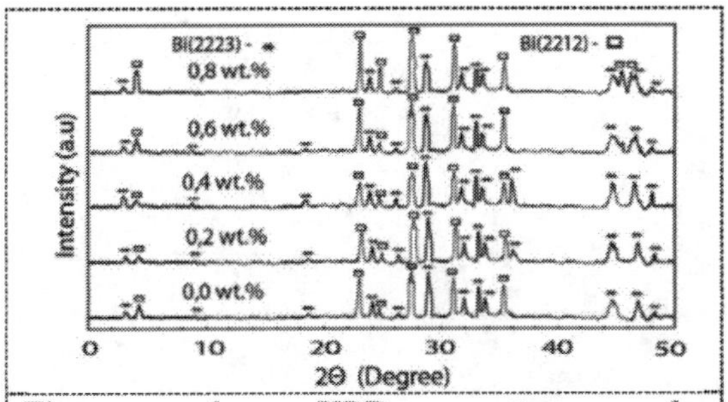

Figure 1. XRD patterns for $Bi_{1.7}Pb_{0.3}Sr_2Ca_2Cu_3O_y$ and Sb-doped samples.

Figure 2 shows the temperature dependences of the real $(-4\pi\chi')$ part of *ac* susceptibility for un-doped and doped samples, measured in zero magnetic fields (H=0) at h=1 Oe, f=20 kHz. The diamagnetic onset temperature of the superconducting transition for 0.0 wt.% sample is about $T\approx 110$ K and T_c monotonically decreases with increasing Sb content. As we see un-doped (0.0 wt.%) and low doped (0.2 wt.%) samples clearly show a two step decrease with T, which reflects the flux shielding from and between the grains. The high doped samples (0.4-0.8 wt.%) do not show two steps and full screening of applied *ac* magnetic fields for samples, whose Sb_2O_3 doping is in the range 0.4 wt.%, is observed at T~92 K.

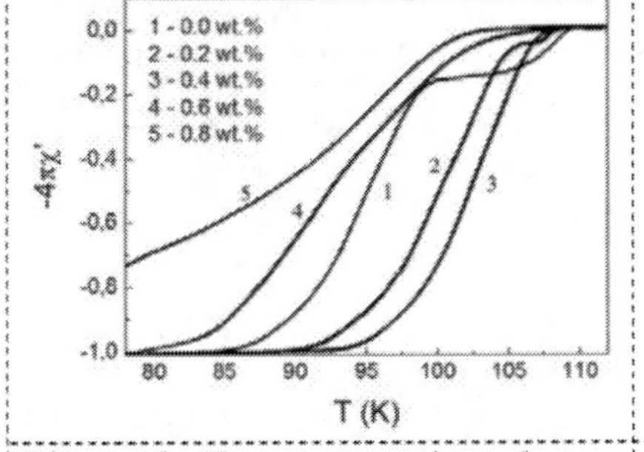

Figure 2. Temperature dependences of the real χ' parts of ac susceptibility.

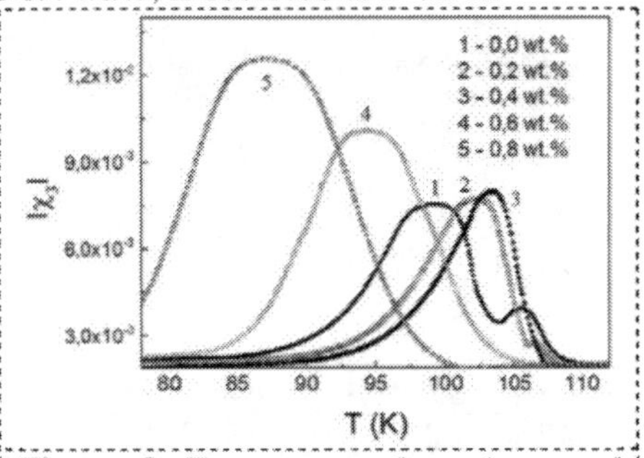

Figure 3. Temperature dependences of the amplitude of the third harmonics.

Measurements of temperature dependence of the magnitude of the third harmonics at h=1 Oe, f = 20 kHz, H = 0 are presented in figure 3. Like$-4\pi\chi'(T)$ dependence for un-doped and 0.2 wt.% doped samples, the $\chi_3(T)$ presents a two-peaks plot. The high-temperature peak could be caused by the penetration of the field into the crystallites and be associated with intragranular critical current density, j_{cg}, while the low-temperature maximum is due to the penetration only into the Josephson medium formed by weak links between the crystallites. The latter is associated with intergranular critical current density j_{cJ}.

According to (1), amplitudes of each harmonics are inversely proportional to the critical current density $b_n \approx 1/j_c$. Thus, $j_c(H)$ function could be determined by studying the dependencies of these harmonics on the magnitude of the constant field H at a fixed amplitude of h [7]. From Figure 4 it follows that for un-doped and doped samples b_3 is very sensitive to the low external *dc* fields and is linearly increasing. The observed behavior of the third harmonic could be explained by Kim-Anderson critical state modal dependences of critical current density on the magnetic induction [10-12] $J_c(B) = J_0/(1 + B/B_0)$. Additionally, it should be noted that we have carried out such experiments in a large temperature range and $b_3(H)$ is not changed for all three samples.

Figure 5 presents dependence of the transport critical current densities on temperature j_{cJ} value calculated from $\chi_3(T)$ measurements using equation (1). As we see j_c value for the un-doped sample is 85 A/cm^2 and the largest value of critical currents observed in sample 0.4 wt.% is equal to 115 A/cm^2.

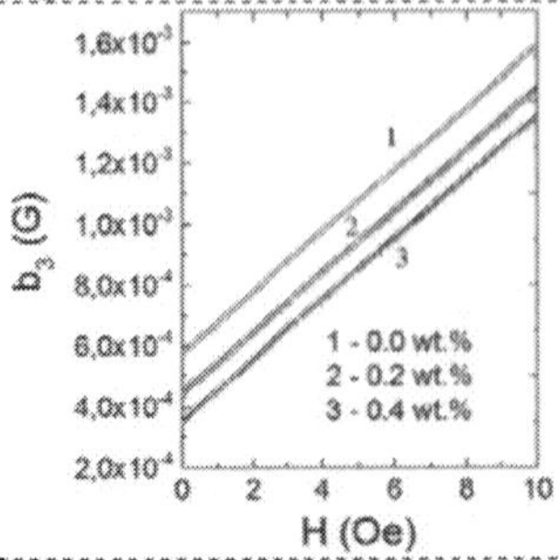

Figure 4. Dependence of the third harmonic on *dc* fields, h=1 Oe, T=78 K

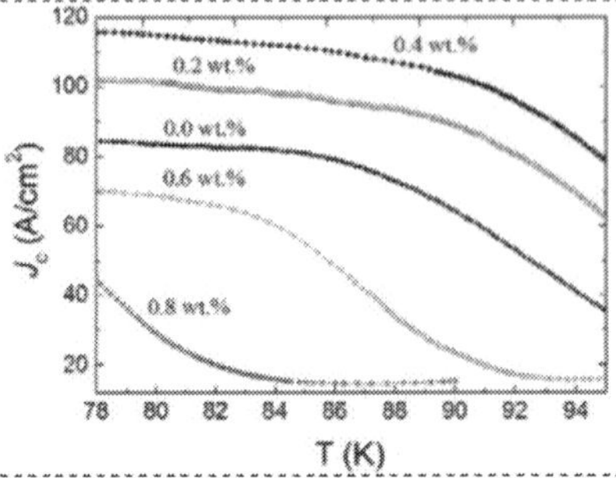

Figure 5. Dependence of the critical current densities on temperature.

4. Conclusion

The effect of Sb_2O_3 doping on Bi-2223 system has been studied by the low fields *ac* susceptibility and the third harmonic response. Temperature dependences of χ' show a two step process, for un-doped and small doped (0.2 wt.%) samples. It is important to note that the high doped samples (0.4-0.8 wt.%) do not show two steps. The absence of two-step behavior could be explained in terms of smaller grain size in the high doping samples, so that grains are fully penetrated at a lower field value. We found that in the low level doped samples, Antimony enhances the value of the transport critical current densities from 85 A/cm^2 (0.0 wt.%) to 115 A/cm^2 (0.4 wt.%). Critical state models are used to explain the nonlinear magnetic response in un-doped and doped samples, with a generalized field dependence of critical current J_c: $J_c(B) = J_0/(1 + B/B_0)$.

5. Acknowledgements

This work has been fulfilled by financial support of the Shota Rustaveli National Science Foundation, Grant No. FR/423/6-260/12. One of the authors (Dr. Ioseb Metskhvarishvili) acknowledges the Shota Rustaveli National Science Foundation (Travel grant #03/137, 01.08.2013) for financial supporting of participation on "11th European Conference on Applied Superconductivity (Eucas-2013)".

References

[1] Kocabas K, Bilgili O andY asar N 2009 J. Supercond. Nov. Magn. **22** 643
[2] Fruth V, Popa M, Ianculescu M A, Stir M, Preda S and Aldica G 2004 J. Eur. Ceram. Soc. **24** 1827
[3] Siddiqi S A and Akhtar B 2003 Mod. Phys. Lett. B **17** 199
[4] Heiras J, Krauss W and Politis C 1990 Int. J. Mod. Phys. B **04** 131
[5] Iqbal M J and Mehmood R 2006 Mater. Sci. Eng. B **135** 166
[6] Hongbao L, Xiaonong Z, Yaozu C et al 1988 Physica C **156** 804
[7] Ginzburg S L, Khavronin V P, Logvinova G Yu et al. 1991 Physica C **174** 109
[8] Metskhvarishvili I R 2009 J. Low Temp. Phys. **155** 153
[9] Metskhvarishvili I R et al 2013 J. Low Temp. Phys. **170** 68
[10] Kim Y B, Hempstead C F and Strnad A R 1962 Phys. Rev. Lett. **9** 306
[11] Anderson P W 1962 Phys. Rev. Lett. **9** 309
[12] Kim Y B, Hempstead C F and Strnad A R 1963 Phys. Rev. **131** 2486

Importance of uniaxial compression for the appearance of superconductivity in $NdO_{1-x}F_xBiS_2$

Omachi A, Hiroi T, Kajitani J, Miura O and Mizuguchi Y

Tokyo Metropolitan University, 1-1, Minami-osawa, Hachioji, 192-0397, Japan

mizugu@tmu.ac.jp

Abstract. We have investigated the crystal structure and superconducting properties of the new layered superconductor $NdO_{1-x}F_xBiS_2$. Bulk superconductivity with a T_c above 4.5 K was observed. It was found that the T_c depended on both F concentration and crystal structure. Uniaxial compression along the c axis upon F substitution seemed to be linked with the appearance of bulk superconductivity. Furthermore, we considered that a higher T_c can be achieved when the c/a parameter was optimized in the $NdO_{1-x}F_xBiS_2$ system.

1. Introduction

Layered materials have been actively studied in the field of superconductivity because superconductors with unconventional paring mechanisms and/or high transition temperature (T_c) had been discovered in layered crystal structures. Recently, we have reported superconductivity in several layered materials possessing a BiS_2-type superconducting layer [1,2]. The crystal structure composed of an alternate stacking of the BiS_2 superconducting layers and blocking layers is quite similar to those of the Cu-oxide and Fe-based superconductors. So far, three structure types of BiS_2-based materials, $Bi_4O_4S_3$, $REOBiS_2$ family (RE = La, Ce, Pr, Nd, Yb) and $SrFBiS_2$, have found to become superconducting upon electron doping into the Bi-6*p* orbitals within the BiS_2 layers [1-9]. Electrical resistivity measurements under high pressure revealed that the T_c of BiS_2-based family was sensitive to application of pressure and can be largely enhanced [10-12] as observed in the Fe-based family [13]. Furthermore, the superconducting properties of $LaO_{1-x}F_xBiS_2$ strongly depended on the sample preparation method; polycrystalline samples prepared with post-annealing at high pressure showed a T_c above 10 K while the sample prepared solid-state reaction in an evacuated quartz tube showed lower T_c of 2.5 K and low shielding volume fraction [14]. These facts indicate that the appearance of a higher T_c requires optimization of local crystal structure in the BiS_2-based superconductors. In this article, we have investigated the crystal structure and superconducting properties of $NdO_{1-x}F_xBiS_2$ synthesized with two preparation temperatures of 700 and 800 °C to clarify crystal structure parameters correlating with the superconducting properties in the BiS_2-based superconductors. We found that uniaxial compression along the c axis is essential for the appearance of bulk superconductivity with a higher T_c in the $NdO_{1-x}F_xBiS_2$ system.

2. Experimental details

The polycrystalline samples of $NdO_{1-x}F_xBiS_2$ (x = 0.1~0.7) were prepared by a solid-state reaction method. Bi grains, Nd_2O_3 powders, Nd_2S_3 powders, BiF_3 powders and Bi_2S_3 powders were used as the starting materials. The Bi_2S_3 powder was synthesized by a direct reaction of Bi grains and

S grains at 500 °C in an evacuated quartz tube. Other chemicals were purchased from Kojundo-Kagaku laboratory. The mixture of the starting materials with nominal compositions of $NdO_{1-x}F_xBiS_2$ (x = 0.1~0.7) was well-mixed, pelletized and sealed into an evacuated quartz tube. The $NdO_{1-x}F_xBiS_2$ pellets were heated at 700 and 800 °C for 15h. The obtained products were ground, sealed into an evacuated quartz tube and heated again with the same heating conditions to homogenize the samples. The obtained samples were characterized by X-ray diffraction using the θ-2θ method. Lattice constants were calculated using the peak positions of the X-ray profile by the least-square method. The temperature dependence of magnetization was measured by a superconducting quantum interface device (SQUID) magnetometer with an applied field of 5 Oe after both zero-field cooling (ZFC) and field cooling (FC).

3. Results and discussion

Figure 1 (a) and (b) show the X-ray diffraction patterns for the $NdO_{1-x}F_xBiS_2$ samples prepared at 700 and 800 °C, respectively. The profiles for the 700 and 800 °C samples exhibit quite similar tendency upon F substitution except for the slight differences in peak shifts corresponding to the changes in lattice constants. For x = 0.2-0.6 the peaks are characterized using the tetragonal space group of *P4/nmm* as indexed with the Miller indices in Fig. 1(a). Almost no impurity peaks are observed for x = 0.2 – 0.5. For $x \geq 0.6$, impurity peaks appear, indicating that the solubility limit of F in the $REOBiS_2$-type structure is near x = 0.5 for this system. Interestingly, the $REOBiS_2$-type structure does not form for $x \leq 0.1$. The estimated lattice constants are plotted in Fig. 2. The a axis does not show a remarkable dependence on F concentration, while the c axis largely decreases with increasing F concentration. Figure 2(c) shows the c/a ratio. It is clear that the uniaxial compression along the c axis is generated upon F substitution.

Figure 3(a) and (b) show the temperature dependence of magnetic susceptibility for $NdO_{1-x}F_xBiS_2$ prepared at (a) 700 and (b) 800 °C, respectively. For the 700 °C samples, superconductivity is observed for $x \geq 0.3$. For the 800 °C samples, superconductivity is observed for $x \geq 0.2$. We estimate two types of transition temperatures, T_c^{onset} and $T_c^{10\,\%}$, to investigate the bulk nature of the samples. The T_c^{onset} is defined as a temperature at which the magnetic susceptibility begins to decrease. The $T_c^{10\%}$ is defined as a temperature at which the value of the ZFC diamagnetic signal corresponds to 10 % of the value at 2 K. This value should represent the bulk properties of the sample because the T_c^{onset} could be affected by local regions with a T_c higher as compared to the bulk T_c.

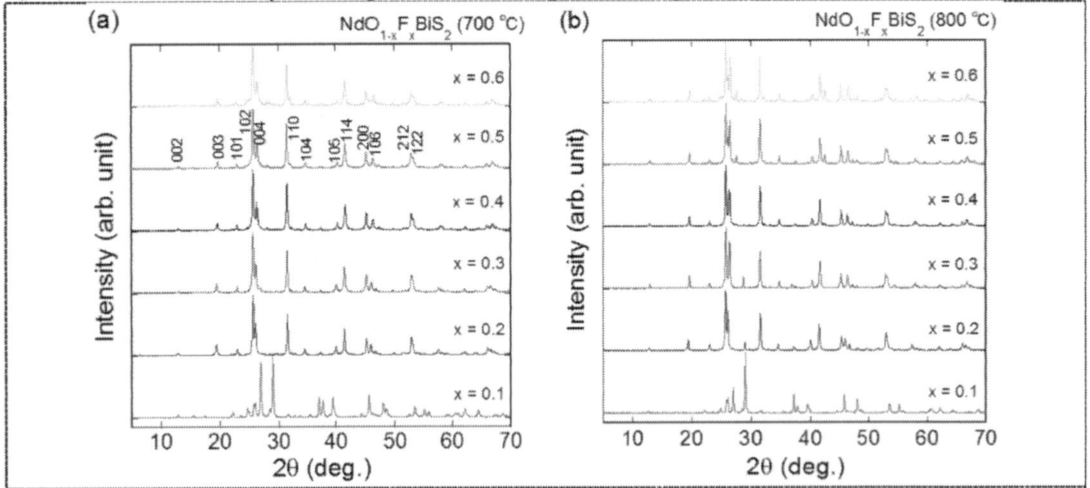

Figure 1. X-ray diffraction patterns for $NdO_{1-x}F_xBiS_2$ prepared at (a) 700 °C and (b) 800 °C. The numbers indicate Miller indices.

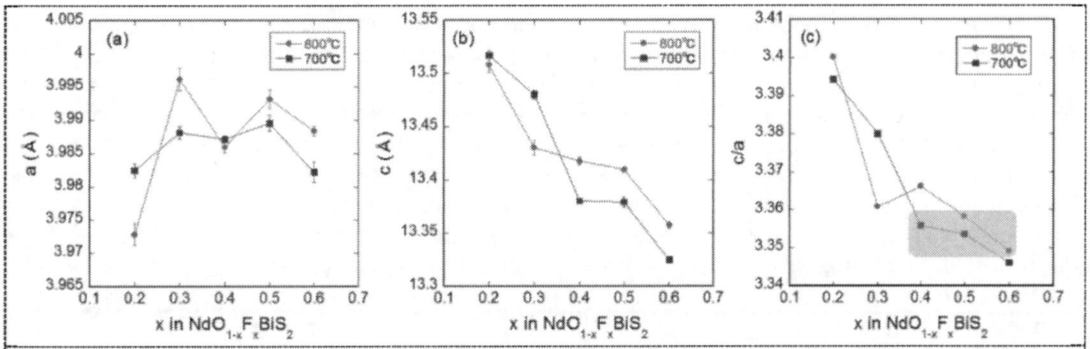

Figure 2. F concentration dependence of lattice constants (a) *a* axis, (b) *c* axis and (c) *c*/*a* ratio for $NdO_{1-x}F_xBiS_2$. The green square indicates the data points whose $T_c^{10\%}$ is around 4.5 K.

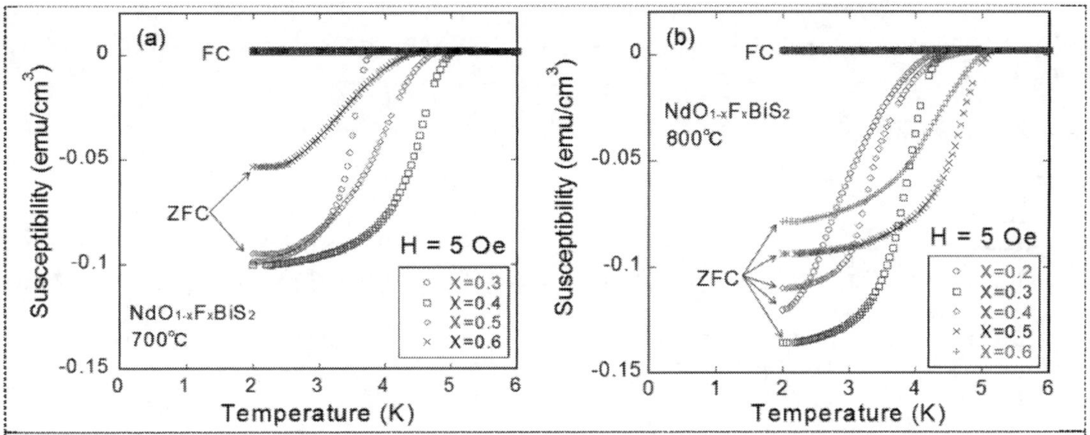

Figure 3. Temperature dependence of magnetic susceptibility for $NdO_{1-x}F_xBiS_2$ prepared at (a) 700°C and (b) 800 °C.

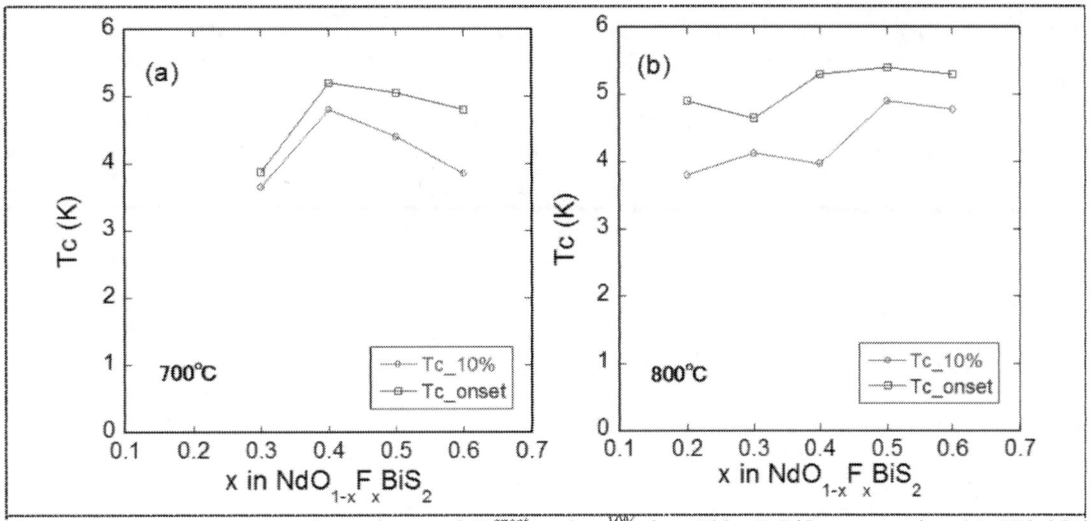

Figure 4. F concentration dependence of T_c^{onset} and $T_c^{10\%}$ for $NdO_{1-x}F_xBiS_2$ prepared at (a) 700 °C and (b) 800 °C.

On the basis of the systematic investigation on the lattice constants and the susceptibility, we find that uniaxial compression seems to be linked with the appearance of superconductivity in $NdO_{1-x}F_xBiS_2$. The higher T_c, with a criterion of $T_c^{10\%} \sim 4.5$ K, is observed for $x = 0.4$ and 0.5 for the 700 °C samples and $x = 0.5$ and 0.6 for the 800 °C samples. We note that the c/a ratio of those higher-T_c samples is located around $c/a \sim 3.35$, as indicated by a green area in Fig. 2 (c). This implies that the optimization of T_c requires the optimization of the c/a ratio in the $NdO_{1-x}F_xBiS_2$ system. Uniaxial compression should affect the local structure parameter such as z coordinate of in-plane S and/or apical S. To analyze which is a local structure which is essential for superconductivity in $NdO_{1-x}F_xBiS_2$, crystal structure analysis using single crystals is required for the next step.

4. Conclusion

We have systematically synthesized the $NdO_{1-x}F_xBiS_2$ polycrystalline samples, and investigated the superconducting properties by X-ray diffraction and the susceptibility measurements. We have found that uniaxial compression should be linked with the appearance of superconductivity in $NdO_{1-x}F_xBiS_2$. Furthermore, it was expected that the optimization of T_c requires the optimization of the c/a parameter, which should be linked to the change (optimization) of the local crystal structure along the c axis.

Acknowledgement

This work was partly supported by Grant-in-Aid for Scientific Research (KAKENHI) and a research fund from The Thermal and Electric Energy Technology Foundation.

References

[1] Mizuguchi Y, Fujihisa H, Gotoh Y, Suzuki K, Usui H, Kuroki K, Demura S, Takano Y, Izawa H and Miura O 2012 *Phys. Rev. B* **86** 220510
[2] Mizuguchi Y, Demura S, Deguchi K, Takano Y, Fujihisa H, Gotho Y, Izawa H and Miura O 2012 *J. Phys. Soc. Jpn.* **81** 114725
[3] Usui H, Suzuki K and Kuroki K 2012 *Phys. Rev. B* **86** 220501
[4] Xing J, Li S, Ding X, Yang H and Wen H H 2012 *Phys. Rev. B* **86** 214518
[5] Demura S, Mizuguchi Y, Deguchi K, Okazaki H, Hara H, Watanabe T, Denholme S J, Fujioka M, Ozaki T, Fujihisa H, Gotoh Y, Miura O, Yamaguchi T, Takeya H and Takano Y 2013 *J. Phys. Soc. Jpn.* **82** 033708
[6] Jha R, Kumar A, Singh S K and Awana V P S 2013 *J. Appl. Phys.* **113** 056102
[7] Lin X, Ni X, Chen B, Xu X, Yang X, Dai J, Li Y, Yang X, Luo Y, Tao Q, Cao G and Xu Z 2013 *Phys. Rev. B* **87** 020504
[8] Yazici D, Huang K, White B D, Chang A H, Friedman A J and Maple M B 2012 *Philosophical Magazine* **93** 673
[9] Lee J, Stone M B, Huq A, Yildrim T, Ehlers G, Mizuguchi Y, Miura O, Takano Y, Deguchi K, Demura S and Lee S H 2013 *Phys. Rev. B* **87** 205134
[10] Kotegawa H, Tomita Y, Tou H, Izawa H, Mizuguchi Y, Miura O, Demura S, Deguchi K and Takano Y 2012 *J. Phys. Soc. Jpn.* **81** 103702
[11] Wolowiec C T, White B D, Jeon I, Yazici D, Huang K and Maple M B 2013 arXiv:1308.1072
[12] Takahashi H, Igawa K, Arii K, Kamihara Y, Hirano M and Hosono H 2008 *nature* **453** 376
[13] Mizuguchi Y, Tomioka F, Tsuda S, Yamaguchi T and Takano Y 2008 *Appl. Phys. Lett.* **93** 152505
[14] Deguchi K, Mizuguchi Y, Demura S, Hara H, Watanabe T, Denholme S J, Fujioka M, Okazaki H, Ozaki T, Takeya H, Yamaguchi T, Miura O and Takano Y 2013 *EPL* **101** 17004

Effects of RE compositional boundaries on fracture strength at 77 K in large single-grained RE-Ba-Cu-O bulk 150 mm in diameter

A Murakami[1], H Teshima[2], M Morita[2], T Kudo[3] and A Iwamoto[4]

[1] Department of Mechanical Engineering, Ichinoseki National College of Technology, Takanashi, Hagisho, Ichinoseki, Iwate 021-8511, Japan
[2] Nippon Steel & Sumitomo Metal Corporation, 20-1 Shintomi, Futtsu, Chiba 293-8511, Japan
[3] Advanced Course of Production Engineering, Ichinoseki National College of Technology, Takanashi, Hagisho, Ichinoseki, Iwate 021-8511, Japan
[4] National Institute for Fusion Science, 322-6 Oroshi-cho, Toki, Gifu 509-5292, Japan

E-mail: amura@ichinoseki.ac.jp

Abstract. Large single-grained RE-Ba-Cu-O high temperature superconducting (HTS) bulk 150 mm in diameter was fabricated by employing the RE compositional gradient technique, where RE denotes rare-earth elements, and fracture strength in the super-large HTS bulk was evaluated at 77 K. No significant difference was observed for the average fracture strength value between the specimens which contained the RE compositional boundaries and the specimens which did not contain the boundaries. The average facture strength values at 77 K were higher than those at room temperature. These results demonstrate that the fracture strengths of the RE compositional boundaries are not degraded by thermal stress in the cooling process.

1. Introduction

Owing to the excellent and unique features such as large current transport capacity in the presence of strong magnetic fields, automatically stable levitation without active control systems and extremely high trapped field ability in compact space, high temperature superconducting (HTS) bulk materials of RE-Ba-Cu-O (RE: Y or rare-earth elements) are highly attractive for the practical application. Recently, super-large single-grained HTS bulk 150 mm in diameter has been successfully produced (Figure 1) [1]. In order to overcome the problem of undesirable nucleation at the position away from the seed for the super-large HTS bulk, a new process based on the difference in the melting temperature between RE elements, that is, the RE compositional gradient technique was employed in the basic melt-growth processing [1,2].

Since HTS bulks are subjected to electromagnetic force and thermal stress in the superconducting devices [3-5], evaluations of the mechanical properties of HTS bulks are important for the development of superconducting devices equipped with HTS bulks. So far, we have investigated the distribution of the mechanical properties in a super-large HTS bulk 150 mm in diameter through three-point bending tests at room temperature for specimens cut from the bulk [6]. No large difference was observed for the mechanical properties between the specimens which contained the RE compositional boundaries and the specimens which did not contain the boundaries [6]. Evaluations of the mechanical properties at cryogenic temperature are informative for the practical application of the super-large HTS bulk. However, the evaluations of the mechanical properties at cryogenic temperature for the super-large HTS bulk have not been carried out extensively [1]. It has not been strictly understood whether the fracture occurs at the RE compositional boundaries or not. In the present study, we investigate the effect of the RE compositional boundaries on the fracture strength at 77 K.

2. Experimental procedure

Evaluations of the fracture strength were carried out through four-point bending tests at 77 K for specimens cut from a super-large single-grained RE-Ba-Cu-O bulk 150 mm in diameter. Specimens were cut from the super-large bulk as shown in Figure 2. The composition ratios of RE-site elements in the precursor of the super-large bulk are also shown. Gd and Dy were chosen as RE. Dy content in the precursor increases with increase of the distance from the seed in order to decrease the peritectic decomposition temperature and suppress the undesirable nucleation at the position away from the seed. The composition ratio of RE-site elements was not changed continuously, but varied step by step, that is, by 5 % in the present study. Molar ratio $REBa_2Cu_3O_x$:RE_2BaCuO_5 of the precursor was 3:1. 10 wt% Ag_2O and 0.5 wt% Pt were added to the precursor. Crystal growth was conducted by using a (Nd,Sm)-Ba-Cu-O seed in air. Two types of specimens with the dimensions of 2.8 x 2.1 x 24 mm^3 were cut from the super-large bulk. Some specimens, denoted as Specimens A1-A5, were cut from the bulk such that the specimens did not contain the RE compositional boundaries between two regions with different Dy content. On the other hand, the other specimens, denoted as Specimens B1-B4, contained the boundaries. These specimens were cut from four layers denoted as 1st-4th layer. Each specimen was placed on the bending test jig and immersed in the liquid nitrogen bath. Four-point bending load was applied as schematically shown in Figure 3 under the crosshead speed of 0.2 mm/min. RE compositional boundary of the Specimens B lies inside the upper loading span where the maximum stress is caused by bending. After the bending tests, RE (Gd and Dy) mapping figures on the 2.1 x 24 mm^2 side surface of

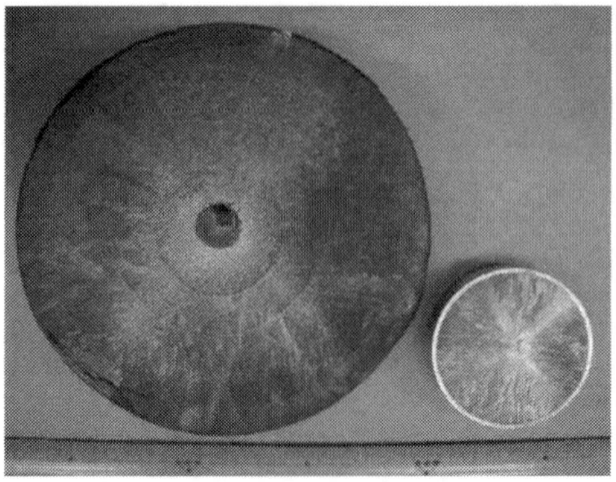

Figure 1. RE-Ba-Cu-O single-grained bulks 150 mm in diameter fabricated by employing RE compositional gradient technique (left) and 60 mm in diameter (right), respectively.

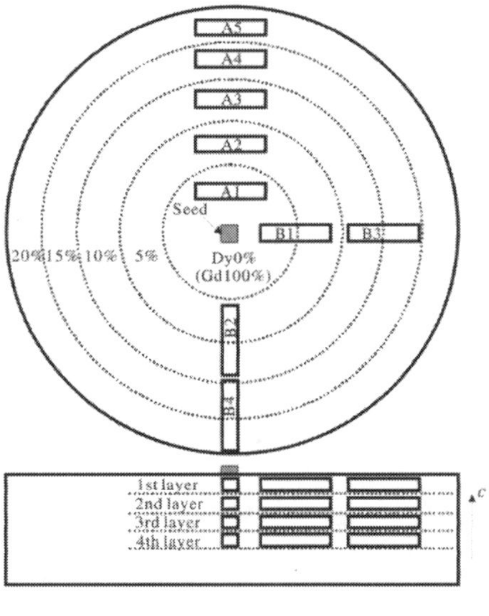

Figure 2. Schematic illustration of cutting out of bending test specimens.

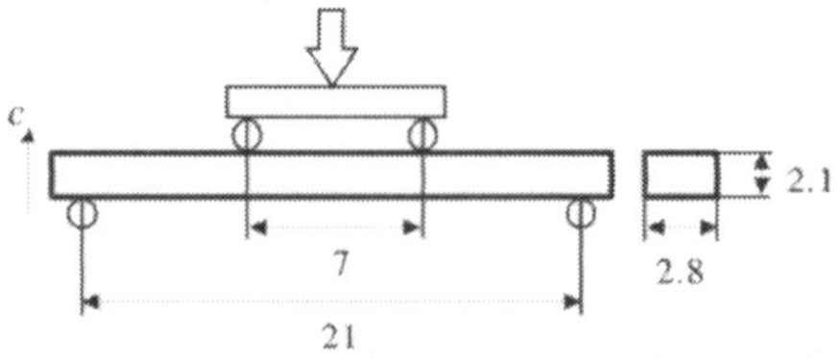

Figure 3. Schematic illustration of four-point bending test.

the fractured specimens were obtained by using micro X-ray fluorescence (μ-XRF) spectrometer.

3. Results and discussion

Fracture strengths of the Specimens A and B evaluated through the four-point bending tests at 77 K are shown in Figure 4. Fracture strengths evaluated at room temperature (RT) are also shown. Specimens for the bending tests at RT were cut from the sites adjacent to the Specimens A1-A3 and B1-B4 of the 2nd and 4th layers. The fracture strength data marked with * was obtained for specimens cut from the 1st layer in our previous study [1]. Few pores were observed for the specimens cut from the 1st layer. On the other hand, the fracture strength data without * was obtained for specimens cut from the 2nd, 3rd and 4th layers in the present study. Porosities of the 2nd, 3rd and 4th layers were 7-14%. The fracture strength values with * are extraordinarily higher than those without *. Higher fracture strength values with * is attributable to the exceptionally low porosity of the specimens cut from the 1st layer. Such a facture strength distribution in the super-large HTS bulk is similar to those in smaller HTS bulks [7,8]. Excluding the exceptional data points with *, the average fracture strength values at 77 K of the Specimens A and B were close to each other. The average fracture strength values at RT of the Specimens A and B were also close to each other. These results are similar to the three-point bending test results at RT obtained in the previous study [6]. Although the crystal growth behavior such as growth rate in the super-large HTS bulk can be changed at the RE compositional boundaries, it was found that the boundaries cause no degradation of the fracture strength. In our previous study, no significant differences in the distributions of RE_2BaCuO_5 particles, Ag particles and pores were observed between the specimens with and without the RE compositional boundaries [6]. Fracture strength values of the super-large HTS bulk at 77 K were higher than those at RT. Such temperature dependence of the fracture strength of the super-large HTS bulk is similar to that of a smaller HTS bulk reported [9]. Improvement of the fracture strength by cooling is consistent with the decrease of inter-atomic distance. These results demonstrate that the fracture strengths of the RE compositional boundaries are not degraded by thermal stress in the cooling process.

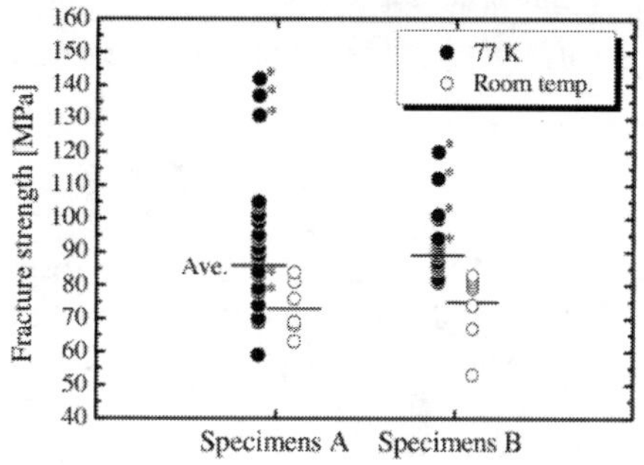

Figure 4. Fracture strength of the Specimens A and B at 77 K and room temperature. Data points for extraordinarily low porosity specimens cut from 1st layer are marked with *. Average values are obtained excluding data points with *.

No degradation of the fracture strength with increase of the distance from the centre or increase of the Dy content was observed for the super-large HTS bulk as shown in Figure 5.

RE (Gd and Dy) mapping figures on the 2.1 x 24 mm^2 side surface of the fractured B1 specimens are shown in Figure 6. Sites where the fracture occurred are marked with arrows. It is observed that the RE compositional boundaries did not always cause the fracture of the specimens, although the fracture of the specimen cut from the 3rd layer occurred at the site adjacent to the boundary. Specimens cut from the 2nd-4th layer contained pores. It is considered that the fracture of these specimens originated from pores or pre-

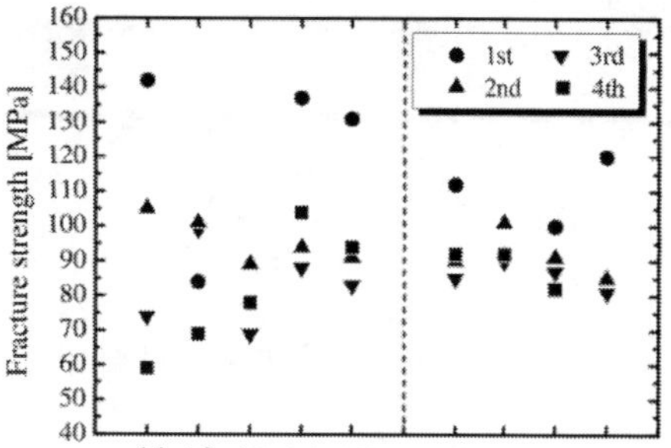

Figure 5. Fracture strengths evaluated through four-point bending tests at 77 K for the Specimens A1-A5 and B1-B4 cut from 1st-4th layer.

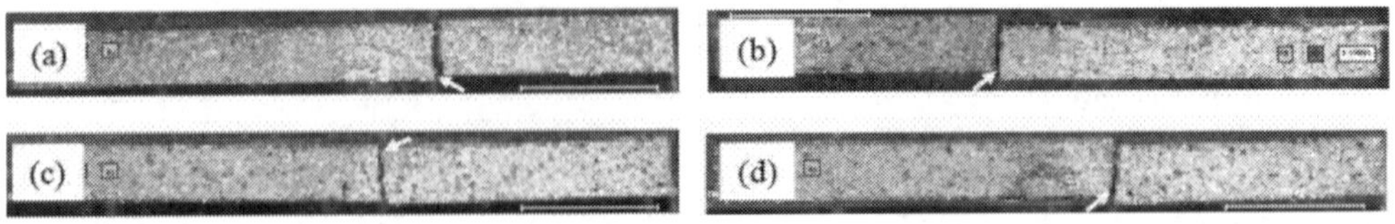

Figure 6. RE (Gd and Dy) mapping figures on the 2.1 mm x 24 mm^2 side surface of the fractured B1 specimens. Red region represents Gd rich and green region represents Dy rich. Sites where the fracture occurred are marked with arrows. Specimens cut from (a) 1st layer (Fracture strength: 112 MPa), (b) 2nd layer (Fracture strength: 90 MPa), (c) 3rd layer (Fracture strength: 85 MPa) and (d) 4th layer (Fracture strength: 92 MPa), respectively.

existing micro-cracks. Since the specimen cut from the 1st layer contained few pores and no significant difference in the fracture strength was observed between the specimens which contained the RE compositional boundaries and the specimens which did not contain the boundaries, it is deduced that the inherent fracture strengths of the RE compositional boundaries and those of non-RE compositional boundary sites are close to each other. While the fractures of the specimens cut from the 1st, 3rd and 4th layers occurred in the Dy rich region, the fracture of the specimen cut from the 2nd layer occurred in the Gd rich region. Although the number of the specimens fractured in the Dy rich region is larger than that in the Gd rich region among the B1 specimens, no degradation of the fracture strength with increase of the Dy content was observed as mentioned above (Figure 5).

4. Summary

Single-grained RE-Ba-Cu-O HTS bulk 150 mm in diameter was fabricated by employing the RE compositional gradient technique. Fracture strength in the super-large HTS bulk was evaluated at 77 K through four-point bending tests for specimens cut from the bulk. Two types of specimens were cut from the bulk: one type was across the boundaries between two regions with different RE compositions and the other was not across the boundaries. There was no significant difference between the average fracture strength values at 77 K of the specimens which had the RE compositional boundaries and specimens which did not contain the boundaries. These fracture strength values at 77 K were higher than those at room temperature. RE mapping revealed that the fracture does not always occur at the RE compositional boundaries. These results demonstrate that the fracture strengths of the RE compositional boundaries are not degraded by thermal stress in the cooling process.

Acknowledgements

We appreciate experimental supports by students of Ichinoseki National College of Technology. This work was supported in part by the NIFS Collaboration Research Program (NIFS13KECA019).

References

[1] Teshima H, Morita M, Okajima N, Terao Y, Sekino M, Ohsaki H and Murakami A 2013 Proc. ICEC 24 - ICMC 2012 879
[2] Morita M, Teshima H and Hirano H 2006 Nippon Steel Tech. Rep. **93** 18
[3] Ren Y, Weinstein R, Liu J, Sawh R P and Foster C 1995 Physica C **251** 15
[4] Miyamoto T, Nagashima K, Sakai N and Murakami M 2000 Supercond. Sci. Technol. **13** 816
[5] Tsuchimoto M and Murata K 2003 Physica C **392-396** 575
[6] Murakami A, Teshima H, Morita M and Iwamoto A 2013 Physica C in press
[7] Nariki S, Sakai N, Murakami M and Hirabayashi I 2004 Physica C **412-414** 651
[8] Fujimoto H and Murakami A 2011 IEEE Trans. Appl. Supercond. **21** 2718
[9] Katagiri K, Nyilas A, Sato T, Hatakeyama Y, Hokari T, Teshima H, Iwamoto A and Mito T 2006 Supercond. Sci. Technol. **19** S545

Journal of Physics: Conference Series **507** (2014) 012035 doi:10.1088/1742-6596/507/1/012035

Scattering theory of the Josephson effect in iron based superconductors

C. Nappi[1,2], S. De Nicola[2], M. Adamo[1,2] and E. Sarnelli[1,2]
[1]CNR-Istituto di Cibernetica "E. Caianiello" - Via Campi Flegrei 34, I80078, Pozzuoli, Napoli, Italy, EU
[2]CNR-SPIN, Complesso Universitario di M. S. Angelo, Via Cintia, I-80126 Napoli, Italy, EU
E-mail: c.nappi@cib.na.cnr.it

Abstract. We study the Josephson effect in $S_{\pm}IS_{\pm}$ junctions made by a two band superconductor with $s_{\pm}$ wave symmetry. We derive the Andreev coefficients for the scattering problem at the junction interface and the temperature dependence of the critical current. We predict various features of the Josephson current for certain values of the second band strengths and tunnel barrier amplitude among which a high temperature π state coupling, and a π to 0 transition as the temperature lowers.

1. Introduction

The Fermi surface of recently discovered iron-based superconductors is composed of several electron and hole sheets. Theory suggests that the superconducting order parameter has opposite signs in the electronlike and holelike bands in these materials ($s_{\pm}$ state or reversed sign s state). Several theoretical models have already analyzed the consequences of such an extendes s-wave symmetry on interface tunneling and also on Josephson junctions [1]. Most of these studies have been focused on junctions between a conventional superconductor and an iron-pnictide with $s_{\pm}$ pairing. Different methods have been proposed to include interband scattering into the microscopic theory of tunneling spectroscopy, which remains still a puzzling issue [2, 3, 4, 5, 6]. Recently, we have studied the Josephson effect in an all-pnictide superconducting junctions ($S_{\pm}IS_{\pm}/S_{\pm}NS_{\pm}$ junctions) [7] by adopting the method used in Ref. [4] for a two-band NS point contact interface. Focusing on the possibility of π-phase shifts formation in the interface of two iron-pnictide intergrains, we obtained results relevant to possible experiments in submicron iron based Josephson junctions and also to possible further theoretical developments. The possibility that the symmetry in these materials be s_{++}, instead of $s_{\pm}$, has not been completely ruled out, at moment. To this end in the present work besides providing details of the calculation method we briefly compare the model predictions for s_{++} and $s_{\pm}$ symmetry material based junctions.

2. A two-band superconductor junction model

We model the $S_{\pm}IS_{\pm}$ (or the $S_{++}IS_{++}$) junction by assuming the following space dependent pair potential for the two bands $j = 1, 2$ along the x coordinate, normal to the junction interface

$$\Delta_j(x) = \Delta_j e^{i\varphi_j}\Theta(-x) + \Delta_j e^{i(\varphi_j+\varphi)}\Theta(x), \quad j = 1, 2 \tag{1}$$

Here $\Theta(x)$ is the Heaviside function, φ is the global phase difference between the two superconductive regions and φ_1 and φ_2 are the 'internal' pair potential phases. In the case of $s_\pm$ wave gap model, $\varphi_2 - \varphi_1 = \pi$, while in the standard two band model, with gaps of the same sign (s_{++} wave, like in MgB_2), $\varphi_2 - \varphi_1 = 0$. We neglect the self-consistency condition for $\Delta_1(x), \Delta_2(x)$, which is equivalent to neglecting the proximity effect. The quasiparticle wave functions, $u_j(x)$ and $v_j(x)$, satisfy the following Bogoliubov de Gennes equations [8]:

$$\begin{aligned}
&\left[-(\hbar^2/2m)(d^2/dx^2) + U(x) - E_F\right] u_j(x) + \Delta_j(x)v_j(x) = Eu_j(x)\\
&\left[(\hbar^2/2m)(d^2/dx^2) - U(x) + E_F\right] v_j(x) + \Delta_j{}^*(x)u_j(x) = Ev_j(x), \qquad j = 1,2
\end{aligned} \qquad (2)$$

where E is the energy of the quasiparticles relative to the Fermi energy E_F. The potentials $U(x) = U_0\delta(x)$ and $\Delta_j(x)$ are responsible for the normal scattering and the scattering of electrons into holes (Andreev scattering), respectively.

3. Wave functions and boundary conditions in the presence of a second band

The two gaps are felt by the quasiparticles as potential wells. To solve the resulting quasiparticle scattering problem, $E > \Delta_1$ (we assume $\Delta_1 < \Delta_2$), in a $S_{+-}IS_{+-}$ or a $S_{++}IS_{++}$ junction, one should write four independent eigenfunctions, one for each elementary scattering process [9], and include the presence of the second band. For instance, for the scattering of electronlike excitations incoming from the left (passage of Cooper pairs to the right), we have [4]

$$\begin{aligned}
\Psi^{S_L}(x) = &\begin{pmatrix} u_1 \\ v_1 e^{-i\varphi_1}\end{pmatrix}\phi_{p_e}(x) + \alpha_0\begin{pmatrix} u_2 \\ v_2 e^{-i\varphi_2}\end{pmatrix}\phi_{q_e}(x) + \\
&a_e\left[\begin{pmatrix} v_1 \\ u_1 e^{-i\varphi_1}\end{pmatrix}\phi_{p_h}(x) + \alpha_0\begin{pmatrix} v_2 \\ u_2 e^{-i\varphi_2}\end{pmatrix}\phi_{q_h}(x)\right] + \\
&b_e\left[\begin{pmatrix} u_1 \\ v_1 e^{-i\varphi_1}\end{pmatrix}\phi_{-p_e}(x) + \alpha_0\begin{pmatrix} u_2 \\ v_2 e^{-i\varphi_2}\end{pmatrix}\phi_{-q_e}(x)\right] \qquad x < 0\\
\Psi^{S_R}(x) = \; &c_e\left[\begin{pmatrix} v_1 \\ u_1 e^{-i(\varphi_1+\varphi)}\end{pmatrix}\phi_{-p_h}(x) + \alpha_0\begin{pmatrix} v_2 \\ u_2 e^{-i(\varphi_2+\varphi)}\end{pmatrix}\phi_{-q_h}(x)\right] + \\
&d_e\left[\begin{pmatrix} u_1 \\ v_1 e^{-i(\varphi_1+\varphi)}\end{pmatrix}\phi_{p_e}(x) + \alpha_0\begin{pmatrix} u_2 \\ v_2 e^{-i(\varphi_2+\varphi)}\end{pmatrix}\phi_{q_e}(x)\right] \qquad x > 0
\end{aligned} \qquad (3)$$

where a_e is the Andreev coefficient, b_e accounts for the normal (specular) reflection of the incoming electronlike excitation, c_e and d_e represent the normal transmission and the anomalous transmission of the excitation in the right electrode as an electronlike excitation or an holelike excitation, respectively. The coefficient α_0 defines the ratio of the probability amplitudes for the incoming electronlike excitation that these four processes occour in the second band q rather than in the first p, with $p \neq q$. The functions $\phi_{p,q}(x)$ represent Bloch waves in the two-band superconductor. The use of Bloch waves, instead of simple plane waves here, is imposed by the obvious circumstance that two plane waves of the same energy would necessarily share the same momentum [4, 7]. u_1 and v_1 and u_2 and v_2 are the standard Bogoliubov coefficients $u_{1,2}^2 = 1/2 + \sqrt{E^2 - \Delta_{1,2}^2/2E}$ and $v_{1,2}^2 = 1/2 - \sqrt{E^2 - \Delta_{1,2}^2/2E}$ while $p_{e,h}$ and $q_{e,h}$ are the Fermi vectors of the two bands (taken up to a reciprocal lattice constant), $p_{e,h}^2 = k_F^2 - k_\parallel^2 \pm 2m\sqrt{E^2 - \Delta_1^2/\hbar^2}$, $q_{e,h}^2 = k_F^2 - k_\parallel^2 \pm 2m\sqrt{E^2 - \Delta_2^2/\hbar^2}$ respectively. The total wave function $\Psi_1 + \alpha\Psi_2$ has to satisfy the following boundary conditions at the interfaces $x = 0$.

$$\Psi_{S_L}(0^-) = \Psi_{S_R}(0^+), \qquad \frac{d\Psi_{S_R}(0^+)}{dx} - \frac{d\Psi_{S_L}(0^-)}{dx} = \frac{2mU_0}{\hbar^2}\Psi_{S_L}(0^-) \qquad (4)$$

Assuming $E > \Delta_1$ [10] and equal band interface velocities, i.e. $v_q = -(i\hbar/m)\phi_q'(0)/\phi_q(0) = v_p = -(i\hbar/m)\phi_p'(0)/\phi_p(0) = v$, [4, 7], plugging (3) in (4) gives the Andreev coefficient a_e (and the coefficients b_e, c_e, d_e). In the same way, consideration of scattering of holelike excitations incoming from the left (passage of Cooper pairs to the left), provides the Andreev coefficient a_h.

4. Josephson current from the scattering coefficients

The total Josephson current, i.e. the Cooper pair flow, is given by [9]

$$I = \frac{e\Delta_1 k_B T}{\hbar} \sum_{\omega_n} \frac{1}{\Omega_{1n}} \left[a_e(\varphi, i\omega_n, \alpha) - a_h(\varphi, i\omega_n, \alpha)\right] \tag{5}$$

Here, $\Omega_{1n} = \sqrt{\omega_n^2 + \Delta_1^2}$, $\omega_n = \pi k_B T(2n+1)$ are the Matsubara frequencies and the detailed balance requires $a_h(\varphi) = a_e(-\varphi)$.

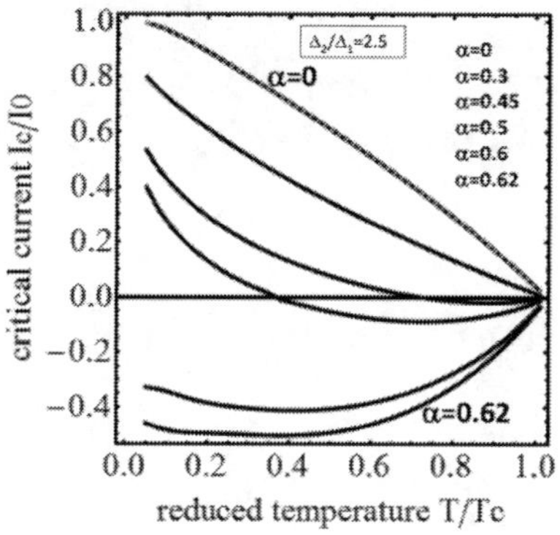

Figure 1. Maximum Josephson current as a function of the temperature for a high transparency ($Z = 0.0001$) $S_{+-}IS_{+-}$ junction. The red line is the clean limit of Kulik-Omelyanchuck.

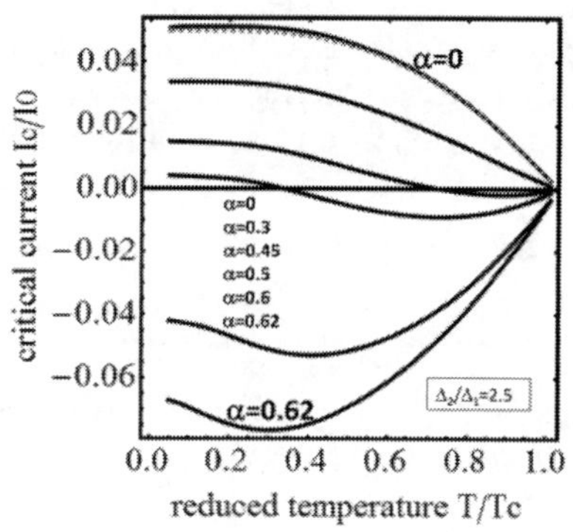

Figure 2. Maximum Josephson current as a function of the temperature for a low transparency ($Z = 3$) $S_{+-}IS_{+-}$ junction. The red line is the Ambegaokar-Baratoff limit (tunnel limit).

4.1. s_{+-} - symmetry

For a $S_{+-}IS_{+-}$ junction, i. e. $\varphi_2 - \varphi_1 = \pi$, Eqs. (3) and (4) provides for a_e the expression

$$a_e(\varphi, E) = \frac{1}{M}\left[A_1 \sin(\varphi/2)\left(A_2 \sin(\varphi/2) - A_3 \cos(\varphi/2)\right)\right] \tag{6}$$

where

$$M = M_1 - M_2 + M_3 \cos\varphi; \quad M_1 = \left(1 + 2Z^2\right)\left(1 + 2\alpha^2\sqrt{(E^2-1)(E^2-r^2)} + r^2\alpha^4\right);$$

$$M_2 = 2E^2\left[1 - \alpha^2 + \alpha^4 + Z^2(1+\alpha^4)\right]; \quad M_3 = 1 + \alpha^2\left(-2E^2 + 2\sqrt{(E^2-1)(E^2-r^2)} + r^2\alpha^2\right);$$

$$A_1 = 2\alpha\left[\sqrt{(E-1)(E+r)} - \sqrt{(E+1)(E-r)}\right] + 2(\alpha^2 r - 1);$$

$$A_2 = E(\alpha^2 - 1); \quad A_3 = i\left(\sqrt{E^2-1} - \alpha^2\sqrt{E^2-r^2}\right);$$

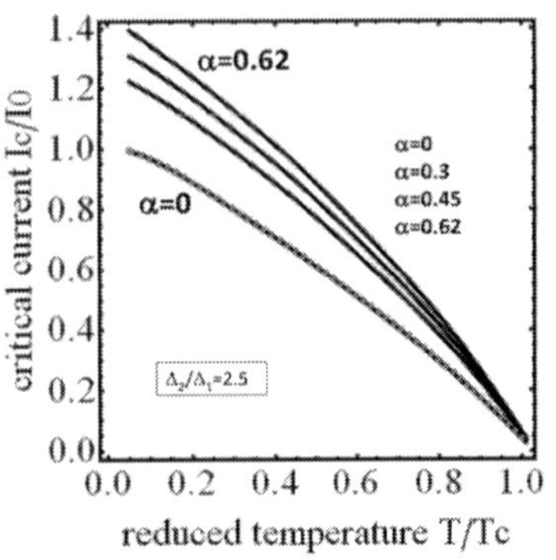

Figure 3. Maximum Josephson current as a function of the temperature for a high transparency ($Z = 0.0001$) $S_{++}IS_{++}$ junction. The red line is the clean limit of Kulik-Omelyanchuck.

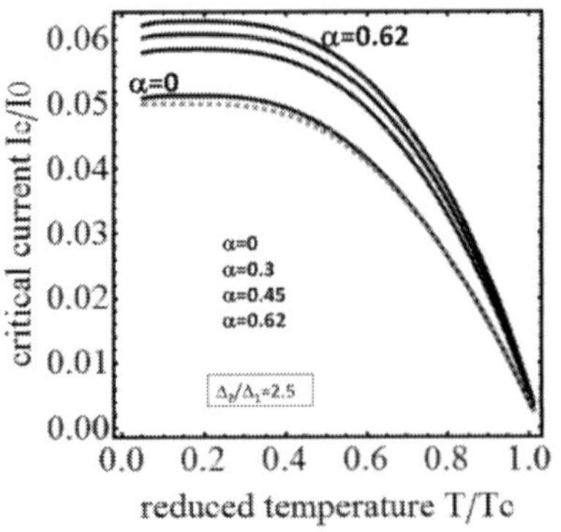

Figure 4. Maximum Josephson current as a function of the temperature for a low transparency ($Z = 3$) $S_{++}IS_{++}$ junction. The red line is the Ambegaokar-Baratoff limit (tunnel limit).

We have introduced here the barrier strength $Z = U_0/\hbar v$ and the second band weight parameter $\alpha = \alpha_0\phi_{q_e}(0)/\phi_{p_e}(0)$ [4],[7]. For values of α such that $0 < \alpha < \sqrt{\Delta_1/\Delta_2}$, a_e has poles which correspond to discrete Andreev states. In this case the supercurrent is essentially determined by the the Andreev discrete spectrum [7]. Figs.(1) and (2) show the maximum Josephson current, in the clean and tunnel limit, respectively, as a function of the reduced temperature T/T_c calculated for different α values in a $S_{+-}IS_{+-}$ junction. The current is in units of $I_0 = e\Delta_1(0)/\hbar$. The gap ratio has been set to $\Delta_2/\Delta_1 = 2.5$. Noteworthy, indipendent from the barrier transparency, negative critical currents appear for sufficiently high values of the α parameter, i.e. when the contribution of the second band has a stronger influence. Negative critical currents indicate the formation of π-state coupling junctions [12, 13, 14].

4.2. s_{++} - symmetry

Figs.(3) and (4) show the model predictions for a $S_{++}IS_{++}$ junction, i.e. $\varphi_2 - \varphi_1 = 0$. In this case too we have assumed the gap ratio $\Delta_2/\Delta_1 = 2.5$. The expression of a_e is more involved and is not reported. Notably, no discrete spectrum exists and only the continuous spectrum contributes to the supercurrent. No negative critical current are predicted for any α value, neither in the clean limit (Fig.(3)) nor in the tunnel limit (Fig.(4)); the inter-band interference effect however makes the current exceed the usual values obtained in a s-wave junction ($\alpha = 0$).

4.3. Conclusions

We have presented a model theory describing the temperature dependence of the Josephson critical current in $S_\pm IS_\pm/S_\pm NS_\pm$ junctions. Starting from the Bogoliubov - de Gennes equations, the Josephso-like current is derived in the presence of double-band superconductor electrodes. Depending on the barrier strength and the range of the second band weight parameter α, a π to 0 crossover is predicted. On the contrary the same analysis for a $S_{++}IS_{++}$ junction points out the total absence of π-state in such structures and no anomalous dependence of the critical Josephson current on the temperature.

Acknowledgments

We thank A. A. Golubov and Y. Tanaka for useful discussions on this topic. The financial contribution of EU NMP.2011.2.2-6 IRONSEA project nr. 283141 is gratefully acknowledged.

References

[1] Seidel P 2011 *Superc. Sci. Technol.* **24** 043001
[2] Burmistrova A V, Devyatov I A, Golubov A A, Yada K, Tanaka Y 2013 *J. Phys. Soc. Jpn.* **82** 034716
[3] Liu C S, Chang J Y, Wu W C, Mou Chung-Yu 2013 J. Phys.: Condens. Matter **25** 365701
[4] Golubov A A, Brinkman A, Tanaka Y, Mazin I I, Dolgov O V 2009 *Phys. Rev. Lett.* **103** 077003
[5] Araújo M A N and Sacramento P D 2009 *Phys. Rev.* B **79** 174529
[6] Sperstad I B, Linder J, Sudbø A 2009 *Phys. Rev.* B **80** 144507
[7] Nappi C, De Nicola S, Adamo M, Sarnelli E 2013 *Europhys. Lett.* **102** 47007
[8] De Gennes J G 1966 *Superconductivity of Metals and Alloys* Benjamin: New York
[9] Furusaki A and Tsukada M 1991 *Solid State Commun.* **78** 299.
[10] No discrete Andreev states exist for $\Delta_1 < E < \Delta_2$. The continuous Andreev spectrum starts with $E > \Delta_1$
[11] Furusaki A 1999 *Superlattices Microstruct.* **25** 809.
[12] Bulaevskii L N, Kuzii V V and Sobyanin A A 1977 *Pis'ma Zh. Eksp. Teor. Fiz.* **25** 314. [JEPT Lett.**25**, 290]
[13] Testa G *et al.* 2004 *Appl. Phys. Lett.* **85** 1202. Testa G, Sarnelli E, Monaco A, Esposito E, Ejrnaes M, Kang D J, Mennema S H, Tarte E J, Blamire M G 2005 *Phys. Rev.* B **71** 134520.
[14] Kashiwaya S, Tanaka Y 2000 *Rep. Prog. Phys.* **63** 1641.

Comparison of critical properties in MgB_2 nanometer films prepared on SiC/Si substrate

A Nishida[1], C Taka[1], S Chromik[2] and R Durny[3]

[1] Department of Applied Physics, Fukuoka University, 8-19-1 Nanakuma, Jonan-ku Fukuoka 814-0180, Japan
[2] Institute of Electrical Engineering, Slovak Academy of Sciences, Dubravska cesta 9, 841 04 Bratislava, Slovak Republic
[3] Department of Physics, Faculty of Natural Sciences, Constantine the Philosofer University, Tr. A. Hlinku 1, 949 01 Nitra, Slovak Republic

E-mail: nishida@cis.fukuoka-u.ac.jp

Abstract. Critical superconducting properties in 100nm MgB_2 film are investigated and examined in comparison with our previous results in 10nm and 50nm films. MgB_2 films were prepared by sequential evaporation of boron and magnesium on SiC-buffered Si substrate followed by in situ annealing. The amount of supplied boron was controlled so as to result in the required MgB_2 film thickness with excess Mg top layer. The superconducting transition temperature T_c is estimated to be 33.1K from the onset of AC diamagnetic susceptibility. The upper critical field H_{c2} is also estimated from the AC susceptibility. Their temperature dependences are almost linear with $|dH_{c2}/dT|$~5.6kOe/K for the parallel magnetic field and $|dH_{c2}/dT|$~3.9kOe/K for the perpendicular field. These values are larger than those in our previous 50nm film (T_c of 34.5K) in spite of a little lower transition temperature. The critical current density J_c and irreversibility field H_{irr} are estimated from DC magnetization hysteresis. Scaling behavior of $H_{irr} \sim [1 - (T/T_c)^2]^n$ is examined in comparision to our previous studies in 10nm and 50nm films with different critical exponents n.

1. Introduction

Investigations of the binary metallic MgB_2 superconductor in the form of thin films are important both from basic as well as application points of view, since the form of thin film can provide variety of different sample conditions and variety of electronics applications [1].

In our previous article [2], nanometer thickness effects were investigated in MgB_2 thin films (10nm thick with the superconducting transition temperature $T_c = 26.0$K and 50nm thick with $T_c = 34.5$K). Under magnetic field perpendicular to the film surface, temperature dependence of the upper critical field H_{c2} was almost linear with $|dH_{c2}/dT|$~3.1kOe/K for the 50nm film, while for the 10nm film overall dependence was about $|dH_{c2}/dT|$~4.8kOe/K yet with apparent positive curvature. The lowest temperature value of H_{c2} for the 10nm film with short coherence length exceeded the one for the 50nm film in spite of much lower T_c. Scaling behavior of the irreversibility field $H_{irr} \sim [1-(T/T_c)^2]^n$ with different critical exponents n was examined in terms of superconducting character.

In this work, we prepare thicker film of 100nm and comparatively investigate superconducting critical properties along with examinations on T_c, H_{c2}, J_c and H_{irr}.

2. Experimental

MgB_2 films were prepared by sequential electron beam evaporation of boron and magnesium on SiC-buffered Si substrate followed by in-situ annealing [3]. The precursor Mg-B bilayers were deposited at room temperature substrate, where the amount of supplied boron was controlled so as to result in film thickness of 100nm. As-deposited film was in-situ heated to 280°C for 30 min in an argon atmosphere at a pressure of 0.06 Pa. Subsequently, the pressure was increased up to 16 Pa and the temperature was increased to maximum temperature of 750°C and kept there for 10 min. The sample was then cooled down to room temperature at 10^3 Pa Ar pressure.

AC and DC magnetizations were measured with magnetic fields parallel and perpendicular to the film using PPMS magnetometer (Quantum Design). The transition temperature $T_{c2}(H)$ was evaluated from the onset of AC diamagnetic susceptibility at each magnetic field H. The critical current density J_c was estimated from DC magnetization hysteresis with the Bean critical state model: $J_c = 30\Delta M/r$, where ΔM is the height of the magnetization loop and r is the sample half-width (about 0.16 cm). Irreversibility field H_{irr} was also estimated from diminishing magnetization hysteresis.

3. Results and discussion

3.1. AC magnetization measurements and upper critical field

Figure 1 shows onsets of AC diamagnetic susceptibility χ' under various magnetic fields perpendicular to the surface of the present 100nm MgB_2 film. The measurements were performed with the field-cooling condition. The superconducting transition temperature T_c is estimated to be 33.1K at minimal magnetic field. This is a little lower than T_c of 34.5K in our previous 50nm film. Decrease of T_c suggests some more impurities or defects mingled by chance into the present 100nm film. From onset transition temperatures $T_{c2}(H)$ as indicated by arrows, we can estimate temperature dependent upper critical fields $H_{c2}(T)$. Diamagnetism was much weaker with magnetic field parallel to the film surface.

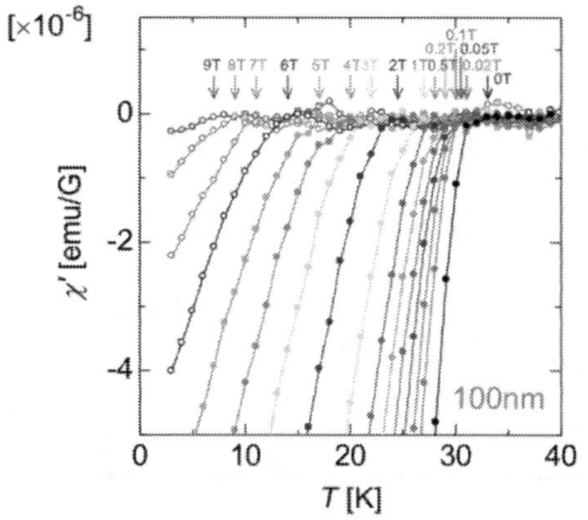

Figure 1. AC susceptibility χ' under various magnetic fields perpendicular to the surface of 100nm MgB_2 film.

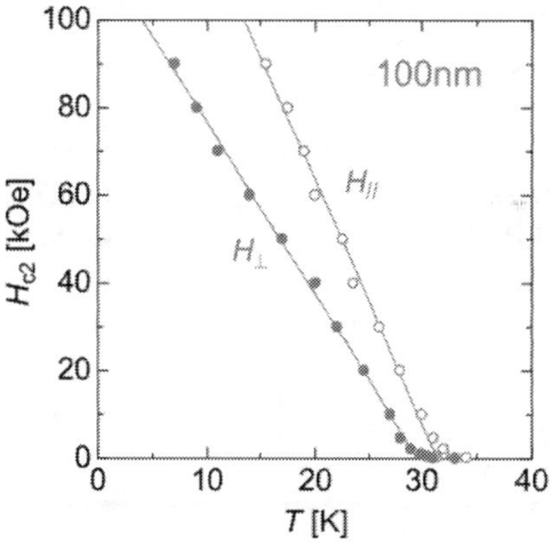

Figure 2. Upper critical field H_{c2} for perpendicular (solid circles) and parallel (open circles) fields.

Thus estimated $H_{c2}(T)$ is indicated in figure 2 as a function of temperature T for the perpendicular (solid circles) and parallel (open circles) magnetic fields. As seen in the figure, temperature dependence is almost linear except for just below T_c. The temperature coefficient $|dH_{c2}/dT|$ is about 5.6kOe/K for the parallel magnetic field and 3.9kOe/K for the perpendicular field. These values are larger than those in our previous 50nm film in spite of a little lower transition temperature for 100nm film. This infers that the coherence length ξ is shorter in the present film, which is possibly due to more impurities or defects reducing mean-free path along with the dirty limit.

Almost linear temperature dependence in $H_{c2}(T)$ is consistent with the behavior in the previous 50nm film and absence of distinct positive curvature indicates that the present 100nm film (as well as 50nm film) is in the category of three-dimensional superconductors (though anisotropic).

The anisotropy parameter γ can be estimated from the temperature coefficient ratio between parallel and perpendicular H_{c2} as $\gamma = \xi_{//} / \xi_{\perp} = |dH_{c2}^{//}/dT| / |dH_{c2}^{\perp}/dT|$. In table 1 we summarize critical superconducting parameters evaluated in our MgB_2 films of 10nm, 50nm and 100nm thickness.

Table 1. Summary of critical superconducting parameters evaluated in our MgB_2 films with different thickness.

Thickness (nm)	T_c (K)	$\lvert dH_{c2}^{//}/dT\rvert$ (kOe/K)	$\lvert dH_{c2}^{\perp}/dT\rvert$ (kOe/K)	γ
10	26.0	5.6	4.8	1.2
50	34.5	4.9	3.1	1.6
100	33.1	5.6	3.9	1.4

The very low T_c of 26.0K as well as the very small value of γ in 10nm film probably reflects anomalous two-dimensional character (granular superconductivity) as discussed previously [2].

A little smaller γ in 100nm film than 50nm can be explained by the presence of more defects or impurities as mentioned above. In our previous report [2], we discussed possibility of limitation of coherence length by the crystal grain size. Because the present films are not single crystal, grain sizes would be less than at least 1/10 of the film dimensions and almost the same between 50 and 100 nm films. Thus, the perpendicular coherence length $\xi_{\perp}$ may be limited by the perpendicular grain size of a few nanometer and almost unchanged, while the parallel coherence length $\xi_{//}$ may not be limited by larger grain sizes in the layer direction and then mingled impurities may result in decrease of $\xi_{//}$ as well as γ in 100nm film.

3.2. Critical current density and irreversibility field

Critical current density J_c in 100nm MgB_2 film is given in figure 3 as a function of magnetic field H perpendicular to the film, where J_c was deduced from DC magnetization hysteresis and Bean model.

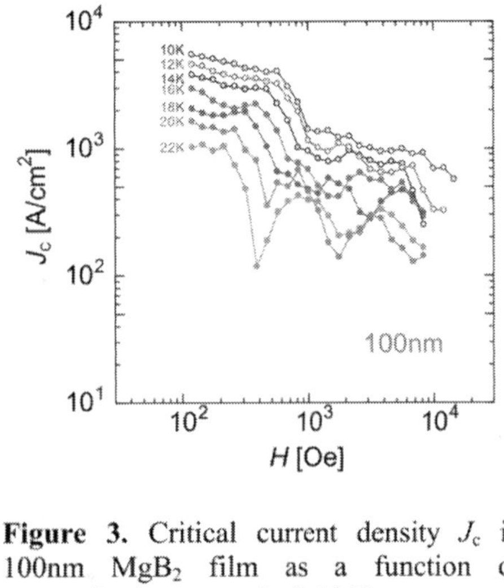

Figure 3. Critical current density J_c in 100nm MgB_2 film as a function of perpendicular magnetic field H.

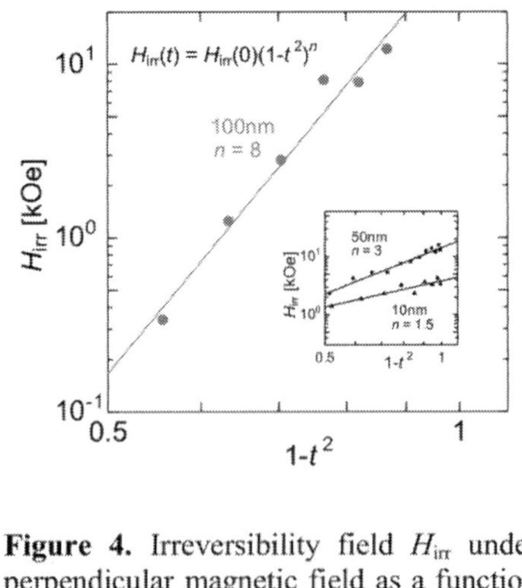

Figure 4. Irreversibility field H_{irr} under perpendicular magnetic field as a function of $1-t^2$ with $t = T/T_c$.

Obtained J_c is rather small compared with maximum values of 0.6MA/cm^2 and 5MA/cm^2 in the previous 10 and 50 nm films, respectively. The reason is not entirely clear at present, though above mentioned defects or impurities may obstruct coupling between crystal grains and reduce J_c values throughout the whole sample dimension.

We next estimate the irreversibility field H_{irr} based on vanishing criterion of J_c=300A/cm^2 (about 1/20 of the maximum observed J_c). Figure 4 indicates variation of H_{irr} in 100nm film under perpendicular magnetic field as a function of 1-t^2, where $t = T/T_c$. In the inset of the figure we also represent the former results in 10 and 50 nm films [2]. For high-T_c superconductors H_{irr} is often examined in relation to the scaling law as $H_{irr}(T) = H_{irr}(0)\ [1-(T/T_c)^2]^n$ [4, 5]. The straight lines in figure 4 correspond to the least square fitting to the data based on this scaling equation. From such fitting, the critical exponent n is deduced to be about 8 for 100nm film in contrast with 3 and 1.5 for 50nm and 10nm films. The n value seems to increase with increasing thickness of the MgB_2 film.

The difference in n values reflects difference in the pinning mechanism. Matsushita *et al.* [6] examined the pinning mechanism in melt-textured Bi-2212 superconductor and discussed that smaller n value corresponds to flux pinning due to non-superconducting inclusions, while larger n value corresponds to the pinning due to small defects such as point defects or dislocations. Thus, the large value of 8 for our 100nm film infers that the main pinning centers in this film are defects or impurities, whose presence was repeatedly suggested in the above discussion in §*3.1*. The irreversibility field at 0K, $H_{irr}(0)$, can also be derived from the least square fitting to the data, the results being 46kOe, 14kOe and 3.8kOe, respectively, for 100nm, 50nm and 10nm films. The largest value of $H_{irr}(0)$ in 100nm film is again explained by the presence of many defects in this film.

4. Conclusion

Thick film of 100nm MgB_2 was prepared and comparatively investigated on various critical superconducting properties along with examinations on T_c, H_{c2}, J_c and H_{irr}. MgB_2 films were prepared by sequential electron beam evaporation of boron and magnesium followed by in-situ annealing.

Comparing our critical parameters in three different MgB_2 films, 100nm thickness resulted in anisotropic three-dimensional superconductor with rather many defects or impurities, showing small J_c and large $H_{irr}(0)$. Such impurities may come from rather long evaporation time needed for accumulating the very thick B and Mg layers. On the other hand, the smallest thickness of 10nm resulted in anomalous two-dimensional superconductivity with the largest H_{c2} but the smallest H_{irr}. After all, as far as we prepared, the 50nm thickness seems to result in the best film, indicating fair values of H_{c2} and H_{irr} along with the highest T_c as well as the largest J_c.

Acknowledgements

This work was supported in part by the Central Research Laboratory, Fukuoka University (Project #115003), and measurements were performed at the Advanced Materials Institute Facilities established at Fukuoka University under the support from the Ministry of Education, Culture, Sports, Science and Technology.

References

[1] Nishida A, Taka C, Chromik S and Durny R 2009 *Journal of Physics: Conference Series* **150** 052186

[2] Nishida A, Taka C, Chromik S and Durny R 2011 *Proceedings of ICEC 23 – ICMC 2010 Wrocław, Poland*, 1249-1254

[3] Chromik S, Nishida A, Strbík V, Gregor M, Espinos J P, Liday J and Durny R 2013 *Applied Surface Science* **269** 29-32

[4] Matsushita T, Nakatani T, Otabe E S, Ni B, Umemura T, Egawa K, Kinouchi S, Nozaki A and Utsunomiya S 1993 *Cryogenics* **33** 251

[5] Nishida A, Taka C, Chromik S and Durny R 2012 *Physics Procedia* **36** 644-648

[6] Matsushita T, Fujiyoshi T, Toko K and Yamafuji K 1990 *Applied Physics Letter* **56** 2039-2041

Iron isotope effect in the iron arsenide superconductor $(Ca_{0.4}Na_{0.6})Fe_2As_2$

Y Tsuge[1], T Nishio[1], A Iyo[2,3], Y Tanaka[2], and H Eisaki[2,3]

[1]Department of Physics, Tokyo University of Science, 1-3 Kagurazaka, Shinjuku, Tokyo 162-8601, Japan
[2]National Institute of Advanced Industrial Science and Technology, Tsukuba, Ibaraki 305-8568, Japan
[3]Transformative Research Project on Iron Pnictides (TRIP), Japan Science and Technology, Chiyoda, Tokyo 102-0075, Japan

E-mail: nishio@rs.kagu.tus.ac.jp

Abstract. We report a new sample synthesis technique for polycrystalline $(Ca_{1-x}Na_x)Fe_2As_2$ $(0<x\leq0.6)$ and measurements of an iron isotope effect in optimally doped $(Ca_{0.4}Na_{0.6})Fe_2As_2$ with three types of iron isotopes (^{54}Fe, natural Fe, and ^{57}Fe). We synthesized isotope samples carefully not to give rise to a difference in the Na content x between different isotope samples, which becomes potentially a factor for an extrinsic difference in the superconducting transition temperature T_c between those samples. No significant difference in lattice parameters between those samples is shown by measurements of powder x-ray diffraction (XRD), implying that the Na content in samples is well-controlled. Our estimate of the iron isotope coefficient α_{Fe} defined by $-d\ \ln T_c/d\ \ln M_{Fe}$, where M_{Fe} is the iron isotope mass, is -0.19. These indicate that in $(Ca_{0.4}Na_{0.6})Fe_2As_2$, the iron isotope coefficient becomes definitely negative. We discuss the implications of this fact, considering previous measurements of an iron isotope effect in different iron-based superconductors.

1. Introduction

The recent discovery of superconductivity at 26 K in $LaFeAsO_{1-x}F_x$ [1] inspired an effort to find a new high-temperature superconductor containing iron, followed by finding of a new class of high-temperature superconductors, the so-called iron-based superconductors. The highest superconducting transition temperature T_c in iron-based superconductors has reached to 56 K until now [2]. Whether the mechanism of superconductivity in iron-based superconductors is identified with the conventional phonon mechanism or not is a matter for argument. Several studies [3-5] suggest that superconductivity at the high T_c cannot be explained by conventional electron-phonon interactions but antiferromagnetic spin fluctuations, which evokes an analogy between cuprates and iron-based superconductors. However cuprates are much more anisotropic and have d-wave gap symmetry, while iron-based superconductors show primarily s-wave gap symmetry (conventional superconductors have typically s-wave gap symmetry) [6-8]. We are still uncertain about the mechanism of superconductivity in iron-based superconductors. In order to investigate the implications of electron-phonon interactions for the mechanism of superconductivity, it is helpful to study an isotope effect on T_c. The isotope coefficient α defined by $-d\ln T_c/d\ln M$, where M is the isotope mass, is 0.5 in conventional superconductors in principle, actually depending on Coulomb interactions between electrons. In many simple metals like Hg, Pb, and Zn, isotope coefficients are found to be nearly 0.5. For a binary alloy superconductor, MgB_2 (T_c=40 K), α of 0.3 is found in a boron isotope effect study [9]. It demonstrates that the electron pairing is primarily phonon-mediated in MgB_2, though the mechanism was argued owing to relatively higher T_c than in conventional superconductors. For iron-based superconductors, several groups have reported on an iron isotope effect. Shirage and co-workers show that α_{Fe}= -0.18 [10] and -0.024 [11] in $(Ba_{0.6}K_{0.4})Fe_2As_2$ (T_c=38 K) and $SmFeAsO_{1-x}$ (T_c=54 K), respectively. Tsuge and co-workers [12] show α_{Fe}= -0.54 in $FeSe_{1-x}Te_x$ (x=0.65), i.e, the so-called inverse isotope effect. On the other hand, Liu and co-workers [13] show that α_{Fe}=0.37 and 0.34 in $(Ba_{0.6}K_{0.4})Fe_2As_2$ (T_c=38 K) and $SmFeAsO_{1-x}F_x$ (T_c=42 K), respectively. Khasanov and coworkers [14] report α_{Fe}=0.81 in $FeSe_{1-x}$. There are non-negligible discrepancies in α_{Fe} values between measurements for compounds of the same type. Since T_c of these superconductors strongly depends on differences in the K content, O content, Te content, or Se content, the reason for the discrepancies might be such a compositional difference between isotope samples synthesized separately. Thus, in different iron-based superconductors, we would like to investigate the iron isotope effect for isotope samples all of which have no difference in the composition.

In this paper, we show a new sample synthesis technique for polycrystalline $(Ca_{1-x}Na_x)Fe_2As_2$ ($0<x\leq0.6$) not to give rise to a difference in the Na content x and measurements of the iron isotope effect on T_c in samples synthesized by this technique. A negative isotope effect has been definitely observed by using several sets (S1, S2, and S3) of three iron isotope samples (^{54}Fe, natural Fe, and ^{57}Fe). We discuss the implications of this fact, considering the previous measurements of the iron isotope effect mentioned above.

2. Experimental

We used three types of Fe: ^{54}Fe enriched powder (ISOFLEX USA, purity 99.99%, M_{Fe}=54.01), ^{57}Fe enriched powder (ISOFLEX USA, purity 99.99%, M_{Fe}=57.00), and natural Fe powder (M_n=55.85, including 5.8% of ^{54}Fe, 91.8% of ^{56}Fe, and 2.1% of ^{57}Fe). In a new technique, precursors of CaAs, NaAs, and As powder were mixed and divided into two or three equal weights, resulting in two or three mixed precursors with the same Na content. Then the mixed precursors were mixed with ^{54}Fe, ^{57}Fe and natural Fe powder respectively. The mixtures were pressed into pellets. The two pellets were sandwiched with other two $(Ca_{0.4}Na_{0.6})Fe_2As_2$ pellets with natural Fe. These pellets were put into a sealed stainless steel tube and then heated at 750℃ for 24 h. The Na content in samples is decreased somewhat by vaporization of Na. However, Na vaporized in outer pellets can compensate the loss of the Na content in samples. Powder x-ray diffraction measurements were carried out by using CuK_α radiation. All the samples have a single phase and differences in lattice parameters between two samples synthesized simultaneously are negligible. The unit cell lattice parameters of both samples

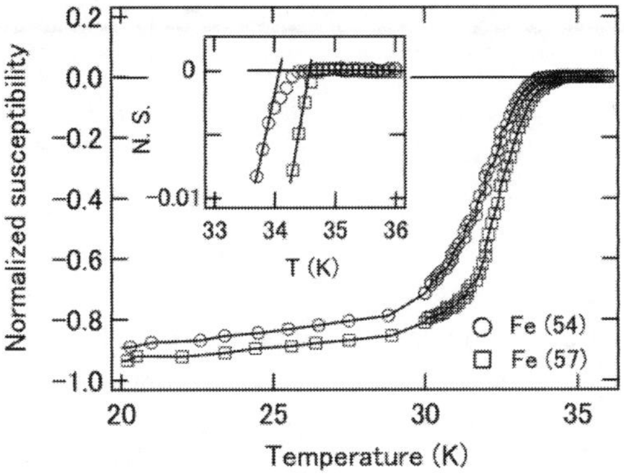

Figure 1 Temperature dependence of magnetic susceptibility for S1 isotope samples (^{54}Fe and ^{57}Fe) of $(Ca_{0.4}Na_{0.6})Fe_2As_2$. The susceptibility is normalized by its value at 5 K. The data of the ^{54}Fe and ^{57}Fe samples are shown by circular and square symbols. The inset shows the determination of T_c. The natural Fe data almost overlaps the ^{57}Fe data, so that it is not shown here.

can be estimated as a = 3.845 Å and c = 12.19 Å, indicating that a Na content difference between the two samples is negligible. The dc magnetic susceptibility was measured by using a SQUID magnetometer (MPMS, Quantum Design) under a magnetic field of 5 Oe for the determination of T_c.

3. Results and discussion

Figure 1 shows temperature dependence of magnetic susceptibility for S1 isotope samples (^{54}Fe and ^{57}Fe) of $(Ca_{0.4}Na_{0.6})Fe_2As_2$. The susceptibility is normalized by its value at 5 K. The susceptibility curves exhibit the sharp superconducting transition, indicating our samples are of high quality. Susceptibility curves similar to those are obtained for S2 and S3 isotope samples. No significant difference in the superconducting volume fraction is observed in each combination. The onset T_c is determined from the definition as shown in the inset of Fig. 1. Isotope shifts in T_c are clearly seen, showing that a sample with larger isotope mass has higher T_c in S1. It has been confirmed that a factor for an extrinsic difference in T_c between different isotope samples was ruled out by powder x-ray diffraction measurements. It is considered that the observed isotope shifts in T_c are intrinsic.

Figure 2 shows T_c as a function of the iron isotope mass for different isotope sample sets. Apparently, a sample containing larger isotope mass has higher T_c in each combination. Small differences in T_c between ^{54}Fe samples as well as ^{57}Fe samples are presumed to be due to some differences in the Na content x, according to our powder x-ray diffraction measurements. Isotope samples synthesized separately must have some differences in the composition, which results in nontrivial changes in T_c. This is an issue of isotope sample preparation for iron-based superconductors. Our results are highly reproducible and evidently demonstrate that α_{Fe} becomes negative. The average of α_{Fe} about all of sample sets is -0.19, which is consistent with α_{Fe} in $(Ba_{0.6}K_{0.4})Fe_2As_2$ of the same type as $(Ca_{0.4}Na_{0.6})Fe_2As_2$ [10].

Figure 3 shows α_{Fe} values as a function of T_c obtained in various measurements [10-14]. There is a remarkable difference in α_{Fe} between samples of the same type: $FeSe_{0.35}Te_{0.65}$ and $FeSe_{1-x}$. It is hard to explain why the α_{Fe} values show such a large difference if both values are intrinsic. We focus on negative values of α_{Fe} obtained by Shirage *et al.* [10, 11] and Tsuge *et al.* [12], since isotope samples which were prepared carefully to rule out any compositional difference were used in their measurements. Apparently, α_{Fe} decreases with decreasing T_c, which means that T_c in low-T_c iron-based superconductors is more isotope mass dependent than in high-T_c ones. Although several factors including anharmonic phonons, a pair breaking effect, and van Hove singularity of the density of states may be speculated as the reason why α_{Fe} becomes negative, we notice a factor other than these: a two-band multi-channel effect [15]. Yanagisawa and co-workers [15] calculate negative α_{Fe} values, using a two-band model where electron-phonon interactions and antiferromagnetic spin fluctuations in each band as well as the interband interactions are included. In the model, electron-phonon interactions affect T_c; T_c decreases and the negative contribution to α_{Fe} increases as the

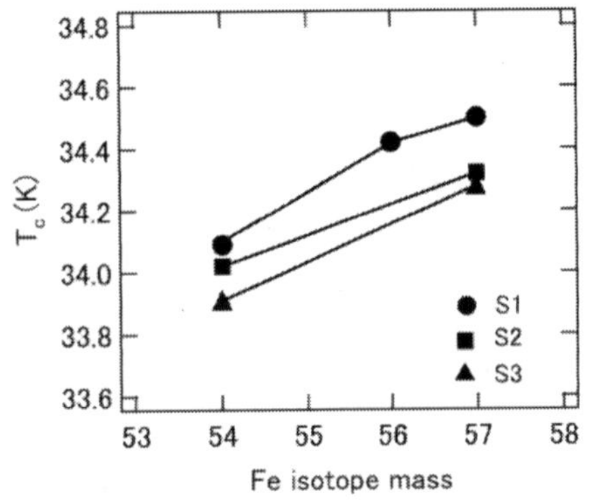

Figure 2 Isotope mass dependence of T_c for the three sets of isotope samples S1, S2, and S3. The average of α_{Fe} is -0.19.

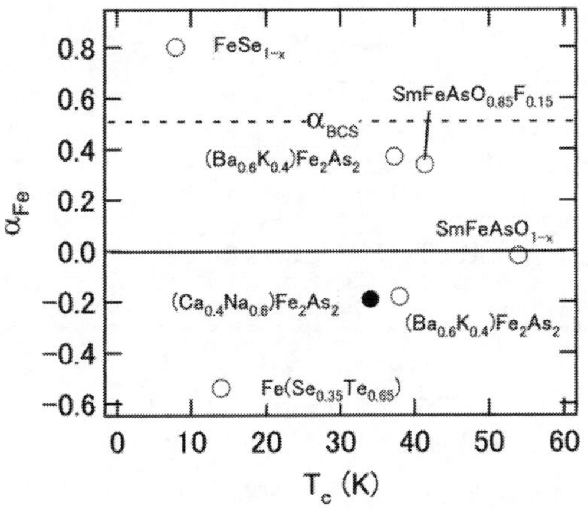

Figure 3 α_{Fe} versus T_c for various measurements : $SmFeAsO_{1-x}$ (Ref. 11), $SmFeAsO_{1-x}F_x$ (Ref. 13), $(Ba_{0.4}K_{0.6})Fe_2As_2$ (Ref. 10 and 13), $FeSe_{0.35}Te_{0.65}$ (Ref. 12), and $FeSe_{1-x}$ (Ref. 14). A value of α_{BCS} is 0.5 for phonon-mediated superconductivity.

intraband electron-phonon interactions become weaker. Actually, band structure calculations [16] show that electron-phonon interactions in the FeSe system are weaker than in the LaFeAsO system. Thus, it is likely that the isotope mass dependence of T_c can be explained by the two-band model.

In conclusion, an iron isotope effect has been observed in $(Ca_{0.4}Na_{0.6})Fe_2As_2$ in which the Na content is well-controlled. The iron isotope coefficient is estimated to be -0.19, which is consistent with a value [10] observed in $(Ba_{0.6}K_{0.4})Fe_2As_2$. Definitely, α_{Fe} becomes negative and decreases with decreasing T_c in iron-based superconductors. It seems like that this behaviour originates from two-band superconductivity in which the electron-coupling is mediated by the multi-channel interaction in each band.

References

[1] Kamihara Y *et al.* 2008 *J. Am. Chem. Soc.* **130** 3296.
[2] Wang C *et al.* 2008 *Europhys. Lett.* **83** 67006.
[3] Boeri L *et al.* 2008 *Phys. Rev. Lett.* **101** 026403.
[4] Christianson A D *et al.* 2008 *Nature* **456** 930.
[5] Imai T *et al.* 2009 *Phys. Rev. Lett.* **102** 177005.
[6] Hashimoto K *et al.* 2009 *Phys. Rev. Lett.* **102** 017002.
[7] Evtushinsky D V *et al.* 2009 *New J. Phys.* **11** 055069.
[8] Nakayama K *et al.* 2009 *EuroPhys.Lett* **85** 67002.
[9] Bud'ko S L *et al.* 2001 *Phys. Rev. Lett.* **86** 1877.
[10] Shirage P M *et al.* 2009 *Phys. Rev. Lett.* **103** 257003.
[11] Shirage P M *et al.* 2010 *Phys. Rev. Lett.* **105** 037004.
[12] Tsuge Y *et al.* 2012 *Physics Procedia* **36** 731.
[13] Liu R H *et al.* 2009 *Nature* **459** 64.
[14] Khasanov R *et al.* 2010 *New J. Phys.* **12** 073024.
[15] Yanagisawa T *et al.* 2009 *J. Phys. Soc. Jpn.* **78** 094718.
[16] Boeri L *et al.* 2009 *Physica C* **469** 628.

Systematic Enhancements of Switching Rate in Intrinsic Josephson Junctions

Y. Nomura[1], T. Mizuno[1], H. Kambara[1], Y. Nakagawa[1], T. Watanabe[2], I. Kakeya[1] and M. Suzuki[1]

[1]Department of Electronic Science and Engineering, Kyoto University, Japan
[2]Graduate School of Science and Technology, Hirosaki University, Japan

E-mail: nomura@sk.kuee.kyoto-u.ac.jp

Abstract. We measure the switching probability distributions of the first and second switching in the stack of $Bi_2Sr_2CaCu_2O_{8+\delta}$ and $Bi_2Sr_2Ca_2Cu_3O_{10+\delta}$ intrinsic Josephson Junctions. The saturation temperature of the first switching in Bi2212 is ~ 2 K. This temperature is almost same in Bi2223. It is a little higher than crossover temperature estimated from thermal activation. On the other hand, in the second switching, the saturation temperature in Bi2212 is ~ 10 K, while it is ~ 2 K in Bi2223. Surprisingly, the high saturation temperature is observed only in Bi2212, although the difference of the two materials is only the thickness of the superconducting layer. This result suggests that the high saturation temperature of the second switching in the stack of $Bi_2Sr_2CaCu_2O_{8+\delta}$ is caused by capacitive coupling effects.

1. Introduction

High-T_c superconductors $Bi_2Sr_2Ca_{n-1}Cu_nO_{4+2n+\delta}$ (BSCCO) are highly anisotropic and their transport properties along the c-axis are described as stacks of weakly coupled Josephson junctions with $(CuO_2)_n$ layers being superconducting electrodes and Bi_2O_2 layers being barrier layers. These systems are referred as intrinsic Josephson junctions (IJJs) in contrast to artificially stacked multi-layered Josephson junctions. In IJJs, inductive (magnetic) coupling between stacked Josephson junction is essential because the separation between Josephson junctions are much shorter than the London penetration depth [1]. Furthermore, capacitive (charge) coupling is relevant because charge conservation within the superconducting layers does not hold due to the short Thomas-Fermi screening length ~ 1 nm being comparable to the interlayer separation [2]. These interactions emerges in phase dynamics in stacked IJJs observed as switching properties of connected IJJs in series.

The current-biased single Josephson Junction can be well described by the RCSJ (resistively and capacitively shunted junction) model, which depicts the phase dynamics as motion of a phase particle in a tilted washboard potential. The macroscopic quantum tunneling (MQT) is described as the quantum tunneling of the phase particle through the potential barrier due to Josephson plasma oscillation. The most sensitive measure for the phase dynamics is the switching probability distribution (SPD), the histogram of current where a Josephson junction switches from superconducting ($R = 0$) to resistive state ($R \neq 0$). MQT in Bi2212 IJJ has been observed as low-temperature saturation of the SPD width below 1 K, which is one order of magnitude higher than that of a single Josephson junction made by metallic superconductors like

Nb/AlO$_x$/Nb [3]. This raised crossover temperature between MQT and thermal activation (TA) has been explained by high critical current density j_c of Bi2212 IJJ. The crossover temperature can be estimated by $\sim \hbar\omega_p/2\pi k_B$, where ω_p is the current dependent plasma frequency which is roughly consistent with $\sqrt{j_c}$, and k_B is the Boltzmann constant.

More interestingly, it has been reported that the SPD width saturates around $\sim$ 10 K for the second switching, where another IJJ included in the stack switches to resistive state subsequent to the first switching [4, 5]. The origin of the further enhancement is possibly attributed to interaction between adjacent IJJs, which can be described by coupled phase particles in a tilted washboard potentials. However, heating artifact cannot be excluded from the possible origins of the enhancement.

In this paper, we present systematic difference on the first and second switchings of the stack of $Bi_2Sr_2Ca_1Cu_2O_{8+\delta}$ (Bi2212) and $Bi_2Sr_2Ca_2Cu_3O_{10+\delta}$ (Bi2223). The thicker superconducting layer in Bi2223 (0.65 nm) than that in Bi2212 (0.3 nm) may cause weaker coupling between adjacent IJJs. The crossover temperature of the first switching in the stack of Bi2212 and the temperature of Bi2223 are almost identical. However, the saturation temperature of the second switching in Bi2212 is much higher temperature than that in Bi2223. The significant difference of the second switching is considered as a direct evidence of the charge coupling with nm-length scale rather than the inductive coupling.

2. Experimental method

We fabricated mesa structures including a few stacked IJJs on Bi2212 and Bi2223 single crystal with the sizes of 1 $\times$ 1 μm^2 by using electron-beam lithography and Ar-ion milling. The details of mesa-fabrication are described elsewhere [6]. The mesa structure has advantages against the Z-type bridge structures in the following reasons. Heating in the IJJs is removed by the thick Ag electrode covering the top of the IJJs. The first switching definitely takes place at the topmost IJJ among the stacked IJJs because the critical current of the topmost IJJ is suppressed due to the proximity effect. The second switching is presumably induced in the adjacent IJJ, the second topmost IJJ in the mesa, when the inter-IJJ is taken into consideration.

Switching probability distributions were measured with following system. First, send a start signal to the counter and the constant current source. Second, the ramp current flows to the IJJs from the zero-voltage to the first or second resistive state. A stop signal is sent when the junction switches. The counter measures the elapsed time from the start signal to the stop signal. Switching current is calculated by multiplying the measured time and the ramp rate of the bias current. To reduce external noise, we used optical fibers which separates the constant current source and IJJs from the AC supply. We prepared home-made constant current source driven by DC stabilized voltage source to avoid external noise and the influence of changing resistance by switching. This current source consists of integrating circuit and voltage controlled current source.

Effective temperature ($T_{\rm eff}$) is a parameter representing fluctuation of the switching and is derived by fitting the TA model to the experimental data. The saturation temperature T_s is the highest temperature of the temperature independent region in the plot of $T_{\rm eff}$ versus bath temperature ($T_{\rm bath}$) measured by temperature sensor.

3. Results and discussion

Figure 1 shows I-V characteristic and $T_{\rm eff}$ vs $T_{\rm bath}$ plots for the Bi2212 mesa. The critical current densities are found to be 2.0 kA/cm^2 and 3.0 kA/cm^2 for the first and second switchings, respectively. The saturation temperature T_s of the first switching in Bi2212 is $\sim$2 K, which is slightly higher than the crossover temperature estimated from j_c, which is $\sim$ 0.72 K in Bi2212. Figure 2 represents the similar plots for the Bi2223 sample. In (a), j_cs are found to be 0.7 kA/cm^2 and 0.8 kA/cm^2 for the first and second switchings, respectively. T_s of the first switching

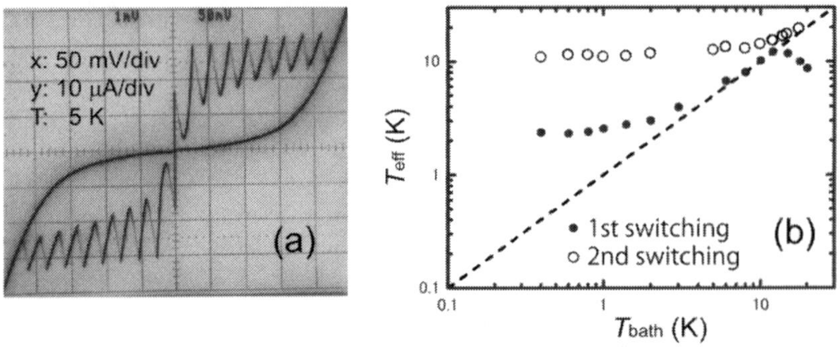

Figure 1. (a) I-V characteristic in Bi2212 at 5 K. (b) Bath temperature (T_{bath}) versus estimated temperature from thermal activation model (T_{eff}) in Bi2212.

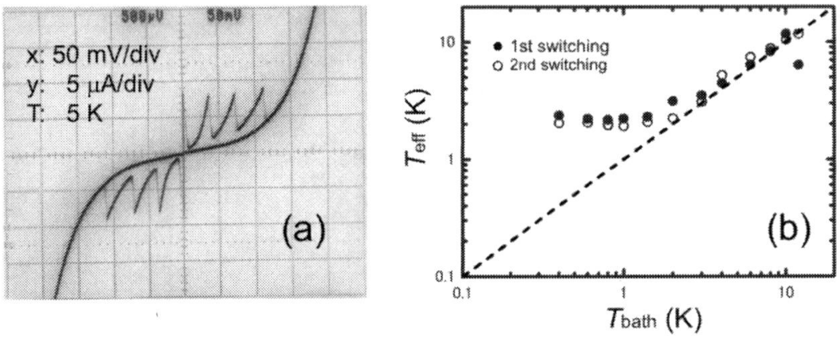

Figure 2. (a) I-V characteristic in Bi2223 at 5 K. (b) Bath temperature (T_{bath}) versus estimated temperature from thermal activation model (T_{eff}) in Bi2223.

is approximately 2 K, slightly higher than the estimated crossover temperature $\sim$ 0.54 K. These results shows that the first switching is roughly described by the single junction model and there exists no essential difference between Bi2212 and Bi2223.

The second switching indicates essential difference between Bi2212 and Bi2223. T_s of the second switching in Bi2212 is $\sim$ 10 K, one order of magnitude higher than T_s of the first switching, as well as the previous reports [4, 5], and T_{eff} starts to merge to the TA model at T_{bath} = 10 K. This is in sharp contrast to the result in Bi2223, in which T_{eff} of the second switching is almost identical to T_{eff} of the first switching except above 10 K, where the phase diffusion takes place for the first switching. Increase in j_c of the second switching in Bi2212 is not sufficient to explain the enhancement of T_s because the crossover temperature is given by $\sqrt{j_c}$ according to the single junction model. Moreover, the T_s enhancement of the second switching in Bi2212 has been reported in stacks with uniform j_cs [5]. The possibility of the heating artifact is also excluded because the self heating and thermal properties are identical for the Bi2212 and Bi2223 mesas. Therefore, the present results clarify that the enhancement of T_{eff} observed in Bi2212 is a result of coupling between IJJs. Switching of the second topmost IJJ of the stack is induced by the running state of the phase particle of the first topmost IJJ.

The enhancement of T_s is observed only in Bi2212 although the difference between the two materials is nothing but the presence of a planar CuO_2 superconducting layer inserted between CuO pyramids. It is surprising that only 0.35 nm increase in the thickness of superconducting electrode completely depresses the enhancement of the second switching in the MQT state.

This drastic change within atomic length-scale suggests that the phase dynamics between two adjacent IJJs are coupled by the capacitive coupling. The enhancement of the second switching is the unique phenomenon in which the capacitive coupling plays dominant role as far as our knowledge.

4. Conclusion

We measured SPDs of the first and second switching in the stack of Bi2212 and Bi2223. The first switching shows the similar behavior in Bi2212 and Bi2223. However the second switching is strongly enhanced in Bi2212 in comparison with the first switching, whereas in Bi2223 the second switching is essentially same as the first switching. It is considered that the enhancement of the second switching is attributed to the capacitive coupling effect because of the drastic depression by the presence of the third CuO_2 layer.

Reference

[1] Sakai S, Bodin P and Pedersen N F 1991 *J. Appl. Phys.* **73** 2411
[2] Koyama T, and Tachiki M 1996 *Phys. Rev.* B **54** 16183
[3] Inomata K, Sato S, Nakajima K, Tanaka A, Takano Y, Wang H B, Nagao M, Hatano H and Kashiwaya S 2005 *Phys. Rev.* Lett. **95** 107005
[4] Kashiwaya H, Matsumoto T, Shibata H, Kashiwaya S, Eisaki H, Ysohida Y, Kawabata S and Tanaka Y 2008 *Phys. Soc. Jpn* **77** 104708
[5] Ota K, Hamada K, Takemura R, Ohmaki M, Machi T, Tanabe K, Suzuki M, Maeda A and Kitano H 2009 *Phys. Rev.* B **79** 134505
[6] Kakeya I, Hamada K, Tachiki T, Watanabe T and Suzuki M 2009 Supercond. Sci. Technol. **22** 114014

Journal of Physics: Conference Series **507** (2014) 012039 doi:10.1088/1742-6596/507/1/012039

Magnetic Flux-Trapping of Anisotropic-Grown Y-Ba-Cu-O Bulk Superconductors during and after Pulsed-Field Magnetizing Processes

T Oka[1], Y Yamada[1], T Horiuchi[1], J Ogawa[1], S Fukui[1], T Sato[1], K Yokoyama[2], M Langer[3]

[1] Niigata University, 8050 Ikarashi-2-nocho, Nishi-ward, Niigata 950-2181, Japan

[2]Ashikaga Institute of Technology, 268-1 Ohmae-cho, Ashikaga, Tochigi 326-8558 Japan

[3]IFW Dresden, Helmholzstr. 20-01069 Dresden, Germany

E-mail: okat@eng.niigata-u.ac.jp

Abstract. The magnetic flux penetration into the melt-textured Y-Ba-Cu-O high temperature superconducting bulk magnets were precisely evaluated during and after the pulsed field magnetization processes operated at 30 K. The bulk magnets were carefully fabricated by the cold seeding method with use of a single and a pair of seed crystals composed of the Nd-Ba-Cu-O thin films. These seed crystals were put on the top surfaces of the precursors to let the large grains grow during the heat treatments. We observed the flux penetrations which occurred in the lower applied-field regions at around 3.1 T for the samples bearing the twin seeds than those of the single-seeded crystals at around 3.8 T. This means that the magnetic fluxes are capable of invading into the twin-seeded samples more easily than the single-seeds. It suggests that the anisotropic grain growths of parallel and normal to the rows of seed crystals affects the variations of J_c values with different distributions of the pinning centers, results in the preferential paths for the invading magnetic fluxes.

1. Introduction

The high temperature superconducting (hereafter abbreviated as HTS) materials fabricated by so-called melt-processes have a unique characteristic to act as the trapped-field magnets or bulk magnets [1-3]. Some investigations including our previous studies have been performed on the pulsed-field magnetization (PFM) methods to develop more compact and easier processes than the field cooling (FC) method [4, 5]. The fact that the field-trapping performance on the PFM process is usually inferior to those by the FC method is attributed to the local heating which is mainly caused by the motion of the applied magnetic flux in the superconducting phases in the sample [6, 7].

The keys for high field-trapping are the temperature controlling [8] and the consecutive magnetic pulsed-field application in the magnetizing processes [9]. Fujishiro *et al.* reported the highest performance of 5.2 T with the precise analysis on the trapped-field distributions throughout the activation processes [10, 11]. The trapped-field distributions which remain in the bulk magnet before the following activation affect the magnetic flux motion in the sample and resultant field-trapping [12].Therefore, it is definitely important to clarify the behaviors of the invading flux into the bulk magnets during the PFM processes in order to enhance their field-trapping ability.

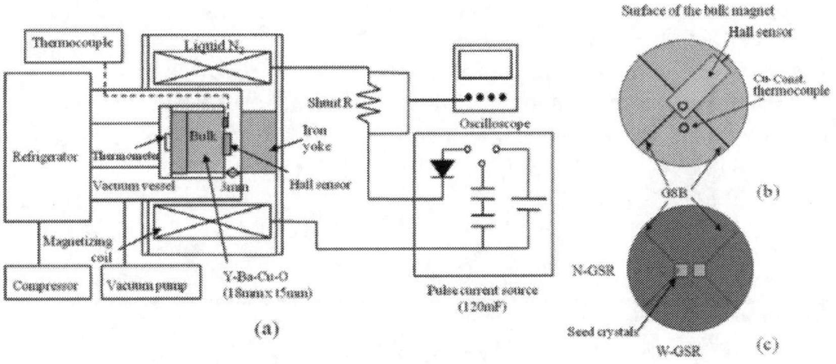

Fig. 1. Schematic illustration of the bulk magnet system (a), the positions of sensors on the sample surface (b), and the positions of EPMA analysis as a function of distance from the centre along the narrow-GSR and wide-GSR directions (c).

The melt-processed bulk magnets show the heterogeneous microstructures according to the presence of non-superconducting Y_2BaCuO_5 (Y211) particles which must contribute to enhance the J_c values [13]. In this paper, the bulk magnets bearing the anisotropic structure were fabricated so as to examine how the field-penetration occurs when the intense magnetic pulsed-fields are applied to the samples in their superconducting states. We attempt to examine the flux motions by discussions on the data of the trapped-flux densities.

2. Experimental

2.1 Sample preparation and structural analysis

The starting powers of $YBa_2Cu_3O_{7-\delta}$ (Y123) and Y211were mixed in the proper ratios with 1 wt% CeO_2 addition and pressed into the pellets. The Y211 nominal content was chosen as 28.5% in molar ratio. The precursors were heat-treated in the cold seeding method with the Nd-Ba-Cu-O thin films on the MgO crystal substrates (THEVA Co.), and heated to 1050 °C and then gradually solidified with the cooling rate 0.25 °C/h. The post annealing was operated in the temperature range of 500 – 300 °C for 167 hours in the flowing pure oxygen atmosphere. The compositional analysis was operated with use of the electron probe micro analyzer EPMA (JEOL Ltd, JXA-8621MX).

2.2 Pulsed-field magnetization procedure

An illustrated structure of the total system is shown in Fig. 1(a) with the electrical circuit which is equipped with a condenser bank of 120 mF (Nihon Denji Sokki Co., SBV-10124), a shunt resister, and an oscilloscope. Figure 1(c) shows the points of EPMA compositional analyses, which were operated along the directions parallel and normal to the row of seed crystals. We defined the direction along the row of seed crystals as narrow side of the growth sector region (abbreviated as n-GSR) and as wide side (w-GSR) which is normal to it.

Sample	#1	#2	#3	#4	#5
Seed crystal	-	Gap 2.5mm	twin	twin	twin
Growth period	80h	80h	80h	80h	60h

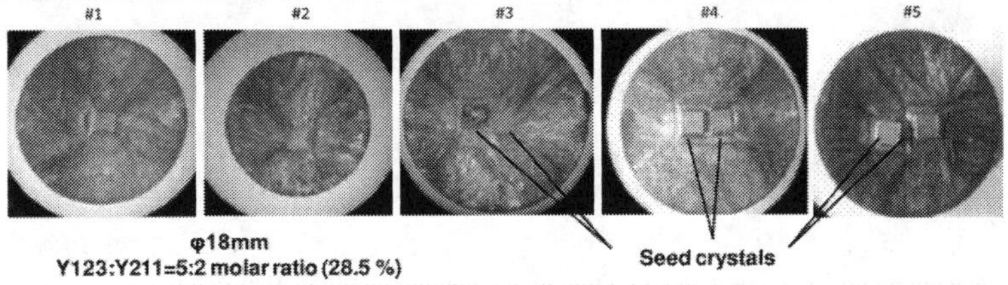

Fig. 2. Anisotropic-Grown Y-Ba-Cu-O Bulk Superconductors with various seed-crystal positions.

3. Results and discussions

3.1 Anisotropic-grown HTS bulk magnets

Figure 2 shows the topside-views of the bulk magnet samples, which were fabricated by the melt-process with varying the positions and numbers of seed crystals. The sample size is approximately 15 mm in diameter. One sees the apparent growth sector boundaries (GSB), which indicate their characteristic "X-shape". We note that #1 sample is the conventional single-seeded sample, #2 is the sample bearing the single-seed whose position was displaced 2.5 mm distance from the center, and #3 – 5 are twin-seeded crystals which exhibited the anisotropic crystal growth. The GSB lines for #3 – 5 samples divide the anisotropic-grown GSR in the narrow and the wide portions along the direction parallel to the rows of seed crystals and normal to them, respectively. Since the GSB lines starts at the corner of the seed crystals, the neighboring corners draw no GSB lines. Therefore, we see the narrow GSRs along the direction parallel to the row of seed crystals and the wide GSRs along the direction normal to it, as shown in Fig. 1(c).

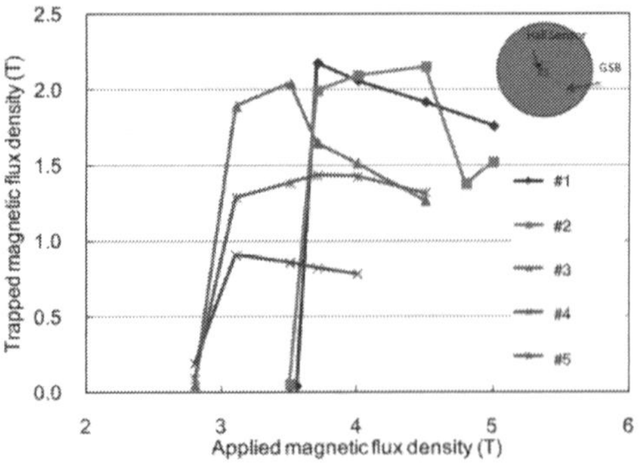

Fig. 3. Experimental results of trapped flux of HTS bulk samples with various seeding patterns after PFM.

3.2 Trapped-fields in the bulk magnets after and during PFM process

The trapped magnetic flux densities which were measured at the center of each pulsed-field application are shown in Fig. 3. One sees that the starting points of the field invasions varied with each samples. We notice that the profiles can be separated in two groups. The magnetic flux did not invade into the center of the sample until when the applied field exceeded over 3.5 T in the single-seeded crystal (#1 and 2), whereas the samples #3 – 5 easily allowed the flux invasion even when the applied field of 3 T was applied. This implies that the anisotropic grain growth structure led by the twin seeds must result in the different flux invasion behaviors into the samples. We should pay attention to the fact that the highest trapped-field of #3 and #1 are comparable and showing a similar degradation behavior caused by the heat generation after showing each peak.

Figure 4 shows the evolutional profiles of magnetic flux invasion in sample #6 twin-seeded crystal as a function of applied field. Although the speeds of the invading fluxes were similar in any magnitudes of fields, the moments when the fields started to invade moved forward with increasing applied fields. Figure 5 shows the times for the fluxes to arrive at 0.5 T from the beginnings of invasion. The profiles of arriving times are roughly separated into two groups, and the magnetic flux invades earlier and in the weaker field ranges in the twin-seeded crystals than single-seeded. This means that the introduction of the flux was promoted in the anisotropic-grown samples.

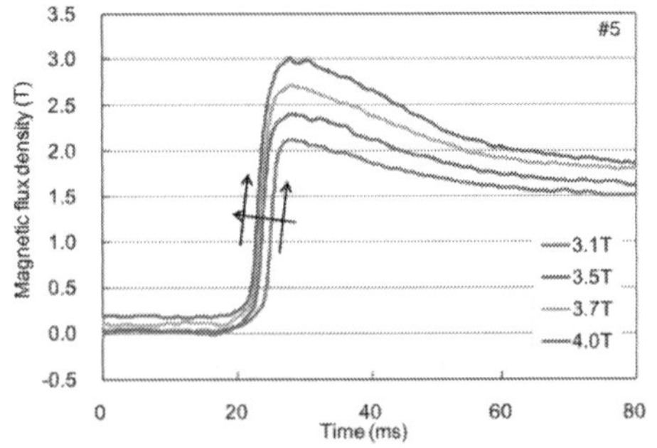

Fig. 4. Time evolutional profiles of magnetic flux invasion as a function of applied field strength.

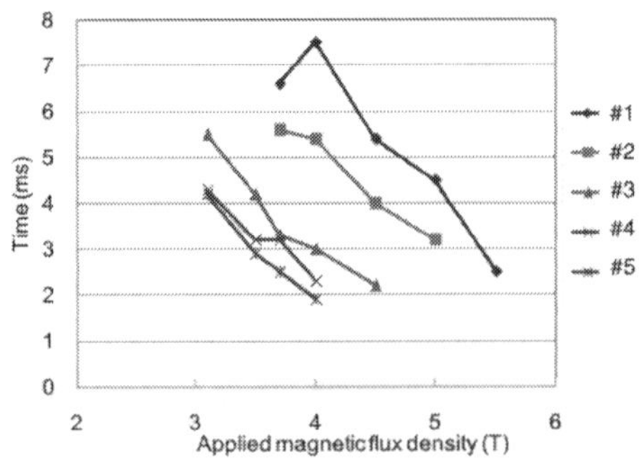

Fig.5. Arriving times of the invading flux at 0.5 T at the center point of the sample.

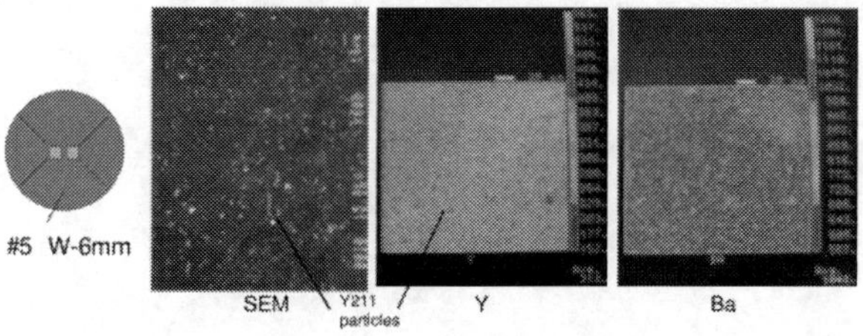

Fig.6. SEM imaging and EMPA mapping of Y and Ba detected at 6 mm distant from the center along w-GSR direction.

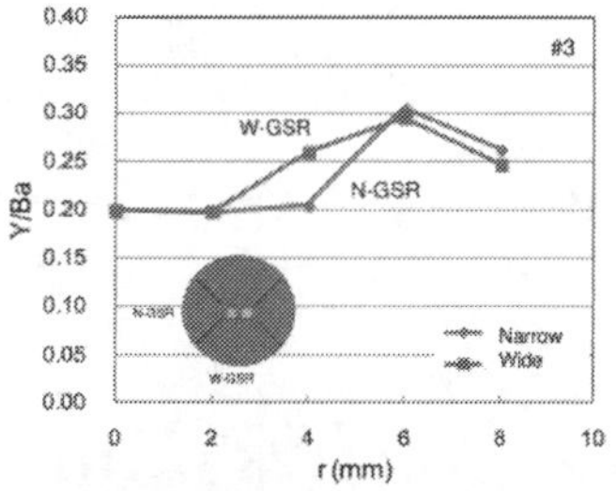

Fig. 7. Distribution of the detected characteristic X-ray intensity ratios Y/Ba along the n- and w-GSR directions.

3.3 Compositional analysis

We operated the compositional analysis on the sample with use of EPMA. Figure 6 shows the SEM image and mapping of compositions Y and Ba at the point 6 mm distant from the center along the w-GSR. One sees the uniformly dispersed Y211 particles and the mapping approves it. It is known that the Y211 particles do not precipitate under the regions where the seed crystals are placed due to the pushing effect at the boundaries of melt and solid phases. The characteristic X-ray signals should exhibit the varieties of their distributions. Figure 7 shows the intensity ratios of Y and Ba measured at various points along the directions n-GSR and w-GSR. We observed a distinct difference in Y/Ba at the position 4 mm distant from the centre point. This implies the anisotropic distributions of Y211, and suggests the different J_c distribution between the directions along n- and w-GSRs.

4. CONCLUSION

The field-trapping behaviors of the anisotropic-grown Y-Ba-Cu-O-based bulk magnets with use of single, displaced, and twin-seeded crystals were evaluated with respect for the trapped-flux density during and after PFM processes. The magnetic flux invades the twin-seeded sample earlier and easier than single-seeded crystals. The inhomogeneous distribution of Y211, which is introduced in the material structure by controlling the crystal growth, affects the flux invasion behaviors. The results suggest the easier, calmer activation method for the HTS bulk magnets than usual PFM process.

References

[1] Weinstein R, Chen In-Gann, Liu J and K Lau 1991 *J. Appl. Phys.* **70** 6501
[2] Krabbes G, Fuchs G, Schatzle P, Gruss S, Park J, Hardinghaus F 2000 *Advances in Superconductivity XII Tokyo Springer-Verlag* 437
[3] Tomita M and Murakami M 2003 *Nature* **421** 517
[4] Oka T, Itoh Y, Yanagi Y, Yoshikawa M, Ikuta H Mizutani U 2000 *Physica C* **335** 101
[5] Ikuta H, Ishihara H. Yanagi Y, Itoh Y, Mizutani U 2002, *Supercond. Sci. Technol.* **15,** 606
[6] Itoh Y, Mizutani U 1996 *Jpn J Appl Phys*, **35,** 2114
[7] Mizutani U, Mase A, Tazoe K, Ikuta H, Oka T, Itoh Y, Yanagi Y, Yoshikawa M 2000 *Physica C* **335** 92
[8] Sander M, Sutter U, Koch R, Klaeser M 2000 *Supercond. Sci. Technol.* **13** 841
[9] Mizutani U, Oka T, Itoh Y, Yanagi Y, Yoshikawa M, Ikuta H 1998 *Applied Superconductivity* **6** 235
[10] Fujishiro H, Tateiwa T, Fujiwara A, Oka T, Hayashi H 2006 *Physica C* **445-448** 334
[11] Fujishiro H, Hiyama T, Tateiwa T, Yanagi Y, Oka T 2007 *Physica C* **463-465** 394
[12] Oka T, Seki H, Ishiduka D, Ogawa J, Fukui S, Sato T, Yokoyama K, Murakami A 2012 *J. Phys. Conf. Ser.* **400** 022089
[13] Oka T, Itoh Y, Yanagi Y, Tanaka H, Takashima S, Yamada Y, Mizutani U 1992 *Physica C* **200** 55

The effect of BZO dopant concentration on magnetically obtained B_{irr} and B_{c2} in YBCO thin films deposited on STO substrates

P Paturi[1], **M Malmivirta**[1], **H Palonen**[1,2] **and H Huhtinen**[1]

[1] Wihuri Physical Laboratory, Department of Physics and Astronomy, University of Turku, 20014 Turku, Finland

[2] National Graduate School on Nanoscience, Finland

E-mail: petriina.paturi@utu.fi

Abstract. The effect of $BaZrO_3$ (BZO) dopant concentration (0-9 wt-%) on magnetically obtained irreversibility field B_{irr} and upper critical field B_{c2} is investigated in $YBa_2Cu_3O_{6+x}$ (YBCO) thin films prepared by pulsed laser deposition on $SrTiO_3$ (001) (STO) substrates. It is found that the magnetic determination of B_{irr} and B_{c2} works well. The results show that BZO-doping increases B_{c2} in all concentrations, implying a change in the coherence length and band structure of the superconductor.

1. Introduction

In preparation of superconducting wires for applications, the flux pinning properties of the superconductor are essential. For most applications, isotropic dependence of critical current density, J_c, on magnetic field angle is desired and the most common way of inducing it, is by addition of non-superconducting nanorods or -particles. In the past years there has been discussion on how the apparently lowered anisotropy should be interpreted: can the Blatter anisotropy scaling [1], which comes from electron mass anisotropy, be used for analysis of $J_c(\theta)$ curves? Most researchers have recently come to the conclusion that the lowered anisotropy of $J_c(\theta)$ curves is due to strong flux pinning by the nanoinclusions, not a change in electron mass anisotropy. The correct way of measuring the electron mass anisotropy, γ, is from high pulsed magnetic field measurements of resistivity and determination of upper critical field, B_{c2}. Such measurements have been recently published with inconsistent results: in $BaZrO_3$-doped (BZO) films with nanorods the anisotropy was seen to lower from 5 of undoped film to 3.4 in 4% BZO-films [2], but in films with BZO-nanoparticles no lowering of anisotropy was seen [3].

In this paper we test the method presented in [4] to determine the irreversibility field, B_{irr} and B_{c2}. These are determined for $YBa_2Cu_3O_{6+x}$ films doped with varying amounts of BZO. We compare the results to those made in high fields for similar samples [2] and also resistive measurements at low fields from the same films [5].

2. Experimental

The YBCO thin films were deposited by pulsed laser ablation from five differently BZO-doped nanograined targets on single crystal $SrTiO_3$ (100) (STO) substrates. The targets were prepared

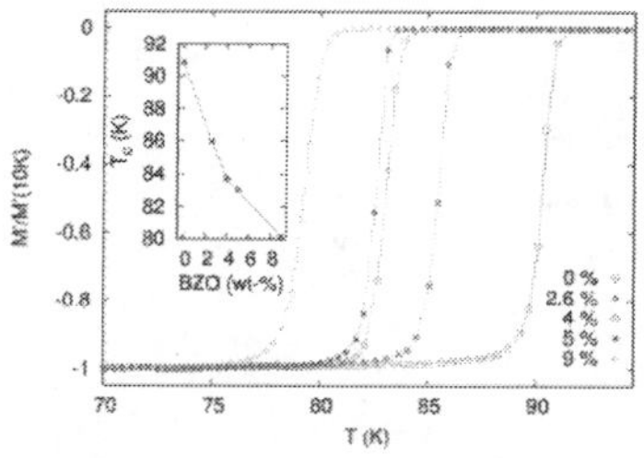

Figure 1. The normalized real parts of the ac-magnetization for all the films. The inset shows the dependence of T_c on temperature.

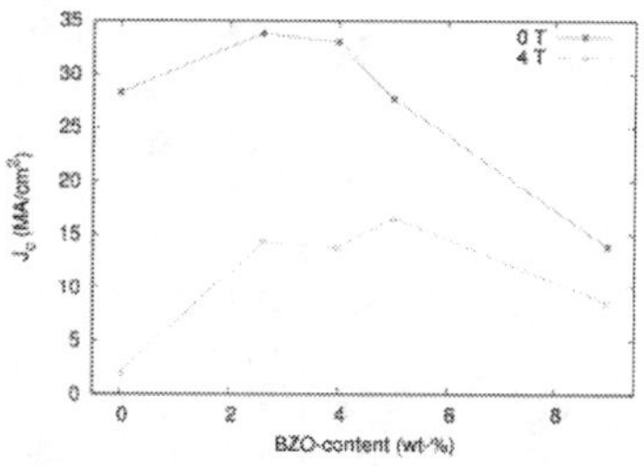

Figure 2. J_c of all the films in 0 T and 4 T determined from hysteresis loops measured at 10 K.

using the citrate-gel method [6] adding the BZO already in the sol-phase. The BZO contents of the targets were 0, 2.6, 4, 5 and 9 wt-%. The used ablation parameters were optimized for J_c and the substrate temperature was 770 °C and the ablation pressure was 23 Pa. The detailed deposition conditions can be found at [2]. At this substrate temperature the BZO forms straight 5-7 nm diameter rods perpendicular to the surface of the substrate [7].

The critical temperatures, T_c, of the films were determined using the ac-magnetometry option in a Quantum Design PPMS with zero dc-field and 0.1 mT ac-field at 113 Hz frequency. The real parts of the magnetizations are shown in Fig. 1. T_c was defined as the temperature where the reduced magnetization is -0.1. The transition widths, ΔT_c were defined as the temperature difference between -0.9 and -0.1 values of the reduced magnetization. For all the samples ΔT_c was between 1.3 and 2 K with no clear dependence on BZO-content. The films show typical decrease of T_c with increasing BZO-content [8], but as the transition widths show, the quality of all the films is good.

The critical current densities, J_c, were determined from magnetization loops measured at 10 K between -8 and 8 T using the Bean critical state model [9]. The obtained critical current densities in zero field and in 4 T are shown in Fig. 2. It can be seen that at zero field the 2.6 % and the 4 % sample give the highest J_c, but at 4 T the highest J_c is obtained with the 5 % sample. This is in accordance with our previous results [10] where it was found that the optimal BZO-content increases with increasing magnetic field. The obtained values of J_c show that the samples are highly optimized also for critical current density.

The upper critical field, B_{c2}, and the irreversibility field, B_{irr}, were determined using the method introduced in [4]. Using the PPMS dc-magnetization option the zero field cooled, ZFC, and field cooled, FC, temperature dependent magnetizations were measured at fields of 0, 0.125, 0.25, 0.5, 1, 2, 4, 6 and 8 T. The irreversibility temperature for the measurement magnetic field, T_{irr}, was determined as the temperature where the ZFC and FC curves deviate from each other (Fig. 3). Similarly, the critical temperature related to the upper critical field, T_{c2}, was defined as the temperature where both the ZFC and FC curves deviate from the constant diamagnetic signal given by the STO substrate. The obtained temperatures define the temperature dependences of B_{c2} and B_{irr}.

3. Results and discussion

The overall shape of the ZFC and FC curves can be explained using the critical state models and vortex compression models [11, 12, 13]. In ZFC the sample is cooled down in zero field and then at low temperature the measurement field is switched on. The vortices penetrate the superconductor according to the critical state model and a large diamagnetic signal is observed

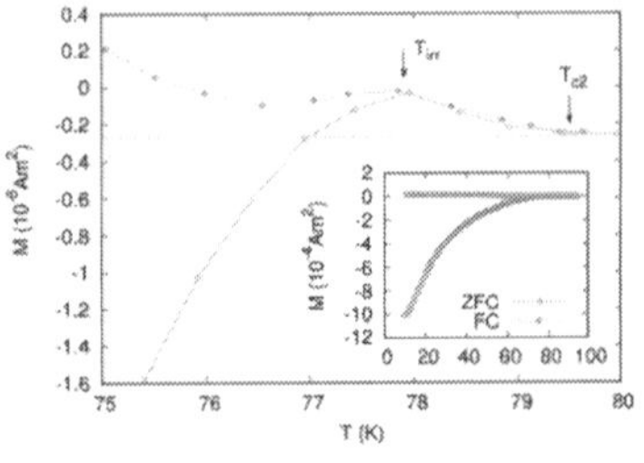

Figure 3. The definition of $T_{\rm irr}$ and $T_{\rm c2}$ for the 5 wt-% BZO sample in magnetic field of 1 T. The temperature independent diamagnetic signal given by the STO substrate shown as a blue line. The inset shows the full ZFC and FC curves.

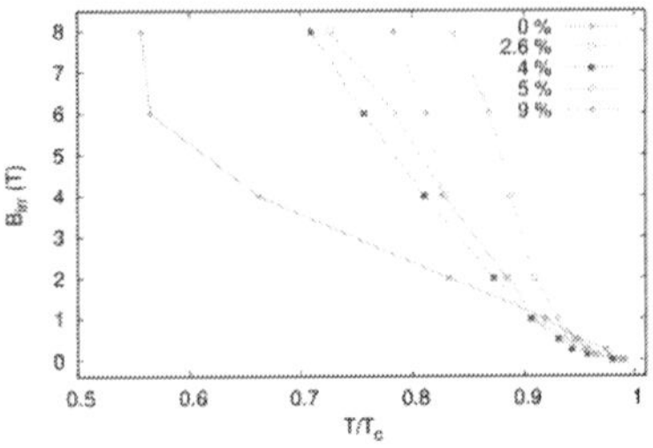

Figure 4. The temperature dependence of $B_{\rm irr}$ for all the samples.

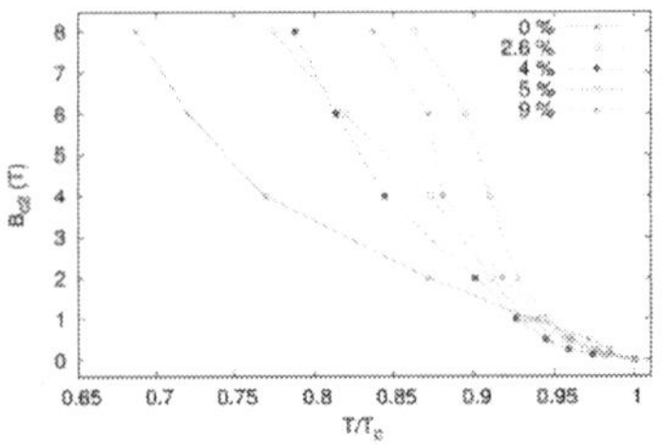

Figure 5. The temperature dependence of $B_{\rm c2}$ for all the samples.

(inset of Fig. 3). As the temperature is increased, the vortices move more into the sample and the diamagnetic signal decreases. Close to the critical temperature, a paramagnetic signal is observed. The origin of this paramagnetic signal has been discussed earlier [11, 12, 13], and in general it is thought to arise from the compression of the vortices away from the surface by the shielding currents. In all of our measurements the ZFC curve has a maximum at $T_{\rm irr}$, *i.e.* where the vortices are not pinned anymore. It is natural that the highest compression and paramagnetic signal is obtained just before the vortices are free to move. Above $T_{\rm irr}$ the magnetization approaches zero, which it reaches at $T_{\rm c}$. The maximum in the ZFC is observed in all our measurements regardless of external field, but the shape of the ZFC curve has been observed to depend on both the external field and the length of the measurement scan [12].

During field cooling the field fully penetrates the sample above $T_{\rm c}$ and upon cooling the vortices are formed and compressed to show a paramagnetic response. At $T_{\rm irr}$ the ZFC and FC curves deviate and the FC curve continues to give a paramagnetic signal. Below $T_{\rm irr}$ the signal slightly increases due to further compression of the vortices, which is caused mainly by the increase of $J_{\rm c}$ with temperature [12].

The determined irreversibility fields for all the samples are shown in Fig. 4. The $B_{\rm irr}(T)$ of all the doped films are fairly close to each other, but the undoped film is clearly different from these. At low temperatures the undoped film has lowest $B_{\rm irr}$ and the 5% doped sample has highest. Around $T = 0.94\ T_{\rm c}$ there is a crossover above which the undoped sample has the highest $B_{\rm irr}$. The result at low temperature is easily understood to be caused by the strong pinning by the BZO-rods. The rods form much stronger pinning sites for the vortices than the dislocations prevalent in the undoped sample, which increases $B_{\rm irr}$. Close to the critical temperature the undoped sample is better probably due to more homogeneous material.

The determined upper critical fields show (Fig. 5) a similar tendency as the $B_{\rm irr}$. $B_{\rm c2}$ of the doped films are clearly higher than those of the undoped film, although the difference is not as large as in $B_{\rm irr}$. This is in line with the previous observations for $BaHfO_3$-doped films [4] as well as data from pulsed high fields [2]. Increase in B_{c2} implies a decrease in the Ginzburg-Landau coherence length, ξ. According to the BCS-theory the ξ depends on the superconducting gap and the Fermi-velocity, *i.e.* BZO-doping causes a change in the band structure of YBCO. The ξ can also change when the superconductor shifts from the clean limit to dirty limit. This effect is most notable at high temperature, where it can explain the increase of B_{c2}, but at lower temperatures, $T < 0.9\ T_c$, the distance between the BZO-rods is much larger than ξ and the decrease of the electron mean free path is not enough to explain the two-fold increase in B_{c2}.

When these data are compared to the resistive data from the same samples [5], we can see that this method of determining the B_{c2} from magnetic measurements is actually closer to the "correct" high field measurements. Both methods give higher B_{c2} for the BZO-doped films, where as the low current and constant field measurements give similar values for all samples except the 9 wt-%, where B_{c2} is lower. This is easily understood as the low current resistive measurements probe only the very best superconducting paths in the samples.

In [2] it was seen that for 4 % -doped YBCO the increase of B_{c2} is connected to lower electron mass anisotropy ($\gamma = 3.4$ for BZO-doped film). Since the BZO-nanorods are only about 20 nm apart, it is plausible that the strain they cause to the YBCO lattice, also would have an effect for the band structure. Naturally, based on data from only one orientation, we cannot say anything about the electron mass anisotropy for these films, but since the results are similar as in [2], it can be assumed that the electron mass anisotropy is lowered also for the other BZO-concentrations.

4. Conclusions

We have magnetically measured thin YBCO films with different BZO-concentrations and determined the B_{irr} and B_{c2} from ZFC and FC measurements. It was found that the method works well, specially for the determination of B_{c2}. It was found that the BZO-doping increases B_{irr} for all the samples as was expected. The B_{c2} was also found to increase for all the BZO-doped films. This implies a change in the band structure of YBCO and is consistent with the results from high magnetic field measurements [2].

5. Acknowledgments

Jenny and Antti Wihuri Foundation is acknowledged for financial support. M. M. acknowledges Finnish Foundation for Technology Promotion and Finnish Cultural Foundation for financial support.

References

[1] Blatter G, Feigel'man M V, Geshkenbein V B, Larkin A I and Vinokur V M 1994 *Reviews of Modern Physics* **66** 1125
[2] Palonen H, Huhtinen H, Shakhov M A and Paturi P 2013 *Supercond. Sci. Technol.* **26** 045003
[3] Llordés A, Palau A, Gázquez J, Coll M, Vlad R, Pomar A, Arbiol J, Guzmán R, Ye S, Rouco V, Sandiumenge F, Ricart S, Puig T, Varela M, Chateigner D, Vanacken J, Gutiérrez J, Moshchalkov V, Deutscher G, Magen C and Obradors X 2012 *Nat. Mater.* **11** 329
[4] Matsushita T, Nagamizu H, Tanabe K, Kiuchi M, Otabe E S, Tobita H, Yoshizumi M, Izumi T, Shiohara Y, Yokoe D, Kato T and Hirayama T 2012 *Supercond. Sci. Technol.* **25** 125003
[5] Malmivirta M, Palonen H, Huhtinen H and Paturi P 2013 *Journal of Physics: Conference Series* **submitted**
[6] Raittila J, Peurla M, Huhtinen H, Paturi P and Laiho R 2003 *Physica C* accepted
[7] Malmivirta M, Yao L, Huhtinen H, Palonen H, van Dijken S and Paturi P 2013 *J. Appl. Phys.* **submitted**
[8] Peurla M, Paturi P, Stepanov Y P, Huhtinen H, Tse Y Y, Bódi A C, Raittila J and Laiho R 2006 *Supercond. Sci. Technol.* **19** 767–771
[9] Wiesinger H P, Sauerzopf F M and Weber H W 1992 *Physica C* **203** 121
[10] Huhtinen H, Irjala M, Paturi P, Shakhov M A and Laiho R 2010 *J. Appl. Phys.* **107** 053906
[11] Koshelev A E and Larkin A I 1995 *Phys. Rev. B* **52** 13559
[12] Luzhbin D A, Pan A V, Komashko V A, Flis V S, Pan V M, Dou S X and Esquinazi P 2004 *Phys. Rev. B* **69** 024506
[13] Moshchalkov V V, Qiu X G and Bruyndoncx V 1997 *Phys. Rev. B* **55** 11793

Magnetic phase diagram and critical current of $BaFe_2As_2$ single crystals with hole- and electron-doping

K. Pervakov[1,2], V. Vlasenko[1], E. Khlybov[2,3], V. Pudalov[1] and Yu. Eltsev[1]

[1]P. N. Lebedev Physical Institute, Russian Academy of Sciences, Moscow 119991, Russian Federation;

[2]International Laboratory of High Magnetic Fields and Low Temperatures, Wroclaw 53-421, Poland;

[3]Institute for High Pressure Physics, Russian Academy of Sciences, Troitsk 142190, Moscow region, Russian Federation;

E-mail: pervakov@sci.lebedev.ru

Abstract.
We report on comparative study of magnetic phase diagram and critical current of the hole- and electron-doped $BaFe_2As_2$ single crystals with close values of superconducting critical temperature, T_c, (slightly underdoped $Ba_{0.64}K_{0.36}Fe_2As_2$ with T_c=25K and optimally doped $BaFe_{1.9}Ni_{0.1}As_2$ with T_c=20K) obtained from measurements of the temperature dependence of ac-susceptibility and isothermal irreversible magnetization loops, M(H), in magnetic fields parallel to the c-axis of the crystal. From ac-susceptibility measurements we get estimation of a slope of the upper critical field, H_{c2}, in dependence on temperature, $dH_{c2}/dT \approx -4.2T/K$ for $BaFe_{1.9}Ni_{0.1}As_2$ single crystal and $H_{c2}/dT \approx -1.75T/K$ for $Ba_{0.64}K_{0.36}Fe_2As_2$ sample that in accordance with Werthamer, Helfand, and Hohenberg (WHH) model gives $H_{c2}(0) = 0.69T_c((dH_{c2})dT) \approx 56T$ for $BaFe_{1.9}Ni_{0.1}As_2$ sample and lower value of $H_{c2}(0) \approx 31T$ for $Ba_{0.64}K_{0.36}Fe_2As_2$ crystal. However, obtained from $M(H)$ measurements temperature dependence of the irreversibility field, $H_{irr}(T)$, for $BaFe_{1.9}Ni_{0.1}As_2$ crystal located below the one for$Ba_{0.64}K_{0.36}Fe_2As_2$ crystal. Furthermore, at T=4.2K and higher temperatures our results for critical current density, J_c, calculated from $M(H)$ curves clearly show slower reduction of J_c with increasing field in even underdoped $Ba_{0.64}K_{0.36}Fe_2As_2$ sample compared to optimally doped $BaFe_{1.9}Ni_{0.1}As_2$ crystal demonstrating higher capacity of K-doped 122 compounds for production of superconducting cables and wires with high critical current in strong magnetic fields.

1. Introduction

At present, about 5 years after discovery of superconductivity in iron based superconductors (FBS) [1] great potential of these compounds for practical applications becomes increasingly evident. To date a few different families of FBS superconductors are known. The highest values of superconducting critical temperature, $T_c \sim 55K$, and upper critical field, $H_{c2}(0)$ exceeding 100T were found in 1111 family (REO(F)FeAs, where RE - rare earth element) [2,3]. However, single phase 1111 compounds may be prepared using extreme growth conditions with high temperature $\sim 1300^oC$ and high pressure ~50 kbar that makes a serious problem to produce practical 1111 superconducting elements with high critical parameters. On the other hand the simplest from the point of view of its chemical composition and synthesis parameters FeSe compound has rather low $T_c \sim 11-12K$ [4]. For this reason, essential efforts of various research groups to produce superconducting wires using FBS were focused on 122 family materials ($BaFe_2As_2$) with hole- or electron-doping, in particular, $Ba_{1-x}K_xFe_2As_2$, $BaFe_{2-x}Co_xAs_2$ and $BaFe_{2-x}Ni_xAs_2$ compounds. Among these candidates for practical use $BaFe_{2-x}Ni_xAs_2$ single crystals with optimal doping ($x \approx 0.1$) have the lowest $T_c \sim 20K$ compared to optimally doped $Ba_{1-x}K_xFe_2As_2$ ($x \approx 0.4$) and $BaFe_{2-x}Co_xAs_2$ ($x \approx 0.15$) crystals with $T_c \sim 38K$ and $T_c \sim 25K$, respectively [5]. At the same time recently very high value of critical current density around $\sim 3 \times 10^6 A/cm^2$ in $BaFe_{2-x}Ni_xAs_2$ crystals at T=4.2K in low magnetic fields exceeding J_c data for other 122 superconductors was reported [6]. In order to estimate potential of $BaFe_{2-x}Ni_xAs_2$ compound for applications in more detail in this paper we report on comparative study of magnetic phase diagram and critical current of optimally doped $BaFe_{1.9}Ni_{0.1}As_2$ and slightly under-doped $Ba_{0.64}K_{0.36}Fe_2As_2$ crystals with close T_c values.

2. Experiment and discussion.

$Ba_{0.64}K_{0.36}Fe_2As_2$ and $BaFe_{1.9}Ni_{0.1}As_2$ single crystals were grown using the self-flux method as described elsewhere [6, 7]. For measurements we used crystals of the in-plane size of 2÷3 mm and thickness around 0.1÷0.2 mm. The mass of $Ba_{0.64}K_{0.36}Fe_2As_2$ and $BaFe_{1.9}Ni_{0.1}As_2$ crystals was 34 mg and 56 mg, respectively. Temperature dependence of ac-susceptibility was measured by the Quantum Design PPMS system. Irreversible magnetization loops we obtained using home-built low-frequency (3.6 Hz) vibrating sample magnetometer with a step motor [8].

High quality of crystals of both compositions was confirmed by X-ray diffraction analysis demonstrating absence of any foreign phases as well as by the results of ac-susceptibility measurements of zero-field superconducting transition showing transition width (10%-90%) of about 1K. As an example in Fig.1 one can see data for $Ba_{0.64}K_{0.36}Fe_2As_2$ sample at H=0.

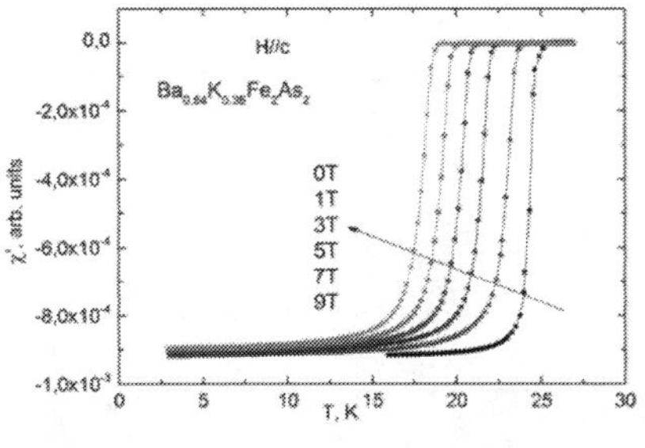

Figure 1. Temperature dependence of ac-susceptibility of $Ba_{0.64}K_{0.36}Fe_2As_2$ sample in the region of superconducting transition in magnetic field up to 9T.

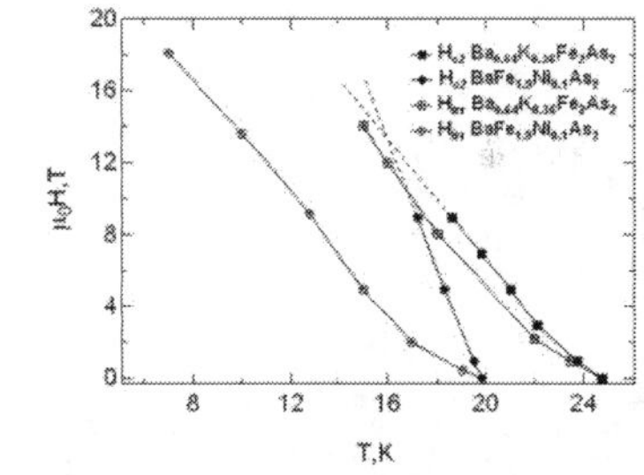

Figure 2. Temperature dependence of the upper critical field and irreversibility field for $Ba_{0.64}K_{0.36}Fe_2As_2$ and $BaFe_{1.9}Ni_{0.1}As_2$ single crystals.

In Fig. 1 we also show superconducting transitions obtained for $Ba_{0.64}K_{0.36}Fe_2As_2$ crystal by ac-susceptibility measurements in magnetic field up to 9T parallel to the c-axis. With increasing field transition curves moved to lower temperatures without any broadening of the transition. Very similar data were found for $BaFe_{1.9}Ni_{0.1}As_2$ crystal [6]. The difference between two sets of data for $Ba_{0.64}K_{0.36}Fe_2As_2$ and $BaFe_{1.9}Ni_{0.1}As_2$ single crystals is the following: due to lower zero-field T_c transition curves for $BaFe_{1.9}Ni_{0.1}As_2$ sample located at lower temperatures while the total temperature shift of the transition by 9T field is larger for $Ba_{0.64}K_{0.36}Fe_2As_2$

crystal signaling lower value of a slope of the temperature dependence of the upper critical field, $H_{c2}(T)$, for $Ba_{0.64}K_{0.36}Fe_2As_2$ crystal compared to $BaFe_{1.9}Ni_{0.1}As_2$ sample. This result is much more clearly seen from Fig. 2 where we plot $H_{c2}(T)$ dependence for $Ba_{0.64}K_{0.36}Fe_2As_2$ and $BaFe_{1.9}Ni_{0.1}As_2$ samples. Superconducting transition point was obtained by extrapolation of a linear part of the transition to zero level. Estimation of the slope of the $H_{c2}(T)$ dependence gives $dH_{c2}/dT \approx -4.2T/K$ and $\approx -1.75T/K$ for $BaFe_{1.9}Ni_{0.1}As_2$ and $Ba_{0.64}K_{0.36}Fe_2As_2$ crystals, respectively. Using these data and Werthamer, Helfand, and Hohenberg (WHH) expression [9] we obtain $H_{c2}(0) = 0.69T_c((dH_{c2})/dT) \approx 56T$ for $BaFe_{1.9}Ni_{0.1}As_2$ sample and lower value of $H_{c2}(0) \approx 31T$ for $Ba_{0.64}K_{0.36}Fe_2As_2$ crystal suggesting crossing of $H_{c2}(T)$ lines for these samples as presented in Fig.2 by linear extrapolation of $H_{c2}(T)$ dependences shown by dotted lines.

Also in Fig. 2 we present temperature dependence of the irreversibility field, $H_{irr}(T)$, for both samples obtained at different temperatures from irreversible magnetization loops, $M(H)$, shown in Fig.3 and Fig.4. To get H_{irr} at fixed temperature for closed $M(H)$ loops we used value of field when $\Delta M(H) = M_{down}(H) - M_{up}(H)$, where M_{up} and M_{down} - magnetization measured with increasing and decreasing field, falls below instrument resolution limit. For opened $M(H)$ loops in order to estimate H_{irr} (somewhat arbitrarily) we employed an empirical extrapolation to zero of a part of $M(H)$ curves at fields lower than the value of the inflection point [6]. As one can see from Fig. 2 in striking contrast to expected crossing behavior of $H_{c2}(T)$ lines $H_{irr}(T)$ line for$BaFe_{1.9}Ni_{0.1}As_2$ crystal located this one for $Ba_{0.64}K_{0.36}Fe_2As_2$ crystal at all fields and temperatures used in our study reflecting higher field and temperature limits of the vortex solid region with non-zero critical current for $Ba_{0.64}K_{0.36}Fe_2As_2$ crystal.

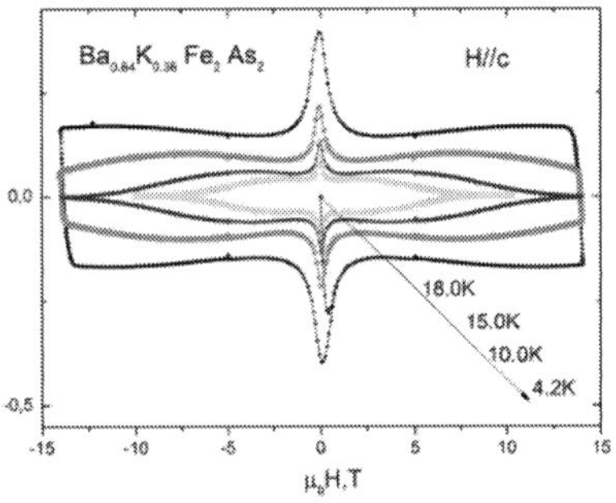

Figure 3. Isothermal magnetization loops in fields up to 14T for $Ba_{0.64}K_{0.36}Fe_2As_2$ single crystal at various temperatures.

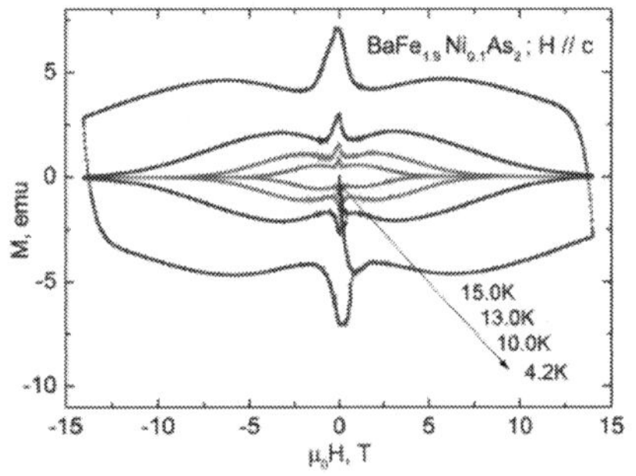

Figure 4. Isothermal magnetization loops in fields up to 14T for $BaFe_{1.9}Ni_{0.1}As_2$ single crystal at various temperatures.

This result finds a support from Fig. 5 where we present field dependence of the critical current density for $Ba_{0.64}K_{0.36}Fe_2As_2$ and $BaFe_{1.9}Ni_{0.1}As_2$ single crystals at different temperatures. J_c value was calculated from irreversible magnetization loops shown in Fig.3 and Fig.4 using the well-known expression $J_c = 20\Delta M/a(1 - a/3b)$ obtained within Bean critical state model [10,11] where a and b ($b > a$) are the in-plane crystal sizes with field applied parallel to the c-axis. Presented in Fig. 5 data show substantially higher J_c values in about one order of magnitude for sample $BaFe_{1.9}Ni_{0.1}As_2$ compared to $Ba_{0.64}K_{0.36}Fe_2As_2$ crystal at fields below 1T. With further increase of field difference between J_c values for these two crystals rapidly decreases. At T=15K and $H > 4T$ as well as at T=10K and $H > 12T$ J_c value for $Ba_{0.64}K_{0.36}Fe_2As_2$ crystal exceeds critical current density for $BaFe_{1.9}Ni_{0.1}As_2$ sample. At T=4.2K higher J_c for $Ba_{0.64}K_{0.36}Fe_2As_2$ crystal may be expected at fields above 15-16T. Thus, these results clearly demonstrate greater critical current potential of even under-doped $Ba_{0.64}K_{0.36}Fe_2As_2$ compound

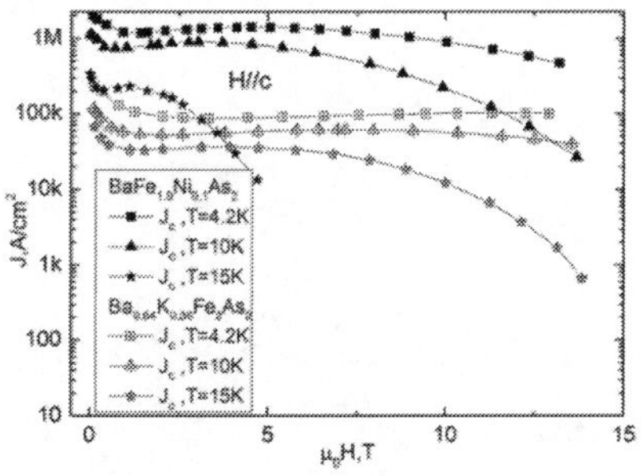

Figure 5. Calculated from M(H) curves field dependence of critical current density for $Ba_{0.64}K_{0.36}Fe_2As_2$ and $BaFe_{1.9}Ni_{0.1}As_2$ single crystals.

compared to optimally doped $BaFe_{1.9}Ni_{0.1}As_2$ sample in high magnetic fields.

3. Conclusions

We have studied magnetic phase diagram and critical currents in $BaFe_{1.9}Ni_{0.1}As_2$ and $Ba_{0.64}K_{0.36}Fe_2As_2$ crystals with approximately the same T_c values. Our results show that vortex solid phase boundary for $Ba_{0.64}K_{0.36}Fe_2As_2$ crystal located at higher temperature and magnetic field compared to $BaFe_{1.9}Ni_{0.1}As_2$ sample. Critical current density of $BaFe_{1.9}Ni_{0.1}As_2$ crystal exceeds J_c for $Ba_{0.64}K_{0.36}Fe_2As_2$ at fields below $\sim 1T$. With increasing field difference between J_c values for $BaFe_{1.9}Ni_{0.1}As_2$ and $Ba_{0.64}K_{0.36}Fe_2As_2$ crystals rapidly decreases, thus, demonstrating higher critical currents in $Ba_{0.64}K_{0.36}Fe_2As_2$ samples in strong magnetic fields above ∼10-15T.

4. Acknowledgements

This study was supported by RFBR (grant N13-02-01180) and by grants of Ministry of Education and Science of the Russian Federation.

References

[1] Y Kamihara, T Watanabe, M Hirano, and H Hosono 2008 *J. Am. Chem. Soc.* **130** 3296
[2] H. Takahashi et al 2008 *Nature* B **130** 376
[3] E P Khlybov et al 2009 *JETP Lett.* **90** 429
[4] F C Hsu et al 2008 *Proc. Natl. Acad. Sci. U.S.A.* **105** 14262
[5] D. L. Sun et al 2011 *Phys. Rev. B* **80** 144515
[6] K S Pervakov et al 2013 *Supercond. Sci. Technol.* **26** 015008
[7] H Luo et al 2008 *Supercond. Sci. Technol* **21** 125014
[8] V I Nizhankovskii and L B Lugansky 2007 *Meas. Sci. Technol.* **18** 1533
[9] N R Werthamer, E Helfand, and P C Hohenberg 1966 *Phys. Rev.* **147** 295
[10] C P Bean 1964 *Phys. Rev. Lett.* **8** 250
[11] C P Bean 1964 *Rev. Mod. Phys.* **36** 31

Journal of Physics: Conference Series **507** (2014) 012042 doi:10.1088/1742-6596/507/1/012042

Hole Concentration Effect on the Microwave Nonlinearity of $Tl_2Ba_2CaCu_2O_{8\pm\delta}$ Superconducting Thin Films

M.M. Bischak[1], **J. Thomas**[1], **R.R. Philip**[2], **and S.K. Remillard**[1,3]
[1] Hope College, 27 Graves Place, Holland, MI 49423, USA.
[2] Thin Film Research Lab, UC College, Aluva, Cochin, Kerala, India.

E-mail: remillard@hope.edu

Abstract. The carrier concentration of $Tl_2Ba_2CaCu_2O_8$ films was modified by annealing in N_2 gas. X-ray analysis of the structure and the oxygen content revealed a correspondence between carrier concentration and oxygen depletion. The T_C and nonlinear surface impedance was measured using a dielectric resonator and the nonlinearity slope parameter $r=\Delta X_S/\Delta R_S$ was found to converge to unity at the critical temperature, indicating a dominance of Josephson fluxon hysteresis on the nonlinearity. Highly inductive nonlinearity was observed in a small range of doping levels between $0.180<p<0.195$ holes/Cu, which does not include the optimal doping level of 0.16 holes/Cu.

1. Introduction

High temperature superconductor (HTS) thin films are used in passive microwave devices such as RF filters[1] and MRI pick-up coils[2]. These applications are best served by materials with little nonlinearity, which is manifest as a dependence of the surface impedance, Z_S, on the microwave magnetic field, H_{RF}, as well as with signal distortion[3]. In this experiment, modification of the hole doping moves the nonlinear response of the thin film to a new quiescent operating point, and permits two key observational opportunities. First is the opportunity to examine nonlinearity at lower power with suppressed carrier concentration as if it were being operated at higher power. Second is the opportunity to directly examine the effect of carrier concentration on the nonlinearity of Z_s.

The starting material for these experiments comes from a large supply of 50 mm diameter $Tl_2Ba_2CaCu_2O_8$ thin films on $LaAlO_3$ substrates, which presents a unique opportunity to conduct these experiments that may otherwise be prohibited by material supply limitations.

2. Microwave Nonlinearity in $Tl_2Ba_2CaCu_2O_8$ Films

A temperature and H_{RF} dependent fit function for Z_S of $Tl_2Ba_2CaCu_2O_8$ was used by Gaganidze, *et al.*[4], with various terms corresponding to the mechanisms behind the loss. At very low H_{RF}, weak links formed by twin and grain boundaries are in the Meissner state and Z_S is dominated by leakage through the weak links. R_S then primarily depends on H_{RF}^2 [5], although at higher H_{RF} the grain boundaries decouple and the dependence becomes weaker. A low field saturating flux-flow contribution varies as $(1+H_0/H_{RF})^{-1/2}$, where the characteristic scaling field H_0 can be especially pronounced in $Tl_2Ba_2CaCu_2O_8$ which has very weak links[6]. Above J_{Cj} Josephson fluxons nucleate

[3] To whom any correspondence should be addressed.

and their dissipation is hysteretic, mostly due to pinning, as characterized by $R_S \propto H_{RF}$ and $X_S \propto H_{RF}$[7]. Although stronger H_{RF} dependence is seen if the fields are enhanced at film edges [7] or at pores and exposed cleavage plane edges, H_{RF}^2 dependence can be attributed to leakage current through weak links[8]. Oxygen deficiency causes enhanced kinetic inductance in weak links in YBCO[9], and other experiments have also found grain boundary transport[10] and pinning[11] to be influenced by doping.

The dimensionless slope parameter ($r=\Delta X_S/\Delta R_S$) of $r\approx 1$ along with $R_S \propto H_{rf}$, or H_{rf}^2 in the case of edge effects, indicates hysteresis loss due to the nucleation, annihilation and pinning of fluxons[4]. However, $r>>1$ corresponds to intrinsic nonlinearity[8], which scales as the H_{C1} of the bulk material. In the case of unaligned small grains, the weakly coupled grain model has been shown to be consistent with an observed negative value of r at high reduced temperature[12]. In these measurements H_{RF} remains under about 1,500 A/m and the reduced temperature remains under $t=0.95$, meaning that H_{C1} is not approached. Thus the nonlinearities in these measurements do not involve the intrinsic effects of Abrikosov fluxons.

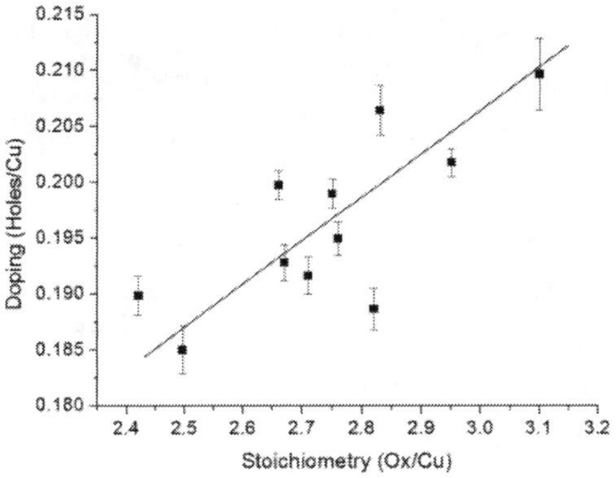

Figure1. Doping from T_C as a function of Oxygen/Copper ratio determined from EDS.

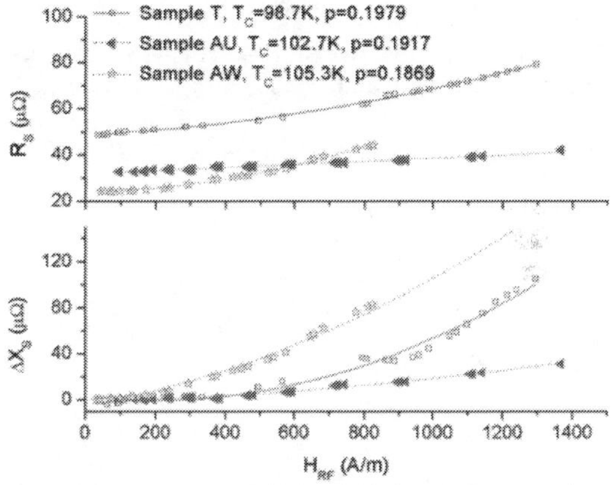

Figure 2. R_S and ΔX_S for films of three doping levels at 5.5 GHz and reduced temperature of 0.79 with 2nd order polynomial fits.

3. Experiment

3.1. The Films and Their Hole Doping

The films were made by DuPont Superconductivity by annealing off-axis magnetron sputtered amorphous BaCaCuO films in a Tl_2O_3 vapor at 1 atm[13]. We confirmed the manufacturer's report that R_S at 8.1 GHz and 77K was less than 200μΩ and that the thicknesses were about d=400 nm.

Carrier density, p, was depleted by annealing for 7 hours in N_2 flowing at 20 cm^3/min between 250°C and 360°C. p is related to $t=T_C/T_{C,max}$ through the phase boundary parabola[14] $p=0.16\pm\sqrt{(1-t)/82.6}$ where $T_{C,max}$ is about 112 K for $Tl_2Ba_2CaCu_2O_8$. The positive sign corresponds to overdoping, and the 2212 phase is over-doped. So N_2 annealing can modify the T_C [15]. Tl becomes mobile only above our anneal temperatures. [16]. Thus, changes in T_C correspond to changes in oxygen. Figure 1 shows holes per Cu, from T_C, plotted against oxygen per Cu, from energy dispersive X-ray spectroscopy. A variation in oxygen results in a scattered variation in doping illustrating the connection between oxygen depletion and carrier concentration.

3.2. Surface Impedance Measurement

The surface impedance in Figure 2 is measured using a polished sapphire/HTS resonator following a well-established 3-sample round-robin [17,18]. $X_S(H_{RF})$ is also measured in the round robin, and corresponds to a variation in effective penetration depth, λ, as $\Delta X_S=(\omega\mu_o)\Delta\lambda$, which comes directly from the resonant frequency, f_o, by $\Delta X_S=2G\cdot(\Delta f/f_o)$. G is the sample geometry factor, in our case 203 Ω, and Δf is the resonant frequency shift.

Thin films require a correction to R_S for the substrate [19]. Surface reactance, also corrected for the film thickness, is $\Delta X_S = \Delta X_{S,eff} \tanh(d/\lambda)$ where $\Delta X_{S,eff}$ is computed from the frequency shift. The effective penetration depth, λ, is a composite of the bulk London depth and field penetration at weak links. Iteration is used to find λ using $\lambda_{ab}(0)$=200 nm [20] applied to the d-wave dirty limit $\lambda(T)=\lambda(0)/\sqrt{1-(T/T_C)^2}$ which is applicable to these DuPont $Tl_2Ba_2CaCu_2O_8$ thin films [21].

4. Results

R_S and ΔX_S of three representative films are shown at t=0.79 in Figure 2. All films exhibit both linear and quadratic terms in $R_S(H_{RF})$, although all films with T_C around 102 K have a much weaker function of $Z_S(H_{RF})$. In all cases, the X_S has both linear and quadratic field dependence. For these films, all examined at 77K and above, including the flux flow term proportional to $(1+H_0/H_{RF})^{-1/2}$ did not improve the fit. Using a wider temperature range, Hein *et al.*[22] scaled the currents at different temperatures and found the H_{RF} dependence of R_S to be independent of Oxygen doping in YBCO.

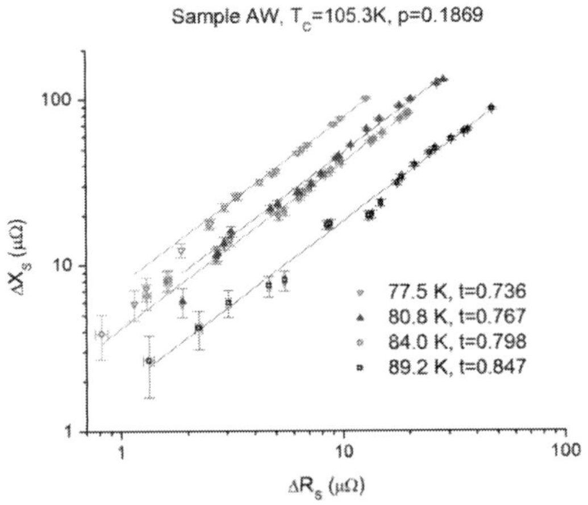

Figure 3. X_S varies nearly linearly with R_S, with H_{RF} as the implicit variable. The slope (corresponding to height on the log graph) decreases with increasing T from 77.5K.

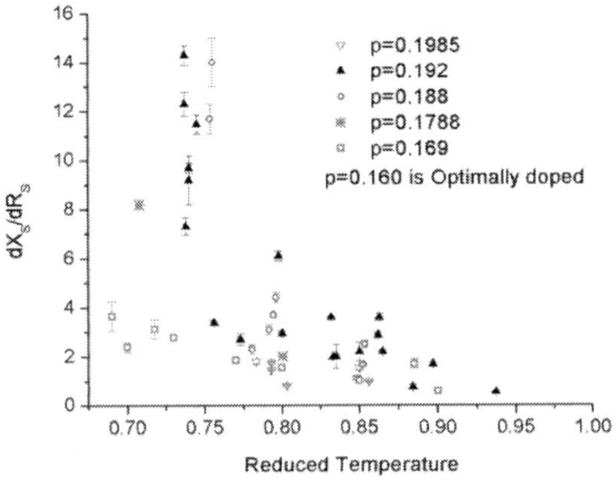

Figure 4. All slopes of 20 samples fall in a band of values that converges to unity at T_C. Binning according to doping level reveals the trends.

X_S and R_S usually have similar dependence on H_{RF} [23], thus producing a linear variation of ΔX_S with ΔR_S, with H_{RF} as the implicit parameter, shown in Figure 3. The slope, $r=\Delta X_S/\Delta R_S$, varies with doping, p, and with reduced temperature, t. Forty five TBCCO/LAO wafers were used in some way throughout this study and the temperature dependence of the slopes fall within a region that converges at unity at T_C and ranges from r=2 to r=15 at t=0.75.

The dependence of r on p at constant t appears to be quite scattered. But by viewing the slope results as in Figure 4, with the slopes binned according to p, the trend stands out. For doping in the range $0.180<p<0.195$ the slope becomes large ($r>10$) at about t=0.75. For films in this range of carrier concentration, the low temperature nonlinearity is dominated by the kinetic inductance. Thus the low temperature inductive nonlinearity survives to the highest temperature for this doping region. The dramatic increase in r does not occur above and below this doping range. For p<0.180 and p>0.195, the slope for $0.75<t<1$ remains in the range of only $1<r<4$ and the surface resistance is linear in H_{RF}, indicating impedance nonlinearity dominated by Josephson fluxon hysteresis.

5. Conclusion

The doping of 400 nm thick $Tl_2Ba_2CaCu_2O_8$ films on $LaAlO_3$ substrates was modified by annealing the films in N_2 gas below 400°C. The nonlinear slope parameter was found to peak for carrier concentrations in the range of $0.180<p<0.195$ holes/Cu indicating the dominance of significant inductive loss in weak links to higher temperatures. Outside this doping range, hysteretic loss from nucleation, annihilation and pinning of Josephson fluxons dominates the nonlinear surface impedance. In continuing work, these results are being applied to the preparation of patterned single resonators in order to investigate the influence of carrier concentration on intermodulation distortion.

6. Acknowledgement

This work was supported by grants DMR/1206149 and PHY-DMR/1004811 from the U.S. National Science Foundation, and by a grant from the Research Corporation for Science Advancement. RRP acknowledges UBCHEA, NY for their support through the United Board Fellows program.

References

[1] Yi H R, Remillard S K and Abdelmonem A 2003 *IEEE Trans. on Microwave Theory and Tech.* **51** 2290-2296
[2] Wosik J, Xue L, Xie L M, Kamel MR, Nesteruk K and Bankson JA 2007 *Appl. Phys. Lett.* **91** 183503
[3] Eben A M, Bunnell V A, Goodson C J, Pease E K, Lee S C and Remillard S K 2011 *IEEE Trans. on Applied Superconductivity* **21** 595
[4] Gaganidze E, Heidinger R, Halbritter J, Shevchun A, Trunin M and Schneidewind H 2003 *J. Appl. Phys.* **93**, 4049-4054
[5] Gaganidze E, Heidinger R, Halbritter J and Schneidewind H, 2002 *Physica C* **372-376** 511-514
[6] Gurevich A, Rzchowski M S, Daniels G, Patnaik S, Hinaus B M, Carillo F, Tafuri F and Larbalestier D C 2002 *Phys. Rev. B* **88** 097001
[7] Halbritter J 1995 *J. Supercond.* **8**, 691
[8] Halbritter J 2006 *J. Superconductivity and Novel Magn,* **19** 547-560
[9] Moeckly BH and Buhrman RA 1995 *IEEE Trans. Appl. Supercond.* **5**, 3414-3417
[10] Guth K, Krebs H U, Freyhardt H C and Jooss C. 2001 *Phys. Rev. B*, **64**, R140508
[11] Gallitto A A, La Duca M, Li Vigni M, Topal U, and Yildiz Ş 2012 *Physica C* **483**, 71-78
[12] Remillard S K, Reeves M E, Rachford F J, and Wolf S A 1994 *J. Appl. Physics* 75, 4103
[13] Holstein W L, Parisi L A, Wilker C and Flippen R B 1992 *Appl. Phys. Lett.* **60**, 2014-2016
[14] Presland M R, Tallon J L, Buckley R G, Liu R S and Flower N E 1991 *Physica C* **176**, 95
[15] Morosin B, Venturini E L and Ginley D S 1991 *Physica C: Superconductivity* **183** 90-98
[16] Raveau B, Hervieu M and Michel C 1994 Extended defects and redox mechanisms in thallium cuprates *Thallium-Based High-Temperature Superconductors* ed A M Hermann and J V Yakhmi (New York: Marcel Dekker) chapter 4 pp. 41-88
[17] Wilker C, Shen Z Y, Nguyen V X and Brenner M S 1993 *IEEE Trans. Appl. Supercond.* **3** 1457
[18] International Standard IEC 61788-7, Superconductivity – Part 7 (2006).
[19] Klein N, Chaloupka H, Müller G, Orbach S, Piel H, Roas B, Schultz L, Klein U and Peiniger M 1990 *J. Appl. Phys*, **67**6940-6945
[20] Cohen L F, Cowie A, Gallop J C, Ghosh I S and Goncharov I N 1997 *J. Supercond.* **10** 85-90
[21] Bonn D A, Liang R, Riseman T M, Baar D J, Morgan D C, Zhang K, Dosanjh P, Duty T L, MacFarlane A, Morris G D, Brewer J H, Hardy W N, Kallin C and Berlinsky A J 1993 *Phys. Rev. B* **47** 11314-11328
[22] Hein M A, Humphreys R G, Hirst P J, Parks S H and Oates D E 2003 *J. Supercond.* **16** 895-903
[23] Velichko A V, Lancaster M J and Porch A 2005 *Supercond. Sci. Technol.* **18**, R24-R49

MgB_2 superconductors with addition of ZrB_2 and different carbon sources

L B S Da Silva[1], E E Hellstrom[2], D Rodrigues Jr[1]

[1]Escola de Engenharia de Lorena, Universidade de São Paulo, Lorena, SP, Brazil.
[2]Applied Superconductivity Center, Florida State University, Tallahassee, FL, USA.

E-mail: lucasarno@usp.br, durval@demar.eel.usp.br

Abstract. MgB_2 has been catching the attention due to the possibility to apply the material in magnets and electronic devices, operating with cryocoolers. In this work, MgB_2 bulks were developed and analyzed with addition of ZrB_2, another diboride with the same C32 hexagonal structure as MgB_2, and simultaneous addition of different carbon sources (SiC, graphite, and carbon nanotubes). The objective of these additions is to modify the Mg planes with the diborides and to dope the material with carbon, improving the upper critical fields. Besides the doping of the material, this method creates crystalline defects in the superconducting matrix, which can act as pinning centers. As a result we could improve the critical current density of the material and estimate the behavior of dopants on the superconducting properties.

1. Introduction

Magnesium diboride (MgB_2) is one of the most promising superconductors for practical applications, due to its properties and characteristics, and to the low price of the precursor elements. The operation temperature of the application can be kept around 20K by cryocoolers. To reduce the problem with weak links and to improve the densification and intergrain connectivity in MgB_2, high energy ball-milling and hot isostatic pressing (HIP) have been used by research groups around the world [1,2].

The intensity that the magnetic flux lines are pinned in the MgB_2 defines the maximum critical current density (J_c) that the material can carry under a certain applied magnetic field (H). These pinning behaviors can be produced by defects in the superconducting matrix or other inhomogeneities in the material [3]. As MgB_2 is a granular material, the principal mechanism of flux pinning is due to the grain boundaries [4]. The processing scheme and the introduction of defects in the superconducting matrix can also influence the flux pinning efficiency and grain connectivity, simultaneously.

MgB_2 doping is another way to change the intrinsic superconducting properties, as the critical temperature (T_c) and upper critical magnetic field (B_{c2}). Research groups around the world have been investigating the doping of MgB_2 with different materials. Metallic elements, such as Li, Mn, Al, Ta, Zn, Cd, and others, are used to modify the Mg planes in the crystalline structure of the MgB_2. Carbon is one of the most effective element for significant enhancement of the critical current density at high magnetic fields, modifying the B planes [5,6]. The addition of materials with AlB_2–type crystalline structure (ZrB_2, TaB_2, VB_2, and AlB_2) has been investigated by our research group [1,5,7]. In these works, the addition of the diborides was efficient, changing the superconducting matrix, improving the magnetic flux pinning and, consequently, improving the transport properties of these superconductors. Simultaneous addition of diborides and carbon has not been systematically explored before [7].

In the present work a methodology to optimize J_c in bulks of MgB_2 superconductors is described. The method uses the addition of another diboride (ZrB_2) and simultaneous addition of a carbon source (SiC, graphite, and carbon nanotubes) to MgB_2 by high energy ball milling. The heat treatments were performed in a Hot Isostatic Press (HIP) to improve the densification and grain connectivity.

2. Experimental Procedure

MgB_2 samples were developed and analyzed with addition of ZrB_2 and simultaneous addition of carbon sources (SiC, graphite, and carbon nanotubes). The samples were prepared using pre-reacted Alfa-Aesar powders. The powder of MgB_2 was mixed to the ZrB_2 following the $Mg_{1-x}Zr_xB_2$ ratio, where x = 0.05 corresponds to the 5at.% concentration of the doping element. In some samples, 10wt.% of SiC, 3wt.% of graphite, and 3wt.% of carbon nanotubes (CNT) were added with the ZrB_2.
The mixing of the powders was performed with a SPEX 8000D high energy ball mill following the ball-to-powder mass ration of 3:1. All steps, including the preparation and mixture of the powders, were realized in a glove-box with argon atmosphere. The powders were compacted using a Cold Isostatic Press (CIP) under pressure of ~30 MPa. The heat treatments were performed in a Hot Isostatic Press (HIP), using slow heat-up with a 0.5°C/min rate to 1000°C, heat treatment for 24h under a pressure of ~30 MPa, and slow cooling-down with a 0.2°C/min rate to room temperature, to recrystallize and form the final phases, as well promoting the densification and the good intergrain connectivity. Table 1 shows the MgB_2 samples produced and analyzed in this work.

Table 1. Composition of the MgB_2 samples

Sample composition	Sample name
MgB_2	MgB_2
MgB_2+5at.%ZrB_2	ZrB_2
MgB_2+5at.%ZrB_2+10wt.%SiC	ZrB_2+SiC
MgB_2+5at.%ZrB_2+3wt.%graphite	ZrB_2+Graph
MgB_2+5at.%ZrB_2+3wt.%CNT	ZrB_2+CNT

X-ray diffractometry (XRD) was used to analyze the samples in an attempt to determine the formation of phases and the structural changes. The analysis were made using Cu-Kα X-ray radiation, scanning between 20° and 80°, and angular step of 0.02° in 2θ with 1s of counting per point, in a Shimadzu XRD-6000 Lab X diffractometer. The crystal structure refinements of all samples were performed from the X-ray diffractograms using the Rietveld method through the FullProf software (freeware). It was possible to extract an estimate of the sample's lattice parameters and composition.
The microstructural and compositional analysis of the samples after heat treatment were made using a scanning electron microscope (SEM) model LEO1450VP with an Oxford INCA energy dispersive spectrometer (EDS). The micrographs were obtained using a backscattered electrons detector. These analyses were important to verify the homogeneity, composition and granularity of the samples. A transmission electron microscope (TEM) model JEOL JEM2100 was used to complete the analysis.
The superconducting characterization was performed using a Quantum Design 9T PPMS. The critical temperatures T_c were extracted as the intersection point between the zero-field cooled (ZFC) and field cooled (FC) branches from the DC magnetization versus temperature curves measured under an applied magnetic field of 30 Oe. DC magnetization loops versus applied magnetic field at 5K were used to extract the $J_c(H,T=5K)$ *vs.* $\mu_0 H$ curves using the follow equation [8]:

$$J_c = \frac{20\Delta M}{a_1\left(1-\frac{a_1}{3a_2}\right)}; a_1 < a_2 \quad (1)$$

where ΔM is the difference of magnetization in a given field, a_1 and a_2 are the width and thickness, respectively, of the rectangular-shaped section extracted from the samples.

3. Results and Discussion

3.1. Crystallography Characterization

Table 2 is a compilation of results from X-ray analysis and the crystal structure refinement using the Rietveld method. All samples showed some MgO contamination and the formation of the MgB_4 phase, despite the careful sample preparation in all steps. The formation of the Mg_2Si phase can also be observed for samples with SiC additions. The formation of this phase can be explained by the Mg reaction with Si, which can be provided by the free Mg generated after its replacement by Zr in the crystalline structure. However, using conventional XRD analysis, there are no evidences of Mg replacement by Zr atoms. It is required a more accurate method to detect the doping. On the other hand, the significant decrease of the MgB_2 *a* lattice parameter indicates some carbon replacement at the boron sites of MgB_2, which is in agreement to the literature when using 10wt.% of SiC [6].

Table 2. Parameters obtained after Rietveld Refinement.

Sample	Phase	Composition (fract. %)	Lattice Parameters a (Å)	b (Å)	c (Å)
MgB_2	MgB_2	63.77	3.0835	3.0835	3.5216
	MgB_4	7.49	5.4543	4.3997	7.4745
	MgO	28.74	4.2138	4.2138	4.2138
ZrB_2	MgB_2	77.62	3.0817	3.0817	3.5259
	ZrB_2	21.24	3.1690	3.1690	3.5303
	MgB_4	0.57	5.6911	4.1949	7.5224
	MgO	0.56	4.2326	4.2326	4.2326
ZrB_2+SiC	MgB_2	44.99	3.0683	3.0683	3.5236
	ZrB_2	13.40	3.1649	3.1649	3.5260
	SiC	2.20	4.3508	4.3508	4.3508
	Mg_2Si	2.98	6.3551	6.3551	6.3551
	MgB_4	29.50	5.4643	4.4283	7.4723
	MgO	6.92	4.2176	4.2176	4.2176
ZrB_2+Graph	MgB_2	64.89	3.0639	3.0639	3.5254
	ZrB_2	19.97	3.1674	3.1674	3.5299
	MgB_4	6.45	5.6815	4.2039	7.5367
	MgO	8.69	4.2205	4.2205	4.2205
ZrB_2+CNT	MgB_2	61.21	3.0599	3.0599	3.5228
	ZrB_2	25.20	3.1669	3.1669	3.5285
	MgB_4	4.20	5.6866	4.2137	7.5739
	MgO	9.38	4.2138	4.2138	4.2138

3.2. Microstructural Characterization

Figure 1 presents the microstructure of the MgB_2 with addition of ZrB_2. The micrographs were obtained for the fractured surfaces by SEM analysis performed using the detector for backscattered electrons. The sample shows good connectivity, indicating that the preparation method and the heat treatment profile were efficient to create a dense MgB_2 superconducting phase. Some small clusters (light gray) were identified as ZrB_2 by EDS analysis. Figure 2 illustrates the TEM analysis of the ZrB_2 grains, which present an average grain size around 1 μm. Part of these secondary phases can act as pinning centers, improving the maximum J_c that the sample is capable to transport in the presence of an applied magnetic field. This effect will be even more pronounced if the secondary phase has dimensions in the order of the MgB_2 coherence length (less than 25nm), as reported in literature [9].

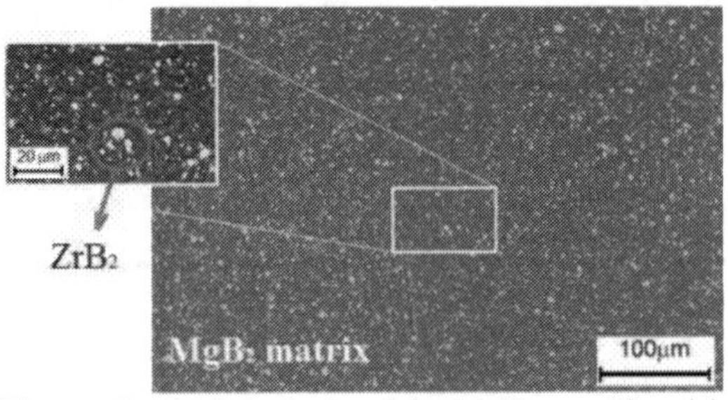

Figure 1. SEM analysis of the MgB_2 sample with ZrB_2 addition (ZrB_2), using backscattered electrons detection mode.

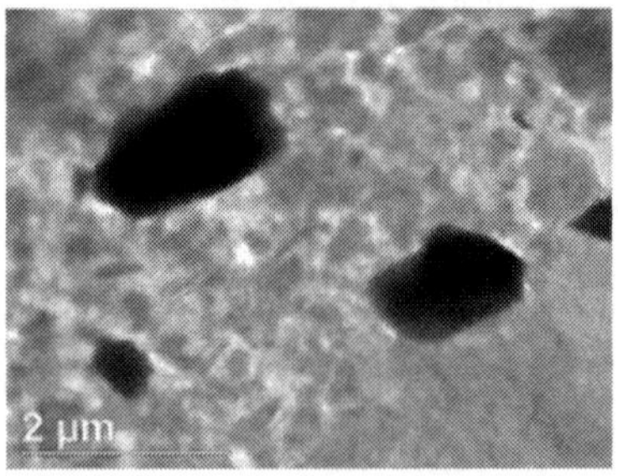

Figure 2. TEM analysis of the MgB_2 sample with ZrB_2 addition (ZrB_2). The grains of ZrB_2 are the dark regions and the light-gray is the MgB_2 matrix.

3.3. Analysis of the Superconducting Properties

Figure 3 shows the normal-to-superconductor transition using the ZFC and FC characterization. The critical temperatures T_c were extracted as the intersection point of the ZFC and FC curves. The pure MgB_2 sample showed T_c around 38.9 K. The critical temperature decreases to 38.2 K for the MgB_2 with ZrB_2 addition, and decreases to 36.2, 31.5, and 29.0 K for the MgB_2 samples with addition of SiC, graphite, and CNT, respectively. T_c decreases mainly due to changes in the MgB_2 matrix, which

is expected for doped samples. However, the values of T_c were maintained high for all samples and the transition widths were relatively narrow, showing the good quality of the superconducting phases.
Figure 4 shows the J_c vs. H curves for the MgB_2 samples at 5K. It can be seen that the samples with ZrB_2 addition showed improvements in J_c in the entire range of applied magnetic field, especially at low magnetic fields. This behavior can be probably explained by the creation of regions acting as effective pinning centers for the flux lines, like unreacted ZrB_2 or phases containing carbon.
On the other hand, simultaneous addition of ZrB_2 and graphite also presents an increase on J_c in the entire range of applied magnetic field, if compared with the pure MgB_2 sample, but more pronounced in high magnetic fields at 5K. The effective doping of the MgB_2 superconducting phase with carbon, as evidenced by the XRD analysis, is probably due to the replacement of boron by carbon in the MgB_2 crystalline structure. This behavior was already reported in the literature for samples mixed with a carbon source [6].

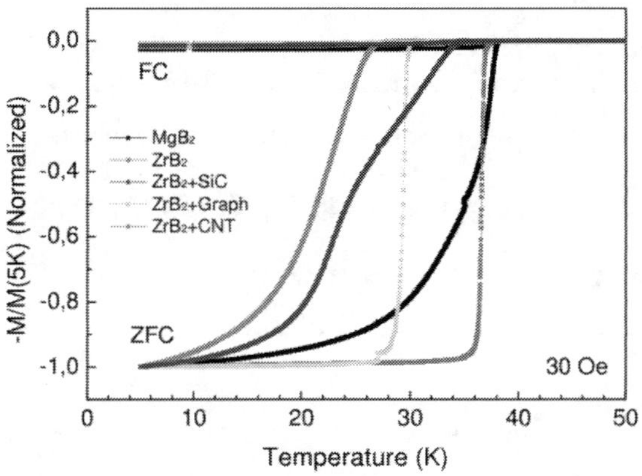

Figure 3. Normalized DC magnetization as a function of the temperature for MgB_2 samples, measured under 30 Oe.

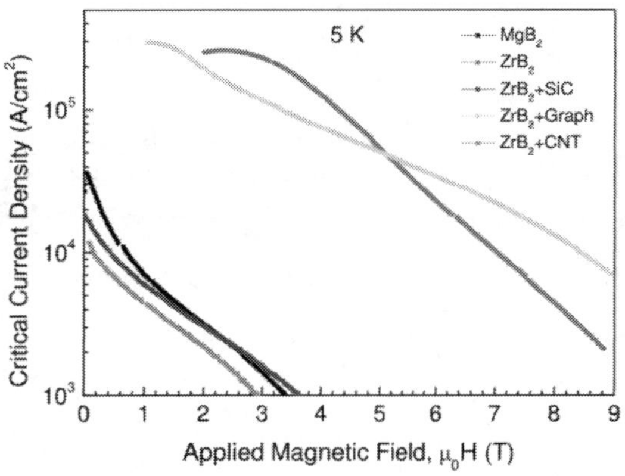

Figure 4. Critical current density as a function of the applied magnetic field for MgB_2 samples, measured at 5 K.

4. Conclusion

This work showed the results and a preliminary discussion of the possible reasons for the improvement of the transport properties of MgB_2 superconducting bulks when other diboride (ZrB_2) and carbon sources are added into the material. The additions of ZrB_2 and ZrB_2+graphite were efficient to improve J_c of the MgB_2 samples in all magnetic fields at 5 K. The combination of ZrB_2 and graphite showed the best transport conditions in high magnetic fields and at 5 K.

References

[1] Rodrigues Jr. D, Jiang J, Zhu Y, Voyles P, Larbalestier D C and Hellstrom E E 2009 *IEEE Trans. Appl. Supercond.* **19** 2797-2801.
[2] Serrano G, Serquis A, Dou S X, Soltanian S, Civale L, Maiorov B, Holesinger T G, Balakirev F and Jaime M 2008 *J. Appl. Phys.* **103** 023907.
[3] Da Silva L B S, Rodrigues C A, Oliveira Jr N F, Bormio-Nunes C and Rodrigues Jr D 2010 *Supercond. Sci. Technol.* **23** 115012.
[4] Buzea C and Yamashita T 2001*Supercond. Sci. Technol.* **14** R115-R146.
[5] Da Silva L B S, Rodrigues Jr D, Serrano G, Metzner V C V, Malachevsky M T and Serquis A 2011 *IEEE Trans. Appl. Supercond.* **21** 2639-2642.
[6] Dou S X, Pan A V, Zhou S, Ionescu S, Liu H K and Munroe P R 2002 *Supercond. Sci. Technol.* **15** 1587.
[7] Rodrigues Jr. D, Da Silva L B S, Metzner V C V and Hellstrom E E 2012 *Physics Procedia* **36** 468-474.
[8] Bean C P 1962 *Phys. Rev. Lett.* **8** 250.
[9] Mazin I I, Andersen O K, Jepsen O, Dolgov O V, Kortus J, Golubov A A, Kuz'menko A B and van der Marel D 2002 *Phys. Rev. Lett.* **89** 107002.

Evidence of a miscibility gap in the $FeTe_{1-x}Se_x$ polycrystalline samples prepared with a melting process

A Sala[1,2], A Palenzona[1,3], C Bernini[3], F Caglieris[2], M R Cimberle[3], C Ferdeghini[3], G Lamura[3], A Martinelli[3], M Pani[1,3], J Hecher[4], M Eisterer[4] and M Putti[2,3]

[1] *Chemistry and Industrial Chemistry Department, University of Genova, via Dodecaneso 31, I-16146 Genova, Italy*
[2] *Physics Department, University of Genova, via Dodecaneso 33, I-16146 Genova, Italy*
[3] *CNR-SPIN-Genova corso Perrone 24, I-16152, Genova, Italy*
[4] *Atominstitut, Vienna University of Technology, Stadionallee 2, A-1020 Vienna, Austria*

alberto.sala@spin.cnr.it

Abstract. The study of overdoped $FeTe_{1-x}Se_x$ ($0.5 < x < 1$) polycrystalline superconductor samples is reported. The samples were prepared using a melting technique previously developed by our group. Increasing the Se content a phase separation related to the formation of FeSe inside the Fe(Se,Te) phase happens, as demonstrated by structural analysis and magnetic characterization. The proposed phase separation picture is likely the fingerprint of a miscibility gap in the Fe(Se,Te) system.

1. Introduction

From the discovery of the chalcogenides superconductors [1], the ubiquitous character of $FeTe_{1-x}Se_x$ compounds to show a non homogeneous microstructure was enlightened [2,3]: a phase separation in the whole sample volume of two isostructural tetragonal phases with different Se/Te ratio was observed. The $FeTe_{1-x}Se_x$ phase diagram with $0 < x \leq 0.5$ has been deeply investigated, instead few studies on the compositions with $0.5 < x \leq 1$ have been performed [4].

For this reason we synthetized $FeTe_{1-x}Se_x$ polycrystalline samples with x = 0.5, 0.6, 0.7, 0.8, 0.9, 1 by a melting process technique [5,6] and we investigated their crystallographic, microstructural and physical properties using: Polarized Optical Microscopy (PLOM), X-Ray Powder Diffraction (XRPD), Scanning Electron Microscope (SEM) coupled with Energy Dispersive X-Ray Spectroscopy (EDS) analysis and DC magnetic susceptibility measurements.

2. Experimental

$FeSe_xTe_{1-x}$ samples (with x = 0.5, 0.6, 0.7, 0.8, 0.9, 1) powders were synthesized using the common solid state reaction method[3,4]. The obtained powders, once pelletized in a hydraulic press, were melted in a high frequency induction furnace in inert gas atmosphere. After the melting process a vacuum annealing treatment at 450° C for 20 days is performed. Phase identification was carried out by X-ray powder diffraction (XRPD–PHILIPSPW1830; Bragg–Brentano geometry; Cu Kα;

range 10°-80° 2θ; counting time 3 s) and the crystal structure indexing was performed using the FullProf software. Microstructure was analysed by a scanning electron microscope (SEM: CAMBRIDGE S360), coupled with an electron dispersive spectrometer for chemical analysis (EDS: Oxford AtztecEnergy X-Max) and PLOM. Magnetic measurements were performed by a dc SQUID magnetometer (MPMS by Quantum Design).

3. Results and discussions

The XRPD analysis reveals that the samples are mainly constituted of tetragonal anti PbO-type structure, with few traces of hexagonal polymorph NiAs-type phase. The diffractograms present a peak broadening of the diffraction lines, due to de-mixing of the main tetragonal phase in two isostructural phases characterized by a slightly different [Se]/[Te] ratio. As the Se content increases a progressive peak splitting is observed (see Figure 1).

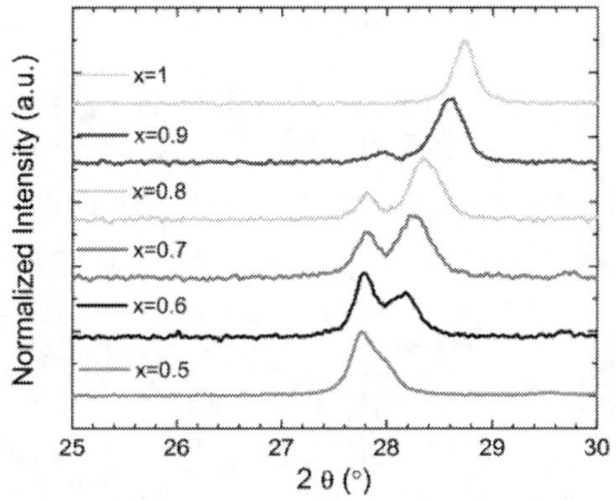

Figure 1. Magnification of the (101) Bragg's peak of the Anti PbO-Type structure, varying the nominal Se content.

Both lower angle peak, related to the Te enriched structure, and higher angle one, related to the Se enriched structure, shift at higher angles with increasing x, indicating that Se enters in both the structures. The calculated volumes for the tetragonal crystal cells are reported in Figure 2. For x that varies from x = 0.5 to x = 1, the percentage volume shrinking is about 8% for the Se enriched structure and about 3.5% for the Te enriched ones.

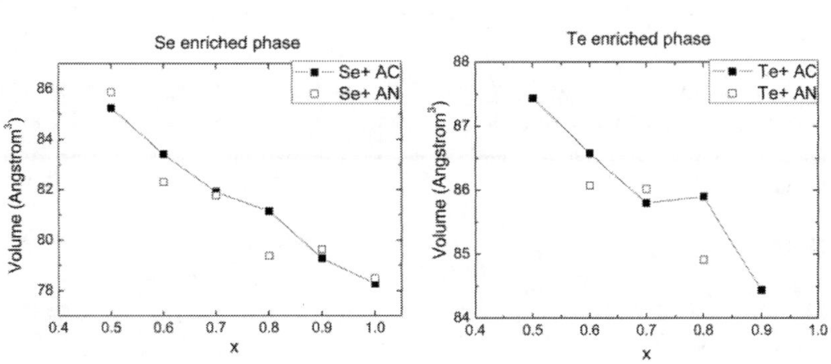

Figure 2. x dependence of the unit cell volume for the Se enriched (left) and Te enriched (right) structures.

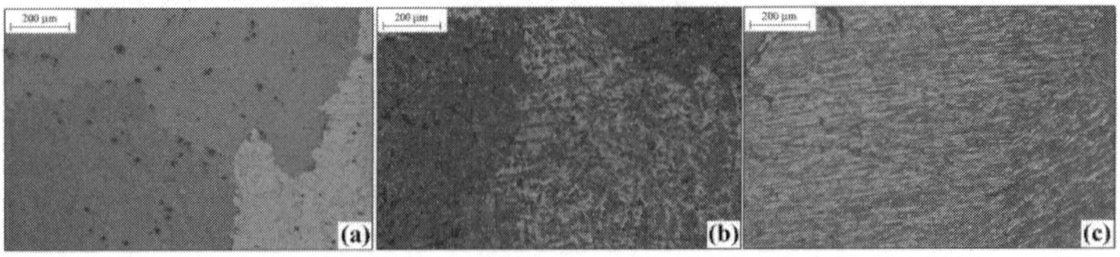

Figure 3. Polarized Light Optical Microscopy images of three samples: (a) x = 0.5, (b) x = 0.7 and (c) x = 1.

The contrast in PLOM imaging technique is related to the grains orientation; with this technique we are able to observe the grains structure in our samples. The PLOM images show an anomalous grain growth, varying the nominal Se concentration (see Figure 3). In the sample with x = 0.5 big oriented grains (~0.5 mm^2 size) are visible, while the FeSe sample (x = 1) presents small grains (~100 μm^2 size). Increasing x small grains with size close to those observed in the sample with x = 1 grow inside bigger ones (see Fig. 3 (b)). For all the samples no significant variations were observed after the annealing process. SEM & EDS analysis shows that the samples are constituted by three phases that differ mainly for the Se/Te ratio, Fig. 4: a light grey phase (B), a middle grey phases (D) surrounded by a dark grey phase (C).

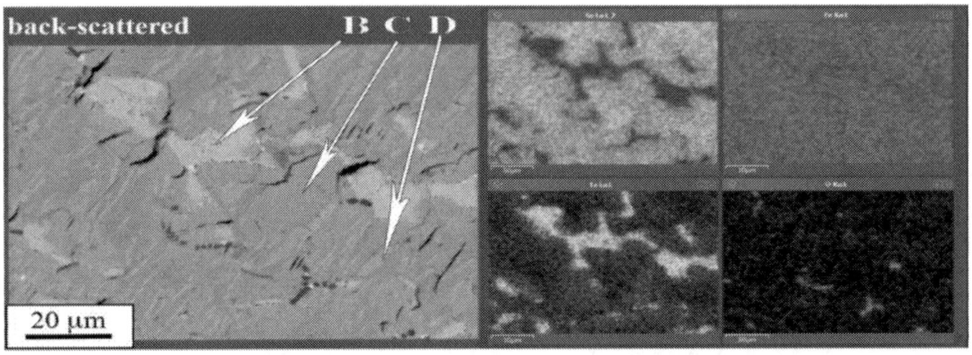

Figure 4. Back-scattered SEM image of the x = 0.9 sample (left), with EDS element mapping (right) for Se (top-left), Fe (top-right), Te (bottom-left) and O (bottom right).

Te is mainly concentrated in the B phase, where the Se/Te ratio varies from ≈ 0.7 to ≈ 1.2 and monotonically increases as x increases. On the contrary, C and D are mainly constituted of Se-rich phase with a Se/Te ratio variation from ≈ 2 up to ≈ 20, that rapidly increases with x. The Fe concentration is homogeneous (around 1) in the B and C phases, while the D phase presents an iron contraction from ≈0.90 to ≈1 that increases with increasing x. All the samples present similar microstructure with different relative quantity of the above cited phases, as reported in Figure 5 for the samples with $0.5 \le x \le 1$.

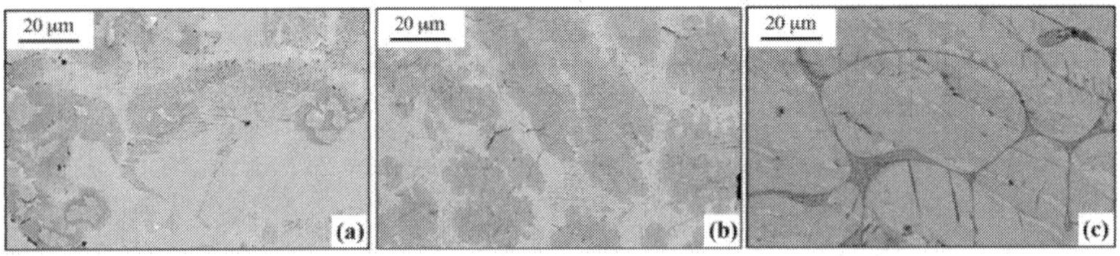

Figure 5. Back-scattered SEM image of three samples: (a) x = 0.5, (b) x = 0.7 and (c) x = 1.

Increasing x, C and D phases grow at the expense of the B one.

ZFC susceptibility measurement at low field (10^{-3} T). Are presented in Fig. 6. The x = 0.5 sample presents a sharp single step transition with Tc-onset ≈ 13.4 K and a full shielding volume fraction at 4.2 K. Double step transitions are observed in x = 0.6, 0.7 samples, with Tc-onset at 13.9 K and 12 K respectively. The sample with x = 1 presents a single step transition with Tc-onset ≈ 7.7 K as expected for FeSe samples. The onset of the second transition of the x = 0.6, 0.7 samples is ≈ 7.5 K, suggesting the presence of FeSe inclusions.

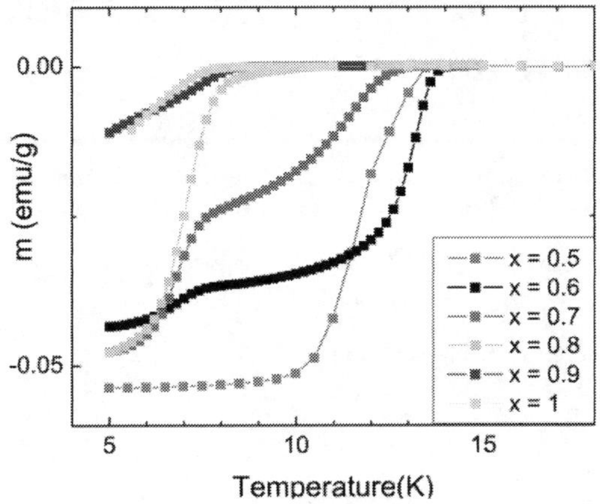

Figure 6. Magnetic susceptibility as a function of the temperature.

4. Conclusions

In this work we studied the Fe(Se_xTe_{1-x}) overdoped region of the phase diagram. Samples with nominal Se content x = 0.5, 0.6, 0.7, 0.8, 0.9, 1 were prepared by a melting process. Form the XRPD analysis we observed two isostructural anti-PbO type structures with different Se/Te ratio. This phase de-mixing increases rapidly with increasing Se content, with a clear bi-phasic behaviour for the x = 0.6, 0.7 as shown from the ZFC susceptibility. PLOM imaging shows an anomalous grain growth: small grains grow inside bigger ones with increasing x. The typical dimension of the small grains is the same as for the x = 1 sample. SEM-EDS analysis confirms a growth of a selenium enriched phase inside the tellurium enriched one. This suggest the existence of a miscibility gap in the FeSe – FeTe pseudo binary phase diagram. We observed that this phase de-mixing bring always to a worsening of the superconducting properties.

References

[1] Hsu F C, et al. 2008 *Proceedings of the National Academy of Sciences* **105** 14262
[2] Bhatia V, et al. 2011 *Chemical Communications* **47** 11297
[3] Wittlin A, et al. 2012 *Superconductor Science and Technology* **25** 065019
[4] Mizuguchi Y, et al. 2010 *Journal of the Physical Society of Japan* **79** 102001
[5] Palenzona A, et al. 2012 *Superconductor Science and Technology*, **25** 115018
[6] Sala A, et al. 2013 *Physica C: Superconductivity* **494** 69

Dependences of microwave surface resistance of HTS thin films on applied dc magnetic fields parallel and normal to the substrate

K. Sato[1], S. Sato[1], K. Ichikawa[1], M. Watanabe[1], T. Honma[1], Y. Tanaka[1], S. Oikawa[1], A. Saito[1] and S. Ohshima[1]

[1]Graduate School of Science and Engineering, Yamagata University, 4-3-16 Jonan, Yonezawa, Yamagata 992-8510, Japan

E-mail: txh11900@st.yamagata-u.ac.jp

Abstract. We investigated the dc magnetic field and temperature dependences of microwave surface resistance (R_s) of high-temperature superconductor (HTS) films. Previously, we reported that the surface resistance $R_s^{(n)}$ under a dc magnetic field applied normaly to the substrate increased when increasing the applied magnetic field. For NMR application, we have to examine the $R_s^{(p)}$ under the dc magnetic field parallel to the substrate. We measured the $R_s^{(p)}$ of the YBCO and DyBCO thin films with a thickness of 500 nm deposited on a MgO (100) substrate using the dielectric resonator method at 21.8 GHz, and a dc magnetic field of up to 5 T. In a zero magnetic field, the values of $R_s^{(n)}$ and $R_s^{(p)}$ were 0.35 mΩ at 20 K. Under the dc magnetic field, the $R_s^{(n)}$ and the $R_s^{(p)}$ also increased with increasing magnetic field, however, the $R_s^{(p)}$ had a lower magnetic field dependence and the value was about 1/10 of that of the $R_s^{(n)}$. The $R_s^{(p)}$ at 16.4 T and at 700 MHz could be estimated by the two-fluid model. The $R_s^{(p)}$ value was about 1/2600 compared with that of copper at 20 K. As a result, we clarified that 500 nm thick YBCO and DyBCO thin films could provide advantages for NMR application.

1. Introduction

The microwave surface resistance (R_s) of high-temperature superconductor (HTS) thin film is approximately two or three orders of magnitude smaller than that of the conducting metals usually used in the microwave frequency region. Because of this characteristic, passive microwave devices, such as filters [1] and antennas [2], have been widely investigated. Applications of HTS films with low R_s under a high magnetic field have been proposed, such as pick-up coils for magnetic resonance imaging and nuclear magnetic resonance (NMR). In order to realize these microwave devices, it is necessary to select and evaluate HTS materials by precise R_s measurement. For example, Honma et al. studied the R_s of YBCO thin films under high magnetic fields up to 12 T [3]. According to their study [3], the R_s of HTS films increases rapidly according to applied magnetic fields and thus it is thought that superconducting NMR pick up coils is difficult. However, in their experiment, a magnetic field was applied normaly to the substrate-plane, defined as $R_s^{(n)}$. In this study, a magnetic field was applied parallel to the substrate, defined as $R_s^{(p)}$. It is important to know how the $R_s^{(p)}$ behaves in a high magnetic region. However, there were few experimentally reports.

In this paper, we measured the $R_s^{(p)}$ of $YBa_2Cu_3O_y$ (YBCO) and $DyBa_2Cu_3O_y$ (DyBCO) superconducting films under high dc magnetic fields up to 5 T and investigated whether or not these materials are valid for NMR pick-up coils.

2. Experimental

We measured the $R_s^{(p)}$ in dc magnetic fields using the dielectric resonator method [4-7] and a superconducting magnet with NbTi (Suzuki Shokan Co., Ltd.: SSR-75). The dielectric resonator method is an international measurement standard for the R_s of HTS superconducting films [8].The R_s measurement system was described in detail in Ref. [7]. The resonator frequency of the TE_{011} mode was set to 21.8 GHz. In the cavity, we set the HTS films as shown in Fig. 1. Fig. 1 (a) represents the measurement cavity for $R_s^{(n)}$, and Fig. 1 (b) represents the measurement cavity for $R_s^{(p)}$. Therefore, the field was (a) normal and (b) parallel to the substrate-plane of the HTS films.

The film thickness and crystalline properties of the measured YBCO and DyBCO thin films are listed in Table 1. These films were deposited on MgO (001) substrates by the thermal co-evaporation method. The crystalline of the films was evaluated by $\theta/2\theta$ and φ-scan X-ray diffraction. The $\delta\gamma$ was the full-width half-maximum (FWHM) of the rocking curve of the (005) peak of the films. The $\delta\varphi$ was the FWHM of the φ-scan of the (102) plane of the thin film. From table 1, the $\delta\gamma$ values of the YBCO and DyBCO films were lower than 1.2 degree and the $\delta\varphi$ values were lower than 1.7 degree, respectively. Therefore, the crystalline properties of the measured YBCO and DyBCO thin films were almost the same.

3. Results and discussion

3.1. Temperature dependence of $R_s^{(n)}$ and $R_s^{(p)}$

Fig. 3 shows the temperature dependence of the $R_s^{(n)}$ and $R_s^{(p)}$ of 500 nm thick YBCO and DyBCO thin films in a zero magnetic field at 21.8 GHz. The $R_s^{(n)}$ [3] and $R_s^{(p)}$ values at 20 K were both 0.35mΩ. Measurement error between each cavity was less than 5 percent.

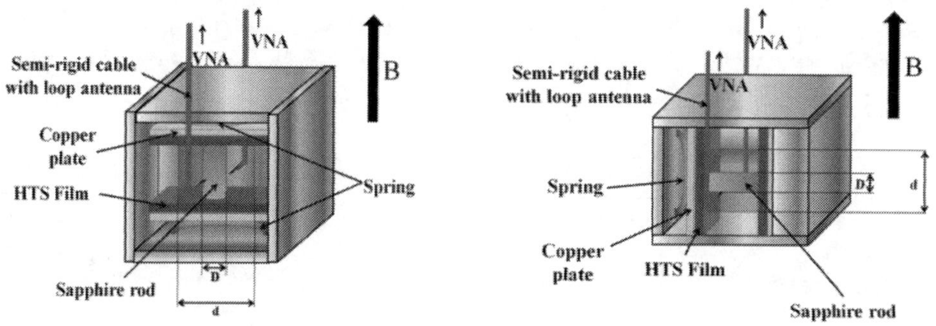

(a) (b)

Fig. 1. Schematic drawing of cavity for (a) $R_s^{(n)}$ and (b) $R_s^{(p)}$

Table 1. Crystalline properties of YBCO and DyBCO thin films (Produced by THEVA)

HTS thin films	$YBa_2Cu_3O_{7-\delta}$	$DyBa_2Cu_3O_{7-\delta}$
Substrate	MgO (0.5 nm thickness)	
Cap layer	CeO_2 (30 nm thickness)	
Thickness [nm]	500	
$\delta\gamma$ [deg.]	1.09	1.17
$\delta\varphi$ [deg.]	1.63	1.3

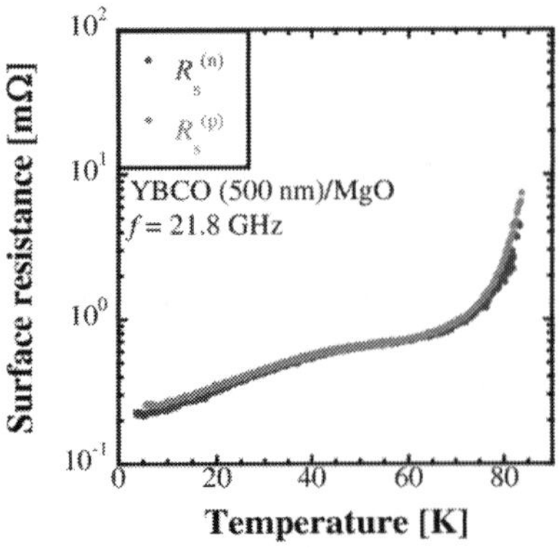

(a)

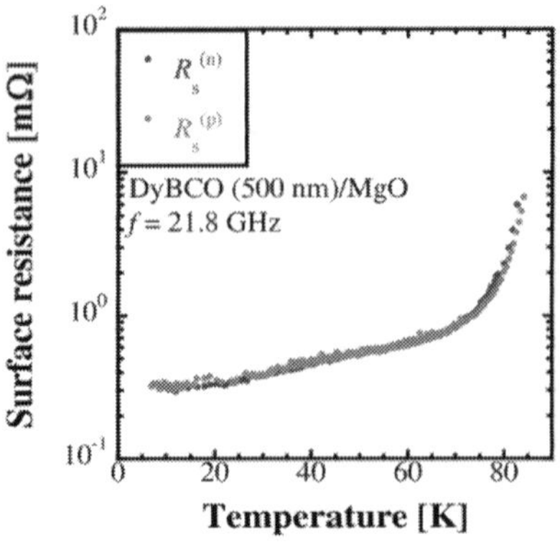

(b)

Fig. 3. Temperature dependence of $R_s^{(n)}$ and $R_s^{(p)}$ for (a) YBCO and (b) DyBCO thin films with a thickness 500 nm in a zero magnetic field.

3.2. DC magnetic field dependence of $R_s^{(n)}$ and $R_s^{(p)}$

Fig. 4 shows the dc magnetic field dependence for the R_s of YBCO and DyBCO thin films with a thickness of 500 nm in a dc magnetic field between 0 and 5 T at 20 K. The $R_s^{(n)}$ and $R_s^{(p)}$ values of the two materials increased with a dc magnetic field up to 5 T. However, the value of $R_s^{(p)}$ had low magnetic dependence compared with $R_s^{(n)}$. The value of $R_s^{(p)}$ was about 1/13 for YBCO and 1/9 for DyBCO compared with the $R_s^{(n)}$. These results could be explained by using the two-fluid model for high frequency and low magnetic field limits [10].

Under a low magnetic field, the R_s is given as follows,

$$R_s(B) = R_s(0\mathrm{T}) + \frac{B\Phi_0}{2\lambda_L \eta_{\mathrm{low}}(B,T,d)} \qquad \left(B \ll \frac{\omega\mu_0\lambda_L^2}{\Phi_0\hat{\mu}} \right) \qquad (1).$$

Under a high magnetic field, the R_s is given as follows,

$$R_s(B) = \sqrt{\frac{\omega\mu_0\Phi_0 B}{2\eta_{\mathrm{high}}(B,T,d)}} \qquad \left(B \gg \frac{\omega\mu_0\lambda_L^2}{\Phi_0\hat{\mu}} \right) \qquad (2).$$

Here, B is the magnitude of the applied dc magnetic field, d is the thickness of films, ω the angular frequency, Φ_0 the flux quantum, μ_0 the vacuum permeability, λ_L the London penetration depth, $\hat{\mu}$ the dynamic mobility and η the viscous drag coefficient which is related to the motion of the magnetic flux. We considered η as a function of the magnetic field and temperature [3]. In Fig. 4, solid lines represent the fitted data of R_s under a low magnetic field calculated using Eq. (1). Dotted lines represent fitted data of R_s under a high magnetic field calculated using Eq. (2). In low magnetic regions, $R_s^{(n)}$ increased linearly with the applied magnetic field. In high magnetic regions, $R_s^{(n)}$ also increased proportional to the square root of the magnetic field. The values for the $R_s^{(n)}$ of YBCO and DyBCO thin films increased linearly with the applied magnetic field up to 5 T respectively. Meanwhile, the $R_s^{(p)}$ of various materials displayed extremely low magnetic field dependences up to 5 T. Therefore, in thin films with strong intrinsic pinning, motion of flux was reduced.

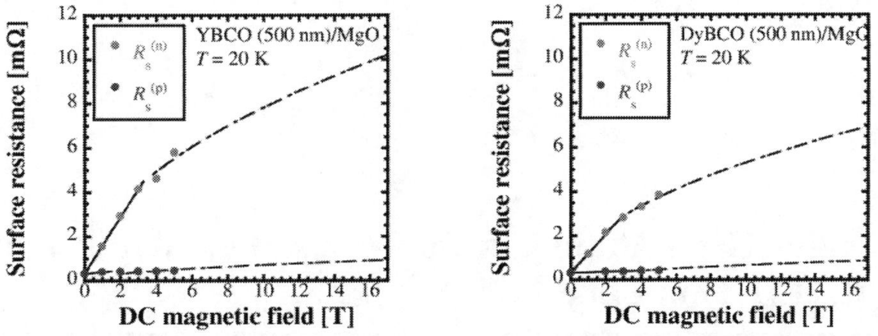

(a) (b)

Fig. 4. The dc magnetic field dependence of $R_s^{(n)}$ and $R_s^{(p)}$ at 20 K

An R_s value of 25 μΩ at f = 700 MHz, T = 20 K, B = 16.4 T was required for NMR pick-up coils. The $R_s^{(p)}$ of YBCO and DyBCO were 0.94 and 0.88 μΩ respectively. These values were about 1/2650 and 1/2840 compared with that of copper. These results show that 500 nm thick YBCO and DyBCO thin films could provide advantages for NMR pick-up coil materials. For NMR application, HTS materials need not only possess low R_s, but also have a high critical current density (J_c). We will continue to consider HTS pick-up coils for NMR application in the future.

4. Conclusion

In this paper, we investigated the dc magnetic field and temperature dependences of microwave surface resistance ($R_s^{(n)}$, $R_s^{(p)}$) in HTS films for NMR application. The $R_s^{(p)}$ value shows low magnetic field dependence compared with $R_s^{(n)}$. This behavior is caused by the layered structure of YBCO and DyBCO thin films. The motion of flux was hindered. The estimated values of 500 nm thick YBCO and DyBCO thin films at f = 700 MHz, T = 20 K, and B = 16.4 T were 0.94 and 0.88 μΩ respectively. These values were lower than 25 μΩ, which was less than 1/1000 of the R_s of copper. From these results, we clarified that YBCO and DyBCO thin films could provide advantages for NMR application.

Acknowledgements

This work was supported by the Strategic Promotion of Innovation Research and Development of JST (Development of the next-generation NMR technology using HTS materials), JSPS KAKENHI Grant Numbers 22560317, 24560393.

References

[1] A. Saito, J. H. Lee, H. Teshima, and S. Ohshima, "Power-Handling Capability of Transmit Filters Using Superconducting Bulk Ring Resonators," *IEEE Trans. Appl. Supercond.*, vol. 19, No. 3, pp.899-902, 2009.

[2] S. Ohshima, "High-Temperature superconducting passive microwave devices, filters antennas," *Supercond. Sci. Technol.*, vol. 103-108, pp.61-66, 2000.

[3] T. Honma, S. Sato, K. Sato, M. Watanabe, A. Saito, K. Koike, H. Kato, S. Ohshima, "Microwave surface resistance of YBCO superconducting thin films under high DC magnetic field," *Physica C*, vol. 484, pp. 46-48, 2013.

[4] Y. Kobayashi, M. Katoh, "Microwave measurement of dielectric properties of low-loss materials by the dielectric rod resonator method," *IEEE Trans. Microwave Theory Tech.*, vol. MTT-33, No. 7, pp. 586-592, 1985.

[5] M. Kusunoki, M. Inadoumaru, S. Ohshima, K. Aizawa, M. Mukaida, M. Lorenz, H. Hochmuth, "Dielectric loss tangent of sapphire single crystal produced by edge-defined film-fed growth method," *Physica C*, vol. 377, pp. 313-318, 2002.

[6] S. Y. Lee, B. J. Soh, J. W. Ahn, J. Y. Cho, B. H. Park, C. S. Jung, V.B. Fedorov, A. G. Denisov, Y. H. Kim, T. S. Hahn, S. S. Choi, B. Oh and S. H. Moon, "Use of a dielectric-loaded cylindrical cavity in measurements of the microwave surface resistances of high-Tc superconducting thin films," *IEEE Trans. Appl. Supercond.*, vol. 7, pp. 2013-2017. 1997.

[7] K. Nakagawa, T. Honma, K. Takeda, S. Ono, H. Kai, A. Saito, M. Mukaida, K. Nakajima, S. Ohshima, "Intrinsic Surface Resistance of YBCO Thin Films under DC Magnetic Field," *IEEE Trans. Appl. Supercond.*, vol. 21, No3, pp. 587-597. 2011.

[8] International Electrotechnical Commission (IEC) 61788-7, second ed., Superconductivity – Part 7 : Electronic characteristic measurement-Surface resistance of superconductors at microwave frequencies, 2007.

Bicrystalline Grain Boundary Junctions of Co-doped and P-doped Ba-122 Thin Films

S Schmidt[1], S Döring[1], F Schmidl[1], F Kurth[2], K Iida[2], B Holzapfel[2], T Kawaguchi[3], Y Mori[3], H Ikuta[3], P Seidel[1]

[1] Friedrich-Schiller-University Jena, Institute of Solid State Physics, Helmholtzweg 5, 07743 Jena, Germany
[2] IFW Dresden, Institute for Metallic Materials, Helmholtzstrasse 20, 01069 Dresden, Germany
[3] Nagoya University, Department of Crystalline Materials Science, Chikusa-ku, 464-8603 Nagoya, Japan

E-mail: stefan.schmidt2@uni-jena.de

Abstract. We prepared GB junctions of $Ba(Fe_{0.9}Co_{0.1})_2As_2$ thin films on bicrystalline [001]-tilt $SrTiO_3$ substrates. The junctions show clear Josephson effects. Electrical characterization shows asymmetric I-V characteristics which can be described within the resistively shunted junction (RSJ) model. A large excess current is observed. Their formal I_CR_N product is 20.2 μV at 4.2 K, which is decreased to 6.5 μV when taking I_{ex} into account. Fabrication methods to increase this value are discussed. Additionally, measurements on GB junctions of $BaFe_2(As_{0.66}P_{0.34})_2$ thin films on LSAT bicrystalline substrates are shown. Their symmetric RSJ/flux flow-behavior exhibits a formal I_CR_N product of 45 μV, whereas the excess corrected value is 11 μV.

1. Thin Film preparation

1.1. Co-doped Ba-122

Grain boundary junctions play an important role regarding applications from superconducting wires to superconducting quantum interference devices (SQUID). Pnictide superconductors in particular have advantageous properties compared to cuprates at large grain boundary (GB) angles [1]. In systematic investigations of suitable substrate materials by Iida et al. [2] and Thersleff et al. [3] STO substrates proved beneficial for the superconducting properties of Co-doped Ba-122. By introducing a buffer layer of crystalline iron the critical current density j_C and the transition temperature T_C can be increased compared to other substrate materials or STO substrates without additional iron buffer layer.

To improve these results further a buffer layer of $MgAl_2O_4$ with a thickness of 10 nm is deposited by Pulsed Laser Deposition (PLD) between the STO [001]-tilted bicrystal substrate with an artificial grain boundary (GB) angle of 30° and the Fe buffer (20 nm). The crystal orientation is rotated by 45° between the substrate and the Fe buffer due to the different lattice constants. The (001) surface plane of crystalline iron (bcc structure) and the distance between two iron atoms within the unit cell of Co-doped Ba-122 deviate by just 2% [3]. By depositing superconducting $BaFe_{1.84}Co_{0.16}As_2$ on the buffer stack the crystal orientation is rotated by another 45° while the artificial lattice mismatch of the substrate is perfectly passed on to the pnictide thin film [4]. The superconducting transition starts at 24 K and reaches zero resistance at 23 K.

1.2. P-doped Ba-122

In contrast to the PLD growth of Co-doped Ba-122 on STO substrates we used Molecular Beam Epitaxy (MBE) to deposit P-doped Ba-122 (figure 1). By using MBE epitaxial growth of optimally doped (34 % phosphorous on the arsenic site) Ba-122 could be achieved without any buffer layers, see figure 2. We want to note that the optimal doping level of P-doped Ba-122 depends on the substrate material.

The [001]-tilted LSAT bicrystalline substrate with a mismatch angle of 45° is heated up to 780 °C in the MBE chamber, where the bulk materials Fe, Ba, As, and GaP are used as precursors. The solids are heated up in their effusion cells to 1065 °C (Fe), 537 °C (Ba), 186 °C (As), and 702 °C (GaP), respectively. The resulting gases Fe, Ba, As_4 and P_2 are deposited stoichiometrically on the LSAT substrate at a very slow growth rate of 100 nm per hour. The superconducting transition of the resulting $BaFe_2As_{1.32}P_{0.68}$ thin films occurs between 34.5 K ($T_{C,90}$) and 30.5 K ($T_{C,10}$).

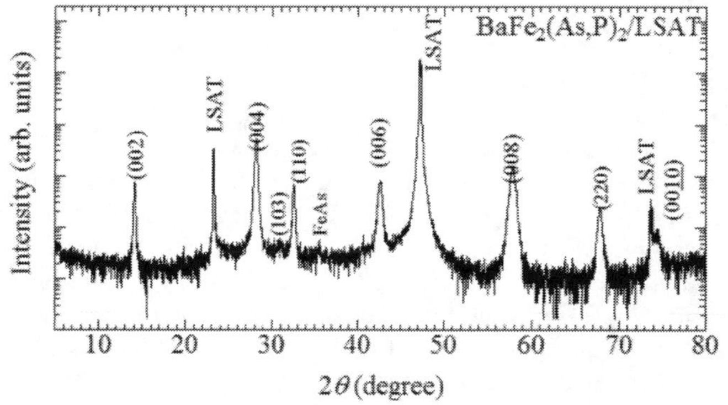

Figure 1. XRD scan of a whole P-doped Ba-122 thin film on LSAT without any additional buffer layers.

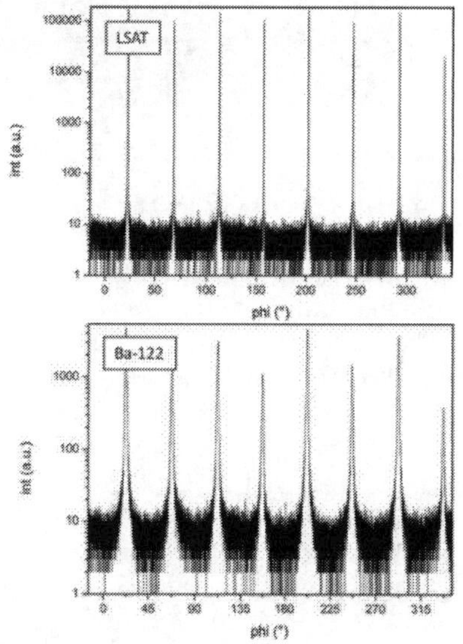

Figure 2. Phi-scans of LSAT (220) and P-doped Ba-122 (103). No signs of twinning or imperfect epitaxy between substrate and thin film are visible.

2. Junction preparation

The pnictide thin films are patterned by Ion Beam Etching (IBE) with a beam voltage of 500 V according to a photomask. This photolithographic layout forms microbridges of Ba-122 over the artificial grain boundary of both the substrate and the pnictide thin film (see figure 3). The bridge widths vary between 2 μm (see figure 4 left) and 10 μm (see figure 4 right).

Prior to the etching process the Co-doped thin film was covered with a gold layer to prevent surface degradation due to the exposure to air and chemicals used in the photolithographic process. Thus, the Co-doped GB junctions are shunted resistively by both the Au cap layer and the Fe buffer layer. The P-doped Ba-122 grown by MBE resulted to be more stable and do not need any cap layer. Not needing buffer layers, the P-doped thin films are not shunted extrinsically.

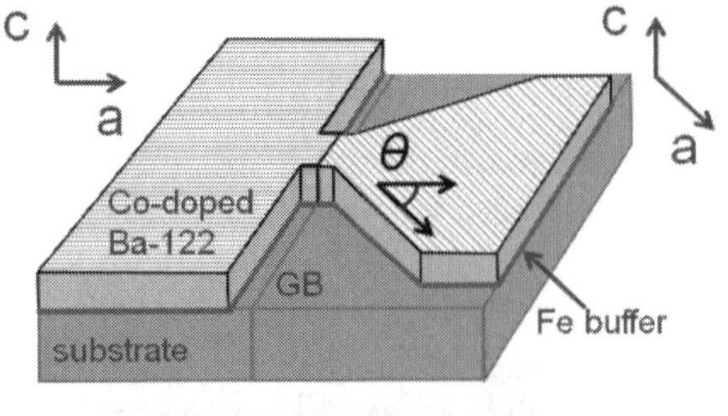

Figure 3. (left) Sketch of a microbridge forming a GB junction on a bicrystalline substrate (blue). The red layer underneath the superconducting thin film (white) denotes the Fe buffer, which is just present at the Co-doped Ba-122 films. The surface of the Co-doped material is capped with an Au layer, additionally.

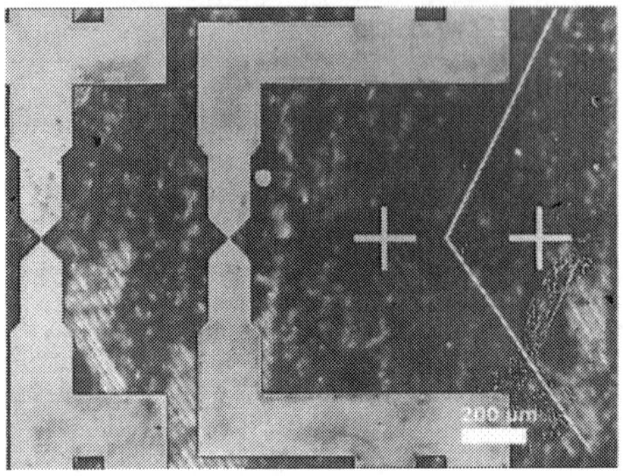

Figure 4. (left) Two microbridges of P-doped Ba-122 over the GB area with widths of 2 μm (left) and 10 μm (right), respectively. The GB area is marked with red lines and is visible as a thin line in the middle, oriented horizontally. The LSAT substrate orientation and its mismatch of 45° are marked by the yellow lines. The crosses are markers for adjusting the photomask more precisely.

3. Electrical measurements on Co-doped Ba-122

Both thin film materials show Josephson-like behaviour when measuring over the GB.

In the case of Co-doped Ba-122, the I-V characteristics can be modelled within an asymmetric RSJ model at low temperatures. A large excess current of 645 μA lowers the formal I_C from 950 μA to 305 μA (see figure 5) which decreases the formal I_CR_N value of 20.2 μV to a corrected value of 6.5 μV [4]. The critical current density across the microbridge is up to 1.4 10^5 A cm^{-2} (formal value), which is also reduced to 1.2 10^4 A cm^{-2} (corrected value). Instead the P-doped Ba-122 microbridges exhibit symmetric I-V characteristics, which can be fitted by a flux-flow extension of the RSJ model as shown by Saitoh et al. [5]. A formal I_C of 440 μA as visible in figure 6 leads to a formal I_CR_N product 45 μV which has to be corrected to a value as low as 11 μV when taking I_{ex} into account.

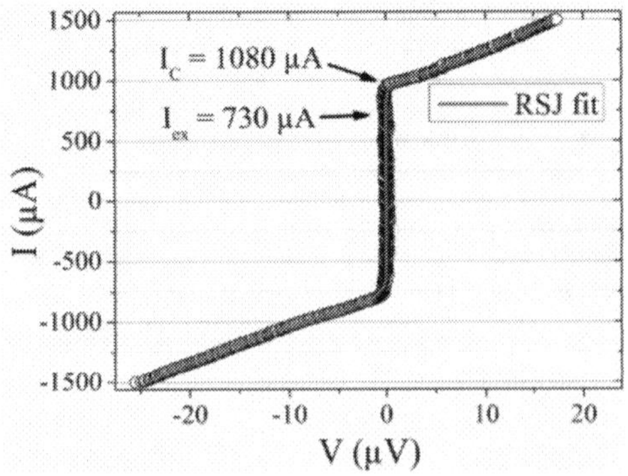

Figure 5. I-V characteristics of a microbridge (d = 7 μm) of Co-doped Ba-122 on a 30° STO bicrystal substrate. The red line shows a classical RSJ fit.

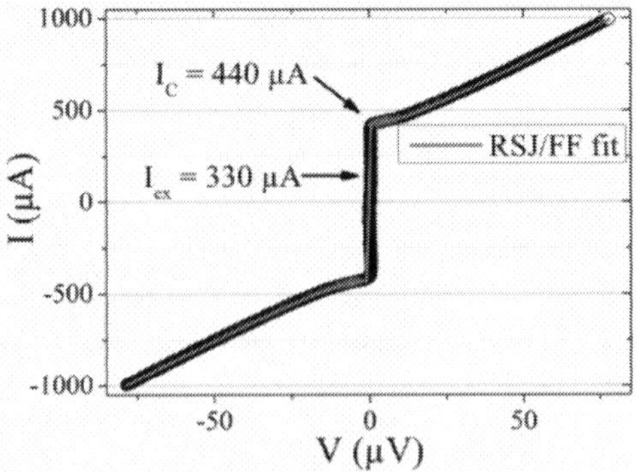

Figure 6. I-V characteristics of a microbridge (d = 10 μm) of P-doped Ba-122 on a 45° LSAT bicrystal substrate. The red line shows a combined RSJ and flux-flow fit as in ref. [5].

The critical currents over the microbridges decay with increasing temperature differently for both materials. While the Co-doped junctions follow a rather linear decline (see figure 7), the P-doped ones behave nonlinear at both low temperatures and near T_C (see figure 8), which may be explained by a more SINIS-like behaviour of the GB [6], but further investigation would be required to verify this assumption. The low critical current of the Co-doped bridges in comparison to the results shown in figure 5 are caused by disturbances of the GB area during an ion beam etching process to remove the resistive shunt of the Au cap layer mentioned in section 2. The general temperature dependence before the cap layer removal was linear as well. During this process not only the critical current drastically decreased, but the normal state resistance increased simultaneously leaving the I_CR_N product unchanged. With a more sensitive way to remove the Au cap layer, the normal state resistivity should be increased by leaving I_C more or less constant. In this way we could tune, i.e. increase the I_CR_N product.

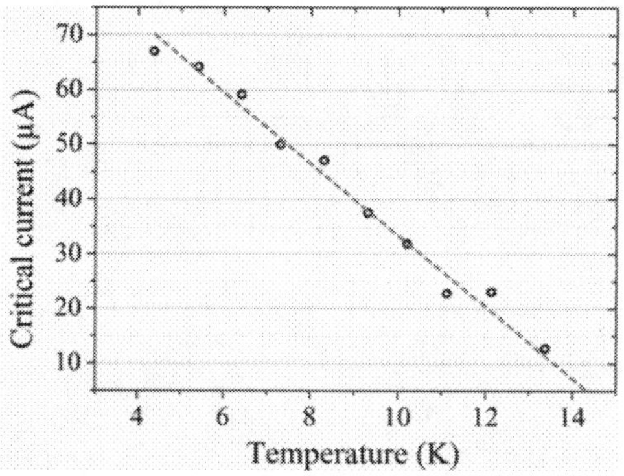

Figure 7. Critical current versus temperature of a Co-doped Ba-122 microbridge. The dashed grey line is a guide for the eye.

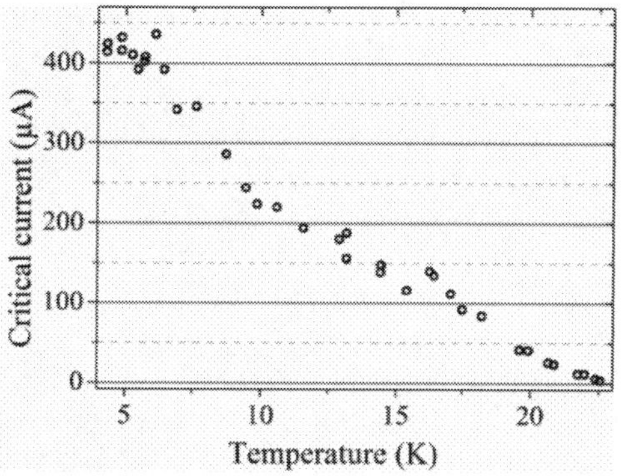

Figure 8. Critical current versus temperature of a P-doped Ba-122 microbridge.

Measurements of the Co-doped GB junctions under microwave irradiation and under influence of an external magnetic field are shown in ref. [4] and suggest that the Fe-buffer shields microwaves, effectively. Additionally, the buffer stack disturbs magnetic field measurements by trapping magnetic flux and making the critical current over magnetic field dependence highly asymmetric. First measurements on P-doped and thus non-buffered Ba-122 GB junctions show clear microwave dependence and a higher stability in magnetic fields instead.

Acknowledgements
This work was partially supported by DFG under project no. SE 664/15-2, the EU under project no. FP7-283141 (IRON-SEA), and the Landesgraduiertenförderung Thüringen.

References

[1] Katase T, Ishimaru Y, Tsukamoto A, Hiramatsu H, Kamiya T, Tanabe K and Hosono H 2011 *Nature Comm.* **2** 409
[2] Iida K, Haindl S, Thersleff T, Hänisch J, Kurth F, Kidszun M, Hühne R, Mönch I, Schultz L, Holzapfel B and Heller R 2010 *Appl. Phys. Lett.* **97** 172507
[3] Thersleff T, Iida K, Haindl S, Kidszun M, Pohl D, Hartmann A, Kurth F, Hänisch J, Hühne R, Rellinghaus B, Schultz L and Holzapfel B 2010 *Appl. Phys. Lett.* **97** 022506
[4] Schmidt S, Döring S, Schmidl F, Tympel V, Haindl S, Iida K, Kurth F, Holzapfel B and Seidel P 2013 *IEEE Trans. Appl. Supercond.* **23** 7300104
[5] Saitoh K, Ishimaru Y, Fuke H and Enomoto Y 1997 *Jpn. J. Appl. Phys.* **36** L272–L275
[6] Golubov A A and Kupriyanov M Y 1989 *Sov. Phys. JETP* **69** 805-812

Journal of Physics: Conference Series **507** (2014) 012047 doi:10.1088/1742-6596/507/1/012047

Growth of a smooth CaF_2 layer on NdFeAsO thin film

N Sumiya[1], T Kawaguchi[1, 5], M Chihara[1], M Tabuchi[2], T Ujihara[3], A Ichinose[4], I Tsukada[4], H Ikuta[1]

[1]Department of Crystalline Materials Science, Nagoya University, Japan
[2]Synchrotron Radiation Research Center, Nagoya University, Japan
[3]Department of Materials Science and Engineering, Nagoya University, Japan
[4]Central Research Institute of Electric Power Industry, Kanagawa, Japan

E-mail: sumiya@iku.xtal.nagoya-u.ac.jp

Abstract. We studied the method to grow a smooth and flat CaF_2 layer on NdFeAsO thin films since CaF_2 is a promising candidate material for the barrier layer of a superconducting junction. When the CaF_2 layer was grown at 800°C, the surface was very rough because {111} facets had grown preferentially. However, when CaF_2 was grown at lower temperatures and post-annealed *in situ* at 800°C for 30 min the facets were eliminated and a CaF_2 layer with a smooth surface was obtained. Fluorine diffusing from CaF_2 into NdFeAsO was observed when CaF_2 was grown at high temperatures, but the diffusion was suppressed by lowering the growth temperature to 400°C.

1. Introduction

The recently discovered iron based superconductors [1] have a high transition temperature (T_c) that benefits greatly to practical applications. Fabrication of Josephson junctions using iron-based superconductors are currently actively studied as it forms the basis of electronic applications [2]. Among the various types of junctions, sandwich-type planar junctions have an advantage of the easiness of integration, and are the most fundamental and elemental type of junctions for superconducting electronics. Fabrication of this type of junctions has been reported for $Ba(Fe,Co)_2As_2$ [3, 4], which is one of the most studied iron-based superconductors. However, a junction of this type using a *Ln*FeAs(O,F) (*Ln*=lanthanide) thin film has yet not been realized, although it has the highest T_c among the iron-based superconductors discovered so-far [5].

Over the past few years, a considerable effort has been devoted to grow high quality *Ln*FeAs(O,F) thin films. Our group has grown NdFeAs(O,F) [6-9] and LaFeAs(O,F) [10] thin films by molecular beam epitaxy (MBE). Other groups have reported on the growth of SmFeAs(O,F) thin films with a similar method [11], and LaFeAs(O,F) thin films by pulsed laser deposition [12]. As the next step toward realizing Josephson junctions, it is necessary to grow a thin and smooth barrier layer on these superconductors. CaF_2 is known as a good insulator with a wide band gap. In addition, the NdFeAs(O,F) thin film exhibited the highest T_c when it was grown on a CaF_2 substrate in our previous study where various substrates were used [9]. Therefore, the second superconducting layer that would be grown on the barrier layer is expected to have a high T_c if CaF_2 were used as the insulator.

[5] Present Address: Institut für Festkörperphysik Friedrich-Schiller-Universität Jena, Germany

For these reasons, we have grown CaF_2 on NdFeAsO thin films in this work. However, as will be shown below, we found that the surface of the CaF_2 layer was very rough because {111} facets had grown preferentially. The surface roughness was too large that the fabrication of a Josephson junction would be difficult. On the other hand, it has been reported that CaF_2 layers grown on Si(100) substrates had similarly rough surfaces, but could be flattened by post-growth annealing at high temperatures [13]. This idea was applied in the present work to study the method of fabricating a flat CaF_2 layer on NdFeAsO thin film.

2. Experiment

NdFeAsO thin films with a thickness of 50 nm were grown on MgO(100) substrates by MBE. The growth method was similar to that reported elsewhere [8], except that the growth temperature was increased to 800°C and the oxygen source was changed to He diluted O_2 gas. In the present study, we focused on fabricating a flat and smooth CaF_2 layer on the NdFeAsO phase. Therefore, fluorine was not doped to the NdFeAsO phase to shorten the fabrication process. After the growth of the NdFeAsO layer, CaF_2 was grown with a rate of 5 nm/min for 5 min at temperatures that will be described later. A Knudsen cell was used to supply CaF_2. After the growth of CaF_2, some of the thin films were annealed at 800°C for 30 min, which was carried out *in situ*.

Reflection high energy electron diffraction (RHEED) was used to characterize the surface during thin film growth. The obtained thin films were characterized by x-ray diffraction (XRD), atomic force microscope (AFM), and transmission electron microscope (TEM), and the composition by electron probe microanalysis (EPMA). Resistivity was measured by a four-probe method.

3. Results

Figure 1(a) shows an AFM image of CaF_2 grown at 800°C on NdFeAsO. Obviously, the surface is very rough and consists of many islands. The root mean square roughness (R_{RMS}) was 17.8 nm, which is too large for junction fabrication. Figure 1(b) shows the RHEED pattern after the growth of CaF_2 for the same sample. While a streaky RHEED pattern was observed during the growth of NdFeAsO that indicates the base layer was flat, the RHEED pattern of the CaF_2 layer was spotty, consistent with the island structures shown in figure 1(a). Weak streaks running between the spots can also be seen in figure 1(b). The angle between the off-angle streaks is 109°, which indicates the existence of {111} facets.

The formation of a faceted surface was also reported when CaF_2 was grown on Si(100) substrates [13,14]. In addition, it was shown that the CaF_2 surface can be flattened by post-growth annealing [13]. Figure 2 shows AFM images of two CaF_2 layers grown on NdFeAsO at 800 and 600°C, both followed by an annealing at 800°C for 30 min. When the growth temperature (T_g) was 800°C, the CaF_2 surface had a morphology similar to figure 1(a), thus the annealing had no effect on the surface roughness. On the other hand, the T_g = 600°C sample had a far smoother surface. Figure 3 indicates the RHEED patterns of the sample shown in figure 2(b). Figure 3(a) was taken right after the growth of CaF_2, which was spotty and indicates an island growth mode. The RHEED pattern changed to a streaky one after annealing as shown in figure 3(b). Thus, the flattening of the surface occurred during the annealing at 800°C.

We have also grown the CaF_2 layer at a lower temperature, T_g = 400°C, and annealed it at 800°C. Figure 4(a) shows the AFM image of this sample. R_{RMS} was 2.38 nm, which is better than the T_g = 600°C sample. Figures 4(b) and (c) show cross-sectional scanning TEM (STEM) images of the T_g = 800°C (figure 2(a)) and T_g = 400°C (figure 4(a)) samples. Consistent with the AFM images, the

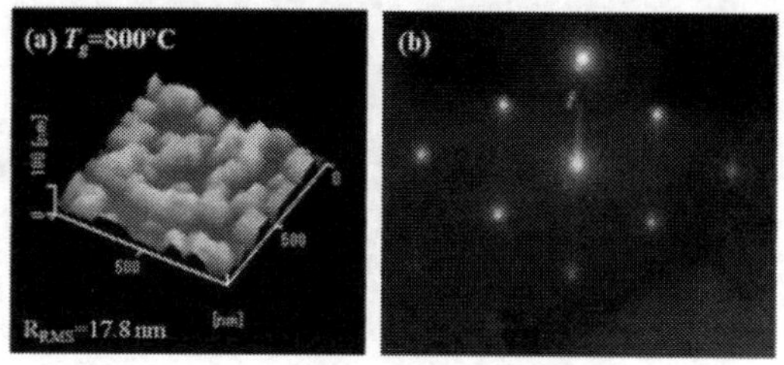

Figure 1. (a) An AFM image of the CaF_2 layer grown at 800°C on NdFeAsO. (b) The RHEED image of the same sample right after the growth of CaF_2.

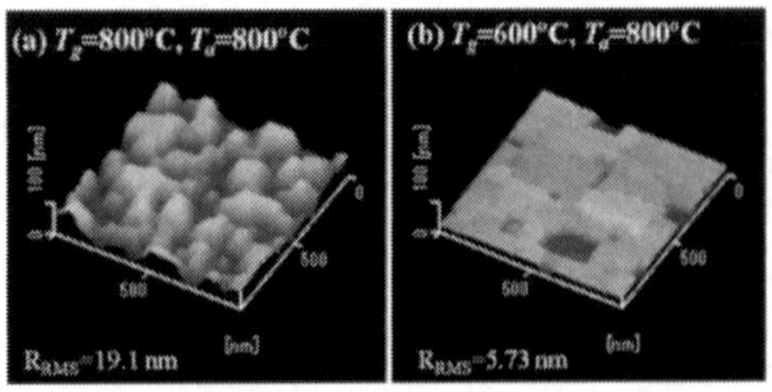

Figure 2. AFM images of the CaF_2 layers grown on NdFeAsO at (a) 800°C and (b) 600°C. Both samples were annealed at 800°C for 30 min.

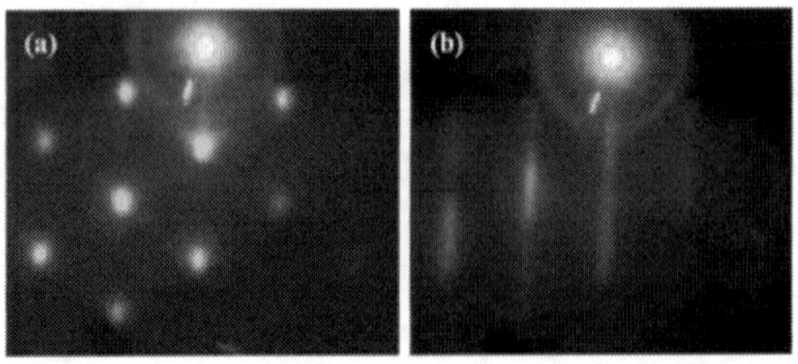

Figure 3. RHEED patterns of the same sample as figure 2(b) taken (a) before and (b) after annealing at 800°C for 30 min.

STEM observation revealed a very rough surface for the T_g = 800°C sample. It should also be noted that NdFeAsO is covered only partially with CaF_2. In contrast, CaF_2 covered completely NdFeAsO when it was grown at 400°C.

4. Discussion

The faceted growth of CaF_2 can be understood by the difference in the free energies of the {111} and {100} surfaces [14]. As mentioned above, Asano *et al.* have shown that a flat CaF_2 layer can be obtained on Si(100) by post-annealing [13]. They have also grown a second CaF_2 layer on the post-annealed (flattened) first layer, and found that the surface was facetted again when the second layer was grown at 600°C but remained flat at 800°C, which led them to suggest that the relative ratio of the free energy might alter at high temperatures. This is consistent with the RHEED patterns shown in figure 3 that suggest the {111} plane of CaF_2 is stable at 600°C while the {100} plane is more stable at 800°C. However, in contrast to the second CaF_2 layer on the flattened first layer in the study by Asano *et al.*, our sample had a rough surface when CaF_2 was grown at 800°C (figure 2(a)). A possible reason might be that the bonding between CaF_2 is stronger than that to NdFeAsO. In that case, island growth of CaF_2 can take place before the NdFeAsO surface is covered if the migration of CaF_2 is sufficiently large at high temperatures. This is probably the reason why part of the surface of NdFeAsO was not covered in figure 4(b). At lower temperatures, on the other hand, CaF_2 covers the surface of NdFeAsO (figure 4(c)) because the migration is suppressed.

Figure 5 shows the temperature (T) dependence of resistivity (ρ) of the thin films prepared with different growth temperatures of CaF_2. As mentioned above, we used non-doped NdFeAsO as the base layer, and a superconducting transition was not expected. However, we observed clear superconducting transitions for the T_g = 600 and 800°C thin films, which suggest that fluorine had diffused from CaF_2 to the NdFeAsO layer. The lower T_c of the T_g = 600°C thin film, together with the behavior of the ρ-T curve, indicates that this film has a lower fluorine content. On the other hand, no superconducting transition was observed for the T_g = 400°C thin film. Hence, the higher the growth temperature, the more fluorine was diffused into the NdFeAsO layer. Note that all three samples were

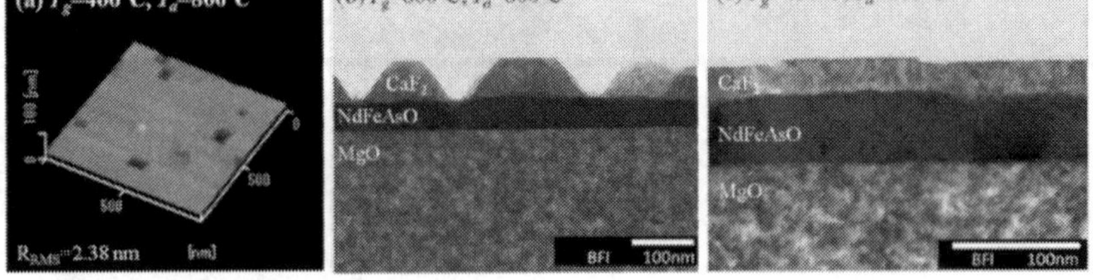

Figure 4. (a) An AFM image of the CaF_2 layer grown on NdFeAsO at 400°C and annealed at 800°C. (b) A cross-sectional STEM image of the sample shown in figure 2(a). (c) A cross-sectional STEM image of the sample shown in figure 4(a).

annealed at 800°C for 30 min after the growth of CaF_2, which is equal to the highest T_g of the three samples. Nevertheless, the amount of diffused fluorine depended on the growth temperature, suggesting that fluorine diffusion did not occur during annealing but the growth process. We speculate that fluorine may easily diffuse from molecular CaF_2 at the arrival at the substrate when it has enough thermal energy, but not after it is crystallized. Regardless of the reason why fluorine diffusion depended on T_g, however, the results of figure 5 suggest that fluorine was not diffused from CaF_2 into the NdFeAsO phase for the T_g = 400°C thin film and we can expect that a good interface is formed between the two layers.

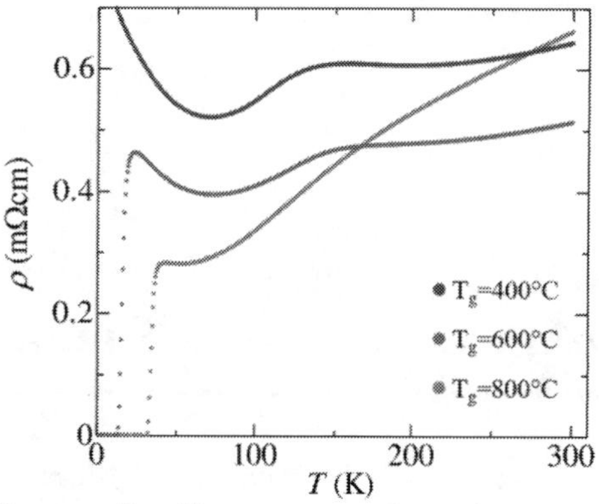

Figure 5. Temperature dependence of resistivity of CaF_2/NdFeAsO/MgO multilayers. CaF_2 was grown at 400, 600 and 800°C, all followed by an annealing at 800°C for 30 min.

5. Conclusions

Aiming the fabrication of sandwich-type planar superconducting junctions, we studied the method to fabricate a flat and smooth CaF_2 layer on NdFeAsO thin films. The as-grown CaF_2 layer exhibited a very rough and faceted surface. The facets were eliminated and a flat surface was obtained by post-growth annealing at 800°C when CaF_2 was grown at lower temperatures. Fluorine diffusing from CaF_2 into NdFeAsO was observed when CaF_2 was grown at high temperatures, but was suppressed at T_g = 400°C.

Acknowledgements

This study was partially supported by Strategic International Collaborative Research Program (SICORP), Japan Science and Technology Agency, and by Grant-in-Aid from MEXT, Japan.

References

[1] Kamihara Y, Watanabe T, Hirano M and Hosono H 2008 *J. Am. Chem. Soc.* **130** 3296
[2] Seidel P 2011 *Supercond. Sci. Technol.* **24** 043001
[3] Schmidt S, Döring S, Schmidl F, Grosse V, Seidel P, Iida K, Kurth F, Haindl S, Mönch I and Holzapfel B 2010 *Appl. Phys. Lett.* **97** 172504
[4] Döring S, Schmidt S, Schmidl F, Tympel V, Haindl S, Kurth F, Iida K, Mönch I, Holzapfel B and Seidel P 2012 *Physica C* **478** 15
[5] Tanabe K and Hosono H 2012 *Jpn. J. Appl. Phys.* **51** 010005
[6] Kawaguchi T, Uemura H, Ohno T, Watanabe R, Tabuchi M, Ujihara T, Takeda Y and Ikuta H 2009 *Appl. Phys. Express* **2** 093002
[7] Kawaguchi T, Uemura H, Ohno T, Tabuchi M, Ujihara T, K. Takenaka, Takeda Y and Ikuta H 2010 Appl. Phys. Lett. **97** 042509
[8] Kawaguchi T, Uemura H, Ohno T, Tabuchi M, Ujihara T, Takeda Y and Ikuta H 2011 *Appl. Phys. Express* **4** 083102
[9] Uemura H, Kawaguchi T, Ohno T, Tabuchi M, Ujihara T, Takeda Y and Ikuta H 2012 *Solid State Commun.* **152** 735
[10] Kawaguchi T, Uemura H, Ohno T, Tabuchi M, Ujihara T, K. Takenaka, Takeda Y and Ikuta H 2011 *Physica C* **471** 1174
[11] Ueda S, Takeda S, Takano S, Yamamoto A and Naito M 2011 *Appl. Phys. Lett.* **99** 232505
[12] Kidszun M, Haindl S, Reich E, Hänisch J, Iida K, Schultz L and Holzapfel B 2010 *Supercond. Sci. Technol.* **23** 022002
[13] Asano T, Ishiwara H and Furukawa S 1988 *Jpn. J. Appl. Phys.* **27** 1193
[14] Fathauer R W and Schowalter L J 1984 *Appl. Phys. Lett.* **45** 519

Size dependence of the trapped magnetic field in large single-grained Gd-Ba-Cu-O bulk superconductors

H Teshima and M Morita

Nippon Steel & Sumitomo Metal, 20-1 Shintomi, Futtsu, Chiba 293-8511, JAPAN

teshima.a5b.hidekazu@jp.nssmc.com

Abstract. Based on careful consideration to crystal growth mechanism, three different sizes of Gd-Ba-Cu-O cylindrical bulk superconductors have been fabricated. The trapped magnetic field proportionally increased with the sample size and the peak heights at 77 K were 0.91-0.95 T, 1.38-1.51 T and 1.91-2.03 T for samples 30 mm, 46 mm and 64.5 mm in diameter, respectively. In addition, macroscopic J_c of approximately 13,000 A/cm^2 was estimated for all samples with different sizes. Such a high value of macroscopic J_c and the liner increase of trapped fields with size can be ascribed to excellent quality of large single-grained Gd-Ba-Cu-O bulk superconductors.

1. Introduction

High temperature superconducting (HTS) bulks of RE-Ba-Cu-O (RE: Y or rare earth elements) are highly attractive from a practical viewpoint because of their strong pinning force and a variety of interesting bulk applications have been proposed. In order to realize bulk applications, large single-grained HTS bulks with high critical current density, J_c, are indispensable. Regarding enlargement of a single-grained HTS bulk crystal, a breakthrough was achieved by Morita et al. shortly after the discovery of Y-Ba-Cu-O: control of nucleation and crystal orientation by seeding [1, 2]. This seeding technique is very useful and has been widely utilized for fabricating single-grained HTS bulks in many laboratories around the world. The values of J_c are usually evaluated for small specimens 1-2 mm in size and these J_c values are defined as microscopic J_c in this paper. Microscopic J_c values higher than 10,000 A/cm^2 at 77 K and 1 T are often reported and in most cases they are for small specimens cut from HTS bulks 20-30 mm in size. It is well-known that, according to the Bean model, the trapped field capability is proportional to the product of their single-domain size and J_c. Assuming that J_c is constant, the peak height of the trapped field should be proportional to the sample size. There were, however, few experimental reports on the size dependence of the trapped field so far. This is probably because it is still difficult to produce large single-grained HTS bulks with high J_c kept constant, due to undesirable nucleation and deterioration, especially for HTS bulks larger than 30 mm.

In this paper, in order to experimentally demonstrate the relationship between the trapped field and the sample size, we have fabricated and investigated three different sizes of Gd-Ba-Cu-O cylindrical bulk superconductors: 30 mm, 46 mm and 64.5 mm in diameter.

2. Crystal growth mechanism of RE123 bulk crystals with much excess RE211

It is very important to understand the complicated crystal growth mechanism of a $REBa_2Cu_3O_x$ (RE123) crystal with much excess RE_2BaCuO_5 (RE211) inclusions finely dispersing in the superconducting matrix, in order to fabricate HTS bulks of good quality, especially for large samples. It is known that the RE123 phase is produced by the following peritectic reaction:

RE211 + L → RE123 + excess RE211

where L denotes a liquid phase consisting $BaCuO_2$ and CuO. Crystal growth is generally limited by interface kinetics, mass transport and heat transport. Since HTS bulk crystals exhibit faceted growth, it is no doubt that the effect of interface kinetics is essential in these materials. Both mass and heat transports are also important for stable growth of the faceted interface.

First, as for mass transport, the supply of RE elements into the growing interface is crucial for steady crystal growth. A model that RE ions dissolving in the liquid phase move toward the growing interface can be considered and this model may be true for crystal growth with no or little excess RE211 phase, but the flux of dissolving RE ions alone is not sufficient for the growth of large HTS bulks since the solubility of RE ions in the melt is very small, a few % or less. In the growth of HTS bulks containing much excess RE211 phase of a few tens %, another model should be proposed that RE211 particles trapped on the growing interface play a much more important role on the supply of RE elements into the growing interface. In this model, RE211 particles trapped on the growing front are decomposed at the three phase interface between the liquid phase, RE211 and RE123 and, as shown in Figure 1, the resultant RE ions move along the interface plane for a certain time determined by interface kinetics. Some of them are incorporated into the growing interface at kinks and others are dissolved into the liquid. According to this model, it is possible to supply a sufficient amount of RE element to the growing interface even for large HTS bulks. This model also predicts that stable faceted interface is difficult to grow in the case that the distribution of excess RE211 particles is not uniform, leading to instability in crystal growth. It is thus particularly important to enhance the uniformity of excess RE211 phase distribution inside the precursor for producing large HTS bulks of good quality.

Next, as for heat transport, the internal temperature gradient at solid/liquid interface is important for stable faceted growth. Compared with the positive temperature gradient where the liquid phase temperature is higher than the solid phase one, it is more difficult to keep the solid/liquid interface flat due to the Mullins-Sekerka instability in the case of negative temperature gradient. Although positive temperature gradient models based on directional solidification are often proposed for HTS bulks, the internal temperature gradient is much different. In the HTS bulk melt-process, the internal temperature gradient is negative rather than positive, because the entire precursor including both the crystallized and liquid parts is slowly cooled during the crystal growth step and the crystallized phase temperature rises due to the latent heat generated at the growing interface. The HTS bulk melt-process therefore requires more careful heat treatments for stable faceted growth, especially for larger samples.

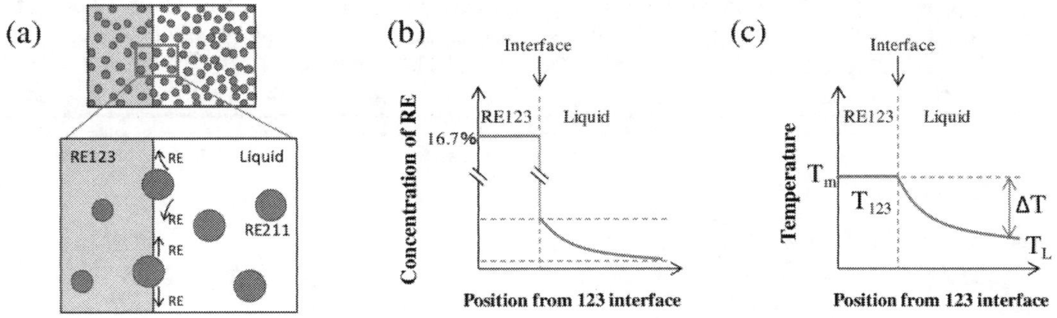

Figure 1. (a) Schematic diagram, (b) RE concentration and (c) temperature near the interface.

3. Sample preparation

Large single-grained Gd-Ba-Cu-O bulk superconductors were prepared by a melt-growth process called the modified QMG method. QMG originally stood for "Quench and Melt-Growth", but in the modified QMG process the quench step is eliminated by employing a small amount of Pt or CeO_2 addition into the starting materials [2] and thus "Q" means "Quality" at present instead of "Quench.

In order to enhance the uniformity of excess RE211 phase distribution within the precursor, very careful attention was paid to the precursor preparation. First, appropriate amounts of high purity Gd_2O_3, BaO_2 and CuO powders were ground and mixed well with 0.5 wt% of Pt. These mixed powders were calcined at about 1173 K and then pulverized. This calcination process was repeated twice to ensure complete reaction. Next, separately-calcined powders of Gd123 and Gd211were ground and blended well in a molar ratio of Gd123:Gd211=3:1 with 10 wt% Ag_2O. This mixture process was also repeated three times to ensure complete mixing. Pt was added for the refinement of Gd211 particles and Ag was for improving the fracture strength properties. Finally, the well-mixed powders were pressed into three different sizes of cylindrical precursors.

Special care was also taken of the heat treatment for crystal growth. First, the precursor was heated to a relatively high temperature, approximately 1443 K, for a certain length of time in order to transform the whole precursor into the partially-melted state completely. If part of the precursor is incompletely melted, undesirable nucleation occurs and the sample becomes poly-grained, not single-grained. Next, the melted precursor was rapidly cooled to about 1310 K and the seed crystal of (Nd, Sm)-Ba-Cu-O was placed on the top of the precursor. The precursor was then cooled to the point just above the peritectic temperature, followed by slow cooling to 1245 K for crystal growth in air. The as-grown samples were machined into three different sizes of Gd-Ba-Cu-O cylindrical bulk superconductors: 30 mm, 46 mm and 64.5 mm in diameter, as shown in Table 1. Finally, the samples were annealed for 100 h at 673 K in flowing oxygen gas

Table 1. Size and trapped field properties of Gd-Ba-Cu-O bulk superconductors.

Sample No.	A-1	A-2	B-1	B-2	C-1	C-2
Diameter (radius)	30 mm (15 mm)		46 mm (23 mm)		64.5 mm (32.25 mm)	
Thickness	10 mm		15 mm		20 mm	
Measured peak height	0.91 T	0.95 T	1.38 T	1.51 T	1.94 T	2.03 T
Corrected for r/t =1.5	0.91 T	0.95 T	1.39 T	1.52 T	1.98 T	2.08 T

4. Results and discussion

Figure 2 shows the appearance and the trapped field distribution at 77 K for three different sizes of Gd-Ba-Cu-O. The samples were field-cooled under an applied field of 3.5 T and the trapped fields were measured by scanning a Hall sensor two-dimensionally at a gap of 0.6 mm above the sample surface. As shown in Figure 2(b), each sample exhibited a symmetric trapped field distribution with a single peak, indicating that the sample was a single crystal with no serious weak-links. In other words, stable crystal growth continued to the sample edge even in large HTS bulks 64.5 mm in diameter. The peak values of the trapped field at 77 K were 0.91-0.95 T, 1.38-1.51 T and 1.91-2.03 T for samples 30 mm, 46 mm and 64.5 mm in diameter, respectively. Figure 3(a) displays the relationship between the measured values of the peak height and the sample diameter, and the peak height increases linearly with the diameter. This result experimentally demonstrates that the trapped field ability proportionally increases with the sample size up to at least 64.5 mm.

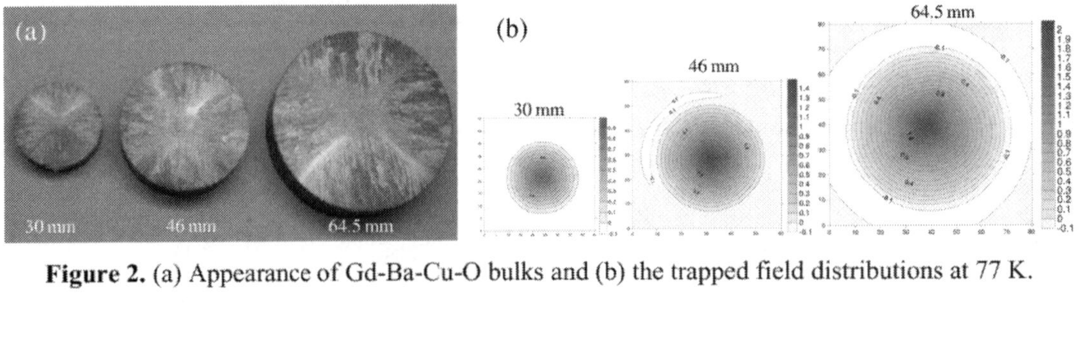

Figure 2. (a) Appearance of Gd-Ba-Cu-O bulks and (b) the trapped field distributions at 77 K.

The Bean model simply predicts that the peak height of the trapped magnetic field is proportional to the sample size in a one-dimensional model, but it is more appropriate to use a two-dimensional model for real HTS bulks. For a cylindrical HTS bulks with a finite thickness assuming a field-independent J_c, the peak height of the trapped field, B_p, is described as follows [3]:

$$B_p = \frac{\mu_0 J_c t}{2} \ln\left\{\frac{r}{t} + \sqrt{1+\left(\frac{r}{t}\right)^2}\right\}$$

where μ_0 is the magnetic permeability in air, r the radius and t the thickness. According to the above equation, the peak height is strictly proportional to the sample size only in the situation where the aspect ratio, r/t, is constant. The aspect ratio of the samples used in this study was not constant as shown in Table 1 and thus we should correct the peak height. Open circles in Figure 3(b) are the corrected values for r/t =1.5 and it seems that the linearity is slightly improved.

In addition, J_c can be estimated from the trapped magnetic field by using the above equation. This J_c value is defined as macroscopic J_c in this paper because it is calculated on the assumption that the superconducting current flows throughout the whole sample. The estimated vales of macroscopic J_c are approximately 13,000 A/cm^2 at 77 K for all samples with different sizes. It is amazing that there are no large differences in macroscopic J_c between different-sized HTS bulks and that these macroscopic J_c value are as high as the microscopic J_c ones reported so far. Such a high value of macroscopic J_c and the liner increase of trapped fields with size can be attributed to good quality of large single-grained Gd-Ba-Cu-O bulk superconductors.

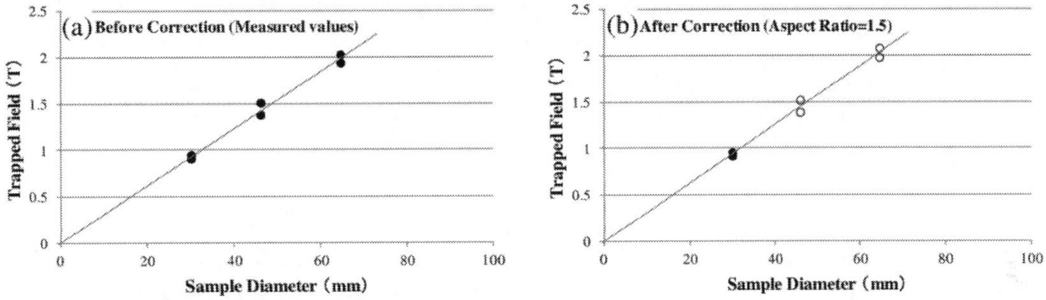

Figure 3. Relationship between trapped field and sample diameter of Gd-Ba-Cu-O bulks: (a) before correction (measured values) and (b) after correction (aspect ratio r/t = 1.5).

5. Summary

We have fabricated three different sizes of Gd-Ba-Cu-O cylindrical bulk superconductors, based on careful consideration to crystal growth mechanism. The trapped field ability proportionally increased with the sample size up to at least 64.5 mm and macroscopic J_c of approximately 13,000 A/cm^2 at 77 K was estimated for all samples with different sizes. High quality of large single-grained Gd-Ba-Cu-O bulk superconductors is responsible for such a high value of macroscopic J_c and the liner increase of trapped fields with size.

References

[1] Morita M, Takebayashi S, Tanaka M, Kimura K, Miyamoto K and Sawano K 1991 *Advances in Superconductivity 3 (ISS1990)*, pp733-736

[2] Morita M, Sawamura M, Takebaayshi S, Kimura K, Teshima H, Tanaka M, Miyamaoto K and Hashimoto M 1994 *Physica C 235-240*, pp209-212

[3] Tsuchimoto M and Takashima H 2001 *IEEE trans. on Appl. Supercond.* vol 11, pp1992-1995

Investigation of the strain-sensitive superconducting transition of $BaFe_{1.8}Co_{0.2}As_2$ thin films utilizing piezoelectric substrates.

S. Trommler[1], J. Hänisch, K. Iida, F. Kurth, L. Schultz[1] and B. Holzapfel[2], R. Hühne

IFW Dresden, Institute for Metallic Materials, P.O. Box 270116, 01171 Dresden, Germany
[1] Technische Universität Dresden, Department of Physics, Institute for Physics of Solids, 01062 Dresden, Germany
[2] TU Bergakademie Freiberg, 09596 Freiberg, Germany

E-mail: s.trommler@ifw-dresden.de

Abstract. The preparation of biaxially textured $BaFe_{1.8}Co_{0.2}As_2$ thin films has been optimized on MgO single crystals and transfered to piezoelectric (001) $Pb(Mg_{1/3}Nb_{2/3})_{0.72}Ti_{0.28}O_3$ substrates. By utilizing the inverse piezoelectric effect the lattice parameter of these substrates can be controlled applying an electric field, leading to a induction of biaxial strain into the superconducting layer. High electric fields were used to achieve a total strain of up to 0.05% at low temperatures. A sharpening of the resistive transition and a shift of about 0.6 K to higher temperatures was found at a compressive strain of 0.035%.

The interplay of structural properties and intrinsic parameters is widely used to study superconducting materials [1, 2, 3]. Traditionally, the lattice mismatch between substrate and film material is used to generate statically strained films. Simultaneously, relaxation effects lead to the implementation of defects during the growth process, which changes the microstructure and thus affect thin film properties like the critical current density and pinning mechanisms. Therefore, the suitability of this method is strongly limited, since strain-driven effects are hard to separate from preparation related changes.

Alternatively, the application of single crystalline piezoelectric $Pb(Mg_{1/3}Nb_{2/3})_{0.72}Ti_{0.28}O_3$ (PMN-PT) substrates has been demonstrated to provide the control of reversible and dynamic strain in thin functional films [4, 5, 6, 7] and thus give the opportunity to study the strain sensitivity of microstructure related properties in superconducting thin films. However, the adaptation of this approach to pnictides requires the application of suitable buffer layers to prevent any oxidation. In this paper we report on the utilization of an Fe/MgO buffer to the model system $BaFe_{1.8}Co_{0.2}As_2$ (Ba-122) which might be also suitable for other Fe based superconductors.

The samples were prepared in ultra-high vacuum (10^{-9} mbar) using pulsed laser deposition and stoichiometric targets. A detailed description of the target and sample preparation was published previously [8, 9, 10]. During deposition, the film evolution is monitored with reflection high energy electron diffraction (RHEED), whereas for structural characterization x-ray $\theta/2\theta$ scans and pole figure measurements in a four circle goniometer were performed. For the application of dynamic strain the sample was split into 1 mm wide stripes and contacted at the Au/NiCr bottom electrode. The top electrode is provided by the film itself and connected to ground level to avoid any field effects. For electrical characterization contacts were attached to the sample surface with silver glue in a four-probe geometry. The PMN-PT was poled with 10 kV/cm at room temperature (RT) and subsequently cooled down at 6.6 kV/cm. The resistive transition was recorded with 10 μA and 1 K/min in a Physical Property Measurement System (Quantum Design). The critical temperature is defined at a criterion of 90% of the normal state resistance, noted here as $T_{c,90}$.

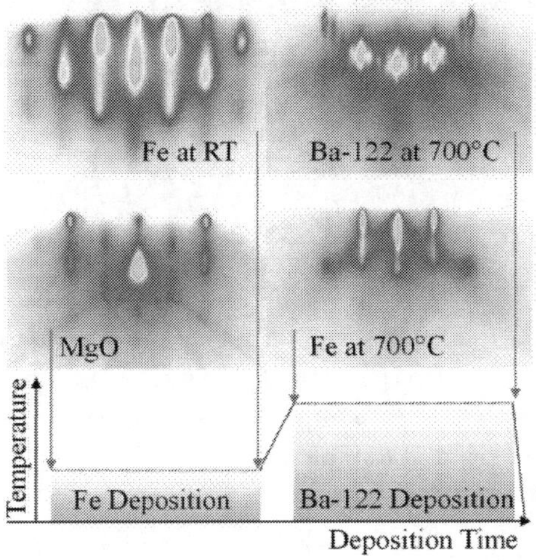

Figure 1. RHEED pattern of selected deposition steps for the preparation of a Ba-122/Fe bilayer on MgO single crystal.

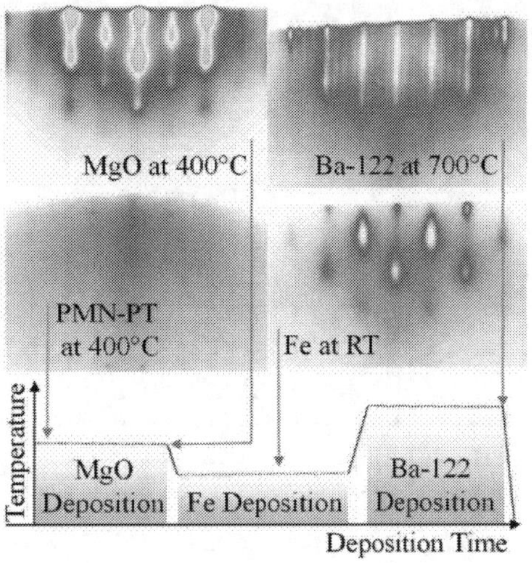

Figure 2. RHEED pattern of selected deposition steps for the preparation of a Ba-122/Fe bilayer on MgO buffered PMN-PT.

In the first step the film preparation was optimized on MgO single crystalline substrates (MgO SC). To improve the epitaxial growth of Ba-122, an artificial Fe buffer layer was introduced resulting in a clean and coherent Ba-122/Fe interface [9, 11]. A typical deposition sequence on MgO with the corresponding RHEED images is given in fig. 1. The MgO diffraction pattern reveals sharp peaks and Kikuchi lines, typical for polished single crystalline substrates. The diffraction pattern of 20 nm Fe, deposited at RT, originates from the transmission of textured 3D structures, indicating island growth. The peaks broadening arises from a a small grain size, which is due to the small lattice coherence. During the heating ramp the Fe diffraction pattern changes drastically, which can be attributed to recrystallization effects leading to a peak sharpening. The arrangement on Laue circles and the occurrence of Kikuchi lines indicate a reduction of the surface roughness and an enhancement of the lattice coherence. Directly after approaching the desired temperature, 50 nm Ba-122 are deposited. The distinct arrangement of the diffraction peaks on Laue circle evidences a 2D-like surface, whereas additional intermediate peaks reveal the formation of a surface reconstruction, which was also observed for previous films [12].

A similar layer architecture was applied to PMN-PT substrates (fig. 2). The low intensity of the diffraction peaks for PMN-PT arises from a higher mosaicity of the crystal. However, the peaks are arranged on Laue circles, indicating a smooth surface. 20 nm MgO are deposited at 400 °C since a diffusion barrier is required to avoid any oxidation of the Ba-122/Fe bilayer. The diffraction pattern is similar to a MgO SC, but with broadened peaks. After cooling down to RT iron is deposited, showing an similar diffraction pattern as the corresponding Fe on MgO SC. In contrast, no significant change is observed during ramping up the temperature. This can be attributed to the transfer of the substrate mosaicity into the subsequent layers, avoiding recrystallization effects. Finally, the Ba-122 film exhibits vertically elongated peaks, which is typical for the transition between layer and island growth. Additional lines connected to a surface reconstruction are observed, which indicate similar surface properties on both substrates.

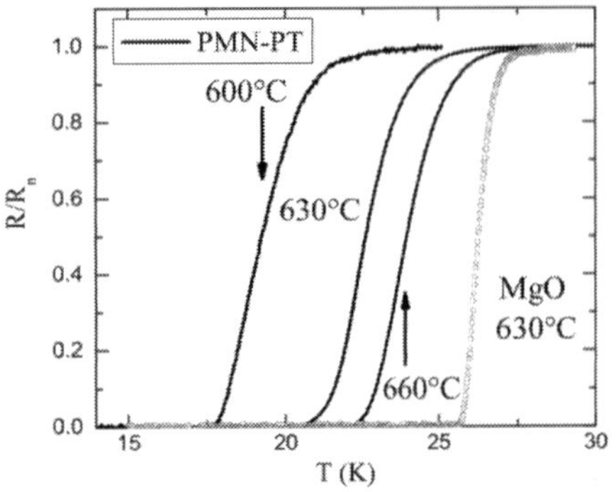

Figure 3. Resistive transition of Ba-122/Fe bilayers on MgO-buffered PMN-PT for different temperatures in comparison to a Ba-122/Fe bilayer on MgO single crystal.

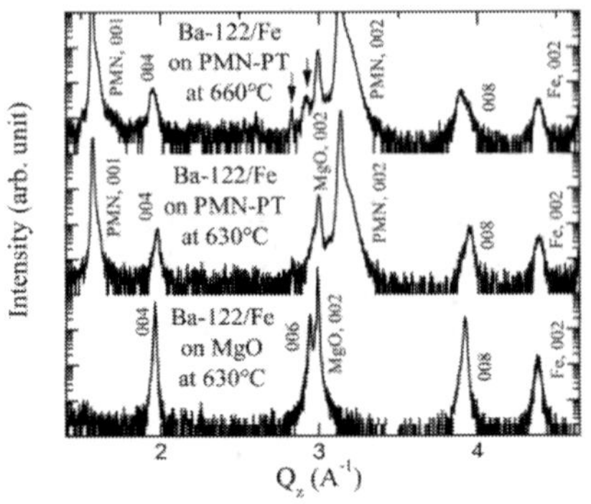

Figure 4. $\theta/2\theta$ scans of Ba-122/Fe bilayers on MgO-buffered PMN-PT for two temperatures in comparison to a Ba-122/Fe bilayer on MgO single crystal.

In general, we observe a a higher T_c with increasing deposition temperature of Ba-122. For MgO substrates the optimal deposition temperature was found to be 630 °C, with a maximum $T_{c,90}$ of 26.9 K and a very small transition width of $\Delta T = 1$ K (see fig. 3). A similar dependency is found for films on PMN-PT substrates, with an optimal deposition temperature of 660 °C resulting in $T_{c,90} = 25.5$ K. The broadening of the resistive transition with $\Delta T = 2.6$ K is found to be independent from the resulting $T_{c,90}$ and might be due to the higher defect density in films on PMN-PT, as discussed above.

Standard $\theta/2\theta$ scans (fig. 4) prove pure c-axis growth of Ba-122 on both MgO and PMN-PT, and the known epitaxial relationship was confirmed by pole figure measurements. With increasing deposition temperature, additional peaks appear (arrows), which are also present for just heated reference substrates and originate from a temperature-induced decomposition of the substrate. To avoid any degradation of the piezoelectric performance of the PMN-PT crystal, the strain experiments were carried out on films grown at 600 °C with $T_{c,90} = 20.5$ K. After poling and cooling down the sample to 10 K the resistive transition was measured for different electric fields (fig. 5). According to previous studies at 20 K [7], the substrate provides about a linear strain of $\epsilon_{ab} = 0.017\%$ for 13.3 kV/cm. Due to the polarity of the substrate, a positive field leads to compressive strain, whereas a negative field induces tensile strain.

An increasing biaxial compression results in a shift of the resistive transition to higher temperatures with a total shift of $T_{c,90}$ of about 0.6 K for $\epsilon_{ab} = 0.035\%$. To illustrate the strain sensitivity of the whole transition slope, a criterion-dependent temperature shift can be defined as $\Delta T(p)$ at the resistance criterion R_p, where $p = R_p/R_{30\mathrm{K}}$. Exemplary, this is displayed for $\Delta E = 26.6$ kV/cm in the inset of fig. 5. For all slopes we find an increasing $\Delta T(\epsilon_{ab})$ with decreasing p. This behavior can be explained taking into account, that the shape of the resistive transition is influenced by different mechanisms, as pinning, thermal activation or fluctuations, which generally have different strain sensitivities. $T_{c,90}$ as well as T_{irr} is highlighted in the inset, where T_{irr} is defined for $p = 10^{-2}$ and exhibits roughly twice the temperature shift compared to $T_{c,90}$.

To get deeper insight, we fitted the slopes according to the vortex-glass theory, where

$R \propto (T - T_g)^\alpha$ [13]. The total shift in dependence of the applied electric field is displayed in fig. 6. For compressive strain we find a linear behavior for both T_g and $T_{c,90}$, where T_g represents the broadening of the resistive transition close to zero resistance (see the inset of fig. 6). In contrast, for tensile strain we observe almost no change in $T_{c,90}$ but a strong broadening of the transition, which might be connected to the formation of cracks. This would lead to a non-uniform strain transfer into the superconducting layer, where the lattice keeps almost unstrained. With increasing tension the cracks will open and reduce the number of superconducting current paths resulting in an increased local current density, which thus increases the resistance and broadens the slope. Due to the low thermal expansion coefficient of PMN-PT [14] the film is tensile-strained at low temperature compared to RT. Therefore, any crack formation may be closed and not noticeable at RT.

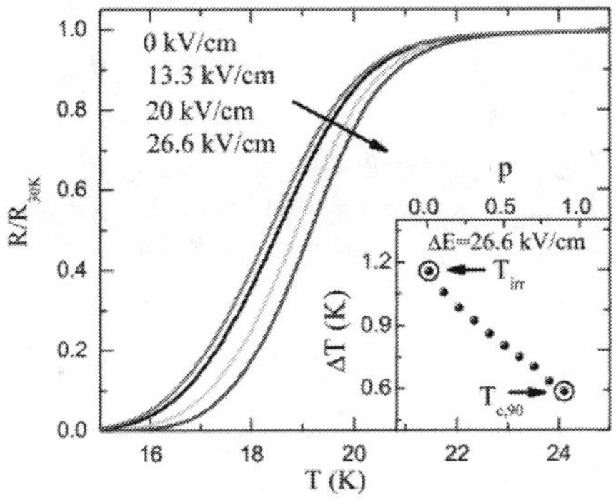

Figure 5. Resistive transition of a Ba-122/Fe bilayer on MgO-buffered PMN-PT for different electric fields, resulting in compressive strain. The inset displays the shift of the transition in dependency of R_p.

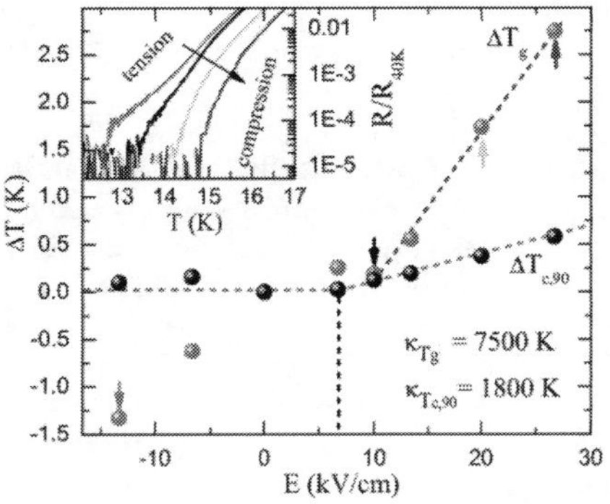

Figure 6. Strain-induced shift of $T_{c,90}$ and T_g. The inset displays the curvature change of the transition close to zero resistance with increasing compressive strain (fields are marked by arrows).

Based on the total differential of the specific resistivity, the strain sensitivity is defined as $\kappa = \Delta T/\epsilon_{ab}$. At compressive strain, the best fit yields a strain sensitivity of 1800 K for $T_{c,90}$ and 7500 K for T_g, respectively. The 4 times larger strain sensitivity of T_g compared to $T_{c,90}$ is related to a transition sharpening with increasing in-plane compression. Therefore, the linear field dependency for T_g has to deviate at high electric fields due to the limited transition width.

[1] L. Gao, Y. Y. Xue, F. Chen, Q. Xiong, R. L. Meng, D. Ramirez et al., Phys. Rev. B **50** (1994), 4260.
[2] H. Takahashi, K. Igawa, K. Arii, Y. Kamihara, M. Hirano and H. Hosono, Nature **453** (2008), 376.
[3] S. Medvedev, T. M. McQueen, I. A. Troyan, T. Palasyuk et al., Nature Materials **8** (2009), 630.
[4] R. Hühne, D. Okai, K. Dörr, S. Trommler, A. Herklotz et al., Supercond. Sci. Technol. **21** (2008), 075020.
[5] A. D. Rata, A. Herklotz, K. Nenkov, L. Schultz, and K. Dörr, Phys. Rev. Lett. **100** (2008), 076401.
[6] A. Herklotz, A. D. Rata, L. Schultz, and K. Dörr, Phys. Rev. B **79** (2009), 092409.
[7] S. Trommler, R. Hühne, K. Iida, P. Pahlke, S. Haindl, L. Schultz et al., New J. Phys. **12** (2010), 103030.
[8] F. Kurth, K. Iida, S. Trommler, J. Hänisch, K. Nenkov et al., Supercond. Sci. Technol. **26** (2013), 025014.
[9] K. Iida, S. Haindl, T. Thersleff, J. Hänisch, F. Kurth et al., Appl. Phys. Lett. **97** (2010), 172507.
[10] S. Trommler, J. Hänisch, V. Matias, R.Hühne, E. Reich et al., Supercond. Sci. Technol., **25** (2012), 084019.
[11] T. Thersleff, K. Iida, S. Haindl, M. Kidszun, D. Pohl et al., Appl. Phys. Lett. **97** (2010), 022506.
[12] K. Iida, J. Hänisch, S. Trommler, S. Haindl, F. Kurth et al., Supercond. Sci. Technol., **24** (2011), 125009.
[13] D.S. Fisher, M.P.A. Fisher and D.A. Huse, Phys. Rev. B **43** (1991), 130159.
[14] H.W. King, S.H. Ferguson, D.F. Waechter and S.E. Prasad, Proceedings of ICONS (2002).

Magnetic measurements based on magneto-optical Kerr effect on pnictide $Ba(Fe_{1-x}Co_x)_2As_2$/Fe thin film

M Valentino[1,2], C Bonavolontà[1,2], C Porto[2], C Delisio[1,2], L Parlato[1,2] GP Pepe[1,2]

[1]CNR-SPIN Napoli, Italy
[2]Department of Physics, University of Naples "Federico II", Italy

E-mail: massimo.valentino@spin.cnr.it

Abstract: We present characterization of $Ba(Fe_{1-x}Co_x)_2As_2$ thin film using a magneto-optical Kerr effect setup based on photoelastic modulation. The magneto optical technique has been used to analyse the thickness and the temperature dependence of the magnetic properties of epitaxial $Ba(Fe_{1-x}Co_x)_2As_2$/Fe thin films grown on MgO substrate. The first harmonics, of the reflected laser beam intensity from pnictide thin films is compared with a reference Fe thin film. The hysteretic loops of such samples at room temperature and at 5K are presented.

1. Introduction

Magneto-optics studies the interaction of light with a medium exerting action of an external magnetic field, including changes in the polarization state and in the propagation direction of light. It greatly contributes to the development of modern optical measurement techniques. For this reasons the magneto-optical Kerr effect MOKE has emerged as a powerful experimental technique to study the magnetic properties of thin films and multilayers.

Compared to other elegant techniques such as superconducting quantum interference device magnetometry, magnetic force microscopy, etc., the advantages of MOKE [1] include high sensitivity down to the monolayer resolution, high temporal and spatial resolution, simplicity, and straightforward in situ implementation all of that at very low costs in comparison with its alternatives. MOKE has been extensively employed to investigate several important phenomena of modern magnetism. Some of them are: determination of electronic structure [2], discovery of the oscillations in exchange coupling between ferromagnetic layers separated by an antiferromagnetic or nonmagnetic layer [3-4], observation of perpendicular anisotropy in ultrathin films [5], test of two-dimensional Ising model in monolayer films [6], spin reorientation transition, and correlation between magnetic anisotropy and lattice symmetry breaking [7]. MOKE measurements can be realized with various setups. Here we focus on the analysis of the widely used modulation technique in longitudinal configurations. The performance of the realised MOKE set up has been tested on Fe- based superconductors thin films. The recent discovery of superconductivity with transition temperature T_c up to 55 K in iron–pnictide systems [8] has attracted enormous interest in this class of materials. Investigations on this new class of superconductors have revealed for instance, a multiband nature, and a short coherence length. A common phase diagram for Fe–pnictides suggests the existence of an antiferromagnetic spin density-wave (SDW) ground state that is stabilized by Fermi-surface nesting [9] as well as by strong antiferromagnetic (AFM) spin interactions along the Fe-square-diagonal [10]. Moreover, in underdoped compounds a coexistence of SDW and superconducting phase can be present. In this work, AFM characterization of the Co-doped Ba-122/Fe thin films with a thickness of 20nm and 40nm is reported. Moreover, the MOKE characterization in terms of hysteretic loop at 300K and 5K has been carried out.

2. Experimental set-up

As reported in figure 1 (right) the components of the MOKE set up include: an He-Ne laser having wavelength = 633nm and an output power of 15 mW, a polarizer *P*, the sample *S* among Helmholtz coils, a photoelastic modulator *PM*, an analyzer *A*, and the photodetector *D*. Since the Kerr rotation angle θ_K and ellipticity θ'_K are typically small, i.e., 10^{-3} rad, optimization of the signal-to-noise S/N ratio is very crucial. In order to achieve an optimum S/N ratio, appropriate placements and orientations of the optical components are crucial.
The laser produces a nearly linearly polarized beam allowing for s-polarized electric field vector oscillating perpendicular to the plane of incidence or p-polarized electric field vector oscillating in the plane of incidence configurations. Subsequently, we discuss s-polarized incoming light only. In our setup, due to geometrical constraints of the Helmholtz coils the laser beam makes only an angle of about 20° with the normal of the sample surface as reported in figure 2 (right).
The laser beam then passes through a Glan–Thompson polarizer (Edmund Optics) with an extinction coefficient of 10^{-6} which produces high degree of polarization. A lens of focal length *f* =300 mm and diameter of *D* =50 mm is used to focus the light beam onto the sample surface.
The reflected beam is periodically modulated between left and right circularly polarized light by the photoelastic modulator *PEM*-90 (Hinds Instruments). Modulation takes place with a frequency of 50 kHz. The modulation signal is used as reference signal for a lock-in amplifier (EG&G 7260 Instruments). The beam then is transmitted through an analyzer and is finally detected by a photosensitive fast responding diode (Nirvana detector-Model 2007) providing the input signal to the lock-in amplifier. An electromagnet, home made, in the Helmholtz configuration, capable of generate a magnetic field of more than 50mT, powered by a bipolar power supply Elind KL 1200 has been used. The intensity of the magnetic field has been measured by a Hall sensor model 516-60, realized by Leybold. The sample was mounted on a cold finger in a temperature controlled liquid helium continuous flow optical cryostat realized by Oxford Instruments, operating between 4.2 and 350 K.

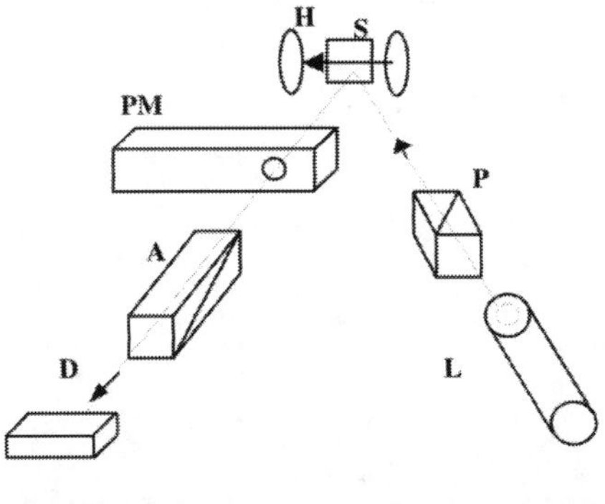

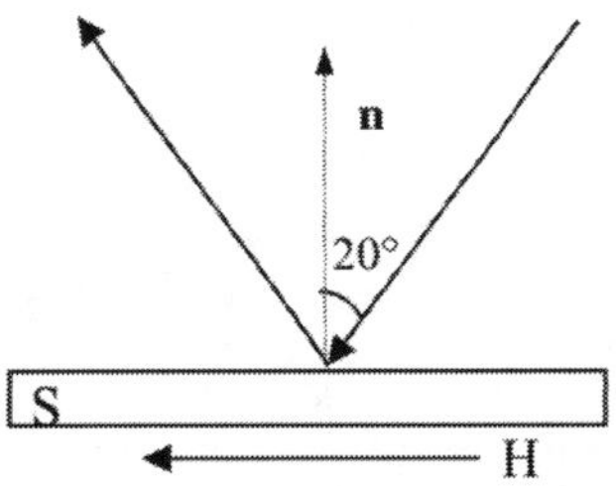

Figure 1. (left) used MOKE set-up: L = He-Ne laser, P = polarizer, S = sample surface, H= magnetic field direction, PM = electro-optical modulator, A= polarizer used as analyzer, D = detector; (right) represents the longitudinal MOKE configuration in which the orientation of the magnetic field H is parallel both to the sample surface and the plane of incident light.

3. Results and discussion

Using the experimental set-up reported in figure 1 different magneto-optical measurements have been performed. The investigated samples are $Ba(Fe_{1-x}Co_x)_2As_2/Fe$ having thickness of 20nm and 40nm where the Fe layer is 20nm and the substrate of MgO is 0.5mm. the content of Co-doping is nominally 8%. Moreover, a Fe 20nm thick thin film has been used as reference. More details about the fabrication of the sample are reported elsewhere [11]. The transport properties of the sample used in this experiment are quite different, in particular the 20nm thin film shows a superconductive behaviour with a critical temperature of T_c about of 9.5K, while the sample of 40nm thick shows a T_c of 16.5K.

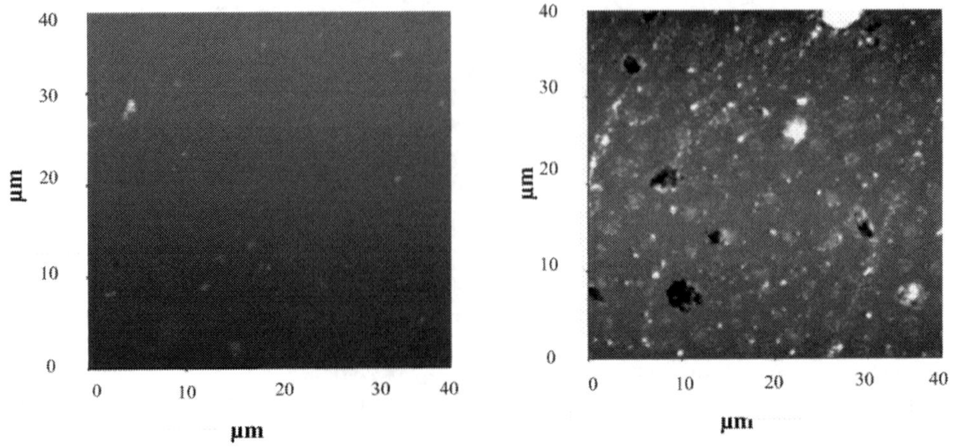

Figure 2. AFM of the samples having thickness of 40 (left) and 20nm (right) respectively.

The laser spot used in the MOKE measurements is about $0.8\ (mm)^2$ where the optical penetration depth is about 15-20nm. It is interesting to note that the surface roughness of the two sample is quite different as reported in the figure 2, in fact the surface of the sample of Ba-122 (20nm) thick is full of FeO based outgrowth. Measurements by a conventional VSM (Vibrating Sample Magnetometer) has been performed on the same sample analysed in this work. The results, not shown here, cannot enable us to detect different behaviours with respect to the sample thickness and cannot be useful for clarify the role of iron layer on the magnetization at room temperature.

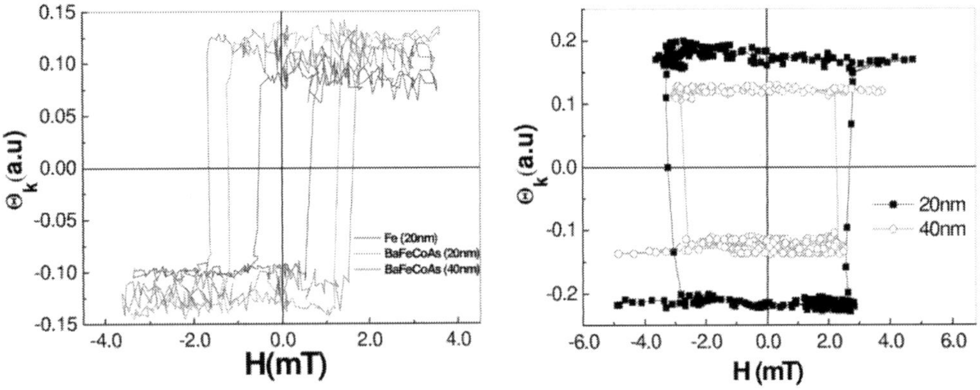

Figure 3. (left) MOKE measurements at room temperature of the 20nm and 40nm Ba-122/Fe are compared with the Fe 20nm thick thin film; (right) comparison of MOKE measurements of the Ba-122/Fe thin film of 20nm and 40nm measured at 5K.

The MOKE measurements at 300K show (figure 3 left) that the coercitivity of the sample increases with the thickness of layer of Ba-122 instead the measurement performed at 5K, reported in the (figure 3 right), indicate an opposite behaviour. In the latter case the coercitivity of the thin film reduces when the thickness of the sample increases. At room temperature we obtain for Ba-122/Fe of 20nm thick a coercitivity of $H_c=(1{,}3\pm0{,}1)$ mT while for Ba-122/Fe 40nm it is $H_c= (1{,}8 \pm0{,} 1)$ mT. Both the samples show an H_c larger than that of the only iron thin film 20nm thick.
This result demonstrates that the MOKE measurements reveal the contribution to the magnetic behaviour mainly due to the Co-doped Ba-122 thin films. Moreover, since the H_c is larger than the only Fe thin film, it is reasonable that the hysteretic loops of Ba-122/Fe of 40nm and 20nm include also the Co magnetic contribution. Then, considering that the Ba-122 Co-doped of 20nm has a thickness comparable to the optical penetration depth, the hysteretic loop of such samples at 300K

includes two contributions: the first due to the FeO outgrowths, and the second due to the interface between the Fe and the $Ba(Fe_{1-x}Co_x)2As_2$ layer. Moreover, regarding the figure 3 (right) the increasing of the H_c at low temperature for both Ba-122 samples could be addressed to the possible coexistence of the SDW and superconducting phase. This statement could be due considering the hypothesis that the samples are not 8% Co-doped but they are underdoped samples since a strong Co diffusion into the Fe buffer can be observed [12]. Moreover, further experiments are currently in progress aiming to better understand the nature of such result.

4.Conclusion

MOKE measurements and AFM have been realized on superconductive $Fe/Ba(Fe_{1-x}Co_x)_2As$ thin film at room and low temperature. The AFM measurements shown that the surface of the thin film 20nm thick is full of outgrowth due to the iron diffusion inside the Ba-122 film. The magnetization of the 20nm thin film is greater than the magnetization of the 40nm $Fe/Ba(Fe_{1-x}As_x)_2Co$ thin film measured at 5K. The MOKE set up shows a good sensitivity and a good signal to noise ratio that enable us to perform hysteretic loop, from room temperature up to 5K, due to the superconducting Ba-122 thin films even if they are grown on Fe buffer layer.

Acknowledgment

The authors acknowledge the financial support of the EU (Iron-Sea under project No. FP7-283141).

Reference

[1] J. Kerr, 1877 *Philos. Mag.* **3**, 321
[2] J. L. Erskine and E. A. Stern, 1973 *Phys. Rev. Lett.* **30**, 1329
[3] P. Grünberg, R. Schreiber, Y. Pang, U. Walz, M. B. Brodsky, and H. Sowers, 1987 *J. Appl. Phys.* **61**, 3750
[4] E. E. Fullerton, M. J. Conover, J. E. Mattson, C. H. Sowers, and S. D. Bader, 1993 *Phys. Rev. B* **48**, 15755
[5] C. Liu, E. R. Moog, and S. D. Bader, 1988 *Phys. Rev. Lett.* **60**, 2422
[6] J. Kohlhepp, H. J. Elmers, S. Cordes, and U. Gradmann, 1992 *Phys. Rev. B* **45**, 12287
[7] J. Chen and J. Erskine, 1992 *Phys. Rev. Lett.* **68**, 1212
[8] Kamihara Y, Watanabe T, Hirano M and Hosono H 2008 *J. Am. Chem. Soc.* **130** 3296
[9] Mazin I I, Singh DJ, Johannes M D and Du MH 2008 *Phys. Rev. Lett.* **101** 057003
[10] Yildirim T, 2008 *Phys. Rev. Lett.* **101** 057010.
[11] K. Iida, J. Hänisch, S. Trommler, S. Haindl, F. Kurth, R. Hühne, L. Schultz, and B. Holzapfel, 2011 *Supercond. Sci. Technol.* **24**, 125009
[12] F. Kurth, K. Iida, S. Trommler, J. H¨anisch, K. Nenkov, J. Engelmann, S. Oswald, J. Werner, L. Schultz, B. Holzapfel, S. Haindl, 2012 arXiv:1205.5219v1

Corrigendum: Magnetic measurements based on magneto-optical Kerr effect on pnictide $Ba(Fe_{1-x}Co_x)_2As_2$/Fe thin film

M Valentino[1,2], C Bonavolontà[1,2], C Porto[2], C Delisio[1,2], L Parlato[1,2] G P Pepe[1,2]

[1]CNR-SPIN Napoli, Italy
[2]Department of Physics, University of Naples "Federico II", Italy

CORRIGENDUM TO: M Valentino et al 2014 *J. Phys.: Conf. Ser.* **507** 012050

We would like to acknowledge the contribution of K Iida and F Kurth to this work by adding their names as co-authors of the published article. The correct list of authors for the paper `Magnetic measurements based on magneto-optical Kerr effect on pnictide $Ba(Fe_{1-x}Co_x)_2As_2$/Fe thin film' is

M Valentino[1,2], C Bonavolontà[1,2], C Porto[2], C Delisio[1,2], L Parlato[1,2], G P Pepe[1,2], K Iida[3] and F Kurth[3]

[1]CNR-SPIN Napoli, Italy
[2]Department of Physics, University of Naples "Federico II", Italy
[3]IFW Dresden, Germany

Experimental study in the development of HTS NMR probe

M Watanebe[1], S Sato[1], K Sato[1], K Ichikawa[1], A Saito[1], S Ohshima[1]
[1]Graduate School of Science and Engineering, Yamagata University, 4-3-16 Jonan, Yonezawa, Japan

E-mail: tsh23840@st.yamagata-u.ac.jp

Abstract. We investigated the effect of introducing artificial pinning centers (APCs) into $YBa_2Cu_3O_{7-\delta}$ (YBCO) thin films to improve the critical current density (J_c) and reduce the microwave surface resistance (R_s) in a high dc magnetic field. Introduction of the APC was realized by dispersing $BaHfO_3$(BHO) particulates in the YBCO thin films. Four types of sintered YBCO ceramics with x wt % (x = 0.5, 1.0, 2.0) BHO were used as laser targets in order to fabricate YBCO thin films with various amounts of BHO and investigate the influence of the BHO doping concentration. J_c was measured by a four probe method. We measured J_c in a magnetic field which was applied parallel to the c-axis (J_c(p)), and J_c in a magnetic field which was applied perpendicular to the c-axis (J_c(n)). The R_s was calculated using a correlating equation of J_c and R_s. The J_c(p) at 5 T and 20 K of 1.5 wt% BHO doped YBCO thin film was the largest at 3.4 MA/cm^2. The J_c(n) of 1.5 wt% BHO doped YBCO thin film was 2.5 MA/cm^2, which is comparable to non-doped. The R_s(n) and R_s(p) of all YBCO thin films was less than 25 μΩ at 500 MHz. Therefore, we clarified the possibility that the most effective way to fabricate thin film for NMR coil application is by doping YBCO with BHO 1.5 wt%.

1. Introduction

High resolution nuclear magnetic resonance (NMR) spectroscopy is an important technique for determining molecular structure and physical properties of liquid samples [1]. A major drawback is its low signal-to-noise ratio (S/N) compared with other spectroscopic methods. (S/N) can be improved by raising the magnetic field strength, lowering the temperature of the sample, or improving the performance of the radio frequency (RF) probe and electronics [2]. RF probes can be improved in three general ways: First, reducing the volume of the detection coil, the signal per nuclear spin can be increased [3]. Second, by cooling the coils and preamplifiers, the noise can be reduced by up to a factor of four [4]. Finally, coils built from low-loss materials such as superconductors can improve the sensitivity [5]. In particular, the R_s of high temperature superconductivity (HTS) thin film is smaller than that of the conducting metals usually used in the microwave frequency region [6], and improvement of sensitivity in nuclear magnetic resonance (NMR) analysis can be expected from the characteristics of the low R_s under a high magnetic field. However, when NMR pick-up coils are applied to high magnetic fields, such as 11.7 T in the case of a 500 MHz NMR, the J_c of HTS films reduce rapidly according to the applied magnetic fields. Furthermore, the pick-up coils are necessary to generate a radio frequency during the excitation phase of a pulsed NMR experiment. Due to these factors, superconducting of NMR pick-up coils are not yet wide spred.

In particular, because the magnetic field is applied perpendicular to the c-axis, it is necessary to improve the J_c and R_s for NMR detection coil applications. It has been reported that to improve the J_c at high magnetic fields, high temperature superconductivity (HTS) thin film doped with $BaMO_3$ (M = Metal, Zr,

Sn) as APCs are effective [7]-[8]. Recently, H. Tobita et al reported that GdBCO coated conductors doped with BHO exhibit superior properties as APC material [9]. Additionally, Y. Ichino et al reported that a J_c of 2 MA/cm^2 at 20 K, 10 T was obtained by doping a 3.7 vol% of BHO to SmBCO films. In this study, we will clarify the doping amount of BHO suitable for NMR pick-up coils by evaluating the J_c of YBCO films doped with BHO at a low temperature and high magnetic field, targeting an increase in J_c and a decrease in R_s for YBCO thin films [10].

2. Experiment

Non-doped and BHO doped YBCO thin films with a thickness of 200 nm were deposited by using a pulsed-laser-deposition (PLD) method which used a KrF excimer laser (λ = 248 nm) on MgO (100) single crystalline substrates. The laser energy and repetition rate were 180 mJ and 1 Hz, respectively. We used commercial 1.5, 2.0 and 3.0 wt% BHO doped YBCO targets to fabricate BHO doped YBCO thin films. For comparison, we also used a non-doped YBCO target. The holder was heated by a SiC heater, and a substrate was attached directly to the holder and was heated by heat conduction. The holder was rotated to form homogeneous films. In this study, all films were deposited at approximately 720 °C with an oxygen pressure of 250 mTorr. The deposited thin films were then annealed at 430 °C with an oxygen pressure of 2000 Pa for 1 hour. For the analysis of the transport properties, UV photolithography and dry etching were used in order to obtain 2 mm long and 100 μm wide strip patterns. Au was deposited in order to reduce contact resistance to the electrode. The patterned samples were then lodged on a sample holder and mounted in cryostat endued with a 5 T superconducting magnet. The standard 1 μV/cm criterion was used to evaluate critical current values from the characteristics collected in the four probe configuration. In this study, since we could not use the 1 μV/cm standard due to influence of noise from the characteristics of *I-V*. We defined the critical current values by using the standard 1 μV/cm^2 after averaging the data from 10 experiments. The R_s values were calculated using a correlation equation with J_c and R_s [11-12].

$$R_s\ [\mathrm{m}\Omega] = 1.8 \times J_c^{-1}\ [\mathrm{MA/cm^2}] \text{ at } 21.8\ \mathrm{GHz} \quad (1)$$

The crystalline of the films was evaluated by X-ray diffraction (XRD) to obtain the $\theta/2\theta$ value and ϕ-scans. The $\delta\omega$ was the full-width half-maximum (FWHM) of the rocking curve of the (005) peak of the films, and the $\delta\phi$ was the FWHM of the ϕ-scan of the (102) plane of the thin films.

3. Result & Discussion

Figure 1 (a) shows the $\theta/2\theta$ results of the XRD measurement of the YBCO superconducting thin films. Both the non-doped and BHO doped samples showed strong YBCO (00l) peaks. Thus, the observations of the YBCO phase suggest that the YBCO films grew with c-axis orientation on the MgO substrate. Figure 1(b) shows X-ray ϕ-scan diffraction patterns of the (102) face of the YBCO. Each of the YBCO (102) peaks was clearly detected at 90° intervals, indicating that thin films have a four-hold symmetry on the MgO substrate. Thus, all thin films were revealed to be epitaxially grown cube-on-cube biaxially oriented thin films. Table 1 shows the $\delta\omega$, $\delta\phi$ and T_c of the non-doped YBCO and BHO doped YBCO thin films. For all BHO doped YBCO thin films other than 3 wt% doped thin film, the value of $\delta\omega$ was less than 1.6 and the value of $\delta\phi$ was less than 1.8. Therefore, no effect on the crystallinity in 2.0 wt% or less for BHO doped YBCO was observed.

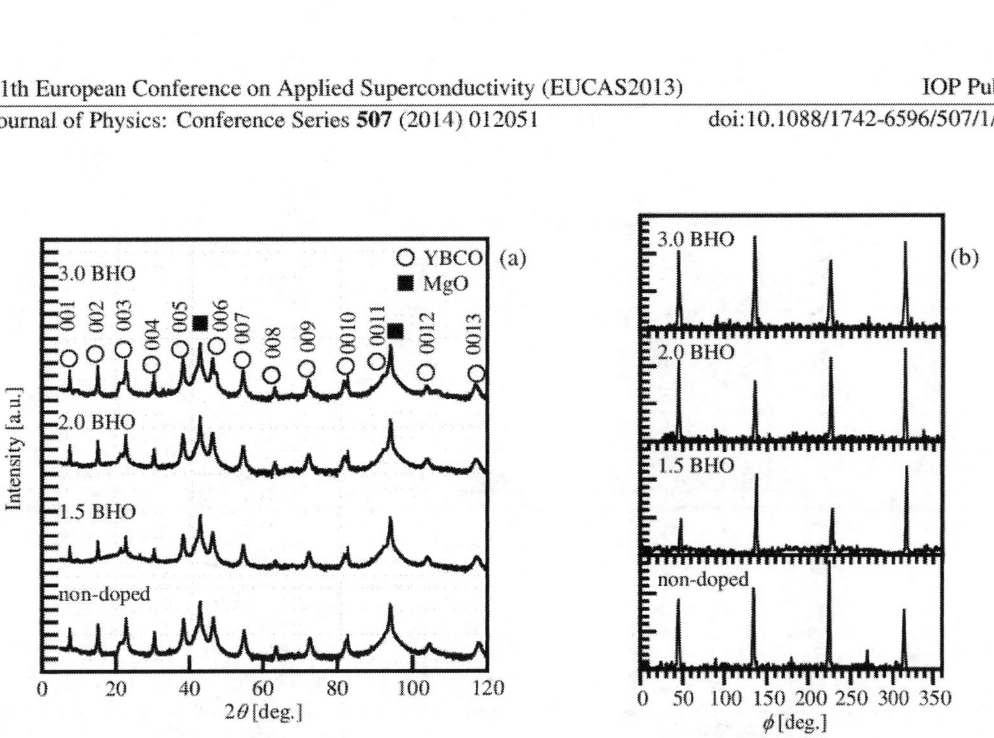

Figure 1. (a): X-ray θ-2θ pattern collected on the non-dope and BHO1.5, BHO2.0 and BHO3.0. (b): φ-scan of the non-dope and BHO1.5, BHO2.0 and BHO3.0.

Table 1. APC material and composition of the thin film. T_c, $\delta\omega$ and $\delta\phi$ of the each thin films.

Sample name	Super conductor	APC material	Additive amount [wt%]	$\delta\omega$	$\delta\phi$	T_c [K]
non-doped	$YBa_2Cu_3O_{7-\delta}$	–	–	1.62	1.79	86.9
BHO1.5	$YBa_2Cu_3O_{7-\delta}$	$BaHfO_3$	1.5	1.34	1.79	82.3
BHO2.0	$YBa_2Cu_3O_{7-\delta}$	$BaHfO_3$	2.0	1.54	1.71	81
BHO3.0	$YBa_2Cu_3O_{7-\delta}$	$BaHfO_3$	3.0	2.23	2.57	83

Figure 2 (a) shows the J_c(n) and J_c(p) measured at 20 K. The dotted lines represent the J_c fitting data calculated though formulas (1), (2) and (3). These fitting data could be explained by using the two-fluid model for high frequency and low magnetic field limits [13]. The R_s is given as (1) in a low magnetic field and (2) in a high magnetic field [14]. Where λ_L is the London penetration depth, B is the dc magnetic field, Φ_0 is the flux quantum, and $\hat{\mu}$ is the dynamic mobility. The η is the viscous drag coefficient which is related to the motion of the magnetic flux.

$$R_s(B) = R_s(0\mathrm{T}) + \frac{B\Phi_0}{2\lambda_L \eta_{low}(B,T,d)}, \quad B \ll \frac{\omega\mu_0\lambda_L^2}{\Phi_0\hat{\mu}} \quad (2) \qquad R_s(B) = \sqrt{\frac{\omega\mu_0\Phi_0 B}{2\eta_{high}(B,T,d)}}, \quad B \gg \frac{\omega\mu_0\lambda_L^2}{\Phi_0\hat{\mu}} \quad (3)$$

Non-doped typically showed J_c = 6.75 MA/cm^2 in zero magnetic field. The J_c of the YBCO thin films doped with BHO were comparable J_c of non-doped in zero magnetic field. However, the YBCO doped with BHO 1.5 wt% was 7.4 % greater than the non-doped YBCO. Improvement of the J_c and R_s by APCs under a zero magnetic field was reported by Matsumoto et al. and Mukaida et al [1-2]. In particular, the J_c(p) of BHO1.5 was 4.0 MA/cm^2 in 5 T, 41 % greater than that of the non-doped. It was possible to improve the J_c(p) by doping the BHO. The J_c(n) of BHO1.5 was 3.4 MA/cm^2, and was almost the same for the non-doped.

In addition, it is necessary to generate a magnetic field of B1 = 3.69×10^{-3} T for the 500 MHz NMR pick-up coil. It is assumed that for the application of NMR to a one-turn coil 1mm in width and 1 μm thick, 18.3 A of current will flow through the coil in order to product B1, J_c(p) 2 MA/cm^2 critical current density is required at 11.7 T. Thus, it was revealved from the fitting data that the J_c(p) at 10 T of BHO1.5 was 2.0 MA/cm^2, and 1.5wt% BHO doped YBCO is the most effective in the application of NMR . Figure 2.(b) shows the R_s(n) and R_s(p) calculated from a correlating equation of J_c. The dotted line is the fitting data of R_s under a weak magnetic field calculated by using equation (1) and under a strong magnetic field calculated by using equation (2). The R_s(p) and R_s(n) of BHO1.5 was two orders of magnitude lower than the R_s of copper, and is thus effective for NMR pick-up coils.

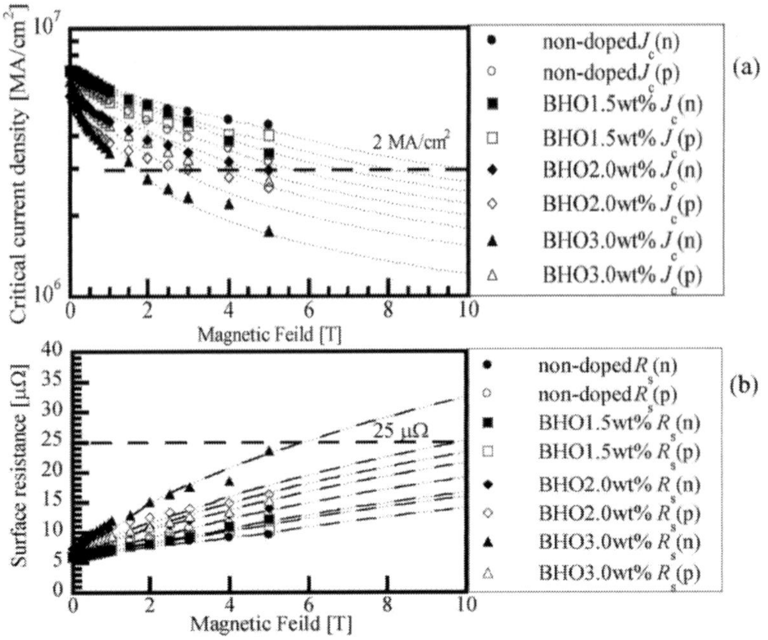

Figure 3. (a): J_c of the YBCO films as a function of magnetic field applied parallel and vartical to the c-axis of the films. (b): Magnetic field dependence of the R_s calculated from the magnetic field dependence of J_c.

4. Conclusion

We measured the J_c of 0, 1.5, 2.0 and 3.0 wt% BHO doped YBCO thin films. In addition, we have calculated the correlating equation of J_c. The J_c(p) of BHO1.5 was 7.4 % bigger than non-doped in a zero magnetic field. The J_c(p) of BHO1.5 in 5 T was 41 % bigger than non-doped. J_c(n) was also comparable to the non-doped. The R_s(n) and R_s(p) of BHO1.5 of 10 T were smaller by approximately two orders of magnitude than the R_s of copper. Therefore, we clarified the possibility that the most effective way to fabricate thin film for NMR coil application is by doping YBCO with BHO 1.5 wt%.

References

[1] Howard D. W. Hill, Varian NMR Instruments, Palo Alto, CA 94304, USA "Improved sensitivity of NMR spectroscopy probes by use of high-temperature superconductive detection coils" IEEE Trans. vol. 7, No. 2, pp. 3750-3755, 1997

[2] Willian W. Brey, Arthur S. Edison, Robert E. Nast, James R. Rocca, Saikat Saha, Richard S. Withers, "Design, construction, and validation of a 1-mm triple-resonance high-temperature-superconducting probe for NMR" Journal of Magnetic Resonance vol. 179, No. 2, pp. 290-293, 2006

[3] D. Hoult, R. Richards, "Singnal-to-noise ratio of nuclear magnetic resonance experiment" J. Magn. Reson. vol. 24, pp. 71-85, 1976

[4] D.L. Olson, T.L. Peck, A.G. Webb, R.L. Magin, J.V. Sweedler, "High-resolution microcoil H-1-NMR for mass-limited, nanoliter-vol Samples" Sience vol. 270, pp. 1967-1970, 1995

[5] W.A. Anderson, W.W. Brey, A.L. Brooke, B. Cole, K.A. Delin, L.F. Fuks, H.D.W.Hill, M.E. Johanson, V. Kotsubo, R. Nast, R.S. Withers, W.H. Wong, "High-sensitivity NMR spectroscopy probe using superconductive coils" Bull. Magn. Reson. vol. 17, pp. 98-102 1995

[6] A. Saito, J. H. Lee, H. Teshima, and S. Ohshima, "Power-Handling Capability of Transmit Filters Using Superconducting Bulk Ring Resonators" IEEE Trans. Appl. Supercond., vol. 19, No. 3, pp.899-902, 2009.

[7] J. L. MacManus-Driscoll, S. R. Foltyn, Q. X. Jia, H. Wang, A. Serquis, L. Civale, B. Maiorov, M. E. Hawley, M. P. Maley and D. E. Peterson, "Strongly enhanced current densities in superconducting coated conductors of $YBa_2Cu_3O_{7-x}$ + $BaZrO_3$" Nat. Mater., vol. 3, pp. 439, 2004.

[8] Chakrapani V Varanasi, P N Barnes, J Burke L Brunke, I Maartense, T J Haugan, E A Stinzianni, K A Dunn and P Haldar, "Flux pinning enhancement in $YBa_2Cu_3O_{7-x}$ films with BaSnO3 nanoparticles" Supercond. Sci. Techonol. vol. 19, L37-L41, 2006

[9] H. Tobita, K. Notoh, K. Higashikawa, M. Inoue, T. Kiss, T. Kato, T. Hirayama, M. Yoshizumi, T. Izumi, Y. Shiohara, "Fabrication of $BaHfO_3$ doped $Gd_1Ba_2Cu_3O_{7-\delta}$ coated conductors with the high I_c of 85 A/cm-w under 3 T at liquid nitrogen temperature (77 K)Supercond. Sci. Technol. vol.25 no. 6, pp. 062002, 2012

[10] Y. Ichino, Y. Sawano, A. Tsuruta, Y. Yoshida, S. Awaji, "Superconducting transport properties in high magnetic fields of nano-structured RE-Ba-Cu-o superconducting films"

[11] S. Ohshima, M. Kusunoki, M. Inadomaru, M. Mukaida, Y. Tanaka, and H. Ihara, "Relationship Between the Surface Resistance and Depairing Current Density of Superconductors" IEEE Trans. on Applied Superconductivity vol. 13, No. 2, pp. 3578-3580, 2003

[12] K. Nakagawa, S. Nakayama, A. Saito, S. Ono, H. Kai, M. Mukaida, T. Honma, S. Ohshima, Physica, "Relationship between intrinsic surface resistance and critical current density of YBCO thin films with various thickness" Physica C vol. 470, pp. 1361-1364, 2010"

[13] M.J. Lancaster, passive Microwave Device Applications of High-Temper ature Superconductors, Cambrige University Press, Cambridge, pp. 36-40, 1997

[14] T. Honma, S. Sato, K. Sato, M. Watanabe, A. Saito, K. Koike, H. Kato, S. Ohshima, "Microwave surface resistance of YBCO superconducting thin films under high DC magnetic field" Physica C vol. 484, pp. 46-48, 2013

The abnormal temperature-dependent rectification effect in $BiFeO_3/YBa_2Cu_3O_x$ heterostructures

Qianqian Yang, Huai Zhang, Qian Dai, Ruijuan Nie, Furen Wang
School of Physics, Peking University, Beijing 100871, People's Republic of China

E-mail: qianyqq@pku.edu.cn

Abstract. We report the fabrication and properties of heterostructures formed by the $BiFeO_3$ (BFO) ferroelectric and $YBa_2Cu_3O_x$ (YBCO) superconductor. Four measurement configurations on BFO/YBCO heterojunctions are proposed and showed different I-V characteristics. Moreover, one type of BFO/YBCO heterostructures which had the strongest rectification effect at 15K, showed the abnormal rectification effect dependent on temperature. The rectifying effect enhanced with decreasing temperature below the superconducting transition temperature of YBCO. YBCO electrode played an important role in increasing the rectifying parameter because of the high density of states for quasi-particles in YBCO in the superconducting state.

1. Introduction

The discovery of high temperature superconductors (HTS) initiated studies about oxide heterostructures. The perovskite oxide heterostructures have attracted much attention owing to their exciting interface-related physics and excellent performances over polycrystalline materials.

Recently, $BiFeO_3$ (BFO) is widely known as a Pb-free ferroelectric and of large remnant polarization of up to 100 $\mu C/cm^2$ in its ground state (rhombohedral phase)[1]. It is found to have the diode effect, where BFO behaves like a semiconductor[2]. Ferroelectric bulk materials are traditionally regarded as insulators, but thin films are always treated as semiconductors as becoming thinner[3, 4]. The Bi-deficiency is the main reason for BFO films to become p-type semiconductor[5, 6] while n-type semiconductor BFO arises from the oxygen vacancy[7, 8]. HTS are expected to apply in electronic devices. However, the ordinary semiconductors such as Si and GaAs are difficult to combine with HTS because of the different crystal structures. This activated our interest in studying the BFO/YBCO heterostructures. In this paper, we propose the fabrication of the BFO/YBCO heterostructures. The properties of this heterostructures are of great potential for the development of new electronic devices such as diodes with very high rectifying parameters and transistors based on the Josephson field effect.

2. Experimental details

We have fabricated 180nm-thick BFO films epitaxially grown on 60nm-thick YBCO films on $SrTiO_3$ (STO) substrates by pulsed laser deposition (PLD). YBCO and Au as top electrodes are fabricated by PLD through a metal shadow mask. Silver (Ag) electrodes (~1 mm^2) were placed on the surface of YBCO, BFO and Au for measuring the electrical characteristics. The detailed growth parameters are summarized in Table 1. The growth rate for YBCO, BFO and Au are 0.067nm/pulse, 0.021nm/pulse, 0.015nm/pulse, respectively.

Table 1. PLD growth parameters of YBCO, BFO and Au thin films.

	Target-to-substrate distance(mm)	Deposition temperature (°C)	Deposition pressure (Pa)	Annealing time (min)	Annealing temperature (°C)
YBCO on STO	50	760	26	30	430
BFO-1	40	650	17	5	430
BFO-2	40	650	6	5	430
YBCO on BFO	50	700	26	30	430
Au	40	Room temperature	$2*10^{-5}$	-	-

3. Results and discussion

Figure 1 shows the typical current-voltage (I-V) characteristics for a dc voltage loop ranging from $0 \rightarrow V_{max} \rightarrow 0 \rightarrow -V_{max} \rightarrow 0V$ of BFO/YBCO heterstructures. Here the Ag/Au and Ag/YBCO interface showed practically ohmic characteristics. Although the I-V curves in figure 1a and figure 1b are asymmetric for the forward and reverse biases, they are in the same order of magnitude and the current at V_{max} is small. The rectification behavior is mostly obvious in Au/BFO-2/YBCO as shown in figure 1d. The oxygen deposition pressure as fabricating BFO-2 is 6 Pa lower than that of BFO-1 with 17 Pa, which will increase the quantity of oxygen vacancies in BFO-2 films. The growth of top YBCO films in YBCO/BFO-2/YBCO (shown in figure 1b) results in the oxygen content of BFO increases and the oxygen vacancies are filled in BFO again, which is discussed in our previously reported paper[9]. The oxygen deficiencies are crucial to the rectification effect because they are the main reason for BFO films to become n-type semiconductors[7], as confirmed from the oxygen pressure experiment and annealing experiment[8].

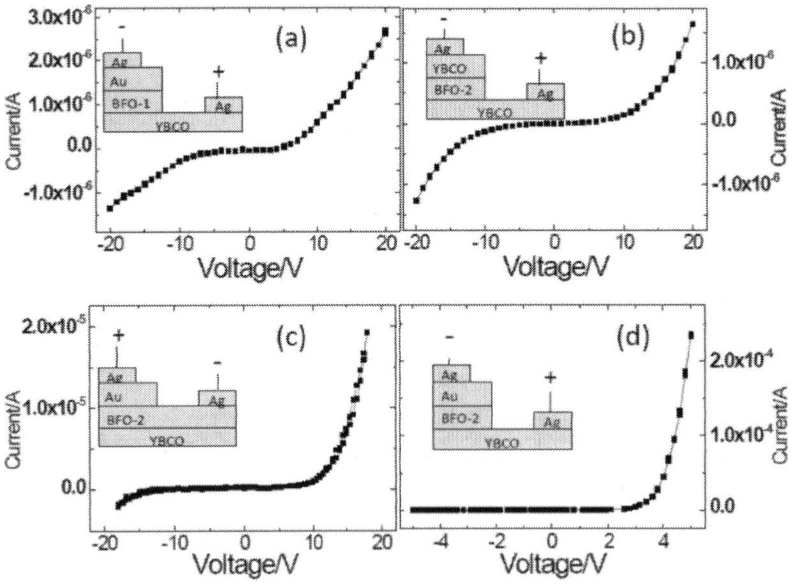

Figure 1: I-V characteristics of BFO/YBCO heterostructures with four measurement configurations at 15K.

The rectifying parameter, r_p, is defined by, $r_p = \frac{I^+_{max}}{I^-_{max}}$, where I^+_{max} and I^-_{max} denote the forward and reverse current at V_{max} and $-V_{max}$, respectively. Although the rectification behaviour is observed in in-plane BFO-2 films (as shown in figure 1c), the rectification parameter is just 10 much less than that of the Au/BFO-2/YBCO with r_p=1100. YBCO electrode played an important role in increasing the rectifying parameter in Au/BFO-2/YBCO.

The conduction mechanism can be inferred by I-V curve dependent on temperature. For Au/BFO-1/YBCO and YBCO/BFO-2/YBCO with no rectification behavior, the I^+_{max} and I^-_{max} increased by increasing temperature, because more thermally activated charge carriers are generated in the film and they participate in the conduction. The thermoionic emission is the most important process determining the current in a Schottky contact as a bias voltage is applied.

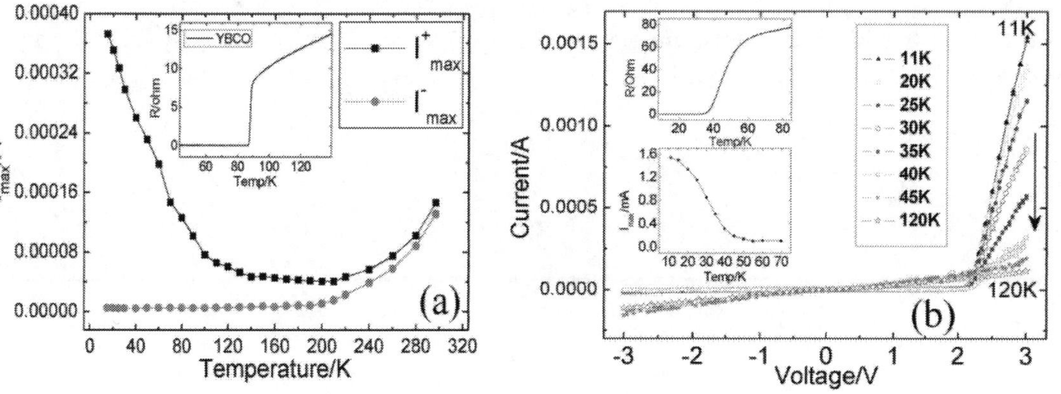

Figure 2: The measurement is carried out in the configuration of figure 1d. (a) I^+_{max} and I^-_{max} versus temperature curves of Au/BFO-2/YBCO heterostructures. The inset shows the resistance versus temperature (R-T) curve of YBCO. (b) I-V curves of Au/BFO-2/ $YBa_2Cu_3O_{x\sim6.43}$ heterostructure. The insets show the R-T curve (top) and I^+_{max} versus temperature curve (bottom) separately.

The ferroelectric polarization of BFO is considered to have an effect on the rectification behavior. In the pseudo cubic crystal structure, the spontaneous polarization direction lies close to (111), and the values measured along a-axis and c-axis are projections of (111) onto these orientations and thus equal[1]. Therefore, the ferroelectric polarization should play the same role in the temperature dependent I-V curves of the in-plane BFO and out of plane BFO (Au/BFO-2/YBCO). I^+_{max} of in-plane BFO-2 films increased with increasing the temperature. The ferroelectric polarization didn't reduce rectification parameter with increasing the temperature. Nevertheless, in Au/BFO-2/YBCO, the rectification parameter decreased as the temperature increased. Figure 2a shows the temperature dependence of the rectifying I-V curve behavior in the out of plane configuration (inset of figure 1d). The superconducting transition temperature (T_c) of YBCO is 89K. There is no rectification effect at room temperature. I^+_{max} and I^-_{max} firstly decrease with decreasing the temperature from room temperature. This is consistent with the change of the resistance of BFO film with the temperature. The rectifying effect is enhanced substantially with decreasing temperature below T_c of YBCO, which is dependent on the T_c of YBCO.

The similar results are observed in the other Au/BFO-2/YBCO heterstructure where T_c of YBCO is about 40K. $YBa_2Cu_3O_x$ shows a significant T_c-x relationship, where the oxygen content x could be determined by the (00l) diffraction peak and controlled by changing the partial oxygen pressure during cooling in the deposition process[10, 11]. Figure 2b shows the I-V curves dependent on temperature for Au/BFO-2/ $YBa_2Cu_3O_{x\sim6.43}$ heterostructure. The resistance versus temperature curve of YBCO film is shown in the inset of figure 2b. We can see the rectification effect also diminishes above the T_c.

Although the rectification effect is recently observed in BFO capacitors, all the heterojunctions showed the rectification parameter decreased with the temperature below T_c [2, 12]. This excluded the impact of Au and BFO interface and BFO films. Therefore, the BFO and YBCO interface is the main reason for the observed increase of the rectifying parameter.

The corresponding schematic energy band diagram at BFO/YBCO interface is shown in figure 3. For BFO, the band gap E_g is 2.8 eV and the electron affinity is 3.3 eV[13], and then the work function of n-type BFO should be less than 4.7 eV. The work function of YBCO is about 6.1 eV[14]. Because the BFO has higher Fermi level than YBCO, the electrons of BFO would move spontaneously into YBCO at the interface. This results in the upward-bend conduction band CB and valence band VB of BFO. At the BFO/YBCO interface, transport current at temperature below T_c is possibly from the correlated charge carriers and quasi-particles. Electrons from the conduction band of the BFO can approach the interface due to the decrease of the band-bending under the forward bias. The electrons and the holes can recombine at the interface. The superconductor opens an energy gap near the Fermi energy below the superconducting transition temperature. Meanwhile, the density of state for quasi-particles of YBCO increases with decreasing temperature below T_c substantially, which directly increased the current.

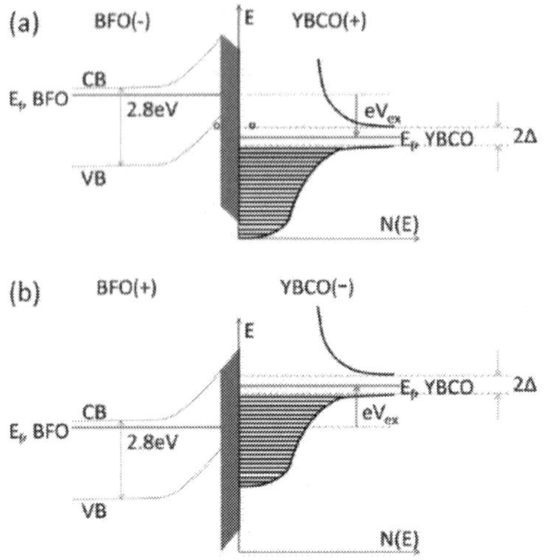

Figure 3: Schematic energy band diagram of BFO/YBCO heterostructures under forward bias (a) and reverse bias (b) at the superconducting state ($T < T_c$).

4. Conclusions

The BFO/YBCO heterostructures are fabricated by PLD on STO substrates. Four measurement configurations on BFO/YBCO heterostructures are proposed and showed different I-V characteristics. The oxygen vacancies in BFO films are crucial to the observation of the rectification effect in BFO. Especially, the rectification effect in Au/BFO-2/YBCO is enhanced with decreasing the temperature below T_c, which is attributed to the BFO/YBCO interface.

5. References

[1] J. Li, J. Wang, M. Wuttig, R. Ramesh, N. Wang, B. Ruette, A. Pyatakov, A. Zvezdin, and D. Viehland, *Appl. Phys. Lett.* **84**, 5261 (2004).
[2] T. Choi, S. Lee, Y. Choi, V. Kiryukhin, and S.-W. Cheong, *Science* **324**, 63 (2009).
[3] Y. Watanabe, *Phys. Rev. B* **57**, R5563 (1998).
[4] M. Dawber, K. Rabe, and J. Scott, *Rev. Mod. Phys.* **77**, 1083 (2005).
[5] Z. Zhang, P. Wu, L. Chen, and J. Wang, *Appl. Phys. Lett.* **96**, 232906 (2010).
[6] T. R. Paudel, S. S. Jaswal, and E. Y. Tsymbal, *Bulletin of the American Physical Society* **57** (2012).
[7] S. Clark and J. Robertson, *Appl. Phys. Lett.* **94**, 022902 (2009).
[8] C.-H. Yang *et al.*, *Nature Mater.* **8**, 485 (2009).
[9] Q. Yang, X. Ma, Q. Dai, H. Zhang, R. Nie, and F. Wang, *Physica C* **492**, 181 (2013).
[10] J. Jorgensen *et al.*, *Phys. Rev. B* **36**, 3608 (1987).
[11] J. Ye and K. Nakamura, *Phys. Rev. B* **48**, 7554 (1993).
[12] H. Li, K. Jin, S. Yang, J. Wang, M. He, B. Luo, J. Wang, C. Chen, and T. Wu, *Journal of Applied Physics* **112**, 083506 (2012).
[13] S. Clark and J. Robertson, *Appl. Phys. Lett.* **90**, 132903 (2007).
[14] T. Hirano, M. Ueda, K.-i. Matsui, T. Fujii, K. Sakuta, and T. Kobayashi, *Jpn. J. Appl. Phys.* **31**, 1345 (1992).

Synthesis of Bi-based superconductor by microwave-assisted hydrothermal method

R. G. Lima[1], V. D. Rodrigues[1], C. L. Carvalho[1], S. R. Teixeira[2], A. E. Souza[2] and R. Zadorosny[1]

[1]Depto de Física e Química, Faculdade de Engenharia de Ilha Solteira, Univ Estadual Paulista - UNESP, Caixa Postal 31, 15385-000, Ilha Solteira, SP, Brazil

[2]Depto de Física, Química e Biologia, Faculdade de Ciências e Tecnologia, Univ Estadual Paulista - UNESP, Presidente Prudente, SP, Brazil

rafazad@yahoo.com.br

Abstract. In this work we studied the synthesis of BSCCO-2212 superconducting phase associating a quite similar method developed by Pechini with the microwave-assisted hydrothermal method. To study the influence of the microwaves on the properties of BSCCO system, we synthesized two samples by such method. For one sample we used carbonates and for the other one we used nitrates as chemical reagents. We also produced a reference sample just using carbonates by Pechini´s method to compare their morphological and superconducting properties. The structural properties of the samples were analyzed by scanning electron microscopy and X-ray diffraction. The Bi-2212 phase is predominant in all samples and despites the nitrates-like sample has a broader distribution of grain size in comparison with the reference sample, its magnetic behaviour is closer to that presented by the reference one.

1. Introduction

Studies of new routes to synthesized oxide superconductors are focused on the production of materials with higher T_c and J_c [1,2]. Chemical methods as sol-gel, Pechini, coprecipitation and hydrothermal have been applied on the synthesis of superconducting materials due to the homogeneity to produce samples. [1,3] Thereby, on the last years, the use of microwaves in the synthesis of a variety of materials has been raised in areas as chemistry, condensed matter physics and materials engineering. In some researches, the use of domestic microwave-oven in scientific activities has been of great interest due to its simplicity and low cost operation [4]. It is interesting to emphasize that several inorganic oxides, including the CuO, absorbs micro-wave radiations as those produced by domestic microwave-oven, i.e., with frequency of 2.45GHz. Baghurst and coworkers [6] were the pioneers on the application of microwaves to synthesize mixtures of metallic oxides with superconducting properties. In such samples, the heating process begins in the inner of the copper oxides and then the heat is transferred to the vicinity. Thus, the crystallization temperature of the material is reached faster than in conventional heating processes, saving time and energy. [5,6]

Particularly, in the microwave-assisted hydrothermal method, MAH, those waves interact with the solution and part of the electromagnetic energy is converted in thermal energy. Thus, this heat is generated in the inner solution and homogeneously propagates to the entire system promoting an extremely fast crystallization kinetics. [7,8] In contrast with the conventional heating process, there are several advantages in use MAH such as the high heating ratios which are reached, reduced processing time, energetic efficiency, formation of nano-sized grains in many cases, and so on. [8]

Thus, in this work we synthesized samples of the BSCCO system, focused on the 2212 phase. [9] We used the polymeric precursor method developed by Pechini, MP, associated to MAH. In one hand we worked with a chemical method which is a good synthesis route to produce homogeneous materials [1,3] and on the other hand we added a step on the synthesis process which consisted in the use of MAH.

2. Materials and methods

We focused on the synthesis of Bi-2212 due to its chemical stability and absence of toxic elements. We had already worked with different kind of chemical reagents, i.e., we used carbonates and nitrates. Thus, the samples were synthesized following PM and an association of PM with MAH. It was used a molar ratio of 3/1 for citric acid/metal and a mass ratio of 40/60 for citric acid/ethilenoglicol. [10].

A reference sample, REF, was synthesized by PM where we used carbonates as $(BiO_2)CO_3$, $SrCO_3$, $CaCO_3$ and $CuCO_3(OH)_2$ [11]. Another sample was also prepared using carbonates however, after the formation of the polymer, the precursor solution was submitted to MAH. The equipment used in this stage was a commercial microwave oven adapted with a Teflon® autoclave in which was inserted a collector cup. Such system is constituted by a thermocouple, a silicone sealing gasket, fixing screws, pressure gauge, safety valve and temperature controller. The synthesis was carried out using a heating rate of 2°C/min and maintained at 140°C for 60 minutes. During the heating process, the registered pressure was 3 bar. After that, the solution was kept in rest for during 24 hours and all the process was repeated once more. Thus, a supernatant material and a dark brown and viscous precipitate were originated. The precipitate was isolated and dried at 70°C during five days in an oven. The resulting sample was labeled as CMAH. A third sample was produced using nitrates as $Bi_5H_9N_4O_{22}$, $Sr(NO_3)_2$, $Ca(NO_3)_2$ and $Cu(NO_3)_2.3H_2O$ for which was followed the same procedure as described above to obtain CMAH. This sample was labeled NMAH.

All powders were calcinated at 200°C/1h and 400°C/2h. After the calcinations, the powder was heat treated at 850°C [12] in two steps, one of them for 2h and the other one by 6h. The heating rate used was 2°C/min. The materials were characterized by x-ray diffraction, XRD, (Shimatzu model XRD-6000) and scanning electron microscopy, SEM, (ZEISS model EVO LS15). The XRD measurements were done in the 2θ ranged from 4° to 60° with 1°/min, steps of 0.02° and λ=1.542 Å with CuKα radiation.

The powders were pelletized and sintered at 845°C for 24h in air. After that, the samples were characterized by XRD, SEM and electric and magnetic measurements. The four-dc probe method was used to do the electric measurements which were carried out in a home-made system consisted of a current source Keithley 228A; a multimeter Keithley 2000, a nanovoltmeter Keithley 2182 and a Dewar with liquid nitrogen. The magnetic measurements were carried out in Quantum Design PPMS model 6000.

3. Results and discussion

3.1. XRD and SEM characterizations

The XRD indicates that REF is a polyphasic sample with the presence of the Bi-2212 phase and, at least, the $(BiO)_2Sr_2CaCu_2O_6$ secondary 2212 phase, as shown in Figure 1(a). For the CMAH sample, the XRD data show two phases, Bi-2212 as major phase and Bi-2201 as second phase (see Figure 1(b)). The sample NMAH has the Bi-2212 phase with traces of Bi-2201 phase and other spurious compositions such as $SrCuO_2$ and $Bi_2Sr_3O_6$. Those analysis were made using JCPDS cards or equivalent.

Figure 2 shows SEM images of the as obtained powders and pellets using REF, CMAH and NMAH processes. In all cases were observed plate-like grains that are characteristic of the BSCCO superconducting system [13]. Figure 2 shows SEM micrographs of the pelletized samples. As can observed, in this figure, the REF sample presents plate-like grains with dimensions higher than 2x10 μm^2 and several small rounded grains with minimum dimensions around 1x2 μm^2; in CMAH sample micrograph could be identified few grains with minor sizes of the order of 1x1 μm^2 even with thickness around 250 nm and the NMAH present plates higher than 2x4 μm^2 and few long needles with dimensions around 1.3x16 μm^2 which indicates that such sample has a broader distribution of grain size.

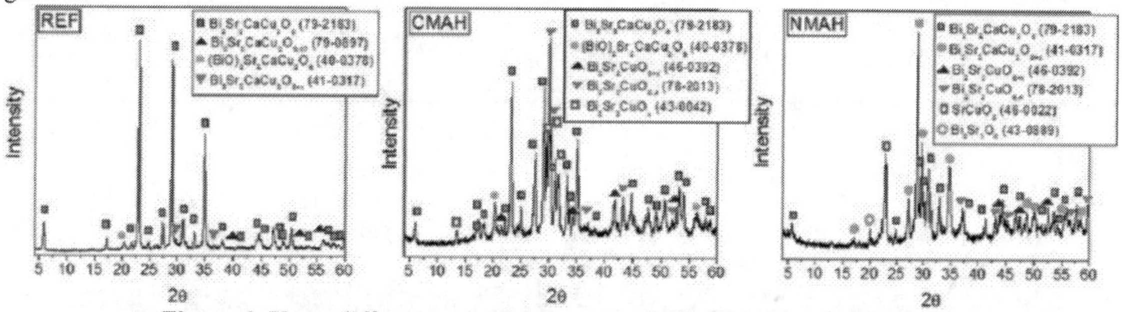

Figure 1. X-ray diffractogram of the samples REF, CMAH and NMAH.

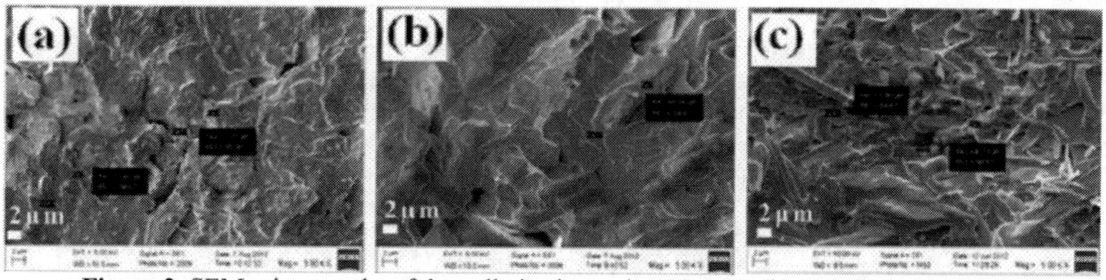

Figure 2. SEM micrographs of the pelletized samples (a) REF, (b) CMAH and (c) NMAH.

3.2. Electric and magnetic characterizations

Figure 3 shows the R(T) curves of the samples. In the worked range of temperatures can note some transitions in REF at 86K, 81K and 78K. The CMAH presents a transition at 81K and the NMAH sample also presents several transitions at 92K, 90K, 85K and 80K. For all samples can note the T_c signature of the Bi-2212, as observed in the XRD analysis. The several transitions presented by all samples could be due to the presence of other superconducting phases with different T_c´s.

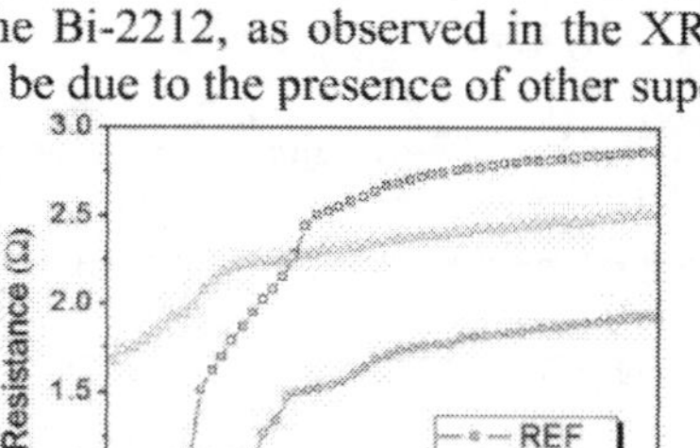

Figure 3. Resistance as a function of the temperature curves of the samples REF, CMAH and NMAH. The several transitions presented by each sample is an indicative of the presence of different superconducting phases.

Figure 4 shows the M(T) curves of the samples. In all curves the T_c is around 80K, characteristic of the Bi-2212 phase. It should be noted that the zero field cooled, ZFC, curve of REF has a greater diamagnetic response, in modulus, than for CMAH. Such behaviour is an indicative that the intergranular current flows through REF easier than in the CMAH. We can also note that the NMAH sample has a response quite different from that presented by REF, probably associated to its large grain-size distribution observed in SEM micrographs. However, the magnitude of its magnetization signal is comparable with that of the REF sample.

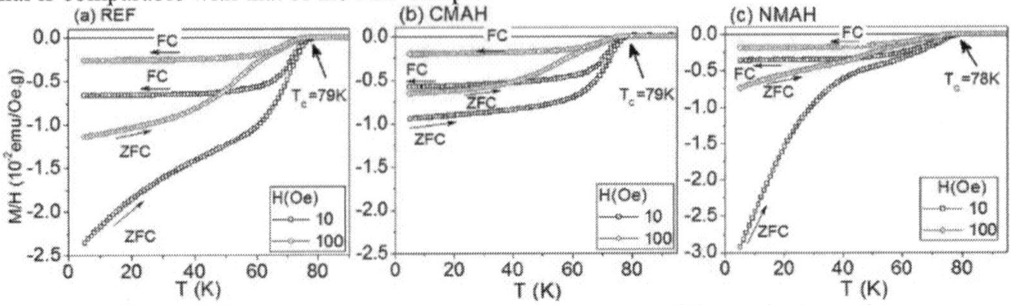

Figure 4. Magnetization versus temperature curves of the studied samples.

As conclusions we can note that both methods used in the synthesis of the samples produced the Bi-2212 phase with segregate phases. SEM micrographs show the presence of at least two type of grains, i.e., a plate-like grains which are characteristic of BSCCO superconducting system and rounded grains. In the electric characterizations, the several transitions presented by some samples could be related to other superconducting phases. The differences between electric and magnetic responses might be related to the fact that while the surface response is predominant in electric characterizations the magnetic ones measure an average response of the bulk sample. The measurements of the magnetization versus temperature followed the ZFC and FC procedures of the REF and NMAH samples corroborates with the SEM observations, i.e., the broader distribution of grain size could be responsible to the different behaviours exhibited under external magnetic fields. Thus, it can be concluded with those analyses that the sample produced only by Pechini's method present a better superconducting response than the samples produced by MAH. However, between MAH samples, the sample produced by nitrates, NMAH, presented a behaviour closer to that presented by REF sample. It is worth to emphasize that this is an initial study of the production of oxide superconductors by the microwave-assisted hydrothermal method.

We thank the Brazilian Agencies CAPES, CNPq, Fundunesp/PROPe and the São Paulo Research Foundation (FAPESP), grant 2013/11114-7, for financial support.

References

[1] Rao, C N R, Nagarajan R. and Vijayaraghaven R 1994 *Superc. Sci. Techn.* **6** 1
[2] Deimling C V, Motta M, Lisboa-Filho P N, Ortiz W A 2008, *J. Mag. Mag. Mat.* **320** e507
[3] Zhang Y, Yang H, Li M, Sun B and Qi Y 2010 *Cryst. Eng. Comm.* **12** 3046
[4] Keyson D, Longo E, Vasconcelos J S, Varela J A, Éber S, Dermaderosian A 2006 *Cerâmica* **52** 321
[5] Volanti D P, Cavalcante L S, Keyson D, Lima R C, Moura A P, Moreira M L, Macario L R, Godinho M, 2007 *Metalurgia & Materiais* **63** 351
[6] Baghurst D R, Chippindale A M and Mingos D M P 1988 *Nature* **332** 311
[7] Souza A E, Silva R A, Santos G T A, Moreira M L, Volanti D P, Teixeira S R, Longo E 2010 *Chemical Physics Letters* **488** 54
[8] Zhu X H and Hang Q M 2013 *Micron* **44** 21
[9] Maeda H, Tanaka Y, Fukutumi M and Asano T 1988 *J. Appl. Phys* **27** L209
[10] Chu C T and Dunn B 1987 *J. Am. Ceram. Soc.* **70** C-375
[11] Kakihana M 1996 *Sol-Gel Sci. Tech.* **6** 7
[12] Peng Z S, Hua Z Q, Li Y N, Di J, Ma J, Chu Y M, Zhen W N,Yang Y L, Wang H J and Zhao Z X 1998 *Journal of Superconductivity* **11** 6
[13] Majewski P 2000 *Journal of Materials Research* **15** 4

Journal of Physics: Conference Series **507** (2014) 012054 doi:10.1088/1742-6596/507/1/012054

Electronic transport properties of superconducting FeSe thin films in a magnetic field

AG Zaitsev[1], R Schneider, D Fuchs, A Beck and R Hott
Karlsruhe Institute of Technology, Institute for Solid-State Physics
D-76021 Karlsruhe, Germany

E-mail: alexander.zaitsev@kit.edu

Abstract. Superconducting c-axis oriented tetragonal FeSe films were grown on (001) MgO substrates. The transition temperature increased with film thickness and reached 10.5 K for a 2 μm film. The upper critical field, B_{c2}, was measured up to 14 T and then extrapolated to lower temperatures by an interpolation scheme based on Werthamer-Helfand-Hohenberg theory. The highest $B_{c2}(0)$ was obtained for the 2 μm film: 28.8 T and 24.7 T in parallel and normal field, respectively. The effect of thermally activated flux-flow on the film resistivity was analysed, and the activation energies (U_0) were extracted. At temperatures not far below T_c and fields above 1 T we observed a clear logarithmic dependence of $U_0(B)$, which is an evidence for the 2D vortex matter. At lower fields and temperatures the vortex matter behaved 3D-like.

FeSe with a tetragonal lattice is one of the simplest iron-based superconductors with respect to its binary chemical composition and crystallographic structure consisting of quasi-2D Fe-Se layers (*a-b* plane) separated by a c-axis lattice spacing of ~5.5 Å[1]. Recently we have shown that the evolution of 2D electronic transport in FeSe thin films gives rise to a superconductor-insulator quantum phase transition [2] as the film thickness was reduced from 2 μm to 1 nm. In the current work we focus in more detail on the properties of superconducting films of this series, especially in dc magnetic fields.

The FeSe films were prepared *in situ* by simultaneous magnetron sputtering of a sintered planar FeSe and a pure Se target onto 5×10 mm^2 (001) MgO substrates [3]. For all examined films the θ-2θ x-ray diffraction (XRD) scans revealed (001) tetragonal FeSe structure with lattice constant c between 5.526 Å and 5.548 Å [3]. Minor amount of (101) tetragonal FeSe phase was often present and the spurious (001) hexagonal FeSe phase was observed in some films. The film lattice exhibited a good off-plane alignment. The (001) rocking curve of a 0.5 μm thick film had a full width at half-maximum (FWHM) $\Delta\omega$ = 0.22°. $\Delta\omega$ increased with the film thickness up to 2.8° for a 2 μm film. The in-plane lattice alignment examined by ϕ-scan XRD revealed an in-plane orientational relationship with the (100) axis of tetragonal FeSe parallel to the (110) axis of MgO, see figure 1. In this case each three unit cells of (001) FeSe ($a = b$ = 3.76 Å) are accommodated diagonally over two unit cells of (001) MgO ($a = b$ = 4.21 Å), see figure 1b, resulting in a reasonably low 5% lattice mismatch.

The dc resistivity (ρ) of the FeSe films was measured using a Keithley 6220/2182 system as a function of temperature (T) in dc magnetic fields (B) up to 14 T directed either parallel or normal to the film surface. The measurements were performed by four-point technique over bridges patterned in the films. Figure 2 shows the whole set of resistivity data measured for a 2 μm FeSe film over a 200 μm

[1] To whom any correspondence should be addressed.

wide and 2 mm long bridge. At zero applied field this film exhibited low normal state resisitvity, $\rho_n(15K,0T) \approx 0.17$ mΩ·cm, which is comparable to bulk material [4]. In increasing field ρ_n grew at a rate of 0.3 μΩ·cm/T in parallel field and 1 μΩ·cm/T in normal field. Similar magnetoresistive behaviour has been reported for other Fe-based superconductors [5]. Above 15 K ρ_n was fairly linear with temperature and we used a linear extrapolation of $\rho_n(T)$ for evaluating $\rho(T)/\rho_n(T)$ ratios at $T < T_c$. At no field applied the T_c onset was defined at 12.9 K ($\rho(T)/\rho_n(T) = 0.9$), the T_c offset at 8.2 K ($\rho(T)/\rho_n(T) = 0.1$) and the midpoint T_c at 10.5 K. These values are quite close to those of bulk material under ambient pressure, indicating (i) high film quality and (ii) that the film lattice was free of considerable mechanical stress. Increase of the magnetic field shifted the $\rho(T)$ curves to lower temperatures but did not affect their shape: neither steps nor other weak-link induced features [6] appeared. Furthermore, the highest applied field of 14 T did not suppress the superconductivity to a remarkable extent. Thus we conclude the FeSe films to be free from distinct inhomogeneities.

Figure 3 shows the upper critical fields B_{c2} extracted from the data of figure 2 using the $\rho(T)/\rho_n(T) = 0.9$ criterion, as well as $B_{c2}(T)$ data obtained for three other FeSe films of various thicknesses. In agreement with our former findings [2] thicker films exhibited higher T_c. B_{c2} also increased with film thickness. $B_{c2}(T)$ of the 2 μm film were very close to those reported for bulk FeSe at ambient pressure [4]. For a more detailed analysis we focused on the 2 μm film, see figure 3b. Most of all we were interested in the development of $B_{c2}(T)$ at low temperatures. Usually, Werthamer, Helfand and Hohenberg (WHH) theory [7,8] is applied to predict B_{c2} at $T = 0$ K from the slope of $B_{c2}(T)$ at T_c according to the WHH result $B_{c2}(0) = 0.7\ T_c\ [-dB_{c2}(T_c)/dT]$. But this approach is only reliable for well-defined T_c and $B_{c2}(T \approx T_c)$ data evaluation which is not really the case for our FeSe films: It leads to $B_{c2}(0) \approx 61$T for both parallel and normal field. However, since $B_{c2}(T)$ data extracted well below T_c are much more reliable we derived an interpolation function $B_{c2}(0K \leq T \leq T_c)$.

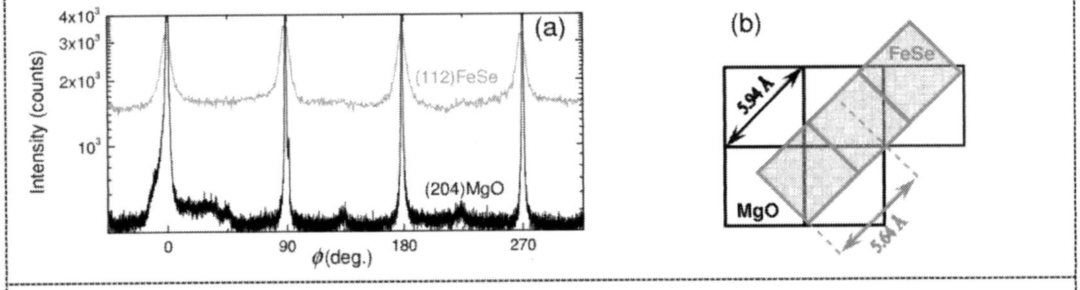

Figure 1. (a) The XRD ϕ-scan on the (112) reflection of the FeSe film and (204) reflex of the MgO substrate and (b) the resulted in-plane orientation of the film on the substrate.

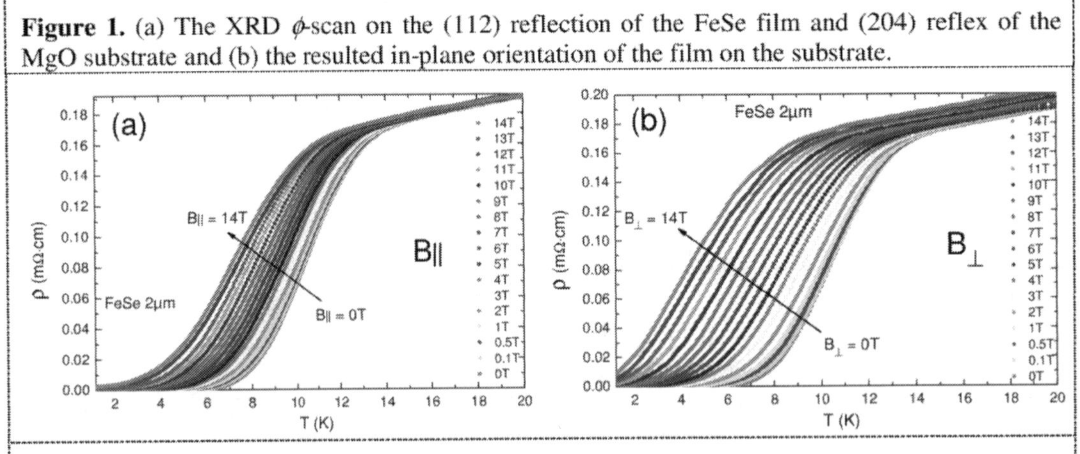

Figure 2. Resistivity of the 2 μm thick FeSe film in dc magnetic field applied parallel (a) and normal (b) to the film surface.

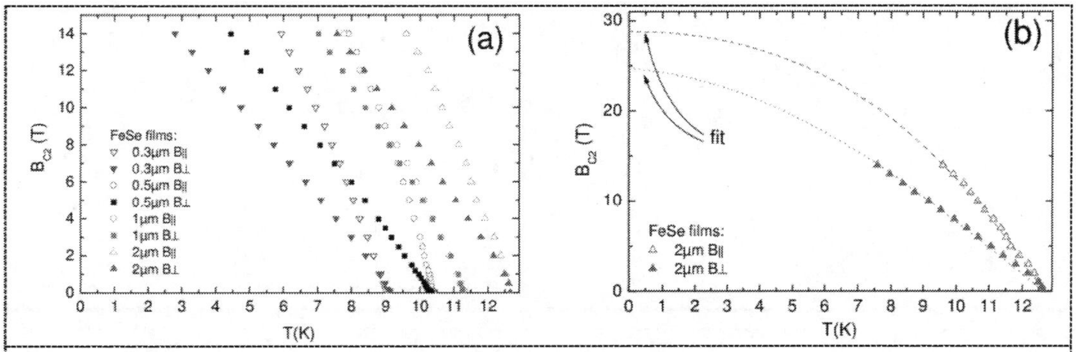

Figure 3. (a) $B_{c2}(T)$ measured for the films of different thicknesses. (b) Data extrapolation to lower temperatures for the 2 μm film: the dashed and the dotted lines are fitted to the experimental data using our WHH functional fit for parallel and normal fields, respectively.

We started from the WHH fixpoint equation $\log[t] = F\,[t, b[t], \alpha]$, where the functional F depends on $t = T/T_c$, $b[t] = B_{c2}(t)/\,B_{orb}(T = 0\text{ K})$ and the Maki parameter, α, quantifying the weakening influence of the Pauli electron spin paramagnetism on the superconducting state. For $\alpha = 0$ this does not play any role and $B_{c2}(T)$ is given as the pure "orbital field limit" $B_{orb}(T)$ due to the supercurrents circulating around the vortex cores. For $\alpha > 0$, the weakening of the superconducting state reduces the maximum magnetic field that the superconductor can tolerate below this orbital limit. In our calculations we first solved the WHH fixpoint equation numerically for the full range of the up to now experimentally reported Maki parameters $0 \leq \alpha \leq 6$ to obtain, for a given α_0, a set of solution points $\{b[t_i], t_i\}$. We then derived an analytic interpolation function $b_{interpol}[t; \alpha]$ which approximates these sets of solution points within an accuracy of ~ 1 % in the temperature range from 0 K to T_c and for $0 \leq \alpha \leq 6$. Our fit procedure allows to readily identify the reliable data range from which first α and then in the second step $B_{orb}(T = 0\text{ K})$ and $B_{c2}(T = 0\text{ K})$ are derived. More details of this evaluation scheme will be explained in a future publication. For $B_{c2}(T)$ of the 2 μm film we derived $\alpha_{||} \approx 1.7$ for parallel and $\alpha_{\perp} \approx 0.4$ for normal field. Such an anisotropy of the Maki parameter and the Pauli limiting field reflects the different paramagnetic influence on the Cooper pairs for the two different field orientations and is even more expressed in FeSe than in $CeCoIn_5$ [9]: In the fields perpendicular to the Fe-Se layers a conventional pure orbital limitation prevails, whereas in $B_{||}$ the Pauli limit is more dominant. The respective fits lead to the prediction $B_{c2||}(0) = 28.8$ T and $B_{c2\perp}(0) = 24.7$ T (see figure 3b). The conventional WHH estimate $B_{c2}(0) \approx 61$T actually has to be compared to our respecttive fit results $B_{orb||}(0) = 50$ T and $B_{orb\perp}(0) = 25$ T, demonstrating the error of the conventional $B_{orb}(0)$ prediction method.

The $\rho(T,B)$ data allow analysing the thermally activated flux-flow, which occurs above the irreversibility line and enforces $\rho = \rho_n{\cdot}\exp(-U_0/kT)$, where U_0 is the activation energy and k is the Boltzmann constant [10]. The slope of the Arrhenius plot, see figure 4, yields U_0 for a given field B. However, for

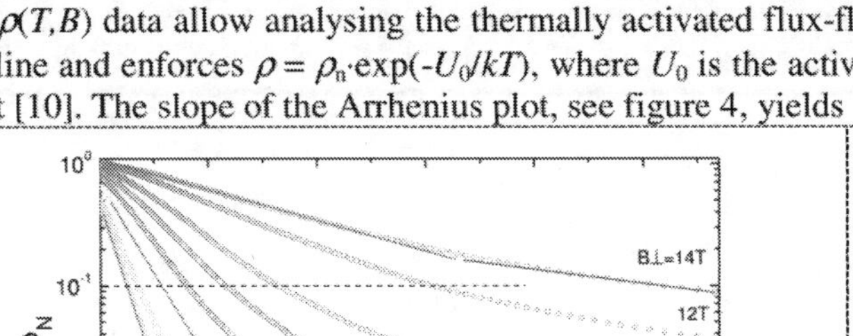

Figure 4. Data of figure 2b arranged in the Arrhenius plot. The dashed line at $\rho/\rho_n = 0.1$ illustrates the inapplicability of this criterion for the extraction of irreversibility fields of FeSe films, since the dissipative TAFF regime extends far below this line.

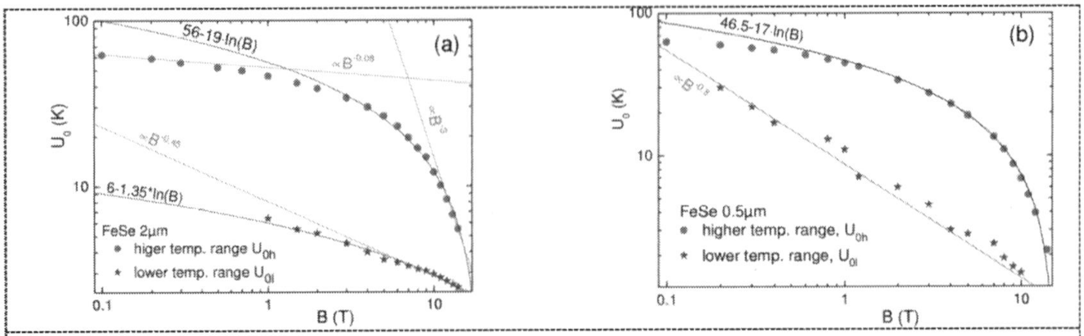

Figure 5. The activation energies extracted form the $\rho(T)$ at various $B_{\perp}$. (a) for the 2 μm thick FeSe film and (b) for the 0.5 μm thick FeSe film along with logarithmic and power-law fits.

layered superconductors, e.g. oxygen-deficient $YBa_2Cu_3O_x$ [10], this slope may not remain constant: each curve is divided in two roughly linear parts with different slopes attributed either to a quasi-2D 'pancake' vortex matter at temperatures not far below T_c with a logarithmic $U_0(B)$ dependence, or to a 'conventional' 3D vortex matter at lower temperatures with a power-law $U_0(B)$. For FeSe films these two ranges can be distinguished as well, see figure 4a. The extracted activation energies are assembled in figure 5a as U_{0h} for the higher temperature range and U_{0l} for the lower one. At higher temperatures the dissipation is likely due to 2D vortices: A logarithmic fit of $U_{0h}(B)$ appears more reasonable than any power-law fit. The classification of the lower temperature range is not unambiguous. For $B > 3$ T both fits describe $U_{0l}(B)$ well. Below ~2 T the logarithmic fit may appear better, though the accuracy of U_{0l} fades due to small resistivities. This unclear distinction between 2D an 3D behaviour can be explained by the lattice disorder accumulated over the 2 μm film growth and distorting the long-range order of FeSe planes (as noted above, the mosaic spread of the 2 μm film reached $\Delta\omega = 2.8°$, compared to $\Delta\omega = 0.22°$ of the 0.5 μm film). For comparison, $U_0(B)$ for the 0.5 μm film is shown in figure 5b. The data of figure 5b also reveal a convincing logarithmic fit of $U_{0h}(B)$ extending to lower fields than in figure 5a, and the power-law fit of $U_{0l}(B)$ becomes distinctive. Apparently, high structural perfection is crucial for clear distinction between 2D and 3D vortex behaviour in FeSe films.

Furthermore, the data of figure 4 imply the inapplicability of the $\rho/\rho_n = 0.1$ criterion for estimating the irreversibility fields for FeSe films since the TAFF region extends to much lower values of ρ/ρ_n.

In summary, superconducting films of (001) tetragonal FeSe were grown on (001) MgO substrates with 5% lattice mismatch. Both T_c and B_{c2} of our films increased with the film thickness d and reached at $d = 2$ μm the values reported for bulk under ambient pressure [4]. The concomitant reduction of the lattice alignment from $\Delta\omega = 0.22°$ for $d = 0.5$ μm to $\Delta\omega = 2.8°$ for $d = 2$ μm likely affected the electronic transport behaviour under application of magnetic field: The higher lattice misalignment smeared the distinction between 2D and 3D regimes of thermally activated flux-flow.

References

[1] Stewart G R 2011 *Rev. Mod. Phys.* **83** 1589
[2] Schneider R, Zaitsev A G, Fuchs D and v Löhneysen H 2012 *Phys. Rev. Lett.* **108** 257003
[3] Schneider R, Zaitsev A G, Fuchs D and Fromknecht R 2013 *Supercond.Sci.Technol.* **26** 055014
[4] Mizuguchi Y, Tomioka F, Tsuda S, Yamaguchi T and Takano Y 2008 *Appl Phys Lett* **93** 152505
[5] Fuchs G et al. 2009 *New J.Phys.* **11** 075007
[6] Haindl S et al. 2010 *Phys. Rev. Lett.* **104** 077001
[7] Werthamer N R, Helfand K and Hohenberg C 1966 *Phys. Rev.* **147** 295
[8] Brison J P et al. 1995 *Phys. C* **250** 128
[9] Agosta C C http://physics.clarku.edu/superconductor/PauliFold/CalculatingHP.html
[10] Qiu X G, Wuyts B, Maenhoudt M, Moshchalkov V V and Bruynseraede Y 1995 *Phys. Rev. B* **52** 559

Rapid thermal decomposition for $YBa_2Cu_3O_{7-\delta}$ films derived by DEA-modified TFA-MOD

Q Q Zhang, S C Zhao, Z Y Liu, R S Rui, W B Qiu, Y Q Guo, M J Li, W T Yang and C B Cai*

Research Center for Superconductors and Applied Technologies, Shanghai University, China

E-mail: cbcai@shu.edu.cn

Abstract. Thermal decomposition of YBa2Cu3O7-δ (YBCO) films derived by diethanolamine (DEA)-modified trifluoroacetic acid-metal organic deposition (TFA-MOD) was investigated with respect to the understanding of the correlation between the stress releasing and rapid decomposition. It is revealed that the evaporation of DEA and the decomposition of precursor films occur simultaneously. A pyrolysis time as 20 seconds is optimal to keep the proper amount of DEA which prevents the films from severe stress during the pyrolysis. Then smooth surface of resultant films appears. In case of a pyrolysis time longer than 40 s, cracks emerge in the films accompanied with complete evaporation of DEA and appearance of Cu-rich particles, while films with pyrolysis time shorter than 10s is excessively soft, with large amount of DEA and TFA remaining in the film, implying insufficient pyrolysis.

1. Introduction

It is well known that TFA-MOD process is very promising for the industrial scaling up for YBCO coated conductors due to low cost. One difficulty in TFA-MOD method is the need of a long pyrolysis step to evaporate the solvent and decompose the trifluroacetates into oxyfluoride precursor, accompanying with great volume shrinkage and out-gassing. During the decomposition process, buckling and cracking generally occur if the film does not maintain sufficient elasticity to endure the strain. The thermal decomposition for conventional TFA-MOD is more than 10h [1], and it is a disadvantage to be applied into commercial-scale production of kilometre-length tapes. In past years, extensive efforts were proposed to reduce the pyrolysis time mainly by three ways, using precursor with low content of fluorine [2~4], additions [5, 6] , or avoiding the appearance of water in the precursor solution [7~9]. Several groups have managed to reduce the pyrolysis time to several hours. As a solution starting method, MOD can be easily modified by additions. Many additives, such as DEA [5], PEG [6], have been demonstrated effective to reduce the pyrolysis time. Of all additives, DEA appears especially effective because of its high boiling- point (268℃). With the addition of DEA in TFA-MOD precursor solution, the pyrolysis time can be reduced as short as 20s [5]. The pyrolysis time greatly affects the development of YBCO phase and morphology of the DEA-YBCO films. Thus, more detailed information regarding DEA-modified pyrolysis process is required to give insight in the possible reasons. In the present work, a series of precursor films quenched at different pyrolysis time are investigated with respect to the morphology and residual particles, to give more understanding of pyrolysis process in the DEA-modified TFA-MOD.

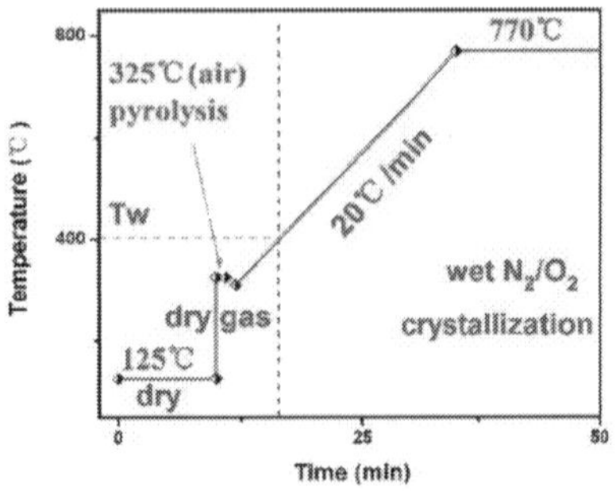

Figure 1. Typical thermal profile for our research

2. Experimental details

In our TFA-MOD process, Y-acetate, Ba-acetate, Cu-acetate are dissolved in TFA with the molar ratios of Y: Ba: Cu =1: 2: 3.15 to prepare the precursor. Diethanolamine (DEA) and methanol are used as solvents, with the ratio of Cu: DEA=1: 2. The molar concentration of the metal ions is 1.5mol/L. Precursor solutions are deposited onto $LaMnO_3$/Epi–MgO /IBAD-MgO /Y_2O_3/Al_2O_3/Hastelloy substrates by dip coating. The coated substrates are then placed at 125℃in a dry N_2/O_2 gas mixture for 10 minutes to get dried. Then they are placed directly to pyrolysis process at 325℃ [5] for different times from 6 to 120 sec in air. The following crystallization carried out at 770℃ for 60min at a heating rate of 20℃/min from 310 ℃ to 770℃ under 150ppm O_2/bal. N_2 atmosphere, and the dry gas is changed to wet gas at 400℃ (Tw). Finally the crystallized films are annealed at 450℃ for 60min in dry O_2 for full oxidation. In this work, we focus on the process of pyrolysis. The Metalloscopy and Scanning Electron Microscopy (SEM) test is used for the morphology observation and Fourier Transform Infrared Spectroscopy (FTIR) is used for the characterization of the component of the films.

3. Results and discussion

In order to further understand the pyrolysis process, the samples quenched after different pyrolysis time (tp) are studied. The pyrolysis process is as the following: The coated films are placed directly to 325℃ for different time in air to pyrolysis, then pulled out of the tube furnace and placed at room temperature. In figure 2, both Metalloscopy and SEM images (after gilded) of samples quenched after different pyrolysis time are obtained.

Figure 2. Series of (a)Metalloscope images(b)SEM images (after Gilded)of pyrolyzed films quenched after different pyrolysis time

According to Metalloscopy image, the surface of the films with pyrolysis time shorter than 10s is smooth [Figure 2(a)]. Increasing tp to 20s, the whole surface of the sample is bestrewn with pits of almost the same size. The bubbled surface looks like gas rush through the films intensively. During 30~40s, the surface of films are generally flat again except a few large bubbles. Further increase tp to 60~120s, more and more particles and cracks appear on the surface. The decomposition temperature range for Cu-fluoroacetate is 240-270℃and 320-400℃for Y- and Ba-fluoroacetates, respectively. The decomposition is runaway exothermic and out-gassing. The films are pyrolyzed at 325℃, a large amount of gas is expected for the evaporation of DEA and the part products of decomposition of organic salts. According to the phenomenon, it is supposed that tp varies at 1~10s, both evaporation and decomposition are mild due to the short time and the surface remains smooth. As tp increases to10~20s, DEA evaporates intensively and pits the surface by rushing through the film. All the pits are shallow which may imply that there is enough DEA remaining in the film, thus the film is still elastic. By increasing tp to 30~40s, the amount of DEA and TFA in the film decreases with the long heat treatment time thus the evaporation and decomposition became mild. A few large bubbles appear because of gas accumulation while other places remain flat. Further lengthening tp to 60~120s, there is nearly no DEA left and the film become rigid. Hence, defects and cracks occur. SEM images show the same trends with the change of pyrolysis time. The particles in the surface of samples pyrolyzed for more than 60s are Cu rich according to the EDS analysis. This may be the result of composition inhomogeneity.

Figure 3 shows the dependence of the size and number of Cu-rich particles on pyrolysis time. Data in figure 3 is acquired by capturing images of the same size (3.9 μ m *3.9 μ m) by visual counting and vernier caliper respectively. The relative size to the average diameter of the particles (330nm) at 60s and particles number, changes from zero to 1.49 and 23 respectively with the increase of pyrolysis time. This phenomemon indicates that the composition homogeneity degrade when the pyrolysis time is longer than 50s. In both figure 2(a) and 2(b), samples pyrolyzed for tp =30 s or 40 s are smoother than those with shorter pyrolysis time except for the big bubble. This may suggest that fluoroacetates decomposition reaction has finished or become slow after 30-40 s for the surfaces reveal a very slow rate of emission of gas.

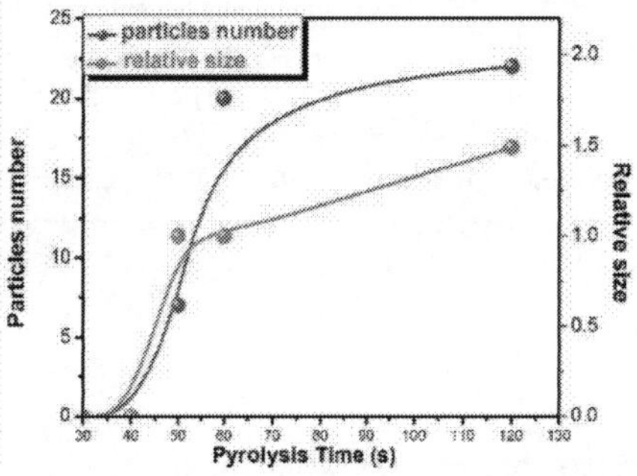

Figure 3. The particle number and relative size at different pyrolysis time in figure 2(b).

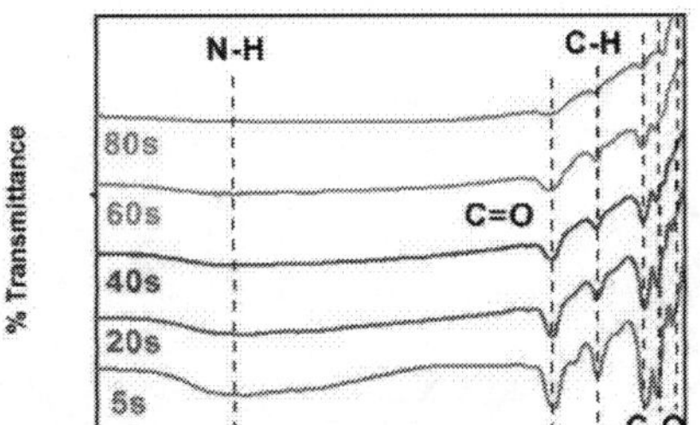

Figure 4. Series of FTIR patterns of pyrolyzed films quenched after different pyrolysis time.

The FTIR patterns of precursor films quenched after different pyrolysis time is measured in figure 4. The strong carbonyl (C=O) absorption appears at 1685 cm-1. Stretching vibrations of C-O absorptions appear at 1209 cm-1, 1145 cm-1, and 1064cm-1. The absorption appears at 1456cm-1 maybe indicating C-H stretching vibration, which needs more evidence. However, N-H (3300-3500cm-1) peak indicates DEA which appears as a small trough. It is reasonable that the intensity of N-H peak is generally mild, while that of C=O is strong. The C=O and C–O bonds are related to carboxyl absorption, which could be ascribed to the TFA species. With the increasing of tp, the peak intensity of C=O and C-O decreased quickly, but the tiny peak at 80s indicates that TFA is not totally decomposed. Meanwhile, the vibration of N-H cannot be observed after pyrolyzing for 40s. It implies that the evaporation of DEA and the decomposition of TFA proceed simultaneously, and once no more DEA remains in the samples which are pyzolyzed for more than 40s, the decomposition continues slowly.

To summary, the morphology diagram of DEA-modified TFA-MOD pyrolysis is given in figure 5. We divide the pyrolysis process into three stages. In the stage I, we achieve smooth surface with appropriate amount of DEA remaining in the film as the pyrolysis time (tp) is in the range of 0~40s. Then in the stage II, Cu-rich particles appear with few DEA remaining and the decomposition proceeding as the tp varies in the range 40~60s. In the stage III, with the tp increasing to 60~120s, more and more Cu-rich particles accompanied by cracks emerge with the absence of DEA.

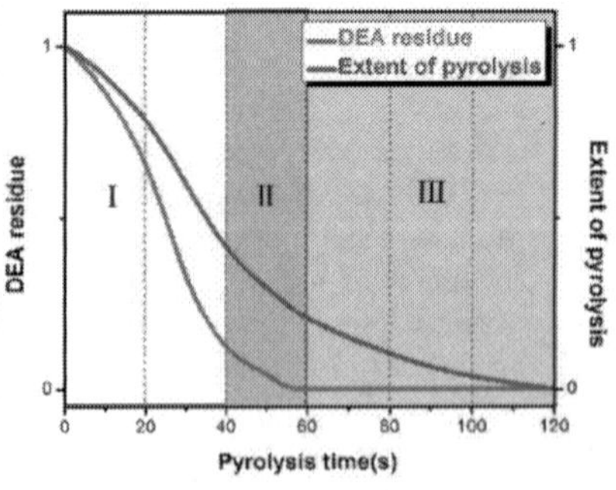

Figure 5. Morphology diagram of DEA-modified TFA-MOD pyrolysis.

4. Conclusions

In this work, the precursor YBCO films quenched after different pyrolysis time derived by DEA-modified TFA-MOD are investigated to give more understanding of pyrolysis process. These films are studied by Metalloscopy, SEM, FTIR, which reveal that a pyrolysis time as short as 20s is optimal. The short pyrolysis time appears necessary to keep the proper amount of DEA, which prevents the films from severe stress, and assure smooth surface of resultant films. According to FTIR, the clusters of C=O, C-O and N-H correspond to TFA acid radical and DEA respectively, which give the evidences that the evaporation of DEA and the pyrolysis process occur simultaneously. And the pyrolysis continues slowly with the absence of DEA after 60s. A pyrolysis time longer than 60s leads to the appearance of cracks and Cu-rich particles without the protection of DEA. In case of excessively short pyrolysis time, namely shorter than 10s, there would be insufficient pyrolysis leaving the surface extremely soft.

Acknowledgments

This work is partly sponsored by the Science and Technology Commission of Shanghai Municipality (10dz1203500 and 11dz1100302), the National Natural Science Foundation of China (10774098 and 5 1202141), the Ministry of Science and Technology of China (973 Projects, 2011CBA00105).

References

[1] Araki T and Hirabayashi I 2003 *Supercond. Sci. Technol.*16 R71

[2] Tokunaga Y, Fuji H, Teranishi R, Matsuda J, Asada S, Kaneko A, Honjo T, Izumi T, Shiohara Y, Yamada Y, Murata K, Iijima Y, Saitoh T, Goto T, Yoshinaka A, and Yajim A, 2004 *Physica C* 412 910

[3] Tokunaga Y, Honjo T, Izumi T, Shiohara Y, Iijima Y, Saitoh T, Goto T, Yoshinaka A and Yajima A 2004 *Cryogenics* 44 817

[4] Ichikawa H, Nakaoka K, Miura M, Sutoh Y, Nakanishi T, Nakai A, Yoshizumi M, Izumi T and Shiohara Y 2009 *Physica C* 469 1329

[5] Dawley J T, Clem P G, Boyle T J, Ottley L M,Overmyer D L and Siegal M P 2004 *Physica C* 402 143

[6] Wei Wu, FengFeng, Kai Shi, Wei Zhai, Timing Qu, Rongxia Huang, Xiao Tang, Xiaohao Wang, Qingyu Hu,Jean-Claude Griveland Zhenghe Han, 2013 *Supercond. Sci. Technol.* 26 055013

[7] Zalamova K, Roma N, Pomar A, Ricart S, Gázquez J, Puig T, Sandiumenge F and Obradors X 2006 *J. Phys.: Conf. Ser.* 43 150

[8] Obradors X, Puig T, Pomar A, Sandiumenge F, Mestres N, Coll M, Cavallaro A, Romà N, Gázquez J, González J C, Castaño O, Gutierrez J, Palau A, Zalamova K, Morlens S, Hassini A, Gibert M, Ricart S, Moretó J M, Piñol S, Isfort D and Bock J 2006 *Supercond. Sci. Technol.* 19 S13

[9] Zalamova K, Roma N, Pomar A, Ricart S, Gazquez J, Puig T,Sandiumenge F and Obradors X 2006 7th *European Conf. on Applied Superconductivity* vol 43 p 150

Phase Evolution of $YBa_2Cu_3O_{7-x}$ films by all-chemical solution deposition route for coated conductors

Yue Zhao[a], Xiao Tang[a], Wei Wu[b], and Jean-Claude Grivel[a]

a, *Department of Energy Conversion and Storage, Technical University of Denmark, Roskilde 4000, Denmark.*

b, *Department of Physics, Tsinghua University, Beijing 100084, People's Republic of China*

zhyu@dtu.dk

Abstract. In order to understand the all-chemical-solution-deposition (CSD) processes for manufacturing coated conductors, we investigated the phase evolution of $YBa_2Cu_3O_7$ (YBCO) films deposited by a low-fluorine metal-organic solution deposition (LF-MOD) method on CSD derived $Ce_{0.9}La_{0.1}O_2/Gd_2Zr_2O_7$/NiW. It is shown that the phase transition from the pyrolyzed film to fully converted YBCO film in the LF-MOD process is similar to that in typical trifluoroacetates-metal organic deposition (TFA-MOD) processes even though the amount of TFA in the solution is reduced by almost one half compared with typical TFA-MOD cases. Moreover, we found that the formation of impurities (mainly $BaCeO_3$, $NiWO_4$ and NiO) is strongly related to the annealing temperature, i.e., the diffusion controlled reactions become intensive from 760 °C, which might be connected with the poor structural and superconducting properties of the films deposited at high sintering temperatures. Based on these results, the optimized growth conditions of YBCO films were established, and a high critical current density (Jc) of about 2 MA/cm^2 (77 K, self field) is achieved in a 200 nm thick YBCO film in the architecture made by our all CSD route.

1. Introduction

Due to the great potential for their use in electric power applications, the YBCO based coated conductors have been extensively studied and developed during the past two decades. Long-length wires with superior performance have been successfully manufactured by several routes [1, 2]. Among those techniques, combining Rolling-Assisted Biaxially Textured Substrates (RABiTS) and chemical solution deposition (CSD) is considered as one of the most promising routes, which could easily be scaled-up and further reduce the costs due to initial investment and production. However, because of the lack of comprehensive understanding of CSD techniques, the industrialization of CSD processes for both buffer layers and superconducting layer (all CSD processes) is still quite challenging and rarely reported [3]. Further investigations of all CSD processes, including compatibility between the materials used in each layer, solution chemistry, as well as nucleation and growth mechanism are needed.

In our previous studies, we reported a CSD process for a full buffer layer architecture [4]. A critical current density (J_c) value higher than 1 MA/cm^2 (77 K, in self-field) was achieved in a YBCO film grown by pulse laser deposition, demonstrating the high quality of this buffer stack for coated conductors. In this work, we focused on an all CSD process, where a low-fluorine metal-organic solution was used for depositing YBCO directly on the CSD derived buffer layer. The low-fluorine metal-organic deposition (LF-MOD) route chosen for this work has the purposes of shortening the pyrolysis process and reducing amount of the TFA in the precursor solution. In this work, the phase evolution of YBCO during the firing processes was investigated in order to establish the heat treatment profiles. We paid additional attention on transition of the intermediate phases, formation of YBCO phase and inevitable impurities ($BeCeO_3$, $NiWO_4$, NiO), all of which have strong influences on the superconducting performance of the final products. With the guideline of phase evolution, the growth conditions of the YBCO films were optimized.

2. Experimental

The propionate-based MOD route for preparing $Ce_{0.9}La_{0.1}O_2$(CLO) and $Gd_2Zr_2O_7$(GZO) buffer layer stack on textured NiW substrates was described previously [4, 5]. With optimized coating and heat treatment conditions, high quality buffered substrates were obtained routinely for further YBCO deposition.

The LF-MOD process for YBCO film deposition was similar to standard TFA-MOD, where the only difference in solution is that the copper salt is replaced by a non-TFA precursor. First, a solution (referred to as solution A) was made by dissolving home-made yttrium and barium trifluoroacetates in methanol with a Y/Ba ratio of 1/2, like in standard TFA-MOD solutions [6]. Secondly, copper acetate was dissolved in a small amount of acrylic acid (10% excess) and de-ionized water. After full reaction and removal of residual solvent and by-products by vacuum evaporation, a dried copper-based powder was obtained. The final precursor solution was made by mixing stoichiometrically solution A and copper-based powder at room temperature. The total cation concentration was 1.5 M by adjusting the amount of methanol. The precursor solution was deposited onto substrates by spin-coating with a rotation speed of 3000 rpm. A typical heat treatment procedure, including pyrolysis, sintering and oxygenation, was used for obtaining superconducting YBCO films. The pyroslysis process started from room temperature to 400 °C with a ramp of 1.7 °C/min in humid oxygen (dew point (D.P.) =20 °C). Then partially reacted samples were quenched from different firing temperatures (400 °C, 600 °C, 700 °C, 740 °C, 780 °C, 820°C) to study the phase evolution. All the firing processes were performed in a humid atmosphere (D.P.=20 °C, with 100 ppm oxygen). For the fully reacted YBCO films, following oxygenation was carried out at 450 °C for two or three hours in flowing oxygen gas.

The phase assembly and crystallographic textures were investigated by means of x-ray diffraction (XRD) using Cu Kα radiation in a four-circle diffractometer (Bruker D8). The surface morphology of the films was observed by scanning electron microscopy (SEM, Supra 35) equipped with an inlens detector. the superconducting transition temperature (*Tc*) was determined by AC susceptibility measurements with AC field amplitude of 0.1 mT and frequency of 21 Hz, while the critical current densities, *Jc*, were calculated based on Bean model [7].

3. Results and discussion

θ-2θ theta scans of the samples quenched from different temperatures are shown in Fig. 1. In the pyrolyzed film, we observed $Ba_{1-x}Y_xF_{2+x}$ (BYF, x = 0.35) and CuO phases, which are basically the same compositions as those in all TFA solutions [8]. When the sintering temperature reaches 600°C, the peak shift of barium oxyfluorides can clearly be seen, which means that yttrium starts to dissolve out of BYF. The peaks corresponding to the BYF and CuO phases become narrow at the elevated temperatures up to 780 °C, indicating the grain growth of these phases. The BYF peaks nearly remain at the same positions after 700 °C, implying that almost all Y was released from the BYF matrix. However, since the peak of the Y_2O_3 the most likely Y phase, is too close to the buffer layer peaks, it is difficult to claim if there is any such phase forming at this stage. However, $Y_2Cu_2O_5$ (211) (2θ = 31.2°) clearly appears at 740 °C, probably due to the reaction between the Y_2O_3 and CuO. This could indirectly support the view of the formation of Y_2O_3 formation at an earlier stage. The first appearance of the YBCO (00l) peaks occurs at temperatures lower than 760 °C, while the trace of random orientation (103)/(013) implies that the epitaxial growth is predominating under these conditions.

Regarding the impurities formation caused by elemental diffusion at the interlayers, we mainly focus on $BaCeO_3$ and $NiWO_4$ which are normally observed when using nickel based substrates. As early as 700 °C, we could observe a small amount of $BaCeO_3$ and $BaCeO_3$ (110) peak became intense with increasing temperature. Compared with other reports where $YBCO^{MOD}$ films were grown on CeO_2 cap layers made by physical deposition technique (i.e., sputtering or PLD) [9, 10], the formation of $BaCeO_3$ in our case takes place much earlier. This is probably due to the microstructure of the CLO^{CSD} cap layer, which leads to its high chemical activity [11]. However, it seems that under an optimized firing profile, $BaCeO_3$ has little negative influence on the superconducting properties in the final YBCO films, as will been shown in the following. The $NiWO_4$ peaks appear at 700 °C as well and keep developing at higher temperatures.

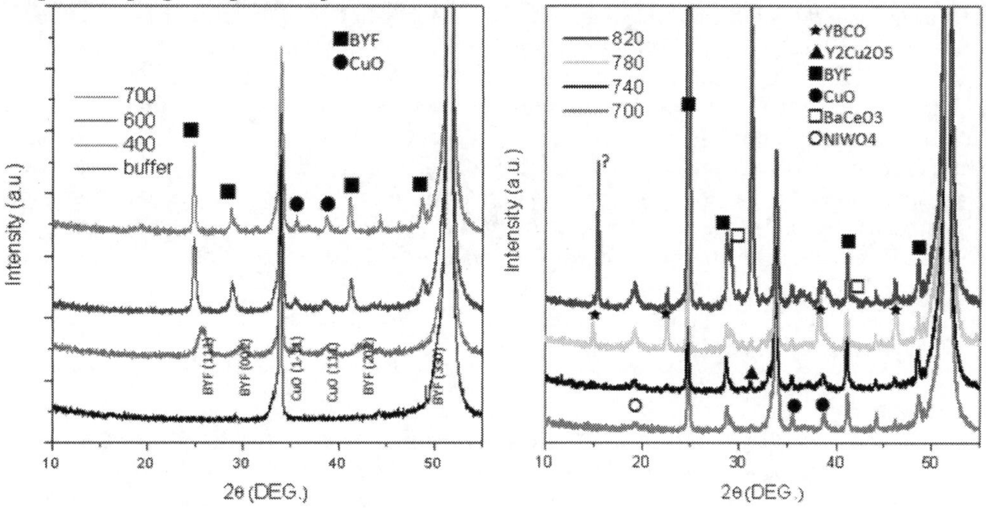

a) b)

Figure 1. *θ-2θ* scans of the buffered substrate and a series of YBCO samples quenched from different temperatures

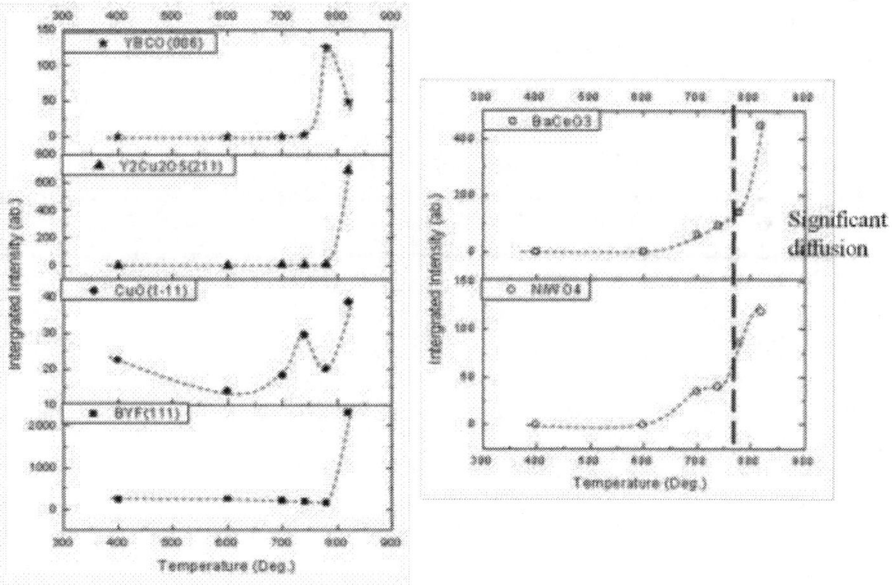

Figure 2 Integrated area evolution versus temperature of XRD peaks of intermediate phases (left) and impurities (right) of films.

Journal of Physics: Conference Series **507** (2014) 012056 doi:10.1088/1742-6596/507/1/012056

The integrated peak area evolution of several phases during the YBCO sintering process are plotted as a function of temperature and presented in figure 2. It allows us to draw the evolution trends when the temperature increases. It can be seen that the $Y_2Cu_2O_5$ and BYF phase are still not converted completely at 820 °C due to the short dwell time. The dramatic increase of BYF (111) peak intensity is due to the texture formation. The drop of the YBCO (006) peak intensity is associated with both unknown phase formation and the complex reactions in the interlayer (as evidence by rapid increase of the $BaCeO_3$ content). Considering these results, we conclude that sintering YBCO films on such CSD-derived substrates at a temperature higher than 780 °C should be avoided. One should notice that the aforementioned phase evolution is based on the kinetics of the firing process up to 820 °C. In the practical cases, longer dwell time will change the starting formation temperatures of the intermediate phases and final products, e.g., YBCO *(00l)* peaks appear as early as 700 °C after 3 hours dwell.

According to the results of phase evolution, several sintering temperatures (from 700 °C to 780 °C with an interval of 20 °C) were chosen to optimize the growth conditions of YBCO films. Larger amount of grains with undesirable orientations can be discerned on the surface of the films sintered at temperatures above 760 °C, suggesting that the elemental diffusion significantly changes the YBCO nucleation and growth. On the contrary, less oxidation of the metallic substrate was obtained at lower sintering temperatures. Since the growth rate at 700 °C is very low, we could not obtain well crystallized YBCO film at this temperature with reasonable dwell time (e.g., 3 hrs). The films sintered at 720 °C and 740 °C exhibit sharp in-plane and out-of-plane texture quality, and J_c values of 2.1 MA/cm^2 and 1.0 MA/cm^2 were obtained, respectively, demonstrating the feasibility of the all-CSD route for coated conductors fabrication.

4. Conclusion

Phase evolution of YBCO grown on $Ce_{0.9}La_{0.1}O_2/Gd_2Zr_2O_7$/NiW substrates by a low-fluorine metal-organic solution deposition method was studied. It is shown that the compositions in the pyrolyzed film are similar to those in typical TFA MOD films. Besides, we noticed that impurity formation (mainly $BaCeO_3$, $NiWO_4$ and NiO) becomes intensive from 760 °C, which might deteriorate the superconducting performance of YBCO layers. According to these results, the optimized growth conditions of YBCO films were established, and a *Jc* of 2 MA/cm^2 (77 K, self field) is achieved in such an all CSD derived architecture.

References

A reference

[1] M.W. Rupich, X. Li, C. Thieme, et.al., 2010, *Supercond. Sci. Technol.*, 23,014015,.

[2] V. Selvamanickam, [Online]. available at: http://www.eucas2011.org/include/content/plenarytalks/selvamanickam_pdf.php

[3] Michael Bäcker, [Online]. available at:http://d-nano.com/en/news/events.html

[4] Y. Zhao, X.-F. Li, A. Khoryushin, et.al., 2012, Supercond. Sci. Technol., 25, 015008.

[5] Y. Zhao, J. C. Grivel. 2013, CrystEngComm, 15, 3816-3823.

[6] A. Gupta, R. Jagannathan, E. I. Cooper, et.al., 1988, Appl. Phys. Lett. 52, 2077-2079.

[7] E. M. Gyorgy, R. B. van Dover, K. A. Jackson, et.al., 1989, Appl. Phys. Lett. 55 283-285

[8] K. Zalamova, A. Pomar, A. Palau, et.al., 2010, Supercond. Sci. Technol., 23, 014012.

[9] F Lu, F Kametani and E E Hellstrom. 2013, Supercond. Sci. Technol., 26, 045016.

[10] I. Nagino, K. Matsumoto. H. Adachi, et.al., 2010, Physica C, 20, 1308 - 1312.

[11] S. Solovyov, H. Wiesmann, presented at the *Superconductivity for Electric Systems 2008 Annual Peer Review*, Washington, D. C., 2008

Journal of Physics: Conference Series **507** (2014) 022001 doi:10.1088/1742-6596/507/2/022001

Analysis of critical current reduction in self- field in stacked twisted 2G HTS tapes

N V Bykovsky[1], S S Fetisov[1], A A Nosov [1], V V Zubko[1] and V S Vysotsky[1]

[1]Russian Scientific R&D Cable Institute (VNIIKP), Moscow, Russia

E-mail: nvbykovsky@gmail.com

Abstract. Twisted stacked-tape cable (TSTC) is one of the methods to produce flexible current carrying elements (CCE) with high current density made of 2G coated conductors. The critical current of CCE in self-field is less than the sum of critical currents of all tapes used to make CCE. This critical current reduction is attributed to self-field influence of critical currents of tapes. To describe the current reduction accurately in TSTC it is necessary to take into account current and field distribution in a stack cross-section using dependencies of critical current on magnetic field for each tape. In other words one has to make self-consistent calculations of current and magnetic field distribution in a stack. In this paper we present the self-consistent analysis of magnetic field and current distribution in TSTC with different number of tapes using the generalized Kim model. We also measured current degradation in TSTC made of from two to ten 2G tapes and compared it with calculations. We found good coincidence of calculated and measured current degradation that confirm the validity of our approach.

1. Introduction

The common way to create HTS current carrying element consists of tapes laid in helix on round core (CORCC) [1] or in making ROEBEL element [2]. Both technologies have advantages such as uniform current distributions between tapes, but also have significant limitations, such as low current density for CORCC and high tapes outgoings for ROEBEL.

One more way to create high current density carrying element is a twisted stack of HTS tapes (TSTC) [3]. TSTC have a high potential for the introduction to DC applications with high current densities, for example for degaussing of ships or for low voltage DC cables for data centers [4]. This technology provides compactness, high current density and negligible tapes outgoings. On the other hand, undesirable feature of twisted stacked tapes is critical current reduction due to self-field effect as it was observed in [3] and [5]. Advanced study of this effect is needed if low field application of TSTC is considered. In this paper we summarize our detailed analysis of current reduction in self-field both numerically and experimentally.

2. Critical currents of TSTC

2.1. Method of calculation of magnetic field in TSTC

In order to calculate magnetic field at arbitrary point (x,y), produced by twisted tape, one can note, that in limit *twist pitch>>tape width (w)>>tape thickness*, Biot–Savart law for straight tape is a fine approach:

$$\vec{b}(x,y,\eta)=\frac{\mu_0}{2\pi}\int_{-w/2}^{w/2}\frac{-\vec{e_x}(y-\eta)+\vec{e_y}(x-\xi)}{(x-\xi)^2+(y-\eta)^2}J(\xi)d\xi \tag{1}$$

where are: (ξ,η) - coordinates of tape, $J(\xi)$ - current density.

Each tape in stack is located in external magnetic field, produced by all another tapes of stack. Thus, for k-th tape in stack, corresponding external magnetic field will be expressed as:

$$\vec{B_k}(x)=\sum_{i\neq k}\vec{b}(x,\eta_k,\eta_i) \tag{2}$$

In HTS layers for $j_c(B)$ Kim's model [6] is used

$$j_c=\frac{j_{c0}}{\left(1+\frac{\sqrt{k^2B_x^{\ 2}+B_y^{\ 2}}}{B_0}\right)^{\beta}} \tag{3}$$

with parameters, which measured at VNIIKP at temperature 77 K for the used in work SuperPower SCS4050 [7] HTS tape: $k=0.5, \beta=0.7, B_0=0.11\ T$.

Then, modified critical current density allows to recalculate magnetic field distributions, therefore looping the calculations. This iterative process runs until consistent magnetic field ⇔ critical current density distributions will be found. It should be noted that process converges very fast – for only several iterations steps.

2.2. Measurements tools

Measurements of each tape critical current reduction in stack for various numbers of tapes in stack and various gaps between tapes was performed on insert (figure 1) at temperature 77 K. Set of shunts (10 long copper strips for each tape) provides very uniform current distribution between tapes (0.5% accuracy) and serves as handy current sensor of each tape. Single tape critical current measurements and calibration of shunts were made preliminarily. Tapes are laying in stack on a length ~50 cm. Each tape has its own voltage taps pair.

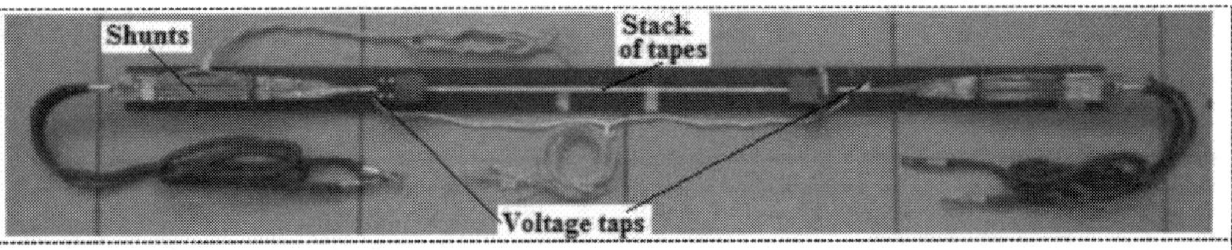

Figure 1. . Optimal configurations of stacks with cross section 4*4 mm^2.

2.3. Experimental and calculated results

Results of each tape's critical current reduction measurements in tight stack of (n=2…10) tapes are represented in figure 2 (number of tapes in stack in graph's legend).

By averaging this data over tapes in stack, one can find average critical current degradation:

$$\mathrm{degr}(n,\Delta)=\frac{\sum_{i=1}^{n}Ic_i(n,\Delta)}{\sum_{i=1}^{n}Ic_{i0}} \tag{4}$$

where are: n – number of tapes in stack, I^i_{c0} – single tape critical current, Δ - gap between tapes in stack.

Both numerical and experimental results for tight (Δ=0) stack of I_c reduction $\mathrm{degr}(n,0)$ function are shown in figure 3 There is good agreement with numerical results. Some deviations can be explained by difference in critical currents of single tapes (ignored in calculations).

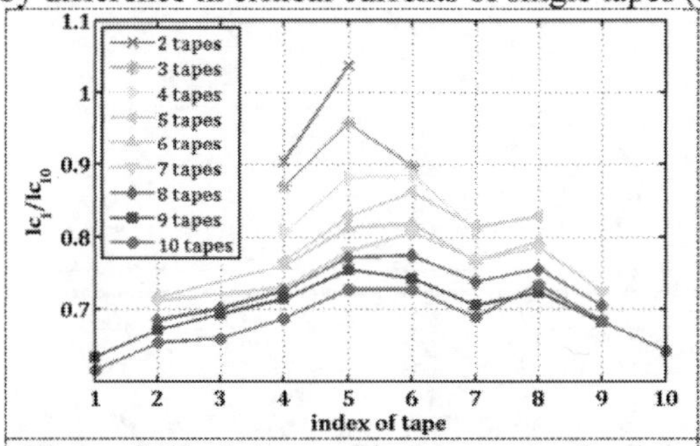

Figure 2. Each tape critical current reduction in tight stack.

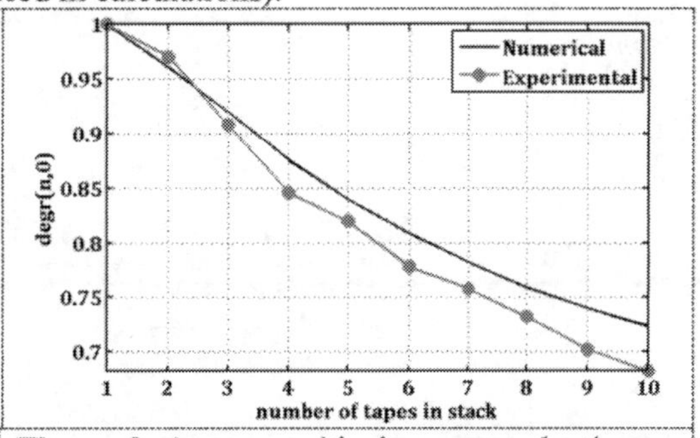

Figure 3. Average critical current reduction as a function of number of tapes in tight stack.

Next part of the experimental work was study influence on critical current reduction of gaps between tapes in a stack. This was done for stack of 7 tapes by inserting of electrical insulating paper between tapes. Thickness of paper is 120 μm, which allows to check multiple gap values: 120, 240, 360, 480 and 600 μm for stack of 7 tapes. Results of measurements are shown in figure 4.

Critical current reduction of each tape averaged over 7-tapes of stack and numerical values of $\mathrm{degr}(7,\Delta)$ function are compared in figure 5.

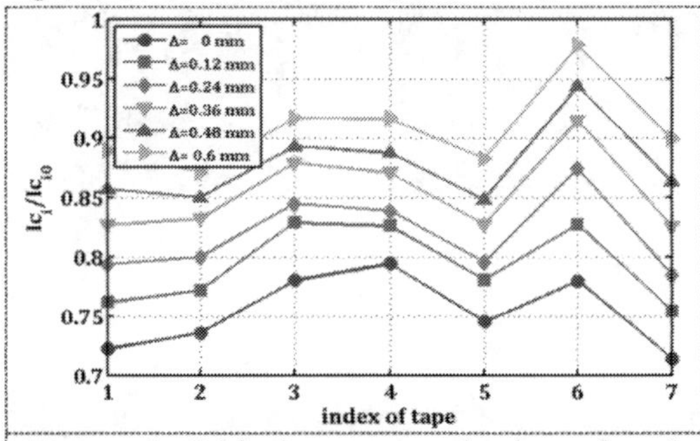

Figure 4. Critical current reduction of each tape in stack of 7 tapes.

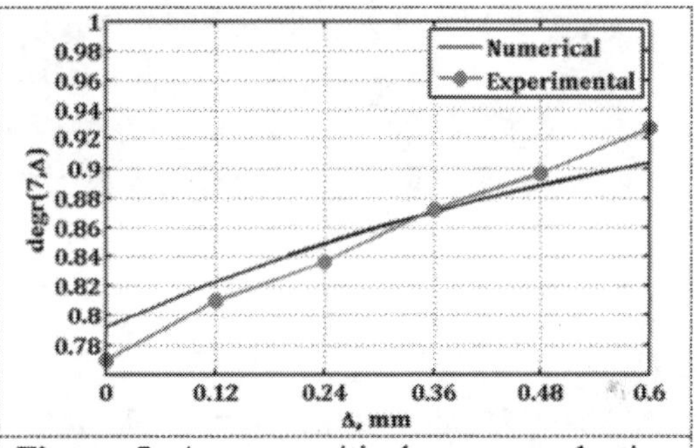

Figure 5. Average critical current reduction as a function of a gap between tapes.

Thus, numerical method was confirmed experimentally. Additional attention should be paid while choosing $j_c(B)$ models (or interpolation data) for numerical evaluation, which plays central role in all calculations.

3. Optimization of tape location (tapes configuration) in a stack with fixed cross-section

With implemented numerical model of mutual tapes influence in stack one can solve optimization problem – find N-tapes positions in a stack with fixed cross-section, corresponding to maximal current density. Chosen cross section configuration for this problem is 4x4 mm^2. This cross-section is filled by N tapes (width-4 mm, thickness-0.1 mm). The remaining space of cross section is filled by additional layers (insulating, stabilizing, etc.). In general, N-tapes problem have N-1 independent parameters of optimization (gaps between tapes). For clarity, there is a simplification about each tape's critical current – all of them are set to be of the same value. This simplification reduces the number of parameters in two times (because of horizontal symmetry of the problem). For example, for stack of 4 tapes only 1 parameter of optimization is needed: if gap Δ between 1 and 2 tape is chosen, then gap between 3 and 4 tape is the same (due to symmetry), and gap between 2 and 3 tape will be equal to 4-4*0.1-2* Δ.

Finally, optimization problem solutions against number of tapes in stack from 3 to 10 are given in table 1 and illustrated in figure 6.

Table 1. Optimal values of gap.

Number of tapes in stack	3	4	5	6	7	8	9	10
gaps, μm	1850, 1850	850, 1900, 850	290,1460, 1460, 290	0, 1060, 1280, 1060, 0	0, 630, 1020, 1020, 630, 0	0, 320, 830, 900, 830, 320, 0	0, 70, 700, 780, 780, 700, 70, 0	0, 0, 500, 650,700,650, 500, 0, 0

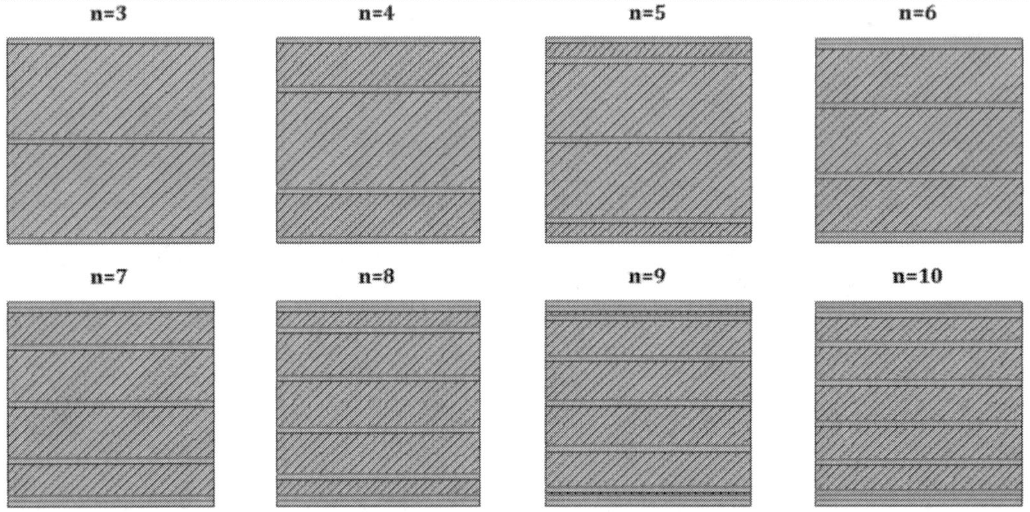

Figure 6. Optimal configurations of stacks with cross section 4*4 mm^2.

4. Conclusion

TSTCs in self-field conditions were experimentally and numerically evaluated. Calculation method of mutual HTS tapes influence in stacked twisted tapes was proposed. This method was confirmed by experimental measurements. Thus, it can be used for analysis of more complicated problems. Solution for stack optimization problem, based on calculation method of this work, was given also.

References

[1] D. van der Laan et. al., "Compact GdBa2Cu3O7–δ coated conductor cables for electric power transmission and magnet applications", Supercond. Science Technology, Vol. 24, N 4, 2011.

[2] W. Goldacker, et al., "Status of high transport current ROEBEL assembled coated conductor cables", Superconductor Science and Technology, Vol. 22, N 3, 2009.

[3] M. Takayasu et al., "HTS twisted stacked-tape cable conductor" Super-conductor Science Technology vol. 25 no. 1, 014011, 2012.

[4] Available at: http://www.amsc.com/library/LVDC_BRO_0412_web.pdf

[5] V. E. Sytnikov, et al, Study of HTS Conductors Made From Combinations of HTS Tapes, J. Phys.: Conf. Ser. 43 1059-1062, 2006 (Proceedings EUCAS – 2005) 2006 J. Phys.: Conf. Ser. 43 1059

[6] F. Gömöry et al., "AC losses in coated conductors" Superconductor Sci-ence Technology vol. 23 no. 3, 034012, 2010.

[7] Available at: www.superpower-inc.com.

Ag Doping Effects of $GdBa_2Cu_3O_{7-\delta}$ Films by Low Fluorine MOD Method

M. J. Sun, W. B. Qiu, Z. Y. Liu, C. Y. Bai, Y. Q. Guo, C. B. Cai *
Research Center for Superconductors and Applied Technologies, 99 Shangda Road , Shanghai University, 200444, Shanghai, China

*E-mail: cbcai@staff.shu.edu.cn

Abstract: A series of Ag-doped GdBaCuO superconducting films were prepared by low fluorine MOD process on buffered Hastelloy substrates. The effects of Ag doping on the microstructures and superconducting properties of GdBCO thin films were investigated. It was found that an optimum addition of about 5 mol % Ag lead to better c-axis orientation and surface microstructure of the resultant films. Meanwhile, the pyrolysis process shortened and the superconducting performance is improved with a wide available window.

1. Introduction

$REBa_2Cu_3O_{7-\delta}$ (REBCO) (RE = Y, Gd, etc.) coated conductors have been intensively focused due to their great potential in practical applications [1-3]. Compared with YBCO, GdBCO superconductors have a higher critical temperature (~90*K*) and can keep relatively good current density value(*Jc*) in high magnetic field[4,5]. The metal–organic deposition (MOD) route is considered as a cost-effective and non-vacuum technique which can produce REBCO coated conductors with high performance [6,7].

In recent years, several groups studied[8-10] the effects of Ag doping in YBCO bulks or films. It is reported that in melt-textured polycrystalline, Ag particles segregate in the grain boundaries of YBCO which results in a thinner effective grain boundary and a higher *Jc*[8]. Enhancement of *Jc* was also found in Ag-doped PLD-YBCO films under low magnetic field [9].

But for MOD process, Ag doping was rarely reported[12]. In this study, a series of Ag doped $GdBa_2Cu_3O_{7-\delta}$ (GdBCO) films are prepared on buffered Hastelloy substrates by low fluorine MOD. A few positive results are achieved, showing that optimum addition about 5 mol % of Ag in GdBCO solutions lead to rapid pyrolysis process, better c-axis orientation as well as surface microstructure and then a higher *Jc* value.

2. Experimental

The precursor solution was prepared by dissolving TFA salts for Gd, Ba and a fluorine free organic salt for Cu with the 1:2:3 cation ratio into propionic acid with the total metal concentration of 1.5 M/L. Silver trifluoroacetate was added with the molar proportions ranging from 2.5 mol % to 20 mol %. A few drops of diethanol amine were added to the solution to increase its stability. The precursor solution was dip coated on $LaMnO_3/MgO/Y_2O_3/Al_2O_3/$ Hastelloy substrates. Gel films formed after dip coating followed by a fast pyrolysis process at 300°C(20K/min) in a humidified oxygen atmosphere. Growth was performed at 770°C for 1.5h in humid N2/O2 atmosphere and then annealed at 450°C for 1h in dry oxygen atmosphere. For reference, a pure GdBCO sample was prepared by traditional low fluorine route as well.

Metallographic optical microscope (MM) was used to observe the surface quality of the pyrolyzed films. The phase purity and the texture of GdBCO films were investigated by X-ray diffraction (XRD) using Cu Kα radiations (Rigaku D/MAX2000PC). The film thickness was measured by a step profiler system. A typical thickness for all present films are around 400nm. The surface morphology of the YBCO film was observed by field emission scanning electron microscopy (SEM). The critical current density was inductively measured by a CryoScan system at 77 K.

3. Results and discussion

Fig. 1 shows the typical metallographic optical microscope(MM) images of GdBCO precursor films quenched at 500°C during temperature ramping with two kinds of solutions(normal low fluorine solution and 5 mol % Ag doped low fluorine solution) pyrolyzed under the same condition (20 K/min) . Typical wrinkles are observed on the pyrolyzed sample using normal low fluorine solution, as shown in Fig. 1(a). The precursor film with 5 mol % Ag doped low fluorine solution shows a smooth and crack-free microstructure in Fig. 1(b). It is demonstrated that the Ag doped low fluorine solution has beneficial effect on the stress relief and minimization of the surface degradation of pyrolyzed films.

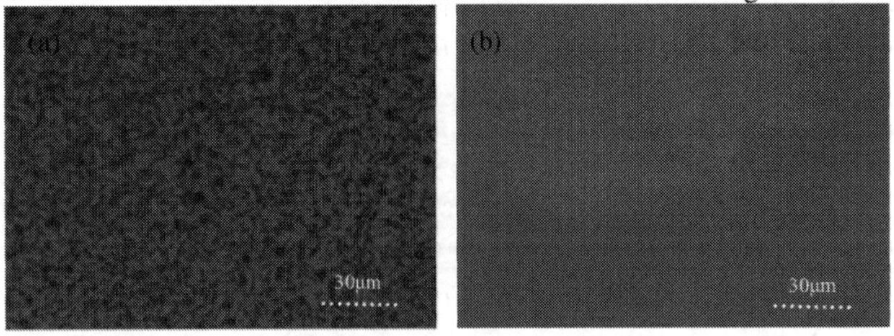

Figure 1. MM images of the samples quenched at 500°C with and without Ag doping (a) normal low fluorine solution without Ag (b) 5mol% Ag doped solution

The XRD θ–2θ patterns of GdBCO-Ag fired at 770°C are shown in Fig. 2. It is seen that all GdBCO-Ag films are pure (00l) orientation without undesirable phases while Ag phase hardly detected. However GdBCO(200) peak is frequently found in pure GdBCO films , which is further confirmed by SEM morphologies. It is believed that although Ag may not occupy the space in lattice of GdBCO, it plays role on the suppression of a-axis of GdBCO films.

It is found that GdBCO(004) peak intensity improve significantly by proper addition of Ag close or equal to 5 mol %. GdBCO films keep good orientation as the percent of Ag doping varing from 2.5% to 10%. However high doping content as 20 mol % may degrade the texture of GdBCO. Obradors et al.[12] believed that the decrease of the peritectic temperature of the Ag–RE–Ba–Cu–O system was the reason for the enhancement of c-axis orientation. Further work is needed to explain the reason why c-axis nucleation becomes easier, since the growth mechanism of thin film and bulk is different.

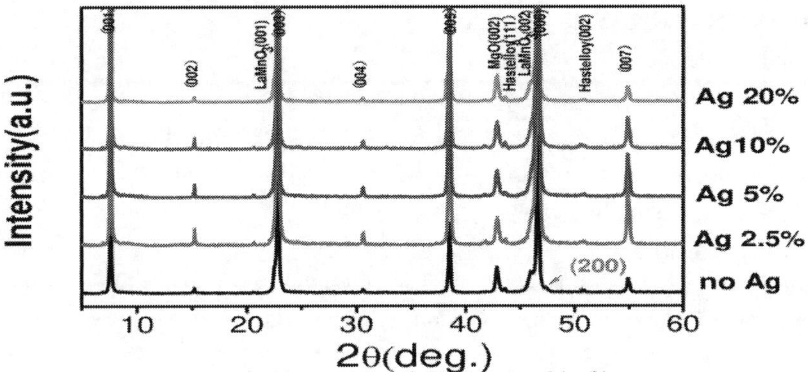

Figure 2. XRDθ–2θ patterns of GBCO-Ag thin films

Fig. 3 shows the SEM morphologies of all derived films. The a-axis oriented grains with needle-shape and more pores are observed in pure GdBCO film. GdBCO films obtained by proper addition of Ag close or equal to 5mol% level show less pores and higher rate of c-axis orientation were. In addition, no distinct Ag was detected on GdBCO matrix except only negligible Ag nanoparticles was detected by EDX, which is consistent with the XRD patterns . As the content of Ag doping increases up to 20% mol, large Ag precipitates with grain size of 100–200 nm emerge as confirmed by EDX, indicating that a large quantity of Ag prefers to gather at film surface rather than segregate in matrix, as shown in Fig. 3.

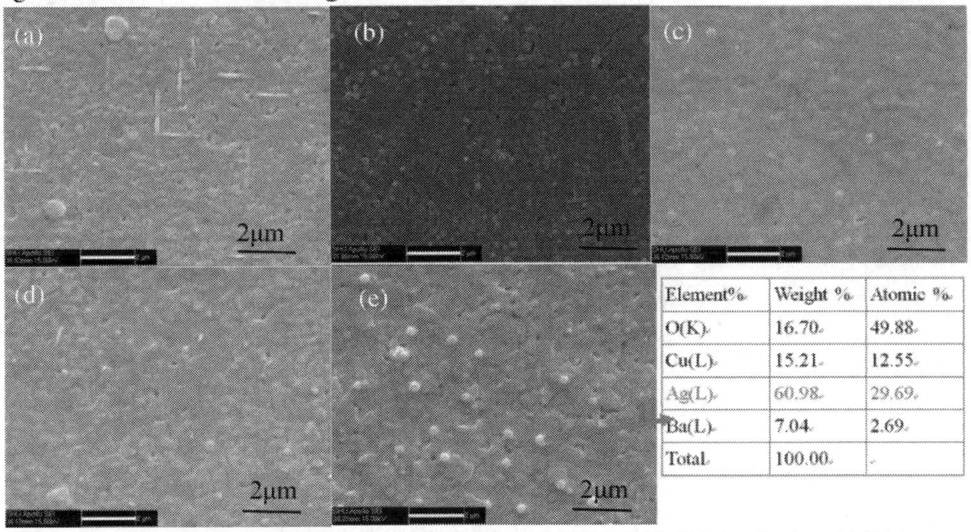

Element%	Weight %	Atomic %
O(K)	16.70	49.88
Cu(L)	15.21	12.55
Ag(L)	60.98	29.69
Ba(L)	7.04	2.69
Total	100.00	

Figure 3. SEM images of GdBCO films (a) pure GdBCO (b)2.5% Ag doping (c)5%Ag doping (d)10% Ag doping(e)20% Ag doping

The highest inductive critical current density(1.5MA/cm^2) at 77 K was obtained by 5 mol Ag doping. It is essential to note that *Jc* values are improved by Ag doping. Although the *Jc* value(0.65MA/cm^2) appear degrading with a high doping content up to 20 mol %, it is still slightly higher than the undoped ones' (0.6MA/cm^2). A possible explanation for the positive result of *Jc* is the enhancement of c-axis nucleation of GdBCO and the improvement of surface quality caused by moderate Ag doping.

In case of 20 mol % Ag addition, the degradation of Jc may be caused by the large amount of impurities, inhomogeneity and decline of texture.

4. Conclusions

In summary, a series of Ag doped GdBCO films were prepared by low fluorine MOD method. The effects of Ag doping on the microstructures and superconducting properties of GdBCO films were investigated. The results showed that optimum addition of Ag close or equal to 5 mol % in the GdBCO solutions lead to better c-axis orientation and surface microstructure and then a higher *Jc*.

Acknowledgment

This work is sponsored by Shanghai Science and Technology Leading Project (No.11dz1100300), by the National Natural Science Foundation of China General Program (No. 11174193) and by the National Key Research Program (973) of China Sub-Project (No.2011CB605905);

References

[1] Miyata S, Matsuse K, Ibi A, Izumi T, Shiohara Y and Goto T 2013 *Supercond. Sci. Technol.* **26** 045020

[2] Iguchi T, Araki T, Yamada Y, Hirabayashi I and Ikuta H 2002 *Supercond. Sci. Technol.* **15** 1420

[3] Rupich M W, Verebelyi D R, Zhang W, Kodenkandath T and Li X 2004 *Mater. Res. Soc. Bull.* **29** 572

[4] Jin L H, Lu Y F, Feng J Q, Zhang S N, Yu Z M, Wang Y and Li C S 2011 *J. Alloys. Comp.* **509** 3353–3356

[5] Chen Y Q , Yan F Y , You C Y , Zhao G Y and Jiao Y 2013 *J. Alloys. Comp.* **576** 265–270

[6] Obradors X, Puig T, Pomar A, Sandiumenge F, Mestres N, Coll M and Ricart 2006 *Supercond. Sci. Technol.* **19** 13–26

[7] Jung-Woo Lee, Yun-Sik Joo, Soon-Mi Choi and Sang-Im Yoo 2013 *IEEE Trans. App. Supercond.* **23** 8002704

[8] Vyacheslav F, Solovyov, Li Qiang, Si Weidong, Maiorov Boris, Timothy J, Haugan Mac, Manus-Driscoll J L, Yao H, Jia Q X and Specht E D 2012 *Phys. Rev. B* **86** 094511

[9] Cheng CH, Zhao Y 2007*Phsica* C **463** 174-177

[10] Pan A V, Pysarenko S V, Wexler D, Rubanov S and Dou S 2007 *IEEE Trans. App. Supercond.* **17** 3585-3588

[11] Ichino Y, Sudoh K, Miyachi K, Yoshida Y and Takai Y, 2003 *IEEE Trans. App. Supercond.* **13** 2735

[12] Obradors X, Martínez-Julián F, Zalamova K, Vlad V R, Pomar A, Palau A, Llordés A, Chen H, Coll M, Ricart S, Mestres N, Granados X, Puig T and Rikel M 2012 *Physica* C **482** 58-67

Crystal structure and critical properties of HTSC 2G prototypes formed by pulsed laser deposition technique on the RABiTS tapes

I Chernykh, A Stroev, M Garaeva, T Krylova, L Klevalina, J Grishchenko, M Presniakov, I Karateev, S Shavkin, M Zanaveskin, A Vasiliev, A Shikov

National Research Center "Kurchatov Institute", Russia, Moscow, 123185, Akademika Kurchatova pl. 1

igor.chernykh@gmail.com

Abstract. HTSC 2G YBCO based short samples were formed by PLD (pulsed laser deposition) on the cube textured Ni-W tapes. The influence of buffer layer architecture on the critical properties of HTS films was investigated with scanning and transmission electron microscopy, X-ray diffraction and four point probe electrical measurement. The best critical properties were found in the HTS grown on tapes with triple buffer layers: Y_2O_3, YSZ and CeO_2. After CeO_2 layer deposition the blister-like defects appeared on the buffers surface. These defects being inherited by YBCO film could lead to degradation of critical properties.

1. Introduction

The formation of second generation of high temperature superconductor cables (HTSC 2G) on RABiTS (rolling-assisted-biaxially-textured substrates) require the intermediate buffer layers [1]. One of the widely used buffer layers architectures for YBCO growth on RABiTS is the 3-layered system consisting of the seed layer Y_2O_3, barrier layer yttrium stabilized zirconia (YSZ) and cap CeO_2 layer. The use of CeO_2 buffer is explained by good match of the crystallographic parameters of (001) CeO_2 and (001) YBCO compounds. In the present work we applied such 3 buffer layers for YBCO film formation on RABiTS. This system demonstrated 145 A/1 cm width current on short samples. The value of the current in our samples could be influenced by specific dome-like defects in YBCO films. Similar defects were found in [1-2], but their nature was not clearly understood. In this study we investigated the mechanisms of formation of such defects and their influence on critical parameters of YBCO films.

2. Experimental details

The EVICO RABiTS tapes with composition Ni + 5 %at W were used for the YBCO based coated conductors. The in-plane texture of substrates determined from the X-ray phi-scans was 6°. All films were grown using PLD (Kr-F, λ=248 nm, energy density 3 J/cm^2). The seed buffer layer Y_2O_3 was deposited in Ar+H_2 (P=1 mTorr) atmosphere to prevent Ni-W substrate from oxidation at initial growth steps. Other buffers were deposited in 1 mTorr O_2 atmosphere. O_2 pressure during YBCO deposition was P=350 mTorr. The deposition temperature of all layers have been controlled by IR-pyrometer and maintained at t=750°C. The sample sizes were 10x10 mm. The microstructure of the systems were characterized using X-ray diffraction (XRD), scanning electron microscopy (SEM), scanning/transmission electron microscopy (TEM/STEM) together with energy dispersive X-ray (EDX) microanalysis. SEM analysis were performed in Helios (FEI, US) dual beam instrument (electron - focus ion beam (FIB)), which was also used for cross-sectional sample preparation. A Pt layer of 2-3 μm thick was deposited on the surface of the sample prior to the FIB cross-sectioning. The protective layer was formed in two steps process: first it was deposited with of Plan-view samples were prepared by conventional method. All specimens were studied using a Titan 80-300 TEM/STEM (FEI, Oregon, US) equipped with a spherical aberration (Cs) corrector (electron probe corrector), a high angle annular dark field (HAADF) detector, an atmospheric thin-window energy dispersive X-ray (EDX) spectrometer (Phoenix System, EDAX, Mahwah, NJ, US), and post-column Gatan energy filter (GIF; Gatan, Pleasanton, CA, US). Measurements of the superconducting properties were made by standard 4 probe technique. Ag was DC sputtered for contacts formation with sequential etching of uncovered film in HNO_3 solution.

3. The buffer layers investigations.

The SEM study demonstrated different morphology of YSZ and CeO_2 surfaces and that is shown in Fig.1 (a) and (b). The blister-like defects appeared on the surface of CeO_2 (three buffer layers) Fig.1 (a). The diameter of blisters was 5-7 μm and that was stable within 10 - 100 nm of CeO_2 film thicknesses. The concentration of the defects was not changed either. However, we found different concentration of the blisters in different areas of the film, which were associated with differently oriented grains in the substrate. In the same time, the surface of YSZ was smooth, without any peculiarities Fig.1 (b). Close inspection of images of YSZ/CEO_2/protective Pt-C layer (Fig. 2 (a) and (b)) demonstrated, that during sample preparation by FIB, amorphization of CeO_2 surface occurred by Ga^+ ions. That caused the formation of additional layer on the top of CeO_2. XRD and TEM data pointed that Ni oxidation occurs after CeO_2 film formation. The NiO layer is visible on cross-section HAADF STEM images on Fig. 2 (a) and (b). The blister area was filled with Ni: the HAADF contrast (Fig 2 (a)) together with EDX microanalysis unambiguously pointed to the absence of W in that area. The NiO layer was not flat (Fig.2 (b)). Surprisingly, that the thickness of NiO layer was higher in the areas, which were close to the perimeter of the blisters and lower in the epicenter. If the ambient atmosphere during cooling was changed to H_2, the blisters were not formed on the surface of CeO_2. All these data indicated that the formation of blister occurred due to oxidation of the Ni-W substrate through the channels, which appeared in the CeO_2/YSZ/Y_2O_3 buffers as a result of unevenness on Ni-W substrate surface from one hand and stress in the system caused by difference in thermal expansion coefficient from the other. The unevenness of the substrate resulted in the formation of grain boundaries in the buffer layers which served as O diffusion paths during the cooling of our system from the growth to room temperature.

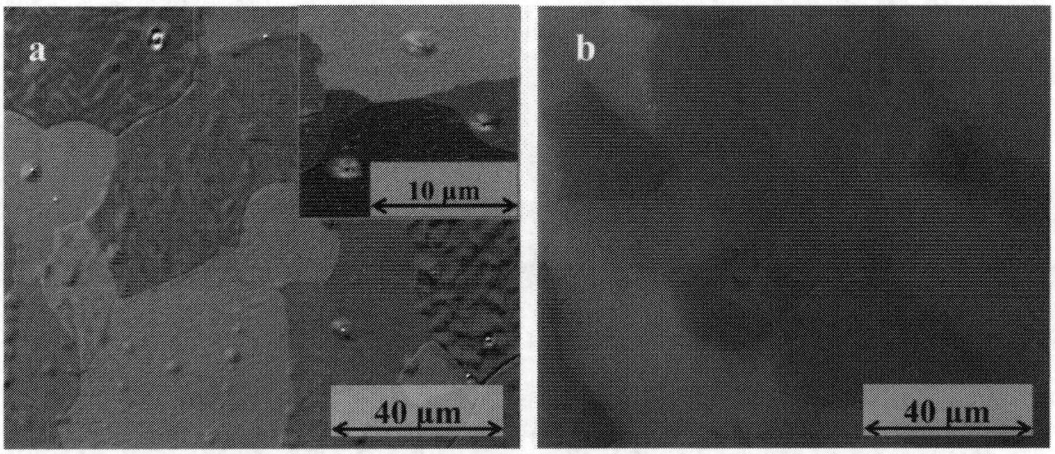

Fig. 1 Secondary electron (SE) SEM images of buffer surfaces. (a) – CeO_2 buffer (3-layer buffer), (b) – YSZ (2-layer buffer).

The stress also triggered on the Kirkendall diffusion of Ni in the oxidized area and that initiated the formation of blisters.

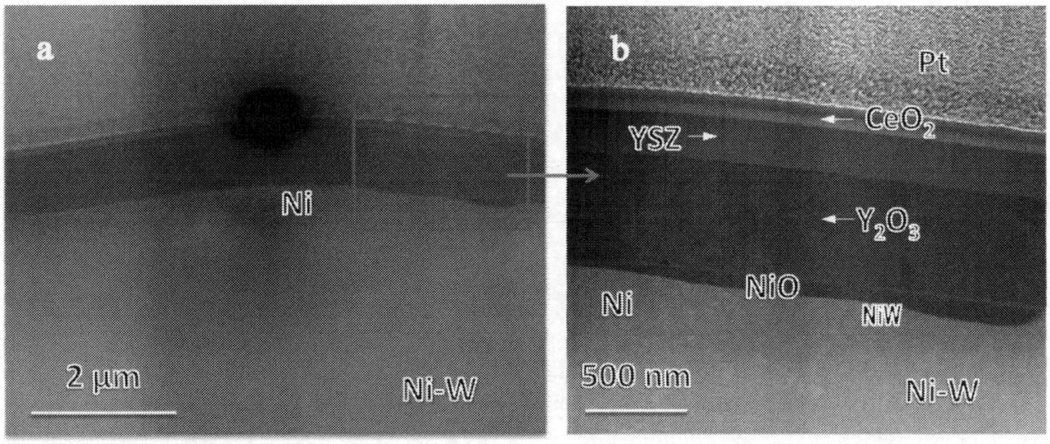

Fig. 2 HAADF STEM cross-section image of the 3 layer buffer system in blister area. (a) – the blister was filled with Ni covered with NiO film. Dark area on the top of the blister corresponds to the hole in the TEM sample. (b) – enlarged image of $NiW/Ni/NiO/Y_2O_3YSZ/CeO_2$ interfaces.

4. YBCO film investigation

The YBCO films were deposited on RABiTS with three and two buffer layers under the same deposition conditions and SEM images of YBCO film surface are shown in Fig.3 (a) and (b) correspondingly. The blisters were found on the surface of $YBCO/CeO_2/YSZ/Y_2O_3/Ni$, however they looked smoother than those on CeO_2 surface. The diameter of blisters was similar to one on CeO_2 surface, namely 5-7 µm and we proposed that these defects were inherited by YBCO film. The SEM study did not reveal similar defects on $YBCO/YSZ/Y_2O_3/Ni$ surface. The SEM shows better epitaxy and better surface morphology of YBCO films on CeO_2: slightly misoriented domains in the film and grain boundaries were clearly visible. Such contrast was not observed in YBCO films grown on YSZ buffer. YBCO films were not uniform in both cases, but overall porosity of YBCO films on YSZ buffer was higher.

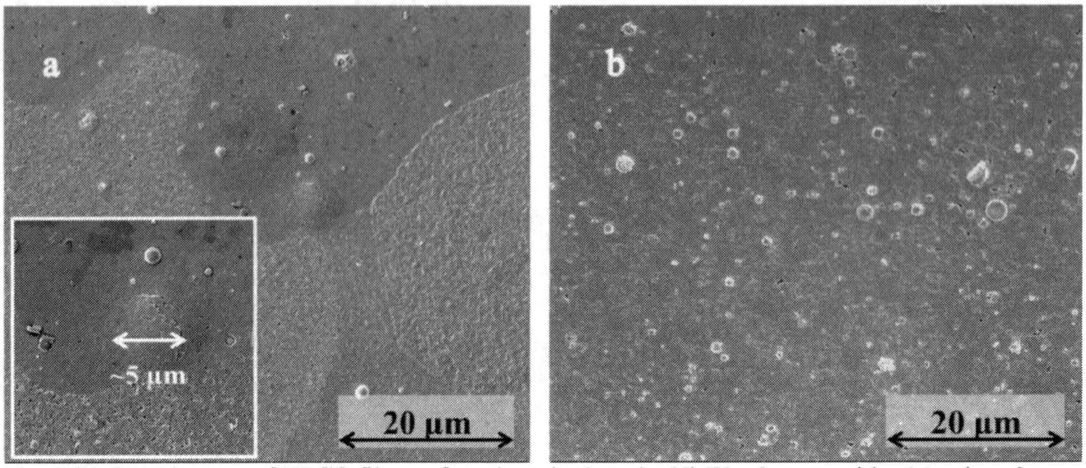

Fig. 3 SE SEM images of YBCO film surface deposited on the Ni-W substrate with : (a) - three layers $CeO_2/YSZ/Y_2O_3$ buffers (enlarged part of the image is in the insert) and (b) - double layers YSZ/Y_2O_3 buffers.

The critical temperature of YBCO on 3 layers buffer was found to be $T_c = 88$ K with the transition width of 2 K and on 2 layers buffer $T_c = 85$ K. However, the transition width in the system with 2 buffer layers was narrower than 1 K. The reason of such critical temperature reduction was caused by changes of oxygen stoichiometry of the film. The X-ray analysis demonstrated that oxygen content in that sample was equal to x = 6.7, while in the samples with 3 buffer it was higher: x = 6.9. Critical current in the sample with 3 buffer layers was $I_c = 14.5$ A/(1 mm width), while in the system with 2 buffer layers $I_c = 7.5$ A, in both cases YBCO film thickness was 1 µm. That also could be explained by the difference in oxygen content. The difference in morphologies of YBCO films on different buffers might result variations in oxygen content.

Conclusion

Different aspects of the influence of buffer layers architecture on microstructure and electro physical properties of YBCO tapes on RABiTS substrate have been illustrated. The mechanism of blisters formation in 3 buffer layers system was proposed. The YBCO films on YSZ buffer layer demonstrated lower critical current density but narrower transition compared to the YBCO films on CeO_2 buffer layer. That was explained by the difference in film oxidation and film morphology.

References

[1] Goyal A, Norton D P, Budai J D, Paranthaman M, Specht E D, Kroeger D M, Christen D K, He Q, Saffian B, List F A, Lee D F, Martin P M, Klabunde C E, Hartfield E, Sikka V K 1996 *Appl. Phys. Lett.* **69** 1795

[2] Sambasivan S, Kim I, Barnett S, Zurbuchen M A, Li J, Kang B W, Goyal A, Barnes P N, Oberly C E 2003 *J. Mater. Res.* **18** 919

[3] Kreiskott S Getta M Moenter B Piel H Pupeter N 2003 *Physica C* **383** 306

Flux pinning distribution and E-J characteristics of 2G YBCO Tapes

Iole Falorio, Edward A. Young, Yifeng Yang
Institute of Cryogenics, Faculty of Engineering and Environment, University of Southampton SO17 1BJ, UK
E-mail: if1c09@soton.ac.uk

Abstract.
E-J characteristics of SuperPower YBCO 2G tapes have been measured in the temperature range 15K-80K. It was found that the E-J characteristics deviate significantly from the standard power-law behaviour with apparent power exponent decreasing continuously at high voltage with increasing current. The deviation of E-J characteristic from the standard power-law was found to be consistent with a Weibull distribution of critical current. The identical scaling of E-J characteristics above 40K suggests a common critical current distribution. At lower temperature the critical current distribution becomes narrower to give higher apparent power exponent at lower temperature. With the critical current distribution model it is shown that the dissipation can be correlated directly to the flux flow resistance of Bardeen. The distribution of the critical current can be associated with a distribution of pinning potential of collective pinning and the headline pinning range is obtained as function of temperature.

1. Introduction

The development of 2G YBCO tapes has led to the availability of practical conductors suitable both for medium/high-field applications at liquid helium temperature and at intermediate temperature range. E-J characteristics of these tapes at such conditions are essential for magnet design, however they are not yet available in the literature due to the difficulties of transient measurement without liquid cooling. A systematic study of E-J characteristics over a wide range of temperatures (80K-40K) has been carried out and presented in previous work [1]. Experimental measurements conducted on YBCO tapes evidenced a deviation of transition characteristics from power-law behavior; this deviation has been interpreted in the frame of a collective pinning model. Although the collective pinning model is believed to describe the depinning mechanism of 2G tapes, the parameters introduced do not permit simple comparisons between conductor performances. In this work measurements have been extended to lower temperatures (80K-15K) by using a semi-transient protocol. A simpler statistic model based on a spatial distribution of critical current is used to describe the non-linear behavior observed.

2. Methods

2.1. Sample holder and semi-transient protocol used for measurements

The sample used in this work was standard 4mm wide Cu stabilised YBCO conductor produced by SuperPower Inc. Measurements at variable temperatures were carried out using a cryostat refrigerator cooled by a Sumitomo cryo-cooler, as described in [2]. Measurements were made on a 1 mm bridged structure, obtained by transversely milling the conductor from either side, as described in [1]; the milling does not cause degradation. A pair of hybrid HTS leads were used to inject current up to 600A to reduce the heat load to the system. The E-J measurements were conducted over more than four decades of voltages by a combination of DC current measurements at low voltages (0.1-10 μV) and of transient current measurements at higher voltages, recorded at 10kHz.

2.2. *E-J characteristics of flux flow due to a Weibull distribution of flux pinning*

It has been reported previously [1] that the E-J characteristic of standard YBCO tape at 77K deviates from the simple power-law, as shown in fig.(1) where the underlying superconductor E-J (•) is also presented with the current sharing by the stabilizing copper removed. The solid line shows that the deviation from power-law (dashed line) can be accounted from a distributed depinning model [3] where the depinned flux lines/bundles contribute to local dissipation in the fully flux flow regime by

$$E_e(J_C, J) = \rho_{FF} \cdot (J - J_C), \qquad J > J_C \tag{1}$$

where ρ_{FF} is the flux flow resistivity and J_C is the spatially distributed critical current. Flux flow resistivity ρ_{FF} dependence on temperature has been calculated according to Bardeen's relation $\rho_{FF} = \rho_n(T) \cdot (B_{sf}(T) + B_{ext})/B_{C2}(T)$ which has been proved to hold for YBCO epitaxial films in the range of temperature 85K-76.5K. B_{ext} is the external field applied whilst B_{sf} is the self field contribution, which has been calculated as a function of the tape geometry according to [6]. The calculation of the flux flow resistivity allows to establish the dissipation level (based on empirical data) giving strong support to the model. The depinning probability of flux bundles has been found to be well described by a Weibull distribution [7], given by

$$P(J_C) = \begin{cases} \frac{m}{J_0} \cdot \left(\frac{J_C - J_{Cmin}}{J_0}\right)^{m-1} \cdot \exp\left[-\left(\frac{J_C - J_{Cmin}}{J_0}\right)^{m}\right] & J_C > J_{Cmin} \\ 0 & J_C < J_{Cmin} \end{cases} \tag{2}$$

where J_{Cmin} is the minimum value of the distribution,J_0 is a scaling parameter which is roughly equal to the width of the distribution and m is the exponential parameter which determines the shape of the distribution. The macroscopic E-J characteristic can be obtained by summing the contribution to dissipation of different domains

$$E(J) = \int_{J_{Cmin}}^{J} P(J_C) \cdot E_e(J, J_C) \cdot dJ_C = E_C \cdot \left(\frac{J - J_{Cmin}}{J_{C0} - J_{Cmin}}\right)^{m+1} \tag{3}$$

This E-J leads to a local power-law index $n_J = d(lnE)/d(lnJ) = (m+1)/(1 - (J_{Cmin}/J))$ which reduces with increasing current and approaches $n = (m+1)$ when $J \to \infty$. The equivalent power law exponent for the commonly used critical current $J_{C0} = J_{Cmin} + [(J_0^m \cdot (m+1) \cdot E_C)/\rho_{FF}]^{\frac{1}{m+1}}$ at $E_C = 1\mu V/cm$ is correlated to the Weibull exponent m and the critical current threshold J_{Cmin} i.e. $n_{J_{C0}} = (m+1)/(1 - (J_{Cmin}/J_{C0}))$. Note that J_{C0} with a smaller $n_{J_{C0}}$ is increased by a wider distribution of a lager J_0 (blue line in fig.(1)).

3. Results and discussion

3.1. *E-J characteristics of 2G YBCO predicted by depinning model*

The E-J characteristics predicted by depinning model are shown and compared with the respective experimental data in fig.(1). The interpolating curve has been obtained by varying the exponent m until a good fitting was observed and by calculating J_{Cmin} as a function of m

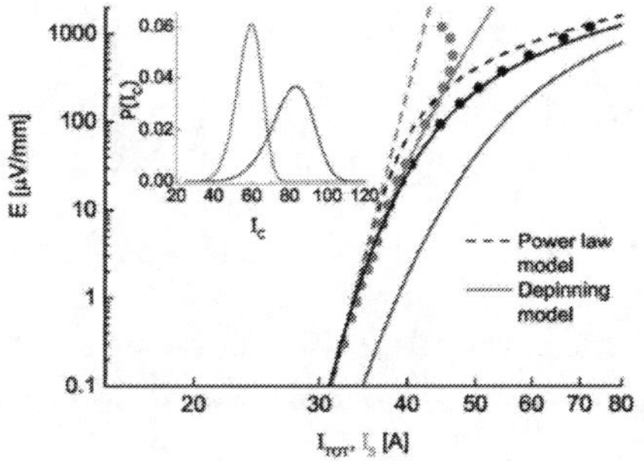

Figure 1. E-J characteristic of bridged 1mm 2G YBCO with (•) and without (•) current sharing contribution compared with power-law (dashed lines)and distributed depinning model (solid lines).

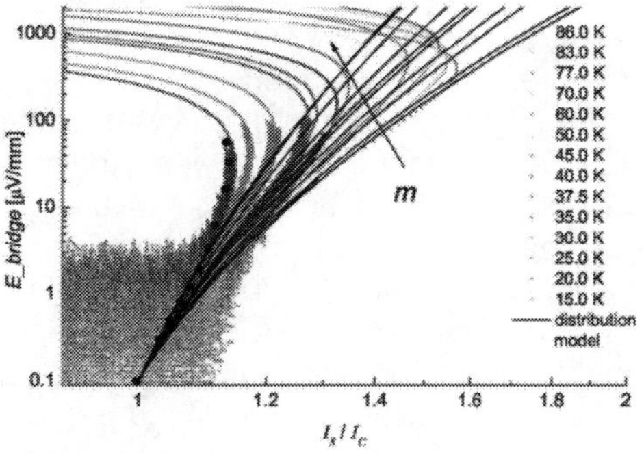

Figure 2. E-J characteristics measured on 1 mm YBCO tape in a range of temperature (86K-15K). Distributed depinning model predictions once current sharing distribution has been removed (black solid lines)

for each attempt according to eq.(3). The fitting parameters obtained are demonstrated to be independent of the engineering current criteria chosen, returning the same parameter m and J_{Cmin}.

Deviation from power-law behaviour has been observed across the temperature range 15K-86K. The equivalent n-number $n_{J_{C0}}$ has been found to remain constant at a value of 30 in the temperature range 40K-86K, and to slowly increase up to 40 at 15K, as observable in fig.(2). The resistance temperature dependence has been taken into account according to [8]. The data set of E-J characteristics can be satisfactory predicted with the depinning model law as shown in fig.(2), although curves at very low temperature are strongly affected by heating at low dissipation levels ($\sim 10\mu V/mm$) making data fitting difficult. The parameter m has been found to have a dependence on temperature qualitatively similar to the n-number as shown in the inset of fig.(3). The respective set of distribution curves, normalised with the pinning range I_0, shows a narrower bell as the lower temperature is approached, as shown in fig(3).

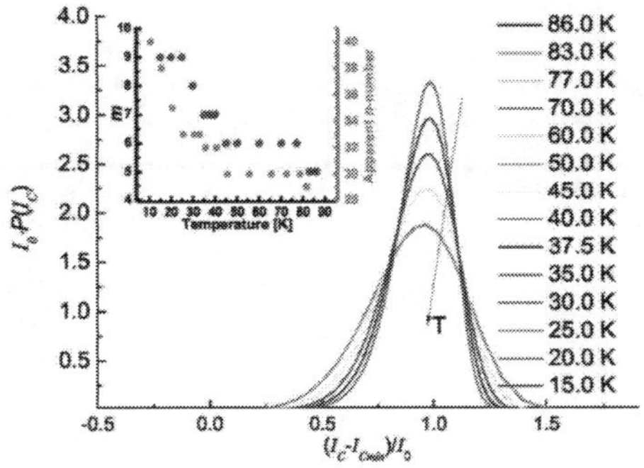

Figure 3. Current distribution of 2G tape in a range of temperature (86K-15K). Inset-Dependence of equivalent n-number $n_{J_{C0}}$ and exponent m on temperature.

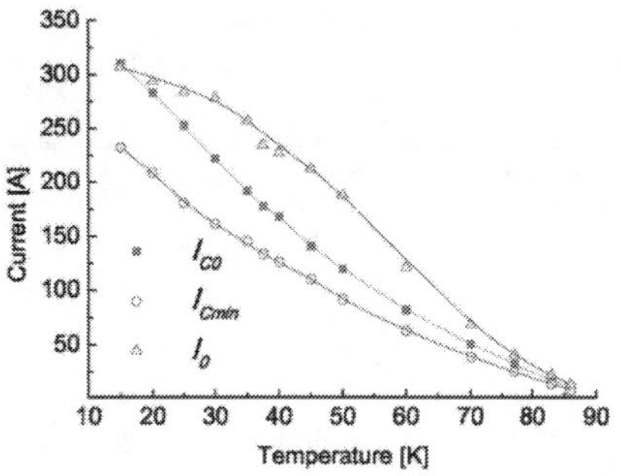

Figure 4. Weibull distribution parameters I_{cmin} and I_0 and critical current determined with the electric field criteria I_{C0} dependence on temperature.

3.2. Characteristics of Weibull distribution parameters

The dependence of the Weibull distribution parameters I_{Cmin} and I_0 on temperature is highlighted in fig.(4) and compared with the critical current I_{C0}.

The pinning threshold I_{Cmin} shows a similar temperature dependence of critical current I_{C0} with an accelerated increase at lower temperatures, which signifies the extended influence of the thermal activated flux flow [9]. In contrast, the pinning range I_0 shows a moderating increase with reducing temperature. Such a behaviour is similar to the depinning current and suggests that the pinning range is related to the underlying pinning strength of the material.

In order to demonstrate the effect that these parameters have on the E-J characteristic, a transition characteristic with same current threshold I_{Cmin} and same Weibull exponent m of the curve which fits the data, but with a greater pinning range I_0, has been included in fig.(1) (solid blue line). In the inset of the figure, the critical current distributions associated with the data fitted to the experimental results and with the data obtained using a greater pinning range, are compared. The bell of the distribution associated with the greater pinning range is broader than the one corresponding to the real data, and corresponds to a conductor with stronger pinning centres, which is thus more stable.

A good quality superconductor is conventionally considered to have a larger n-number, however this comparison shows that this criteria is artificial because the more stable conductor presents an E-J characteristic with a higher I_{C0} and in turn smaller $n_{I_{C0}}$, as expressed in eq.(3). Surprisingly, the distributions curves observed in the data set presented in fig.(3) are narrower at low temperatures. This phenomenon is attributed to a strong dependence of I_{Cmin} on temperature, when compared to I_0. Furthermore, an increment in temperature has a faster effect on I_{Cmin} rather then on I_0, as shown in fig.(4), which shows that I_0 tends to saturate at low temperatures. The latter observation indicates the presence of strong pinning centres that are not particularly affected by temperatures, whilst weak pinning centres quickly appears and contribute in dissipation as the temperature is increased.

4. Conclusion

Transport measurements at different temperatures on 2G YBCO conductor revealed an underlining E-J characteristic described by inhomogeneous distribution of critical current. The model does not furnish macroscopic information about the depinning mechanism, which is still believed to be described by collective pinning of small flux bundles, but it is of practical utility since it introduces parameters that can be related to tape performance such as J_{Cmin} and J_0 and permits comparisons between conductors with different critical current densities. Future studies will relate the model presented in this work to the collective pinning model.

References

[1] Falorio I. and Young E. A. and Yang Y. 2012 *Physic Procedia* **36** 1462
[2] Young E. A. and Bailey W.O.S. and Al-Mosawi M.K. and Beduz C. and Yang Y. and Chappell S. and Twin A. 2011 *Physic Procedia* **36** 1343
[3] Kiss T. and Nakamura T. and Takeo M. and Kuroda K. and Matsumoto Y. and Irie F. 1995 *IEEE Tans. App. Superconductivity* **5** 1363
[4] Bardeen J. and Stephen M. J. 1965 *Physical Review* **140** 1197
[5] Kunchur M. N. and Christien D. K. and Phillips J. M.1993 *Physical Review Letters* **70** 998
[6] Yang Y. and Hughes T. and Beduz C. and Spiller D.M. and Scurlock R.G. and Norris W.T. 1996 *Physica C* **256** 378
[7] Irie F. and Tsujioka Y. and Chiba T. 1992 *Supercond. Sci. Technol.* **5** 379
[8] Young E. A. and Chappell S. and Falorio I. and Yang Y. 2011 *IEEE Tans. App. Superconductivity* **21** 3062 Chappell, S. ; Falorio, I. ; Yang, Y.
[9] Ihara N. and Matsushita T. 1996 *Physical Review* **257** 223
[10] Kiss T. and Inoue M. and Nishimura S.and Kuga T. and Matsushita T.and Iijima Y. and Kakimoto K. and Saitoh T. and Awaji S. and Watanabe K. and Shiohara Y 2002 *Physica C:Superconductivity* **382** 57

Development Status of AMSC Amperium® Wire

S Fleshler, K DeMoranville, J Gannon Jr, X Li, E Podtburg, M W Rupich, S Sathyamurthy, C L H Thieme, D Tucker and L Whitman

AMSC, 64 Jackson Road, Devens, MA USA

E-mail: Steven.Fleshler@amsc.com

Abstract. AMSC produces Second Generation (2G) HTS wire for utility power applications as well as coil, motor and generator solutions. In this paper, various types of AMSC's Amperium® wire suitable to power cables, fault current limiters and coils are reviewed. In addition, recently developed performance-improvements in amperage, reduced ac power loss and mechanical properties are summarized. The introduction of thicker HTS layers coupled with optimized heat treatments to enhance critical current density dramatically improve both cable and coil wire current-carrying capability. A non-magnetic RABiTS™ substrate has now been developed to the point where it is compatible with the manufacturing process and capable of sustaining large critical currents. Finally, the ability of Amperium® wires to withstand cable-winding stresses, and to exhibit the high transverse c-axis strength critical to the reliability of the wire in coils, are discussed.

1. Introduction

AMSC manufactures and sells high temperature superconductor (HTS) wire for use in high power density transmission cables, fault current limiters (FCL's), distribution power cables with integrated FCL capability, and light-weight high power motors and generators. AMSC's wire manufacturing is based on the Rolling Assisted Biaxially Textured Substrate/Metal Organic Deposition (RABiTS/MOD) processes described in detail elsewhere [1,2]. The final wire consists of either 4mm or 10mm HTS 'insert', slit from wider strips, laminated between two metal stabilizer strips. The width of the stabilizer is designed wider than the HTS insert so that the regions at the edges are filled with solder (referred to as solder fillets) supplying mechanical strength and hermeticity to the wire.

In this paper, different types of AMSC's Amperium® wire architectures, and the ongoing development efforts to improve their performance, are described. The paper is arranged in three sections, the first section covers I_c and ac loss for wire used in ac applications operating in low magnetic field at temperatures accessible with liquid nitrogen (LN_2), the second section describes I_c of wires designed for coil-based devices operating at low temperature and high field, and the last section highlights the tolerance of the laminated wire architectures to mechanical stresses.

2. Wire for LN_2 AC Applications: I_c and AC Loss

Targeted applications that fall into this category include resistive FCL's and power cables (transmission and distribution). For resistive FCL's, the wire architectures have been designed with 75μm thick stainless steel (SS) lamina, 12mm in width, around a 10mm HTS insert. Two variants are available: a three layer version with a single HTS insert between the two SS lamina and a four layer option with two (double) HTS inserts and between the lamina [3]. The ability to assemble one or two HTS inserts into a single wire allows for a wide range of wire currents. The typical ranges for the minimum critical current for the 3-layer and 4-layer wires (with a 0.8 μm HTS layer) are 225-275A and 475-550A, respectively. Figure 1 shows plots of I_c versus position for the 3-layer and 4-layer wire options over representative 500m lengths. The minimum I_c's are 262A and 538A with a standard deviations of 1.3% and <1% for the 3-layer and 4-layer variants, respectively.

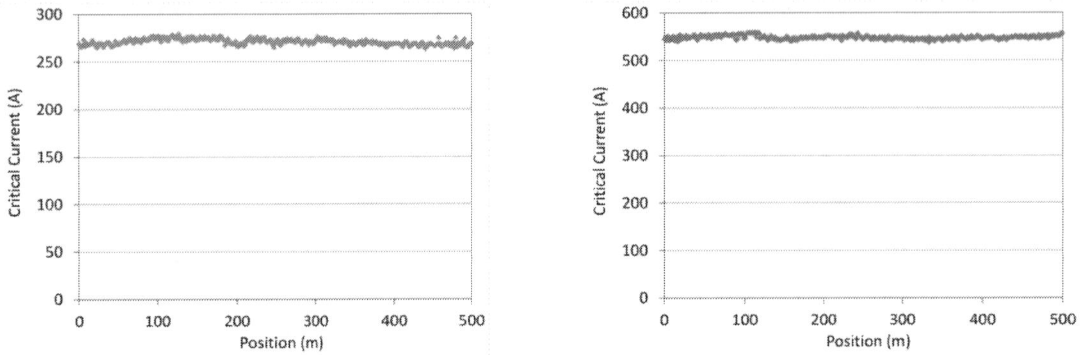

Figure 1. I_c displayed for each meter of 3-layer and 4-layer SS wires for resistive FCL application.

Brass cable wire was designed to be robust in cable stranding operations and resistant to buckling under the compressive and bending loads experienced in transmission cables with thick dielectric layers needed for high voltage operation. The wire has a 3-layer composite structure consisting of a 4mm HTS insert laminated between two 4.4mm wide, 150μm thick brass stabilizers. As previously reported, nominal values of the self-field, 77K I_c's ranging from 140-200A have been achieved in brass cable wire by incorporating a 1.2 μm HTS layer optimized for use at LN2 temperatures and low fields [3].

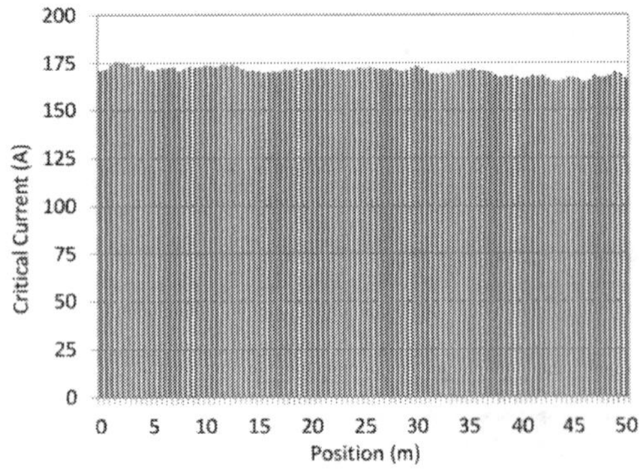

Figure 2. I_c map of 50m long brass cable wire with NMS and 1.2μm HTS layer using one meter gauge-length. The minimum I_c is 165A.

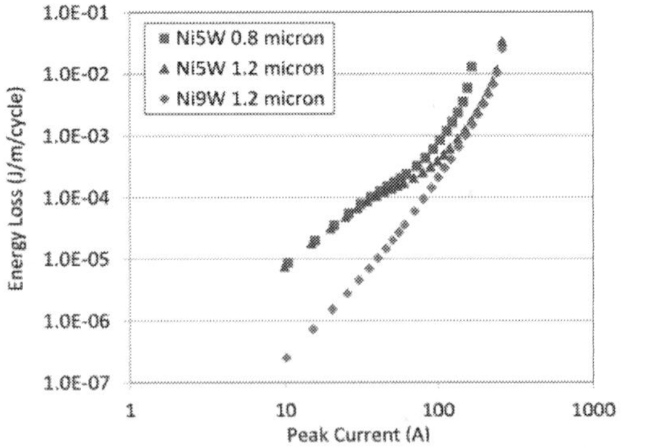

Figure 3. AC loss per cycle vs. peak current for different substrates and HTS thicknesses measured at 50 Hz and temperature ~75K.

In addition to increasing current capacity, development work at AMSC has been focused on reducing the ac loss and magnetic permeability associated with the RABiTS substrate. The Ni5W alloy (atomic percent) presently used in the production wire is slightly ferromagnetic, resulting in a hysteretic energy loss and magnetic susceptibility at low currents (and magnetic fields). The first demonstration of cable wire with a non-magnetic substrate (NMS) using Ni9W RABiTS with good local metrics for texture and I_c on par with the standard wire was reported in reference [3]. Once the feasibility of the NMS technology was established in short lengths with the 0.8μm HTS layer manufacturing process, 50-70m long strips were produced to verify compatibility with the 1.2μm HTS coating and modified heat treatments designed for higher I_c. Brass cable wire incorporating NMS has achieved I_c ranging from 165A to 176A over 50m as exhibited in Figure 2. The wire with NMS is in the same range as its long-length counterpart with standard Ni5W substrate reported in reference [3].

Figure 3 compares the transport ac loss of a section of NMS cable wire with wires containing 0.8μm and 1.2μm HTS layers on standard Ni5W substrate measured at Los Alamos National Laboratory. The results are as expected, at low currents, the FM contribution dominates the loss of the wire with the standard Ni5W substrate independent of the coating thickness (and I_c) while the loss of the wire with NMS is greatly reduced. At higher currents, the loss is dominated by the HTS contribution and the wires with 1.2μm HTS layer have nearly the same loss regardless of the type of substrate, which is considerably lower than the 0.8μm wire possessing much lower I_c.

3. Wire for Coil Applications: Ic at Low Temperature, High Magnetic Field

Amperium wire designed for coil-based applications is laminated with 50μm thick copper stabilizers and is available in 4.8mm and 12mm widths with 4mm and 10mm HTS inserts, respectively. These wires have wider solder fillets, providing additional strength against delamination when subjected to high transverse c-axis stresses.

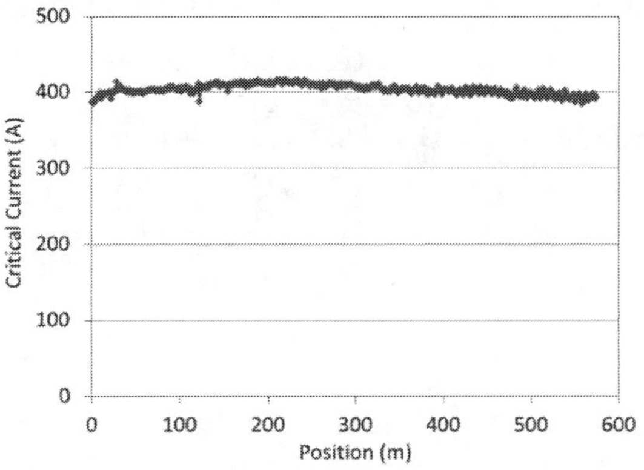

Figure 4. I_c over length for qualification run of 12mm copper coil wire with 1.2μm HTS layer optimized for low temperature operation.

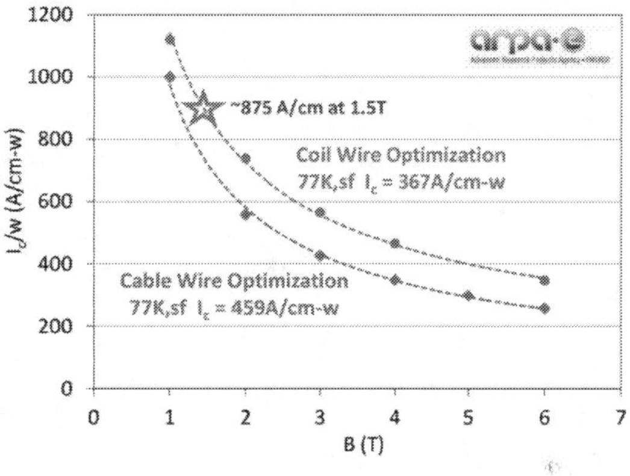

Figure 5. I_c vs. perpendicular magnetic field at 30K for wires with heat treatments optimized for cable (red) and coil (blue) applications.

Figure 4 shows the I_c versus length for an initial qualification run of a 12mm wide coil wire processed with a 1.2μm HTS layer optimized for coil operating regimes. The minimum I_c for this wire is 385A with a standard deviation less than 1.7% over nearly 575m. Although the 77K, self-field critical current density for coil wire is lower than that for cable wire with the same HTS thickness, its absolute critical current at coil operating conditions (i.e. low temperatures and high fields) is higher. Figure 5 shows a comparison of the critical current in perpendicular field (oriented along the c-axis) at 30K of two wires with 1.2μm HTS layer optimized for the cable or coil operating regimes. The coil wire, optimized for operation at low temperature, outperforms the cable wire by about 20% even though the 77K, self-field I_c of the cable wire is about 20% higher.

4. Key Mechanical Properties: Cable-Winding Tolerance and C-Axis Strength

To date, development and qualification of 4.4mm brass and 12mm copper wires containing the 1.2μm HTS layer have met all existing Amperium wire specifications (see www.amsc.com). Herein one key mechanical test is highlighted for each application.

Sections of 1.2μm HTS-NMS brass cable wire were subjected to a reel-to-reel cable winding test to characterize its integrity in its intended use. The test consists of sequentially going through reverse double bends around six inch rollers, a 360° twist and helical winding around a 19mm diameter former at 10 lb tension. The I_c was measured every one meter before and after the test. The test was carried out twice, once with the wire oriented so that the substrate faced radially inward to the former and once with the substrate oriented radially outward. The results of the cable-winding test, shown in Figure 6, confirm the wire maintains its I_c after repeated tests.

For 12mm copper coil wire, the c-axis strength or transverse delamination strength (breaking strength) was tested since this is an essential property that insures the reliability of the wire in coils, especially when considering stresses endured in epoxy-impregnated coils during cool-down to operating temperature. Figure 7 shows the minimum and maximum c-axis strength at 77K for seven wires with three or four samples taken per wire. The lowest value for this sampling is 40MPa, well-above requirements for in-house coil-designs. The distribution in strength reflects variation in centering the HTS insert within the laminated structure. Future improvement in the alignment will raise the minimum values. Naturally more statistics will be assembled as the qualification progresses.

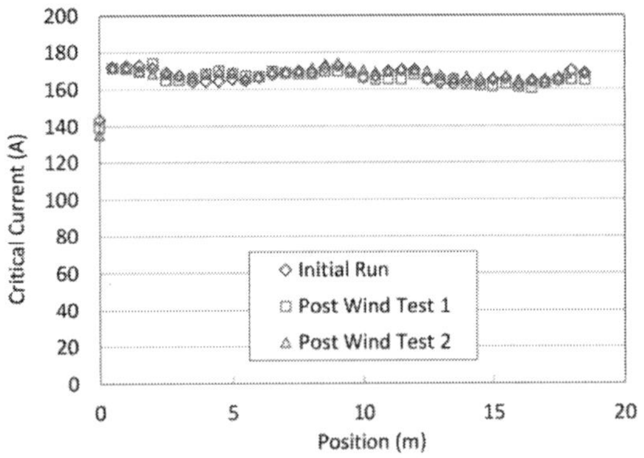

Figure 6. I_c over length is retained after simulated cable-winding test relative to the initial run. Tests were run with substrate oriented both in and out.

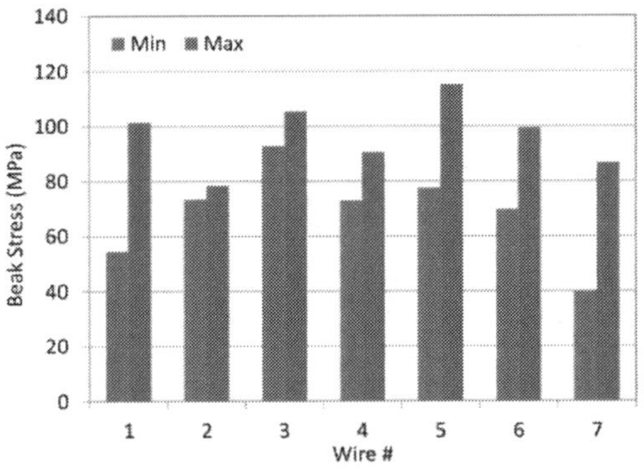

Figure 7. Minimum and maximum values of the c-axis strength at 77K for seven 12mm copper wires with 1.2μm HTS layer.

5. Conclusion

Increased performance of Amperium wire with 1.2μm thick HTS layers provides opportunity to reduce the cost of both LN_2 ac applications and low temperature coil applications by requiring less wire to build devices. Enhanced performance in robust brass laminated cable wire has been demonstrated on a non-magnetic RABiTS substrate over moderate lengths which will lower the cost of refrigeration. Initial qualification results for copper laminated over long length show promise of high current with enhanced field performance in a durable wire architecture capable of withstanding large c-axis stresses and resilient to delamination for coil applications.

References

[1] Fleshler S, Buczek D, Carter W, Cedrone P, DeMoranville K, Gannon Jr J, Inch J, Li X, Lynch J, Otto A, Podtburg E, Rupich M, Sathyamurthy S, Schreiber J, Thieme C, Thompson E, Tucker D, Nagashima K, and Ogata M, 2009 *Physica C* **469** 1316.

[2] Rupich M, Li X, Thieme C, Sathyamurthy S, Fleshler S, Tucker D, Thompson E, Schreiber J, Lynch J, Buczek D, DeMoranville K, Inch J, Cedrone P,and Slack J, 2010 *Supercond. Sci. Technol.* **23** 014015.

[3] Rupich M. Li X, Sathyamurthy S, Thieme C, DeMoranville K, Gannon J, and Fleshler S, 2013 *IEEE Trans. Appl. Supercond.* **23**, 6601205.

11th European Conference on Applied Superconductivity (EUCAS2013) IOP Publishing
Journal of Physics: Conference Series **507** (2014) 022006 doi:10.1088/1742-6596/507/2/022006

Operation and experience of a 2 km coated conductor REEL – to – REEL copper pulse plating facility

U Floegel- Delor, T Riedel, D Wippich, R Rothfeld, P Schirrmeister, R Koenig, F N Werfel , A Usoskin [1], A Rutt[1]

Adelwitz Technologiezentrum GmbH (ATZ), Naundorfer Str. 29, 04860 Torgau, Germany
[1]Bruker HTS, Siemensstr. 88, 63755 Alzenau, Germany

E-mail: floegel-delor@t-online.de

Abstract. Bruker HTS manufactures YBCO based superconducting wires of the second generation on low- cost Stainless Steel substrate (100 µm thick). With 250 – 500 A/cm@77 K, SF, 650 MPa tensile strength and 6 mm bending radius excellent electrical and mechanical properties are achieved. As complementation of the 2G fabrication technology an automated 2 km copper pulse plating facility has been installed in 2012. We report here the operation requirements and the experiences of the copper plating technique.

1. Introduction

Copper plating technique enables a fast and economic fabrication of a hard bonded copper layer on the superconductor surface. Thereby, as for LTS, metallic copper is the obvious metallic partner for the second generation 2G conductor. In addition, copper is one of inherent chemical constituents of all HTS. The copper shunt on the conductor gives a number of electric, thermal and mechanical advantages if power electric applications are considered. The developed physics and chemistry of pulse plating has been studied in the last years and transformed into the REEL – REEL technique for continuous deposition a 10 – 20 µm thick copper layer. The key issues of $CuSO_4$ plating and the governed electrochemical parameters following the Nernst equation for the Cu^{2+} dissociation are reported earlier [1-3].

For the purpose of the copper deposition on the surface of superconducting wires a fine- tuned electroplating technology using an aqueous $CuSO_4$ solution is found. Both acidic copper sulphate and alkaline solutions has been tested to optimize the plating process and deposition speed. An automated 12-chamber plating facility with up to 2000 m conductor length in REEL - to - REEL mode has been installed in 2012. We review and discuss here the wire plating concept, and the operation experiences with the REEL –to –REEL unit.

2. YBCO –Copper Hybrid wire

The basic wire architecture of the Bruker HTS wire is displayed in Fig. 1. It consists of the nonmagnetic 100 µm Nickel – Chromium Stainless Steel substrate, an YSZ/CeO_2 double buffer structure, and the pulsed laser deposited 1.2 µm thick Y-123 high-temperature superconductor on top. Finally, 0.5 µm thick Ag cap layer protects the wire in the initial processes. After this stages, the ready - processed tape is transferred into the copper plating facility, inserted into the REEL dynamics, and are subsequently covered with an appropriate metallic Cu layer. The advantage of the last process is the non-vacuum technology and the relatively low processing costs. From the plating process point of view the geometrical parameters of the conductor can be varied in a wide window. The electrochemical parameters follow the optimum current density per square centimeter of the electrode.

Further advantage of the Cu plating is the ability to deposit metallic copper both on Ni-W substrate directly as well as on Stainless Steel / Hastelloy surfaces (with a seeding layer between).

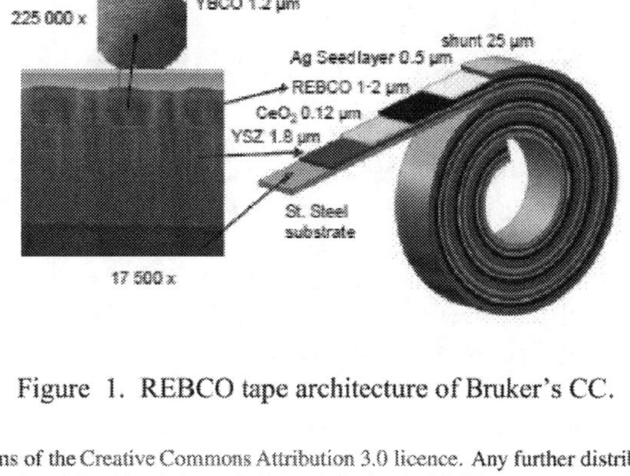

Figure 1. REBCO tape architecture of Bruker's CC.

In Fig. 1 the quite sophisticated wire structure consists of altogether 5 layers of a total thickness of about 3 µm. The ABAD / IBAD template possess the advantage of a rather perfect texture transformation and avoid imperfect Y123 layer growth. In Fig. 1 electron microscopy demonstrates the perfection the REBCO layer. The high-resolution picture shows in addition the well oriented lattice structure in c-axis with the smooth surface. The copper layer of typically 20 µm thickness in one- side or double –sided geometry on top, that is was the wire user sees finally from the conductor.

3. Automated copper plating facility

3.1. Design and construction

The automated 6 m long modular plating facility is shown in Fig. 2, consisting of the central 12 chamber plating modules with Teflon slits for the moving tape cathode throughout. On both sides the coil bobbin reels are located capable to operate a REEL of a maximum diameter of 33' (838 mm). The electrical insulated REELS are connected with driving stepper motors on a shaft for a continuous tape speed through the plating chambers. The tape way is carefully aligned to avoid edgewise strains during the tape movement.

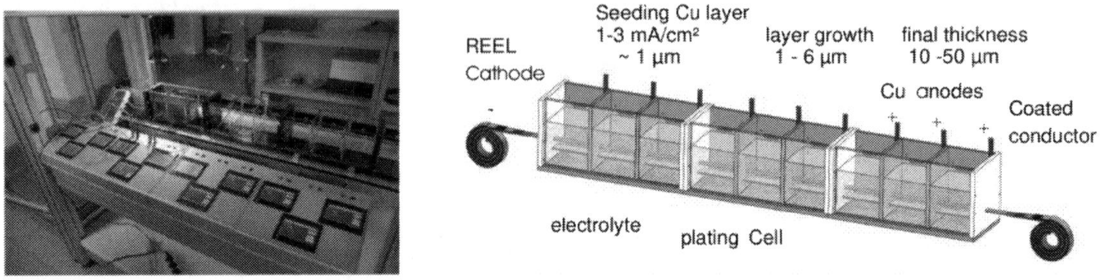

Figure 2. Automated continuous copper plating machine with individual electrolytic cells.

The REEL-to-REEL transport system of the equipment can handle tape length up to 4 km which has been tested for all speeds with NiCr Stainless Steel tape. The control of the speed torque and force reduces the handling and fits the plating process cycle time by virtue it's closed –loop constant movement and constant tape tension control.

The plating cells are positioned in the centre of the facility, neighboured by the two bobbin REELS. The conductor is run through slits from one cell to the next driven by two plating propulsion motor units at a speed of up to 100 m/h. The complete machine is designed and constructed according to the technique of semiconductor processing machines. The modular character with the plating stations provides an ease of process modification, variation and line expansion. The electric plating parameters are delivered by individual current voltage power supplies with a maximum current per cell of 3 Amperes. The cell voltage is a sensitive function of the actual substrate and electrolytic solution conductivity. The tape conductivity in the process is a strong function of the growing thickness of the Cu layer. At a copper thickness of a few micrometers the cell voltage is a measure of the $CuSO_4$ concentration of the plating solution and provides a control of the copper yield per cell.

3.2. Mechanical tape properties

Coated conductors produced by Bruker HTS are optimized for application at 77 K, where most of the applications are performed. For low temperature and high- field applications the wires are usually tested at 4.2 K showing much higher I_c values. A 4 mm PLD wire test sample with a length of about 6 m with 2 x 45 µm thick copper layers could carry at liquid Helium temperature (4.2 K) currents of almost 0.5 kA and 2.0 kA at a external field of 18 Tesla by field orientation B||c and B||a, b, respectively. In general, the wire manufacturers'effort is aimed to improve the electrical parameters I_c (B) at 77 K.

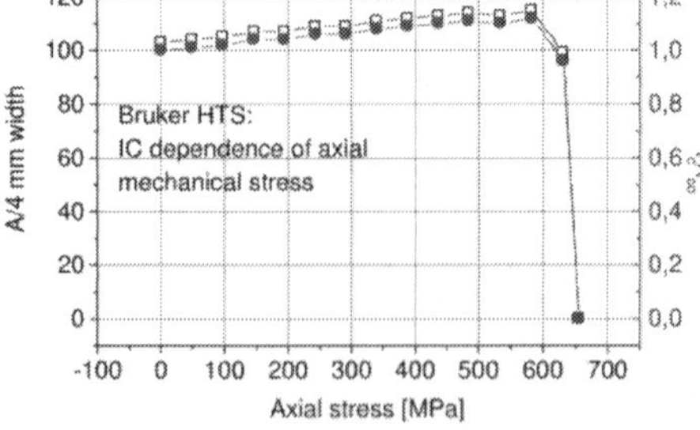

Figure 3. Critical current as a function of axial strain at 77 K.

In Fig. 3 the critical current J_c for the applied axial force of a 4 mm wire is plotted in absolute Ampere values and relative to the zero force data I_{c0}. It is demonstrated that the current increases slightly up to a critical stress of about 550 MPa. Above 600 MPa it drops down within 50 MPa to zero indicating the situation that the generated strain is fully transferred to the YBCO layer.

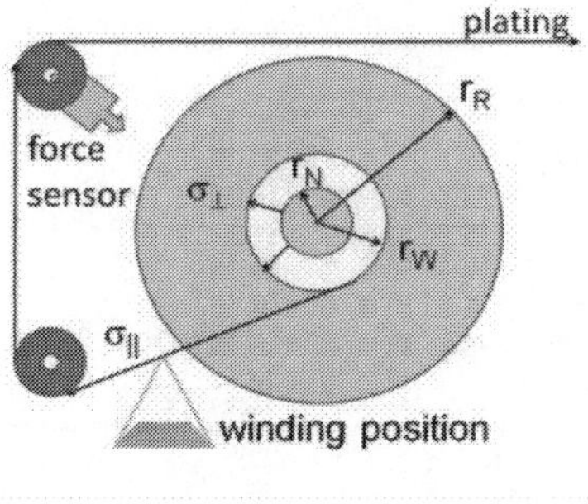

Figure 4. Force speed and winding control of the large reel in the plating apparatus.

The design of the Cu reel-to-reel plating facility considers this strain sensitivity. Thereby, the transport and the plating operation implicate mechanical forces on the tape. The tape has to be kept in a defined and constant distance to the separate anodes to ensure an efficient electrolytic field. The 33 inch REELS are driven of variable speed and electronic control for the transport through the cells. Because of the high sensitivity of the fully processed and therefore valuable coated conductor a careful operating procedure during the plating process is ultimate.

To control and to limit the possible forces and torques on the processing tape a mechatronics part and sensors are adapted to the REEL system. Fig. 4 displays the principal control detectors of the dynamical force and torque. In addition, for large tape lengths on has to limit the pre- tension on the tape. Applying pre-tension to the already processed tape results often in too strong radial forces that keep the inner layers on the REELS under unwanted high compressive stress.

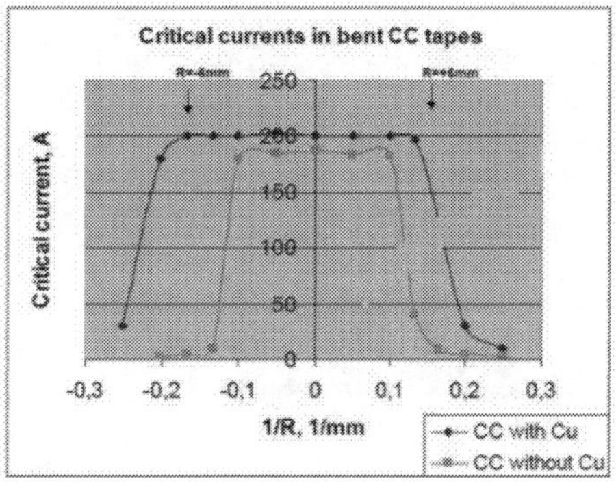

Figure 5. Critical bending behavior on the current density with and without Cu layer.

Figure 4 shows the winding and tension control parts on both REEL chambers. In a compromise between force and the necessary tape stiffness in the cells, the control system reduces the torque of the powered motor drive continuously with increasing diameter of the winded tape on the REEL. For a tape length of up to 2000 m the REEL control reduces the winding force in three steps from 12 N to 10 N, and finally to 8 N for the outermost layers.

Fig. 5 displays the measurement of the influence of the tape bending on the critical current density J_c. In extreme assembling situation of coil winding the pure tape shows a tolerance in J_c up to bending radius of 10 mm. The copper layer on the wire improves further the J_c tolerance level up to very narrow bend radius of 6 mm.

3.3. Acidic versus alkaline Cu solution

The $CuSO_4$ plating process is most effective and non-poisonous for copper plating. The power supplies for the individual galvanic cells are designed and manufactured for low HTS wire conductivity (Hastelloy). It allows the current to flow constant, reverse or in a wide ratio of pulse plating. The typical electrolytic current distribution is shown in Fig. 2, where the starting current density is small, and few mA per square centimeter wire area. A copper seed is the primary target here followed by higher currents in the next cells. All plating parameters are determined and governed by the Nernst equation. The cell potential E of a $CuSO_4$ solution relative to the standard potential E_0 is given by,

$$E = E_0 - \frac{RT}{nF} \ln \frac{C_1}{C_2}$$

R and F are the common Gas constant and the Faraday constant, respectively. C_1 and C_2 ($C_1 > C_2$) denote the electrolytic concentrations in the concentric and dilute status for n electron oxidation or reduction process, respectively. The standard potential for the reduction process at the cathode Cu^{2+} (aqua) + 2 e^- → Cu (s) is E_0 = 0.337 V. For the two electron process Cu (s) → Cu^{2+} + 2 e^- at the anode, the above equation at room temperature becomes $\Delta E = 0.0295\ V * \log_{10} [C_1 (Cu^{2+}{}_{conc.}) / C_2 (Cu^{2+}{}_{dilute})]$, whereby $\log_{10}$ is the common logarithm to the bases 10. The copper mass plated via $CuSO_4$ can be calculated then by,

$$m_{Cu} = I\,t / (n\,F)$$

Faraday constant F = 96485 As (C)
n =2 electron process
1mol Cu = 63.54 g

At an electrolytic current of 1.5 A applied for 2 minutes the theoretical galvanic deposited mass is 59 mg copper. This plated mass gives a thickness of 33 µm on 10 cm long and 0.4 cm width HTS 2G wire.

The Cu^{2+} concentration profile must adjust to match the mass transfer and reaction rate within the cell geometry.

An alternative way of plating is the use of a non-aqueous alkaline solution instead of acidic $CuSO_4$ aqueous electrolyte. The alkaline electrolytic solution is a mixed copper complex salt and non-poisonous in contrast to the $Cu(CN)_2$ route. We utilized again pulse plating technology to protect the tape. The applied chemical current density in the alkaline complex salt solution is by a factor of 2-3 smaller compared to $CuSO_4$. The Cu layers become fine grained and are mechanically stable on the surface. A comparison of the two electrolytes is given in table 1. While the aqueous Cu sulfate electrolyte shows a 100% electrolytic current transformation the alkaline non-aqueous solution is less effective. On the other hand, the alkaline plating way displays a dense fine grained layer on the coated conductor wire.

Table 1: Acidic Cu sulfate versus alkaline Cu plating electrolyte

Solution	$CuSO_4+H_2SO_4+NaCl$	Cu complex+corrector+KOH
Composition	200 + 10-30+ 0.1 g/l	30 Vol % + 6.5 Vol % + add.
pH value	3.4	9.3
Anode material	Cu + 0.03% P	Cu 99.9 + 0.03 P
Currrent density	40 – 50 mA/cm²	5 – 15 mA/cm²
El. Conductivity	100 mS/cm	30 -50 mS/cm
Deposition rate	0.8 µm@40 mA/cm²	0.25 µm@15 mA/cm²
Current yield	100%	80- 95%

3.4. Operation experience

Dependent on the wire substrate, non- textured Ni-Cr Stainless Steel or textured Ni-W5% tape, the plating tape speed with the 12 electrolytic cells is varied between 5 and 30 m/h. Within a program the current density is increased from 1-3 mA/cm² up to maximum values of 30 – 50 mA/cm² getting copper layer thicknesses of 20 – 30 µm. The total current of each cell should limited to less than 3 Amperes preventing the summarized current on the tape of less than 30 A.

In principle, the plating facility is capable to operate tape length up to 4000 m. The tests of a SS substrate showed a safe and controlled REEL dynamics with speeds up to 100 m/h. The pulse plating technique can be programmed and selected in a wide variety of pulse durations and a fine tuned electrolytic load on the different tape positions.

4. Summary

After more than one year operation the experiences of automated REEL-to-REEL copper galvanic deposition equipment for 4 mm 2G PLD conductors is reported here. The plating unit is tested with speeds up to 100 m/h, and long- length SS substrate of 4 km was operated safe and controlled. The maximum axial stress of the tape is 600 MPa and the narrowest bend radius was 6 mm without degradation of J_c. Both the standard aqueous Cu sulfate electrolyte and an alkaline non- aqueous solution has been used for continuous and pulse plating. The properties and the behavior of the different solutions are compared and evaluated.

References

[1] Floegel- Delor U, Rothfeld R, Wippich D, Goebel B, Riedel T, Werfel, F N, Metallic Cu Coating on HTS Surfaces using Electrochemical preparation, *IEEE Trans. Appl. Supercond. Vol. 11(2001)2838*

[2] Floegel-Delor U, Riedel T, Wippich D, Goebel B, Rothfeld R, Schirrmeister P, Werfel F N, Usoskin A, Rutt, A (2011) Reel-to-Reel Copper Electroplating on Pulse Laser Deposition, *IEEE Trans. Appl. Supercond. 21, 2984*

[3] European *Patent No. 131385, US Patent 7,048,840*

Journal of Physics: Conference Series **507** (2014) 022007 doi:10.1088/1742-6596/507/2/022007

In-field critical current property of IBAD/PLD coated conductors

S Fujita[1], M Daibo, M Igarashi, R Kikutake, K Kakimoto, Y Iijima, M Itoh and T Saitoh

Fujikura Ltd., 1440, Mutsuzaki, Sakura, Chiba, 285-8550, Japan

shinji.fujita@jp.fujikura.com

Abstract. $REBa_2Cu_3O_y$ (RE : rare earth elements, RE123) coated conductors are expected to show high performance in superconducting applications, due to their high mechanical strength and high current density in magnetic fields. Fujikura has developed ion-beam-assisted-deposition (IBAD) and pulsed-laser-deposition (PLD) technique, and today we routinely manufacture coated conductors with length over 500 m and I_c over 500 A/cm-width at 77 K, self field. We have also been able to fabricate long conductors with higher I_c by thickening RE123 layer. Although I_c performances at 77 K, self field are important, but performances in magnetic fields at lower temperatures below 77 K are more important for coil applications. In this work, we evaluated coated conductors with thick RE123 layer in magnetic fields at low temperatures. All samples are fabricated with long length. We measured samples by four-probe transport method at wide temperature range from 77 K to 10 K by using cryo-cooled cryostat. A conductor with 5.5 μm thick RE123 exhibits high I_c values of 937 A/cm-width at 77 K, self field, 637 A/cm-width at 50 K, 5T and 976 A/cm-width at 40 K, 5T.

1. Introduction

High performance $REBa_2Cu_3O_y$ (RE : rare earth elements, RE123) coated conductors(CCs), such as those over a few hundred amperes per centimeter width at 77 K, self field(s.f.) with a few hundred meters length, have become commercialized. Fujikura has been developing the combination of ion-beam-assisted-deposition (IBAD) and pulsed-laser-deposition (PLD) technique and succeeded in fabrication of an over 800 m-length CC with critical current (I_c) over 570 A/cm-width [1]. Recently, quite uniform I_c over 500 A/cm-width with length over 500 m are routinely produced. On the other hand, we have successfully obtained a very high I_c value of over 1000 A/cm-width in a short sample by thickening RE123 layer [2]. Furthermore, not only short samples, 800 A/cm-width-class CC with over 100 m length has also been developed [1]. These high I_c properties, however, are those of at 77 K,s.f. and in-field properties, especially at low temperatures of under 50 K, are actually more important in many applications. In order to prove the practical utility of our high performance CCs, various CCs sampled from long length CCs with various RE123 layer thickness including over 5 μm-thick are evaluated in magnetic fields at low temperatures in this report.

1 shinji.fujita@jp.fujikura.com

2. Experiments

The samples evaluated here are 1 cm-width CCs with configuration of /Ag (6~13 μm) / $GdBa_2Cu_3O_x$: GdBCO (2.0~5.5 μm) / CeO_2 (400 nm) / MgO (5 nm) / Y_2O_3 (20 nm) / Al_2O_3 (150 nm) / Hastelloy(100 μm or 75 μm) /. Ag, Y_2O_3, Al_2O_3 layers were deposited by sputtering and MgO layers were formed by IBAD and CeO_2, GdBCO layers were deposited by PLD respectively. As shown in table 1, all the CCs are fabricated with many hundreds or tens of meters, including some production CCs with over 300 m length (sample A and B), and they have I_c values of over 500 A/cm, max value 937 A/cm in a 5.5 μm-thick GdBCO sample.

Table 1. Specifications of evaluated samples.

Sample designation	GdBCO thickness [mm]	I_c [A/cm-width][a]	Deposition length [m]
A	2.0	513	350
B	2.4	600	500
C	3.0	658	150
D	3.5	656	50
E	4.1	724	20
F	4.5	692	50
G	5.1	801	50
H	5.5	937	110

[a] The I_c values are measured by four-probe transport method at 77 K,s.f. with 1 cm-width short samples cut out from long length CCs.

We made microbridges with ~400 μm-width and ~1 cm-length stripe on each sample by wet etching for I_c measurements at low temperatures. Four-probe transport *I-V* measurements were performed in magnetic fields up to 3 T and temperatures ranging from 77.3 K to 10 K using a 3 T cryo-cooled NbTi superconducting magnet with 200 mm warm bore and a cryo-cooled cryostat with rotatable sample stage; some of the measurements were conducted with a 5T cryo-cooled RE123 magnet we have developed [3]. The electric field criterion for I_c determination was 1 μV/cm. As to sample A, we measured in high magnetic fields up to 17 T using a superconducting magnet at Tohoku University and then made microbridge with ~30 μm-width and ~1 mm-length by photolithograph technique. Measurements of in-high-field were conducted at the temperatures from 20 K to 4.2 K by means of gas or liquid helium cooling. In order to derive I_c per cm-width from the actual measured microbridge $I_c(T,B,\theta)$, we multiplied the ratio of $I_c(T,B,\theta)$ to I_c(77.3 K,s.f) of the microbridge by I_c(77 K,s.f.;1 cm-width); that is the I_c values of table 1, where T is the sample temperature, B and θ are the magnitude and the angle of magnetic fields respectively.

3. Results and discussion

3.1. In-field properties of a production CC

I_c-B characteristics for sample A which is a typical production CC at various temperatures is shown in figure 1. The field angle is parallel to the c-axis. I_c decreases with increasing field, but I_c in low temperatures nonetheless reaches 461 A/cm-width at 50 K,3 T and 1017 A/cm-width at 20 K,3 T. figure 2 shows I_c-θ characteristics for the same sample at 3 T. The field angle is defined as $\theta = 0^\circ$ at B//ab and $\theta = 90^\circ$ at B//c. Focusing on the peak of I_c around B//ab at temperatures below 50 K, the height of peak become higher and the width of peak become broader with decreasing temperature.

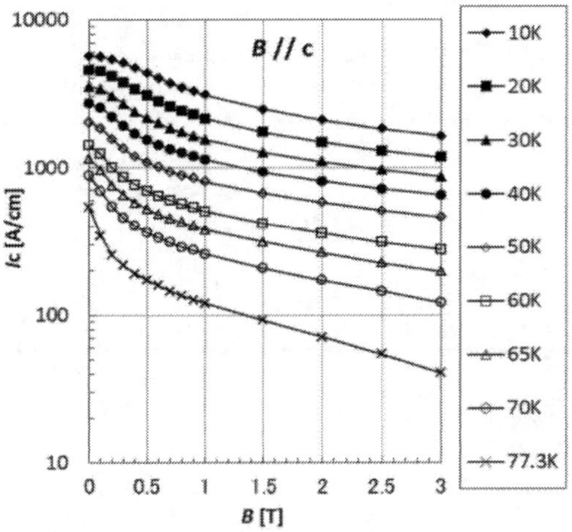

Figure 1. Magnetic field dependence of I_c per cm-width at B//c for sample A.

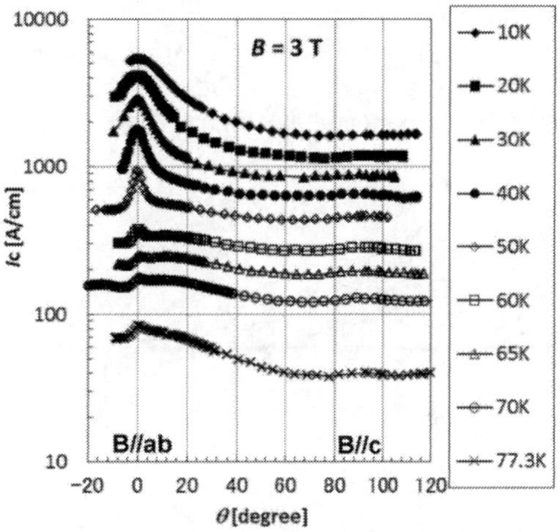

Figure 2. Magnetic field angle dependence of I_c per cm-width at 3 T for sample A.

I_c-B characteristics and I_c-θ characteristics in the high magnetic fields for sample A have successfully been obtained as shown in figure 3 and figure 4 respectively. The I_c values at 17 T, B//c indicate 310 A/cm-width at 20 K and 518 A/cm-width at 4.2 K. The peaks of I_c around B//ab are extremely sharp and the ratio of I_c at B//ab to B//c is approximately 8 irrespective of temperature at least these temperatures.

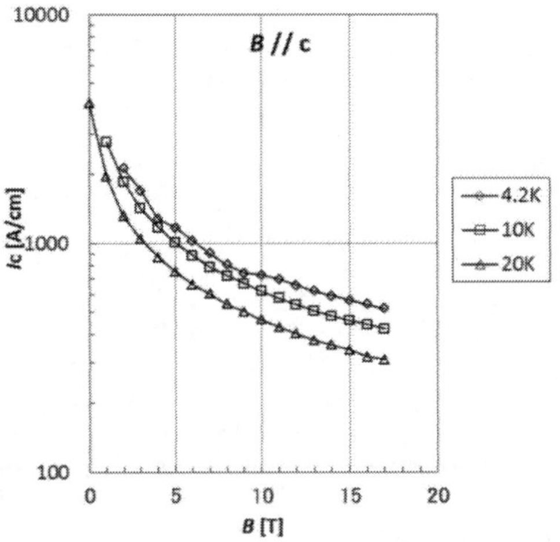

Figure 3. Magnetic field dependence of I_c per cm-width at B//c up to 17 T for sample A.

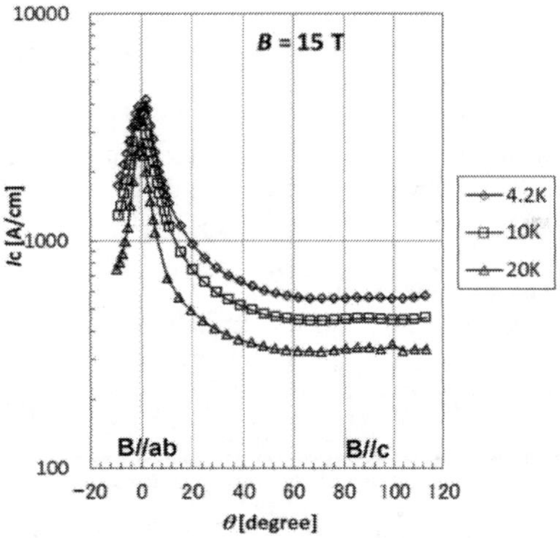

Figure 4. Magnetic field angle dependence of I_c per cm-width at 15 T for sample A.

3.2. In-field properties of a thick-GdBCO CC

The magnetic field dependence of I_c per cm-width for sample H which is the thickest sample in this work, 5.5 μm-thick, is shown in figure 5. This sample exhibits high I_c values of 637 A/cm-width at 50 K, 5 T and 976 A/cm-width at 40 K, 5 T. Furthermore, it shows 2440 A/cm-width at 20 K, 3 T and this is beyond twice the value of sample A.

3.3. GdBCO thickness dependence of in-field I_c

I_c at 20 K, 3 T of all the samples are plotted as a function of GdBCO thickness in figure 6. I_c at 20 K, 3 T tends to be proportional to GdBCO thickness unlike at 77.3K, s.f. . Therefore, it is indicated that in-field I_c at low temperatures increase linearly along with the GdBCO thickness.

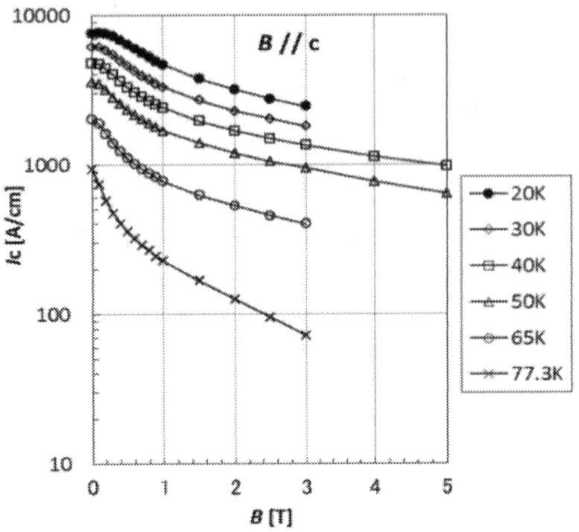

Figure 5. Magnetic field dependence of I_c per cm-width in B//c for sample H.

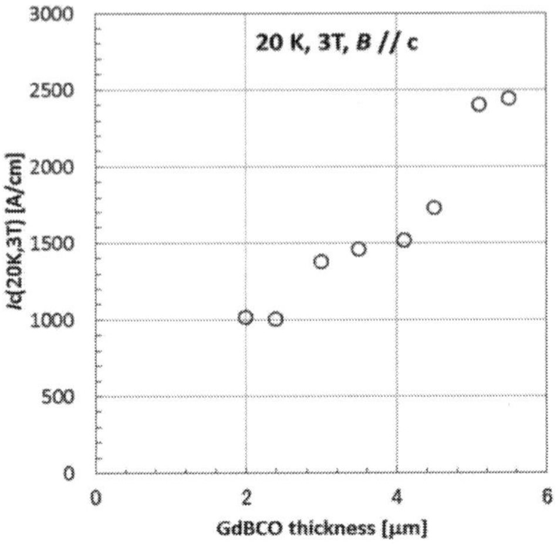

Figure 6. I_c at 20 K, 3T as a function of GdBCO thickness.

4. Summary

We have obtained in-field property of a typical production CC over a wide range of temperatures and magnetic fields. These results are useful for design of practical applications. Moreover, we have measured a very high I_c CC with the thickness of 5.5 μm and obtained high I_c values of 637 A/cm-width at 50 K, 5 T and 976 A/cm-width at 40 K, 5 T. From the measurement results of various GdBCO thicknesses, it is confirmed that I_c at low temperatures tends to be proportional to GdBCO thickness.

5. Acknowledgement

This work includes data measured at the High Field Laboratory for Superconducting Materials, Institute for Materials Research, Tohoku University.

References

[1] M. Igarashi, *et al.*, *Physics Procedia* **36** (2012) 1412.
[2] K. Kakimoto, *et al.*, *Supercond. Sci. Technol.* **23** (2010) 014016.
[3] M. Daibo, *et al.*, *IEEE Trans. Appl. Supercond.* **23** No. 3 (2013) 4602004.

Journal of Physics: Conference Series **507** (2014) 022008 doi:10.1088/1742-6596/507/2/022008

Hysteretic Dependence of Magnetic Flux Density on Primary AC Current in Flat-Type Inductive Fault Current Limiter with YBCO Thin Film Discs

Masayuki Harada, Yasunobu Yokomizu and Toshiro Matsumura

Department of Electrical Engineering and Computer Science, Nagoya University, Furo-cho, Chikusa-ku, Nagoya 464-8603, Japan

E-mail: masayu_h@echo.nuee.nagoya-u.ac.jp

Abstract. This paper focuses on a flat-type inductive superconducting FCL (FIS-FCL) consisting of a pancake coil and a YBCO thin layer disc. AC current injection experiments and magnetic field analysis were carried out for two kinds of FIS-FCL, single-disc model and double-discs model. In the former, the pancake coil was putted on the YBCO disc. In the latter, the pancake coil was sandwiched with two YBCO discs. The double-discs model cancels out the magnetic flux density more effectively than the single-disc model. In the double-discs model, the superconducting state period is longer than in the single-disc model. Thus, it may be concluded that the double-discs model is considered to be suitable for FIS-FCL.

1. Introduction

Recently, by the spread of distributed power supplies, fault current tends to increase, which may cause power system equipment to break. Therefore, the development of fault current limiters (FCL) are expected. We have proposed a flat-type inductive superconducting fault current limiter (FIS-FCL) consisting of a pancake coil with copper wire as the primary winding and a superconducting thin film discs as secondary windings[1]. The small FIS-FCL modules, which were constructed by using discs with yttrium barium copper oxide (YBCO) thin film layer, have been experimentally examined[2]. The FIS-FCL module might be scaled up by piling up for use in practical applications. When a normal load current flows through the primary winding, the magnetic flux generated by primary current is canceled out by that produced by the shielding current induced in the superconducting disc. Thus, the inductance of the FCL is suppressed to be almost zero. On the other hand, when over current such as short circuit current flows through the primary winding, the prospective magnitude of the shielding current induced in the superconducting disc increases more than the critical current level, and then the shielding current has to be limited within critical level. Therefore, the magnetic flux density generated by primary current is no longer canceled out so that the inductance appears in the FCL.

In this paper, in order to understand the periodic behavior of magnetic flux density arising in the FIS-FCL, over current injection experiments were carried out for two different configuration of the FIS-FCL. Furthermore, the magnetic field analysis was performed by using the finite element method to find out the shape of the shielding current induced in the YBCO layer.

2. Experiment

2.1. *Experimental condition*

Fig. 1 shows schematic arrangements of two different structure of the FIS-FCL and the pick-up coil for measuring magnetic flux density. A pancake coil with copper wire as primary winding is putted on the YBCO disc #1 in figure (a) (single-disc model) and sandwiched by two YBCO discs #1 and #2 in figure (b) (double-discs model). The discs #1 and #2 work as secondary windings. In both discs, YBCO thin layer of 185 nm in thickness was coated over sapphire substrate discs of 2 mm in thickness by metal-organic deposition (MOD) method[3].

Fig. 2 shows the experimental circuit. In this paper, AC current of 15 A in peak value was injected into the pancake coil by adjusting the output voltage v_s to be 30 V in peak value. The electro-motive force e was induced in the pick-up coil by AC magnetic flux generated in the FCL. The magnetic flux density B_z at the center of the FCL can be estimated from e as follows:

$$B_z = -10.8 \int edt \tag{1}$$

The coefficient -10.8 was derived by another experiment. In our experiment, waveforms of current i injected into the primary winding and electromotive force e induced in the pick-up coil were recorded with a digital oscilloscope.

2.2. *Measurement result and magnetic flux density*

The over current injection experiments were carried out for two kinds of single-disc model with disc #1 or #2 and double-discs model. The measured waveforms of i, e and B_z are illustrated in Fig. 3. In all models, e hardly appears after the peak point of the injected current. This implies that the change of primary magnetic flux is canceled out by secondary current flowing YBCO layer. The peak value of B_z in the double-discs model is lower than that in the single-disc model.

Fig. 4 (a) and (b) show the instantaneous magnitude of B_z against those of i in both single-disc model and double-discs model. In this figure, the B_z-i characteristic of a standalone primary winding without the YBCO thin film disc is also plotted for the peak value of v_s was 30 V. The gradient of this characteristic represents the inductance of the FIS-FCL. In each model (single-disc model with #1 or #2 and double-discs model), the B_z-i characteristic exhibits hysteresis. The high gradient before current peak is similar to that of the standalone primary winding. Thus, the YBCO layer is considered to be in the resistance-generation state. After

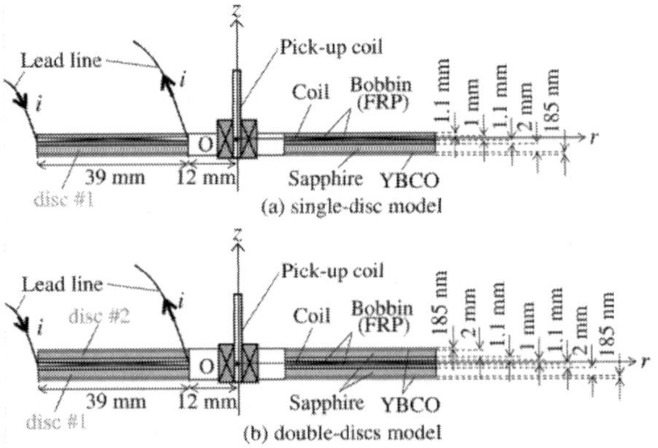

Figure 1. Schematic arrangement of FIS-FCL and pick-up coil.

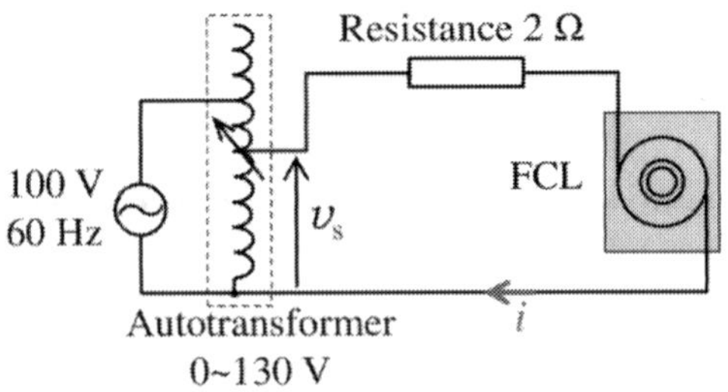

Figure 2. Experimental circuit.

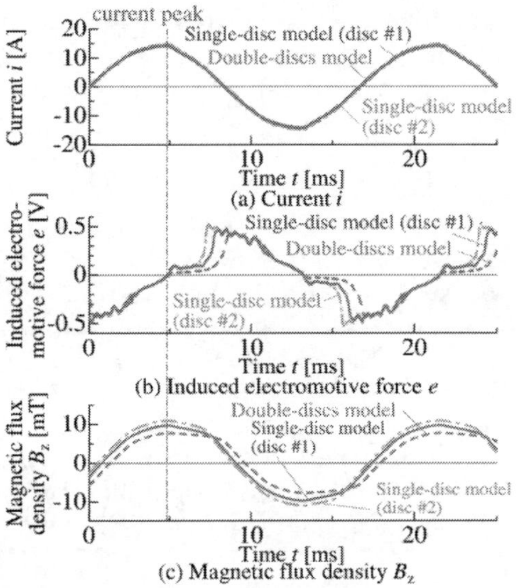

Figure 3. Measured waveforms in over current carrying experiments.

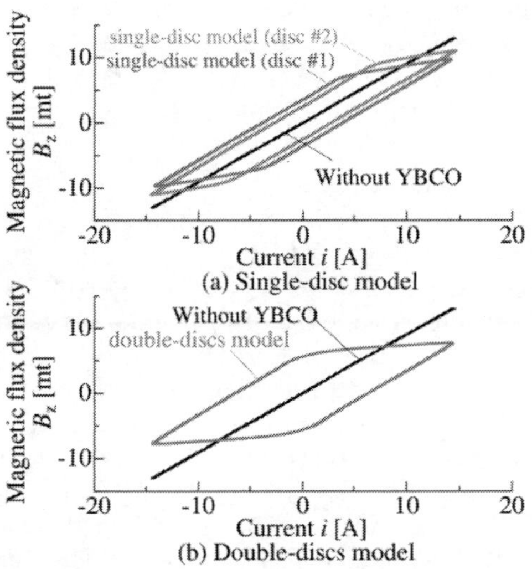

Figure 4. Hysteresis relationship between the magnetic flux density and injected current.

current peak, in contrast, the B_z-i characteristic has a very low gradient. This implies that the inductance of FIS-FCL is much smaller than that of standalone primary winding because YBCO layer is superconducting state so that the change of the primary magnetic flux is canceled out by the shielding current induced in YBCO layer. Comparing these low gradients in all models, it is found that the double-discs model has lower gradient than single-disc model. This means that the double-discs model suppress the inductance to small when the YBCO layers are superconducting state. Moreover, in the double-discs model, the period when the YBCO keeps superconducting state is longer than that in single-disc model.

3. Numerical calculation

3.1. Magnetic field analysis

Magnetic field analysis was carried out for a simplified FIS-FCL with a 10-turn pancake coil (Fig. 5). The governing equation is given from Maxwell's equations and Ohm's law as follows:

$$\frac{\partial}{\partial z}\left(\frac{1}{\mu}\frac{\partial}{\partial z}A_\theta\right)+\frac{\partial}{\partial r}\left\{\frac{1}{\mu}\frac{1}{r}\frac{\partial}{\partial r}(rA_\theta)\right\}+J_0-\frac{1}{\rho}\frac{\partial A_\theta}{\partial t}=0, \tag{2}$$

where J_0, ρ, μ and A_θ are the forced current density, resistivity, magnetic permeability and rotation-direction magnitude of vector potential in the element, respectively. In our calculation, $\mu = 1.25 \times 10^{-6}$ H/m for all elements, $\rho = 0$ Ωm for the elements corresponding to the primary winding, and $\rho = 1.00 \times 10^{18}$ Ωm for the elements corresponding to the spacer and LN_2. The resistivity of YBCO layer ρ_{sc} was determined according to the following equation in the n-value model[4]:

$$\rho_{sc}=\frac{E_c}{J_c}\left(\frac{|J_{sc}|}{J_c}\right)^{n-1}, \tag{3}$$

where J_{sc}, n and J_c is the eddy current density induced in the YBCO layer, n-value and critical current density of YBCO, respectively. For another YBCO sample, n and J_c were measured

40.4 and 9.4 kA/mm^2, respectively. Then these values of n and J_c were used in this calculation. The vector potential A_θ is obtained by solving the governing equation at each element. In this paper, the AC current $i_s = I_m \sin(120\pi t)$ was injected into the primary winding, where peak value I_m was adjusted to 7 A.

3.2. Calculated waveform

Fig. 6 shows the injected current flowing through the primary winding i, total shielding current induced in each YBCO layer i_{sc} and magnetic flux density B_z, respectively. The magnitude of the critical current in the adopted YBCO layer, I_c, is estimated to be 18.8 A, which is the critical current value of the single-disc model. The total critical value of the double-discs model is $2I_c = 37.6$ A. Before current peak, both i_{sc} are limited at the critical current I_c or $2I_c$ and both B_z changes according to i.

After current peak, both i_{sc} begin to change against i and the instantaneous absolute value of i_{sc} is under the corresponding critical current. It means that the YBCO layer recovers to the superconducting state. In double-discs model, as shown in Fig. 6 (b), the saturation is reached for a higher applied field than in single-disc model. As seen in Fig. 6 (c), in double-discs model, B_z is limited smaller than in the single-disc model.

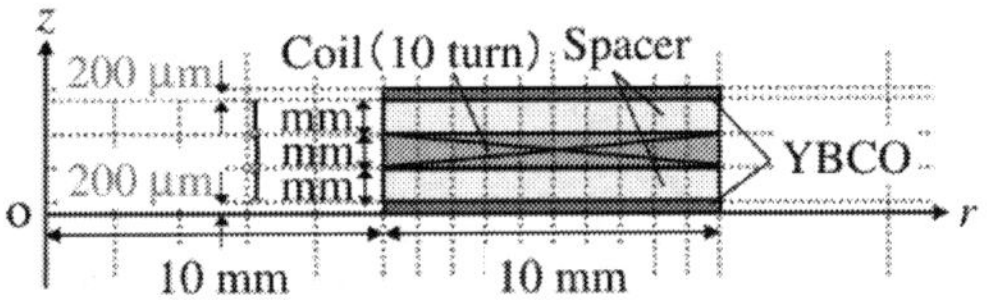

Figure 5. Simplified FCL model for numerical calculation.

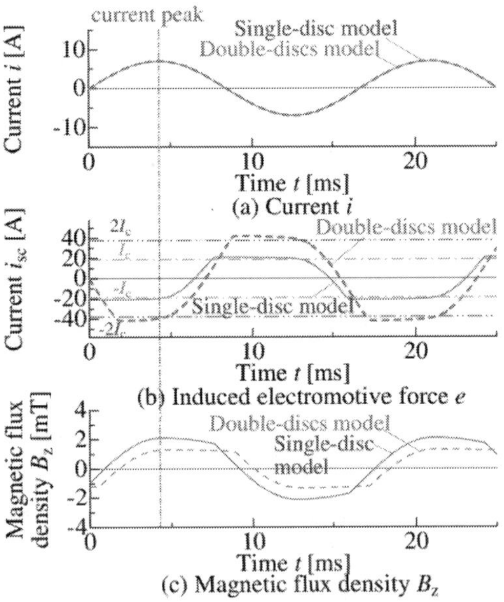

Figure 6. Calculated waveforms of injected current i, shielding current induced in YBCO layer i_{sc} and magnetic flux density B_z.

Acknowledgements

We thank the National Institute of Advanced Industrial Science and Technology (AIST) for producing the YBCO thin film disc adopted in experiments.

References

[1] Matsumura T, Sugimura M, Yokomizu Y, Shimizu H, Shibuya M, Ichikawa M and Kado H 2005 *IEEE Trans. Appl. Supercond.* **15** 2015-18

[2] Higuchi K, Guan Y, Yokomizu Y and Matsumura T 2012 *Physics Procedia* **36** 1254-57

[3] Manabe T, Ahn Jun H, Yamaguchi I, Sohma M, Kondo W, Tsukada K, Mizuta S and Kumagai T 2006 *IEICE Trans. Electron.* **E89-C** 186-90

[4] Ishiyama A, Nakatsugawa J, Noguchi S, Kado H and Ichikawa M 1999 *IEEJ Trans. Power and Energy* **119** 1201-09

Effect of boron particle size on microstructure and superconducting properties of in-situ Cu addition MgB_2 multifilamentary wire

Y. Hishinuma[1], A. Kikuchi[2], Y. Shimada[3], S. Hata[3], T. Takeuchi[2], S. Yamada[1] and A. Sagara[1]

[1] National Institute for Fusion Science, 322-6, Oroshi-cho, Toki, Gifu, 509-5292 Japan
[2] National Institute for Materials Science, 1-2-1, Sengen, Tsukuba, Ibaraki, 305-0047 Japan
[3] Kyusyu University, 6-1, Kasugakoen, Kasuga, Fukuoka, 816-8580 Japan

E-mail: hishinuma.yoshimitsu@nifs.ac.jp

Abstract. In previous studies, the secondary (impurity and non-reactive) phase and voids were observed in MgB_2 matrix after the heat treatment, and then these are the lowering factors of critical current density (J_c) property. In order to improve J_c property by microstructure control of MgB_2 matrix, the fine elemental boron powder as the raw material was carried out using the high-speed vibrated milling with tungsten carbide (WC) jar. The average particle size of metal boron powder was decreased from 1.14 μm to 0.20 μm by the high-speed vibrated milling. The various fine particle boron powders as the function of milling time were also prepared, and in-situ Cu addition MgB_2 multifilamentary wires using these fine boron powders were fabricated. Critical transition temperature (T_c) value of Cu addition MgB_2 wire using fine boron powder obtained to about 37 K. No change of the T_c property by the different particle sized boron powders was confirmed. In this paper, the comparisons of microstructure and superconducting properties between the different boron particle sizes were investigated.

1. Introduction

Discovering of MgB_2 superconducting compound in 2001 [1], many R&D activities of MgB_2 tape and bulk materials were carried out for the practical applications at liquid hydrogen temperature. Because MgB_2 has higher critical transition temperature (T_c) of 39 K compared with the other metallic superconducting compounds. Furthermore, the features of MgB_2 compound are not only higher T_c but also binary chemical composition, lower specific gravity and lower cost. These features will be contributed to the some future electrical power applications. In addition, it is well known that MgB_2 compound is one of the lower induced radio-activation superconducting materials, so-called "Low activation superconductor". Noda *et al.* reported that induced radio-activity of MgB_2 compound remarkably lower than that of Nb-based superconducting materials, and the half-life period of MgB_2 compound was estimated within 1 year based on International Thermonuclear Experimental Reactor (ITER) design [2]. MgB_2 compound will be one of the attractive superconducting materials to use under the higher neutron irradiation and flux environment applications such as future fusion reactor and high energy accelerator device, and then it will be alternative material of Nb-Ti wire in these field

[1] To whom any correspondence should be addressed.

applications. In previous study, remarkable critical current density (J_c) under the magnetic field at 4.2 K was improved by the Cu addition using Mg_2Cu compound and "Low temperature diffusion process" [3]. As the view point of advanced fusion application, microstructure and superconducting properties of Cu addition MgB_2/Ta multifilamentary wire using boron-11 isotope powder, which isotope was stable against neutron irradiation, for the restraint of the nuclear heating and transformation by the nuclide conversion were investigated [4]. T_c value of MgB_2 wire using ^{11}B powder was obtained to be 37 K, which was similar value compared with the MgB_2 wire using natural boron powder. However, we confirmed that J_c value of MgB_2 wire using ^{11}B isotope powder was lower than that of the sample via natural boron powder. In MgB_2 core on lower J_c wire, the much increase of non-reactive boron particle fraction and decrease of current path were observed. A part of J_c-B enhancement on MgB_2 wire by the microstructure control, we performed the boron powder refining using high-speed vibrated milling in order to increase MgB_2 volume fraction with decreasing of non-reactive boron particles. The milling technique is one of the smart methods to improve the J_c-B performance due to the fine particle MgB_2 grain formation [5-7]. In this paper, the comparisons of microstructure and superconducting property on Cu addition MgB_2 wire between the different raw boron particle sizes were investigated.

2. Sample preparations and experimental procedure

Precursor mixture powders were made by metal Mg powder (99.9%), Mg_2Cu compound and natural boron powder. The Cu additional composition of precursor powder was fixed to the optimum 3 at% from the previous results of J_c-B performances [3, 4, 8, 9].

Boron powder refining was carried out using the high-speed vibrated milling. Boron powder and tungsten carbide (WC) ball were set into the WC jar. When boron powder was set into the WC jar, liquid nitrogen was also used in order to drive off the air gas. The milling time was 1, 3 and 5 minutes. The particle size of refined boron powder was measured using the laser diffraction typed particle size distribution meter (HORIBA LA-920), and average particle size was defined as 50 % of a passing portion cumulative distribution (D_{50}). W contamination and B oxidation were evaluated by the Inductively Coupled Plasma-mass spectrometry (ICP-mass) and X-ray diffraction (XRD) analysis.

The precursor mixture powders were packed into metal Ta tubes. Precursor mono-cored wires were fabricated through the PIT process. Wire drawing was carried out using grooved-roller and hexagonal cassette-roller dies, and the precursor mono-cored wires had a hexagonal diameter of 1.90 mm. The prepared mono-cored wire was cut to short piece wires, and 19 pieces of mono-cored wires were stacked into metal Cu tube. The stacked composite was wire drawn to a final diameter of 1.04 mm. Prepared precursor wires were heat treated using "Low temperature diffusion process" which is various lower temperatures (450-550 ℃) during 200 hours under Ar atmosphere [8-10].

T_c value was estimated from magnetization as a function of the temperature using a Quantum Design SQUID magnetometer. The magnetization hysteresis curves as a function of magnetic field under the various temperatures from 4.2 K to 30 K were also measured. The magnetic field from -5 T to +5 T applied to the wire samples. J_c values estimated by the magnetization, so called "Magnetization J_c (J_{cm})", of all samples were calculated by the "Bean-model" based on the critical state model [11]. The longitudinal microstructure was observed using SEM-EDX and TEM-EDX (Lorenz electron microscope (TECNAI-20F)).

3. Results and Discussions

3.1. Change of particle size distribution on the refined boron powder using high-speed vibrated milling

The particle size distributions of the boron powder after various milling time are shown in figure 1. The particle size distribution of boron powder without milling had two main peaks around 1.5 and 0.5 μm, and average particle size was obtained to 1.07 μm. In the case of the refined boron powder using vibrated milling during 1 minute, the particle size distribution was also shown two main peaks around 0.5 and 0.2 μm, and average particle size was obtained to 0.22 μm, which value was corresponded to about 1/5 compared with boron powder without milling. These mentioned that the particle size of boron powder could be reduced by the high-speed vibrated milling. According to the XRD analysis of all refined boron powders, the diffraction peaks of B_2O_3 phase were not observed at all. The liquid N_2 into WC jar was effective to restrain the boron oxidation. Figure 2 shows that change of average particle size and W contamination as a function of the vibrated milling time. The average particle size was reduced remarkably during short time (1min). However, average particle size did not change on milling time (3 and 5 min). This suggested that boron refining did not depend on the milling time and shorter milling time promoted to refine boron powder due to the low temperature brittleness by liquid N_2. On the other hands, W contamination into refined boron powder was increased with increasing of vibrated milling time from ICP-mass results. From the TEM-EDS mapping, however, W concentration was not clearly observed in powder core filament matrix. The high-speed vibrated milling with liquid N_2 was preferable method to refine boron powder.

3.2. Effect of boron particle size on superconducting properties and microstructure of MgB_2 wires using atomized boron powder

Figure 3 shows that magnetization curves of MgB_2 wires using refined boron powder as a function of the vibrated milling time. These wires were sintered at 525 ℃ for 200 hrs. The magnetic moment was normalized by the total volume of MgB_2 core filaments. T_c values of MgB_2 wire using refined boron powder were almost obtained to 36.5 K and they were same to the MgB_2 wire using commercial boron powder. On the other hands, magnetization moment which was normalized by the total core volume was increased by the boron powder refining. We found that the volume fraction of MgB_2 phase was increased by the boron powder refining. This tendency was confirmed on each sintering temperatures (450-550 ℃), and the maximum magnetization moment was obtained to the sintering temperature of 525 ℃. Furthermore, we confirmed that magnetization property was improved by the refined boron powder and the optimum milling time, which was obtained to the maximum magnetization, was 3 min. Based on these M-H hysteresis curves, the comparisons of J_{cm}(B) performances as a function of the temperature was shown in figure 4. In each temperature (5-30 K), the J_{cm} properties of MgB_2 wire using refined boron powder were higher than that of commercial MgB_2 wire (not refined B). H_{c2} values estimated by Kramer plot of refined boron wire were improved.

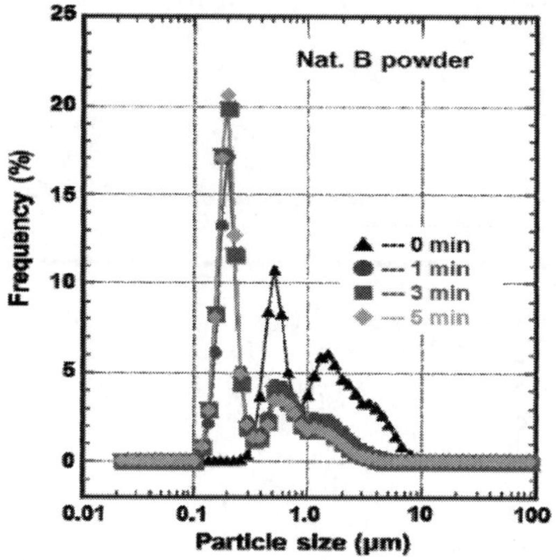

Figure 1. The particle size distributions of the boron powder after various vibrated milling time.

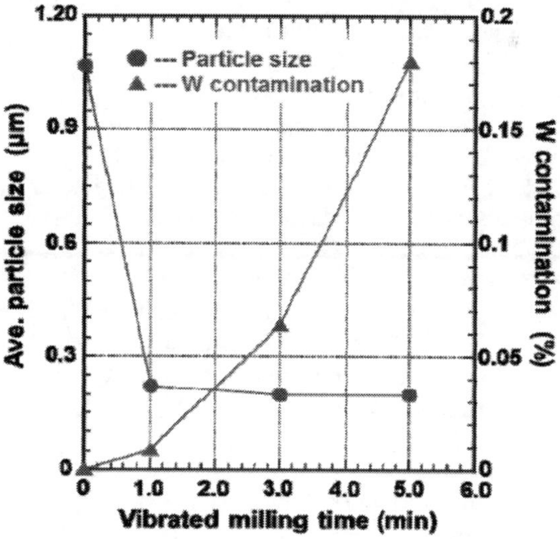

Figure 2. Change of average boron particle size and W contamination as a function of the vibrated milling time.

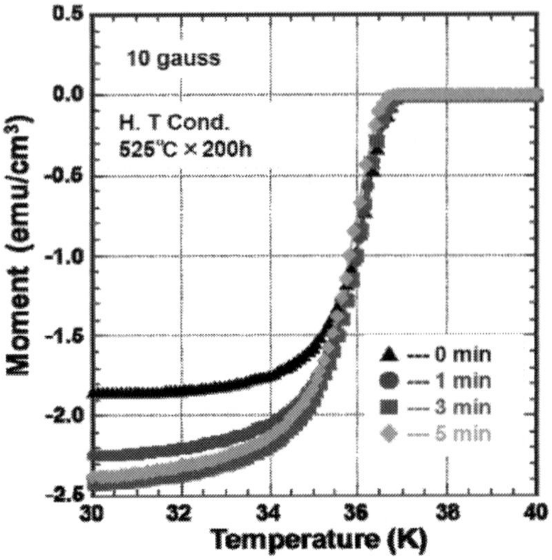

Figure 3. The magnetization curves of MgB_2 wires using refined boron powder as a function of the vibrated milling time.

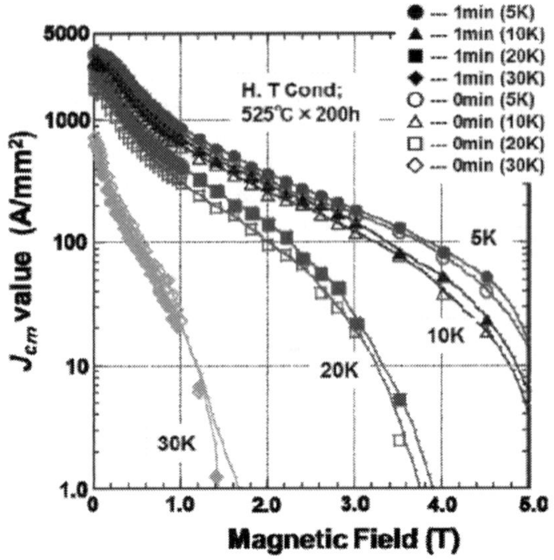

Figure 4. Comparisons of J_{cm}(B) performances of MgB_2 wires as a function of the temperature.

We thought that J_{cm} and H_{c2} properties enhancements were caused by the increase of volume fraction of MgB_2 phase due to the boron powder refining.

4. Conclusions

The high-speed vibrated milling with liquid N_2 was suitable to refine boron powder, and particle size was reduced to 1/5 during short time such as 1 minute. And liquid N_2 was also effective to restrain the boron oxidization and promote boron powder refining due to the low temperature brittleness. The J_{cm} and H_{c2} properties of MgB_2 wire were improved by the boron refining. These were caused by the increase of MgB_2 volume fraction.

Acknowledgments

This work was supported by the Fusion Engineering Research project (UFFF036) in NIFS and JSPS KAKENHI Grant number 25420892. A part of this work was conducted in Institute for Molecular Sciences, supported by Nanotechnology Platform Program (Molecule and Material Synthesis) of MEXT, Japan.

Refferences

[1] Nagamatsu J, Nakagawa N, Zenitani Y and Akimitsu J 2001 *Nature* **410** 63
[2] Noda T, Takeuchi T and Fujita M 2004 *J. Nucl. Mater.* **329-333** 1629
[3] Hishinuma Y, Kikuchi A and Takeuchi 2007 *Supercond. Sci. and Technol.* **20** 1178
[4] Hishinuma Y, Kikuchi A, et al, 2012 Physics Procedia **36** 1486
[5] Xu X, Kim J, et al, 2006 *Supercond. Sci. Technol.* **19** L47
[6] Vignolo M, Romano G, et al, 2008 *IEEE Trans. Appl. Supercond.* **18** 1175.
[7] Senkowicz B J, Giencke J E, et al, 2005 *Appl. Phys. Lett.* **86** 202502
[8] Hishinuma Y, Kikuchi A, et al, 2006 *Supercond. Sci. Technol.* **19** 1269
[9] Hishinuma Y, Kikuchi A, et al, 2007 *IEEE Trans. Appl. Supercond.* **17** 2798
[10] Hishinuma Y, Kikuchi A, et al, 2006 *Fusion. Eng. and Des.* **20-22** 2467
[11] C. P. Bean, 1964 *Rev. Mod. Phys.* **36** 31

Journal of Physics: Conference Series **507** (2014) 022010 doi:10.1088/1742-6596/507/2/022010

Inkjet printing of multifilamentary YBCO for low AC loss coated conductors

S C Hopkins[1], D Joseph[1], T B Mitchell-Williams[1], A Calleja[2,3], V R Vlad[2,3], M Vilardell[2,3], S Ricart[3], X Granados[3], T Puig[3], X Obradors[3], A Usoskin[4], M Falter[5], M Bäcker[5], B A Glowacki[1]

[1]Department of Materials Science and Metallurgy, University of Cambridge, 27 Charles Babbage Road, Cambridge, CB3 0FS
[2]Institut de Ciència de Materials de Barcelona (ICMAB), Barcelona, Spain
[3]OXOLUTIA SL, Parc de Recerca de la UAB, Barcelona, Spain
[4]Bruker HTS GmbH, Ehrichstraße 10, 63450 Hanau, Germany
[5]Deutsche Nanoschicht GmbH, Heisenbergstraße 16, 53359 Rheinbach, Germany

E-mail: sch29@cam.ac.uk

Abstract. Considerable progress has been made with the development of REBCO coated conductors in recent years, and high performance conductors are available commercially. For many applications, however, the cost remains prohibitive, and AC losses discourage their selection for higher frequency applications. Chemical solution deposition (CSD) methods are attractive for low-cost, scalable preparation of buffer and superconductor layers, and in many respects inkjet printing is the method of choice, permitting non-contact deposition with minimal materials wastage and excellent control of coating thickness. Highly textured coatings of YBCO and Gd-doped CeO_2 have previously been reported on buffered metal substrates. Inkjet printing also introduces the possibility of patterning – directly depositing two and three dimensional structures without subtractive processing – offering a low-cost route to coated conductors with reduced AC losses. In this contribution, the inkjet deposition of superconducting YBCO tracks is reported on industrially relevant buffered metal substrates both by direct printing and an inverse patterning approach. In the latter approach, ceria tracks were printed reported, which are a candidate both for resistive filament spacers and buffer layers. TFA-based precursor solutions have been printed on SS/ABAD-YSZ/CeO_2 and Ni-W/LZO/CeO_2 RABiTS substrates, and the resulting multifilamentary samples characterised by microscopy and scanning Hall probe measurements. The prospects for future inkjet-printed low AC loss coated conductors are discussed, including control of interfilamentary resistivity and bridging, transposed filamentary structures and stabilisation material.

1. Introduction

Whilst YBCO and related rare earth (REBCO) coated conductors with high J_c are now available commercially, high cost and AC losses prevent their use in many applications. Hysteresis losses scale linearly with width [1], so for AC applications there is a strong motivation to divide the REBCO layer into filaments. This is often accomplished by laser striation or chemical etching, which adds an additional processing step and is intrinsically wasteful. Striation alone is not sufficient to maximize performance because of electrical and magnetic coupling losses. The former can be addressed by increasing the interfilamentary resistance, encouraging stabilization striation and the use of resistive filament spacers. Ideally, however, full transposition of filaments is also required [2]. As twisting is impractical for coated conductors, this is often addressed at the cabling stage, e.g. with the CORC or Roebel architectures, but a tape-level solution would clearly be beneficial [3].

Chemical solution deposition (CSD) methods are attractive for low-cost, scalable preparation of the buffer and superconductor layers of REBCO coated conductors; and inkjet printing is a particularly versatile approach. Drop-on-demand inkjet printing offers excellent coating thickness control and uniformity, reduced materials wastage and, by virtue of a sealed ink reservoir, fewer issues with solution purity and solvent vapour containment than alternatives such as dip and slot die coating. Furthermore, the high resolution patterning capability exploited for graphics printing could allow superconducting patterns and structures to be deposited directly without subtractive processing [3,4].

Piezoelectric printheads are widely used for both graphics and industrial applications. The small nozzle (10–60 μm) through which ink is ejected is uncapped, so the ink is kept under negative pressure. Very small drops can be produced (1–100 pL) at rates up to 50 kHz, and printheads are available with more than 1000 nozzles, so this technology is suited to high throughput, high resolution deposition. In the low resolution limit, electromagnetic printheads based on solenoid valves is also available. The jetting rate (<1 kHz) and nozzle count are much lower, but the drop volume is larger (>1 nL), so similar throughputs can be achieved. The large (~100 μm) orifice size and positive ink pressure can increase the robustness and maintainable lifetime of the printheads.

The TFA MOD approach has been very widely investigated for YBCO, and it has been successfully implemented by inkjet printing for both continuous and mutifilamentary patterns [3-5]. Oxides suitable for resistive spacers and buffer layers have also been printed, including Gd-doped ceria by the present authors, both as highly textured films [6] and randomly-oriented thicker films.

This article demonstrates two inkjet approaches to producing multifilamentary YBCO coatings on industrially relevant substrates. Firstly, initial tests with direct printing of YBCO filaments on ABAD substrates are reported. Secondly, an inverse approach (based on a previous patent by some of the present authors [7]) is demonstrated on RABiTS, in which untextured CeO_2 filaments are used to template an over-printed YBCO film. The future prospects are then discussed.

2. Methods

2.1. Directly printed YBCO filaments

For direct YBCO printing, a 4 mm wide stainless steel/ABAD YSZ/PLD CeO_2 tape (Bruker) was used. TFA solutions in methanol with a total metal ion concentration of 0.5 mol dm^{-3} (ICMAB) were printed in Cambridge using a 60 μm piezoelectric dispenser (MJ-AT, Microfab). A bipolar drive waveform was designed using quantitative drop visualisation system (Cambridge) to achieve optimal drop formation. Samples were printed on the ABAD substrate after cleaning with acetone and methanol. The printed coatings were processed immediately using a two-stage heat treatment.

2.2. Inverse printing of YBCO filaments

For the inverse printing approach, a 10 mm wide Ni-W/LZO/CeO_2 RABiTS tape was used (Deutsche Nanoschicht). CeO_2 was chosen as the templating material, and a propionic acid and acetate based ink (0.5 mol dm^{-3}) was prepared in Cambridge [6]. The same Microfab dispenser was used to print longitudinal tracks on the substrate following ultrasonic cleaning. A short low temperature heat treatment (1 h, 650 °C) was applied in pure Ar to dry and convert the filaments without crystallising well-textured CeO_2, which is expected to require a higher temperature [6]. Another TFA solution (~0.9 mol dm^{-3}, Deutsche Nanoschicht) was then printed using a 100 μm electromagnetic nozzle (SMLD 300, Gyger). The substrate was fully coated, including the CeO_2 filaments. A two-stage heat treatment was applied, and the resulting sample was characterised by scanning Hall probe microscopy.

3. Results and discussion

3.1. Directly printed YBCO filaments

Optimisation of the Microfab drive waveform for the YBCO ink produced a single drop 97 pL in volume with a velocity of 1.5 m s^{-1} within 0.5 mm of the orifice (figure 1(a)). Track widths varied significantly between substrates cleaned by wiping and ultrasound, figure 1(b) and (c); and tracks one drop wide could be as thin as 20 nm after heat treatment. Multiple depositions (5–10) were necessary for an acceptable thickness, and straight tracks ~200 μm wide could then be achieved, figure 1(d), although occasional breaks along the filament length occurred due to variations in substrate wetting.

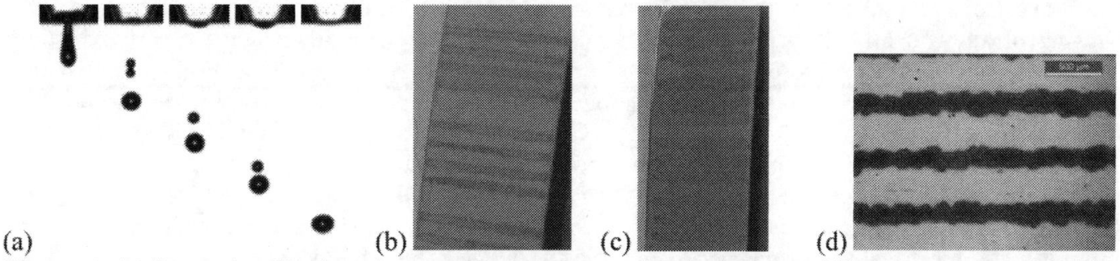

Figure 1. (a) Drop visualisation showing the formation of a single drop of YBCO precursor ink (80 μs intervals); transverse tracks with various drop spacings after (b) wiping and (c) ultrasonic cleaning; and (d) longitudinal tracks with 5 superimposed depositions.

3.2. Inverse printed YBCO filaments

Optimisation of the Microfab drive waveform for the CeO_2 precursor ink resulted in a drop volume of 80 pL and a velocity of 2 m s^{-1}. On the ultrasonically cleaned substrate, the average drop deposited in isolation reached a diameter of 0.25 ± 0.02 mm. Six tracks were printed with a drop spacing of 0.1 mm, producing straight-edged uniform filaments approximately 1.3 mm in width (figure 2).

Figure 2. Photograph of inkjet printed CeO_2 filaments on a RABiTS substrate.

After heat treatment of the CeO_2 precursor filaments, the sample was over-printed with YBCO precursor ink. The electromagnetic jetting parameters were optimised at 400 mbar and 400 μs, producing ~15 nL drops, and a continuous coating was produced with a 1 mm drop spacing. Assuming a dense coating of uniform thickness, the expected thickness after heat treatment is ~230 nm.

A Hall probe scan clearly showed distinct superconducting filaments (figure 3), correlating in position with the channels between CeO_2 filaments, with an approximate width of 0.5 mm.

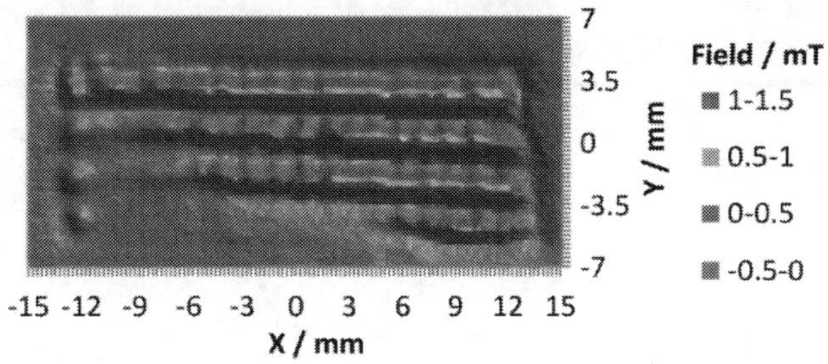

Figure 3. Hall probe scan of an inverse printed YBCO sample showing clearly resolved superconducting filaments.

Estimating the critical current density, J_c, assuming a uniform thickness of 230 nm, yields a value of 0.6 MA cm^{-2} [8]. There are also spatial variations in performance, but in one filament a continuous length of 18 mm exceeds 50% of the maximum J_c for this sample. Whilst the maximum J_c and spatial variations are not yet acceptable for applications, this is a promising result for an initial study.

3.3. Discussion and future prospects

It has been demonstrated that continuous filaments in an acceptable width and thickness range can be produced on ABAD substrates by direct printing of YBCO precursor inks. The large number of deposition stages required is problematic, so future studies should target a higher ink concentration: 1.5 mol dm^{-3} should be attainable. The sensitivity to substrate preparation indicates that further wetting optimisation is required. Crystallographic and superconducting characterisation is now in progress.

The inverse printing approach is promising and versatile. A detailed crystallographic study of the YBCO and CeO_2 regions is required, but the results suggest that locally inhibiting the growth and/or texturing of subsequently deposited YBCO is sufficient to achieve a filamentary superconducting structure. An ink could therefore be selected for templating solely based on rheological and jetting properties. CeO_2 printing is particularly interesting because of the prospect of adjusting texture locally through heat treatment variations and/or selective Gd substitution. Deposited on either the substrate or YBCO, this could form part of the production process for filament transposition structures, and/or for local interfilamentary bridges for reliability in long conductors (at the expense of increased losses) [9]. Electrical transport measurements are in progress for the present samples, but there is clearly potential for tailoring the electrical and/or magnetic properties of inkjet-printed filamentary spacers.

In an all-inkjet RABiTS coated conductor, textured filamentary CeO_2 could also be printed as part of the buffer stack, essentially the inverse of the present approach and analogous to previous reports of striating the buffer stack [10]. Widespread activities in printed electronics also suggest that inkjet printing should be readily applicable to producing filamentary silver stabilisation layers.

4. Conclusions

Filamentary YBCO structures have been prepared by inkjet printing on ABAD and RABiTS substrates. Direct printing on ABAD substrates can produce fine filaments with multiple depositions, but more uniform wetting and an increased ink concentration are required. The feasibility of the inverse printing approach has been demonstrated using printed CeO_2 filaments, and Hall probe characterization has confirmed the production of superconducting YBCO filaments. Both methods could form part of an inkjet route to low AC loss coated conductors and merit further investigation.

Acknowledgements

The authors from the University of Cambridge, ICMAB, OXOLUTIA and Bruker gratefully acknowledge funding from the European Union's Seventh Framework Programme (FP7/2007-2013) under the EUROTAPES project (grant agreement no. 280432).

References

[1] Brandt E H and Indenbom M 1993 *Phys. Rev. B* **48** 12893
[2] Mawatari Y 1996 *Phys. Rev. B* **54** 13215
[3] Glowacki B A and Mosiadz M 2009 *J. Sol-Gel Sci. Technol.* **51** 335-347
[4] Van Driessche I, Feys J, Hopkins S C, Lommens P, Granados X, Glowacki B A, Ricart S, Holzapfel B, Vilardell M, Kirchner A, Baecker M 2012 *Supercond. Sci. Technol.* **25** 065017
[5] Duckworth R C, Paranthaman M P, Bhuiyan M S, List F A and Gouge M J 2007 *IEEE Trans. Appl. Supercond.* **17** 3159
[6] Mosiadz M, Tomov R I, Hopkins S C, Martin G, Hardeman D, Holzapfel B and Glowacki B A 2010 *J. Sol-Gel Sci. Technol.* **54** 154
[7] Bäcker M, Glowacki B A, Falter M and Engell J, Patent DE102006030787A1
[8] Vandaele K, Mosiadz M, Hopkins S C, Patel A, Van Driessche I, Glowacki B A 2012 *Mater. Res. Bull.* **47** 2032
[9] Takacs S 2013 *Supercond. Sci. Technol.* **26** 055022
[10] Majkic G, Kesgin I, Zhang Y, Qiao Y, Schmidt R and Selvamanickam V 2011 *IEEE Trans. Appl. Supercond.* **21** 3297

Measurement of local temperature in tapes and bulks by the Double Thickness Method

[1]V Korotkov, [1,2]E Krasnoperov, [1,3]D Yashkin, and [1]A Kartamyshev

[1]National Research Center Kurchatov Institute, Kurchatov sq., 1, 123182, Moscow, Russia
[2]Moscow Institute of Physics and Technology, 141700, Dolgoprudny, Russia
[3]Moscow State Engineering Physics Institute (MEPhI), Kashirskoe sh. 31, 115409, Moscow, Russia

E-mail: **vasmephi@mail.ru**

Abstract. The average temperature of the heated metal conductor are usually measured by the hot wire method. This method is not applicable to measure the temperature of superconductor. As an alternative we proposed the Double Thickness Method. In DTM two identical elements of superconductor are glued together and miniature thermometer is placed between them. This method allows to measure temperature on superconducting tapes and bulks under the condition of intense cooling (for example, in liquid nitrogen). DTM can be used in AC, DC and pulse heating regimes.

1. Introduction

The heating of superconductor by various thermal disturbances is the most important parameter in the problem of thermal and electromagnetic instability and in the relaxation process. Under adiabatic conditions the heating can be reliably measured by the thermometer attached to the surface. Under liquid and gaseous environment this technique indirectly characterizes the heating due to presence of the temperature jump on the interface with the coolant. This problem can be overcome using specially selected thermal insulating element, which covers the thermometer [1].

In this work the double thickness method (DTM) for the temperature measurement is proposed. In DTM miniature thermometer is placed between two identical elements and it is thoroughly isolated from liquid coolant. If the size of the thermometer, and accordingly the gap between the elements is much smaller than the distance from the thermometer to the cooling medium, the actual internal temperature is maximal in the central region, where there is no temperature gradient.

2. Experimental setup. DTM test experiment

The DTM was tested on the stainless steel tape (width = 6 mm, thickness = 0.18 mm). Scheme of measurements is shown in Fig.1.

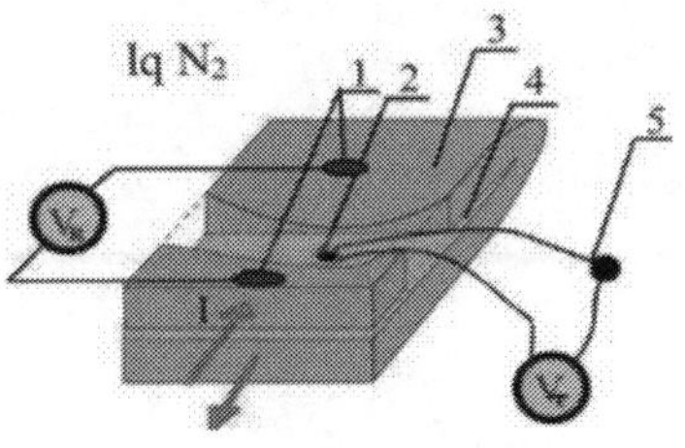

Fig.1. Scheme of measurements by the double thickness method.

The tape resistance was studied preliminary in the temperature range 50-90 K. It is well-known that such calibration allows to use conductor as internal resistive thermometer (hot wire method) [2]. Measured T values were compared to the data of the thermocouple placed between the tapes. The thermocouple was made of copper-constantan wires of 50 μm thickness. The active junction – 2 was set between the tapes and isolated by mica strips of 0.05 mm thickness. Reference joint of the thermocouple – 5 was placed in the liquid nitrogen back from the tapes. The space between the layers - 4 was filled with vacuum sealant Apiezon, which prevents the penetration of nitrogen to the thermocouple. At T = 78 K this design remained stable in currents up to 600 A. The

Journal of Physics: Conference Series **507** (2014) 022011 doi:10.1088/1742-6596/507/2/022011

tape heating was determined by the 4-contact method (voltmeter – V_R).

Because of the weak dependence of the stainless tape resistance on the temperature, the heat dissipation can be considered uniform over the cross section. It is known from the solution of the heat transfer equation that distribution of temperature is parabolic with a maximum θ_0 in the middle between tapes [3] :

$$\theta = \theta_0 - qx^2/2k \qquad (1)$$

Where θ =T - 78 is the relative temperature, k is thermal conductivity (k = 8 W/(m·K) for stainless steel), q – power heat released in unit volume. Boundary conditions $\theta(-d)= \theta(d)= \theta_b$ determine the maximal temperature $\theta_0 = \theta_b+qd^2/2k$. Temperature measurements by both methods during DC heating of the tape are shown in Fig. 2.The internal thermometer registers the average temperature $<\theta>$, but thermocouple - the maximum temperature θ_0 between the tapes. Difference between the readings of

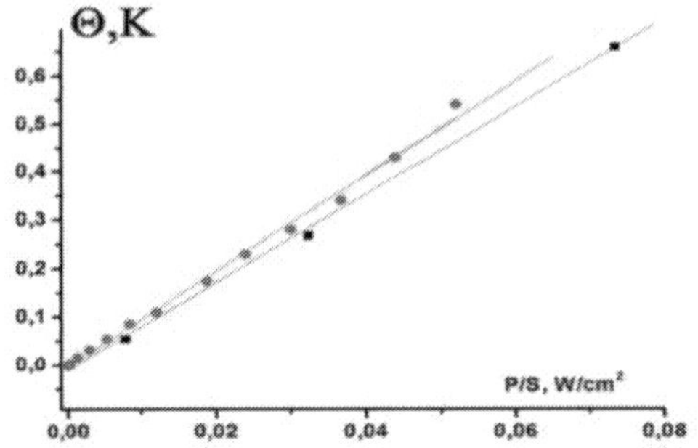

Fig.2. The temperature of the stainless tape under DC heating in liquid nitrogen vs the dissipated power P/S. Circles (red) - thermocouple, squares (black) – internal thermometer.

thermometers increases with increasing heat dissipation $\theta_0 - <\theta> = qd^2/3k$, and for P/S=5 W/cm² (P/S=q*d) it is $\theta_0 - <\theta> =$ 0.2 K. In the region of low heat generation, as in Fig. 2, difference in the readings is not noticeable.

3. Temperature measurements on stainless tape during pulse heating

Low heat capacity of the thermocouple and short measurement time of modern voltmeters allows using DTM in the pulse measurements. In particular, cooling process of stainless steel tape and heat capacity were measured. Current pulses of duration 3 ms were realized by discharging the condenser bank with capacity of 10 mF, the temperature of the thermocouple was measured by the nanovoltmeter Keithley 2181. Fig. 3 shows the temperature measurements by DTM after the current

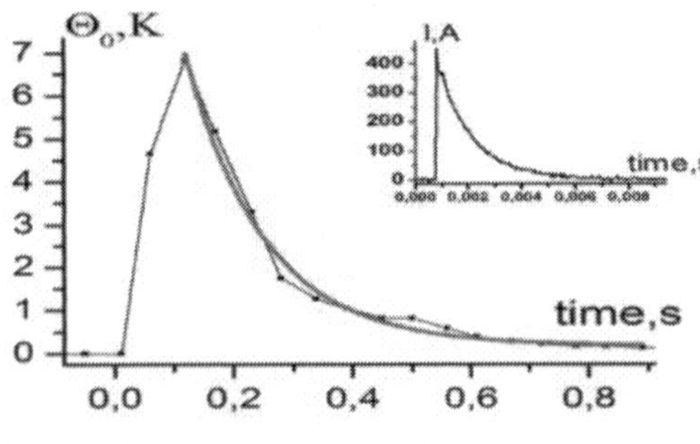

Fig. 3. Temperature in the gap between tapes after the pulse heating. Current pulse (in the inset) at t=0 has a duration of 3 ms and indistinguishable in sec time scale.

pulse passing. Some time after the pulse (less than 0.1 seconds) the graph characterizes the heating of the gap between tapes, including a thermocouple, and voltmeter's time recovery.

The falling part of the curve describes the tape's cooling by the liquid nitrogen. Reduction of the temperature can be approximated by an exponential law $\theta(t) = \theta_0 \cdot \exp(-t/\tau)$, where θ_0 – temperature jump in the gap, which is obtained by extrapolation to t = 0, τ – cooling time parameter. In Fig. 3 it is seen that cooling time on 2-3 orders of magnitude more than the pulse duration. Therefore, one can conclude that the heating happens under adiabatic conditions. In this case temperature jump and heating energy Q are related by

$$\theta_0 = Q/C \qquad (2)$$

where C - specific heat of the tape. This ratio was used for determine C. In Fig. 4 the dependence of temperature jump on the pulse energy is shown. In the initial part of the curve, where this function is linear, heat capacity, calculated from (2) as $C = \lim (\Delta Q/\Delta\theta_0)$ is equal C = 0.5 J/K.

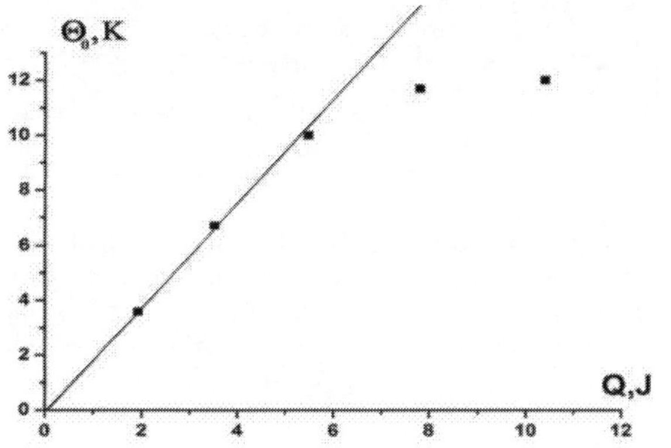

Fig . 4 Maximal temperature of pulse heating as a function of dissipated energy.

This value on 25 % higher than data published in [5]. Difference arises due to the heat capacity of the gap filler, which has a thickness comparable with the thickness of the tape. DTM is very promising for quick and easy estimating heat capacity of tape.

4. Double thickness method for bulk superconductors

The measurement of the local temperature is important during the process of pulsed field magnetization of bulk superconductors [4]. DTM was applied to temperature distribution measurements on the superconducting annuli in the process of PFM in liquid nitrogen. As in the case of tapes, the measurement joint was placed between two melt-grown annuli. Magnetization was realized by the pulsed method using copper coils placed outside the annuli. The pulses duration was of 40 ms, pulse amplitude of H_a = 0.4 T, 0.7 T, 0.8 T.

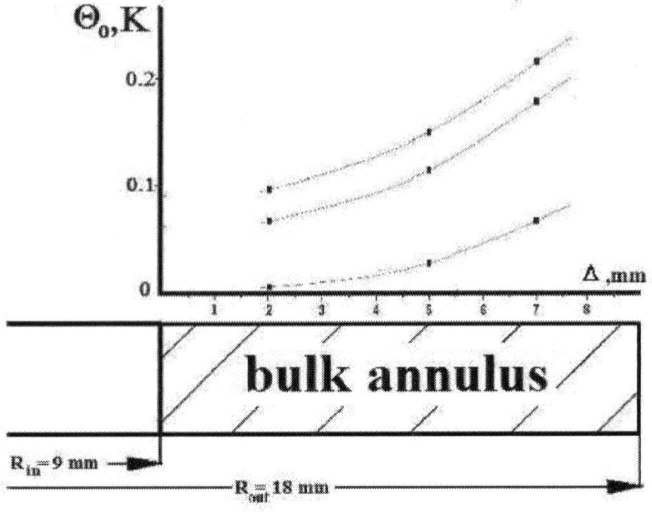

Fig. 5. Temperature distribution along the radius of superconducting annuli. Pulse amplitude from top to bottom: H_a = 0.8 T, 0.7 T and 0.4 T.

Temperature distribution along the radius of the annulus for different H_a is shown in Fig. 5. The maximum temperature difference between the outer and inner edge of the annuli is approximately 0.25 K (at H_a = 0.8 T). The average heating temperature at H_a = 0.8 T is equal to 0.17 K. This value is consistent with the data in [6], for tablet of the same diameter after pulse heating with field amplitude H_a = 1.0 T and pulse duration of 40 ms.

5. Conclusion

The Double Thickness Method is a promising method of temperature measurement on bulk and tape superconductors. The DTM was compared with hot wire method on the stainless steel tape. In pulse heating regime the DTM can be applyed to estimate heat capacity. With DTM the temperature gradient in pulsed magnetized superconducting annuli was measured in liquid nitrogen.

6. References

[1] Ashworth S P, Suenaga M 2001 *Cryogenics* **77** p 412
[2] Healy J, de Groot J and Kestin 1976 *Physica B+C* **82** (Issue 2) pp 392–408
[3] Lienhard J H V and Lienhard J H IV 2011 *"Heat Transfer"*, Dover Publications, p 23
[4] Fujishiro H, Naito T *et al* 2010 *Physica C* **470** pp 1181–1184
[5] Iwasa Y 2005 *"Case studies in superconducting magnets"*, p 634
[6] Yanagi Y *et al* 2005 *Supercond. Sci. Technol.* **18** p 839

The effect of Y_2O_3 and $YFeO_3$ additions on the critical current density of YBCO coated conductors

M Lao,[1] M Eisterer,[1] O Stadel,[2] A Meledin,[3] G van Tendeloo[3]

[1]Atominstitut, Vienna University of Technology, Stadionallee 2, 1020, Vienna, Austria
[2]PerCoTech AG, Bienroder Weg 53, 38108 Braunschweig, Germany
[3]EMAT Research Group, University of Antwerp, Groenenborgerlaan 171, B-2020, Antwerp, Belgium

E-mail: mlao@ati.ac.at

Abstract.
The pinning mechanism of MOCVD-grown YBCO coated conductors with Y_2O_3 precipitates was investigated by angle-resolved transport measurement of J_c in a wide range of temperature and magnetic fields. Aside from the Y_2O_3 nanoprecipitates, *a*-axis grains and threading dislocation along the *c*-axis were found in the YBCO layer. The Y_2O_3 precipitates are less effective pinning centers at lower temperature. The tapes with precipitates show a higher anisotropy with larger J_c at $H \parallel ab$ than $H \parallel c$. This behavior was attributed to the preferred alignment of the nanoprecipitates along the *ab*-plane.

1. Introduction

The second generation high-temperature superconducting wire based on YBCO ($YBa_2Cu_3O_{7-\delta}$) grown-epitaxially on metallic substrates has shown a great promise for high-current transmission applications. The presence of vortex pinning sites either in the form of dislocations, twin boundaries, stacking faults or the addition of artificial pinning centers such as nanoprecipitates and columnar defects was shown to greatly enhance J_c especially at high magnetic fields. The addition of random pinning centers has attracted much attention since they are sources of isotropic pinning in the superconductor. One of the most common forms of random pinning center are Y_2O_3 precipitates which greatly enhance J_c of YBCO tapes [1, 2, 3, 4]. Some studies explored the effect of adding magnetic nanoparticles such as $YFeO_3$ that also enhanced J_c without degrading the superconducting characteristics of the films [5]. Another important aspect is the deposition process for the production of YBCO coated conductors, such as pulsed laser deposition (PLD), chemical solution deposition (CSD) and metal-organic chemical vapor deposition (MOCVD). Each deposition process leads to a particular formation of pinning centers which cause variations in the behavior of J_c at different temperatures and fields. The actual effects of the nanoprecipitates and correlated defects formed through these deposition processes are still to be determined in a broader range of temperatures and fields.

The MOCVD technique offers a high potential for large-scale production of superconducting tapes. The cheapest possible way to produce high-efficiency wires has to be found. A combination of the MOCVD and CSD approach to grow YBCO and oxide buffer layers on textured metal substrates such as NiW is one of the low-cost alternatives to high-vacuum deposition techniques [6].

In this study, the effect of Y-based precipitates on J_c of MOCVD-grown YBCO on a MOCVD $(Y,Ce)O_{1.5+x}$ and CSD-grown $La_2Zr_2O_7$ (LZO) buffer layer on a magnetic Ni5%W substrate will be investigated. The microstructure formed in the superconducting layer will be related to angle-resolved measurements of J_c at fields up to 5 T and temperatures down to 20 K to provide a more concise view of the pinning contributions in these specific tapes.

2. Experimental details

The oxide buffer layer $(Y,Ce)O_{1.5+x}$ and YBCO layer were grown via a MOCVD reel-to-reel system in PerCoTech [10] on a Ni5%W substrate with a 140 nm CSD-grown LZO buffer layer. The YBCO layer of the first tape was deposited in two-pass process to form a total thickness of 700 nm. A 10 mol% of Y_2O_3 was added to the second tape and the third tape contains 1 mol% of $YFeO_3$. Both tapes were grown in a single-pass process. The estimation of the $YFeO_3$ and Y_2O_3 additions are based on the Fe and Y content in the precursor solution. The tapes will be denoted as YBCO, YO and YFO, respectively. The critical temperature of the YFO tape was found to be 1.3 K lower than YBCO which indicates that iron was incorporated in the superconducting layer. The characteristics of the tapes are summarized in Table 1. All the tapes have a width of 10 mm.

Table 1. Summary of the PerCoTech tape characteristics.

Name	Addition	Thickness (nm)	$T_{c,onset}$ (K)	I_c (A), 77 K 10mm-width	Bridge dimension (mm^2)
YBCO	-	700	90.5	85	0.45×2.40
YO	10 mol% Y_2O_3	380	89.7	48	0.65×3.25
YFO	1 mol% $YFeO_3$	450	89.2	47	0.55×3.65

Bridges were patterned on the tapes for measurements at lower temperature. The dimensions are given in Table 1. Transport measurements were made at magnetic fields of up to 5 T. The measurements at 77 K were made with the 10-mm tapes and the bridged tapes were used for measurements at temperatures from 55 K down to 20 K. The four-probe method was used with 1 $\mu V\ cm^{-1}$ criterion. All J_c-anisotropy measurements were made under maximum Lorentz force configuration.

The microstructure of the tape with $YFeO_3$ addition was investigated by TEM using a FEI Tecnai G2 electron microscope operated at 200 kV, a Philips CM30 electron microscope operated at 300 kV and a FEI Titan electron microscope operated at 120, 200 and 300 kV.

3. Results and Discussion

The self-field J_c of the three tapes is 1.21, 1.26 and 1.04 MA cm^{-2} for the YBCO, YO and YFO tape, respectively. Figure 1a shows angle-resolved measurements of J_c at 77 K for the YBCO tape which shares a similar behavior with the other two tapes. An *ab*-peak is observed and a small *c*-axis peak appears at intermediate fields. Comparing the $J_c(\theta)$ of the three tapes in figure 1b, the YO tape has the highest J_c along $H \parallel ab$ followed by YFO. Interestingly, the J_c of the tapes with the additions is only large at $H \parallel ab$ but the values of the J_c converge at $H \parallel c$. This is quite unexpected since previous studies have found that Y_2O_3 additions are isotropic pinning centers in PLD-grown [2] and MOD (Metal Organic Deposition)-grown [1] YBCO tapes. Therefore, a decrease in anisotropy and increase in J_c should be found in the whole angular range.

According to STEM data, the YBCO layer of the YFO sample contains dislocations parallel to the *c*-axis such as the areas marked by the horizontal arrows in figure 2a-b. Such dislocations commonly form in MOCVD-grown tapes [7, 8]. We also observe YBCO *a*-axis grains marked by

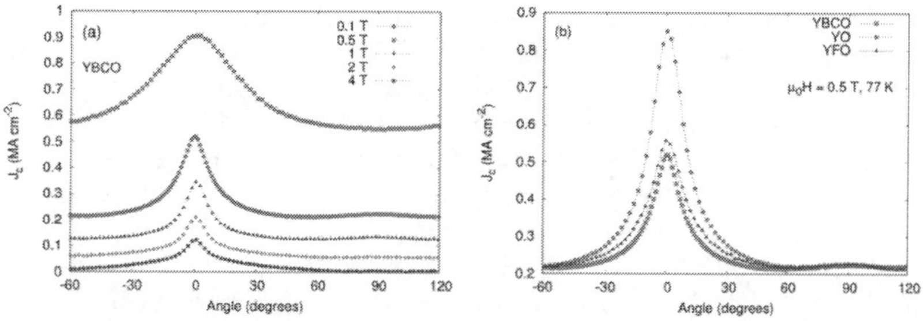

Figure 1. (a) Angular dependence of J_c at 77 K for different fields. (b) Comparison of the three tapes at 0.5 T.

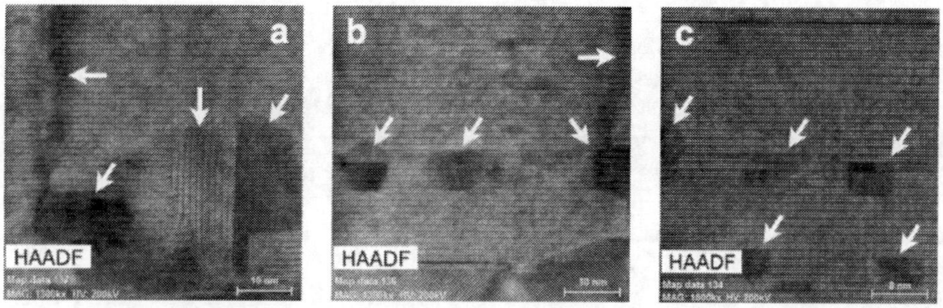

Figure 2. HAADF-STEM images of various parts of the YBCO layer showing dislocations, Y_2O_3 nano precipitates and a-axis grains.

the vertical arrow in figure 2a and Y_2O_3 nanoprecipitates with a diameter of 8-10 nm. Larger particles such as CuO and YCuO grains (100-200 nm in size) were also found by TEM. According to the EDX (Energy-dispersive X-ray spectroscopy) maps, $YFeO_3$ precipitates were not formed but the iron is homogeneously distributed within the YBCO layer and the precipitates consists again of Y_2O_3. This indicates that Cu was partially substituted by Fe in YBCO and the excess Cu formed the CuO and YCuO grains. Therefore, we can deduce that the same type of precipitates is common to YO and YFO tapes and they only vary in amount since the YFO tape contains 1 mol% addition while 10 mol% Y_2O_3 was added to the YO tape.

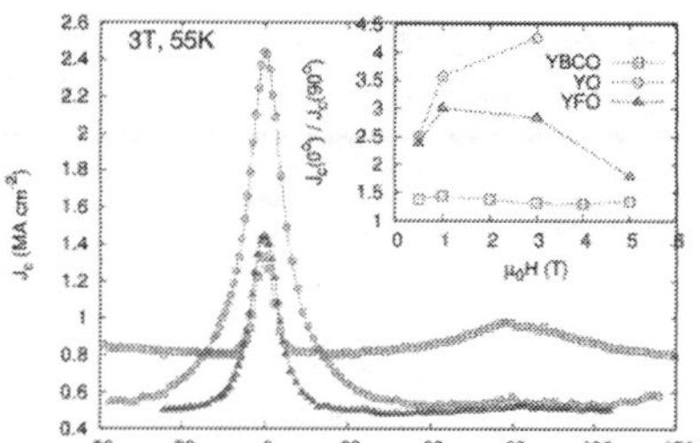

Figure 3. Comparison of $J_c(\theta)$ of the three tapes. The inset shows the ratio of J_c between $H \parallel ab$ and $H \parallel c$.

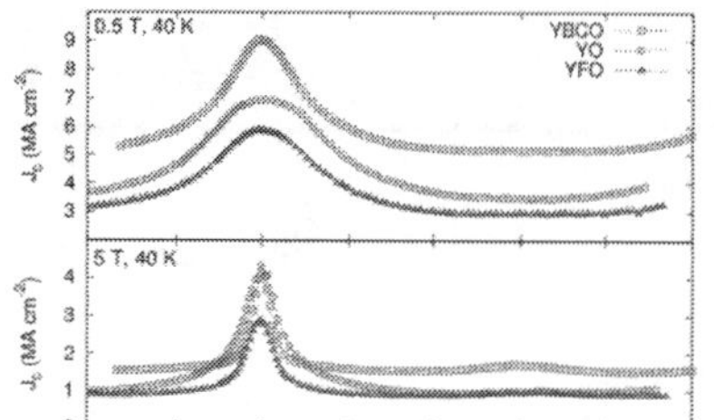

Figure 4. Angular dependence of J_c in the three tapes at 40 K, 0.5 T (top) and 5 T (bottom).

In the measurement at 55 K shown in figure 3, the c-axis peak of the standard YBCO was found to be more prominent than in YO and YFO tapes. We conclude from the TEM data that the appearance of the c-axis peak may be related to the vortices pinned by the threading dislocations oriented along the c-axis. The standard YBCO was deposited by a two-pass process, however, previous studies showed that the c-axis oriented dislocations are formed continuously over the whole layer thickness in multi-pass processes of the MOCVD technique [8]. The YBCO is also the least anisotropic of the three tapes as shown by the inset in figure 3 in contrary to the unexpectedly strong anisotropy of YFO and YO tapes. The large difference between J_c at $H \parallel ab$ and $H \parallel c$ of the two tapes may be related to the aligned formation of the Y_2O_3 nanoprecipitates parallel to the ab-planes as shown in the TEM image in various parts of the YBCO layer (figure 2b-c). Thus, pinning by these precipitates is strongest along the ab-planes. In $H \parallel c$, only a small section of a vortex is pinned by a precipitate. Also, extra energy arises from the meandering of the vortex lines between the precipitates in other direction not parallel to the ab-planes, thus, pinning is weaker along these directions. The formation of these precipitates is not random in contrast to the precipitates formed by the PLD and MOD techniques. This correlated formation of the nanoparticles was also observed for MOCVD-grown (Y, Sm)$_1$Ba$_2$Cu$_3$O$_y$ films where the (Y, Sm)$_2$O$_3$ particles are arranged in layers and tilted by around 7° from the interface [7, 8]. Although the tilt was not observed in our data, this preferrential formation of precipitates was found to be related to MOCVD deposition process.

The Y_2O_3 nanoprecipitates are no longer effective pinning sites over the whole angular range at 40 K and low fields (3 T and below). The same is also observed at 20 K where $J_c(\theta)$ of YFO and YO tapes is much smaller than in sample YBCO at all fields. Also at this temperature, intrinsic pinning dominates at $H \parallel ab$ in all the tapes, as expected at low temperatures.

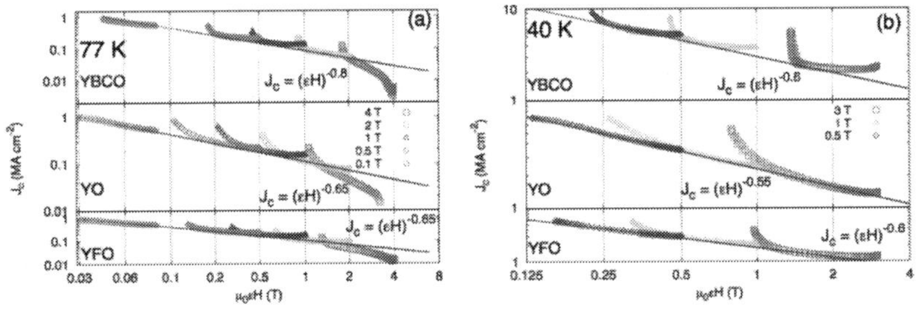

Figure 5. Anisotropic scaling analysis of the three tapes at (a) 77 K and (b) 40 K.

The effective anisotropy γ obtained from anisotropic scaling [9] is 2.2, 3.8 and 3.1 for the standard YBCO, YO and YFO tapes, respectively. The values for γ are smaller than the mass anisotropy in YBCO (γ = 5-7) but close to findings for MOCVD grown tapes [7]. Using $J_c = [\varepsilon(\theta)H]^{-\alpha}$, the determined α decreases at lower temperature. The decrease in α indicates an increase in the density of pinning sites, as smaller defects start to serve as pinning centers when their pinning energy exceeds the thermal energy at low temperature.

4. Conclusion

Y_2O_3 nanoprecipitates were successfully introduced into the YBCO layer of MOCVD grown YBCO tapes with MOCVD-grown (Y, Ce)O$_{1.5+x}$ and CSD-LZO buffer layers on a magnetic Ni5%W substrate. Aside from the Y_2O_3 precipitates, threading dislocations along the c-axis and a-axis grains of YBCO were found by TEM. The Y_2O_3 precipitates contribute to pinning at higher temperatures and hence enhance J_c. However, their effectivity decreases at lower temperatures and lower fields. These precipitates contribute significantly to pinning when $H \parallel ab$ and less for $H \parallel c$. This behavior was attributed to the preferred alignment of the precipitates parallel to ab-planes. Threading dislocations along the direction of the c-axis in these MOCVD-grown tapes are related to the c-axis peak in $J_c(\theta)$. Intrinsic pinning at $H \parallel ab$ becomes dominant at lower temperature.

Acknowledgment

The research leading to these results has received funding from the European Union Seventh Framework Programme [FP7/2007-2013] under grant agreement n° NMP-LA-2012-280432.

References

[1] Yamasaki H, Yamaguchi I, Sohma M, Kondo W, Matsui H, Manabe T and Kumagai T 2012 *Physica C* **478** 19-28
[2] Mele P, Matsumoto K, Horide T, Ichinose A, Mukaida M, Yoshida Y and Horii S 2007 *Supercond. Sci. Technol.* **20** 616-20
[3] Sparing M, Backen E, Freudenberg T, Hühne R, Rellinghaus B, Schultz L and Holzapfel B 2007 *Supercond. Sci. Technol.* **20** S239-46
[4] Haugan T, Barnes P, Wheeler R, Meisenkothen F and Sumption M 2004 *Nature* **430** 867
[5] Wimbush S, Durrell J, Tsai C, Wang H, Jia Q, Blamire M and MacManus-Driscoll J 2010 *Supercond. Sci. Technol.* **23** 045019
[6] Muydinov R, Stadel O, Falter M and Bäcker M 2012 *Physics Procedia* **36** 1468-74
[7] Chen Z, Kametani F, Chen Y, Xie Y, Selvamanickam V and Larbalestier D 2009 *Supercond. Sci. Technol.* **22** 055013
[8] Holesinger T, Maiorov B, Ugurlu O, Civale L, Chen Y, Xiong X, Xie Y and Selvamanickam V, 2009 *Supercond. Sci. Technol.* **22** 045025
[9] Civale L, Maiorov B, Serquis A, Willis O, Coulter J, Wang H, Jia Q, Arendt P, MacManus-Driscoll J, Maley M and Foltyn S 2004 *Applied Physics Letters* **84** 12
[10] Stadel O, Muydinov R 2009 *IEEE/CSC & ESAS European Superconductivity News Forum (ESNF)* 9

Analysis of the anisotropic critical current behaviour of HTS coated conductors

P M Leys, M Klaeser, F Schleissinger and Th Schneider

Karlsruhe Institute of Technology, Institute for Technical Physics, Hermann-von-Helmholtz Platz 1, 76344 Eggenstein-Leopoldshafen, Germany

E-mail: pauline.leys@kit.edu

Abstract. In order to generate fields of more than 20 T using pure superconducting magnets, the implementation of HTS coated conductors (CCs) is an option. Amongst the characteristics of the REBCO is its intrinsic anisotropy which means its critical current, I_c, varies as a function of its orientation with respect to the external magnetic field, Φ, as well as being dependent on external field, B, and temperature, T. For the design of the high field insert coils for the experimental magnet facility HOMER II, the behaviour of I_c at fields of about 24 T and at Φ approximately 80 ° is required. Within our JUMBO facility it is possible to measure the $U(I)$ behaviour of commercial CCs at liquid helium and liquid nitrogen temperatures. The free bore available is 100 mm at 4.2 K; the applied magnetic field can be varied up to 10 T, and Φ can be set between 0 and 180 °. Measurements were carried out on commercially available CCs with emphasis on Φ values near to 90 °. The critical current was determined by means of a power law function fitted to the resulting $U(I)$ curves using specially written Matlab® programs. A number of different three dimensional functions were fitted to the measurement data and compared with each other.

1. Introduction

REBCO high temperature superconductor (HTS) coated conductors can be used for energy applications at 77 K and, due to their excellent low temperature current carrying capabilities at high fields, they are ideal candidates for operation at 4.2 K. Having studied the dependence of the critical current, I_c, of an HTS on the angle, Φ (defined as the angle between the magnetic field and the normal vector to the conductor tape) at different magnetic fields and temperatures in a previous paper [1], we wanted to further develop our analysis of mathematical models or functions that would fit the three dimensional surface plot of $I_c(B, \Phi, T)$ measurement data at both 4.2 K and 77 K.

All high field insert coils have a maximum radial field, B_r, component at their outer edges where the angle Φ is greatest. For an optimal design we need to know the I_c anisotropy for each conductor to be used in developing our HOMER II facility, NMR projects and all other high field applications at 4.2 K. Anisotropy is also an important feature at 77 K. Here the B_r component must be considered in project design as well, but in this case, due to the strong $I_c(B)$ decrease, only low fields are considered. Therefore the aim of this paper is to present an analysis and comparison of the critical current data for the HTS samples at both liquid helium and liquid nitrogen temperatures.

2. Experimental Details

The conductor used in the experiments was a coated conductor from the company SuNAM [2]. We tested two conductor types: type I with copper plating and type II with copper plating and brass lamination. The conductor width was approximately 4.1 mm and its thickness 0.11 mm (type I) or

0.19 mm (type II). The I_c values are 130 - 190 A at 77 K and 1160 – 1230 A at 4.2 K self-field, dependent on conductor type. Measurements were made across the complete conductor width of 4.1 mm for the two conductor samples at 4.2 K and at 77 K and the angle Φ was adjusted between 0 – 180 °. At both temperatures the experiments followed the same procedure as described in [1] with more measurements at Φ = 76 -104 ° in steps of $\geq$ 2 °, as this range is most relevant to magnet design. Each set of measurements was made with a different conductor sample.

At 4.2 K we measured the voltage-current (U(I)) characteristics [1] for the samples at fields from 3 - 10 T. Measurements at fields below 3 T were not made due to the high critical current at low fields, and also our area of interest at this temperature is high field operation. At 77 K we measured U(I) in fields from 0 – 4 T with steps of 0.1 T being made between 0 and 1 T in consideration of the usual operational range at this temperature.

3. Theoretical Background

The intrinsic anisotropy of HTS CC's leads to an angular dependency of the critical parameters, in particular I_c and n [3]. $I_c(B, \Phi, T)$ was determined from measured $U(I)$ curves by means of a power law fit with E_c defined as 0.1 μV/cm.

The first mathematical model considered for the dependence of I_c on B was that according to [3] and combined with anisotropy from [4] as already used in our previous paper [1], given as

$$I_c(B,\Phi) = I_{c0} \cdot \left[1 + \varepsilon(\Phi) \cdot \left(\frac{B}{B_0}\right)\right]^{-\gamma} \tag{1}$$

This was adapted to form the following equation:

$$I_c(B,\Phi) = I_{c0} \cdot \left[1 + \varepsilon(\Phi)^{\alpha} \cdot \left(\frac{B}{B_0}\right)^{\beta}\right]^{-1} \tag{2}$$

A second model considered was the so-called Percolation model according to [5] which states that the current flows by percolation through the weak-link network of the conductor. Combining this model with the anisotropy gives:

$$I_c(B, \Phi) = I_{c0} \cdot \left\{\delta + \exp\left[-\varepsilon(\Phi)^{\alpha} \cdot \left(\frac{B}{B_0}\right)^{\beta}\right]\right\} \tag{3}$$

A third function considered is given in [6] where the magnetisation of superconducting wires was studied and a remarkably good fit with the following function (which has been expanded to include the anisotropy and further mathematical fit parameters) was found:

$$I_c(B, \Phi) = I_{c0} \cdot \exp\left[\lambda - \varepsilon(\Phi)^{\alpha} \cdot \left(\frac{B}{B_0}\right)^{\beta}\right] \tag{4}$$

In each case, the anisotropy function $\varepsilon(\Phi)$ was defined as

$$\varepsilon(\Phi) = \sqrt{g^2 \cdot \cos^2(\Phi) + \sin^2(\Phi)} \tag{5}$$

with the parameters to be fitted by the Matlab® software grouped into the physical fit parameters I_{c0}, B_0, δ, λ and g and the mathematical fit parameters α,β and γ

4. Analysis of the results

The measurements showed that the angle at which I_c has its maximum value deviated slightly from 90 ° at both 4.2 K and 77 K. Therefore an extra variable must be included in the anisotropy function $\varepsilon(\Phi)$ in equation (5) and Φ becomes $\Phi + \Delta\Phi$. One theory is that this off-set arises due to the crystal structure of the HTS and is specific to the conductor sample tested. In the above equations I_{c0} was determined by extrapolating the value of $I_c(B)$ at the angle at which the maximum measurement values were found to B=0. Once $\Delta\Phi$ was determined by equation (1), it was fixed at this value (see table 1).

Taking the normal operating range of the conductor into consideration, the fits were made for the liquid nitrogen data at 0 - 1 T. For the liquid helium data the fits were made over the range 3 – 10 T as a basis for extrapolation to B > 20 T. The best fit was determined by comparing the values of the Mean value of the Absolute Difference (MAD) between the I_c values from our measurement data and the corresponding I_c values determined by the fit function as a percentage of the average measured I_c

value. The three dimensional surfaces resulting from the best of the fit models at 77 K and 4.2 K are shown in figures 1 and 2 respectively for the conductor type II and the fit parameters in table 1.

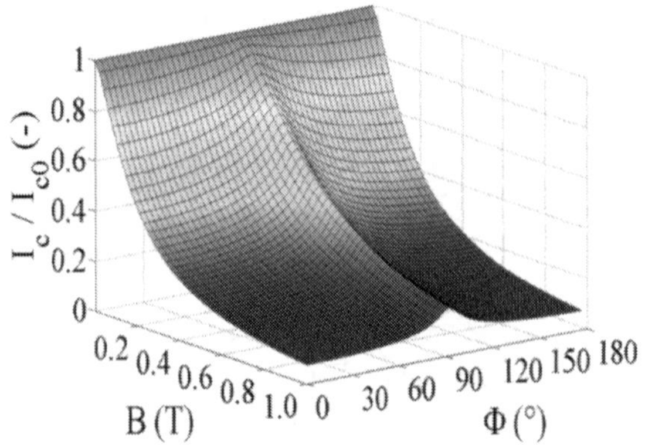

Figure 1. $I_c(B, \Phi, T$=77 K$)/I_{c0}$ surface approximation from equation (1) for CC type II (normalised).

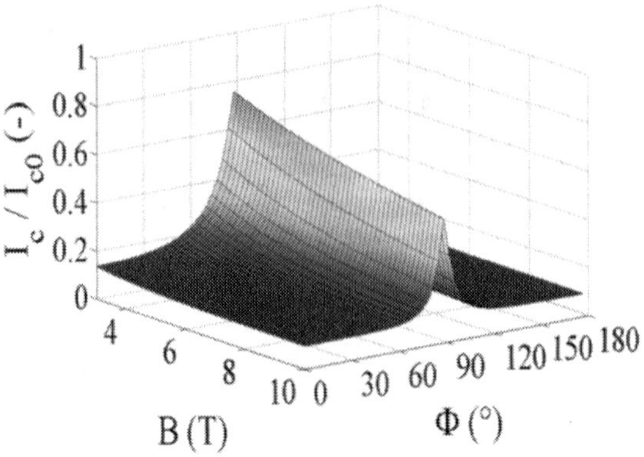

Figure 2. $I_c(B, \Phi, T$=4.2 K$)/I_{c0}$ surface approximation from equation (3) for CC type II (normalised).

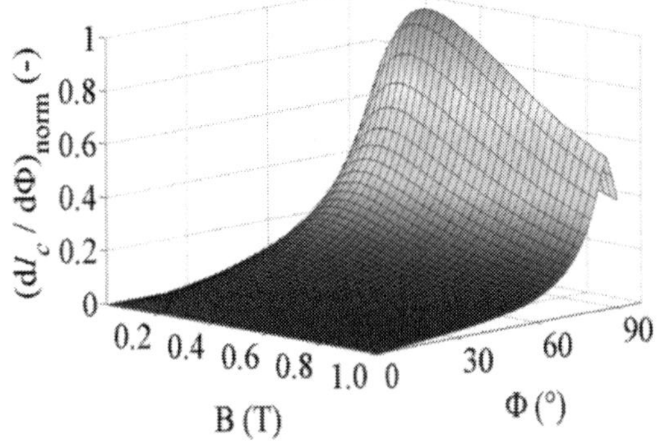

Figure 3. Partial derivative of $I_c(B, \Phi, T$=77 K$)$ fit with respect to Φ for CC type II (normalised).

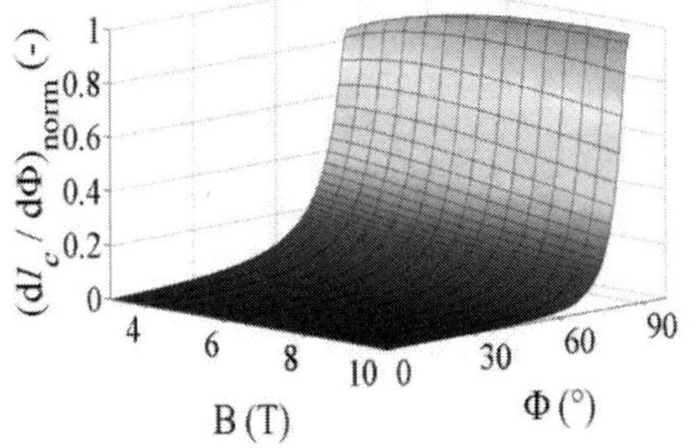

Figure 4. Partial derivative of $I_c(B, \Phi, T$=4.2 K$)$ fit with respect to Φ for CC type II (normalised).

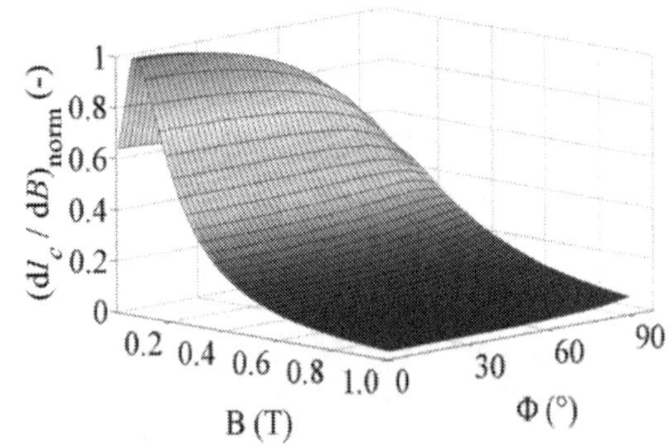

Figure 5. Partial derivative of $I_c(B, \Phi, T$=77 K$)$ fit with respect to B for CC type II (normalised).

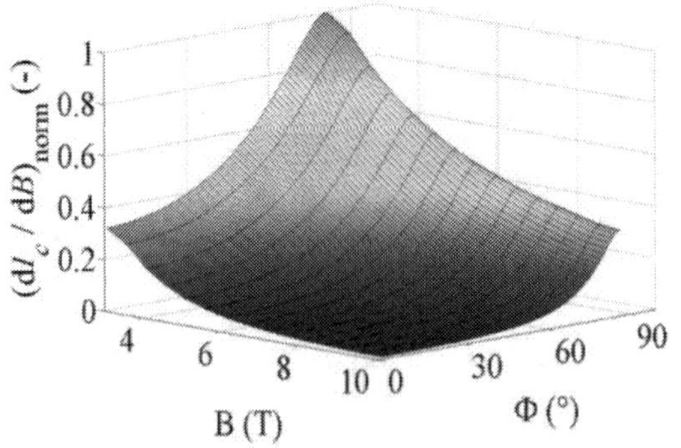

Figure 6. Partial derivative of $I_c(B, \Phi, T$=4.2 K$)$ fit with respect to B for CC type II (normalised).

In order to better illustrate the dependence of I_c on Φ and B individually, figures 3 - 6 show the normalised partial derivatives of $I_c\ (B,\Phi,T)$ for the functions at 77 K and 4.2 K up to the maximum I_c value. From figure 3 it can be seen that at 77 K varying Φ has no effect on I_c at zero field, as expected, but rises quickly to a maximum at around 90 ° as soon as an external field is applied. Figure 5 clearly shows that at extremely low fields a slight increase in the external field results in a rapid decrease in I_c at 77 K which is more apparent when the field is perpendicular (Φ = 0 °) to the conductor. At 4.2 K, figure 4 demonstrates that when the field is at approximately 90 ° to the conductor any change in Φ dramatically reduces I_c which is even more apparent as the field strength increases. This means that the peak in figure 2 becomes increasingly sharper at higher fields. Figure 6 illustrates that at 4.2 K I_c changes more rapidly with B when the field lies parallel to the conductor (Φ = 90 °) than when it is perpendicular (Φ = 0 °).

All fits made using equations (1) – (4) gave an MAD value of less than 10 %. Equation (2), the modified version of equation (1), improved the MAD value of equation (1) by almost 50 %. Table 1 summarises the results of the best fit functions for the individual sets of measurement data. The MAD percentage error value shows a good correlation with the R^2 value generated by the Matlab® software.

Table 1. Summary of Analysis

Temp. (K)	CC-type	ΔΦ (°)	Equation	Fit Parameters I_{c0} (A)	B_0 (T)	g(-)	α(-)	β(-)	δ(-)	MAD (%)
4.2	II	1.84	3	1168	12.70	30.82	0.59	0.60	0.10	3.52
4.2	I	5.10	3	1222	20.28	53.96	0.65	0.77	0.15	3.02
77	II	-1.64	2	138	0.38	13.03	0.42	1.46	-	3.89
77	I	3.28	2	187	0.34	14.32	0.38	1.47	-	4.01

5. Conclusion

We have made $I_c\ (B,\Phi,T)$ measurements at 4.2 K and 77 K for CC's from SuNAM and analysed a number of possible three dimensional fit functions. The variation in $\Delta\Phi$ was possibly due to pinning structure variations between the individual samples. We found that all functions studied gave a suitable fit of the measurement data and that whilst the fit function based on [3] used in our previous paper [1] was still valid (equation (1)), we were able to reduce the percentage error of the function in some cases by up to 50% by introducing new parameters. The best fits gave an error of less than 5 %. At both liquid nitrogen and liquid helium temperatures the error was minimised in all our fits due to the large number of I_c measurements made in the range 76 -104 ° and, in the case of 77 K, the consideration of the magnetic field range at which the HTS would be likely to be operating. At 4.2 K the functions used can be extrapolated to higher fields, as required.

6. References

[1] Leys P, Klaeser M, Schleissinger F and Schneider Th 2013 Angle-dependent U(I) measurements of HTS coated conductors IEEE Transactions on Applied Superconductivity **23** 3 3 8000604

[2] www.i-sunam.com

[3] Kim Y B, Hempstead C F and Strnad A R 1962 Critical persistent currents in hard superconductors Phys. Rev. Letters **9** 7 306-9

[4] Blatter G, Feigelman M V, Geshkenbein V B, Larkin A I and Vinokur V M 1994 Vortices in high-temperature superconductors Rev. Mod. Phys. **66** 1125

[5] Doyle T B 1993 Morphology-dependent intergranular behaviour in high-Tc superconducting Y-Ba-Cu-O Phys. Rev. B **47** 8111-8

[6] Fietz W A, Beasley M R , Silcox J and Webb W W 1964 Magnitization of Superconducting Nb-25%Zr wire Phys. Rev. **136** 2A 335-345

Journal of Physics: Conference Series **507** (2014) 022014 doi:10.1088/1742-6596/507/2/022014

New progress of ITER-PF strand production in WST

J.F. Li[1,2], W.T. Liu[1,2], L. X. Yan[1,2], J. H[1,2], H. X. Gao[1,2], J. W. Liu[1,2], S.J. Du[1,2], X.H. Liu[1,2], Y. Feng[1,2], P.X. Zhang[1,2,3], S. Liu[4] ,H.W. Li[4], E.W. Niu[4]

[1]Western Superconducting Technologies Co.,Ltd., Xi'an 710018, China
[2]National Engineering Laboratory for Superconducting Materials Preparation, Xi'an 710018, China
[3]Northwest Institute for Nonferrous Metal Research, P.O. Box 51, Xi'an 710016, China
[4]China International Nuclear Fusion Energy Program Execution Center, Beijing, 100862, China
E-mail: Ljfwst@gmail.com

Abstract. ITER Poloidal Field (PF) systems consist of 6 independent coils with different dimensions and require NbTi superconductor and copper strands. Western Superconducting Technologies Co.,Ltd.(WST) will supply PF2-5 NbTi strand for ITER, and over 14,000 km of NbTi strands have been produced in the past two years. Main performance of NbTi strands, including critical current, n value, wire diameter, Cu/non-Cu ratio, hysteresis loss and RRR are reported and analysed in this paper.

1. Introduction

The ITER magnet coils are wound from Cable-In-Conduit Conductors (CICC) made up of superconducting and copper strands assembled into a multistage, rope-type cable inserted into a conduit of butt-welded austenitic steel tubes. The conductors for the Toroidal Field (TF) and Central Solenoid (CS) coils require about 500 tons of strands while the Poloidal Field (PF) and Correction Coil (CC) conductors need around 250 tons of NbTi strands [1].

CN contribute to ITER project includes PF conductors, TF conductors and CC/Feeder which requires about 170 tons NbTi strand and about 30 tons Nb_3Sn strand. In the past two years, WST has produced over 50 tons NbTi strand for PF conductor.

2. Strand Performance

Figure 1 shows a cross section microphotograph of NbTi wire with 0.73mm and the enlarged microphotograph of NbTi filaments area. The niobium barrier is used between NbTi filament and copper matrix to avoid intermetallic of Cu and Ti which can't be co-deformed with the strands or stabilizer matrix [2].

The main specifications of PF NbTi strand including critical current (5T, 4.2K), n values, strand diameter, Cu/ non-Cu ratio, hysteresis loss and RRR values are listed in Table 1. The main strand performance of WST is shown in figure 2 to 7, and satisfied the ITER requirements which marked as dash line in each figure.

The implementation of Statistical Process Control (SPC) calls for the identification of key process parameters and of sound methods for relevant statistical sampling in order to reduce and control manufacturing process variability, prevent defects and ensure suitable and reproducible performance.

SPC methods rely on the monitoring of mean values of key process parameters, and the evolution of these mean values, to detect deviations in production and assess their potential impact on performances.

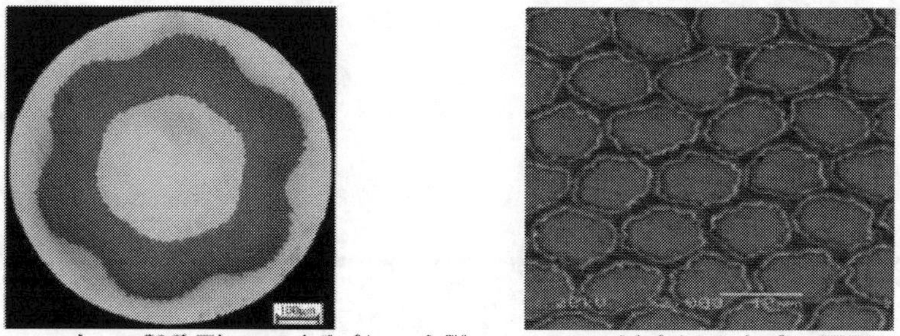

Figure 1 Cross section of NbTi strand (left) and filament area (right) made for ITER-PF conductor

Table 1. Main Strand Specification

	NbTi strand
Minimum critical current	339A
n-Value[a]	> 20
Strand diameter	0.730 ± 0.005 mm
Cu-to-non-Cu volume ratio	2.25- 2.45
Maximum Hysteresis loss	45 mJ/cm3
Residual Resistivity Ratio of plated strand	> 100

[a] This value is defined in the 0.1-to-1 μV/cm range at 4.22 K and B_{ref} = 5.0T for minimum critical current

The critical current and n value has the similar trend and a large distribution along with the mass production yield increased.

The diameters of nickel plated NbTi strand data which shown in figure 4 are with small range and well controlled. The Cu/non-Cu ratio points are spread from lower limits to upper limits and within the specification and some of the data near lower limits because of the intended cuts.

The hysteresis loss and RRR values appeared within specification, and large variation of RRR values compared with other parameters was primary attributed to different oxygen free copper suppliers and temperature control.

3. Production Yield

Figure 8 shows the overall production yield of PF NbTi strand up to now. A stable production yield has been achieved for PF NbTi production. The highest average piece length of batch is around 90000 meters which above 300 kg.

4. Qualification Test

Full-size ITER conductor qualifications are carried out at the SULTAN facility, located at CRPP, in Villigen, Switzerland [3]. Nb-Ti conductor samples are single lengths mounted in a hairpin configuration at the bottom. SULTAN samples are instrumented with voltage taps and temperature sensors at various locations. The measurements are carried out at fixed temperature and field (B_{ref}) and increasing current to obtain critical current (I_c) or at fixed current and field and increasing temperature

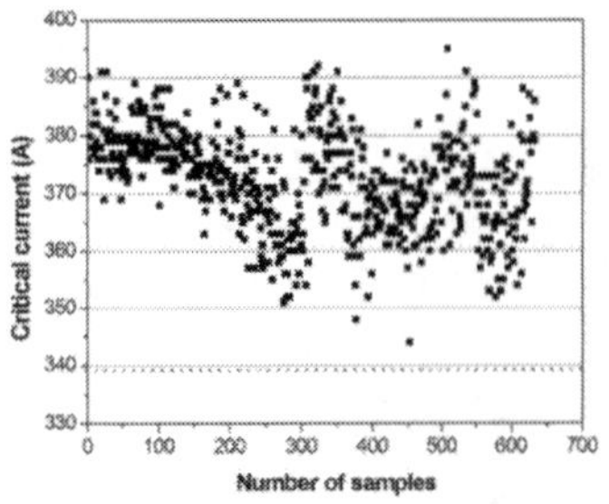

Figure 2. The critical current of PF NbTi strand

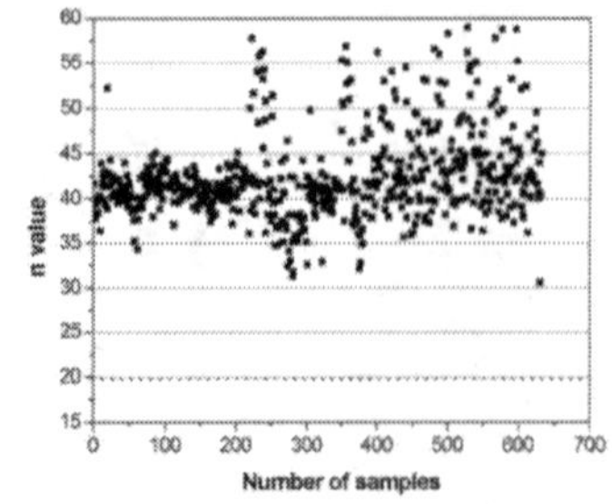

Figure 3. The n value of PF NbTi strand

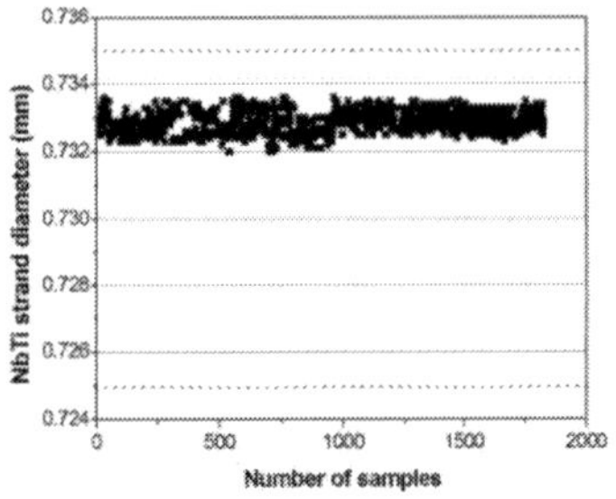

Figure 4. The diameter of PF NbTi strand

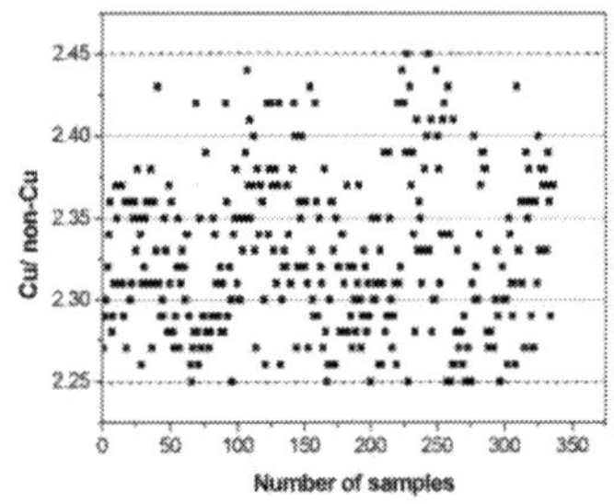

Figure 5. The Cu/non-Cu ratio of PF NbTi strand

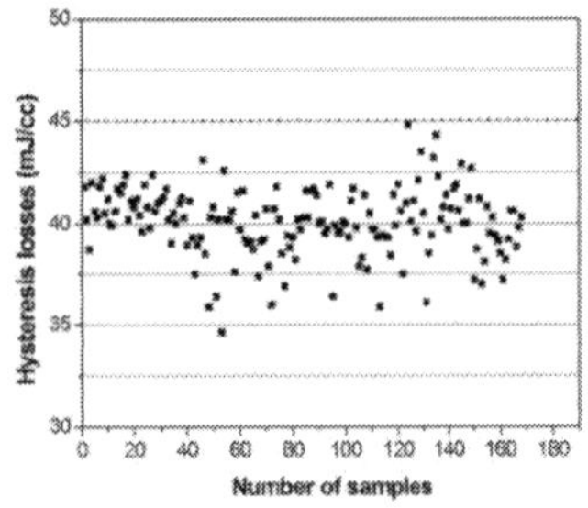

Figure 6. The hysteresis loss of PF NbTi strand

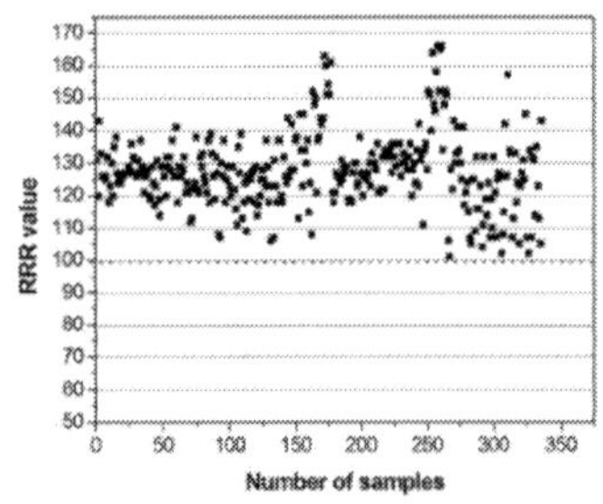

Figure 7. The RRR value of PF NbTi strand

to obtain critical sharing temperature (T_{cs}). The runs are the most relevant to assess conductor performance.

The required values of I_c, B_{ref} and $T_{Cs,min}$ for the various PF coils are summarized in Table 2. For a strand design to be qualified its current sharing temperature at I_{op} and B_{ref} must be greater than the value specified in Table 2.

Two sultan samples named PFCN1 (PF2-4) and PFCN2 (PF5) have been tested in CRPP. The test results of PFCN1 and PFCN2 samples showed that the performance of the PF conductor satisfy the ITER requirement. The test results of the samples are shown in the Fig 9 (a) and (b).

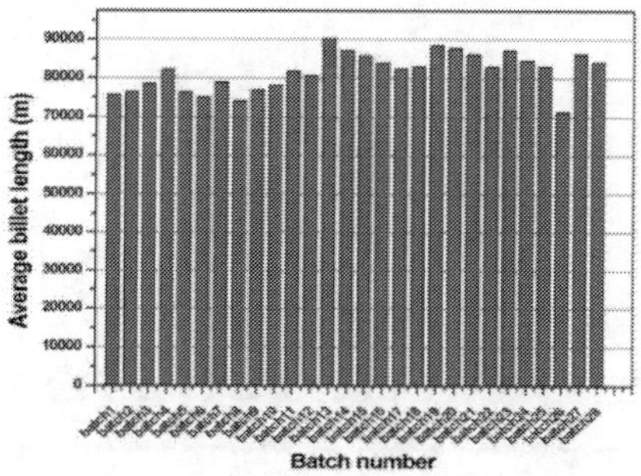

Figure 8 The average piece length of PF NbTi strands

Table 2. Minimum current sharing temperature of ITER PF Strands

Coil	Current (A)	Field (T)	Min. Current Sharing Temp.(K)
PF1&6	33	6.4	5.9
PF2-4	69	5.0	6.1
PF5	45	5.7	6.1

5. Conclusion

WST have produced PF2-5 NbTi strand successfully and will finish the PF strand production in the next year.

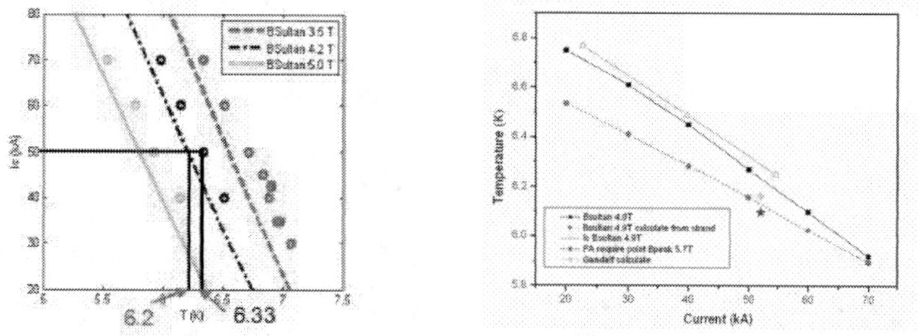

Figure 9 Test results of PFCN1(a, left) andPFCN2(b, right) sultan samples(Courtesy of Vostner A, IO)

References

[1] Devred A, Backbier I, Bessette D, Bevillard G, Gardner M, Jewell M, Mitchell N, Pong I and Vostner A, 2012, Status of ITER Conductor Development and Production, *IEEE Transactions on Applied Superconductivity*, vol. 22,p4804909

[2] Cooley L, Lee P J and Larbalestier D C, 2003, B3.3.2 *Conductor processing of low Tc materials: the alloy Nb-Ti of the handbook on superconducting materials*.pp603-639

[3] Bruzzone P, Anghel A, Fuchs A and Pasztor G, Upgrade of operating range for SULTAN test facility, 2002, *IEEE Transactions on Applied Superconductivity*,, vol. 12, no. 1,pp520-523

Effects of Laser Repetition Rate on the Structural and Superconducting Properties of YBCO films Deposited by PLD on NiW Tapes

Linfei Liu, Guina Xiao, Da Xu, Ying Wang, Xiang Wu, Qiang Luo, Yijie Li

Key Laboratory of Artificial Structures and Quantum Control (Ministry of Education), Department of Physics and Astronomy, Shanghai Jiao Tong University, 800 Dongchuan Road, Shanghai 200240, People's Republic of China

E-mail:yjli@sjtu.edu.cn

Abstract. $YBa_2Cu_3O_{7-x}$ (YBCO) films were fabricated at various laser repetition rates on $CeO_2/YSZ/CeO_2$ buffered Ni-W tapes by pulsed laser deposition (PLD). The dependence of structural and superconducting properties on the laser repetition rates was investigated. The microstructure and surface morphology of YBCO films were characterized by X-ray diffraction and atomic force microscopy. And the critical current (I_c) of YBCO films was measured by the conventional four-probe method. The results show that with increasing laser repetition rate the island density decreased and the island size increased. It was found that the texture and I_c of YBCO films were largely dependent on the laser repetition rate. And furthermore it was observed that the film thickness was not in directly proportional to laser repetition rate. Under optimum experimental condition, high quality YBCO films with critical current I_c above 500 A/cm at 77 K and 0T were obtained.

1. Introduction

$YBa_2Cu_3O_{7-\delta}$ (YBCO) is one of the most widely studied superconductors because of its significant potential for practical applications. YBCO have been fabricated by various techniques, such as metal organic deposition (MOD) [1], pulsed laser deposition (PLD) [2], and metal-organic chemical vapor deposition (MOCVD) [3,4]. Since PLD technique was widely used in preparing thin films with precise control of films reproducibility, uniformity and simplicity, it has been regarded as one of the most promising methods [5]. The effects of texture, deposition temperature and thickness on the growth of epitaxial YBCO film with well c-axis orientation have been extensively studied [6]. However the effect of laser repetition has rarely been studied. Therefore, in order to achieve YBCO films with high critical current, it is necessary to understand the relationship between the laser repetition rate and performance of YBCO film.

The aim of the present work is to prepare the YBCO films on buffered Ni-W tapes at different laser repetition rates and hence to investigate the influence of repetition rate on epitaxial growth, surface morphology and superconducting property.

2. Experimental

YBCO films were deposited on Ni-W tapes by PLD. The in-plane and out-of-plane textures of Ni-W tapes used in this study were evaluated by the full width at half maximum (FWHM) values of the X-ray diffraction (XRD) φ-scan and ω-scan as $\Delta\varphi$ = 8-9° and $\Delta\omega$ = 6-7°, respectively. Prior to the fabrication of YBCO films, the buffer layers of cerium oxide (CeO_2) and yttria-stabilized zirconia (YSZ) were well prepared as described in reference [7]. The thicknesses of CeO_2-seed layer, YSZ layer and CeO_2-cap layer were approximately 50 nm, 100 nm and 250 nm, respectively. During YBCO film deposition, a KrF excimer laser with a wavelength of 248 nm was used and the incident angle between the laser beam and the target surface was 45° . The distance between the target and the substrate was about 3 cm. The energy of the laser was 200 mJ and the oxygen pressure was 200 mTorr. The substrate temperature was 760-800 ℃ and the repetition rate of the laser was range from 30-180 Hz. After deposition, Ag films were deposited using DC sputtering system and then the tapes were annealed at 500℃ in O_2 pressure for one hour.

The structure of the film was studied by a general area detector diffraction system (D8 Discover with GADDS, Bruker Advanced X-ray Solutions, Inc.) with Cu Kα radiation. The generation voltage is 40 KV and tube current is 40 mA. The surface morphology and root mean square (RMS) surface roughness of the film were examined with atomic force microscope (AFM, BioScope, Veeco Instruments, Inc.) in tapping mode. The thickness of the film was measured by a stylus-type profilometer (Dektak XT). The critical current (I_c) was measured in liquid N_2 (77K) using a DC four-probe method with the criterion of 1μv/cm.

3. Results and discussion

3.1. The structure of YBCO films deposited at different laser repetition rates

Table 1. Optimum substrate temperature of YBCO film deposited at various repetition rates.

Repetition rate (Hz)	**30**	**60**	**90**	**120**	**150**	**180**
Substrate temperature (℃)	760	770	780	790	800	800

First we fabricated the YBCO films at different laser repetition rates and substrate temperatures. It was found that the optimum substrate temperature was dependent on the laser repetition rate. Table 1 shows the optimum substrate temperature of YBCO film deposited at different laser repetition rates. As shown in table 1, with increasing laser repetition rate the optimum substrate temperature first increased and then kept constant when the laser repetition rate increased to 150 Hz.

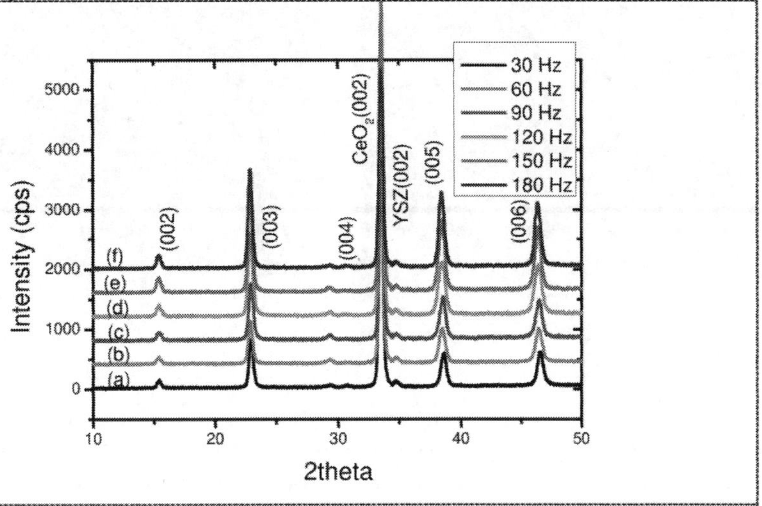

Figure 1. XRD θ-2θ scans of YBCO films deposited at different laser repetition rates. (a) 30 Hz, (b)60 Hz, (c) 90 Hz (d) 120 Hz, (e) 150 Hz,(f) 180 Hz

X-ray diffraction (XRD) θ-2θ scans of the YBCO films were carried out to characterize the structure of the YBCO films deposited at optimum substrate temperature and different laser repetition rates. The XRD pattern of the YBCO films was shown in Fig.1. The appearance of peaks (00l) is observed, which shows that the YBCO film with a c-axis orientation was preserved. The peak intensity of the films increased and full width at half maximum (FWHM) decreased with increasing laser repetition rate. The crystallite size of the films was calculated using the Scherrer formula, and we found that the crystallite size increased with increasing laser repetition rate.

3.2. The surface morphologies of YBCO films deposited at different laser repetition rates

Fig.2 show the AFM images of the YBCO films deposited at various repetition rates. The images show the uniform distribution of crystallites with dense structure. The crystallite size, the particles and surface roughness increased with increasing laser repetition rate and the result is confirmed with our XRD pattern.

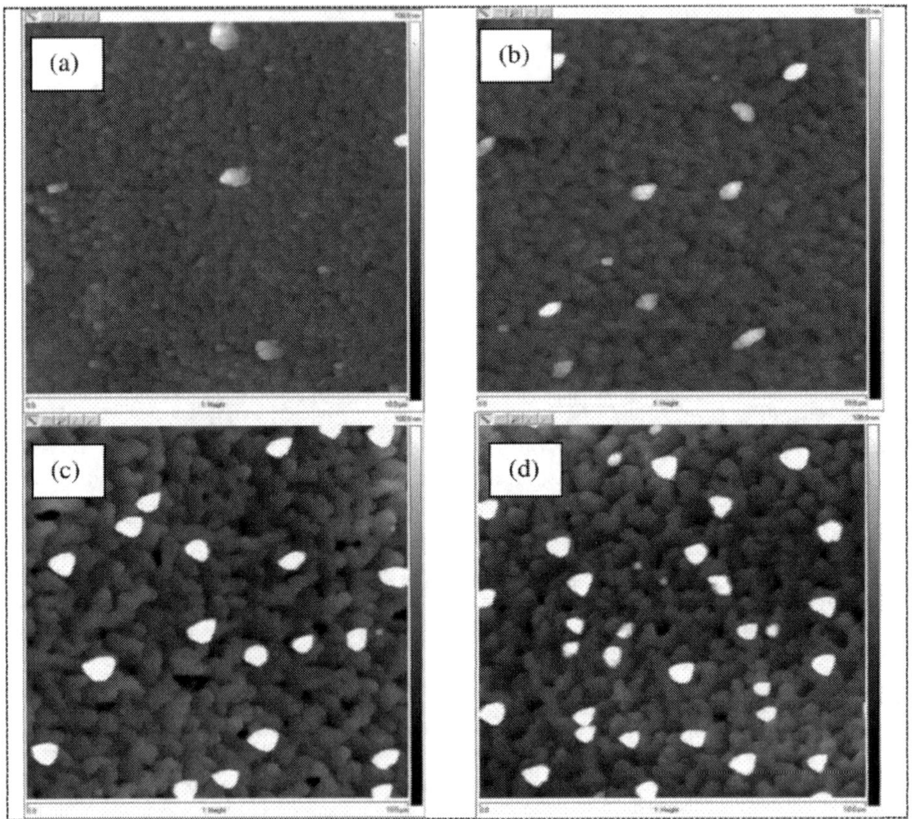

Figure 2. AFM images of YBCO films deposited at various laser repetition rates. (a) 30 Hz, (b)60 Hz, (c) 150 Hz,(d) 180 Hz

3.3. The superconducting properties of YBCO films deposited at different laser repetition rates

The thickness and I_c of YBCO film deposited at different laser repetition rates were measured by a stylus-type profilometer and the conventional four-probe method without microbridge patterning at 77 K and self field, respectively. Then we calculated the deposition rate and the critical current density J_c of the YBCO film, which were shown in table 2. As can be seen from table 2, with increasing laser repetition rate, the deposition rate decreased, and the J_c first increased and then decreased. The highest

Jc of 5.1× 10^6 A/cm^2 was obtained for the YBCO film deposited at 150 Hz. Then we fabricated 2 μm thick YBCO film at optimum condition, which has the highest I_c of 520 A/cm and J_c of 2.6×10^6 A/cm^2 as shown in Fig.3.

Table 2. The dependence of deposition rate and J_c of YBCO films on laser repetition rate

Repetition rate (Hz)	**60**	**90**	**120**	**150**	**180**
Deposition rate (nm/pulse)	0.7	0.63	0.63	0.56	0.53
J_c(MA/cm^2,at 77K,0T)	3.9	4.1	4.3	5.1	4.7

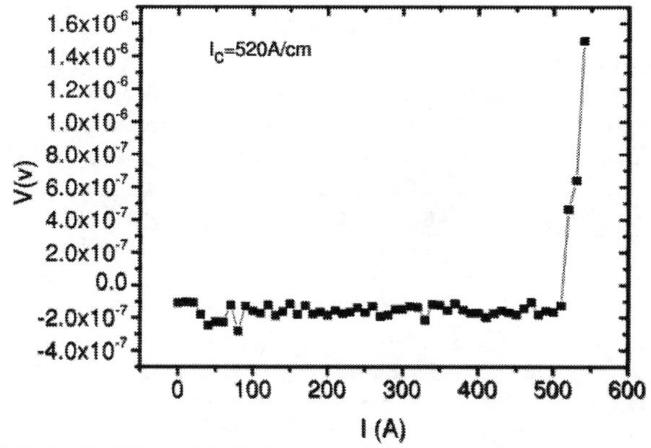

Figure 3. End-to-end I-V characteristic of 2.0 μm thick YBCO film deposited at optimum experimental condition, tested at 77 K and self field.

4. Conclusion

YBCO films were grown on buffered Ni-W tapes by PLD at different laser repetition rates. As a result, the microstructure and the performance of YBCO were dependent on the laser repetition rate. With increasing laser repetition rate, the optimum deposition temperature, particles, surface roughness and grain size increased, but the deposition rate decreased. The J_c first increased and then decreased with increasing repetition rate. The highest J_c of 5.1× 10^6 A/cm^2 was obtained at 150 Hz. High quality 2 μm thick YBCO film was obtained, which had I_c of 520 A/cm and J_c of 2.6 MA/cm^2 at 77 K and self field.

References

[1] Izumi T, Matsuda J S, Nakaoka K, Kitoh Y, Sutoh Y, Nalkanishi T,Yoshizumi M, Yamada Y and Shiohara Y 2007 *IEEE Trans. Appl. Supercond.* **17** 3329-31.
[2] Yamada Y, Miyata S, Yoshizumi M, Fukushima H, Ibi A, Kionoshita A, Izumi T, Shiohara Y, Kato T and Hirayama T 2009 *IEEE Trans. Appl. Supercond.* **19** 3236-9.
[3] Molodyk A, Novozhilov M, Street S, Castellani L and Ignatiev A 2011 *IEEE Trans. Appl. Supercond.* **21** 3175-8.
[4] Selvamanickam V et al 2011 *IEEE Trans. Appl. Supercond.* **21** 3049-54.
[5] M. Fukutomi, S. Aoki, K. Komori, R. Chatterjee and H. Maeda 1994 *Physica C* **219** 333.
[6] Y. Li, Z. Zhao, L. Liu, Q. Ye and H. Zheng 2009 *IEEE Trans. Appl. Supercond.* **19** 3295.
[7] L. Liu, Z. Zhao, H. Liu and Y. Li 2010 *IEEE. Trans. Appl. Supercond.* **20** 1561.

A Method to Estimate the Necessary Twist Pitch in Multi-filamentary Superconductors

S Lindau[1], N Magnusson[2], H Taxt[2]

[1] Norwegian University of Science and Technology, Department of Electrical Power Engineering, NO-7491 Trondheim, Norway
[2] SINTEF Energy Research, NO-7465 Trondheim, Norway

E-mail: niklas.magnusson@sintef.no

Abstract. Twisting of multi-filamentary superconductors is an important step in the development of wires with AC losses at an acceptable level for AC applications. The necessary twist pitch depends on wire architecture, critical current density, matrix material, and external factors such as temperature, frequency and applied magnetic field. The development of an AC optimized MgB_2 superconductor would be facilitated by a fast method to set the requirements for the twist pitch. A problem often encountered when comparing wires with different twist pitches is the degradation in critical current occurring at small twist pitches due to mechanical deformation. In this work we propose to use a non-twisted conductor to estimate the influence of twisting on the AC losses. A long superconductor is cut into smaller lengths, each simulating one third of the twist pitch, and the AC losses due to applied magnetic fields are compared between samples of different lengths. With this method, the effect of reducing the size of the loop of the coupling currents is studied without changing the superconducting parameters. AC loss measurement results are presented for a round titanium matrix MgB_2 wire with simulated twist pitches between 9 mm and 87 mm.

1. Introduction

Medium- and high-temperature superconductors on or close to the market have the technical performances necessary for consideration in several large-scale applications, like DC [1] or low-field AC [2]-[3] cables, generator rotors [4], and induction heaters [5]. These applications set only minor requirements on the wire's behavior under AC operation. To broaden the possible range of applications to include components like transformers, reactors, and generator stators, where the AC magnetic field is significant, low AC loss conductors are required.

For YBCO coated conductors there is ongoing research aiming at reducing the AC losses by e.g. transverse cuts [6]-[7]. For the multi-filamentary conductors, BSCCO and MgB_2, the key is to manufacture wires with thin, electromagnetically decoupled filaments. For BSSCO conductors the research toward such wires was intensive in the late 1990s [8]-[11], without reaching loss levels which lead to a commercial breakthrough, mainly due to difficulties to obtain a sufficiently short twist pitch without degradation of the superconducting properties.

For MgB_2 conductors the AC optimization step is yet to be taken. The route towards a low AC loss wire includes twisting of the filaments, use of high resistivity (and non-magnetic) matrices, and reduction of filament size. To significantly reduce the losses compared to the losses in conventional copper conductors, the AC losses need to be about or below 0.1 W/kAm at an operating temperature of MgB_2 of 20 K. To reach that target the filaments need to be fully decoupled and their size to be of the order 10 μm [12], although a larger filament size can be accepted for applications with other benefits than only loss reduction.

When twisting the filaments, the conductor may suffer from degradation in the critical current density, due to the mechanical deformation [13]. The effect of the twisting with respect to reduction of AC losses may therefore be difficult to assess, since the AC losses are highly dependent on the critical current density [14]. However, in the conductor development phase, the knowledge of the necessary twist pitch for different wire architectures is crucial to set the development target, and to balance the efforts of inserting resistive barriers and applying strong filament twisting.

Some guidance of the necessary twist pitch can be obtained by equation (1) defining a critical length L_c of the twist pitch [15],

$$L_c = 4\sqrt{\frac{I_c \rho}{2\pi f B}}, \quad (1)$$

where I_c is the critical current, ρ the electrical resistivity of the matrix, and f and B are the frequency and amplitude of an applied sinusoidal magnetic field. At twist pitches longer than L_c, the filaments are fully coupled, at shorter twist pitches the filaments become first partly decoupled, and at twist pitches much shorter than L_c, fully decoupled.

In real situations, the matrix architecture may be complicated and ρ difficult to predict. Furthermore, one would not only be interested in the twist pitch where decoupling starts, but also of the twist pitch leading to sufficient decoupling for a given wire architecture.

In this work we propose an experimental technique to estimate the necessary twist pitch in multi-filamentary conductors by measurements on non-twisted wires.

2. Simulating the twist pitch by cutting non-twisted wires

When a multi-filamentary wire is exposed to a time-varying magnetic field, currents are induced in a loop consisting of two filaments and bridges in the metal matrix between them. These currents contribute to the total losses by an increase in the superconductor hysteresis losses and by the ohmic losses caused by the coupling currents crossing the metal matrix.

In a non-twisted multi-filamentary superconducting wire, the loop becomes infinitely long, and the filaments are fully coupled, i.e. the filaments act as one bundle leading to larger hysteresis losses than if they were acting separately. To reduce the loop and thereby the losses, the wire can be twisted, as in figure 1 (right). The enclosed magnetic flux (and change in flux) in the loop becomes smaller with a smaller twist pitch, and hence the driving voltage for the coupling currents becomes smaller. With strong enough twisting, the driving voltage becomes lower than necessary for the currents to cross the metal matrix and the filaments become decoupled and the losses thereby reduced.

To study the effect of twisting we propose to cut a non-twisted multi-filamentary wire into shorter lengths each approximately representing one third of the twist pitch. Compare the drawings in figure 1. For the twisted wire to the right, the area of an enclosed loop for half a twist pitch is $2/\pi$ of the area of the loop for the cut non-twisted wire to the left (assuming the loop is closed at the ends). Hence, the behavior of a twisted wire, when exposed to an AC magnetic field, can approximately be studied considering short non-twisted wires. The lengths of the samples correspond to one third of the twist pitch of twisted wires.

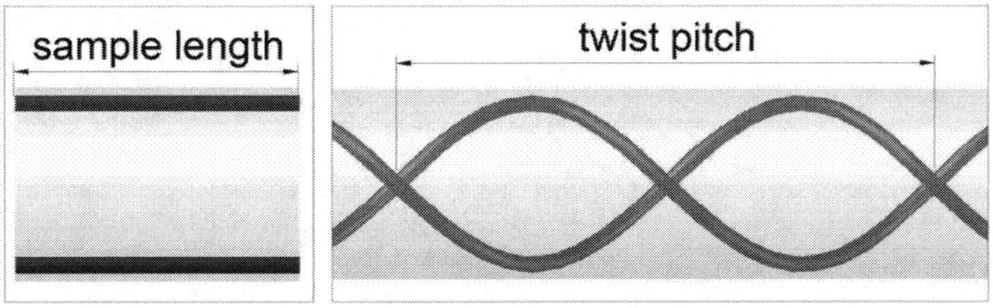

Figure 1. Schematic drawings of a twisted wire (right) and a non-twisted sample cut to a length corresponding to one third of the twist pitch. Only two filaments are shown in each sample.

3. Experimental

3.1 Sample and sample preparation

A 37 filament MgB_2 non-twisted round wire with a titanium matrix was studied, see figure 2 (left) for a cross-section photo. The wire was manufactured by Columbus Superconductors for research purposes. The wire diameter was 1.14 mm, the total cross-sectional area 1.01 mm^2, and the superconducting part 0.25 mm^2.

The sample wire was cut into six pieces. To each piece a thin copper wire (used for temperature measurements) and a high resistance wire (used as heater) were attached, and the assembly was placed in a mold and casted in Stycast 2850 epoxy, see figure 2 (right).

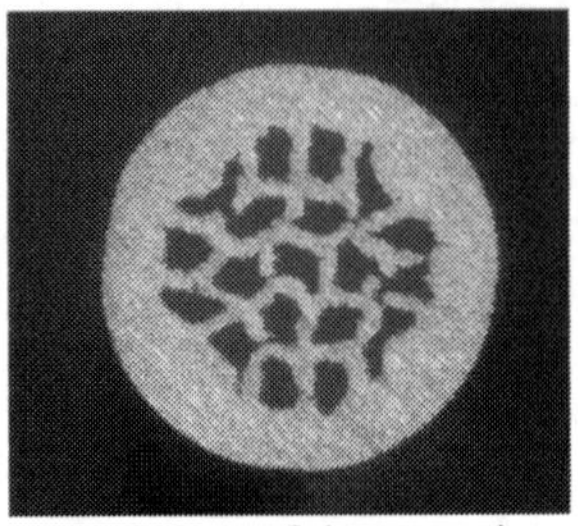

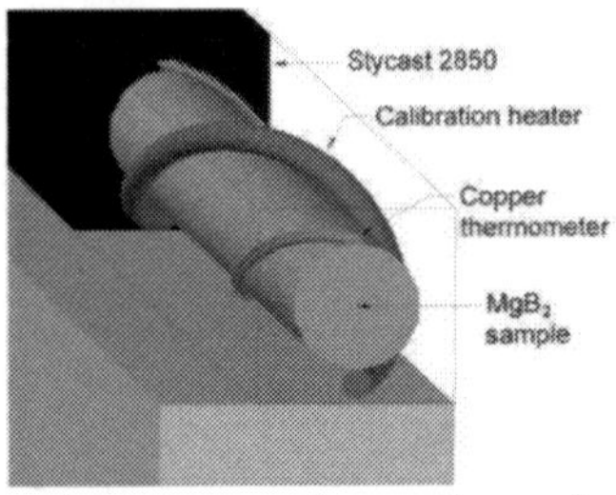

Figure 2. Cross-section of the sample wire (left) and the principle sample configuration (right) [16].

The six pieces were then carefully sanded from the ends and inwards about 4 mm to remove parts at the ends potentially mechanically damaged during cutting. The remaining sample lengths were 3, 3, 4, 7, 12, and 29 mm.

3.2 Measurement method

The measurement method was calorimetric. The system used is presented in detail elsewhere [16], here a short summary is given. The samples were glued with Stycast 2850 to the top of the sample holder. The samples and the sample holder were placed in vacuum and were in thermal contact with the cold head of a cooling machine. In a nitrogen bath just outside the cryostat, a set of magnetizing copper coils was placed, generating a magnetic field over the samples. The exposures of time-varying magnetic fields resulted in temperature increases in the samples due to AC losses. These increases were calibrated against the temperature increases (determined by measuring the change in resistance in a copper wires glued to each sample) due to the known power dissipation in heaters wound around the samples and fed by a low DC current. The samples of different lengths were placed after each other and were thereby simultaneously exposed to practically identical magnetic fields.

4. Results and discussion

Measurements were performed at 35 K. At this for MgB_2 conductors relatively high temperature, the critical current and hence the penetration field is low, allowing us to study the losses in the for medium field applications interesting region above the penetration field. Figure 3 shows the losses per unit length at 50 Hz as function of magnetic field for different sample lengths. Clearly the losses per unit length are reduced for sample lengths of 7 mm and below, whereas the losses for the 12 and 29 mm samples show only minor differences. The loss reduction can be attributed to the reduction in hysteresis losses due to the decoupling of filaments with reduced sample length. From the results it can be seen that the decoupling starts between sample lengths of 7 mm and 12 mm, corresponding to twist pitches of 21 to 36 mm, and decoupling further reduces losses down to the shortest sample length of 3 mm (corresponding to a twist pitch of 9 mm).

It is important to notice that the indicated L_c of about 30 mm is valid for this wire alone. The wire architecture, including the effective resistivity of the matrix, will strongly influence L_c as well as the loss reduction for twist pitches below L_c.

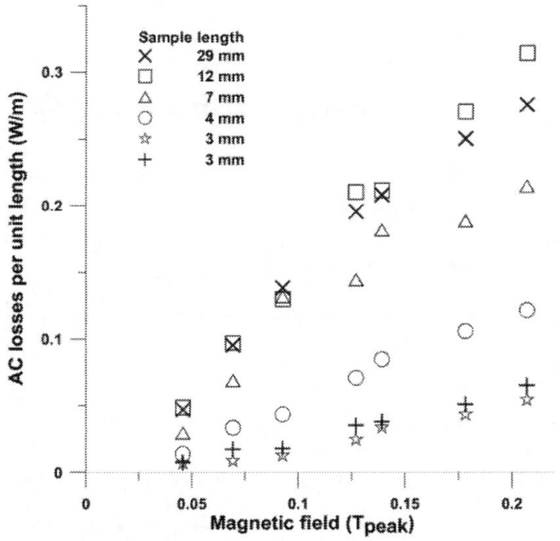

Figure 3. AC losses per unit length as function of applied magnetic field for non-twisted samples cut to different lengths. The temperature was 35 K and the frequency 50 Hz.

5. Conclusions

It has been presented how the necessary twist pitch in multi-filamentary superconductors can be predicted by studying short non-twisted samples. For an MgB_2 superconductor with a titanium matrix, the decoupling was estimated to start at a twist pitch corresponding to about 30 mm.

Acknowledgment

This work has been supported by the Nowitech programme and the Research council of Norway.

References

[1] Runde M 1995 *IEEE Trans. Appl. Supercond.* **5** 813

[2] Maguire J F, Yuan J, Romanosky W, Schmidt F, Soika R, Bratt S, Durand F, King C, McNamara J and Welch T E, 2011 *IEEE Trans. Appl. Supercond.* **21** 961

[3] Honjo S, Mimura T, Kitoh Y, Noguchi Y, Masuda T, Yumura H, Watanabe M, Ikeuchi M, Yaguchi H and Hara T, 2011 *IEEE Trans. Appl. Supercond.* **21** 967

[4] Gamble B, Snitchler G and MacDonald T 2011 *IEEE Trans. Appl. Supercond.* **21** 1083

[5] Runde M, Magnusson N, Fülbier C and Bührer C 2011 *IEEE Trans. Appl. Supercond.* **21** 1379

[6] Suzukia K, Yoshizumia M, Izumia T, Shioharaa Y, Iwakumab M, Ibic A, Miyatac S and Yamadac Y 2008 *Phys.* C **468** 1579

[7] Zhang Z, Duckworth R C, Ha T T, List F A, Gouge M J, Yimin C, Xuming X, Selvamanickam V and Polyanskii A 2011 *IEEE Trans. Appl. Supercond.* **21** 3301

[8] Christopherson C J and Riley Jr G N 1995 *Appl. Phys. Lett.* **66** 2277

[9] Goldacker W, Eckelmann H, Quilitz M and Ullmann B 1997 *IEEE Trans. Appl. Supercond.* **7** 1670

[10] Sugimoto M, Kimura A, Mimura M, Tanaka Y, Ishii H, Honjo S and Iwata Y 1997 *Phys.* C **279** 225

[11] Yang Y, Hughes T J, Beduz C and Darmann F 1998 *Phys.* C **310** 147

[12] Taxt H, Magnusson N, Runde M and Brisigotti S 2013 *IEEE Trans. Appl. Superconductivity* **23** 8200204

[13] Malagoli A, Bernini C, Braccini V, Fanciulli, Romano G, Vignolo M 2009 *Supercond. Sci. Technol.* **22** 105017

[14] Bean C P 1964 *Rev. Mod. Phys.* **36** 31

[15] Wilson M 1983 'Superconducting Magnets' *Clarendon Press, Oxford* 175

[16] Taxt H, Magnusson N and Runde M 2013 *Cryogenics* **54** 44

Ferromagnetic nanoparticles as efficient bulk pinning centers in HTSC

V A Kashurnikov[1], **A N Maksimova**[1] **and I A Rudnev**[1,2]
[1]National Research Nuclear University, Moscow Russia
[2]International Laboratory of High Magnetic Fields and Low Temperatures, Wroclaw Poland

Abstract. The self-consistent interaction of Abrikosov vortex with both point and extended ferromagnetic defects was taken into account in the 2D model of layered HTSC. Magnetization curves for HTSC with magnetic nanoparticles as bulk pinning centers were obtained by Monte-Carlo method. The nonlinearity of the interaction of the superconductor with ferromagnetic defects and irreversibility of initially reversible magnetization processes due to nonlinearity were demonstrated.

1. Introduction

The effect of ferromagnetism on the superconducting state attracted considerable attention in recent years. Ferromagnetic impurities in high-temperature and low-temperature superconductors were investigated to improve current-carrying ability, and the interaction of ferromagnetic phases with superconductivity has been used for solve the problem of the coexistence of superconductivity and ferromagnetism in a mixed vortex state. An enhancement of vortex pinning by a magnetic nanoparticle is expected due to an energy of particle's magnetic moment in the field of vortex in addition to conventional, nonmagnetic part of interaction.

For example, Fe particles in Hg-13%In were investigated and the magnetization curves were obtained for the parallel and antiparallel directions of the residual magnetization of the impurities and the external field [1]. The enhancement of vortex pinning with an increase of the impurity concentration was demonstrated. The interaction of a vortex with a ferromagnetic particle was described theoretically as an energy of point dipole in the field at the particle produced by vortex. The force of interaction of a single vortex line with a spherical magnetic nanoparticle of arbitrary radius was calculated [2]. Analysis of the behavior of ferromagnetic Gd in Nb [3] showed that the pinning force is due to the hysteretic loss for magnetization reversal.

A considerable part of investigations were associated with the vortex structures appearing in thin superconductor films in the vicinity of ferromagnetic nanoparticles. The enhancement of pinning of vortices in a thin film with ferromagnetic nanodots was detected experimentally, various configurations of one, two or three vortices around a magnetic dot were observed [4], the structures formed by vortices around magnetic dipole were investigated numerically [5].

Thus, analysis of the interaction of the vortex state with the ferromagnetic phase and correct estimation of the influence of its magnetic moment on the vortex dynamics in the superconductor is one of the important problems of superconductivity, it is especially important for an HTSC material due to its high critical parameters.

The aim of this paper is a correct description of magnetic particles and a superconductor as a self-consistent system and an exact calculation of magnetization reversal processes in a layered HTSC material with ferromagnetic nanoparticles as bulk pinning centers. Here, we represent the results of calculation of the magnetization reversal loops by using the Monte-Carlo method. Orientation of ferromagnetic nanoparticles' easy axes and values of magnetic anisotropy parameter were taken into account, the case of periodic lattice of elongated ferromagnetic defects was also analyzed.

2. Model

To calculate an equilibrium distribution of vortices, we numerically minimized the Gibbs thermodynamic potential of the 2D vortex system with a varying number of pancake vortices; this potential has the form (see [6,7], where the model and the algorithm are described in details):

$$G = sN\varepsilon + \sum_{i<j} U_{in-plane}(r_{ij}) + s\sum_{i,j} U_p(r_{ij}) + \sum_{i,j} U_{surf}(r_{ij}^{(im)}) \quad (1)$$

where ε is a self-energy of vortex, s is a thickness of a superconducting layer and N is the number of vortices in the system, the second term describes pair interaction of vortices, the third term takes into account the interaction of vortices with pinning centers and the fourth term describes the interaction of vortices with the surface and the external field. Pancakes in a plane interact with the long-range potential in the following form:

$$U_{in-plane}(r_{ij}) = s\frac{\Phi_0^2}{8\pi^2\lambda^2} K_0(\frac{r_{ij}}{\lambda}), \quad (2)$$

where r_{ij} is a distance between vortices and $\lambda = \lambda(T) = \lambda(T=0)/\sqrt{1-(T/T_c)^{3.3}}$ is the magnetic field penetration depth, $K_0(x)$ is a Bessel function and T_c is a superconducting critical temperature.

Then we take into account an ensemble of ferromagnetic nanoparticles in the bulk of superconductor. Gibbs potential of the ferromagnetic particle has a form:

$$U_p = U + U_{pm} + U_{pn}, \; U_{pm} = -\mu H_v, \quad (3)$$

μ is the projection of the magnetic moment onto the direction of the external field, H_v is the field produced by the vortex at the point of location of the particle. In our calculations, we choose $\mu \approx 10^4 \mu_0$. U_{pn} is the nonmagnetic part of the interaction, U is a part of particle's energy, not associated with vortices. Ferromagnetic defects may be represented as an assembly of magnetic moments with fixed absolute value. For each magnetic moment,

$$U = KV\sin^2(\varphi - \theta) - \mu H_M \cos\varphi, \quad (4)$$

θ is the angle between the easy axis and the external magnetic field, φ is the angle between the particle magnetic moment and the external field, KV term represents the energy of magnetic anisotropy [8], H_M is the magnetic field by Meissner current. In an assembly of ferromagnetic nanoparticles, the elementary process of a change in the magnetization is the rotation of the magnetic moment of a particle. Figure 1 represents the calculated magnetization loops of such an assembly at different values of θ: $\theta = 0$ (easy axes parallel to an external field), $\theta = \pi/2$ (perpendicular), and the last curve is for the case of ferromagnetic nanoparticles with randomly oriented easy axes.

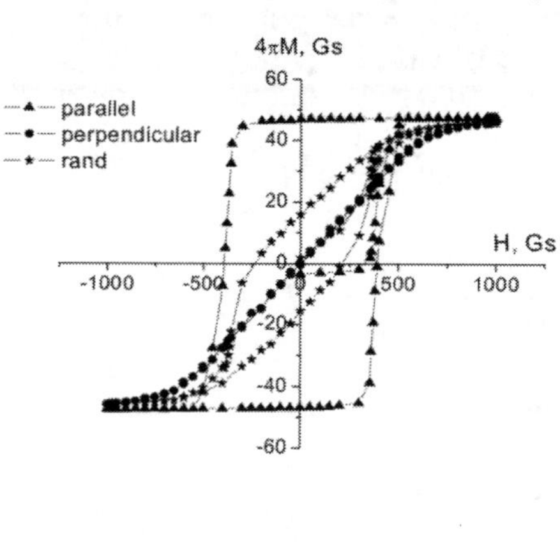

Figure 1. The magnetization loops for ferromagnetic nanoparticles magnetized with an external field with different easy axes orientation. $T = 10K$, much smaller than $T_c = 84K$ and Curie temperature.

3. Calculation results

Consider now the magnetization process for the case of ferromagnetic particles randomly distributed in the bulk of the superconductor. The magnetization data for HTSC sample and ferromagnetic nanoparticles are replotted in figures 2 and 3. As shown in figure 2, the magnetization loop for the particles with easy axes parallel to an external field in HTSC is wider then the same curve for isolated particles, and in figure 3 (easy axes perpendicular to an external field) an initially reversible magnetization process become irreversible for particles in HTSC. This effect is due to the nonlinearity of the interaction of the superconductor with ferromagnetic defects (see figure 4): the value of magnetic field at the location of each particle is $\sim n\Phi_0$ (n is vortex concentration), which do not coincide with the value of an external field.

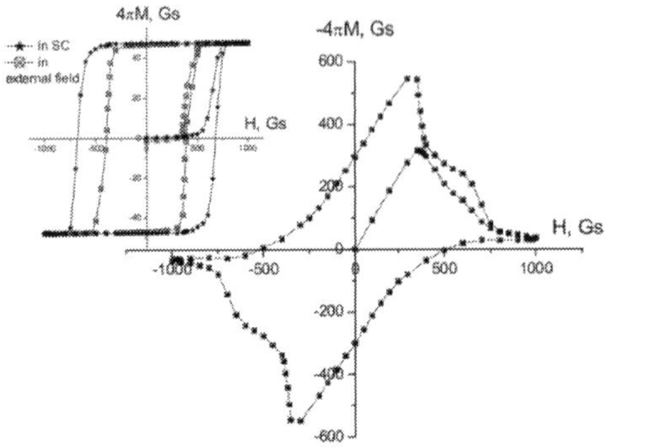

Figure 2. Magnetization curves for the case of particles' easy axes are parallel to the external field. Inset: magnetization curves for nanoparticles only. T=10K, particles' concentration everywhere $c = 1{,}1 \cdot 10^9 cm^{-2}$.

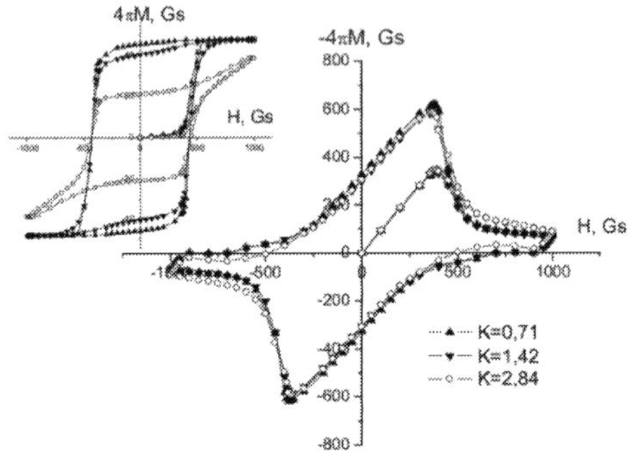

Figure 3. Magnetization curves for the case of particles' easy axis are perpendicular to the external field. Inset: magnetization curves for nanoparticles in SC. K value is given in units of $\varepsilon_0 = \Phi_0^2 s/(4\pi\lambda)^2$. T=10K.

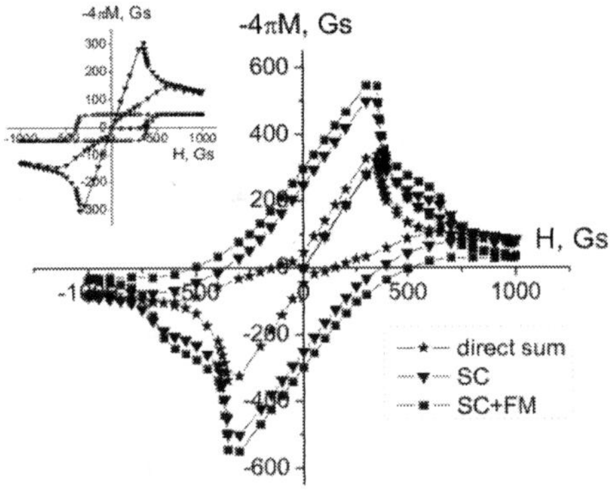

Figure 4. Nonlinear interaction of a ferromagnet with a superconductor: the magnetization curves for nanoparticles only and a pure superconductor (inset) and comparison of the result of exact calculation for superconductor with ferromagnetic defects with the magnetization loop obtained by direct summation of the curves in the inset.

Consider now a periodic lattice of extended ferromagnetic defects. In our calculations, an extended defect is represented as a cluster of isolated magnetic moments, the distance between moments in the cluster is $\sim \lambda$. The results of calculations for the case of elongated defects are parallel and perpendicular to the direction of vortex entrance, are represented in figure 5. As shown, the linear size of defect does not influence on the magnitude of remanent magnetization in the parallel case and strongly influence on the remanent magnetization in the perpendicular case (see inset in figure 5). There is also a peak near H_{c1} on the curve for perpendicular case. Both effects are due to so-called screening of regions near surface [9, 10]. Magnetic flux, pinned on elongated defects, does not let next vortices enter the sample.

Interaction between magnetic moments is strong enough only for particles in the same cluster, when distance between particles is less than λ and we may disregard screening effects. The interaction energy is assumed to be an interaction energy between two point dipoles and reads:

$$U_{\text{int}} = -\frac{\mu_1 \mu_2}{R_0^3}\left[2\sin\varphi_1 \sin\varphi_2 - \cos\varphi_1 \cos\varphi_2\right], \quad (5)$$

where φ_1 and φ_2 are the angles between magnetic moment dipoles μ_1 and μ_2 and an external field. The magnetization data for clusters of ferromagnetic nanoparticles are shown in figure 6.

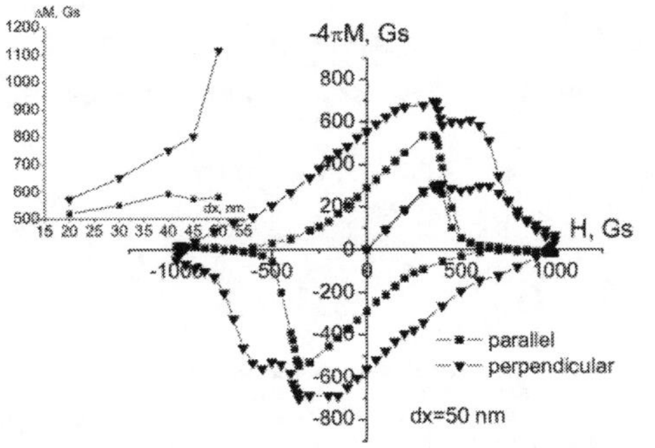

Figure 5. Magnetization curves for HTSC with elongated magnetic defects with fixed distance dx between magnetic moments in cluster. Inset: dependence of residual magnetization on the dx value.

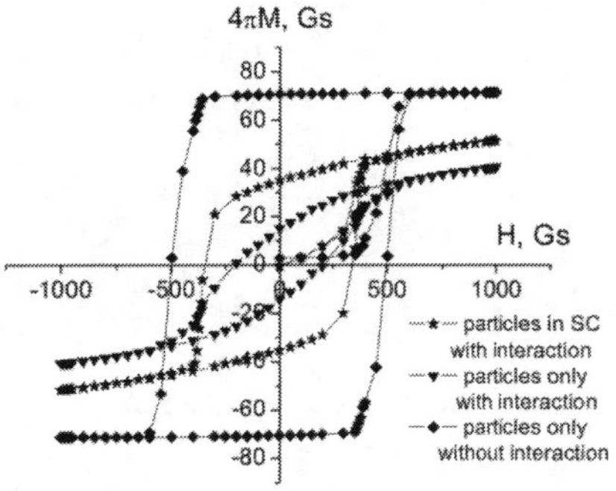

Figure 6. Magnetization curves for ferromagnetic nanoparticles with interaction.

4. Conclusion

In summary, the magnetization reversal loops of 2D layered HTSC were calculated by Monte-Carlo method for both point and extended ferromagnetic defects. Orientation of ferromagnetic nanoparticles' easy axes and interaction between magnetic moments in extended defect were taken into account. It was shown, that interaction between magnetic moments within extended defect decreases the area of magnetization loop, coercive force and, therefore, hysteresis energy losses.

Acknowledgements

The work was supported by RFBR, Grant **12-02-00561**.

[1] Alden T H, Livingston J D 1996 *J. Appl. Phys.* **37** 3551
[2] Snezhko A, Prozorov T, Prozorov R 2005 *Phys. Rev. B: Condens. Matter* **71** 024527
[3] Palau A, Parvaneh H, Stelmashenko N A *et al* 2007 *Phys. Rev. Lett.* **98** 117003
[4] Shapoval T, Metlushko V, Wolf M, Holzapfel B *et al* 2010 *Phys. Rev. B: Condens. Matter* **81** 092505
[5] Milosevic M V, Yampolskii S V, Peeters F M 2002 *Phys. Rev. B: Condens. Matter* **66** 174519
[6] Lawrence W E, Doniach S 1970 in *Proceedings of LT 12* ed E Kanda, p 361
[7] Kashurnikov V A, Rudnev I A, Gracheva M E, Nikitenko O A 2000 *JETP* **90** (1) 173
[8] Prozorov R, Yeshurun Y, Prozorov T, Gedanken A 1999 *Phys. Rev. B* **59** 6956
[9] Zyubin M V, Rudnev I A, Kashurnikov V A 2003 *JETP* **96** (6) 1065
[10] Kashurnikov V A, Rudnev I A, Zyubin M V *Supercond. Sci. and Techn.* 2001 **14**(9) 695

The role of Pd over layer thickness on PLD YBCO coated conductor properties

A Mancini[1], F Rizzo[1], A Rufoloni[1], F Fabbri[1], A Augieri[1], A Vannozzi[1], G Thalmaier[2], T Petrisor[2], A Angrisani Armenio[1], V Galluzzi[1], and G Celentano[1]

[1]ENEA Frascati Research Centre, Via E. Fermi 45, I-00044 Frascati, Italy
[2]Technical University of Cluj, 3400 Cluj-Napoca, Romania

E-mail: antonella.mancini@enea.it

Abstract. Varying the thickness of the Pd layer, the role of the Pd films deposited on the cube-textured Ni-based substrates, was assessed in order to investigate the issues related to substrate oxidation phenomena and their impact on the quality of the buffer layer structures. Using a low temperature annealing, 600 °C, a threshold for the Pd film thickness, beyond which the Pd layer hinders the oxides formation on the substrate, can be defined. When the Pd-Ni interdiffusion is complete, oxide phases of Ni and of Ni-W coexist in the substrate. Besides, at temperatures as high as 800°C, the $NiWO_4$ formation is favoured as the thickness of the Pd layer increases. Despite of these phenomena, development of a buffer layer template suitable for YBCO growth are reported with a thickness for the Pd layer as low as 20 nm.

1. Introduction

For the development of superconducting tapes based on $YBa_2Cu_3O_{7-x}$ (YBCO) films, the beneficial role of a Pd layer between the Ni- or Cu-based substrates and the buffer layer structures has been widely demonstrated [1-5]. On a cube-textured Ni-W substrates buffered with a 200 nm thick Pd layer and a standard buffer-layer architecture, CeO_2/YSZ/CeO_2 (YSZ is Y_2O_3 stabilized ZrO_2), YBCO films have been obtained with J_C up to 2.2 MA/cm^2 in self-field. These results were obtained for YBCO films grown both by pulsed laser deposition (PLD) or by metal organic decomposition (MOD) techniques. The Pd has different fundamental properties. First of all, the lattice parameter and the coefficient of thermal expansion for the Pd have an intermediate value between the one of the substrate and those of some materials typically used as buffer layer, eg CeO_2 or MgO. This contribute to relieve the misfit stress and the thermal stress of the buffer layer structures grown on cube-textured metal tape. Therefore, the eteroepitaxial growth at relatively low temperatures, as low as 350 or 500°C, and a higher stability against the delamination are assured. The Pd layer shows an orientation distribution in the out-of-plane direction up to 60% sharper with respect to the substrate [5]. This characteristic can be related to the formation of a Ni-Pd solid solution at the interface that can alleviate the high lattice mismatch, about 10%, between the Pd layer and the Ni-W substrate [6]. Besides, it was observed that the quality of the CeO_2/substrate interface after an oxidation process is largely improved by the presence of the Pd layer [7].

In the present work, a study regarding the effects of the Pd layer thickness on the substrate oxidation process and on the properties of the CeO_2/YSZ/CeO_2 buffer layer structure is reported. The obtained results indicate that a Pd film thickness as low as 20 nm can promote the development of high quality buffer layer structures for the development of coated conductors based on YBCO film.

2. Experimental details

The Pd films were deposited by e-beam technique on the cube textured $Ni_{95}W_5$ (Ni-W) tapes. Details on the substrate preparation and on the Pd film growth are reported elsewhere [8, 5].

The buffer layer architecture, CeO_2/YSZ/CeO_2, and the YBCO film have been grown by Pulsed Laser Deposition (PLD) in two different vacuum chambers: one chamber is equipped with a Nd:YAG laser in the 4th harmonic configuration and used for the deposition of the buffer layer structures, and the other chamber equipped with a XeCl excimer laser and used for the YBCO film deposition. In the case of the buffer layer structure, the starting materials were high density sintered targets with purity of 99.99 %. The seed layer of CeO_2 and the intermediate layer of YSZ were grown at 500 °C in vacuum and in 10 mTorr of oxygen flowing atmosphere, respectively. Subsequently, the temperature was raised up to 750 °C in vacuum at a rate of 10°C/minute, then a CeO_2 cap layer was deposited in a 10 mTorr atmosphere of flowing oxygen. The laser parameters were: fluence of about 2.2 J/cm^2 and frequency of 3 Hz. The deposition time is adjusted in order to obtain the following thickness for the deposited layers: 15 nm for the CeO_2 top layer, 110 nm for the intermediate YSZ layer and 45 nm for the CeO_2 seed layer. The growth rates are 2.5 Å/s and 0.7 Å/s for CeO_2 and YSZ, respectively. Details on YBCO film preparation are reported elsewhere [9].

Structural and morphological properties were analysed by means of X-ray diffraction (XRD) and Scanning Electron Microscopy (SEM). Standard X-ray θ-2θ and ω-scans were performed with the CuKα radiation using a Rigaku Geigerflex diffractometer equipped with a graphite monochromator on the diffracted beam, or with a 9 kW rotating anode Rigaku SmartLab diffractometer equipped with a Johansson monochromator on the incident beam. SEM investigations have been done using a LEO 1525 field emission high resolution scanning electron microscope. The zero resistance temperature transition, T_C(R=0), was derived from the temperature dependence of the electrical resistance measured by a four-point method.

3. Results and discussion

3.1. Pd thickness and substrate oxidation mechanism

X-ray spectra of Pd/Ni-W samples subjected to annealing process at two different temperatures in 10 mTorr oxygen pressure for 30 minutes are reported in Figure 1. The thicknesses of Pd films is ranging from 0 to 100 nm. In the case of the uncoated substrate, the formation of a NiO layer for both the annealing temperatures, 600°C and 800°C, is observed.

At 600°C, Figure 1a, the complete interdiffusion between the Pd film and the substrate occurs for a Pd film thickness of 15 nm, and the formation of NiO as well as of $NiWO_4$ in the substrate is observed. In the case of thicker Pd films, a part of the Pd film resists against the interdiffusion and the $NiWO_4$ phase is not detectable.

At the higher annealing temperature, 800°C, Figure 1b, even the thicker Pd layer interdiffuses almost entirely in the substrate, and oxide phases of Ni and of Ni-W coexist. In Table 1, the integrated

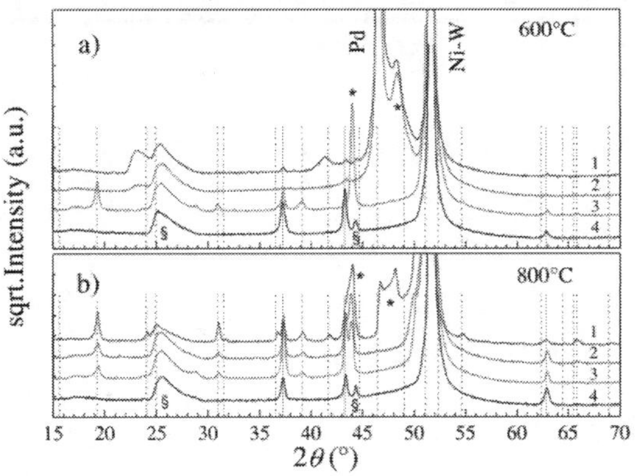

Figure 1. XRD spectra acquired on Ni-W tapes coated with a Pd layer of thickness 1) 100 nm, 2) 50 nm, 3) 15 nm and 4) without the Pd layer. Samples were annealed in a low oxygen pressure (10 mTorr), at temperature of: (a) 600 °C and (b) 800 °C. Symbols * and § indicate features related to the interdiffusion process and to the substrate, respectively [5]. Dashed lines mark $NiWO_4$ stronger peaks, continuous lines mark NiO peaks.

Table 1 Integrated intensity of NiO and $NiWO_4$ XRD peaks for Pd/Ni-W samples annealed at 800°C in 10mTorr of O_2 atmosphere.

sample	Pd (nm)	NiO	$NiWO_4$
1	100	40.2	96.2
2	50	165.4	52.1
3	15	312.7	37.4
4	0	103.7	0

intensities of the diffraction peaks of NiO and $NiWO_4$ are listed. The greater the thickness of the Pd layer, the easier the formation of $NiWO_4$. This reveals that an increasing amount of W is involved in the oxidation process. It is possible to speculate that the Pd presence in the Ni-W matrix, promotes the W migration in a region where the oxidation is easier, e.g. on the surface, or even favors the oxygen diffusion inside the substrate, for example due to an increase of the lattice disorder.

Thermo-Gravimetric analyses, TGA, performed in air on the bare substrate and on the Pd coated substrate are reported in Figure 2. The sample with the Pd coating shows an increase mass higher if compared with the uncoated one. Moreover, the cross sections, analyzed by SEM, reveal different thicknesses for the oxide penetration in the bare and in the covered substrate, of about 5 and 10 μm, respectively. This remarkable thickness difference together with TGA results confirm the greater ease to absorb oxygen for the coated sample.

3.2. Pd thickness and coated conductors properties

In Figure 3a, the XRD spectra of CeO_2/YSZ/CeO_2 buffer layer structures deposited on Pd/Ni-W substrates, with different thickness of the Pd layer, are reported. The intense (00ℓ) peaks reveal a good epitaxial growth for each layer, with the $[00\ell]$ direction perpendicular to the substrate surface. As shown in Figure 4 panel a) for the sample grown on a 20 nm thick Pd layer, despite the oxidation and the interdiffusion processes, the buffer layer structures surfaces are smooth and continuous.

In the Figure 3b the spectrum for the YBCO sample deposited on the buffer layer structure with a 110 nm thick Pd layer is reported. The YBCO film is completely *c*-axis oriented and a full width at half maximum (FWHM) for the (006) YBCO peaks ω-scan of about 2.4° in the rolling direction was

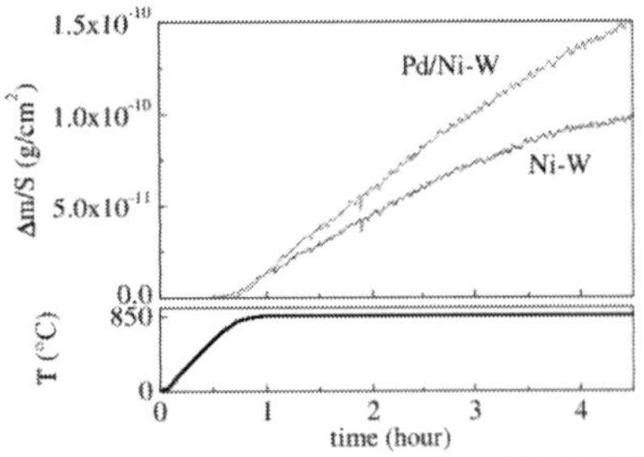

Figure 2 Mass gain profile and temperature, from in air TGA test on an uncoated Ni-W substrate and on a 200 nm thick Pd buffered substrate.

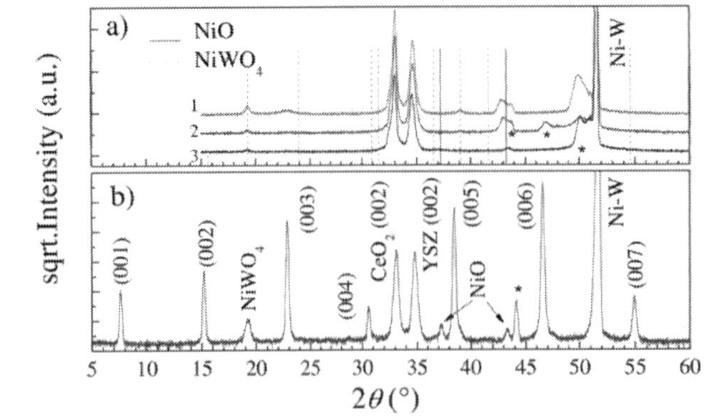

Figure 3. a): XRD spectra for CeO_2/YSZ/CeO_2 buffer layer structure deposited on Ni-W tape coated with a Pd layer of thickness (1) 165 nm, (2) 110 nm and (3) 20 nm. Marks are the same as figure 1. b): XRD spectrum for the YBCO film deposited on CeO_2/YSZ/CeO_2 buffer layer structure with a 110 nm Pd over layer.

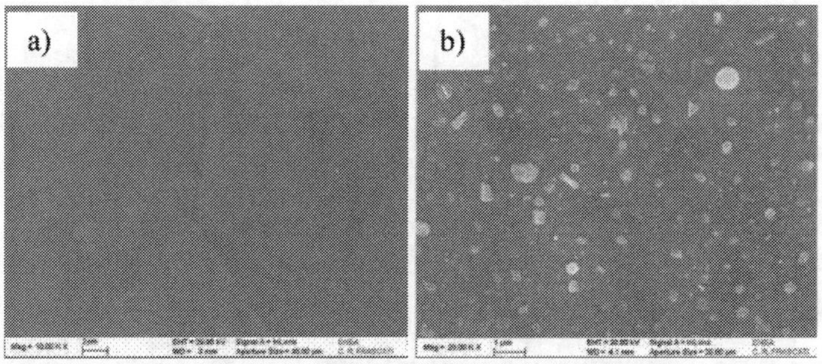

Figure 4. SEM surfaces images of: a) the CeO_2/YSZ/CeO_2 buffer layer structure deposited on Ni-W tape coated with a 20 nm thick Pd layer; b) YBCO film deposited on the buffer layer structure with a 110 nm thick Pd layer.

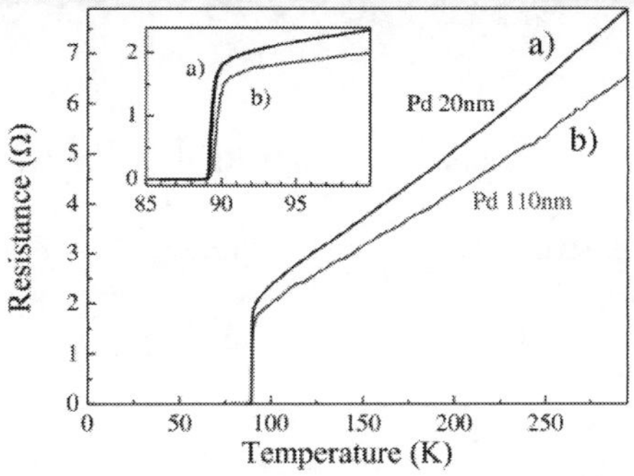

Figure 5. Resistance versus temperature dependence for YBCO films deposited on CeO_2/YSZ/CeO_2/Pd/Ni-W with a Pd layer of about: (a) 20 nm and (b) 110nm, respectively. In the inset, the detail of the transition region is shown.

detected, indicating a sharp out-of-plane orientation. The morphological characterization of the YBCO surface, Figure 4b, reveals that the film is compact and dense. Some features typical of the YBCO film grown by PLD, like droplets and out-growth, are present. Similar characteristics are obtained in the YBCO films grown exploiting Pd layers of about 20 and 165 nm thick.

Regardless of the thickness of the Pd film, the dependence of the electrical resistance with the temperature shows similar behavior, as reported in Figure 5. YBCO films exhibit a narrow superconducting transition and zero resistance critical temperature values, T_C, of about 89 K. As a result, Pd/Ni-W structures with a Pd layer 20 nm thick, have been validate as suitable for the development of the YBCO based coated conductor.

4. Conclusions

The role of the Pd layer thickness on the substrate oxidation process in the case of the Pd/Ni-W structure has been investigated. Results show that if a complete Pd-Ni interdiffusion occurs an easier $NiWO_4$ formation is observed. Besides, the oxidation process is progressively enhanced as the thickness of the Pd layer increases. However, at temperature as low as those used for the deposition of the first layers of the buffer structure, 500°C, a part of the Pd film is not involved by the interdiffusion. As a result, the epitaxial growth of a high quality buffer structure and YBCO film is allowed even exploiting a Pd layer with thickness as low as 20 nm.

References

[1] He Q *et al.* 1997 *Physica* C **275** 155–61.
[2] Celentano G *et al.* 2005 *IEEE Trans. Appl. Supercond.* **15** 2691–4.
[3] Ciontea L *et al.* 2008 Journal of Physics: Conference Series **97** 012302.
[4] Vannozzi A *et al.* 2011 *IEEE Trans. Appl. Supercond.* **21** 2908-11.
[5] Mancini A *et al.* 2008 *Supercond. Sci. Technol.* **21** 015003.
[6] Je J H *et al.* 2002 *Physica* C **383** 241–6.
[7] Mancini A *et al.* 2011 *IEEE Trans. Appl. Supercond.* **21** 2891-95.
[8] Varesi E *et al.* 2002 *Physica C* **372-376** 763.
[9] Varesi E *et al.* 2003 *Supercond. Sci. Technol.* **16** 498-505.

Influence of middle-energy ion-irradiation on the flux pinning properties of YBCO films: Comparison between different synthesis methods

H Matsui, H Ogiso, H Yamasaki, M Sohma, I Yamaguchi, T Kumagai, and T Manabe
National Institute of Advanced Industrial Science and Technology (AIST), Tsukuba 305-8565, Japan

E-mail: h.matsui@aist.go.jp

Abstract. We studied effects of 3-MeV gold-ion irradiation on critical current density (J_c) in $YBa_2Cu_3O_7$ films prepared by three different deposition methods: fluorine-free metal-organic deposition, thermal co-evaporation, and metal-organic chemical-vapor deposition. The three films showed strikingly distinct responses to the extrinsic pinning-center introduction in their J_c values in magnetic fields, namely, about 300%, 70%, and no enhancement in the above order. Transmission electron microscopy results suggest that this characteristic process dependence is closely related to the microstructure of the as-deposited films.

1. Introduction

Recently, we observed a 4-fold enhancement of critical current density (J_c) in magnetic fields by 3-MeV gold ion irradiation in $YBa_2Cu_3O_7$ (YBCO) films prepared by fluorine-free metal-organic deposition (FF-MOD) [1]. This result was remarkable because it evidenced that ion irradiation in the so-called nuclear stopping regime (0.1-5 MeV in YBCO), which creates point-like defects in a target, can provide additional pinning force as strong as those by irradiation in the electronic stopping regime (> 100 MeV), which creates columnar defects [2]. Properties and the mechanism of this strong point-like pinning-center have not been elucidated, especially how a microstructural character before irradiation plays a role in the remarkable J_c enhancement. To clarify this point, we performed 3-MeV-gold-ion irradiation on YBCO films prepared by thermal co-evaporation (TCE) and metal-organic chemical-vapor deposition (MOCVD) and compared their results with FF-MOD.

2. Experiments

Basic properties of YBCO films used in this study are summarized in Table 1. We prepared FF-MOD films [3], and purchased TCE and MOCVD films from CERACO Ltd [4] and Superpower Ltd [5], respectively. The MOCVD films were dry-etched before irradiation to remove a silver protection layer and to make the film thickness comparable to that of the other film types. The dry-etch was done by conventional argon-ion milling (Hakuto Co. Ltd, 3-IBE) with accelerating voltage 300 V and ion-current density 1 mA cm^{-2}. The irradiation experiments for extrinsic pinning-center introduction were performed using a 1MV tandem-type Van-de-Graaff accelerator (NEC3SDH-4) in a vacuum of 10^{-5} Pa. The incident angle of ions was set normal to the film surface. J_c data was collected by a non-destructive ac inductive method at the liquid nitrogen temperature [6]. All J_c data of FF-MOD films was presented in our previous publication [1]. Transmission electron microscope (TEM) measurements were performed using TOPCON EM-002B operated at 120 kV.

Table 1. YBCO films used in this study.

Process	Substrate	Thickness (μm)	T_c (K)	J_c^{self} (MA/cm^2)
FF-MOD	$SrTiO_3$	0.70	91.0	2.6
Thermal co-evaporation	$SrTiO_3$	0.65	87.9	2.8
MOCVD	Hastelloy	0.67	90.5	3.0

3. Results and discussion

Figures 1(a)-1(c) show dependences of J_c on magnetic field and on irradiation dose (B_Φ) in YBCO films prepared by the different processes. Magnetic field was directed normal to the film surface. We note that B_Φ does not represent an areal defect density as in the case of columnar defects created by sub-GeV ion irradiation [2]. The dose-dependence of J_c at a representative field-strength (1 T) is shown in figure 1(d). We find a striking deposition-process dependence in the width of J_c enhancement, namely, about 300%, 70%, and 0% in FF-MOD, TCE, and MOCVD films, respectively. We also find that J_c drops at the highest dose in all film types due to the universal "excess-irradiation" effect, namely, a drop in both the condensation energy density and the percolation path of supercurrent associated with an extensive destruction of superconducting matrix. This effect seems to be dominant irrespective of B_Φ in MOCVD films resulting in no J_c enhancement by the irradiation.

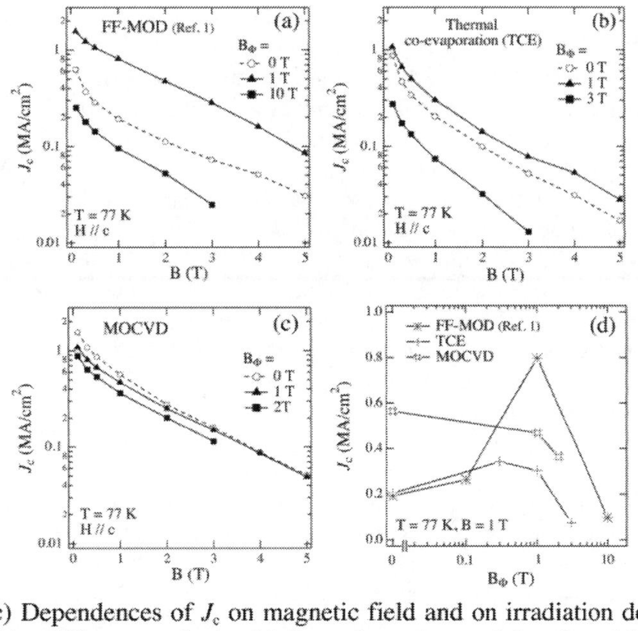

Figure 1. (a-c) Dependences of J_c on magnetic field and on irradiation dose B_Φ in YBCO films prepared by different methods. (d) Dose dependence of J_c at B = 1 T.

To study the dependence of the irradiation effect on magnetic-field-angle θ, in figure 2, we show $J_c(\theta)$ curves normalized to H // *ab* obtained before and after irradiation (B_Φ = 1 T). Essentially the same results were obtained at 0.1 and 1 T where we performed θ-dependent measurements. As we reported previously [1], MeV-ion irradiation on a FF-MOD film enhances J_c more significantly at H // *c* than H // *ab* [figures 2(a) and 2(d)]. This basic anisotropy trend is also found in figures 2(b) and 2(e), suggesting that the same type of pinning centers were introduced by the irradiation in FF-MOD and TCE films, although the width of J_c enhancement is largely different between the two. In figures 2(c) and 2(f) showing J_c anisotropies in MOCVD films, we find no significant irradiation effect indicating an isotropic suppression of J_c by the physical treatment. This can be understood by considering the above-mentioned excess-irradiation effect independent of the direction of magnetic field.

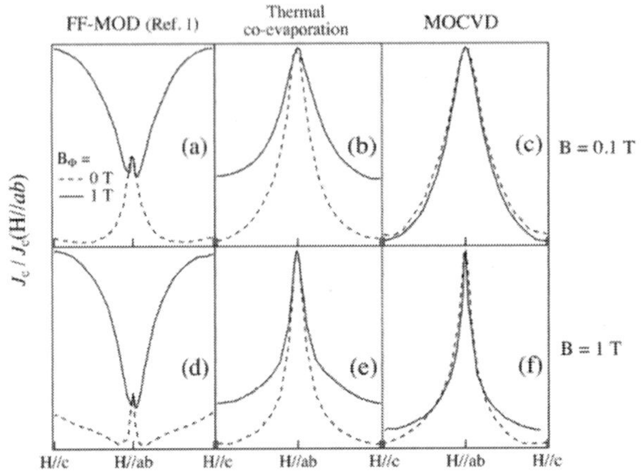

Figure 2. Magnetic-field-angle dependence of J_c normalized to H // *ab* obtained before and after irradiation (B_Φ = 1 T). The data were collected at 0.1 and 1 T.

Next, we argue the origin of the strong deposition-process dependence observed in figures 1 and 2. Since we set both irradiation conditions and film thickness almost identical among the three types of YBCO films, it is unlikely to attribute the dependence to the total energy transferred from incident ions to the target films. Instead, the dependence may be explained by a difference in the microstructural characters of the as-deposited films, because J_c is determined not only by the property of normal-state cores, but also by that of the superconducting matrix surrounding the cores through the condensation energy density and the percolation path of supercurrent. Namely in the simplest picture, given the same amount of additional pinning centers, a larger J_c enhancement is expected in a system with a larger fraction of unperturbed matrix. To evaluate this simplest explanation of figures 1 and 2, we performed cross-sectional TEM measurements on as-deposited films prepared by the three processes (figure 3). In consistent with our previous report [7], figures 3(a) and 3(d) show that the microstructure of a FF-MOD film is characteristic in its extremely small density of defects in nm-scale. In striking contrast to this relatively "clean" microstructure provided by FF-MOD, the microstructure of a TCE and MOCVD film are characterized by dense nm-scale defects. Indeed, we find in figures 3(b) and 3(e) that the TCE film is filled with stacking faults parallel to the *ab*-plane, and in figures 3(c) and 3(f) that the MOCVD film includes *c*-axis-correlated defects identified by moiré patterns and the stacking faults with considerable density. Although it is difficult to quantitatively correlate the results in figure 3 with those in figures 1 and 2, we can find a qualitative consistency between them in the context of the above-mentioned simple scenario. Therefore, the present results suggest that in YBCO films, an as-deposited microstructure plays a central role in how J_c in magnetic field responses to the extrinsic pinning-center introduction by middle-energy (0.1 – 5 MeV) ion irradiation.

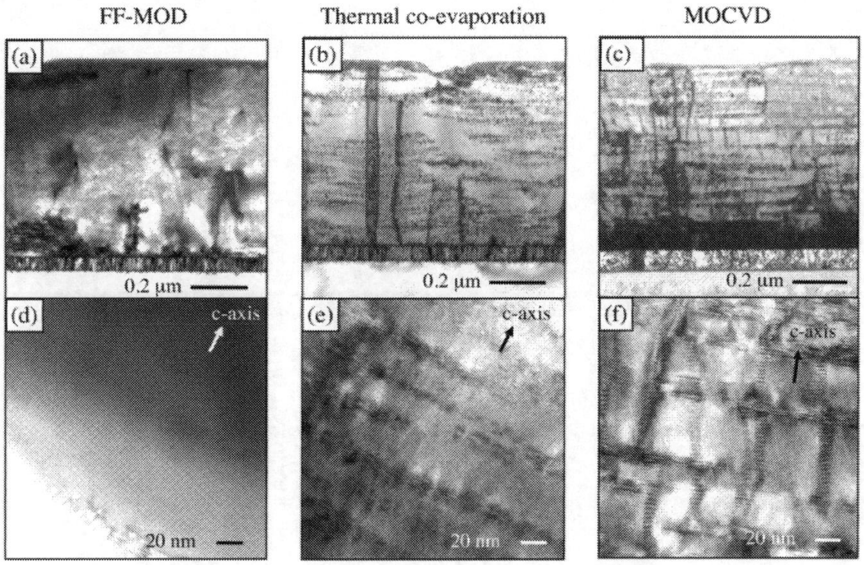

Figure 3. Cross-sectional TEM images of as-deposited YBCO films.

4. Conclusion

We studied effects of 3-MeV Au^{2+}-ion irradiation in YBCO films prepared by three different deposition processes: FF-MOD, thermal co-evaporation, and MOCVD. The response of in-field J_c values to the extrinsic pinning-center introduction was found strongly dependent on the film-deposition processes namely, about 300%, 70%, and no enhancement in the above order. TEM results indicated that this characteristic process dependence is closely related to the microstructure of the as-deposited films.

Acknowledgement

This research was partially supported by TEPCO Memorial Foundation.

References

[1] Matsui H, Ogiso H, Yamasaki H, Kumagai T, Sohma M, Yamaguchi I and Manabe T 2012 *Appl. Phys. Lett.* **101** 232601
[2] Civale L *et al.* 1991 *Phys. Rev. Lett.* **67** 648
[3] Matsui H, Tsukada K, Tsuchiya T, Sohma M, Yamaguchi I, Manabe T and Kumagai T 2011 *Thin Solid Films* **519** 8063
[4] Utz B, Semerad R, Bauer M, PrusseitW, Berberich P and Kinder H 1997 *IEEE Trans. Appl. Supercond.* **7** 1272
[5] Zie Y Y *et al.* 2005 *Physica C* **426–431** 849
[6] Yamasaki H, Mawatari Y, Nakagawa Y, Manabe T and Sohma M 2007 *IEEE Trans. Appl. Supercond.* **17** 3487
[7] Matsui H, Tsukada K, Tsuchiya T, Sohma M, Yamaguchi I, Manabe T and Kumagai T 2011 *IEEE Trans. Appl. Supercond.* **21** 2933

Bi-2212 round wire development for high field applications

H Miao, Y Huang, S Hong, M Gerace and J Parrell
Oxford Superconducting Technology, Carteret, NJ 07008, USA

E-mail: hanping.miao@oxinst.com

Abstract. Oxford Superconducting Technology (OST) has been continuously improving Bi-2212 round wire performance because of its potential for application in high-field magnets (> 25 T). We focused on Bi-2212 wire configuration design, filament densification and reducing carbon and hydrogen contamination to improve the engineering critical current density (J_E). Several wire configurations have been developed to meet different wire diameter and operating current requirements. The swaging, cold isostatic pressing (CIP) and over-pressure heat treatment processes have been demonstrated to effectively increase Bi-2212 filament mass density in the final wire and result in high performance over long length. The J_E values exceeding 550 A/mm^2 at 4.2 K, 15 T have been achieved on the CIPed 1 m long sample using a 10 bar over-pressure (OP) heat treatment. The twisted Bi-2212 wire significantly reduced ac loss without the critical current degradation.

1. Introduction

It is well known that Nb_3Sn superconducting magnets cannot be applied in the magnetic field higher than 25 T because of the limit of its upper critical field (B_{c2}). High T_c superconductor $Bi_2Sr_2CaCu_2O_{8-x}$ (Bi-2212) wire has demonstrated many attractive features, such as high irreversibility field (B_{irr} > 100 T), isotropic electromagnetic performance, solenoid layer winding, and compatibility with conventional cabling and strand insulation methods. All of these advantages make it easy to adapt the well established Nb_3Sn technology and make Bi-2212 one of most promising materials to enable superconducting magnets > 25 T range [1-8].

One of the largest challenges for large scale application of Bi-2212 wires is the current-limited by filament porosity. Therefore, OST focuses on improving J_c and J_E in long length Bi-2122 round wires by exploring wire processing conditions to reduce the porosities. In this paper, we report our recent results from densifying Bi-2212 core filaments, developing new wire configurations and applying the twist on Bi-2212 round wires.

2. Experimental

Bi-2212/Ag round wires were fabricated by the traditional powder-in-tube and multifilament restack techniques and the detail is described in our earlier papers [3, 7]. In this work, Nexans granulate powder with the optimized composition of $Bi_{2.17}Sr_{1.94}Ca_{0.89}Cu_{1.98}Ox$ was used for all wire fabrication. Bi-2212/Ag wires were heat treated using a partial melt-solidification process in 1 atmosphere flowing oxygen. The over pressure heat treatment was performed at Applied Superconductivity Center of National High Magnetic Field Lab (ASC-NHMFL) as described in other publication [8].

The critical current, I_c was measured using the four-probe transport method with a 0.1 μV/cm criterion at 4.2 K under applied field up to 15 T. The barrel sample is typically 1.2 m in wire length and the voltage tap spacing is 72 cm. Bi-2212 core density was calculated by measuring sample weight, volume and HTS core fill factor. Wire diameter was measured with a laser micrometer. Microstructures were examined using optical and scanning electron microscopies. The quench experiments and detailed microscopy study were performed at ASC-NHMFL [5, 6]. Hysteresis losses were measured in a quantum design physical property measurement system, equipped with a vibrating sample magnetometer at National Institute of Standards and Technology (NIST).

3. Results and Discussion

3.1. Optimum wire configuration and size

One purpose of the present work is development of various sized wires with optimum J_E/J_c values for satisfying the different magnet design and operating current requirements. As well known, the peak J_c and J_E in Bi-2212/Ag wires depend on the final filament size and the best J_E falls in the filament size ~15 μm, demonstrated in our early study using 0.8 mm 85x7 configuration wires [3]. Therefore, in order to produce a large diameter wire, we have to increase the filament number to make the filament size close to 15 μm. The 19, 37, 85 or 121 mono-core hex rods were stacked in the sub-billets. 18 or 36 sub-billet hexes plus 1 Ag central rod hex were bundled and restacked into the AgMg alloy tubes and drawn to a range of final wire diameters from 0.8 to 1.5 mm (referred as 19x36, 37x18, 85x18, 121x18 wires) [9]. Table 1 presents optimized wire size for each configuration. It can be seen that the configurations of 19x36 and 37x18 are most suitable for wire ~0.8 mm, 85x18 for wire ~1.2 mm, and 121x18 for wire ~1.4 mm, respectively. The best J_E values (4.2 K and 15 T) of wires with different configurations versus number of filaments of sub bundle are also listed in table 1. All the samples with the similar fill factor of ~25% were made by OST standard process [3] without further densification and heat treated on barrel (1.2 m in sample length). It can be observed that the as-drawn wire performance strongly depends on the number of filaments in sub elements. It is probably due to their difference in the bridges between filaments within the sub elements. By microstructure study, the filaments within a sub bundle normally merge together after final heat treatment, while not linking across sub bundles [10]. In the as-drawn wires without core densification, because of a large amount of porosity, the bridges between filaments may play an important role for filament continuity, resulting in a better wire performance for higher filament number wire configuration.

Table 1. Bi-2212 round wire configurations, optimum wire diameters and their peak J_E values in as-drawn wires. All samples were heat treated using OST standard HT profile in 1 atmosphere oxygen.

Wire configuration (filament number in sub x number of sub bundle)	Fill factor (%)	Optimized wire diameter (mm)	Peak J_E value (A/mm^2, 4.2 K, 15 T) in as-drawn wires
19 x 36	24.0	0.8	195
37 x 18	24.8	0.8	240
85 x 18	25.2	1.2	315
121 x 18	25.4	1.4	320

3.2. Core densification by swaging, cold isostatic pressing and over pressure heat treatment

In the as-drawn wire, Bi-2212 filament density is about 70% of its theoretical density. The residual 30% void space is distributed within the filaments. During melting treatment, the powder porosity agglomerates in the liquid Bi-2212 of filaments. The C and H from powder and sheath could lead to gas-filled bubble formations [5, 11]. The large bubbles significantly reduce Bi-2212 filament effective cross-section and degrade the long length wire performance. In order to densify Bi-2212 filaments in the as-drawn wire prior to melt heat treatment, swaging or cold isostatic pressing (CIP) processes were applied. By 20% area reduction swaging, the Bi-2212 filament density before heat treatment is increased to nearly 90% of the theoretical density. The cold isostatic pressing at pressures up to 1400 MPa has been used to increase the powder core density. Figure 1 shows the Bi-2212 core density corresponded with CIPing pressure. It is clear to see that the filament density is increased from 71% in the as-drawn wire to > 85% by CIPing with pressure of 650 MPa and > 90% with the pressure of 1200 MPa. As shown in the cross-sections of quenched samples, many large (filament sized) bubbles are observed in as-drawn wire samples quenched from the melt. After 650 MPa CIP, there are fewer bubbles evident during the melting process. The quench study on 0.8 mm 37x18 wire confirms the result [6], showing reduced void fraction in the melted filaments and doubled critical current density.

Figure 2 shows the J_E progress in the recent billets after the core densification. While J_E of ~320 A/mm^2 was obtained in the control as-drawn wires, J_E values of ~470 A/mm^2 at 4.2 K, 15 T were achieved in the core densified wires after swaging and CIPing processes. Further improvement was realized by using the recently developed over-pressure heat treatment by ASC/NHMFL [8]. This new process prevents wire swelling during heat treatment and improves wire performance significantly. J_E of 550 A/mm^2 at 4.2 K and 15 T has been achieved on a 650 MPa CIPed, 1.2 m long sample using a 10 bar over-pressure (OP) heat treatment. Higher performance has been achieved by 100 bar OP process on 5 cm long samples [8]. From this figure, it is also shown highly reproducible Bi-2212 wire performance from billet to billet.

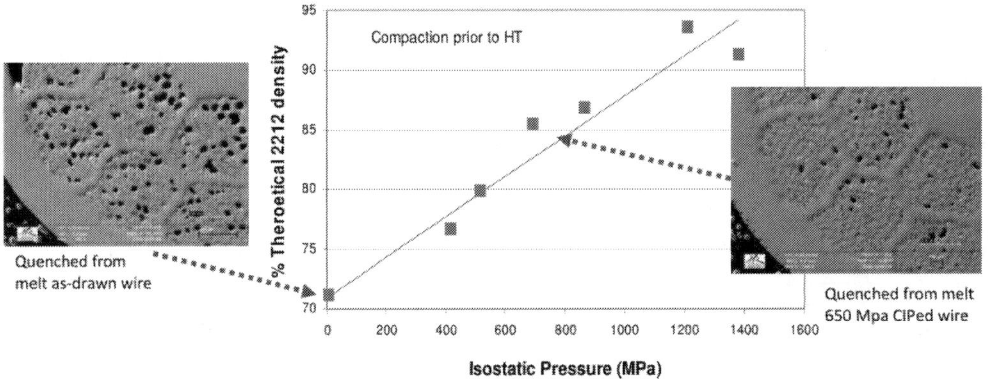

Figure 1. Bi-2212 core density as a function of CIPing pressure. The left picture shows the quenched cross-sections from melt samples in as-drawn and the right picture is by 650 MPa CIPing.

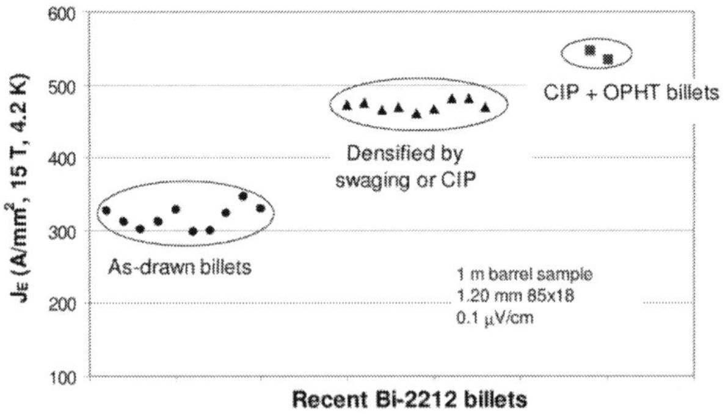

Figure 2. J_E(B) values at 4.2 K, 15 T for 1 m barrel samples obtained in the recent 85x18 wire billets made at OST by the condition of as-drawn, core densified and over-pressure heat treated at 10 bar, respectively.

3.3. Applying twist on Bi-2212 wires

At OST, we recently applied the twist process on the Bi-2212 wires and found that the twisting process does not deteriorate wire performance even twisting to 12 mm in twist pitch length [9]. Figure 3 shows the direct comparison of the hysteresis losses for the untwist sample and twisted sample with 12 mm twist pitch length of 1.2 mm 85x18 wires. It is clear to see that ac loss is reduced to 50% on the twisted Bi-2212 wire.

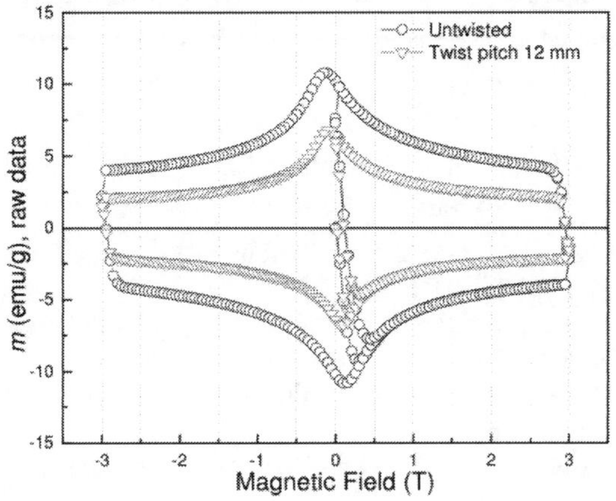

Figure 3. The direct comparison of the hysteresis losses between the untwisted and twisted sample with twist pitch length of 12 mm. The samples were made by the same spiral size and heat treatment.

4. Conclusions

Bi-2212 round wire performance has been significantly improved recently through filament densification. Filament densification achieved via swaging and cold isostatic pressing (CIP) before the melting process, and over pressure heat treatment during the melting process, have been shown to reduce the porosity and enhance the performance of the Bi-2212/Ag wire. The continued improvements in Bi-2212/Ag wire have enabled demonstration of J_E of over 550 A/mm^2 at 4.2 K, 15 T in the 1 meter length wire barrel samples, which provide a great potential for future high field magnet applications. Several filament configurations have been developed to meet different operating current requirements. The twisted Bi-2212 wire significantly reduced ac loss without the critical current degradation.

References

[1] Weijers H W, Trociewitz U P, Marken K R, *et al* 2004 *Supercond. Sci. Technol.* **17** 636
[2] Godeke A, Cheng D, Dietderich D, *et al* 2009 *IEEE Trans. Appl. Supercond.* **19** 2228
[3] Miao H, Marken K R, Meinesz M, *et al* 2005 *IEEE Trans. Appl. Supercond.* **15** 2554
[4] Friend C, Miao H, Huang Y, Melhem Z, *et al* 2010 *IEEE Trans. Appl. Supercond.* **20** 583
[5] Kametani F, Shen T, Jiang J, *et al* 2011 *Superconductor Science and Technology* **24** 075009
[6] Jiang J, Starch W L, Hannion M, *et al* 2011 *Supercond. Sci. and Technol.* **24** 082001
[7] Miao H, Huang Y, Meinesz M, Hong S and Parrell J A, 2012 *Adv. in Cryogenics Eng.* **58** 315
[8] Larbalestier D C, Jiang J, *et al* submitted to *Nature Materials* April 2013, arXiv 1305.1269
[9] Huang Y, Miao H, Hong S and Parrell J A, 2013 submitted to MT-23
[10] Shen T, Jiang J, Kametani F, *et al* 2010 *Superconductor Science and Technology* **23** 025009
[11] Malagoli A, Kametani F, *et al* 2011 *Superconductor Science and Technology* **24** 075016

Journal of Physics: Conference Series **507** (2014) 022021 doi:10.1088/1742-6596/507/2/022021

Improvement of critical current densities in $SmBa_2Cu_3O_y$ films with $BaHfO_3$ nano-rods using low temperature growth technique

Shun Miura[1], Yutaka Yoshida[1], Yusuke Ichino[1], Ataru Ichinose[2], Kaname Matsumoto[3] and Satoshi Awaji[4]

[1]Department of Energy Engineering and Science, Nagoya University, Furo-cho, Chikikusa-ku, Nagoya 464-8603, Japan
[2]Electric Power Engineering Research Laboratory, Central Research Institute of Electric Power Industry, 2-6-1 Nagasaka, Yokosuka, Kanagawa 240-0196, Japan
[3]Department of Materials Science and Engineering, Kyushu Institute of Technology, 1-1 Sensui-cho, Tobata-ku, Kitakyushu 804-8550, Japan
[4]Institute for Materials Research, Tohoku University, Katahira 2-1-1, Aoba-ku, Sendai 980-8577, Japan

E-mail: miura-syun12@ees.nagoya-u.ac.jp

Abstract. We studied growth substrate temperature dependent morphologies of $BaHfO_3$ (BHO) nano-rods for $SmBa_2Cu_3O_y$ (SmBCO) films with BHO fabricated by pulsed laser deposition (PLD) method. The morphology of BHO nano-rods within the SmBCO matrix could be controlled by changing substrate temperatures adoping a Low Temperature Growth (LTG) technique. Using the LTG technique, we could fabricate SmBCO films showed *c*-axis orientation even at low substrate temperature during deposition without a degradation of superconducting properties, which is one of the most advantages of the LTG technique. In this study, we found an optimal morphology of the BHO nano-rods and the morphology was fine, straight and various growth directions. The BHO nano-rods were more effective to improve the J_c in a magnetic field than other morphologies of the BHO nano-rods.

1. Introduction

$REBa_2Cu_3O_y$ (RE: Y and rare-earth elements, REBCO) superconductors are very promising materials, because they have high critical temperatures (T_cs). Research and development of electrical power applications, such as cables, superconducting magnetic energy storage and magnetic resonance imaging, using the REBCO films have been actively carried out [1]. For the applications, it is necessary to improve the critical current density (J_c) of REBCO films in magnetic fields. In order to improve the J_c of REBCO films in magnetic fields, it is effectively to introduce a non-superconducting second phase into a REBCO matrix. Size of the non-superconducting second phase is ideally hoped to be similar to vortex core (~10 nm, for *ab* plane at 77 K). BMO_3 (M: Zr, Sn and Hf, BMO) is very famous vortex pinning centres for *c*-axis of REBCO matrix. The BMO formed "nano-rods" in REBCO matrix which fabricated by vapour phase deposition method such as pulsed laser deposition (PLD) and chemical vapour deposition (CVD). The BMO nano-rods become correlated pinning centres for *c*-axis of a REBCO matrix. Furthermore, morphologies of the BMO nano-rods could be controlled by deposition condition such as substrate temperatures, laser repetition rates and BMO content during deposition [2-5]. In order to control flux pinning properties of BMO nano-rods in REBCO matrix, it is necessary to control the morphologies of the BMO nano-rods. In this paper, we focus on growth substrate temperatures during deposition. The morphology of BHO nano-rods within the SmBCO matrix was controlled by changing substrate temperatures adoping a Low Temperature Growth (LTG) technique. Using the LTG technique, we could fabricate SmBCO films which showed *c*-axis orientation even at low substrate temperature without a degradation of superconducting properties. In this work, we carried out to control morphologies of BHO nano-rods and compared the flux pinning properties of the BHO nano-rods in SmBCO films.

2. Experimental details

All films in this study were deposited on a $LaAlO_3$ (100) single-crystal substrates by PLD method using a KrF (λ = 248 nm) excimer laser at a repetition rate of 10 Hz. The laser energy density and distance between substrate and targets were 2.0 J/cm^2 and 45 mm, respectively. All the films were fabricated by LTG technique. The LTG technique consists of the following two steps. Firstly, a thin SmBCO seed layer was deposited at the relatively high substrate temperature, in order to form a purely *c*-axis oriented SmBCO film on a substrate. Secondly, a SmBCO upper layer was homoepitaxially grown on the seed layer. A SmBCO seed layer was deposited at 850 °C with a thickness of about 50 nm on the substrate, and then an upper layer of BHO doped SmBCO was grown at 700 ~ 840 °C with a thickness of about 500 nm on the seed layer. Then, we define substrate temperatures of upper layer as T_s^{upper}. We applied an alternating targets (ALT) technique to introduce BHO into the SmBCO upper layer. The ALT technique is a way to obtain films with a designed composition by exchanging two different targets alternately. In this study, we used a pure SmBCO target and a pure BHO target to deposit the BHO doped SmBCO films via the ALT technique. The number of total irradiation times with pulse laser of upper layer of the SmBCO target and the BHO target are 24000 (= 30 pulses × 800) and 1600 (= 2 pulses × 800), respectively. O_2 partial pressure during the deposition was 400 mTorr.

The crystalline orientation of the films was evaluated by X-ray diffraction (XRD) analysis and microstructure of the films was observed by high-resolution transmission electron microscope (TEM). The BHO contents were measured by inductively coupled plasma-atomic emission spectroscopy (ICP-AES). In this paper, BHO contents of the SmBCO films were 2.7 vol.%. The resistivity and critical current at various magnetic fields were measured by standard four-probe method with a physical properties measurement system (PPMS : Quantum Design). A T_c was defined from a temperature dependence of resistivity by using the resistivity criterion of 0.1 μΩcm and the probe current was fixed to about 25 A/cm^2. The J_c was evaluated from current vs. voltage curves with electric field criterion of

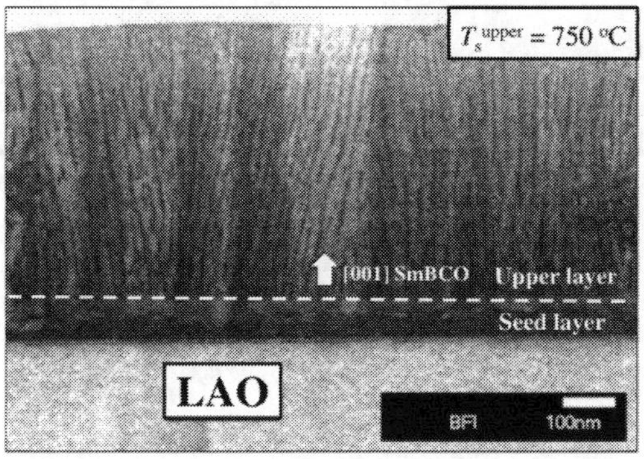

Figure 1 A cross-sectional TEM image of 2.7 vol.% $BaHfO_3$-doped $SmBa_2Cu_3O_y$ film fabricated at T_s^{upper} of 750 °C.

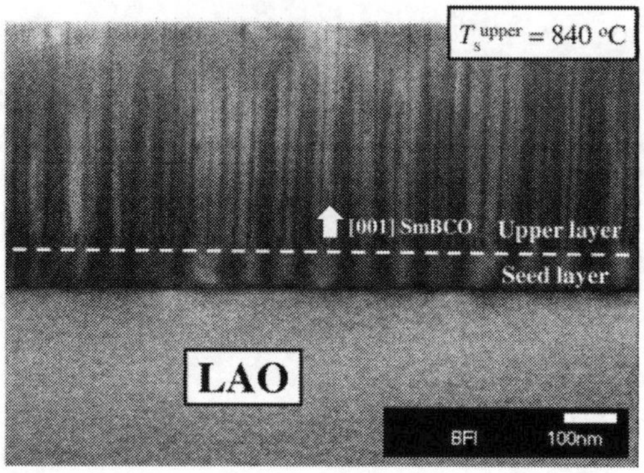

Figure 2 A cross-sectional TEM image of 2.7 vol.% $BaHfO_3$-doped $SmBa_2Cu_3O_y$ film fabricated at T_s^{upper} of 840 °C.

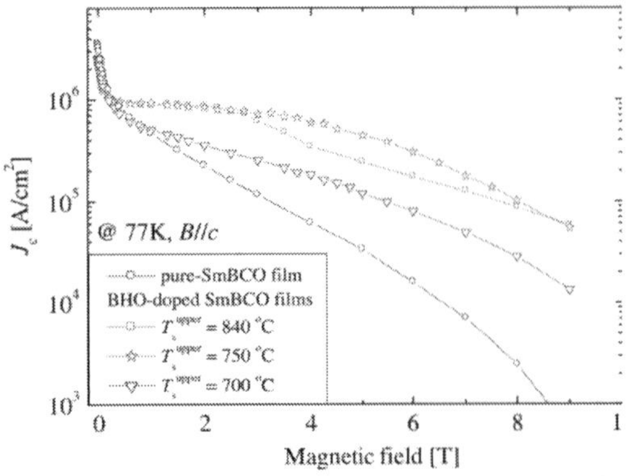

Figure 3 Magnetic field dependence of J_cs of $B//c$ for BHO-doped SmBCO films and a pure-SmBCO film.

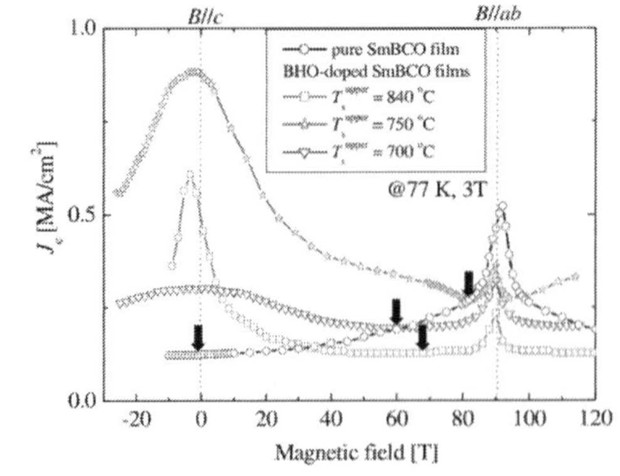

Figure 4 Field angular dependence of J_cs for the BHO-doped SmBCO films and a pure-SmBCO film.

1μV/cm at 77 K. Thicknesses of the films were measured by the TEM and ICP-AES. From the measurements, thicknesses of the films were about 500 nm.

3. Results and discussion

Firstly, in order to investigate morphologies of BHO nano-rods within a SmBCO film fabricated by the LTG technique at various substrate temperatures, microstructures in the LTG-SmBCO + BHO films were measured by TEM images. Figure 1 shows a cross-sectional TEM image of the BHO-doped LTG-SmBCO film fabricated at relatively low substrate temperature, T_s^{upper} = 750 °C. The BHO formed fine and long nano-rods with the diameter and the number density of about 5.7 nm and 2300 /μm^2, respectively. Then the matching field (B_φ) calculated by the number density was 4.7 T. Another important feature is that some BHO nanorods inclined for *c*-axis of the SmBCO. On the other hand, morphologies of BHO nano-rods within a SmBCO film fabricated at relatively high substrate temperature (T_s^{upper} = 840 °C) were different from that of the T_s^{upper} = 750 °C sample. Figure 2 shows a cross-sectional TEM image of the BHO-doped LTG-SmBCO film fabricated at T_s^{upper} = 840 °C. The BHO formed fat nano-rods and grown straightly along with the *c*-axis of SmBCO. The diameter and the number of the BHO nano-rods were about 8.5 nm and 1440 /μm^2, respectively. Then the matching field (B_φ) calculated by the number density was 3.0 T. Based on these, we successfully controlled morphologies of BHO nano-rods in SmBCO films by changing substrate temperature.

We discuss flux pinning properties of the BHO nano-rods in SmBCO films fabricated at T_s^{upper} = 700, 750 and 840 °C. We show magnetic field dependence of J_cs at 77 K of $B//c$ for the BHO-doped SmBCO films and a pure-SmBCO film in Figure 3. The J_c of all the BHO-doped films in high magnetic field were higher than that of the pure-SmBCO film. T_s^{upper} = 750 °C and 840 °C samples showed plateau regions in the J_c-B curves. In addition, the ending points of these plateau regions are about B_φ. In the plateau regions, fluxes motion in the films was reduced because BHO nano-rods are occupied by magnetic fluxes. Thus, in this magnetic field range, J_c is not affected to applied magnetic field. On the other hand, there was no plateau region on the J_c-B curve of T_s^{upper} = 700 °C sample. It was likely to be due to becoming the BHO to particles or very short nano-rods.

We show the applied field angular dependences of J_c at 77 K in 3 T for the BHO-doped SmBCO films fabricated at various T_s^{upper} and the pure-SmBCO film in Figure 4. We can confirm the J_c peak for $B//c$ of T_s^{upper} = 750 °C sample was the highest in this study. Furthermore, the J_c value in applied magnetic field angular of a wide range (0 ~ 80 °) was higher than that of the pure-SmBCO film. Then, we focuses on the detailed shapes of the J_c peaks for $B//c$, the widths of the peaks became broader with T_s^{upper} decreasing. The T_s^{upper}-dependent change could occur as the result of a change in BHO morphologies with T_s^{upper} . In other word, the vortex pinning by inclined the BHO nano-rods could lead to a broader J_c peak for *B//c*. In Figure 4, the minimum J_c values (J_c^{min}) are indicated by arrows. The J_c^{min} of T_s^{upper} = 750 °C sample was higher than other samples, the J_c^{min} value was 0.27 MA/cm^2 (77 K, 3 T). It was confirmed that various growth directions of the BHO nano-rods was effective to improve the J_c^{min}. Superior properties were also found in the angular dependence of J_c of the T_s^{upper} = 750 °C sample compared with the other samples remarkably in this study. Because, maybe the BHO nano-rods became high number density and grew to the various directions.

4. Conclusion

The relationships between morphologies of BHO nano-rods and flux pinning properties were clarified in the SmBCO films with BHO nano-rods. We succeeded in controlling the morphologies of the BHO nano-rods by growth substrate temperatures. It could be confirmed that BHO nano-rods of T_s^{upper} = 750 °C sample were inclined for *c*-axis of SmBCO matrix. In addition, these BHO nano-rods formed fireworks structures. In the future, understanding the mechanism of the formation of the fireworks structures will help us to control morphologies of BHO more effectively. In this study, we found an optimal morphology of BHO nano-rods which was fine, straight and various growth directions. The BHO nano-rods were more effective to improve the J_c^{min} than other morphologies of BHO nano-rods.

Acknowledgment

This work was partly supported by Grants-in-Aid for Scientific Research (19676005, 23226014 and 25289358) from the Japan Society for the Promotion of Science (JSPS).

References

[1] Shiohara Y, Fujiwara N, Hayashi H, Nagaya S, Izumi T and Yoshizumi M 2009 *Physica C* **469** 863

[2] Ozaki T, Yoshida Y, Ichino Y, Takai Y, Ichinose A, Matsumoto K, Horii S, Mukaida M and Takano Y 2010 *J. Appl. Phys.* **108** 093905

[3] Haruta M, Ichinose A, Fujita N, Saura K, Maeda T and Horii S 2013 *IEEE Trans. Appl. Supercond.* **23** 8000904

[4] Ichinose A, Naoe K, Horide T, Matsumoto K, Kita R, Mukaida M, Yoshida Y and Horii S 2007 *Supercond. Sci. Technol.* **20** 1144

[5] B Maiorov, S A Baily, H Zhou, O Ugurlu, J A Kennison, P C Dowden, T G Holesinger, S R Foltyn and L Civale 2009 *Nature Materials* **8** 398

[6] A Ichinose, P Mele, K Matsumoto, H Kai, M Takamura, K Yamada, M Mukaida, S Horii, S Funaki, Y Ichino, Y Yoshida and R Kita 2009 *Phys. C* **469** 1374

Journal of Physics: Conference Series **507** (2014) 022022 doi:10.1088/1742-6596/507/2/022022

Effect of Gadolinium substitutions in La2Zr2O7 films grown on textured Ni-5at.%W substrate by chemical solution deposition

H. Muguerra[1], A-C Pescheux[1], Jean-Louis Soubeyroux[1]

[1] Univ. Grenoble Alpes, Inst NEEL, F-38042 Grenoble, France

CNRS, Inst NEEL, F-38042 Grenoble, France
E-mail : herve.muguerra@grenoble.cnrs.fr

Abstract

In the $YBa_2Cu_3O_{7-x}$ coated conductor, the buffer layer is crucial for the texture transfer from the Ni-5%W substrate to the superconducting layer. We developed original $La_{2-x}Gd_xZr_2O_7$ buffer layers ($x \leq 1$) by chemical solution deposition to decrease the structure mismatch between the substrate and the buffer layer. They show a strong cube texture up to the surface, a crack free surface and a very low roughness. In conclusion, the $La_{2-x}Gd_xZr_2O_7$ layers with $0.4 < x < 0.9$ are promising and should be employed in the future as buffer layer in coated conductor to replace $La_2Zr_2O_7$ buffer layer.

Introduction

The second generation of high temperature superconductor tapes, called coated conductors, presents a considerable interest in the field of high power applications with low energy consumption [1]. Among the different architectures, the association of a rolling-assisted biaxially textured substrate (RABITS) and Chemical Solution Deposition methods (CSD) are considered as one of the most promising approach for the cost-effective processing of long lengths of coated conductors [2]. These coated conductors are built with a $YBa_2Cu_3O_{7-x}$ (YBCO) layer deposited on a Ni-5%W bi-axially textured substrate coated with intermediate buffer layers. In this architecture the superconducting layer is cube-textured along its principal axes, [001] and [100], to overcome weak superconducting coupling in high angle grain boundaries [3]. The buffer layer is an essential ingredient because it acts as crystallographic template for the growth of YBCO and chemically insulates the YBCO layer from the metal substrate. Therefore, the candidates for buffer materials must be selected based on: i) similar lattice parameters with both substrate and YBCO to support epitaxial growth; ii) the efficiency of the barrier against oxidation of the metallic substrate during YBCO film deposition and metal diffusion from the substrate to the YBCO. $La_2Zr_2O_7$ (LZO) has emerged as a promising buffer layer on Ni-5%W and has been obtained by CSD method by a four steps process: 1) synthesis of the precursor solution; 2) deposition of the precursor film on a substrate by dip-coating; 3) low temperature pyrolysis phase to remove the organic matter (typically 400°C); 4) high temperature heat-treatment for nucleation, crystallization and densification of the buffer layer (typically 1000°C under reducing atmosphere). We employed for the first time the CSD method to substitute Gd for La in $La_2Zr_2O_7$. The objectives were to modify the lattice parameter of the buffer layer in order to decrease the structural mismatch between the buffer layer and the substrate and then to improve the texture transfer. Indeed the structural mismatch NiW/$La_2Zr_2O_7$ is 7.5% and NiW/$Gd_2Zr_2O_7$ 5%. In the present work, we report the synthesis of $La_{2-x}Gd_xZr_2O_7$ buffer layers (LGZOx with $x \leq 1.2$). We have employed X-ray diffractions, and microscopies (Atomic Force Microscopy, Scanning Electron Microscopy, Transmission Electron Microscopy) to study the influence of the gadolinium substitution on the texture and the microstructure.

Experimental

The precursor $La_{2-x}Gd_xZr_2O_7$ solutions ($0 \leq x \leq 1$) are produced by mixing a stoichiometric amount of lanthanum (III) 2,4-pentanedionate hydrate, gadolinium (III) 2,4-pentanedionate hydrate and zirconium (IV) 2,4-pentanedionate with propionic acid under a continuous stirring at 50°C for 15 minutes. The final solution concentration is 0.6 mol/L. Biaxially textured Ni-5at.%W tapes are used as substrates (Evico, Dresden). Before coating they are cleaned in an ultrasonic bath for 15 minutes in acetone and for 15 minutes in methanol and then are heated up to 800°C for 1 hour in a reducing atmosphere under a continuous flow of Ar-5%H_2 gas to restore the surface properties [02]. The substrates are dip-coated at 60 mm/min and then the precursors layers are dried for 2 minutes at 60°C using infrared lamps. Finally, the coatings are heated at 450°C for 1 hour under primary vaccum and then annealed at 1000°C for 30 minutes under Ar-5%H_2 gas flow. The heat treatment is the same for the different $La_{2-x}Gd_xZr_2O_7$ samples ($0 \leq x \leq 1.2$). X-ray data were collected using a Seifert four circles diffractometer equipped with a Cu K_α source at room temperature in air. Transmission electron microscopy (FEI Titan 50-80), scanning electron microscopy (FEG-SEM, Zeiss Ultra +) and atomic force microscopy (AFM, Veeco D3100) were used to determine the thickness, the morphology and the roughness of the buffer layers.

Results and discussion

X-ray diffraction Θ-2Θ scans for $La_{2-x}Gd_xZr_2O_7$ buffer layers are shown in Figure 1a (LGZOx with $0 \leq x \leq 1.0$). The crystal structure of LGZO buffer layers is described as the $La_2Zr_2O_7$ (LZO) pyrochlore structure. For the different samples only the (004) peak is observed showing a very strong c-axis preferred orientation of the buffer layer on the Ni-5%W substrate. Undesirable (222) peak was not detected (Figure 1a) while it was always observed for the $Gd_2Zr_2O_7$ buffer layers obtained by chemical solution deposition [4-5]. The lattice parameter decreases almost linearly with increasing gadolinium fraction x (Figure 1b), as expected from the smaller ionic radius of Gd^{3+} (1.00 Å) compared to La^{3+} (1.03 Å). Consequently, the structural lattice mismatch between the buffer layer and the substrate also decreases as a function of the Gd fraction, from 7.5% for LZO to 6.25% for LGZO0.9. However at gadolinium contents higher than 0.9, the intensity of the (004) peak decreases. We have increased the annealing temperature up to 1020°C nevertheless we observe no improvement. We concluded that the upper limit of the solid solution in the $La_{2-x}Gd_xZr_2O_7$ system seem to be close to x = 0.9. A higher Gd content seems disturb the crystal growth of the layer, a similar feature has been reported for cerium substitution in $La_2Zr_2O_7$ layer [02].

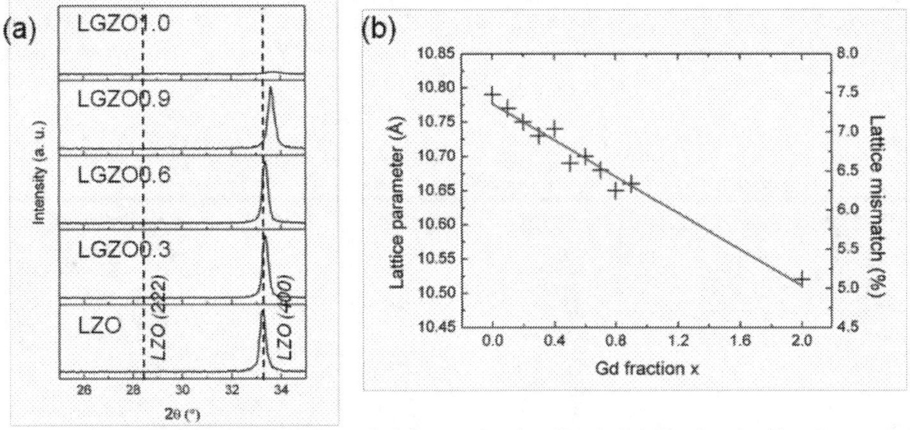

Figure 1 : (a) X-ray diffraction Θ-2Θ scans for $La_{2-x}Gd_xZr_2O_7$ buffer layers, (b) Variation of lattice parameters and structure lattice mismatch LGZO/Ni-5at.%W across the $La_{2-x}Gd_xZr_2O_7$ series as a function of x.

We have determined the in-plane and out-of-plane texture of the LGZO layers by φ-scans of the LGZO(222) reflection and ω-scans of the LGZO(400) reflection in the rolling direction (RD) and transverse direction (TD) of the Ni-5%W tapes. Figure 2 represents the φ-scan and the ω-scans of LGZO0.5 and summarizes the average full width at half maximum (FWHM) measured for the $La_{2-x}Gd_xZr_2O_7$ series as a function of x. In the in-plane direction, the FWHM values are constant as a function of Gd content and remain close to 6.2° which is very close to the values reported for LZO and GZO buffer layers [05]. These values indicate a successful transfer of the substrate biaxial texture to the buffer layer. In the out-plane direction the FWHM values increase almost linearly with increasing gadolinium percentage, from 8° to 9.5° in the RD and from 5.1° to 6.0° in the TD directions. However, they are below the FWHM values reported for $Gd_2Zr_2O_7$ buffer layers obtained by chemical solution deposition, 10.2° in the RD and 6.9° in the TD direction [5], and are in the appropriate range for the deposition of YBCO layers with high Jc [6].

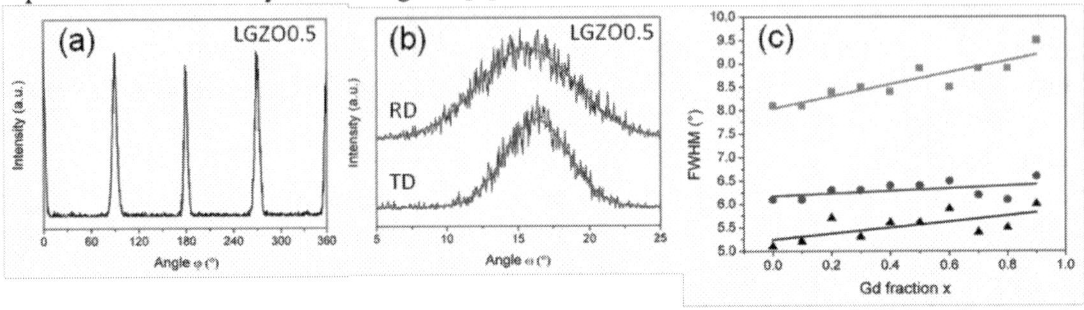

Figure 2 : (a) φ-scan and (b) ω-scans of the LGZO0.5(222) and LGZO0.5(400) reflections in the rolling direction (RD) and transverse direction (TD) of the Ni-5at.%W tapes. (c) Variation of the FWHM value across the $La_{2-x}Gd_xZr_2O_7$ series as a function of x. (●) = φ-scan (■) = ω-scans TD, (▲) = ω-scans RD.

For the different buffer layers, homogenous and crack free surfaces are observed by SEM analysis, as shown in Figure 3 for LGZO0.8. No modification of the grain size is observed as a function of Gd content. The grain diameter is close to 30 nm and is in agreement with the values reported in the literature for LZO and GZO thin layers [3-4]. The thickness of the buffer layers has been estimated by TEM cross-section analysis to be between 60 and 80 nm (Figure 3). Dark regions with a rectangular shape are observed on the TEM images of the different LGZO layers. They are homogenously distributed throughout the bulk of the materials and their diameter is between 10 and 20 nm. In the literature they are described as nanovoids and are observed for LZO and GZO layers [7]. The origin of the nanovoid formation mechanism is not clearly established and different explanations are proposed [7]: i) the escaping of carbon-rich gases creates the nanovoids and they are intrinsic to chemical solution deposition method; ii) the nanovoid appearance is intrinsic to pyrochlore structure and has no relation with the deposition technique.

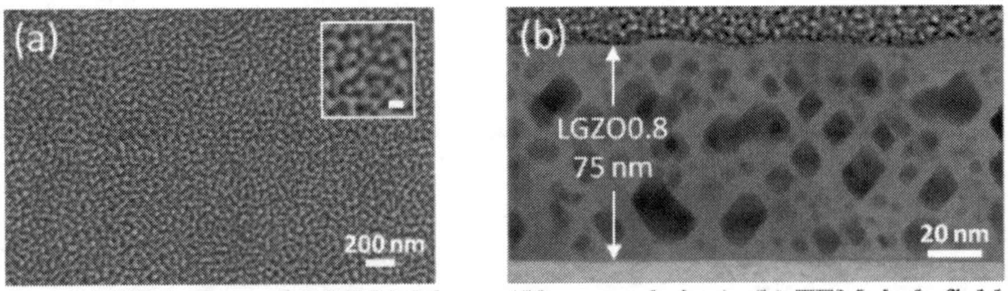

Figure 3 : (a) SEM images of LGZO0.8 layer (50 nm scale bar), (b) TEM dark-field image of LGZO0.8 in cross-section.

AFM investigations of the surface structure of the LGZO samples were carried out and are illustrated in Figure 4. The different pictures confirm the homogenous grain size distribution observed by the SEM analysis. Nevertheless the root mean square (RMS) roughness of the LGZO buffer layers decreases as a function of Gd ratio, from 2.4 to 1 nm measured on $4\mu m^2$ square samples (Figure 4c). Indeed, we observe also a decrease of the grains size as a function of Gd substitution, from 45 nm for LGZO0.1 to 25 nm for LGZO0.9. The RMS roughness of LGZO layers is lower than the values reported for LZO layer, 2.81 nm [03]. A smooth surface is absolutely necessary for the successful epitaxial growth of YBCO layer on the top of the buffer layer.

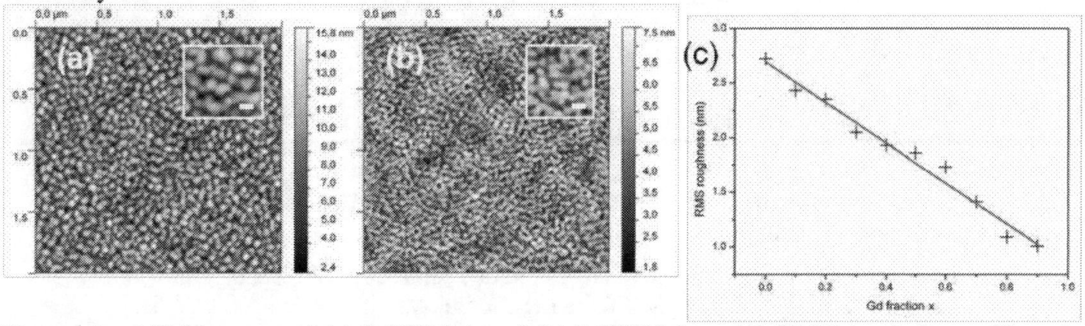

Figure 4 : AFM images of (a) LGZO0.1 and (b) LGZO0.9 on Ni-5%W (scale bar = 50 nm). (c) Variation of the RMS roughness across the $La_{2-x}Gd_xZr_2O_7$ series as a function of x.

Conclusions

This work presents the synthesis and characterization of $La_{2-x}Gd_xZr_2O_7$ layers ($0 < x < 1$) on Ni-5%W. The structural mismatch between the buffer layer and the substrate decreases linearly as a function of the Gd fraction. Nevertheless, we observe no improvement in the texture of the buffer layers. The SEM and AFM analysis revealed a dense morphology with no reduction of porosity as a function of Gd content. The roughness of the $La_{2-x}Gd_xZr_2O_7$ layers decreases linearly as a function of the Gd percentage. Their surfaces are very smooth which is very important for the epitaxial growth of the YBCO layer. The $La_{2-x}Gd_xZr_2O_7$ layers with x between 0.4 and 0.9 have all the characteristics to be promising buffer layers and their integration in coated conductor will be described in a future article focused on superconducting properties.

Acknowledgments

This work has been done under European Integrated Project EUROTAPES. The authors also thank for transmission electron microscopy Alexander Meledin and Stuart Turner of EMAT laboratory, Antwerp, Belgium.

References

[01] A. P. Malozemoff, S. Fleshler, M. Rupich, C. Thieme, X. Li, W. Zhang, A. Otto, J. Maguire, D. Folts, J. Yuan, H.-P. Kraemer, W. Schmidt, M. Wohlfart, H.-W. Neumueller, Supercond. Sci. Technol. 21 (2008) 034005.

[02] V. Narayanan, I. Van Driessche, Progress in Solid State Chemistry 40 (2012) 57.

[03] D. Dimos, P. Chaudhari, J. Mannhart, F. K. LeGoues, Physical Review Letters 61 (1998) 219.

[04] Y. Zhao, J.-C. Grivel, M. Napari, D. Pavlopoulos, J. Bednarčík, M. von Zimmermann, Thin Solid Films 520 (2012) 1965.

[05] Y. Zhao, X.-F. Li, A. Khoryushin, D. He, N. H. Andersen, J. B. Hansen, J.-C. Grivel, Supercond. Sci. Technol. 25 (2012) 015008.

[06] D. Dimos, P. Chaudhar, J. Mannhart, Phys. Rev. B 41 (1999) 4038.

[07] V. Cloet, P. Lommens, R. .Hühne, K. De Buysser, S. Hoste, I. Van Driessche, Journal of Crystal Growth 325 (2011) 68.

Influence of laser striations on the properties of coated conductors

R Nast[1], M Vojenčiak[1,2], E Demenčik[1], A Kario[1], B Ringsdorf[1], A Jung[1], B Runtsch[1], F Grilli[1], W Goldacker[1]

[1]Karlsruhe Institute of Technology, Institute for Technical Physics, Hermann-von-Helmholtz-Platz 1, D-76344 Eggenstein-Leopoldshafen, Germany
[2]Institute of Electrical Engineering, Slovak Academy of Sciences, Dúbravská cesta 9, 841 04 Bratislava, Slovakia

E-mail: rainer.nast@kit.edu

Abstract. Due to their high current carrying capability, coated conductors are regarded as the most promising high-temperature superconductor tapes for power applications. However, their high aspect ratio causes too high magnetization losses. To reduce the ac loss, one way is to striate the wide tapes into filaments. We used a picosecond laser for the structuring of (*RE*)BCO coated conductors. The laser allows to burn 18 μm to 21 μm wide grooves (Ag-cap) with a depth between 0.5 μm to more than 100 μm into the coated conductors, with negligible heat effects at the edges of the structures. Different numbers of filament were structured in Cu- and Ag-cap coated conductors. Patterns with up to 120 parallel filaments in 12 mm wide conductor were made. The critical current and the total ac-magnetization loss were measured as a function of the filament count. With an increasing number of filaments I_c degradation occurs. This current reduction has two contributions, the removed HTS material and current inhomogeneities within the superconductor for instance defects along the tape causing secondary phases. For 120 filaments Ag-cap tapes the hysteresis loss reduction is about two orders of magnitude, as expected. The observation of some remaining filament coupling was investigated.

1. Introduction

Because of their high critical current density $(RE)Ba_2Cu_3O_{7-x}$ (*RE* is a rare earth, usually Y, Gd or Sm) coated conductors are the most promising high-temperature superconducting tapes for power applications. However, for applications involving windings, like transformers, generators and motors, the AC loss is too large. This is because of the perpendicular field component causing important magnetization AC loss on flat tapes with high aspect ratio [1].

Filamentization of the superconductor reduces the magnetization AC loss and many techniques have been evaluated [2-12]. All these techniques differ in the groove width and depth, the degree of I_c degradation, the amount of coupling losses and the stage of complexity which is directly coupled with the cost. The used picosecond laser provides a reliable one step process with minimized I_c degradation scalable to a reel to reel process. However, these striations are only effective if the filaments are transposed, either by twisting the tape or winding it helicoidally, making composite cables, as it was proven experimentally [10]. This transposition is required because of the coupling currents in the metal connecting the filaments to each other, either at the current leads or along the length of the tape.

But in case of a technical conductor some filament coupling is desirable for a good current sharing and stabilization.

2. Experimental

The structuring was made on 12 mm wide SuperPower tapes with and without Cu stabilization. We used an infrared Nd:YAG-laser with a wavelength of 1030 nm which was pulsed with a frequency of 400 kHz. The pulse duration is less then 10 ps. The maximum power of the laser is 25 W, with maximum pulse energy of 125 μJ. For the structuring of Ag-cap conductors we only used 20 % of the power (5 W) with 10 repetition cycles for each single groove. For Cu-cap conductors we used 50 % of the power (12.5 W) with 50 repetition cycles for each single groove. For the scribing process a combination of a scanner optic with a speed of 90 m/min and a moving table with 22.5 m/min was used. The width of laser grooves is between 18-21 μm for Ag-cap and 37-42 μm for Cu-cap conductors. The resolution of the repeatable track is 0.5 μm.

Samples with up to 120 filaments on 20 cm long tapes were prepared. For the current injection 2.5 cm at both ends of the tapes were not structured. After measuring the U-I curves the ends were cut by laser. These ends were used to investigate the microstructure of the grooves in a scanning electron microscope (SEM). The remaining parts were used for AC loss measurements. For this we used the calibration-free method [13]. In order to understand how the coupling currents flow inside the coated conductor, we measured the transverse resistance between different filaments. All presented measurements were performed at 77 K.

3. Results and Discussion

3.1. Striations

Figure 1 shows top view of samples with 2 μm Ag-cap layer (a-b) and with 20 μm Cu-cap layer (c-d).

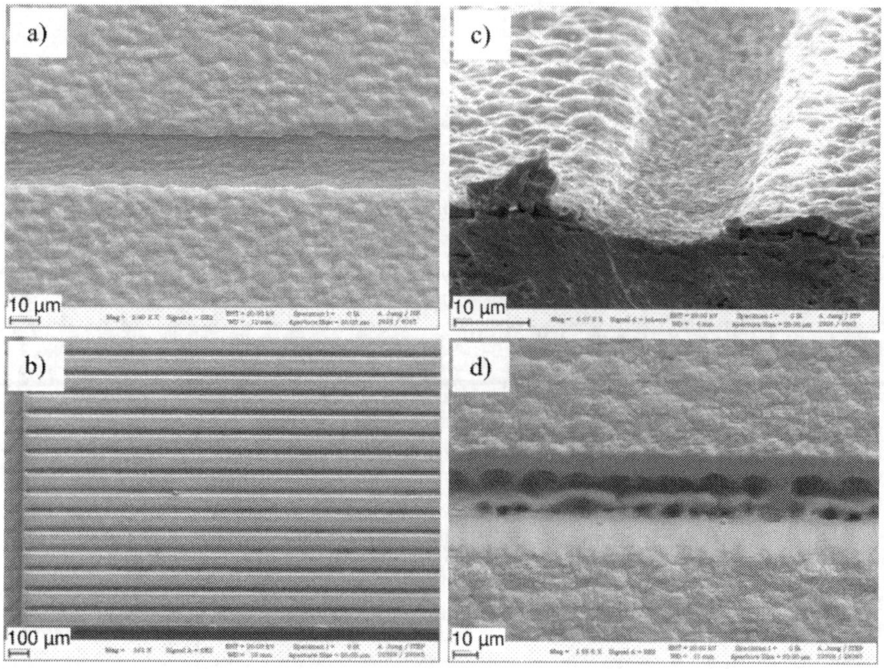

Figure 1. SEM images of striated tapes, a-b) Ag-cap conductor, c-d) Cu-cap conductor with 120 filaments.

For the structuring of Cu-cap conductors in comparison to Ag-cap conductors one has to use higher laser power (12.5 W) and more repetition cycles (50) to go through the Cu. In this case also the hastelloy substrate will be strongly affected by the laser beam, resulting in an inhomogeneous groove edge and bottom. Mainly rows of holes in the hastelloy tape will be seen. For the Ag-cap tapes the structuring is more homogeneous and mainly stops at the hastelloy tape (fig. 1 a)). Nevertheless the I_c measurements together with SEM and FIB investigations indicate that the striation process had negligible heat effects at the edges of the grooves.

3.2. DC measurements
The DC measurements of the critical current as a function of the number of filaments are shown in figure 2.

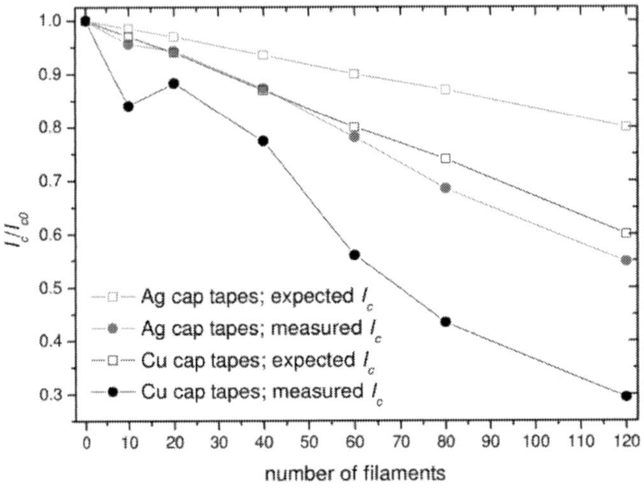

Figure 2. Normalized critical current as a function of the number of filaments.

The open symbols in figure 2 represent the expected I_c values solely caused by the loss of superconducting material. Filled symbols represent the measured I_c values. The additional degradation of I_c occurs from statistically distributed inhomogeneities. The influence of inhomogeneities on the current carrying capability increases with decreasing filament width (increasing number of filaments). The Ag-cap sample with 120 filaments has an I_c of 187.5 A, which represents 55% of the critical current of the original tape (342 A). Taking into account that one loses 20 % of superconducting material it reached 68% of the expected value (274 A). The Ag-cap samples with 10 and 20 filaments are basically unaffected by the striation process. They retain 97 % of their current carrying potential. These findings indicate that the striation process does not strongly affect the critical current of the samples. Even the sample with 80 filaments still retains 79 % of its potential. For Cu-cap samples the reduction of I_c is steeper. Here we cut off the edges of the tapes to reduce coupling losses, but this additionally affects and hinders the current rearrangement and percolation, decreasing I_c.

3.3. AC measurements and transverse resistance
We measured the losses in dependence of the filament number for different frequencies. The loss reduction with the filament number was slightly less than expected (figure 3), which indicates some degree of coupling between the filaments. We are able to calculate the losses properly by finite-element method (FEM), assuming fully uncoupled filaments [14]. The agreement between simulations

and experiments worsens with increasing number of filament (figure 3). This indicates that the coupling effect is more important for increasing number of filaments.

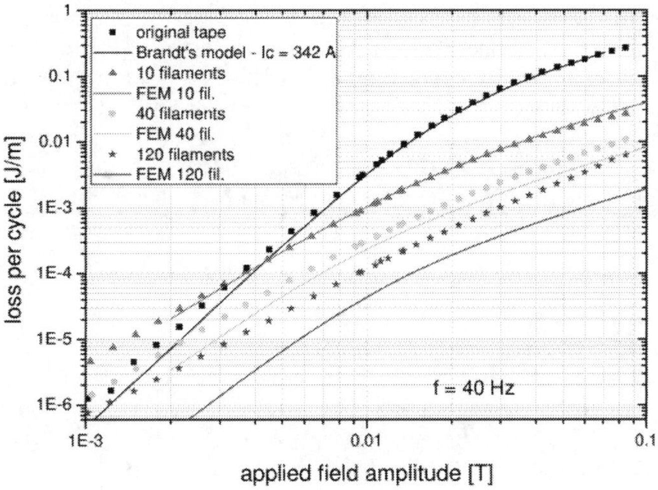

Figure 3. Magnetization AC losses in dependence on the number of filaments for Ag-cap samples.

We tried to influence the coupling losses by changing the laser parameter, like varying the power and the number of runs. But there are no clear dependencies. Additionally we oxidized the Ag-cap samples after the standard laser process at 500°C for 1 h in pure oxygen to see if the resistance of the groove increases [12]. With this process the groove resistance (measured with four point method) increases about two orders of magnitude up to 650 mΩ compared to the grooves after the standard laser process with 3 mΩ. In figure 4 the losses normalized by the square of the applied field amplitude for standard laser processed tapes and additional oxidized tapes at 130 Hz are shown.

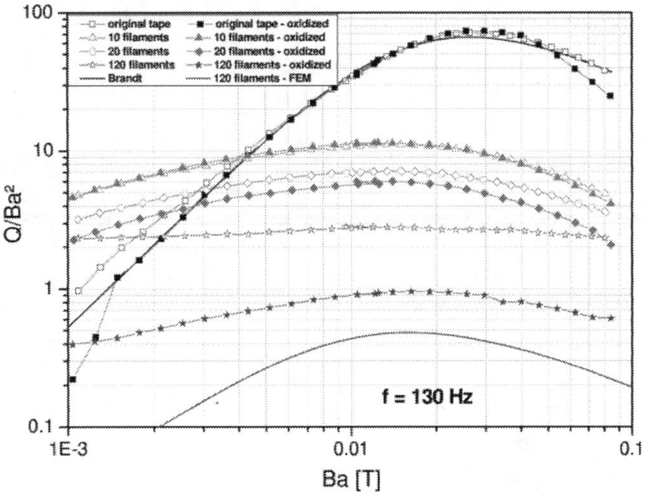

Figure 4. AC losses normalized by the square of the applied field amplitude for standard and oxidized Ag-cap tapes. Blue line: FEM simulation for totally uncoupled filaments.

In figure 4 it is clearly visible that the oxidation treatment is a quite effective tool to reduce the coupling loss by increasing transverse resistivity. For the sample with 10 filaments nearly no effect of the oxidation is seen, because even in the standard sample the coupling loss is very low. With increasing number of filaments the loss reduction increases. For the 120-filament tape a large loss reduction occurs. At 130 Hz the standard 120-filament tape is almost flat. Because the coupling loss is proportional to Ba^2, the loss of a fully coupled tape should be constant in the whole range of the field amplitude. The oxidized tape with 120 filaments is in between the two extreme cases, the FEM simulation with no coupling and the standard 120-filament tape fully coupled. In the case of Cu-cap samples one frequency was measured. The loss is about 6 times higher than in Ag-cap samples. Cutting off the Cu edges from the samples reduces the loss about a factor of 2.

4. Summary

Ag-cap and Cu-cap tapes from SuperPower were structured with an Nd:YAG picosecond laser system. We made up to 120 filaments in 12 mm wide tapes. Because of using higher power for the Cu-cap tapes, the substrate tape is also affected resulting in row of holes inside the hastelloy tape. Nevertheless in both Ag- and Cu-cap tapes the laser did not strongly affect the material beneath the edges of the grooves. DC measurements showed a decreasing I_c with increasing filament numbers. This is caused by the loss of superconducting material and by statistically distributed inhomogeneities affecting the rearrangement and percolation of the current. The striated tapes show some filament coupling, which is more important as the number of filaments increases. The groove resistance could be increased of about two orders of magnitude during a heat treatment at 500°C in pure oxygen for Ag-cap samples. The AC loss measurements of these samples indicate a clear decrease of the coupling loss, especially for tapes with high filament numbers.

References

[1] Brandt E H and Indenbom M 1993 *Physical Review B* **48** 12893–906
[2] Sumption M D, Collings E W and Barnes P N 2005 *Supercond. Sci. Technol.* **18** 122
[3] Duckworth R C, List F A, Paranthaman M P, Rupich M W, Zhang W, Xie Y Y and Selvamanickam V 2007 *Physica C* **463-465** 755
[4] Majkic G, Kesgin I, Zhang Y, Qiao Y, Schmidt R and Selvamanickam V 2011 *IEEE Trans. Appl. Supercond.* **21** 3297
[5] Marchevsky M, Zhang E, Xie Y, Selvamanickam V and Ganesan P G 2009 *IEEE Trans. Appl. Supercond.* **19** 3094
[6] Suzuki K, Matsuda J, Yoshizumi M, Izumi T, Shiohara Y, Iwakuma M, Ibi A, Miyata S and Yamada Y 2007 *Supercond. Sci. Technol.* **20** 822
[7] Tsukamoto O and Ciszek M 2007 *Supercond. Sci. Technol.* **20** 974
[8] Lee J K, Kim W S, Park C, Byun S, Han B W, Lee S, Park S and Choi K 2009 *Physica C* **469** 1432
[9] Terzieva S, Vojenčiak M, Grilli F, Nast R, Šouc J, Goldacker W, Jung A, Kudymow A and Kling A 2011 *Supercond. Sci. Technol.* **24** 045001
[10] Šouc J, Gömöry F, Kováč J, Nast R, Jung A, Vojenčiak M, Grilli F and Goldacker W 2013 *Supercond. Sci. Technol.* **26** 075020
[11] Kopera Ĺ, Šmatko V, Prusseit W, Polák M, Semerad R, Štrbík V and Šouc J 2008 *Physica C* **468** 2351
[12] Kesgin I, Majkic G and Selvamanickam V 2013 *Physica C* **486** 43
[13] Šouc J, Gömöry F and Vojenčiak M 2005 *Supercond. Sci. Technol.* **18** 592
[14] Brambilla R, Grilli F and Martini L 2007 *Supercond. Sci. Technol.* **20** 16

Journal of Physics: Conference Series **507** (2014) 022024 doi:10.1088/1742-6596/507/2/022024

Optimisation of composite superconducting bulks made from (RE)BCO coated conductor stacks using pulsed field magnetization modelling

A Patel[1] and B A Glowacki1[1,2,3]
[1]Applied Superconductivity and Cryoscience Group, Department of Materials Science and Metallurgy, University of Cambridge, Pembroke Street, Cambridge, CB2 3QZ, UK
[2] Institute of Power Engineering, ul. Augustówka 6, 02-981 Warsaw, Poland
[3] Department of Physics and Energy, University of Limerick, Plassey, Ireland

E-mail: ap604@cam.ac.uk

Abstract. Coated conductors, although designed to carry transport current, are ideally suited to carrying persistent current and can therefore be cut and stacked to form a type of composite bulk which has superior thermal properties compared to existing bulks despite having less than 2 % superconductor by volume. The magneto-thermal modelling reported follows on from previous experimental work on pulsed magnetization of a 12 mm square tape stack. The magnitude of the applied field has a strong effect on the trapped field and flux. The optimum applied field depends on sample height and the maximum trapped field and flux saturates as the height reaches the diameter of the stack. The nature of a composite bulk made from a stack of tapes gives complete control over the height of the stack which needs to be optimised for pulsed magnetization.

1. Introduction

The ability of superconducting bulks to trap high fields once magnetized allows them to act as powerful permanent magnets thanks to persistent currents circulating in the bulks. However, for high trapped fields and temperatures below 77 K, bulks suffer from poor thermal stability [1] and lack mechanical strength. A stack of tapes created from commercially available 12 mm wide tape produced by SuperPower Inc has already been shown to trap up to 2 T when pulse magnetized [2] and over 7 T between two stacks when field cooled [3]. Given the stack size and I_c properties, these results show great potential for a stack of tapes to be used as a trapped field magnet. Optimisation of the stack parameters is now needed including critical state modelling of applied field and sample height as reported here.

2. Modelling framework and thermal properties

The modelling of pulsed field magnetization was carried out using the H-formulation for magnetic fields, coupled with heat transfer in the finite element modelling package COMSOL Multiphysics 4.2a. The framework used the E-J power law and is the same as that used in [4]. The applied pulsed field was half a sinusoid. The full mathematical description of heat generation and the temperature and field dependent critical current density for the model can all be found in [4]. The total cooling power applied to the cold head (see figure 1) used in the simulation is described by the following function, equation (1), which gives realistic cooling saturating at a maximum of 30 W for a temperature rise greater than

15 K. These parameters approximately reflect the real cooling power in cryocoolers, such as the one used in [5].

$$\begin{array}{ll} P_0 = 0 & T < T_0 \\ P_0 = Q_c(T - T_0) & T_0 \le T \le (T_0 + 15) \\ P_0 = 15Q_c & T > (T_0 + 15) \end{array} \quad (1)$$

The simulation was split into two time domains, $0 < t < 0.1$ s (which includes the 28 ms pulsed field) and $0.1 \le t < 30$ s, with the second time domain having a higher n value to give realistic flux creep rates as described in [4]. The equilibrium temp, T_0, was 38K with a starting temp of 40 K. All trapped field and flux values were evaluated at 0.8 mm above sample to match previous experiments.

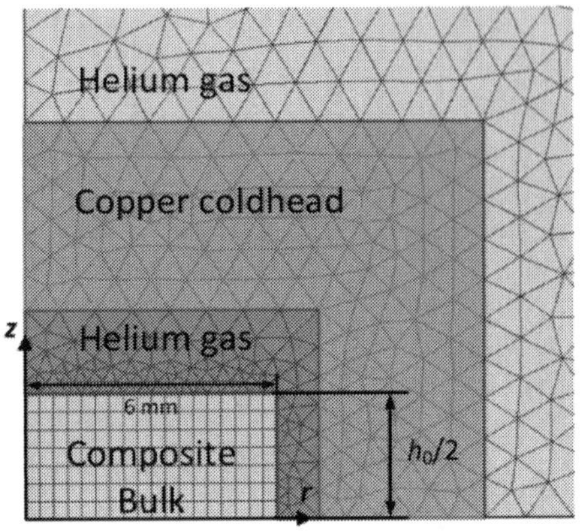

Figure 1. Geometry and mesh used in model for a 12 mm diameter stack with variable sample height h_0.

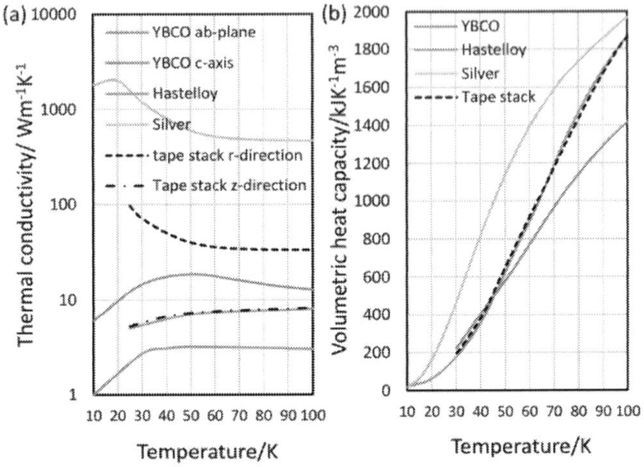

Figure 2. (a) Effective thermal conductivity of the tape stack in the radial and azimuthal directions compared to constituent materials. (b) Effective volumetric heat capacity of tape stack compared to constituent materials.

Modelling of individual stack layers is a very challenging task given the number of mesh elements required due to thin layers. An approximation can be made for the thermal properties of the stack which are highly anisotropic. The effective thermal conductivity parallel and perpendicular to layers of a 3 layer structure can be shown analytically to be:

$$\overline{\kappa_{\parallel}} = f_1\kappa_1 + f_2\kappa_2 + f_3\kappa_3, \quad \overline{\kappa_{\perp}} = \frac{\left(\frac{\kappa_1}{f_1}\right)\left(\frac{\kappa_2}{f_2}\right)\left(\frac{\kappa_3}{f_3}\right)}{\left(\frac{\kappa_1}{f_1}\right)\left(\frac{\kappa_2}{f_2}\right)+\left(\frac{\kappa_2}{f_2}\right)\left(\frac{\kappa_3}{f_3}\right)+\left(\frac{\kappa_1}{f_1}\right)\left(\frac{\kappa_3}{f_3}\right)} \quad (2)$$

where f_n is the volume fraction of each layer. Applying this equation to the stack of tapes magnetized in [2], gives an enhanced radial thermal conductivity as shown in figure 2(a), but the azimuthal conductivity is dominated by Hastelloy as well as the average volumetric heat capacity (figure 2(b)).

3. Optimisation for pulse magnetization of a 12mm diameter stack

3.1. Optimisation of applied field

The applied pulsed field was varied from 0 to 10 T for three different sample heights. The trapped field and flux results are shown in figure 3. Trapped flux needs to be considered as much as trapped field when optimising the sample as flux is as important for some applications such as motors and generators. Significant relaxation has occurred 30 s after the pulse but the general dependence of trapped field and flux on applied field is similar to 0.1 s. There exists a height dependent optimum applied field as too low an applied field failed to induce current deep enough into the sample, whereas too high an applied field generates excessive heat leading to a larger temperature rise. This matches experimental behaviour for trapped flux in a bulk [6]. Interestingly, the optimum applied field did not fully penetrate any of the heights modelled and less so for the largest height. This is a clear indication that trying to saturate the sample is not beneficial for pulse magnetization at lower temperatures (< 50 K). Experimental applied fields need to be optimised depending on the sample height.

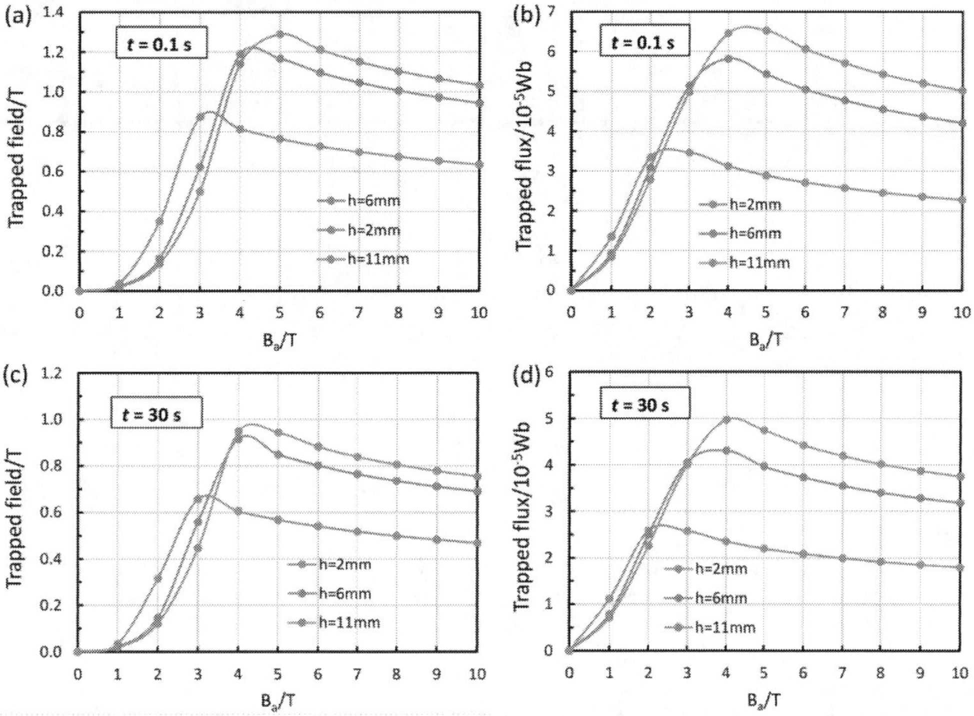

Figure 3. Effect of applied field magnitude on trapped field and flux shortly after pulsed field (a) and (b), and after relaxation (c) and (d), for different sample heights.

3.2. Optimisation of sample height

The optimum applied field dependence on height was estimated by interpolating the optimum fields determined for the three heights modelled in figure 3. The trapped field data in figure 3(c) was used for the optimisation of applied field to account for relaxation effects. Once the dependence of applied field on height was determined, 11 different sample heights were modelled from 2 mm to 12 mm as shown in figure 4. The dotted lines show the expected dependence if the samples have constant uniform current density. Therefore the dotted lines show the form of behaviour resulting purely from geometric scaling of the sample. In order to make a comparison to the full thermal critical state model, the uniform current density for this approximation was chosen to be that of the $h_0 = 6$ mm critical state model. Looking at the dotted lines, it is clear that the trapped field and flux above a sample saturate for larger heights, because for the point above a sample, the contribution to field from far away layers is small. The trapped field and flux is dominated by the top few mm of the sample. The full thermal critical state results using optimised applied field have remarkably close behaviour to the uniform J_c scaling curves. This correlation is related to the average temperature rise (figure 4(c)) being the same for all the heights modelled. If the same applied field was used for all the heights, there would be a large difference in temperature rise between small and large heights resulting is greater differences in trapped field and flux. The surface area to volume changes significantly with height for small heights (inset of figure 4 (a)), but surprisingly, for the optimum applied field used for each height, the enhanced surface cooling expected for the smallest sample heights is not so apparent in the trapped field and flux results. The overall result from the data is that large sample heights close to the sample diameter (12 mm) can be used to maximise trapped field. There does not appear to be a maxima for the trapped field and flux for the sample size and cooling power modelled. In practice, to minimise cost whilst still having high performance, a sample height of approximately 8 mm would be a good compromise which corresponds to 145 layers of the tape tested in [2].

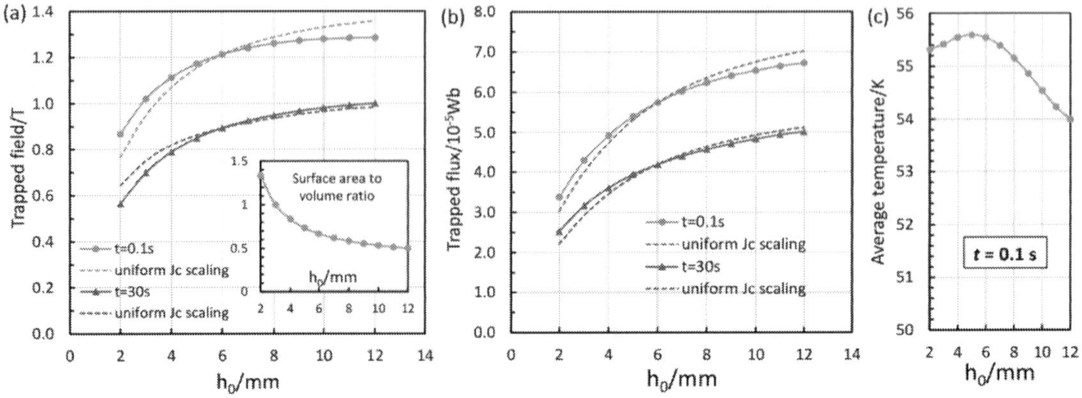

Figure 4. Effect of sample height on trapped field (a) and flux (b), where the applied field used for each height was optimised to give maximal trapped field. (c) The average sample temperature shortly after the pulse for each height simulated.

4. Conclusions

A thermal critical state model was applied to a 12 mm diameter stack of tapes assuming homogeneous material properties. The effective thermal conductivity was calculated based on Hastelloy, YBCO and silver contributions and found to be highly anisotropic compared to bulk YBCO. There exists a height dependent optimum applied field for pulsed magnetization of the samples. When using optimum applied fields for various sample heights, the maximum trapped field and flux possible varies almost as expected from geometric scaling of a sample with uniform constant persistent current density. The conclusion from this behaviour is that increasing the sample height always increases trapped field and flux for the sample modelled but there is saturation as the height approaches the sample diameter. Given materials cost, the results suggest that there is no benefit is creating a stack with height greater than or equal to its width or diameter. The model made a number of assumptions which do not completely match experimental reality. In future, to make the model more realistic, multiple pulses will be modelled as well as eddy current heating in the silver layer which was not taken into account. Future modelling will also consider magnetization of larger sized stack currently under experimental investigation.

References

[1] Fujishiro H, Naito T, and Oyama M, 2012 *Physics Procedia.* **36** 687-692
[2] Patel A, Hopkins S C, and Glowacki B A, 2013 *Supercond. Sci. Technol.* **26** 032001
[3] Patel A, Filar K, Nizhankovskii V I, Hopkins S C, and Glowacki B A, 2013 *Appl Phys Lett.* **102** 102601-5
[4] Patel A and Glowacki B A, 2012 *Supercond. Sci. Technol.* **25** 125015
[5] Patel A, Giunchi G, Albisetti A F, Shi Y, Hopkins S C, Palka R, Cardwell D A, and Glowacki B A, 2012 *Physics Procedia.* **36** 937-942
[6] Mizutani U, Oka T, Itoh Y, Yanagi Y, Yoshikawa M, and Ikuta H, 1998 *Appl Supercond.* **6** 235-246

Journal of Physics: Conference Series **507** (2014) 022025 doi:10.1088/1742-6596/507/2/022025

The study of relaxation characteristics of stack of HTS tapes for use in levitation systems and trapped flux magnets

S Pokrovskiy, N Mineev, A Sotnikova, Y Ermolaev, I Rudnev

National Research Nuclear University "MEPHI", Russia, Moscow
International Laboratory of High Magnetic Fields and Low Temperatures, Wroclaw, Poland

sergeypokrovskii@gmail.com

Abstract. We present experimental data and analysis of application of coated conductors (CC) for levitation and trapped field magnets. The magnetization loops and magnetization relaxation curves were obtained in fields up to 14 T and temperatures from 4.2 K to 77.4 K. The levitation force and its relaxation were measured for the stack of superconducting tapes at T=77 K. The measurements of levitation force for different stack sizes demonstrate the linear growth of the force amplitude and the improvement of the relaxation characteristics as a number of tapes in a stack increases.

1. Introduction

The recent development in the high temperature superconductor (HTS) tapes production expands their application area to magnetic systems. In this paper we represent the study of properties of CCs allowing to use them in trapped flux magnets and magnetic levitation systems [1-3]. It is known that the Bi-based tapes show a large critical current degradation under applied magnetic field as compared to the YBCO-based coated conductors. In turn, the modern CC samples maintain the critical transport current at several amperes at 77 K in the fields up to 8 T. This is the main reason to choose coated conductors for magnetic applications.

The bulk YBCO superconductors were widely used for levitation systems over the last decades. However they have some obvious weaknesses such as heterogeneity of superconducting properties through the sample volume, frailness, the problem of manufacturing of the extended samples, etc. In addition, there is a question about the performance of the superconducting material: large part of a bulk superconductor (up to 30-40 % [4]) is normally shielded from an external magnetic field and does not contribute to the levitation force. On the other hand the industrial high quality coated conductors have a good uniformity and texturing properties across the length. It also exhibits the extremely high critical current densities. The geometric parameters of such tapes provide flexibility in the formation of stacks of tapes of different sizes and structures.

2. Experimental details

Firstly, we studied the behavior of individual coated conductors samples in a wide range of magnetic fields and temperatures. The experiments were carried out on the SuperPower tapes by using the vibrating magnetometer [5]. The external magnetic field was perpendicular to the 4x4 mm sample large surface. Under these conditions, the magnetization loops $M(H)$ were measured in the fields up to 14 T and in a temperature range from 4.2 K to 77.4 K. Also the relaxation curves were obtained after finishing each loop measurement in the zero magnetic field.

Journal of Physics: Conference Series **507** (2014) 022025 doi:10.1088/1742-6596/507/2/022025

The setup for measuring of levitation force (Figure 1) provides an ability to determine the levitation force over the gap between the superconducting sample and permanent magnet [6]. The coated conductors of 12 mm wide and 15 mm long were used for measurements. Stack size was varied from 3 to 10 tapes and the minimum levitation gap was 2 mm. The dependencies of the levitation force on the time were measured in a similar way as the magnetic moment relaxation.

3. Results

The dependencies of magnetization on applied magnetic field (magnetization loops) obtained at different temperatures were used to calculate the critical current over the temperature relationship (Figure 2). A strong relationship between magnitude of the critical current density and levitation force points out to a proportional increase in levitation force with decreasing temperature. Also the obtained relaxation curves allow to calculate the magnetization change time constants for a wide range of temperatures. An analysis of dynamic magnetization properties demonstrated that the range of possible optimal temperatures for a steady magnet system is 4.2 – 20 K.

The dependence of the levitation force F_z on the levitation gap z for CC stacks was measured at 77 K in two modes: zero field cooling (ZFC), Figure 3a, and cooling in the field of permanent magnet (FC), Figure 3.b. The obtained results and its comparison with ones for bulk samples show that the main difference between a stack of tapes and the bulk superconductors is a strong hysteretic behavior (the appearance of the attractive force) of the stack, in ZFC as well as in FC mode, which is confirmed in [2]. This may be due to easier magnetic field penetration to the HTS film.

As was mentioned above, the number of tapes in the stack was varied from 3 to 10. The maximum repulsive force (for levitation gap of 2 mm), and maximum attractive force were chosen for the comparison of values for various number of tapes (Figure 4). Resulting curves show a linear rise with increasing of tapes numbers in a stack. It is expected that a further increase in the number of layers in the HTS composite will cause the rise of the force with subsequent saturation, similar to bulk superconductors [7].

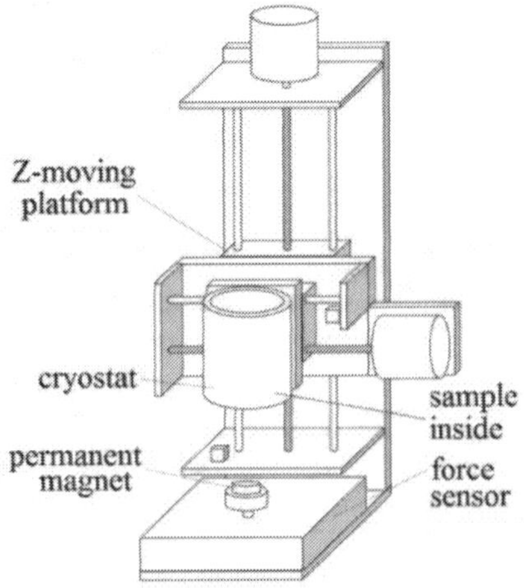

Figure 1. Scheme of the levitation force measuring setup.

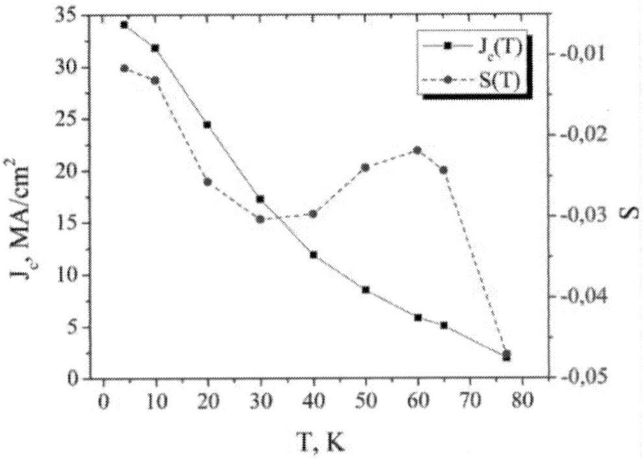

Figure 2. Temperature dependences of critical current density J_c and relaxation constant S in zero magnetic field.

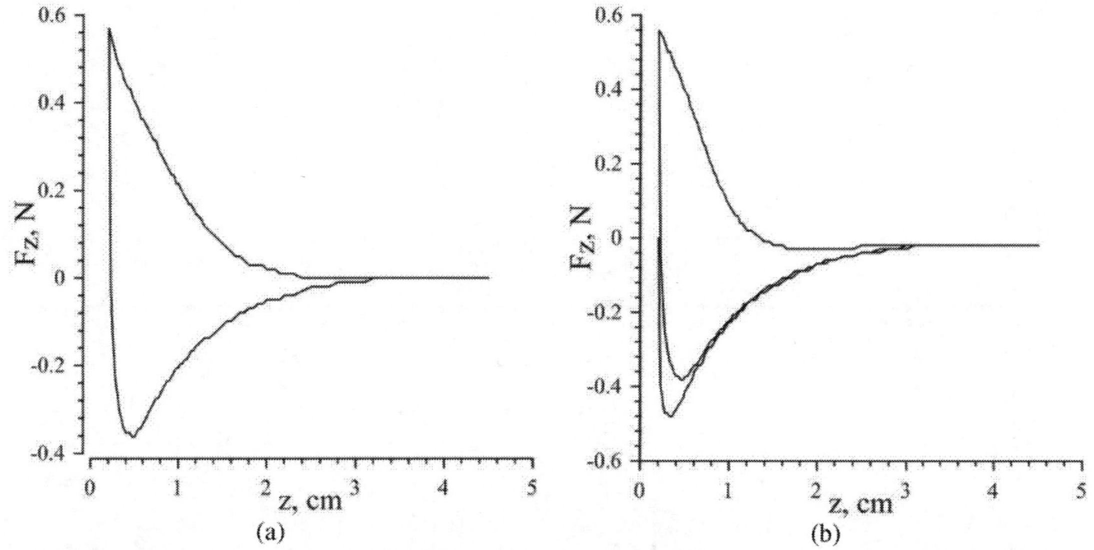

Figure 3. Levitation force for 10 CC tapes stack at T=77.4 K in two modes: (a) ZFC, (b) FC.

Figures 5 and 6 present the comparisons of the repulsive force relaxation curves for the stack of tapes and a bulk superconductor. In contrast to the bulk superconductor, the decrease in the force at the initial stage is more abrupt for the stack. However, after approximately of 50 seconds the decrease gets slower, and the value reaches a plateau earlier than one for the bulk sample. Also a residual repulsive force for a tapes stack maintains higher. Comparison of relaxation of the normalised magnetization (single tape) and the levitation force (for a stack of 10 layers) leads to the conclusion that the stack is more stable in time interaction.

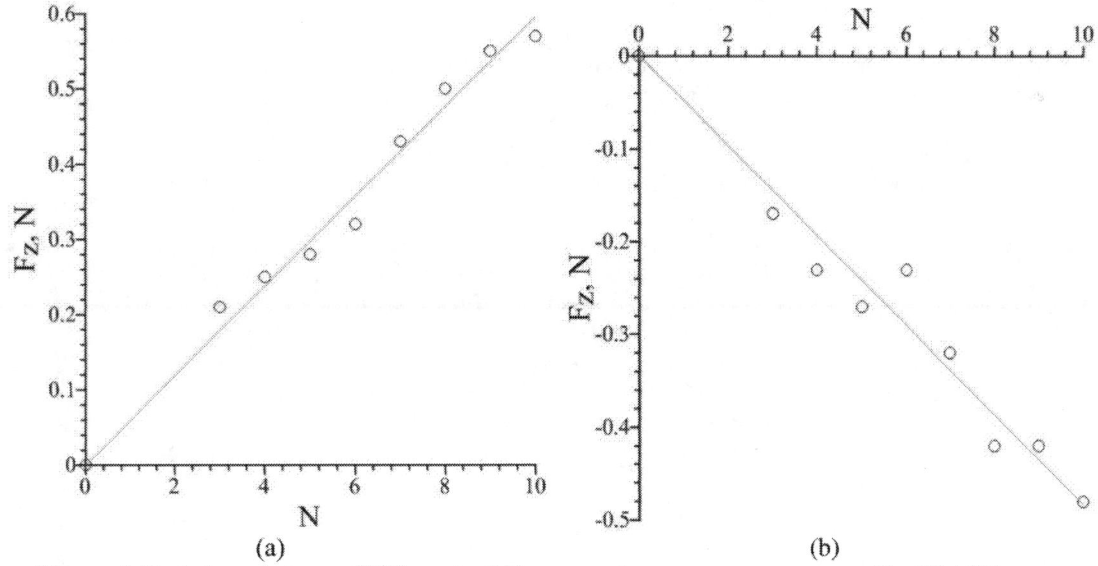

Figure 4. Peak force between HTS stack of N tapes and a permanent magnet at T = 77.4 K in two modes: (a) repulsive force, (b) attractive force.

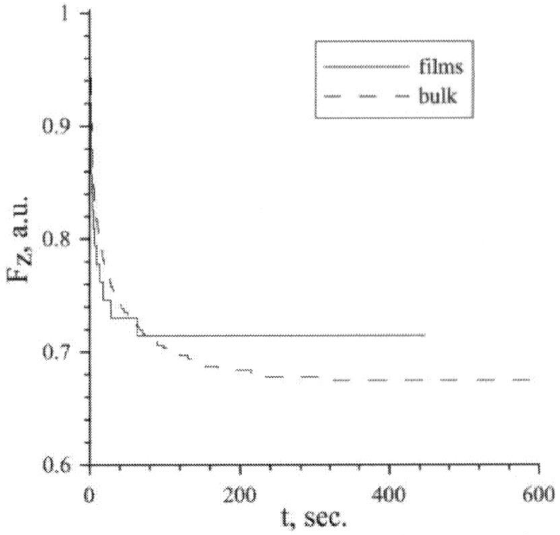

Figure 5. Comparison of relaxation curves for 10 tape stack and a bulk sample at T=77.4 K.

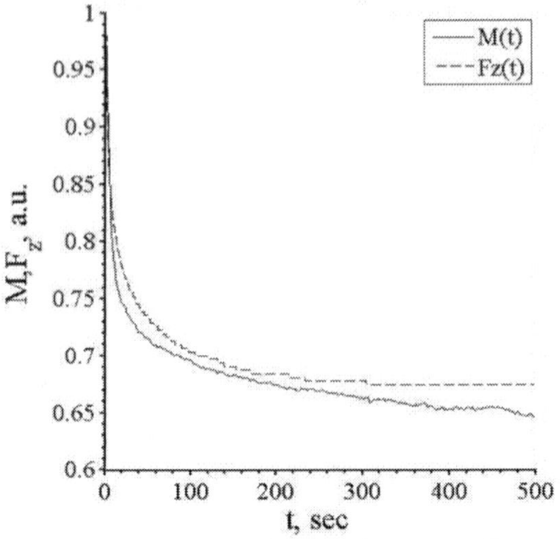

Figure 6. Compared relaxation curves for levitation force (10 tape stack) and magnetization of a single tape at T=77.4 K.

4. Conclusions

The main aspects of possible realization of an effective levitation systems or trapped field magnets by using CC tapes were discussed: working temperature, relaxation and stack size. The amplitude of the tape critical current and magnetization decreases dramatically as a temperature rises, what can be used by applying modern cryocooling systems and shifting the working temperature down to 20 K. On the other hand, the relaxation constants of individual tape shows a local minimum of the relaxation speed at the temperature of 60 K, even lower speed is reached only at 15 K.

Tapes stack looks more stable than a bulk sample and addition of more tapes in a stack increases the magnitude of the levitation force and the time stability. Making of a multilayer tape would be very useful for applications in levitation systems.

Acknowledgments

The work was supported by Russian Ministry of Education and Science, Grant 14.A18.21.0349.

References

[1] Patel A, Hopkins S C and Glowacki B A 2013 *Supercond. Sci. Technol.* **26** 032001
[2] Sassa F, Dias D H, Sotelo G G and de Andrade R Jr 2012 *Physics Procedia* **36** 1008-13
[3] Sassa F, Dias D H, Sotelo G G and de Andrade R Jr 2013 IEEE *Trans. Appl. Supercond.* **23**
[4] Rudnev I and Ermolaev Yu 2006 *J. Phys.: Conf. Ser.* **43** 983-6
[5] Nizhankovskii V and Lugansky L 2007 *Meas. Sci. Technol.* **18** 1533-37
[6] I A Rudnev and Yu S Ermolaev 2008 *J. Phys.: Conf. Ser.* **97** 012006
[7] Leblond C, I Monot I, D Bourgault D and Desgardin G 1999 *Supercond. Sci. Technol.* **12** 405-10

Journal of Physics: Conference Series **507** (2014) 022026 doi:10.1088/1742-6596/507/2/022026

Evolution of superconducting critical temperature in asymmetrical $F_1/F_2/S$ trilayers due to singlet-triplet correlations and external magnetic field

Yu.N. Proshin and M.V. Avdeev

Kazan Federal university, Kremlevskaya, 18, Kazan, Russia

E-mail: yurii.proshin@ksu.ru

Abstract. We theoretically explore a competition between magnetism and superconductivity for the three-layered ferromagnet/superconductor (F/S) structures. The critical properties of $F_1/F_2/S$ trilayers in an external parallel magnetic field **H** are investigated. The different mutual orientations of the F layer magnetizations are examined. Under this condition the triplet component of the superconducting condensate is arisen. Assuming that all layers are dirty we solve boundary problem for the Usadel function. To find the critical parameters for trilayer as function of the F layers thicknesses we use Gor'kov's self-consistency equation. The results of numerical calculations for critical temperature T_c and critical magnetic field H_c at various parameters of F/S structure show that the external magnetic field and triplet correlations can essentially influence on the critical properties of considered systems. Unusual dependencies are discussed.

1. Introduction

It is well known that the singlet superconductor (S) and ferromagnet (F) order parameters are fundamentally incompatible and their coexistence in homogeneous materials is difficult to realize. However in the artificial layered heterostructures consisting of alternating F and S layers this is possible [1,2] due to the proximity effect [3]. In the F/S structures the singlet superconducting correlations can penetrate from the S layer into the F layer on the depth of order of the coherence length $\sim \sqrt{a_F l_f}$ ($a_F = v_F/2I$ is spin stiffness length, I is exchange field value, velocity v_F is Fermi velocity and l_f is free path length in ferromagnet). Note that the penetration depth into the F is much smaller than that in the normal metal (N), since condition $I \gg T_c$ is satisfied in conventional ferromagnets. The features of interplay of the S and F ordering phases leads to interesting physical phenomena [1,2,4].

Recently, it was demonstrated that not only the singlet component of the superconducting condensate (correlations between quasiparticles with parallel spins) but also the triplet ones can be induced in the ferromagnet [5–9]. The triplet superconducting component is generated in case of nonhomogeneous magnetization of F layers. For the F/S structures the noncollinear orientation of magnetizations in the F layers may also give rise to the triplet superconductivity, in this case each magnetization may be homogeneous [8,9]. The triplet component penetrates into F layer over more long distance compared to the singlet component. This circumstance may add new features to the proximity effect in the S/F structures.

Journal of Physics: Conference Series **507** (2014) 022026 doi:10.1088/1742-6596/507/2/022026

Reach physics of the F/S system makes them attractive for technical applications. Thus the spin valve device based on the three layered FS systems switched by weak external magnetic field was proposed in works [10–12].

We study the $F_1/F_2/S$ structures in the presence of an external parallel magnetic field $\mathbf{H}$. The different mutual orientations of the F layer magnetizations are examined. The triplet superconducting component is generated at noncollinear magnetizations of the F layers. Assuming that all S and F layers are dirty we solve boundary problem for the Usadel function. Then we use Gor'kov's self-consistency equation to find the critical parameters at various parameters F/S structure. The results of numerical calculations for critical temperature T_c and critical magnetic field H_c as function of thicknesses of both F_1 and F_2 layers are presented.

2. Theoretical background

The critical temperature T_c at the second order transition is obtained from the self-consistent equation for the superconducting gap $\Delta(\mathbf{r})$ [13]

$$\Delta(\mathbf{r}) \ln t = \pi T_c \sum_{\omega>0}^{\infty} \mathrm{Tr}\left(\hat{F}(\mathbf{r},\omega) - \frac{\Delta(\mathbf{r})}{\omega}\right), \tag{1}$$

where $t = T_c/T_{cs}$ is the reduced critical temperature (T_{cs} is the superconducting critical temperature for the bulk material without spin exchange interaction), Tr means the trace function, the summation is performed over repeated spin indices, ω is the Matsubara frequency.

In presence of the external magnetic field $\mathbf{H} = \mathrm{rot}\mathbf{A}$ the pair amplitude $\hat{F}$ satisfies the Usadel-like equations [14–16]

$$\begin{gathered}\left[|\omega| - \mathrm{i}I(\hat{\sigma}_3)_{\alpha\beta} - \frac{1}{2}D_{\alpha\beta}\hat{\mathcal{L}}^2\right] F_{\alpha\beta}(\mathbf{r},\omega) = \delta_{\alpha\beta}\Delta(\mathbf{r}),\\ D_{\alpha\beta} = \frac{D}{1-\mathrm{i}2I\tau_f(\hat{\sigma}_3)_{\alpha\beta}},\quad \hat{\mathcal{L}} = \nabla - \frac{2\pi\mathrm{i}}{\Phi_0}\mathbf{A},\end{gathered} \tag{2}$$

where Φ_0 is the magnetic flux quantum, D is the diffusion constant and τ_f is the elastic scattering time of non-magnetic impurities in F layer. Note that the diffusion constant $D_{\alpha\beta}$ in (2) is complex due to the presence of exchange interaction [1]. In case of the magnetization in F layer has the form $\mathbf{M}(0, M\sin\phi, M\cos\phi)$ then the solution of the (2) is $\hat{F}' = \hat{U}^\dagger \hat{F}\hat{U}$. Here $\hat{U}$ is unitary matrix, which has the form

$$\hat{U} = \begin{pmatrix} \cos\frac{\phi}{2} & \mathrm{i}\sin\frac{\phi}{2} \\ \mathrm{i}\sin\frac{\phi}{2} & \cos\frac{\phi}{2} \end{pmatrix}.$$

We use the boundary conditions derived by microscopic approach in the work [17]. For the SF interface, where F and S layers occupy the regions $x < 0$ and $x > 0$, respectively, they have the form

$$\frac{4D_s}{\sigma_s v_F^s}(\hat{\mathcal{L}}\mathbf{n})F^s_{\alpha\beta} = \frac{4D_f}{\sigma_f v_F^f}(\hat{\mathcal{L}}\mathbf{n})[\hat{U}^\dagger \hat{\mathcal{F}}^f \hat{U}]_{\alpha\beta} = F^s_{\alpha\beta} - [\hat{U}^\dagger \hat{F}^f \hat{U}]_{\alpha\beta},\ \mathcal{F}^f_{\alpha\beta} = \frac{F^f_{\alpha\beta}}{1-\mathrm{i}2I\tau_f(\hat{\sigma}_3)_{\alpha\beta}}. \tag{3}$$

The boundary conditions at the outer surfaces have the form

$$(\hat{\mathcal{L}}\mathbf{n})F^{s,f}_{\alpha\beta} = 0. \tag{4}$$

Here $\mathbf{n}(1,0,0)$ is the unit vector normal to the S, F_1, F_2 planes; the indices s and f denote S and F layers; σ_s and σ_f are the transparencies from the S and F side, respectively [1]. Note that the boundary conditions at the interface (3) take into account the fulfillment of the condition of detailed balance

$$\sigma_s v_F^s N_s = \sigma_f v_F^f N_f, \ \sigma_{f1} v_F^{f1} N_{f1} = \sigma_{f2} v_F^{f2} N_{f2}$$

where $N_{s,f}$ is the density of states on the Fermi surface for the S and F layers, respectively.

The set of the equations (1) and (2) and the appropriate boundary conditions (3), (4) are sufficient to calculate critical temperature T_c of the SF layered structures.

3. Results and discussion

We consider a $F_1/F_2/S$ system in presence of the external magnetic field $\mathbf{H}$ applied parallel to the plane of the contact. We choose the coordinate system so that the S layer occupies the region $0 < x < d_s$, F_1 and F_2 layers occupy the regions $-d_{f1} - d_{f2} < x < -d_{f2}$ and $-d_{f2} < x < 0$, respectively. The magnetization vector of the F_1 layer may be rotated $\mathbf{M}_1(0, M_1 \sin\phi, M_1 \cos\phi)$, while the magnetization of the F_2 layer is fixed $\mathbf{M}_2(0, 0, M_2)$. Following the work [18] we perform the numerical calculations of T_c for the $F_1/F_2/S$ trilayer in similar manner.

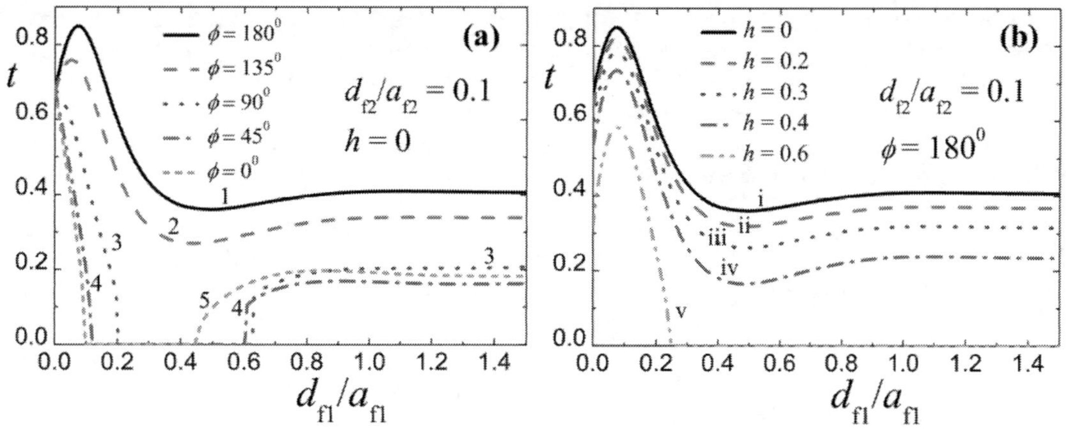

Figure 1. (Color online) (a) The phase diagram $t(d_{f1})$ at fixed thicknesses $d_{f2} = 0.1a_{f2}$ without magnetic field ($H = 0$) at various values of the angle ϕ: (1) $\phi = 180^0$, (2) $\phi = 135^0$, (3) $\phi = 90^0$, (4) $\phi = 45^0$, (5) $\phi = 0^0$;
(b) The phase diagram $t(d_{f1})$ for the antiparallel state ($\phi = 180^0$) at various values of the h: (i) $h = 0$, (ii) $h = 0.2$, (iii) $h = 0.3$, (iv) $h = 0.4$, (v) $h = 0.6$;
Other parameters of the system are: $d_s/\xi_{s0} = 0.65$, $l_s/\xi_{s0} = 0.3$, $\sigma_s = 3$, $n_{sf} = 0.75$, $l_{f1}/a_{f1} = l_{f2}/a_{f2} = 0.25$, $\sigma_{f1} = \sigma_{f2} = 5$, $I_1/\pi T_{cs} = I_2/\pi T_{cs} = 15$, $d_{f2}/a_{f2} = 0.1$.

In figure 1(a) the reduced critical temperature t as function of the reduced F_1 layer thicknesses at fixed thicknesses $d_{f2} = 0.1a_{f2}$ without magnetic field ($H = 0$) at various values of the angle ϕ is displayed. So we have an unusual dependence when a maximum corresponding to symmetrical case with $d_{f1} \approx d_{f2}$. It is caused by relatively small values of the $F_{1,2}$ layer thicknesses, when the antiparallel magnetizations compensate each other in most complete manner as in clean case [19], in which the solitary re-entrant superconductivity is predicted for clean $F_1/S/F_2$ trilayer. In common case a position of this peak strongly depends from the $F_1/S/F_2$ parameters (transparencies, etc). It is quite probably that the solitary re-entrant superconductivity may be realized in the dirty $F_1/S/F_2$ trilayer too.

The cases for $\phi = 180^0$ and $\phi = 0^0$ correspond to the antiparallel and parallel mutual orientation of the magnetizations respectively. It is seen, that the change of the angle from the

$\phi = 180^0$ to $\phi = 0^0$ leads to the suppression of the superconductivity. This is easy to understand. When the mutual orientation of the magnetization is antiparallel, the effective exchange field in the F layers is decreased and hence the critical temperature is maximal. Note that changing the angle ϕ varies qualitatively the dependence $t(d_{f1})$: a change of the mutual orientation of the magnetizations leads to the re-entrant superconductivity appearance.

The phase diagrams with the magnetic field influence are plotted in figure 1(b). The $t(d_{f1})$ curves for the antiparallel mutual orientation of the magnetizations are plotted under certain values of reduced magnetic field $h = H/H_c$ (H_c is critical field for isolated S metal). The thickness $d_{f2} = 0.1a_{f2}$ is fixed. The magnetic field variation leads to qualitative change of the $t(d_{f1})$ dependence. Without field this dependence has a maximum at the value $d_{f1} \approx 0.1a_{f1}$ and then it reaches a plateau. The magnetic field monotonically depresses the critical temperature, but it may also lead to qualitative change of dependence $t(d_{f1})$. Note that the dependence $t(d_{f1})$ at the value $h = 0.6$ has a clear pronounced peak and further ($d_{f1} > 0.1a_{f1}$) drops monotonically to zero.

4. Conclusions

In this work we consider the asymmetrical $F_1/F_2/S$ trilayers. Our theoretical approach is valid in the external magnetic field presence as in real experimental setup for the FS spin valve explorations [20, 21]. It is shown that the external magnetic field and triplet correlations can essentially influence on the critical properties of considered systems. In particular, the change of the magnetic field and mutual orientation of the magnetizations can lead to a qualitative modification of the phase diagrams of $F_1/F_2/S$ system.

Acknowledgments

The work is partially supported by Ministry of Education and Science of the Russian Federation and Russian Fund of Basic Research (No 13-02-01202).

References

[1] Izyumov Y A, Proshin Y N and Khusainov M G 2002 *Physics-Uspekhi* **45** 109–148
[2] Buzdin A 2005 *Rev. Mod. Phys.* **77** 935–976
[3] de Gennes P G 1964 *Rev. Mod. Phys.* **36** 225–237
[4] Efetov K B, Garifullin I A, Volkov A F and Westerholt K 2008 *Magnetic Heterostructures (Springer Tracts in Modern Physics* vol 227) (Springer Berlin / Heidelberg) chap Proximity Effects in Ferromagnet/Superconductor Heterostructures, pp 251–290
[5] Bergeret F S, Efetov K B and Larkin A I 2000 *Phys. Rev. B* **62**(17) 11872–11878
[6] Bergeret F S, Volkov A F and Efetov K B 2001 *Phys. Rev. Lett.* **86** 4096–4099
[7] Volkov A F, Bergeret F S and Efetov K B 2003 *Phys. Rev. Lett.* **90**(11) 117006
[8] Fominov Y V, Golubov A A and Kupriyanov M Y 2003 *JETP Letters* **77** 510–515
[9] Fominov Y V, Golubov A A, Karminskaya T Y, Kupriyanov M Y, Deminov R G and Tagirov L R 2010 *JETP Letters* **91** 308–313
[10] Oh S, Youm D and Beasley M 1997 *Appl. Phys. Lett* **71** 2376–2378
[11] Tagirov L R 1999 *Phys. Rev. Lett.* **83** 2058–2061
[12] Buzdin A I, Vedyayev A V and Ryzhanova N V 1999 *EPL (Europhysics Letters)* **48** 686
[13] Abrikosov A A and Gorkov L P 1961 *Sov. Phys. JETP* **12** 1243
[14] Usadel K D 1970 *Phys. Rev. Lett.* **25** 507
[15] Radović Z, Dobrosavljević-Grujić L, Buzdin A I and Clem J R 1988 *Phys. Rev. B* **38** 2388–2393
[16] Krunavakarn B and Yoksan S 2006 *Physica C: Superconductivity and its Applications* **440** 25–34
[17] Proshin Y N and Khusainov M G 1998 *JETP* **86** 930–942
[18] Avdeev M and Proshin Y 2012 *Journal of Physics: Conference Series* **394** 012015
[19] Khusainov M, Khusainov M, Ivanov N and Proshin Y 2009 *JETP Letters* **90** 359–364
[20] Leksin P V, Garif'yanov N N, Garifullin I A, Schumann J, Vinzelberg H, Kataev V, Klingeler R, Schmidt O G and Bchner B 2010 *Appl. Phys. Lett.* **97** 102505
[21] Leksin P V, Garif'yanov N N, Garifullin I A, Fominov Y V, Schumann J, Krupskaya Y, Kataev V, Schmidt O G and Büchner B 2012 *Phys. Rev. Lett.* **109**(5) 057005

Development of the paramagnetic RABiTS tapes for coated conductors

A E Vorobieva[1], I M Abdyukhanov[1], D N Rakov[1], V V Samusevich[1], A V Nikolaev[1], A V Borisov[1], Yu N Belotelova[1], V A Drobyshev[1], M M Potapenko[1], L M Kryukova[1], M V Polikarpova[a], V V Gur'ev[2], A K Shikov[2] and S V Shavkin[2]

[1] Bochvar Institute of Inorganic Materials (JSC VNIINM), 123098, Rogova 5a, Moscow, Russia
[2] National Research Center «Kurchatov Institute», 123182, Kurchatov sq.1, Moscow, Russia

E-mail: vniinmbochvar@yandex.ru

Abstract. RABiTS production process was developed for tapes from Ni-W alloys with W content in the range of 4-10.5 at.%. Tapes' composition, structure and magnetic properties were investigated. High purity Ni-W ingots were smelted and tapes with thickness of 0.06-0.09 mm and various total accumulated deformations in the range of 90-99.2% were fabricated. Texture quality dependence on accumulated deformation in the tapes with W various content was studied. X-ray analysis data including pole figures show that high quality cube texture can be formed in the tapes with W content up to 9 at.% inclusive, but not formed in the tapes with 10.5 at.%. According Curie temperature values and saturation magnetization measurements of alloys with W content more than 8 at.% are paramagnetic at liquid nitrogen temperature. Summarizing it can be stated that RABiTS tapes with 8-9 at.% W content have optimal combination of structure and magnetic properties for using in coated conductors.

1. Introduction

Substrate materials are the key element of the HTSC-2G coated conductors. There are two types of substrate materials: textured RABiTS Ni-W tapes and non-textured tapes based on Hastelloy alloys or stainless steel. The main requirement for the RABiTS tapes is the presence of the sharp biaxial cube texture [1]. Usually for the fabrication of such tapes the alloys on the base of Ni-W systems are used. These alloys have the FCC crystallographic structure and enable to form the desired sharp biaxial cube texture with {100} direction perpendicular to the tape's surface. Such texture is formed as a result of the appropriate thermomechanical treatment in the process of thin tape formation which consist of cold rolling with large amount of total accumulated deformation (usually 95% and more) and subsequent annealing in vacuum or in reduction atmosphere in the temperature range of 1050 – 1250 °C [2].

Applications of HTSC-2G in different superconducting devices, which work in alternating current mode, demand the minimization of electromagnetic hysteresis losses values which depend on the magnetic properties of the substrate material at working temperature (77 K and lower) and are the function of the value of mass magnetization (M_{sat}) of the chosen alloy. Nickel is a ferromagnetic metal with Curie temperature 631 K. The doping of Ni with W leads to decrease of the Curie temperature and M_{sat} value [2]. Curie temperature of these alloys diminishes lower than 77 K with the increase of

W content up to 8 at.% [3]. But the drawback of high W content is related with the enhanced difficulty to form the sharp texture, especially with increase of W content more that 5 at.%. Recently it was shown that the cube texture forming is also possible in the Ni-W alloys with W content up to 9 at.% but with much difficulties at tapes fabrication and the part of cube texture orientation in {100} direction was only 97-98.5% [4].

In the previous work [5] we have reported about possibility of the forming sharp cube texture in Ni-W tapes with W content up to 7.7 at.% (20.7 wt.%). M_{sat} values at 77 K and at 4.2 K were as small as 7 and 10 emu/g correspondently, that is by factor of 9 and of 6 lower than of pure Ni. Although such M_{sat} value decrease is favorable for using such alloy as HTSC-2G substrate but nevertheless the Curie temperature of this alloy was 97 K and so it remains ferromagnetic at all possible working temperatures of HTSC. So the main objective of present work was to increase the W content in alloy up to 10.5 at.% and to elaborate paramagnetic Ni-W tapes in combination with sharp cube texture.

In our previous work 100% cube texture in tapes with W content in the range of 5 – 7.7 at.% was obtained by thermomechanical treatment that consisted of tapes rolling with 99% of total accumulated deformation and subsequent texturing vacuum annealing at 1150 °C [5]. The texture quality of various alloys at lower amounts of accumulated deformation wasn't investigated. In the present work we have investigated the effect of W content in the range of 4 – 10.5 at.% and total accumulated deformation during cold rolling in the range of 90 – 99.2% on texture quality of Ni-W tapes.

2. Experimental part

At Ni-W alloys preparation we proceeded from the assumption that the most part of metal impurities have negative effect on the texture quality and so for ingots melting only high purity Ni and W were used. Also we avoided any doping elements (e.g. Zr, Y, Al) which are usually added for improvement of the homogeneity of the ingot.

The set of 5 ingots with the weight of ~ 7 kg each has been fabricated with the following W contents: 4.8, 6.0, 7.7, 9.0 and 10.5 at.% (13.6, 16.6, 20.7, 23.5, 27.0 wt.% correspondently). After appropriate thermomechanical treatment tapes with the thickness of 60 – 90 μm were obtained. The total accumulated deformation of these tapes before the texturing annealing was varied from 90 up to 99,2% with 0.5 – 1.0% interval.

According to the metallographic investigations after cold rolling all samples had similar homogeneous structure. Grains which were formed at earlier stages of thermomechanical treatment were completely eliminated.

Mechanical properties of tapes with various W content differed from each other. For example figure 1 shows the dependence of microhardness as function of total accumulated deformation for various alloys. The microhardness of the alloy with 4.8 at.% W at final stage of the cold rolling was only 370 Hv, whereas for alloys with 7.7 and 9.0 at.% its value exceeds 500 Hv.

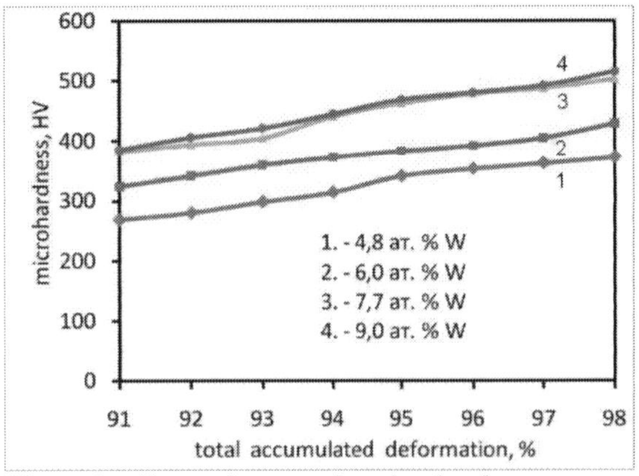

Figure1. Dependence of microhardness as a function of total accumulated deformation for alloys with various W content.

After texture annealing (1150 °C, 15 min) the texture quality was studied by X-ray analysis. It was carried out at $Co_{K\alpha}$-irradiation by 3 steps. At the first stage the general survey was made in the 2Θ angles range from 30 up to 120 degrees and integrated intensity ratio between the main Ni line (002) and secondary ones (111), (220) and (113) were determined. At the second stage, for samples in which integrated intensity of the secondary lines were less than 1% of the intensity of the main Ni line (002) the full width at half maximum of Ni (002) peak were determined. These measurements were carried out in two directions – along and across the direction of cold rolling. And at the third stage for complete texture characterization pole figures were created for some selected samples.

The magnetic properties in relation with the concentration of W in Ni-W alloys were investigated by the measurements of the Curie temperatures on the tape samples by the method of magnet shielding at the frequency of 1000 Hz in temperature range 330 – 4.2 K.

3. Results and discussion

X-ray investigations have shown that the tapes with W content 4.8 at.% have the best structure. Only Ni (002) peaks were presented at diffractogram. For samples with total accumulated deformation from 92% up to 99.2% the full width at half maximum along and across the direction of cold rolling was 4.4 – 4.8 and 7.3 – 7.6 degrees accordingly. Only in samples with 90 and 91% deformation the full width at half maximum increased up to 9.6 degree and so they had somewhat worse texture. Hence the best texture quality in tapes Ni-4.8 at.% W can be obtained only in the case of cold rolling with total accumulated deformation no less than 92%.

Contrary to this required texture wasn't obtained in any sample with W content 10.5%. At all diffractogram the Ni (002), (113), (220) and (111) peaks were presented. Intensity of the last 3 peaks decreased with the increasing of total accumulated deformation value but even at 99% deformation they haven't disappeared. The integrated intensity ratio of (002) and (113) lines was no more than 9:1.

The sharp cube texture was achieved in tapes of 3 others compositions. The texture quality depends on W content in tape and on total accumulation deformation value during cold rolling. For Ni-6.0 at.% W tapes after deformations 93 – 99% only Ni (002) lines were presented at diffractogram. At lower deformations traces of Ni (113) and (220) peaks were also presented. The minimum widening of the texture maximum with the full width at half maximum values less than 5 and 8 degrees in two perpendicular directions was obtained only after no less than 96% deformation.

The conditions of obtaining reasonable texture in tapes Ni-7.7 at.% W were more strict. Quality texture was achieved only after deformation 98 – 99%. Pole figures confirmed that in tapes with W content 4.8 – 7.7 at.% the part of cube texture exceeds 99%.

The most difficult was the texture formation in tapes with W content 9.0 at.%. Although the integrated intensities of Ni (113), (220) and (111) secondary peaks decreased greatly with increase of total accumulation deformation value but nevertheless even at 99.2% deformation the trace of (113) (the main among secondary peaks) was presented at the diffractogram. The integrated intensity ratio of (002) and (113) intensities was 400:1 (figure 2). Such high ratio was only in the case of deformations no less than 99.2%. At lower deformations secondary peaks were more noticeable. For example at 97.5% deformation the ratio of (002) and (113) was only 100:3. The full width at half maximum in two perpendicular directions in samples with 99% deformation was 4.5 and 7.5 degrees (figure 3) which is not worse than that in tapes with W content 4.8 at.%.

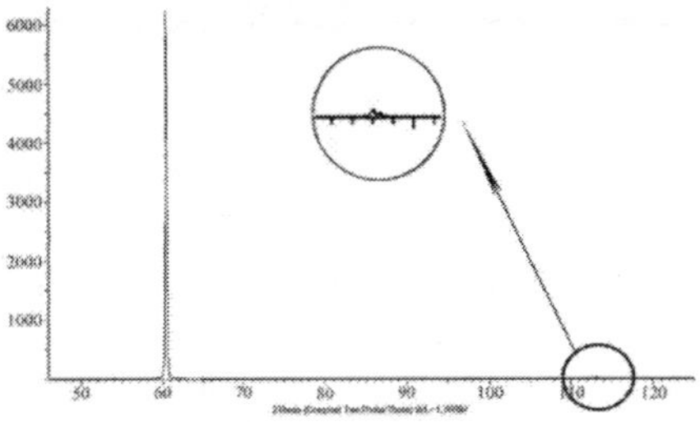

Figure 2. X-ray diffractogram of the Ni - 9.0 at.% W tape, 99.2% total accumulates deformation ($Co_{K\alpha}$ irradiation)

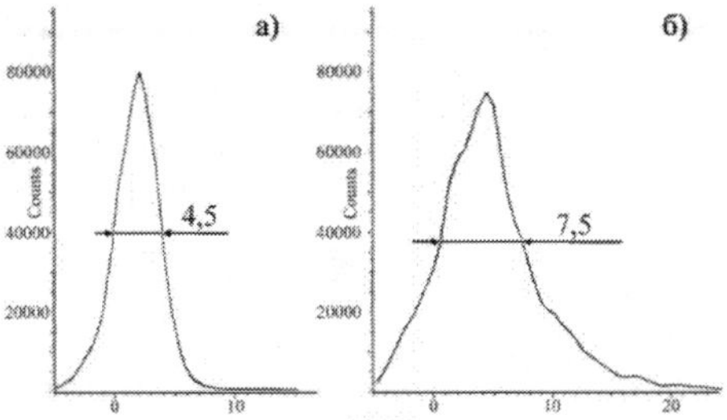

Figure 3. Tilt curves for Ni (002) line along (left) and across (right) direction of cold rolling. The full width at half maximum is shown by arrows. The sample Ni-9.0 at.%W, 99% total accumulated deformation

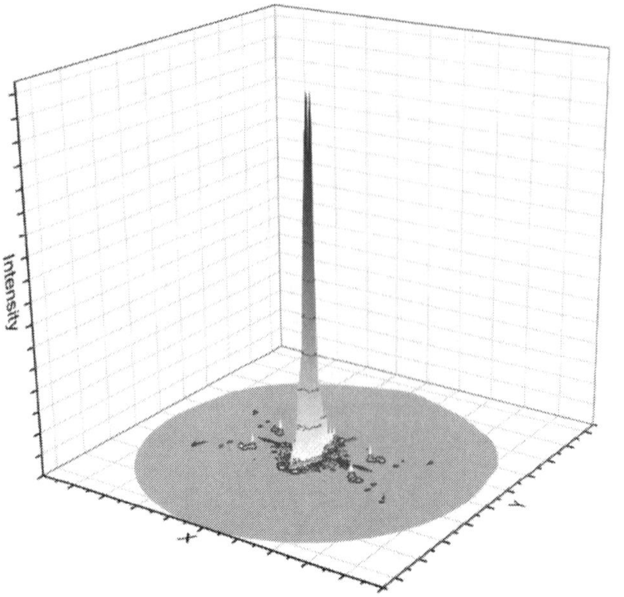

Figure 4. 3-D projection of {100} pole figure. Ni-9.0 at.%W tape.

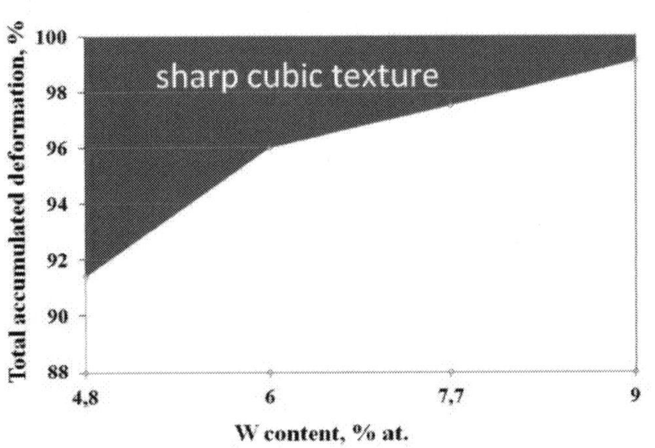

Figure 5. Border conditions for quality texture formation in Ni-W tapes

According to the pole figure (figure 4), the portion of cube texture {100} orientation in this sample was more than 97%. It is somewhat worse than that in tapes with lower W content but available for HTSC-2G production.

Thus it can be stated that desired texture was achieved in tapes with W content up to 9.0 at.% inclusive. The texture quality depends on total accumulation deformation. The tapes with higher W content demands more deformation to obtain desired quality. The border conditions for achieving available texture are shown on figure 5.

Previously we have reported [5] that Curie temperature linearly decreased with increase W content and drop from 300 K for alloy Ni-5.5 at.% W down to 97 K for alloy with 7.7 at.% W. Present work measurements confirmed these data and also had shown that tapes with 9.0 at.% W are paramagnetic in whole investigated temperature range.

4. Conclusion

In Ni-W tapes biaxial cube texture quality dependence on W content and total accumulated deformation value at cold rolling was investigated. It was shown that acceptable texture can be formed in tapes with W content up to 9.0 at.%. At the same time tapes with such high W content are paramagnetic. Summarizing it could be stated that tapes Ni-9 at.% W are favorable for the HTSC-2 coated conductors to be used in ac applications.

5. References

[1] Goyal A, Norton D P, Budai J D et al. 1996 *Appl. Phys. Lett.* **69**(12) 1795

[2] Specht E D, Goyal A, Lee D F et al. 1998 *Supercond. Sci. Technol.* **11** 945

[3] Sakamoto H, Nagasu Y, Ohashi Y, Nakasaki R, Mimura M, Nakai A 2007 *Physica C* **463–465** 600–603

[4] Rupich M W, Xiaoping Li, Sathyamurthy S et al. 2013. *IEEE Transactions on Applied Superconductivity*, vol: 23, issue: 3, part: 3. Article# 6601205

[5] Pantsyrny V I, Vorobieva A E, Abdyukhanov I M et al. 2012 *Proc. of ICMC-2012 Conf. (Fukuoka, Japan, 14-19 May 2012)* pp.895-898

Current Instability of High Temperature Superconducting Tapes in the AC Modes

V Romanovskii [a], ***IEEE Member*, K Watanabe** [b] **and S Awaji** [b]
[a] NBIKS-Centre, National Research Centre "Kurchatov Institute", Moscow, 123182, Russia
[b] High Field Laboratory for Superconducting Materials, Institute for Materials Research, Tohoku University, Sendai, 980-8577, Japan

E-mail: vromanovskii@netscape.net

Abstract. The basic physical reasons rooted in the current instability phenomenon of AC regimes are discussed. To explain their stable formation, the conduction-cooled high-temperature superconductors (HTS) are investigated studying mutually dependent thermal and electrodynamics states of a tape. It is proved that there exist characteristic times defining time windows of stable development of overloaded AC regimes. They lead to the states for which the limiting peak values of charged current and induced electric field before instability onset much more than the corresponding critical values of a superconductor. Moreover, the limiting stable peak values of the temperature, electric field and current are higher than the corresponding quench ones calculated at DC modes. It is shown that the stability boundary of overloaded AC modes slowly depends on the current frequency. As a result, the high-T_c superconducting magnets may stably operate under very high AC losses even at the conduction-cooled conditions. The obtained results should be taken into consideration when the range of stable losses is defined in HTS during AC regimes.

Introduction

The boundary of the current instability onset is an important factor, which determines the limitation in the possible application of superconductors. To design a high-temperature superconducting magnet with high operational properties, it is possible to use the regimes when a charged current is larger than the critical current of a superconductor. For example, the HTS operating regimes are stable in the overcurrent pulses [1]-[5]. At the same time, the mechanisms, which lead to the existence of the stable overloaded AC regimes, are not discussed. The presented below investigation allowed to formulate the thermo-electrodynamics mechanisms underlying the stable formation of overloaded AC regimes in HTS tapes.

2. Alternating current model of HTS tapes

Let us investigate the formation of AC regimes in a cooled superconducting tape ($-a < x < a$, $-\infty < y < \infty$, $-b < z < b$, $b >> a$) placed in a fully penetrated DC external magnetic field B. Assume that a superconducting core is uniformly distributed over a cross section of a tape with the volume fraction coefficient η; the sinusoidal current I having peak value I_m is charged in the y-direction and its self-field is greatly less than the external magnetic field. To make the analysis without excessive computations, let us accept that the temperature and electric field are uniform over the cross section of a tape for which the superconductor's voltage-current characteristic is described by a power low with the linear temperature dependence of the critical current density J_c. In accordance with these assumptions, the thermal and electric states of a superconducting tape depend on the currents flowing in the superconducting core J_s and the normal conducting sheath J_n. Therefore, the following equations

$$C(T)dT/dt = -hp(T - T_0)/S + E(t)I(t)/S, \quad T(0) = T_0 \qquad (1)$$

$$E(t) = E_c\left[J_s(t)/J_c(T,B)\right]^n = J_n(t)\rho_n(T,B), \quad E(0) = 0, J_c = J_{c0}(T_{cB} - T)/(T_{cB} - T_0) \qquad (2)$$

$$I(t) = \eta J_s(t)S + (1-\eta)J_n(t)S, \quad I(t) = I_m \sin(2\pi f t) \qquad (3)$$

may be used to describe the dynamics of the temperature and electric field in zero-dimensional approximation according to the continuous medium model. Here, C is the specific heat capacity of tape, h is the heat transfer coefficient, S is the cross section of tape, p is the cooled perimeter, E_c is the voltage criterion used in the J_c definition, n is the power exponent of the E - J curve, J_{c0} and T_{cB} are the critical parameters of a superconductor at the given external magnetic field, f is the current frequency.

As an example, we take the tape to be Ag/Bi2212 during the calculations. The sheath residual resistivity ratio RRR=$\rho_n(273\ K)/\rho_n(4.2\ K)$ was varied assuming that $\rho_n(273\ K)=1.48\times10^{-6}$ Ω·cm according to [6]. The sheath resistivity as a function of temperature and magnetic field was approached by the relations proposed in [6, 7]. The following parameters a = 0.019 cm, b = 0.245 cm, $S/p = a$, η = 0.2, n = 10, $E_c = 10^{-6}$ V/cm, $J_{c0} = 1.5\times10^5$ A/cm^2, T_{cB} = 26.1 K were set. The simulation was made at operating temperature T_0=4.2 K assuming that the external magnetic field is equal to B=10 T and the heat removal conditions on the surfaces of the tape take place at h=10^{-3} W/(cm^{2}K). The values of the peak current and frequency were varied. The heat capacities of the superconductor C_s and the sheath C_n are taken into account to calculate the heat capacity of the tape in the form $C(T)=\eta C_s(T)+(1-\eta)C_n(T)$ in compliance with the additive law. The temperature dependence of the heat capacity of a superconductor was given as in [8]. The heat capacity of silver was calculated in accordance with [6].

3. Results and discussion

Figures 1 and 2 prove the existence of the stability boundary of overloaded AC regimes. Note that the critical current of the tape equals 566 A in the case under consideration. Besides, the calculations, which were made according to the current instability theory discussed in [9], depict that the DC-quench values of the electric field, current and temperature equal $E_q = 2.9\times10^{-6}$ V/cm, I_q = 573 A, T_q = 5.9 K at RRR=10, respectively. The presented results show the following features of the AC formation regimes.

There are the stable and unstable regimes, which are similar to regimes observed in superconductor near the DC stability boundary. Thereby, the limiting peak current exists, which defines the stability boundary of the overloaded AC regimes in HTS tapes. At the same time, the limiting stable peak values of charged current and induced electric field are greater than the critical ones. Moreover, these values also exceed the corresponding DC-quench values. In particular, the stable peak value of the electric field is over 10 times greater than a priori chosen critical electric field in the both operating regimes depicted in figures 1b and 2b. As a result, this feature leads to the high level of the stable AC losses. As the calculations show, the peak value of the heat generation power (G=EJ) exceed 0.4 W/cm^3 during stable overloaded state at f=10 Hz and RRR=100. It is a very important result for a practical superconductivity application because these intensive losses do not destroy the superconducting properties of HTS tapes.

Figures 1b and 2b allow to understand the features leading to such overloaded AC regimes. They demonstrate the characteristic stages of their evolution, which, first of all, depend on frequency.

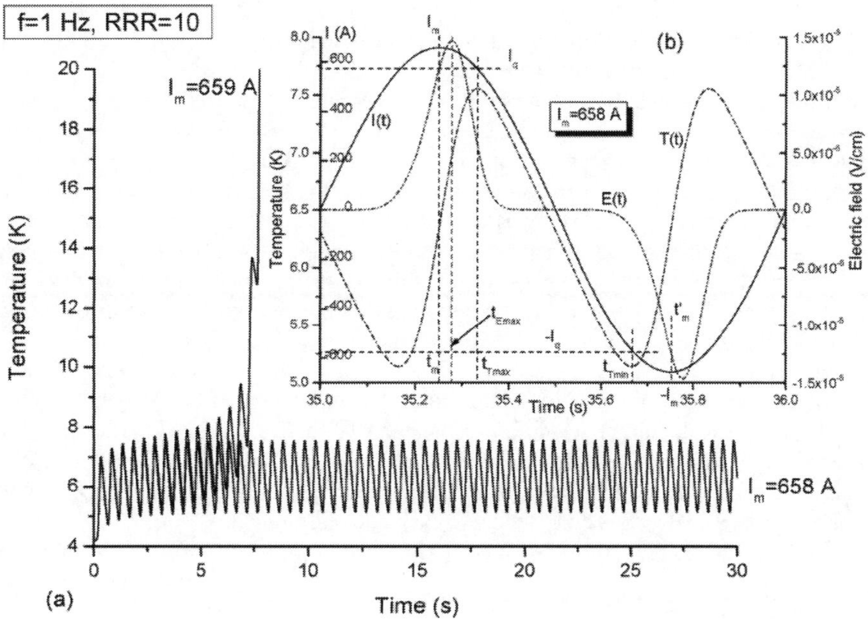

Figure 1. Stable and unstable temperature evolution (a) and typical stable stages (b) of the overloaded AC regime at $f = 1$ Hz at high resistivity of a silver sheath for Ag/Bi2212.

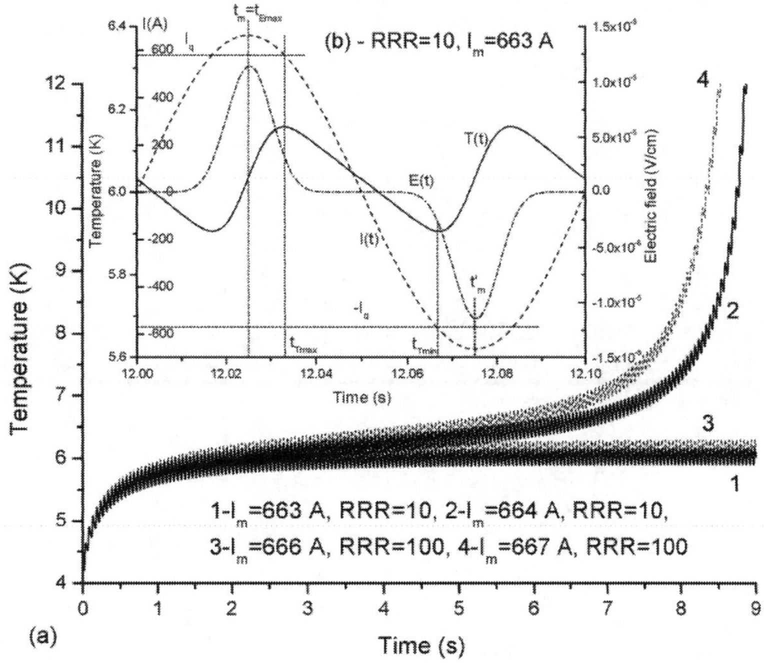

Figure 2. Stable and unstable temperature evolution (a) and typical stable stages (b) of the overloaded AC regime at $f = 10$ Hz and different resistivity of a silver sheath for Ag/Bi2212.

Journal of Physics: Conference Series **507** (2014) 022028 doi:10.1088/1742-6596/507/2/022028

They can consist of four or three time windows in the first half of charging cycle. First of all, let us discuss the formation peculiarities of the regime that take place at f=1 Hz (figure 1b). During this current load, the temperature still increases although the charged current decreases in the initial stage that starts after $t = t_m$. This stage is in the time interval $t_m < t < t_{Emax}$ where t_{Emax} is the time when the electric field has a maximum. As calculations show, the heat generation in a tape ($G=EJ$) exceeds the heat removal ($W=h[T-T_0]/a$) in this time interval. However, the heat removal has maximum at $t = t_{Emax}$. Therefore, the induced electric field starts to decrease at $t > t_{Emax}$. This is second stage when decreasing current passes into the corresponding value of the DC-quench current ($I(t) < I_q$). This stage exists in the time interval $t_{Emax} < t < t_{Tmax}$ where t_{Tmax} is the time when the temperature of a tape has a maximum. Therefore, the temperature of the superconductor continues to rise in this stage. The joint reduction of the electric field and current leads to the third stage at $t > t_{Tmax}$. It is characterized by decreasing temperature. Thereby, the stable decrease of the current, electric field and temperature take place at $t > t_{Tmax}$. The end of this stage is determined by t_{Tmin} after which the heat generation starts to exceed the heat removal, as the calculations show. Therefore, it is the fourth stage when the temperature of a tape starts to increase. In the end of this stage, the electric field and the current equal the corresponding peak values $-E_m$ and $-I_m$, respectively. The next mode of the AC regime evolution repeats the above-discussed stages and the stable cycle development goes on.

Figure 2b demonstrates that the first stage existing after I_m may be absent when the frequency increases. As calculations depict for the tapes under considerations, the current and the electric field decrease simultaneously in the first stage when the current frequency is more than 3 Hz. But the remaining stages repeat the above-discussed ones. As a whole, the calculations indicate that:

– the higher the current frequency or the lower the critical current density of a superconductor, the lower the first time window;

–the higher the sheath resistivity, the lower the first time window;

– the higher the heat capacity, the lower the first time window.

All of these peculiarities lead to the fact that the limiting instability currents of high-T_c superconducting tapes depend weakly on the current frequency during AC regimes.

4. Conclusion

The mechanisms underlying the existence of stable overloaded AC regimes in HTS tapes are investigated. It is shown that formation of them has the stages, which are defined by characteristic time windows depending on frequency, properties of a Bi2212 superconductor and a silver sheath. This feature defines the existence of the limiting value of a peak current of stable overloaded AC regimes. It is higher not only the critical current of a superconductor but also the corresponding quench current defining the current stability boundary in the DC regimes. Besides, stable peak values of the electric field and temperature are also higher than the related DC-quench values. These results demonstrate that the application of overloaded AC regimes would be very successful for many superconducting electro- power devices.

References

[1] Seong-Woo Y *et al* 2003 *IEEE Trans. Appl. Supercond.* **13** 2968

[2] Ishiyama A *et al* 2005 *IEEE Trans. Appl. Supercond.* **15** 1879

[3] Lue J W, Gouge M J, Duckworth R C 2005 *IEEE Trans. Appl. Supercond.* **15** 1835

[4] Ahn M C, Yang S E, Park D K 2007 *Cryogenics* **47** 425

[5] Fetisov S S, Vysotsky V S, Sytnikov V E 2009 *IEEE Trans. Appl. Supercond.* **19** 2411

[6] Dresner L 1993 *Cryogenics* **33** 900

[7] Lim H, Iwasa Y 1997 *Cryogenics* **37** 789

[8] Muller K-H, Andrikidis C 1994 *Phys. Rev.* B **49** 1294

[9] Romanovskii V R, Watanabe K 2009 *Superconducting Magnets and Superconductivity* ed H Tovar and J Fortier (Nova Science Publishers Inc., New York, USA) chapter 6 pp 293-399

Contactless measurements of local transport characteristics of coated conductors under the bending strain

I Rudnev[1,2], A Mareeva[1], N Mineev[1,2], S Pokrovskiy[1,2], A Sotnikova[1,2]

[1] National Research Nuclear University MEPHI, Moscow Russia
[2] International Laboratory of High Magnetic Fields and Low Temperatures, Wroclaw Poland

E-mail: iarudnev@mephi.ru

Abstract. The influences of bending strain on the critical current density J_c in coated conductor (CC) tapes were investigated at 77 K by contactless method based on scanning Hall probe magnetometry (SHPM). In order to clarify the reasons of variations in critical current for different bending radius we carried out local studies of critical currents in CC tapes before and after bending strain. By using SHPM we defined two-dimensional maps of critical current distribution at various bending radiuses. It was demonstrated that the superconducting layer in CC tape cracked at critical bending radius into periodical regions with unchanged and decreased transport properties. Appearance of local cracks with reduced value of the critical current leads to degradation of the critical current of the whole tape. It is found that the both critical bending radius and kind of tape cracking depend on the type of metallic substrates. The results of contactless studies are in good agreement with direct transport measurements.

1. Introduction

Advances in the fabrication processes of coated conductor tapes based on high temperature superconductors (HTSC) have lead to real possibility to use CC tapes in electrical application such as superconducting electric power cables, motors, wind generators, magnets, etc. During fabrication and practical operation of superconducting devices the coated conductor experiences various stresses. One of them is the bending stress during the winding of coils. So it is necessary to investigate the mechanical bending strain property of $YB_2Cu_3O_{7-x}$ (YBCO)-coated conductors with high J_c. For YBCO CC tapes, some electro-mechanical properties have been investigated by various research groups which include tensile, transverse compressive and bending strain dependence of the J_c [1–8]. Under bending deformation, when a strain was over 0.5%, cracking was observed using a scanning electron microscope (SEM). The generation of crack appeared to be the primary reason for J_c degradation [2,6]. In the above papers the critical current was measured by transport methods, and the observation of cracks by using SEM was carried out after chemical removing of the protective layers that could result in loss of integrity of HTSC composite. It should also be noted that the transport methods do not reflect the local distribution of critical current and do not allow to observe the dynamics of appearance of local areas with reduced critical current.

In this report we present new results of visualization of critical current distribution in CC tapes under bending deformation. We demonstrate that the superconducting layer in CC tape cracks at some bending radius into periodical regions with unchanged and decreased critical current.

2. Materials and methods

For studies we selected two types of tapes produced by SuperPower (SP) and American Superconductor (AMSC), which differ in the substrate and internal architecture. The tapes were wound on cylinder with diameter d , cooled, warmed and then the measurements of trapped magnetic flux were done at T= 77 K. Whole procedure was repeated for several cylinders with diameters d from 50 mm to 4 mm. For measurements of trapped magnetic field we used home-made SHPM equipment. Two-dimensional distributions of the critical current as well as the distribution of the longitudinal and transverse J_c components were obtained from the local magnetic induction data according to algorithm described in Ref. [9, 10]. The studies were performed for two variants of tapes winding: the superconducting layer was inside or outside the bend of tapes. The first case corresponds to compression strain in superconducting layer, the second one – to tensile strain.

3. Results and discussion

Figure 1 shows the two-dimensional distributions of the trapped magnetic induction $B_z(x,y)$ for the cases of tensile and bending strain. In both cases the surfaces $B_z(x,y)$ are given for the three bending diameters d = 46 mm, 6 mm and 4 mm. From these figures one can see that for large values d= 46 mm distribution $B_z(x,y)$ are sufficiently homogeneous. Distributions $B_z(x,y)$ are approximately the same for the cases of tension and compression. A sharp discontinuity $B_z(x,y)$ is observed at d = 6 mm for the case of tensile bending strain. This discontinuity $B_z(x,y)$ indicates that there is a separation of the sample to the superconducting and non-superconducting regions. Perhaps it means the appearance of mechanical cracks in HTSC layer like it was shown in [2, 6]. A further decrease in diameter down to 4 mm leads to disappear of superconductivity almost completely. In the case of compressive bending strain a different scenario takes place. At d = 6 mm and d = 4 mm the superconducting layer does not crack, and the distributions $B_z(x,y)$ are sufficiently uniform. However, decrease in diameter d leads to gradual reduce of trapped flux. From the matrix of values $B_z(x,y)$ we calculated spatial distribution of longitudinal $J_x(x,y)$ and transverse $J_y(x,y)$ components of the critical current density. Component $J_x(x,y)$ formally characterizes the transport critical current as it is directed along the tape, while $J_y(x, y)$ reflects the presence of current flow heterogeneity, which causes a deviation of the current lines along the direction of the tape. Figure 2 shows the longitudinal components $J_x(x,y)$ obtained from $B_z(x,y)$ data. In the case of tension the appearance of regions with sharply reduced J_c for d = 6 mm and a full J_c loss at d = 4 mm can be clearly seen. Figure 3 shows the dependencies of the maximum and minimum critical current on bending diameters d obtained from values $J_x(x,y)$ for different d.

To compare the different types of tapes (see Figure 4) the experimental data were reconstructed as the dependence of the critical current on bending strain ε, which was defined as $\varepsilon = t / (2R + t) \times 100\%$, where t is the thickness of tape and R is bend radius.

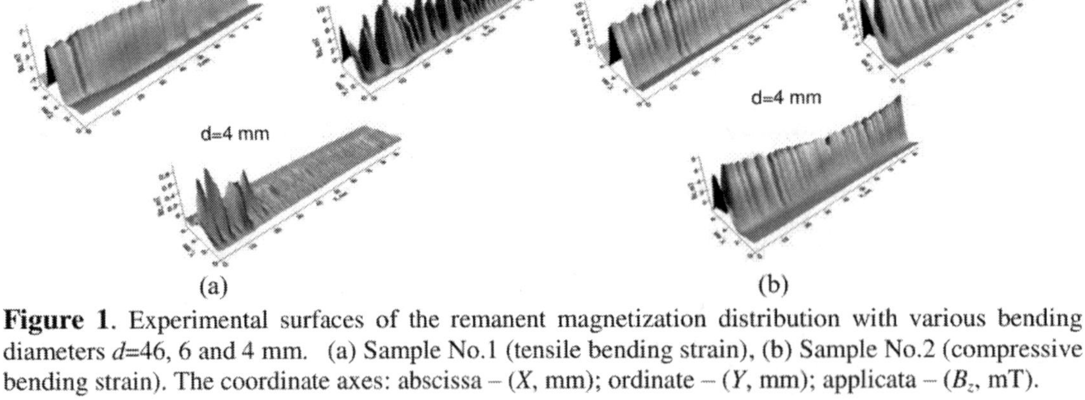

Figure 1. Experimental surfaces of the remanent magnetization distribution with various bending diameters d=46, 6 and 4 mm. (a) Sample No.1 (tensile bending strain), (b) Sample No.2 (compressive bending strain). The coordinate axes: abscissa – (X, mm); ordinate – (Y, mm); applicata – (B_z, mT).

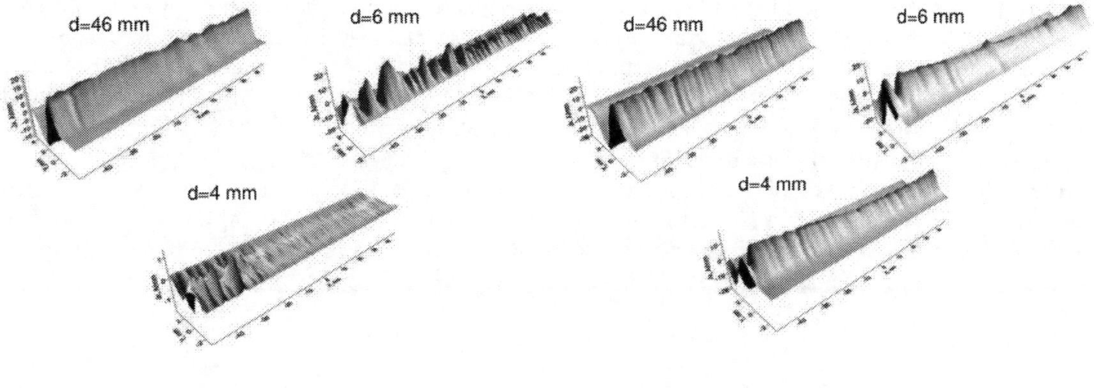

(a) (b)

Figure 2. Distribution of shielding currents along the tapes with various bending diameters d=46, 6 and 4 mm. (a) Sample No.1 (tensile bending strain), (b) Sample No.2 (compressive bending strain). The coordinate axes: abscissa – (X, mm); ordinate – (Y, mm); applicata – (J_x, A/mm).

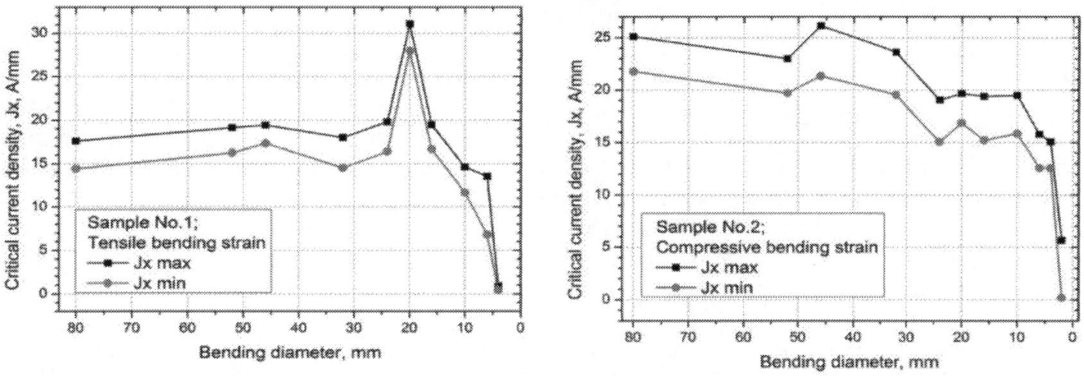

Figure 3. The dependence of the maximum and minimum values of the longitudinal component of the critical current density on the bending diameter for tensile bending strain (left) and compressive bending strain (right).

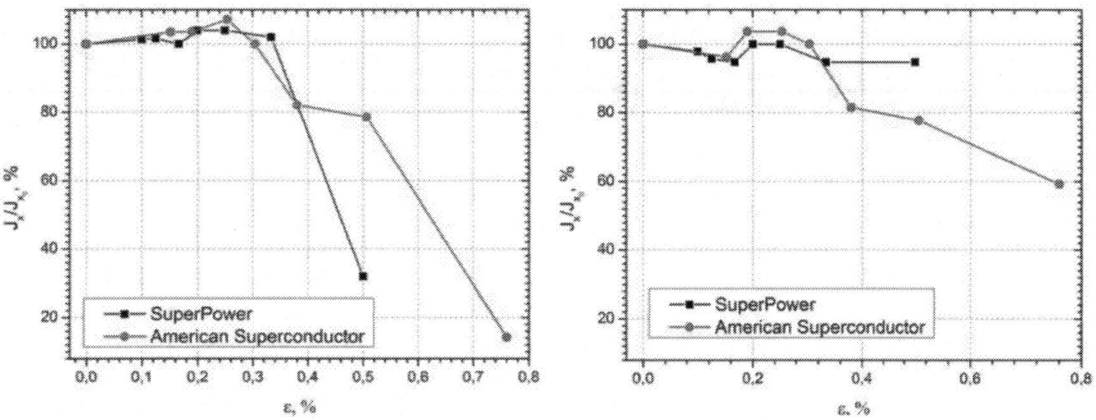

Figure 4. The dependence of normalized critical current density on the bending strain for SP and AMSC tapes. Left - tensile bending strain, right - compressive bending strain.

Calculations of bending strain ε show that the superconducting layer in the AMSC tapes is subjected to the high mechanical stresses compared to SP tapes for the same radius of curvature due to position of the neutral axis of strain through the thickness of the substrate. SP samples do not experience significant changes in the critical current of up to 0.5% strain, then in the case of stretching the critical current drops sharply by 70%. The strains 0.5% corresponds to d=10 mm.

Degradation of the critical current of AMSC tapes clearly observed from ε=0.38% strain, which corresponds to the bending diameter 25 mm. At 0.5% strain (d=15 mm) J_c is reduced in 20%. When the deformations reach 0.76% (10 mm) J_c decreases in 40% for the case of compressive strain and in 85% for the case of tensile strain. The difference in the data for different types of tapes is related to the difference of mechanical properties and structural features of substrates.

4. Conclusion

The influence of bending strain on critical current density in YBCO coated conductors was investigated by local magnetic method based on scanning Hall probe techniques. The obtained results are summarized as follows:

(1) In the region of small deformation (i.e. small strain) HTSC layer deforms elastically and critical current density does not change. At strain 0.3% – 0.5% plastic flow of the substrate materials begins that leads to degradation of HTSC critical current.

(2) Behavior of the critical current is different for the cases of the tensile or compressive bending strain. Namely, for the tensile strain samples crack at bending diameter 10 mm while in the second case no occurrence of cracks is observed down to d=4 mm.

(3) The behavior of the critical current density in the bending strain is different for SP and AMSC tapes. Plastic flow of AMSC tapes begin at lower values of strain, but cracking of HTS layer occurs first in the SP samples. The observed differences may be due to several reasons: different positions HTS layer relative to the neutral axis, the different mechanical properties of the substrate materials, various mechanisms of stress for these materials.

It should also be noted that we observed a decrease in J_c after eight weeks at a constant exposure to bending strain. Value reduction ranged from 8% to 25% for different diameters of the bend. This phenomenon may be due to materials creep.

Acknowledgments

The work was supported by Russian Ministry of Education and Science, Grant 14.A18.21.0349.

References

[1] Cheggour N, Ekin J W, Clickner C C, Verebelyi D T, Thieme C L H, Feenstra R, Goyal A and Paranthaman M 2003 *IEEE Trans. Appl. Supercond.* **13** 3530
[2] Park C, Norton D P, Budai J D, Christen D K, Verebelyi D, Feenstra R, Lee D F, Goyal A, Kroeger D M and Paranthaman M 1998 *Appl. Phys. Lett.* **73** 1904
[3] Shin H S, Kim K H, Dizon J R C, Kim T H, Ko R K and Oh S 2005 *Supercond. Sci.Technol.* **18** S364
[4] Sugano M, Osamura K, Prusseit W, Semerad R, Itoh K and Kiyoshi T 2005 *Supercond. Sci. Technol.* **18** 369
[5] Cheggour N, Ekin J W, Thieme C L H, Xie Y Y, Selvamanickam V and Feenstra V 2005 *Supercond. Sci. Technol.* **18** S319
[6] Sutoh Y, Kakimoto K, Kaneko N, Iijima Y and Saitoh T 2005 *Physica C* **421–430** 933
[7] Shin H S, Dedicatoria M J, Dizon J R C, Ha H S and Oh S S 2009 *Physica C* **469** 1467
[8] Sugano M, Machiya S, Sato M, Koganezawa T, Shikimachi K, Hirano N and Nagaya S 2011 *Supercond. Sci. Technol.* **24** 075019
[9] Pokrovskiy S V, Rudnev I A and Podlivaev A I 2009 *J. Phys. Conf. Ser.* **150** 052211
[10] Rudnev I A, Pokrovskiy S V and Podlivaev A I 2012 *IEEE Trans. Appl. Supercond.* **22** 9001304

Diagnostics of target inhomogeneity and influence of YBCO target oxygen content on properties of HTSC 2G samples fabricated by PLD technique

S V Shavkin[1], A K Shikov[1], I A Chernykh[1], V V Guryev[1], E S Kovalenko[1], E V Yakovenko[1], M L Zanaveskin[1], D N Rakov[2] and A E Vorobieva[2]

[1] National Research Centre "Kurchatov Institute", 1, Kurchatov square, Moscow, 123182, Russia
[2] Bochvar All-Russian Scientific Research Institute for Inorganic Materials (VNIINM), 5a, Rogova street, Moscow, 123098, Russia

E-mail: shavkin@mail.ru

Abstract. The pulsed laser deposition (PLD) targets used for HTSC 2G fabrication should be uniform and chemically suitable to make long tapes with reproducible critical properties. We investigated inner density distribution of flat YBCO targets using non-destructive neutron introscopy and tomography techniques. The methods are based on absorption of monochromatic neutron beam which is going through the specimen. In first set of targets we measured variance of material density (up to 10-25%) with maxima density located in central layer of the target body. After updating of target fabrication technology no significant density inhomogeneity were detected. The dependence of critical current of superconducting films with thickness of 300 nm deposited on sapphire substrate with appropriate buffer layers using PLD YBCO targets with various oxygen content was also investigated. Oxygen index of YBCO targets vary from 6.12 to 6.85. No significant influence of target oxygen index on critical current density of fabricated films was detected.

1. Introduction

The method of Pulsed Laser Dispersion (PLD) allows manufacturing of long length second generation (2G) HTS conductors with extremely high current-carrying capacity. The main advantage of the PLD method is stoichiometric transfer of a material from the target to the substrate that defines high requirements to quality of the targets to be sprayed. The target for PLD has to be uniform by density to keep conditions of laser sputter plume formation for a long time and to provide fabrication of long tapes with reproducible critical characteristics. In this paper the application of nondestructive methods of a neutron radiation introscopy and tomography for testing of uniformity of PLD target density.

Besides, practical interest is the influence of an oxygen index of PLD target on properties of the fabricated coating. The influence of number of PLD target parameters such as grain size, density, cationic stoichiometry (Y, Ba and Cu), composition and percentage of the impurities (acting as the pinning centers) on critical characteristics of fabricated HTSC coatings was under intensive investigations [1-5]. However influence of an oxygen index of the target on characteristics of formed epitaxial layers remains insufficiently studied. It is well known that critical characteristics of both bulk HTSC and films depend on the content of oxygen (index) in the unit cell of $YBa_2Cu_3O_x$ [6, 7]. In

YBCO the superconducting orthorhombic phase with a critical temperature of about 90 K exists at 6.8<x<7.0, the phase with T_C about 60 K corresponds to 6.6<x<6.8 but when oxygen concentration x<6.5 there is a transition to a tetragonal not superconducting phase. Possibility of formation of coatings with high value of critical temperature from a target with the low oxygen content of x=6.55 was demonstrated earlier [8].

It is credible that during long-duration PLD process the temperature of the target surface could reach up to several hundred centigrade due to heat radiation from substrate. As the deposition process of all epitaxial layers (buffer and superconducting) of HTSC tapes is carried out in high vacuum or in inert gas atmosphere or in the oxygen environment there can be a considerable variations of HTSC target oxygen index due to diffusion of oxygen from/to the target. So the influence of such target surface oxygen index variation on critical current of fabricated epitaxial HTSC films is very important.

2. PLD target preparation

All targets were fabricated in Bochvar Institute. The basic manufacturing techniques are earlier described in paper [9]. Briefly, the technology includes preparation of the mix of nitrate solutions of yttrium and copper and barium nitrite solution; carrying out joint sedimentation of oxalates; drying and processing of an oxalate deposits; pyrolysis of oxalate powder; powder processing after pyrolysis; pressing of targets, heat treatment of targets.

The first set of targets (denoted below as #216, 217, 259, 261, 262, 264) were made with dry powder pressing technique without using of binding substance. Diameter of these targets was about 20 mm. Mean powder size was about 10 microns. Pressing power varies from 0.5 to 5 t/cm^2. The average density of targets was from 5.3 to 5.95 g/cm^3.

The updated target fabrication technology includes powder pressing technique with specific organic binder. These are targets denoted below as #282, 284, 352-1. Diameter of targets was about 65 mm. Average density of targets was 5.7 – 5.75 g/cm^3.

3. Diagnostics of target inhomogeneity by neutron introscopy and tomography techniques

Advantage of neutron methods of research in comparison with x-ray is large penetration length. The investigations of YBCO PLD targets were carried out on DICTOR installation of neutron research reactor IR-8 in Kurchatov Institute. Determination of macroscopic non- uniformity of the object was done by means of a transmission of a monochromatic thermal neutrons beam through a sample with using of absorbing and diffraction contrast. Output beam from reactor is monochromatic (with wavelength of 1.52 Å) with 40 × 50 mm^2 in size which passes through the sample located on remotely operated motorized goniometer with a vertical axis of rotation. The neutron radiation is registered by the detector consisting of the scintillation screen, the mirror, a lens and remotely operated video camera with Peltye cooled CCD matrix of 4004 x 2671 pixels in size. Spatial resolution of the image is limited by angular divergence of incident neutron beam (about 2.5 deg.) and is about 400 microns on the sample. The sample image on the detector was formed mainly due to absorbing contrast that allows to visualize distribution of absorbing ability and substance density. In the presence of large crystalline grains there can be an additional diffraction contrast Program-controlled registration of images by sample rotation round an axis of the goniometer allows to carry out tomographic shooting of a sample in an automatic mode.

In case of a homogeneous uniform sample the distribution of intensity of transferred beam on introscopy image will be uniform. Appearance of contrast areas on the image denote to heterogeneity of a chemical composition and/or local density of the sample.

More detailed information on density distribution in PLD targets can be received by carrying out of full tomographic shooting of targets with small steps of scanning angle of a target regarding to vertical axis. Total duration of shooting of introscopy experiment is about of 2 minutes, in case of a tomography experiment it is about 12 hours. Tomographic shooting consists of 360 projections with a steps of 0.5 deg., and empty beam registration with dark current of the detector. Preprocessing of projections (cleaning of noise, the accounting of a background and non-uniformity of the beam, image normalization), and also decomposition of virtual cross-section on projections was done by means of ImageJ program macros. Advantage of neutron tomography is that one can easy look on shape of any inhomogeneity areas in the sample and can make any cross-sections inside the sample.

Typical introscopy images and results of interpretation for target #261 are shown on figure 1. Both neutron beam projections (full-face and side-face) allow estimate the measure of inhomogeneity of target. On side-face image the narrow dark area exists in the central part of target. Taking into account intensity of transferred neutron beam (see intensity cross-section profiles marked by 1, 2 and 3 for side-face) we can calculate appropriate variance of density (see corresponding density line for profile 2 which goes though target geometrical center). Maximum variation of density of target is about 26%.

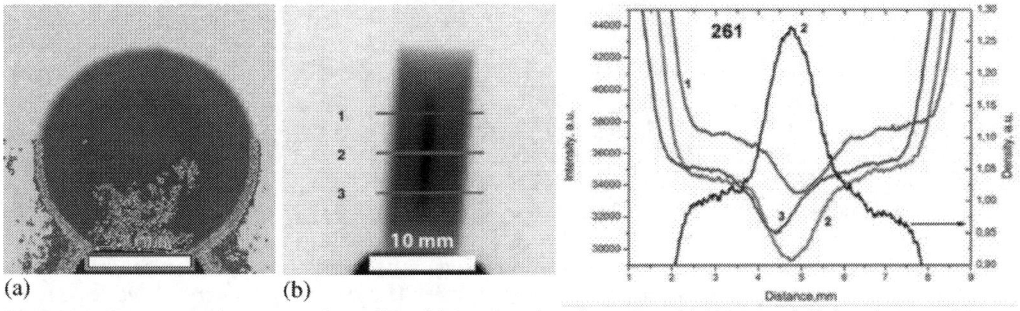

Figure 1. Introscopy images and results of interpretation for target #261. (a) - full-face image, (b) - side-face image. Right graph shows side-face intensity profiles (1, 2 and 3) and density profile 2.

Calculated maximum variations of density for all PLD targets from first set (dry pressed) are presented in table 1. All targets have area with increased density in the central layer of target body.

Table 1. Calculated maximum density variation for PLD targets.

Target #	262	216	259	261	264
Maximum density variation (%)	12	10	15	26	15

In contrast both tomographic and introscopy images of targets fabricated with organic binder show no variations of density (also in central layer). Figure 2 shows tomographic side-face cross-section of three new targets (#282, 284, 352-1) and of one dry pressed target (#262) for comparison. Maximum density variation for PLD targets does not exceed 3%.

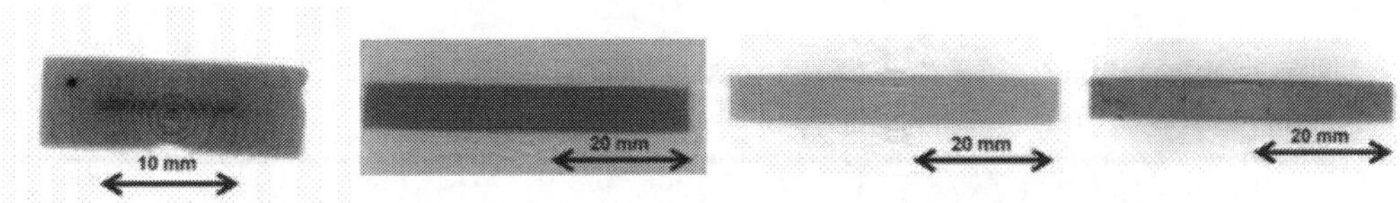

Figure 2. Tomographic side-face target cross-section. From left to right: #262 (dry pressed, strong inhomogeneity and dense inclusion inside); #282, 284 and 352-1 (all pressed with organic binder, homogenous). Visible concentric circles are artifacts due to tomographic image recovery procedure.

4. Influence of an oxygen index of PLD targets on properties of HTSC coating

We used the YBCO target #217 with diameter of 25 mm and 10 mm in height with average density of 5.95 g/cm^3 to fabricate HTSC films. Estimation of the oxygen index variation of the target was carried out by means of x-ray diffraction method (using "oxygen index" vs "*c*- lattice parameter" dependence [10]) and by changing of the target weight (to within 0.1 mg) after each deposition experiment. Initial oxygen index of the target was x=6.85. All HTSC films with thickness of 300 nm was deposited by excimer (KrF) laser on single crystal sapphire Al_2O_3 (r-plane) with CeO_2 buffer layer. Substrate temperature was 760°C, oxygen pressure in the chamber was 350 mTorr. Each deposited film had been subjected to post-growth annealing in the oxygen atmosphere with a pressure of 650 Torr during one hour. Critical temperature of films was measured by inductive method. Critical current was measured using the four-probe method on bridges with 1 mm width (77K, self-field). Temperature of the target during process was measured in-situ by thermocouple. It has been found that the target surface temperature reached the maximum of about 500°C in 30 minutes during deposition process.

At the first experiment the influence of target heating on its oxygen index was examined. Target was subjected to warming up within 30 minutes inside deposition chamber, thus temperature of a target reached 500°C. The oxygen index of a target before and after radiation heating has changed from x=6.85 to x=6.60. To simulate influence of buffer layer (CeO_2) growth process on the target oxygen index the vacuum annealing of the target was done with temperature of 500°C during 3 hours. As the result the target oxygen index decreased from x=6.60 to x=6.12. To recover the target oxygen index to an initial x=6.85 the annealing in pure oxygen with a pressure of 600 Torr was carried out with temperature of 500°C during 3 hours. Thus, target oxygen index can vary within 6.85-6.12 during standard deposition process and post-growth annealing.

At the second experiment we deposited HTSC films using the same target but various starting target oxygen index (x=6.12 and x=6.85). Measured oxygen index of both HTSC films was x=6.9 independently on starting target oxygen index. Critical temperature of films was 88-89 K, critical current and critical density were 9.4 A (3.2 MA/cm^2) and 10.4 A (3.5 MA/cm^2) respectively. Thus, critical characteristics of both samples are almost the same. This can be explained by two factors: 1) oxygen environment in the chamber can re–saturates target with oxygen, and 2) post-growth oxygen annealing of film can re–saturates film with oxygen.

5. Conclusions

In this paper unique nondestructive neutron introscopy and tomography techniques were developed and successfully applied to investigate the PLD target inhomogeneity. The PLD targets fabricated by updated technology with pressing with organic binder are uniform and homogenous as distinct from dry pressed targets which demonstrate up to 25% density variation.

The oxygen index of the PLD target varies in the chamber during deposition process (within limits 6.12-6.85) but the variation does not influence on current-carrying capacity of fabricated HTSC films.

Acknowledgments
Work is supported by Rosatom State Corporation (contract H.4g.43.90.13.1164).

References

[1] Paturi P, Schlesier K and Huhtinen H 2009 *Physica C* **469** 839
[2] Irjala M, Huhtinen H, Jha R, Awana V P S and Paturi P 2011 *IEEE Transactions on Applied Superconductivity* **21** 2762
[3] Wee S H, Specht E D, Cantoni C, Zuev Y L, Maroni V, Wong-Ng W, Liu G, Haugan T J and Goyal A 2011 *Phys. Rev.* B **83** 224520
[4] Peurla M, Huhtinen H, Paturi P, Stepanov Yu P, Raittila J and Laiho R 2005 *IEEE Transactions on Applied Superconductivity* **15** 3050
[5] Develos-Bagarinao K, Yamasaki H, Nakagawa Y and Endo K 2004 *Physica C* **412–414** 1286
[6] Maiti K, Fink J, Jong S, Gorgoi M, Lin C, Raichle M, Hinkov V, Lambacher M, Erb A and Golden M S 2009 *Phys. Rev.* B **80** 165132
[7] Mori Z, Doi T and Hakuraku Y 2010 *J. Appl. Phys.* **107** 023903
[8] Huhtinen H, Paturi P, Lahderanta E and Laiho R 1999 *Supercond. Sci. Technol.* **12** 81
[9] Vorobieva A E, Abdukhanov I M, Rakov D N, Kotova E V and Shikov A K 2012 *VANT. Materialovedenie i nivie materialy (in Russian)* **73** 108
[10] Benzia P, Bottizzoa E and Rizzia N 2004 *J. of Crystal Growth* **269** 625

Cooling Stability Test of MgB2 Wire Immersed in Liquid Hydrogen under External Magnetic Field

Yasuyuki Shirai*, Kyosuke Hikawa*, Masahiro Shiotsu*, Hideki Tatsumoto, Yoshihiro Naruo***, Hiroaki Kobayashi***, Yoshifumi Inagaki*****

*Kyoto University, Graduate School of Energy Science, 606-5801 Kyoto, Japan、
**JAEA, J-Park Center, Ibaraki, Japan, ***JAXA, Sagamihara, Japan

shirai@energy.kyoto-u.ac.jp

Abstract. Liquid hydrogen (LH_2), which has large latent heat, low viscosity coefficient, is expected to be a candidate for a cryogen for superconducting wires, not only MgB_2 but also other HTC superconductors. LH_2 cooled superconducting wires are expected to have excellent electro-magnetic characteristics, which is necessary to be clear for cooling stability design of LH_2 cooled superconducting device, however, due to handling difficulties of LH_2, there are only few papers on the properties of LH_2 cooled superconductors, especially under external magnetic field. We designed and made an experimental setup which can energize superconducting wires immersed in LH_2 with the current of up to 500A under the condition of external magnetic field up to 7 T and pressure up to 1.5 MPa. In order to confirm experimental method and safety operation of the setup, over current tests were carried out using MgB_2 superconducting wires under various external magnetic field conditions. Critical current of the test wire at the temperature 21, 24, 27, 29 K under external magnetic fields up to 1.2 T was successfully measured. The resistance of the wire also was measured, while the transport current exceeded the critical current of the wire.

1. Introduction

Generally, the high-Tc superconductors, such as YBCO and BSCCO are cooled by liquid nitrogen (77K), however, it is considered that the excellent electro-magnetic properties of such materials are achieved with temperature of 20 – 40 K. Liquid hydrogen (LH_2) has excellent properties as a coolant, such as large latent heat, low viscosity coefficient and so on [1]-[3]. The characteristics of Jc-B (critical current density and critical magnetic flux density) of high-Tc superconductors cooled by LH_2 are fairly good compared with those by LN_2, while the heat capacity of the materials in LH_2 temperature is still large for better cooling stability. Especially, LH_2 is expected to be the coolant for MgB_2 superconductor whose critical temperature is 39 K.

Cooling stability in steady state and transient state of superconductors and their coils cooled by LH_2 in a certain magnetic field is important issue for design of LH_2 cooled superconducting power apparatus. We have developed a thermal-hydraulics experimental system for liquid hydrogen [4]-[6] in order to investigate heat transfer characteristics for wide ranges of subcoolings, pressures up to supercritical and flow velocities in forced flow cooling. The experimental set-up for investigating electrical properties of liquid hydrogen cooled superconductors under external magnetic field was designed and made [7]. In this paper, over-current test of an MgB_2 short sample (supplied by Hyper Tech) under magnetic field in order to evaluate the experimental set-up was described.

2. Experimental set-up[7]

Experimental set-up for investigating LH_2 cooled superconductors under magnetic field was designed and made as shown in Fig.1 (cross-sectional view). The LH_2 cryostat has 61 L volumetric capacity of LH2 with 309 mm inner diameter. The insert of LH_2 cryostat has 3 current leads (~ 500 A), then two test sample can be set at once. A superconducting magnet which can produce up to 7 T in experimental space (~300 mm diameter and 400mm height) lower part of the LH_2 cryostat is set in the liquid helium (LHe) cryostat. Major specification of the test facility is listed in Table 1.

Test superconductors, which are set immersed in LH_2, can be tested with excitation of various current patterns (~500 A) in a certain magnetic field (~7 T). Test pressure can be set up to 2.0 MPa. Test temperature can be set 20 ~32 K using sheath heater placed at the bottom part of the cryostat.

The LH_2 cryostat is equipped though the bore area of the magnet. Maximum current sweep rate is 0.09 A/s. It takes about one hour to excite the magnet up to 7 T.

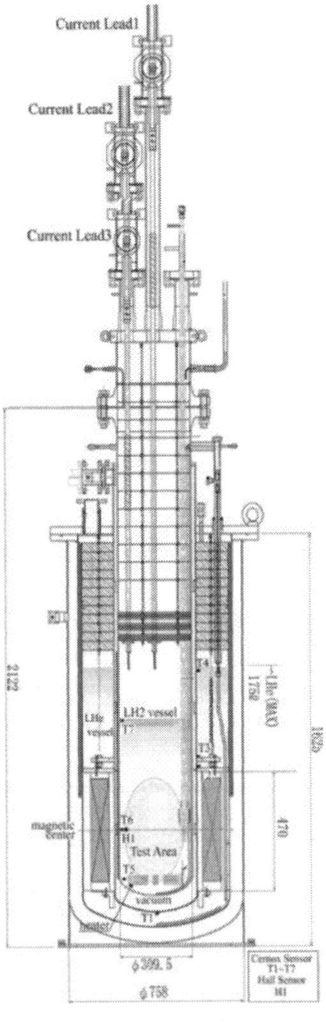

Fig. 1 Cross-sectional view of cryostats for electro-magnetic property tests of liquid hydrogen cooled superconductors.

TABLE 1

SPECIFICATION OF CRYOSTAT AND MAGNET

Liquid Hydrogen Cryostat	
Inner diameter	309.5 mm
Height (bottom to top flange)	2218 mm
Volumetric capacity for LH_2	61 L max
Liquid Helium Cryostat	
Inner diameter	350 mm
Outer diameter	630 mm
Height (bottom to top flange)	1625 mm
Volumetric capacity for LHe	175 L max
Superconducting Magnet	
Material	NbTi
Inductance	112.36 H
Rated current	175A
Max. magnetic field (center)	7 T

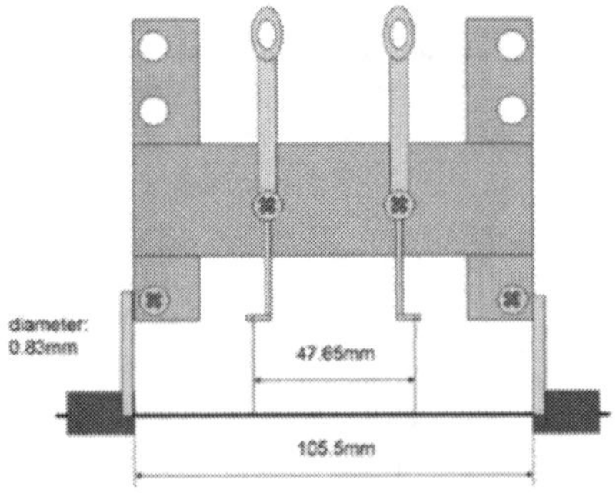

Fig. 2 Illustrated test sample of MgB2 short wire and set-up.

3. Test insert

Test section is inner-bottom area (309 mm diameter and 400 mm height) of the LH_2 cryostat where the external magnetic field is generated by the superconducting magnet immersed in LHe (see Fig. 1). The test sample, that is, short MgB_2 wire as illustrated in Fig. 2 was set orthogonal oriented to the field and at the center in height of the magnetic field profile.

The sample wire is monolith MgB_2 wire of 0.83 mm diameter, 160mm length. Configuration of the wire is 20 % MgB_2, 41.8% Copper and 38.2 % Nb barrier.

4. Critical current and Electrical resistance of the wire

4.1 Quasi steady state heat transfer experiment

The current through the MgB_2 short sample was controlled to meet the exponential heat inputs $Q=Q_0\exp(t/\tau)$ with various periods τ in order to evaluate the heat transfer property from heated wire surface to LH_2. The pressure of the main tank was set to 1.1 MPa. The bulk-liquid temperature (subcooling temperature) was set 21, 24 27 and 29 K by use of the sheath heater equipped at the bottom of the cryostat. The heat input was calculated by the current and the tap voltage of the test wire.

One of the experimental results with quasi-steady state heat input (τ = 2.0 s) is shown in Fig 3. The AC loss of the wire during current increasing was considered to be negligible. The bulk temperature of the liquid hydrogen was 21 K (subcooled by 11 K) and the pressure was 1.1 MPa. As the transport current I increased, at around 5.2 s, the wire resistance R appeared and the heat input Q to the wire started to increase. At about 8.0 s, as the nucleate boiling began, cooling property was improved. The current was successfully shut down preventing from burning out of the wire.

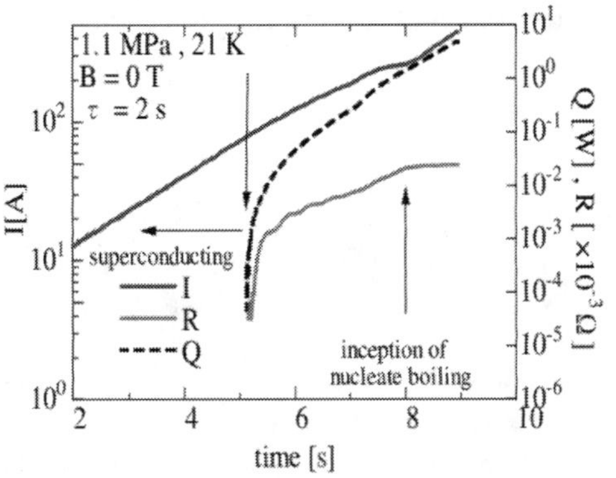

Fig. 3 Quasi-steady state heat input test (τ= 2.0 s, T_B= 21 K, subcooled by 11 K, and the pressure was 1.1 MPa).

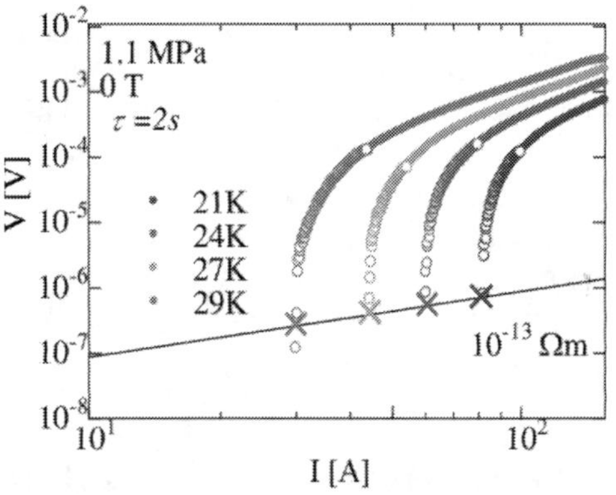

Fig. 4 Tap voltage versus transport current during quasi-steady heat input (τ= 2.0 s) under 21, 24 ,27 and 29 K without magnetic field. The critical current was evaluated for each temperature.

4.2 Critical Current of the test sample

Fig 4 shows an experimental data of the tap voltage versus transport current at the appearance of the resistance of the test wire with the quasi-steady state heat input (τ = 2.0 s) under the bulk liquid temperature of 21, 24 ,27 and 29K without magnetic field. The intersection point of each line with the constant resistivity line of 10^{-13} Ωm is indicated the critical current. The critical current decreases as higher temperature. The gradient of the tangential line at the intersection point which is related to the n-value of the wire is evaluated to become smaller as the higher temperature.

4.3 Critical Current under magnetic field

The same experimental data of the tap voltage versus transport current with magnetic field of B=0.41 T is shown in Fig. 5. The critical current was degraded with external magnetic field at each temperature, such as from 82 A to 27 A at 21K. It can be observed the n-value becomes smaller with higher magnetic field.

The same experiments were carried out with external magnetic field of 0.205, 0.41, 0.82 and 1.25 T. The characteristics of Ic(B) was evaluated as shown in Fig. 6 with temperature as a parameter.

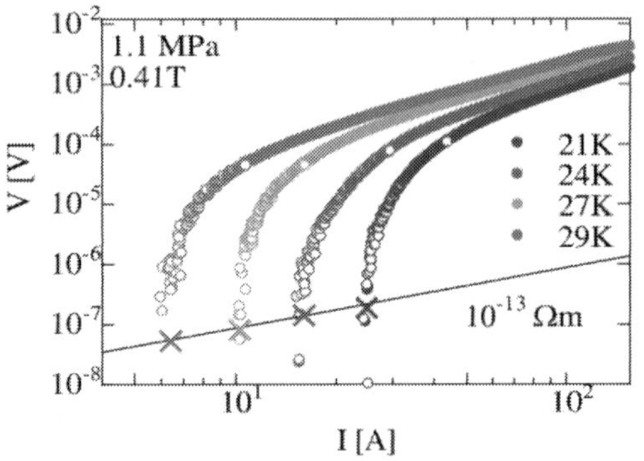

Fig. 5 Tap voltage versus transport current during quasi-steady heat input (τ= 2.0 s) under 21, 24, 27 and 29 K with external magnetic field of 0.41 T. The critical current was evaluated for each temperature.

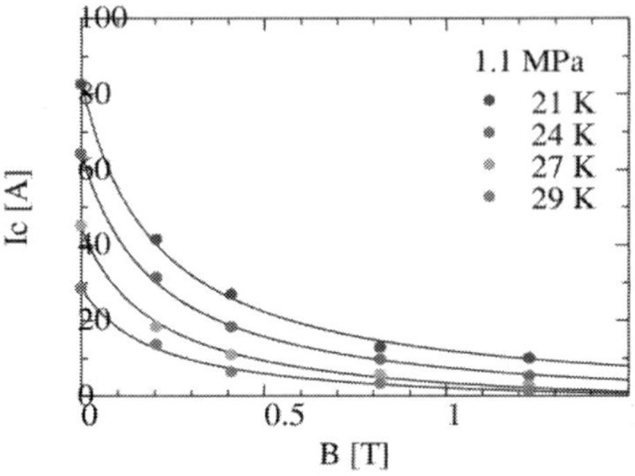

Fig. 6 Critical current of the test wire versus external magnetic field under 21, 24, 27 and 29 K.

5. Conclusion

We designed and made an experimental setup which can energize superconducting wires immersed in LH_2 with the current of up to 500 A under the condition of external magnetic field up to 7 T and pressure up to 1.5 MPa. Using this experimental set-up, over-current tests were carried out using MgB_2 superconducting short wire under various external magnetic field conditions. Critical currents of the test wire at the temperature 21, 24, 17, 29K under external magnetic field up to 1.25 T were measured. All the tests were performed with remote-operation system in a safe manner.

Acknowledgments
This research was supported in part by JST, ALCA. The authors thank technical staffs of JAXA. The authors also thank Hyper Tech Research, Inc. for providing the MgB_2 short wire samples

References

[1] Steward, W. G., Adv. in Cryogenic Eng. 35, Plenum, New York, 1990, pp.403-412.
[2] Graham, R. W., et.al., Adv. in Cryogenic Eng. 10, Plenum Press, New York, 1965, pp. 342-352.
[3] Class, C. R., et.al., Adv. in Cryogenic Eng. 5, Plenum, New York, 1960, pp. 254-261.
[4] Shirai Y., et.al., Adv. in Cryogenic Eng. 55A, AIP, New York, 2010, pp.337-344.
[5] Shirai, Y., et.al., Cryogenics, 50, pp.410-416 (2010).
[6] Yasuyuki Shirai, et.al., Cryogenics Vol..51(2011) 295–299
[7] Yasuyuki Shirai, et.al., IEEE Trans. on ASC, 10.1109/TASC.2012.2234820

Pulsed laser deposition of thick $BaHfO_3$-doped $YBa_2Cu_3O_{7-\delta}$ films on highly alloyed textured Ni-W tapes

M Sieger, J Hänisch, K Iida, U Gaitzsch, C Rodig, L Schultz, B Holzapfel, R Hühne

Institute for Metallic Materials, IFW Dresden, PO Box 270116, 01171 Dresden, Germany

m.sieger@ifw-dresden.de

Abstract. $YBa_2Cu_3O_{7-\delta}$ (YBCO) films with a thickness of up to 3 µm containing nano-sized $BaHfO_3$ (BHO) have been grown on Y_2O_3/Y-stabilized ZrO_2/CeO_2 buffered Ni-9at% W tapes by pulsed laser deposition (PLD). Structural characterization by means of X-ray diffraction confirmed that the YBCO layer grew epitaxial. A superconducting transition temperature T_c of about 89 K with a transition width of 1 K was determined, decreasing with increasing BHO content. Critical current density in self-field and at 0.3 T increased with increasing dopant level.

1. INTRODUCTION

The high-temperature superconductor $YBa_2Cu_3O_{7-\delta}$ (YBCO)[1] might be used for technical applications such as superconducting transmission cables, motors and electrical power components due to its superior current transport capability in high magnetic fields.[2] The critical current I_c can be increased by improving the critical current density J_c for a constant film thickness t or by increasing t for a constant J_c. Artificial pinning centres as nanoparticles like $BaHfO_3$ (BHO)[3,4] form effective immobilization points for magnetic flux lines,[5] thereby increasing $J_c(\mu_0 H)$. One of the main approaches to realize long length coated conductors is to use Ni-W tapes with a strong cube texture, formed by heavy cold rolling and annealing.[6] Recently, Ni-9at% W (Ni9W) tapes were realized, which are non-magnetic at 77 K and therefore reduce AC losses of the conductors.[7] A suitable buffer layer architecture as for example Y_2O_3/Y-stabilized ZrO_2 (YSZ)/CeO_2[8] ensures epitaxial growth to prevent high-angle grain boundaries, which are highly detrimental to the transport current.[9] Here, we demonstrate thick YBCO films grown on Ni9W. Furthermore, BHO pinning centres were incorporated into the YBCO matrix using PLD targets from mixed powders to increase the current transport capability in magnetic fields.

2. EXPERIMENTAL DETAILS

A *Lambda Physics LPX305i* KrF excimer laser (λ = 248 nm, t_{pulse} = 25 ns) with an energy density of 1.6 J/cm² at the surface of the PLD target was used to grow superconducting films on 10 mm × 10 mm × 1 mm single crystalline (100) oriented $SrTiO_3$ (STO) substrates as well as on 10 mm × 10 mm × 80 µm sized biaxially textured Ni9W tapes with a Y_2O_3/YSZ/CeO_2 buffer layer stack on top. The details of the tape preparation and buffer layer deposition may be found in Ref. [8]. An oxygen partial pressure of 0.4 mbar was maintained during the deposition process. The deposition temperature was set to 810°C for STO substrates and to 830°C for Ni9W tapes, which were found as optimum conditions for high critical temperature and current density in previous experiments. Undoped as well as 2, 4 and 6 mol% BHO-doped YBCO films were deposited from PLD targets which were prepared from mixed powders by a conventional solid state reaction. In-situ oxygen loading of the YBCO films took place in 400 mbar O_2 during cooling at 20 K min^{-1} in order to obtain superconducting YBCO.

The surface morphology was studied using a *Philips XL 20* scanning electron microscope, and the film thickness was measured on cross sectional cuts prepared by a focused gallium ion beam in a *FEI Helios 600i*. The crystal structure was determined in a *Bruker D8 Advance* X-ray diffractometer in Θ-2Θ geometry. Pole figures were taken by a *Philips X'Pert PW3040* on the YBCO (102) and BHO (110) planes. Superconducting properties, i.e. critical temperature $T_{c,50}$ (measured at midpoint of transition), the transition width ΔT_c, and the critical current densities J_c^{sf} and $J_c^{0.3T}$ (J_c in self-field and in a magnetic field of 0.3 T parallel *c*), were measured inductively.

3. RESULTS AND DISCUSSION

3.1. Growth of thick YBCO films on STO single crystalline substrates

The X-ray diffraction patterns of three different samples on STO are shown in figure 1(a). The deposition frequency was increased with increasing film thickness to keep the growth time nearly constant. All samples show strong (00ℓ) YBCO peaks, which is a clear sign of *c*-axis oriented growth on the STO substrates. Additional peaks are assigned as Y_2O_3 inclusions originating from excess yttrium in the target. The thickest sample shows additional *a*-axis oriented YBCO grains due to the presence of (*h*00) reflections. The (102) YBCO pole figures for the 1 μm and 3 μm thick films are given in figures 1(b) and 1(c). Whereas an undisturbed epitaxial growth was observed for the 1 μm thick film, a significant amount of *a*-axis oriented grains is present in the 3 μm thick film, which is visible at an angle ψ of approximately 33°. It should be noted that the *a*-axis oriented growth was prevented in a 3 μm thick film by employing a laser frequency of 10 Hz.

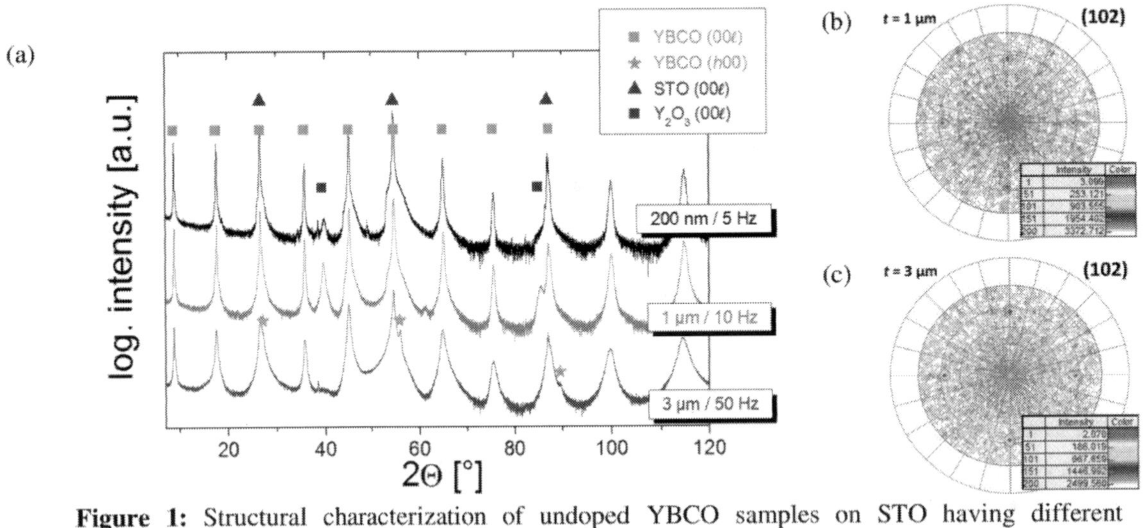

Figure 1: Structural characterization of undoped YBCO samples on STO having different thicknesses: X-ray diffraction pattern (a); YBCO (102) pole figure of a 1 μm (b) and 3 μm (c) thick layer

Table 1 summarizes the superconducting properties for this sample series. The critical temperature $T_{c,50}$ of the 1 μm thick sample is slightly increased compared to the 0.5 μm thin film. However, the critical current density J_c^{sf} is strongly reduced from 2.7 MA/cm² (0.5 μm) to 1.8 MA/cm² (1 μm). The

3 μm thick sample shows a significant drop in both, T_c and J_c^{sf}, which may have several reasons. On the one hand, a relatively long *c*-axis, compared to the other samples, indicates a lower oxygen content leading to a reduced critical temperature. On the other hand, high-angle grain boundaries due to *a*-axis oriented grains might be responsible for lower J_c values. A change to rougher morphologies with increasing thickness may also contribute to the J_c drop.[10]

Table 1: $T_{c,50}$ and critical current density in self-field, J_c^{sf}, at liquid nitrogen temperature for undoped YBCO samples with different thicknesses grown on STO

thickness	f_{Dep}	$T_{c,50}$	ΔT_c	J_c^{sf}
[μm]	[Hz]	[K]	[K]	[MA/cm²]
0.5	5	90.9	1.2	2.71 ± 0.12
1	10	92.4	1.0	1.83 ± 0.62
3	50	86.4	1.8	0.38 ± 0.11

3.2. BHO-doped YBCO samples on highly alloyed textured metal tapes

Thicker YBCO layers were grown on Ni9W in the next step, using a laser repetition rate of 10 Hz. Figure 2(a) shows a top view of a 1.8 μm thick YBCO film on a buffered Ni9W tape. The grain structure, which is transferred from metal tape via buffer layers, is clearly visible. Only a small number of Ni recrystallization twins, which are detrimental to *c*-axis aligned growth of YBCO, are present. The higher magnification view (figure 2(b)) shows the small YBCO islands which coalesce during film growth.[11] The cross sectional view in figure 2(c) shows the complete layer architecture with the metal tape, the threefold buffer layer stack and YBCO. Additionally, nickel oxide (NiO) and nickel tungsten oxide ($NiWO_4$) are found at the tape/buffer interface.

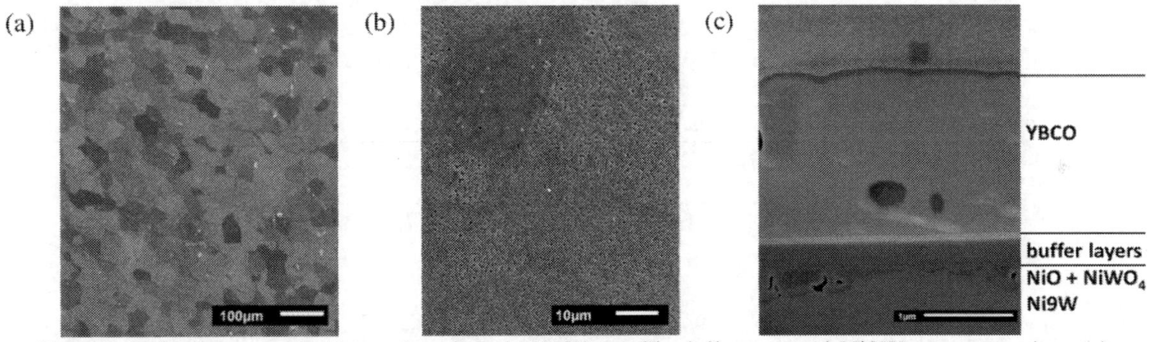

Figure 2: Surface morphology of undoped YBCO on biaxially textured Ni9W tape: top view (a), higher magnification (b), lateral view of a cross section (c)

The epitaxial growth on the biaxially textured metal tape is proven by the presence of (00ℓ) peaks of the buffer layers and YBCO in figure 3(a) as well as by pole figure measurements. The YBCO (102) peaks in the pole figure in figure 3(b) show the typical broadening, already present in the Ni9W substrate. The average (4 peaks) in-plane full width at half maximum is 6.2° for YBCO, which is similar to the value of 6.4° for the substrate. Figure 3(c) shows that BHO is incorporated into the films with a biaxial, cube-on-cube relationship. The additional strong peaks at $\Psi = 55°$ arise from the Y_2O_3 (222) planes, which have almost the same 2Θ value as the BHO (110) plane.

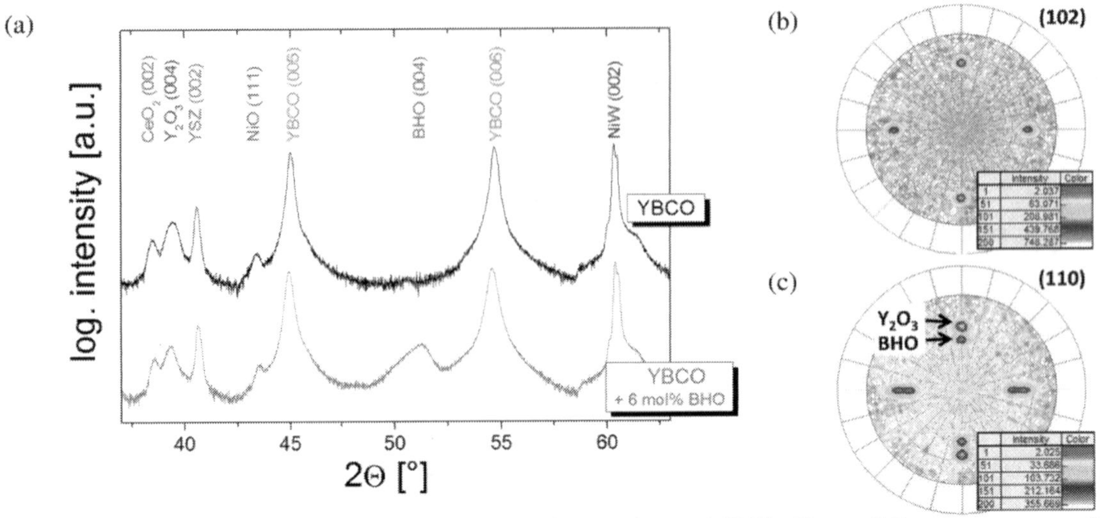

Figure 3: Structural characterization of YBCO samples on Ni9W: X-ray diffraction pattern of 1.8 µm thick undoped and 6 mol% BHO-doped YBCO (a); with respective YBCO (102) pole figure (b); and BHO (110) pole figure (c)

Whereas the critical temperature of the YBCO films on Ni9W decreases slightly with increasing BHO concentration, the critical current density increases in self-field and even more pronounced at 0.3 T applied parallel to the *c*-axis of the films (see table 2). Pinning effects in higher fields will be investigated by detailed transport J_c studies in the near future.

Table 2: Critical temperature and critical current density of 1.8 µm thick undoped and BHO-doped YBCO samples on biaxially textured Ni9W tapes deposited with a frequency of 10 Hz

BHO content	$T_{c,50}$	ΔT_c	J_c^{sf}	$J_c^{0.3T\|c}$
[mol%]	[K]	[K]	[MA/cm²]	[MA/cm²]
0	89.1	1.0	0.38 ± 0.01	0.15 ± 0.01
2	88.1	0.8	0.33 ± 0.01	0.19 ± 0.01
4	88.4	0.7	0.60 ± 0.04	0.32 ± 0.01
6	87.7	0.9	0.73 ± 0.02	0.33 ± 0.02

The authors gratefully thank J. Scheiter, U. Fiedler and M. Kühnel for technical assistance and acknowledge the financial support from EUROTAPES, a collaborative project funded by the European Community's Seven Framework Programme (FP7 / 2007 - 2013) under Grant Agreement no. 280432.

References

[1] Wu M *et al* 1987 *Phys. Rev. Lett.* **58** 908-910
[2] Foltyn S *et al* 2007 *Nat. Mater.* **6** 631-642
[3] Hänisch J *et al* 2006 *Supercond. Sci. Technol.* **19** 534-540
[4] Engel S *et al* 2000 *Appl. Phys. Lett.* **90** 102505
[5] Matsushita T 2000 *Supercond. Sci. Technol.* **13** 730-737
[6] Eickemeyer J *et al* 2010 *Supercond. Sci. Technol.* **23** 085012
[7] Gaitzsch U *et al* 2013 *Supercond. Sci. Technol.* **26** 085024
[8] Hühne R *et al* 2007 *Supercond. Sci. Technol.* **20** 709-714
[9] Hilgenkamp H and Mannhart J 2002 *Rev. Mod. Phys.* **74** 485-549
[10] Foltyn S *et al* 1999 *Appl. Phys. Lett.* **75** 3692-3694
[11] Hühne R *et al* 2010 *Supercond. Sci. Technol.* **23** 034015

Pulsed Laser Deposition of Thin YBCO Films on Faceted YSZ Single Crystal Fibers

O. Snigirev, [1] M. Chukharkin, [1, 2] N. Porokhov, [1] S.Y. Rusanov, [3] V.V. Kashin, [3] V.B. Tsvetkov, [3] A. Kalabukhov, [1, 2] and D. Winkler [2]

[1]*M.V. Lomonosov Moscow State University, Moscow, Russian Federation*
[2]*Chalmers University of Technology, Department of Microtechnology and Nanoscience – MC2, Gothenburg, Sweden*
[3]*Institute of General Physics Russian Academy of Sciences, Moscow, Russian Federation*

E-mail: maxim.chukharkin@gmail.com

Abstract. Flexible rods of single crystals of 9% Y_2O_3-stabilized ZrO_2 (YSZ) were used as substrates for deposition of high-critical temperature superconducting (HTS) thin films. YSZ fibers were prepared by mini-pedestal method with laser heating and had average diameter of 300 micrometers and 30 mm length. X-ray diffraction analysis demonstrated high crystalline quality of obtained fibers and also indicated the presence of 15° deviation of the fiber axis from the [001] YSZ direction. Thin $YBa_2Cu_3O_{7-x}$ films were grown by pulsed laser deposition on YSZ rods using CeO_2 buffer layer. Films have shown high critical temperature of 90 K with sharp superconducting transition. Critical current density was estimated to about $3x10^4$ A/cm^2 at 80 K. Temperature dependence of critical current density suggests granular structure of films with grain size about several microns. Our results demonstrate feasibility of flexible YSZ fibers coated by HTS thin films for practical use.

Keywords: high-temperature superconductors, YBCO thin films, flexible substrates

PACS numbers: 74.78.-w, 74.72.-h, 74.25.Sv.

1. Introduction.

A significant progress in fabrication of high-quality high-critical temperature superconductors (HTS) power cables and ribbons could allow making HTS flux transformers. However, metal tapes used as substrates for fabrication of HTS cables are magnetic and conductive that are not compatible with superconducting quantum interference devices (SQUIDs) due to inserted magnetic noise and problem with realization of superconducting contacts [1, 2] to HTS thin-film multiturn input coil placed on the SQUID chip [3]. Therefore, flexible and non-magnetic non-conductive substrates should be used. Another important issue is providing a good superconducting contact between the HTS wire and the SQUID pick-up loop, if needed. Both problems have not been solved yet and represent a significant technological challenge.

To solve aforementioned problems, flexible single crystal fibers of different materials have been used as substrates for deposition of HTS thin films. Recently, $YBa_2Cu_3O_{7-x}$ (YBCO) thin films with small width-to-thickness aspect ratio fabricated on sapphire faceted fibers by pulsed laser deposition (PLD) were reported [4]. Besides the sapphire, yttrium-stabilized zirconia (YSZ) could be used as a possible flexible substrate material for the HTS wires. The flexibility of the YSZ polycrystalline substrates in the form of thin sheets with thickness close to 50 μm has been shown previously [5]. Due to the cubic crystallographic structure it is much easier to obtain facets with correct orientation suitable for the YBCO thin film deposition. However, only few reports on fabrication of YBCO thick films by liquid phase epitaxy [4] and hot-wall type metal-organic chemical vapor deposition [6] on YSZ fibers have been published.

2. Fabrication details.

High-quality single-crystal fibers were fabricated using mini-pedestal method with laser heating [7, 8]. Several fibers with length of about 30 mm and average diameter of about 300 μm were fabricated using this method and had at least one facet. Image of central part of one obtained faceted fiber is shown in figure 1 (a), and face image of the end of fiber is shown in figure 1 (b).

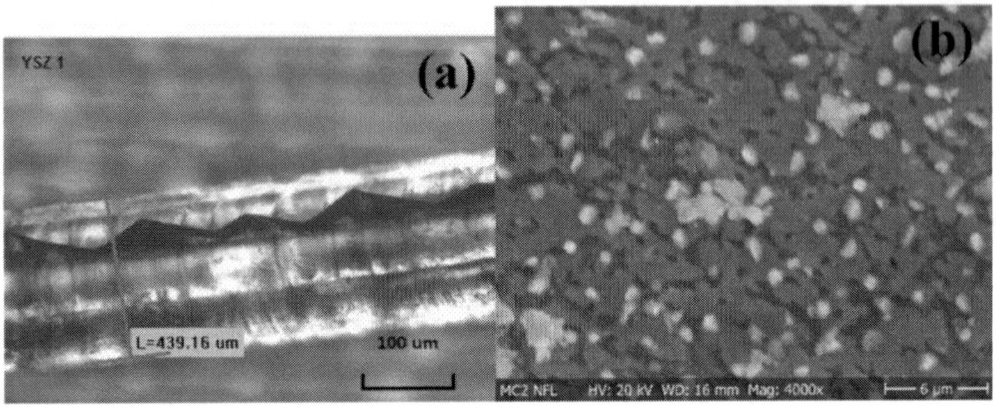

Figure 1.The images of the obtained YSZ faceted fiber. The long single-side facet is visible in fiber face – (a). Scanning electron microscopy images of the central part of the fiber with deposited film and silver-paint contacts placed on the holder - (b).

YSZ fibers were placed directly into the sapphire sample holder with 10×10 or 5×5 mm^2 open windows and covered by another 0.5 mm thick YSZ single crystal substrate to prevent deposition of film material on the heater. During the deposition, substrates were heated by SiC radiation heater from the back side.

25 nm thick CeO_2 buffer layers were deposited using RF sputtering at the temperature of 780 °C. The layer was deposited with an RF source power of 100 W in a mixed argon-oxygen atmosphere (60% O_2 + 40% Ar) at a partial pressure of 0.1 mbar. Distance between target and sample was 30 mm. Sample was slowly (10 °C/min) cooled down to a room temperature under 500 mbar oxygen pressure. After CeO_2 deposition, sample was transferred into the PLD chamber without breaking the vacuum.

YBCO films were deposited using pulsed laser deposition from a 3 inch stoichiometric target at the temperature of 880 °C. Oxygen pressure during deposition was set to 0.6 mbar, repetition rate 10 Hz and laser energy density at the target 1.25 J/cm^2.The substrate-to-target distance was set to 60 mm. 7500 laser pulses approximately correspond to YBCO thickness of about 360 nm.

The surfaces of the obtained YBCO films were investigated using scanning electron microscopy (SEM). Figure 1 (b) shows SEM image of the central part of the YSZ fiber with the deposited YBCO film. The size of grains was estimated about 5 μm.

3. Electrical characterization of the YBCO films.

Electrical measurements of deposited YBCO films were performed in the PPMS system (Quantum Design) in the temperature range 5 - 300 K and magnetic fields up to 8 T. Four electrical contacts were made along the fiber using gold bonding wire and additional silver glue reinforcement. Distance between contacts was about 1 mm.

Critical temperature, T_C of two deposited films was measured at zero magnetic field. Sharp superconducting transition at temperature about 90 K was observed in both samples. Residual resistance ratio between 300 K and 100 K was about 2. A small residual resistance below superconducting transition of the order 5 Ω was observed in both samples. Estimated normal state resistivity at 300 K, ρ_n is close to 500 μΩcm.

Magnetic field dependence of resistance, R, for one of the samples is shown in figure 2a. Obtained $R(H,T)$ dependence is typical for the case when magnetic field is oriented parallel to $(a$-$b)$ planes of the YBCO [9, 10]. Increasing the magnetic field does not only shift critical temperature but makes the transition wider. Broadening of the transition at high magnetic field can be explained by thermal activated flux-flow [11]. There is no shoulder in transition curve that is usually attributed to a lattice melting transition or giant thermal activated flux creep dissipation [10].

Temperature dependences of critical current for this sample are shown in figure 2b. $I_c(T)$ curve for the sample measured near the critical temperature is well-fitted to the Ambegaokar-Baratoff model corresponding to Josephson tunneling between film grains that is well approximated by $(1-T/T_c)$ dependence near T_c. In the case of low critical temperature granular thin film [12 - 15] such behavior corresponds to situation when supercurrent critical density is low and did not suppress the grain's gap parameter.

Due to the round shape of fiber and films it is rather difficult to estimate cross-section area of the YBCO film and corresponding critical current density. A reasonable assumption is that film covers uniformly facet surface with a width about 1/3 of fiber diameter. For the YBCO film thickness of about 300 nm, the critical current density, J_c can be estimated as 3×10^4 A/cm^2 at 80 K for both samples.

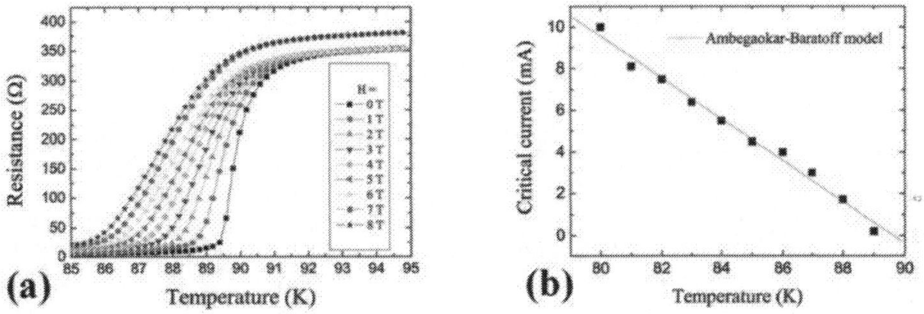

Figure 2. Temperature dependencies of YBCO film resistance at different magnetic fields (a). Experimental temperature dependence of critical current of YBCO films deposited on 10 mm long YSZ fiber (b).

4. Discussion

Our results show that YSZ faceted fibers can be used as substrates for pulsed laser deposition of YBCO thin films with good superconducting parameters (T_c about 90 K, reasonable critical current density). The residual resistivity below critical temperature can be explained by the presence of steps at the facet surface due to the presence of 15° deviation of the fiber axis from the [001] YSZ direction shown by XRD measurements. Large superconducting grains connected by normal metal (or semiconducting) inclusions could be formed here. This is further supported by the fitting of $I_c(T)$ curves that have behavior characteristic for granular films.

We note that these results have been obtained for fibers that did not have facets with proper crystallographic orientation suitable for epitaxial growth of YBCO films. Deviation of the fiber axis from the crystallographic [001] YSZ direction may be the main reason for granular structure of the YBCO films, formation of intergrain boundaries, and lower values of critical current densities. To fulfill the conditions of the epitaxial growth of the CeO_2 and YBCO layers the axis direction of the fiber should correspond to the <001> family of directions. In this case there should be the areas on the flank of the fiber corresponding to the directions [100], [010], [-100], [0-10] due to the cubic structure of the YSZ.

We used rather thick YSZ fibers that have been earlier demonstrated flexibility with typical bending diameter of 40 mm [16]. For practical applications it is important to grow fibers with smaller diameter of about 50 um that can significantly reduce the bending diameter to about 10 mm. It is also important to achieve proper facet formation conditions that would allow for well-oriented surface of the fiber and provide better conditions for the epitaxial growth of HTS films. We are currently working in this direction.

Acknowledgments

The work was supported in part by Russian Ministry of Education and Science (Grant 8244), the Swedish Research Council, Swedish Institute Visby program, and Knut and Alice Wallenberg foundation. We also acknowledge support from Swedish national research infrastructure for micro and nano fabrication (Myfab).

References

[1] Divorne H, Scols J, Fermon C, Jacquinot J F, Pannetier-Lecceur M 2008 *Rev. Sci. Instrum.* **79** 025107-025111

[2] Divorne H, Guerrero R, Fermon C, Pannetier-Lecceur M, Jacquinot J F 2009 *IEEE Trans. Appl. Supercond.* **19** 761-763

[3] Chukharkin M, Kalabukhov A, Schneiderman J, Öisjöen F, Snigirev O V, Lay Z, and Winkler D 2012 *Appl. Phys. Lett.* **101** 042602-(1-5)

[4] Yamada Y, Kawashima J, Niiori Y, Masegi T, Hirabayashi I 1997 *Advances in Superconductivity IX. Proc. of the 9th International Symposium on Superconductivity (ISS'96)* **2** 653-5

[5] Divorne H, Scols J, Fermon C, Jacquinot J F, Pannetier-Lecceur M 2008 Rev. Sci. Instrum. **79** 025107-025111

[6] Yang C S, Moeckly B 2010 *IEEE Trans. Appl. Supercond.* **21** 107-110

[7] Hasegawa M, Yoshida Y, Ito Y, Iwata M, Kawashima J, Takai Y, Hirabayashi I 1998 *Advances in Superconductivity X. Proc. of the 10th International Symposium* **2** 623-6

[8] Bufetova G A, Kashin V V, Nikolaev D A, Rusanov S Ya, Seregin V F, Tsvetkov V B, Scherbakov I A, Yakovlev A A 2006 *Quantum Electronics* **36** 616-619

[9] Bufetova G A, Kashin V V, Nikolaev D A, Papin Yu M, Rusanov S Ya, Seregin V F, Scherbakov I A, Tsvetkov V B, Yakovlev A A 2007 *Laser Phys. Lett.* **4** 440-444

[10] Ott H R. *High-Tc superconductivity* Springer, Zurich, p. 797.

[11] Palstra T T M, Batlogg B, Van Dover R B, Schneemeyer L F, and Waszczak J V 1988 *Phys. Rev. Lett.* **61** 1662

[12] Tinkham M 1995 *Introduction in superconductivity* 2nd ed. (Singapure) McGrow-Hill Ch. 9

[13] Campwirth R T, Gray K E 1981 *IEEE Trans. on Magn.* **MAG-17** 1 565-568

[14] Clem J R, Bumble B, Raider S I, Gallagher W J, and Shin Y C 1987 *Phys. Rev. B* **35** 13 6637-6642

[15] Mannhart J, Chaudhari P, Dimos D,Tsuei C C, and McGuire 1988 *Phys. Rev. Lett.* **61** 21 2476-2479

[16] Kawashima J, Kitamura T, Taniguchi S, Niiori Y, Shiokara Y, Hirabayashi I, Tanaka S 1996 *Advances in Superconductivity VIII. Proc. of the 8th International Symposium on Superconductivity (ISS'95)* **2** 767-70

Journal of Physics: Conference Series **507** (2014) 022034 doi:10.1088/1742-6596/507/2/022034

AC loss properties of single-layer CORC cables

M Solovyov[1], J Šouc[1], F Gömöry[1]

[1] Institute of Electrical Engineering Slovak Academy of Sciences
Dubravska cesta 9, 84104 Bratislava, Slovakia

E-mail: mykola.solovyov@savba.sk

Abstract. In this work, an experimental and theoretical study of the transport and magnetization AC losses of a single layer *Re*BCO cable is presented. Examined cable consisted of five coated conductor tapes wound in parallel helically on the round core with 40 mm pitch and six full transposition resulting in the 24 cm total cable length. Transport currents up to 600 A amplitude with frequencies 36 Hz and 72 Hz were used in the transport loss measurement. AC fields up to 50 mT were applied during the magnetization loss measurement at 36 Hz and 72 Hz, respectively. Very good agreement between the experimental results and the simulations for magnetization loss was achieved. However, the experimental results for transport loss are markedly greater than the prediction reached from simulation. Nevertheless, they are lower than the transport AC loss theoretically assumed for a straight superconducting tape with the identical critical current.

1. Introduction

Helical layout of the 2nd generation of high-temperature superconducting coated conductor (CC) tapes on a flexible round core (called CORC cables - Coated Conductors on Round Core) represents a promising option for achieving round conductor elements suitable for the winding of magnet coils. The advantage is its manufacturing by classical cabling procedures. The main drawback is rather low fill factor. However, it can be improved by increasing the number of layers. In this way also the critical current can be enhanced. For a pulse or AC magnets or power transformers the energy dissipation resulting in AC loss is of prime importance. The CORC cables [1-3] may be good alternative to Roebel cables [4-6] for this kind of applications.

2. Cable model and experiments

2.1. Cable model

In this work we present the study of AC losses of single-layer CORC cable (figure 1), consisting in five CC tapes. The summary of cable design parameters is shown in table 1.

Figure 1. The CORC cable from five CC tapes.

Table 1. Parameters of the CORC cable.

CC tape	*SuperPower®*, SCS4050
Critical current of one tape (77 K, self-field)	113A
Length of one cable pitch	4 cm
Number of transpositions	6
Cable length including current terminals	30 cm
Core material and dimensions	fiberglass former, ϕ 8 mm
Current terminals	tapes soldered at the cable ends

2.2. Transport loss measurement

The cable transport loss was measured by the standard lock-in technique at transport current increasing with defined steps. Sinusoidal transport current with amplitude in the range 60 - 600 A and frequencies 36 Hz and 72 Hz was used. Cable was cooled by liquid nitrogen to 77K. To avoid the ambiguity of the voltage taps placing directly on some of the CC tapes, they were located on the current terminals – figure 2. In such configuration the voltage drop on resistive joints, U_R, adds to the cable loss voltage, U_{SC}, resulting in $U_{TOT} = U_{SC} + U_R$ measured in experiment. The joint resistance (about $2 \cdot 10^{-7}$ Ω) was estimated from U_{TOT} measurement at low AC currents (from 1 to 5 A) when U_{SC} is negligible. It was afterwards used to obtain U_{SC} from U_{TOT} by subtracting U_R during the data processing.

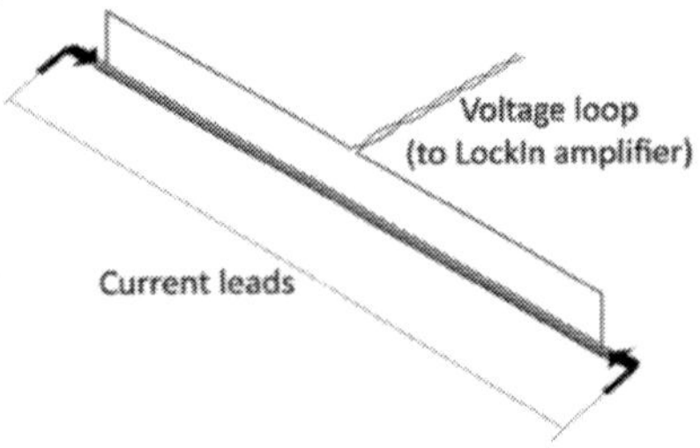

Figure 2. Experimental setup for transport measurement.

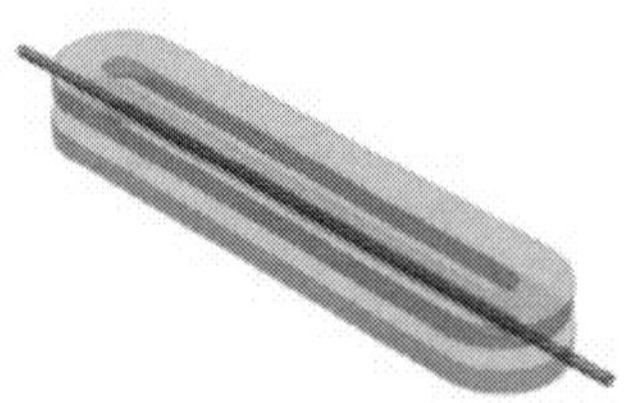

Figure 3. Experimental setup for magnetization measurement.

2.3. Magnetization loss measurement

The calibration free method [7] combined with the lock-in technique was utilized for the magnetization loss measurement. Briefly, two identical Cu magnets generating AC magnetic field and each involving pick-up coils form the base of this method. If no sample is placed in one of the magnets, a voltage equilibrium exists on pick-up coils connected in series with opposite polarity. When the sample is located in one of the magnets, the voltage proportional to the sample loss will appear.

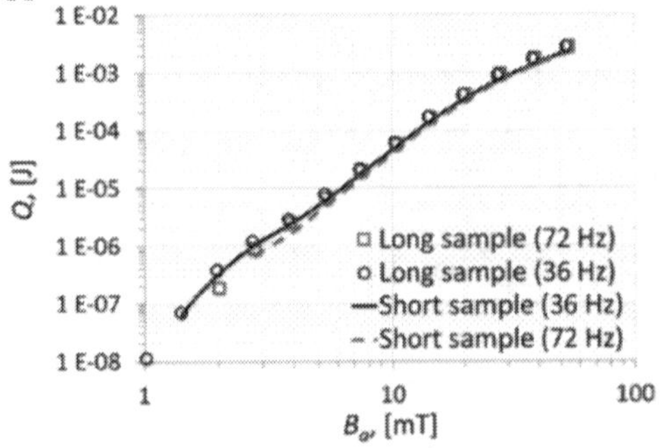

Figure 4. Magnetization AC losses in single CC tape, measured on short sample and long sample (with respect to magnet bore).

The cable magnetization loss was measured at AC magnetic field increasing with defined steps. The frequencies of applied magnetic field were 36 Hz and 72 Hz, respectively. The length of the cable was longer than the magnet bore to exclude any influence of the massive normal metal current terminals in which eddy currents could appear – see figure 3. In this condition the measured AC magnetization loss represents only the dissipation generated in the part of cable inside the magnet bore. This approach provides relevant data because the magnetic field outside the bore is much lower. Such conclusion was experimentally tested by the magnetization loss measurement of two samples from a straight CC tape. "Short" sample's length corresponded to the bore dimension and the "long" sample exceeding the magnet winding on both sides in the same way as shown in figure 3. The result of this comparison is shown on figure 4 confirming that the loss is generated only in the "active" part of the tape inside the magnet bore.

3. Numerical simulations

Distributions of the current density and the magnetic field were calculated in two-dimensional approximation - in the cable cross-section. The finite element technique of Comsol Multiphysics® in vector potential formulation was used [9]. For each current amplitude the AC loss was obtained by numerical integration of the calculated I-Φ hysteresis loop formed by 40 points.

The geometry of cable used for numerical simulations in shown on the right side of the figure 5. This geometry assumes the same cross-section as the real cable and the same critical current of the tapes (I_c). Inclination because of lay angle is reflected by increasing the tapes width from 4 mm to 4.66 mm and reduction of the critical current density (J_c) by 16.5 %.

Figure 5. Real geometry of the CORC cable (left) and geometry used for numerical simulations (right).

It is known that the loss is influenced by a non-uniform J_c distribution across the tape width [10]. Therefore we determined this non-uniformity profile experimentally from the magnetic field mapping and inversion procedure [11]. For the purpose of numerical calculation the trapezoidal approximation shown in figure 6 was used. The $J_c(B)$ dependence was not used in this simulation.

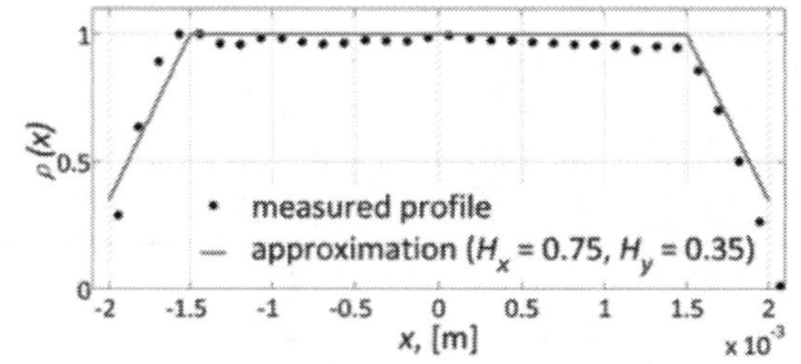

Figure 6. Non-uniformity profile across the tape width (circles) and its trapezoidal approximation used in simulations (full line).

4. Results and discussion

The measured magnetization losses is shown in the figure 7a. There was no frequency dependence observed. Remarkable is the agreement between numerical simulations and experiment.

In the case of transport AC losses the measured data are about four times higher than the numerical prediction (figure 7b). This experiment was repeated several times with different configurations of the voltage taps and wiring, with the same result. Therefore we search to find the cause of such discrepancy in the assumptions implemented in the calculations. For example, the irregularity of gaps between the tapes (see figure 1), which have high influence on transport AC losses in the CORC cable [11-12] has not been considered in the calculation. Another possible reason could be a non-uniform current distribution in parallel superconducting tapes. Better understanding of these factors needs a separate study.

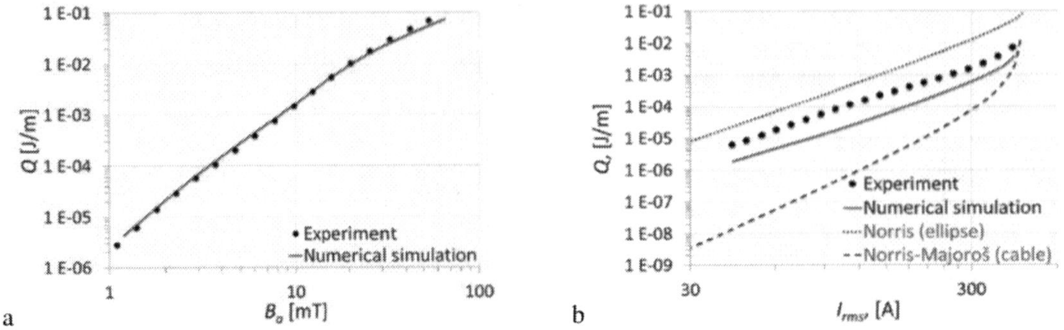

Figure 7. Measured and calculated magnetization (a) and transport (b) AC losses of CORC cable (only losses at 36 Hz are shown). For sake of comparison the transport losses calculated for the cable from uniform tapes using the Norris-Majoroš formula [12] as well as the prediction assuming a round cylinder with the same critical current [13] are shown.

In conclusion we have still some problems in the prediction of transport AC losses in our CORC cable. On the other hand, for the application of these cables in large magnets the magnetization loss is of prime concern. Here we get very good agreement between numerical simulations and experiment.

5. References

[1] van der Laan D, Noyes P, Miller G, Weijers H and Willering G 2013 *Superconductor Science and Technology* 26 045005
[2] Šouc J, Gömöry F, Kováč J, Nast R, Jung A, Vojenčiak M, Grilli F, Goldacker W 2013 *Superconductor Science and Technology* 26 075020
[3] Lee J, Kim Y, Lee S, Kim W, Park S, Park C and Choi K 2013 *IEEE Transactions on Applied Superconductivity* 23 5900804
[4] Grilli F, Zermeño V, Vojenčiak M, Pardo E, Kario A, and Goldacker W 2013 *IEEE Transactions on Applied Superconductivity* 23 5900205
[5] Badcock R, Bumby C, Jiang Z, Long N 2012 *Physics Procedia* 36 1159-64
[6] Pardo E and Grilli F 2012 *Superconductor Science and Technology* 25 014008
[7] Šouc J, Gömöry F, Vojenčiak M *Superconductor Science and Technology* 18 (2005) 592-5
[8] Li Q, Amemiya N, Nishino R, Nakamura T, Okuma T, *Physica C* 484 (2013) 217-22
[9] Gömöry F, Vojenčiak M, Pardo E, Solovyov M and Šouc J 2010 Superconductor Science and Technology 23 034012
[10] Solovyov M, Šouc J, Gömöry F 2012 *Physics Procedia* 36 617-22
[11] Fukui S, Hatakeyama S, Nanayama J, Maruko A, Ogawa J, Sato T, Oka T, and Tsukamoto O 2013 *IEEE Transactions on Applied Superconductivity* 23 5900105
[12] Majoroš M 1996 *Physica C* 272 62-4
[13] Norris W T 1970 *Journal of Physics D: Applied Physics* 3 489

Acknowledgments
Financial support of the European Commission for the project NMP3-LA-2012-280432 "EUROTAPES" and from the Slovak Research and Development Agency under the contract No. DO7RP-0003-12 is acknowledged

Critical current density of Nb_3Sn wires after irradiation with 65MeV and 24GeV protons

T. Spina, C. Scheuerlein, D. Richter, L. Bottura, A. Ballarino, R. Flükiger
CERN, TE-MSC, Geneva, Switzerland

tiziana.spina@cern.ch, christian.scheurlein@cern.ch, david.richter@cern.ch, luca.bottura@cern.ch, amalia.ballarino@cern.ch, rene.flukiger@cern.ch

Abstract. Industrial Nb_3Sn wires with Ti and Ta additives (RRP process) and with Ta additives (PIT process) with a diameter of 1 mm have been irradiated at room temperature with protons of 65 MeV and of 24 GeV at various fluences up to 1×10^{21} p/m^2. A steady increase of J_c vs. fluence was observed for all the wires up to the highest fluence. The observed increase of J_c at 4.2K in all wires was quite similar in spite of the very different proton energies. With increasing fluence, the radiation induced pinning force was found to increase, the enhancement J_c/J_{co} after 5.04×10^{20} p/m^2 reaching 1.4 for Ta and 1.8 for Ti alloyed wires at 10T. The present results were quantitatively analysed by assuming a radiation induced point pinning mechanism in addition to grain boundary pinning. The results are compared with those of an ongoing neutron irradiation study undertaken on the same Nb_3Sn wires in collaboration with the Atominstitut Vienna. Proton irradiation was found to produce considerably higher damage than neutron irradiation.

1. Introduction
The High Luminosity Large Hadron Collider upgrade (HL-LHC) aims at reaching an integrated luminosity of 3000 fb^{-1} with a peak luminosity of 5×10^{34} cm^{-2}s^{-1}. The present NbTi inner triplet in the interaction regions will be replaced by larger aperture quadrupoles based on Nb_3Sn conductors [1]. RRP or PIT Nb_3Sn wires will be used, the requirement for the critical current density being $J_{c(non\text{-}Cu)}$ = 1500 A/mm^2 at 4.2 K and 15 T. The radiation issue is very complex, the quadrupoles being simultaneously exposed to various high energy radiation sources [2]. The aim of the present work is to study the effect of proton irradiation on current carrying capability of RRP and PIT Nb_3Sn wires with Ti and Ta additives during the LHC lifetime. At the same time, the effect of neutron irradiation on the same wires was studied at the Atominstitut Vienna in the frame of a collaboration.

2. Experimental

2.1. Proton energies, proton irradiated Nb_3Sn wires and the two facilities (IRRAD1 and UCL)
The maximum radiation load acting on the quadrupole Q2a over the whole operation time has recently been calculated by F. Cerutti at CERN [2]. At the inner winding of the Q2a quadrupole, where the maximum damage is expected, the radiation spectrum will consist of photons, electrons, protons and pions, with energies ranging from 0.1 MeV to 10^3 GeV (with maxima around 1GeV for electrons and positrons and around 200 MeV for protons and pions) and neutrons with an energy peak value around 1 MeV. The maximum fluence at 3000 fb^{-1}, calculated for the case without W shielding, is: $\sim1.5\times10^{21}$ neutrons m^{-2}, $\sim 2\times10^{20}$ protons m^{-2} and $\sim 2\times10^{20}$ pions m^{-2} [2]. The present room temperature proton irradiation study was performed at two different energy values: 65 MeV and 24 GeV at the Université Catholique de Louvain-La-Neuve (UCL) and at CERN (IRRAD1) respectively. The corresponding

proton fluxes were 8×10^{13} p/m^{2}s at IRRAD1 and 2×10^{16} p/m^{2}s at the UCL. The set-up for irradiation facilities has been described in detail by Otto et al. [3]. The proton fluences were chosen based on earlier works at proton energies E ≤ 3 MeV [4, 5]. However the present proton energies being substantially higher, work at higher fluences has still to be undertaken to reach the maximum of J_c/J_{c0}. Three types of Nb_3Sn industrial multifilamentary wires with a length of approximately 5 mm and a mass of ~ 20 mg were irradiated up to 1×10^{21} p/m^2: two RRP (or Restacked Rod Process) wires from Oxford Superconducting Technology (Ta alloyed, billet #7419, and Ti alloyed, billet #11976) and a PIT wire from Bruker EAS (Ta alloyed, billet #0904). These wires have been characterized in a recent article (Flükiger et al. [6]).

2.2. Measurements

The magnetic moment of the Nb_3Sn wires was measured in a Vibrating Sample Magnetometer (VSM) at 4.2K. The measurements, performed in He gas flow, were accomplished by sweeping the applied field perpendicular to the wire axis, between 0 and 10.5 T with steps of 250 mT while increasing and decreasing the field (rate of 500 mT/min for each step). The temperature was measured by means of three Cernox thermometers. The values of J_c were determined from magnetization under the assumption that each sub-element can be treated as an independent superconducting hollow cylinder, as recently reported by Baumgartner et al. [7]. A conventional AC magnetic susceptometer (2 mT and 40.8 Hz) was used to determine the variation of the critical temperature T_c vs. proton fluence.

3. Results and discussion

The experimental values of T_c after proton irradiation are reported in table 1. The observed decrease of T_c due to proton irradiation is higher with respect to that obtained after neutron irradiation [6].

Table 1. Variation of T_c after proton irradiation for the RRP Ti alloyed (#11976) and PIT Ta alloyed (#0904) wires at 24GeV and 65MeV respectively.

Φt (p/m^2)	T_c(K) #11976	T_c(K) #0904
1.08×10^{20}	16.90	-------
2.00×10^{20}	-------	17.10
5.04×10^{20}	16.85	-------
1.00×10^{21}	-------	16.95

"---" no data available for these fluences

The variation of the ratio $J_c(\phi t)/J_{c0}$ (J_{c0} is the unirradiated value) with proton fluence at different energies is shown in figure 1.

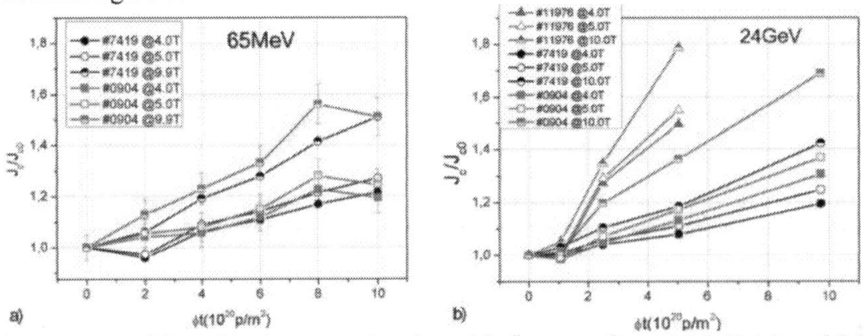

Figure 1 Enhancement of the critical current density with fluence after 65 MeV (a) and 24 GeV (b) proton irradiation. The small decrease of J_c/J_{c0} up to 2×10^{20} p/m^2 was discussed by Brown et al. [10]

Up to the maximum applied fluence, J_c/J_{c0} of all irradiated wires was found to increase with fluence, the enhancement being more important at higher magnetic fields. The small decrease of J_c/J_{c0} at low fluences (discussed earlier by Brown et al. [10]) was not observed in all wires. The present results allow two main observations: (a) The enhancement of J_c in all wires is quite similar, in spite of the very different proton energies. This could be explained on the basis of the PHITS calculation of Fukahori et al. [8] on Cu, our wire diameter being small with respect to the proton penetration depth at these energies, (b) Both Ta alloyed wires exhibit a similar enhancement, regardless of the fabrication technique. The nature of the additive seems to be important, J_c/J_{c0} of the Ti alloyed wire (#11976) exhibiting a larger enhancement than the Ta alloyed wires #7419 and #0904: at the fluence 5×10^{20} p/m^2 and 10 T, the ratio J_c/J_{c0} reaches 1.8, while both Ta alloyed wires reach values between 1.2 and 1.4. The variation of J_c/J_{c0} for the Ta alloyed PIT wire irradiated at 65 MeV (figure 1b) seems to indicate a peak around 8×10^{20} p/m^2, within the error bar. A conclusive answer is expected after ongoing irradiations at higher fluences.

3.1. Contribution of radiation induced defects to the volume pinning force

As shown in figure 2, the enhancement of J_c for the irradiated wires is higher at small reduced fields b thus reflecting enhanced pinning after irradiation. This behaviour is not affected by the uncertainty in the extrapolated $B_{c2(Kramer)}$ (see table 2).

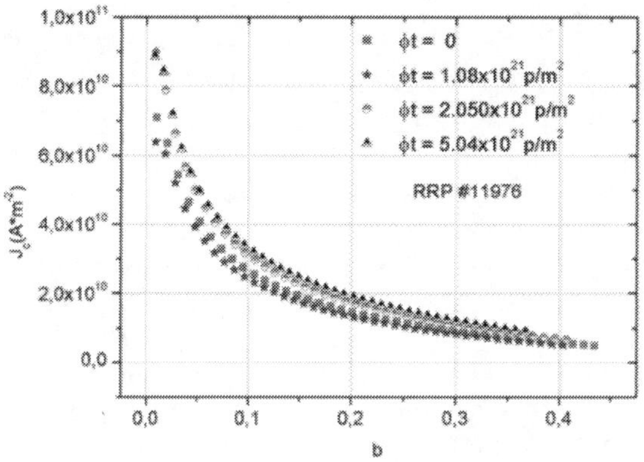

Figure 2 Behaviour of J_c vs. b for various fluences for RRP Ti alloyed (#11976) wires at 24 GeV.

Table 2 B_{c2} values calculated with the Kramer extrapolation and scaling fields B_{c2}^* (fitting parameter) at 4.2 K for RRP Ti alloyed wires (#11976) at 24GeV. These values are only indicative, the highest applied field being 10 T.

ϕt (10^{20}p/m^2)	**B(4.2K)$_{c2(Kramer)}$ (T)**	**B*(4.2K)$_{c2}$ (T)**
1.08	25.9	26.2
2.5	25.7	25.8
5.04	28.5	28.2

As proposed by various authors, the enhancement of J_c after irradiation with neutrons [7,9,10,11] or protons [4,5,12] is attributed to the radiation induced nanosize defect clusters acting as new pinning centers. Küpfer et al. [11] proposed a two-mechanism model which was recently improved by Baumgartner [13]: $f(b) = \alpha\, b^{p_1}(1-b)^{q_1} + \beta\, b^{p_2}(1-b)^{q_2}$, with $\alpha + \beta = 1$ and $b = B/B^*_{c2}$ (B^*_{c2} is a scaling field parameter, see table 2). The first contribution is due to grain boundary pinning and does not change after irradiation (p_1, q_1 are fixed at the values found for unirradiated wires), while the second contribution is due to defect clusters, various combination of the fitting parameters yielding values close to $p_2 = 1$ and $q_2 = 2$, thus suggesting point defects. The two contributions of the volume pinning force are shown in figure 3. For the RRP Ta alloyed wire at 24 GeV the field $(b_{max})_{Defect\ Cluster}$ increases from 0.17 (unirradiated) to 0.24 for 1×10^{21} p/m^2, while $(b_{max})_{Grain\ Boundairies}$ remains unchanged at ~ 0.17.

3.2. Comparison between the damage caused by neutrons and by protons

The enhancement of radiation induced pinning force can be represented by the variation of the factor β in the formula given above for the normalized pinning force $f(b)$ (figure 4).

Comparing the present results with neutron irradiation data obtained on the same wires by Baumgartner [13], it follows that for the same fluence, the damage produced by proton irradiation is considerably higher than the damage due to neutron irradiation. The same value of β (for example 0.5) obtained after proton fluences between 4 and 8×10^{20} p/m^2 (figure 4), is obtained after a neutron fluence of ~1×10^{22} n/m^2 [13], i.e. a factor > 10 higher.

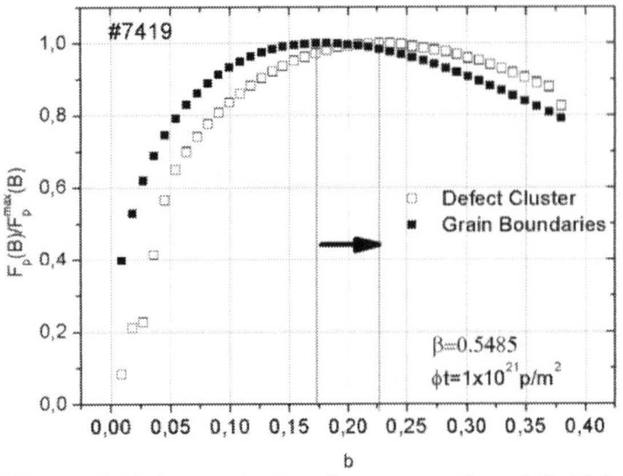

Figure 3 Volume pinning force vs. reduced field b for Ta alloyed wire #7419 irradiated with 65 MeV protons at 1×10^{21} p/cm^2.

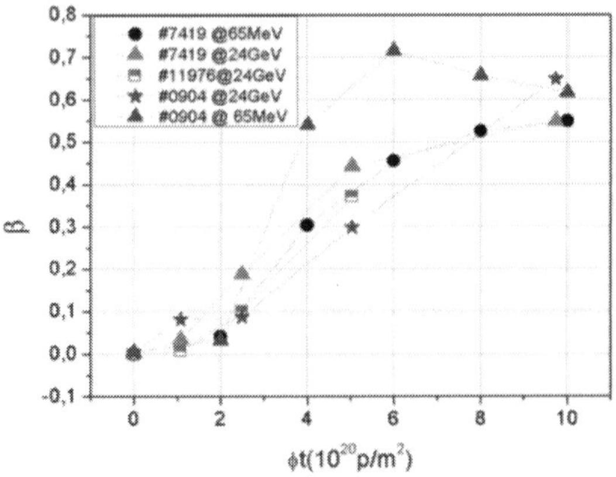

Figure 4 Increase of the proton radiation induced pinning force term β with fluence. (β uncertainty: ~ 10%)

Conclusion

Ternary alloyed RRP and PIT Nb_3Sn wires with high J_c values have been irradiated at the UCL and IRRAD1 reactors up to ϕt = 1×10^{21}p/m^2. The enhancement of J_c/J_{c0} in all irradiated wires is quite similar, in spite of the very different proton energies. Both Ta alloyed wires (RRP and PIT) exhibit a similar enhancement regardless of the fabrication technique, while J_c/J_{c0} of the Ti alloyed wire exhibits a stronger increase than the Ta alloyed wires. With the present fluences the expected maximum in J_c/J_{c0} was not yet achieved. The variation of J_c after proton irradiation was quantitatively analysed by assuming a radiation induced point pinning mechanism in addition to grain boundary pinning. A comparison with the results obtained on the same wires after neutron irradiation [6,13] shows that proton irradiation causes the same damage at fluences being one order of magnitude lower.

References

[1] Todesco E, Allain H, Ambrosio G, Borgnolutti F, Cerutti F, Dietderich D, Esposito L S, Felice H, Ferracin P, Sabbi G, Wanderer P, Van Weelderen R 2013 *Trans. App. Sup.* **23** 3
[2] Cerutti F, Lechner A, Mereghetti A, Brugger M, 2012 *WAMSDO workshop* CERN (CH)
[3] Otto T, Scheurlein C, Catherall R, Glaser M, Militaru O, Flukiger R, Ballarino A and Bottura L, 2013 *IEEE Trans. Appl. Supercond.*, **23** 6000504
[4] Bode H J and Wohlleben K 1966 *Phys. Letters* **24A** 1
[5] Voronova I V, Mihailov N N, Sotnikov G V and Zaikin V Ju 1978 *J. Nucl. Materials* **72** 129
[6] Flükiger R, Baumgartner T, Eisterer M, Weber H W, Spina T, Scheuerlein C, Senatore C, Ballarino A, Bottura L 2013 *Trans. App. Sup.* **23** 3
[7] Baumgartner T, Eisterer M, Weber H W, Flükiger R, Bordini B, Bottura L and Scheuerlein C, 2012 *Trans. Appl. Supercond.*, **22** 6000604
[8] Fukahori T, Iwamoto Y, 2012, *IAEA/TM on Primary Radiation Damage*, Vienna, Austria
[9] Fähnle M 1977 *Phys. Stat. sol.* **(b) 84** 245-251
[10] Brown B S, Blewitt T H, Wozniak D G and Suenaga M 1975 *J. App. Phys.* **46** 5163
[11] Küpfer H, Meier-Hirmer R, and Reichert T 1980 *J. App. Phys.* **51** 1121
[12] Snead C L, Jr 1978 *J. Nucl. Materials* **72** 190-197
[13] Baumgartner T, 2013, *PhD thesis, Vienna University*, Austria

Analysis of transient state in HTS tapes under ripple DC load current

M Stepien and B Grzesik

The Silesian University of Technology, Faculty of Electrical Engineering,
B. Krzywoustego 2, 44-100 Gliwice, Poland

E-mail: mariusz.stepien@polsl.pl

Abstract. The paper concerns the analysis of transient state (quench transition) in HTS tapes loaded with the current having DC component together with a ripple component. Two shapes of the ripple were taken into account: sinusoidal and triangular. Very often HTS tape connected to a power electronic current supply (i.e. superconducting coil for SMES) that delivers DC current with ripples and it needs to be examined under such conditions. Additionally, measurements of electrical (and thermal) parameters under such ripple excitation is useful to tape characterization in broad range of load currents. The results presented in the paper were obtained using test bench which contains programmable DC supply and National Instruments data acquisition system. Voltage drops and load currents were measured vs. time. Analysis of measured parameters as a function of the current was used to tape description with quench dynamics taken into account. Results of measurements were also used to comparison with the results of numerical modelling based on FEM. Presented provisional results show possibility to use results of measurements in transient state to prepare inverse models of superconductors and their detailed numerical modelling.

1. Introduction

Development of high temperature superconductors (HTS) results in continuous progress in manufacturing of HTS wires and tapes. Application of new structure of wire in power devices needs its characterization, especially for quench and stability. The most simple are tests of wire under DC load to measure the critical parameters. Two main drawbacks of this type of measurements are important. The first one, the high power is needed. The second drawback is possible destruction of wire due to long-time quench. These can be eliminated using measurements of short time pulse. This method is well known and used for testing both LTS and HTS wire [1], [2].

The authors decided to carry out the second method as they wanted to make the measurement with various shapes of ripple. Measurement under pulse load allows to determine an influence of AC current on additional AC losses. Since superconductors is loaded with DC current with AC ripple component are widely used, e.g. SMES coils or DC transmission cable authors of this paper present results of HTS tapes such type of load.

An important reason to undertake measurements of HTS tapes under such loads is the need of its parameters for modelling. Detailed analysis of processes occurring inside of HTS tape during its operation, particularly during a quench, can be described by FEM modelling [4].

The main difficulty during development of the numerical models is to determine the material parameters. Correctness of introduced material parameters influences directly on accuracy of the results of numerical analysis. Measurements can be used to improve numerical modelling based on direct models. The result of measurements can be also used to develop inverse numerical models.

2. Measuring and control system

Presented in the paper results were obtained using measuring system shown in Fig. 1, for the sample tape (2) depicted in Fig. 2. The results presented in the paper are carried out for superconducting 1G Bi-2223 sample tape (2) made as two bifilarly arranged tapes with the length of 1 m each. The current was measured by means of the shunt resistor (3). Signal of the current taken from the shunt was sent to DAQ SCXI (5) via terminal (4). The voltage signal was taken from the sample tape and sent to DAQ SCX converter (5) via SCXI terminal (4).

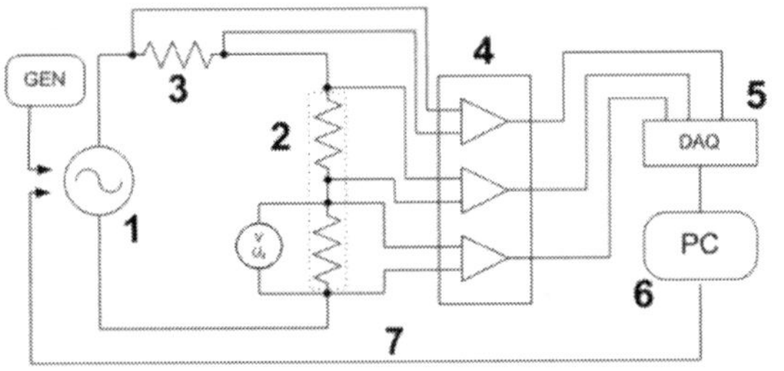

Figure 1. Block scheme of measuring system, 1 - programmable DC current supply, 2 - bifilar HTS tapes, 3 - shunt resistor, 4 - SCXI terminal, 5 - DAQ SCXI converter, 6 PC, 7 - RS 485 link.

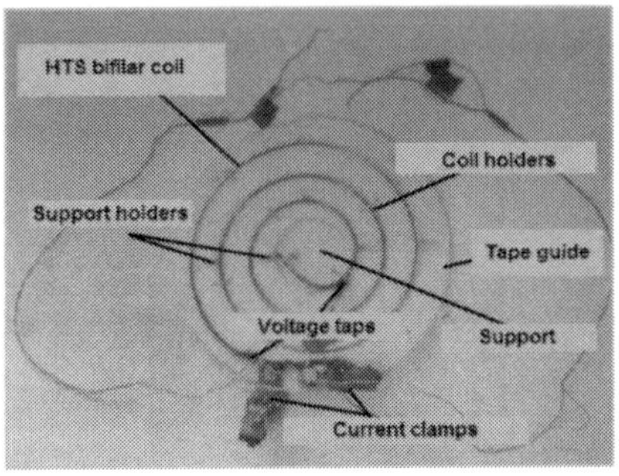

Figure 2. Photograph of bifilar HTS tape shaped as spiral.

The second part of the measuring system was a control system of programmable DC current supply. The supply (1) was controlled by LabView user interface using RS-485 interface (7). The DC current supply has over-voltage protection and limited output power. When the voltage along the HTS tape increased above the limit, the output current was reduced. The output power was limited to 1.2 kW. Photograph of the whole system is shown in Fig. 3.

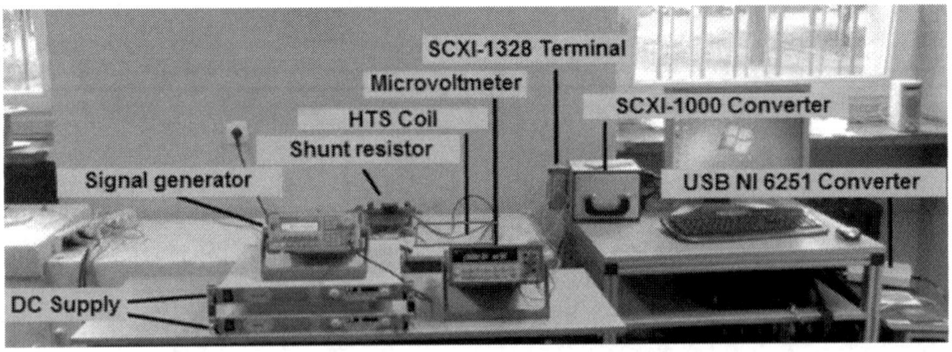

Figure 3. Photograph of the laboratory set-up for control and measurements (HTS tape inside of cryostat immersed in LN2).

3. Results of measurements for sine and triangle ripple of current components

The superconducting tape was immersed in LN2. The voltage drop along the tape was measured versus time. The measurements were carried out for the non-zero positive current in the tape (it means a peak-to-peak amplitude no higher than a twice of DC component).

The comparative measurements were carried out for different maximum current (the minimum value was fixed as a constant), different frequencies (0.5 Hz, 1 Hz and 2 Hz) and different ratio of rise to fall time at triangle ripple. Results of measurements under sine and triangle pulse of current are presented in Fig. 4. Maximum and minimum value of current of both shapes are the same ($I_{max} = 180A$, $I_{min} \approx 10A$).

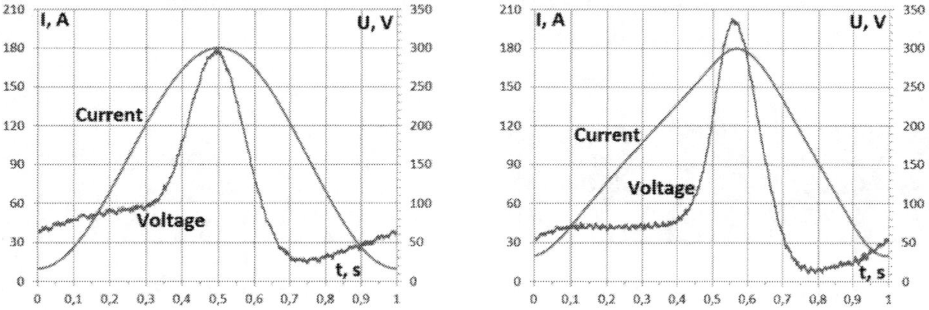

Figure 4. Quench measured under sine (left) and triangle ripple (right) at 1 Hz.

As one can observe the voltage at rising sine current is different than during falling one (the symmetrical pulse of current results with unsymmetrical voltage), what is the influence of a local temperature on the critical current density (and equivalent resistivity of the tape). The local resistivity is an important parameter of the tape used for numerical (FEM) modelling [3]. Since resistivity is a function of voltage and current the results of the measurements were presented in a form of voltage vs. load current for sine (Fig. 5) and triangle (Fig. 6) current pulse.

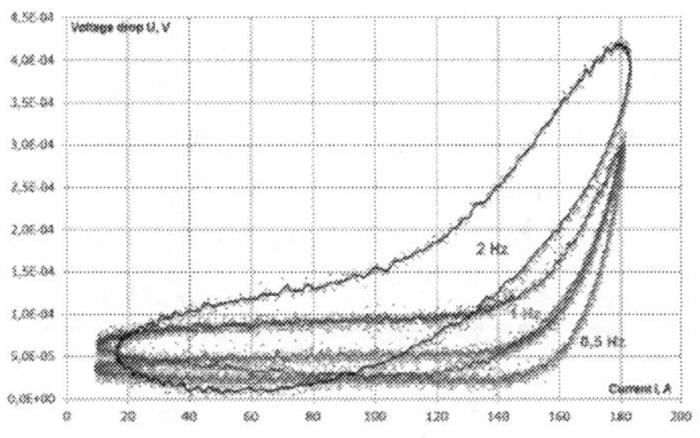

Figure 5. Voltage drop at sine ripple current of 0.5 Hz, 1 Hz and 2 Hz.

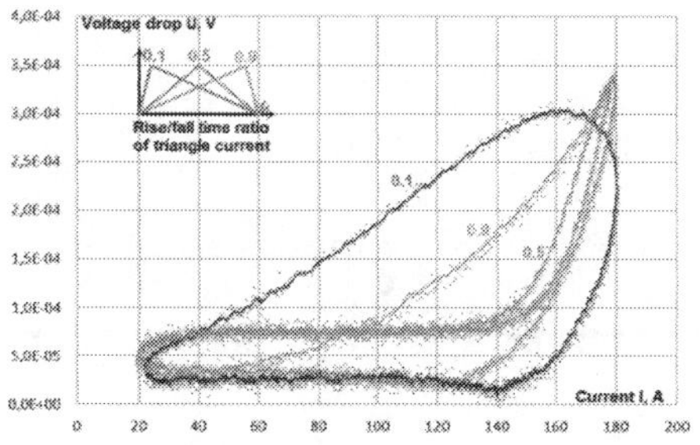

Figure 6. Voltage drop vs. current with triangle ripple, var. rise/fall time rat., 1 Hz.

The results in Fig. 5 shows non-linear dependence between voltage and current (an equivalent resistivity of the tape varies with the current). The resulting voltage vs. current forms a loop during rise and fall of the current. It is due to thermal capacity of the tape. The results given in Fig. 6 shows that equivalent resistivity of the tape depends additionally on current rate of rise. At the same frequency and different rise/fall ratio the voltage is different at the same current. The best results (the smallest loop area) were obtained at rise time equal to fall time.

4. Numerical modelling and future directions of research

The authors developed numerical FEM modelling based on ANSYS software [4]. Transient state modelling allows to calculate spatial and terminal parameters of superconductor (current density, temperature, voltage drop). Example geometry of 1G wire and resulting spatial distribution of the power losses is shown in Fig. 7. The voltage generated under the sine current pulse, shown in Fig. 8 is similar to the waveforms presented in Fig 4. In order to obtain results of numerical modelling reflecting real operation of given structure of HTS tape, the input parameters (material properties) must be corrected.

The results of measurements presented in above sections allows to calculate selected equivalent parameters of the tape, particularly resistivity and thermal capacity. It allows to obtain proper results of numerical modelling for tested HTS tape. Note, that numerical modelling allows to obtain more detailed results than measurements.

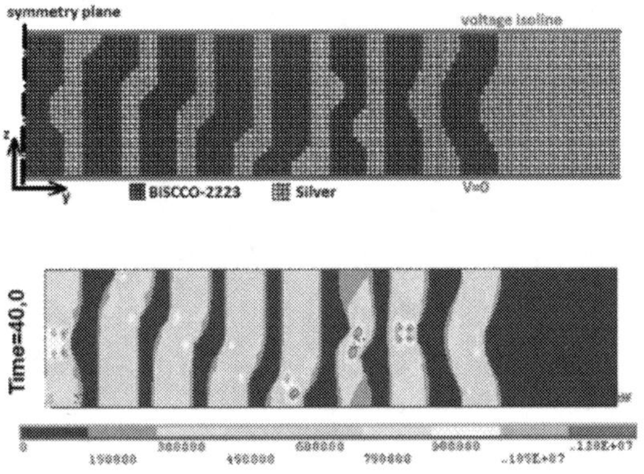

Figure 7. 2D mesh of FEM model (top) and distribution of power losses (bottom)

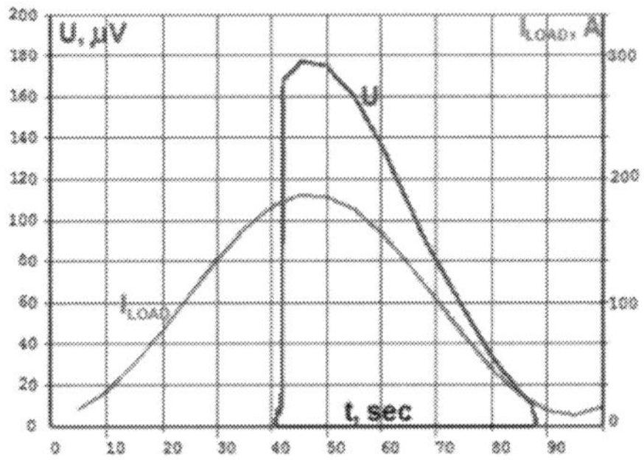

Figure 8. FEM results - voltage under sine pulse of current; quench at 42 s

5. Conclusions

Presented in the paper results were obtained for HTS 1G tape carrying DC current with sinusoidal and triangle ripple. The results were presented averaging for several periods of current. Measured voltage drop under ripple load allows to calculate equivalent parameters of HTS tape. Results of measurements at rising and falling current allows to estimate parameters for inverse FEM model of superconductors, also in the part of thermal parameters. Future work will be directed to prepare FEM computational models of superconductors taking results of measurements (measured equivalent parameters) into account.

Acknowledgments

Presented work is supported by National Science Centre (Poland) under research project number DEC-2011/03/B/ST8/04908.

References

[1] Rogacki K et al. 2002 *Superconductor Sci. Technol.* **15** 1151.
[2] Woniak M, Hopkins S C, Glowacki B A and Janowski T 2009 *Przeglad Elektrotechniczny (Electrical Review)* **85** (5) 198-199
[3] Stepien M 2012 Transient state modelling in HTS using ANSYS APDL *Proc. of 3rd Intl Workshop on Numerical Modelling of HTS*, Barcelona, Spain
[4] Stepien M, Krosny S and Grzesik B 2010 *J. Phys.: Conf. Series* **234** 022036

Journal of Physics: Conference Series **507** (2014) 022037 doi:10.1088/1742-6596/507/2/022037

Cryogen-free 1kA-class I_c measurement system featuring an 8 T HTS magnet

N M Strickland[1], C Hoffmann[1,3], S C Wimbush[1], D M Pooke[2], T Huang[2], Z Lazic[2], V Chamritski[2], E F Talantsev[1], N J Long[1] and J L Tallon[1]

[1]Callaghan Innovation, PO Box 31310, Lower Hutt 5040, New Zealand
[2]HTS-110 Ltd, PO Box 31310, Lower Hutt 5040, New Zealand

E-mail: nick.strickland@callaghaninnovation.govt.nz

Abstract. We have developed a cryogen-free critical-current (I_c) measuring system comprising a conduction-cooled 8 T HTS magnet and convection-cooled sample, both cooled by commercial cryocoolers. The sample can be rotated and transport currents of up to 800 A delivered with less than 0.5 K temperature rise during the I_c measurement. The system is automated with respect to variations in temperature (30–90 K), field (0–8 T), and field angle (0–360°). We have used this system to measure HTS wire samples, concentrating on metal-organic deposited YBCO on RABiTS substrates. Particular emphasis is given to the evolution of I_c anisotropy with temperature, and the dangers of extrapolating from 77 K to 30 K.

1. Introduction

The critical current, I_c, is one of the principal metrics for the quality of high-temperature superconducting (HTS) wires. For convenience, I_c values are often specified only at liquid-nitrogen temperature, 77 K, however this simple metric ignores the fact that I_c is a complex function of external parameters temperature, T, magnetic field, B, and magnetic-field angle, θ. This $I_c(T,B,\theta)$ variation arises from a combination of intrinsic and extrinsic properties and can therefore vary significantly for wires produced by different fabrication methods. Even quite similarly fabricated wires can have different behaviors and the implication of this for applications is that $I_c(T,B,\theta)$ should be measured over the parameter space that will be encountered in an application. The paucity of such data in the literature for T<77 K is testament to the challenge that arises in performing these measurements. Magnetic fields beyond 2 T and sample temperatures below 65 K are typically obtained in specialized systems in which both sample and magnet are cooled with liquid-helium [1]–[3][3], an inconvenient, expensive and non-renewable resource. For this reason, cryogen-free systems for cooling both the sample and the magnet are of interest [4],[5]. The transport measurements themselves are also difficult, particularly when a commercial conductor is to be measured in which case currents of several hundred amps must be delivered to the sample.

With these demands and difficulties in mind, we have developed a bespoke short-sample I_c measurement system incorporating a conduction-cooled HTS magnet and gas convection-cooled variable-temperature sample insert both cooled by cryocoolers. Sample temperatures down to 30 K, magnetic fields up to 8 T, full field-angle dependence and transport currents up to 800 A have been

[3]Current address: Ceraco Ceramic Coating GmbH, Ismaning, Germany

achieved. In this paper we give a brief description of the principal parts of the system and illustrate its performance with some example data obtained on commercial HTS wire.

2. I_c measurement system

A photograph of the I_c measurement system is shown in figure 1. The sample loading and cooling subsystem is contained within the vacuum chamber mounted on top of the magnet, while the magnet conduction cooling system is partially obstructed from view behind the magnet. The HTS sample is mounted on a top-loading cold-swappable sample rod and is typically 10–40 mm long.

Figure 1. Photograph of the I_c measurement system, including 8 T HTS magnet (diameter 525 mm) and cryogenic vacuum chamber.

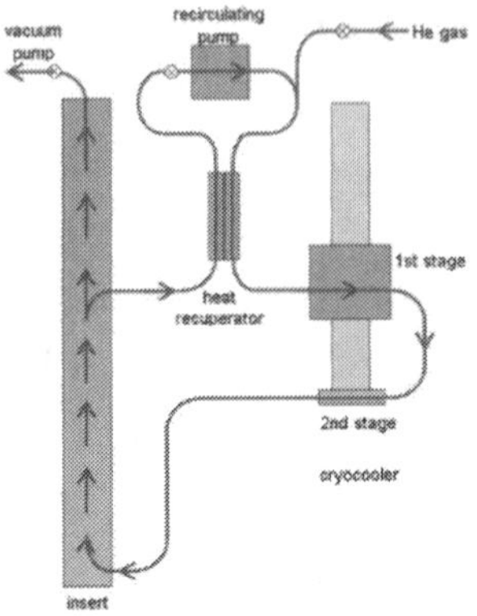

Figure 2. Schematic diagram of the closed-cycle gas-flow system for convection cooling the variable-temperature insert.

The 8 T magnet at the heart of this system was designed and constructed by HTS-110 Ltd, New Zealand. This magnet is a horizontal-field split-pair dipole with a vertical, rectangular 30 mm × 80 mm room temperature bore. With this configuration the sample rod can be conveniently rotated by a computer-controlled rotation stage to explore field-angle dependences. The coils are wound from BSCCO HTS wire and are conduction cooled to below 20 K by a SHI Cryogenics CH-208 cryocooler. The full field of 8 T is achieved with a current of 200 A. The ramp time from 0 to 8 T is 9 minutes.

Gas convection cooling of a variable-temperature insert was chosen to ensure a consistent sample temperature as well as to provide direct cooling of the sample current leads; a schematic of the cooling cycle is shown in figure 2. Helium gas flows in a closed cycle transferring heat from the sample to a cryocooler (SHI Cryogenics CH-204) via heat exchangers mounted on both the first and second stages of the cold head, shown in place within the vacuum chamber in figure 3(a). The first-stage heat exchanger has a helical path while the second-stage heat exchanger has a serpentine path. Circulation pressure is provided by a room-temperature diaphragm pump with a counter-flow heat recuperator used to pre-cool the gas to about 100 K before returning to the cold head. The design of heat recuperator chosen is a tube-in-tube counter-flow type formed into a helix to minimise space, and is shown in figure 3(b). A gas over-pressure of about 300 mbar is maintained, with about 100 mbar differential across the pump, and with a flow rate of 5–10 L/min. A heater and heat exchanger on the variable-temperature insert provides temperature control of the gas stream.

Temperature control is provided by two feedback loops: one controlling the temperature of the gas stream and one controlling a heater on the sample mount. A cernox temperature sensor soldered directly onto the sample provides the sample temperature and feedback to the sample-mount heater.

To measure HTS wires over a wide range of conditions, high currents must be delivered to the sample without unacceptable heating. We have demonstrated up to 800 A as in the *E(I)* curve shown in figure 4(a). *E(I)* measurements are made in a ramp mode with current ramped quasi-continuously under computer control to above I_c while the current and voltage are monitored. A standard 4-terminal configuration is used with voltage taps spaced 5–10 mm apart and an electric-field criterion for I_c of 1 μV/cm.

The current leads are a composite of stainless steel, copper and HTS wire. The stainless steel provides strength and thermal stability, the HTS wire carries high currents at the cold end and the copper carries current at the warm end of the current lead or whenever the local I_c of the HTS is exceeded. The HTS wire terminates before the sample to avoid distorting the applied field.

Heating originates from three sources: the sample itself when currents exceed I_c, contact resistances, and more remotely in the current leads. Heating on and adjacent to the sample is of greatest concern as this can rapidly raise the sample temperature. Remote heating can manifest as slower temperature excursions coming well after the measurement. Both of these types of temperature rise are illustrated in figure 4(b), where the sample temperature was monitored during a series of I_c measurements. In this series the I_c increased as the field angle approached the *ab*-peak up to the 800 A measurement shown in figure 4(a). Measurements up to 400 A could be made with less than 0.2 K immediate temperature rise, and up to 800 A with less than 0.5 K. The maximum current was limited by the available power supply; currents up to and above 1000 A would otherwise be manageable with no further changes provided temperature rises of 0.5–1.0 K were deemed acceptable. Both immediate and delayed temperature rises can be reduced by limiting the duration of the current ramp; in a typical measurement the ramp time is limited to 5–10 seconds. The delayed temperature rises can also build up if many measurements are made in rapid succession, as in the example here. This effect can be minimised by allowing time between measurements for heat to be removed.

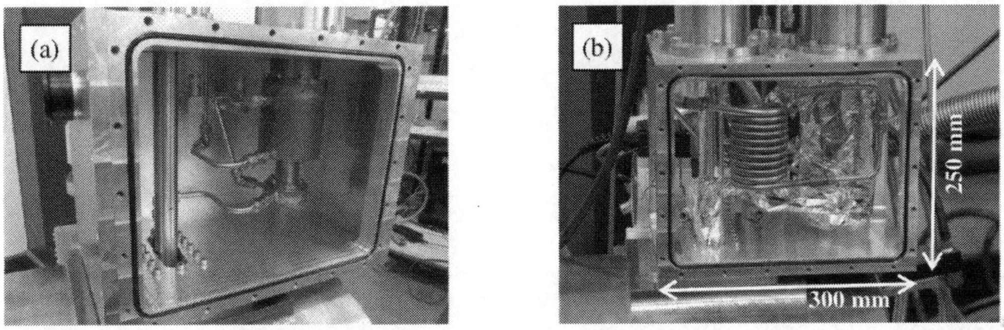

Figure 3. Vacuum chamber housing sample-cooling cryogenics: (a) the variable-temperature insert on the left and heat exchangers on the cryocooler on the right, and (b) the helical counter-flow heat recuperator, with other components now wrapped in super-insulation.

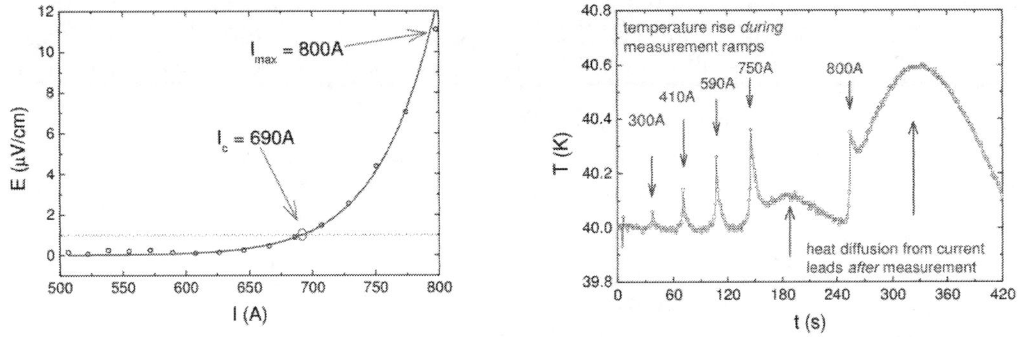

Figure 4. Example I_c measurement: (a) *E(I)* curve with current ramp up to 800 A with power-law fit giving an I_c of 690 A and (b) temperature excursions measured on the sample during a sequence of I_c measurements. The temperature spikes arise from heating on or near the sample, while slower excursions arise from heat diffusion to the sample from further up the current leads.

3. I_c measurements of HTS wires

With temperature, magnetic field, sample orientation, current ramping and current–voltage measurements under computer control, $I_c(T,B,\theta)$ data collection is readily automated. In figure 5 we compare I_c data for two wire samples fabricated by the MOD-TFA method on RABiTS substrates by AMSC [6]. The processing conditions were modified slightly to impact on the pinning landscape. This is evident in comparing the traditional 77 K, 1 T angle dependences in figure 5(a) and 5(b), where a flat background and prominent *ab*-peak in sample 1 contrasts with a broad *c*-axis peak and negligible *ab*-peak in sample 2. However when we compare the temperature series of angle dependences taken at $B = 6$ T in figure 5(c) and 5(d) we find that contrasting shapes at 60–70 K give way to similar shapes at lower temperatures. We draw the conclusion that extrapolations of I_c from 77 K to 30 K, or vice versa, are not robust and that conductor characterisation should therefore routinely cover the parameter space relevant to the application of interest. The data is fitted here with maximum-entropy functions [7] which describe how $I_c(\theta)$ arises from averaging over defect populations; the details will be described in a future report.

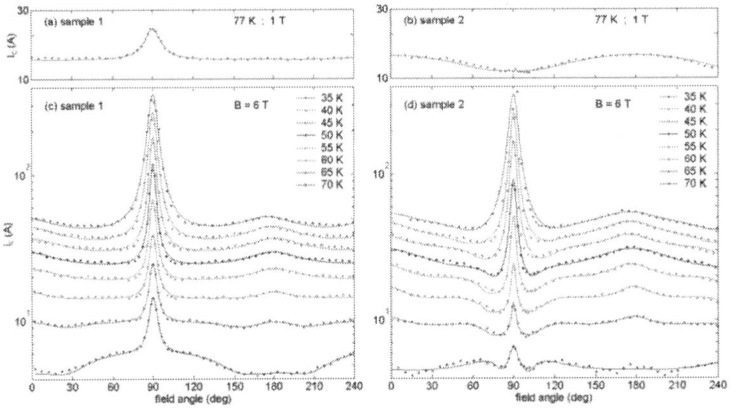

Figure 5. The angle dependence of I_c in different temperature and field regimes. (a) Sample 1 at 77 K, 1 T, (b) Sample 2 at 77 K, 1 T, (c) Sample 1 temperature variation at 6 T, (d) Sample 2 temperature variation at 6 T.

4. Conclusions

We have constructed and demonstrated a cryogen-free critical-current measurement system covering a parameter space of 30–90 K, 0–8 T, and full field-angle dependence, automated with respect to these parameters and $E(I)$ data collection. Critical currents up to 800 A can be measured with an immediate temperature rise of less than 0.5 K, allowing for extensive measurements of commercial HTS wires. Extension to higher currents would be possible with an appropriate power supply.

With this instrument under computer control, a large amount of data can be conveniently collected covering the stated parameter space. We have illustrated this with examples showing that $I_c(\theta)$ curves can evolve differently as functions of temperature and field even in quite similarly fabricated samples.

References

[1] Xu A *et al.* 2010 *Supercond. Sci. Technol.* **23** 014003.
[2] Xu A *et al.* 2012 *Phys. Rev. B* **86** 115416.
[3] Herzog R and Evetts J E 1994 *Rev. Sci. Instrum.* **65** 3574.
[4] Sohn M H *et al.* 2011 *Physica C* **471** 1449.
[5] Wang Q *et al.* 2009 *IEEE Trans. Appl. Supercond.* **19** 2325.
[6] Rupich M W *et al.* 2013 *IEEE Trans. Appl. Supercond.* **23** 6601205.
[7] Wimbush S C and Long N J 2012 *New J. Phys.* **14** 083017.

11th European Conference on Applied Superconductivity (EUCAS2013) IOP Publishing
Journal of Physics: Conference Series **507** (2014) 022038 doi:10.1088/1742-6596/507/2/022038

Critical current measurement of HTS tape relating with cable structure for a DC power cable

Jian Sun[1], Hirofumi Watanabe[1], Makoto Hamabe[1], Toshio Kawahara[1] and Satarou Yamaguchi[1,2]

[1]Center of Applied Superconductivity and Sustainable Energy Research, Chubu University, 1200, Matsumoto-cho, Kasugai, Aichi 4878501, Japan

[2]Department of Electrical Engineering, Chubu University, 1200, Matsumoto-cho, Kasugai, Aichi 4878501, Japan

E-mail: j_sun@isc.chubu.ac.jp

Abstract.
In the 200 m high temperature superconducting (HTS) cable test facility at Chubu University, a coaxial power cable is used and composed of two BSCCO (Bi-2223) superconducting layers. The tapes are wound closely to reduce effects on the critical current of BSCCO at self-field. Accordingly, each superconducting layer has a different number of BSCCO tapes. Previously, we have investigated dependence of the critical current (I_c) on the gap in order to optimize the HTS DC cable design. We have been studying the effect on the performance of HTS tapes for the superconducting DC power cables by critical current measurements. In the present experiments several HTS tapes are used and set as a similar structure in the cable with a two-layer structure. The critical current of HTS tapes are measured against the gap between the tapes in the same layer. The experiments show the improvement of the critical current by optimizing the tape arrangements due to magnetic field interaction between the tapes. We will present the experimental results and discuss the design of the HTS DC cable.

1. Introduction

DC superconducting power transmission system has been studied at Chubu University by using HTS tapes [1]. In the 200 m HTS cable system, called as CASER-2, a coaxial cable is used as shown in Fig. 1, which is made of BSCCO tapes by Sumitomo Electric Industries (SEI) [2]. There are 23 tapes with a two-layer structure for the inner HTS conductor and 16 tapes with a mono-layer structure for the outer one in consideration of the gap effect on the critical current of the tapes in the cable, which affects the rated current of the cable. In order to optimize the structure of a HTS power cable, we have been studying the effects of tape arrangements on the critical current to improve the superconductivity characteristics of HTS tape in the cable. Previous study shows strong dependence of the critical current of BSCCO tape on the gap. The critical current of the middle tape in three straight tapes in parallel increases by 10% for a two-layer structure [3].

We continue to measure the critical currents of BSCCO tape due to HTS tape arrangements to study the effect on the performance of BSCCO tapes for the superconducting DC power cables relating with cable structure [4]. As shown in Fig. 1b, the tapes are wound crossly and there are small gaps due to the shape of the former which may degrade the performance of the

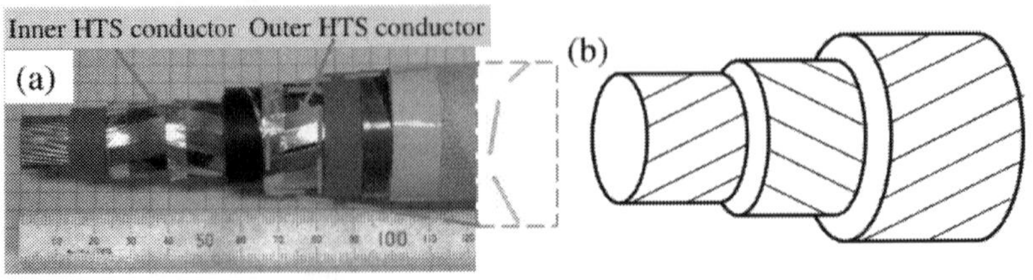

Figure 1: (a) The coaxial HTS power cable for CASER-2 and (b) A diagram of the tapes winding arrangement in the CASER-2 power cable. The insert of (a) shows gaps between HTS tapes.

tape. In this work more HTS tapes are set as a similar structure in the cable with a two-layer structure by considering the winding directions. This paper presents the I_c measurements of BSCCO tape relating with the cable structure by varying the transport current independently.

2. Experiments

BSCCO tapes are used with a cross section of 4.5 mm wide and 0.35 mm thick by SEI [5],[6]. The I_c of BSCCO tape is about 160 A at 77 K, self field. Figure 2a shows the tape arrangements with a two-layer structure for different winding directions. d is the lateral space gap between the tape edges in the same layer. The gaps are set to be 0.4 mm and 2.0 mm. Five tapes are used with a length of 27 cm and are insulated from each other by an insulating tape. Figure 2b shows a current loop applied to the five tapes conductor. Different currents are applied to the tapes by two power-supplies. The critical current of the middle tape #3 in the lower layer is measured by changing the lateral space gaps for different transport current in the neighboring tapes.

The experimental setup for the critical current measurement is similar to previous experiment [3] in which the tapes are immersed into liquid nitrogen at 77 K. Three voltage taps are attached on BSCCO tapes with distance of 8 cm and 10 cm for parallel tape conductor and 2 cm and 3 cm for cross tape conductor. The transport current is measured by a current shunt resistor. The voltage signals are measured by a KEITHLEY 2700 digital multimeter. The V - I characteristics of HTS tapes are obtained by the four-probe method.

3. Results and discussion

Figure 3 shows E-I curves of middle tape #3 for single and five parallel tape conductors with different gaps by dividing the measured voltages to the distance between the voltage taps. The

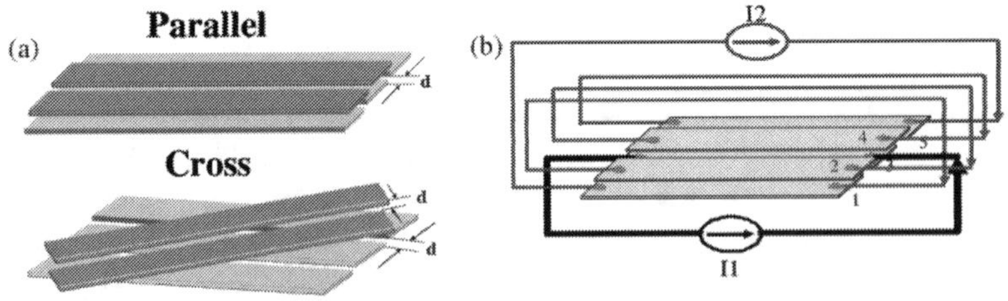

Figure 2: (a) A schematic diagram of the experimental setup for the parallel and cross tape conductors and (b) Circuit loop for the current applied to five tapes conductors. Power supply I1 is applied to tape #3 and I2 to other tapes in series mode.

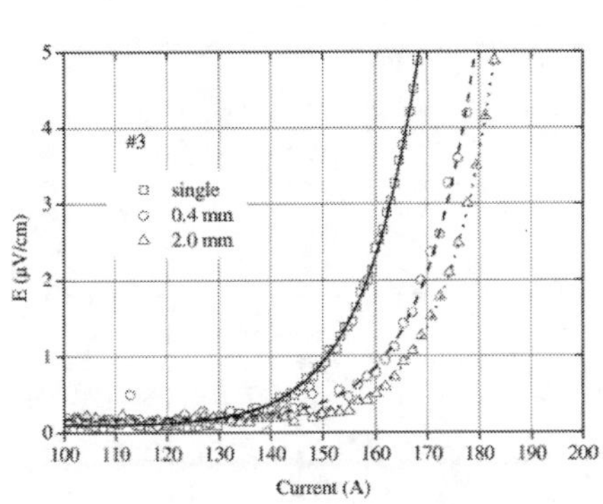

Figure 3: *E-I* curves of middle BSCCO tape in five tapes conductor with different gaps. The neighboring current *I*2 is 120 A.

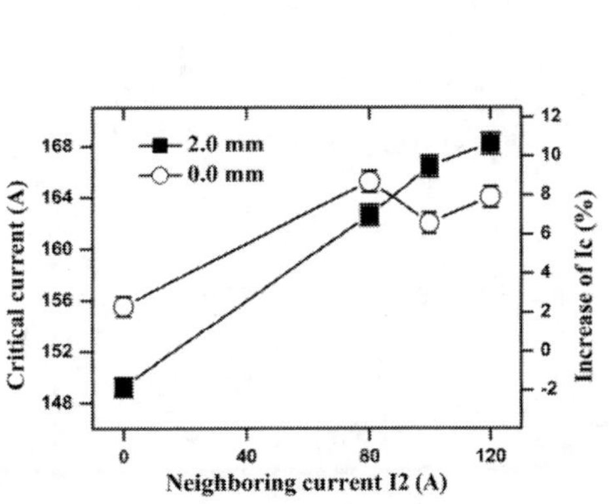

Figure 4: Dependence of the critical current of middle tape #3 in five parallel tapes conductor on the neighboring current *I*2.

transport currents in the neighboring tapes are 120 A. *E-I* curves are quite different between small and large gaps and the critical current of tape #3 in the five tapes is larger than that of a single tape. The critical current is determined on the criterion of 1 μV/cm. Figure 4 presents the dependence of the critical current of middle BSCCO tape #3 in five parallel tapes conductor on the neighboring current *I*2. When the neighboring current is larger than 80 A, the critical current becomes larger than that of a single tape. When the neighboring current is larger than 100 A, the critical current for large gap of 2.0 mm becomes larger than that for small gap of 0.4 mm. The measured critical currents of middle tape #3 increases 10% for 2.0 mm gap in the five parallel tapes arrangement and thus the critical current of BSCCO tape is improved when there are gaps between the tapes in the same layer.

We calculate the magnetic field distribution by the commercial finite element method code (ANSYS) [4, 7, 8]. Figure 5a-f present the magnetic flux lines around the five tape conductors with increasing gaps between the tapes. The transport current of 160 A is assumed to be uniformly distributed in the BSCCO filaments area. The cross section of the transport current area is assumed to be 4.5 mm × 0.25 mm for BSCCO tape by subtracting the reinforcing copper layer [6] and thus the current density is 142 A/mm^2. The magnetic flux density in the BSCCO filament area is reduced by increasing the gaps between the tapes. When the gap is 2.0 mm, about half of the tape width, neutral zone is formed in the intermediate tape in the second layer as in Fig. 5f. Therefore, the critical current in the five tape conductors is improved when there exists a gap of 2.0 mm as shown in Fig. 3 and 4.

As shown in Fig.1b, the tapes for HTS conductor are wound crossly between each layer. Five long tapes are wound around a supporter with a similar pitch as the CASER-2 power cable. Figure 6 presents the dependence on the gaps of the increase of I_c and the n-value of BSCCO tape in the five tapes conductor for different neighboring currents by comparison of the parallel and cross arrangements. The I_c of BSCCO tape is measured to be 161 A. The I_c and the n-value of BSCCO tape become larger than those of the single one. The increased I_cs of BSCCO tape by about 10% is measured for the parallel arrangement. However, the cross arrangement leads to the decrease of the I_c and the n-value of BSCCO tape. Therefore, the performance of HTS power cable can be improved by the parallel arrangement.

4. Conclusion

We investigated the critical current of BSCCO tape relating with the cable structures of gaps and winding methods for the five tapes conductor with a two-layer structure. The critical current

Figure 5: Magnetic flux lines around five tape conductors with increasing gaps between 0.4 mm (a) and 2.4 mm (f) by a step of 0.4 mm.

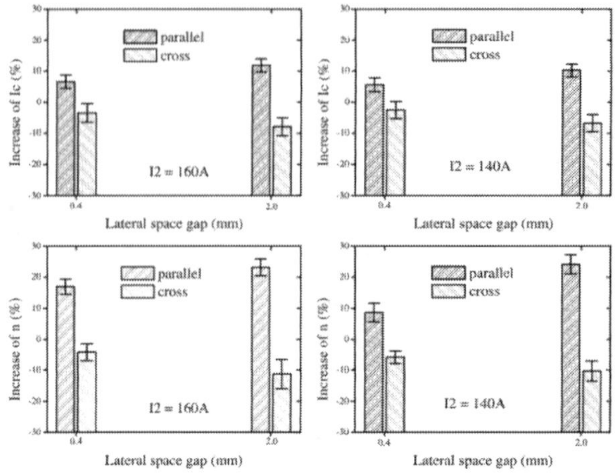

Figure 6: The dependence of increase of I_c and n-value of BSCCO tape on the gaps between the tapes for parallel and cross tapes conductor. Left: $I2 = 160$ A; Right: $I2 = 140$ A.

of BSCCO tape is improved when gaps exist for the five parallel tape conductor. The critical current and the n-value of BSCCO tape for the parallel arrangement become larger than those for the cross arrangement. Therefore fewer BSCCO tapes can be wound in parallel with gaps for a DC power cable with a two-layer structure.

Acknowledgments

The authors thank Dr. Atsuo Iiyoshi, Chancellor of Chubu University for his continuous encouragement through this work. We acknowledge Dr. Hiroshi Fujiwara for his support on the construction of 200 m superconducting facilities of CASER-2. Parts of this work were supported by a grant of Strategic Research Foundation Grant-aided Project for Private Universities, by Ministry of Education, Culture, Sports, Science and Technology (MEXT).

References

[1] Yamaguchi S, Hamabe M, Yamamoto I, Famakinwa T, Sasaki A, Iiyoshi A, Schultz J, Minervini J, Hoshino T, Ishiguro Y and Kawamura K 2008 *J. Phys.: Conf. Series* **97** 012290

[2] Yamaguchi S, Kawahara T, Hamabe M, Watanabe H, Ivanov Y, Sun J and Iiyoshi A 2011 *Phyica C: Superconductivity* **471** 1300–3

[3] Sun J, Yamauchi S, Sugino M, Watanabe H, Hamabe M, Kawahara T and Yamaguchi S 2011 *Physica C: Superconductivity* **471** 1313–6

[4] Sun J, Watanabe H, Hamabe M, Kawahara T, Iiyoshi A and Yamaguchi S 2012 *Phys. Proc.* **36** 372–5

[5] Kikuchi M, Ayai N, Fujikami J, Kobayashi S, Yamazaki K, Yamada S, Ishida T, Kato T, Hayashi K, Sato K, Hata R, Iihara J and Yamaguchi K, Kumakura H, Kitaguchi H, Osamura K and Shimoyama J 2008 *AIP Conf. Proc.* **986** 407–15

[6] Ayai N, Yamazaki K, Kikuchi M, Osabe G, Takaaze H, Takayama H, Kobayashi S, Fujikami J, Hayashi, Sato K, Osamura K, Kitaguchi H, Matsumoto S, Kiyoshi T and Shimoyama J 2009 *IEEE Trans. Appl. Supercond.* **19** 3014–7

[7] ANSYS® Academic Teaching Advanced, Release 12.1 ANSYS Inc. www.ansys.com

[8] Farinon S, Fabbricatore P and Gömöry 2010 *Supercond. Sci. Technol.* **23** 115004

Journal of Physics: Conference Series **507** (2014) 022039 doi:10.1088/1742-6596/507/2/022039

Development of cube textured Ni-W alloy substrates used for coated conductors

Hongli Suo[1], Lin Ma[1,2], Mangmang Gao[1], Yichen Meng[1], Yi Wang[1], Min Liu[1], Yue Zhao[2] and Jean-Claude Grivel[2]

[1] College of Materials Science and Engineering, Beijing University of Technology, Pingleyuan 100, Chaoyang District, 100124 Beijing, People's Republic of China.

[2] Department of Energy Conversion and Storage, Technical University of Denmark, Frederiksborgvej 399, 4000 Roskilde, Denmark

E-mail: honglisuo@bjut.edu.cn

Abstract. It is considered as a challenge for RABiTS route to get cube textured Ni-W alloy substrates with high mechanical and magnetic properties for coated conductors. The works of our group in recent years are summarized about different Ni-W substrates with high W content and composite tapes made by RABiTS technique. The fabrication process and the mechanism of cube texture formation in above different tapes are studied systematically. Compared with commercial Ni-5 at.%W substrate, these alloy substrates show a comparable texture quality and improved mechanical properties as well as reduced or zero magnetization especially in the novel composite substrates.

1. Introduction

Due to the excellent performance, High Temperature Superconductors (HTS) are generally considered the most promising functional materials for many potential applications such as electrical generators and transformers. For the high current purpose in long length coated conductor tapes, highly cube textured metal template tapes are required with not only well mechanical but also lower AC loss. Some works for decreasing magnetic losses are studied in Ni-W substrates [1].

Ni-5 at.%W (Ni5W) alloy tapes, a metallic substrate with highly cube texture, has been fabricated in worldwide [2]. However, it is unsuitable for applications due to the high Curie temperature (~335K) and the low yield strength (~160 MPa) [3]. Increasing the W content leads indeed to a decrease of the Curie temperature and an increase of the yield strength of Ni-W alloys, but also to a prompt decrease of the cube texture unfortunately [4]. Therefore, it is considered as a main challenge for the RABiTS route to obtain a sharp cube textured Ni-W alloy substrates with high W content. Moreover, a kind of composite substrates with sandwich-like structure have been proposed, in which the outer layers possess a sharp cube texture, and the inner layer such as Ni-9.3 at.%W (Ni9.3W) [5] supplies both high mechanical and magnetic properties. By either Powder In Tube (PIT) or co-rolling routes, many groups in the world took more efforts to obtain strong interface bonds between the outer layer and the inner layer as well as a highly cube textured surface on the composite tapes.

This paper summarizes the works about the fabrication process and the mechanism of cube texture formation in different Ni-W alloy substrates prepared by our group in recent years, including both the high W content Ni-W alloy substrates and the composite substrates fabricated by RABiTS technique. A special attention is given to the novel composite substrates made by a "composite ingot" method.

2. Experimental

Ni-7 at.%W (Ni7W) and Ni9.3W alloys were prepared by mixed powders with Ni (99.9%) and W (99.9%), which were pressed under 200~250 MPa by Cold Isostatic Pressing (CIP) after ball milling, then were sintered at around 1100°C for 10 h to obtain the initial Ni-W alloy ingots. Furthermore, the as-obtained ingots underwent heavy cold rolling down to 65 μm thickness with about 3~5% deformation per pass until more than 99% total reduction reached. The optimized multi-step annealing procedures were selected to get the required cube texture in a flowing Ar (4 vol.% H_2) atmosphere [6].

"Composite ingot" method was proposed for the development of multi-layer composite substrates. This kind of composite ingots were made "layer by layer", in which the Ni-W alloy with low W content as an outer layer being in charge of the formation of highly cube texture after rolling and recrystallization annealing processes, meanwhile the one with high W content as an inner layer being responsible for high mechanical and magnetic properties of the whole composite substrate. In this route, composite ingots were designed to different types such as welding type bonding [7] and diffusion type bonding for a good interface joint between the outer layer and the inner layer. After heavy deformation, as-fabricated composite ingots still keep the original sandwich-like structure, and show a good ductility during the following cold rolling process. Optimized heat treatments were performed as same as above-mentioned conditions.

Pole figures, (111) phi-scans and (002) rocking curves were measured by XRD with CuK_α radiation. EBSD was used to analyse the local orientation and obtain GBs information. Besides, the mechanical and magnetic behaviours were tested by mechanical testing machine and PPMS, respectively.

3. Results and discussion

3.1. Textured Ni-W substrates with high W content

3.1.1. Formation of cube texture in Ni7W alloy tape. Compared with one-step annealing process to heat up to final recrystallization temperature directly, in a two-step annealing process, the tapes were held at the initial recrystallization temperature for a certain time prior to full recrystallization at higher temperature. After different heat treatment processes, it was found that the two-step annealing processes are more beneficial to the formation of low angle GBs (LAGBs) and cube texture in Ni7W alloy tapes prepared by powder metallurgical route. The result is similar to that by casting route [4]. A sharp and pure {100}<001> cube texture has been obtained in the Ni7W tape, while the FWHM values of X-ray (111) phi-scan and (002) rocking curve are 6.33° and 4.3°, respectively. Both 99.8% (≤10°) biaxial cube grains and 94.6% LAGBs (2°~10°) are calculated from the EBSD orientation map detected on the surface of Ni7W alloy tape. This is probably due to the rotation of grains from off-cube to cube easily and the merger of annealing twins by cube grains during two-step annealing process. In consideration of influence on the epitaxially deposited superconducting layer, the grains connected with LAGBs will be conducive to transport current in YBCO film [8].

3.1.2. Formation of cube texture in Ni9.3W alloy tape. Due to the lower Stacking Fault Energy (SFE) in Ni9.3W alloy tape, a recovery annealing stage with lower temperature was added prior to the two-step annealing process in order to obtain a sharper cube texture, so-called three-step annealing process. It could be found that both the cube texture content and GBs quality in Ni9.3W alloy tape were improved by this heat treatment method. 88.8% (≤10°) cube texture was obtained in a Ni9.3W alloy tape prepared like Ni7W tapes from EBSD maps of the surface, which is much higher than that in a Ni9.3W alloy tape prepared by melt casting route in our work. Recently a new higher value, 96.0% (≤10°) cube texture in Ni9.5W, is reported by IFW Dresden [9]. In their work, intermediate heat treatments were also introduced many times before the two-step annealing procedure. Therefore, it is necessary to take recovery annealing processes in preparation procedures. However, there are a little of off-cube grains, including annealing twins and rotated cube grains along the rolling direction. More studies on the heat treatment profiles of substrates with high W content should be further optimized.

3.2. Interface characterization and surface texture analysis in composite substrates

3.2.1. Welding type composite substrate. Compared with a single layer substrate, a composite one has advantages combining a sharp cube texture with high mechanical strength as well as low magnetism. Welding type composite tape was prepared from a composite ingot with a total stacking sequence Ni5W/Cu/Ag/Cu/Ni9.3W/Cu/Ag/Cu/Ni5W, where Cu/Ag/Cu layers act as solder between the outer Ni5W layer and the inner Ni9.3W layer. Well welding bonds can be obtained between the interfaces of the tri-layers by forming a CuNi solid solution and AgCu eutectic phases [10] after sintering at 900°C.

The content of cube texture was 98.6% (≤10°) detected by XRD after heat treatment process, and the FWHM values of (111) phi-scan and (002) rocking curve are 7.6° and 6.1°, respectively. The texture relationship between outer layer and inner layer is investigated in an EBSD map of cross section. The outer layer has formed a preferred orientation, while the inner one presents a random orientation with highly misoriented grains. There is no diffusion between each layer from the EDS analysis results. The reason is that the Ag layer prevents the diffusion of W element from the inner to outer layers during the heat treatment process.

3.2.2. Diffusion type composite substrate. Different from welding type composite substrates, diffusion type ones form not a welding bond between the different layers but a W distribution gradient by diffusion from inner layer to outer layer of the composite ingot during the sintering process. Both CIP [11] and Spark Plasma Sintering (SPS) [12] methods can be used to obtain this kind of initial ingots. The details for preparing ingots are given elsewhere [13]. In this part of works, we made Ni12W as the inner layer and chose different Ni-W alloy as the outer layer such as Ni5W, Ni7W, Ni8W and Ni9W.

For example, the Ni5W/Ni12W/Ni5W composite tape has a good connectivity and clear boundaries, with a thickness about 75 μm consisting of a 25 μm/30 μm/20 μm sequence after heavy rolling. But the clear boundary becomes vague due to W gradient diffusion and continuous distribution after heat treatment. The permeation of W results in a broader diffused interface width and a W content slightly higher at the surface. According to a macro-texture measurement, a highly oriented cube texture was obtained on the Ni5W surface, which can also be verified by the ODF section. There is only a little effect on the cube texture formation by the diffusion of W element in the surface of composite tape. To evaluate the local micro-texture, EBSD is employed on the Ni5W surface, and the percentage of cube texture exceeds 98.8% (≤10°). The off-cube grains include a twin grain type and an equiaxial grain type. Equiaxial grains can be eliminated by improving the temperature and prolonging time. The number fraction of cube grains exceeds 67% (≤7°) and the maximum appears at 6°. These results confirm the highly cube texture in the outer layer, despite of W element diffusion from the inner layer.

In the reel-to-reel process, for epitaxial growth of buffer layers and YBCO film at 800°C to 1100°C, the substrate will endure a certain strain. It was reported [14] that the substrate does not withstand strains above 0.2% under tensile stress. Some micro-texture studies were performed by *in-situ* EBSD analysis on the Ni5W surface with tension by a tensile tester accessory at heat temperature. A strain was put along the sample's rolling direction, and the yield strength ($\sigma_{0.2}$) exceeds 240 MPa [15] which can meet the demands for practical applications [1]. The percentage of cube texture is 98.7% (≤10°) at 2% elongation, which is similar as 99.8% (≤10°) at original stage. These results indicate that this kind of composite substrate can be used in the reel-to-reel method with a stable and high quality of cube texture. In addition, it was found that some of the grains have been rotated by a small angle relative to cube grains although the content of cube grains does not change obviously. However, the evolution of misorientation angle is due to complex reasons but not simply related to the increase of elongation.

Similar works were also proceeding on other Ni-W alloy composite substrates. A special attention is given to the Ni8W/Ni12W/Ni8W composite tape especially to its 96.5% cube texture and non-ferromagnetic. The total thickness of this kind of substrate is 80 μm finally and the ratio of thickness in each layer is 1:2:1. Moreover, a detailed optimization for increasing cube texture in Ni8W surface is under study and the results will be reported elsewhere. The key questions are focused on the formation of highly cube texture in the surface of outer Ni-W alloy layers with high W content.

3.3. Magnetic property and yield strength in the as-obtained substrates

Table 1 summarizes the results obtained on cube texture content, mechanical and magnetic properties for different kinds of single component Ni-W alloy tapes and composite tapes. As-obtained substrates combine an excellent cube texture, enhanced mechanical properties and reduced magnetization. Due to the Ni12W inner layer, this kind of composite substrates as alternatives are very competitive to Ni-W alloy ones with single component for the potential applications. Furthermore, the successful deposition of high quality epitaxial $La_2Zr_2O_7$ buffer layer by MOD method directly on several high W content Ni-W alloy substrates and composite substrates, proves that these advanced substrates are appropriate for the further development of the HTS coated conductors [15].

Table 1. Cube texture content, mechanical and magnetic properties of various Ni-W alloy tapes.

Ni-W alloy tapes	Cube texture	$\sigma_{0.2}$ (MPa)	M (emu/g)	Ni-W alloy composite tapes	Cube texture	$\sigma_{0.2}$ (MPa)	M (emu/g)
pure Ni	/	63	57.1	Ni5W/Ni9.3W/Ni5W	98.6%	180	16.7
Ni5W	98.9%	176	24.8	Ni5W/Ni12W/Ni5W	99.0%	240	14.9
Ni7W	99.8%	216	10.7	Ni7W/Ni12W/Ni7W	99.6%	333	5.0
Ni8W	90.2%	/	/	Ni8W/Ni12W/Ni8W	96.5%	/	0
Ni9.3W	88.8%	220	0	Ni9W/Ni12W/Ni9W	72.3%	/	0
commercial Ni5W	98.0%	170	27.5				

4. Conclusions

We have described the development and summarized the properties of various advanced substrates. High W content Ni-W alloy substrates and a series of Ni-W composite substrates have been prepared with a sharp cube texture, high mechanical and magnetic properties. It can be concluded that, by designing innovative configurations and optimizing the preparation parameters, the RABiTS technique can remarkably improve the properties of the substrates. We expect that these advanced Ni-W alloy substrates have a potential for the large scale production of HTS coated conductors in the near future.

Acknowledgements

The authors would like to sincerely acknowledge Prof. Flükiger from University of Geneva for his great support and helpful comments. This work is supported by NSFC (51171002), Beijing NSF (2132011), RFDP (20121103110012), Beijing NSF B Type (KZ201310005003), Beijing PCIT&TCD (IDHT20130510), CFSD Special Funds and 211 Program of Beijing UT.

References

[1] Suenaga M, Iwakuma M, Sueyoshi T, et al. 2008 *Phys. C* **468** 1714-1717
[2] Zhao Y, Suo H L, Liu M, et al. 2006 *Phys. C* **440** 10-16
[3] Ijaduola A O, Thompson J R, et al. 2003 *Phys. C* **43** 163-171
[4] Subramanya V S, Eickemeyer J, et al. 2004 *Scripta Mater.* **50** 953-957
[5] Suo H L, Zhao Y, Liu M, et al. 2007 *IEEE T. App. Supercon.* **17** 3420-3423
[6] Zhao Y, Suo H L, Liu M, et al. 2007 *T. Nonferr. Metal. Soc.* **17** 208-211
[7] Zhao Y, Suo H L, Liu M, et al. 2007 *Acta Mater.* **55** 2609-2614
[8] Zhao Y, Suo H L, Zhu Y H, et al. 2009 *Acta Mater.* **57** 773-781
[9] Gaitzsch U, Hänisch J, Hühne R, et al. 2013 *Supercond. Sci. Technol.* **26** 085024
[10] Suo H L, Zhao Y, Liu M, et al. 2008 *Acta Mater.* **56** 23-30
[11] Suo H L, Zhao Y, Liu M, et al. 2008 *Supercond. Sci. Technol.* **21** 025006
[12] Liu M, Suo H L, Zhao Y, et al. 2007 *Scripta Mater.* **56** 129-131
[13] Zhao Y, Suo H L, Liu M, et al. 2008 *Supercond. Sci. Technol.* **21** 075003
[14] Goyal A, Norton D P, Kroeger D M, et al. 1997 *J. Mater. Res.* **12** 2924
[15] Gao M M, Suo H L, Zhao Y, et al. 2010 *Acta Mater.* **58** 1299-1308

Investigation of REBCO Twisted Stacked-Tape Cable Conductor Performance[1]

Makoto Takayasu[1], Luisa Chiesa[2], and Joseph V. Minervini[1]

[1]MIT Plasma Science and Fusion Center, Cambridge, MA 02139, U.S.A.
[2]Mechanical Engineering Department, Tufts University, Medford, MA 02155 U.S.A.

Email: takayasu@psfc.mit.edu

Abstract. Performances of Twisted Stacked-Tape Cables (TSTC) made of YBCO tapes were discussed for self-field and high-field operations. Power transmission cables of a single stack cable and a 3-channel conductor were evaluated by self-field distributions. A new concept of a double coaxial cable was discussed to obtain a high performance, high current density power cable. High field test results of 50-tape and 40-tape TSTC conductors were summarized, and their degradations were evaluated with single-tape data measured.

1. Introduction

The second generation high temperature superconductor (HTS) REBCO tapes such as coated YBCO tapes are very attractive to various low field and high field applications. For high current applications of power transmission cables and various large magnets, a few cabling methods of REBCO tapes such as Roebel, CORC and TSTC have been developed [1]-[3]. We have been developing one of the cabling method; TSTC (Twisted Stacked-Tape Cable) [3]. The TSTC method consists of stacking flat tapes and twisting them along the stack axis. Compact cabling technique of a TSTC for REBCO coated tapes is very useful for both power transmission and magnet conductors. For a power cable, the presence of a self-field significantly affects and degrades performance especially at around 77 K [3]. In order to obtain a high performance from the conductor, it is important to improve the magnetic self-field effect.

Studies of field orientation effects for NbTi and Nb_3Sn wires were investigated [4], [5]. The critical currents of Nb_3Sn increased by a factor of two with the longitudinal field at high magnetic fields [6]. Similarly the longitudinal field enhancement of critical current performances for HTS tapes has been investigated, and Matsushita, et al. has developed an advanced power cable structure of a traditional HTS power cable to improve its performance using the longitudinal magnetic field effect [7]. We are developing a double coaxial cable of a new compact high performance HTS cable composed of two types of cables: an inner cable of TSTC and an outer wrapping-wire cable like a CORC conductor. In this paper we will first present magnetic field orientation dependences measured for a single YBCO tape, and discuss performances of a single stack and 3-channel TSTC conductors including a newly developed double coaxial cable taking into account self-field effects. We will then discuss the degradations of TSTC conductors tested at 20 T in liquid helium.

[1] This work supported by the U. S. DOE, Office of Fusion Energy Science under Grants: DE-FC02-93ER54186, and partially DE-SC0004062 and Supercon DOE STTR Phase I DE-SC0007722 and Phase II DE-SC0004269.

Journal of Physics: Conference Series **507** (2014) 022040 doi:10.1088/1742-6596/507/2/022040

2. Low Field Performance

2.1. Field orientation effect on critical current of a single tape

Critical currents of single tapes (4 mm width, 0.1 mm thick YBCO (SuperPower SCS4050-AP) were measured at low magnetic fields at 77 K in two different configurations: transverse and longitudinal field orientations. Four samples were mounted 90° apart on a sample holder, so that all samples were tested at once in the different orientations. In the transverse orientation, the transverse field B_T was rotated from the x-axis and the y-axis for Samples T_{ab} and from the y-axis and the x-axis for Sample T_c in the xy plane shown in Figure 1(a), while in the longitudinal orientation, the longitudinal field B_L was rotated for Samples L_{ab} and L_c from the z-axis (the tape axis) to the x-axis and the y-axis, respectively. The applied fields were varied between 0 and 0.5 T.

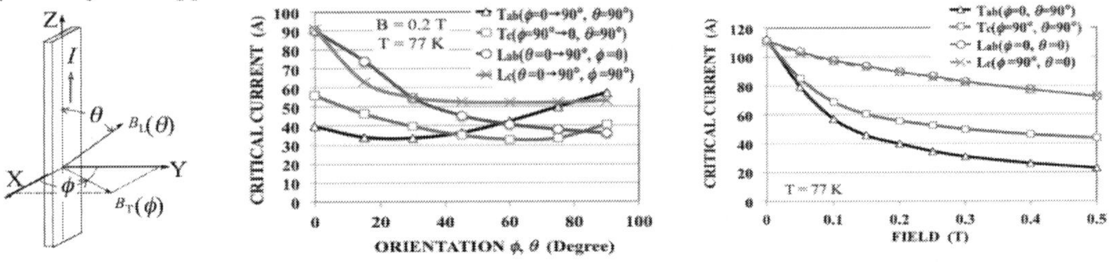

(a) (b) (c)

Figure 1. (a) Geometric illustration of a single tape test, and Field orientation effects on critical current I_c measured: (b) I_c vs. field orientation, and (c) I_c vs. magnetic field.

Measured critical currents are shown in Figures 1(b) and (c). In Figure (b) field orientation dependences of critical currents measure at 0.2 T are shown as a function of the field orientation given by the angles ϕ and θ. The angle dependences of critical currents for Samples T_{ab} and T_c are approximately mirror symmetry of each other as expected from the geometrical symmetry of the sample orientation. Note that the critical current of Sample L_c decreased more sharply than that of L_{ab}. The similar behaviors were confirmed at the applied field of 0.1 T and 0.5 T. Figure 1(c) shows the critical currents vs. the field intensity in transverse fields parallel to the ab plane (Sample T_{ab}) and parallel to the c-axis (T_c), and in longitudinal fields (L_{ab} and L_c). Longitudinal field affects the critical current much less than transverse fields. These data are used in the following analyses. The critical current of the tested YBCO-AP tape in the c-axis field was larger than that in the in-plane field.

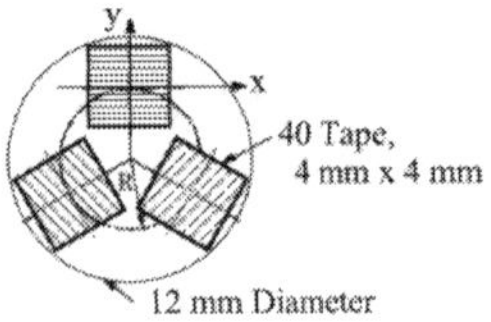

Figure 2. 3-channel conductor of three 40-tape, 4 mm x 4 mm, placed 120° apart shown in the inset (R=3.5 mm).

2.2 Self-field analysis of various TSTC conductors

Performance of a 32-tape single-stack TSTC conductor of YBCO tapes at 77 K has been characterized by the self-field effect [8]. Self-field of an untwisted stacked-tape conductor was analytically calculated for an infinitively long conductor with a uniform current density over the cross-section [3]. Using a similar method self-fields of a 3-channel conductor have been analytically calculated by superimposing vector fields of three conductors. Self-field distributions over the cross-sections of a single stack and 3-stack-in-a-round-rod (3- channel) conductor were calculated with a tape current of 44.5 A. Each stack (4 mm width, 4 mm height) was composed of 40 YBCO tapes (0.1 mm thick, 4 mm width). The 3-channel conductor was comprised of three stacks 120° apart which locate circumferentially as shown in Figure 2. The distance between the stack center and the conductor axis, R, was 3.5 mm. The self-field distributes more uniformly in a single stack conductor than in the 3-channel conductor where the field distributions of each stack affect other stacks while the in-plane field components are enhanced. The average self-field over the cross-section for the 3-channel conductor is B_{ave}= 137 mT and the average orientation ϕ_{ave}=38°, while for a single stack conductor B_{ave}= 108 mT and the average orientation ϕ_{ave}=45°.

Critical currents of a TSTC conductor can be dominated by the overall minimum tape critical current $I_{c\text{-min}}$ [A]. It has less orientation dependences for a transverse field angle $\phi \approx 20°$ as seen in Figure 1(b). From the experimental data for fields between 0.1 T and 0.5 T, the overall minimum tape current $I_{c\text{-min}}$ was found to be given by $I_{c\text{-min}} = 13.08 B^{-0.562}$ (B in unit of T) in a limited field range between 0.08 T and 0.55 T for the tested YBCO tape. From this equation and the average field B_{ave} the cable critical currents of a single stack cable and 3-channel conductors were evaluated at 77 K in Table 1. In the 3rd column the estimated overall cable diameters are also given. Measured critical current of a 40-tape single stack conductor was 1.8 kA which agreed well with the calculated value of 1.78 kA.

Table 1. Calculated critical currents of various TSTC conductor in self-field at 77 K.

Conductor	Tapes	R (mm)	Cable OD (mm)	B_{ave} (mT)	Tape $I_{c\text{-min}}$ (A)	Cable I_c (kA)
A. Single stack	40 x 1	3.5	6.0	108	44.5	1.78
B. 3-channel	40 x 3	3.5	12	128	41.7	5.00
C. 3-channel	20 x 3	3.5	10	94	50.0	3.00
D. 3-channel	20 x 3	2.5	8.2	102	48.0	2.88

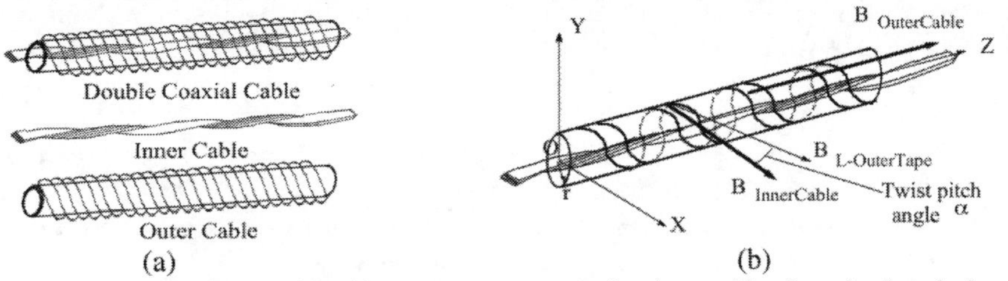

Figure 3 (a) An illustration of a double coaxial cable concept, composed of an inner cable of a twisted stacked-tape cable (TSTC) and an outer cable of a helical winding wire cable on a round surface over the inner cable. (b) Magnetic filed arrangement along a double coaxial cable.

2.3 Double coaxial cable

Double coaxial cable is composed of an inner cable and an outer cable which have two different type cables mounted co-axially, as shown in Figure 3(a). For example, the inner cable is made of a twisted stacked-tape cable (TSTC) and the outer is a helical winding tape type cable like a CORC conductor on the round surface of the inner cable. The magnetic field produced by the inner cable current creates the circumferential field $B_{InnerCable}$ longitudinal to the tapes of the outer cable, and the magnetic field by the outer cable currents creates the axial field $B_{OuterCable}$ longitudinal to the tapes of the inner cable, as shown in Figure 3(b). The longitudinal magnetic fields of these inner and outer cables enhance each other performances. It is confirmed from the experimental data in Figure 1 that the longitudinal fields do not degrade the cable performance if the cable is designed properly.

For example, the minimum tape current $I_{c\text{-min}}$=41.7 A of Conductor B in Table 1 (40-tape, 3-channel TSTC conductor) creates the average transverse field B_{ave}= 128 mT. The inner cable can reach the total field of 200 mT with the longitudinal angle θ=33° if the outer cable applies a longitudinal field of 154 mT. From L_c in Figure 1(b) the longitudinal field increases the tape critical current to 55 A. If the inner cable current increases the transverse self-field also increases, therefore a precise analysis requires iteration. By iteration the inner cable current was obtained to be 6.2 kA with the inner tape current of 52 A under the total field of 200 mT with the longitudinal angle θ=52° if the outer cable applies a longitudinal field of 121 mT, which is produced by an outer cable carrying 4 kA, (10-helically wound cable, each subcable carrying 400 A). The inner cable current can be increased to more than 6 kA (from to 5.0 kA) through the longitudinal field of the outer cable. The inner cable can be enhanced by 20%. This double coaxial cable will carry total 10 kA at 77 K, composing the inner cable current of 6 kA and the cable outer current of 4 kA. The diameter of this double coaxial cable will be less than 15 mm. If the outer cable is made of a 6 kA cable of 5-helically windings (each subcable of 1.2 kA), this outer cable generates the longitudinal field of 370 mT, which will allows the inner cable current 6 kA by the longitudinal field of 400 mT with the longitudinal angle θ=23°. This double coaxial cable would carry a total of 12 kA or ±6 kA in a cable diameter of 18 mm at 77 K.

3. High Field Performance

For testing a TSTC cable in a high field magnet, we have been developing a new magnet winding method called the Stacked-Tape Twist-Wind (STTW) technique, suitable for the fabrication of complicated 3D windings of 2G HTS tape cables [9]. Two samples of YBCO tapes were fabricated and tested at 4.2 K using a 20 T, 195 mm warm bore Bitter magnet at NHMFL, Florida State University [9]. Both tested cables were 2.3 m length composed of 50-stacked tapes (SuperPower, 0.1 mm thick and 4 mm width). They were wound on a 165 mm diameter pentagonal cylinder by the STTW method. The test result of the first sample showed that the critical current at the criterion of 100 μV/m was 4.0 kA [9]. The cable quenched at 4.7 kA. It was degraded. The cable itself was soldered between the stacked tapes but it was not well supported in the round groove. The sample could be degraded by the Lorentz force. The second sample was not soldered but the conductor was supported with Stycast epoxy in the groove. The critical current of the second sample was measured at various fields from 4 T to 20 T. The sample carried 10 kA at 4 T without developing any voltage at the start of the test. However the sample gradually degraded showing noticeable resistive component as the tests progressed. After the whole test, it was noticed that a high current spike (few times of the critical current) was produced at the end of each shot when the sample current was dumped. This over-current could be degrading the sample performance.

A 40-tape TSTC conductor soldered in a copper tube of 9.5 mm OD was tested at the KIT FBI facility [10]. A cycle test was performed between 6 T and 12 T. The test results showed degradations between the first measurements and the second measurements of about 11% at 6 T and 4% at 12 T. After the first test no further degradation was confirmed [10].

From these three sample tests it was found that in high magnetic field that TSTC cables showed about 50% degradation compared with the strand data obtained in perpendicular field (parallel to the c-axis). Transverse load tests of YBCO single tapes and TSTC conductors have not shown such large critical-current degradations observed in the cable samples [11]. In order to understand the high field degradations further electrical and mechanical work is required.

4. Conclusions

Performances of TSTC conductors for power cable applications have been evaluated based on self-field and field orientation data measured at low fields up to 0.5 T. It will be possible to obtain a 3-channel YBCO TSTC conductor of 5 kA at 77 K. A double coaxial cable composed of a TSTC inner cable and a wrapped wire type outer-cable can provides conductors of 12 kA or ±6 kA. Longitudinal field of a double coaxial cable can enhance inner cable performance by more than 20%. A double coaxial cable can provide a high current compact power cable with a low heat loss. High-field performance of TSTC conductors tested at up to 20 T have revealed significant degradations. Its origin has not yet been understood. Further electrical and mechanical investigations will be required for a twisted stacked coated-tape conductor.

References

[1] Goldacker W et al. 2007 *IEEE Trans. Appl. Supercond.* **17** 3398
[2] van der Laan D C 2009 *Supercond. Sci. Technol.* **22** 065013
[3] Takayasu M, et al. 2012 *Supercond. Sci. Technol.* **25** 014011
[4] ten Haken B 1992 *IEEE Trans. on Magnetics* **28** 755
[5] Matsushita T 2012 *Japan. J. Appl. Phys.* **51** 010111
[6] Takayasu M, et al. 1998 *Adv. Cryog. Eng.* **44** 1035
[7] Matsushita T 2012 US Patent application, US 2012/0214676
[8] Takayasu M 2012 *Adv. Cryo. Eng.* Plenum, N.Y. **57** 273
[9] Takayasu M, et al. 2013 *IEEE Trans. Appl. Supercond.* **23** 4800104
[10] Barth C et al. presented at ASC 2012
[11] Chiesa L 2013 presented at MT23

Journal of Physics: Conference Series **507** (2014) 022041 doi:10.1088/1742-6596/507/2/022041

Comparison of J_c characteristics in PIT wires based on $BaFe_2As_2$ with different substitutions

T Tamegai[1], S Pyon[1], Q P Ding[1], H Inoue[1], H Kobayashi[1], Y Tsuchiya[1], Y Sun[1], H Kajitani[2], and N Koizumi[2]

[1]Department of Applied Physics, The University of Tokyo, Hongo, Bunkyo-ku, Tokyo 113-8656, Japan
2 Fusion Research and Development Directorate, JAEA, 801-1 Mukoyama, Naka-shi, Ibaraki 311-0193, Japan

E-mail: tamegai@ap.t.u-tokyo.ac.jp

Abstract. Three kinds of superconducting wires based on $BaFe_2As_2$ with different substitutions are fabricated using powder-in-tube method and characterized including magneto-optical imaging. In the case of $(Ba,K)Fe_2As_2$ wires processed by hot isostatic press, critical current density (J_c) of 32 kA/cm^2 has been achieved at 4.2 K under self-field. Wires fabricated in a similar way with $Ba(Fe,Co)_2As_2$ resulted in much lower J_c of 7.8 kA/cm^2 at 4.2 K under self-field. In the case of $BaFe_2(As,P)_2$, J_c of ambient-pressure processed wire has J_c of only 1.0 kA/cm^2 at 4.2 K under self-field. Origins of these differences are discussed.

1. Introduction

Extensive research works on iron-based superconductors (IBS) in the last five years made it clear that these superconductors have fascinating fundamental properties for applications, especially for high field use [1]. The upper critical field reaches over 50 T even for the weak direction and the anisotropy parameter is much smaller than cuprate superconductors. In addition, high critical current density, J_c, is demonstrated in most of IBS single crystals [2-5] and irradiations of heavy-ions or protons can enhance it [6-8]. Among various IBS, $BaFe_2As_2$ system is most widely studied because high-quality single crystals can be easily obtained. In $BaFe_2As_2$ system, superconductivity can be induced by replacing elements in three different sites. Superconducting wires of $BaFe_2As_2$–based materials have been fabricated by several groups [9-13]. Among them, a very large transport J_c of ~1x10^5 A/cm^2 has been achieved in $(Ba,K)Fe_2As_2$ wires processed at low-temperature combined with mechanical alloying and densification using hot isostatic press (HIP) process [12]. In the present study, we fabricate superconducting wires using powder-in-tube (PIT) technique for three different $BaFe_2As_2$ systems with different substitutions and compare their J_c properties using transport and magneto-optical (MO) measurements. We chose relatively high temperature process without mechanical alloying, but applied HIP process to know the effectiveness of high-pressure densification.

2. Experiments

We have prepared $(Ba,K)Fe_2As_2$, $Ba(Fe,Co)_2As_2$ and $BaFe_2(As,P)_2$ wires by the PIT method. To prepare polycrystalline materials of $(Ba_{0.6}K_{0.4})Fe_2As_2$, we used Ba(2N) pieces, K(2N5) pieces, and FeAs powder as starting materials and reacted in an evacuated quartz tube. Polycrystalline materials of $Ba(Fe_{0.93}Co_{0.07})_2As_2$ were prepared in a similar way by using Ba pieces, FeAs, and CoAs powders as starting materials. FeAs and CoAs powders were prepared using Fe(Co) powder and As(7N) grains.

$BaFe_2(As_{0.65}P_{0.35})_2$ polycrystals were prepared from Ba pieces, Fe powder, As grains and P(6N) grains. After suitable soakings at intermediate temperatures, final reactions were conducted for 5(24) hours at 1100°C(900°C) for $(Ba,K)Fe_2As_2$ ($Ba(Fe,Co)_2As_2$, $BaFe_2(As,P)_2$). Synthesized materials were ground in a globe box, filled in a Ag tube (4.5 mm OD, 3.0 mm ID), and drawn into wires of about 1 mm diameter. In case of $Ba(Fe,Co)_2As_2$, we add 15%(weight) of Ag. The drawn wires were cut into a few centimeter length and inserted into 3 mm OD Cu tube and further drawn into a diameter of 1mm. After this process, both ends of Cu tube were welded in the arc furnace for HIP process. Densification using HIP was conducted at 120 MPa and 600°C for 4 hours. Qualities of fabricated wires were evaluated using powder X-ray diffraction, magnetization, and current-voltage characteristic measurements. We also conducted magneto-optical (MO) imaging of the wire to characterize the grain connectivity.

3. Results and Discussion

3.1. $(Ba,K)Fe_2As_2$ wire

Figure 1(a) shows temperature dependence of magnetization for the as-grown material and sintered (HIP) wire of $(Ba,K)Fe_2As_2$. Usually, as-grown sample show the onset of superconductivity close to 37 K, while T_c is suppressed after sintering and transition becomes broader. The same trend of T_c is observed also in $Ba(Fe,Co)_2As_2$ as shown in Fig. 1(b). In the latter case, we confirm that Co-rich (Fe,Co)As impurities increase after sintering and the Co content of the matrix decreases.

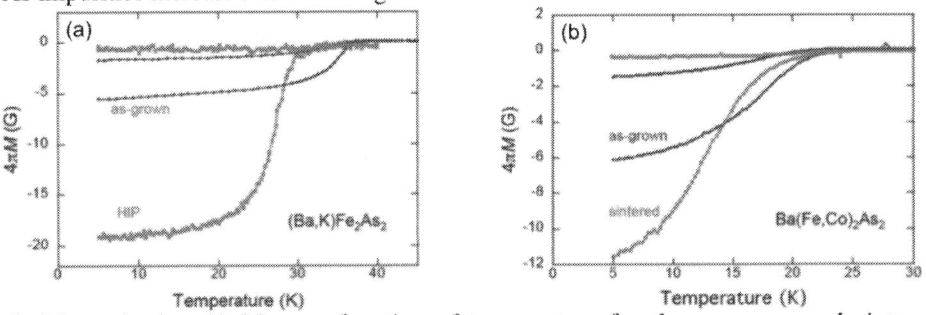

Figure 1. Magnetization ($4\pi M$) as a function of temperature for the as-grown and sintered (HIP processed) samples of (a) $(Ba,K)Fe_2As_2$ and (b) $Ba(Fe,Co)_2As_2$.

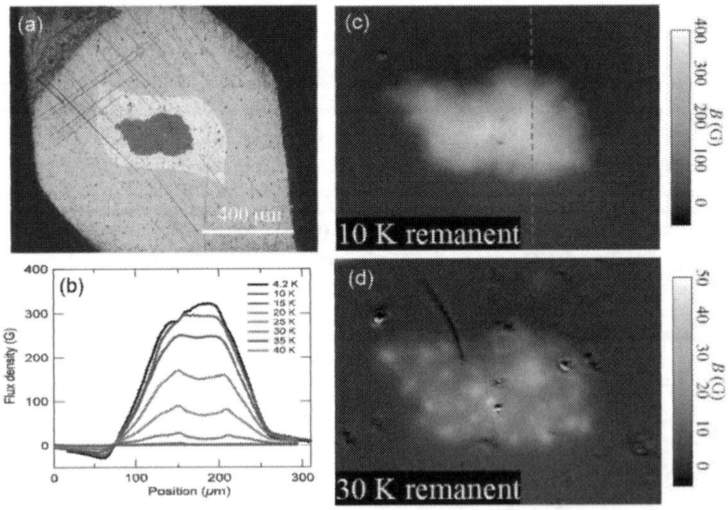

Figure 2. (a) Optical micrograph of the cross section of $(Ba,K)Fe_2As_2$ wire. The dark core is surrounded by Ag and Cu sheaths. MO images of the wire core in the remanent state at (c) 10 K and (d) 30 K. (b) Temperature dependence of local magnetic induction profile along the dotted line in (c).

Figure 2(a) shows an optical micrograph of the cross section of $(Ba,K)Fe_2As_2$ wire. MO image at 10 K in the remanent state shown in Fig. 2(c) indicate a strong grain coupling, although grains become visible at higher temperature as shown in Fig. 2(d). Temperature dependence of the magnetic induction profile along the dotted line in Fig. 2(c) are shown in Fig. 2(b), from which we can evaluate the intergranular critical current density, J_c^{inter}~58 kA/cm^2 at 5 K.

3.2. $Ba(Fe,Co)_2As_2$ wire

Figure 3(a) shows an optical micrograph of the cross section of $Ba(Fe,Co)_2As_2$ wire. MO image at 5 K in the remanent state (Fig. 3(c)) indicate a modest grain coupling, although grains become visible at 15 K (Fig. 3(d)). Temperature dependence of the magnetic induction profile along the dotted line in Fig. 3(c) are shown in Fig. 3(b), from which we can evaluate J_c^{inter}~7.3 kA/cm^2 at 5 K.

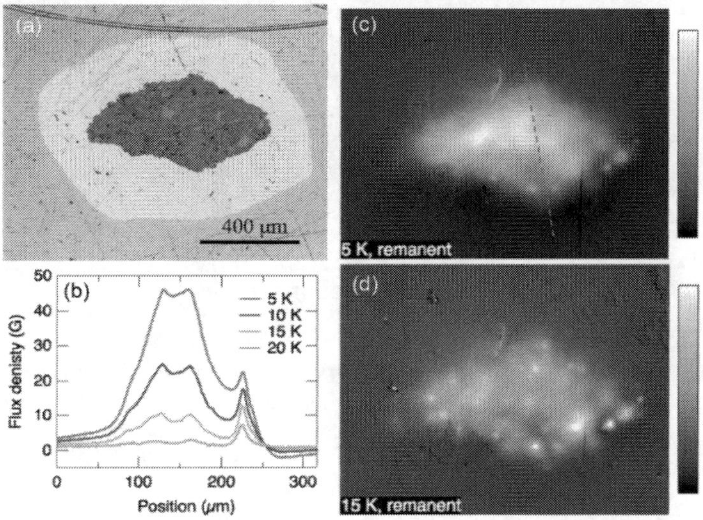

Figure 3. (a) Optical micrograph of the cross section of $Ba(Fe,Co)_2As_2$ wire. The dark core is surrounded by Ag and Cu sheaths. MO images of the wire core in the remanent state at (c) 5 K and (d) 15 K. (b) Temperature dependence of local magnetic induction profile along the dotted line in (c).

3.3. $BaFe_2(As,P)_2$ wire

Figure 4(a) shows an optical micrograph of the cross section of $BaFe_2(As,P)_2$ wire. Since the transport J_c of this wire was very low as described in next section, we made a polycrystalline pellet of this material (Fig. 4(b)) and made MO observations. The MO image shown in Fig. 4(c) is taken at 5 K in the remanent state. It is obvious that grains are not coupled strongly. Some of the bright spots remains even above T_c, which we identify originated from magnetic Fe_2P impurity phase.

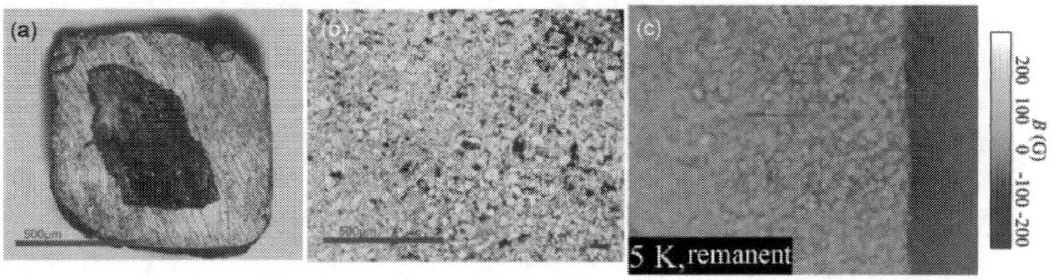

Figure 4. (a) Optical micrograph of the cross section of $BaFe_2(As,P)_2$ wire. The dark core is surrounded by Ag sheath. (b) Optical micrograph of the polycrystalline $BaFe_2(As,P)_2$. (c) MO image of the polycrystalline sample at 5 K in the remanent state.

3.4. Transport critical current densities

By using standard d.c. transport measurement, we evaluate the transport critical current density, J_c, with a criterion of E=1 μm/cm. The resulting J_c as a function of magnetic field is summarized in Fig. 5. HIP processed wires of (Ba,K)Fe_2As and Ba(Fe,Co)$_2As_2$ achieved reasonably high J_c at 4.2 K under self-field, 32 kA/cm^2 and 7.8 kA/cm^2, respectively. More importantly, they do not decrease strongly with magnetic field. As a reference we also plot J_c from ref. [13] processed at ambient conditions and J_c from ref. [12], where mechanical alloying and HIP process are combined. Although value of J_c with only HIP process is lower than those in ref. [12], our data definitely proof the effectiveness of densification by HIP process.

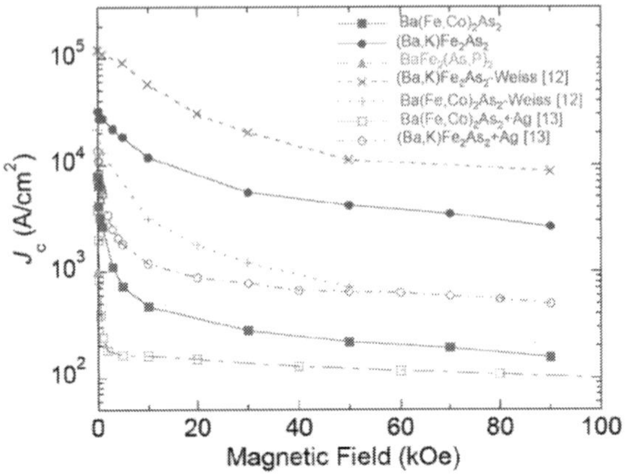

Figure 5. J_c as a function of magnetic field for (Ba,K)Fe_2As,P_2 (circles), Ba(Fe,Co)$_2As_2$ (squares), and $BaFe_2$(As,P)$_2$ (triangles) wires. As a references, similar data from ref. [13] on wires processed at ambient conditions (Ba,K)Fe_2As_2 (open ciecles) and Ba(Fe,Co)$_2As_2$ (open squares), and from ref. [12] on (Ba,K)Fe_2As_2 (crosses) and Ba(Fe,Co)$_2As_2$ (pluses) wires fabricated by HIP process are also plotted.

4. Summary

Three kinds of superconducting wires based on $BaFe_2As_2$, namely, (Ba,K)Fe_2As_2, Ba(Fe,Co)$_2As_2$, $BaFe_2$(As,P)$_2$ wires have been fabricated using powder-in-tube method combined with hot isostatic press (HIP) process. J_c at 4.2 K under self-field for (Ba,K)Fe_2As_2, Ba(Fe,Co)$_2As_2$, and $BaFe_2$(As,P)$_2$ wires are 32 kA/cm^2 (HIP), 7.8 kA/cm^2 (HIP), and 1.0 kA/cm^2 (ambient), respectively. Magneto-optical images clearly show the presence of strong intergranular current in (Ba,K)Fe_2As_2, while grains are almost decoupled in $BaFe_2$(As,P)$_2$. Impurity phases such as (Fe/Co)As and Fe_2P are responsible for the low transport J_c in Ba(Fe,Co)$_2As_2$ and $BaFe_2$(As,P)$_2$ wires, respectively.

References

[1] Gurevich A 2011 *Rep. Prog. Phys.* **74**, 124501
[2] Nakajima Y, Taen T and Tamegai T 2009 *J. Phys Soc. Jpn.* **78**, 023702
[3] Taen T, Tsuchiya Y, Nakajima Y and Tamegai T 2009 *Phys. Rev. B* **80**, 092502
[4] Moll P J W, Puzniak R, Balakirev F *et al.* 2010 *Nat. Mater.* **9** 628
[5] Sun Y, Taen T, Tsuchiya Y, Shi Z X and Tamegai T 2013 *Supercond. Sci. Technol.* **26**, 015015
[6] Nakajima Y, Tsuchiya Y, Taen T, Tamegai T *et al.* 2009 *Phys. Rev. B* **80** 012510
[7] Tamegai T, Taen T, Yagyuda H, Tsuchiya Y *et al.* 2012 *Supercond. Sci. Technol.* **25** 084008
[8] Taen T, Nakajima Y, Tamegai T and Kitamura H 2012 *Phys. Rev. B* **86** 094527
[9] Togano K, Matsumoto A and Kumakura H 2011 *Appl. Phys. Exp.* **4** 043101
[10] Ma Y W 2012 *Supercond. Sci. Technol.* **25** 113001
[11] Ding Q P, Prombood T, Tsuchiya Y *et al.* 2012 *Supercond. Sci. Technol.* **25** 035019
[12] Weiss J D, Tarantini C, Jiang J *et al.* 2012 *Nat. Mater.* **11** 682
[13] Tamegai T, Ding Q. P., Inoue H *et al.* 2013 *IEEE-Trans. Appl. Supercond.* **23** 7300304

Journal of Physics: Conference Series **507** (2014) 022042 doi:10.1088/1742-6596/507/2/022042

Development of an experimental system for characterization of high-temperature superconductors cooled by liquid hydrogen under the external magnetic field

H Tatsumoto[1], Y Shirai[2], M Shiotsu[2], Y Naruo[3], H Kobayashi[3] and Y Inatani[3]

[1]J-PARC Center, Japan Atomic Energy Agency, Tokai, Ibaraki, JP
[2]Department of Energy Science & Technology, Kyoto University, Kyoto, JP
[3]Institute of Space and Astronautical Science, JAXA, Kanagawa, JP

E-mail: tatumoto@post.j-parc.jp

Abstract. An experimental system has been developed to investigate electro-magnetic properties of high-Tc superconductors cooled by liquid hydrogen under the external magnetic field of up to 7 T. A LH_2 cryostat is concentrically mounted on the inside of a LHe cryostat to cool a NbTi superconducting magnet. The experimental system is installed in an explosion-proof room. Explosion proof electrical devices are used and current leads are covered with an enclosure filled with nitrogen gas. A remote control system has been developed. Furthermore, the effects of stray magnetic field on the existing and the new devices are investigated and electro-magnetic shielding panels and enclosure made of iron were designed. It is confirmed through the cryogenic test that the experimental system meets the design requirements.

1. Introduction

Liquid hydrogen (LH_2) is expected as a coolant of high-Tc superconductors because of the excellent cooling properties and the spectacular improvement of the electromagnetic characteristics when the temperature is decreased from 77 K to 20 K. However, there have been few experimental data for the cooling properties of LH_2 because of its explosive properties. For the first step, we have developed an experimental system to study cooling properties of liquid and supercritical hydrogen for wide ranges of pressures, temperature and flow rates [1]. Furthermore, over-current characteristics of MgB_2 immersed in LH_2 have been studied under no external magnetic field [2]. It is necessary to understand cooling stability of the superconductors cooled by LH_2 under some external magnetic field.

In this study, an experimental system has been developed to study electromagnetic characteristics of high-Tc superconductors cooled by liquid and supercritical hydrogen under magnetic field up to 7 T. The cryogenic test has been conducted so as to verify the performance.

2. Design of experimental system

2.1. Design details

Figure 1 shows a schematic of the developed experimental system, which has been designed based on the existing thermal-hydraulic one [1]. They are installed in an explosion-proof room. The design pressure of a LH_2 cryostat is 2.0 MPa. The hydrogen inventory is 61 L. There are three current leads (up to 500 A) in enclosures where nitrogen gas purges with a pressure of 5 kPaG, in order to set two test pieces at a time. The liquid hydrogen can be pressurized to a desired value by a pure hydrogen gas (99.999%) that is adjusted by a dome-loaded gas regulator. The bath temperature is controlled by a sheathed heater with a capacity of 500 W. Five cernox temperature sensors are vertically located at the liquid levels of 13, 16, 20, 40 and 50 L. Hydrogen gas is temporarily supplied to the bottom so as to uniform the bath temperature.

A LHe bath, whose design pressure is 0.04 MPa, is concentrically mounted on the outside of the LH_2 bath through a vacuum layer. A Nb-Ti superconducting magnet (7 T) with a height of 406.4 mm, an inner diameter of 400.1 mm and an outer diameter of 558.8 mm is cooled. The superconducting magnet has an inductance of 112.4 H and a rated current of 175.27A. The current leads are also covered with the GN_2 enclosures, which are isolated from those for the LH_2 cryostat. It takes about an hour to achieve the excitation of 7 T due to the maximum current sweep of 0.09 A/s. The inventory of LHe is determined to 100 L above the top of the magnet. The allowable heat inleak need to be reduced below 18 W so as to get experimental duration of more than three hours. The heat inleak was estimated 12 W except for that the power lead. The cooling design for the power leads was carried out so as to reduce the heat inleak through them down to 2.5 W using the CFD code, STAR-CCM+.

When quench occurs, the stored energy of 1.71 MJ at the excitation of 7 T is used by the evaporation of LHe and the temperature rise of the magnet. The blow off rate is estimated to be 397 g/s based on the authors' data of the critical heat flux in a pool of saturated LHe (5 x 10^3 W/m^2) [3]. A flange insert valve whose size is 50A is installed not to exceed the allowable pressure in the LHe bath.

Two hydrogen gas detectors (RIKEN KEIKI, GD-A8-16), which are metal oxide and catalytic based sensor, are located on the ceiling with a height of 4 m. Explosion proof electrical devices are used such as a solenoid valve, a pressure transmitter and so on. The power leads are covered with the enclosure mentioned above. A remote operational system has been established and we maintain a safe distance from the apparatus during the experiment. If an off-normal event such as hydrogen leakage occurs, hydrogen is automatically released to the outside through the vent line and the power sources are shut off. However, the magnet is degaussed at the allowable speed because it is in helium atmosphere and the power leads are isolated by the GN_2 enclosure.

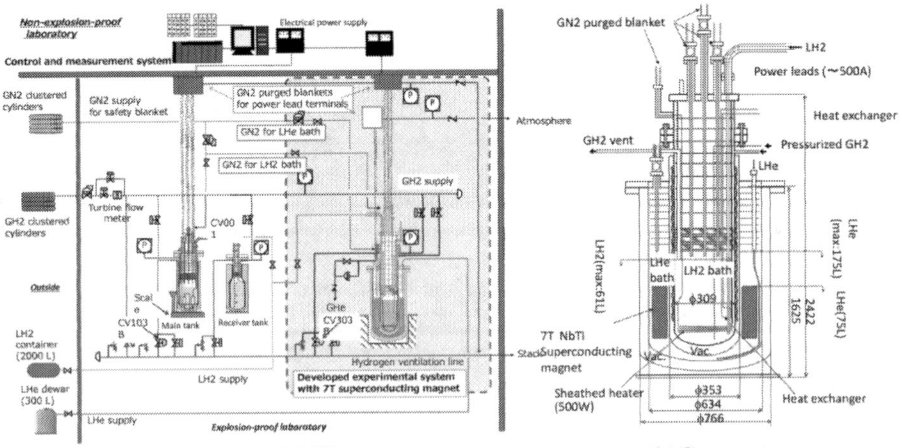

(a) P&ID (b)Cryostat

Figure 1. Schematic of an experimental system.

2.2. Countermeasure against stray magnetic field

The stray magnetic field is analyzed under the excitation of 7 T using ANSYS. The solenoid coil is located at the center of a cubic 14 m on a side in the analytical model. The current density of 1.08 A/m^2 is applied, which corresponds to the magnetic field of 7 T at the center of the coil. As shown in Fig.2, the analytical result indicates that the stray magnetic fields are estimated to be 30 G and 10 G at the horizontal distances of 3.2 m and 4 m away from the center of the magnet, respectively. It is necessary to clarify the effect of the stray magnetic field on the existing and the newly-installed devices: the hydrogen gas detector, the explosion proof solenoid valve (ASCO, JE4) for on-off valves, that for pneumatic control valves (TOKO VALEX, T-8810), the hydrogen gas detector, pressure transmitter (NAGANO KEIKI, KH41-173), the scale corrected by electromagnetic force (Mettler Toledo, WMHC 300s) and an air-conditioning control panel located at the next control room. The effect of the magnetic field on the solenoid valves, pressure transmitter and the hydrogen gas detector were investigated in the magnetic field up to 300 G using a copper coil magnet with an inner diameter of 165 mm and a height of 30 mm. The solenoid valve functioned normally below 210 G because of the explosion-proof enclosure. The hydrogen detector and the pressure transmitter were not affected.

The solenoid valves were moved to 4-m away from the magnet where the magnet field is below 10 G. The magnetic field is less than 20 G at the location of the hydrogen detectors. The electro-magnetic shielding made of iron was designed for the control valves, the scale and the control panel. As shown in Figs.3 to 5, it is clarified that the stray magnetic field can be reduced below 25 G by using the shielding panel and enclosure with the thickness of 3.2mm and 7 mm, respectively.

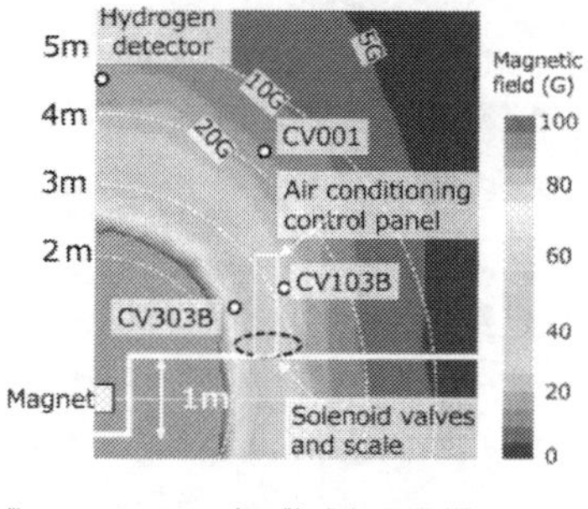

Figure 2. Stray magnetic field at 7 T.

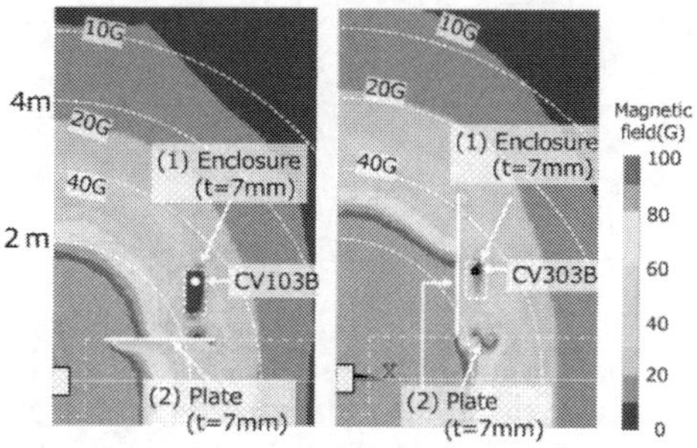

Figure 3. Countermeasures for CV103B and CV303B.

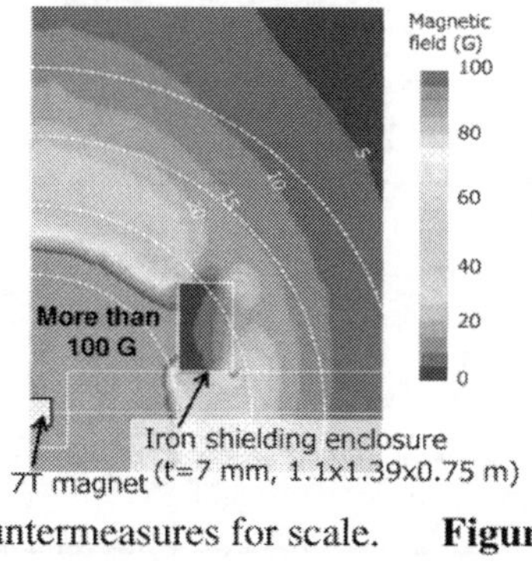

Figure 4. Countermeasures for scale.

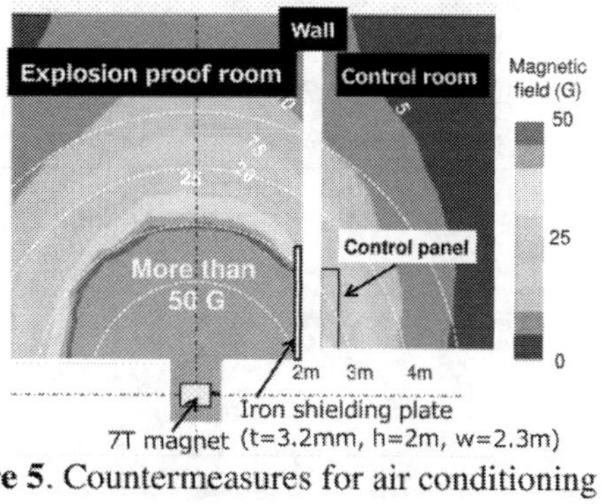

Figure 5. Countermeasures for air conditioning control

2.3. Commissioning

Figure 6 shows the first excitation test result. As shown in Fig.6, the heat inleak into the LHe bath is 15.5 W given by the evaporation of LHe and agrees with the design value, when the evaporation gas flows only to the cooling channels for the power leads. When the evaporation gas flows not only to the power leads but also along the wall of the cryostat, the heat inleak is decreased down to 5.3 W. The experimental duration of more than 3 hours can be achieved for the heat inleak of 15.5 W. Figure 7 shows one of over current test results using a short MgB_2 wire with a diameter of 0.51 and a length of 130 mm (@HyperTech), which was immersed in subcooled LH_2 at the bath temperature of 21 K and the pressure of 1.1 MPa. The current was controlled to meet the exponential heat inputs $Q=Q_0 \exp(t/\tau)$ with the period τ of 5 s. As the transport current I increased, the wire resistivity R appeared at $I = 65$ A, which corresponds to a critical current, and the heat input Q to the wire also started to increase. It was confirmed that the over current characteristic test was successfully carried out using the MgB_2 wire. Figure 8 shows the measured magnetic field distribution at the ground level for 6.4 T. The magnetic field at the devices can be mitigated below 20 G, which is lower than the allowable value. It is confirmed that the experimental system satisfies the design requirement.

3. Conclusion

An experimental system has been designed and fabricated to investigate electro-magnetic properties of high-Tc superconductors cooled by liquid and supercritical hydrogen under the external magnetic field up to 7T. The countermeasures against the explosion proof and the stray magnetic field have been conducted. It was confirmed through the commissioning that the developed experimental system satisfied the design requirements.

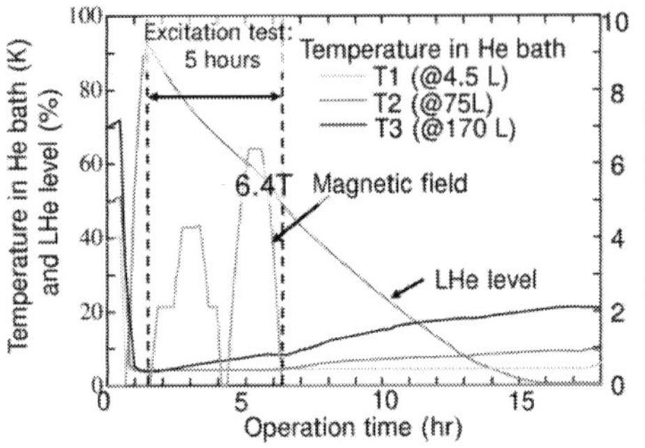

Figure 6. First excitation test result.

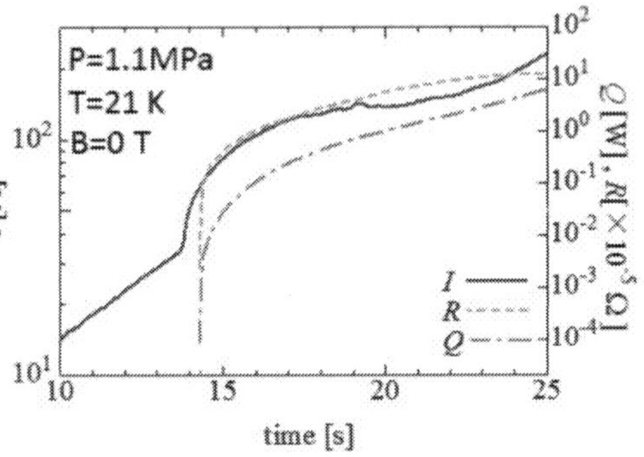

Figure 7. Over current test result.

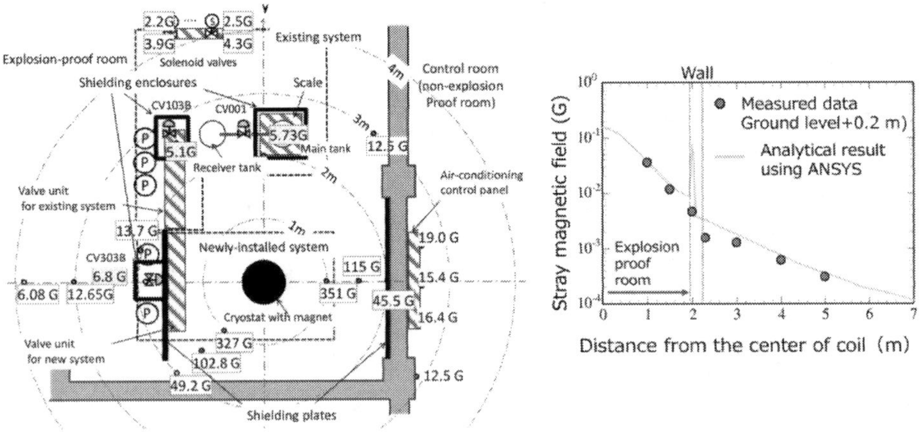

Figure 8. Measured stray magnetic field at the excitation of 6.4 T.

References

[1] H. Tatsumoto, Y. Shirai, M. Shiotsu, K. Hata, H. Kobayashi, Y. Naruo, Y. Inatani, Journal of Physics: Conference Series. (2010)234,032056A

[2] K. Hikawa, Y. Shirai, T.Takegami, M. Shiotsu, H. Tatsumoto, K. Hata, H. Kobayashi, Y. Naruo, Y. Inatani, N. Narita, "Over-Current Characteristics of MgB2 Wire Cooled by Liquid Hydrogen", Proc. of the 24th ICEC, pp.203-206, 2012.

[3] K. Hata, H. Nakagawa, H. Tatsumoto, Y. Shirai and M. Shiotsu, "Critical heat flux on a flat plate in a pool of Msubcooled liquid helium", Advances in Cryogenic Engineering, 47B, pp.1460-1467,2002.

$BaMO_3$ (M=Zr, Hf, Sn) material dependence of T_c reduction in $BaMO_3$-doped $SmBa_2Cu_3O_y$ films

Akihiro Tsuruta[1], Yutaka Yoshida[1], Yusuke Ichino[1], Ataru Ichinose[2], Kaname Matsumoto[3] and Satoshi Awaji[4]

[1]Department of Energy Engineering and Science, Nagoya University, Furo-cho, Chikusa-ku, Nagoya 464-8603, Japan
[2]Electric Power Engineering Research Laboratory, Central Research Institute of Electric Power Industry, 2-6-1 Nagasaka, Yokosuka, Kanagawa 240-0196, Japan
[3]Department of Materials Science and Engineering, Kyushu Institute of Technology, 1-1 Sensui-cho, Tobata-ku, Kitakyushu 804-8550, Japan
[4]Institute for Materials Research, Tohoku University, Katahira 2-1-1, Aoba-ku, Sendai 980-8577, Japan

E-mail: tsuruta-akihiro11@ees.nagoya-u.ac.jp

Abstract. For the improvement of superconducting properties of $BaMO_3$ (M=Zr, Sn, Hf)-doped $REBa_2Cu_3O_y$ films, we focused on the T_c reduction caused by lattice misfit between $BaMO_3$ and $REBa_2Cu_3O_y$. We compared the influence on the T_c reduction of $BaZrO_3$, $BaSnO_3$ and $BaHfO_3$-doped $SmBa_2Cu_3O_y$ (SmBCO) films with various BMO contents. From the correlation between configuration of nanorods and *c*-axis expansion, it was suggested that the lattice strain working on the SmBCO matrix became stronger in the order of $BaHfO_3$, $BaZrO_3$ and $BaSnO_3$. Therefore, the combination of SmBCO and $BaHfO_3$ is better regarding the suppression of the T_c reduction compared with other combinations of SmBCO and $BaMO_3$.

1. Introduction

For the realization of the high performance superconducting applications such as superconducting magnetic energy storage (SMES) and magnetic resonance imaging (MRI) using $REBa_2Cu_3O_y$ (REBCO: RE = rare earth) coated conductors, they must have a high critical current density in the applied magnetic fields $J_c(B)$ [1]. In order to improve the in-fields J_c, it is well known that the introduction of the artificial pinning centers (APCs) is effective. So far, various kinds of APCs had been reported [2-4]. Among them, $BaMO_3$ (BMO: M = Zr, Sn, Hf) nanorods are known as strong *c*-axis correlated pinning centers in REBCO films fabricated by vapor phase epitaxial method such as a pulsed laser deposition (PLD) and a chemical vapor deposition (CVD) methods [5-8]. However, the introduction of BMO nanorods also causes a reduction of the critical temperature (T_c) in the REBCO films, because the REBCO lattice is distorted by the lattice misfit between REBCO and BMO. The *c*-axis of REBCO matrix is extended by the lattice strain and the T_c decreases. The suppression of the T_c reduction is considered to be effective for the increase in J_c of BMO-doped REBCO films. Therefore, the lattice strain in the REBCO matrix and its influence on the T_c reduction should change with the kind of BMO materials due to their different lattice constants.

In this study, we aimed to suppress the T_c reduction of the BMO-doped REBCO films such as $BaZrO_3$ (BZO), $BaSnO_3$ (BSO) and $BaHfO_3$ (BHO). We have investigated the correlations between T_c reduction and several BMO materials for BMO-doped SmBCO films with various BMO contents.

2. Experimental details

All the BMO-doped SmBCO films were deposited on $LaAlO_3$ (100) single crystalline substrates by the conventional PLD method using a KrF excimer laser (λ = 248 nm) at a repetition rate of 10 Hz. The laser energy density, distance between substrate and targets, and O_2 partial pressure during the deposition were 1.7 J/cm^2, 70 mm, and 400 mTorr, respectively. The BMO-doped SmBCO films were deposited by an alternating-targets technique (ALT-PLD) [9, 10]. We fixed substrate temperatures during the deposition of BZO-doped films (940°C), BSO-doped films (840°C), and BHO-doped films (960°C). These temperatures are the optimum temperatures for the straightly growth of each BMO nanorods. We changed the BMO contents in each BMO-doped SmBCO film from 1.0 to 8.3 vol.% in order to change the nanorod density and diameter. All samples were taken O_2 post-annealing at 350°C for 1.5 hours after the deposition. The crystalline orientation of the films was evaluated by x-ray diffraction (XRD) analysis and the microstructure of the films was investigated using high-resolution transmission electron microscope (TEM). The diameter and the number densities of nanorods were measured from the cross-sectional and plan-view TEM images. The BMO content in each film were measured by energy dispersive x-ray spectroscopy (EDX). The resistivity was measured with a physical property measurement system (PPMS: Quantum Design) by standard four-probe method. The T_c was defined from the temperature dependence of the resistivity using a resistivity criterion of 0.1 μΩ cm and the probe current was fixed to about 25 A/cm^2.

3. Results and discussion

Figure 1 (a), (b) and (c) show the cross-sectional TEM images of the 3.0 vol.% BZO-doped, 4.8 vol.% BSO-doped and 4.9 vol.% BHO-doped SmBCO films, respectively. From these TEM images, it was observed that all BMO materials formed nanorods and all nanorods grew straightly and continuously. In this study, we confirmed that nanorods grew straightly and continuously in most of all films with different BMO contents by TEM images and magnetic field angular dependence of J_c (data not shown). Therefore, it is not necessary to consider the influence of nanorods formation on the lattice strain, c-axis expansion and T_c reduction.

Figure 2 shows the correlation between T_c and c-axis length of each BMO-doped SmBCO films with various BMO contents. In all kinds of BMO-doped SmBCO films, the T_cs decreased with increasing of c-axis length. Additionally, these tendencies quantitatively agree for all BMO material. The correlation between T_c and c-axis length of the BMO-doped SmBCO films does not depend on the kind of BMO material.

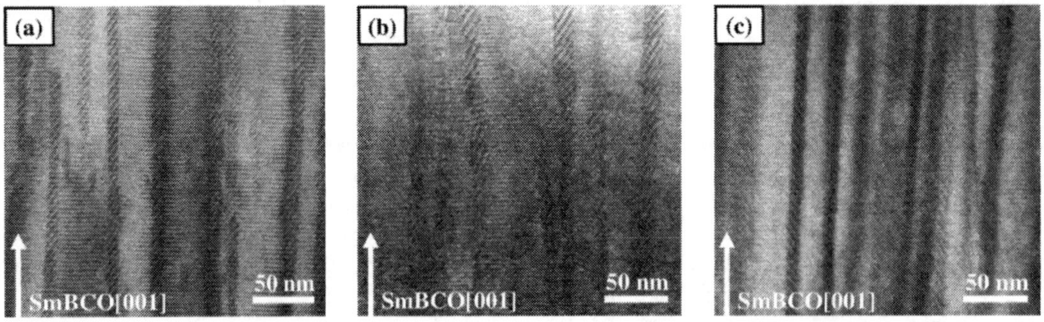

Figure 1. Cross-sectional TEM images of (a) 3.0 vol.% $BaZrO_3$-doped $SmBa_2Cu_3O_y$ film, (b) 4.8 vol.% $BaSnO_3$-doped $SmBa_2Cu_3O_y$ film, and (c) 4.9 vol.% $BaHfO_3$-doped $SmBa_2Cu_3O_y$ film.

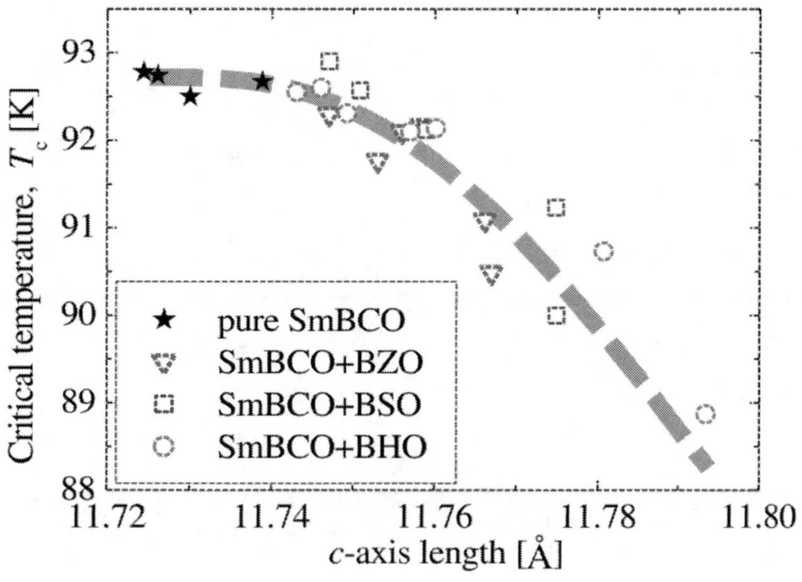

Figure 2. Correlation between T_c and c-axis length of each BMO-doped SmBCO film with various BMO contents.

Furthermore, we investigated the correlation between configuration of nanorods and lattice strain. Figure 3 shows the c-axis lengths versus the interfacial area between BMO nanorods and SmBCO matrix in unit volume and figure 4 shows the c-axis lengths versus the distance between adjacent nanorods interfaces. The c-axis length increases with increasing of interfacial area (fig. 3) or decreasing of distance between nanorods (fig. 4). In addition, the c-axis lengths become longer in the order of BHO-doped films, BZO-doped films and BSO-doped films at the same interfacial area or nanorods distance. From this result, it is considered that the lattice strain in the SmBCO matrix become stronger in the order of BHO, BZO and BSO. The lattice strain should be stronger with increase of lattice misfit $((3c_{BMO}-c_{SmBCO})/c_{SmBCO})$ in the order of BSO (5.4%), BHO (6.7%), and BZO (7.2%). However, these orders were different with our results. We have not clarified the reason of this difference between the order of c-axis expansion and lattice misfit, yet. We anticipate that the lattice strain is composed of the tensile stress by the lattice misfit and stress relaxation by the dislocations. From the results of this study, BHO might be better for the stress relaxation by the dislocations than BZO and BSO in the SmBCO matrix.

4. Conclusion

In order to suppress the T_c reduction of the BMO-doped REBCO films, we have investigated the correlations between T_c reduction and different kinds of BMO material in SmBCO films with various BMO contents. From the comparisons between T_c and c-axis length of each BMO-doped SmBCO film, it was revealed that the correlation between T_c and c-axis length is not depending on the kind of BMO material. Additionally, we investigated the correlation between configuration of nanorods and c-axis lengths. From the results, lattice strain became stronger in the order of BHO, BZO and BSO in both cases focusing on interfacial area and nanorods distance, and the order differed from the order of lattice misfit. In this study, it was considered that the combination of SmBCO and BHO are better for the suppression of the T_c reduction compared with other BMO-doped SmBCO films.

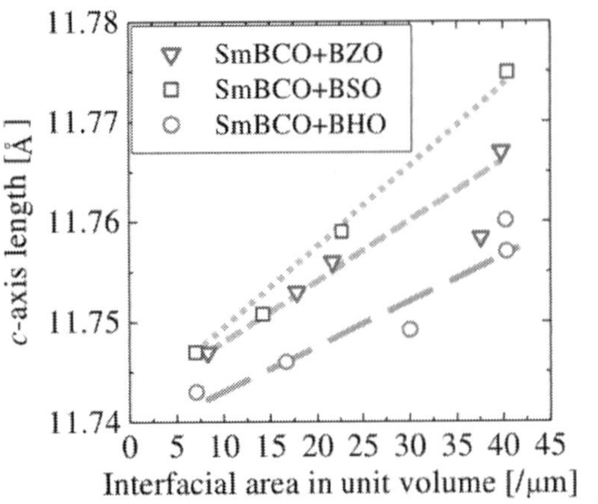

Figure 3. *C*-axis lengths versus the interfacial area between BMO nanorods and SmBCO matrix in unit volume for BMO-doped SmBCO films with various BMO contents.

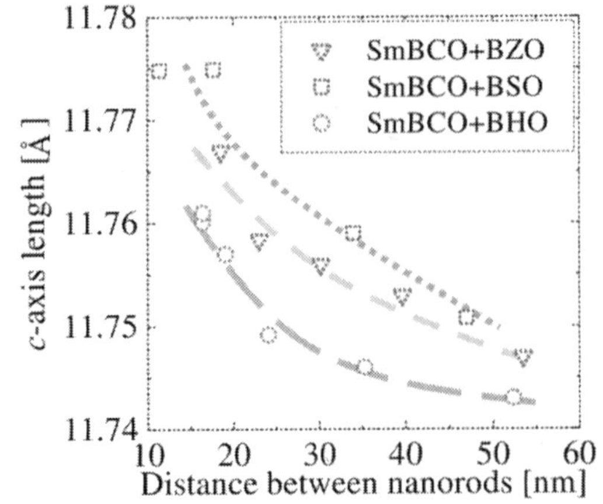

Figure 4. *C*-axis lengths versus the distance between adjacent nanorods interfaces for BMO-doped SmBCO films with various BMO contents.

Acknowledgment

This work was partly supported by Grants-in-Aid for Scientific Research (19676005, 23226014 and 25289358) and Grant-in-Aid for JSPS Fellows (25002829).

References

[1] Larbalestier D, Gurevich A, Feldmann D M and Olyanskii A 2001 *Nature* **414** 368
[2] Broussard P R, Cestone V C and Allen H R 1995 *IEEE Trans. Appl. Supercond.* **5** 1222
[3] Roas B, Hensel B, Saemann-Ischenko G and Schultz L 1989 *Appl. Phys. Lett.* **54** 1051
[4] Miura M, Kato T, Yoshizumi M, Yamada Y, Izumi T, Shiohara Y and Hirayama T 2008 *Appl. Phys. Exp.* **1** 51701
[5] Macmanus J L, Foltyn S R, Jia Q X, Wang H, Serquis A, Civale L, Maiorov B, Hawley M E, Maley M P and Peterson D E 2004 *Nature Materials* **3** 439
[6] Yamada Y, Takahashi K, Kobayashi H, Konishi M and Watanabe T et al. 2005 *Appl. Phys. Lett.* **87** 132502
[7] Mele P, Matsumoto K, Horide T, Ichinose A, Mukaida M, Yoshida Y, Horii S and Kita R 2008 *Supercond. Sci. Technol.* **21** 32002
[8] Tobita H, Notoh K, Higashikawa K, Inoue M, Kiss T, Kato T, Hirayama T, Yoshizumi M, Izumi T and Shiohara Y 2012 *Supercond. Sci. Technol.* **25** 062002
[9] Haugan T, Barnes P N, Maartense I and Cobb C B 2003 *J. Mater. Res.* **18** 2618
[10] Tsurta A, Yoshida Y, Ichino Y, Ichinose A, Matsumoto K and Awaji S 2013 *IEEE Trans. Appl. Supercond.* **23** 8001104

11th European Conference on Applied Superconductivity (EUCAS2013) IOP Publishing
Journal of Physics: Conference Series **507** (2014) 022044 doi:10.1088/1742-6596/507/2/022044

Effect of heat treatment on superconducting properties of FeSe wire fabricated by PIT method

V. Vlasenko[1], K. Pervakov[1], V. Pudalov[1], V. Prudkoglyad[1], V. Berbentsev[2], S. Gavrilkin[1], and Yu. Eltsev[1]

[1]P. N. Lebedev Physical Institute, Russian Academy of Sciences, Moscow 119991, Russian Federation;

[2]Institute for High Pressure Physics, Russian Academy of Sciences, Troitsk 142190, Moscow region, Russian Federation;

E-mail: vlasenkovlad@gmail.com

Abstract.
We report on successful fabrication of superconducting FeSe wire using hot gas extrusion (HGE) ex-situ PIT (powder in tube) method. Length of the obtained wire was about 60cm with cross-sectional superconducting core area nearly 2.5×10^{-3} cm^2. For the wire sample we observed superconducting transition temperature, $T_c^{onset} \approx 11$ K, about 1.2K lower compared to preliminary prepared FeSe powder. Heat treatment in argon atmosphere at 350°C resulted in transition width decrease from $\triangle T(10\% - 90\%) \approx 1.75$ K in sample without heat treatment down to $\triangle T \approx 0.9$ K in annealed samples. Estimated derivative of the upper critical field as a function of temperature of the sample annealed during 72h in argon atmosphere at 350°C is $dH_{c2}/dT \sim -2.9$ T/K. Applying WHH theory to our data allows to define $H_{c2}(0K) = 0.69T_c \times (dH_{c2}/dT) \approx 19.8T$. The untreated wire shows critical current density, $J_c \sim 75$ A/cm^2 at T=4.0 K in zero field. Increasing annealing time up to 72 hours at 350°C in argon atmosphere gives rise to J_c increase of about 60% approaching 120 A/cm^2 at T=4.0K and H=0T . Also J_c measurements were made in magnetic fields up to 9T. Our results show applicability of the HGE PIT method for fabrication of superconducting wires based on FeSe compound. Long-range heat treatment is necessary to improve superconducting properties of the samples.

1. Introduction

Recently discovered iron-based superconductors have superconducting critical temperature (T_c) approaching 55K [1] that is below liquid nitrogen temperature in contrast to high-T_c cuprate superconductors (HTSC) with T_c up to about 130K. However, already earliest studies of iron-based superconductors have shown great potential of these compounds for high-current applications in high magnetic fields. Obvious advantages of iron-based superconductors are the very high upper critical field (H_{c2}) of about 150-200T [2–4], low anisotropy in comparison with HTSC, and high values of the critical current density (J_c), exceeding 10^6 A/cm^2 at 4.2K in zero field [5–8]. FeSe is the simplest iron-containing superconductor from the point of view of its chemical composition. It has tetragonal structure formed by alternating iron and selenium layers [9,10]. T_c of FeSe compound is near 12K at normal conditions [11,12] and increases up to 37K under external pressure of about 5 GPa [13]. FeSe may be easily synthesized compared to other families of iron-based superconductors and, furthermore, does not contain toxic arsenic.

For this reason FeSe may be considered as the model iron-based superconductor to check the possibility to produce superconducting wires and tapes using various methods. Experiments with this compound are useful to get necessary technical experience and also could help to understand the main factors influencing superconducting properties of wires and tapes made using iron-based superconductors. The basic requirement to a quality of the superconducting wire is T_c, J_c, and H_{c2} values close to values of these parameters obtained in bulk samples. The main technique of manufacturing of superconducting wire is the PIT (powder in the tube) method. This method has two different modifications known as in-situ and ex-situ [14]. The first method suggests producing of the superconducting wire from the mixture of initial components, when superconductive phase is obtained inside the wire as a result of heat treatment. The second method requires preparing of preliminary synthesized polycrystalline superconductor. To date, there are already some reports concerning manufacturing of superconducting wires and tapes using FeSe [9, 12, 15]. The highest J_c values up to 10^3 A/cm^2, in 7-core wire were obtained using gas diffusion technique [15]. About two times lower J_c values were reported for 3-core wire produced by chemical transformation PIT process [9].

In this work the single-core superconducting wire on the basis of FeSe was made by ex-situ HGE method. We report on production of the FeSe superconducting wire and its transport properties in fields up to 14T. Our results indicate that the critical current density J_c approaches 120 A/cm^2(T=4.0K, H=0T) after low temperature (350°C) heat treatment in argon during 72h.

2. Experiment and discussion.

The FeSe wire was prepared using the ex-situ HGE PIT technique with steal C22(DIN) sheath in several intermediate steps. Firstly, stoichiometrical mixture of Se and Fe pieces was sealed into quartz tube and synthesized during 120h at temperature near 900°C. The phase composition of the synthesized polycrystalline FeSe was defined by X-ray powder diffraction method. The results of X-ray studies showed that the tetragonal FeSe is a main phase with amount above 90%. Then superconductive material was regrounded, packed into a steel tube of 9.3 mm outer diameter and 3 mm wall thickness and plugged as shown in Fig.1a and Fig.1b by a steel fitting. After packing, the tube with FeSe powder was drawn to wires of 2.2 mm in diameter and 600mm in lengths using HGE method (Fig.1c and Fig.1d) at argon pressure of about 330-340 MPa and temperature of about 950°C. Prepared wire (Fig. 2a) was cut into 1 cm lengths specimens (Fig. 2b), and finally, a few batches of these samples were subjected to heat treatment in argon atmosphere at 350°C during 24, 48, 72, and 96 hours. Fig. 3 shows temperature dependence of zero field resistance of polycrystalline samples without heat treatment and annealed at 350°C during 24, 48, 72, and 96 hours. One can see that the superconducting transition of preliminary prepared polycrystalline sample starts at $T_c^{onset} \approx 12.1K$ with transition width (10%-90%) of about $\triangle T \approx 3^oC$. This value is similar to data obtained in other studies [10, 11]. For our wire samples superconducting transition starts at lower temperature while the transition width gradually decreased to the value below 1K. It should be also noted that with increasing temperature resistive response for all samples with heat treatment up to 72 hours appears at the same temperature $\sim$ 9.5K indicating stability of FeSe phase in our wire samples. Higher heat treatment time resulted in displacement of the resistivity onset to lower temperatures.

Further, we measured temperature dependence of the resistance of our wire samples near superconducting transition in magnetic fields up to 14T (up to 9T using PPMS system and up to 14T with Oxford magnetic system). As an example in Fig.4 we show resistive superconductive transitions of FeSe wire sample annealed during 24h with probing current of 5 mA. Similar results without any systematic dependence on annealing time were obtained for the samples subjected to the heat treatment up to 96 hours at 350°C in argon atmosphere. These measurements clearly show presence of superconductivity in our wire samples in magnetic fields up to 12T at liquid helium temperature 4.2K. From the linear extrapolation of the $H_{c2}(T)$ dependence

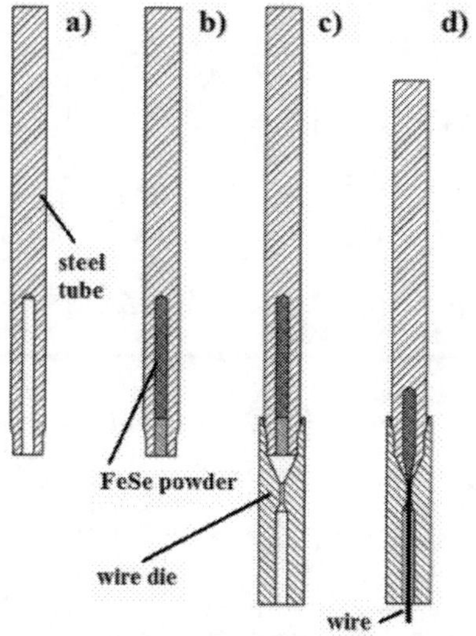

Figure 1. (a) original steel tube; (b) tube with FeSe powder; (c) setup for HGE; (d) HGE process.

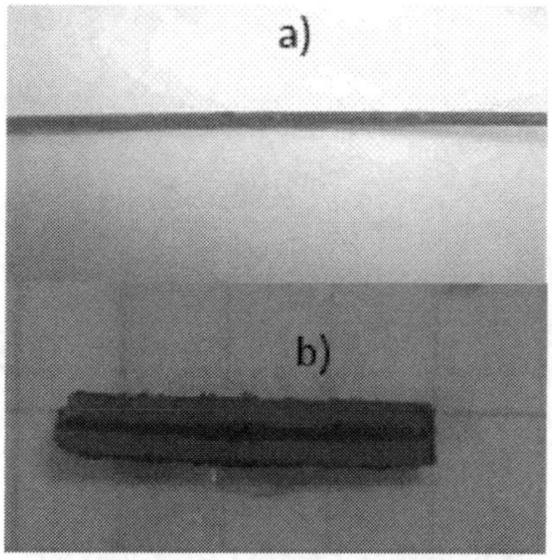

Figure 2. (a) superconducting FeSe wire of 2.2 mm diameter obtained by gas extrusion method; (b) longitudinal cross-section of the FeSe wire.

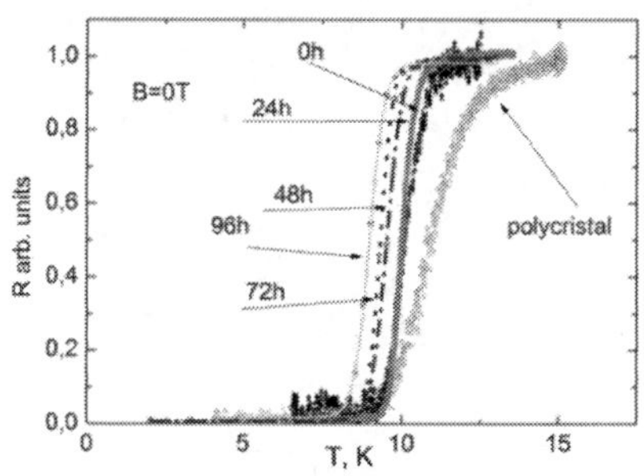

Figure 3. Zero field R(T) dependence for wire FeSe samples with different heat treatment in the region of superconducting transition.

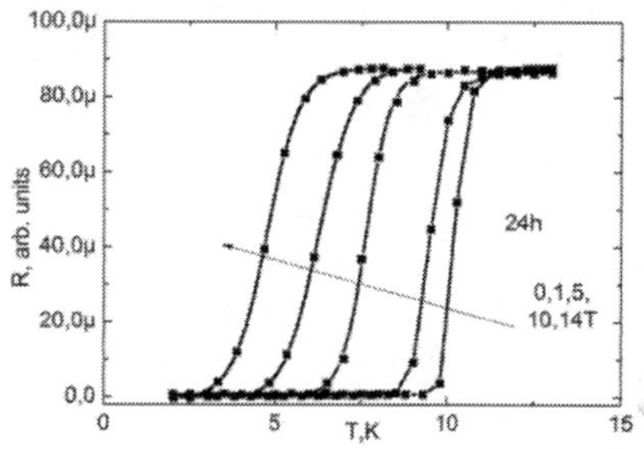

Figure 4. Temperature dependence of resistance below 14K for the as-prepared FeSe wire under magnetic fields up to 14 T.

constructed using mid-point of the resistive superconducting transition $H^*_{c2}(0)$ value nearly 31T may be obtained. Applying the Werthamer, Helfand, Hohenberg (WHH) expression $H_{c2}(0K) = 0.69T_c \times (dH_{c2}/dT)$ [16] we get the value of the upper critical field $H_{c2}(0) = 19.8T$ for our FeSe wire samples.

Transport critical current (I_c) was measured on 1 cm length samples by a standard four-probe resistive method at different temperatures below T_c and applied magnetic fields up to 9T using a Physical Property Measurement System (PPMS). The magnetic field was applied perpendicularly to the wire axis. I_c value was defined with 1 μV/cm criterion. To get J_c, I_c values were normalized to the smallest cross-sectional area of 1 cm sample found after inspection under optical microscope of the cross-section of the wire along its length. In Fig. 5 and Fig. 6 we show $J_c(T)$ dependence, for obtained in our experiment FeSe wires in zero field and 9T field. One can clearly see essential increase of J_c for samples with increasing heat treatment time.

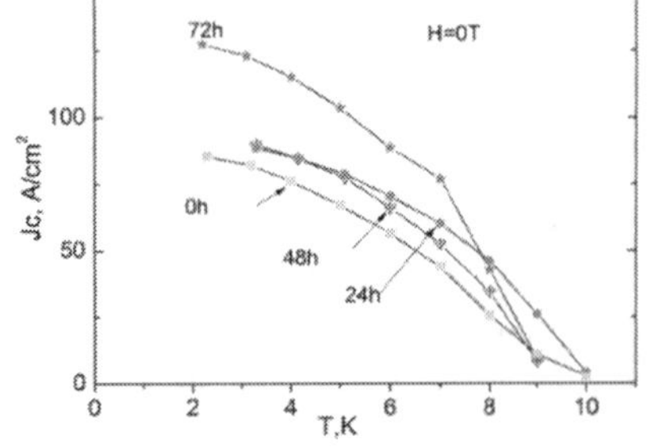

Figure 5. Temperature dependence of critical current density for the obtained single core FeSe wires at zero applied magnetic field.

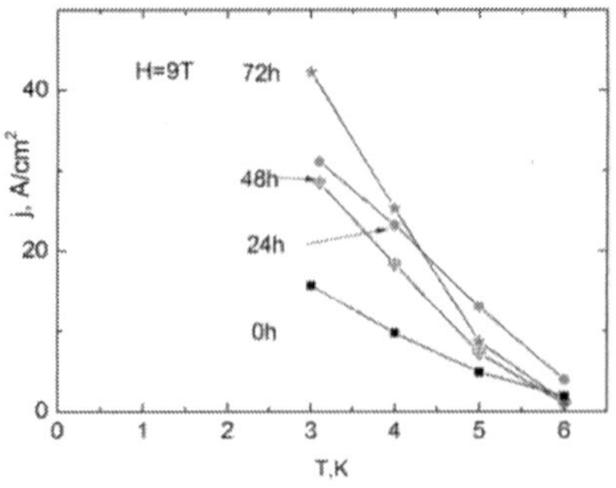

Figure 6. Temperature dependence of critical current density for the obtained single core FeSe wires in magnetic field 9T

In particular, enhancement of J_c about 60% after 72h annealing (T=4.0K, H=0T) indicates important role of heat treatment at 350°C in argon atmosphere to increase critical current density of FeSe wires prepared using HGE method. Our data are below results reported by other groups [9, 12, 15]. The probable reason is porous microstructure of our wires observed under high resolution optical microscope.

In conclusion, in our study we show that PIT technique with hot gas extrusion procedure allows to produce FeSe wires with reasonable superconducting parameters. Long time (up to 72 hours) heat treatment at 350°C in argon atmosphere results in essential increase of J_c of FeSe wire prepared by this method.

3. Acknowledgements

This study was supported by RFBR (grant N 13-02-01180) and by grants of Ministry of Education and Science of the Russian Federation.

References

[1] Y Kamihara, T Watanabe, M Hirano, and H Hosono 2008 *J. Am. Chem. Soc.* **130** 3296
[2] J Jaroszynski et al 2008 *Phys. Rev.* B **78** 174523
[3] Chunlei Wang et al 2011 *Supercond. Sci. Technol.* **24** 065002
[4] E P Khlybov et al 2009 *JETP Lett.* **90** 429
[5] T Katase et al 2011 *Appl. Phys. Lett.* **98** 242510
[6] Y Nakajima, Y Tsuchiya, T Taen, T Tamegai, S Okayasu, M. Sasase 2009 *Phys. Rev.* B **80** 012510
[7] N D Zhigadlo et al 2008 *J. Phys. Condens. Matter* **20** 342202
[8] K S Pervakov et al 2013 *Supercond. Sci. Technol.* **26** 015008
[9] Y Mizuguchi, H Izawa, Ti Ozaki, Y Takano and O Miura 2011 *Supercond. Sci. Technol.* **24** 125003
[10] S I Vedeneev, B A Piot, D K Maude, A V Sadakov 2013 *Phys. Rev.* B **87** 134512
[11] B H Mok et al 2009 *Cryst. Growth Des.* **9** **(7)** pp 32603264
[12] Toshinori Ozaki et al 2012 *J. Appl. Phys.* **111** 112620
[13] Satoru Masaki et al 2009 *J. Phys. Soc. Jpn.* **78** 063704
[14] Yanwei Ma 2012 *Supercond. Sci. Technol.* **25**
[15] Zhaoshun Gao et al 2011 *Supercond. Sci. Technol.* **24** 065022
[16] N R Werthamer, E Helfand, and P C Hohenberg 1966 *Phys. Rev.* **147** 295

Journal of Physics: Conference Series **507** (2014) 022045 doi:10.1088/1742-6596/507/2/022045

Current-carrying capacity of HTS DC cables with the reduced Lorentz force

V.S. Vyatkin, K. Tanabe, E.S. Otabe, M. Kiuchi, T. Matsushita

Faculty of Computer Science and Systems Engineering,
Kyushu Institute of Technology, 680—4 Kawazu, Iizuka, Fukuoka 820—8502, Japan

E-mail: vyatkin@aquarius10.cse.kyutech.ac.jp

Abstract. The structure of superconducting DC cable made with high temperature superconducting tapes is designed to improve the current-carrying capacity by reducing the Lorentz force acting on flux lines. For this purpose an axial magnetic field is produced by the outer shield conductor. The analysis was performed for the optimum structure for various superconducting tapes with different $J_c(B,\theta)$ characteristics. It is found that the current-carrying capacity increases appreciably when the difference in the critical current density of the tape is large enough between normal and parallel magnetic fields.

1. Introduction

It is known that the critical current density dramatically increases when a metallic superconductor is placed in a parallel magnetic field. This is the longitudinal magnetic field effect. However, the critical current density does not appreciably increase in a parallel magnetic field in high temperature superconductors (HTS) due to weak links at boundaries of inhomogeneous distribution of tilted grains. Nevertheless, the critical current of HTS tapes in the parallel magnetic field is higher than that in the normal magnetic field [1-4].

Then, we proposed to use this superior property in a parallel magnetic field for superconductivity DC cables [5]. The optimum structure was investigated based on the critical current property of a typical superconducting tape. However, the optimum structure is different from tape to tape. In this paper we calculated the current-carrying capacity of the cable made with some commercial HTS tapes to investigate the influence of superconducting properties of HTS tapes.

The layout of the tapes in the inner conductor of the cable is shown in Fig.1 [6]. The outermost layer of the cable has the maximal inclination of the superconducting tapes. The axial magnetic field produced by the outer layers makes the magnetic field at the inner layers more axial, resulting in a smaller Lorentz force acting on the flux lines [3]. The current-carrying capacity was calculated by the iteration method similarly to the previous analysis [6].

2. Method of calculation

It is assumed that the external axial magnetic field B_{ext} is applied to the inner conductor of the cable by the current flowing back the outer shield layer. For the estimation of the total current-carrying capacity, the critical current density of each layer, which depends on the local magnetic field, must be determined. That is, the critical current density J_{ci} of i-th layer is formally given by

$$J_{ci} = f(B_i, \theta_i), \qquad (1)$$

where B_i is the strength and θ_i is the angle of the magnetic field from the current direction on the i-th layer. These variables, B_i and θ_i, are functions of the current flowing in each layer. Thus, these must be solved consistently. For this purpose, the iteration method was used for the set of equations (1). The details of the analysis are given in [6].

One of the parameters of the cable is the angle θ_{max}, the maximal inclination of the superconducting outermost layer. The innermost layer is approximated to be parallel to the cable axis, for simplicity. The distribution of the angle θ_i is assumed to be linear in the region between the two layers. It is found that any variation from this linear dependence leads to a decrease in the current-carrying capacity. It is means that the linear dependence is optimal.

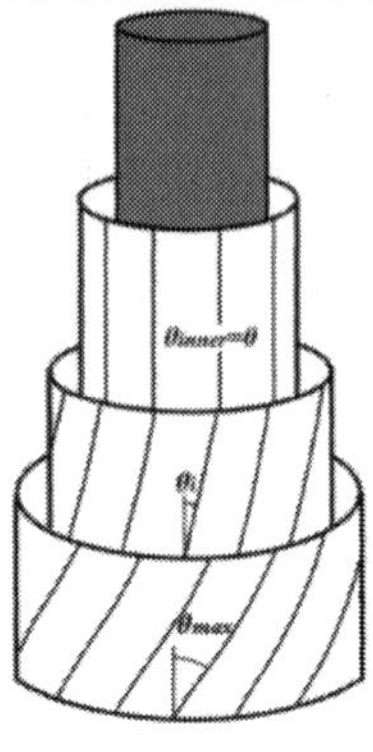

Figure 1. The layout of superconducting tapes in the cable with reduced Lorentz force.

3. Calculated results

We examined 3 superconducting tapes, the critical current properties of which are shown in the leftmost column in Fig.2. Sample C shows a gradual increase in J_c with increasing parallel magnetic field, while J_c's in samples A and B decrease monotonically. The middle and right columns show the calculated current-carrying capacity (I_c) as a function of θ_{max} under various external magnetic fields B_{ext} for the cable of 40mm diameter with 6 and 13 layers, respectively.

In the case of 6-layer cable made with sample A, for example, I_c of the cable increases with increasing θ_{max}, because the external layers create the axial magnetic field for inner layers. When θ_{max} becomes too large, I_c decreases because of a worse efficiency of the transport current, resulting in some maximum capacity at a medium value of θ_{max}. The peak value of the capacity tends to increase with increasing B_{ext} first, and then decreases. The first increase is caused by the increase in the critical current density of the tape due to the change in the field direction, and the latter decrease comes from the decrease in the critical current density in a higher magnetic field. If we compare the graphs in the middle column to the corresponding graphs in the rightmost column, it is found that I_c of the cable enhances when the number of the layers is increased. It can be also shown that the ratio of the current-carrying capacity of the present cable to that of the conventional cable increases with increasing layer number.

4. Discussion

The performance of the proposed cable was also investigated for superconductors with different critical current densities. We focus sample C, because its critical current density increases with increasing parallel magnetic field. Hence sample C seems to be suitable for application to the present cable. Nevertheless, the obtained I_c is not so large. In particular, I_c for this sample decreases with increasing θ_{max} for $B_{ext} = 0$, while those for other samples increase. We want to clarify the reason for this. For this purpose, we approximate $J_c(B)$ characteristics by lines as in Fig. 3:

$J_{cLon}(B) = 7.173 \times 10^9 + a \times 1.232 \times 10^{10} \times B$ A/m^2 (3),

$J_{cAB}(B) = 7.173 \times 10^9 - b \times 2.132 \times 10^9 \times B$ A/m^2 (4),

where $J_{cLon}(B)$ and $J_{cAB}(B)$ are the critical current densities in a parallel and normal magnetic field B. The coefficients a and b are the multiplication factor, and a=b=1 represents the observed results. We assume a cable with 7 layers. Then, the magnetic field inside the cable does not exceed 0.1T under the condition of $B_{ext} = 0$, and the linear approximation holds safely.

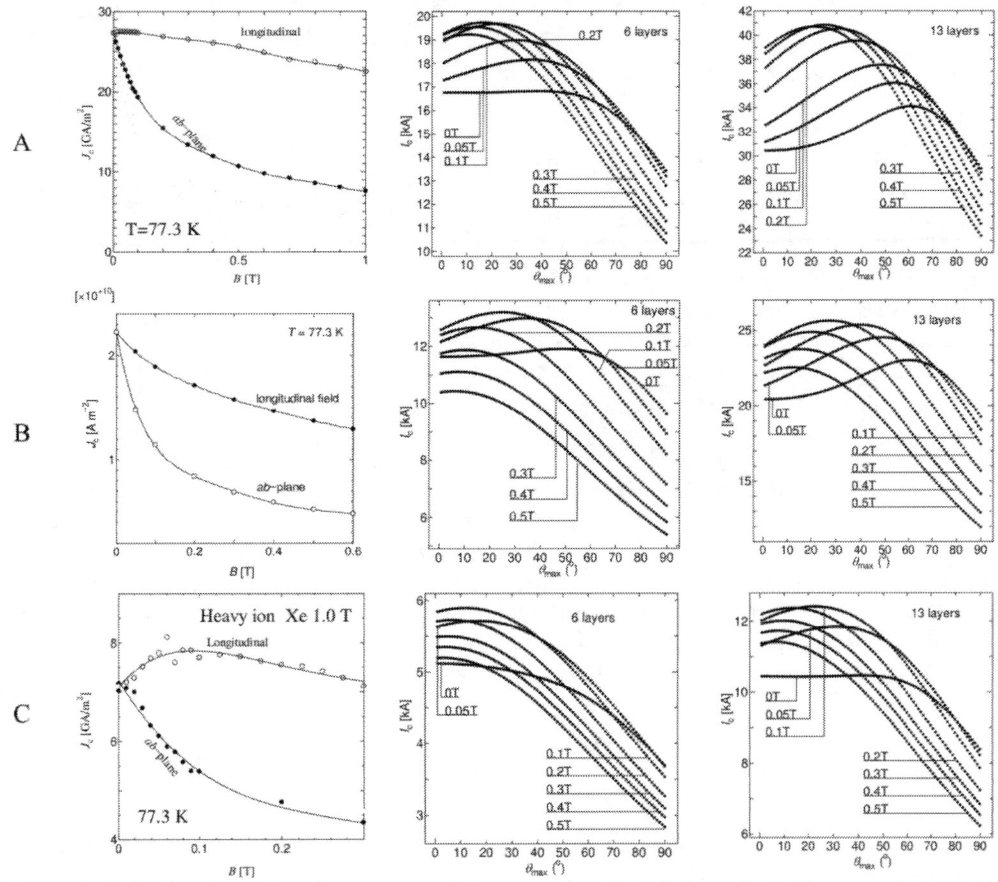

Figure 2. Calculated results of current-carrying capacity (I_c) of the reduced Lorentz force cable using the various superconducting tapes. Left column - J_c vs. parallel (upper line) and normal (lower line) field. Middle and right columns - the current-carrying capacity vs. angle θ_{max} for 6 and 13 layers cable in an external axial magnetic field of 0T -0.5T

It is possible to obtain different characteristics for other samples by changing a and b. Figure 4 shows I_c vs. θ_{max} under $B_{ext} = 0$ for various values of a and b. In Fig. 4 the boundary is also shown between the areas showing positive and negative variation rate of I_c with respect to θ_{max} at small values of θ_{max}. It is found from Fig. 4 that the positive variation rate, a more efficient property, is obtained when the sum of a and b is greater than some value about 4. When this sum becomes large, i.e., when the difference in the two $J_c(B)$'s become large, the variation rate increases appreciably. When a is sufficiently large, the axial magnetic field produced by helical windings sufficiently increases the critical current density of the tape, resulting the positive variation rate. When b is sufficiently large, the magnetic azimuthal field produced by the inner layers decreases. If the outer layer is twisted, the field angle at outer layer decreases, and the critical current increases. As a result, the variation rate becomes positive. Thus, I_c vs. θ_{max} property $B_{ext} = 0$ is not simply determined only by the $J_{cLon}(B)$. In sample C, although a is not large, b is not large. Thus, the variation rate is not positive in sample C.

In a practical cable the optimal structure must be sought under the avail magnetic field B_{ext}.

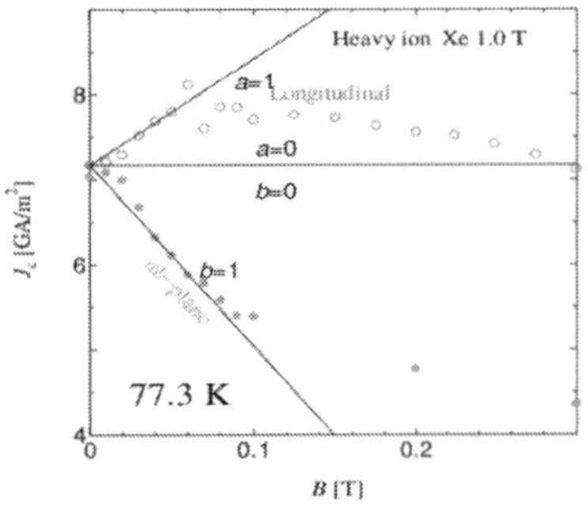

Figure 3. The Fig. 2c dependence with interpolating lines for low magnetic field. The parameters a and b, the increments of $\Delta J_c/\Delta B$ of parallel and normal application of magnetic field will be modified.

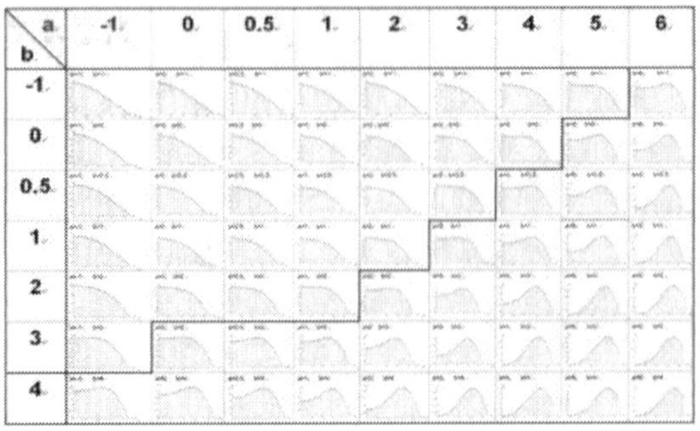

Figure 4. Influence of the inclination of the dependence $J_c(B)$ (parameters a and b) for the efficiency of the cable with reduced Lorentz force.

5. Conclusion

In the present work the new design of HTS cable is investigated. It is demonstrated that I_c of the cable gradually increases with increasing layer number and external magnetic field, and takes a maximum at some external magnetic field. It was found that the characteristics of a cable made with superconducting tape with J_c that increases with parallel magnetic field are not good under the condition of $B_{ext} = 0$. It was shown that the difference of J_c between parallel and normal magnetic fields must be sufficiently large to get good characteristics of current-carrying capacity. In the next investigation the optimum condition including the axial magnetic field B_{ext} must be sought.

6. Acknowledgements

This work is supported by Advanced Low Carbon Technology Research and Development Program (ALCA) of Japan Science and Technology Agency (JST).

References

[1] Sekula S T, Boom R W and Bergeron C J 1963 *Appl. Phys. Lett.* **2** 102
[2] Cullen G W, Cody G D and McEvoy J P Jr 1963 *Phys. Rev.* **132** 577
[3] Cullen G W and Novak R L 1964 *Appl. Phys. Lett.* **4** 147
[4] Bychikov Yu F *et al.*, 1969 *JETP Lett.* **9** 404
[5] T. Matsushita, M. Kiuchi and E.S. Otabe 2012 *Supercond Sci. Technol.* **25** 125009
[6] V.S. Vyatkin *et al.*, 2013 Physica C 494, pp. 135-139.

Effects of YSZ Buffer Layer Surface Morphology on Superconducting Performance of YBCO films Deposited by Pulsed Laser Deposition on NiW Tapes

G N Xiao, L F Liu, D Xu, X Wu, Q Luo and Y J Li[1]
Key Laboratory of Artificial Structures and Quantum Control, Ministry of Education, Department of Physics and Astronomy, Shanghai Jiao Tong University, Shanghai 200240, P.R. China

E-mail: yjli@sjtu.edu.cn

Abstract. YSZ buffer layers and YBCO superconducting layers were grown on biaxially textured NiW tapes by a reel-to-reel pulsed laser deposition (PLD) technique. It was found that orientation and surface morphology of YSZ buffer layers were very sensitive to deposition parameters, especially to oxygen pressure and laser repetition rate. Relationships between microstructure, surface morphology of YSZ buffer layer and PLD deposition parameters were systematically studied using X-ray diffraction (XRD) and scanning electron microscopy (SEM). The effects of surface morphology of YSZ buffer layer on the superconducting performance of YBCO layer were investigated. As increasing oxygen pressure from 0.5 to 10 mTorr, the surface morphology of YSZ buffer layer became better first and then worse, however, it deteriorated with the increase of laser repetition rate. It was found that the surface morphology of YSZ buffer layer had great influence on the superconducting transport properties of YBCO layer. Under optimized experimental condition, high quality YBCO layer with critical current density J_c above 4×10^6 A/cm^2 was obtained at 77 K and self field.

1. Introduction

Second generation high temperature superconductors (2G-HTS) have attracted much interest worldwide due to its potential application prospects in the fields of transportation, power transmission, medical equipment, new energy, and so on. In the past years, great progress has been made in the fabrication of $REBa_2Cu_3O_{7-x}$ (REBCO) tapes with remarkable superconducting performance [1, 2]. Three main technical routes have been developed for the fabrication of REBCO tapes, including inclined substrate deposition (ISD), ion beam assisted deposition (IBAD), and rolling-assisted biaxially textured substrates (RABiTS). Among them, the RABiTS approach has been demonstrated to be one of the most promising routs for industrially producing HTS tapes that it is fast, cost-effective, and easier to control buffer layers growth.

So far, various oxide materials have been employed as a single buffer layer for REBCO coated conductors, for example, $La_2Zr_2O_7$, $LaMnO_3$, and so on [3, 4]. However, two kinds of multilayer buffer architectures of CeO_2/YSZ/CeO_2/Ni and CeO_2/YSZ/Y_2O_3/Ni are still the most stable and widely used buffer-layer architectures. It has been reported that the superconducting performance of YBCO layer was not only dependent on the pure (001) orientation of CeO_2 seed layer, but also very sensitive

[1] To whom any correspondence should be addressed.

to its surface morphology [5]. In this paper, we fabricated a series of YSZ barrier layers on RABiTS tapes by PLD method under different oxygen pressures and laser repetition rates. Relationships between surface morphology of YSZ barrier layer and superconducting transport properties of YBCO layer were studied.

2. Experimental

The substrates used in this study were rolling assisted biaxially textured (RABiTS) Ni-5%W tapes with 10 mm width and 50 μm thickness from EVICO GmbH in Germany. RABiTS tapes were annealed in an atmosphere of forming gas (3%H_2 + 97%Ar) to prevent oxidation of the nickel substrate prior to deposition. Then CeO_2 seed layer, YSZ barrier layer, and CeO_2 cap layer were prepared on RABiTS tapes in a multi-target reel-to-reel pulsed laser deposition (PLD) system, which were reported in previous paper detailedly[6].

Ceramic YSZ target with desired composition of ZrO_2 doped with 8 mol% Y_2O_3 was used for ablation. Laser ablation was performed using a Lambda Physik LPX 220 KrF excimer laser with the wavelength and pulse duration of 248 nm and 20 ns. The incident angle of laser beam was 45°. The laser energy and substrate temperature were 280 mJ and 750 °C. The target-substrate distance and tape speed were 6.5 cm and 6 m/h. The oxygen pressure was varied from 0.5 mTorr to 10 mTorr. The laser repetitio n rate was varied from 80 Hz to 140 Hz.

After the deposition of buffer layers, the YBCO films with thickness of 400 nm were then fabricated on the buffer layers with different YSZ layers at 800 °C under a pure oxygen pressure of 200 mTorr. The laser energy of 350 mJ and laser repetition rate of 60 Hz was used. The YBCO films were quickly cooled to room temperature.

The orientation and texture of the YSZ and YBCO films were analyzed by a Bruker D8 Discover X-ray Diffractometer with a General Area Detector Diffraction System (GADDS), with Cu Kα radiation operated at 40 kV and 40 mA, including θ-2θ scan, in-plane ϕ-scan and out-of-plane ω-scan rocking curve measurements. The surface morphology of the films was determined by a high-resolution field emission scanning electron microscope (FESEM), FEI Sirion 200 with 5 kV accelerating voltage applied to the electron beam.

The critical current (I_c) values were measured at 77 K by the conventional DC four-probe method with the electric field criterion of 1μV/cm. The thickness of YBCO film was measured by a Bruker DektakXT stylus step-profilometer.

3. Results and discussion

Figure 1(a) shows the X-ray θ-2θ scans for YSZ films deposited at different oxygen pressures. All the YSZ films were deposited at the same laser repetition rate of 120 Hz. The diffraction peaks at 33.10° was attributed to the (002) planes of CeO_2 seed layer films. It was observed that all the YSZ films exhibited two diffraction peaks at 2θ equals to 29.78° and 34.74°, which were corresponding to (111) and (002) planes of YSZ cubic structures (JCPDS 30-1468), respectively. As the oxygen pressure increased from 0.5 mTorr to 10 mTorr, the intensity of (002) peak decreased sharply, by contrast, the intensity of (111) peak increased first and then decreased. The in-plane full width at half maximum (FWHM) values of (002) peak were 5.68°, 5.70°, 7.05°, and 7.41° for YSZ films deposited at oxygen pressure of 0.5, 1, 5, and 10 mTorr, respectively.

Figure 1(b) shows the X-ray diffraction θ-2θ scans for YSZ films deposited at different laser repetition rates. All the YSZ films were deposited at the same oxygen pressure of 0.1 mTorr. It can be seen that pure (002) oriented YSZ films were fabricated at the whole laser repetition rate range of 80 Hz to 140 Hz, and the intensity of (002) peak increased with the increase of laser repetition rate. Compared figure 1(a) with (b), it was found that the structure of YSZ was very sensitive to the oxygen pressure. At low oxygen pressure of 0.1 mTorr, the YSZ films had a pure (002) orientation. When the oxygen pressure was above 0.5 mTorr, the YSZ films had mixed (111) and (002) orientations.

Figure 1(c) shows a typical XRD pattern of 400 nm thick YBCO film deposited on the optimized YSZ buffer layer. In the XRD pattern, only (00*l*) peaks from YBCO film were observed, which indicated that the YBCO film had a pure c-axis orientation and high epitaxial quality. The in-plane FWHM value for YBCO (103) was 6.12°.

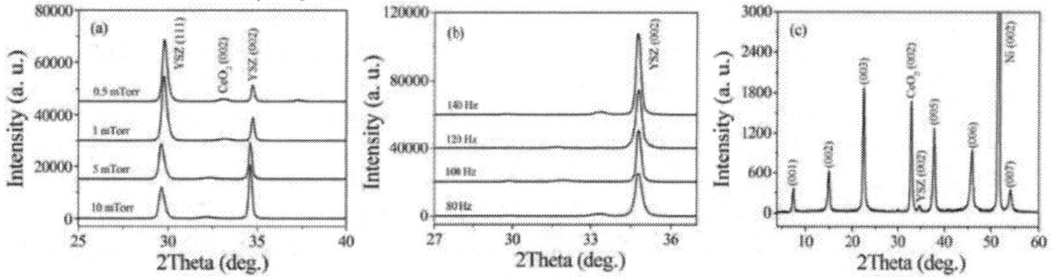

Figure 1. X-ray θ-2θ scans of YSZ films deposited at different (a) oxygen pressures and (b) laser repetition rates, and (c) YBCO film deposited on the optimized YSZ buffer layer.

Figure 2 exhibits SEM images of YSZ films deposited at different oxygen pressures. When the oxygen pressure was 0.5 mTorr, several particles (acicular and hill-and-valley-like structures), as well as cracks were formed on the film surface. The smoother and crack-free surface with little particles developed at 1 mTorr (figure 2b). However, as the oxygen pressure increased above 5 mTorr (figure 2c), rough surface with crack appeared again. Further increase of oxygen pressure to 10 mTorr, the number of particles increased greatly (figure 2d). The rough surface was probably due to the deposition of large ionic clusters. At high oxygen pressure, zirconium ions in the plasma gas can easily react with oxygen molecules because of reduced mean free path of gas particles, which could induce the increase of the number of large ionic clusters.

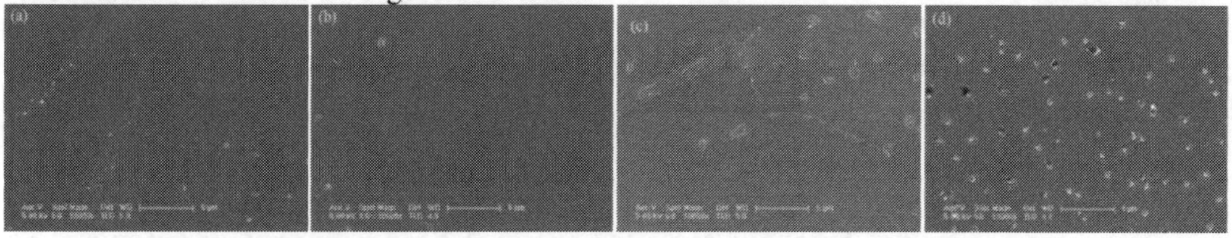

Figure 2. SEM images of YSZ films deposited at different oxygen pressures. (a) 0.5 mTorr, (b) 1 mTorr, (c) 5 mTorr, and (d) 10 mTorr.

Figure 3 exhibits SEM images of YSZ films deposited at different laser repetition rates. It can be seen that all the YSZ films were dense without cracks. However, particles appeared in all YSZ films. With increasing the laser repetition rate from 80 Hz to 140 Hz, the number of the particles increased, and large YSZ particles were easily formed along the Ni grain boundaries. These large particles generally resulted from splashing taking place during the PLD processing [7]. The rapid local heating of the target by the laser beam caused large particulates to splash onto the substrate.

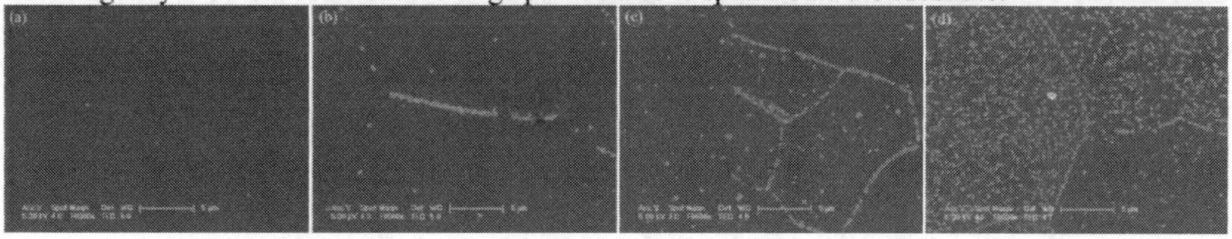

Figure 3. SEM images of YSZ films deposited at different laser repetition rates. (a) 80 Hz, (b) 100 Hz, (c) 120 Hz, and (d) 140 Hz.

YBCO films with thickness of 400 nm were deposited on different YSZ buffer layers under the optimized condition. The critical current (I_c) values of YBCO films were measured at 77 K and self-field by the conventional four-probe method. The critical current density (J_c) was calculated from the cross sectional area of the YBCO layer. I-V curves of the fabricated YBCO films were shown in figure 4. As shown in figure 4a, I_c values of 60 A/cm, 70 A/cm, 65 A/cm, and 36 A/cm were measured for YSZ films deposited at oxygen pressure of 0.5 mTorr, 1 mTorr, 5 mTorr, and 10 mTorr, respectively. As oxygen pressure increased from 0.5 mTorr to 1 mTorr, the J_c increased from 1.5×10^{-6} A/cm^2 to 1.75×10^{-6} A/cm^2. When oxygen pressure was above 5 mTorr , the J_c decreased to 9×10^{-5} A/cm^2 finally. With increasing the laser repetition rate of YSZ buffer layer from 80 Hz to 140 Hz, the I_c increased from 60 A/cm to 180 A/cm, and the corresponding J_c increased from 1.5×10^{-6} A/cm^2 to 4.5×10^{-6} A/cm^2. These results indicated that the superconducting transport properties of YBCO films were largely dependent on the surface morphology of YSZ buffer layers.

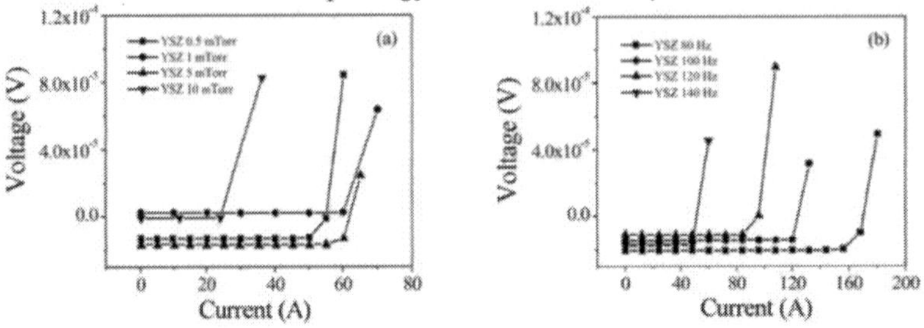

Figure 4. End-to-end I-V characteristics of the YBCO films deposited on YSZ buffer layers with different (a) oxygen pressures and (d) laser repetition rates.

4. Conclusions

In this work, we have investigated the influence of oxygen pressure and laser repetition rate on the structure and surface morphology of YSZ films deposited on RABiTS by PLD. We found that the orientation and in-plane texture of the YSZ films were very sensitive to the oxygen pressure. Pure (002) oriented YSZ films were obtained at 0.1 mTorr. The in-plane texture of YSZ films became better as the oxygen pressure decreased. It was found that the superconducting transport properties of YBCO films were dependent on not only structure but also surface morphology of YSZ buffer layer. As the surface morphology of YSZ buffer layer improved, the I_c and J_c of YBCO films increased. On the optimized YSZ buffer layer with pure (001) orientation and smooth surface, the YBCO film exhibited high c-axis oriented and excellent superconducting properties with J_c of 4.5×10^{-6} A/cm^2 under 77 K and self field.

5. Acknowledgements

This research was sponsored by the Ministry of Science and Technology of China ITER Project (Grant 2011GB113004), the Youth Fund of Natural Science Foundation of China (Grant 11204174), and the Shanghai Science and Technology Committee (Grants 09DZ206000 and 11DZ1100402).

References

[1] Igarashi M, Kakimoto K, Hanyu S, Tashita C, Hayashida T, Hanada Y, Fujita S, Morita K, Nakamura N, Sutoh Y, Kutami H, Iijima Y and Saitoh T 2010 *J. Phys.: Conf. Ser.* **234** 022016
[2] Ko K P, Moon S H, Park C and Yoo S I 2011 *IEEE Trans. Appl. Supercond.* **21** 2965
[3] Caroff T Morlens S, Abrutis A, Decroux M, Chaudouët P, Porcar L, Saltyte Z, Jiménez C, Odier P and Weiss F 2008 *Supercond. Sci. Technol.* **21** 075007
[4] Wang S F, Zhou Y L, Zhao S Q, Fu G S, Jin K J, Cheng B L, Chen Z H, Lu H B and Yang G Z 2006 *J. Supercond. Nov. Magn.* **19** 485
[5] Li Y J, Liu L F, Liu H R, Sun X K, Hong D, Xu D and Wang Y 2011 *IEEE Trans. Appl. Supercond.* **21** 2924
[6] Li Y J Zhao Z C, Liu L F, Ye Q H and Zheng H 2009 *IEEE Trans. Appl. Supercond.* **19** 329
[7] Chrisey D B and Hubler G K 1994 *J. Wiley* pp 14-15

Inductance and Current Distribution Analysis of a Prototype HTS Cable

Jiahui Zhu[1,2], Zhenyu Zhang[2], Huiming Zhang[2], Min Zhang[2,3], Ming Qiu[1], and Weijia Yuan[2,4]

[1] China Electric Power Research Institute, China,
[2] Department of Electronic and Electrical Engineering, the University of Bath, UK,
[3] Department of Engineering, the University of Cambridge, UK
[4] Corresponding author: E-mail: **w.yuan@bath.ac.uk**

This project is partly supported by NSFC Grant 51207146, RAEng Research Exchange scheme of UK and EPSRC EP/K01496X/1.

Abstract: Superconducting cable is an emerging technology for electricity power transmission. Since the high power capacity HTS transmission cables are manufactured using a multi-layer conductor structure, the current distribution among the multilayer structure would be non-uniform without proper optimization and hence lead to large transmission losses. Therefore a novel optimization method has been developed to achieve evenly distributed current among different layers considering the HTS cable structure parameters: radius, pitch angle and winding direction which determine the self and mutual inductance. A prototype HTS cable has been built using BSCCO tape and tested to validate the design the optimal design method. A superconductor characterization system has been developed using the Labview and NI data acquisition system. It can be used to measure the AC loss and current distribution of short HTS cables.

Key words: BSCCO, HTS cable, current distribution, structure parameters, self and mutual inductance, optimal design
(Some figures in this article are in colour in the electronic version)

I. Introduction

High temperature superconducting (HTS) cables are an emerging and attractive technology for modern power grid transmission due to the rapid decline in available underground space for conventional cable and higher power quality requirements of a growing digital economy nowadays, which encourages utilities to seek for HTS cables as an alternative solution.

An important factor in the design of the HTS cable is the AC loss reduction. Since the high power capacity HTS transmission cables are manufactured in the structure of multi-layer conductors for a large current-carrying capability, the current distribution among the multilayer structure is non-uniform without proper optimization design. The AC loss will be increased if the current is not evenly distributed among the HTS conducting layers. Because of the resistance of the HTS conductors is negligible, the current distribution is dominated by the inductance of each conducting layer. For this reason, the self and mutual inductance of HTS cable are investigated in terms of the cable structural parameters: radius, pitch angle and winding direction.

In this paper, a 1 m long, 110 kV/3 kA cold dielectric (CD) HTS cable given in figure 1 has been proposed and mathematically investigated the inductance varying with the radius, pitch angle and winding direction. An optimized short prototype HTS cable has been constructed and current distribution has been tested using Rogowski coils.

II. Inductance calculation method of HTS cable

The structure of the prototype CD HTS cable is consisted of copper former, four HTS conductor layers, electrical dielectric layer and two HTS shielding layers, as shown in the figure 1. Each conductor layer of CD HTS cable is helically wound around on the copper former. Figure 2 gives a schematic diagram of one CD HTS conductor layer in where r is the radius of conductor layer, P is the winding pitch for a single HTS tape, β is the pitch angle of HTS tapes, and w is the width of the HTS tape. From the schematic, it can be seen that the relationship of the winding angle and winding pitch is shown in (1):

$$P = \tan\frac{2\pi r}{\beta} \tag{1}$$

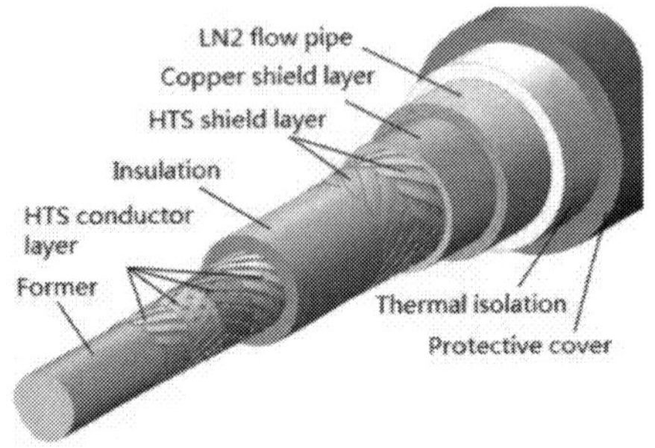

Figure 1. The configuration of CD HTS cable

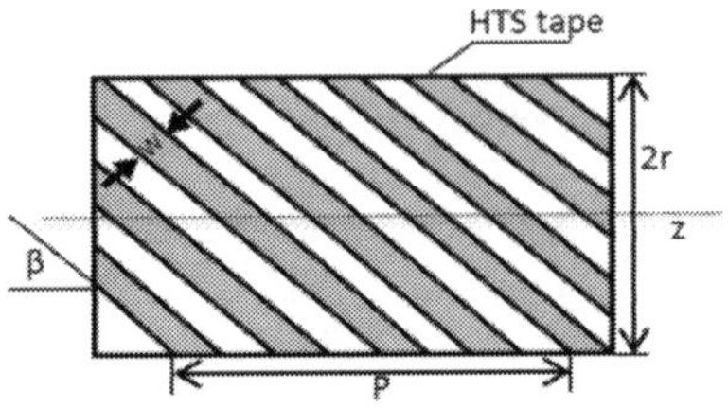

Figure 2. Schematic diagram of CD HTS cable conductor layer

This kind of structure will cause the inductive characteristics for HTS cable varying with cable geometry. The resistance of each HTS conductor layer can be ignored at operating temperature (77 K for YBCO tapes), and the contact resistance in series with each HTS layer is also ignored in the analysis, hence the inductance of each layer can be considered to be the key factor for determining the HTS cable current distribution.

Each HTS conductor layer has a self-inductance as well as a mutual inductance for multi-layer HTS cable. The formulas of self and mutual inductance can be derived in terms of the Ampere's law and enclosed magnetic field energy, which has been published by Olsen[1]. The self-inductance is given by (2)

$$L_s = \mu_0\frac{\pi r_i^2}{P_i^2} + \frac{\mu_0}{2\pi}\ln\left(\frac{D}{R_i}\right); \tag{2}$$

mutual inductance is given by (3)

$$M_{ij} = M_{ji} = \frac{\mu_0}{2\pi}\left(\frac{\alpha_i\alpha_j}{2}\frac{r_i}{r_j}\tan(\beta_i)\tan(\beta_j) + \ln(\frac{D}{r_j})\right); \quad r_i < r_j; \tag{3}$$

Where: L_s is the self-inductance of the layer (H/m); M_{ij} is the mutual inductance between layer i and layer j (H/m); μ_0 is the permeability of free space ($4\pi\times10^7$ T-m/A); R_i is the radius of ith conductor layer; β_i is the pitch angle of ith conductor layer; D is the radius of the shielding layer; i, j is layer number, $i, j = 1, 2, 3, 4, \cdots$.

In the self and mutual inductance formulas, the thickness is considered to be infinitesimal due to the aspect ratio of HTS tape is more than 100. The gap between adjacent filaments of each layer is ignored and a current sheet is assumed. It can be seen that the inductance characteristics are only varied with radius, pitch angle and winding direction.

III. The inductance varying analysis of the HTS cable

A. Radius

For safety reason, the radius of the former can be varied from R_f to 1.2 R_f, and the radii of the other HTS layers are adjusted accordingly. The minimum radius of former is initially set to be 12.25 mm. With an increment step of 0.2 mm, the maximum radius of former is 14.05 mm. The inductance variation of each HTS layer and total inductance of the HTS cable in terms of radius is shown in the figure 3 and figure 4, respectively.

The results show that the self and mutual inductance of each HTS layer are decreasing when the cable radius increases. The total inductance of the HTS cable is linearly decreasing with the cable radius increasing.

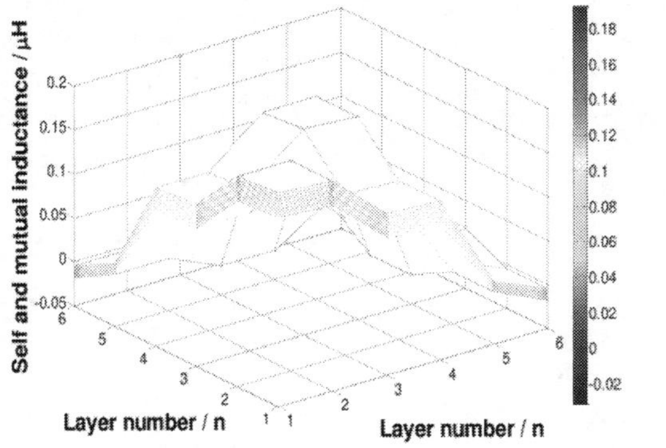

Figure 3. Self and mutual inductance varying with HTS layer radius

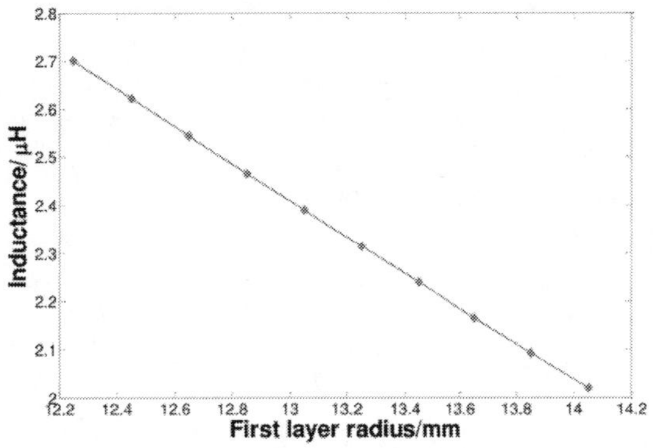

Figure 4. Total inductance of HTS cable varying with radius

B. Pitch angle

The pitch angle of every HTS layer is initially set to be 27^0 while the maximum pitch angle is 81^0 with an increment step of 6^0 applied to every layer accordingly, so as to investigate the inductance variation in terms of pitch angle at fixed radius and winding direction. Figure 5 gives the result of the inductance of first HTS layer varying with pitch angle. It can be showed that the self and mutual inductances of each HTS layer is increasing with increased pitch angle (The negative mutual inductances are due to the opposite winding direction between the two relevant layers, which results into a negative current component induced in the first layer). The remaining layers have the similar results. Figure 6 shows that the total inductance of HTS cable is increased as approximately quadratic function varied with pitch angle.

C. Winding direction

Table 1 shows the 12 groups of winding direction combination for each HTS layer of the cable ('+1' means clock-wise direction and '-1' means anticlockwise direction). The equation (3) shows that only the mutual inductance varies with different winding directions.

To minimize the cable axial flux, the total HTS layer number should be even and the sum of the winding direction coefficient should be equal to zero. The total cable inductance varying with winding direction shows that adjacent layers with opposite winding directions can minimize the mutual inductance, as shown in the figure 7.

IV. The current distribution optimization for HTS cable

The current distributions in each superconducting layer are calculated. Based on the equivalent circuit [2] of HTS cable in figure 1, the mathematical matrix model of the CD HTS cable is

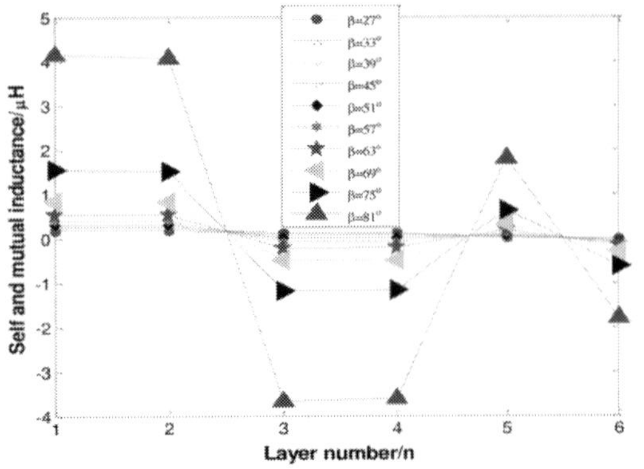

Figure 5. Self and mutual inductance varying with first HTS layer pitch angle

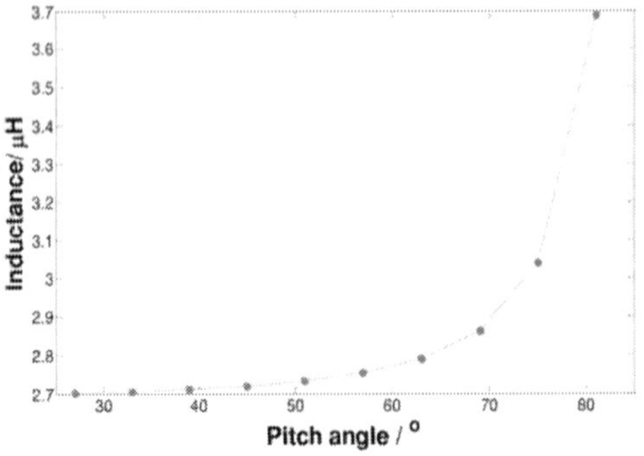

Figure 6. Total inductance of HTS cable varying with pitch angle

Table 1. Winding direction of CD HTS cable

Group no. \ Layer no.	1	2	3	4	5	6
1	1	1	-1	-1	1	-1
2	1	1	-1	-1	-1	1
3	-1	-1	1	1	1	-1
4	-1	-1	1	1	-1	1
5	1	-1	1	-1	1	-1
6	1	-1	1	-1	-1	1
7	-1	1	-1	1	1	-1
8	-1	1	-1	1	-1	1
9	1	-1	-1	1	1	-1
10	1	-1	-1	1	-1	1
11	-1	1	1	-1	1	-1
12	-1	1	1	-1	-1	1

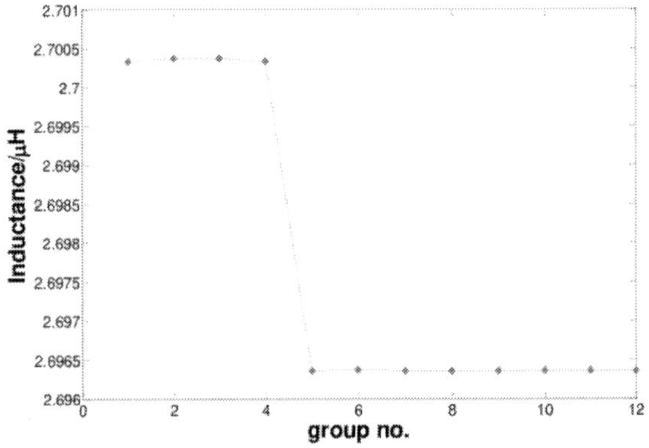

Figure 7. Total inductance of HTS cable varying with winding direction

established in (4) and (5). Where ω is angular frequency, I_{op} is the RMS values of the total operation current and U is the voltage drop across each conductor layer.

The current distribution can be optimized by taking pitch angle and radius as variables to obtain a minimum objective function[3]. The designed variables are presented in (6). The objective function for minimization is presented by (7), which represents the current difference in different conductor layers, where: X is a vector containing the pitch angles β and radii r of each HTS conductor layer, n is the number of total layers, $i,j=1,2,\dots n$, $G(X)$ is the objective function. Detail of the optimized algorithm can be found in [3].

V. Experimental test of current distribution of CD HTS cable

A 0.2 m, 132 kV/1.2 kA prototype HTS two-layer cable has been optimally designed for each of its HTS layer to carry the same transport current, as shown in figure 8. The cable is composed of two BSCCO HTS conducting layers wrapped around a copper former with 9 mm in radius. The pitch angle for each layer is 20^0 and 11^0, respectively. The current of each HTS layer can be measured by Rogowski coils wrapped around the copper current leads connected with each HTS layer, shown in the figure 9. The data are recorded by a high accuracy NI acquisition and Labview system. The measurement result is presented in figure 10. [4]

$$\begin{bmatrix} \mathbf{U} \\ \vdots \\ \mathbf{U} \\ 0 \\ \vdots \\ 0 \end{bmatrix} = j\omega \begin{bmatrix} L_1 & \cdots & M_{1,m} & M_{1,m+1} & \cdots & M_{1,n} \\ \vdots & \ddots & \vdots & \vdots & \ddots & \vdots \\ M_{m,1} & \cdots & L_m & M_{m,m+1} & \cdots & M_{m,n} \\ M_{m+1,1} & \cdots & M_{m+1,m} & L_{m+1} & \cdots & M_{m+1,n} \\ \vdots & \ddots & \vdots & \vdots & \ddots & \vdots \\ M_{n,1} & \cdots & M_{n,m} & M_{n,m+1} & \cdots & L_n \end{bmatrix} \begin{bmatrix} \mathbf{I}_1 \\ \vdots \\ \mathbf{I}_m \\ \mathbf{I}_{m+1} \\ \vdots \\ \mathbf{I}_n \end{bmatrix} \tag{4}$$

$$\mathbf{I}_1 + \cdots + \mathbf{I}_m = -\left(\mathbf{I}_{m+1} + \cdots + \mathbf{I}_n\right) = \mathbf{I}_{op} \tag{5}$$

$$X = [\beta_1, \beta_2, \dots, \beta_n, r_1, r_2, \dots, r_n] \tag{6}$$

$$G(X) = \min F(X) = \min\left(\sum_{i=1}^{n-1} \sum_{j=i+1}^{n} \left|I_i(X) - I_j(X)\right|\right) \tag{7}$$

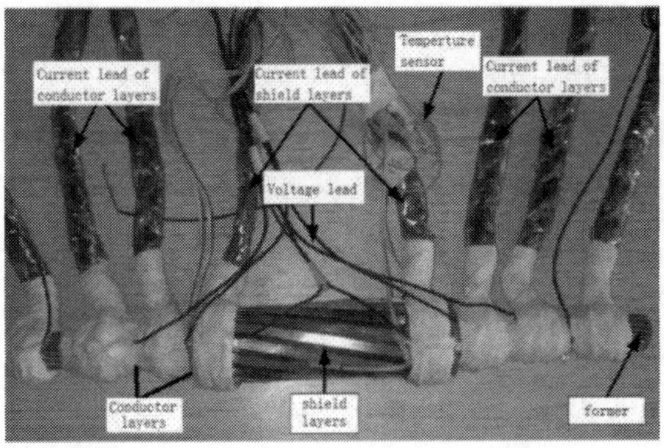

Figure 8. 0.2 m, 132 kV/1.2 kA prototype HTS cable

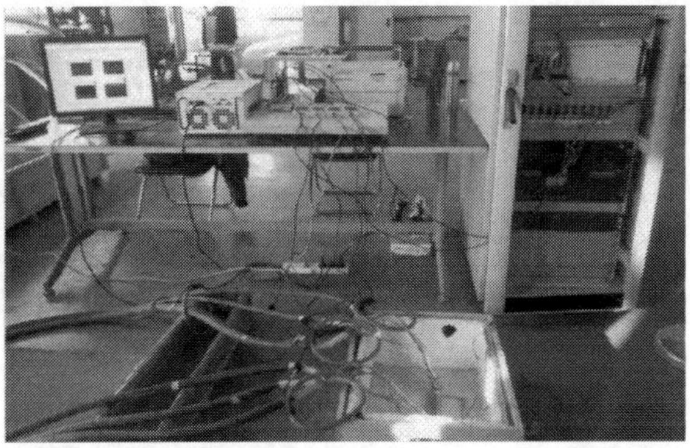

Figure 9. Current distribution measured by Rogowski coils

The testing current of each layer is not exactly evenly distributed partly due to the inaccurately controlled pitch angle during the cable constructing process, and the optimizing procedure also introduce errors. The optimized algorithm is based on the self and mutual inductance formulas which the gaps between filaments of each layer is ignored in the inductance calculation.

The contact resistance between the copper current leads and HTS layers is comparable to the inductance of HTS layers in the real constructed cable, hence the contact resistance also causes current distribution unequally.

VI. Conclusions

The self and mutual inductances of multi-layer HTS cables have been investigated in terms of the radii, pitch angles and winding directions of each HTS layer and inductance variation is summarized. It is found that the geometry of the cable is not fully considered regarding to the gaps between filaments, which will introduce errors for accurately optimizing current distribution. Hence, this method can be only used in the approximate current distribution analysis.

Furthermore, a 0.2 m-long, 132 kV/1.2 kA prototype two-layer HTS cable has been optimally designed and constructed. The current distribution of the two-layer HTS BSCCO prototype cable has been measured to validate the optimal design method. The current is almost evenly distributed as in the design. This experiment can be further improved by removing the effect of contact resistance between the copper current leads and HTS layers.

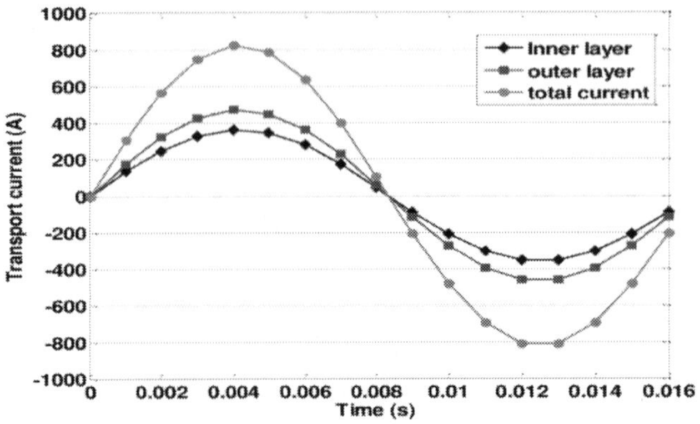

Figure 10. Current distribution testing results of HTS cable at 60 Hz

References

[1] S. Kriiger Olsen, C. Trieholt, A. Kiihle, 0. Tsnnesen,"Loss and Inductance Investigations in a 4-layer Superconducting Prototype Cable Conductor," *IEEE Trans. Appl. Supercond.*, vol.9, no.2, pp 833-836, 1999

[2] Zhixuan Wang, et. al., "Design of CD HTS power cable," *International Conference on Trans. Appl. Supercond. And Electromagnetic Devices. 2009:64-67*

[3] Jiahui Zhu, Xuzheng Bao, Lijie Guo, Zhanjun Xia, Ming Qiu, and Weijia Yuan, "Optimal Design of Current Sharing in Transmission Conductors of a 110 kV/3 kA Cold Dielectric Superconducting Cable Consisted of YBCO Tapes," IEEE Trans. Appl. Supercond., vol.23, no.3, p 5402505, 2013

[4] Jiahui Zhu, Zhenyu Zhang, Huiming Zhang, Min Zhang, Ming Qiu, and Weijia Yuan, "Electric Measurement of the Critical Current, AC Loss, and Current Distribution of a Prototype HTS Cable," *IEEE Trans. Appl. Supercond.*, VOL. 24, NO. 3, June 2014

On the use of copper-based substrates for YBCO coated conductors

A Vannozzi[1,1], F Fabbri[1], A Augieri[1], A Angrisani Armenio[1], V Galluzzi, A Mancini[1], F Rizzo[1], A Rufoloni[1], J A Padilla[2], E Xuriguera[2,3], D De Felicis[4], E Bemporad[4] and G Celentano[1]

[1] ENEA Frascati Research Centre, Superconductivity Laboratory, Fusion Technical Unit, Via E. Fermi 45, I-00044 Frascati, Rome, Italy
[2] Department of Materials Science and Metallurgical Engineering, Universitat de Barcelona, Martí i Franquès 1, E-08028 Barcelona, Spain
[3] La Farga Lacambra, Ctra. C17z km 73.5, E-08508 Les Masies de Voltregà, Barcelona, Spain
[4] University "Roma Tre", Engineering Department, Via della Vasca Navale 79, I-00146 Rome, Italy

E-mail: angelo.vannozzi@enea.it

Abstract. It is well known that the recrystallization texture of heavily cold-rolled pure copper is almost completely cubic. However, one of the main drawbacks concerning the use of pure copper cube-textured substrates for YBCO coated conductor is the reduced secondary recrystallization temperature. The onset of secondary recrystallization (i.e., the occurrence of abnormal grains with unpredictable orientation) in pure copper substrate was observed within the typical temperature range required for buffer layer and YBCO processing (600–850 °C). To avoid the formation of abnormal grains the effect of both grain size adjustment (GSA) and recrystallization annealing was analyzed. The combined use of a small initial grain size and a recrystallization two-step annealing (TSA) drastically reduced the presence of abnormal grains in pure copper tapes.
Another way to overcome the limitation imposed by the formation of abnormal grains is to deposit a buffer layer at temperatures where secondary recrystallization does not occur. For example, $La_2Zr_2O_7$ (LZO) film with a high degree of epitaxy was grown by metal-organic decomposition (MOD) at 1000 °C on pure copper substrate. In several samples the substrate underwent secondary recrystallization. Our experiments indicate that the motion of grain boundaries occurring during secondary recrystallization process does not affect the quality of LZO film.

1. Introduction

The use of pure copper biaxially textured substrate for $YBa_2Cu_3O_{7-\delta}$ (YBCO) coated conductor was studied by some authors during the last ten years [1–3]. In fact, pure copper shows some desirable characteristics such as an almost perfect cube texture development, a high electrical and thermal conductivity and a diamagnetic behaviour [2, 4, 5]. Moreover, copper is not expected to poison YBCO

[5] To whom any correspondence should be addressed.

Journal of Physics: Conference Series **507** (2014) 022048 doi:10.1088/1742-6596/507/2/022048

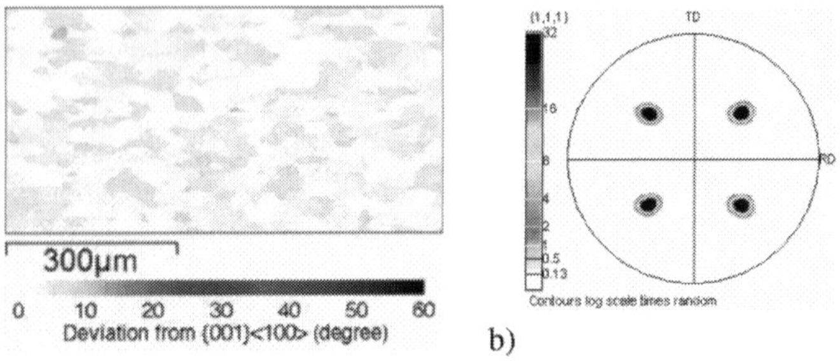

Figure 1. EBSD misorientation map (a) and (111) pole figure (b) of ETP Cu sample (batch B1) recrystallized for 1 h at 900 °C.

during film deposition. On the other hand, the use of a pure copper substrate for YBCO coated conductor is challenging due to the occurrence of abnormal grain growth at low temperature, the poor mechanical properties and the ease of oxidation. While the last two issues are dealt with by the cited authors, there is no investigation concerning the former, which will be addressed in the present contribution.

2. Experimental details

High purity copper bars (either electrolytic tough pitch, ETP or oxygen-free high conductivity, OFHC) were annealed for 1 h in high vacuum at 500 °C. Grain size adjustment (GSA) was performed by cold-rolling with a deformation degree above 50% and annealing in high vacuum for 10 min in the temperature range 400–500 °C. The final cold-rolling was performed using a 4-high rolling mill equipped with DLC-coated steel cylinders with a radius R=24 mm as working rolls, reaching a deformation degree higher than 97% and a final thickness in the range 50–100 μm. Cube texture was obtained by annealing as-rolled sample either in high vacuum or in reducing atmosphere. $La_2Zr_2O_7$ (LZO) film was grown on cube-textured copper substrate by metal-organic decomposition (MOD) as described by Augieri *et al.* [6], using a temperature plateau at 1000 °C.

Structural and morphological properties of the samples were analysed by means of X-ray diffraction and scanning electron microscopy (SEM). The X-ray θ-2θ and ω-scans were performed using a Rigaku Geigerflex diffractometer working with Cu-Kα radiation. A LEO 1525 field emission, high-resolution SEM equipped with an Oxford INCA Crystal electron backscattering diffraction (EBSD) system was used for microstructural investigation. No tilt correction was applied. The area fraction of cube-oriented grains A_{cub} was evaluated from EBSD maps by counting the points falling within 12° from the ideal (001)[100] orientation. Cross sections were obtained by focused ion beam (FIB) technique using a FEI Helios NanoLabTM 600. A protective platinum layer (1 μm thick) was deposited before sample ion-milling.

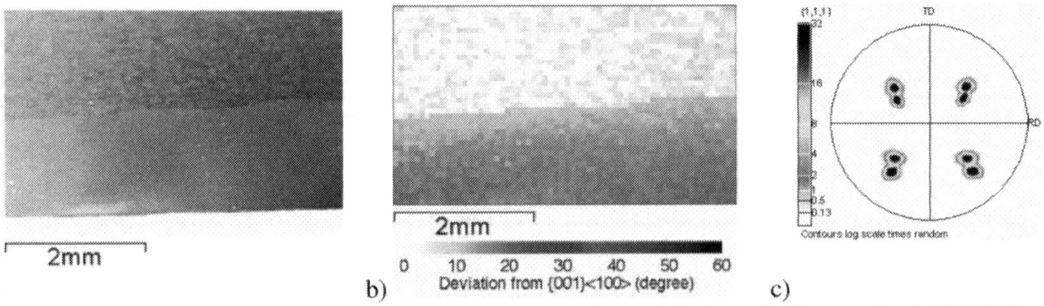

Figure 2. FS-SEM image (a), EBSD misorientation map (b) and (111) pole figure (c) of OFHC Cu sample (batch A) recrystallized for 1 h at 850 °C. The lower half of the sample consists of a single, abnormal grain. In the pole figure, lower poles refer to that grain.

Table 1. Onset temperature of secondary recrystallization T_{2R} observed in different high-purity copper batches.

Batch	Type	T_{2R} (°C)
A	OFHC	850
B1	ETP	> 900
B2	ETP	600
B3	ETP	850
C	OFHC	700

3. Results and discussion

Cold-rolled pure copper is known to develop an almost perfect cube texture as primary recrystallization texture. In particular, it is found that a purity as low as 99.9% is good enough to obtain a sharp cube texture (Figure 1). However, abnormal grain growth caused by secondary recrystallization occurring at high temperature are often observed. The orientation of abnormal grains is in general unpredictable and they can grow as large as several mm, and even though in pure copper such grains often show an almost {100}<001> orientation, as shown in Figure 2, they are sometimes twinned, thus making the tape useless as oriented substrate for YBCO coated conductor. The onset temperature at which abnormal grains develop was observed is as low as 600 °C, depending on the batch, and does not seem to be correlated with the purity of raw material (Table 1). Rather, experiments indicate that the development of abnormal grains can be related to the thermomechanical process followed. In particular, it seems to depend on both the initial grain size and the thermal treatment, namely the use of a single or two-step (TSA) annealing [7]. Two different grain sizes were obtained by annealing a pre-rolled ingot for 10 min at either 400 or 500 °C (GSA). As can be seen in Table 2, the combination of low-temperature annealing of pre-rolled ingot, i.e. smaller initial grain size, and TSA with high final temperature permits to obtain fully cube textured substrates with no abnormal grain development.

On the other hand, the formation of abnormal grains seems to prevent the use of chemical solution deposition (CSD) method for buffer layer deposition on pure copper substrate, since the conversion temperature of most of the suitable buffer layer materials is typically quite high. However, the nucleation temperature of CSD oxide films is generally much lower. As an example, $La_2Zr_2O_7$ (LZO) deposited by MOD nucleates in the temperature range 700–750 °C [8]. As a result, LZO was successfully grown on copper substrate using batches A, B1 and B3 (Table 1, Figure 3a).

Table 2. Influence of the thermal treatment used in both grain size adjustment (GSA) and recrystallization two-step annealing (TSA) on the final texture of Cu OFHC (batch C) samples. A_{cub}=fraction of cube-oriented area within 12° from the ideal cube orientation; AG= occurrence of abnormal grain.

GSA	TSA	A_{cub} (%)	AG
No	800 °C - 60 min (single step)	–	Yes
500 °C - 10 min	800 °C - 60 min (single step)	98.2	No
400 °C - 10 min	300 °C - 20 min + 800 °C - 40 min	>99	No
400 °C - 10 min	250 °C - 10 min + 700 °C - 50 min	>99	No
500 °C - 10 min	250 °C - 10 min + 700 °C - 50 min	98.7	No
400 °C - 10 min	250 °C - 10 min + 600 °C - 50 min	97.1	No
500 °C - 10 min	250 °C - 10 min + 600 °C - 50 min	95.2	No
400 °C - 10 min	250 °C - 10 min + 550 °C - 50 min	89	No
400 °C - 10 min	350 °C - 10 min + 550 °C - 50 min	84.9	No

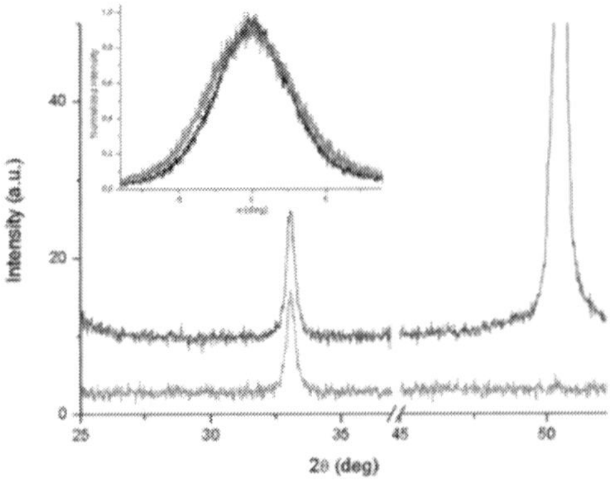

a)

b)

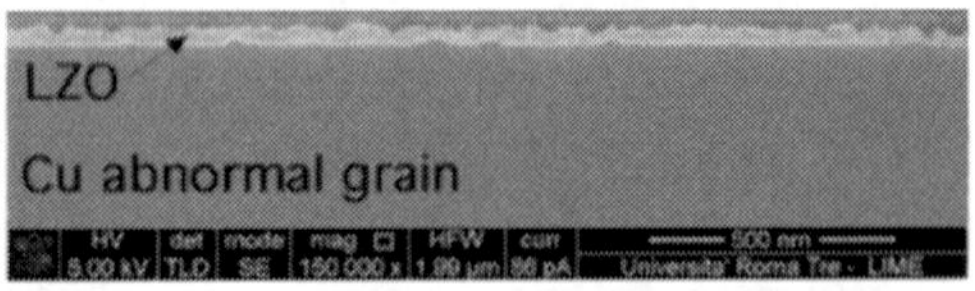

Figure 3. a) X-ray θ-2θ spectrum of MOD LZO films grown on copper substrate at 1000 °C, with (bottom, red) or without (top, black) secondary recrystallization in the copper substrate; in the inset, the respective rocking curves in transverse direction of (004)LZO peak; b) STEM cross-section of the former sample. Recrystallization was performed by annealing copper substrate (batch A) for 1 h at 800 °C.

Remarkably, in several LZO/Cu samples the substrate underwent secondary recrystallization. This can be deduced by the absence of (200)Cu peak in the θ-2θ spectrum. Nevertheless, the structural properties of LZO film are not altered by the secondary recrystallization occurred in the substrate. In fact, in the two cases (400)LZO ω-scans are quite similar. Moreover, both the formation and the migration of secondary recrystallization grain boundary do not affect the LZO film, since neither cracks nor delamination could be observed (Figure 3b). This result indicates that the occurrence of abnormal grains can be neglected if the first buffer layer is deposited below the onset temperature of secondary recrystallization.

Conclusion

The stability of primary recrystallization in pure copper substrate was studied. It is shown that the drawbacks due to the occurrence of abnormal grains at high temperature can be avoided either by secondary recrystallization suppression by controlling the thermomechanical process, or by the use of buffer layer grown at low temperature, i.e. when secondary recrystallization will not occur. Experiments performed with MOD LZO show that grain boundary motion occurring during secondary recrystallization does not cause any delamination or crack of the buffer layer.

Acknowledgments

This work was partially supported by the MIUR through the FIRB project SURE:ARTYST and by the European Community's Seven Framework Programme through the project EUROTAPES. We are grateful to F Maierna and S Chiarelli for technical assistance. We are also grateful to A Concustell and M Tristany for their initial work.

References

[1] Aytug T *et al.* 2003 *J. Mater. Res.* **18** 872
[2] Cantoni C *et al.* 2003 *J. Mater. Res.* **18** 2387
[3] Kashima N *et al.* 2009 *IEEE Trans. Appl. Supercond.* **19** 3299
[4] Varanasi C V *et al.* 2006 *Supercond. Sci. Technol.* **19** 85
[5] Kim K *et al.* 2006 *Supercond. Sci. Technol.* **19** R23
[6] Augieri A *et al.* 2013 *IEEE Trans. Appl. Supercond.* **23** 6600505
[7] Subramanya Sarma V *et al.* 2004 *Scripta Mater.* **50** 953
[8] Knoth K *et al.* 2007 *Acta Mater.* **55** 517

Design of an MgB_2 race track coil for a wind generator pole demonstration

Asger B. Abrahamsen[1], Niklas Magnusson[2], Bogi B. Jensen[3], Dong Liu[4] and Henk Polinder[4]

[1]Department of Wind Energy, Technical University of Denmark, Denmark
[2]SINTEF Energy Research, Norway
[3]Department of Electrical Engineering, Technical University of Denmark, Denmark
[4]Electrical Power Processing Group, Delft University of Technology, Netherlands

Corresponding author's e-mail[1]: asab@dtu.dk

Abstract. An MgB_2 race track coil intended for demonstrating a down scaled pole of a 10 MW direct drive wind turbine generator has been designed. The coil consists of 10 double pancake coils stacked into a race track coil with a cross section of 84 mm x 80 mm. The length of the straight section is 0.5 m and the diameter of the end sections is 0.3 m. Expanded to a straight section of 3.1 m it will produce about 1.5 T magnetic flux density in the air gap of the 10 MW 32 pole generator and about 3.0 T at the edge of the superconducting coil with an operation current density of the coil of 70 A/mm^2.

1. Introduction

Offshore wind power demands large turbines to drive down the cost of energy. In this respect, superconducting direct drive generators may be advantageous due to a better size scaling compared to permanent magnet technology for ratings above 10 MW [1]. Several superconductors are considered for use in the generator rotor, e.g. high-temperature superconducting (HTS) YBCO tapes (coated conductors) [2] and low-temperature superconducting (LTS) NbTi [3]. With a critical temperature of 39 K, the superconductor MgB_2 can be operated in the10-20 K range, well above the usual operation temperature of 4.2 K given by the boiling point of liquid helium for the LTS, and below the operation window of 25-27 K for cooling with liquid neon for the HTS. MgB_2 tapes and wires are produced by the Powder-In-Tube (PIT) method, which is simpler and cheaper than the thin film technology used for the coated conductors. Furthermore, the up-scaling to single piece lengths in the order of km is easier than for the coated conductors. This makes MgB_2 wires and tapes candidates for the field coils of wind turbine generators, although the large scale and in-field performance of the MgB_2 technology still needs verification.

2. Demonstration coil design

We have used MgB_2 engineering current densities of a tape from Columbus Superconductors in the order of 100-200 A/mm^2 in a magnetic field of 2-3 T, corresponding to operation temperatures of 10-20 K [4], to design a race track coil with a straight section of 0.5 m and bending diameter in the end section of about 0.3 m. The coil is targeted at a demonstration in a Ø 1.2 m cryostat equipped with a cryocooler providing a base temperature of about 10 K [5]. Conduction cooling is employed to extract

Table 1. Layout of the MgB_2 race track coil.

Length strait section: L_{strait} [m]	0.5	Opening inside pancake: W_{coil}[m]	0.3
Radius of end winding: R_{end}[m]	0.15	Turns in pancake layer: N	100
Thickness of tape: $t_{SC,\,tape}$ [mm]	0.7	Width of tape: $W_{SC,\,tape}$ [mm]	3.0
Insulation thickness: t_{insul} [mm]	$2\cdot 0.07 = 0.14$	Space between pancakes: W_{insul} [mm]	1
Coil winding thickness: t_{coil}[mm] [a]	84	Double pancake height: H_{DPcoil}[mm] [b]	7
Tape in pancake: $L_{singlepancake}$[m] [c]	220.8	Tape in double pancake: L_{DP} [m]	441.6
Pancake spacing: $L_{coilspace}$ [mm]	1	No. of double pancakes: $N_{pancakes}$	10
Field coil height: H_{coil} [mm] [d]	80	Total tape usage: $L_{tapetotal}$ [m]	4416
Field coil width: W_{coil} [mm]	84	Coil filling factor: f_{coil} [%] [e]	62.5

[a] $t_{coil} = N\cdot(t_{SC,tape} + t_{insul}) = N\cdot t_{wire}$
[b] $H_{DPcoil} = 2\cdot W_{SC,tape} + W_{insul}$
[c] $L_{singlepancake} = N\cdot[2\cdot L_{strait} + 2\pi R_{end}] + 2\pi\cdot t_{wire}\cdot N(N+1)/2$
[d] $H_{coil} = N_{pancakes}\cdot[H_{DPcoil} + L_{coilspace}]$
[e] $f_{coil} = A_{SC}/A_{coil}$

the heat from the coil. Table 1 lists a coil layout based on a series of stacked double pan-cake coils insulated with Kapton tape and impregnated with Stycast 2850. The amount of tape needed for the demonstration is about 4.5 km.

3. Direct drive 10 MW generator

A study of a corresponding 10 MW direct drive wind turbine generator has been conducted by keeping the cross section dimension of the coil constant, but to expand the length and increase the number of poles in the generator to comply with the specifications of the 10 MW INNWIND.EU reference turbine [6]. Figure 1 shows the magnetic flux density distribution in the cross section of the pole at a coil current density, J_{coil}, of 70 A/mm^2. (Note that J_{coil} is the current density of the wire times the coil fill factor). The generator consists of a non-magnetic support of the race track coils, a cryostat wall, an air-cored copper stator at ambient temperature and back-iron to contain the flux. The magnetic flux density at the coil will reach about 2.9 T in the straight section and about 3.0 T at the end section as shown in figure 2. The corresponding load line of both the straight and the end section of the coil are shown in figure 3 along with engineering current densities of the tape at different temperatures.

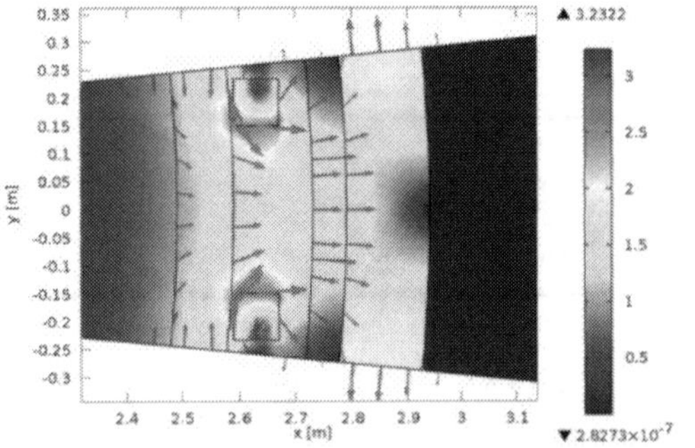

Figure 1. Modeled magnetic flux density in the cross section of a pole of a 10 MW direct drive generator with a coil engineering current density of 70 A/mm^2.

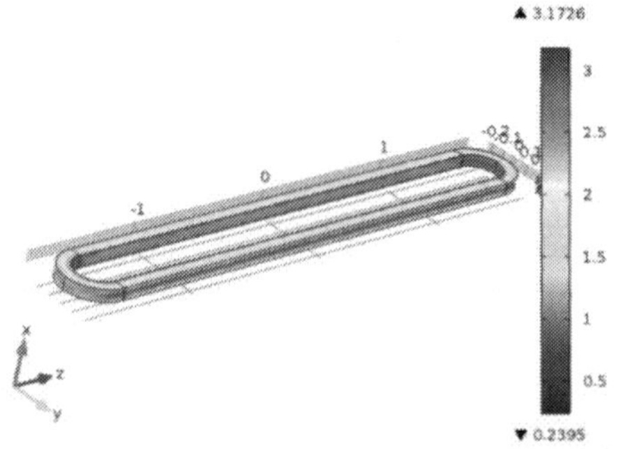

Figure 2. Modeled magnetic flux density of a race track coil in a 10 MW direct drive generator. This coils have the same cross section as the demonstration coil, but the length is 3.1 m.

4. Discussion

The load line reveals that operation at 20 K is not possible with the current design and the wire considered. However, at 10-15 K a sufficient margin to the critical current density of the tape appears attainable. The main properties of the generator are shown in table 2. The diameter and length of the generator are 6.0 m and 3.1 m. The magnetic flux density in the air gap is 1.5 T, notably higher than for a typical permanent magnet direct drive generator, but considerably lower than the LTS proposal of [3]. The usage of tape for a double pancake, a race track and the entire 32 pole generator are 1.5 km, 15 km and 474 km, respectively. Thus such a generator can be produced using the current piece length from the MgB_2 tape and wire production. The first estimate of the cost of the machine has been done on the basis of a tape cost of 4 €/m, which is expected to decrease to approximately 1 €/m after up-scaling. By additionally including the cost of active materials like copper and steel laminates, the cost of the active materials are estimated at 2.3 M€ or 226 €/kW (using 4 €/m). This is one of the main performance indicators, which are evaluated in the INNWIND.EU project for a 10-20 MW offshore turbine including blades, nacelle tower and offshore foundation. The cost per capacity of the direct drive generator should be below 20% of the total offshore turbine cost of 1.5 M€/ to be competitive with the gearbox solution. Hence, when comparing the 226 €/kW obtained for the modelled 10 MW generator with the 300 €/kW requirement, it can be seen that the MgB_2 based generator passes the initial cost criterion already at today's cost level of MgB_2 wire. Further work need to evaluate the cost of the cryostat and cooling system as well as finding the optimal amount of iron minimizing the cost (taking the weight into account). Additionally it should be mentioned that the development of MgB_2 wires is intensive, and a significantly better wire may be anticipated[8]. Looking at figure 3, it is notable that a 50 % increase of the critical current would allow for operation at 20 K and 3 T in an otherwise identical design[8].

Table 2. Properties of a 10 MW direct drive superconducting 32 pole generator based on MgB_2 race track coils with cross sectional dimensions equal to the small scale demonstration coil in table 1.

$R_{Fe\ out}$ [m]		2.94	Torque [MNm]		10.6
$R_{Armature\ out}$ [m]		2.79	Speed [rpm]		9.65
$R_{Armature\ in}$ [m]		2.73	Poles [2p]		32
$R_{Supercond\ out}$ [m]		2.69	Frequency [Hz]		2.57
$R_{Supercond\ in}$ [m]		2.59	$B_{air\ gap}$ [T]		1.5
$L_{generator}$ [m]		3.1	Arm. loading [A/m]		10^5
R_{End} [m]		0.15	Arm. Fill [%]		50
W_{coil} [mm]		84	Shear stress [kN/m^2]		75
H_{coil} [mm]		80	Efficiency [%]		97.7
$L_{SC\ single\ pancake}$ [m]		740.9	J_{coil} [A/mm^2]		70 @ 3 Tesla
$L_{SC\ double\ pancake}$ [m]		1481.7	J_{tape} [A/mm^2]		113 @ 3 Tesla
$L_{SC\ Race\ track\ coil}$ [km]		14.82	M_{Cu} [kg]		19415
$L_{SC\ total}$ [km]		474.2	M_{Fe} [kg]		24998
Tape unit cost [€]	4	($\rightarrow$ 1)	M_{active} [kg]		52331
SC cost [k€]	1897	($\rightarrow$ 474)	Cost Cu [€]		291234
$M_{Superconductor}$ [kg]		7918.1	Cost Fe [€]		74994
$M_{cryostat+cooler}$ [kg]		TBD	Cost total [k€]	2263	($\rightarrow$ 840)
Cost cryostat		TBD	Cost / cap. [€/kW]	226	($\rightarrow$ 84)

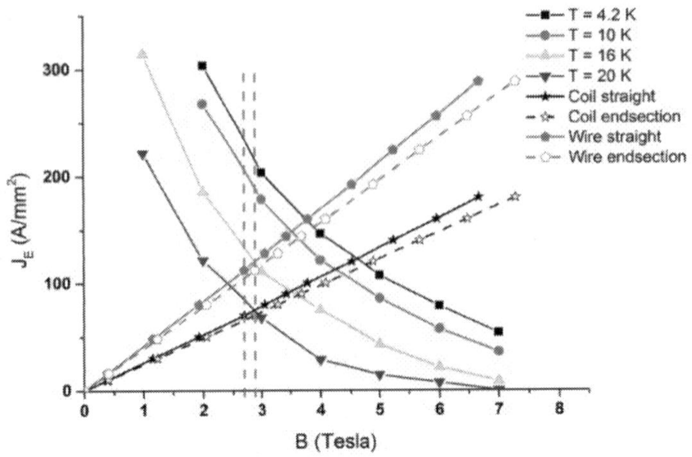

Figure 3. Load line of MgB_2 pole coil positioned in a 10 MW direct drive wind turbine generator. A coil J_{Ecoil} = 70 A/mm^2 is desirable and corresponds to a J_{Ewire} = 112 A/mm^2 taking insulation and spacers into account (crossing of vertical dashed line). The J_E of the wire includes the MgB_2 tape (3.0 mm x 0.5 mm) and a Cu strip for stabilization (3.0 mm x 0.2 mm) [7]. Operation at 2 T seems possible at 20 K whereas T = 10-15 K is needed for 3 T operation.

5. Conclusion

We have designed an MgB_2 race track coil intended for demonstration of a pole of a 10 MW direct drive wind turbine generator. The coil consists of 10 double pancake coils stacked into a race track coil with a cross section of 84 mm x 80 mm. The length of the straight section is 0.5 m and the diameter of the end sections is 0.3 m. It will produce magnetic flux density of about 1.5 T in the air gap of the 10 MW 32 pole generator and about 3.0 T at the edge of the superconducting coil with an operation current density of the coil of 70 A/mm^2. The load line proposal indicates that the operation temperature will be 10-15 K.

Acknowlegdement
This work is done as part of work package 3 "Electro-mechanical conversion" of the INNWIND.EU project funded under FP7. The work is part of the task 3.1. "Superconducting generators" - See www.innwind.eu for more information.

References

[1] Abrahamsen A B, Jensen B B, Seiler E, Mijatovic N, Rodriguez-Zermeno V M, Andersen N H and Østergård J, 2011 *Physica* C **471** 1464
[2] Snitchler G, Gamble B, King C and Winn P 2011*IEEE Trans. Appl. Supercond.* **21** 1089
[3] Fair R 2012 "Superconductivity for Large Scale Wind Turbines", *DOE report* DE-EE0005143
[4] Mine S, Xu M, Buresh S, Stautner W, Immer C, Laskaris E T, Amm K and Grasso G 2013 IEEE Trans. Appl. Supercond. **23** 4601404
[5] Sætre F, Hiltunen I, Runde M, Magnusson N, Järvelä J, Bjerkli J and Engebrethsen E 2011 *Supercond. Sci. Technol.* **24** 035010
[6] DTU 10 MW reference turbine, http://www.hawc2.dk/ September 2013
[7] G. Grasso, private communication
[8] Rindfleisch M et. al.,2013, EUCAS conference 1M-WT-I1 , & Brisgotti S et. al., 2013, EUCAS conference 1M-WT-O1

Design and market considerations for axial flux superconducting electric machine design

M D Ainslie[1], A George[2], R Shaw[2], L Dawson[3], A Winfield[3], M Steketee[4] and S Stockley[5]
[1] Bulk Superconductivity Group, Department of Engineering, University of Cambridge, Trumpington Street, Cambridge CB2 1PZ, UK
[2] Department of Materials Science & Metallurgy, University of Cambridge, Pembroke Street, Cambridge CB2 3QZ, UK
[3] Department of Physics, Cavendish Laboratory, University of Cambridge, J J Thomson Avenue, Cambridge CB3 0HE, UK
[4] Department of Chemical Engineering and Biotechnology, University of Cambridge, New Museums Site, Pembroke Street, Cambridge CB2 3RA, UK
[5] Judge Business School, University of Cambridge, Trumpington Street, Cambridge CB2 1AG, UK

E-mail: mark.ainslie@eng.cam.ac.uk

Abstract. In this paper, the authors investigate a number of design and market considerations for an axial flux superconducting electric machine design that uses high temperature superconductors. The axial flux machine design is assumed to utilise high temperature superconductors in both wire (stator winding) and bulk (rotor field) forms, to operate over a temperature range of 65-77 K, and to have a power output in the range from 10s of kW up to 1 MW (typical for axial flux machines), with approximately 2-3 T as the peak trapped field in the bulk superconductors. The authors firstly investigate the applicability of this type of machine as a generator in small- and medium-sized wind turbines, including the current and forecasted market and pricing for conventional turbines. Next, a study is also carried out on the machine's applicability as an in-wheel hub motor for electric vehicles. Some recommendations for future applications are made based on the outcome of these two studies. Finally, the cost of YBCO-based superconducting (2G HTS) wire is analysed with respect to competing wire technologies and compared with current conventional material costs and current wire costs for both 1G and 2G HTS are still too great to be economically feasible for such superconducting devices.

1. Introduction

Over many years of research, various superconducting machines have been shown to be technically feasible over a wide range of power ranges. For low temperature superconducting (LTS) materials, in particular, the complexity and cost of 4 K cryogenics hindered the commercial development of LTS machines [1], although there were a number of successful technical feasibility demonstrations [2]. The discovery of high temperature superconducting (HTS) materials in 1987 renewed enthusiasm for

[1] To whom any correspondence should be addressed.

applied superconductivity research with the expectation that these materials could be exploited at 77 K, the boiling point of liquid nitrogen. Since then, a number of projects around the world have demonstrated the technical feasibility of HTS machines in various forms, including 5 MW and 36.5 MW motors for ship propulsion [3,4] by American Superconductor (now AMSC); a 380 kW motor [5], which was later developed into a 4 MW project machine [6], by Siemens; a 1.7 MW hydroelectric power generator by Converteam [7]; different HTS induction/synchronous motors at Kyoto University [8,9]; a 30 kW motor for an electric passenger car by Sumitomo Electric [10]; a 1 MW class synchronous motor for industry by KERI and Doosan Heavy Industries [11]; a 1 MW class podded ship propulsion motor by Kawasaki Heavy Industries [12]; and a sub-megawatt class propulsion system by Kitano Seiki [13]. However, both economic and technical challenges have meant that so far none of these machines have been commercialised.

In this paper, the authors investigate a number of design and market considerations for an axial flux superconducting electric machine design that uses high temperature superconductors. The axial flux motor design (for examples, see [14,15]), in general, provides higher torque/power density than other motor designs [15], and the use of superconductors is expected to improve these advantages even further [13]. In this study, the axial flux machine design is assumed to utilise high temperature superconductors in both wire (stator winding) and bulk (rotor field) forms, to operate over a temperature range of 65-77 K (liquid nitrogen temperatures), and to have a power output in the range from 10s of kW up to 1 MW (typical for axial flux machines), with approximately 2-3 T as the peak trapped field in the bulk superconductors. In Section 2, the applicability of this type of machine as a generator in small- and medium-sized wind turbines and as an in-wheel hub motor for electric vehicles is investigated. In Section 3, the cost of YBCO-based superconducting (2G HTS) wire is analysed with respect to competing wire technologies and compared with current conventional material costs.

2. Application Studies

2.1 Small- & Medium-sized Wind Turbines

One potential application for this type of superconducting machine is as a generator in wind turbines, and the assumed power rating (up to 1 MW) puts this machine in the small- to medium-sized turbine category. The increased power density of a superconducting machine would mean a lighter generator and permanent magnet direct-drive (PMDD) wind turbines have received significant interest because elimination of the gearbox and slip rings can result in reduced down-time associated with maintenance and replacement, resulting in reduced cost-of-ownership and increased reliability. The low operating speed of a turbine generator require a high torque for a given power output, resulting in a physically larger machine [16]. However, the use of bulk HTS superconductors is expected to improve the machine performance even further, with trapped fields recorded as high as 17 T at 29 K [17] and up to 3 T at 77 K [18], allowing the removal of iron and an air-core design.

The wind turbine industry continues to grow and is rapidly expanding – the global wind turbine market saw 11% growth in capacity in 2010 with 39 GW delivered worldwide [19]. It is expected that by the end of 2016, global wind capacity will be close to 500 GW, in comparison with 237.7 GW at the end of 2012 [20]. Figure 1 shows the wind turbine market segmented by power output in terms of the expected amount of capacity installed each year from 2010 to 2025 [19]. In 2010, new turbines less than 1 MW accounted for 7% of the market (2.73 GW) and those greater than 3 MW accounted for 13%. In 2018, turbines greater than 3 MW are expected to account for 40% of the market, while turbines less than 1 MW are expected to shrink to 4% [19]. The wind turbine market up to 1 MW is quite mature and there are a large number of companies producing conventional turbines of a high standard Although the wind turbine market on the whole is growing, new installations within the specified power range are expected to decrease, and the 1.5-2.49 MW class of wind turbine will dominate the market for the next five years [19].

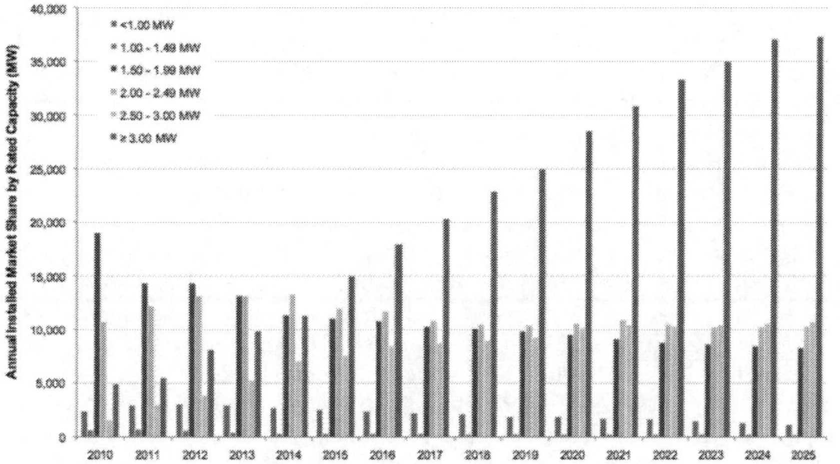

Figure 1. Predicted global annual wind turbine installations by rated output power range, 2010-2025 [19].

2.2 In-wheel Hub Motor for Electric Vehicles

In-wheel hub motors are a concept gaining momentum for use in electric vehicles. This type of motor is commonly found on electric bicycles, and the idea is to place an electric motor into the unused space inside of a wheel to drive the wheel directly. There are a number of advantages: 1) manufacturers can remove the conventional engine bay, which allows for new and creative car designs, 2) removal of much of the powertrain (transmission, differential) results in a significant weight saving and reduced losses in and deterioration of mechanical transmission components, and 3) directly driving each wheel may improve car safety and dynamic drivability. Figure 2 shows the electric vehicle global sales forecast for 2012-2017. For electric vehicles, a new component market is developing rapidly with growing electric motor sales [21].

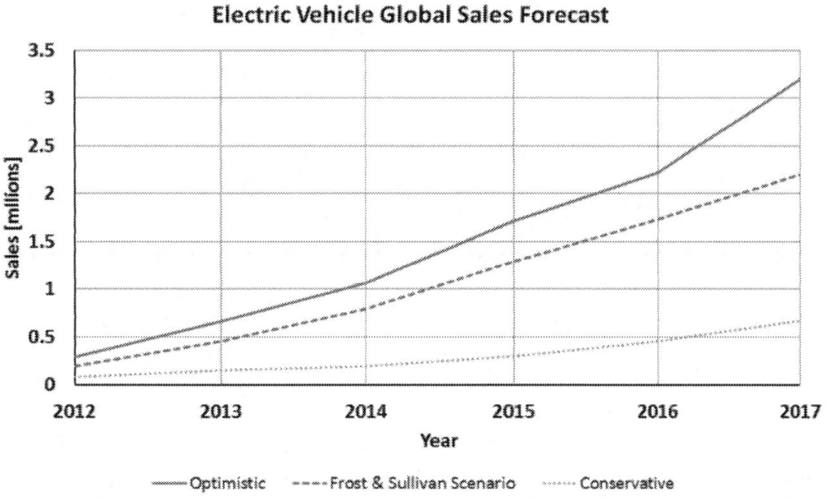

Figure 2. Electric vehicle global sales forecast for 2012-2017 [22].

There are several firms developing in-wheel hub motors at various stages of commercialisation and three companies are highlighted here. YASA Motors [23] is a spin-out from the Engineering

Department at the University of Oxford, who are manufacturing high-torque axial flux motors and have just received £1.45m of funding from a private investment firm, whilst also securing their first contracts with major automotive OEMs [23]. Protean Electric is a US company manufacturing in-wheel motors, which is planning to have their latest prototype ready to present to OEMs by 2013, ready for volume production in 2014. They have just received US$84m in funding to proceed with building new manufacturing facilities in China [24]. In terms of superconducting motors for electric vehicles, the closest to commercialisation is Sumitomo Electric, but this is not an in-wheel hub motor. Sumitomo first demonstrated the use of a superconducting motor to power a passenger vehicle in 2008 [10]. However, due to the likely cost and issues with reliability and cooling for this application, they have now shifted focus to buses, small trucks and forklifts, suggesting heavy duty usage may be a more profitable sector within the electric vehicle market for superconducting motors. Therefore, suggested potential applications would be heavy duty vehicles, such as trucks and plant machinery, and buses. The desire for higher power/torque densities for these functions means that cost and cooling system requirements become less of an issue. Although still many years away, electric aircraft may be an attractive application in the long term, since the power densities required can only be achieved with HTS motors [25].

3. HTS Wire Cost

For a superconducting machine to be competitive with conventional machines, its cost-performance (C-P) ratio needs to be similar to or lower than that of conventional alternatives. The C-P ratio for wire is commonly expressed in US dollars per kiloamp metre ($/kAm), which takes into account the raw wire cost (including labour and manufacturing) in addition to any improved performance under certain operating conditions. Competitive HTS wire requires a C-P ratio of approximately $20/kAm, based on the current C-P ratio of copper wire. 1G (BSCCO) and 2G (YBCO) HTS wire is compared with copper in Table 1.

Table 1. Summary of C-P ratio for HTS and conventional wires.

Wire Type	C-P Ratio
Competitive Point	Approx. 20
Copper	15 – 55
1G (Di-BSCCO)	180 – 230
2G (YBCO, current, SuperPower)	450 [26] – 500
2G (YBCO, projected, SuperPower)	175 [26]

It is important to note that the superconductor C-P ratios above are for performance at 77 K in self-field (i.e., with no externally applied field), and the trapped field in the bulk superconductors of 2-3 T would further reduce the performance of the superconducting wire and increase the C-P ratio. The in-field performance of BSCCO, which has a much lower irreversibility line, is inferior to that of YBCO, so the relative increase in its C-P ratio would be much greater. The general consensus is that the current generation (2G) of HTS wire will not be competitive for some time, if at all, for the desired operating temperature range. However, this machine topology would be preferable cost-wise to a wound DC rotor topology (for example, [3,4]) because significantly less wire would be used in the overall design. The limiting factors still remain very much the same as those raised 15 years ago [27] and reasons include the inherent complexity of the manufacturing process, demand not meeting projections, and the rising raw material costs of metals such as copper (used in both BSCCO and YBCO) and silver (used in Di-BSCCO), both of which have seen their prices rise by about 500% over the past decade.

4. Conclusions & Recommendations

In this paper, a number of design and market considerations have been investigated for an axial flux superconducting electric machine design that uses high temperature superconductors in bulk and tape form. Although the wind turbine market on the whole is growing, new installations within the specified power range are expected to decrease to less than 4% by 2018. The likely cost and issues with reliability and cooling means it is unlikely a superconducting motor would be feasible for electric passenger vehicles, but a suggested potential application would be heavy duty vehicles, such as trucks and plant machinery, and buses. The general consensus is that the current generation (2G) of HTS wire will not be competitive for some time, if at all, for the desired operating temperature range. However, this machine topology would be preferable cost-wise to a wound DC rotor topology because significantly less wire would be used in the overall design.

Acknowledgements

Dr Mark Ainslie would like to acknowledge the support of a Royal Academy of Engineering Research Fellowship. This work was carried out as part of the University of Cambridge's Centre for Entrepreneurial Learning ETECH Project programme, designed to accelerate entrepreneurship and diffusion of innovations based on early stage and potentially disruptive technologies from the University.

References

[1] Tixador P 2010 *Physica C* **470** 971–979
[2] Shimada T, Shibuya M, Takahashi R, Imai Y, Kusafuka H, Shiobara R, Yamaguchi K, Takahashi M, Suzuki K and Miyaike K 1999 *IEEE Trans. Appl. Supercond.* **9** 1189–1192
[3] Snitchler G, Gamble B and Kalsi S S 2005 *IEEE Trans. Appl. Supercond.* **15** 2206–2209
[4] Gamble B, Snitchler G and MacDonald T 2011 *IEEE Trans. Appl. Supercond.* **21** 1083–1088
[5] Nick W, Nerowski G, Neumuller H W, Frank M, van Hasselt P, Frauenhofer J and Steinmeyer F 2002 *Physica C: Superconductivity* **372–376** 1506–1512
[6] Neumuller H W, Nick W, Wacker B, Frank M, Nerowski N, Frauenhofer J, Rzadki W and Hartig R 2006 *Supercond. Sci. Technol.* **19** S114–S117
[7] Fair R, Lewis C, Eugene J and Ingles M 2010 *J. Phys.: Conf. Ser.* **234** 032008
[8] Nakamura T, Nagao K, Nishimura T and Matsumura K 2009 *Supercond. Sci. Technol.* **22** 045022
[9] Nakamura T, Matsumura K, Nishimura T, Nagao K, Yamada Y, Amemiya N, Itoh Y, Terazawa T and Osamura K 2011 *Supercond. Sci. Technol.* **24** 015014
[10] Sumitomo Electric Industries, Ltd. – Superconductivity Web Site EV Motor [Online]. Available: http://global-sei.com/super/magnet_coil_e/evmotor.html
[11] Kwon Y K, Kim H M, Baik S K, Lee E Y, Lee J D, Kim Y C, Lee S H, Hong J P, Jo Y S and Ryu K S 2008 *Physica C* **468** 2081–2086
[12] Umemoto K, Aizawa K, Yokoyama M, Yoshikawa K, Kimura Y, Izumi M, Ohashi K, Numano M, Okumura K, Yamaguchi M, Gocho Y and Kosuge E 2010 *J. Phys.: Conf. Ser.* **234** 032060
[13] Miki M, Tokura S, Hayakawa H, Inami H, Kitano M, Matsuzaki H, Kimura Y, Ohtani I, Morita E, Ogata H, Izumi M, Sugimoto H and Ida T 2006 *Supercond. Sci. Technol.* **19** S494–S499
[14] Hendershot J R and Miller T J E 2010 *Design of Brushless Permanent Magnet Motors* (2nd ed.) Motor Design Books LLC
[15] Gieras J F, Wang R-J and Kamper M J 2008 *Axial Flux Permanent Magnet Brushless Machines* (2nd ed.) Springer
[16] Cao W 2011 High-Temperature Superconducting Wind Turbine Generators, Wind Turbines, Dr. Ibrahim Al-Bahadly (Ed.) *InTech Open*
[17] Tomita M and Murakami M 2003 *Nature* **421** 517-520
[18] Nariki S, Sakai N and Murakami M 2005 *Supercond. Sci. Technol.* **18** S126

[19] Emerging Energy Research, 2011. Global Wind Turbine Market and Strategies: 2011-2025 [Online]. Available: http://www.emerging-energy.com/uploadDocs/Excerpt_GlobalWindTurbineMarketsandStrategies2011.pdf
[20] Global Wind Energy Council (GWEC), Global Wind 2011 Report [Online]. Available: http://www.gwec.net
[21] Frost & Sullivan Automotive & Transportation Group 2011 360 Degree Perspective of the Global Electric Vehicle Market [Dec 2011]
[22] ENEVATE project survey, 2012. The development of e-mobility supply chains in Europe. Kassel: ENEVATE.
[23] YASA Motors [Online] Available: http://yasamotors.com/about-us
[24] Protean Electric 2012 Protean Electric Announces $84 Million in New Funding: New Manufacturing Facility in China [Online] Available: http://www.proteanelectric.com/?page_id=14&post=2212
[25] Masson P J, Brown G V, Soban D S and Luongo C A 2007 *Supercond. Sci. Technol.* **20** 748-756
[26] Hazelton D W 2012 2G HTS Conductors at SuperPower *presented at Low Temperature High Field Superconductor Workshop 2012, Napa, USA* [6 Nov 2012]
[27] Grant P M and Sheahen T P 2002 arXiv:cond-mat/0202386 [cond-mat.supr-con]

The first Italian Superconducting Fault Current Limiter: Results of the field testing experience after one year operation

L. Martini[1], M. Bocchi[1], M. Ascade[1], A. Valzasina[1], V. Rossi[1], C. Ravetta[2] and G. Angeli[1]

[1] RSE S.p.A., via Rubattino 54, 20134 Milano, Italy
[2] A2A Reti Elettriche S.p.A., via Ponte Nuovo 100, 20128 Milano, Italy

E-mail: luciano.martini@rse-web.it

Abstract. Ricerca sul Sistema Energetico S.p.A. (RSE) has been gaining a relevant experience in the simulation, design and installation of resistive-type Superconducting Fault Current Limiter (SFCL) devices for more than five years in the framework of a R&D national project funded by the Ricerca di Sistema (RdS). The most recent outcome of this research activity is the installation of a resistive-type BSCCO-based 9 kV / 3.4 MVA SFCL device in a single feeder branch of the Medium Voltage (MV) distribution network managed by A2A Reti Elettriche S.p.A (A2A) in the Milano area. This installation represents the first SFCL successfully installed in Italy. In this paper, we report on the main outcomes after a more than 1-year long steady-state field testing activity. The design of an upgraded device to be installed in the same substation has already been initiated: the new SFCL will allow to protect four different feeders, therefore implying a device upgrade up to 15.6 MVA.

1. Introduction

Superconducting Fault Current Limiter (SFCL) devices enable us to manage the ever-increasing short-circuit currents in distribution grids. For this reason, SFCLs are highly desirable for utilities [1]-[2]-[3]. The historical trend of SFCL designs and applications was already summarized in several works [4]-[5], therefore we hereby focus the attention on the first Italian SFCL device successfully installed in a real distribution grid. This project has been carried-out by Ricerca sul Sistema Energetico S.p.A. (RSE) in the framework of a R&D national project funded by RdS: the main goal of the project is to conceive a first generation (1G) BSCCO-based resistive-type SFCL for Medium Voltage (MV) distribution grids. The main outcome consists in a 9 kV / 3.4 MVA SFCL 3-phase device nowadays permanently installed as a single-feeder fault protection in the S. Dionigi substation (Milan) belonging to A2A Reti Elettriche S.p.A. (A2A), the second largest Italian utility.

The first steps of the project were already summarized [6]-[7]. This paper is aimed at describing the results gathered during the first year of steady-state operation at the S. Dionigi substation.

2. SFCL basic aspects

2.1. SFCL specifications and tests

Table 1 reports on the network requirements characterizing the installation site. According to A2A, the most important specification is that the SFCL device has to guarantee a Limitation Factor (LF) comprised between 1.7 and 2, being LF the ratio between the prospective short-circuit current I_{SC} and the limited short-circuit current I_{Lim}. Besides the grid requirements, Table 1 provides also a comparison with the short-circuit tests parameters.

Table 1. Comparison between grid requirements and short-circuit test parameters

Parameter	A2A Requirements	Short-Circuit Test
Rated voltage V_{nom}	9 kV_{rms}	12 kV_{rms}
Rated current I_{nom}	220 A_{rms}	220 A_{rms}
Prospective short-circuit current I_{SC}	12.3 kA_{rms}	12.3 kA_{rms}
Maximum prospective short-circuit current I_{SCP}	30 kA_p	33.2 kA_p
Ungrounded short-circuit duration t_{fault}	400 ms	300 ms
Limitation factor LF = (I_{SC} / I_{Lim})	1.7 - 2	1.83

All the design details are described in [6]-[7], therefore it is worthwhile here just mentioning that the SFCL device is made of a HTS BSCCO-2223 tape total length of less than 2 km (1880 m) with self-field end-to-end critical current I_C of 180 A at 77 K; in addition, each phase is shunted by an air-core reactor and a closed-circuit Liquid Nitrogen (LN) re-liquefier is used. As described in [6], dielectric and short-circuit tests validated the SFCL readiness for the grid installation.

2.2. *SFCL installation*

The whole system is composed by several parts. In Figure 1 and Figure 2 are shown, respectively, a picture of the installed SFCL system and a general description of the main SFCL system components. A detailed description of components and of the acquisition system is given in [6]-[7].

Figure 1. SFCL ready for in-field activity

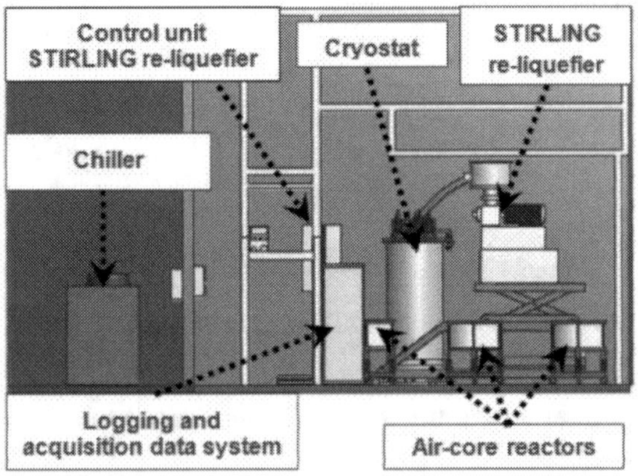

Figure 2. Components description of the installation

3. Field-testing activity of the 9kV/220A SFCL

After having definitely fixed the final connections (Figure 1), the device was officially energized in March 2012, launching a more than one year long in-field testing activity.

The SFCL device worked until the end of June 2013 featuring a 100% availability in the last ten months. At the end of June, the refrigeration system routine maintenance was mandatory due to the achievement of the maximum amount of operation hours foreseen by the manufacturer, therefore the device was switched-off. This temporary outage offered the chance to improve at the same time some features of the acquisition system, hence the SFCL was "re-energized" in the beginning of October 2013 for other six months of in-field testing activity.

The first preliminary results of the field testing activity were already shown in [7]; here we step forward with the discussion of the most relevant data gathered throughout the whole experimentation. Figure 3(a) is referred to June 2013 and shows the time-behavior throughout 24 hours of both LN pressure and motor speed (i.e. the motor driving the LN re-liquefier), along with the temperature of the

LN bath at two different levels (T_{btm} at the cryostat bottom and T_{middle} at the cryostat medium height), whereas Figure 3(b) provides a graphical representation of the probes position for pressure, temperature and LN level measurement. Figure 4(a) emphasizes the time period from April 2012 to May 13th 2012 characterized by high load conditions (up to 120 A_{rms}) and Figure 4(b) shows the trend of SFCL currents from October 2012 until June 2013.

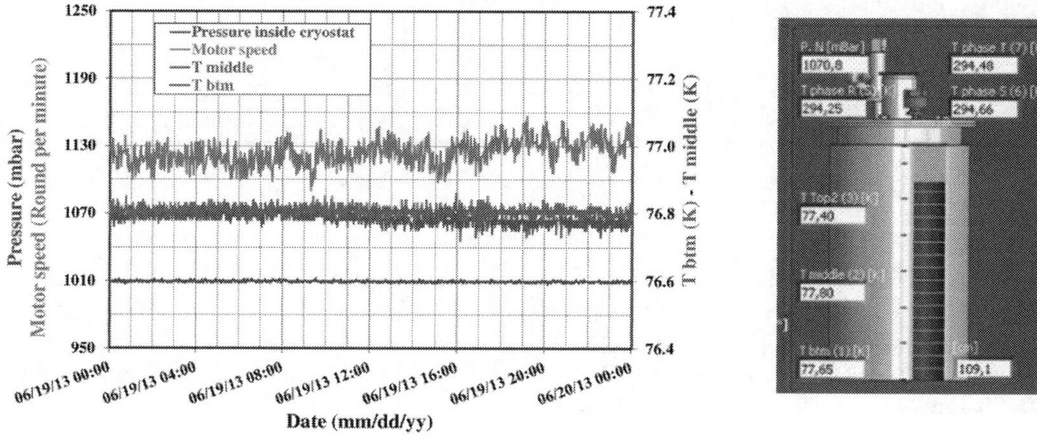

(a) (b)

Figure 3. (a) Time evolution throughout 24 hours of the motor speed, pressure inside cryostat and liquid nitrogen temperatures at two different levels. (b) Synoptic scheme showing the measurement probes position for pressure, temperature and LN level

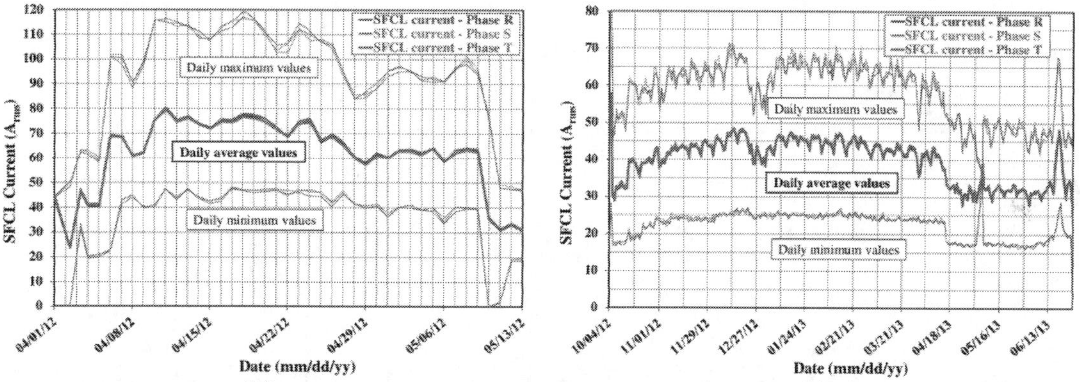

(a) (b)

Figure 4. (a) Transport current in the period from April 2012 to May 13th 2012 that corresponds to the condition of higher load; (b) Trend of currents flowing through the SFCL device in a selected period of time (October 2012 - June 2013)

The current level was in agreement with the one A2A foresaw for this first phase of in-field activity and in Figure 5 a zoom of the current level in the third week of June 2013 can be appreciated. One of the most interesting features of the testing activity occurred in January 2013, when an unexpected increase in the LN bath level was noticed. Since the checks on the LN level sensors did not point-out any anomalies, it was assumed that some air could have entered the cryostat, being the pressure inside slightly lower than outside, and could have been liquefied by the refrigeration system. In agreement with this assumption, it was decided to slightly increase the pressure inside the cryostat from 990 mbar, lower than the room pressure and corresponding to a liquid nitrogen temperature of 76.8 K, to 1010 mbar, higher than the room pressure and corresponding to a liquid nitrogen temperature of 77.1 K. In Figure 6 the stabilizing effect of the pressure enhancement on nitrogen level is shown.

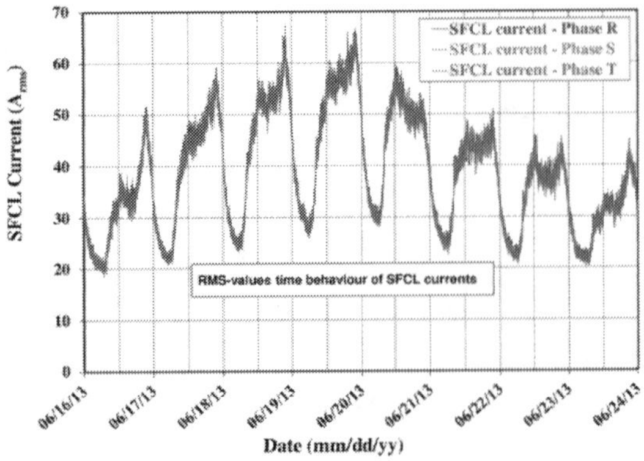

Figure 5. Current RMS-values trend inside the SFCL during the third week of June 2013

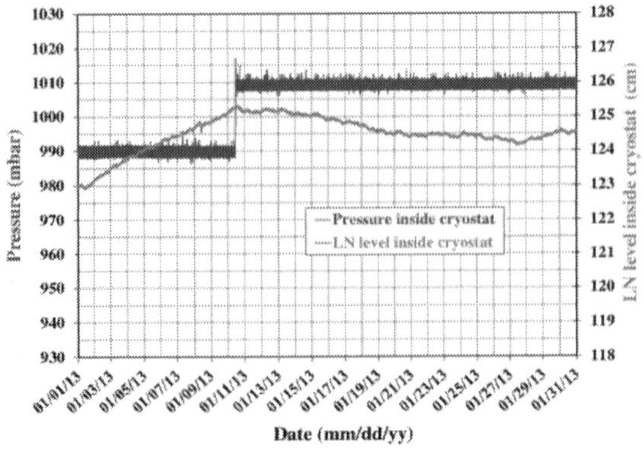

Figure 6. Enhancement in cryostat pressure and effect on liquid nitrogen level

4. Conclusions

The first Italian SFCL device has been installed in a real distribution grid. This project has been carried-out by RSE in the framework of a R&D national project and the result is a resistive-type BSCCO-based 9 kV / 3.4 MVA SFCL installed as a single-feeder fault protection in a A2A substation of the Milan area. The SFCL was commissioned in March 2012, thereby launching a more than one year long field-testing activity. The results gathered in the first phase of field-testing activity confirmed that the SFCL worked as expected in steady-state conditions. After this first success, we will proceed with the analysis of the SFCL behaviour during real fault conditions in the distribution grid; afterwards, the design of the upgraded SFCL (9 kV / 15.6 MVA) to be installed as transformer protection in the same substation has been already initiated. In the upgraded installation, the device will protect 4 feeders where the probability of faults is much higher.

Acknowledgements

This work was supported by the Research Fund for the Italian Electrical System under the Contract Agreement between RSE and the Ministry of Economic Development.

References

[1] Noe M. and Steurer M. 2007 High temperature superconductor fault current limiters: concepts, applications and development status *Supercond. Sci. Technol.* **20** R15-29

[2] Martini L., Ravetta C. and Bocchi M. 2011 Development and testing of innovative fault current limiters for distribution system applications *Proc. 21th Int. Conf. on Electricity Distribution (Frankfurt)* Paper 0339

[3] Martini L., Noe M., Pascal T., Hobl A. and Bocchi M. 2011 The ECCOFLOW project: design and simulation results of a superconducting fault current limiter for operation in electricity networks *Proc. 21th Int. Conf. on Electricity Distribution (Frankfurt)* Paper 1162

[4] Bock J., Breuer F., Walter H., Elschner S., Kleimaier M., Kreutz R. and Noe M. 2004 CURL 10: development and field-test of a 10 kV /10 MVA resistive current limiter based on bulk MCP-BSCCO 2212 *IEEE Trans. Appl. Supercond.* **15** p.1955

[5] Llambes J., Hazelton D. and Weber C. 2009 Recovery under load performance of 2nd generation HTS superconducting fault current limiter for electric power transmission lines *IEEE Trans. on Appl. Supercond.* **19** pp. 1968-1971

[6] Martini L., Bocchi M., Ascade M., Valzasina A., Rossi V., Angeli G. and Ravetta C. 2012 Development, testing and installation of a Superconducting Fault Current Limiter for medium voltage distribution networks *Physics Procedia* **36** pp. 914 – 920

[7] Martini L., Bocchi M., Ascade M., Valzasina A., Rossi V., Ravetta C. and Angeli G. 2013 Live-grid installation and field testing of the first Italian Superconducting Fault Current Limiter *IEEE Transactions on Applied Superconductivity* **vol. 23 no. 3**

Development and test of resistive superconducting fault current limiter; acting time and its recovery conditions.

Carlos A. Baldan[1,2], Luciano C. Guedes[1], Jérika S. Lamas[3], Carlos Y. Shigue[1], and Ernesto Ruppert[4]

[1]School of Engineering of Lorena, University of Sao Paulo, EEL-USP, Lorena - SP, Brazil

[2]Department of Electrical Engineering, State University of Sao Paulo, UNESP, Guaratinguetá – SP, Brazil

[3]Department of Solid State Sciences, Ghent University, Ghent, Belgium

[4]Faculty of Electrical and Computing Engineering, State University of Campinas, UNICAMP, Campinas – SP, Brazil

E-mail: cabaldan@demar.eel.usp.br

Abstract. Resistive-type of superconducting fault current limiters (RSFCL) have been developed for medium voltage class aiming to operate at 1 MVA power capacity and short time recovery (< 2 s). A RSFCL in form of superconducting modular device was designed and constructed using 50 m-length of YBCO coated conductor tapes for operation under 1 kV / 1 kA and acting time of 0.1 s. In order to increase the acting time the RSFCL was combined with an air-core reactor in parallel to increase the fault limiting time up to 1 s. The tests determined the electrical and thermal characteristics of the combined resistive/inductive protection unit. The combined fault current limiter reached a limiting current of 583 A, corresponding to a limiting factor of 3.3 times within an acting time of up to 1 s.

1. Introduction

Superconducting fault current limiters (SFCLs) are current limiting devices which present advantages of very low losses in steady state operation, high limiting impedance under fault conditions, reliable operation, very short reaction times to fault currents and an automatic response feature without the requirement of an external trigger mechanism. The resistive SFCLs present the electrical behavior near the ideal such as low impedance under steady operation and fast actuation. Nevertheless under the fault condition the maximum allowed electrical field, E = 50 V/m must be obeyed in order to withstand the critical current degradation [1-3]. The RSCFL actuation is attained with the insertion of a fast transition resistance of 0.354 Ω/m during 5 cycles, thus limiting the fault current value up to 100 ms.

In order to increase the acting time during a fault current event, a resistive SFCL was combined with a magnetic circuit using a short-circuited transformer with saturated core to analyze the coupling effect with the grid line [4]. More recently, a resistive SFCL was tested using an air-core reactor connected in parallel, but with the superconducting device acting only during the first 100 ms, being them disconnected by a circuit breaker, leaving only the air-core reactor limiting the current up to a total elapsed time of 1 s [5].

In the design of superconducting devices several restrictions must be taken into account for the thermal and electrical protection, such as the maximum allowable electrical field and the energy loss per volume, respectively E = 50 V/m and 1,200 J/cm^3 for the YBCO CC tape [5]. For the medium voltage class applications using the maximum electrical field, continuous long lengths of YBCO CC tapes are needed (e.g., for 15 kV it will need 300 m/phase). Also in the SFCL design several modular units should be considered in a series and/or parallel electric connection arrangements. The modular units permit to adjust the maximum current and voltage during limiting time, usually less than 100 ms, to a recovery under load using an external shunt protection. Fast and safe operation of the SFCL can prevent damage to the circuit components after 50 ms before circuit breaker actuation without air-core reactor and 1 s with it connected in parallel. The combined SFCL has the potential to reduce the fault current level by a factor of 3 to 10 times turning them into an potential candidate of the future smart electric grid.

The architecture of the commercial YBCO CC tapes, such as stabilizer thickness, high resistivity substrate and use of stainless steel tape as reinforcement, with a linear electric resistance of 0.354Ω /m, requires the pulsed current-voltage (I-V) measurements using short samples for determining the peak current limiting period, the maximum allowable temperature of 350 K and the recovery time from full normal state transition without any irreversible tape degradation. An efficient cooling scheme should also be considered to expose the broad tape surface to the liquid nitrogen coolant; in this condition the maximum allowed power dissipation per surface area is 32 W/cm^2 [6].

In this work, a modular SFCL was built using 2 x 25 m length of YBCO CC tapes in parallel, with a shunt protection with equivalent resistance R_{sh} = 0.078 Ω per sector, or R_{eq} = 1.86 Ω for the whole modular device with 24 sectors, without electrical joints (Fig. 1). This configuration provides a homogeneous quench behavior of the HTS tapes, with a copper contact in each 20 cm-length of YBCO CC tapes, working as a thermal sink (barrier) thus contributing for limit ing the maximum temperature within the tape (reducing the length to recovery) and acting as stabilizer in the device for decreasing the recovery time.

2. FCL design

The SFCL was constructed using the 344S type American Superconductor's YBCO CC tape (stainless steel tape reinforcement) with 4.4 mm-width, 0.15 mm thickness, and critical current, I_c = 72±2 A (equivalent to 163 A/cm-width). A short sample test was carried out for the I-V curve characterization applying the peak current value under non-fault condition to carry currents above I_c without quenching during 0.1 s; the current peak was limited to 2.5 times I_c.

The geometry of the modular superconducting device (Fig. 1) aims to expose all the surface of YBCO CC tape to the coolant with a pair of straight and parallel tapes wound on G10 tube and soldered at their ends to the copper terminals. The modular unit constituted by 24 sectors in series was designed for operating at 761 V under a steady current of 150 A.

The nominal power per area of the SFCL can be calculated multiplying U_{nom}/L by I_c/w, where U_{nom} is the nominal voltage, I_c the critical current, L and w are the conductor length and width, respectively, giving the nominal power of 96 kVA for this unit, corresponding to a rms power value 57 VA/cm^2 for fault duration varying from 50 up to 100 ms [3]. The power dissipation can achieve 73.3 kW (SFCL voltage 256 V and limited current 573 A) during 100 ms; the energy density of about 488 J/cm^3 is lower than the critical value of 1,200 J/cm^3 for YBCO tapes [6].

3. Electrical performance

3.1. Pulsed Current Characterization

A controlled DC power supply (10V/1000A) was used to characterize the stack with 2 x 344SS (I_c=72A each) in parallel tapes (0.1 m-length) soldered with Sn-In alloy, and protected by a shunt resistance R_{sh}= 0,17Ω/m. The test was carried out applying pulsed current during 100 ms up to a maximum value of 350 A, and a maximum electric field, E = 1 V/cm. During the transition and the recovery process for current I = 310 A, the elapsed time to recovery the superconducting properties is lower than 1 s, under load steady current of 144A. From the E-t curve (Fig. 2) the power loss can be determined at the maximum value for I = 310 A (I_{peak} =500 A) when reaches 700 W or 109 W/cm^2, corresponding to the energy density 1,548 J/cm^3 (774 J/cm^3 per tape).

Fig. 1. Modular superconducting device constituted by 24 turns of a pair of YBCO CC tapes in parallel.

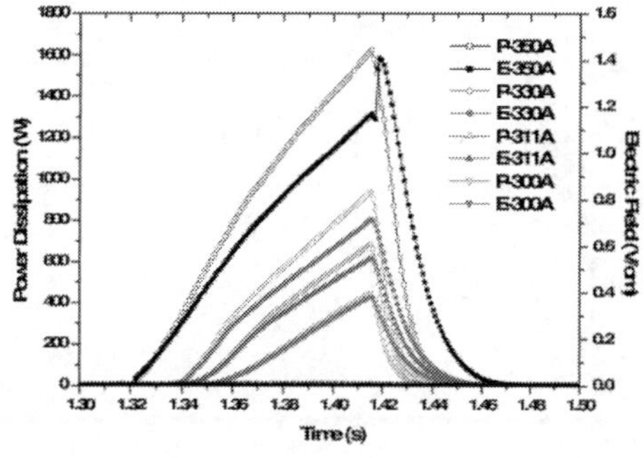

Fig. 2. Power loss and electric field as function of time and applied current peak.

3.2 Fault Current Test of Resistive Modular Device

The modular superconducting device unit (MSD) constituted by 24 sectors in series was designed for operation at 761 V and steady current of 150 A. During the fault current test carried out using a 96 kVA motor-generator the device was connected in series between phase to ground, with a steady rms current of 70 A. The fault is induced when a resistive load (Y-connection) is short-circuited (single phase) using a fast static switch. The electrical circuit for testing was presented elsewhere [7].

The prospective fault current with this configuration can reach 1.9 kA with adjustable time to operate from 1 to 5 cycles. Fig. 3 shows the prospective fault current reaching 1.9 kA, after 2 ms the SFCL acts for limiting the current to 573 A, corresponding to a limiting factor of 3.3 for the first peak and 6.9 for the fifth peak. The measured voltage in the SFCL was low, V_{lim} = 256 V, when compared with the design value of 761 V, indicating that transition happens only in part of superconducting tape length. Fig. 4 shows the equivalent resistance increasing in RSFCL after transition (partial length) for the normal conductor reaching 0.9 Ω compared with the designed value of 1.86 Ω.

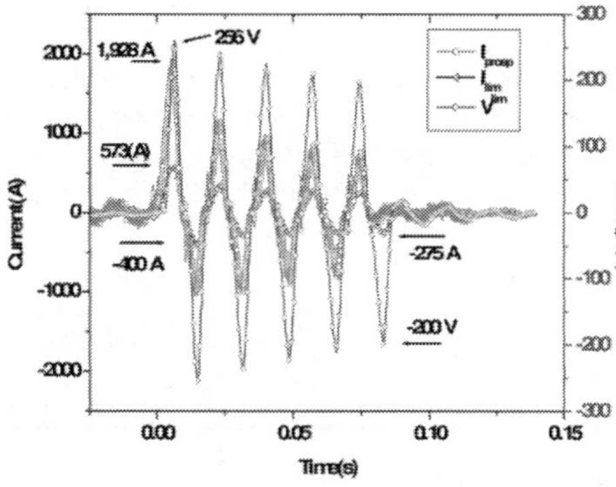

Fig. 3. Prospective fault current of 1.9 kA limited to 573 A with maximum voltage of 256 V

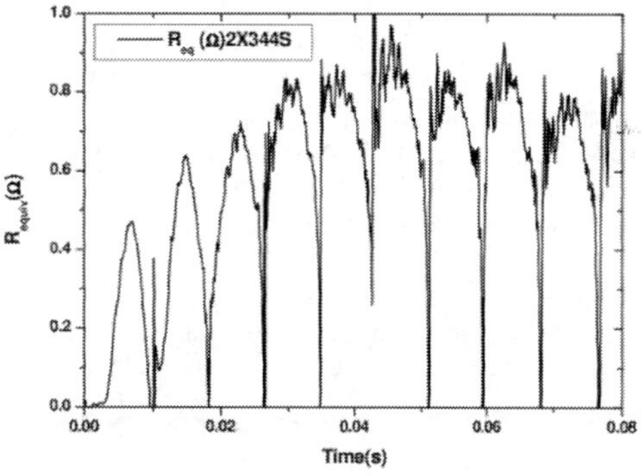

Fig. 4. Equivalent resistance of SFCL

3.3 Fault Current Test using Superconducting Device in parallel with an Air Core Reactor.

The basic concept for combining a magnetic circuit with the modular superconducting device is to get some benefits such as increasing the acting time to 1 s. Compared with the operation of conventional air-core reactors the advantages are the reduction of the Joule losses during steady current operation, because the current flows through superconducting device, and the reduction of the recovery time of the air-core reactor from several hours to several minutes.

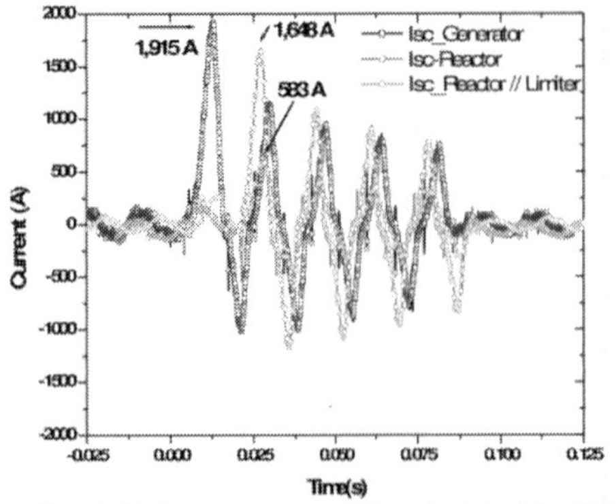

Fig. 5. Fault current values with and without the SFCL in parallel to the air-core reactor

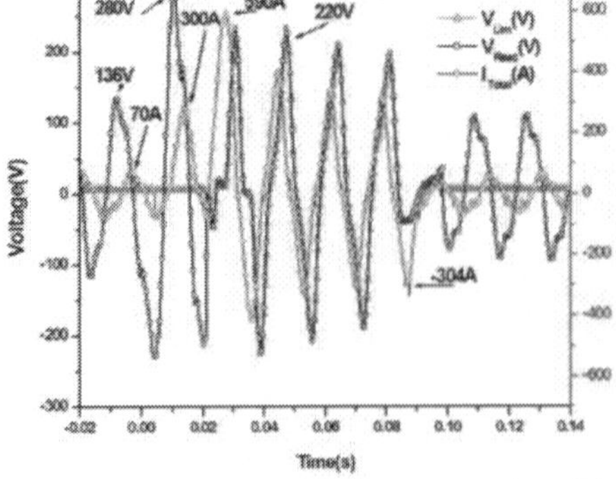

Fig. 6. Voltage and current waveforms for the SFCL in parallel to the air-core reactor

The current waveforms showed in Fig. 5 are the prospective fault current, the limited current by the air-core reactor and combined effect of SFCL with the air-core reactor connected in parallel. Together they can limit the prospective fault current value from 1.9 kA to 583 A. The waveforms showed in Fig. 6 present the voltages within the reactor and in the superconducting module acting together to limit the fault current value by a factor of 3.3. During next fault current test the limited current values reaches 590 A in the first peak and 304 A in the fifth peak, with very low voltage developed within the RSFCL, V_{lim} = 230V when compared with the design value of 761 V.

After measuring the voltage in each sector the results showed that the transition occurred in all of the 24 sectors, what suggests that all of them have the same characteristics and similar critical current values. The total electrical field per element is 0.12 V/cm.

4. Conclusions

The performance of resistive model of SFCL built with 344S YBCO tapes confirmed the limiting of the prospective fault current peak from 1.9 kA down to 583 A, acting just after 2 ms. The limiting factor was of 3.3 (first peak) and 6.9 (fifth peak). The maximum voltage measured in the SFCL reached 256 V (E= 10.2V/m) being a low value compared to the design value of 761 V, indicating partial transition of the YBCO tape length. The combined device with SFCL in parallel with the air-core reactor, to enhance the acting time from 0.1 s to 1 s, were tested and the result showed the limiting factor measured during the fault current test was 3.3 times, using an air core reactor with X/R= 33.46 (L=4.58 mH, X = 1.73 Ω). The equivalent impedance during 5 cycles when became fully resistive, reached $|Z|$ = 0.87 Ω. The test results of the combined SFCL with the air-core reactor presented increased acting time concomitant with the reduction of the fault current level.

References

[1] W. Schmidt et al., "Investigation of YBCO coated conductors for fault current limiter applications," IEEE Trans. Appl. Supercond., vol. 17, pp. 3471-3474, 2007.

[2] K. Nam et al., "Thermal and electrical analysis of CC under AC overcurrent," IEEE Trans. Appl. Supercond., vol. 17, pp. 1923-1926, 2007.

[3] M. J. Kim et al., "Determination of maximum permissible temperature rise considering repetitive over-current characteristics of YBCO coated conductors," IEEE Trans. Appl. Supercond., vol. 18, pp. 660-663, 2008.

[4] F. Moriconi et al., "Development and Deployment of Saturated-Core Fault Current Limiters in Distribution and Transmission Substations", IEEE Trans. Appl. Supercond., vol. 21, pp. 1288-1293, 2011

[5] A. Hobl et al., "Design and production of the ECCOFLOW resistive fault current," IEEE Trans. Appl. Supercond., vol. 23, n.3,p. 5601804. Jun. 2013.

[6] C. Schacherer et al., "Dissipated energy as a design parameter of coated conductor for their use in resistive fault current limiters," Journal of Physics Conf. Series, vol. 97, 012193, 2008.

[7] C. A. Baldan et al., "Test of a modular fault current limiter for 220V line using coated conductor tapes with shunt protection," IEEE Trans. Appl. Supercond., vol. 21, pp. 1242-1245, 2011.

Investigation of 2G coil SCFCL - Modeling and Testing

W.T.B. de Sousa [1,2], A. Polasek[3], R. Dias[3], C.F.T. Matt[3], F.A. da Silva[3], and R. de Andrade Jr.[2]

[1] KIT - Karlsruhe Institute of Technology, Institute for Technical Physics, 76344 Eggenstein-Leopoldshafen, Germany

[2] LASUP - Laboratory for Applied Superconductivity, Federal University of Rio de Janeiro, P.O. 68553, 21945-970, Rio de Janeiro, Brazil

[3] CEPEL - Electric Energy Research Center (ELETROBRAS System), Av. Horácio Macedo, 354, Ilha do Fundão, 21941-911, Rio de Janeiro, Brazil

E-mail: wescley.sousa@partner.kit.edu

Abstract. A non-adiabatic model for SCFCL has been developed. The model is based on the superconductor EJ curve. Heat transfer is modeled by means of a thermal electrical analogy, making the simulations easier to be implemented, because the coupling between electrical and the heat transfer differential equations becomes a straightforward task. Simulation results agree well with experimental testing.

1. Introduction

In spite of the recent advances achieved with superconducting fault current limiters (SCFCL's) [1, 2], simulation and tests of such devices are still important issues. There are different approaches for modeling these devices, whereas testing procedures have been under discussion aiming at standardization [3]. In the present work, a non-adiabatic model is used to simulate the thermal and electrical behavior of a commercial resistive SCFCL component (R-SCFCL) based on second generation superconducting tapes (2G tapes). The model takes into account the characteristic E-J curve of the HTS material as well as the strong coupling between electrical and thermal phenomena. A commercial SCFCL component was tested in the High Power Facilities of CEPEL (Electric Power Research Center). This component was manufactured by Nexans SuperConductors GmbH. Similar components were employed in the ENSYSTROB project, which demonstrated a 2G coil based R-SCFCL in a medium voltage grid [4].

2. Short-Circuit Tests

The R-SCFCL component consists of parallel 2G tapes arranged in two shunted coils which are mounted in an anti-parallel configuration, resulting in a negligible inductance [4]. This component can transport a nominal current of 600 A_{rms} and withstand a maximum voltage of 400 V_{rms} ($\approx$ 0.5 V/cm). It was designed with a high I_c (1200 A, 77 K), in order to withstand large inrush currents, since the SCFCL should not actuate (quench) under inrush currents. The R-SCFCL component was cooled into LN_2 bath (77 K) and tested in the High Current Laboratory of CEPEL. The test circuit is described in [5]. Prospective current was measured without the presence of SCFCL (an asymmetrical prospective current of 53 kA_{peak} (20 kA_{rms})). Fault current limiting test was carried out by connecting the SCFCL component to the test circuit. The fault current limiting test was repeated under the same conditions, showing high repeatability, with negligible variations in limited currents and voltage values.

3. Model

A 1D model was developed where each 2G tape layer is represented by a resistor. Buffer layers are not considered, since they are much thinner (submicrometric range) than the other layers. The HTS layer is modeled as a variable non-linear resistor connected in parallel to ohmic resistors (substrate, silver layer and external shunt). The HTS layer is based on the characteristic EJ curve of the 2G tapes employed in the component (non stabilized SuperPower tapes, according to the SCFCL manufacturer). This tape has an Ag cap layer (2 μm) and an additional Ag layer (1 μm) surrounding the tape [6]. For the present simulations we consider an Ag top layer (3 μm) above the HTS layer and a bottom Ag layer (1 μm) below the substrate. Each layer is considered as a homogeneous material with spatially uniform properties. A MATLAB Script-File (m-file) was used to build the model and solve the electrical and thermal differential equations.

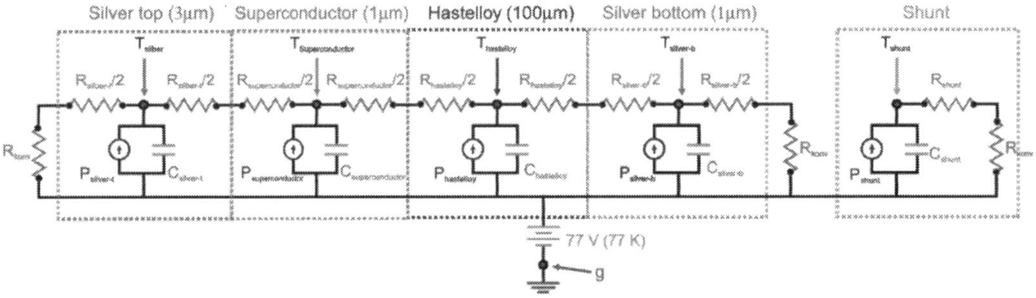

Figure 1. Thermal-electric circuit to solve the thermal behavior.

The one dimensional time-dependent heat transfer across the layers is modeled by a thermal electrical analogy (figure 1). Deep details about thermal electrical analogy can be found in [7, 8]. In the equivalent RC circuit shown in figure 1, resistances correspond to heat conduction of each material, while capacitors correspond to respective thermal capacity. The convective heat transfer between 2G tapes and LN_2 is modeled as a convective resistance. Internal heat generation at each layer is represented by current sources. A DC voltage source (77 V) simulates the LN_2 bath (77 K). The temperature rise of each layer is calculated as a voltage drop between $T_{material}$ and g (figure 1). The use of a thermal-electrical analogy makes the simulations easier to be implemented, because the coupling between the electrical circuit and the heat conduction differential equations becomes a straightforward task. Such analogy is possible, once the temperature is assumed to be spatially uniform within each layer during the transient. This assumption is reasonable since the thicknesses of the layers are very small and the Biot number is lower than 0.1 [9].

4. Results and Discussion

Testing results are shown in figure 2. An asymmetrical prospective current of 53 kA_{peak} (20 kA_{rms}) was limited to 4.6 kA_{peak} (first peak) and the follow current decayed from 4 kA_{peak} to 3 kA_{peak}.

Simulation and testing results agree well (figure 3). Slight differences between simulation and test curves may be attributed to model assumptions, specially the assumption of a HTS

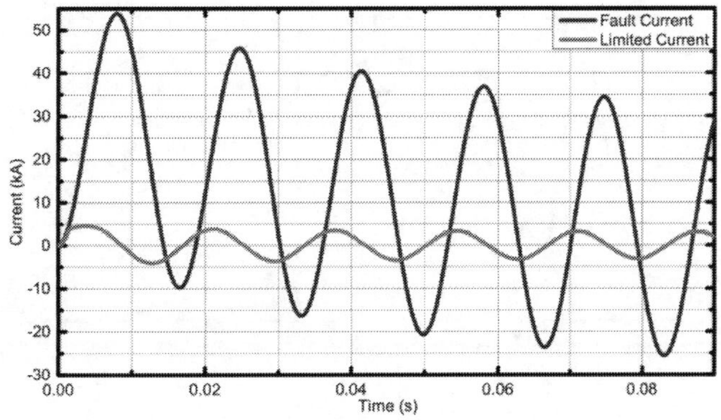

Figure 2. Test result: prospective and limited current.

homogeneous layer, with no local variations of I_c along 2G tapes. In fact, I_c may vary locally along the tapes due to inhomogeneity in the HTS material [10].

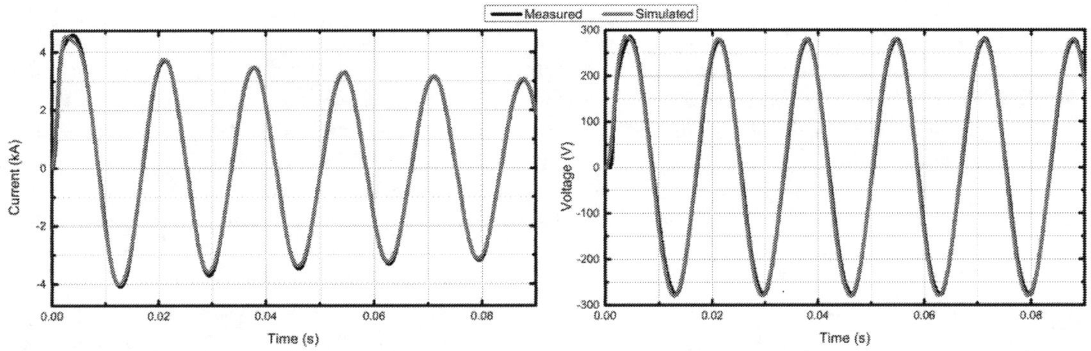

Figure 3. Experimental and simulated results: (a) limited current, (b) voltage; prospective current = 53 kA_{peak} (20 kA_{rms}).

Figure 4a shows the simulated current passing through each layer and external shunt. The current in HTS layer attains a maximum of about 1740 A in 1.5 ms and goes to almost zero in about 4 ms. This actuation time with a relatively low quench current agrees well with the expected behavior of 2G tapes. The HTS layer attains a relatively high resistance value and the current is diverted to the external shunt and silver layers. Since the thickness of Ag top layer is higher than the thickness of Ag bottom layer, the current flowing in the top layer is higher than that of the bottom layer. Just a small current flows in the Hastelloy substrate due to its high resistivity.

The slow decay of follow current presented in figure 3a can be understood by looking at the behavior of follow currents in shunt and silver layers (figure 4a). The resistance of these layers rises considerably with temperature, lowering the follow current, mainly in the case of silver layers. Figure 4b depicts the calculated temperature rise in each layer as well as in the external shunt. Quenching of HTS layer ($T_c \approx 92$ K) takes place in about 2 ms; after quenching, the temperature of HTS layer exhibits a quasi monotonic increase. The other layers present a similar behavior, excepting the fast initial transient of the HTS layer (see the insert in figure 4b). Tape layers attain practically the same temperature due to their low thickness and consequent

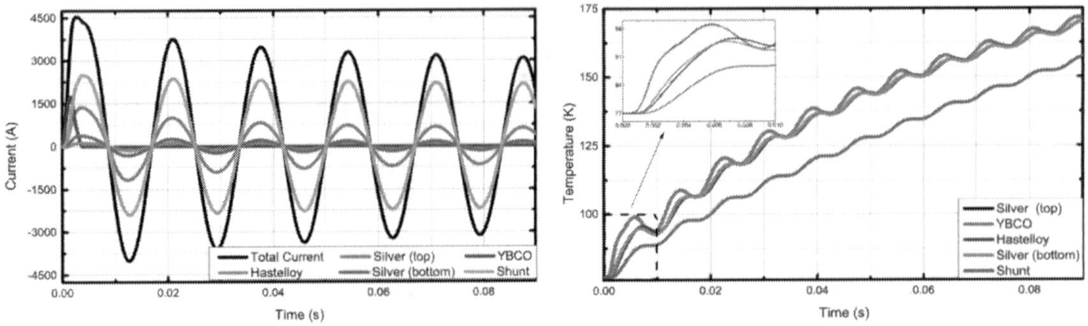

Figure 4. Simulated current in each SCFCL tape layer and shunt (a); temperature rise in tape layers and shunt (b).

fast interlayer heat transfer. Since the metallic shunt is not in contact with 2G tape, it presents a different temperature.

5. Conclusions

In conclusion, the 2G coil R-SCFCL presented an optimal current limiting performance, in good agreement with manufacturer specifications. Simulation and test results agree very well. In addition, simulations provided results of hard experimental access such as the currents flowing in each subcomponent as well as the thermal evolution of each subcomponent. The use of a thermal-electrical analogy showed to be an effective and straightforward way to deal with the strong coupling between electrical and heat conduction phenomena. Further improvements may be achieved by considering the presence of possible inhomogeneities in the HTS layer, since these inhomogeneities may affect material properties, especially local I_c values.

Acknowledgments

The authors are thankful to the personnel of the High Current Laboratory (CEPEL) for the fault current limiting tests and technical support and also to the agencies CNPq and DAAD for the financial support. A special thank to Prof. Dr.-Ing. Mathias Noe for very important advices about the SCFCL modeling.

References

[1] Noe M and Steurer M 2007 *Supercond. Sci. Technol.* **20** 15–29
[2] Morandi A 2013 *Phys. C Supercond.* **484** 242 – 247
[3] Schmitt H, Filho J A, Adapa R, Braun D, Brissette Y, Buchs G, Cvoric D, Darmann F, KEdwards, Fernandez P, Folts D, Hartung K, Hyun O, Jäger J, Iioka D, HKameda, Kim Y, Kleimaier M, Lambert F, Martini L, Noe M, Park K, Rasolonjanahary J L, Steurer M and van der Burgt J 2012 Application and Feasibility of Fault Current Limiters in Power Systems Tech. Rep. 497 Cigré Working Group A3.23
[4] Elschner S, Kudymow A, Brand J, Fink S, Goldacker W, Grilli F, Noe M, Vojenciak M, Hobl A, Bludau M, Jänke C, Krämer S and Bock J 2012 *Phys. C Supercond.* **1** 1–7
[5] de Sousa W T B, Polasek A, Silva F A, Dias R, Jurelo A R and de Andrade Jr R 2012 *IEEE Trans. Appl. Supercond.* **22** 5600106
[6] Park D K, Yang S E, Kim Y J, Chang K S, Ko T K, Ahn M C and Yoon Y S 2009 *Cryogenics* **49** 249 – 253
[7] Robertson A F and Gross D 1958 *J. Res. Natl. Bur. Stand.* **61** 105 – 115
[8] Lawson D I and McGuire J H 1956 *P. I. Mech. Eng.* **167** 275 – 290
[9] Ozisik M N 1993 *Heat Conduction* 2nd ed (John Wiley & Sons)
[10] Gannon J J, Malozemoff A P, Diehl R C, Antaya P and Mori A 2013 *IEEE Trans. Appl. Supercond.* **23** 8002005

Discharge mechanisms in liquid nitrogen with thermally induced gas bubbles

M Blaz and M Kurrat
University of Braunschweig, Institute for High Voltage Technology and Electrical Power Systems, Schleinitzstraße 23, 38106 Braunschweig, Germany

E-mail: m.blaz@tu-bs.de

Abstract. Discharge voltages of a sphere-plane electrode arrangement in subcooled liquid nitrogen stressed with impulse voltages were investigated. Tests were performed with and without thermally generated bubbles between the electrodes. Polarity effects could be observed at higher gaps and lower utilization factors. These effects are explained by the occurrence of positive and negative streamers, which have different propagation mechanisms. The measured discharge voltage with thermally generated bubbles is compared to calculated values using Paschen's law.

1. Introduction

Liquid nitrogen (LN_2) is often used as coolant and electrical insulation of superconducting fault current limiters (SFCL) [1]. During the current limitation phase of a resistive SFCL heat is transferred into the LN_2. This heat input can produce bubbles, which weaken the electrical insulation. In pure insulation liquids it is assumed that the discharge starts due to emission currents from the electrodes [1-2]. With thermally induced bubbles the initiation process of the discharge changes due to the lower dielectric strength of the gas. In [3] the discharge in LN_2 with thermally induced bubbles was investigated in a homogeneous field. At higher pressures and gap distances a polarity effect was observed due to the bubble on the electrode. In this paper the influence of bubbles in the inhomogeneous field on the discharge voltage is analysed.

2. Experimental setup

The measurements were performed with a sphere-plane electrode arrangement, immersed in LN_2 at 78.5 K and 5 bar_{abs}. The setup can be seen in figure 1. The electrodes are made of stainless steel. The sphere (a) has a diameter of 12 mm and the plane (b) of 100 mm. The bobbin of the resistive heater (c) is within the lower part of the sphere electrode and thermally connected to the sphere via a copper rod (d). The heater was supplied with a constant voltage of 11.2 V and a current of 3.6 A. The gap distance can be adjusted by moving the sphere electrode arrangement via a steel cable. The insulation was tested with 1.2/50 lightning impulses using the up and down method. A precise description of this method is given in [3].
The shape of the gas bubbles is dependent on the gap as can be seen in figure 2. Until a gap of 2 mm the bubbles coalesce and reach from the top of the sphere electrode to the surface of the plane electrode (figure 2, 2 mm). At higher gaps discrete bubbles are formed between the electrodes. These bubbles have a diameter of approximately 0.5 mm and are located at the top of the sphere electrode (figure 2, 7.5 mm).

Journal of Physics: Conference Series **507** (2014) 032006 doi:10.1088/1742-6596/507/3/032006

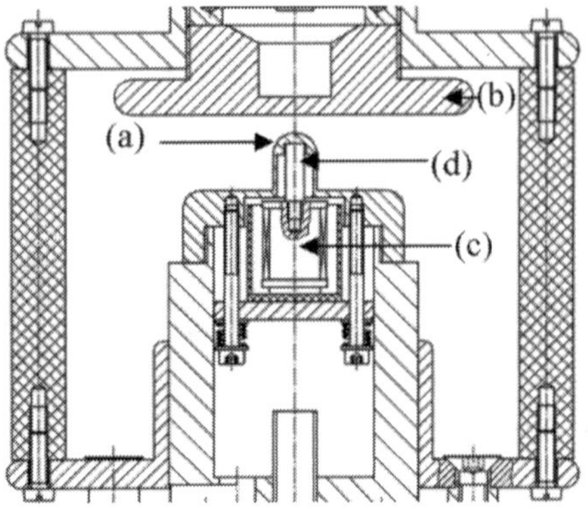

Figure 1. Sphere-plane electrode setup with heater bobbin in the lower electrode.

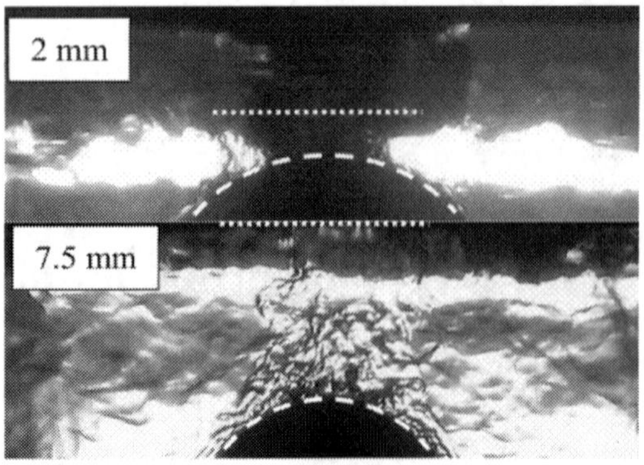

Figure 2. Bubbles at different gaps. Dotted line surface of plane electrode above the sphere, dashed line surface of sphere electrode

3. Results and discussion

The measured discharge voltages without bubbles can be seen in figure 3. If the discharge is triggered by emission currents from the cathode, then the discharge voltage should yield a constant field strength at the sphere surface for different gaps. Up to 3 mm gap, the discharge voltage can be calculated by assuming a constant maximum field at the sphere tip of 82.5 kV/mm and is in good agreement with the measurements (figure 3, dot-and-dash line). But above 3 mm gap the measured values differ from the calculated ones. Additionally the graphs show a polarity dependent change in the discharge voltage, which could be explained by the appearance of positive streamers for the positive sphere and negative streamers for the negative sphere [2]. Positive streamers are faster than negative streamers and can bridge large gaps easier [1]. Therefore the formation of a breakdown at large gaps is more difficult for negative streamers yielding higher discharge voltages [1-2]. It was observed that the negative streamer needs a constant mean field for propagation [4, 5]. For the measured values in figure 3 it is in the range of 50 kV/mm (figure 3, dashed line).

The results of the lightning surge measurements with thermally generated bubbles at an ambient pressure of 5 bar_{abs} can be seen in figure 4. Due to the gas phase, the discharge voltage of these bubbles can be calculated with Paschen's law (1) [3, 6].

$$U_d = \frac{B \cdot p \cdot g}{ln((A \cdot p \cdot g)/k)} \cdot \eta \qquad (1)$$

with p for the ambient pressure, g for the gap, the constant k and the constants for nitrogen gas $A = 510$ bar^{-1} mm^{-1} and $B = 19.65$ kV bar^{-1} mm^{-1} for $E/p = 2 - 8.48$ kV bar^{-1} mm^{-1} and 293 K [3]. If the discharge mechanism represents the Townsend mechanism the constant k is between 2.5 and 18 [3]. In (1) the influence of the temperature on the discharge field strength is not considered. But the density of the gas increases with decreasing temperature. To take the higher density at lower temperatures into account an equivalent pressure is used in (1). This equivalent pressure represents a gas pressure at 293 K with the same gas density as the gas in the bubble at cryogenic temperature and ambient pressure [7]. As a first assumption, the temperature inside the gas bubble is equal to the boiling temperature at ambient pressure. With this temperature for 5 bar_{abs} ambient pressure an equivalent pressure of 17.9 bar_{abs} can be determined [8]. Due to the creation of bubbles at the tip of the sphere, it is assumed that the discharges will start at this point. Therefore the electric field inhomogeneity is taken into account via multiplication with the utilization factor η.

With equation (1) the breakdown voltage can be calculated up to a gap of 2 mm (figure 4, solid line). In this area the breakdown will take place completely in the gas phase. The solid line until 2 mm in

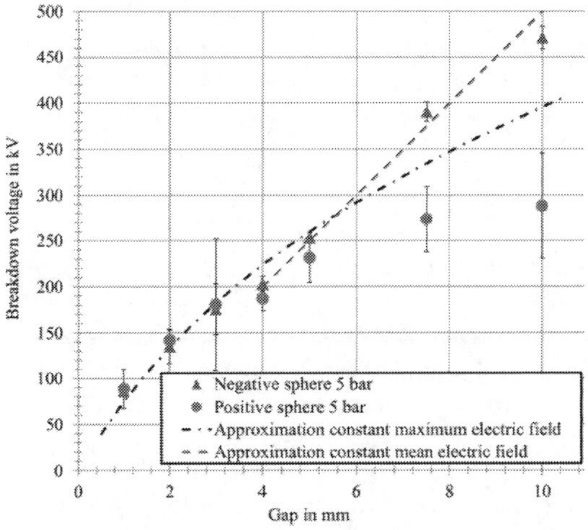

Figure 3. Discharge voltages without bubbles.

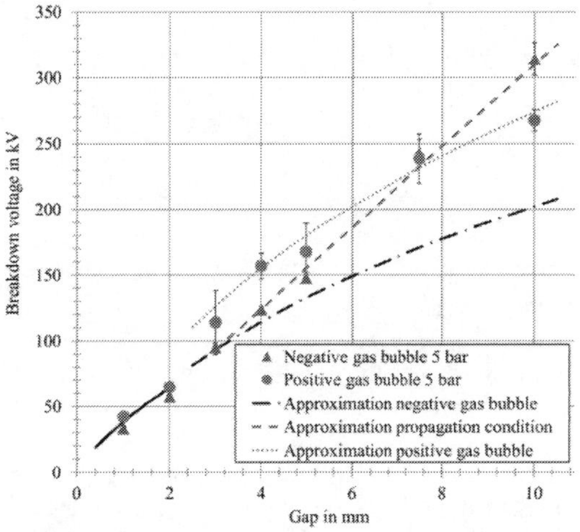

Figure 4. Discharge voltages with bubbles.

figure 4 represents the calculated discharge values. The calculations and measurements show a good agreement. Above a gap of 2 mm discrete gas bubbles are formed and the discharge cannot take place in a complete gas phase anymore. But the gas bubble still represents the weak point of the insulation system. It is assumed, that the complete breakdown occurs, if the discharge in the bubble takes place. The discharge field strength is calculated with Paschen's law again. In contrast to equation (1) the first term of equation (2) calculates the breakdown field strength within a gas bubble with the diameter d_b (0.5 mm). Due to different permittivity of liquid and gas, the field enhancement in the bubble has to be taken into account [3]. The field enhancement in a dielectric sphere located in a homogeneous field can be included with the third term in equation (2).

$$U_d = \frac{B \cdot p}{ln((A \cdot p \cdot d_b)/k)} \cdot \eta \cdot \frac{3 \cdot \varepsilon_{rF}}{\varepsilon_{rG} + 2 \cdot \varepsilon_{rF}} \cdot g \qquad (2)$$

The second term still represents the inhomogeneity. The multiplication of the first three terms yields the mean electric field strength of the arrangement, if the breakdown field strength in the bubble is calculated with Paschen's law. Therefore multiplication with the gap g leads to the discharge voltage across the electrodes. With equation (2) and the Townsend criterion (k=2.5) the dot-and-dash line in figure 4 can be calculated. It is obvious that the measured values do not fit to this calculation.

The discharge voltages for the positive gas bubble on the sphere show a similar tendency as the calculations, but the values are higher. An explanation for this phenomenon could be the required propagation condition for the positive streamer together with the discharge conditions in the bubble on the electrode. If an initiation electron is present in the gas bubble, it will be accelerated towards the anode (figure 5, I). On its way to the anode it will ionize gas molecules. The created ions move to the cathode. At the gas liquid interface the ions can pass into the liquid, but due to their low mobility in the liquid [9] they will accumulate at the interface and reduce the electric field in the bubble but enhance it in the liquid. Additionally the liquid phase on the cathode will hinder the creation of secondary electrons. It is assumed, that the positive streamer propagates through the liquid due to ionization processes in the liquid phase [2]. If the field strength in the liquid is too low for propagation of a positive streamer no breakdown will occur. For the formation of a discharge the charge at the liquid gas interface, produced by an initiation electron, has to be high enough to enable ionization processes in the liquid (figure 5, II). As first hypothesis this situation is considered by applying the streamer criterion (k=18) in equation (2). These calculated values are represented in figure 4 by the dotted line and are in good agreement with the measured values.

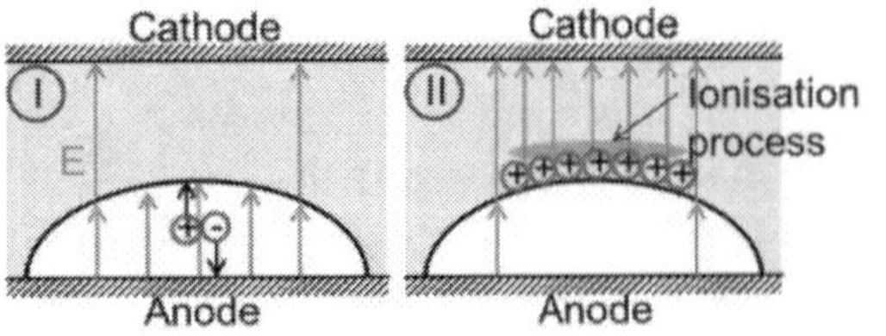

Figure 5. Discharge mechanism in positive bubble

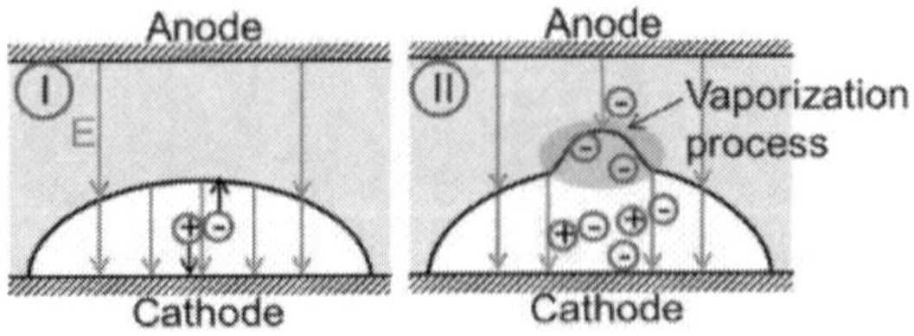

Figure 6. Discharge mechanism in negative bubble

The discharge in the bubble on the negative sphere can still occur after the Townsend mechanism, because secondary electrons can be produced on the surface of the cathode due to its location in the gas phase (figure 6, I). But the measured discharge voltages for the negative gas bubble are also higher than the calculated ones with equation (2) and show a linear dependence of the gap. This effect is similar to the previously investigated propagation condition for negative streamers in LN_2 without gas bubbles (figure 3). The mean field strength, required for propagation of the streamer through the liquid phase, in the measurements with thermal bubbles, is around 31 kV/mm. The propagation of the negative streamer is due to vaporization of the liquid and discharges in the gas phase [1, 2]. In contrast to the ions the electrons have a higher mobility in the liquid. Therefor they will not accumulate at the interface but move through the liquid and will heat it up during their path to the anode (figure 6, II). In the generated gas voids new discharges can occur.

4. Conclusion

In this paper the discharge voltages of a sphere-plane electrode arrangement with and without thermally generated bubbles are investigated. It is shown, that without bubbles, the discharge voltage for small gaps depends on the maximum electric field at the sphere electrode. At higher gaps the field inhomogeneity increases and only streamers with the same polarity as the sphere occur. This yields a polarity effect. Positive streamers have lower discharge voltages than negative ones. The discharge voltage of negative streamers increases linearly with gap due to its propagation condition.

With gas bubbles between the electrodes, the discharge voltage for small gaps can be calculated with Paschen's law considering the Townsend criterion. But at higher gaps polarity effects arise due to different propagation conditions for the positive and negative streamers. For gas bubbles on the positive sphere, the discharge voltage can still be calculated with Paschen's law if the streamer criterion is used. For bubbles on the negative sphere again a linear dependence of the discharge voltage on gap was observed but with a lower slope than for the measurements without bubbles.

5. References

[1] Frayssines P E, Lesaint O, Bonifaci N and Denat A 2002 *IEEE Trans. on Dielectrics and Electrical Insulation* **9** 899-909
[2] Yoshino K, Shiraishi M, Ohseko K and Inuishi Y 1981 *Tech. Repts. Osaka Univ.* **31** 271-6
[3] Blaz M and Kurrat M 2013 *IEEE Trans. on Appl. Superconductivity*, **23** 7700804
[4] Rzad S J, Devins J C, Schwabe R J, 1979 *IEEE Trans. on Dielectrics and Electrical Insulation* **14** 289-96
[5] Lesaint O and Tobazéon R 1988 *IEEE Trans. on Electrical Insulation* **23** 941-54
[6] Atrazhev V M, Vorob'ev V S, Timoshkin I V, MacGregor S J, Given M J, Wilson M P and Wang T 2012 *IEEE Trans. on Dielectrics and Electrical Insulation* **19** 1552-8
[7] Hara M, Suehiro J, Matsumoto H and Kaneko T 1989 *IEEE Trans. on Dielectrics and Electrical Insulation* **24** 609-617
[8] Jacobsen R T, Stewart R B and Jahangiri M 1986 *J. Phys. Chem. Ref. Data* **15** 735-908
[9] Gee N, Floriano A, Wada T, Huang S S S and Freeman G R 1985 *J. Appl. Phys.* **57** 1097-101

Journal of Physics: Conference Series **507** (2014) 032007 doi:10.1088/1742-6596/507/3/032007

Investigation of the cooling conditions for the Fast Ramped Superconducting Magnets of the SIS100 Synchrotron

A Bleile[1], **E Fischer**[1], **H Khodzhibagiyan**[2], **A Mireau**[1] **and P Schnizer**[1]
[1] GSI, Germany
[2] JINR, Russian Federation

E-mail: a.bleile@gsi.de

Abstract. The dynamic losses in fast ramped super ferric magnets will create the main part of the expected heat losses in the superconducting synchrotron SIS100 – the primary accelerator in the Facility for Antiproton and Ion Research (FAIR). The data measured on dipole models at the cryogenic magnet test facility at GSI allow to predict the static and dynamic losses in the superconducting magnets for different operation modes of the SIS100 synchrotron. The calculations of the expected mass flow rates in the cooling channels of the superconducting magnets together with the analysis of the cooling system for different operation modes of the synchrotron will be presented.

1. Introduction

The Facility of Proton and Ion Research (FAIR) will construct as a set of accelerators and storage rings at GSI. The aim of the project is to provide high intensity primary and secondary beams of ions and antiprotons for experiments in nuclear, atomic and plasma physics. The SIS100 synchrotron with a circumference of 1100 meters and a rigidity of 100 Tm will provide the primary beams of different species of ions from protons to uranium and will enable a parallel operation of up to four scientific programs. A big variety of different operation cycles of the SIS100 synchrotron with a repetition rate up to 1 Hz requires fast-ramped superconducting magnets with high dynamic heat release.

The design of the superconducting magnets and of the cooling system of the SIS100 is based on the Nuclotron synchrotron at JINR in Dubna, which was constructed on the basis of the fast-cycling superferric magnets [1]. During an intensive R&D phase the magnet field quality, AC losses and long term mechanical stability have been improved and were experimental tested on the short models and on the full size prototypes [2], [3].

2. Superconducting magnets for SIS100

2.1. Design aspects

The superferric dipoles and quadrupoles for the SIS100 synchrotron (Figure 1) consists of an iron yoke and a superconducting coil made from the Nuclotron type cable where the superconducting NbTi wires are wrapped around a CuNi tube which acts also as a cooling tube. Both the superconducting coil and the yoke are cooled by two phase helium flow. The inner diameters of the cooling tubes in the coils are 4.7 mm for the main dipoles and quadrupoles and 4.0 mm for the corrector magnets. The small inner diameter and the big length of the cooling tube (about 108 m for the coil and bus bars of the main dipoles) limit the mass flow rate of the two phase helium up to 1- 2 g/s.

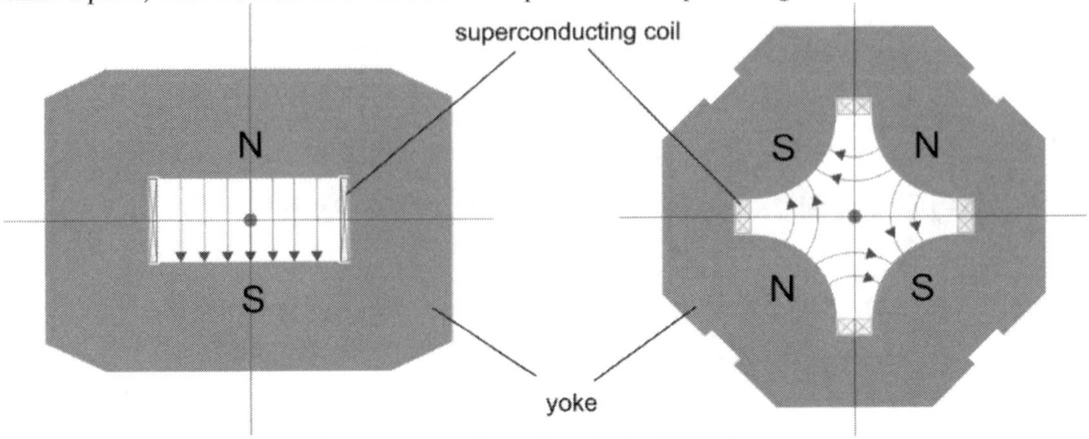

Figure 1. Cross sections of super-ferric dipoles and quadrupoles for SIS100

2.2. Cooling scheme

The cooling scheme of the SIS100 dipole magnet is shown on the Figure 2. The helium in the supply line is sub-cooled and has a temperature of about 4.5 – 4.6 K at P = 1.5 – 1.6 bars. The helium cools successively the magnet bus bars, the coil (Points 1 -2) and the iron yoke (Points 2 -3). Due to the pressure drop in the coil and bus bars as well as due to the static and dynamic heat load the helium in the Point 2 is in the two phase state at $P_2 = P_3 = 1.1 - 1.2$ bar and $T_2 = T_3 = 4.3$ -4.4 K (the subscripts 1,2 and 3 denote the points on the Figure 2). The major part of dynamic heat losses is created in the iron yoke, so the helium in the Point 3 can be either two phase or vapor, depending of the operation mode of the SIS100 synchrotron. All magnets in one sector are connected in parallel (Figure 2, right). The recoolers attached to the supply header (Points 2-2') transfer the heat to the two phase helium and keep the helium in the supply header in the single phase state over the whole length of the SIS100 sector. The cooling scheme of the quadrupole doublets is shown on the Figure 3. A quadrupole doublet module has two parallel flow channels denoted on the Figure 3 as M1 and M2 for the cooling of both quadrupole units. A unit consists of a quadrupole magnet switched hydraulically in series with a corrector magnet. A corrector magnet can be a steerer, chromaticity sextupole, a multipole corrector or a γt-jump quadrupole. All corrector magnets except for the normal conducting γt-jump quadrupoles have the cooling tubes with an inner diameter of 4.0 mm and with the length of 15 – 43 m. The helium flow after each quadrupole unit cools also the common girders – the support structure for the both units.

Since the fast ramped magnetic field generates the heat losses also in the beam vacuum chambers, the dipole and quadrupole magnet modules have a dedicated cooling circuit for the beam vacuum chambers and for the cryo catcher chambers (channel V on the Figure 3).

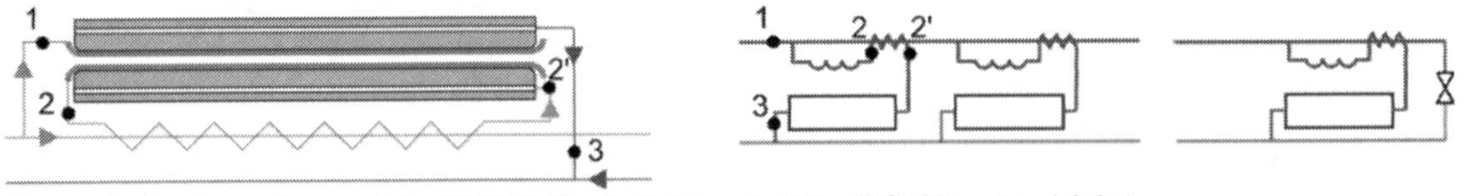

Figure 2. Cooling scheme of the SIS100 dipole (left) and of the SIS100 sector (right)

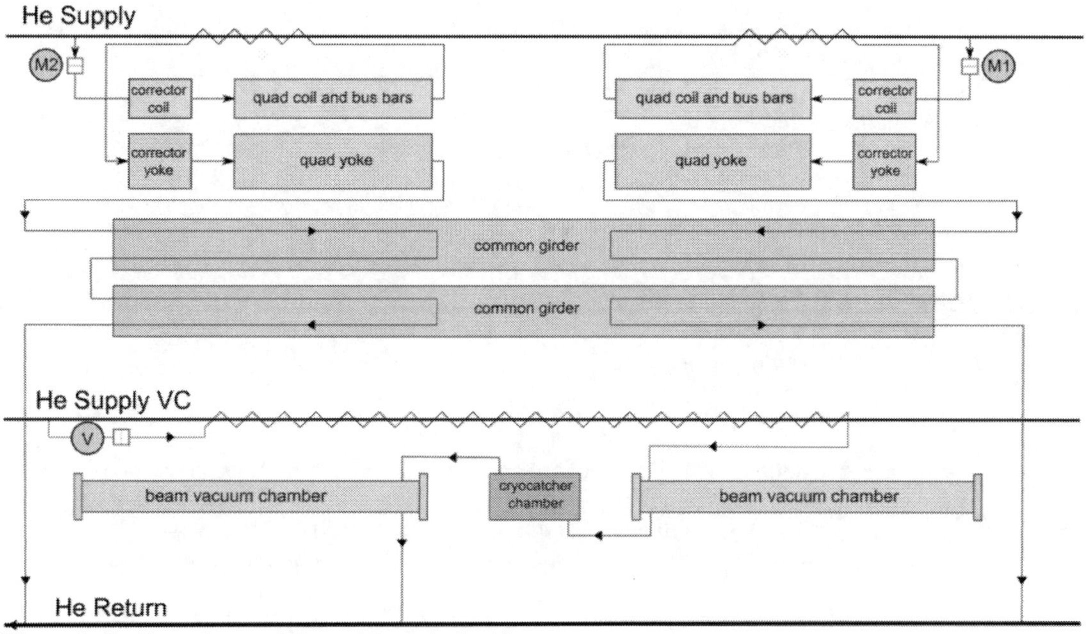

Figure 3. Cooling scheme of the quadrupole doublet

3. Cooling conditions for different operation modes of SIS100

The big variety of operating cycles of the SIS100 causes the wide range of the dynamic heat losses. The dynamic loses for dipole magnets have been measured on prototype test facility at GSI and vary from 0 W in the stand-by mode to 70 W in the operating cycle with B_{max} = 1.9 T at repetition rate of 1 Hz (so called triangular cycle). However, this cycle with a one-turn injection is not foreseen for the standard operation of the SIS100. The inner diameter of the cable for the main dipoles and quadrupoles has been optimized for the operation mode with a multi-turn injection witch causes the dynamic head load of about 50% of the maximum. In this mode the two phase helium at the output of the main dipoles has a vapor quality x~0.9. The estimated values of dynamic losses in the quadrupole magnets are about 35 W. These losses as well as the losses in the corrector magnets will be verified on the first prototype of the quadrupole doublet.

The calculated mass flow rates as well as the conditions on the outputs from the cooling channels are presented in the Table 1 for two cases: 50% and 100% of the maximal dynamic load. The calculations have been performed under following assumption:

- Supply header: P = 1.6 bar, T = 4.5 K, helium is sub-cooled by 0.2 K
- Return header: P = 1.2 bar
- Homogenous model for the two phase flow with the viscosity of the liquid phase ($\eta = \eta_l$)
- Distribution of dynamic losses between coil and yoke: 30% / 70%
- Static losses: 5 W for the dipole modules, 6 W for the quadrupole doublets
- Corrector magnets have a HTS current leads with the heat load < 0.5 W/pair at 4.6 K

The calculations show that all configurations of the cooling channels will differ in the helium mass flow rates and in the vapour quality i.e. in the temperature on the output from the cooling channel. At maximal load the helium on the output from the iron yokes of the dipole modules will have a temperature of about 7.3 K. The operation mode with output temperature of the iron yoke of 7-8 K was verified during the cold testing of the dipole prototypes with two-layer coils [4]. The cold tests of the first series dipole with a single layer coil are scheduled to the end of 2013. The losses in the quadrupole units as well as the reliable operation at the maximal load will be verified within the cold

Table 1. Mass flow rate and output conditions for different configurations of the cooling channels in one sextant of the SIS100

Configuration of the cooling channel	number of channels	50% load		100% load	
		m, g/s	x_{out}	m, g/s	T_{out}, K
Dipole	18	2.4	0.9	2.0	7.3
Quadrupole	10	2.5	0.3	2.0	4.4*
Quadrupole + Sextupole	3	2.6	0.6	2.3	4.7
Quadrupole + Steerer	7	2.3	0.6	2.2	4.6
Quadrupole + Steerer + γt-jump quadrupole	2	2.3	0.6	2.2	4.6
Quadrupole + Steerer + Sextupole	4	1.9	0.9	1.7	6.5
Quadrupole + Multipole Corrector	1	2.0	0.8	1.8	5.7
Quadrupole + Multipole + Steerer	1	1.5	1.0	1.3	8.0

*) two phase helium, x = 0.7

test of the first quadrupole doublet. Especially for the configuration Quadrupole + Steerer + Multipole a dedicated third cooling channel for the multipole corrector may be required because of the long coil of the multipole (43 m cable with inner diameter of 4.0 mm) and thus the low helium mass flow rate.

On the other hand, the mass flow rates through the channels with small hydraulically resistance will be higher than required for the reliable operation, especially at the operation modes with low dynamic losses. Therefore the additional flow impedances (flow restrictors) will be placed on the inputs to the cooling channels M1, M2 and V in the quadrupole modules (Figure 3). The flow restrictors will be sized after the firs tests of each type of the quadrupole unit in order to adjust the vapour quality of about 0.9 for all cooling channels at the operation mode with 50% dynamic load.

4. Conclusions

The big variety of operation modes of the SIS100 synchrotron and different configuration of the parallel cooling cannels require the adjustment of the hydraulic resistances of all cooling channels to one common level at one of the standard cycles of the SIS100. The adjustment will be done by combination of quadrupole magnets with corrector magnets as well as by additional flow impedances on the inputs to the cooling channels.

References

[1] Baldin A.M. et al. 1995, Superconducting fast cycling magnets of the Nuclotron, *IEEE Trans.of Appl.Supercond.*, Vol.5, pp. 875-877

[2] Fischer E et al., Fast Ramped Superferric Prototype Magnets of the FAIR Project, First Test Results and Design Update, *Proc. of the Particle Accelerator Conference PAC'09*, Vancoover 2009

[3] Fischer E, Khodzhibagiyan H, Kovalenko A, Schnizer P, Fast Ramped Superferric Prototypes and Conclusions for the Final Design of the SIS100 Main Magnets, *IEEE Trans.of Appl.Supercond.*, Vol.19, pp. 1087-1091, June 2009

[4] Bleile A, Fischer E, Khodzhibagiyan H, Mierau A, Schnizer P, Measurement of Dynamic Heat Losses in the Fast Ramped Superconducting Magnets for the SIS100 Synchrotron, *Proc. of the 12th International Conference Cryogenics 2012*, pp. 59 – 64, Dresden 2012

Operational experience in the use of 18 kA HTS current leads for Edipo

S March[1], R Wesche[1] and P Bruzzone[1]

[1]EPFL-CRPP Fusion Technology, CH-5232, Villigen PSI, Switzerland

Email: Pierluigi.Bruzzone@psi.ch

Abstract. In spring 2013, the Edipo facility of CRPP was commissioned. The dipole is powered via two 18 kA HTS current leads, designed and manufactured at CRPP. As part of the Edipo commissioning framework, the operational parameters of the leads were implemented in the control system. The in-situ tests were found to be in good agreement with the tests performed without a background field in 2011. The leads consist of a conduction cooled HTS module, made of AgMgAu/Bi-2223 stacks, and a wire bundle heat exchanger. The heat exchanger is cooled by forced flow helium gas, the inlet temperature of which was measured to vary between 65 K and 85 K. During operation with field, the mass flow rate is a function of current (2.05 g/s per lead at full field, 12.35 T, 17.2 kA). Reduced cooling investigations showed that 0.31 g/s per lead is suitable for overnight standby and 0.2 g/s per lead for longer periods. For detection of and protection against quench in the HTS module, a threshold of 10 mV was found to be appropriate. The heat exchanger has a voltage protection threshold of 120 mV. The temperatures of the heat exchanger, the HTS, and the helium inlet temperature were monitored in order to provide a further layer of protection.

1. Introduction

The European Dipole (Edipo) is a new facility, hosted by CRPP, for the testing of high current force flow conductors in a high field [1]. In order to reduce the cryogenic load, the dipole is powered by two CRPP designed HTS current leads [2]. The conduction cooled HTS module consists of AgMgAu/Bi-2223 stacks soldered on a stainless steel support, similar designs have been successfully used for particle accelerators [3] and magnets for fusion [4]. The resistive part of the lead is a wire bundle heat exchanger that is cooled by forced flow helium. The helium inlet temperature is 75 K, which was measured to vary between 65 K and 85 K. The leads were commissioned in 2011 without a background field [5]. The first cool-down of the Edipo coil assembly lasted 50 days with a subsequent cool-down lasting only 24 days. In both cases the leads were cooled down together with the dipole and did not limit the overall cool-down rate. The leads are actively cooled once the dipole temperature is less than 10 K. No problems with the leads were observed during cool-down.

2. Operation with field

During operation with field, the leads are independently operated with the mass flow rate ($\dot{m}$) controlled as a function of current (figure 1). A minimum mass flow rate of 0.38 g/s per lead is activated at the same time as the power supply in order to prevent quenches at low current. The control system has an option to automatically increase the flow and ramp down the field in the event of too high a temperature. Furthermore, the temperature differences between redundant sensors in the same locations are also over-watched by the control system. Temperatures and flow rates for a ramp to the full Edipo field of 12.35 T, corresponding to 17.2 kA, are shown in figure 2. The cooling of the leads was found to be sufficiently stable both during current ramps and flat tops to allow successful operation with field. Variation of the helium inlet temperature did not cause any problems.

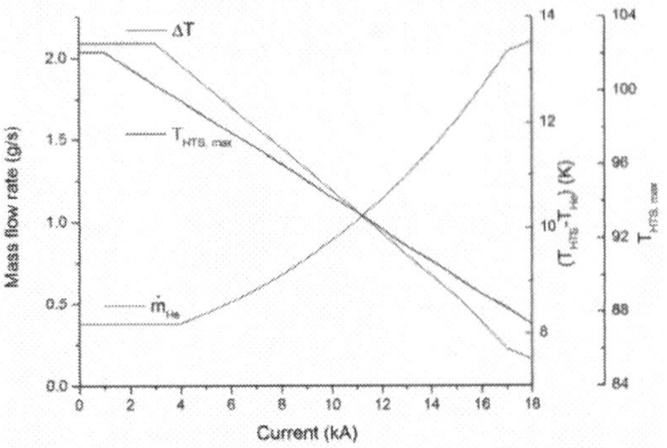

Figure 1. Mass flow rate as a function of current used to control the HTS leads. Also shown are the maximum allowable HTS temperature ($T_{HTS,\ MAX}$) and the maximum allowable temperature difference between HTS and He inlet (T_{HTS} - T_{He}).

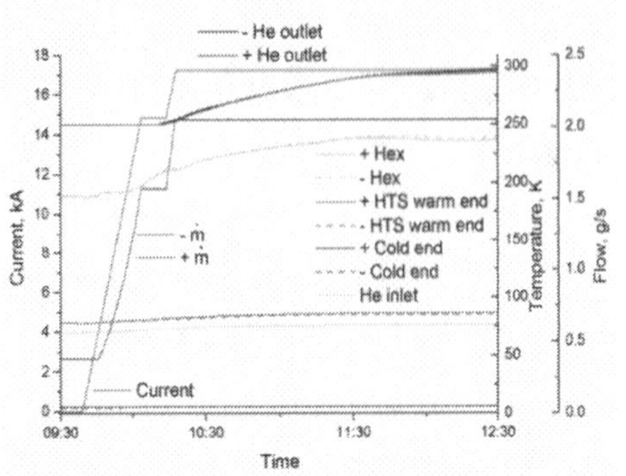

Figure 2. Temperatures and flow rates for the two leads during a current ramp and subsequent stable operation at 17.2 kA. Above 4 kA mass flow is a function of current. The thermocouples for the helium outlets have a lower temperature limit of 250 K.

During commissioning of the leads in 2011, the mass flow was set as a function of the temperature difference (ΔT) between the HTS (T_{HTS}) and the helium inlet (T_{He}). However, it was found that ΔT control was not suitable for use with field due to variations in T_{He}. Control with $\dot{m}(I)$ necessitated the use of a higher flow than with $\dot{m}(\Delta T)$. After commissioning the temperature sensors had 0.4 mm thick glass/epoxy insulation on both sides, previously they were not insulated. Overall, the temperatures during operation were in agreement with the prior measurements (table 1).

Table 1. Comparison between commissioning and operational values. With field; control was $\dot{m}(I)$ and the temperature sensors were insulated Commissioning; control was $\dot{m}(\Delta T)$, there was no insulation.

Parameter	**Commissioning (no field)**	**Operation (12.35 T)**
I (kA)	17	17.2
$\dot{m}$ (g/s)	1.84	2.01
T_{HTS} (K)	84	85-86
T_{He} (K)	76.2	76.4
ΔT (K)	7.8	8.6-9.6

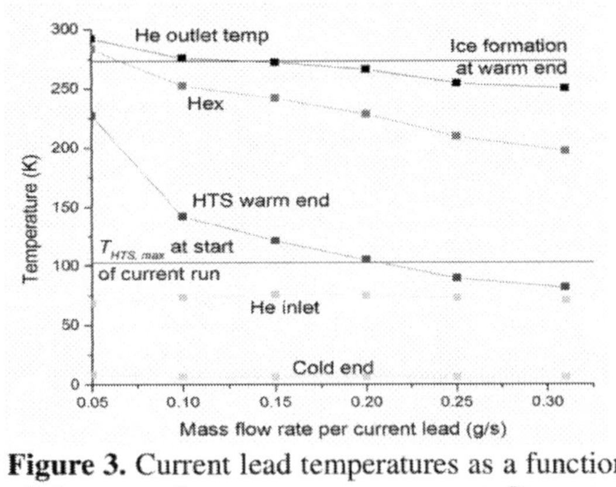

Figure 3. Current lead temperatures as a function of the mass flow rate at zero current. Presented temperatures are an average of all sensors installed in each location and each lead. The allowable $T_{HTS,\ MAX}$ to start current from zero is 102 K. Ice formation line is at 273 K.

3. Standby operation

The nominal standby mass flow rate is 0.31 g/s per lead, when the power supply is not energized. This cools the leads sufficiently enough that operation with current can start immediately. During commissioning, the helium outlet temperature was ~270 K. However, once installed with the complete facility, and with longer periods of standby, this was measured to sink to ~250 K resulting in the formation of ice on the room temperature end of the leads. The variation of the leads' temperatures, at zero current, with different mass flow rates is shown in figure 3. For longer standby the helium mass flow rate is reduced to 0.1 g/s per lead. A mass flow rate of 0.1 g/s per lead was found to be appropriate to avoid ice or moisture formation. However, this leads to a temperature ~130 K at the warm end of the HTS module and so a subsequent re-cooling is required before operation with current. The re-cooling duration is less than one hour. For longer periods of standby (weekends or more) a standby $\dot{m}$ of 0.1 g/s is used, for overnight standby 0.31 g/s is retained with the resulting ice being none or minimal. Fans were installed to enhance air circulation over the leads and other feed throughs.

4. Helium inlet temperature stability

The ~30 year old CRPP cryoplant [6] provides supercritical helium at ~4.5 K and helium gas with a nominal temperature of 75 K. The cryoplant utilizes a 750 kW base load motor and an additional 250 kW motor to produce 1.2 kW at 4.4 K. The cryoplant is shared with the Sultan [7] test facility and there is sufficient cooling power for one facility to be in standby whilst a high field test is performed in the other facility. In standby only one compressor is required; if either facility is operating with field then both compressors are needed.

A particular feature of this configuration is the 75 K helium gas that is used for cooling the current leads and the thermal shields. This temperature was measured to vary between 65 K and 85 K and the temperature is weakly linked to pressure (figure 4), but there is a large amount of scatter. The variation is unavoidable and is due to a variety of factors, including pressure drop over the filters, external temperature, and number of compressors in use. The high helium inlet temperature results in an HTS warm end temperature of ~80 K, significantly higher than that used for other HTS leads. For comparison the LHC leads have an HTS warm end temperature of 48-52 K [8] [9]. This high temperature with a potentially large variation necessitated a large safety factor to be used for the specification of the amount of HTS material [2]. However, the resulting robust design has been proved to allow safe operation of the leads over a variety of helium inlet conditions.

5. Operational aspects of quench and thermal runaway

A voltage based system is used to protect the HTS module against quench and the resistive heat exchanger against thermal runaway (e.g. in case of loss of flow). Protection of the current leads is integrated within the complete Edipo quench detection system (QDS), the development and commissioning of which is reported elsewhere [10] [11]. For detection of quench in the HTS module, the proposed threshold of 10 mV was found to be appropriate. This was a compromise between the relatively high noise (+/- 3 mV) and a high HTS temperature if too high a threshold is specified (~50 mV at 17 kA: ~127 K). Whilst a constant voltage threshold is generally suitable for protecting a superconducting component against quench; it is less suitable for protecting the resistive heat exchanger against thermal runaway- at low current a much higher temperature rise is experienced before the voltage is over-threshold. A temperature based protection system could be used, but in practice a significant lag time between thermal runaway and a sensor's response arises due to the electrical insulation. The heat exchanger is adequately protected at full current with threshold of 120 mV, lower than the 150 mV used for commissioning at 18 kA. To improve the protection of the heat exchanger at low currents; a current dependent voltage threshold could be used, or the heat exchanger temperature could be integrated into the QDS.

During commissioning of the dipole the negative current lead was quenched at low current (720 A). The quench voltages are shown in figure 5. The quench occurred ~72 s after the start of the current ramp and was due to an insufficiently cooled lead. Although the HTS temperature was decreasing, the rate of decrease was not sufficient when compared to the rate of current increase (10 A/s). The control logic was subsequently updated so that the minimum cooling when the power supply was energized was raised from 0.31 g/s to 0.38 g/s per lead; i.e. constant mass flow rate beneath 4 kA and $\dot{m}(I)$ at higher currents. There were no further low current quenches.

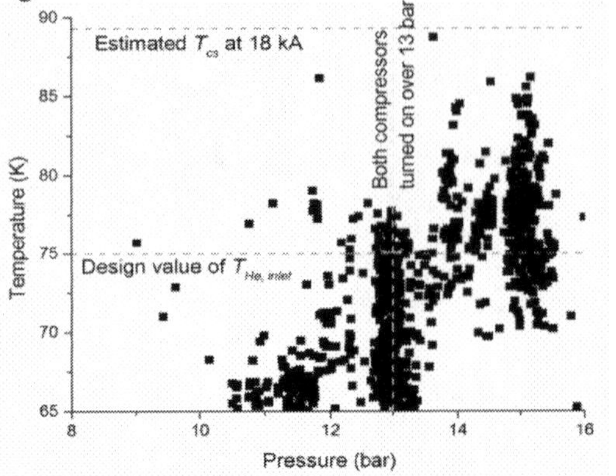

Figure 4. Variation of the helium inlet temperature as a function of the pressure. Above 13 bar, both of the cryoplant compressors are required. The estimated T_{cs} was presented in [5].

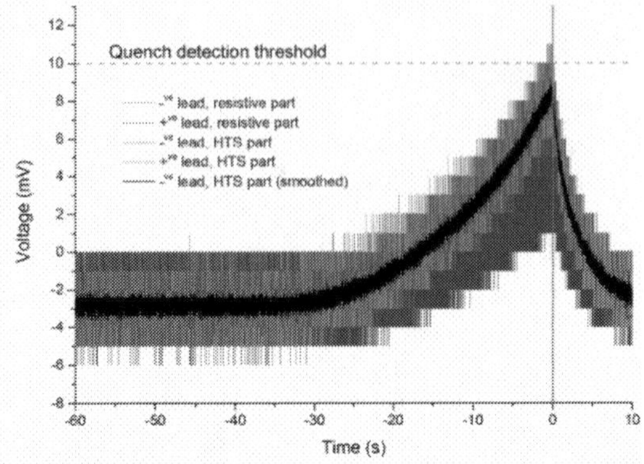

Figure 5. Quench voltages at ~720 A. Quench was detected first on the HTS part of the negative lead; a 5 point smoothing of this signal is shown.

6. Conclusion

The current leads functioned well during the commissioning of the Edipo facility and allowed operation at high field (12.35 T, 17.2 kA) for extended periods of time. During the several weeks of operation with and without field the operational parameters of the leads were refined and were implemented in the control system. Reduced cooling parameters were established for long standby periods in order to prevent ice formation at the warm end of the leads. Due to a robust current lead design, the high helium inlet temperature (75 K) and its large variation (+/- 10 K) did not lead to problems during operation. The quench detection system was shown to adequately protect the leads.

References

[1] Bruzzone P *et al* 2013 Commissioning of the main coil of the Edipo test facility *IEEE Trans. Appl. Supercond.* (in press)
[2] Wesche R *et al* 2009 Design and manufacture of 17 kA HTS current leads *EUCAS 9 (Dresden)*
[3] Ballarino A 2008 Large-capacity current leads Physica C: Superconductivity **468** 2143–8
[4] Wesche R *et al* 2007 Design of high-temperature superconductor current leads for ITER *Fusion Engineering and Design* **82** 1385–90
[5] Wesche R *et al* 2011 Test results of the 18kA Edipo HTS current leads *Fusion Engineering and Design* **86** 1414–7
[6] Arpagaus M *et al* 1984 Helium refrigerator for SULTAN *Sulzer Technical Review* **3** 15–17
[7] Bruzzone P *et al* 2002 Upgrade of operating range for SULTAN test facility *IEEE Trans. Appl. Supercond.* **12** 520–3
[8] Ballarino A 2013 Operation of 1074 HTS current leads at the LHC: Overview of three years of powering the accelerator *IEEE Trans. Appl. Supercond.* **23** 4801904–4801904
[9] Ballarino A *et al* 2008 Commissioning of the LHC current leads *EPAC 11 (Genoa)*
[10] Croari G *et al* 2012 Edipo Quench detection system: concept and development *IEEE Trans. Appl. Supercond.* **22** 4702604–4702604
[11] Croari G and Bruzzone P 2013 From conception to commissioning of Edipo and SULTAN quench detection systems *Magnet Technology 23(Boston)*

Strengthening future electricity grid of the Netherlands by integration of HTS transmission cables

Roy Zuijderduin[1], Oleg Chevtchenko[1], Johan Smit[1], Gert Aanhaanen[2], Rob Ross[2]

[1]Delft University of Technology, The Netherlands
[2]TenneT, Utrechtseweg 310, Arnhem 6800 AS, The Netherlands

Corresponding author's e-mail address: o.chevtchenko@tudelft.nl

Abstract. The electricity grid of the Netherlands is changing. There is a call of society to use more underground cables, less overhead lines (OHL) and to reduce magnetic emissions. At the same time, parts of the future transmission grid need strengthening depending on the electricity demand in the coming decades [1]. Novel high temperature superconductor (HTS) AC transmission cables can play a role in strengthening the grid. The advantages as compared to alternatives, are: economic, underground, higher power capacity, lower losses, reduced magnetic field emissions in (existing) OHL, compact: less occupation of land and less permits needed, a possibility to keep 380 kV voltage level in the grid for as long as needed. The main obstacles are: the relatively high price of HTS tapes and insufficient maturity of the HTS cable technology. In the paper we focus on a 34 km long connection in the transmission grid (to be strengthened in three of the four of TenneT scenarios [1]), present the network study results, derive the requirements for corresponding HTS transmission cable system and compare HTS system to the alternatives (OHLs and XLPE cables).

1. Introduction

In the year 2030 the losses in the Dutch AC grid will be 10 TWh when only copper conductors are used. The lost amount of electricity is comparable to that generated by all Dutch offshore wind parks in the same year (in the moderate scenario). Many grid components are at the end of their lifetime and a replacement wave is expected. Since a regular replacement can take 10 years, new solutions need to be studied timely. In the framework of the national project SuperNet, we study how to strengthen the Dutch grid of 2030 with HTS cables. The study uses four scenarios of the grid development provided by the Dutch Transmission System Operator (TSO) TenneT.

In [2] general trends of Dutch transmission grid were explained using the scenarios of TenneT illustrated with the example of network study of the future grid strengthening with HTS cables. However, economic (and other) aspects of such step were not considered. These are very important, since many of HTS cable demo-projects stop as soon as a subsidy is off. Instead, one can look for economically sound sustainable areas where HTS cables can be applied (without compromising e.g., system reliability) and a reasonable profit can be made at the same time.

In this paper we present results of the techno-economic study performed in this direction for HTS transmission cables, compare HTS cable to available alternatives (OHL and XLPE cable) focus on one selected connection and define areas in the grid where the sustainable approach is valid. Based on the analysis, two new scenarios for integration of HTS cables in Dutch transmission grid are proposed.

HTS cable economy is sensitive to the price of HTS tape. Our approach is that higher voltage allows using less HTS tapes for the same transmitted power, leading to a competitive HTS cable.

Journal of Physics: Conference Series **507** (2014) 032009 doi:10.1088/1742-6596/507/3/032009

2. Dutch transmission grid and the connection Krimpen-Geertruidenberg

2.1. Dutch grid today and in the year 2030

The Dutch AC grid is shown in figure 1a. It consists of the 380 kV ring and several 380 and 220 kV connections. The total (single circuit, 380 kV) length amounts 2127 km. Even though XLPE cable outperforms OHL in public acceptance, rights of way, efficiency, magnetic fields, visual and acoustic impacts, TV and radio interference, electrocution risk, impact on aviation, on building planning, on flora/fauna, impacts of weather and catastrophic events, over 95% of the AC grid is made of OHLs. The total electrical energy produced (consumed) in year 2010 was 124 (109) TWh respectively. Example of the Dutch transmission grid in year 2030 is shown in figure 1b [1]. In this scenario, according to TenneT, business will be as usual. The grid will be made of OHLs where possible and of XLPE cables (where OHLs are impossible to use), with no HTS cables foreseen. Total electrical energy consumed in year 2030 will be 150 TWh and the losses are 1.5-2 TWh.

Four scenarios of possible grid development in year 2030 are provided by TenneT [1] and further elaborated in [2]. The important conclusion for this study was that in three of the four TenneT scenarios, the connection Krimpen-Geertruidenberg K-G (see figure 1a) will need strengthening.

2.2. Connection Krimpen-Geertruidenberg

A closer look at what this 380 kV connection is today and will be in year 2030, exposes the following. Dutch regulation states that the total length of OHL cannot be increased. At present two 2.5 kA OHLs form this 33.7 km long connection, each OHL will be upgraded to 5 kA later on. This however is not sufficient, and in the year 2030 the following alternatives are considered for *n*-1 redundancy: option 1, one new 5 kA OHL will be added. This would require more rights of way and contradicts with the Dutch regulation; option 2, two 5 kA HTS cables in parallel to the two OHL (using the same rights of way); option 3, one 5 kA HTS cable (in series with a flexible alternating current transmission system FACTS) in parallel to the two OHLs (using the same rights of way); option 4, two 5 kA XLPE cables added in parallel to the two OHLs. XLPE cables will need additional right of way, though less than one additional OHL (in option 1).

Reliable transmission system needs n-2 redundancy (n-1 under maintenance). Corresponding options were included in our study, but are out of scope of this paper.

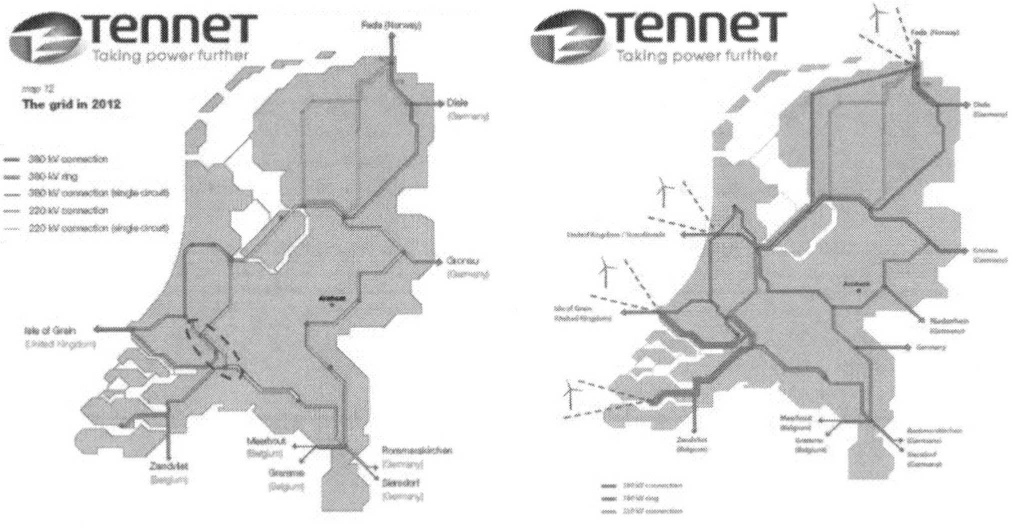

a b

Figure 1. Dutch transmission grid (380 kV and 220 kV are shown in orange and green respectively): a – existing grid in year 2012 [1], the connection K-G is indicated by the dashed ellipse; b – in year 2030, example scenario - business as usual - of TenneT [1].

2.3. Network study

Example of the study is presented in Table 1. Here (for *n*-1 redundancy) from left to right column are shown (calculated values of): transmitted power flow *P*, reactive power Q_1 and Q_2, for respectively the source and the load sites, receiving end current I_2, voltage drop ΔU, the loss per parallel branch and the total loss per connection in GWh/year. The last columns show averaged over the lifetime of 40 years losses (assuming a load growth of 2 % per year). The assumed power load (transport current) in year 2030 and at the end of the lifetime is 7 GW (10 kA) and 10 GW (15 kA), respectively. Under the *n*-1 redundancy the OHL can carry double rated current for days. Therefore, e.g., for option 1, three (not four) OHLs are needed to transmit the 10 GW. While for option 1 the currents are equally distributed over the OHLs, in option 2 most of the current flows in HTS cables. However, when e.g., one OHL fails, too much current will flow in one HTS cable; this is why namely two HTS cables are needed. In option 3 one of the HTS cables is replaced with FACTS (in series with the HTS cable) distributing the load current properly between the cable and the OHLs in all situations. For more details (including other assumptions of the study) see [2].

2.4. Economics of the available options

For options 1, 4 economics is evaluated using TenneT procedures. For options 2, 3 a model was constructed including the technical specifications, genuine design of all HTS cable components and systems, realistic costs of these (terminations, HTS cable core, electrical insulation, cryostat, cooling system, etc.), laying, dismantling, maintenance and other costs allowing to assess the investment (IC) and the lifetime costs (LTC) of the connection. The YBCO tape price was estimated to decline from present 350 down to 25 €/kAm within next 10 years [3], since just 10 km of such HTS cable would require 5 Mm of such tape.

Performed analysis shows that option 3 as compared to option 1 has the same IC and same LTC at the HTS tape price-performance ratio of 103 and 241 €/kAm respectively. At the tape price of 25 €/kAm the savings on IC and LTC are 227 and 625 M€ respectively. As compared to 4, option 3 has the same IC and the same LTC at the HTS tape price of 283 and 356 €/kAm respectively. At the tape price of 25 €/kAm the savings on IC and LTC are 746 and 957 M€ respectively.

Option 2 as compared to 1 has same IC and same LTC at the HTS tape price of 36 and 135 €/kAm respectively. At the tape price of 25 €/kAm savings on IC and LTC are 63 and 611 M€ respectively. As compared to 4, option 2 has the same IC and LTC at the tape price of 135 and 195 €/kAm respectively. At the price of 25 €/kAm the savings on IC and LTC are 582 and 944 M€.

Notably, a hybrid connection (options 2-4) allows to expand the grid underground, as such connection is more reliable than that made of cables only (since if a cable is out of service, remaining OHLs transmit full load current for as long as needed e.g., to repair a cable). Besides, options 2, 4 reduce magnetic emissions from existing OHLs drastically (see table1), a feature desirable for a TSO. Moreover, when HTS cables are used instead of XLPE, substantial cost savings are expected (see text above), connection is more compact and its length has no practical limit in a country like the Netherlands (25 km for option 4) [2]. Other applications of HTS cables include e.g., replacement of aged OHLs and partial undergrounding.

Table 1. Example of the grid study to strengthen the 380 kV connection K-G with OHLs (5 kA each), XLPE cables (2.5 kA each) and HTS cables (5 kA each).

Option	Nr. of circuits x Type	Nr. of cores x rated current, *kA*	P_1 *MW*	Q_1^{**} *MVA_r*	Q_2 *MVA_r*	I_2 *kA*	ΔU *kV*	Loss *GWh/ year*	Total Loss *GWh/year*
1	3xOHL	9 x 5	6644	812	0	10.1	5.7	290	290
2	2xOHL 2xHTS	6 x 5 6 x 5	1436 5366	-65 -50	82 -82	2.1 8.1	0.0	32 29	61
3	2xOHL 1xHTS*	6 x 5 3 x 5	3631 3084	243 255	102 -282	5.5 4.7	-1.3	108 26	134
4	2xOHL 2xXLPE	6 x 5 12 x 2.5	1799 4962	81 219	-30 30	2.7 7.5	1.7	52 61	112

*FACTS in series with HTS cable; ** calculated for cos φ = 1 of the load

3. Future Dutch transmission grid

Therefore, in addition to the scenario(s) of TenneT [1], we propose the following.

Scenario 1 (years 2025-2050, vision of Supernet project): OHLs and HTS cables co-exist, future grid is made of OHLs and HTS cables connected in parallel and using the same rights of way. Expansion of the grid is entirely underground; strengthening of the grid is made with HTS cables. The grid transformation is made at the same or lower IC and LTC. Extrapolating results obtained for connection K-G (option 3 against option 1) on the entire 380 kV grid gives savings of 5 and 13 B€ respectively. Total electrical energy consumed in year 2030 can be as high as 200 TWh and estimated losses are below 0.7 TWh. Almost the same reliability of the grid is expected, the voltage level of 380 kV can be kept forever and there is no need to introduce 750 kV level (as foreseen by ENTSOE).

Scenario 2 (years beyond 2050, vision of Supernet project): HTS cables rule, aged OHLS are replaced with HTS cables, the future grid is made entirely with HTS cables. The cable cores are periodically upgraded with modern HTS tapes, the voltage level is 380 kV and future needs for more electrical energy are met by changing the current capacity of HTS cores.

In order to arrive at scenario 2, HTS cables (as compared to OHLs) have first to show superior behavior in respect to: reliability, power capacity (including circuit breakers and other switching material), maturity, availability, failure rates, maintenance, time of repair, connection length and life expectancy (this requirements are softened in scenario 1 by the presence of the OHLs). A transition from scenario 1 to 2 is achieved by replacing aged OHLs with HTS cables.

4. Conclusions

Our study concludes that reliability of a transmission connection with OHLs alone is the highest, with cables in parallel to OHLs is close to that, and with cables alone is lower. Since reliability, power capacity, maturity, availability, failure rates, maintenance, time of repair, connection length and life expectancy of HTS cables in respect to OHLs are to be demonstrated, until then we propose to use HTS cables in parallel to OHLs. This way the issues of HTS cables can be tolerated in the grid and the cables can be used as soon as their assets are at the level of XLPE cables. HTS cables can be cheaper, more compact and reduce magnetic emissions in OHLs helping TSOs to treat this problem.

From economic point of view it is attractive for connection K-G to replace option 4 with 3 (or 2) at the HTS tape price from 356 to 283 (195 to 135) €/kAm and below. And, it makes sense to replace option 1 with 3 (or 2) at the HTS tape price from 241 to 143 (135 to 36) €/kAm and below. The tape prices and the volumes are within reach in 3-12 years. This allows to start expansion of the Dutch grid underground in a cost competitive way. Results for 220 and 150 kV grids will be treated separately.

Our assessment is based on realistic expectations for HTS cables. New advances can make HTS cables even more attractive: better ReBCO tapes, electrical insulation and cryostats, more efficient cooling using a cold from liquefied natural gas and gas-driven compressors at the cable cooling stations; the latter contributing to a climate in nearby buildings (with heat pumps fed by cable cooling losses), etc.

Acknowledgement

The Supernet project is funded by the AgentschapNL.

References

[1] TenneT: "Vision 2030 national electricity transmission grid", May 2010 and February 2008, Arnhem, the Netherlands url: http:/www.tennet.org/english/tennet/publications/technical_publications/visie2030/visie2030.aspxreference; Tennet Annual report 2012

[2] Zuijderduin R., Chevtchenko O. A., Smit J.J., Aanhaanen G., Melnik I., Geschiere A., 2012, Physics Procedia, 36, 890; Zuijderduin R., Chevtchenko O. A., Smit J.J., Aanhaanen G., Melnik I., Geschiere A., "Sustainable Future Dutch Grid by Integration of High Temperature Superconducting Cables", paper accepted to Applied Superconductivity and Electromagnetics, July 2013, 4.

[3] Matias V., Hammond R.H., Physics Procedia, 2012, 36, 1440.

Fabrication of the Superferric Cyclotron Gas-stopper Magnet at NSCL at Michigan State University

S. S. Chouhan[1], G. Bollen[1], J. DeKamp[1], M. A. Green[1], D. Lawton[2], C. Magsig[1], D. J. Morrissey[2], J. Ottarson[2], S. Schwarz[2] and A. F. Zeller[1]
[1]Facility for Rare Isotope Beams, MSU, East Lansing, MI 48824
[2]National Superconducting Cyclotron Laboratory, MSU, E. Lansing, MI 48824
E-mail: chouhan@frib.msu.edu

Abstract. The magnet for the cyclotron gas stopper is a newly designed, large warm-iron superconducting cyclotron sector gradient dipole. The maximum field in the centre (gap = 0.18 m) is 2.7 T. The outer diameter of magnet yoke is 4.0 m, with a pole radius of 1.1 m and $B^*\rho$ = 1.8 T m. The fabrication and assembly of the iron return yoke and twelve pole pieces is complete. Separate coils are mounted on the return yokes that have a total mass of about 167 metric tons of iron. This paper illustrates the design and the fabrication process for the cyclotron gas-stopper magnet that is being fabricated at MSU.

1. Introduction

Rare Isotope Beams (RIBs) are routinely produced from heavy-ions accelerated up to ~150 MeV/u by the coupled superconducting cyclotrons (K500 and K1200) at National Superconducting Cyclotron Laboratory (NSCL) by reactions in a thin production target (projectile fragmentation). The RIBs are separated in-flight by the A1900 Fragment Separator, and delivered to various experimental areas [1]. Increasing demands for high-quality, low-energy RIBs for nuclear physics and nuclear astrophysics programs have resulted in the construction of a fragment thermalization system coupled to re-accelerator facility, "ReA", to provide exotic low energy RIBs (at keV-energies) [2]. The stopped radioactive ions will also feed an ongoing program centred on precision mass measurements of exotic nuclei and laser spectroscopy in addition to delivery to ReA for reacceleration (presently ~2 and eventually ~15 MeV/u). The fast beams have been slowed down in solid degraders and thermalized in linear helium-filled gas cells. A novel concept to stop the fast beams in a gas-filled reverse cyclotron magnet is being pursed that will avoid many limitations of linear gas cells [3]. The fabrication and assembly of the high performance superferric cyclotron gas stopper magnet for this project with a central field of 2.7 T and a strong focusing gradient with field index below the critical value of k= -0.2 is nearly complete and described in this paper.

2. Magnet Design Overview

The cyclotron gas-stopper magnet is a vertical superconducting split-solenoid configured into a focusing dipole with 120° radial symmetry. The magnet poles are divided into three hill and three valley regions. The magnet field profile along a given radius is dominated by the shape of the pole face. The coil cross-section, iron yoke and size were optimized for an operating field of 2.7 T field at an exciting current of 200 A. Table 1 provides the design parameters of the magnet. The coils were wound and shimmed into two separate helium vessels made out of 304 stainless steel and attached to separate yokes. The two coils and yoke assemblies can be opened and will have a warm electrical

Journal of Physics: Conference Series **507** (2014) 032010 doi:10.1088/1742-6596/507/3/032010

connection. Because the magnet has to operate on a high-voltage platform at 60 kV (to extract thermalized RIBs), each coil is cooled by three two-stage pulse-tube cryo-coolers rated at 1.5 W at 4.2 K. The intermediate heat shield surrounding the cold vessel is cooled by a separate liquid nitrogen supply. An active quench protection system will be used to protect the magnet with an external fast dump resistor circuit. Each coil is also equipped with a cold quench protection circuit using diodes and a resistor assembly (at 4K) to protect the coils against a failure of the HTS current leads. The support system for each cold mass consists of six axial compression links (60 degree apart) and three radial tension links that carry the weight of the cold mass. One side of the magnet structure is fixed and the other is placed on a rail system. The rail system for the moveable half provides access to the beam chamber for service purposes. The detailed magnet design has been published previously [4], [5].

Table 1. Design parameters for the cyclotron gas-stopper magnet.

Parameter	Units	Value
Pole Diameter	m	2.20
Return Yoke Outer Diameter	m	4.00
Minimum vertical Gap	mm	180
Maximum Field	T	2.70
Coil Current Density	A/mm^2	55.0
Peak Field in the Coil	T	2.05
Peak Stored Energy at Operating Current	MJ	3.56

3. Fabrication and assembly

3.1. Pole, Return Yoke And Supporting Structure

The yoke and the pole tips are made up of low carbon iron (AISI 1018 and AISI 1006), respectively. After machining the flat backs of the twelve pole tip blocks, a gradient surface was cut with a CNC milling machine at the NSCL. Before removal from the machine a coordinate measuring machine (CMM) was used to compare every other pole piece to the design specifications. For measurement purposes, each pole was divided into sections as shown in figure 1, and a series of measurements were performed. Figure 2 shows the results of the measurement of Section -2 of Hill-1. A total of 248 measurements points were made of this section. The calculated mean-difference and its standard deviation are 0.008 mm and 0.033 mm, respectively. Ninety-nine percent of the data points lie within a normal distribution and fall well-within the specified machining tolerance (± 0.201 mm), as shown by the red lines in figure 2. The six poles for one yoke of the sector magnet are shown hanging from a mounting fixture in figure 3.

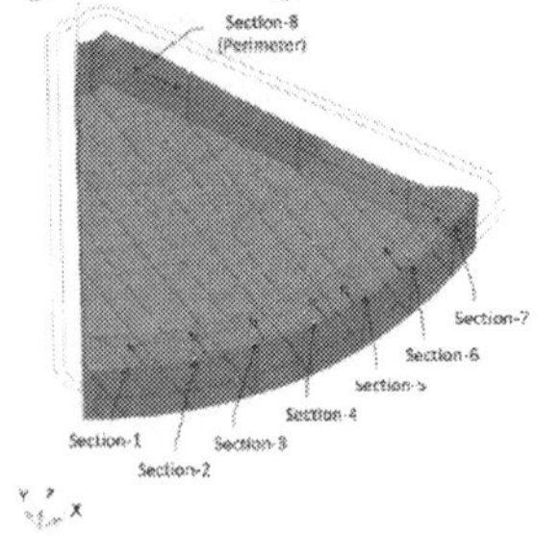

Figure 1. Each pole profile is divided into sections along the radial axis.

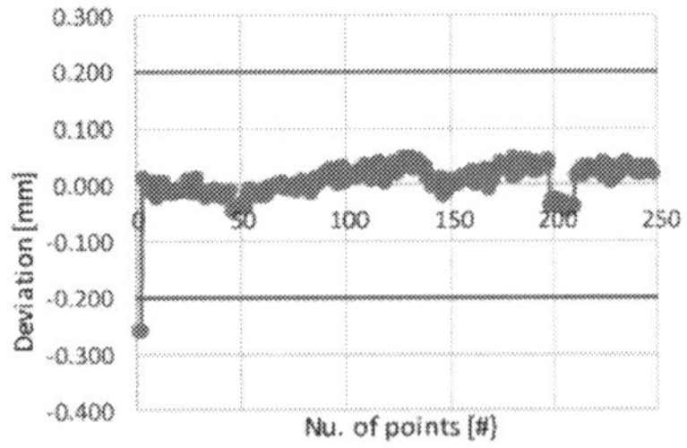

Figure 2. CMM measured data along the radial axis, red horizontal lines indicate the specified tolerance.

Each half of the return yoke consists of major three parts: a lower crescent, a middle slab and an upper crescent, and six return-yoke inserts near the median plane. These pieces were machined at the supplier from cast iron to the final shape and size. Inspections with a pulsed ultrasonic reflection instrument (Olympus-NDT, model EPOCH-XT) and per ASTM A609, Criteria level III were performed to assure that there weren't any voids in the material. All pole and return yoke pieces were completed by the end of 2012. An end-view of the assembled return yoke structure is shown in figure 4. The entire system is mounted on high voltage stand-off posts. The structure has been hi-potted to the desired level (60 kV) without breakdown. The moveable half of the magnet (~100 tons) is on cradle-shaped support system mounted on six Hilman Rollers® rated at 37.5 tons each [5] that move along hardened steel rails. A small 1 HP (1750 RPM) motor with a small torque of only 2 N-m is used to open and close the whole system.

Figure 3. Machined and assembled hill and valley poles attached to a mounting fixture.

Figure 4. Assembled structure of the return yoke on G-10 posts and the moving-track system.

3.2. Coil/ Cryostat And Cooling System

The cold mass that encloses each superconducting coil has been assembled, tested and welded shut. Three coils were wound, with the third one as a spare. The average diameter of the full coil is 2470 mm and the cross-section is 80 mm by 80 mm. The cross-section of the Formvar™-insulated superconductor in the coil is 1.25 mm by 2.5 mm and the copper-to-superconductor ratio is 4.5. There are 31 turns per layer and 57 layers. The total number of turns in each coil is 1760 and the length of the conductor per coil is 13.8 km. These coils were wet wound with fibre-glass cloth (0.2 mm thick) placed between the layers and baked after winding. The insulation of the conductor is sufficient to withstand the calculated 20 V potential that would occur between layers in a quench. A photograph of two finished coils is shown in figure 5. Each coil was shimmed into the helium vessel with wedge-shape stainless steel and G-10 spacers on the outer radius to supply pre-load. The total stored energy in the magnet is 3.56 MJ and an active quench protection system will be used to protect the magnet.

Figure 5. Completed pair of solenoid coils after removal from the winding form and baking.

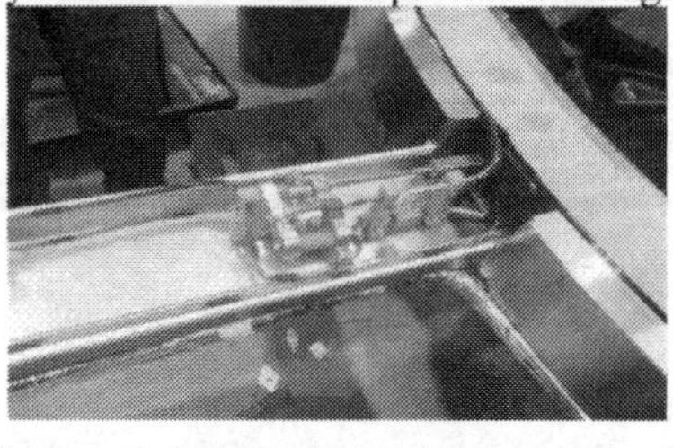

Figure 6. Detail of the shimmed coil in the helium vessel and the diode-resistor assembly.

Each coil is also protected from any failure of current leads by back-to-back cold diodes and a resistor assembly across the coil. The cold diode assemble is located in the neck of the helium vessel near the coil and the resistor is a Kapton™-wrapped 6061 aluminium conductor (1 mm by 3 mm cross-section) that is wound on the outer radius of the coil. Measurements of the diode assembly show a forward voltage at a liquid helium temperature of 10 V. To further protect the magnet, an external slow dump (6 V) circuit in the power supply will be used to discharge the stored energy if the power supply fails.

Figure 6 shows the region where the installed superconducting coil is connected to the back-to-back cold diode assembly and aluminium resistor. A stainless steel cryostat provides the required vacuum and cryogenic environment for each coil. The helium vessel is sufficiently stiff to prevent deflection of cold mass if there are unbalanced magnetic loads due to off-centered radial displacement of the coil. The tension links are made of titanium alloy (Ti 6Al4V ELI) and, typical tension loads on the links are calculated to be 5000 kN plus the weight of the cold mass (1600 kN). Figure 7 shows one of the completely assembled cryostats with the support system and the service box of the magnet. The axial magnetic forces during operation will be directed towards the return yoke and carried by six E-Glass fibre composite tubes in compression on each coil. A series of tests were performed on the tubes and the measured failure load was found to be 270 kN giving a factor of safety of 2.4.
The two superconducting coils will be cooled by six pulse tube cryocoolers. The three coolers on each cryostat operate in parallel and also function as helium re-condensers. Each provides ~1.4 W cooling power at 4.2 K and ~40 W at 40 K. The cold heads and ballast tanks are installed on the cryostat tower as shown in figure 8. The connection between rotary valves and cold heads is made with insulators so that the rotary valves, the high-pressure helium lines, and the compressors can be left at ground potential. The intermediate shield is cooled by flowing nitrogen through a line attached to the copper shield. An extra thermal intercept is added at 40K inside the service box to reduce the heat load to manageable level (<15 W per cooler), connected to the first stage of the cooler. The cold heads are mounted at the top of the cryostat tower in the operating configuration as shown in figure 8 as compared to horizontally during assembly as shown in figure 7.

Figure 7. One cryostat assembly shown in the horizontal position upon completion with labels on axial, radial support links and service box.

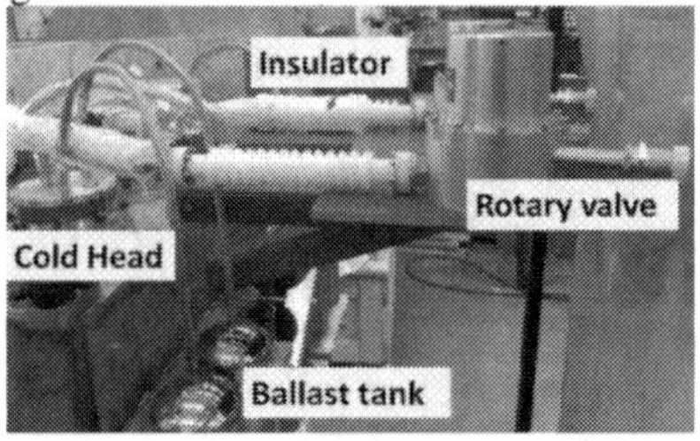

Figure 8. Top of the service box tower showing cold heads and connections during assembly.

4. Summary
Construction of the magnet is in a very advanced state. Fabrication, assembly of the iron structure and both coil cryostats is complete. Extensive testing of the components and instrumentation was carried out during assembly. The initial cool-down and cryogenic testing will be performed once the coil cryostats are mounted in the magnet's iron structure. The operation of the magnet at peak field and field measurements are expected to occur in the first half of 2014.

5. References
[1] Morrissey DJ et al 2003 Nuclear Instruments and Methods in Physics Research B **204** 90.
[2] Kester O *et al* 2009 *Proc. Int. Conf. RF Superconductivity* (Berlin, Germany) MOOCAU05.
[3] Bollen G *et al* 2005 Nuclear Instruments and Methods in Physics Research A **550** 27.
[4] Chouhan S *et al* 2013 *IEEE Trans. Appl. Supercond.* **23** 4101805.
[5] Chouhan S *et al* 2007 *Proc. IEEE Particle Accelerator Conference* **PAC-07** 524.
[6] Hilman Rollers 21150 Butteville Rd NE Donald OR 97020 [Online]. Available: www.hilmanroller-mf.com

Acknowledgments: This work was supported in part by National Science Foundation under grant PHY-0958726 and by Michigan State University.

Journal of Physics: Conference Series **507** (2014) 032011 doi:10.1088/1742-6596/507/3/032011

Synchronous motor with HTS-2G wires

D Dezhin[1], R Ilyasov[1], S Kozub[2], K Kovalev[1], L Verzhbitsky[3]

[1] Moscow Aviation Institute, Dept. 310, Volokolamskoe shosse, 4, Moscow, 125993 Russian Federation
[2] Institute for High Energy Physics, Ploshad' Nauki, 1, Protvino, Moscow region, 142281, Russian Federation
[3] JSC "Scientific-Research Institute of Electromechanics", Panfilova street, 11, Istra, Moscow region, 143502, Russian Federation

E-mail: dezhind@gmail.com

Abstract. One of the applications of new high-temperature superconductor materials (HTS) is field coils for synchronous electrical machines. The use of YBCO 2G HTS tapes (HTS-2G) allows increasing of magnetic flux density in the air gap, which will increase the output power and reduce the dimensions of the motor. Such motors with improved characteristics can be successfully used in transportation as traction motor. In MAI-based "Center of Superconducting machines and devices" with the support of "Rosatom" has been designed and tested a prototype of the 50 kW synchronous motor with radial magnetic flux from a field-coils based on HTS-2G tapes. The experimental and theoretical results are presented.

1. Introduction

Analysis of existing developments showed that the world is gradually moving from internal combustion engines to the use electric motors as traction drive for transport applications. This is due to high efficiency of electric motors, reduction of the size of the traction drive, absence of harmful emissions etc.

We can make a prediction that the next transition in the application of traction motors will use superconducting motor. This is evidenced by numerous examples of the use of superconducting motors for large-scale, powerful traction drives, and, for example, in the automobile industry.

They may provide a number of advantages, such as reducing the size and increasing the efficiency of up to 99% own efficiency (excluding cooling) for large machines.

Many years of experience in the field of creation of cryogenic electrical machines was put in a basis of development of the electric motor with 2G HTS tapes for transport applications.

Currently, the most promising are the following areas of application of powerful high-performance superconducting electric motors as drive:

- Propellers large vessels;
- The leading wheel pairs of electric locomotives;
- Cargo electric cars and electric buses;
- Propellers helicopters and hovercrafts etc.

Preferred area of application for future HTS motor was selected the electric transport. In future will planed to install a new HTS motor on the electrical bus.

2. A small test model of HTSC-2G motor with power of 50 kW has been produced

To gain experience with work with new HTS-2G tapes was designed and produced a small scale model of a synchronous motor. Its power is equal to 50 kW (see figure 1).

Designed synchronous motor has 10 poles, 3 phases, a nominal phase voltage 220 V and nominal current about of 80 Amps. Nominal rotation speed of this electrical motor is 300 rpm. Its calculated efficiency with account for losses in the stator winding, losses in steel and mechanical losses amounted a near to 97% (without cooling).

Figure 1. Experimental model of 50 kW HTS motor.

Field coils were made of SuperPower HTS-2G tape (4 mm, I_c = 100 A SF) at the Institute for High Energy Physics (Protvino). HTS coils were made according to the technology of "double pancake". At the one pole were located two coils with 120 turns in each. For the convenience of winding and further use, winding was made on the steel core (see figure 2). Then, it was fastened to the iron yoke of rotor. The electrical ends of coils were brought out and soldered to the copper wire (see figure 3).

Figure 2. The winding process of HTS-2G tapes in IHEP (Protvino).

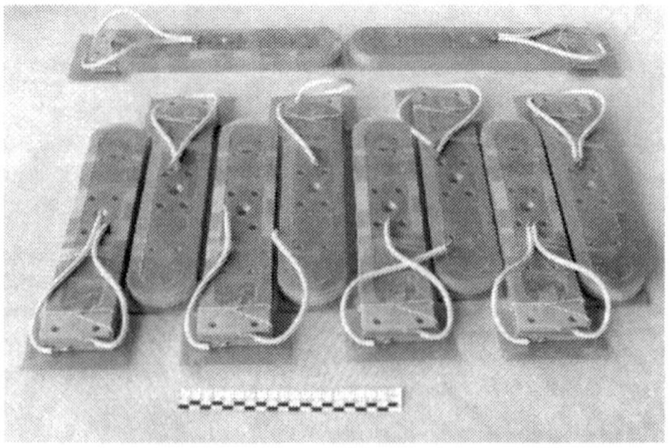

Figure 3. Field-coils was made of HTS-2G tapes with iron core inside.

Cooling of synchronous 50 kW motor was performed by the use of liquid nitrogen. The housing of the motor was a cryostat. Thermal insulation was mounted to the outside of the motor. For the protection from the leakage of liquid nitrogen between the housing and the shaft were used special o-rings.

Electrical power for excitation of field coils was transmitted to the rotor by the use of the slip rings. To prevent the formation of frost, nitrogen vapors were filed to the place of the slip rings.

3. Experimental studies of 50 kW HTS-2G motor were conducted

Experimental studies of 50 kW synchronous motor with HTS-2G field-coils were held on summer 2013. They performed on the test bench of Research Institute of Electromechanics (Istra).

Starting and speedup of the motor were performed by the use of frequency inverter. The load was performed by the using of a high-speed DC machine with gearbox.

Field coils were powered from a DC power supply. Current in the field-coils for safety reasons (related with the critical current), do not exceed 22 A. Due to the high inductance HTS-2G coils, input current increased with the rate of ~1 A/sec. HTS field coils were are not specially shielded from alternate magnetic fields of stator winding.

Some results of the performed test of the 50 kW HTS-2G motor are shown on figure 4 and figure 5. It is seen that due to the limitations of the current in the excitation winding of rotor, voltage in the stator winding was slightly below than calculated. Maximum output power was 42 kW at the speed of motor 300 rpm. Tested electrical motor showed high efficiency (above 96 %) and high power factor in the wide range of loads.

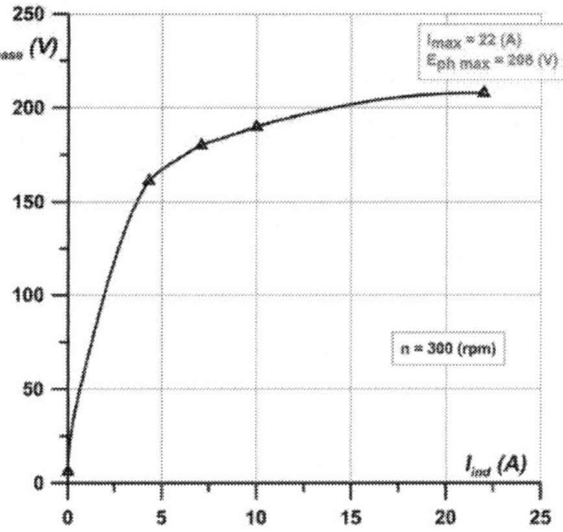

Figure 4. No-load characteristic for the 22 A of excitation current and 300 rpm.

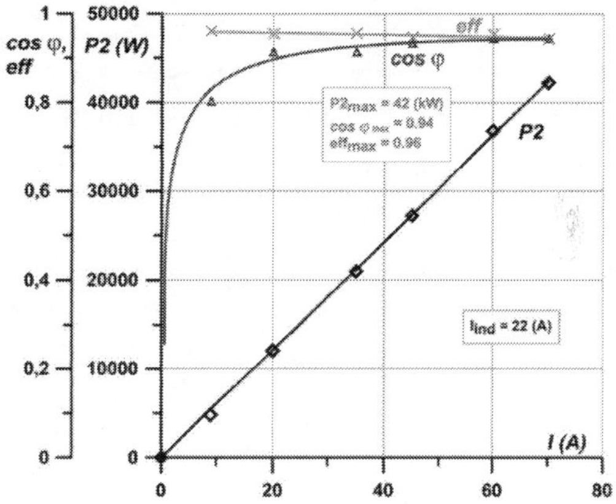

Figure 5. Load characteristics of the 50 kW HTS-2G motor.

The gained successful experience in developing of 50 kW motor with HTS-2G tapes was used to design a new traction motor for transport.

4. Currently in development 200 kW HTS-2G motor for transport applications

At present in the "Russian Center of cryogenic electrical machines" (MAI, Moscow) 200 kW synchronous motor with HTS-2G field-coils is developing. Future motor is designed for installation on electrical bus.

For the decrease of cooling power of motor, its stator will be located at ambient temperature. Field-coils located on the rotor will be made from HTS-2G tapes. Rotor will be placed in a rotating cryostat for cooling of HTS-2G excitation coils. In this regard, the air gap of HTS motor will be increased for placing thermal insulation of the cryostat.

Increased air gap demanded increase of magneto motive force of excitation winding. For these purposes the number of double-pancake HTS coils will be increased up to 3 on one pole.

Stator of 200kW HTS motor will be cooled with water. It will provide to increase of stator current density and reduce the thickness of stator yoke.

Currently checking calculations of the magnetic state of stator are carried out, choosing the most rational scheme for cooling HTS windings, investigate the possibility of the contactless current transfer into the excitation winding.

Main parameters of future 200 kW HTS motor shown in Table 1.

Production of 200 kW HTS motor planned on 2014-2015. The first tests must be conducted in 2015.

Table 1. Parameters of future 200 kW synchronous HTS motor

List of Parameters	Identification	Value
Nominal Output Power (kW)	P	200
Phase voltage (V)	U	450
Nominal Speed (rpm)	n	1500
Maximal Speed (rpm)	n_{max}	4000
Nominal moment (N·m)	M	1300
Maximal moment (N·m)	M_{max}	2600
Phase number	m	3
Pole number	p	6
Nominal current (A)	I	165
Inner diameter of stator (mm)	D	340
Active light (mm)	L	220
Material of stator winding	–	Cu
Operating temperature of stator winding (K)	T	~ 300
Material of field-coils	–	HTS 2G tape
Operating temperature of field-coils (K)	T_{FC}	77
Nominal current in field-coils (A)	I_{FC}	34
Number of turn per one pole	w_{FC}	252
Operating Power factor	cos φ	0.95 – 0.99
Full load efficiency (%)	η	96.3

Acknowledgments

This scientific and research work is carried out under support by State Atomic Energy Corporation "ROSATOM".

Journal of Physics: Conference Series **507** (2014) 032012 doi:10.1088/1742-6596/507/3/032012

Mechanical characterization of journal superconducting magnetic bearings: stiffness, hysteresis and force relaxation

Cristian Cristache[2], Ignacio Valiente-Blanco[2], Efren Diez-Jimenez[1], Marco Antonio Alvarez-Valenzuela[1], Nelson Pato[3] and Jose Luis Perez-Diaz[1]

1 Departamento de Ingeniería Mecánica. Universidad Carlos III de Madrid. Spain.

2 Instituto Pedro Juan de Lastanosa. Universidad Carlos III de Madrid. Spain.

3 Faculdade das Ciencias da Uniersidade de Lisboa. Portugal

E-mail: ediez@ing.uc3m.es

Abstract. Superconducting magnetic bearings (SMBs) can provide stable levitation without direct contact between them and a magnetic source (typically a permanent magnet). In this context, superconducting magnetic levitation provides a new tool for mechanical engineers to design non-contact mechanisms solving the tribological problems associated with contact at very low temperatures. In the last years, different mechanisms have been proposed taking advantage of superconducting magnetic levitation. Flywheels, conveyors or mechanisms for high-precision positioning. In this work the mechanical stiffness of a journal SMBs have been experimentally studied. Both radial and axial stiffness have been considered. The influence of the size and shape of the permanent magnets (PM), the size and shape of the HTS, the polarization and poles configuration of PMs of the journal SMB have been studied experimentally. Additionally, in this work hysteresis behavior and force relaxation are considered because they are essential for mechanical engineer when designing bearings that hold levitating axles.

1. Introduction

Superconducting magnetic bearings (SMB) are used in applications such as in flywheels [1] or Maglev devices among other devices [2–7]. The lack of contact, the extremely low friction and energy losses and the absence of wear are some of the outstanding characteristics that make this kind of bearings interesting for the mechanical engineer [8].

The mechanical interaction between magnets and superconductor has been previously deeply analyzed [9–12]. There are some models that are useful to describe this interaction and that can be applied in finite elements programs as it is commonly used in mechanical engineering [13,14]. Although these models can be very useful for static loads calculations, they are still limited when dynamics and transient effects are taken into account. Hence the real experimentation of SMBs design is still necessary.

Among superconducting bearings, journal bearings have been typically considered in order to bear high axial and radial loads. For most configurations in the literature, a superconducting journal bearing is composed by a ring permanent magnet (PM) levitating over a disk superconductor (SC).

In the FP7 MAGDRIVE project, superconducting magnetic bearings will be used in order to support a novel magnetic harmonic drive where there is no contact between any of the moving parts. As part of this research, a study of different journal superconducting bearings under field cooling conditions has been carried out. Relevant parameters such as the stiffness, the hysteresis or the force relaxation has been characterized using a specifically designed and built test bench.

2. The MAGDRIVE FP7 project

The objective of this FP7 project is to design and build a harmonic drive able to work under cryogenic conditions, with an extremely low friction, long life time and no wear, figure 1. The drive will not use any lubrication and will be free of wearing. This harmonic drive will achieve a great reduction ratio, being based on a non-contact interaction between permanent magnets and soft magnetic materials. In addition, the whole prototype will be supported by High-Temperature SMB intended not only to bear the weight but also to radially stabilize the magnetic harmonic drive core.

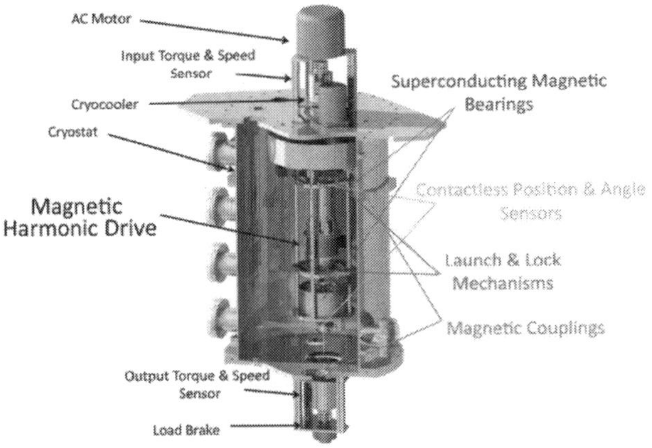

Figure 1. MAGDRIVE parts scheme.

3. Experimental set-up

In order to characterize the axial and radial stiffness of the SMBs, a dedicated test bench has been designed and built. The bench, shown in figure 2, is mainly composed of a metallic structure made of aluminum (1) which fixes the PM in position. Due to the magnetic properties of the aluminum alloy used, low magnetic interaction with the PM and the SC is assured. In addition, the superconductors are placed and fixed inside a LN2 vessel (2) and the last assembled onto a motorized lab-jack from Thorlabs, model L490MZ (3). This lab-jack stand is intended to modify the gap (Z axis) between the PM and the SC with extreme accuracy (about 5μm) within a stroke about 55 mm. The lab-jack stand is assembled onto a linear slider (4) from Igus able to modify the radial distance between the centers of the SC and the PM (X axis). This slider is controlled in a close-loop using the position signal from a laser triangulator from Microepsilon model ILD 1402-50. In summary, overall repeatability in the radial and axial position has been estimated to be better than 100 μm.

The radial and axial forces between the PMs and the SCs have been measured by using two load cells (5) from SENEL, model SX-1, C3 precision category. They were installed at either side of the T bar that supports the PMs as shown in figure 2. Finally, all the electronic systems are connected to a PC and data acquisition is synchronized using Labview.

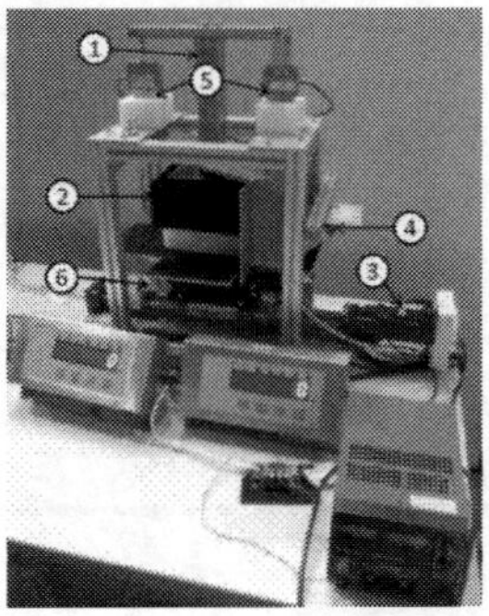

Figure 2. Test bench. 1) T bar, 2) LN2 vessel, 3) lab-jack stand, 4) linear slider 5) laser position sensor and 6) load cells.

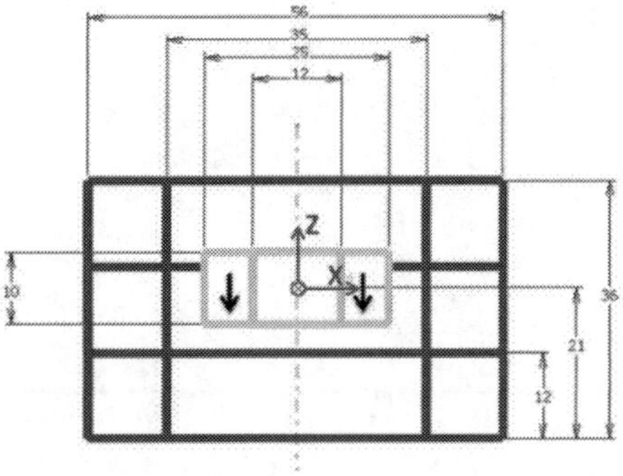

Figure 3. Journal bearing design: grey lined PM and black lined SC.

4. Results and discussion

The superconductors were field cooled using liquid nitrogen (LN2) at ambient pressure (≈ 77 K). The permanent magnets were also immersed in LN2 to assure little variation of the forces arising as a consequence of changes in the magnetization of the permanent magnets due to variations of their temperature (see force relaxation chart). After cooling down the bearing, figure 3, the relative axial and radial position of the permanent magnet with respect to the superconductor is modified.

4.1. Levitation and radial force: Hysteresis

Hysteresis in SMB is a very relevant parameter that must be taken into account when designing a SMB. It not only affects the levitation but also the radial force. Hysteresis for both motion directions in a simple journal bearing is plotted in figure 4 and 5. Additionally, note that the levitation force start to decrease when the axial displacement is about half the height of the PM, figure 4 and 5.

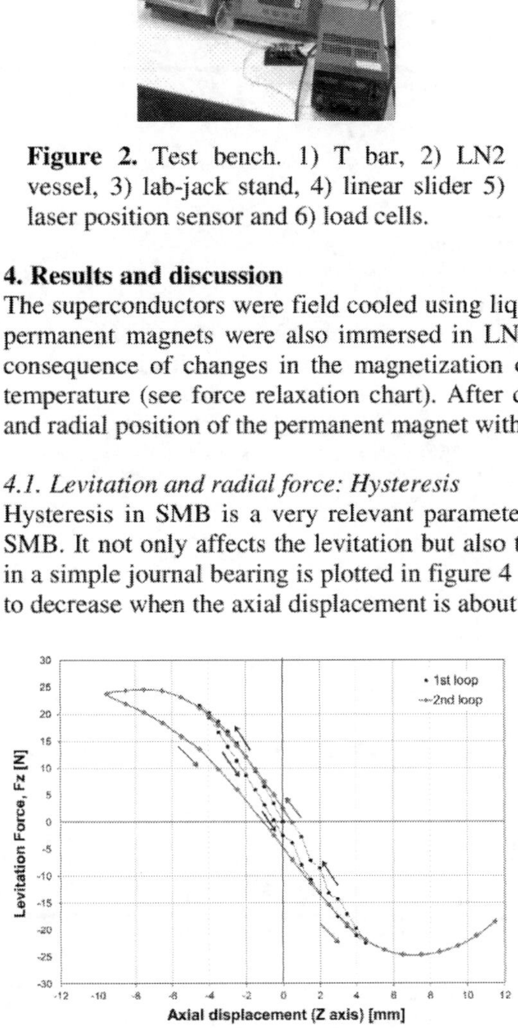

Figure 4. Lift force versus axial displacement.

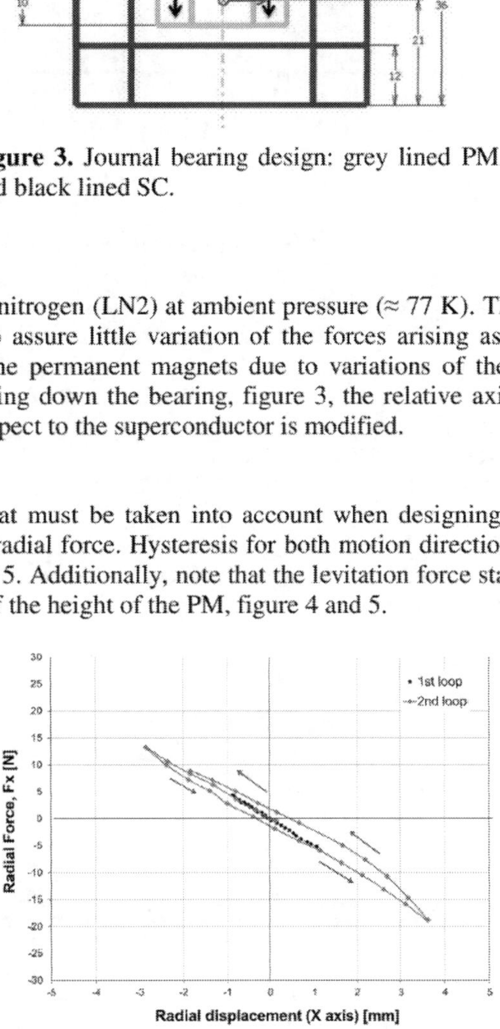

Figure 5. Radial force versus radial displacement.

4.2. Force relaxation

The force relaxation is the drop in the levitation or radial force with time after a relative motion between a permanent magnet and a field cooled superconductor happens. It must be carefully considered when designing a SMB. The relaxation of the levitation force for an axial displacement of - 4 mm from the FC position is shown in figure 6. For this design, the stable final lift force is 4% smaller than the initial one.

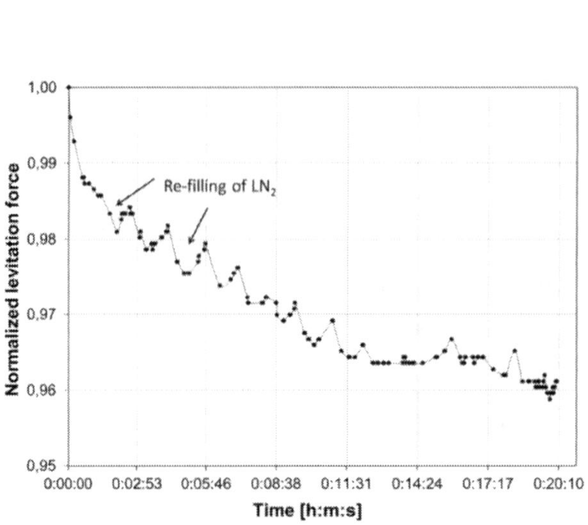

Figure 6. Levitation force in respect to time.

5. Conclusions

According to the experimental results obtained from the tests developed in MAGDRIVE FP7 project, a journal SMB will be designed to bear the weight and to compensate the radial instability of the MAGDRIVE core. Important parameters such us the hysteresis and the force relaxation have been analyzed. Finally, thanks to the relevant tribology advantages of SMB and due to the lack of contact they provide, the viability of a cryogenic-suitable magnetic harmonic drive with absolutely no contact between moving parts is expected to be demonstrated.

6. Acknowledgements

The research leading to these results has received funding from the European Community's Seventh Framework Programme ([FP7/2007-2013]) under grant agreement n° 263014.

7. References

[1] Werfel F N, Floegel-Delor U, Rothfeld R, Riedel T, Goebel B, Wippich D and Schirrmeister P 2012 S *Supercond. Sci. Technol.* **25** 014007

[2] Maillard T, Claeyssen F, LeLetty R, Sosnicki O, Pages A and Vazquez Carazo A 2009 *Proc. SPIE 2009* **7331** 1–9

[3] Perez-Diaz J L and Valiente-Blanco I 2013 10.1109/TMECH.2013.2250988

[4] Iizuka T, Maeda Y, Aihara K and Fujita H 1994 *IEEE Symp. Emerg. Technol. Fact. Autom.* 62–7

[5] Pérez-Díaz J-L, García-Prada J C, Diez-Jimenez E, Valiente-Blanco I, Sander B, Timm L, Sánchez-García-Casarrubios J, Serrano J, Romera F, Argelaguet-Vilaseca H and González-de-María D 2012 *Mech. Mach. Theory* **49** 308–14

[6] Serrano-tellez J, Romera-juarez F, González-de-maría D and Lamensans M 2012 *SPIE Modern Technologies in Space- and Ground-based Telescopes and Instrumentation V.II*

[7] Morales W, Fusaro R and Kascak A 2003 Applications *Tribol. Trans.* **46** 460–4

[8] Theiler G, Gradt T and Klein P 2002 *Tribol. Int.* **35** 449–58

[9] Valiente-Blanco I, Diez-Jimenez E and Perez-Diaz J L 2011 *J. Appl. Phys.* **109** 07E704

[10] Diez-Jimenez E and Perez-Diaz J L 2011 *Superconductivity - Theory and Applications* pp 153–72

[11] Perez-Diaz J L, Diez-Jimenez E, Valiente-Blanco I and Herrero-de-Vicente J 2013 *J. Appl. Phys.* **113** 063907

11th European Conference on Applied Superconductivity (EUCAS2013) IOP Publishing
Journal of Physics: Conference Series **507** (2014) 032013 doi:10.1088/1742-6596/507/3/032013

Development of 16 kA HTS current leads for 40 T hybrid magnet application

Y Bi, K Ding, H Feng, C Liu, X. Liu, K Lu, H Wu, Y Song, C Xu, B Ye, T Zhou

Institute of Plasma Physics, Chinese Academy of Science

E-mail: kzding@ipp.ac.cn

Abstract. A pair of 16 kA high temperature superconducting current leads which will serve a 40 T hybrid magnet at the CHMFL was designed, manufactured and performed the factory acceptance test at the ASIPP site. The major design includes the LN_2-cooling copper heat exchanger part as well as the HTS module made of BSCCO 2223/AgAu matrix tapes in parallel which allows the operation of the current lead at a minimum coolant mass flow. The paper briefly describes the design of the 16 kA HTS current leads. And the cold test results of current leads are also presented in the paper.

1. Introduction

The 40 hybrid magnet project which was approved in 2008 and will be made by Chinese Magnetic field Laboratory, Chinese Academy of Sciences comprised of a 29 T resistive insert and a 11 T superconducting outsert. The final design[1] was updated in 2013. The insert contain six Florida-Bitter-type coils, which will contribute 33-34 T in a working bore of Φ32 mm. The outsert consists of three Nb_3Sn-based CICC solenoid coils, which is cooled by forced-flow supercritical helium of 4.5 K. The superconducting outsert will generate 11.5 T in a room-temperature aperture of Φ800 m. The final hybrid magnet is expected to achieve the magnet field of 44-45 T. A helium refrigerator system (360 W @ 4.5 K in refrigeration mode, 100L/hr in liquefaction mode) is used to cool the superconducting magnet and feeder system including a pair of 16 kA high temperature superconducting current leads which are applied to electrically connect superconducting busbar and room temperature busbar of 16 kA power supply. The HTS leads are assembled at the vacuum cryostat which is remote from magnet distance about 4 meter.

During the past 10 years, the Institute of Plasma Physics, Chinese Academy of Sciences (ASIPP) has developed high temperature superconducting current leads cooled by forced-flow SHe and LN2 in the current range from 10 kA to 90 kA for EAST[2], ITER[3-4], Accelerator magnet[5] and so on. Based on the framework of cooperation between Institute of Plasma Physics, Chinese Academy of Science and Chinese Magnetic field Laboratory, two HTS leads rated at 16 kA current were designed, fabricated and cold tested at ASIPP. The paper briefly describes the design and test result of the 16 kA HTS leads.

2. Design

Total length of the current lead is ~ 2 m. The current feeds from the water cool copper terminal through a 0.4 mm thickness copper foils stacked heat exchanger that is cooled by a counter-flow of liquid nitrogen and vaporing nitrogen. The heat exchanger is connected to an intermediate copper block, and the HTS section. The HTS section consists of 80 stacks of Bi-2223/AgAu matrix soldered on a stainless steel cylinder that provides mechanical support for the tape stacks and an electrical shunt. The cold end of the HTS section is connected to Nb-Ti strands dis-assembled from the CICC cable. More similar design is in detail described in the [5]. Table 1 shows design requirement of 16 kA current leads. Figure 1 is a design overview of 16 kA current leads.

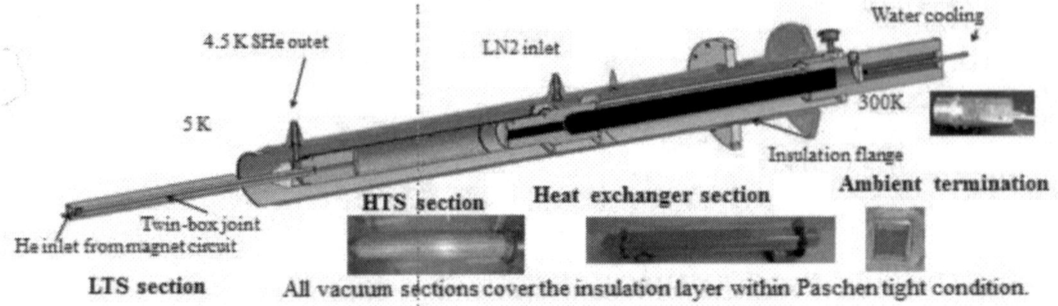

Figure 1 overview of 16 kA HTS current leads design

Table 1 design requirement of 16 kA current leads

Parameters	Values
current	16 kA
HTS Temperature	5~78 K
HEX Temperature	78~300 K
HTS hotspot	200 K
Time constant	10 s
HEX's inlet	LN2 and vapor
Heat load to 5K	< 5W

3. Cold test

3.1. Instrumentation

Figure 2 is a photo of leads before moving into the vacuum cryostat. Figure 3 show the test scheme of HTS tests. Two EAST joints are combined and electrically connected to the cold end of current leads, which are both cooled by liquid helium. Room temperature U-shape copper terminal is connected to a 20 kA power supply. The main HEX is in parallel feed by a 500 L liquid N2 tank.

A temperature sensor is nakedly mounted on the HEX top to monitor and record the temperature profile of the warm terminal. Two temperature sensors (including a backup) mounted at HTS top are used to log the data and monitor the HTS conductor temperature to prevent the HTS quench. In the test, a special temperature sensor is assembled on the HTS at the 90% along the cold end to monitor the quench progress and avoid overheat of the HTS module.

A calibrated CERNOX is assembled in the hole of copper cone of the HTS bottom to monitor and record the temperature profile. 5 pairs of voltage taps are soldered on the surface of the lead to get the detail voltage drop including HEX part, HTS module and joint resistance.

Figure 2 photo of 16kA HTS leads

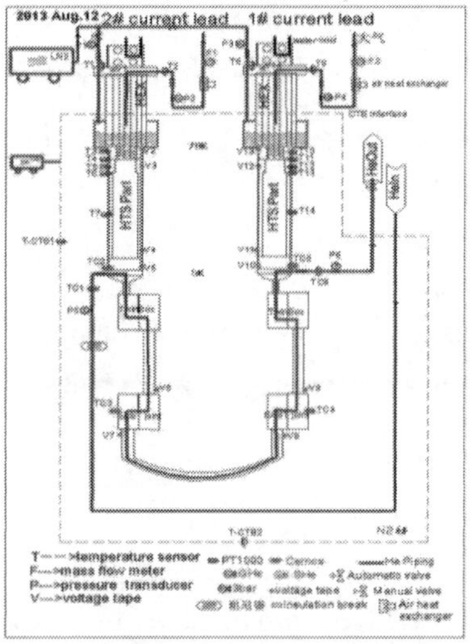

Figure 3 test scheme of HTS leads

3.2. *Test results*

According to the design requirement, multi-measurements containing joint resistance, conduction heat input to 5 K and loss of flow accident of main HEX are performed after HTS leads is cooled down. Table 2 summarizes the test results of 16kA current leads. Here is mainly focus on the results discussion of loss flow accident test. Figure 4 shows detail data of the temperature/voltage drop response after closing the supply valves of the liquid nitrogen tank for the two leads. It indicated that the current lead can bear more than 6 minutes of holding time of loss of coolant in an accident. The current leads have enough margin of time to remind the operator to take action. The hotspot of the HTS module happens at the 95 % location along the cold end which is similar to the past. In figure 4, we can find that the voltage drop of HTS module goes up to milli-volt level quickly. It indicates that quench current starts to shunt from the superconductor to the matrix or stainless steel. When the HTS 95 % temperature along the cold end is more than 200 K, the power supply is interlocked to cut off. Therefore for the required 10 s time constant of holding time of loss of coolant in an accident, both current leads can meet them. The temperature of room temperature terminal becomes much stable because of water-cool design in the terminal.

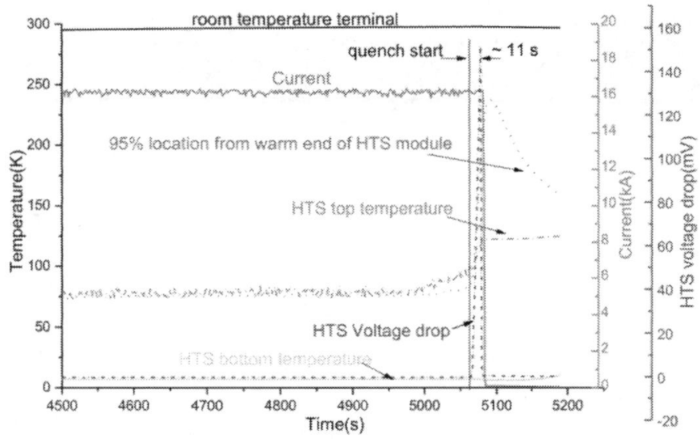

Figure 4 Temperature and voltage evolution of 16 kA operation and LOFA test

Table 2 Summary of test result of 16kA HTS leads

Test items	Test results
maximum current	16kA@81 K
heat load to 5 K end	CL1: 3.6 W, CL2: 3.4 W
Joint resistance of HTS module warm end	CL1: 21.5 nΩ, CL2: 1.3 nΩ
Joint resistance of HTS module cold end	CL1: 10 nΩ, CL2: 13.6 nΩ
LOFA time by cutting supply of LN2 tank.	6 mins without quench.
	11 seconds to reach the hotspot of 200 K after quench

4. Conclusion

A pair of 16 kA HTS current leads was designed and manufactured and performed the factory acceptance test at the ASIPP site. The fruitful development of this two leads provides a significant experience for the next fabrication of 12 kA current leads for the accelerator magnet.

These two current leads will be delivered to the end user soon. On-site acceptance test will be performed with the superconducting magnets.

Acknowledgment

Author thanks the contribution of Hefei KeJuDiwen Inc. including but not limited to Chunlei Xu, Bi Ye, Xiang Liu, and Hefei Keye Inc. for manufacturing.

References

[1] Chen C, Tan Y, Chen Z, Zhu J, Chen Z, Pan Y, Wang F, Huang P, and Kuang G, 2013, *IEEE Transactions on Applied Superconductivity*, **23**, 4300404

[2] Bi Y, 2007, Chinese Cryogenic Physics Journal, **29**, 131

[3] Ding K, Bi Y, Feng H, Liu C, Wu H, Song Y, Zhou T, Huang X, Lin X, Ni Q, Shen G and Yu J, 2012 *Physics Procedia*. **36**, 931

[4] Bauer P, Bi Y, Cheng Y, Devred A, Ding K, Huang X, Liu C, Lin X, Mitchell N, Sahu A, Shen G, Song Y, Wang Z, Zhang H, Yu J and Zhou T, 2010, *IEEE Transactions on Applied Superconductivity*, **20,** 1718.

[5] Ding K, Bi Y, Feng H, Liu C, Wu H, Song Y and Zhou T. 2010,. 936*IEEE Transactions on Applied Superconductivity*, DOI:10.1109/TASC.2012.2230675

A Conduction Cooled High Temperature Superconductor Quadrupolar Superferric Magnet, Design and Construction

I. Dobrin[1], A.M. Morega[2,3], A. Nedelcu[1,2], M. Morega[2], J. Neamtu[1]

[1]National Institute of Electrical Engineering, ICPE-CA, Bucharest, Romania

[2]University POLITEHNICA of Bucharest, Faculty of Electrical Engineering, Bucharest, Romania

[3]"Gheorghe Mihoc – Caius Iacob" Institute of Statistical Mathematics and Applied Mathematics, Romanian Academy, Bucharest, Romania

E-mail: ion.dobrin@icpe-ca.ro

Abstract. The paper presents the prototype of an YBCO superferric quadrupolar magnet for high gradient magnetic field generation, design and construction. The temperature of the superconducting coil has to be kept within safe limits or the HTS would exit the superconductive state. Of particular concern is the „warm" beam tube that passes through the magnet. Cryogenic conduction cooling with a closed cycle G-M Cooler may ensure the removal of the ambient heat influx. Numerical simulation results on the magnetic field and heat transfer problems are then discussed. The computational domain is abstracted out of the CAD design of the system. The design solution is presented and compared with the numerical simulations results.

1. Introduction

Recently, superferric magnets are considered for replacing the copper coils with superconductor windings, a technology is used more and more in the fabrication of high-energy particle accelerators such as dipoles and quadrupoles especially. In particular, quadrupoles have a specific task in shrinking the particles beam size. Usually either NbTi or Nb3Sn superconductors are used, but they have to be cooled by liquid helium to 4.2 K or even lower. The advantages of the HTS coated conductors, like YBCO tape, are the high critical current density at higher magnetic field while operating at a higher temperature (30-77 K). These features recommend their usage in the construction of magnets for either high magnetic field density or high magnetic field gradients.

On the other hand, the increased performances of the closed cycle cryocoolers make them sound solutions for the cooling of the superconducting field windings even the entire magnet, avoiding thus cryogens consumption. In this respect, the HTS coil – cryocooler combinations may provide for a new promising architecture of the magnets for particle accelerators.

From the designer perspective, the use of high temperature superconductors (HTS) in particle accelerators of higher energy raises concerns addressable, in the conception phase, through mathematical modeling and numerical simulation. This approach may unveil the magnetic field quality and the heat transfer paths within the structure. Along this line, the present work presents a quadrupole superferric magnet (QSM) prototype for high gradient magnetic field (~30 T/m in the central zone). The HTS coils are made of coated YBCO tape in double pancake design. The end parts of the coils are racetrack shaped. The profile of the iron yoke is optimized for the magnetic field quality at required field gradient. The superconducting coils are cooled with a Gifford – McMahon cryocooler (GMC) [5]. Numerical simulation is used to solve for the magnetic field and the thermal transfer under steady state operation. The main outcomes are the magnetic field profile and quality and to evaluate accurately the thermal loads for the cryocooler. At this stage we report only design results used in the construction phase. Experimental validation makes this object of a future report.

2. The Quadrupolar Superferic Magnet System – Design and Construction

The first design concern is the magnetic field gradient. Mathematical modeling and extended numerical simulations were performed to optimize the pole profile with the aim to produce high quality field gradient in the central bore of the magnet. The magnetic field is produced by two HTS double pancake racetrack shaped coils per pole. The second objective is the thermal optimization to ensure adequate working conditions for the superconducting coils [1, 2]. A thorough thermal optimization of the system is needed because of the limited power of the cryocooler.

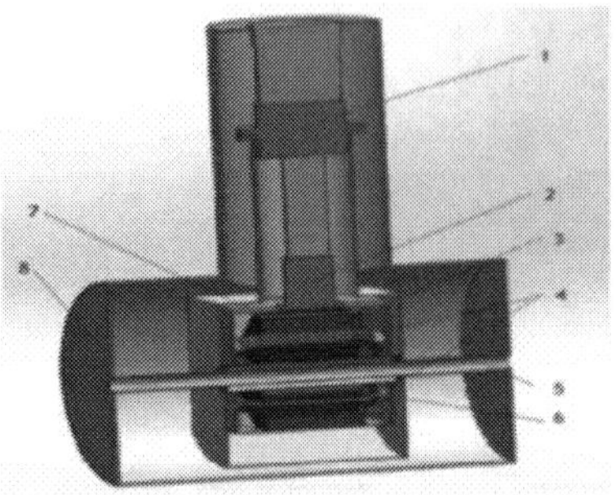

Figure 1. CAD view of the QSM: 1,2 – cooler heads; 3 – iron yoke; 4 – HTS coils; 5 – warm bore; 6 – thermal shield; 7 – thermal shunt; 8 – cryostat.

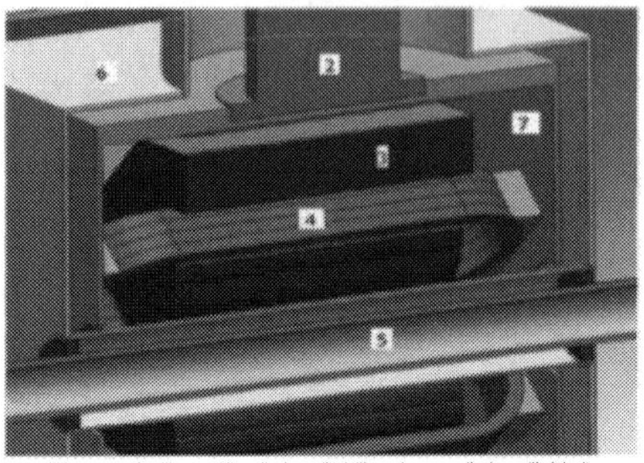

Figure 2. Detail of the CAD view of the QSM.

A simplified CAD geometry (Figures 1, 2) was used as computational domain. The HTS coils are made of YBCO tape [3]. Table I summarizes the main parameters of the tape and the magnet system.

Table I. The main parameters of the wire and QSM system

HTS tape	Value	Units	QSM system	Value	Units
Width	4	mm	Overall size	280 × 450 × 630	mm
Thickness	0.11	mm	Warm bore diameter	30	mm
Critical current	120	A	Magnetic field gradient	33.97	T/m
Critical bend diameter	11	mm	Coils dimensions:	144 × 54 × 12	mm
			- turns per pancake	30	-
			- field current	90	A

The current leads – a copper / HTS combination – are thermally connected to the first stage of the cryocooler (50 K) [2, 8]. The estimated cryogenic heat loads are summarized in Table II.

Table II. Estimated heat load

Heat flux rates [W]	1st stage	2nd stage
Conduction heat transfer in the current leads	18.5	1.26
Heat conduction through measurement wires	0.31	0.12
Heat conduction through mechanical supports	2.0	3.5
Heat transfer through radiation	16.5	0.1
Total thermal load	37.31	4.98

The magnetic yoke of the magnet is made of bulk iron (99.08 % iron content by atomic spectroscopic analysis), machined to the shape shown in Figure 4. The inner diameter is 40 mm, the outer diameter is 160 mm, and it weights ~10 kg. Total losses for this type of iron were 0.34 W/kg measured at 2 Hz, so this contribution to the heat load of the cryocooler is small enough. However, in this study we are concerned with the steady state only (DC operation of the magnet), which is consistent with a slow ramping operation. The cryostat, made of stainless steel, is *T*-shape to meet two main conditions: magnetic field structure and access to the warm bore, and magnet thermal stability.

Figure 3. HTS coils.

Figure 4. Iron yoke.

Figure 5. The QSM assembly.

Figure 6. The QSM inside the cryostat.

Figure 7. The QSM system.

Cooling is by thermal contact using a copper thermal shunt, which connects the coils to the second stage (4.2 K) of the GMC. Another thermal shunt connects the iron yoke to the same cold head. The cold mass is shielded with a thermal radiation screen, which is cooled by the first stage (50 K) of the cryocooler (figures 1-2). This solution provides for 50 K thermal shielding. No MLI (multi layer insulation) was used. Figures 3-7 show the parts of the cold mass and the assembled magnet.

3. Mathematical and Numerical Modelling of the Magnetic Field and Heat Transfer

In the superconducting state the Joule effect is null therefore the magnetic field and heat transfer problems are uncoupled, and they may be solved separately.

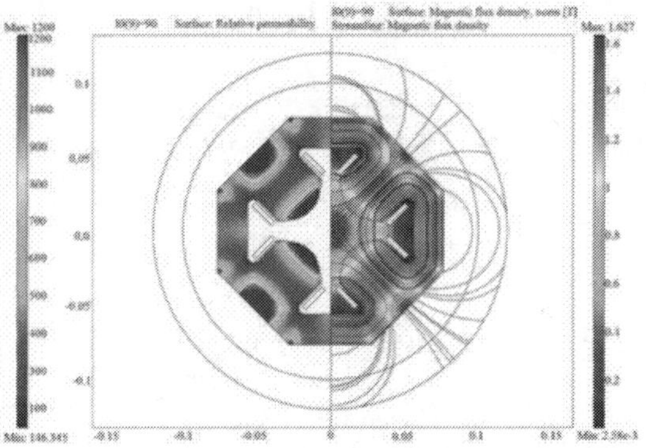

Figure 8. Relative permeability of the iron (left, colour map) and magnetic flux density (right, colour map and field lines).

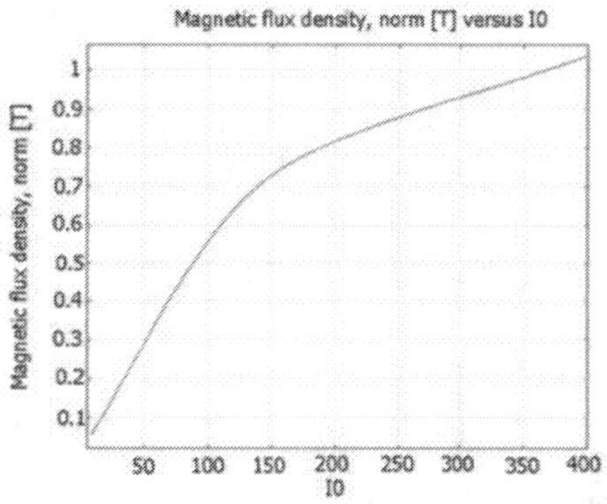

Figure 9. Magnetic flux density versus I_0.

The magnetic field problem is defined through [6] $\nabla\times\left(\mu_0^{-1}\mu_r^{-1}\nabla\times\mathbf{A}\right)=\mathbf{J}_z^e$, where $\mathbf{A}$ [T/m] is the magnetic vector potential, $\mathbf{J}_z^e$ [A/m^2] is the external electrical current density (in the windings), $\mu_0 = 4\pi\times10^{-7}$ H/m is magnetic permeability of vacuum, and μ_r is the relative permeability. Magnetic insulation boundary conditions (BCs) close the problem. The problem was FEM solved [6]. The total current, I_0, is an input parameter (10…400 A). Figure 8 shows the magnetic flux density and the relative permeability in the iron yoke for the optimal external current (I_0 = 90 A). Figure 9 depicts the magnetic flux density versus I_0, 15 mm away from the center.

Conduction heat transfer is governed by [2, 7] $\nabla\cdot\left(-k\nabla T\right)=0$, where T [K] is the temperature, and k [W/(m·K)] is the thermal conductivity. The BCs are as follows: $-\mathbf{n}\cdot\left(-k\nabla T\right)=h\left(T_{inf}-T\right)$, where h [W/m^{2}K] is the convection heat transfer coefficient for the shell (convection to the ambient) and T_{inf} [K] is the ambient temperature; $-\mathbf{n}\cdot\left(-k\nabla T\right)=0$ for the symmetry plane (adiabatic); $T=T_0$, $\left(1-\varepsilon\right)G=J_0-\varepsilon\sigma T^4$, where ε is the surface emissivity [1, 2, 7], for the shields, T_0 [K] is the mutual surface temperature, σ = 5.67·10^{-8} W/(m^{2}K^4), G [W/m^2] is the incoming radiation heat flux, or irradiation, and J_0 [W/m^2] is the surface radiosity. Surface-to-surface radiation conditions were set for different temperature surfaces that can "see" each other. Insulation conditions were set for negligibly small areas, *e.g.* the rims. The temperature of the cooling heads was set at 65 K for the first stage and 10.2 K for the second stage, in compliance with the cold head capacity map of the crycooler.

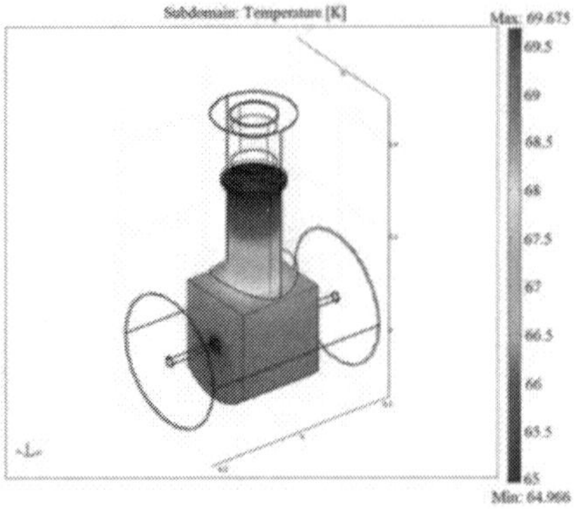

Figure 10. Temperature distribution of the thermal shield.

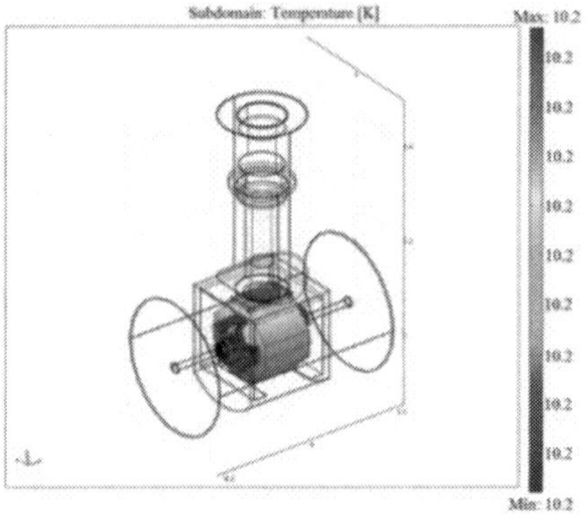

Figure 11. Temperature distribution of the magnet.

The heat transfer problem was solved by the FEM technique too. The temperature field on shield and iron yoke is shown in Figure 10 and Figure 11, respectively.

4. Results and Conclusions

Numerical simulations reveal the magnetic field and the heat transfer paths within the quadrupolar HTS-YBCO superferric magnet for particle accelerators prototype that we designed and fabricated. The results compare well with preliminary available experimental data: the magnetic flux density gradient obtained for optimal current, of 90 A, is 30 T/m, and the peak value in the iron yoke is 1.6 T; the heat loads to the G-M crycooler resulting from numerical simulations are ~5 W for the second stage and ~37 W for the first stage. It may be concluded that the cooling system will keep the coils at 10 K and the thermal shield at 65 K. In this stage the magnet is built. Cooling and functional tests are planned in the next period of time.

Acknowledgments

I. Dobrin, A. Nedelcu, and J. Neamtu acknowledge the project funded by the Romanian National Authority for Scientific Research through the "Core Research" Program under the Contract PN 0935102/2009, and the Grant PCCE-ID_76/2010. A.M. Morega and M. Morega acknowledge the technical support offered by the Laboratory for Multiphysics Modeling at UPB.

References

[1] A.M. Morega, I. Dobrin, and M. Morega, "Thermal and magnetic design of a dipolar superferric magnet for high uniformity magnetic field", *Proceedings of The 7th International Symposium on Advanced Topics in Electrical Engineering – ATEE*, May 12-14, Bucharest, pp. 589-592, 2011, IEEE no: CFP1114P-PRT, ISBN 978-1-4577-0507-6.

[2] A. M. Morega, I. Dobrin, A. Nedelcu, M. Morega, „A Quadrupolar Superferric Magnet", *Proceedings of The 13th International Conference on Optimization of Electrical and Electronic Equipment* (OPTIM), 24-26 May 2012, Brasov, Romania, ISBN 978-1-4673-1653-8/12. IEEE Xplore.

[3] SuperPower Inc., http://www.SuperPower-Inc.com

[4] J. Shi, Y. Tang, et al., "Development of Conduction Cooled HTS SMES", *IEEE Trans. on Applied Superconductivity*, vol. 17, pp. 3846-3850, 2007.

[5] Sumitomo Inc., http://www.shicryogenics.com

[6] Comsol Multiphysics AB, Sweden, v. 3.5a (2010), v. 4.2a (2012), v. 4.3 (2013).

[7] A. Bejan, *Heat Transfer*, John Wiley & Sons, Inc., N.Y., 1993.

[8] HTS-110 Co, http://www.hts-110.nz/current_leads

Electric and magnetic properties measurement and analysis of a conventional and a superconducting power transformer

P Figueira[1], **A G Pronto**[2], **N Vilhena**[1], **J M Pina**[2]

[1] Departamento de Engenharia Electrotécnica, Faculdade de Ciências e Tecnologia, FCT, Universidade Nova de Lisboa, 2829-516 Caparica, Portugal
[2] CTS, Uninova, Departamento de Engenharia Electrotécnica, Faculdade de Ciências e Tecnologia, FCT, Universidade Nova de Lisboa, 2829-516 Caparica, Portugal

E-mail: paulorsfigueira@gmail.com

Abstract. Power transformers based on High Temperature Superconductors (HTS) technology have revealed potential for several practical applications, offering economic, environmental and operational benefits. In this work, two 650 VA single-phase transformers prototypes were developed, tested and characterized: a conventional one, using copper windings, and another with the same primary copper winding, but with a secondary winding made of HTS BSCCO tape. The two prototypes were compared regarding magnetic properties, losses, electric parameters and efficiency, and the results are presented and interpreted. Also, several measures to determine AC critical current of the HTS tape were made. The results are compared with DC critical current for the same tape.

1. Introduction

High temperature superconductors (HTS) power transformers technology have revealed potential for energy applications offering economic, environmental and operational benefits. Among them, perhaps volume and weight reduction, comparing with similar conventional power transformers, is one of the most interesting aspects associated with this technology. Nevertheless, electric and magnetic losses minimization is also very important for transformers' efficiency and HTS transformers could also contribute to it improvement, especially through electric losses reduction related to the use of superconducting materials.

This work presents two 650 VA single phase transformer prototypes, a conventional one using copper windings and another one with the same primary copper winding but with an HTS BSCCO tape on secondary. Several tests were made to measure electric and magnetic properties of constitutive transformer materials, namely to determine AC critical current of BSCCO tape, and to characterize and compare operation performance of both transformer prototypes.

The final objective is to contribute to the analysis of superconducting power transformer potentialities in energy systems.

2. Characterization of the superconducting tape

Before the assemblage of the HTS transformer prototype, the tape was tested in order to measure its AC critical current. In practice, it does not exist a detailed regulation for AC critical criterion, whereas DC critical current measurement is regulated by IEC 61788-3 international standard. Table 1 summarizes the characteristics of the superconducting Bi-2223 tape. Data is provided by the manufacturer and concerns to DC.

Table 1. Innost Bi-2223 technical characteristics

Critical current	90 A
Engineering critical current density	93 A/mm2
Width	4,2 (±0,2) mm
Thickness	0,23 (±0,02) mm
Minimum curvature radius	30 mm (5% I_c degradation)

To determine AC critical current of the tape, two configurations were analyzed: one, where the sample is straight and another where the tape is wound in a coil (30 turns). Measurement procedure followed reference [1]. This method is applied to determine critical current on both configurations (tape and winding) and the results are indicated in figure 1.

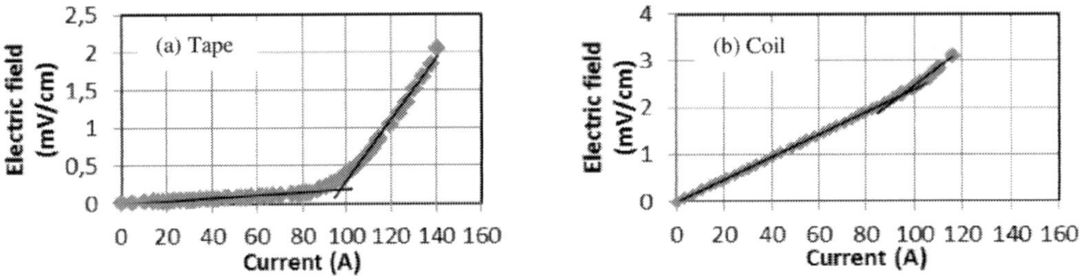

Figure 1. AC critical current approximation. (a) Tape (b) Coil

Following [1], the critical current is defined by the intersection point of the two tangents represented in figure 1. It is 97.3 A for the straight tape, and 95.7 A for the coil. The difference observed is due to the magnetic induction field normal to coil winding, which partially destroys superconductivity and therefore reduces critical current. These values are slightly higher than the manufacturer's indication (90 A in DC).

3. Comparison between conventional and HTS transformer

This section presents the development and comparison between the two built prototypes, one conventional transformer (CT) using copper windings and working at room temperature (298 K) and another with the same primary copper winding, but with secondary one made of HTS BSCCO tape immersed in liquid nitrogen (77 K). For simplicity, the last is called superconducting transformer (SCT) despite it does not have both HTS windings.

3.1. Design characteristics

Main design characteristics of iron core, HTS and copper windings and cryostat, of both transformers are presented in this section. Table 2 and 3 resume the main geometric characteristics and specification of the CT e SCT prototypes.

Table 2. Main geometries of the designed transformers

Iron core	Diameter	62,8 mm
	Filling factor	86,6%
	Laminated factor	95%
	Cross-section area	2550 mm^2
Windings	Voltage per turn	0,6 V/turn
	High voltage turns	30
	Low voltage turns	347
Cryostat	Internal radius	65,3 mm
	External radius	159,0 mm

Table 3. Specifications of the designed transformers

Capacity	650 VA
Frequency	50 Hz
Voltage ratio (prim. / second.)	208,2 V / 18,0 V
Current (prim. / second.)	3,1 A / 36 A

3.2. Core characterization

Due to the fact that SCT has its secondary winding at 77 K, it was decided to characterize the iron core in two different conditions: one, with the core at room temperature and another with the core cooled in a non-uniform way due to the proximity of HTS winding cryostat (in figure 2, 3 and 4 referred as (CT) and (SCT), respectively). The achieved results obtained for magnetic characteristics are represented in figures 2, 3 and 4.

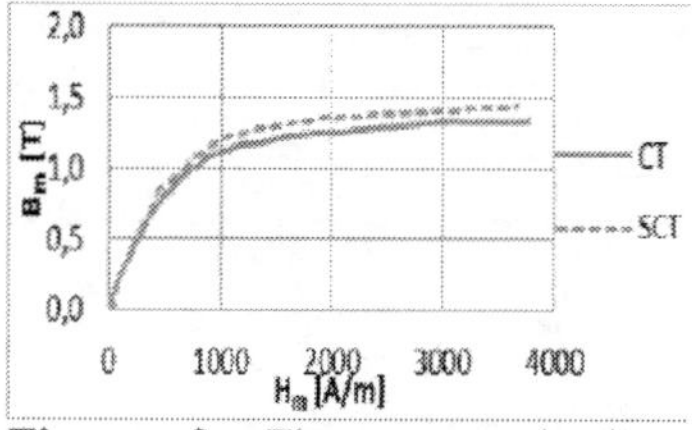

Figure 2. First magnetization curves for ferromagnetic core material in CT and SCT

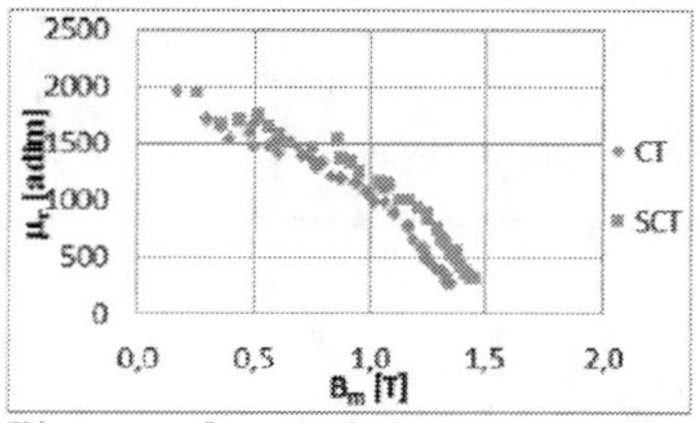

Figure 3. Relative magnetic permeability for CT and SCT

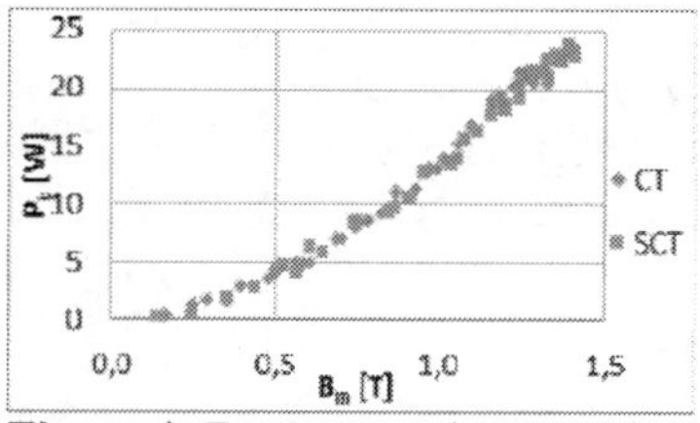

Figure 4. Total magnetic losses for CT and SCT

It can be concluded that:

- The ferromagnetic core exhibits higher saturation flux density on SCT.
- Higher magnetic permeability is observed for SCT core for all range of B_m. These results are justified by the temperature decrease in the cryostat column [2]. According to [3,4], this decrease causes a reduction in thermal agitation of the atomic structure which contributes to decrease the electrical resistivity and to increase magnetic permeability of the ferromagnetic material.
- According to figure 4, total magnetic losses could be considered similar in both transformers.

3.3. Steinmetz equivalent circuit parameters

Open and short circuit tests were performed, in order to characterize and compare both transformers and to establish Steinmetz equivalent circuit parameters. Figure 5 shows the equivalent circuits obtained for CT and SCT.

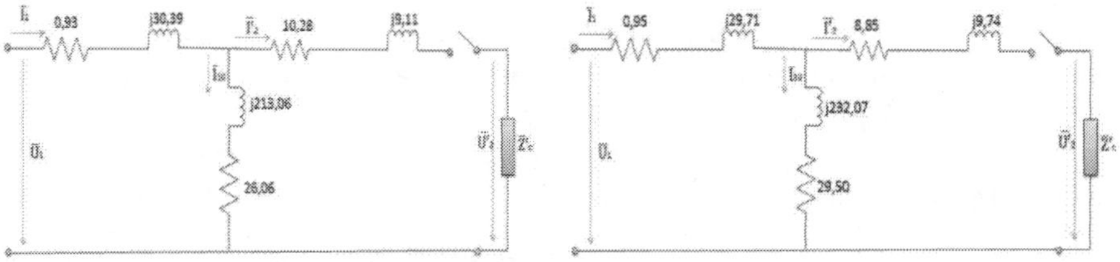

(a) (b)

Figure 5. Steinmetz equivalent circuit (a) for CT (b) for SCT (values in ohm)

Comparing the obtained equivalent circuits, it can be concluded:

- Primary electrical resistance and leakage reactance, are similar for both prototypes (2% variation).
- In SCT the tape exhibits a decrease of 14% in eletric resistance when compared with CT cooper winding. It was expected even a lower value but the resistive tape terminal contacts and the electrical conductor used in the short circuit test have a strong influence on secondary resistence.
- The leakage reactance increases 7% in SCT due to the distance between the HTS coil and the leg of the core, created by the cryostat.
- SCT equivalent magnetizing reactance (232.07 Ω) and core losses equivalent resistance (29.05 Ω) increase 9% and 13%, respectively, compared with CT parameters. It means that, ferromagnetic core has higher permeability and lower resistivity when cooled, which is according with [3,4]. Lower temperatures reduce thermal agitation and, consequently, reduce electrical resistivity.This increases eddy currents leading to higher core losses [3]. Another possible justification for this increasing is, LN_2 evaporation and subsequent water condensation along magnetic flux path, causing short-circuits and increasing eddy currents.

3.4. Transformer load tests

In order to evaluate transformer's behavior, both transformers were tested for different loads. Figure 6 shows real power (P_1), primary voltage (U_1), apparent power (S_1) and efficiency versus current per unit, i.e. $I_2/I_{2n} \cdot 100\%$, for both prototypes. Analyzing the graphics it can be concluded:

- For equal load, real power P_1 is the same for CT and SCT transformers (figure 6a)
- The superconducting transformer shows better voltage regulation (figure 6b). Nevertheless, in both transformers voltage regulation values are higher than desirable: 45% to SCT and 50% to CT.
- SCT requires a lower primary apparent power (figure 6c), to supply the same load current.
- SCT has slightly lower efficiency when compared to CT, a difference of 0.47% (figure 6d).

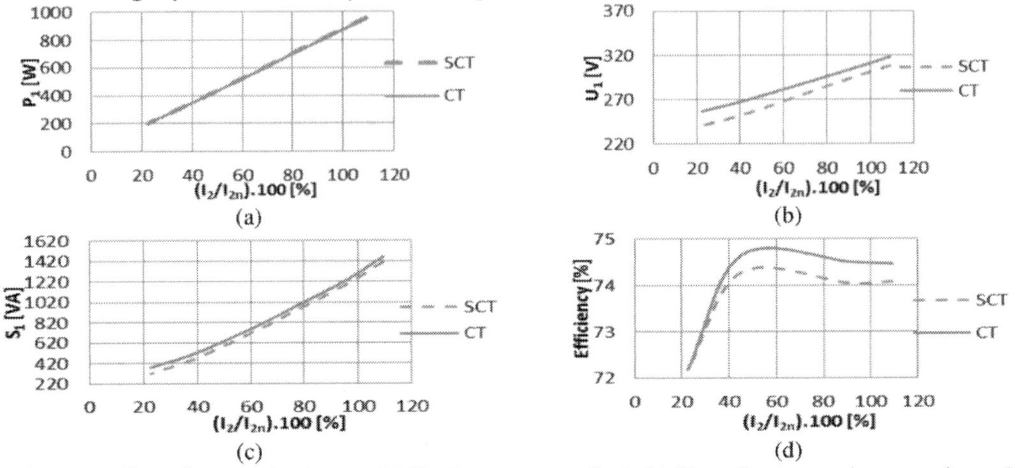

Figure 6. Transformers load tests (a) Real power supplied (b) Transformers primary voltage (c) Transformers primary apparent power (d) Transformers efficiency

4. Conclusions

The use of HTS tape in transformers' secondary winding reduces electrical resistance although less than expected. This is due to electrical contacts made to solder the HTS tape to copper wires outside the cryostat. Superconducting windings impose the existence of a cooling system which increase the distance between transformer leg and the winding, leading to increase leakage reactance, and to decrease the working temperature of the core. This last aspect contribute to increase magnetic losses probably caused by condensation along magnetic flux path which forms short-circuits, increasing eddy currents. Superconducting transformer shows better voltage regulation, lower apparent power but also lower efficiency. For a deeper study of the real impact and potential of using HTS tape on power transformers, it would be relevant to have both windings made of superconducting material.

Acknowledgements

This work was supported by FCT - Fundação para a Ciência e Tecnologia (CTS multiannual funding) through the PIDDAC Program funds.

References

[1] Thakur, K., Raj, A., Brandt, E., et al. (2011, April) Frequency-dependent critical current and transport ac loss of superconductor strip and Roebel cable. Superconductor science and technology, 24, 6, 11pp.

[2] Miyagi, D., Otome, D., Nakano, M., et al. (2010, February), Measurement of Magnetic Properties of Nonoriented Electrical Steel Sheet at Liquid Nitrogen Temperature Using Single Sheet Tester. IEEE Transactions on Magnetics, 46, 2, pp 314-317

[3] A. Gonçalves Pronto, M. Ventim Neves, A. Leão Rodrigues (2011, January), Measurement and Separation of Magnetic Losses at Room and Cryogenic Temperature for Three Types of Steels Used in HTS Transformers, Journal of Superconductivity and Novel Magnetism, vol. 24, n. 1-2, Springer.

[4] A. G. Pronto, M. V. Neves, A. L.Rodrigues (2012), A Possible Solution to Reduce Magnetic Losses in Transformer Cores Working at Liquid Nitrogen Temperature, Phys. Procedia, 36, pp 1103-1108, Elsevier.

Trapped magnetic field between double stacked MgB_2 bulks magnetized by pulsed field

H Fujishiro[1], T Ujiie[1], T Naito[1], A Figini Albisetti[2] and G Giunchi[3]

[1] Iwate University, 4-3-5 Ueda Morioka, Japan
[2] EDISON S.p.A., Foro Buonaparte 31, 20121 Milano, Italy
[3] Materials Science Consultant, via Teodosio 8, 20131 Milano, Italy

E-mail: fujishiro@iwate-u.ac.jp

Abstract. Pulsed field magnetization (PFM) was performed at T_s=14 K for the double stacked MgB_2 bulks (bulk pair) 55 mm diameter fabricated by a reactive liquid Mg infiltration (Mg-RLI) method, compared with that for the single bulk. The trapped field of B_z=0.80 T was achieved between two bulks and B_z at the bulk surface was enhanced from 0.42 T to 0.50 T by stacking of the bulks. The trapped field characteristics of the bulk pair can be qualitatively explained by the model analyses.

1. Introduction

MgB_2 bulk magnet has attractive natures such as low cost, light-weight, and weak-link-free homogeneous current flow, which are clear contrast with REBaCuO (RE=rare earth element) superconducting bulk magnets. To magnetize the superconducting bulks, a pulsed field magnetization (PFM) is an inexpensive and mobile experimental setup with no need of a superconducting magnet, as a substitute for field-cooled magnetization (FCM). We have investigated the PFM procedure for the REBaCuO bulks to enhance the trapped field [1-3]. Considering the obtained experimental results, we proposed a new PFM technique, in which multi-pulses are applied under different temperatures, and realized over 5 T on a GdBaCuO bulk, which is a record-high value by PFM to date [3]. For MgB_2 bulks, the results of the trapped field by FCM have been mainly reported. The higher trapped field was reported to be 2.25 T at 15 K on the single MgB_2 bulk [4] and 3.14 T at 17.4 K in the double stacked MgB_2 bulks [5]. We have performed the PFM procedure for the single MgB_2 bulk fabricated by a reactive liquid Mg infiltration (Mg-RLI) [6] and by a capsule method [7] with 20-50 mm in diameter at 16-30 K, where the trapped field of 0.47 T at 23 K and 0.71 T at 16 K were realized, respectively. To analyze the flux dynamics during PFM, we performed a numerical simulation of PFM for the MgB_2 bulk [8], which reproduced qualitatively the experimental results. The enhancement of the bulk thickness can achieve the higher trapped field because of the elimination of the demagnetization factor.

In this study, we investigated the trapped field characteristics by PFM for large MgB_2 of single bulk and double stacked bulks (bulk pair) fabricated by Mg-RLI method. The magnetic flux intrusion and flux trap during PFM are discussed using a numerical simulation.

2. Experimental Procedure

The thick MgB_2 bulk disk of 55 mm in diameter, which was fabricated by the Mg-RLI technique [9], was sliced and two bulk plates (A and B) were prepared. The thickness of bulk-A and bulk-B was 15.3

and 10.3 mm, respectively. The superconducting transition temperature T_c and the trapped field B_z of the bulk-B magnetized by FCM at 20 K were, respectively, confirmed to be 38 K and 2.0 T. Each bulk was mounted in stainless steel (SUS) ring of 8 mm thickness filled with a stycast 2850GT resin. Figure 1 shows the experimental setups for PFM around the bulk. For the bulk pair as shown in figure 1(a), the bulk-A was set on a soft iron yoke cylinder and tightly anchored onto the cold stage of a Gifford–McMahon cycle helium refrigerator. The bulk-B was also mounted on the bulk-A with the insertion of the brass plate 2 mm in thickness. The initial temperature T_{sA} of the bulk-A was set to 14 K. A magnetizing solenoid copper coil (94 mm i.d., 153 mm o.d., and 50 mm height), which was dipped in liquid nitrogen, was placed outside the vacuum chamber. A magnetic pulse $B_{ex}(t)$ with a rise time of 0.012 s and a duration time of 0.15 s was applied to the bulk by flowing the pulsed current. The time evolutions of the local fields $B_L^C(t)$ and the subsequent trapped fields B_z between two bulks and at the center of the surface were, respectively, monitored by Hall sensors of HS-a and HS-b (F W Bell, BHT 921 and BHA 921). Two-dimensional trapped field profiles of B_z(1 mm) were mapped at a distance of z=1 mm above the bulk surface. During PFM, the time dependence of temperature $T(t)$ of each bulk was measured at each SUS ring using Cernox[TM] thermometers of TM-a and TM-b. The similar experiments were performed for the single bulk-A as shown in figure 1(b).

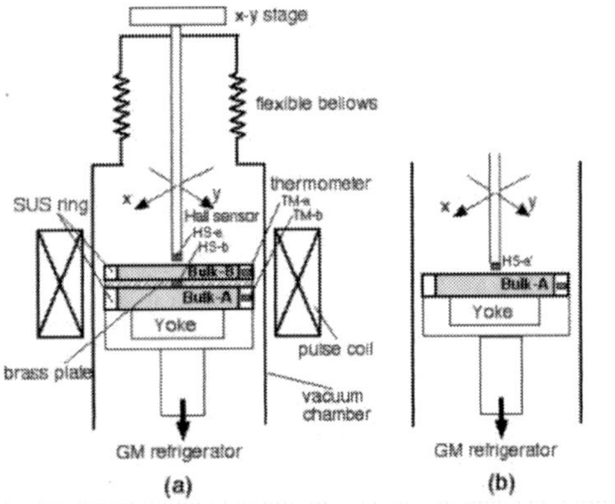

Figure 1. Experimental PFM setups for (a) the bulk pair and (b) the single bulk-A.

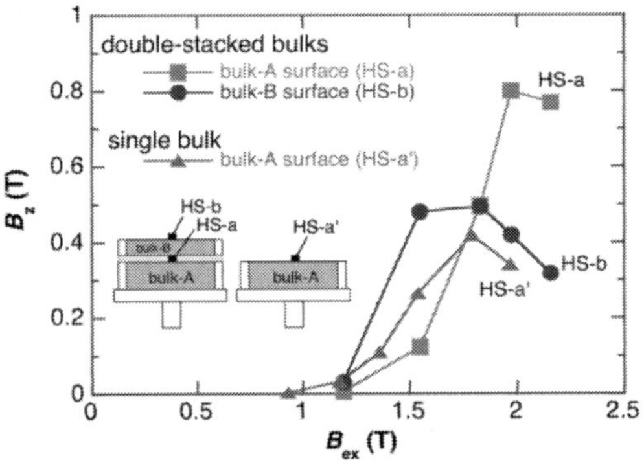

Figure 2. Trapped field B_z for the bulk pair (HS-a and HS-b) and the single bulk (HS-a'), as a function of the applied pulsed field B_{ex}.

3. Results and discussion

3.1. Experimental results

Figure 2 shows the trapped field B_z for the bulk pair and the single bulk-A, as a function of the applied pulsed field B_{ex}. For the bulk pair, B_z at the surface (HS-b) increases, takes a maximum of 0.50 T at B_{ex}=1.79 T and then decreases with increasing B_{ex}. B_z in the bulk pair (HS-a) also increases with increasing B_{ex} and takes a maximum of 0.80 T at B_{ex}=1.98 T, which is the highest B_z value by PFM to date. The B_z vs B_{ex} curve in the bulk pair shifts the higher B_{ex} side, compared with that for the surface because the demagnetization factor between the bulks was eliminated and the shielding behavior of the stack is more effective. For the single bulk-A, the maximum B_z was 0.42 T. The maximum of total trapped flux Φ for the bulk pair and single bulk-A was, respectively, 0.88 mWb for B_{ex}=1.55 T and 0.69 mWb for B_{ex}=1.36 T, which were obtained at B_{ex} lower than that, at which B_z took a maximum.

Figure 3 presents the trapped field profiles on the MgB_2 bulks for B_{ex}=1.55 and 1.98 T. The profiles for the bulk pair are shown in figures 3(a) and 3(b). For B_{ex}=1.55 T, the trapped field profile is concave because of the lower applied field than the optimum one, where the right lower position indicated by the arrow is difficult to trap the magnetic flux, which may reflect the inhomogeneity of critical current density J_c in the bulk-B. On the other hand, for B_{ex}=1.98 T as shown in figure 3(b), the magnetic flux was intruded into the central region of the bulk and the trapped field profiles are nearly the conical one. The profiles on the bulk pair may be influenced by the J_c distribution in the bulk-A. The profiles for the single bulk-A are shown in figures 3(c) and 3(d). For B_{ex}=1.55 T as shown in figure 3(c), the magnetic flux was intruded into the central region of the bulk with a small inflection around the arrows. The conical B_z profile can be obtained B_{ex}=1.79 T and, for B_{ex}=1.98 T, the conical profile maintains with the decrease in the maximum value, as shown in figure 3(d). The trapped field profiles of the bulk pair are not necessarily the same as that for each bulk. The complicated flux intrusion and trap can be considered for the stacking, even though J_c distribution seem to be identical.

Figure 4 shows the time evolution of the applied field $B_{ex}(t)$ and local fields $B_L^C(t)$ at the center of the bulk surface for B_{ex}=1.55 and 1.98 T. For each B_{ex}, $B_L^C(t)$ at the bulk surface (HS-b) first starts to increase and takes a maximum, and then decreases to a final value due to the flux flow. $B_L^C(t)$ in the bulk pair (HS-a) shows the similar time dependence with a time delay because of the strong shielding current in the bulk center. For lower B_{ex}=1.55 T, the final B_z at HS-a is smaller than that at HS-b. On the other hand, for higher B_{ex}=1.98 T, the magnitude relation of the final B_z was reversed.

Figure 5 shows the time dependences of the temperatures $T(t)$ of the bulk pair at TM-a and TM-b for B_{ex}=1.98 T. The temperatures increase abruptly just after the pulse application, and then quickly decrease with increasing time, which are clear contrast with those for the REBaCuO bulk [1]. The sharp temperature changes result from the low specific heat and high thermal conductivity of the MgB_2 bulk. It should be noted that the early temperature profiles are quite different; $T(t)$ at TM-b shows the spike-like behavior, but $T(t)$ at TM-a changes moderately. The difference may come from that of the thermal environment for each bulk. Furthermore, the initial temperature of the bulk-B at TM-b was 17 K, which was 3 K higher than that of the lower bulk-A at TM-a. The result suggests that the thermal contact of the bulk-B is not sufficient to the bulk-A and/or the thermal radiation induces from the surface of the bulk-B. The temperature gradient may exist along the thickness direction in the conduction-cooled bulk, even in the single bulk. In the inset, the applied field dependence of the maximum temperature T_{max} was also shown. T_{max} monotonically increases with increasing B_{ex}.

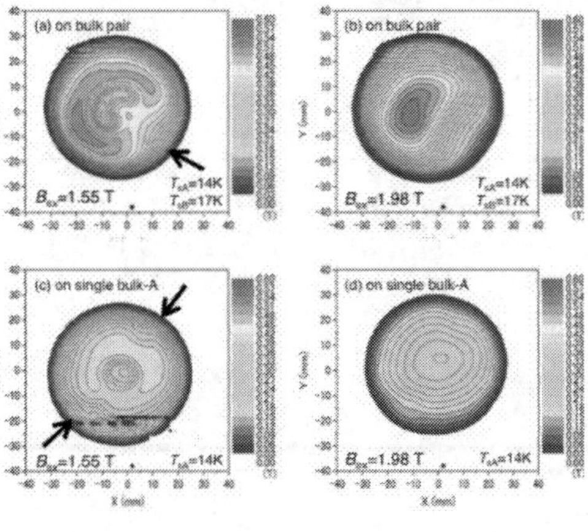

Figure 3. Trapped field profiles on the bulk pair ((a) and (b)) and the single bulk-A ((c) and (d)) for B_{ex}=1.55 and 1.98 T.

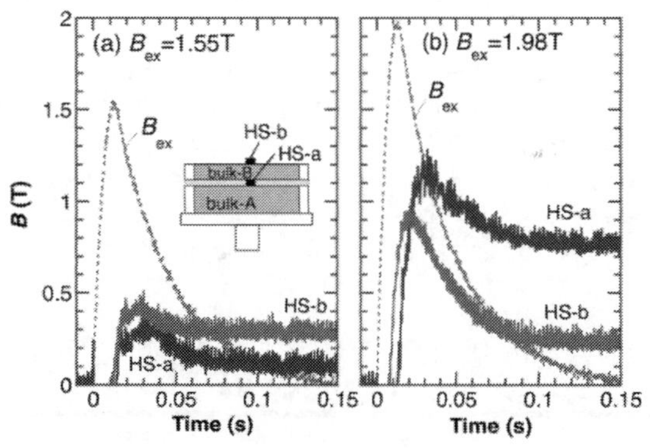

Figure 4. Time evolution of the applied field $B_{ex}(t)$ and local fields $B_L^C(t)$ at the centre of the bulk surface for B_{ex}=1.55 and 1.98 T.

3.2. *Results of numerical simulation and discussion*

In order to understand the experimental results of the PFM for the bulk pair, we performed the numerical simulation. Physical phenomena during PFM were described using electromagnetic and thermal fields. The power-n model (n=50) was supposed to describe the nonlinear E-J characteristic in the bulk. The temperature and magnetic field dependence of the critical current density $J_c(T, B)$ was defined. The details of the simulation are described elsewhere [2, 8].

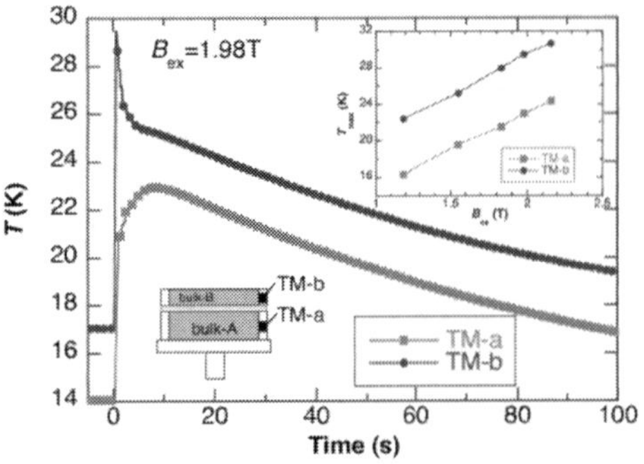

Figure 5. Time dependence of the temperatures of the bulk pair at TM-a and TM-b after applying the magnetic pulse of B_{ex}=1.98 T.

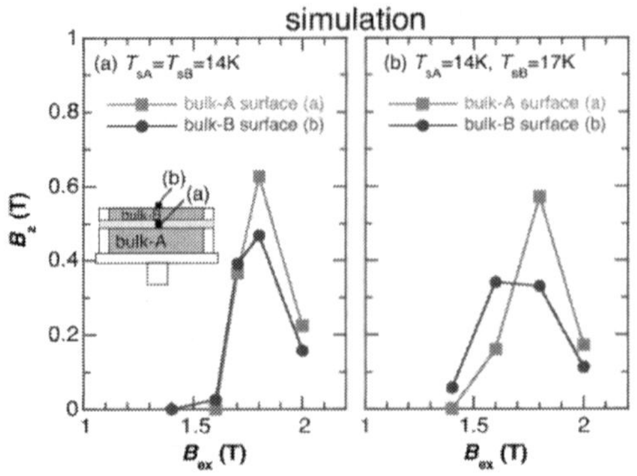

Figure 6. Results of the simulation of the trapped field B_z of the bulk pair at (a) T_{sA}=T_{sB}= 14 K and (b) T_{sA}=14 K, T_{sB}=17 K.

Figure 6(a) shows the results of the simulation of the trapped field B_z of the bulk pair at T_{sA}=T_{sB}=14 K, as a function of B_{ex}. B_z in the bulk pair (HS-a) and on the bulk surface (HS-b) show the similar B_{ex} dependences; both of B_z take a maximum at B_{ex}=1.8 T and then decrease with increasing B_{ex}. B_z at the position (b) is very slightly higher than that at the position (a) for B_{ex} lower than 1.7 T. These results reproduce the experimental ones qualitatively (see Fig. 2). Figure 6(b) shows the similar results of the simulation of the trapped field B_z at T_{sA}=14 K and T_{sB}=17 K, which are the similar condition to the experimental results as shown in figure 5. Both absolute B_z values at the positions of (a) and (b) decreased, compared with those in figure 6(a) and the B_z vs B_{ex} profile at the position (b) shifts lower B_{ex} side. The J_c value in the bulk-B decreases and the magnetic flux is easy to penetrate in the bulk-B. As a result, the magnetic flux is easy to penetrate also into the bulk-A. The temperature rise increases and the trapped field of both bulks decreases. The relation shown in figure 6(b) is more qualitatively similar than that in figure 6(a) to that obtained by the experimental results in figure 2.

In summary, pulsed field magnetization was performed at T_s=14 K for the double stacked MgB_2 bulks fabricated by a reactive liquid Mg infiltration (Mg-RLI) method. The trapped field of B_z=0.80 T was achieved between two bulks and B_z at the bulk surface was enhanced from 0.42 T to 0.50 T by stacking of the bulks. The trapped field characteristics such as the B_z vs B_{ex} relations at the bulk surface and in the bulk pair can be qualitatively explained by the model analyses.

References

[1] Fujishiro H, Oka T, Yokoyama K, Kaneyama M and Noto K 2004 *IEEE Trans. Appl. Supercond.* **14** 1054
[2] Fujishiro H and Naito T 2010 *Supercond. Sci. Technol.* **23** 105021
[3] Fujishiro H, Tateiwa T, Fujiwara A, Oka T and Hayashi H 2006 *Physica C* **445–448** 334
[4] Yamamoto A *et al.* 2011 *Abstracts of The 15th Japan-US Workshop on Advanced Superconductors,* arXiv:1208.1854
[5] Durrell J H, Dancer C E J, Dennis A, Shi Y, Xu Z, Campbell A M, Hari Babu N, Todd R I, Grovenor C R M and Cardwell D A 2012 *Supercond. Sci. Technol.* **25** 112002
[6] Fujishiro H, Tamura T, Arayashiki T, Oyama M, Sasaki T, Naito T, Giunchi G and Albisetti A F 2012 *Jpn. J. Appl. Phys.* **51** 103005
[7] Fujishiro H, Naito T, Sasaki T and Arayashiki T 2013 *Proc. ICEC24-ICMC2012* 571
[8] Fujishiro H, Naito T, Oyama M, Arayashiki T, Tamura T, Sasaki T and Giunchi G' 2013 *IEEE Trans. Appl. Supercond.* **23** 6800804
[9] Perini E, Giunchi G, Saglietti L, Albisetti A, Matrone A, and Cavaliere V 2011 *IEEE Trans. Appl. Supercond.*, **21** 2690

11th European Conference on Applied Superconductivity (EUCAS2013) IOP Publishing
Journal of Physics: Conference Series **507** (2014) 032017 doi:10.1088/1742-6596/507/3/032017

MagLev Cobra: Test Facilities and Operational Experiments

G.G. Sotelo[1], D.H.N. Dias[1], R.A.H. de Oliveira[2], A.C. Ferreira[2], R. De Andrade Jr.[2], R.M. Stephan[2]
1- Fluminense Federal University – UFF
Escola de Engenharia, Rua Passo da Pátria, 156, Bl. E, Sala 431. CEP.: 24210-240, Niterói-RJ, Brazil.
2- Federal University of Rio de Janeiro – UFRJ
Centro de Tecnologia, Cidade Universitária, PO box 68553, 21945-970, Rio de Janeiro, Brazil
E-mail: gsotelo@id.uff.br, dhndias@id.uff.br, raholiveira@ufrj.br, ferreira@ufrj.br, rubens.andrade@ufrj.br, richard@dee.ufrj.br

Abstract. The superconducting MagLev technology for transportation systems is becoming mature due to the research and developing effort of recent years. The Brazilian project, named MagLev-Cobra, started in 1998. It has the goal of developing a superconducting levitation vehicle for urban areas. The adopted levitation technology is based on the diamagnetic and the flux pinning properties of $YBa_2Cu_3O_{7-\delta}$ (YBCO) bulk blocks in the interaction with Nd-Fe-B permanent magnets. A laboratory test facility with permanent magnet guideway, linear induction motor and one vehicle module is been built to investigate its operation. The MagLev-Cobra project state of the art is presented in the present paper, describing some construction details of the new test line with 200 m.

1. Introduction

The requirement of efficient transportation systems, silent, non-polluting and with competitive construction costs is one of the priorities of crowded cities, where nowadays great part of the world population lives. One potential solution for this problem is the superconducting MagLev technology, which is based on the diamagnetic and flux pinning properties of High Temperature Superconductor (HTS) bulk blocks of $YBa_2Cu_3O_{7-\delta}$ (YBCO) in the interaction with Nd-Fe-B permanent magnets. Recently, commercial cryostats with HTS were designed for levitation, contributing for the evolution of the superconducting MagLev technology [1]. Also, the development of high quality Nd-Fe-B and their optimized arrangement in a permanent magnet guideway (PMG) made possible to build linear superconductor magnetic bearings (LSMB) able to levitate high level loads, weighting more than 1 ton [2]. Groups from Brazil [2-3], Germany [4] and China [5] have given important contributions in this field. In special, the Brazilian technology, known as MagLev-Cobra, proposes a full scale vehicle with multiple short units, allowing curves of 50 m radius, ramps of 10% and velocities up to 70 km/h. The MagLev-Cobra can run silently inside cities on elevated structures. For the vehicle traction, a linear induction motor (LIM) is used, whose geometry also contributes to increase the levitation force. Some experiments for the designed LIM are presented here. Finally, this paper shows construction details of the new full vehicle prototype and the 200 meters long test line.

2. Vehicle design and construction

The MagLev-Cobra vehicle design consists of small modular units having 1.5 meters length and 2.4 meters width. These modules can be arranged to fit any specific necessity. The new test line is 200 m long and it is under construction to connect two buildings at the Technology Center (CT) inside the university campus. Figure 1 shows the foundations of a terminal station and Figures 2 (a) and (b) present architectural drawings of this line.

Figure 1. Foundations of a terminal station

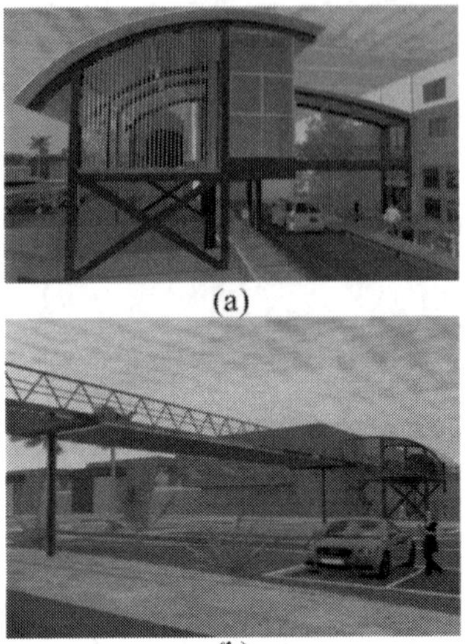

(a)

(b)

Figure 2. Architectural drawings

The vehicle will have four modules. The perspective view of the designed vehicle is presented in Figure 3, and the actual stage of the vehicle under construction with light fiber reinforced plastic materials is presented in Figure 4. This prototype was planned to carry comfortably 30 passengers.

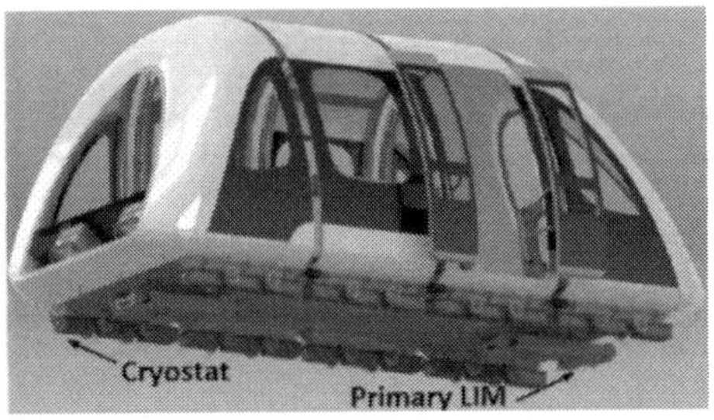

Figure 3. Designed MagLev-Cobra vehicle prototype.

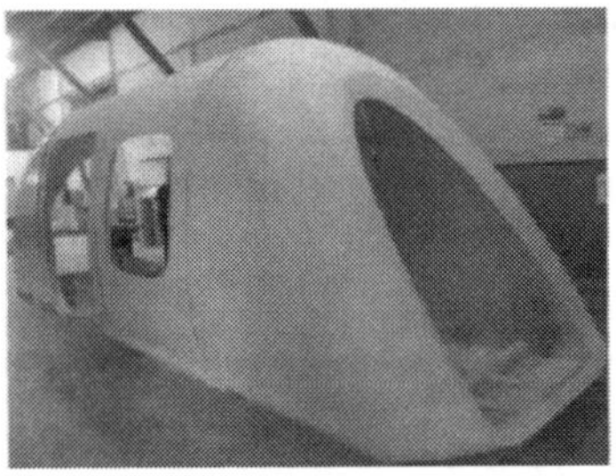

Figure 4. Actual stage of construction of the vehicle.

3. Linear Induction Motor

The designed LIM windings were designed to form a double inverted C configuration (Figure 5). This configuration contributes to the levitation force, since there is an attraction between the stationary and mobile parts besides the propulsion force. The main parts of a LIM are the primary (containing the 3 phase windings) and the secondary (made by laminated iron and short-circuited conducting bars, similar to a squirrel cage rotor). Each vehicle has two short primary windings coupled to the vehicle and the secondary is distributed along the way (long secondary). This solution was chosen because of the lower manufacture cost. The two primary modules work independently, but they are synchronized by sensors installed at the train that gives the control signals for each motor drive. To power supply the primary collecting brushes are used.

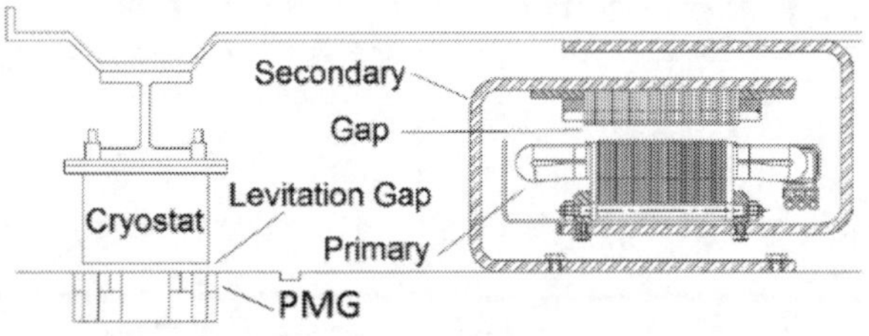

Figure 5. Cross section of vehicle bottom showing the levitation system (cryostat plus PMG) of one side and the inverted C geometry LIM.

The LIM develops a traction force (F_x), responsible for the movement, and a normal force (F_n). The traction force F_x is generated by the interaction between the induced current in the secondary with the travelling field in the air gap. The normal force existing in the LIM is the result of the magnetic flux crossing the air gap and has two components. The first component F_{na} is the electromagnetic attractive force between the primary and the secondary iron core and the second component is the electrodynamics repulsion force F_{nr} between the moving primary and secondary FMM. The repulsive force F_{nr} can be disregarded, since the vehicle operates at low speed. So, F_n improves the levitation gap during the acceleration due the inverted C configuration.

The attraction force was measured by installing a load cell between the ground and the primary of the linear motor. For this test, the longitudinal movement is blocked by a steel cable anchored in the wall. The test adopted the air gap of 8 mm. The frequency varied between 1 Hz to 5 Hz and V/f ratio was kept constant. The attractive force reaches approximately 3000 N, as presented in Figure 6.

Figure 7 presents the operational test of one vehicle module. To evaluate the levitation gap fluctuation during this operational test, ultrasonic position sensors were installed on the vehicle base. Those sensors can give the gap variation with a precision of 0.1 mm. Also, an optic sensor was used measure its velocity. In this test the vehicle was accelerated from standstill and it was brought back again to standstill. For the acceleration, the supply frequency was increased from 0 Hz to 5 Hz with a constant V/Hz relationship. The braking procedure used a combination of regenerative braking with 0.5% direct current component. The acceleration of the vehicle produces an attractive force between the secondary and primary of linear motor as expected.

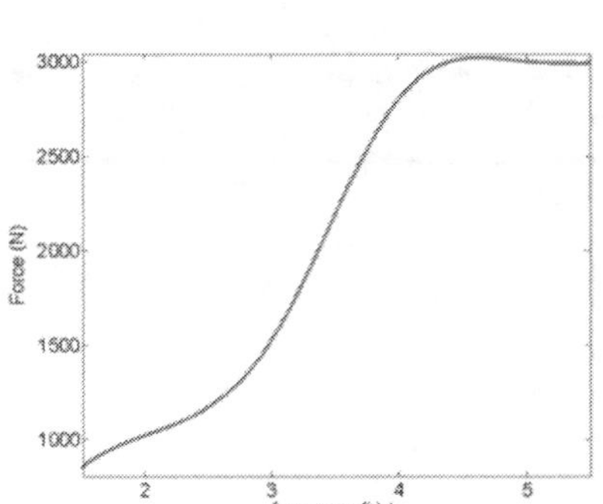

Figure 6. Static force measurement as a function of supply frequency.

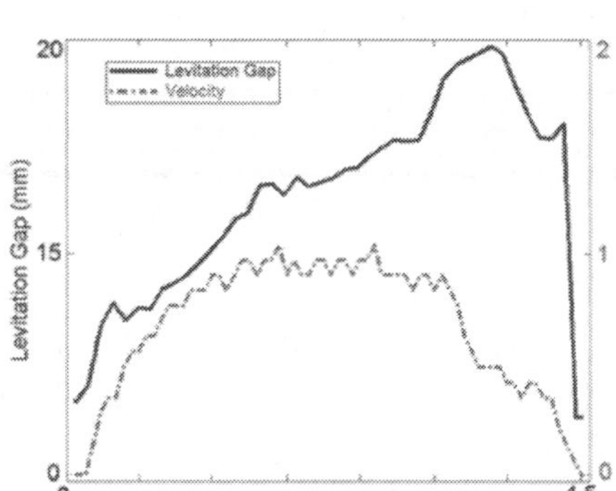

Figure 7. Levitation gap and velocity results during operational tests.

The laboratory tests have the objective to quantify the generated energy during regenerative braking. The tests were realized in a 6 m length ramp. The results were obtained for 1000 kg mass and the inclination of 15 %. The air gap for this experiment is 8 mm. Deceleration is obtained changing the frequency to lower values until stop. A frequency pattern was imposed for the operation as shown in Figure 8. These tests allow the determination of the system regenerative capacity. The DC link voltage and motor AC rms current are presented in Figure 9. It is possible to observe an increase in the DC link voltage during deceleration. The regeneration can also be observed by the DC link current, whose signal is negative during regeneration, as presented in Figure 10, that shows the negative current (in red) has low values of amplitude and duration (~7.8 s $\leq$ t $\leq$ 8.3 s) due to the short test track (6 m).

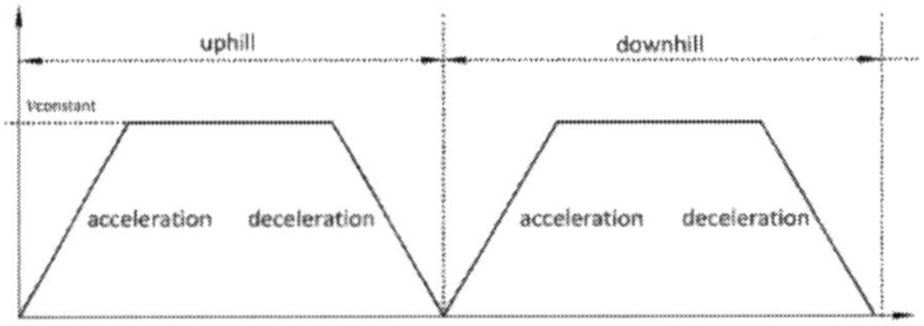

Figure 8. Trapezoidal trajectory used in the experiments.

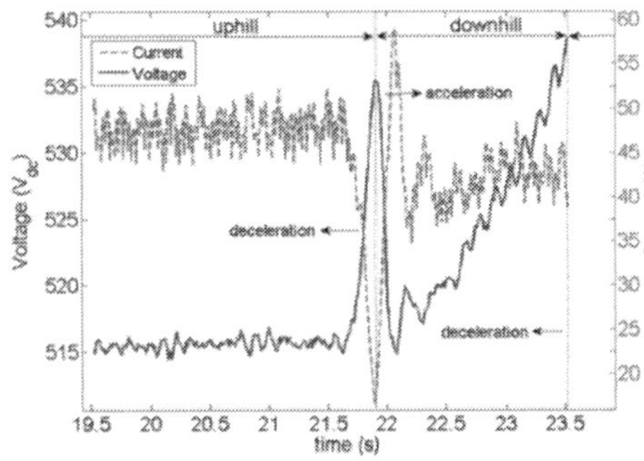

Figure 9. DC link voltage and motor current during a test with 1000kg and 15% ramp.

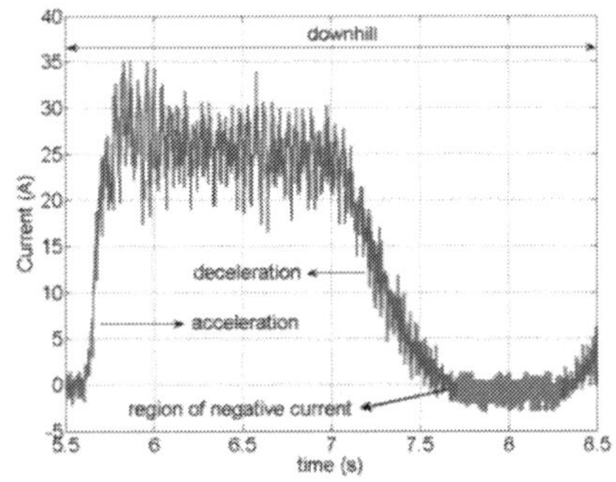

Figure 10. DC link current during regeneration.

4. Conclusions

This paper reported the state of the art of the first full scale superconducting magnetic levitation train prototype, named Maglev-Cobra. The prototype linear induction motor was tested, presenting regenerative braking and levitation gap increase during operation. The daily operation of this system in a 200 m test line is expected to start in September 2014, by the time of the 22nd International Conference on Magnetically Levitated Systems and Linear Drives.

5. References

[1] F. N. Werfel et. al., Supercond. Sci. Technol., vol. 25, pp. 014007, 2012.
[2] G.G. Sotelo et al., IEEE Trans. on Applied Superc.; 2011, vol. 21, p. 1464-1468.
[3] G.G. Sotelo et. al., IEEE Transac. on Appl. Supercond., v. 23 (3), pp.3601204,3601204, 2013.
[4] L. Kuehn et. al., IEEE Trans. Appl. Supercond., v. 17 (2), pp.2079-2082, 2007.
[5] D.H. Jiang et. al., IEEE Trans. Appl. Supercond., v. 22 (3), pp. 3600304, 2012.

Acknowledgments

Authors wishing to acknowledge financial support from the Brazilian agencies: CAPES, CNPq, FAPERJ and BNDES.

Magnetic properties measurement and discussion of an amorphous power transformer core at room and liquid nitrogen temperature

A G Pronto[1], A Maurício[2], J M Pina[1]

[1] CTS, Uninova, Departamento de Engenharia Electrotécnica, Faculdade de Ciências e Tecnologia, FCT, Universidade Nova de Lisboa, 2829-516 Caparica, Portugal
[2] Departamento de Engenharia Electrotécnica, Faculdade de Ciências e Tecnologia, FCT, Universidade Nova de Lisboa, 2829-516 Caparica, Portugal

E-mail: amg1@fct.unl.pt

Abstract. In energy generation, transmission and distribution systems, power transformers are one of the most common and important components. Consequently, the performance of these transformers is crucial to global efficiency of the systems. To optimize transformers efficiency, the selection of an adequate ferromagnetic material is very important. For example, the use of amorphous ferromagnetic materials in transformer cores, replacing crystalline electrical steels, decreases total magnetic losses of the device. Other possible solution to increase energy systems efficiency, is the installation of high temperature superconducting power transformers (HTS transformers), normally cooled by liquid nitrogen at 77 K. In order to contribute to HTS transformer efficiency improvement, a 562.5 VA transformer with an amorphous ferromagnetic core was designed and built. For this core, the most important magnetic properties are measured at room and cryogenic temperature, and then compared with those of a typical crystalline grain-oriented electrical steel. Amorphous material magnetic losses (static and dynamic) at room and 77K are also presented and discussed.

1. Introduction

Amorphous steels used in power transformer cores allow easier magnetization processes and magnetic losses about 60 to 70% lower, when compared to conventional crystalline silicon-steel transformer cores [1, 2]. This loss reduction is very important to optimize conventional transformers efficiency and could became even more important in the optimization of high temperature superconducting power transformers (HTS transformers), where loss reduction has also a direct impact in the cryogenic system.

In order to contribute to HTS transformer efficiency improvement, a 562.5 VA transformer with an amorphous ferromagnetic core (2605SA1 alloy) and cooper windings was designed and built. The magnetic properties was measured at room and liquid nitrogen temperature. This paper presents total magnetic losses curves, magnetization curves and hysteresis loops obtained for both temperatures. The experimental results are analysed and compared to others obtained in similar tests made in a typical crystalline grain-oriented electrical steel.

2. Experimental procedure

The 562.5 VA single phase transformer with an amorphous ferromagnetic core (2605SA1 alloy) was designed and built. Table 1 shows the main characteristics of the developed prototype.

Table 1. Specifications of the designed amorphous core transformer

Capacity (VA)	Frequency (Hz)	Voltage (V) (U_1/U_2)	Turns (N_1/N_2)	Current (A) (I_1/I_2)	Core dimensions (mm) (Heigh×Width×Depth)
562.5	50	112.5 /56.25	203/102	5 /10	300×188×60

The measurements were made under sinusoidal induction waveform at 50 Hz through the application of an AC voltage to the primary of the transformer, leaving the secondary in open circuit. The maximum magnetic induction B was controlled through the variation of the applied voltage U_1, so each voltage imposed a constant B_{max}. The corresponding waveforms of magnetic induction B and applied magnetic field H were registered in a digital oscilloscope. The magnetic field H is calculated from the voltage drop measurement in a calibrated resistance (which is an image of the primary electric current i_1(t)), connected in series with the primary circuit, and the magnetic induction B is obtained from the integration of the open circuit secondary voltage of the transformer, u_2(t). The measurements were made at 298 K for several values of maximum magnetic induction B. Then the transformer were bathed by liquid nitrogen in order to repeat the procedure for the same values of B_{max}.

To compare the magnetic properties at 77 K and 298 K of the amorphous core with a typical grain-oriented core, a Fe-Si transformer designed and built by A. Pronto [3,4], was chosen because it has a core with the same dimensions. In this paper, the amorphous core transformer will be designated by 'NA' and the grain-oriented core transformer by 'AP'. Table 2 shows the main characteristics of the conventional grain-oriented Fe-Si alloy used in AP.

Table 2. Specifications of silicon-iron sample used to build AP core

Class of silicon-iron	Thickness of laminations (mm)	Grade AISI	DIN 46400	% wt Si	Density ($g \cdot cm^{-3}$)
CGO	0.35	M6	VM111-35N	3.9	7.05

3. Experimental results and discussion

The first magnetization curves for the two magnetic cores, NA and AP, at room and cryogenic temperatures, are plotted in figure 1.

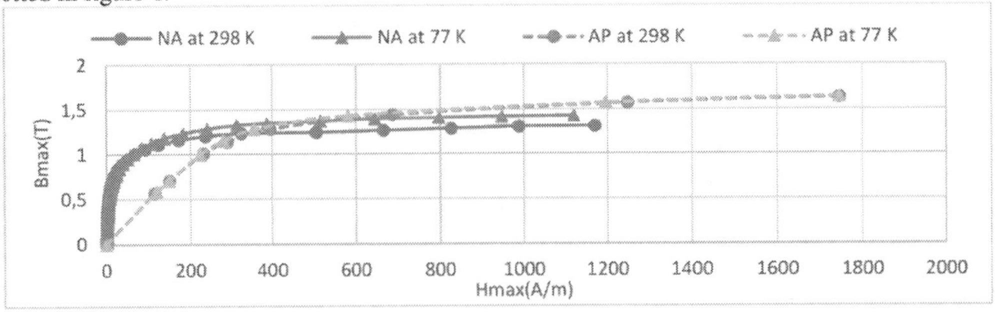

Figure 1. First magnetization curves at 298 K and 77 K for NA and AP (adapted from [3])

Comparing the measured first magnetization curves of NA and AP it can be concluded that:

- For NA core and for maximum magnetic induction B up 1T, the magnetization curves are very similar at 298 K and 77 K. The saturation flux density, B_{sat}, is approximately 0.65 T. Above 1T, and for the same value of magnetic field H, the corresponding maximum magnetic induction B is approximately 10% higher at 77K. This means that, in this range the same magnetomotive force produce higher induction fields.
- For AP core, B_{sat} is approximately 1.2 T, which is about 85% higher than the corresponding value for NA core. This difference is due to the composition of the amorphous materials, namely the smaller quantity of iron (80%), when compared with crystalline Fe-Si steels, added to the presence of silicon (20%), which leads to a decrease in saturation induction, when compared with other ferromagnetic materials [3,5].

- Amorphous material has a much higher magnetic permeability. For $B \leq 0.6$ T (non-saturated region) the relative magnetic permeability of the amorphous material is $\mu_r = 42952$, approximately eleven times higher than that of the silicon-iron for the same region, which is $\mu_r = 3922$.

The hysteresis loops obtain under dynamic conditions for a maximum magnetic induction $B_{max} = 0.65$ T at 298 and 77 K, are represented in figure 2.

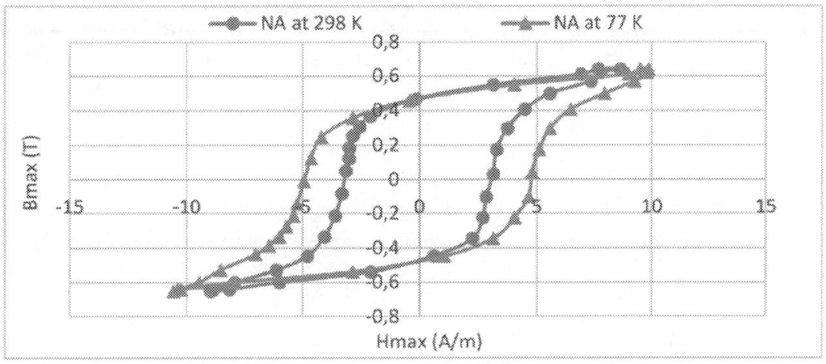

Figure 2. Hysteresis loops at 298 K and 77 K for NA

As can be observed in figure 2, the hysteresis loop obtained at room temperature is narrower, leading to lower total magnetic losses, p_{mag} (0.056 W/Kg at 298 K and 0.084 W/Kg at 77 K). Despite the value of remanence, B_r, remains constant for both temperatures, the coercive magnetic field H_c increase approximately 50% at liquid nitrogen temperature, which corresponds to higher magnetic losses at 77 K.

In order to analyse total magnetic losses behaviour several hysteresis loops were obtained as function of B_{max}, for both temperatures. These measurements allows the determination of total magnetic losses, $p_{mag,t}$, through hysteresis loop area calculation. Figure 3 shows specific total magnetization losses curves obtained for NA and also compared them with total losses obtained for AP [3].

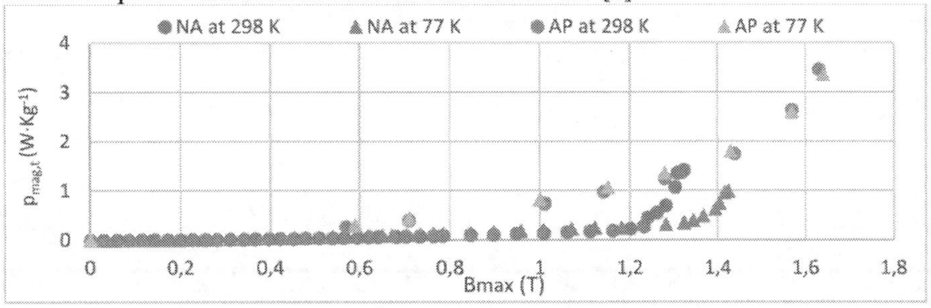

Figure 3. Total specific magnetic losses, $p_{mag,t}$ at 298 K and 77 K for NA and AP (adapted from [3])

Regarding total specific magnetic losses in NA and, for example, for $B = 1.2$ T, magnetic losses increase at 77K up to 8%. According with the results shown in figure 3, and comparing both materials, total specific magnetic losses obtained in AP tests are much higher. This increase is very significant, reaching values 7-8 times higher than the ones obtained for NA. Nevertheless, for NA, when the magnetic material is strongly saturated ($B_{max} > 1{,}2$ T) an inversion on losses behaviour was detected, identically to what happened in AP crystalline core for induction fields equal or higher than 1.5 T: total magnetic losses are lower at 77 K. **T**his difference is much more notorious in ferromagnetic amorphous material than in AP crystalline core. The

Journal of Physics: Conference Series **507** (2014) 032018 doi:10.1088/1742-6596/507/3/032018

loss reduction at low temperature could have its origin in several factors. First, at low temperature, thermal vibration of lattice decreases, which leads to decreasing electric resistivity but, at same time, to increase magnetic permeability of the material. When the material is heavily saturated, the rotation of magnetic domains is the dominant mechanism and it is facilitate by higher permeability, probably reducing hysteresis losses. On the other hand, domain rotation causes a lower variation of dB/dt than that associated with Bloch wall movements, the dominant magnetization mechanism outside saturation region, which contribute to decrease induced electromotive force with corresponding reduction of eddy-currents in amorphous material, and consequent decreasing in excess losses [3,4]. This means that in saturated region, the decreasing in electric resistivity is counterbalanced by easier domain rotation processes which finally results in lower total magnetic losses at cryogenic temperature. [3,4]. For HTS power transformers, this loss reduction would be useful if happens in non-saturated region of the ferromagnetic material.

4. Conclusions

According with the achieved experimental results, it is possible to conclude that:

- amorphous steel has lower saturation flux density than the conventional silicon-steel used in AP, as expected; for this parameter, there are no relevant differences at room temperature and liquid nitrogen temperature for both materials.
- total magnetic losses are much lower in amorphous material when compared with losses obtained in AP core. For example, for a value of B = 1 T, the total magnetic losses in AP are 4 times higher at room temperature and 5 times higher at liquid nitrogen temperature, when compared to NA. In transformer useful operation region (outside saturation), there is an increase in losses at 77 K, for both ferromagnetic core materials, which will correspond to a reduction in transformer efficiency if cores are immersed in liquid nitrogen.
- nevertheless, when the core material are strongly saturated, total magnetic losses at 298 K become larger than losses at 77K. This behaviour could be very useful for HTS transformers if occurs at non-saturated region. A possible explanation is presented above, but further investigation is needed in order to understand the origin of this behaviour and if it's possible to shifted it to normal operation region.

Acknowledgements

The authors would like to thank to Dr. Ryusuke Hasegawa and Metglas Inc. for gently built up and offer the amorphous core used in this power transformer prototype.

This work was supported by FCT - Fundação para a Ciência e Tecnologia (CTS multiannual funding) through the PIDDAC Program funds.

References

[1] DeCristofaro N 1998 Amorphous Metals in Electric-Power Distribution Applications, *Materials Research Society Bulletin*, **23**, n. 5, pp. 50-56.

[2] Hasegawa R 2008 Present status of amorphous soft magnetic alloys, *J. of Magnetism and Mag. Materials* **215-216**, pp 240-245.

[3] Gonçalves Pronto A, Ventim Neves M and Leão Rodrigues A 2011 Measurement and Separation of Magnetic Losses at Room and Cryogenic Temperature for Three Types of Steels Used in HTS Transformers, *J. of Superconductivity and Novel Magnetism*, **24**, n. 1-2.

[4] Gonçalves Pronto A, Ventim Neves M and Leão Rodrigues A 2012 A Possible Solution to Reduce Magnetic Losses in Transformer Cores Working at Liquid Nitrogen Temperature, *Physics Procedia* **36**, pp. 1103-1108.

[5] Luciano A, Benedito S and Kiminami Claudio 1997 An Amorphous core transformer: design and experimental performance, *Materials Science and Engineering* **A226-228**, pp.1079-1082.

Tryout of the vacuum impregnation procedure for the ITER PF1 coil on the VPI mold made of plastic shell

S Grigoriev[1], I Rodin[1], V Tanchuk[1], S Korban[1], A Bursikov[1], A Mednikov[1] and A Pugachev[2]

[1]D.V. Efremov Scientific Research Institute of Electrophysical Apparatus (NIIEFA)
[2] "Forss Consulting" Ltd

E-mail:sergey.grigoriev@sintez.niiefa.spb.su

Abstract. The PF1 Coil as part of the ITER superconducting magnet system is intended for positioning and shaping of the magnetic poloidal field. The technical specification for the PF1 coil requires a full-scale simulation of the basic technological processes including vacuum pressure impregnation (VPI). Usually, a VPI mold is made of massive stainless steel walls to get the required quality of the monolithic structure of the pancake insulation by hot-curing compound. As a result, we obtain a high-cost furnace with a considerable amount of steel inside to be heated up to the specified temperature; excessive energy consumption for the furnace heating system; problems with the vacuum tightness of the VPI mold. The impregnation procedure using a VPI "plastic" mold was proposed so as to avoid the above mentioned shortcomings associated with the use of the furnace made of stainless steel.

1. Introduction

The PF1 Coil (figure1), being part of the ITER magnet system, is designed to create the poloidal magnetic field of a specified value with the aim to control the plasma shape and position [1]. Structurally, the coil is a circular cylindrical solenoid about 9 m in outer diameter, 1 m in height and about 225 t in weight assembled of eight double pancakes.

The mechanical and operation properties of the pancakes, as well as the properties of the entire coil depend to a great extent on the quality of the monolithic structure resulting from vacuum impregnation of the composite pancake insulation by hot-curing compound. The required quality of the monolithic structure is attained, in its turn, by strict adherence to the temperature scenario specified for impregnation in compliance with stringent requirements for temperature non-uniformity of the whole coil structure.

One of the methods of fabrication of electric insulation is the VPI method using, as a rule, metal vacuum volume. The focus of this paper is on fabrication and application of alternative plastic volumes for VPI.

The calculation models and test mock-up (figure 2) of the plastic VPI mold were prepared for tryout of this vacuum impregnation technology. The analysis demonstrates that the proposed method provides the required temperature scenario for the PF1 test mock-ups and the acceptable temperature non-uniformity at a lower power consumption. The qualification sample of the PF1 coil was manufactured and tested to verify the plastic VPI mold technology to be used for the coil vacuum impregnation procedure.

Journal of Physics: Conference Series **507** (2014) 032019 doi:10.1088/1742-6596/507/3/032019

Figure 1. General view of the coil PF1

Figure 2. 3x3 PF1 coil section mock-up

2. Fabrication of fiberglass mold

Solid fiberglass mold for VPI is made of glass-cloth materials and impregnated with cold-curing compound by the infusion method. The proposed fabrication procedure for such molds includes laying of reinforcing material (glass cloth) and wrapping a sample with vacuum film, arrangement of vacuum channels and channels for resin to penetrate, pressurization of the fabricated vacuum volume, infusion of cold-curing resin into the glass cloth with subsequent polymerization at room temperature and removal of secondary materials from the surface of the finished volume. A fiberglass mold is formed around the main insulation of the coil winding and involves the following basic procedures:

- formation of the inner vacuum volume by the plastic film around the sample insulation for VPI with hot-setting compound. At the same time, this film prevents penetration into the main insulation of the sample of the cold-curing compound used for formation of the outer fiberglass mold;
- formation of the outer fiberglass mold body from glass-cloth materials for the infusion with cold-curing compound;
- formation of the vacuum volume with a plastic film wrapped around the outer fiberglass mold body;
- degassing of the inner vacuum volume to a residual pressure of about 10 Pa. In this case, the vacuum volume shell compact the main sample insulation by an air pressure of 1 kg/cm^2;
- degassing of the outer vacuum volume to a residual pressure of about 10 Pa ensuring compaction of glass-cloth materials of the mold body over the outer surface of the sample;
- impregnation of glass-cloth materials of the outer mold body with cold-curing compound followed by polymerization at room temperature.

The above method was used for fabrication of the test sample of the ITER PF1 superconductive coil. The sample was made as stack of 3×3 1.6m-long bars, its cross-section with turn and main insulation is shown in figure 2. The sample was fabricated by the process accepted for manufacture of the PF1 coil [2].

Fabrication procedure of the test assembly for impregnation is shown in figures 3 - 6.

Figure 3. Formation of the turn insulation

Figure 4. Assembly of the 3x3 PF1 coil turns

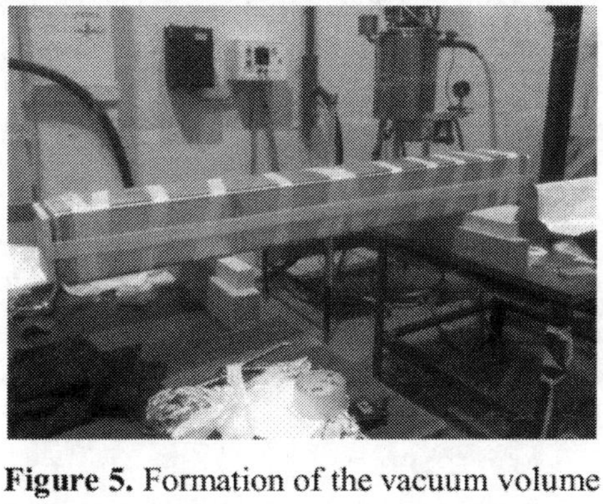

Figure 5. Formation of the vacuum volume

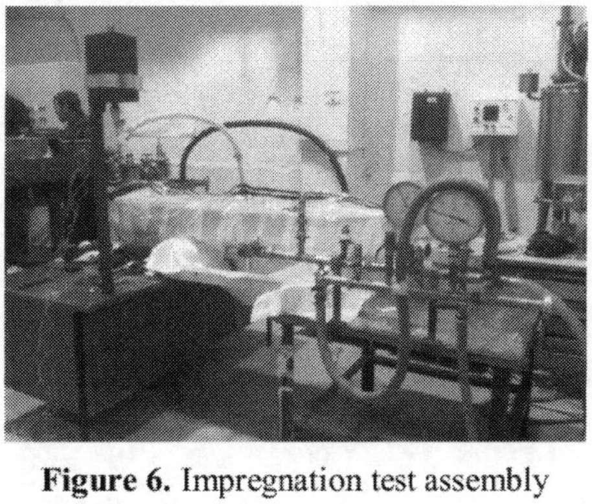

Figure 6. Impregnation test assembly

3. Calculation model

The main purpose of the thermal analysis is to set up a numerical experiment for simulation of the required mode of heat loads providing the required temperature scenario of sample impregnation. The calculation model was developed for the prototype (figure 7) and its temperature state was numerically simulated in compliance with the recommended scenario of the optimal impregnation technology (figure 8).

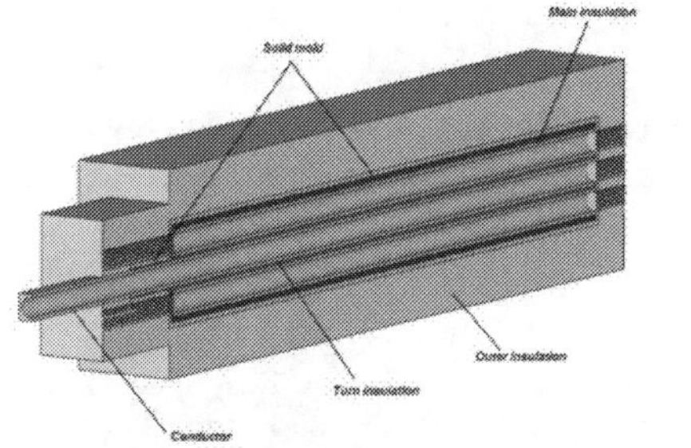

Figure 7. 3DANSYS model of the test sample

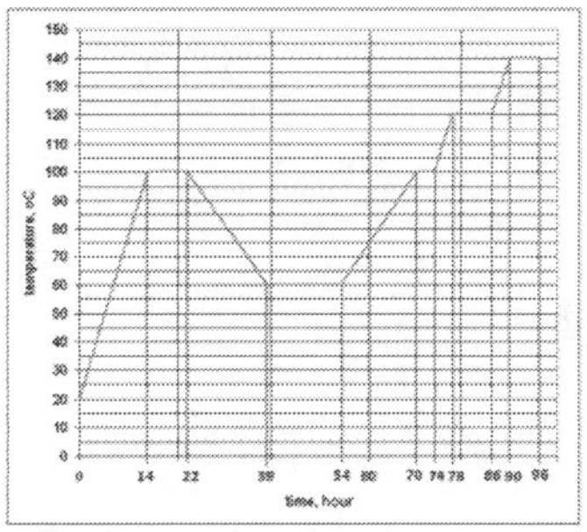

Figure 8. Recommended temperature scenario of VPI process

The calculated powers of the external heaters are shown in figure 9, the temperature distributions over the insulation at the last polymerization stage (140°C level) are given in figure 10.

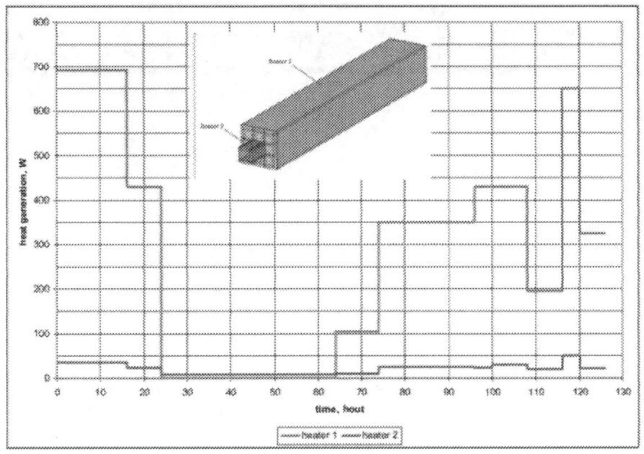

Figure 9. Power on the outer heaters

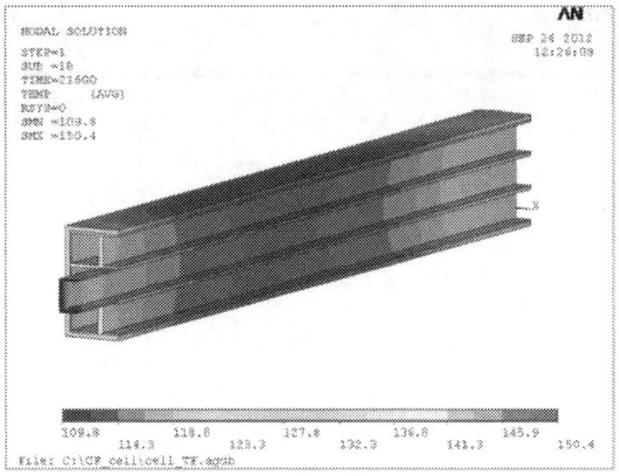

Figure 10. Temperature on the insulation

On the basis of the obtained results the curves of the minimum and maximum temperature were plotted for the whole insulation layer (figure11) and for the central insulation zone, i.e. excluding local angular areas of deteriorated heating (figure12).

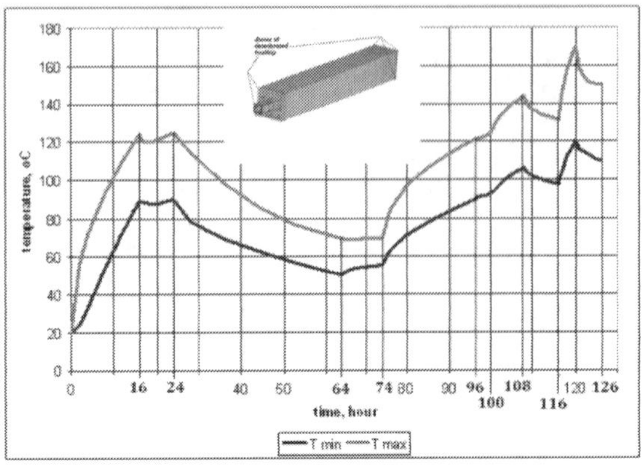

Figure 11. Maximum and minimum temperature on the whole insulation of the sample

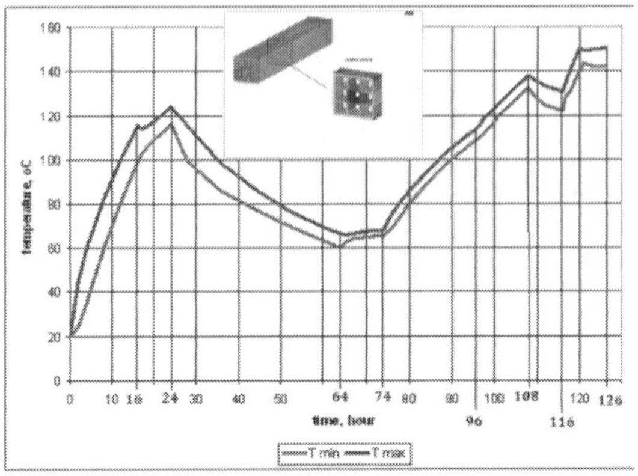

Figure 12. Maximum and minimum temperature in the central zone of sample insulation

4. Impregnation procedure

Impregnation of the inner volume of the sample insulation starts after fabrication of the solid fiberglass mold in accordance with the recommended temperature scenario and calculated power of the heaters (figure 13 shows the outer layer of thermal insulation of the solid fiberglass mold). Figure 14 demonstrates the PF1 coil test mock-up after VPI, polymerization and mold dismounting.

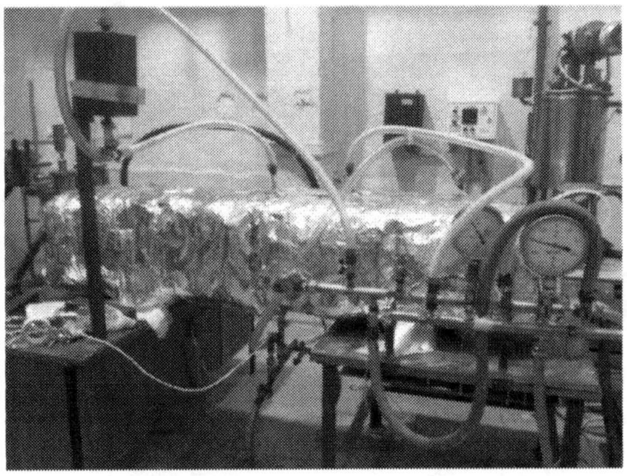

Figure 13. Solid fiberglass mold

Figure 14. Ready-to-deliver mock-up

Conclusions

VPI parameters were calculated by the developed numerical 3D model prior to fabrication of the mock-up and its compounding.VPI process was successfully tried out.

References

[1] Magnet Design Description Document (DDD1.1-1.3), August 2004, ITER internal report
[2] Technical Specification (ANNEX B) to the procurement arrangement of the PF coils of the ITER magnet system 1.1.P3A.RF.01.

Long term operation in 200 m class superconducting DC power transmission test facility in Chubu University

M Hamabe, H Watanabe, J Sun, T Kawahara and S Yamaguchi
Center of Applied Superconductivity and Sustainable Energy Research, Chubu University, Kasugai, Aichi, 487-8501, Japan

E-mail: hamabe@isc.chubu.ac.jp

Abstract. We constructed the 200 m class superconducting DC power transmission test facility (CASER-2) in 2010, and have carried out the cooling down and operation test. The 5th cooling down and operation test was carried out from August to November 2012. Long term current feeding was tested for a month with various currents and temperatures in the 5th test. From the long term current feeding test, the LN2 circulation was clearly affected by the operation of the cryogenic system and the atmosphere, not only by the operation DC current. It was also confirmed that the Peltier current leads worked effectively for the reduction of heat leak at the cable terminal.

1. Introduction

Recently, superconducting DC (SC-DC) power transmission is one of the attractive themes in commercial use of superconductivity, since the SC-DC power transmission is theoretically free from the AC losses, and is able to design the very large current cable for the power transmission. As a practical use, a 380 m/10 kA SC-DC cable was installed in the aluminum electrolysis system of Henan Zhongfu Industrial Co., in China[1]. The Electric Power Research Institute (EPRI) finished a design study of 10 GW SC-DC power transmission over 1000 km in U. S.[2]. Railway Technical Research Institute (RTRI) in Japan started a project to develop SC-DC power transmission system for railways[3].

In Chubu University, the first SC-DC power cable test facility (CASER-1) with a cable length of 20 m was constructed in 2006[4]. Since the first cooling down in October 2006, six cycles of cooling-down were carried out in CASER-1. The performance of the SC-DC cable showed no degradation through these cycles in that the critical currents of the HTS tapes in CASER-1 cable were in the same temperature dependence[5]. Based on the experiments in CASER-1, Chubu University constructed a new test facility of 200 m SC-DC power transmission (CASER-2) in 2010 to establish a power transmission system of a low total cryogenic energy loss[6]. This CASER-2 also has experienced cooling cycles repeatedly, and the 5th cooling cycle of the CASER-2 was carried out from August to November 2012. In this paper, we will describe the results of long term current feeding test for a month with various currents and temperatures in the 5th cooling cycle.

2. Specification of CASER-2

Table 1 shows the electric properties of CASER-2, and Table 2 shows the cryogenic properties to circulate liquid nitrogen (LN2). Figure 1 shows a photograph of a cut-model of a SC-DC cable core fabricated by Sumitomo Electric Industries, Ltd The cable consisted of co-axial three layers of DI-

BSCCO® HTS tapes; one outside layer of 16 tapes and two inside layers of 23 tapes were insulated by polypropylene laminated paper (PPLP®) for bipolar current feed with an insulation voltage of 20 kV in LN2. All HTS tapes were also insulated by PPLP® from the copper former and shield in the earth potential with an insulation voltage of ±10 kV. Designed total current of the cable was 2 kA at 78 K.

One of the features of CASER-2 is that the cable core was placed in the smooth cryogenic pipe with vacuum insulation[6]. Compared with the corrugated pipe, which is widely used for the cryogenic pipes of superconducting cables, the smooth pipe has the advantage to reduce the pressure drop of LN2 flow[7,8]. Moreover, we employed a zinc-coated carbon-steel pipe (216 mm in outside diameter and 5.8 mm in thickness) as the outside pipe, since the outside pipe is used in the atmosphere to keep the vacuum insulation and an expensive stainless-steel pipe is not necessary. As the inside pipe, a stainless-steel pipe (60.5 mm in outside diameter and 1.65 mm in thickness) was employed from the viewpoint of the low-temperature brittleness in a LN2 temperature. Figure 2 shows a photograph of the cryogenic system of CASER-2. Due to the employment of the smooth cryogenic pipe, low pressure drop along the 200 m cable was achieved; consequently, one Stirling type refrigerator and one LN2 pump are sufficient in CASER-2 to circulate the LN2 along the 200 m cable length.

Table 1. Electric specification of CASER-2

	Specification
Current	DC 2 kA at 78 K
Voltage	DC ± 10 kV
Cable structure	Coaxial, bipolar current feed
SC tape	39 DI-BSCCO® HTS tapes (23 tapes in inside HTS layer, and 16 tapes in outside HTS layer)
Current leads	72 Peltier current leads (46 leads in inside HTS layer and 26 leads in outside HTS layer), and 6 copper leads (all in outside HTS layer)

Table.2 Cryogenic specification of CASER-2

	Specification
Cryogenic system	Coolant is subcool LN2 LN2 was circulated by 1 LN2 pump and cooled by 1 Stirling type refrigerator
Cooling power	1 kW @ 77 K (COP=0.083 @ 77K)
Cryogenic pipe	Vacuum insulated double pipe Outside pipe: Zn-coated steel pipe / Inside pipe: stainless-steel pipe Smooth pipes with several bellows

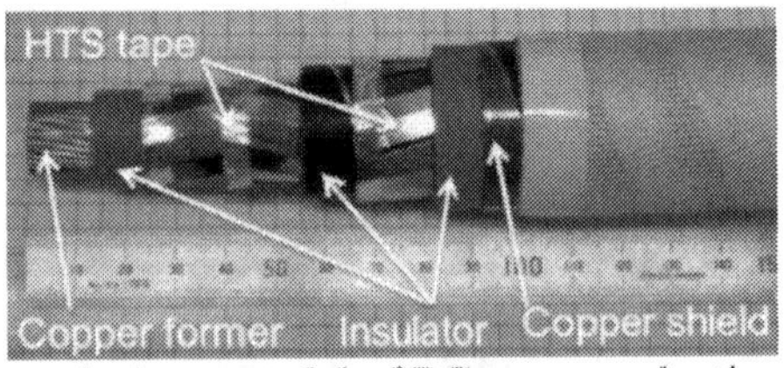

Figure 1. A cut-model of DC superconducting power cable core of CASER-2

Figure 2. Cryogenic system in CASER-2

3. Experimental results in long term operation test

We carried out the 5th cooling down test from August to November in 2012. During the 5th cooling down, we continuously fed the current to the SC cable from September 20th to October 22nd as a long term operation test of CASER-2. We varied operation current and temperature setting of refrigerator as parameters, and kept the speed of the LN2 pump constant, during this long term operation test. Note that the current was limited to 1200 A for the inside layer of the coaxial SC cable and to 600 A for the outside layer, because of the trouble in the connection between the current lead and the power supply.

Figure 3 shows the time evolution of (a) operation current, (b) LN2 temperatures in cryogenic system for the various temperature settings of refrigerator, (c) LN2 flow rate, (d) temperature rise along the 200 m cryopipe as cable line, (e) heat leak on the 200 m cryopipe, (f) temperature rise in and out of the cable terminal B, and (g) heat leak on the terminal B.

From the comparison of figures 3 (a)~(c), the LN2 temperature decreased due to the decrease of refrigerator setting, and increased due to the increase of the operation current. The decrease of LN2 temperature leads to the decrease of the viscosity of LN2. Therefore, the LN2 flow rate also varied following the variation of the LN2 temperature, despite of the constant LN2 pump speed. When the heat leak is stable, the temperature rise along the cryopipe and the cable terminal is inversely proportional to the LN2 flow rate. In figures (d) and (f), the temperature rises shows the undulating

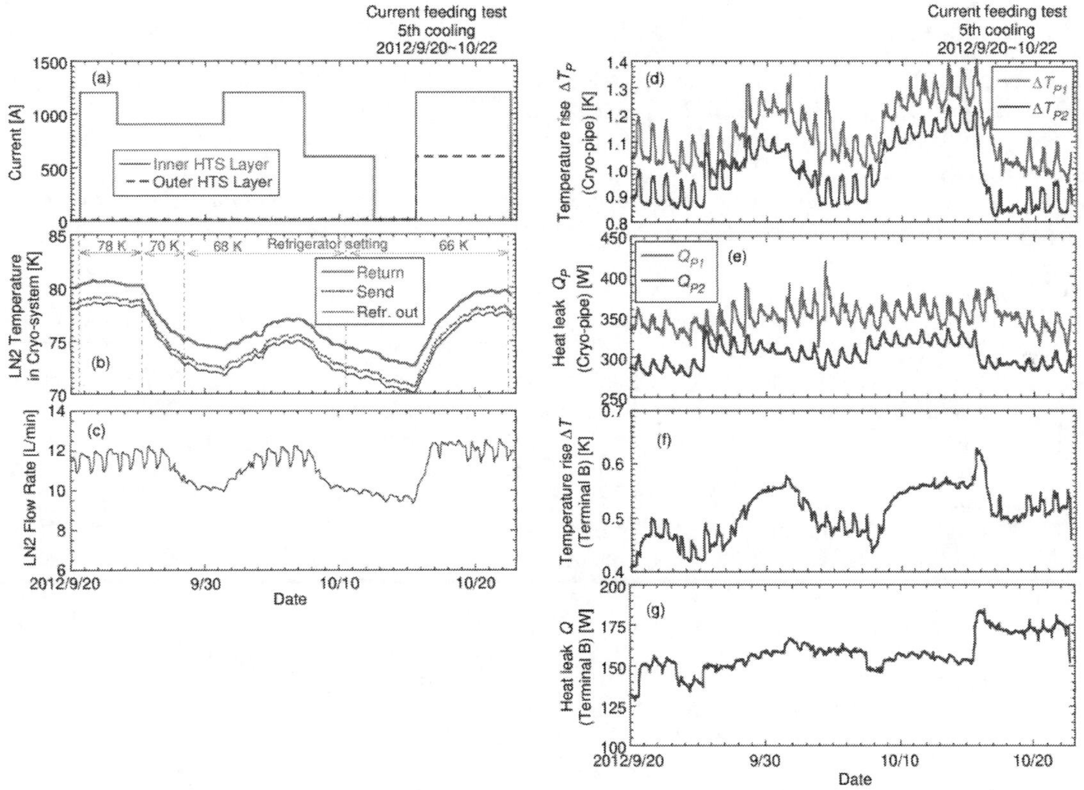

Figure 3. Time evolution of (a) operation current, (b) LN2 temperatures in cryogenic system (refrigerator out, send into and return out of the cable) for the various temperature settings of refrigerator, (c) LN2 flow rate, (d) temperature rise and (e) heat leak along the 200 m cryopipe as cable line, (f) temperature rise and (g) heat leak on the terminal B. In (d) and (e), suffixes 1 and 2 show the data derived from the temperatures of the top and bottom of the cryopipe, respectively.

time evolutions inversely following the variation of the LN2 flow rate.

Figures 3 (c) ~ (g) also shows the spike-like variations with the cycles of 1 day, in particular for (d) and (e). In CASER-2, 80% of the cryopipe was placed outside of the building, and then the temperature of the cryopipe surface was affected by temperature of atmosphere. This variation of the cryopipe temperature led to the daily variation of the heat leak into the cryopipe, in particular as the radiation heat, and also affected to the temperatures of terminals and the LN2 flow rate; consequently, they showed the spike-like daily variations.

The heat leak in the terminal B (figure 3 (g)) slightly increased when only the inside HTS layer operated with a current to 1200 A. This is because the Peltier heat of the Peltier current leads (PCLs) [9] in the cable terminal (Table 1) suppressed the Joule heating due to the operation current. Contrastingly, the heat leak in the terminal B clearly increased when a 600 A current was fed to the outside HTS layer. This was because the current leads in the outside HTS layer included 6 copper leads and Joule heating of copper leads led to the increase of the heat leak in the cable terminal.

4. Conclusion

We carried out the 5th cooling down in 200 m class SC-DC power transmission test facility (CASER-2) from August to November in 2012. During the 5th cooling down, a long term current feeding test was continued for 1 month with varying the operation current and the temperature setting of the refrigerator in the cryosystem. This long term operation clearly showed the relationship of the operation current, the operation temperature of the cryogenic system and the LN2 circulation. In addition, the LN2 circulation was affected by the atmosphere, in particular in the heat leak at the cryopipe; therefore, the LN2 circulation showed the daily variations. It was also confirmed that the PCLs worked effectively to suppress the heat leak at the cable terminal.

Acknowledgement

Parts of this work were supported by a grant of Strategic Research Foundation Grant-aided Project for Private Universities, by Ministry of Education, Culture, Sports, Science and Technology (MEXT).

References

[1] Liang XM, Dai ST, Gao ZY, Song NH, Wang YS, Zhang D, et al., 2010 Design of a 380 m DC HTS power cable. *IEEE Trans. Appl. Supercond.* **20** 1259–62.

[2] Hassenzahl W, Daneshpooy A, Eckroad S, Grant P, Gregory B, Nilsson S 2009 Program on Technology Innovation: A Superconducting DC Cable. *EPRI Report 2009*; 1020458.

[3] Tomita M, Suzuki K, Fukumoto Y, Ishihara A, Muralidhar M 2011 Next generation of prototype direct current superconducting cable for railway system. *J. Appl. Phys.* **109** 063909.

[4] Yamaguchi S, Hamabe M, Yamamoto I, Famakinwa T, Sasaki A, Iiyoshi A, et al., 2008 Research activities of DC superconducting power transmission line in Chubu University. *Jour. of Phys.: Conf. Ser.* **97** 012290.

[5] Hamabe M, Fujii T, Sugino M, Sasaki A, Sugimoto T, Watanabe H, et al., 2010 Cooling cycle test of DC superconducting power transmission cable. *Jour. of Phys.: Conf. Ser.* **234** 032019.

[6] Yamaguchi S, Kawahara T, Hamabe M, Watanabe H, Ivanov Yu, Sun J, et al., 2011 Design and construction of 200-meter high temperature superconducting DC power cable test facility in Chubu University. *Proc. the 23 Int. Cryogenic Eng. Conf. and Int. Cryogenic Mat. Conf. (Wroclaw, Poland, 18-23 July 2010)*, (Wrocław: Oficyna Wydawnicza Politechniki Wrocławskiej) 1041-47.

[7] Weisend II JG, Van Sciver SW 1990 Pressure drop from flow of cryogens in corrugated bellows. *Cryogenics* **30** 935-941

[8] Sasaki A, Hamabe M, Famakinwa T, Yamaguchi S, Radovinsky A 2008 A Numerical Analysis in LN2 Channel for DC-SC Power Transmission Line *Adv. in Cryogenic Eng.* **53A** 75-82

11th European Conference on Applied Superconductivity (EUCAS2013) IOP Publishing
Journal of Physics: Conference Series **507** (2014) 032021 doi:10.1088/1742-6596/507/3/032021

A Novel Electrical Insulating Material for 275 kV High-Voltage HTS Cable with Low Dielectric Loss

N Hayakawa[1], S Nishimachi[1], O Maruyama[2], T Ohkuma[2], J Liu[3] and M Yagi[3]

[1] Nagoya University, Chikusa-ku, Furo-cho, Nagoya 464-8603, Japan
[2] ISTEC-SRL, 3-2-1, Sakado, Takatsu-ku, Kawasaki-shi, Kanagawa 213-0012, Japan
[3] Furukawa Electric Co., LTD., 6, Yawatakaigan-dori, Ichihara, Chiba 290-8555, Japan

E-mail: nhayakaw@nuee.nagoya-u.ac.jp

Abstract. In the case of high temperature superconducting (HTS) power transmission cables at high voltage operation, the electrical insulation technique in consideration of the dielectric loss reduction becomes crucial. In this paper, we focused on a Tyvek/polyethylene (PE) sheet, instead of the conventional polypropylene laminated paper (PPLP). We obtained the dielectric characteristics (ε_r, tanδ) and partial discharge inception strength (PDIE) of PPLP, Tyvek and Tyvek/PE. We pointed out that the dielectric loss of 275 kV HTS cable with Tyvek/PE insulation will be reduced to 21 % of that with PPLP, and the total electrical loss including the AC loss will be reduced to 41 %.

1. Introduction

High temperature superconductivity (HTS) cables have been investigating in the world for realization of high density, large capacity, and low loss transmission of electric power [1]-[3]. Electrical insulation structure of the HTS cables generally has used liquid nitrogen (LN_2) / laminated tapes composite insulation system. Due to the occurrence local partial discharge (PD) in this system, dielectric strength will deteriorate and breakdown (BD) will occur. Therefore, electrical insulation at cryogenic temperatures is a key and common technique for the practical development of HTS cables [4]-[10]. Especially, in the case of HTS cables at high voltage operation, the electrical insulation technique in consideration of the dielectric loss reduction becomes crucial. For example, the dielectric loss of electrical insulation layer for 275 kV HTS cable (Figure 1) in Japanese M-PACC (Materials and Power Applications of Coated Conductors) project is estimated to be three times larger than the AC loss of HTS conductors [1]. Since the dielectric loss depends on the relative permittivity (ε_r) and dielectric dissipation factor (tanδ) of the electrical insulating material, a novel electrical insulating material with high electrical insulation strength and low dielectric loss is required for the high-voltage HTS cables.

From the above background, we have been investigating the feasibility of a novel insulating material for high-voltage HTS cables. In this paper, we focused on a Tyvek/polyethylene (PE) sheet, instead of the conventional polypropylene laminated paper (PPLP), and evaluated the PD inception and dielectric characteristics for the insulation design of 275 kV HTS cables with Tyvek/PE sheet.

Figure 1. Design of 275 kV-3 kA HTS cable (M-PACC)

2. Applicability of novel electrical insulating material (Tyvek/PE)

Tyvek/PE is a synthetic paper which is bonding PE-sheet on Tyvek by heat. Tyvek/PE has a low ε_r and tanδ and is also expected to improve the poor electrical insulation performance of Tyvek due to its porous structure. Figure 2 shows the experimental setup and configuration of cable sample for PD measurement. Six cable samples (inner diameter: 20.4 mm, insulation thickness: 1 mm, effective length: 100 mm) were immersed in LN_2 in the inner vessel of the cryostat. Sample numbers were 4 for PPLP (C-Type), 10 for Tyvek and 4 for Tyvek/PE, respectively. AC high voltage of 60 Hz was applied to the cable sample and generated PD. PD inception voltage (PDIV) was measured 10 times for different LN_2 pressures P (0.1-0.3 MPa) at 77 K. PD inception strength with 0.1 % probability (0.1 % PDIE) at high voltage (HV) electrode surface was calculated by the Weibull analysis. In addition, the dielectric characteristics (ε_r, tanδ) of each sample were measured using schering bridge (300 V/mm).

Figure 3 shows 0.1 % PDIE and relative standard deviation of each insulating material. In Figure 3 (a), 0.1 % PDIE of Tyvek/PE at P = 0.3 MPa is 25.5 kV_{rms}/mm, which is the same level as that of PPLP and about 15 % higher than that of Tyvek. In Figure 3 (b), relative standard deviation of Tyvek/PE is smaller than that of Tyvek.

Figure 4 shows relative permittivity (ε_r) and dielectric dissipation factor (tanδ) of each insulating material (relative ratio against the value of PPLP). Experimental results revealed that the dielectric loss (ε_r * tanδ) of Tyvek and Tyvek/PE was about 20 % of that of PPLP, which means that the dielectric loss of HTS cables insulated by Tyvek or Tyvek/PE will be reduced to 20 %, compared with that with PPLP insulation. These results suggest the technical feasibility of Tyvek/PE as the insulating material of high-voltage HTS cables.

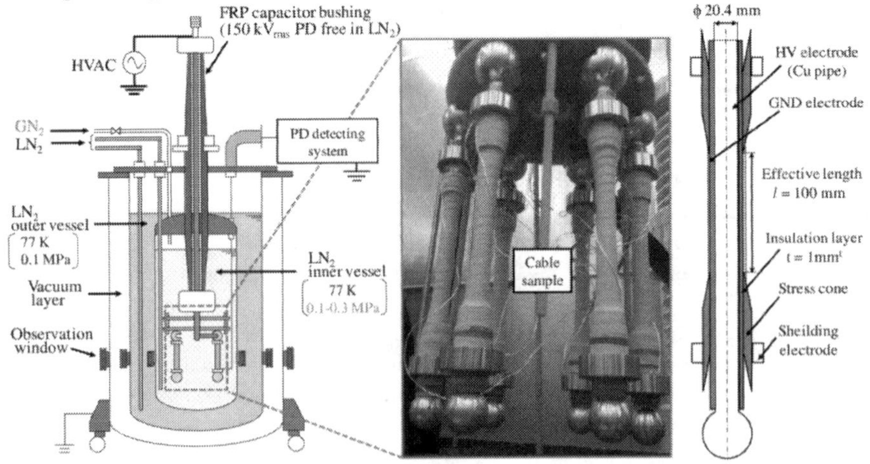

Figure 2. Experimental setup and configuration of cable sample

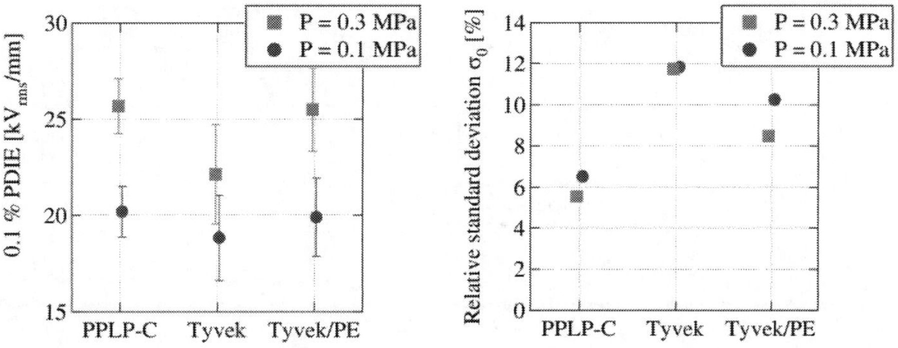

(a) 0.1 % PDIE (b) Relative standard deviation

Figure 3. Partial discharge inception characteristics of each insulating material

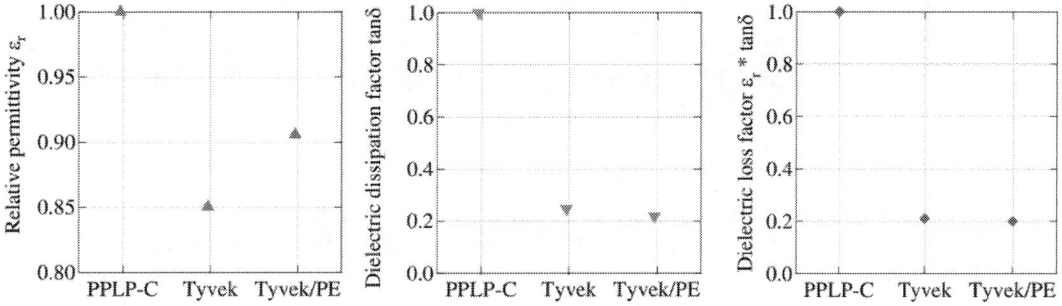

(a) Relative permittivity (b) Dielectric dissipation factor (c) Dielectric loss factor

Figure 4. Dielectric characteristics of each insulating material

3. Electrical insulation and power loss characteristics of 275 kV HTS cable

Using the obtained dielectric and insulation characteristics of Tyvek/PE shown in Table 1, we designed the 275 kV HTS cable in the M-PACC project. Including the evaluation of a composite insulation system with PPLP-C and Tyvek/PE (from the viewpoint of potential distribution, PPLP-C with the higher ε_r is arranged inside), we evaluated PDIV and electrical loss as a function of the radius c of PPLP-C layer (c = a: All Tyvek, c = b: All PPLP-C) as shown in Figure 5.

Figure 6 shows the results of PDIV and electrical loss of 275 kV HTS cable. In Figure 6, the dielectric loss of 275 kV HTS cable with Tyvek/PE insulation will be reduced to 21 % (from 0.60 W/m to 0.125 W/m) of that with PPLP, and the total electrical loss including the AC loss (0.20 W/m) will be reduced to 41 % (from 0.80 W/m to 0.325 W/m), where PDIV based on the 0.1 % PDIE exceeds the testing voltage (310 kV) for 275 kV HTS cables.

Table 1. Insulation material data for 275 kV HTS cable

Condition: LN_2 (T = 77 K)	PPLP-C	Tyvek/PE
0.1 % PDIE at P = 0.3 MPa [kV$_{rms}$/mm]	25.7	25.5
ε_r	1.90	1.73
tanδ	0.00058	0.00013

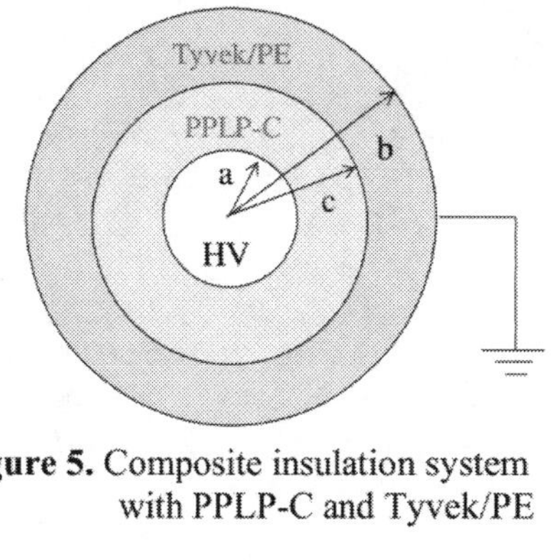

Figure 5. Composite insulation system with PPLP-C and Tyvek/PE

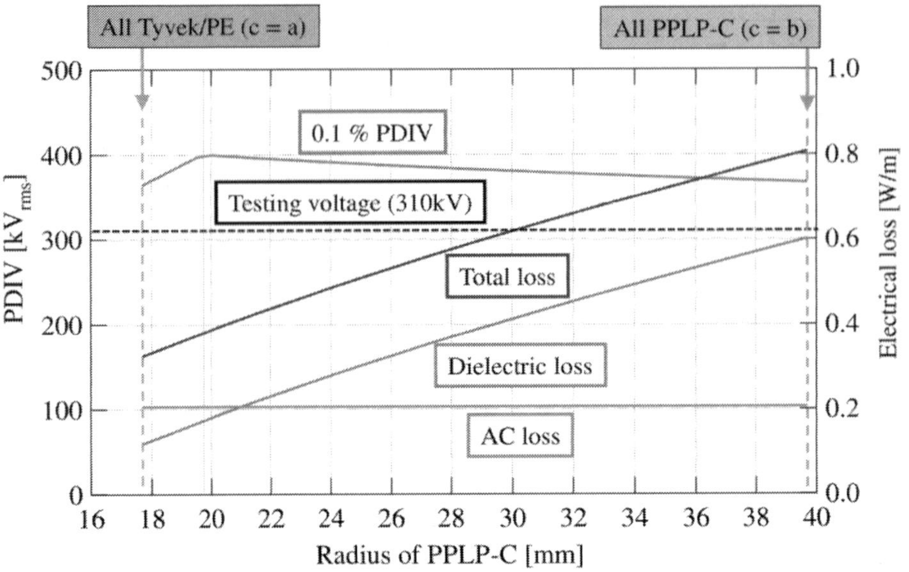

Figure 6. Electrical insulation and power loss characteristics of 275 kV HTS cable

Consequently, we could obtain the consistency between the high electrical insulation performance and the low dielectric loss performance of 275 kV high-voltage HTS cables by using Tyvek/PE as the electrical insulating material.

4. Conclusions

We investigated the electrical insulation and power loss characteristics of a novel electrical insulating material (Tyvek/PE) for high-voltage HTS cables. The main results in this paper can be summarized as follows:

1. 0.1 % PDIE of Tyvek/PE is the same level as that of PPLP.
2. Dielectric loss factor $\varepsilon_r * \tan\delta$ of Tyvek/PE is 20 % of that of PPLP.
3. Total electrical loss of 275 kV HTS cable in Japanese M-PACC project will be reduced to 41 % by using Tyvek/PE insulation, compared with the conventional design with PPLP insulation.

The above results suggested the technical feasibility of Tyvek/PE as the insulating material of high-voltage HTS cables with the coordination between high insulation performance and low dielectric loss.

Acknowledgement
This work has been supported by the New Energy and Industrial Technology Development Organization (NEDO), Japan.

References

[1] Mukoyama S. et al., 2013 *IEEE Trans. on Appl. Supercond.* **23** 5402804
[2] Maguire F. J. et al., 2011 *IEEE Trans. on Appl. Supercond.* **21** 961-966
[3] Choi W. J. et al., 2010 *IEEE Trans. on Appl. Supercond.* **20** 1280-1283
[4] Sauers I. et al., 2011 *IEEE Trans. on Appl. Supercond.* **21** 1892-1895
[5] Blaz M. et al., 2013 *IEEE Trans. on Appl. Supercond.* **23** 7700804
[6] Hayakawa N. et al., 2007 *IEEE Trans. on Appl. Supercond.* **17** 1660-1663
[7] Okubo H. et al., 2008 *IEEE Trans. on Dielectr. and Electr. Insul.* **15** 647-654
[8] Hayakawa N. et al., 2008 *J.Physics: Conference Series* **97** 12053
[9] Hayakawa N. et al., 2010 *J.Physics: Conference Series* **234** 032020
[10] Hayakawa N. et al., 2012 *Physics Procedia* **36** 1305-1308

5 MJ flywheel based on bulk HTS magnetic suspension

V Poltavets [1], K Kovalev [1], R Ilyasov [1], A Glazunov [2], V Maevsky [3], L Verzbitsky[4], V Akhmadyshev [5], A Shikov [6]

[1] Moscow Aviation Institute, Dept. 310, Volokolamskoe shosse, 4, Moscow, 125993 Russian Federation
[2] JSC "Centrotech - SPb", Stachek, 47, St. Petersburg, 198097, Russian Federation
[3] JSC "Gorizont", Omskaya str. 2 J, Rostov-on-Don, 344068, Russian Federation
[4] JSC "NIIEM", Panfilov str., 11, Istra, Moscow region, 143502, Russian Federation
[5] JSC "VPO" Tochmash", North, 1-a, Vladimir, 600007, Russian Federation
[6] NRC "Kurchatov Institute", Academician Kurchatov Square, 1, Moscow, 123182, Russian Federation
E-mail: ilyasov@mai.ru

Abstract. Nowadays the flywheel energy storage systems (FES) are developed intensively as uninterruptible power supply (UPS) devices for on-land and transport (especially airborne) applications worldwide. This work is devoted to the FES with magnetic suspension on the base of bulk HTS YBCO elements and permanent magnets. The developed FES is intended to be used as UPS in Russian atomic industry in case of an emergency. For the successful design of the FES the following questions should be solved: design of the motor/generator, design of the rotor (flywheel), design of the bearing system, design of the control system and system of power load matching, design of the cooling system. The developed small-scale FES with the stored energy 0.5 MJ was used to solve these basic questions. The elaborated FES consists of the synchronous electric machine with permanent magnets, the solid flywheel with axial magnetic suspension on the base of YBCO bulks and permanent magnets, the system of control and power load matching, and the system of liquid nitrogen cooling. The results of theoretical modeling of different schematics of magnetic suspension and experimental investigations of the constructed FES are presented. The design of the future full-scale FES with the stored energy ~5 MJ and output power up to 100 kW is described. The test results of the flywheel rotor and HTS magnetic suspension of 5 MJ FES are presented. This work is done under support of Rosatom within the frames of Russian Project "Superconducting Industry".

1. Past achievements. Moscow Aviation Institute (MAI) has some experience in analysis, design and fabrication of magnetic suspension on the base of bulk HTS elements. In previous years a small model of HTS suspension for high speed transport was built, as well as a model with a transport capacity of 600 kg.

The small-scale model was demonstrated to the President Dmitry Medvedev at the exhibition in 2009 (fig. 1).

2. Description of the design. 0.5 MJ kinetic energy storage was developed and successfully tested in 2010, in MAI. It uses magnetic suspension based on HTS bulk elements (fig. 2). It has 100 kg steel rotor (fig. 3) and iron-free lossless synchronous motor-generator with permanent magnets. Two rotors with fiberglass bandage for test model of flywheel were manufactured, designed and produced for a speed of 26000 rpm. Rotor was simulated by JSC "Centrotech – SPb" (St. Petersburg) and produced by JSC "VPO" Tochmash" (Vladimir).

Figure 1. The President of Russian Federation Dmitry Medvedev in front the small-scale model of bulk HTS suspension for Hi-speed transport

Figure 2. Stator of 0.5 MJ flywheel with bulk HTS elements

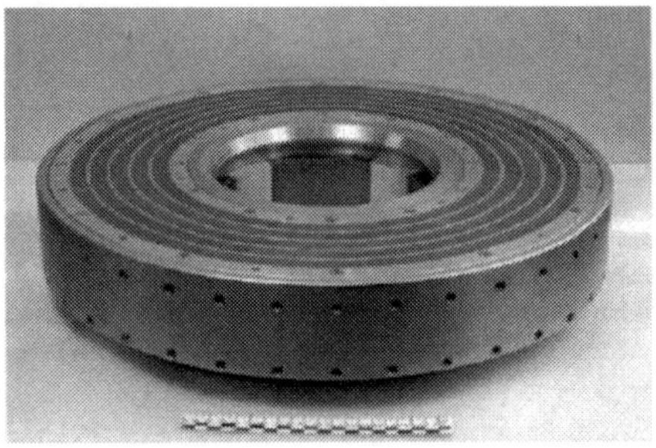

Figure 3. Rotor of small-scale model 0.5 MJ flywheel with PM

In order to understand of the strength properties of fiberglass bandage were made two flywheels (fig. 4, 5) with different fiberglass shell thickness (15 and 25 mm). Sketch of the flywheel and the results of the numerical simulation using finite elements method are shown in Figures 6, 7. Tests have shown that this solution can be successfully applied for the future 5 MJ energy storage.

To reduce the magnetic losses, motor-generator was decided to use the iron-free stator (fig. 8).

Magnetic suspension is generated by of HTS bulk elements on the stator and permanent magnets on the rotor. For axial unloading applied ring-shaped permanent magnets (PM) thrust bearing.

Dependence of the stored energy from the rotor speed of the 5MJ flywheel is shown in fig. 9.

The parameters of 5 MJ flywheel are shown in Table 1.

Figure 4. Two rotors with fibreglass bandage for test model of flywheel. Calculated speed 26000 rpm

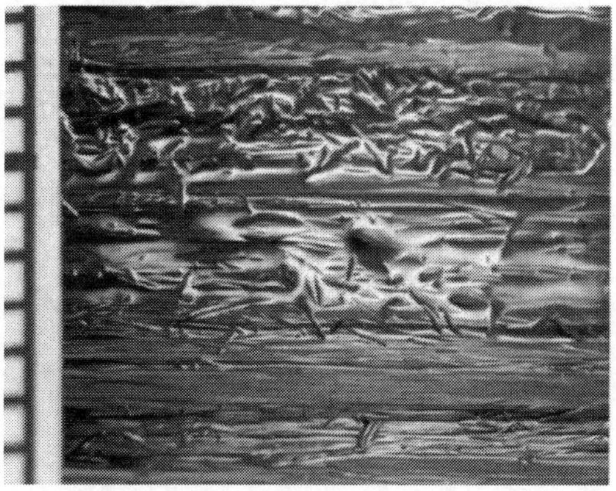

Figure 5. Close-up of fibreglass bandage

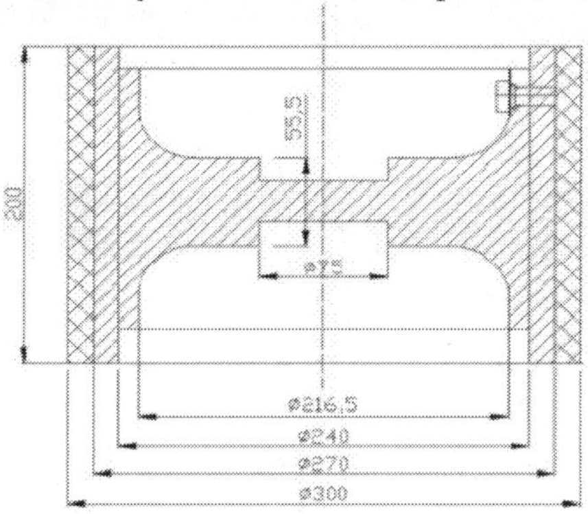

Figure 6. Sketch of flywheel

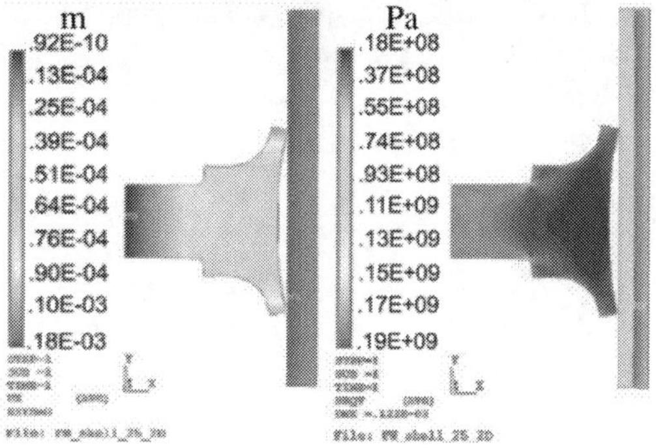

Figure 7. Results of FEM analysis of the strength properties of flywheel 5MJ energy storage

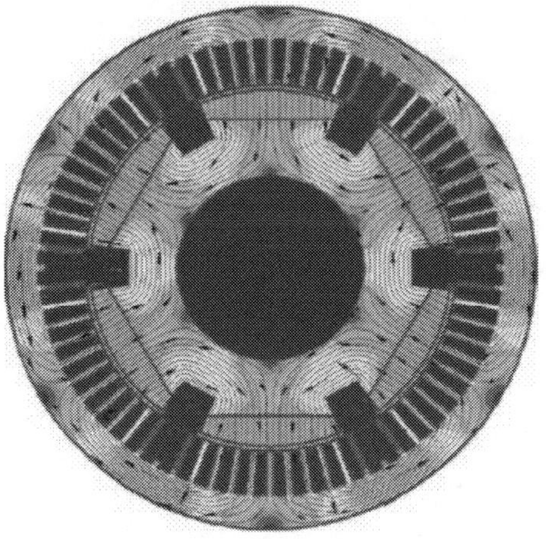

Figure 8. Result of FEM analysis of iron-free teeth of stator of motor-generator for flywheel energy storage

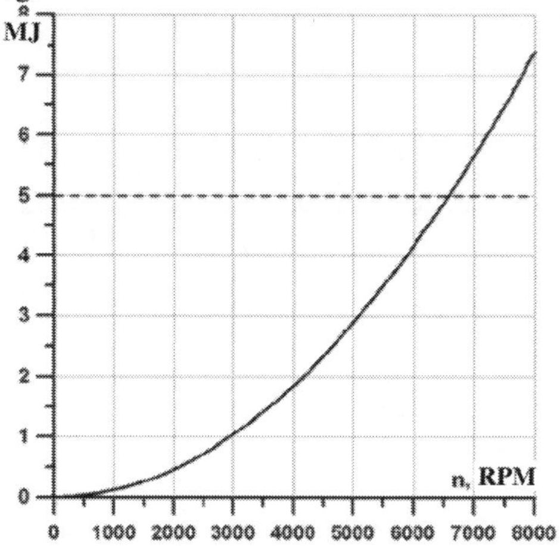

Figure 9. Dependence of the stored energy from the rotor speed of the 5MJ flywheel

Table 1. PARAMETERS OF 5 MJ FLYWHEEL

List of Parameters	units	value
Stored energy	MJ	5
Maximum rotation speed	rpm	8000
Output power of the generator	kW	150
Nominal charging time	sec	300
Minimum discharge time	sec	25
Mass of flywheel	kg	350
Full mass for magnetic suspension	kg	500

Currently the project is in the stage of production of technical documentation. Production of 5 MJ flywheel energy storage will be done in 2014-2015. The first tests are expected in 2015.

Acknowledgment
This work is done under support of Rosatom within the frames of Russian Project "Superconducting Industry".

Synchronous Generator with HTS-2G field coils for Wind-mills with output power 1 MW

K Kovalev[1], L Kovalev[1], V Poltavets[1], S Samsonovich[1], R Ilyasov[1], A Levin[2], M Surin[3].

[1] Moscow Aviation Institute, Dept. 310, Volokolamskoe shosse, 4, Moscow, 125993 Russian Federation
[2] OAO "Aviatsionnoye oborudovaniye, Gogol Boulevard, 21, Moscow, 119991, Russian Federation
[3] NRC "Kurchatov Institute", Academician Kurchatov Square, 1, Moscow, 123182, Russian Federation

E-mail: ilyasov@mai.ru

Abstract. Nowadays synchronous generators for wind-mills are developed worldwide. The cost of the generator is determined by its size and weight. In this deal the implementation of HTS-2G generators is very perspective. The application of HTS 2G field coils in the rotor allows to reduce the size of the generator is 1.75 times. In this work the design 1 MW HTS-2G generator is considered. The designed 1 MW HTS-2G generator has the following parameters: rotor diameter 800 mm, active length 400 mm, phase voltage 690V, rotor speed 600 min^{-1}, rotor field coils with HTS-2G tapes. HTS-2G field coils located in the rotating cryostat and cooled by liquid nitrogen. The simulation and optimization of HTS-2G field coils geometry allowed to increase feed DC current up to 50A. Copper stator windings are water cooled. Magnetic and electrical losses in 1 MW HTS-2G generator do not exceed 1.6% of the nominal output power. In the construction of HTS-2G generator the wave multiplier with ratio 1:40 is used. The latter allows to reduce the total mass of HTS-2G generator down to 1.5 tons. The small-scale model of HTS-2G generator with output power 50 kW was designed, manufactured and tested. The test results showed good agreement with calculation results. The manufacturing of 1 MW HTS-2G generator is planned in 2014. This work is done under support of Rosatom within the frames of Russian Project "Superconducting Industry".

1. The conceptual approach to the selection of the main dimensions

Usual speed of propeller wind-mill with output power of about 1 MW is 7-15 RPM [1]. Electric generator with direct drive and same rotor speed will be the volume of active area about 14 m^3. Generator with HTS field coils is 1.75 times smaller. Increasing of rotor speed to 30 RPM further reduces the size twice. Use of a multiplier and HTS field coils allows reduce the volume to 52 times (figure 1). Figure 2 show that a further increase in speed does not have such a significant impact on reducing the amount of volume generator.

Load characteristics of generator are shown in Figure 3. The horizontal axis is AC phase current, the vertical axis is the AC phase voltage, and series of curves are DC in HTS field coils. Increasing the concentration of curves occurs due to the saturation of the stator teeth. To prevent saturation of tooth, its width has been increased. No-load characteristics are shown in Figure 4. The horizontal axis is DC current in HTS field coils.

This expected reduction DC due to the external magnetic field down to 40 A (at temperature 77 K) to maintain the same magnetomotive force, it is necessary to increase the number of turns per pole up to 400.

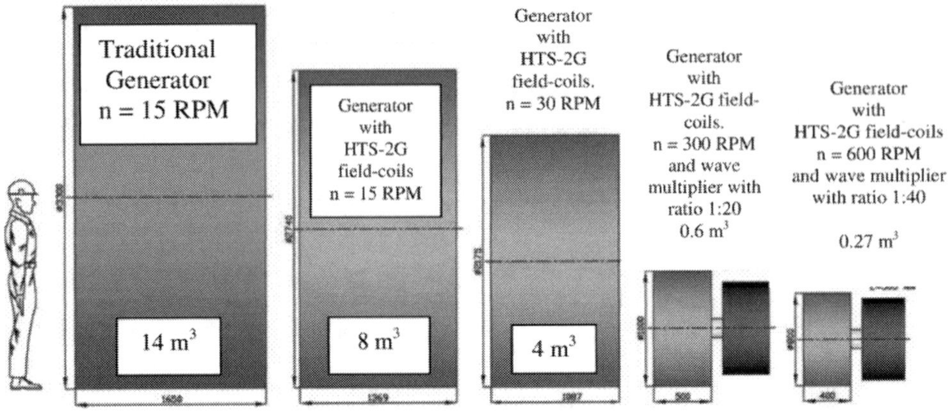

Figure 1. Comparison of the sizes of traditional direct-drive generators and superconducting generators.

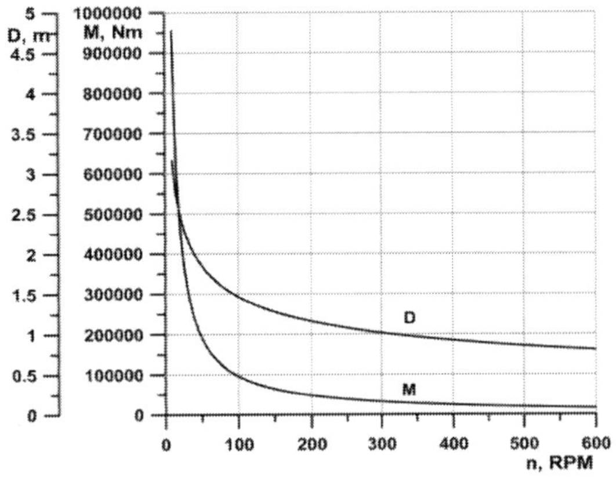

Figure 2. Stator bore diameter (D) and torque dependence (M) on rotor speed

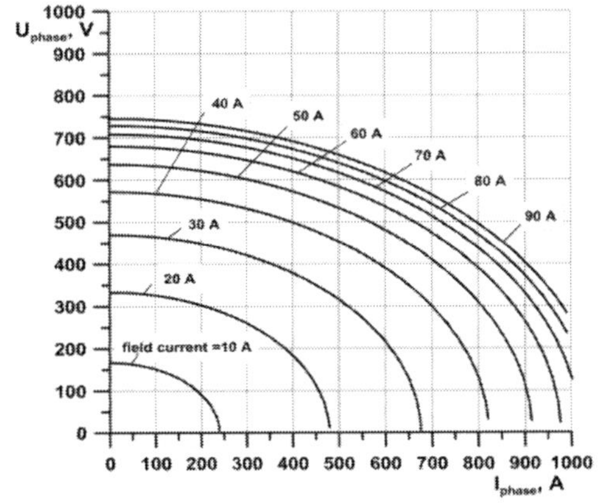

Figure 3. Load characteristic phase voltage on phase current of generator

The calculated characteristics of output power, power factor and phase current for motor mode has been shown figure 5. The horizontal axis is load angle between EMF and voltage.

Series of finite element analysis of magnetic flux distribution were conducted using the software Elcut® Professional [2]. A result has been shown on figure 6. Modeling has allowed revealing rational design of the pole and stator yoke. Fourier analysis of magnetic flux in the gap has been done (fig. 7). In order to increase the EMF it is possible to increase overlap factor the pole up to 0.8 and more. In the case of frequency converter at the output, non-sinusoidal output voltage of the generator may be ignored.

Figure 8 shows the distribution of the magnetic field excitation. As can be seen, the magnetic induction in the corners of the upper HTS-2G field coils near pole zone reaches a value up to 0.5 T. The latter is the reason of the reduction of the critical current due to the magnetic field.

Journal of Physics: Conference Series **507** (2014) 032023 doi:10.1088/1742-6596/507/3/032023

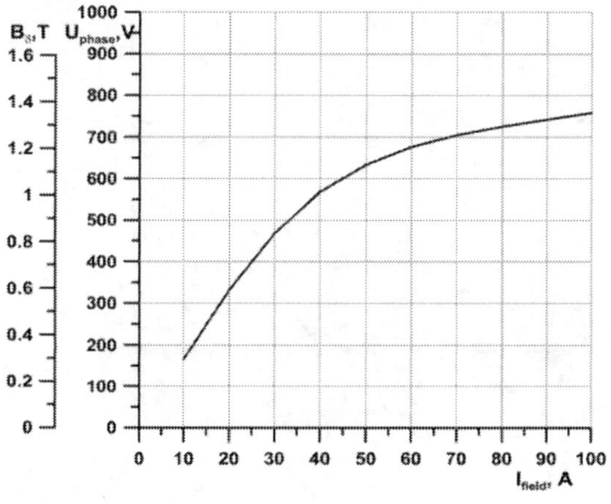

Figure 4. No-load characteristic

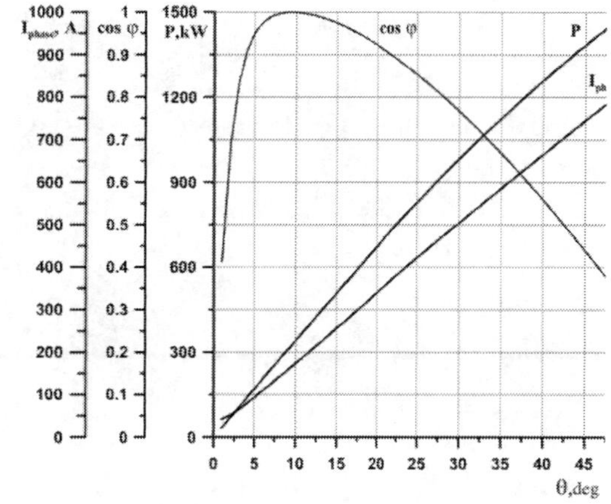

Figure 5. Characteristics of output power

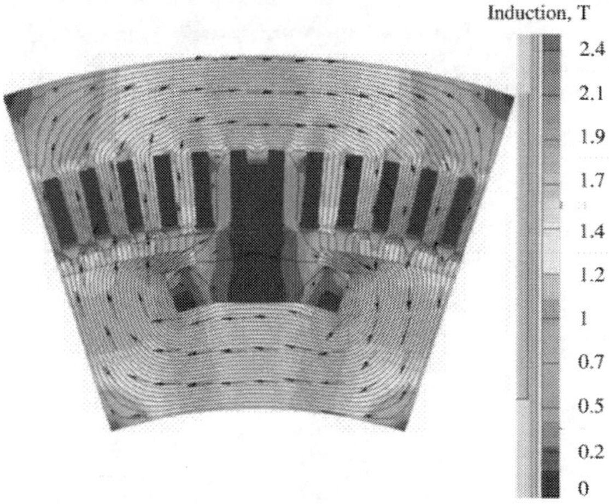

Figure 6. Segment of 1/10 model map of magnetic field

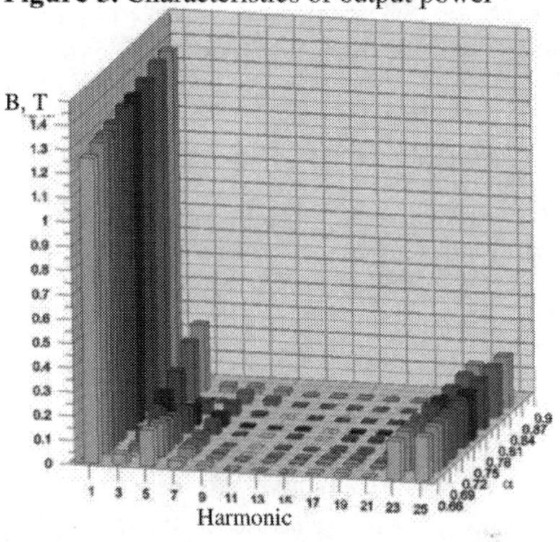

Figure 7. Results of Fourier analysis of magnetic flux in gap

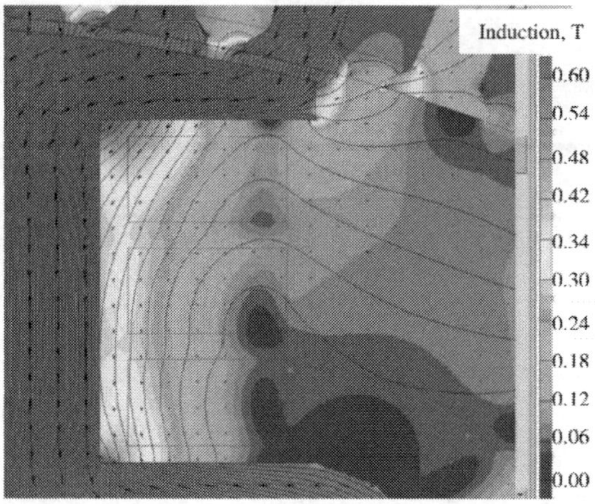

Figure 8. Map of magnetic field (field coils fragment).

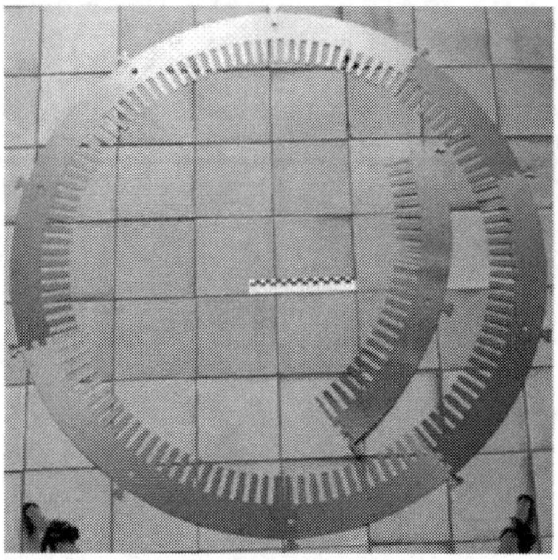

Figure 9. Photo of stator segments.

Figure 10. Pole with HTS-2G tape for 50 kW small-scale model of generator

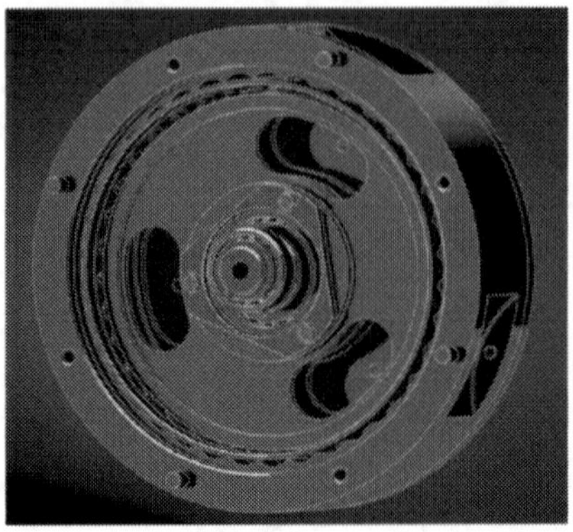

Figure 11. Solid model of wave multiplier.

The external diameter of the yoke of power generators is typically longer than the standard size of electrical steel sheet. In this deal the stator lamination could be fabricated from segment splicers (fig. 9).

For experimental studies the possibility of using HTS-2G winding, was created and successfully tested first in Russia small-scale model of the generator 50 kW with HTS 2G tapes in field coils. Its pole is shown in fig. 10.

The model of wave multiplier is shown in Figure 11. Calculated parameters of the designed 1MW synchronous generator are shown in Table 1.

Table 1. PARAMETERS OF 1 MW SYNCHRONOUS HTS-2G GENERATOR

List of Parameters	units	value
Nominal Output Power	kW	1000
Phase voltage	V	690
Nominal Speed	rpm	600
Pole number	-	10
Nominal current	A	500
Inner diameter of stator	mm	800
Active light	mm	400
Material of stator winding	-	Cu
Operating temperature of stator winding	K	300
Material of field coils	-	HTS-2G tape
Operating temperature of field coils	K	77
Nominal current in field coils (in 0.5 T field)	A	40
Number of turn per one pole	-	400
Operating Power factor	-	0.95 – 0.99
Full load efficiency (without the cryocooler)	%	97 – 98

Manufacture of multiplier with ratio 1:40 and 1 MW generator is planned for 2014 year, the tests for 2015 year.

Acknowledgment

This work is done under support of Rosatom within the frames of Russian Project "Superconducting Industry".

References

[1] L.K. Kovalev, K.L. Kovalev, E.E. Tulinova and N.S. Ivanov, Vultipole Permanent-Magnet Synchronous Generators for Wind Power Installation//The Electricity, #8, 2013

[2] http://elcut.ru/

Partial Discharge Characteristics in Composite Insulation Systems with PPLP for HTS Cable

Y Kikuchi [1], K Yamashita [1], A Kumada [1], K Hidaka [1], K Tatamidani [2] and T Masuda [2]

[1] Department of Electrical Engineering and Information Systems, The University of Tokyo, Japan

[2] Sumitomo Electric Industries, Ltd. Osaka Japan

Email:kikuchi@hvg.t.u-tokyo.ac.jp

Abstract. The electrical insulation system of high-temperature superconducting (HTS) cable consists of liquid nitrogen (N_2(l)) and polypropylene laminated paper (PPLP). Partial discharge (PD) may occur in butt gaps of the insulation layers and its characteristics imply the insulation performance of HTS cables. N_2(l) cooling system is installed in the power system and N_2(l) will flow through the cables during the system operation. Filling the HTS cable with N_2(l) in order to perform pre-shipment inspection is time-consuming and costly for cable manufacturers. Therefore, they are trying to find a cost effective method for pre-shipment inspections. One alternative is to use high pressure gaseous nitrogen (N_2(g)) instead of N_2(l). This article investigates PD characteristics such as PD inception electric field (PDIE) and PD extinction electric field (PDEE) in butt gaps of HTS cables in 0.1 to 0.3 MPa N_2(l) and 0.1 MPa to 1.0 MPa N_2(g) environments. For assessing the surface/volume effects, PD characteristics are measured with changing the size of butt gaps. It turns out that PDIE and PDEE in N_2(g) are linearly correlated with those in N_2(l) at any gas pressure in our testing, and PDIE in 1.0 MPa N_2(g) is almost 30% of that in 0.2 MPa N_2(l). It suggests that PD characteristics in N_2(l) can be extrapolated from those in N_2(g).

1. Introduction

In order to cope with the power demand in urban areas, high temperature superconducting (HTS) cable has attractive performance. In Japanese urban areas the electricity demand is high, therefore HTS cable is an outstanding way to increase the transmission capacity of the electricity, as well as providing a substantial reduction in transmission losses, resulting in a lower CO_2 emission. Moreover the reactance of HTS cable is smaller due to its outer HTS conductor for magnetic shield. This small reactance contributes to the improvement of voltage and synchronous stability of power system [1].

In insulation design of HTS cable, it is inevitable to insert small gaps which are called butt gaps through PPLP® layer. If these gaps had not been inserted properly through PPLP®, the cable insulation layer would have torn by bending the cable. Butt gap should be considered as the weakest points in electrical insulation because PD may be generated in butt gap and it leads to degradation of electrical insulation performance. N_2(l)/PPLP® composite insulation is adopted to the HTS cable because of their high dielectric strength and low dielectric loss. N_2(l) coolant will be circulated in the cable after it is installed in power system. A pre-shipment insulation test is usually performed in the manufacturer before the shipment of cables. The pre-shipment inspection of HTS cable with whole length cooled in N2(l) is not economically efficient and it would be preferable to find a more cost effective method for the pre-shipment inspection.

From the above background, the authors have been investigating the PD characteristics of N_2(l)/PPLP® composite insulation system for HTS cable under ac-voltage application [2]. The authors have focused on the alternative inspection method conducted in N_2(g) at room temperature instead of in N_2(l) for finding easy inspection method. In this paper, therefore, the PD characteristics not only in N_2(l)/PPLP® insulation system but also in N_2(g)/PPLP® insulation system are investigated. Through the comparison of the PD characteristics under these environments, the appropriate pre-shipment test method will be proposed.

2. Materials and methods

Figures 1 and 2 show two types of samples for PD measurement used in this research. The sheet sample shown in Figure 1 consists of PPLP® layers with a hole simulating a butt gap. Samples which have 3, 5, 7, and 9 PPLP layers were used to assess surface/volume effects. To suppress surface discharge from the edge of the high voltage electrode, it was rounded and molded with silicon rubber. The detection electrode was divided from the grounded plate electrode with 1.5 mm thick insulator to measure the PD current. The cable sample shown in Figure 2 is a more practical model, which consists of a high voltage electrode as an inner cylinder, inner semi-conductor layers, PPLP® layers, outer semiconductor layers, and a grounded electrode as an outermost sheath. Samples which have 3, 9, 15 PPLP layers were tested, and the width of the detection electrode was 20 or 140 mm. The detection electrode of each cable sample was also separated to measure PD current. The surface area and the volume of butt-gaps in samples are listed in tables 1 and 2.

Figure 3 shows the experimental setup for measuring PD characteristics from 0.1 MPa to 1.0 MPa pressure N_2(g) at room temperature. AC 50Hz high voltage was applied to the high voltage electrode of the sample. PD current was measured through measuring resistance (50Ω). In case of PD measurement in N_2(l), thermal insulated tank was used, and PD current of a sample was measured by a current transformer (SE Technology Limited SCT-8). The measuring circuit is shown in Figure 4.

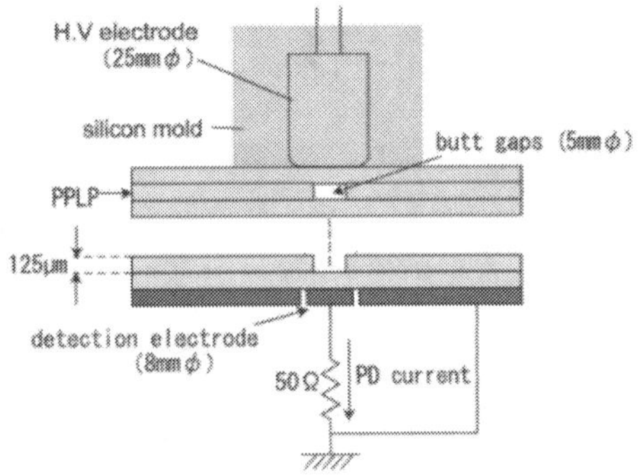

Figure 1. Configuration of sheet sample.

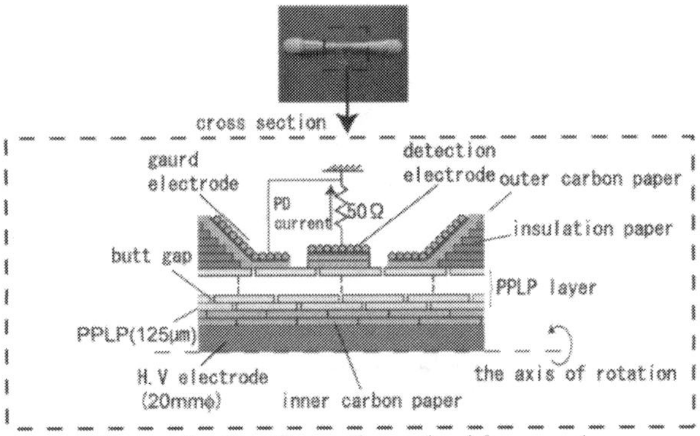

Figure 2. Configuration of cable sample.

Table 1. Surface area and volume of sheet model.

	Number of PPLP layers	Surface area (mm²)	Volume (mm³)
A	3	39.3	2.5
B	5	78.5	4.9
C	7	117.8	7.4
D	9	157.1	9.8

Table 2. Surface area and volume of cable model.

	Number of PPLP layers	Width of detection electrode (mm)	Surface area (mm²)	Volume (mm³)
E	3	20	400.7	25
F	15	20	2286.5	142.9
G	9	140	9009.3	563.1
H	15	140	16005.1	1000.3

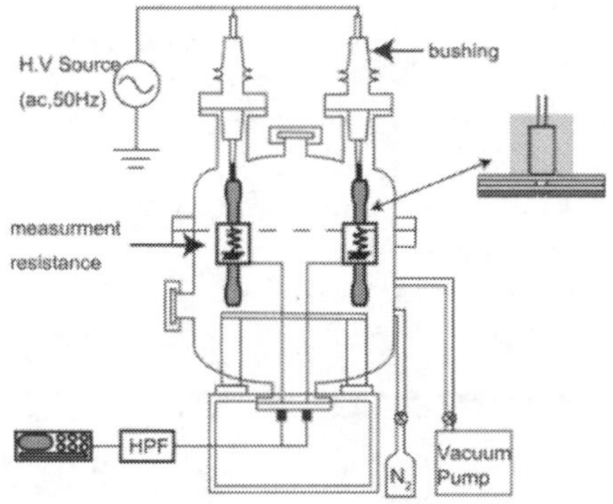

Figure 3. Experimental setup for N_2(g).

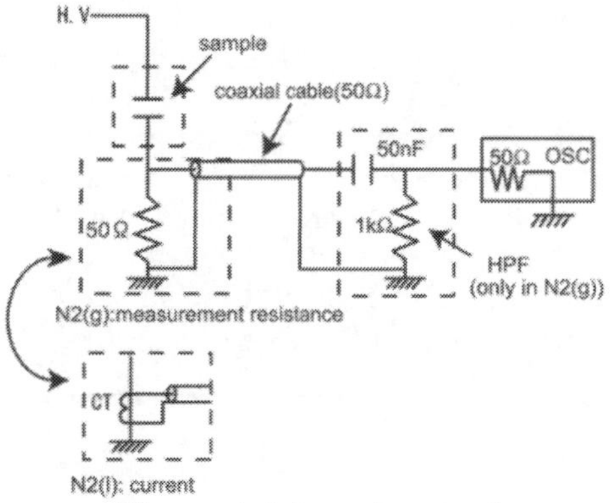

Figure 4. Measuring circuit.

3. Results and discussion

PD inception electric field (PDIE) and PD extinction electric field (PDEE) in N_2(l) and N_2(g) were measured. PDIE was obtained from *Q-E* characteristics *i.e.* the relationship between the PD charge amount (*Q*) and the applied electric field, with increasing applied electric field (*E*). In the same manner, PDEE was obtained with decreasing applied electric field. In this paper, PDIE and PDEE were defined as the electric field where *Q* takes threshold charge of 25 pC in *Q-E* curve. The charge amount of PD in N_2(g) was obtained by integrating PD current waveform monitored by an oscilloscope. On the other hand, that in N_2(l) was derived by the peak value of the current waveform, which was calibrated by measuring a mock PD-current waveform generated by a pulse generator (Nihon Keisokuki Seizosho co,ltd NPG-2).

Typical PD current waveforms in 1.0 MPa N_2(g) are shown in Figure 5. The pulse width of the PD current waveform becomes shorter with the increase of N_2(g) pressure, and the pulse width of PD current waveform in 1.0 MPa N_2(g) is 5 – 10 ns. Figure 6 shows PD characteristics in 1.0 MPa N_2(g). The electric fields in butt gap, E_{gap}, and the averaged applied electric field to the sample, E_{ave}, are expressed in this figure. E_{gap} [kVp/mm] were calculated from the peak applied voltage taking the relative permittivity of PPLP and N_2 into account, and E_{ave} [kVrm/mm] were obtained by simply dividing applied voltage by the sample thickness. As shown in Figure 6, *Q-E* curves are linear in semi logarithmic scale graphs, *i.e.*, the PD charge amount changes exponentially with the applied electric field. This tendency is observed both in N_2(l) and N_2(g) for every tested pressure with any samples.

Figure 7 shows the pressure dependence of PDIE and PDEE. Any significant difference between PDIE and PDEE aren't observed neither in N_2(l) nor N_2(g). PDIE and PDEE in N_2(g) increase lineally with the gas pressure, on the other hand, PDIE and PDEE in N_2(l) are saturated with the pressure at 0.3 MPa.

Figure 8 shows the PDIE with various volume and surface area of butt gap. In 1.0 MPa N_2(g), PDIE decreases linearly with the volume V of butt gaps for $V < 1000$ mm^3 and almost constant for $V > 1000$ mm^3. In Figure 8, PDIE in 0.2 MPa N_2(l) is also plotted. According to the paper by Hayakawa et al. [3] , PDIE in N_2(l) decreases for $V < 100$ mm^3 and keeps constant for $V > 100$ mm^3. The broken line in Figure 8 is drawn based on this report.

The converged values of PDIE in N_2(g) with sufficiently large-volume butt gaps are compared to those in N_2(l). Figure 9 shows the ratio of PDIE in N_2(g) to that in 0.2 MPa N_2(l), hereafter which is denoted as k, as a function of pressure of N_2(g). The suffix "*ave*" and "*gap*" denote the values calculated with PDIEs expressed in the averaged field intensity across the sample and the field intensity in the butt gap, respectively. As shown in this figure, k increases with the gas pressure and k_{ave} reaches 0.3 in 1.0 MPa N_2(g). This means that, subject to E_{ave} is comparable to the PDIE of N_2(l), a PD shall occur at the gap in N_2(g) at room temperature when 0.3 E_{ave} is applied.

4. Conclusions

In this paper, PDIE and PDEE of N_2(l)/PPLP and N_2(g)/PPLP composite insulation systems are investigated under 0.1 MPa to 0.3 MPa N_2(l) and 0.1 MPa to 1.0 MPa N_2(g) environment. The main conclusions are summarized as follows.

(1) There is little difference between PDIE and PDEE in N_2(l) and N_2(g).

(2) In N_2(g), PDIE decreases linearly with the volume of butt gaps while the volume V is smaller than 1000 m^3, and keeps almost constant when V is larger than 1000 mm^3.

(3) The ratio of PDIE in N_2(g) to that in 0.2 MPa N_2(l) increases with the gas pressure and the ratio of E_{ave} reaches 0.3 in 1.0 MPa N_2(g).

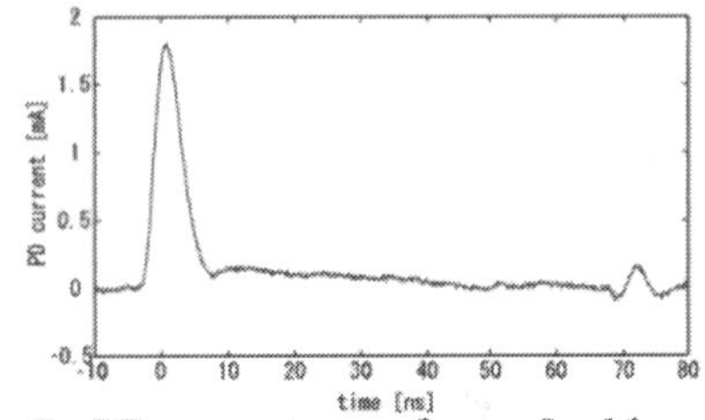

Figure 5. PD current waveform of cable model E in 1.0 MPa N_2(g).

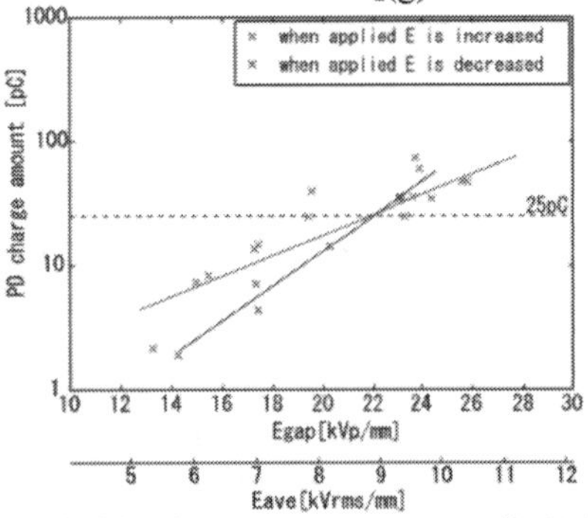

Figure 6. PD charge amount vs. applied electric field of cable model E in 1.0 MPa N_2(g).

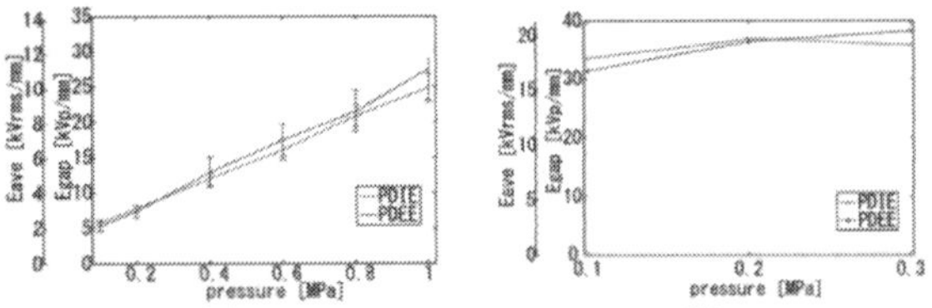

(a) (b)

Figure 7. Plots of PDIE and PDEE of cable model E as a function of pressure (a) N_2(g), (b) N_2(l).

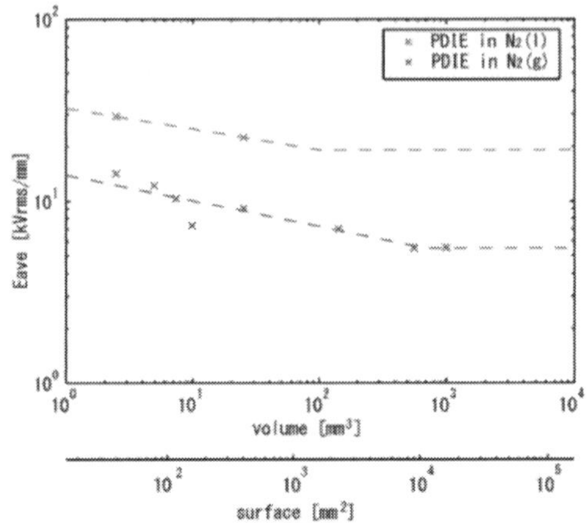

Figure 8. Volume/surface effect of PDIE in N_2(l) and N_2(g).

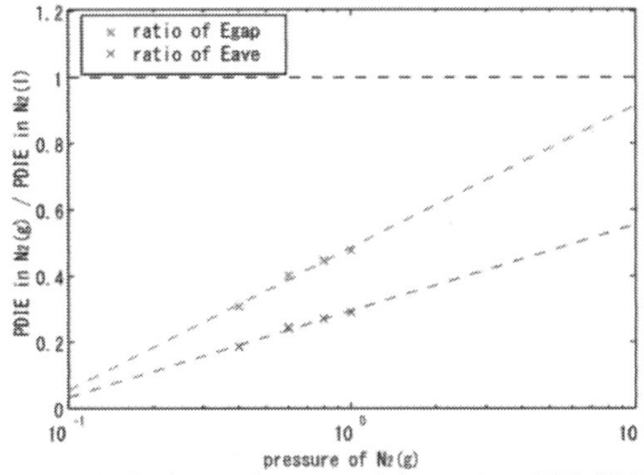

Figure 9. Estimated curve of the ratio of PDIE in N_2(g) to PDIE in 0.2 MPa N_2(l) as a function of gas pressure.

References

[1] Ueda K, Tsukamoto O, Nagaya S, kimura H and Akita S 2003 *IEEE Trans. Applied Superconductivity.* **13** NO.2 1946-51

[2] Rezaeifar F, Suzuki Y, Kumada A, Hidaka K, Masuda T and Nishimura T 2010 *IEEE Trans. Dielectrics and Electrical Insulation.* **17**, No. 6 1747-53

[3] Hayakawa N, Yamaguchi R, Kojima H, Endo F and Okubo 2008 *IEEE Inter, Confe. On. Dielecric Liquids. ICDL 2008.* pp.1-4

Insulation Characteristics of Bushing Shed at Cryogenic Temperature

W J Kim[1], Y J Kim[2] and S H Kim[1]

[1] Department of Electrical Engineering, Gyeongsang National University and ERI, Jinju, Korea

[2] Department of Electric Apparatus Testing, Korea Electrotechnology Research Institute, Ansan, Korea

E-mail: shkim@gnu.ac.kr

Abstract. In the development of high-T_c superconducting(HTS) devices, the bushing for HTS devices (HTS bushing) is the core technology, the need to because of supply high voltage to the cable or the winding of the transformer. The lower part of the bushing is exposed to the liquid nitrogen (LN_2), and it has many sheds. In particular, the insulation body with sheds and electrical insulation at cryogenic temperature have attracted a great deal of interest from the view point of the size, weight and efficiency of bushing. This study has mainly investigated the shed and insulation body by comparing glass fiber reinforced plastics (GFRP) in LN_2. We investigated the surface discharge characteristics according to insulating materials, width and height of the shed.

1. Introduction

The HTS power equipment [1-3] is environment-friendly equipment which exceeds the performance limits of conventional equipment in terms of size, weight, and efficiency. The HTS bushing is core technology which must be developed to apply power to the inside of the equipment in order to operate superconducting power machine. However, HTS bushing is now in a development stage around the globe. It is not yet commercially available.

In general, a conductor is wrapped by insulators in bushing [4]. The insulation factors of HTS bushing consist of an upper insulating unit for air surface discharge, thickness for puncture insulation, and a lower insulating unit for surface discharge in LN_2 at the bottom of the flange. In particular, in the lower insulating unit, material technology and insulation technology at cryogenic temperature are essential because of the existence of LN_2 and cryogenic nitrogen gas. Also, there is a high possibility of bubble generation due to heat penetration or quenching. As a result, a decrease in insulation performance is anticipated [5]. Therefore, this study has investigated the electrical and mechanical characteristics of shed materials in LN_2. In addition, the shape, arrangement of shed have been studied.

2. Experimental Apparatus

The insulation materials used for bushing include GFRP, silicon rubber, ethylene propylene rubber (EPDM)/Si, aromatic polyamide (Nomex), and polytetra-fluoro ethylene (Teflon).

Figure 1 (a) and (b) show the electrode system for surface discharge of planar and cylindrical GFRP samples, respectively. As shown in Figure 1(a), the electrode system has used aluminum needle-plane electrode. Surface length(l) is 20, 30 and 40mm, and body diameter(D) is 20, 50, 95 and

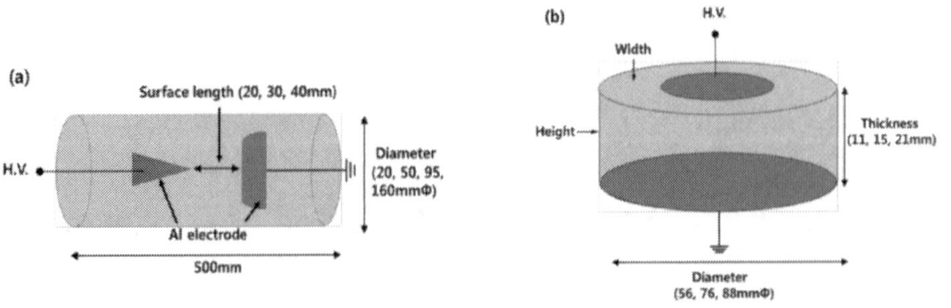

Figure 1. Electrode system for surface discharge; (a) cydrical GFRP sample with electrodes mounted in the side of the cylinder. (b) planar electrodes mounted on top and bottom of circular GFRP sample.

160mm. As shown in figure 2(b), the aluminum circular electrodes were mounted on the upper (diameter: 46mm) and lower (diameter: 76mm) parts each. In addition, the dependence of width was used with cylindrical GFRP shed 15mm in thickness and 56, 76 and 88mm in diameter. The dependence of height was used with cylindrical GFRP shed 76mm in diameter and 11, 15 and 21mm in thickness. The GFRP samples were designated as two different directions which are the longitudinal (parallel to glass-fiber direction) and lateral direction (cross to glass-fiber direction).

The electrode-installed sample was washed with ethyl alcohol and dried in a drying machine to remove humidity. The electrode-installed sample was fixed to the electrode holder. After arranging the electrode holder in a cryogenic container and filling it with LN_2 inside, a high voltage was applied. The high-voltage source was AC (100kV)and the standard impulse (1.2/50μs, 400kV) power supplier. Considering insulation design, impulse breakdown voltage was measured by focusing on the positive impulse voltage with low breakdown voltage.

3. Results and Discussion

Figure 2 shows the characteristics of imp(+) breakdown of various insulating materials under the same conditions. The insulating materials used as planar samples. As shown in this figure, it was the highest in GFRP, followed by in Nomex, Teflon, silicon rubber, and EPDM/Si. Just like the characteristics of AC insulation breakdown, there were good insulation materials with almost no difference except for EPDM/Si and silicon rubber. Also, GFRP was higher than other materials in terms of tensile load, tensile strength, yield stress, and elastic modulus [6].Therefore, this study has investigated the surface discharge characteristics of GFRP which have superior mechanical and electrical characteristics in LN_2.

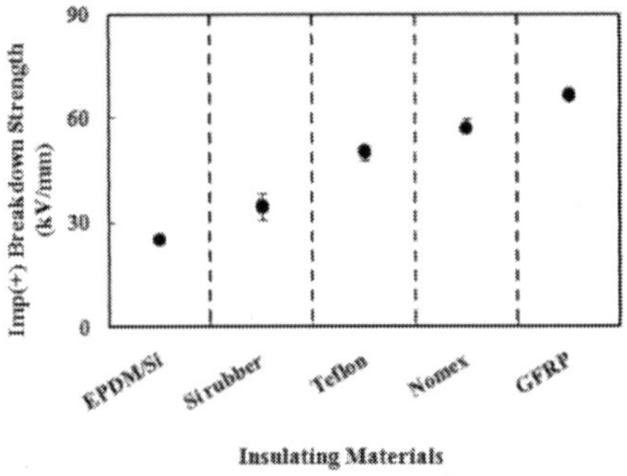

Figure 2. Imp(+) breakdown characteristics of various insulating materials in LN_2.

Figure 3 shows the surface discharge characteristics according to the lateral and longitudinal directions of cylindrical GFRP. The cylindrical GFRP was 160mm in diameter and 40mm in surface length. Then, AC and imp(+) voltage were applied. As shown in this figure, there is no big difference in voltage. However, the surface discharge voltage in longitudinal direction was slightly higher than that in lateral direction in any cases. In case of HTS bushing, the production in lateral direction is inevitable.

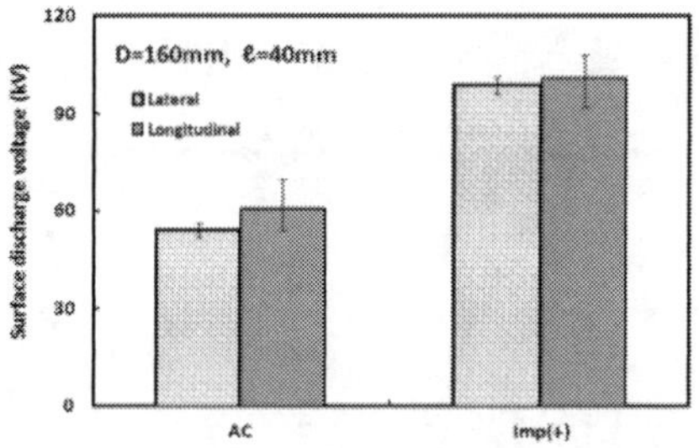

Figure 3. Surface discharge characteristics according to lateral and longitudinal direction in LN_2 for a cylindrical sample of diameter *D* and separation *l* between electrodes (Fig. 1(a)).

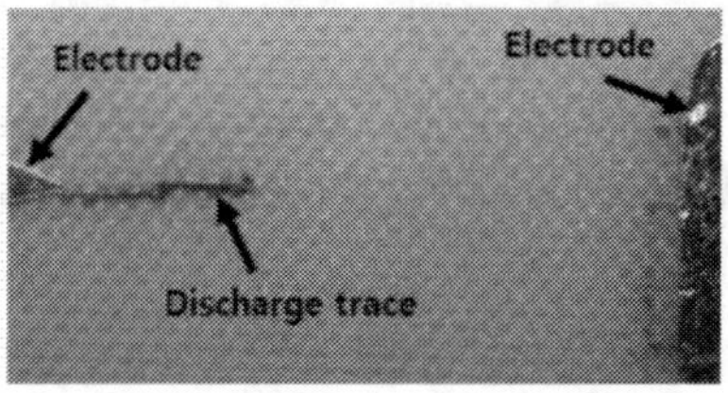

Figure 4. Surface photograph of cylindrical GFRP sample according to the longitudinal direction after impulse surface flashover.

Figure 4 shows the surface photograph of cylindrical GFRP in a longitudinal direction taken under impulse voltage after surface discharge. In the case of a lateral direction, slightly degraded tracking was observed on the surface. In the case of a longitudinal direction, on the contrary, the discharge path developed in zigzag pattern. The surface discharge voltage of longitudinal direction was higher than that of lateral direction because discharge developed toward glass fiber at a right angle. This study has concentrated on the characteristics of a longitudinal direction.

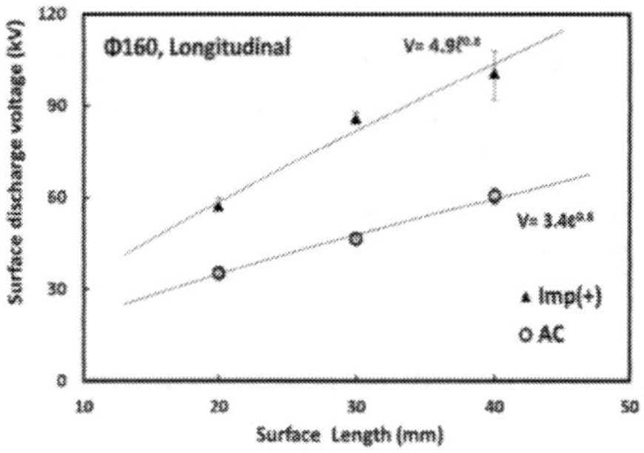

Figure 5. Surface discharge characteristics according to surface length in LN_2 for cylindrical sample.

Figure 5 shows the surface discharge characteristics according to surface length when the cylindrical diameter of GFRP was 160mm. As shown in this figure, the surface discharge of AC voltage is lower than that of impulse voltage in GFRP. The equation between impulse voltage (V_S) and surface length (ℓ) is $V_S=4.9\ell^{0.8}$.

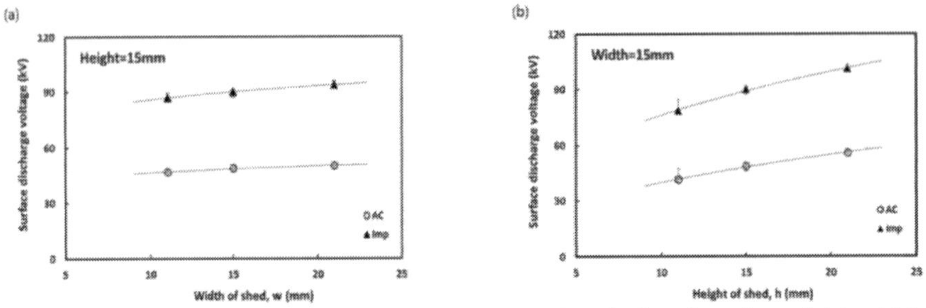

Figure 6. Dependence of the surface discharge voltage of a shed as function of (a) width w and (b) height h.

Figure 6(a) and (b) show that the dependence of surface discharge voltage on the width and height of GFRP shed. Keeping the height and width of shed at 15mm, the width and height were adjusted. As shown in this figure, despite increase in the width of insulation shed, no dramatic increase in surface discharge voltage occurred. As the height of insulation shed increased, on the contrary, surface discharge voltage increased. It's been confirmed that increase in the height of insulation shed instead of increase in the width in suppressing the progress of surface discharge. Therefore, when the shed for HTS bushing is designed, it is more effective to increase height of GFRP shed.

4. Conclusion

In terms of insulation materials for superconducting bushing, GFRP—which is superior at cryogenic temperature electrically and mechanically—is desirable. In addition, in terms of surface discharge voltage, longitudinal direction is slightly higher than lateral direction because of the effect of glass ber direction in GFRP. AC surface discharge voltage is lower than impulse voltage in GFRP. The positive impulse voltage (V_S)-surface length (ℓ) equation can be stated as '$V_S=4.9\ell^{0.8}$.'

Acknowledgments

This research was supported by Basic Science Research Program through the National Research Foundation of Korea (NRF) funded by the Ministry of Education, Science and Technology (No. 2011-0013667).

References

[1] Masuda T, Yumura H, Ohya M, Ashibe Y, Watanabe M, Minamino T, Ito H, Honjo S, Mimura T, Kitoh Y and Noguchi Y 2011 *IEEE Trans. Appl. Supercond.* **21** 1030

[2] Tixador P 2010 *Physica C* **470** 971

[3] Lapthorn A C, Chew I, Enright W G and Bodger P S 2011 *IEEE Trans. Power Delivery* **26** 394.

[4] Yamagiwa T, Kamata Y, Yoshioka Y and Hosokawa M 1981 *IEEE Trans. Power Appar. and Systems*, **PAS-100** 2746

[5] Choi J H, Choi J W, Baek S M and Kim S H 2009 *IEEE Trans. Appl. Supercond.*, **19** 1972.

[6] Ekin J 2006 *Experimental techniques for low-temperature measurements* (Oxford: Oxford University Press)

Influence of longitudinal temperature distribution on current limiting function of Superconducting Fault Current Limiting Cable (SFCLC)

H Kojima[1], T Osawa[1], N Hayakawa[1], M Hanai[1], H Okubo[2]

[1] Nagoya University, Nagoya, Aichi 464-8603, Japan
[2] Aichi Institute of Technology, Toyota, Aichi 470-0392, Japan

E-mail: h-kojima@esi.nagoya-u.ac.jp

Abstract. We have proposed a Superconducting Fault Current Limiting Cable (SFCLC), which is an HTS cable with fault current limiting function. SFCLC is expected to limit the fault current and also immediately recover the cable function after the fault clearance. In the SFCLC operation, a longitudinal temperature distribution will exist due to heat penetration, AC loss, dielectric loss and the performance of cryocooling system, which will influence its current limitation characteristics. In this paper, we investigate the influence of the longitudinal temperature distribution on current limiting function and temperature rise after the current limitation of SFCLC. We suggested the effective measures of parameter control, i.e. decreasing the critical current (I_c@77K), n value at flux flow region ($n_{1\text{-}0}$), increasing the coefficient of longitudinal temperature gradient (α), inflow temperature (T_{in}) to achieve both the higher current limiting function and the lower temperature rise.

1. Introduction

Superconducting technology is expected as next generation electric power technology with higher capacity, efficiency and also higher stability by protection of electric power apparatus and systems from a large fault current, namely as the HTS fault current limiter. In recent years, various researches and development are performed about superconducting power apparatus with these features. Then, we have proposed a Superconducting Fault Current Limiting Cable, abbreviated to "SFCLC" [1]. In the steady state, SFCLC operates as an HTS power cable with high capacity and low loss power transmission. Under the fault condition, SFCLC is expected to generate resistance as a fault current limiter. Additionally, after the fault clearance, SFCLC should recover to the superconducting state with low temperature rise in the fault current limitation.

Excellent fault current limiting function of SFCLC is produced by the generation of uniform resistance along the cable length. However, SFCLC have a longitudinal temperature distribution due to heat penetration, AC loss, dielectric loss and the performance of cryocooling system [2 - 4], which will influence its current limitation characteristics and may induce the fatal temperature rise by the local generation of limiting resistance. In this paper, by using a numerical simulation of SFCLC behaviour, we evaluate the influence of longitudinal temperature distribution on the current limitation characteristics of SFCLC.

2. Simulation method for behaviour of SFCLC

2.1. Simulation model

The numerical model for calculation of current limitation characteristics of SFCLC is developed, where SFCLC is introduced in a distribution system composed of a generator, a transformer and a load [5]. The voltage and length of SFCLC is 77 kV and 10 km, and the prospective fault current I_{PRO} is 20 kA_{peak}. The details of the structure and YBCO tapes of SFCLC can be referred to [1]. The temperature dependence of critical current $I_c(T)$ is shown in figure 1(a).

Assuming a longitudinal temperature distribution of SFCLC, an equivalent circuit of the model system is developed using the circuit equation (1) and the heat equation (2),

$$V = (L_{Tr} + L_{SFCLC})\frac{dI}{dt} + (R_{Tr} + R_{SFCLC})I + R_L I \tag{1}$$

$$c(T)\frac{\partial T}{\partial t} = R_{SC}(x)I^2 - Q_{in} - Q_{out} + S\frac{\partial}{\partial x}\left(\kappa \frac{\partial T}{\partial x}\right) \tag{2}$$

where V is the system voltage, I is the current of SFCLC, T is the temperature of YBCO conductor layer, L_{Tr}, R_{Tr}, L_{SFCLC} and R_{SFCLC} are the inductance and resistance of the transformer and SFCLC, l is the cable length, $R_{SC}(x)$ is the resistance of SFCLC at the position x ($0 \le x \le l$), R_L is the load resistance, $c(T)$ and κ are the thermal capacity and thermal conductivity of the YBCO conductor layer with the copper former, Q_{in} and Q_{out} are the cooling power by the inner and outer cooling layer, S is the cross-sectional area of YBCO conductor layer. $R_{SC}(x)$ fulfills the *E-I-T* characteristics of the YBCO tape as shown in figure 1(b), where the resistive voltage E per unit length of YBCO tape in superconductivity region is based on N value model, and that in flux flow region is based on the experimental results [1]. N value in each domain of figure 1(b) is set to n_0 at superconductivity region, $n_1(T)$ at flux flow (1) region, $n_2(T)$ at flux flow (2) region, respectively, where $n_1(T)$ and $n_2(T)$ are expressed by the following equations:

$$n_1(T) = 4.5 + n_{1\text{-}0} \times exp(-0.14T) \tag{3}$$

$$n_2(T) = 1 + n_{2\text{-}0} \times exp(-0.068T) \tag{4}$$

2.2. Longitudinal temperature distribution

We assumed that the longitudinal temperature distribution of SFCLC in operation is expressed by the following equation:

$$T_{OP}(x) = T_{in} + \Delta T \frac{1 - exp(-\alpha x)}{1 - exp(-\alpha l)} \tag{5}$$

where $T_{OP}(x)$ is the temperature of SFCLC at the position x, α is the coefficient of longitudinal temperature gradient, T_{in} is the inflow temperature of SFCLC, and ΔT is the difference in temperature along the SFCLC (figure 1(c)).

In this paper, we examined the effects of 7 parameters (ΔT = 0.01 - 6 K, α = 0.0001 - 0.005 m^{-1}, T_{in} = 71 - 76 K, I_c@77K = 1.9 – 3.8 kA, n_0 = 1 - 40, $n_{1\text{-}0}$ = 1.5 × 10^5 - 4 × 10^5, $n_{2\text{-}0}$ = 330 - 900) on SFCLC behaviour.

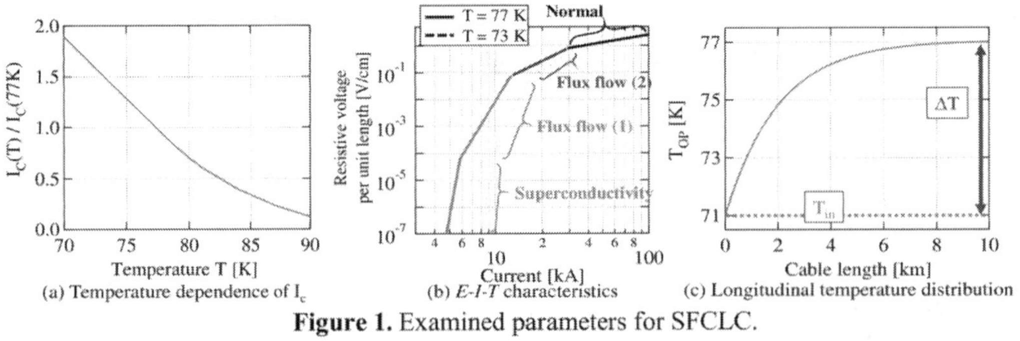

Figure 1. Examined parameters for SFCLC.

3. Results and discussions

3.1. Current limiting characteristics

Figure 2 shows the simulation results for different values of α at $\Delta T = 6$ K, $T_{in} = 71$ K, I_c@77K = 3.1 kA, $n_0 = 36.5$, $n_{1\text{-}0} = 2.5 \times 10^5$, $n_{2\text{-}0} = 460$. Figures 2(a) - (d) show longitudinal temperature and critical current distribution, current waveform, temperature distribution after 5 cycle (T_{5cycle}), and resistance waveform of SFCLC, respectively. The limited current (I_{limit}) at the 1st half cycle in figure 2(b) is 7.2 kA$_{peak}$ (36% of I_{PRO}) by the resistance in figure 2(d). At each longitudinal temperature distribution in figure 2(a), heat by the generated resistance concentrates at $x = 9 - 10$ km of SFCLC in figure 2(c), where T_{5cycle} exceeds 100 K and reaches 300 K at $\alpha = 0.0001$ m^{-1}. α has little influence on I_{limit} in figure 2(b), because the lowest I_c, which is fixed in figure 2(a), is decisive for the initiation of fault current limitation. However, we should note that α has a strong influence on T_{5cycle}, because the cable length contributing to the current limiting resistance is attributed to the longitudinal temperature distribution, i.e. the cable length with the almost lowest I_c. In other words, for the temperature distribution at $\alpha = 0.001$ m^{-1}, SFCLC has the longer length with the almost lowest I_c, and shares the resistance generation by the longer region of the cable, which can prevent the local heat generation.

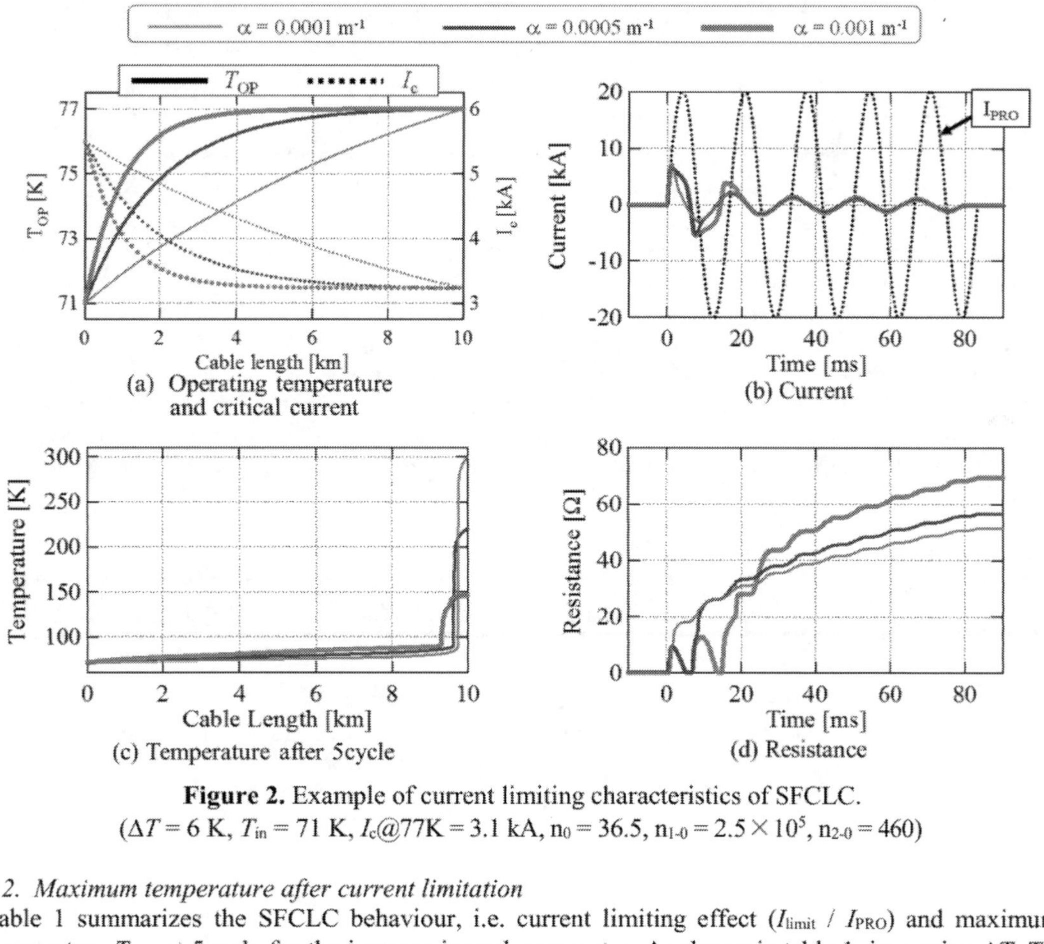

Figure 2. Example of current limiting characteristics of SFCLC.
($\Delta T = 6$ K, $T_{in} = 71$ K, I_c@77K = 3.1 kA, $n_0 = 36.5$, $n_{1\text{-}0} = 2.5 \times 10^5$, $n_{2\text{-}0} = 460$)

3.2. Maximum temperature after current limitation

Table 1 summarizes the SFCLC behaviour, i.e. current limiting effect (I_{limit} / I_{PRO}) and maximum temperature T_{max} at 5cycle for the increase in each parameter. As shown in table 1, increasing ΔT, T_{in}, n_0 and decreasing I_c@77K are expected for the effective current limitation (lower I_{limit} / I_{PRO}). For the lower T_{max}, increasing α, T_{in}, decreasing I_c@77K, n_0, $n_{1\text{-}0}$ and optimum ΔT are expected. Therefore, as the effectiveness for the total SFCLC function in table 1, SFCLC can achieve both a higher current limiting function and a lower temperature rise for the subsequent recovery function, by decreasing I_c@77K and $n_{1\text{-}0}$ from the viewpoint of HTS tape specifications as well as by increasing α, T_{in} from the viewpoint of temperature distribution in operation. These results will contribute to the design and operation of SFCLC.

Table 1. SFCLC behaviour for the increase in each parameter.

		Current limiting effect (I_{limit} / I_{PRO})	Maximum temperature (T_{max} at 5cycle)	Effectiveness for total SFCLC function
α ↗		—	↘	Effective
ΔT ↗		↘	⌒	Effective
T_{in} ↗		↘	↘	More effective
I_c@77K ↗		↗	↗	More effective
N value	n_0 ↗	↘	↗	Less effective
	$n_{1\text{-}0}$ ↗	—	↗	Effective
	$n_{2\text{-}0}$ ↗	—	—	Non-effective

↗ : increase ↘ : decrease — : almost no change ⌒ : have a maximum

4. Conclusion

This paper described the influence of the longitudinal temperature distribution of SFCLC in operation on its current limiting characteristics. Simulation results revealed that the resistance and temperature rise of SFCLC during the current limitation may concentrate on the region with the almost lowest I_c. We demonstrated the critical parameter for current limiting effect and maximum temperature. Then, we suggested the effective measures of parameter control, i.e. decreasing I_c@77K and $n_{1\text{-}0}$ from the viewpoint of HTS tape specifications as well as increasing α, T_{in} from the viewpoint of temperature distribution in operation, in order to achieve both higher current limiting function and lower temperature rise for the subsequent recovery function. These results will contribute to the design and operation of SFCLC.

Acknowledgement

This work was supported by JSPS KAKENHI Grant Number 24760228.

References

[1] Kojima H *et al.* 2012 *IEEE Trans. Applied Superconductivity* **22** 4800105
[2] Melnik I *et al.* 2011 Superconductivity Centennial Conference 2-LA-O8
[3] Maguire J *et al.* 2011 *IEEE Trans. Applied Superconductivity* **21** pp 961-966
[4] Lindsay D *et al.* 2009 Superconductivity for Electric Systems U.S. Department of Energy Annual Peer Review
[5] Kojima H *et al.* 2012 *Phys. Procedia* **36** pp 1296-1300

2T/5T Two-Axis Cryogen Free Superconducting Vector Magnet With Variable Temperature Space

E I Demikhov[1,2], T E Demikhov[1,2], E A Kostrov[1,2,3], V V Lysenko[1,2], and N A Piskunov[1,2]

[1]Lebedev Physics Institute of RAS, 119991 Moscow, Leninskiy prospekt 53, k8

[2]RTI, Cryomagnetic Systems 119333 Moscow Leninskiy prospekt 53, k8

E-mail: kostrov@lebedev.ru

Abstract. A conduction cooled 2T / 5T superconducting vector magnetic system with a variable temperature space was developed and tested. The system is based on a commercial two-stage 4 K Gifford-McMahon cryocooler with the cooling power of 1.5 W at 4.2 K. The cool down time of the magnet from room temperature to 3.2 K is 17 hours. The system provides sample temperature range of 6.0-300 K. The clear diameter of variable temperature space is 39 mm. A 5 T solenoid generates magnetic field in the vertical axis and a 2 T split coil generates field in the horizontal axis. The magnets are made of niobium–titanium wire wound on a copper former. A PC controlled rotary drive is applied to rotate a sample holder around the vertical axis. Thus the measured sample can be exposed to the magnetic field in any desired direction. A helium gas gap heat switch is used as a controllable thermal link between the variable temperature space and the 2nd stage to avoid overheating of the magnet at high temperatures of the sample. The system design, manufacturing and test results are presented.

1. Introduction

Study of the angular and temperature dependence of electro-transport properties in external high magnetic field requires a special magnetic system. One of the possible ways to build such a system based on a closed cycle cryocooler and a superconducting magnet is to use a heat switch. The heat switch is a cryogenic device that allows to make or to break a thermal contact. The temperature of the NbTi magnet should be always near 4 K while the temperature of a sample varies from 6 K to 300 K. Thus the thermal link between the sample chamber and the 2nd stage of a cryocooler should be closed to reach low temperatures and opened at high temperatures of the sample.

The possibility to create a conduction cooled superconducting magnetic system with a variable temperature insert based on a single two-stage Pulse tube cryocooler and a helium actuated gas gap heat switch was experimentally shown in our previous work [1]. The helium actuated gas-gap heat switch is a reliable and simple device with no moving parts [2], [3].

The goal of this work was to design and manufacture a superconducting vector magnetic system with a variable temperature space for measurements of electro-transport properties. The PC controlled rotary drive is used to achieve any desired direction of magnetic field respectively to a sample.

[3] To whom any correspondence should be addressed.

2. General description

The two-axis superconducting cryogen free magnetic system is based on a commercial Gifford-McMahon cryocooler SRDK-415 by Sumitomo [4]. Figure 1 shows the outside view of the magnetic system. Figure 2 shows the schematic representation of the cryomagnetic system.

Figure 1. Outside view of the cryostat.

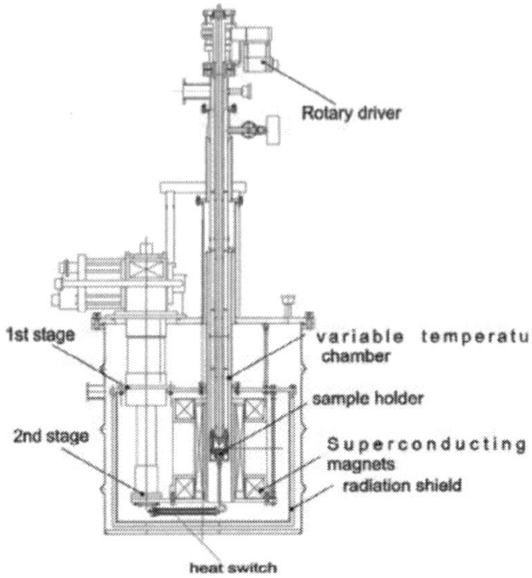

Figure2. Schematic representation of the cryomagnetic system

The superconducting magnet, the 2nd stage of the GM cooler, and the heat switch are placed inside the thermal radiation shield. An additional shield is installed in the magnet bore to reduce the heat flux from the variable temperature insert. Both thermal radiation shields are cooled by the 1st stage of the GM cryocooler.

The helium gas gap heat switch consists of two concentric copper tubes and a stainless steel housing. The internal copper tube is thermally connected to the variable temperature chamber and the external copper tube is thermally connected to the 2nd stage of the GM cooler. The thermal conductivity of the heat switch depends on the helium pressure in the gas gap between the two copper tubes. The detailed description of the heat switch integrated into magnetic system is given in [1].

To achieve low temperatures of a sample (<50 K) the heat switch should be filled with helium gas, otherwise it should be pumped. In addition, for the temperature range of a sample under 15 K the variable temperature chamber should be filled with low pressure helium gas.

The magnet current leads are composed of resistive and HTS parts. The resistive part extending from the room temperature flange to the 1st stage of a cooler is manufactured from brass stripes. The HTS part is made of Coated Conductor tapes (by SuperPower) and connects the magnet to the resistive part. The joints of the HTS part to the resistive part and the NbTi wire are thermally anchored to the 1st and the 2nd stages of the GM cooler respectively. The critical current of the HTS tapes is 110 A at 77 K.

3. Magnetic system

The superconducting magnet consists of one NbTi solenoid coil and NbTi split magnet. The split magnet consists of two coils. All coils are wound on copper formers.

Three coils are assembled on the copper plate directly connected to the 2nd stage of GM cryocooler. Additional brass studs and stainless steel cylinders are used to fix the split coils together and to increase its strength against the magnetic force. Thin indium foil is used to reduce the contact thermal resistance. Figure 3 shows a photograph of the superconducting magnet.

Figure 3. Superconducting magnet (disassembled).

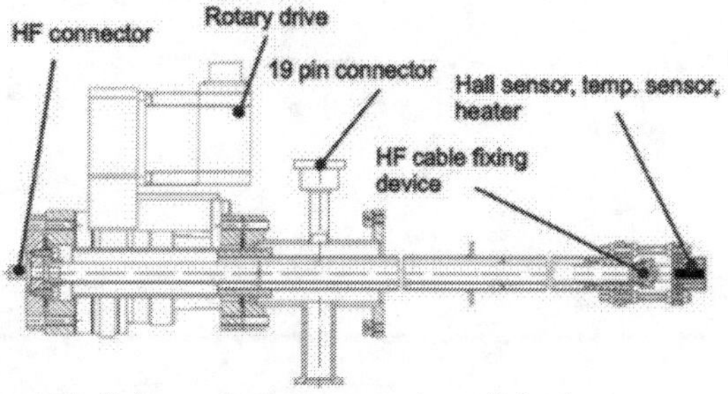

Figure4. Schematic representation of the insert

All coils are fabricated with Ø 0.6 mm (insulated) NbTi/Cu wire and are impregnated with Stycast epoxy resin. The copper to superconductor ratio of the wire is 1.3. The number of the filaments is 630. The main parameters of the magnet are shown in Table 1.

Table 1. The main parameters of the magnet

Characteristic	Solenoid	Split
Winding ID, mm	74.5	129
Winding OD, mm	107	213.9
Field inhomogeneity in 1cm Diameter Sphere Volume	<0.1 %	<0.5%
Max sweep rate	10 mT/s(0 to 3 T)	10 mT/s (0 to 1.5 T)
Combination of the magnetic field	>2 T (when split and solenoid generate equal magnetic field) >4 T (when the magnetic field generated by the split magnet is ten times less than the one generated by the solenoid)	
Operating current	89.1 A@5T	65.8A@2T
Total mass	32 kg	

4. Insert with sample holder

The insert provides measurements of electro-transport properties by means of six low frequency wires and has all necessary connections to install an additional high frequency cable (DC-60Ghz). Figure 4 shows the schematic representation of the insert.

Two coaxial thin stainless steel tubes are connected to the rotary drive (MD35-LB from the UHV Design [6]). The sample holder is mounted on the external tube and is able to rotate in the range of ±190 degrees. The static internal tube is intended for installation of the high frequency (HF) cable. The resistive heater and the temperature sensor mounted on the insert are used to control temperature together with the PID temperature controller. Three components of the magnetic field are measured by means of the commercially available 3-axis cryogenic Hall probe [6].

5. Test results

The magnet is cooled down from room temperature to 3.2 K in 17 hours without any pre-cooling and the insert is cooled down to 6 K in 18.5 hours. The temperature sensor TVO [7] is located on the copper plate which connects the split magnet and solenoid with cryocooler. The temperature measured by this sensor is marked as "Temperature of magnet". The temperature of the insert was measured by Cernox 1050-AA-1.4L by Lakeshore [8]. Figure 6 shows the profiles of temperatures and currents in the split magnet and the solenoid during one of the tests.

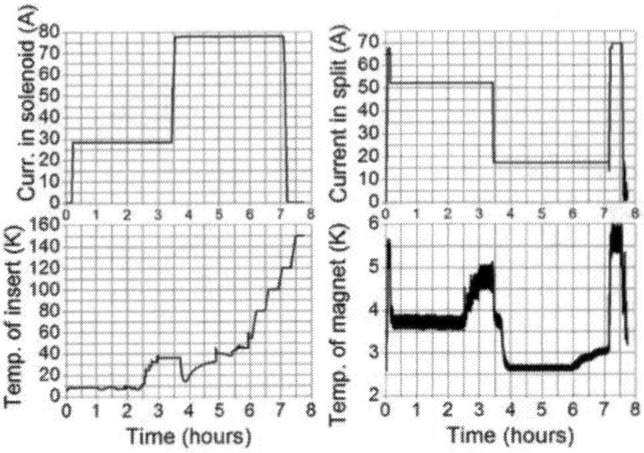

Figure 6. The profiles of temperatures and currents during the test.

The stationary temperature of the magnet was 2.7 K before the current charging. The split magnet was charged up to 67.6 A (2.05 T) with a ramp rate of 0.33 A/s (10 mT/s) and after 5 minutes its current has been decreased to 51.2 A to generate magnetic field of 1.6 T to test simultaneous operation of the split magnet and solenoid, the latter was charged up to 28.7 A (1.6 T) with a ramp rate of 0.2 A/s (11 mT/s). In this mode an overall magnetic field of 2.26 T was generated. Subsequent test shows that the whole system is able to generate 4.4 T provided the magnetic field generated by the split magnet is ten times less than the one generated by the solenoid.

During this test the temperature of insert was stabilized at various setpoints from 6 to 150 K by means of the PID controller. The heat switch was pumped at the temperature of the insert of 36 K to avoid the overheating of the magnet. The system has been tested in the whole range of temperatures of the insert (6-300 K). The typical stability of the sample holder temperature depends on the temperature range and is ±2 mK in range 6-100 K, and ±20 mK in range 100-300 K.

6. Summary

A conduction cooled 2T / 5T superconducting vector magnetic system with a variable temperature space for electro-transport properties measurements was developed and tested. The system generates designed magnetic field for the whole range of sample temperatures (6-300 K).

7. References

[1] Demikhov E I, Kostrov E A, Lysenko V V, Piskunov N A, and Troitskiy V F 2010 8 T Cryogen Free Magnet With a Variable Temperature Insert Using a Heat Switch *IEEE Trans. Appl. Supercond.* vol. 20 No. 3 612-616

[2] Bywaters R P and Griffin R A 1973 A gas-gap thermal switch for cryogenic applications *Cryogenics* vol. 13 6 344-349

[3] Donabedian M 2003 *Spacecraft Thermal Control Handbook* Vol II: Cryogenics The Aerospace Press El Segundo CA

[4] http://www.shicryogenics.com

[5] http://www.arepoc.sk/?p=30

[6] http://www.uhvdesign.com

[7] http://cryo.ru/index.php?lang=en

[8] http://www.lakeshore.com

Acknowledgment

The authors thank prof. Bulat Rameev (Gebze Instute of Technology) for his assistance in the insert design.

Improving homogeneity of the magnetic field by a high-temperature superconducting shield

E Kulikov[1], N Agapov[1], V Drobin[1], A Smirnov[1], G Trubnikov[1], G Dorofeev[2] and H Malinowski[3]

[1] Laboratory of High Energy Physics, Joint Institute for Nuclear Research (JINR), Dubna, Russian Federation
[2] NRC Kurchatov Institute, Moscow, Russian Federation
[3] Laboratory of High Energy Physics, Joint Institute for Nuclear Research (JINR), Dubna, Poland

E-mail: kulikov_box_work@mail.ru

Abstract. The shielding opportunity of the magnetic field perpendicular component by the high-temperature superconducting tape (HTS) is shown experimentally for the first time. The tapes are laid closely to each other with a shift of pieces from layer to layer equal to a half of the tape width at the experimental set-up. This multilayer cylindrical structure inserted into the solenoid is similar to the unclosed shield from a uniform piece of the superconducting foil with the corresponding current-carrying capacity. It has been found that the maximum shielding field is proportional to the number of layers and a half of the full magnetization field of one tape for the regular multilayer structure of the HTS segments. The obtained results are necessary to construct systems with the high magnetic field homogeneity, in particular, for the electron cooling system of charged particle beams at the new accelerator complex which is being developed at JINR in Dubna, Russia.

1. Introduction

Forming of the highly homogeneous magnetic field is the main requirement to construct an electron cooling system for charged-particle beams, as a part of NICA accelerator complex [1, 2]. For this purpose, the magnetic field homogeneity must be at least 10^{-5} with the length of the cooling system solenoid of about 6 m and the magnetic field of up to 0.2 T. However, it is very expensive due to the complexity of the precision winding. Moreover, multiple sections will inevitably lead to the non-uniform field at solenoid junctions. Generation of the highly homogeneous magnetic field using the unclosed superconducting shields is the most promising direction to solve this problem. It is known that low-temperature superconductors can be used for these purposes [3, 4]. On the other hand unique properties of the HTS make this idea even more attractive. The purpose of this paper is to demonstrate an opportunity of using the existing HTS tape to increase the magnetic field homogeneity in the laboratory magnetic system.

2. Planar multilayer structure in the magnetic field perpendicular to the sample

The 2-G HTS Y123 ceramic tape produced by Super Power Company was used as a basic structure element. The tape was 12 mm wide and 50 μm thick. At the first stage the impulse method was used to estimate a simple planar structure shielding ability. Samples of multilayer structures were measured: one was built of stacked HTS 24 mm tape pieces; the other one having the same type tape parts was combined as a „tile". The magnetic field impulse, with the amplitude up to 0.2 T and width of 5 ms, was delivered upon the sample. The electromotive force (EMF) in the measuring coil, placed in the middle of the sample, and current in the energizing coil were recorded. The magnetic field in the middle of the sample was obtained through the integration of the EMF. The full magnetization field of this multilayer structure was defined using the fact that this field depends on the applied field (namely on the current in the energizing coil).

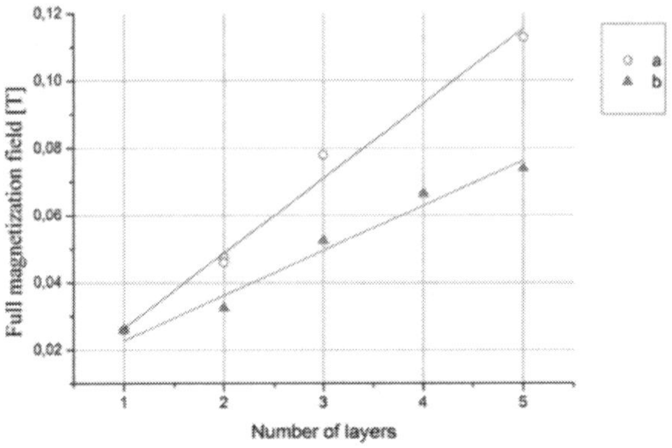

Figure 1. Full magnetization field vs. number of layers. a- structure as a stack of the tapes, b– structure as a "tile" of the tapes.

Figure 1 shows that the full magnetization field of one tape is about 25 mT (curve a). Each next "tile" layer increases the shielding field by an average of 12.5 mT (curve b). So, the shielding ability is proportional to the number of layers and a half of the full magnetization field of one tape for the regular multilayer structure of the HTS segments.

3. Multilayer structure in the form of the lengthwise shield

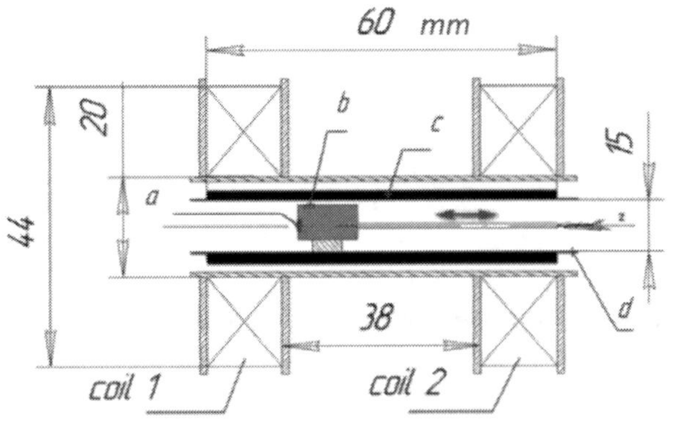

Figure 2. Experimental set-up. a- Hall sensor to measure the longitudinal field component; b– Hall sensor to measure the radial field component; c – HTS lengthwise shield; d - non-magnetic steel tube.

The lengthwise shield is made of the same tape as the samples in the previous experiments. The tapes are laid closely to each other on the tube along its generatrix. The following layers were put with a shift of pieces from layer to layer equal to a half of the tape width to form the "tile".

The magnetic field was measured by Hall sensors (PXA_602817A type) with sensitivities 45 mV/T - for longitudinal component and 245 mV/T - for radial component correspondingly. Liquid nitrogen was used to cool the experimental set-up. Longitudinal (B_z) and radial (B_r) components of the magnetic field near the inner surface of the shield depending on the position along Z axis are given in the figures below.

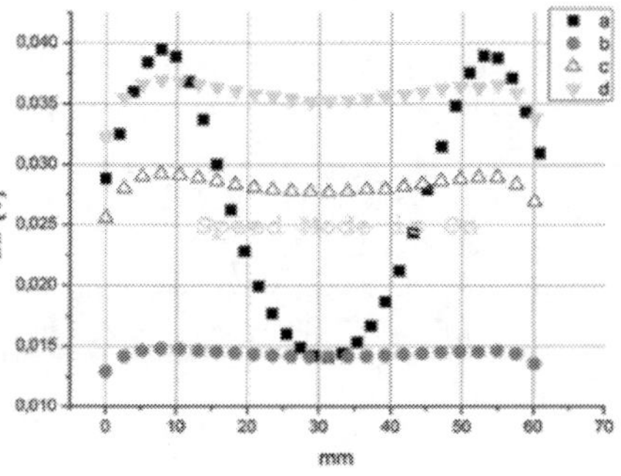

Figure 3. B_Z of the 4-layer shield vs. Z-coordinate for different currents in the coils: a- without shield, 1A; with shield: b- 0.5A; c- 1 A; d- 1.5A.

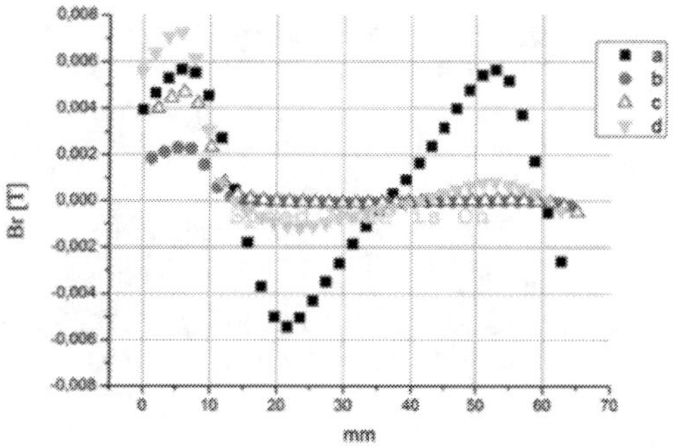

Figure 4. B_r of the 4-layer shield vs. Z-coordinate for different currents in the coils: a- without shield, 1A; with shield: b- 0.5A; c- 1 A; d-1.5A.

The radial component demonstrates the quality of the shielding (See figure 4). Shielding at any current in the coils is observed in figure 3. However, figure 4 shows that the shield performance deteriorates at the current of 1.5 A! The congruence of figure 4 and figure 5 allows to say that the 4-layer shield performs more efficiently than the 2- layer shield.
The comparison of the qualitative results with those of the papers [3-5] have shown that the introduced "tile" lengthwise shield inserted into the solenoid is similar to the unclosed shield from a uniform piece of the superconducting foil with the corresponding current-carrying capacity.

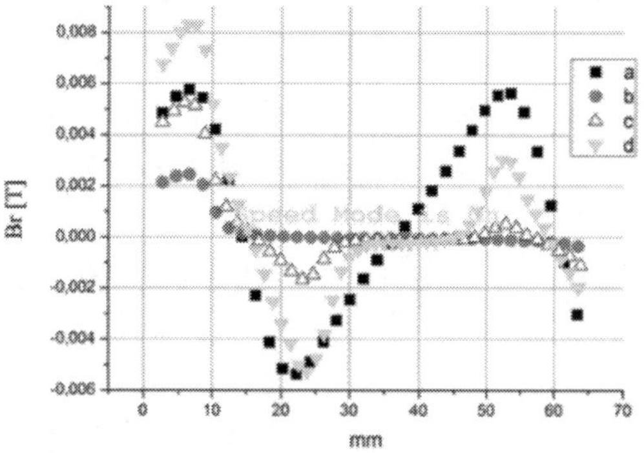

Figure 5. B_r of the 2-layer shield vs. Z- coordinate for different currents in the coils: a- without shield, 1A; with shield: b- 0.5A; c- 1 A; d-1.5A.

The accuracy of the magnetic field measurements was about $\Delta B/B \sim 10^{-2}$, which was basically determined by the stability of the DC power supply.

4. Conclusions

1. The "tile" structure performs as a uniform piece of the superconducting foil and can be used to construct the unclosed shield.
2. The shielding ability of this lengthwise construction is proportional to the number of layers and a half of the full magnetization field of one tape for the regular multilayer structure of the HTS segments.
3. The developed method makes it possible to construct the unclosed superconducting shields using the latest 2-G HTS tape to increase the magnetic field homogeneity of the large magnetic systems, such as the electron cooling system for the NICA complex.

References

[1] Meshkov I N *et al* 2009 *Proc.COOL'09 Workshop on Beam Cooling and Related Topics* (*Lanzhou, China, August 31 – September 4 2009*) p16

[2] Sidorin A O, Meshkov I N, Seleznev I A, Smirnov A V, Syresin E M and Trubnikov G V 2006 *J. Nucl. Instrum. Methods* A **558** 325-28

[3] Agapov N N *et al* 2012 *J. Phys. Part. Nucl. Lett.* **9** 422-24

[4] Bychkov Yu F, Komarov A O and Khlopkov A K 1980 *Metallurgy and Metal Science of Pure Metals* vol 14, ed V Emelyanov and A Evstyuhin (Moscow: Atomizdat) pp 53-63

[5] Kulikov E *et al* 2012 Superconducting unclosed shields for improving homogeneity of the magnetic field in magnetic systems *Proc. 12th IIR Int. Conf. on Cryogenics 2012* (*Dresden, Germany, 11-14 September 2012*) p 351

NbTi Strands Verification for ITER PF CICC Process Qualification of CNDA

F Liu[1], H Liu[1], S Liu[2], B Liu[1], L Lei[1], Y Wu [1]
[1] Institute of Plasma Physics , Chinese Academy of Sciences(ASIPP), Hefei, China.
[2] ITER China, Beijing, 100038, China

E-mail:fangliu@ipp.ac.cn

Abstract. China is in charge of most of Poloidal Field (PF) conductors production for the International Thermonuclear Experimental Reactor (ITER). The execution for PF conductors shall be in three main phases. According to ITER Procurement Arrangement (PA), the Domestic Agency (DA) shall be required to verify the room and low temperature acceptance tests carried out by the strand suppliers. As the reference laboratory of Chinese DA (CNDA), the superconducting strands test laboratory of Institute of Plasma Physics, Chinese Academy of Sciences (ASIPP) was undertaking the task of strands verification for ITER conductors. The verification test includes: diameter, Nickel plating thickness, copper-to-non-copper volume ratio, twist pitch direction and length, standard critical current (I_C) and resistive transition index (n), residual resistance ration (RRR), and hysteresis loss. 48 NbTi strands with 7 billets were supplied for the PF Cable-In-Conduit Conductor (CICC) process qualification. In total, 54 samples were measured. The verification level for PF CICC process qualification was 100%. The test method, facility and results of each item are described in detail in this publication.

1. Introduction

The CICCs for the ITER PF Coils uses twisted, multifilament, Nickel-plated, NbTi based composite strands [1, 2]. China is in charge of most of the ITER PF conductors, including all the Unit Lengths (ULs) for PF2 through PF5. According to ITER PA, the execution for PF conductors shall be in three main phases: phase 1: call for tender, phase 2: supplier disclosure and process qualification and phase 3: production of all conductor ULs. In addition to Cu dummies, process qualification calls for the manufacturing of NbTi conductor qualification lengths.

At the start of process qualification, DA is required to conduct a benchmarking of the room and low temperature acceptance tests. As the reference laboratory of CNDA, ASIPP is undertaking the task of superconducting strands verification for ITER conductors, which has successfully passed the world-wide benchmarking for NbTi strands [3, 4].The verification for the room and low temperature acceptance tests carried out by the strand suppliers is being organized. All the strands used for PF CICC process qualification have been verified. The test method, facility and results of each item are described in this paper.

2. Samples and sample preparation

This work was partially supported by the Equipment Function Developing and Technical Innovation project of the Chinese Academy of Sciences under Grant Y25ZB13291.

For the PF CICC process qualification of CNDA, 2 ULs were required to manufacture, include 1 UL with 425m for PF2-4 and 1 UL with 100 m for PF5. 48 NbTi strands with 7 billets were used. All the strands were fabricated and supplied by the Western Superconducting Technology Co. Ltd (WST). And all the strands are the same type. A micrograph of the transverse cross-section of the WST NbTi strand is shown in figure 1.

The sampling requirements for the strands suppliers are detailed in Table 1. The strand supplier shall ensure that all acceptance tests are carried out at the proper frequency and with the proper procedures and shall compare the test results with the acceptance criteria. According to the test requirements of CNDA, the verification level for each test item was 100%.

The maximum field considering the Lorenz force direction calculated with a 2-D model was around 0.146T when the applied current was 306A for the I_C test barrel we used. The *RRR* testing sample was cut from the straight strand. The hysteresis loss sample shape was a spiral with ~ 5 mm inner diameter and ~1 mm pitch. At least six turns (~100 mm) were required. The strand was wound on a bolt. A section of the spiral was cut and used as hysteresis loss specimen after fixing on the bolt for at least 2 days. A few parameters were required to be measured at room temperature.

Table 1. Acceptance requirements for Nickel-plated, NbTi-based strands.

Verification item	Sampling
Strand diameter	every piece length and over entire length
Twist pitch and twist direction	at minimum: point or tail of every piece length
Ni plating thickness	at minimum: start and end of every Cr plating process
Cu/non-Cu Ratio	at minimum: point and tail of every billet
RRR	at minimum: point and tail of every billet
I_C and n	phase 2: point and tail of every billet plus one sample per breakage
Hysteresis loss	every billet

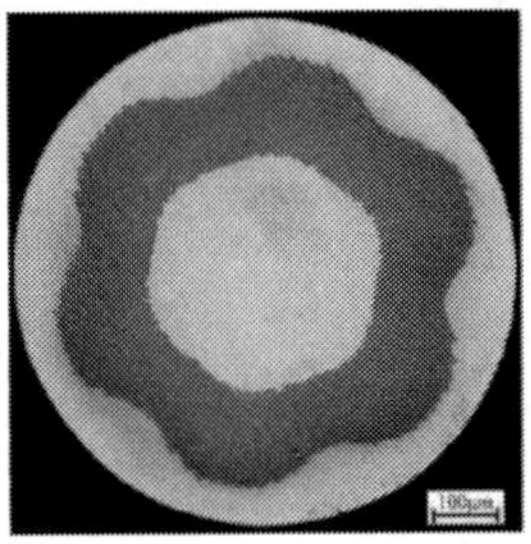

Figure 1. Transverse cross-section micrograph of the NbTi strand used for ITER PF2-5 conductor.

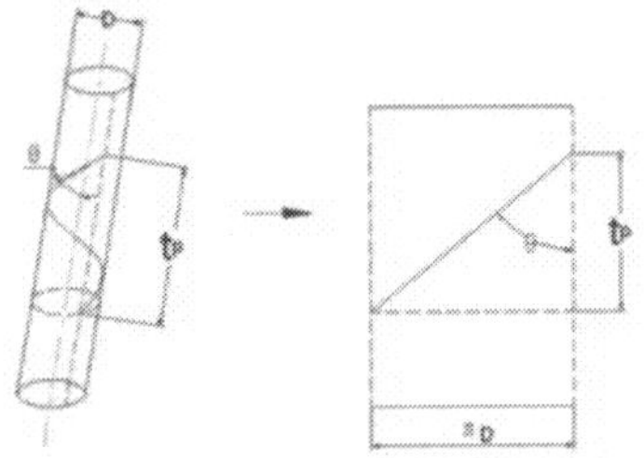

Figure 2. Schematic diagram of twist pitch measurements

3. Experimental methods and test facility

3.1. Room temperature Tests

The room temperature tests include: diameter, Nickel plating thickness, copper-to-non-copper volume ratio, twist pitch direction and length.

Diameters of strands were measured by micrometer, which has been calibrated with the accuracy 0.001mm.

Plating thickness was measured by metallographic techniques. For each cross section, the thickness of the Ni layer was measured with the scanning electron microscope (SEM) at six locations, at 60° each, with an accuracy better than 0.1μm.

The copper-to-non-copper volume ratio *Rcu-ncu* was measured by chemical etching and weighting, and calculated by equation 1.

$$R_{cu-ncu} = V_1/V_2$$
$$V_1 = (D/2)^2 \times \pi \times L \qquad (1)$$
$$V_2 = (W_1 - W_2)/\rho_{Cu}$$

where, L, V and W were the sample length, volume and weight, respectively. The subscript 1 and 2 expressed the sample before and after etching, respectively. D was the diameter of the strand while ρ_{Cu} was the density of Copper which was designated to be 8.93g/cm^3.

Twist pitch measurement was by chemically etching away the stabilizing Cu with the two ends of samples fixed before etching. The photos were getting by SEM. The twisting angle θ was gotten by the image editing software based on the SEM photos. The schematic diagram was shown in figure 2.

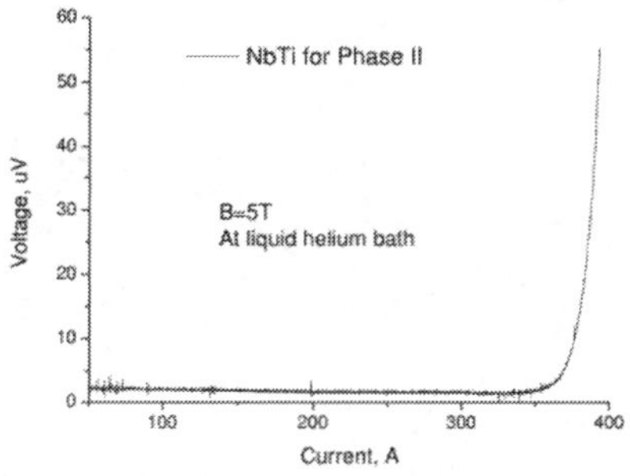

Figure 3. *V-I* curve of one NbTi sample with voltage tap length 50cm.

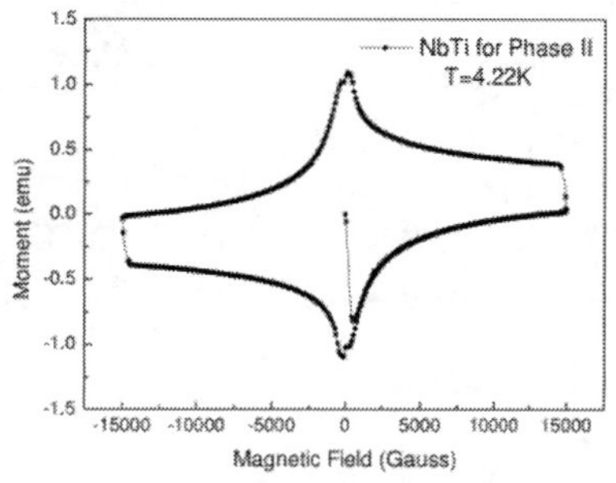

Figure 4. Magnetization curve of one NbTi sample

3.2. Ic, n

For NbTi strand, I_C measurements were required at 5.0 T and 4.22K. The electric field criterion we used was 0.1uV/cm. A voltage-current (*V-I*) curve of one NbTi sample with voltage tap length 50cm was shown in figure 3.

The samples were measured at liquid helium. For liquid helium, when the vapor pressure is around 101.325kPa (1atm), the temperature/pressure relation can be approximated with a linear function [5].

ITER Organization (IO) applied temperature correction using equation 2 [6], since temperature accuracy is important in the *B-T* range concerned.

$$\frac{I_c}{I_m} = \frac{T_c(B) - 4.22}{T_c(B) - T_m} \qquad (2)$$

where $T_c(B)$ is the critical temperature at the specified magnetic field

$$T_c(B) = 9.2 \times (1 - \frac{B}{14.5})^{0.59} \qquad (3)$$

3.3. RRR

RRR sample is mounted on a thermal substrate. The *RRR* of the composite wire was obtained in equation 4 below, where the resistance at 273K (R_{273K}) was divided by the resistance at specified temperature (R_L).

$$RRR = \frac{R_{273k}}{R_L} \qquad (4)$$

RRR for NbTi strands, whose T_C at zero field is approximately 9 K, is defined as the ratio of resistance at 273 K to 10 K. The temperature accuracy at 10 K was less critical, albeit important, than it was at 20 K, because the copper resistivity dependency on temperature was much smaller at 10 K.

3.4. Hysteresis loss
The hysteresis losses were measured by VSM in Ningbo Institute of Material Technology & Engineering Chinese Academy of Sciences (IMTE).

For NbTi strand, the field range requested was ±1.5 T at 4.2 K. The field ramp rate was 50Gs/s. The integration of the magnetization loop was performed and no smoothing of the curves was applied. The magnetization curve of one NbTi sample for PF CICC process qualification was shown in figure 4.The hysteresis loss density Q_{hyst} may be given by the area under the magnetization curve.

4. Test results and conclusions
The verification results of NbTi strands for PF conductors process qualification are summarized in Table 2.

Table 2 Summary of verification results for the NbTi strands with 7 billets

	PA requirement	Mean	Max	Min	σ_s
Diameter, mm	0.73±0.005	0.731	0.733	0.730	0.0006
Twist Pitch, mm Cu/	15±2	14.66	15.22	14.01	0.309
non-Cu Ratio Plating	2.35±0.1	2.316	2.405	2.265	0.0394
thickness, μm	2+0-1	1.5	2.0	1.1	0.22
Ic(5T,4.22K), A	≥339	366.6	379.6	356.2	5.83
n(5T,4.22K)	≥20	43.4	51.6	36.2	3.76
Hysteresis Loss ,mJ/	<45	40.5	42.2	39.3	0.97
cc RRR	>100	157.8	162.0	150.1	3.36

As shown in Table 2, the average diameter after plating of all the NbTi strands which have been supplied is 0.731mm with standard deviation (σ_s) of 0.0001mm. And the difference between maximum and minimum (Δ) is 0.003mm. The average plating thickness is 1.5μm with σ_s of 0.22μm and Δ of 0.9μm. The average Copper to non-Copper ratio is 2.316 with σ_s of 0.0394 and Δ of 0.140. And the average twist pitch is 14.66mm with σ_s of 0.309 mm and Δ of 1.21mm.

The average I_C and n value of all the NbTi strands are 366.6A with σ_s of 5.83A and 43.4 with σ_s of 3.76. The difference between maximum and minimum (Δ) of I_C is 23.4A while Δ of n is 15.4.

The average hysteresis loss is 40.5mJ/cc with σ_s of 0.97mJ/cc and Δ of 2.9mJ/cc.

The average *RRR* of all the NbTi strands is 157.8 with σ_s of 3.36 and Δ of 11.9.

All the data for the strands fit well the requirements for ITER PA. From the feedback from CNDA, most of the results of the strands verification for process qualification showed high agreement with the data from the strand supplier except the twist pitch. The twist pitch verification results showed systematic shift to higher values. That may be caused by the systematic error for the twist pitch method for NbTi strands. A new method was used and verified, and better results were obtained in phase 3.

Acknowledgment
The tests were carried out in ASIPP, the author gratefully acknowledges the ASIPP colleagues for their great help. The author also thanks Huixian Gao at WST for many useful discussions.

References
[1] Shimomura Y 2005 *Fusion Eng. Des.* **74** 9-16.
[2] Holtkamp N 2009 *Fusion Eng. Des.* **84** 98-105.
[3] Pong I Jewell M C Bordini B Oberli L R Liu S Long F Boutboul T Readman P Park S H Park P Y Pantsyrny V Tronza V Martovetsky N Lu J and Devred A 2012 *IEEE Trans. Appl. Supercond.* **22** 4802606.
[4] Liu F Long F Liu B Lei L Wu Y Liu H 2013 *Fusion Eng. Des.* **88** 17-22.
[5] Donnelly R J and Barenghi C F 1998 *J. Phys. Chem. Ref. Data* **27** 1217-1274.
[6] Lubell M S *IEEE Trans. Magn.* 1983 **MAG-19** 754-757.

Demountable Toroidal Field Magnets for Use in a Compact Modular Fusion Reactor

F. J. Mangiarotti*, J. Goh, M. Takayasu, L. Bromberg, J. V. Minervini and D. Whyte

Plasma Science and Fusion Center, Massachusetts Institute of Technology, Cambridge, MA 02139, USA

*E-mail: fjm@mit.edu

Abstract. A concept of demountable toroidal field magnets for a compact fusion reactor is discussed. The magnets generate a magnetic field of 9.2 T on axis, in a 3.3 m major radius tokamak. Subcooled YBCO conductors have a critical current density adequate to provide this large magnetic field, while operating at 20 K reduces thermodynamic cooling cost of the resistive electrical joints. Demountable magnets allow for vertical replacement and maintenance of internal components, potentially reducing cost and time of maintenance when compared to traditional sector maintenance. Preliminary measurements of contact resistance of a demountable YBCO electrical joint between are presented.

1. Introduction

The recent development of YBCO superconducting tapes and cabling methods could be a revolutionary development for magnetic fusion. Subcooled YBCO has larger critical current density at high magnetic fields (more than 20 T) than low temperature superconductors (LTS) such as Nb_3Sn and NbTi. Also, YBCO can be operated at higher temperatures than LTS, reducing the thermodynamic cooling cost and allowing resistive loss in the coil joints.

An innovative concept of fusion reactor has been developed in 2012 by the MIT Fusion Reactor Design course, partially inspired by the Vulcan concept design [1]. A comprehensive report of the characteristics of the reactor is expected to be published later this year. The reactor is 3.3 m in major radius and operates in steady state. The magnetic field on axis is 9.2 T. The tritium breeding blanket is entirely made of liquid FLiBe, and the toroidal field (TF) coils are demountable, allowing for vertical replacement of the internal components of the reactor. This maintenance scheme is much faster, easier and cheaper than sector maintenance, and the replaceable parts such as the vacuum vessel can be fabricated off-site with lower tolerances.

We developed a concept of demountable toroidal field coils for a fusion reactor, with preliminary stress and heating simulations. We have designed demountable electrical joints for YBCO stacked tape conductors for the TF coil system, and tested a small scale prototype of the electrical joint.

2. Toroidal field magnets concept and stress analysis

The TF coil system is composed by 18 demountable coils. The shape of the coils is based on the constant tension D-shape. The coil is divided in two parts, as shown in figure 1: a removable upper leg and a stationary lower leg. The legs are joined in the outer midplane, and in the top of the coils. The structure of the coils is made of stainless steel 316LN.

The forces in the outer joint are supported by bolts. A steel tension ring supports the upper leg on the top joint. The forces in the central column are supported by wedging of the TF coils against each other, and by buckling against the central solenoid. The central solenoid is filled with a fiberglass/epoxy plug to reduce the maximum stresses in this region.

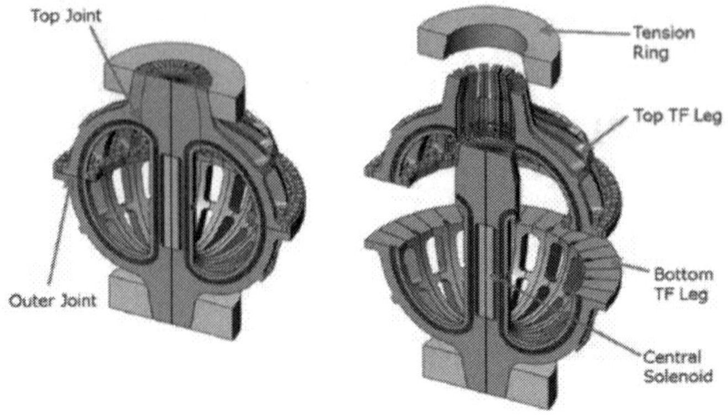

Figure 1. Schematic design of the TF coil system. (Left) assembled; (right) disassembled. The structure is made of stainless steel 316LN, the cables are YBCO conductors. The central solenoid is filled with a fibreglass/epoxy plug to take the compressive loads on the inside.

A finite element method (FEM) simulation of the Lorentz loads in the TF coils was made to estimate the stresses in the structure. The model is half a single TF coil. The load in the coils is due to the Lorentz force in the conductors due to the toroidal magnetic field generated by all 18 coils. Roller boundaries simulate the rotational symmetry, and the contact between the two legs. The simulation results are shown in figure 2. The maximum stress in the structure is approximately 700 MPa and the yield strength of stainless steel 316LN is larger than 1000 MPa [2], which gives a safety factor of approximately 1.4.

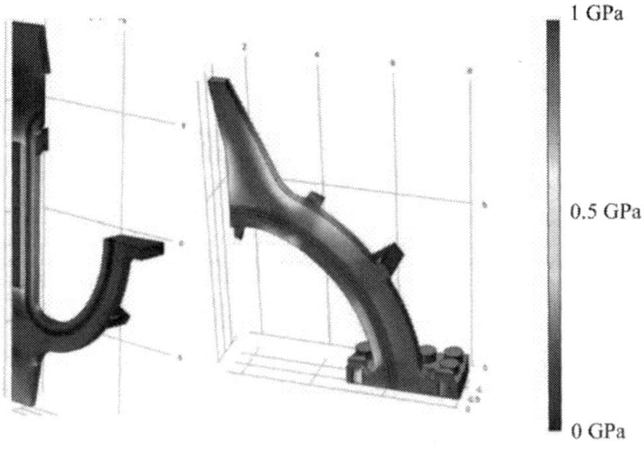

Figure 2. Results of FEM simulation of the stresses in the TF coils. (Left) bottom leg; (right) top leg. The colors represent the von Mises stress. The safety factor is approximately 1.4 in the worst areas.

To provide the 9.2 T on axis, a current of 8.4 MA turn is required in each TF coil. The current is carried by 70 kA YBCO cables; each cable composed by 250 YBCO tapes (25 mm width) assembled using the Twisted Stacked Tape Conductor method [3]. The superconductors are cooled with liquid hydrogen at 20 K. The maximum magnetic field the superconductors are exposed to is approximately 25 T, which gives a critical current margin of approximately 50% (extrapolated from single tape measurements at 25 T and 4.2 K [4]). The cable length is 19 m for the bottom leg, and 7.5 m for the top leg, which allows selecting the segments of each YBCO spool with better performance for the cables located in higher magnetic field areas.

A total of 240 joints in series will be required in each TF coil. The concept for electrical joint between superconductors will be discussed in the next section. Because of the nature of the joint, it is undesirable to have relative motions of its elements. Thus, the structure has to be sufficiently stiff to prevent "scuffing". Some relative motion can be tolerated in the direction of the conductors, but the joint should be designed to prevent having substantial loads perpendicular to them.

3. Demountable YBCO electrical joints concept and preliminary measurements

A demountable "comb teeth" type joint geometry has been developed. The conceptual drawing is shown in figure 3. This geometry offers a good balance between contact area and joint space. A small scale prototype of joint structure for 4 mm wide tape was designed and fabricated using EDM process for the intricate details. This prototype allows joining up to four pairs of YBCO tapes. Figure 4 shows the prototype comb teeth joint with two pairs of YBCO tapes.

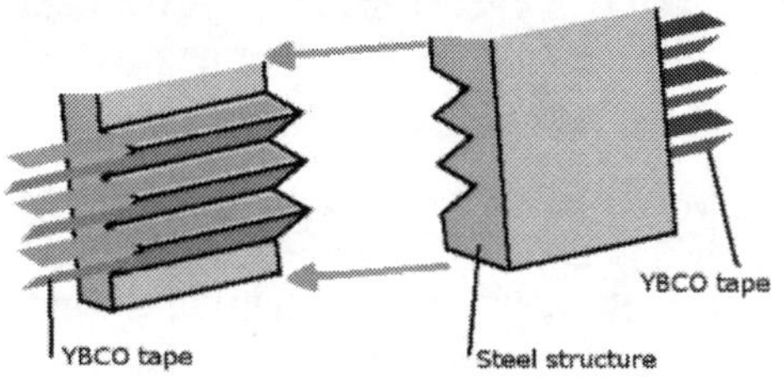

Figure 3. "comb teeth" joint concept. YBCO tapes sit in "comb teeth" in both structures. When pressed together, the YBCO tapes form a "lap" joint.

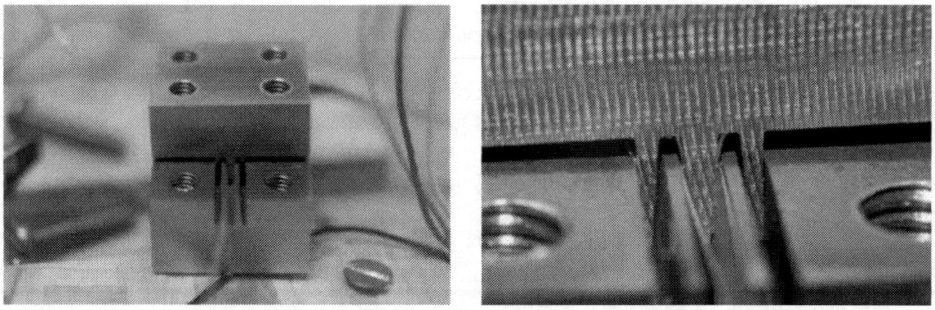

Figure 4. Comb teeth demountable joint device for YBCO tapes. (Left) photograph of a "comb teeth" type joint fabricated by EDM; (right) close-up photograph of the assembled joint, showing details.

Figure 5 shows preliminary results obtained by comb teeth demountable joint devices for a 4-tape device and also a 8-tape device at 77 K. 4 mm SuperPower SCS-4050-AP YBCO tape was used in all measurements. The joint resistances per tape are plotted as a function of the reciprocal of joint length, since the contact resistance is proportional to the reciprocal of the contact surface area. As seen in figure 5 the contact resistances linearly changed with the inverse of joint length; the lowest joint resistance of 25 nΩ was achieved for a 40 mm long joint. The average joint resistance is 30 $\mu\Omega \cdot mm^2$. As seen in the figure it is noted that the uniformity of contact joint resistances is better for shorter length of joints. In this device the very high contact pressure can be obtained by the wedge effect, therefore the low joint resistance could be obtained. However, it was difficult to adjust the pressure and obtain a uniform pressure for all tapes. More work is required to improve the demountable joint device and to scale up.

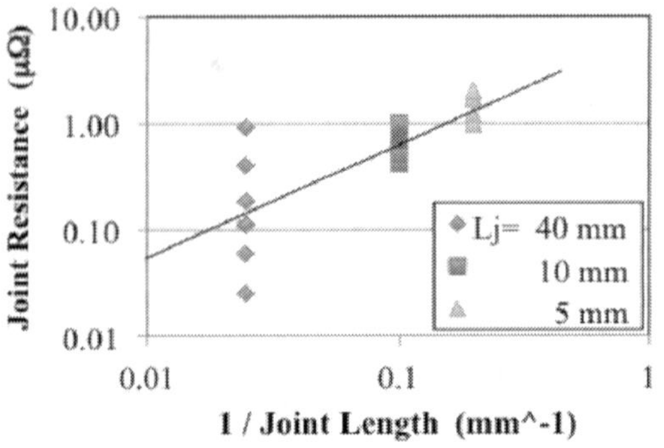

Figure 5. Preliminary results obtained by comb teeth type demountable joint device. Joint resistance plotted as a function of the inverse of joint length. The average contact resistance is 30 $\mu\Omega\cdot mm^2$.

Extrapolating the preliminary results to the TF magnet system described in the previous section, 25 mm wide tape and 10 long joint, a total of 11 kW of heating power is generated in the joints of the entire TF system at the low temperature. To remove that power approximately 800 kW of electricity is required.

To keep the temperature controlled in the joint, each "comb tooth" has a cooling channel in its base, with liquid hydrogen flowing through it. For a 1 mm wide cooling channel in a 2 mm wide comb tooth, with 1 m/s liquid hydrogen flow at 20 K, a 2D FEM simulation showed that the peak temperature is less than 21.3 K in the superconductor.

4. Conclusions

Demountable toroidal field magnets can have a large impact in reducing fusion reactors design limitations. Internal components can be built and assembled off-site and then shipped to the reactor site, allowing for much lower tolerances than on-site building of the first wall, vacuum vessel and other internal components. A maintenance scheme based on vertical replacement of internal components can be much faster, easier and cheaper than sector maintenance.

High field, demountable superconducting TF magnets have been designed for an innovative Fusion Reactor. The magnets would produce a magnetic field of 9.2 T on axis, in a 3.3 m major radius tokamak. A small scale device of its electrical joints between YBCO tapes has been tested. The preliminary measurement of joint resistance is 30 $\mu\Omega\cdot mm^2$ in average.

Further research about the performance of TSTC cables at high magnetic fields (approximately 25 T) and low temperatures (4.2 to 20 K) and large scale measurements of the electrical joints is required to proceed with the development of this magnet system design.

Acknowledgements

We would like to thank B. Sorbom, J. Ball, J. Sierchio, C. Kasten, T. Palmer, D. Sutherland, H. Barnard, C. Haakonsen and the rest of the Spring 2012 MIT Fusion Reactor Design course for their contributions and feedback.

References

[1] Olynyk G *et al* 2012 *Fusion Eng. Des.* **87** 224
[2] Nyilas A *et al* 2004 *Adv. Cryog. Eng. Mat.* **50** 176
[3] Takayasu M *et al* 2012 *Supercond. Sci. Technol.* **25** 014011
[4] Lehner T and Zhang Y 2011 Development, manufacturing and applications of 2G HTS wire at SuperPower *Center for Emergent Superconductivity, Fall 2011 Workshop* (Urbana, USA)

Modeling and measurements of circular and trapezoidal shape HTS coils for electrical machines applications

G. Messina[1], **L. Morici**[1], **U. Besi Vetrella**[1], **G. Celentano**[1], **M. Marchetti**[1], **R.Viola**[1] **and P. Sabatino**[2]
[1] EURATOM-ENEA Association on Fusion, via Enrico Fermi 45, 00044 Frascati, Roma, (Italy).
[2] Università di Salerno, Fisciano (Italy)

Abstract. Axial Flux Electrical Machines (AFEM) with good power-to-weight and diameter-to-length ratio and high efficiency are very attractive for most industrial and power applications. Investigations with both theoretical and experimental methods of ac losses are important for a reliable prediction of dissipation mechanisms in AFEM. In this paper, simulated and measured results for both critical current (I_c) and transport current losses (P_{loss}), obtained on HTS coils, are reported. To investigate shape effects, double pancake coils with variable turns and shapes have been manufacted. Commercial grade $ReBa_2Cu_3O_{7-x}$ (Re = Y or rare earths, ReBCO) tape and epoxy resin has been used for coil winding. A magneto-static 2D finite element model (FEM) for the coils cross section, and a lumped model for AC losses estimations, have been implemented. The agreement among measured and simulated results are satisfactory.

Keywords: HTS (High Temperature Superconductors), AFPM (Axial Flux Permanent Magnets), I_c (critical current), sc (superconductors), HTS 2G (second generation HTS)

1. Introduction

Y-based second generation high temperature superconductors HTS are interesting candidates for AFPM electrical machines applications. To increase electrical loading in the design of Axial Flux Permanent Magnets (AFPM), the use of HTS coils for armature winding is a reasonable choice. HTS coils are typically placed in the stator disk and, in order to maximize the magnetic flux coupling, usually have trapezoidal or circular shapes according to the rotor disk permanent magnet's. HTS tapes AC losses have a significant impact in these applications, and an their estimation gives a better understanding of the overall output power. The Superconductivity Group at ENEA Research Center of Frascati is currently investigating the electromagnetic properties of HTS coils internally manufactured at LN_2 temperature with a particular focus on the AC losses. In this paper we describe a DC FEM code model and a lumped constant AC model for HTS coils. Experimental data have been obtained by electrical methods, and a good agreement with the numerical simulations has been achieved for both DC and AC models.

[1] giuseppe.messina.fra@enea.it, luigi.morici@enea.it.

2. HTS coils preparation

Two different types of HTS coils, A and B (Fig. 1a), have been assembled in double pancake configuration using YBCO Copper laminated certified straight tape Amperium® (4.8 mm x 0.2 mm and self-field critical current Ic>100 A at 77 K). Each coil has been wound around a PVC core side by side with one continuous tape under tension control (2 N); starting at mid-section and so avoiding internal coil electrical joints. Both coils, A and B, have been wet-wound impregnated by diluting Stycast resin in ethyl alcohol 2:1 but with different turns number (10 and 46) and shape (circular and trapezoidal) respectively [1].

Figure 1 (a). HTS coil A and coil B

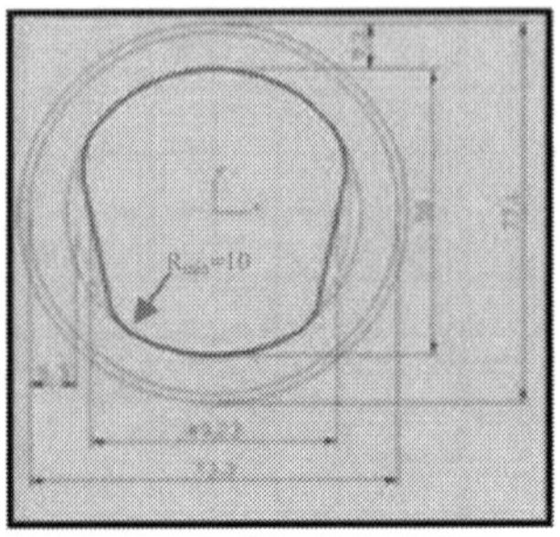

Figure 1 (b). All dimensions in mm. Coil B - height = 11 mm, maximum bending = 10 mm .

The material used for impregnation has been treated according to its own preparation process, high temperature curing has not been performed, all samples being dried at room temperature. During HTS tape winding, the YBCO tape has been outward oriented inter-layering an epoxy resin (Stycast) for electric insulation and mechanical robustness. Coil A configuration parameters are: inner radius (R_i = 20 mm), outer radius (R_{out} = 22.3mm), height (h = 11 mm); coil B dimensions are reported on Fig. 1 (b).

3. DC and AC measurements

To estimate the critical current, I_c, of each HTS coil, current–voltage characteristics at 77 K and self-field condition has been measured with electrical method and the probes placed before any electrical joint. I_c = 100 A and I_c = 75.6 A for coil A and B respectively, have been obtained. AC transport current losses (P_{Loss}) versus rms current (I_{rms}) measures, for each coil at 77 K and various frequencies, have been carried out with the electrical method and are reported in Fig. 2 (a), 2 (b). In particular, P_{Loss} has been obtained multiplying the real rms voltage V_{coil} component, from the lock-in amplifier, with the supplied I_{rms}. An outlook of the obtained results, at a meaningful frequency of 50 Hz, is shown in Fig. 3 (b).

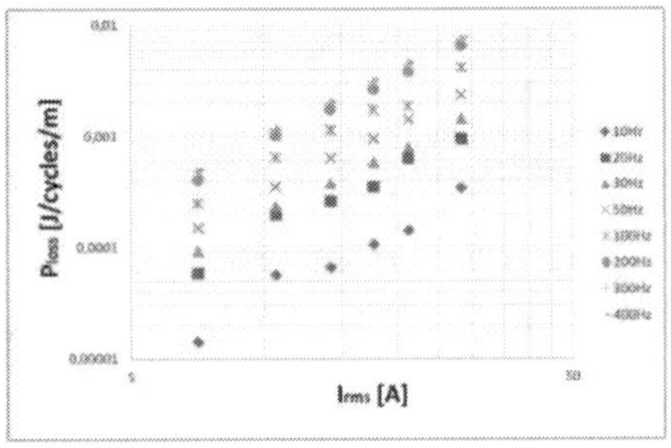

Figure 2 (a). P_{loss} vs I_{rms} @ 77 K and various frequencies for Coil A

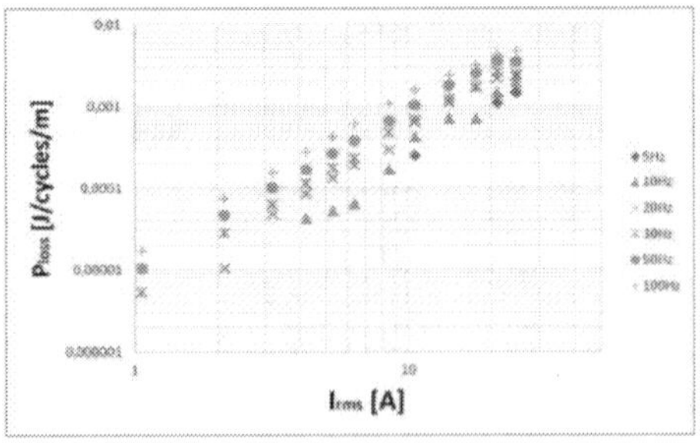

Figure 2 (b). P_{loss} vs I_{rms} @ 77 K and various frequencies for Coil B

4. AC losses lumped model

A simple model for AC losses estimation [2] can be used to explain our measurement results. Hysteretic losses, both of the superconductor and the Ni-W substrate, are modeled by a frequency linearly dependent resistance $R_{sc} = R_0 \cdot f$. The losses on the copper based alloy matrix stabilizer are modeled by a resistance R_{Cu} and the lumped model circuit is shown in Fig. 3 (a). $R_0 = 5 \cdot 10^{-5}$ [Ω/Hz] is the only data fitting parameter in our model and a good agreement with the measured data has been achieved (Fig.3 (b)).

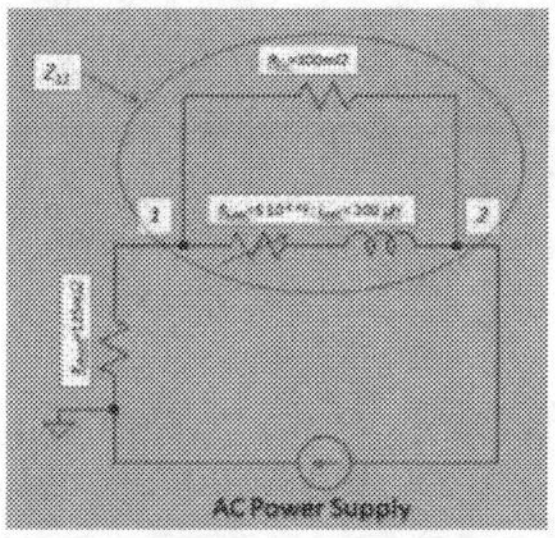

Figure 3 (a). Lumped equivalent circuit for AC losses estimation

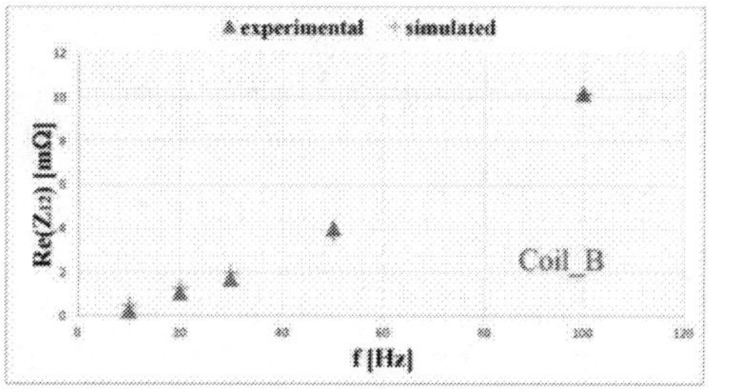

Figure 3 (b). $Re(Z_{12})$ vs frequency @ 77 K and $I_{rms} = 35$ A for Coil B

5. FEM model: 2D approximation

An YBCO double pancake model is implemented using a finite element software (COMSOL Multiphysics 4.3). A 2D axial symmetrical H-formulation has been applied as in Ref.[1]. Tape characteristic properties have been used to improve experimental and data agreement [1].

An elliptical composition of parallel and perpendicular behavior, both of the critical current and the *n*-value, have been introduced for magnetic anisotropic YBCO tapes modeling:

$$I_c(B,\theta) = \sqrt{I_{c\perp}(B)^2\cos^2\theta + I_{c\parallel}(B)^2\sin^2\theta}$$

and

$$n(B,\theta) = \sqrt{n_{\perp}(B)^2\cos^2\theta + n_{\parallel}(B)^2\sin^2\theta}$$

$\theta \equiv$ angle between *c*-axis (i.e. current direction) tape and magnetic induction B. *n*-value, self-field and temperature dependencies of critical current have not been modeled.

The 2D axial symmetrical geometry ((*r*, *z*) cylindrical coordinate system describing coil cross section (Fig.4) has coil current I_ϕ parallel to the electric field $E_\phi = \rho J_\phi$. The parameter ρ is a position function given by (3) inside the material, while on the outside $\rho = 10^5$ Ωm (to speed up numerical calculations, air resistivity [3] was chosen lower than the real value and much higher than the HTS resistivity in dissipative regimes). The YBCO tape *E-J* power law is:

$$\rho(B,\theta) = E_c/J_c(B,\theta) \times \left(J_\phi/J_c(B,\theta)\right)^{n(B,\theta)-1} + \rho_0 \quad (1)$$

where a ρ_0 value of 10^{-16} Ωm, i.e. the Thermally Activated Flux Flow resistance (TAFF) at 77 K of HTS 2G tape [6] has been used in our DC simulations. A code computed I-V curve of double pancake coil has been carried out to test the model. In the 2D simulation, the current is ramped up to a value I and then kept constant for 1 s (in order to relax the calculated voltage). As can be seen in Fig.3, the experimental data $I_c = 75.6$ A are roughly consistent with simulation data $I_c = 78.6\,A$; discrepancies are due to the flattening of the coil section along the x-coordinate (Fig. 1(b)) producing a strength magnetic field higher than the corresponding cylindrical configuration used in the simulation. In any case, the approximation being quite good (only a 4% error on I_c) validates our model and opens the possibility for AC losses estimation. Finally, the total magnetic field B, when the supplied current is equal to the critical value of coil_B, is shown in Fig.4.

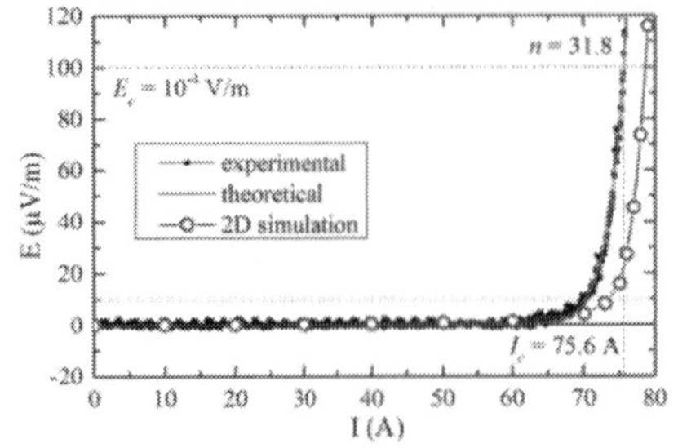

Figure 3. Comparison between the I-V curve measured, theoretical and simulated in 2D for coil_B at T = 77 K

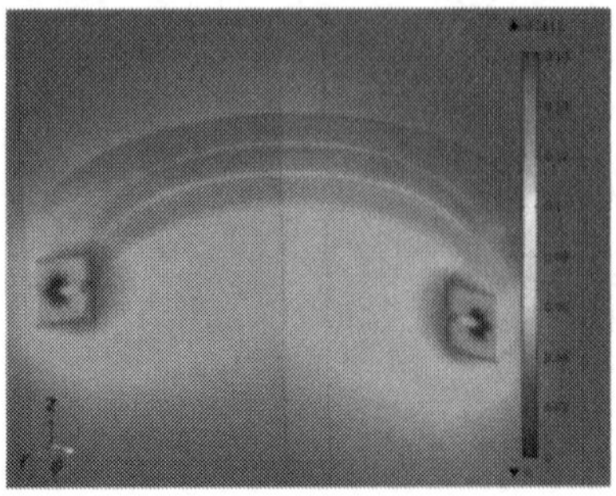

Figure 4. Snapshot of the magnetic field intensity B (T) simulated for coil_B at $I=I_c$

6. Conclusion

In this paper, an improved 2D finite element model, based on the H-formulation and implemented in COMSOL Multiphysics, and a lumped model have been described. The models were used to investigate the electromagnetic properties and calculate the AC losses in different shape HTS coils Stycast wet-wound impregnated. The transport current losses and the critical current density of the coils were measured. Experimental and simulation results are in good agreement. Consequently, the results can be used for AC losses computation in cases of AFPM electrical machines with HTS armature winding.

7. References

[1] U.Besi Vetrella, G. Celentano, M. Marchetti, G. Messina, L. Morici, P. Sabatino, R. Viola and A. della Corte, "HTS coils fabrication and commercial 2G YBCO tapes: measurements and simulation", *IEEE Trans. Appl. Supercond.* vol24, no.3, pp. 4600204-1-4600204-4, June 2014

[2] G. Messina, L. Morici, U.Besi Vetrella, G. Celentano, M. Marchetti, P. Sabatino, R. Viola, "Transport current losses in HTS coils for Axial Flux Electrical Machines applications", *IEEE Trans. Magn.* to be published.

[3] M. Zhang, J. Kim, S. Pamidi, M. Chudy, W. Yuan, T. Coombs, "Study of second generation, high-temperature superconducting coils: Determination of critical current", J. Appl. Phys., vol. 111, 083902 (8pp), April 2012.

[4] E. Pardo, M. Vojenčiak, F. Gömöry, J. Šouc, "Low-magnetic-field dependence and anisotropy of the critical current density in coated conductors", Supercond. Sci. Technol., vol. 24, 065007 (10pp), April 2011.

[5] K. Berger, J. Lévêque, D. Netter, B. Douine, A. Rezzoug, "Influence of Temperature and/or Field Dependences of the E–J Power Law on Trapped Magnetic Field in Bulk YBaCuO", IEEE TAS, vol. 17, pp.3028-3031, June 2007.

[6] S. Stavrev, F. Grilli, B. Dutoit, N. Nibbio, E. Vinot, I. Klutsch, G. Meunier, P. Tixador, Y. Yang, and E. Martinez, "Comparison of numerical methods for modeling of superconductors," *IEEE Trans. Magn.*, vol. 38, no. 2, pp. 849–852, Mar. 2002.

Armature reaction effects on a high temperature superconducting field winding of an synchronous machine: Experimental results

Nenad Mijatovic and Bogi Bech Jensen

Department of Electrical Engineering, Technical University of Denmark, Lyngby 2800 DK

E-mail: nm@elektro.dtu.dk and bbj@elektro.dtu.dk

Abstract. This paper presents experimental results from the Superwind laboratory setup. Particular focus in the paper has been placed on describing and quantifying the influence of armature reaction on performance of the HTS filed winding. Presented experimental results have confirmed the HTS field winding sensitivity to both armature reaction intensity and angular position with respect to the HTS coils. Furthermore, the characterization of the HTS field winding has been correlated to the electromagnetic torque of the machine where the maximal I_c reduction of 21% has been observed for the maximum torque.

1. Introduction

Superconducting (SC) machines have been proposed as direct drive generators for large wind turbines due to their high torque density and very limited, if any, dependance on rare earth elements [1, 2, 3]. Dating from 1950's, the number of large scale application which have tried to exploit potential benefits of SC machines have risen, a few of which are utility generators, ship propulsion drives and industrial motors. For each of the mentioned application a particular SC machine topology has been mostly cited and regarded as an optimal [4]. It seems that multi-pole wound field synchronous machine with SC field winding and conventional copper armature has been the most consistent topology proposed for wind turbine direct drive generators [5]. Since SC field winding is in its essence an electromagnet, rules and guidelines used for SC magnet design have often been the starting point of SC machine design process. Design would involve following steps: required MMF for a given magnetic flux density in air (a load line) is plotted against $I_c(B,T)$ for specific SC after which magnet designer would determine operating current and temperature of SC winding based on the thermal stability and other constrains of the design. If this approach is likewise followed in the SC machine design, magnetic field coming from the AC winding (also known as armature reaction) could compromise stability of the SC field winding which often is not emphasize strongly enough.

The choice of operating current presents a tradeoff between thermal stability of SC and the cost optimization of the system. On the other hand, thermal stability of the SC field winding, or how much SC coils are pron to a quench among other parameters can be correlated to I/I_c ratio and to an appropriate critical current evaluation of SC coils [6]. 30-60% value of I_c is usually regarded as a safe operating current with sufficient current margin for SC coils [7], depending on which criteria I_c has been derived from. Thus, proper I_c evaluation of SC coils in the context of the application is of great importance. Having that in mind we have used the Superwind setup [8, 9], to evaluate the impact of armature reaction on I_c of a HTS field winding derived from voltage measurements, both quantitatively and qualitatively.

a)

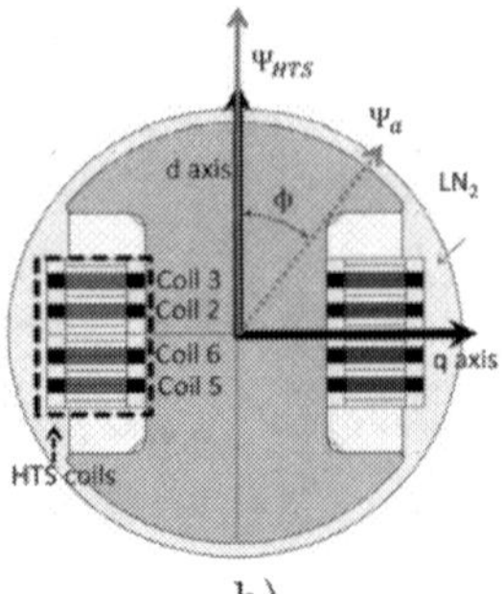

b)

Figure 1. (a) 3D CAD illustration of rotor locking add-in in Superwind setup. ;**(b)** Illustration of the 2D cross section of the HTS field winding assembly with field flux axis colored in blue and armature flux axis in red. Coil ID numbers are identified.

2. Materials and Methods - Superwind setup

Superwind setup [9], has been constructed with the intention to test and characterize verity of HTS coils in field winding of synchronous machine under various conditions. Six epoxy impregnated HTS coils have been constructed and tested at liquid nitrogen (LN_2) boiling temperature (77K). Three coils have been wound with BSCCO HTS and other three coils with YBCO HTS tape [3]. However, due to availability, in the experiment described here only four coils have been used, three BSCOO and one YBCO. Focus of the experimental investigation was to determine the influence of armature reaction on operating condition of HTS winding. Hence, a static locked rotor experiment has been performed. In the experiment, magnetic field of the armature winding is controlled with phase currents. A DC current flows through two phases, phase A and B, connected in series with third phase disconnected. In this way, armature reaction is aligned with disconnected phase while the value of it can be correlated with three phase operation making it fully defined. If the DC current in experiments is $I_{DC} = \sqrt(2) \cdot I_{RMS} \cdot sin(\frac{\pi}{3})$, where I_{RMS} is a RMS value of three phase operation, the armature reaction will have the same value. Field flux has been controlled by the field current control of 4 serial connected HTS coils. The condition monitoring of the HTS field winding was performed by measuring the voltages of individual coils and coil segments. In addition to HTS diagnostics, the armature winding was equipped with an absolute position sensing and torque measurement at the cryostat. Hence, the state of HTS field winding could be correlated to the mechanical output of electrical machine, i.e. electromagnetic torque. The illustration of Locked rotor experimental setup is shown in 1(a) and 1(b) where armature position is controlled by linear adjustment of a threaded rod. Point A is the point of rest while point B can move along the rod resulting in a controlled angular displacement of the armature. The 1(a) illustrates an angular shift of $\approx 45^o$.

3. Experimental results

In the set of HTS coils, two BSCCO coils, Coil 3 and Coil 5, placed as shown in 1(b), have exhibited the lowest critical current from the stack of HTS coils, as presented in 2(a). The critical current of the field winding was determined by comparing the average electrical field of

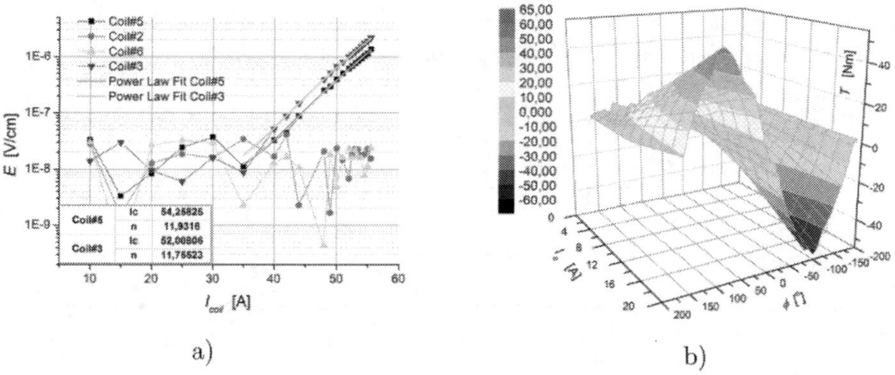

Figure 2. **(a)** IV curves of 4 HTS coils with no armature current. The critical current of the field winding is 52A and is set by Coil 3; **(b)** 3D plot of a measured electromagnetic torque as a function of armature current and angle between field and armature fields while current of HTS field winding is 50A.

coils to $E_0 = 1\mu V/cm$. Hence, the critical current of the field winding in the No load (absence of the armature reaction) is 52A and is set by Coil 3.

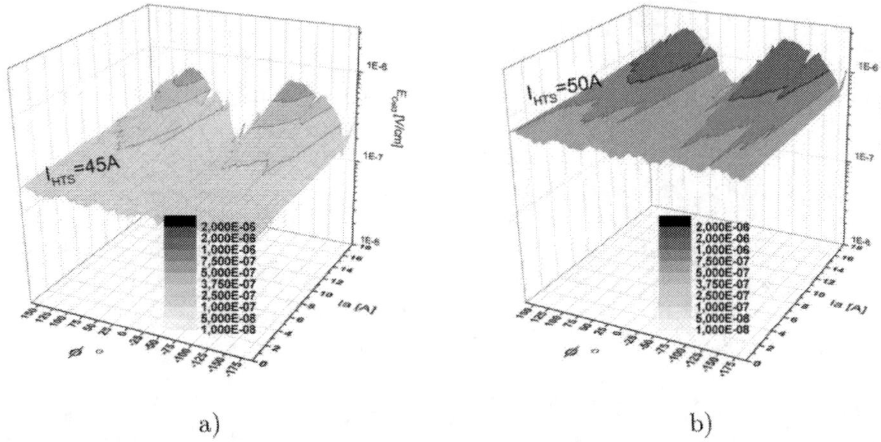

Figure 3. Average electric field of the Coil 3 as a function of armature current I_a and angle between field and armature flux axes, ϕ while current of HTS field winding is **(a)** $I_{HTS} = 45A$; **(b)** $I_{HTS} = 50A$

With the increase of armature reaction, the field condition of the HTS field winding will change which is expected to result in the change of operating current. Indeed, experimental results shown in 3, where the average electric field of HTS coil is shown as a function of both armature reaction intensity (value of the DC current) and the angle between the armature reaction flux and field flux for a constant current of field winding, validate the expectations. Average electrical field of the coils is not necessarily increasing with the increase of armature reaction but has two maximums at dedicated angles $\phi \approx \pm 90^o$, where the electromagnetic torque has its extremes likewise, as seen in 2(b). Having this in mind, we can conclude that the largest reduction of critical current occurs when machine develops maximal electromagnetic torque per Amp of armature current. The I_c for loaded machine (for maximal torque, i.e. $\phi = 90^o$) is presented at 4. For rated armature reaction, which corresponds to I_{DC}=35 A, the critical current for the maximal developed torque is 40 A, which is 20% lower value compared to a No Load.

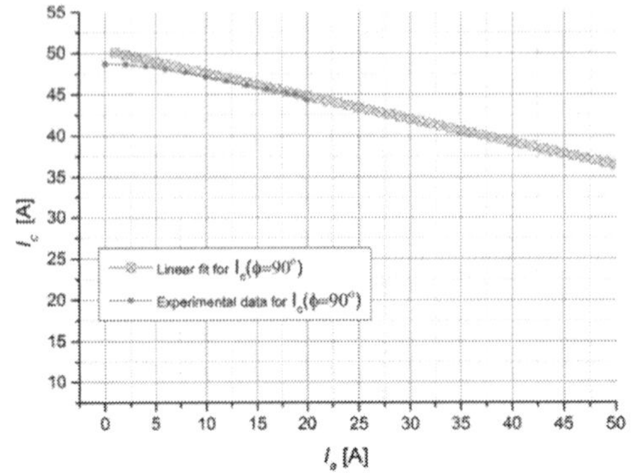

Figure 4. The critical current of the field winding as a function of armature current I_a for $\phi = 90^o$

4. Conclusion

Armature influence on operating performance of SC field winding stems from magnetic field sensitivity of the HTS tape. Based on experiments shown in this paper, magnetic anisotropy was dominant source of armature reaction impact to the HTS field winding since the value of I_c was not affected by the armature reaction at $\phi = 0^o$, as seen at 3. The reduction of critical current without and with the armature reaction can be significant, and in the case of the Superwind setup was found to be as much as 20%. It should be noted that the performed experiments were static, which implies absence of AC loss in HTS winding from slot harmonics or other sources, all of which would additionally decrease the I_c. Thus, the impact of armature reaction on SC field winding can be very significant and should be taken into account in design phase of SC machine.

References

[1] Abrahamsen A B, Jensen B B, Seiler E, Mijatovic N, Rodriguez-Zermeno V M, Andersen N H and Østergaard J 2011 *Physica C: Superconductivity* **471** 1464–1469
[2] Abrahamsen A B, Mijatovic N, Seiler E, Sorensen M, Koch M, Norgard P, Pedersen N F, Træholt C, Andersen N H and Ostergard J 2009 *Applied Superconductivity, IEEE Transactions on* **19** 1678–1682
[3] Abrahamsen A B, Mijatovic N, Seiler E, Zirngibl T, Træholt C, Nørgård P B, Pedersen N F, Andersen N H and Østergaard J 2010 *Superconductor Science and Technology* **23** 034019
[4] Kalsi S S, Weeber K, Takesue H, Lewis C, Neumueller H W and Blaugher R D 2004 *Proceedings of the IEEE* **92** 1688–1704
[5] Jensen B B, Mijatovic N and Abrahamsen A B 2013 *Journal of Renewable and Sustainable Energy* **5** 023137
[6] Wilson M N and Wilson M 1983 *Superconducting magnets* vol 198 (Clarendon Press Oxford)
[7] Fair R 2012 *Superconductivity for Large Scale Wind Turbines* URL `http://www.osti.gov/scitech/servlets/purl/1052970`
[8] Mijatovic N, Jensen B B, Traholt C, Abrahamsen A B, Rodriguez-Zermeno V M, Petersen N F, Henriksen M, Seiler E, Andersen N H and Ostergard J 2011 High temperature superconductor machine prototype *Electrical Machines and Systems (ICEMS), 2011 International Conference on* (IEEE) pp 1–6
[9] Mijatovic N, Jensen B B, Træholta C, Abrahamsen A B, Zermeno V M and Pedersen N F 2012 *Physics Procedia* **36** 771–776

Space Radiation Superconducting Shields

R.Musenich[1], V.Calvelli[1], S.Farinon[1], W.J.Burger[2], and[1] R.Battiston[3]
[1]INFN, via Dodecaneso 33, 16146 Genoa, Italy
[2]Dipartimento di Fisica, Università di Perugia, Perugia, Italy
[3]INFN-TIFPA and Dipartimento di Fisica, Università di Trento, Trento, Italy

E-mail: riccardo.musenich@ge.infn.it

Abstract. The interest on shields to protect astronauts I long term missions against GCR has recently grown and several projects have been funded. Due to their large mass, passive shields for large volume habitable modules are no longer an option and the attention is focused on the more complex, technologically challenging active systems. Among the possible solutions, the most promising is based on huge superconducting coils having a bending power sufficient to deflect out of the habitat charged particles with kinetic energy in the order of 1 GeV. Toroidal magnet systems based wound with Ti clad MgB_2 conductor is proposed and described.

This work is supported by INFN (SR2S-RD) and co-funded by the EU FP7 SR2S Project.

1. Introduction

It is well known that deep space is a hostile environment for human beings: although the spacecraft habitable modules are designed to supply astronauts with all the necessary for their life and to protect them by the sun light and micrometeorites, shielding of the cosmic rays is still an open problem. Cosmic rays, high energy, massive particles of solar or galactic origin, represent a serious threat for the health of crews of future, long term missions in the deep space. The adsorbed dose during a solar particle event (SPE) would be fatal by acute radiation syndrome while the continuous flux of galactic cosmic rays (GCR) exposes the crew to a high risk of develop a cancer [1]: the measurements taken during the voyage of the Mars Science Laboratory show that the dose, due to the GCR, adsorbed by an astronaut into a non-shielded spacecraft is 1.84 mSv/day [2]. SPE are very intense, short bursts of relatively low energy (1-100 MeV) particles, mainly protons, so that it is possible to protect the astronauts recovering them in a passive shelter, made by a material of sufficient thickness, for the duration of the event. A protection from CGR is much more difficult to design as they are composed by protons and ions having a wide energy spectrum with its maximum around 500 MeV/nucleon. Moreover, as the flux is continuous, a CGR shield must surround the whole habitat. As long trips require large habitable volumes, the mass of passive shields is very high and (several hundred tons), so the attention is focused on active systems, which require electric or magnetic field to deflect the charged particles [3], [4].

[1] To whom any correspondence should be addressed.

Among the possible solutions, the most promising is based on huge toroidal magnets having a bending power sufficient to deflect out of the habitable module charged particles with kinetic energy in the order of 1 GeV. The operation environment requires a system fulfilling peculiar characteristics: first of all it must be light and power saving, so it is mandatory to use superconducting coils. Moreover, it must be stable and must safely guarantee an endurance of at least 2 years. These requirements orient the choice of the conductor towards HTS coated conductors or Titanium clad MgB_2.

2. The Superconducting Magnet System

The Space Radiation Superconducting Shield is composed by a large barrel toroid surrounding the habitable module as schematically shown in figure 1. The rear side of the spacecraft is passively shielded by the engine and propellant tanks. The front can be shielded either passively by a module containing equipments and supplies or actively by an end cap toroid.

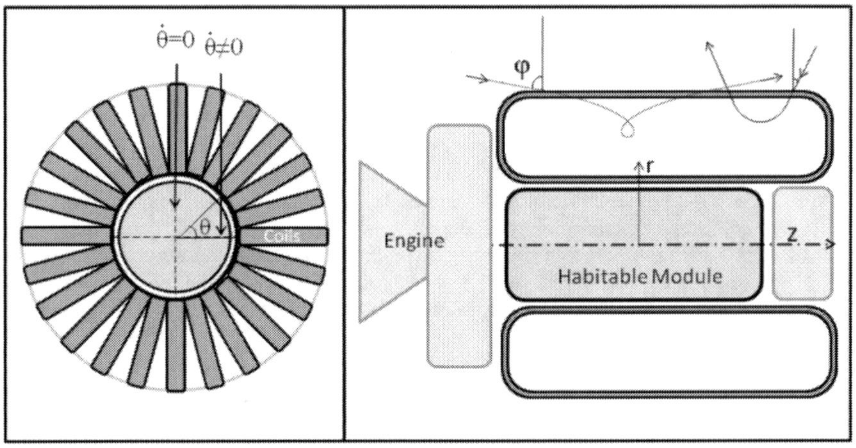

Figure 1. Schematic view of the Space Radiation Superconducting Shield. The trajectories of two particles with different angle of incidence φ are shown.

In the case of a particle moving in the r-z plane, i.e. having zero angular velocity, the motion equations in an ideal toroidal field have an analytical solution [5]. It can also be demonstrated that a particle with zero angular speed is the most penetrating one, so the shielding power of the toroid can be written as:

$$\Xi = \int_{R_i}^{R_e} B_\vartheta \, dR = \frac{\mu_0 I}{2\pi} \ln \frac{R_e}{R_i}. \quad (1)$$

The shielding power can also be written as a function of the particle properties and of the angle of incidence φ:

$$\Xi = \frac{m_0}{q} c\sqrt{\gamma^2 - 1}(1 - \sin\varphi), \quad (2)$$

where m_0 is the rest mass, q the charge and γ is the Lorentz factor. The analytical solution allows calculating the cut-off energy that, in the worst case, i.e. when $\varphi = -\pi/2$, is:

$$K^* = \frac{m_0}{\eta} c^2 \left(1 - \sqrt{\frac{q\Xi}{m_0 c^2 (1 - \sin\varphi)}}\right), \quad (3)$$

where η is the number of nucleons. Assuming an isotropic flux and using the spectra of the CREME database [6], an ideal, infinitely long toroid, with Ξ=5 T·m is able to shield almost 80% of the particle. In case of a real toroid, we have to take into account the azimuthal ripple of the magnetic field.

However, if the number of coils is large enough (≥12 if the internal radius is less than 3 m), the shielding power is not affected by the ripple and the approximation of ideal toroid can be used [5].
Starting from a cylidrical habitat 4 m diameter, 10 m long, our choice is oriented toward a magnet composed of 24 racetrack coils having a winding pack 0.55 m wide and 0.05 m thick as shown in figure 2. The dimensions have been chosen in order to minimize the stray field in the cabin and the maximum field at the conductor.

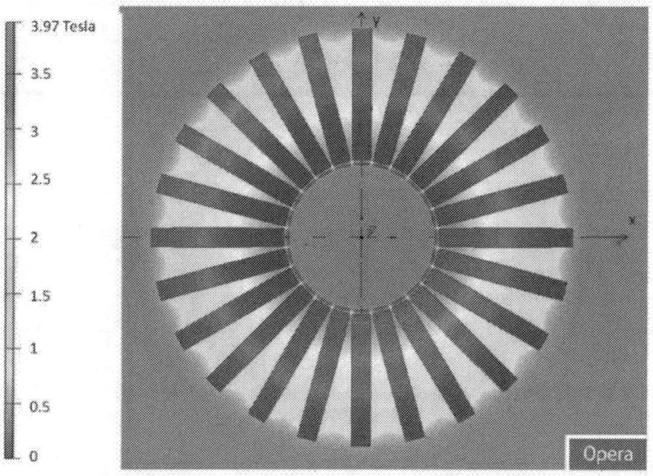

Figure 2. Field map. The toroid is designed to keep the peak field at the conductor less than 4 T and to minimize the stray field in the habitable module.

The outer radius is determined by the cargo launcher size. A heavy launcher with a 9.2 m diameter, 17 m long cylindrical hold, called SLS block II is planned to be operative in the thirties [5]. This opens up two possible scenarios: a compact magnet to be sent in a single launch and a magnet divided in sectors to be separately launched and assembled in orbit. As shown in figure 3, dividing the magnet in 4 sectors the outer radius will be 6.3 m, while a division in 6 sectors leads to 9.5 m.

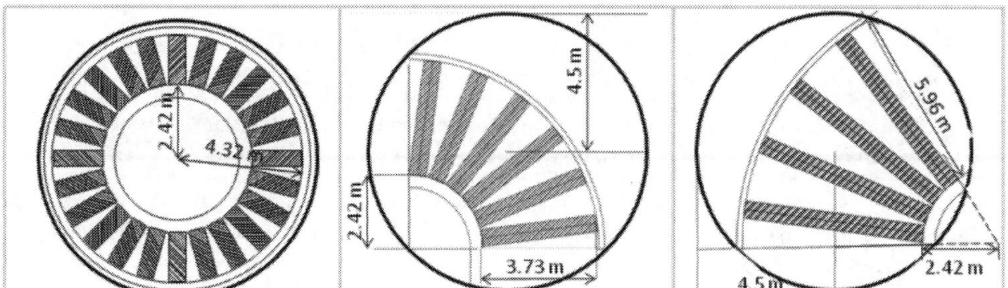

Figure 3. Scheme of the magnet inside the launcher hold (bold circle). Left to right: entire magnet, a quarter and one sixth.

A further subdivision increases the complexity of the system without a significant gain in terms of shielding power. Table 1shows the shielding power and the cut-off energy of the 3 different solutions obtained limiting to less than 4 T the maximum flux density at the conductor.

Considering only the deflection of the incoming particles due to the magnetic field, the shielding power of the compact magnet seems to be sufficient to protect a crew for more than two years, however, monte carlo simulation have shown that a large part of dose is due to the secondary particle generated by the interaction of CGR with the materials [8].

Table 1. Characteristics of the different magnet solutions.

Sectors No.	Outer diameter (m)	Shielding power (T·m)	Cut-off energy (MeV)
1	8.6	5.4	300
4	12.3	8.3	645
6	16.8	11.4	1000

3. Conductor

The conductor is a Ti clad MgB_2 tape, bonded with pure aluminum strips, having 10 mm^2 section area (15% MgB_2, 40% Ti, 45% Al). Such a conductor is light, 3400 Kg/m^3, and allows avoiding helium cryogenics. Solid hydrogen contained in tanks connected with the coils can be used both as cryogen and as passive radiation shield. Conductor prototypes are at present under development at Columbus Superconductors in Genoa, Italy.

The current density being limited to $J=70$ A/mm^2, the total conductor length ranges between 1479 Km and 1770 Km, while the masses of the windings are 49 to 58 tons including the insulation depending on the configuration.

4. Conclusions.

In the perspective of interplanetary manned missions in the near future, superconducting magnets are a serious option to protect crews from cosmic rays. Possible solution have been devised based on MgB_2 based conductors.

5. References

[1] F.A.Cucinotta and M.Durante, "Cancer Risk from Exposure to Galactic Cosmic Rays: Implications for Space Exploration by Human Beings" *Lancet Oncol.* **7**, 431-435, 2006.

[2] C. Zeitlin et al. "Measurements of Energetic Particle Radiation in Transit to Mars on the Mars Science Laboratory", *Science* **340**, 1080-1084, 2013.

[3] Hoffman, J.A., Fisher, P. and Batischev, O., "Use of Superconducting Magnet Technology for Astronaut Radiation Protection", NIAC Phase I Final Report, Contract CP 04-01, 2005.

[4] P.Spillantini et al., "Shielding from Cosmic Radiation for Interplanetary Missions: Active and Passive Methods", *Radiation Measurements,* **42**, 4 – 23, 2007.

[5] R.Battiston, W.J.Burger, V.Calvelli, V.I.Datskov, S.Farinon, R.Musenich, "Superconducting Magnets for Astroparticle Shielding in Interplanetary Manned Missions" *IEEE Trans. on Appl. Supercond.* **23** (3), Art. No. 4101604, 2013.

[6] Cosmic Ray Effects on Micro Electrons, https://creme.isde.vanderbilt.edu/CREME-MC.

[7] http://www.nasa.gov/pdf/664158main_sls_fs_master.pdf

[8] W.J.Burger et al., "Active Magnetic Shielding for Long Duration Manned Space Missions". to be published in the Proceedings of the 6th IAASS Conference, Montreal, 2013.

Cryogenic test facility of superconducting magnets for the accelerator complex NICA

D Nikiforov, N Agapov, H Khodzhibagiyan, N Emelianov and V Korolev
Veksler and Baldin Laboratory of High Energy Physics, Joint Institute for Nuclear Research, Dubna, 141980, Russia

Abstract. The new accelerator complex NICA is being constructed on the basis of the existing superconducting synchrotron Nuclotron at JINR. NICA will consist of 3 rings of superconducting magnets: the Nuclotron, a booster synchrotron and a two-aperture collider ring. Superconducting dipole and quadrupole magnets for the NICA complex will be manufactured and tested at the new test facility. All the magnets contain a cold iron yoke and saddle-shaped superconducting windings made of the hollow NbTi composite superconducting cable refrigerated by two-phase helium flow at 4.5 K. The cryogenic system of the new facility is based on helium refrigerators with excess return flow. The scheme of the cryogenic testing area of the test facility is presented. A mathematical model of the cooling-down processes of the magnets was developed based on the experimental data for the prototypes of the booster and collider superconducting magnets. The calculated parameters allow us having a shorter time of cooling-down and warming-up of the magnets, thus reaching the necessary productivity of the facility.

1. Introduction

A superconducting accelerator Nuclotron has been operated at the Joint Institute for Nuclear Research (JINR) since 1993 up to now. The project of the new superconducting accelerator complex NICA (Nuclotron-based Ion Collider fAcility) is being developed on the basis of the Nuclotron and planned to have been built by 2017 at the Veksler and Baldin Laboratory of High Energy Physics (LHEP) of JINR [1]. To create this project it is necessary to produce and to test 266 superconducting magnets. The new test bench will be created for these purposes.

2. New test facility

The main purpose of the new test bench is to organize the conveyer production of the quadrupole and dipole superconducting magnets for booster and collider accelerator rings. The magnet design allows having a high rate of cooling and a relatively small distribution of temperature gradient in the magnet yoke. The first dipole magnets for the NICA booster was manufactured and tested in 2011. These magnets are 2.2 m long, weight 1020 kilogram and have a radius of curvature of about 14 m. A full-scale model of the quadrupole magnet for the NICA booster was manufactured at the LHEP in December 2011 [2]. The cryogenic tests of the twin-aperture dipole magnet for the NICA collider were successful performed in June 2013.

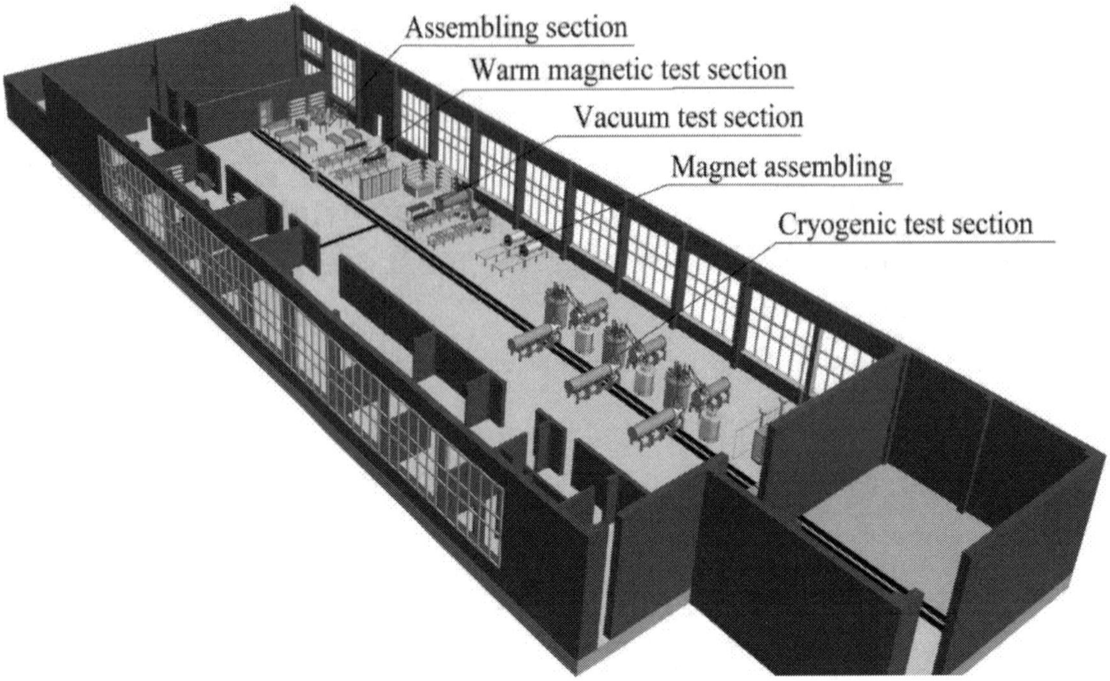

Figure 1.The model of the new test facility.

The new test facility consists of the following sections: assembling section, warm magnetic test section, vacuum test section and cryogenic test section (see Figure 1.) The first section is intended for assembling iron yoke with superconducting coils. The cooling tubes are soldered to the yoke workpiece. The yoke workpieces of the magnets undergo warm magnetic measurements at the next section**.** The main goal of the "warm" magnetic measurements is the assessment of the quality of the magnetic field in the gape of the magnet. At the vacuum section the magnet is inserted into the vacuum shell and the mentioned assembly is vacuum – checked. In addition, during the vacuum tests the hydraulic resistance of the cooling channels is tested. After the magnet has been tested, it goes to the cryogenic section. These tests are carried out at the operating temperature and the maximum currents of the superconducting magnet.

3. Cryogenic test bench

The cryogenic system of the new test stand should provide high reliability of production and uninterrupted operation and implement the necessary performance of the stand. The system was developed based on satellite helium refrigerators for the low-temperature tests of the superconducting magnets. The throttle-type satellite helium refrigerator is one of the most reliable cryogenic plants [3]. Therefore, they have been selected for the cryogenic system of the new test facility. The low-temperature test bench includes: satellite helium refrigerator (1), compressed helium pipeline (5), return helium pipeline (6), liquid nitrogen pipeline from the cryogenic storage tank (7), liquid helium pipeline from the Dewar vessel (2), current leads cryostats (3), and superconducting magnet cryostat (4) (see Figure2.).

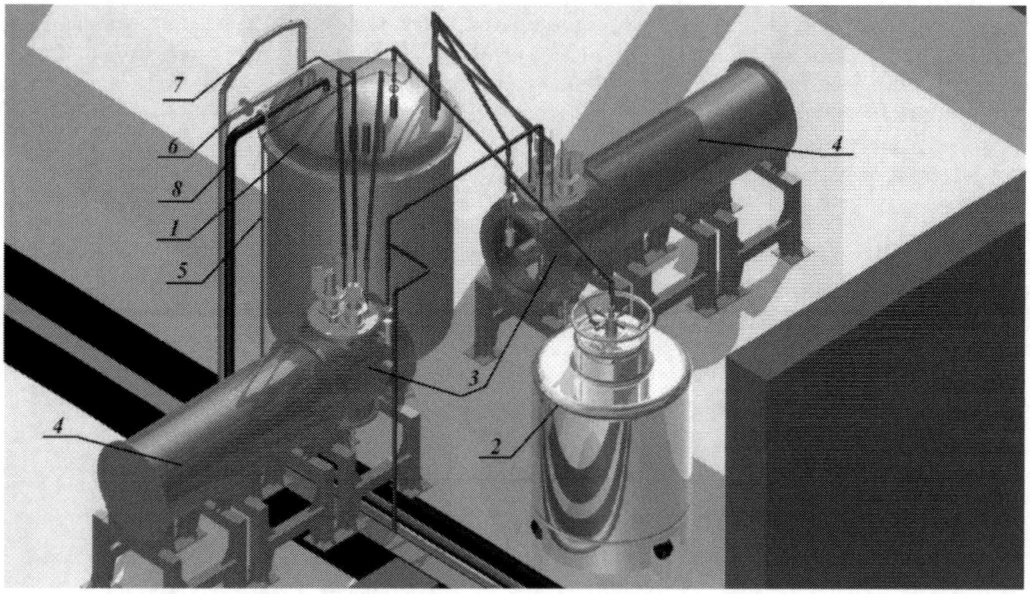

Figure 2.Scheme of cryogenic testing area: 1 – satellite helium refrigerator; 2 – Dewar vessel; 3 – current leads cryostats; 4 – superconducting magnet cryostats; 5 – compressed helium pipeline; 6 – return helium pipeline; 7 – liquid nitrogen pipeline from the cryogenic storage tank; 8 – return nitrogen pipeline.

The satellite helium refrigerator cools down the superconducting magnet from 300 K to 4.5 K, provides refrigerating at the operating temperature of about 4.5 K and warms up from 4.5 K to 310 K. The cooling capacity of satellite helium refrigerator is 100 W. The cooling of superconducting magnet proceeds in two stages. The first stage is cooling of the magnet by helium flow at the temperature of 80 K. At the first stage the cooling capacity is provided by the nitrogen bath of the satellite. The nitrogen bath is fueled with liquid nitrogen from a cryogenic storage tank. The second stage begins when the output temperature of the helium from the magnet is 120 K. At this stage of the magnet cooling the helium flow temperature is 4.5 K. The second stage of cooling is continued until the temperature of the iron yoke is 4.5 K. Cryostatting is to ensure a constant temperature of the magnet at different heat loads during operation. After the test, the magnet should be warmed up to a temperature of about 310 K.

The new test bench will use high temperature superconducting (HTS) current leads. These HTS current leads can be operated at high intensity current up to 18 kA. They will be operated in pulsed mode. HTS current leads have three cooling stages. The top of the HTS current lead is cooled with water. This is to maintain the positive temperature of the top of the HTS current lead. The middle of the current lead is cooled by liquid nitrogen from the satellite helium refrigerator. The cold gaseous nitrogen flowing to the top heats up and its temperature reaches about 290 K. The lower part of the HTS current lead is cooled by liquid helium.

4. Mathematical model

The mathematical model simulated the thermo-physical processes occurring during the cooling of the superconducting magnets from 300 K to 4.5 K. For modeling, it is necessary to know the design and materials of the magnet and thermo-physical properties of the coolant.. As a result, we obtain the evolution of the coolant and material magnet temperatures in each cross-section with a time interval of one second.

The superconducting dipole magnet of the booster was tested. This magnet is 2.2 m long and has a radius of curvature 14 m and weighs about 1 ton. The experimental temperatures of helium in the three main points during the cooling of the dipole magnet are shown in Figure 3. (points). The simulated processes are shown by lines.

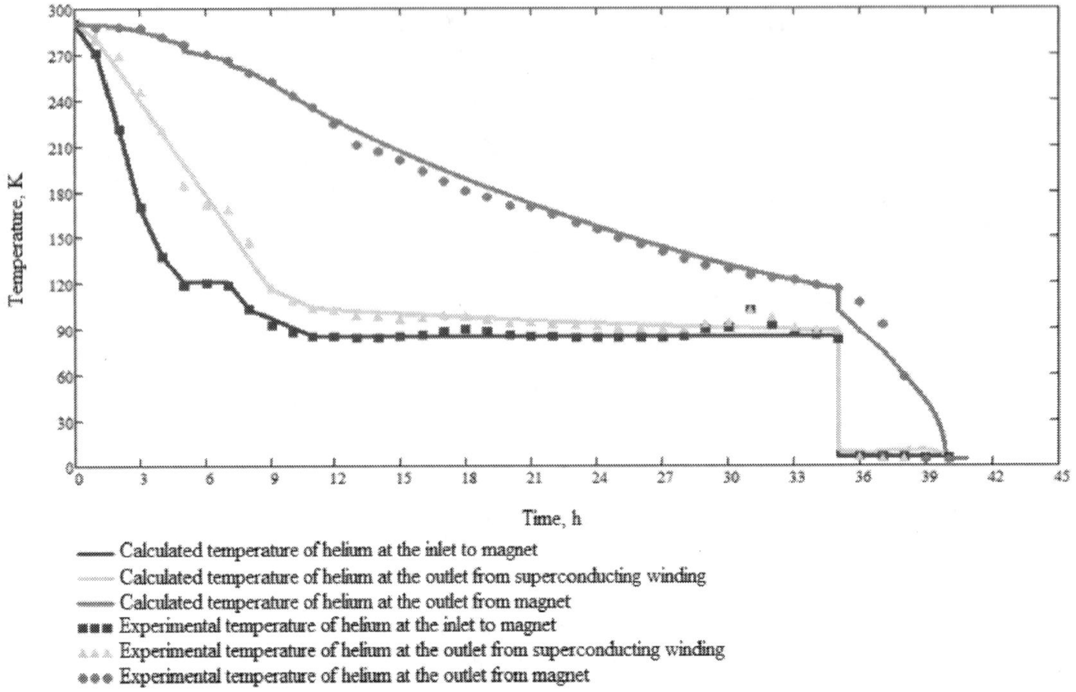

Figure. 3. Cooling process of the dipole magnet booster

While modeling for the new cryogenic test bench the cooling period of the booster dipole magnet was 18 hours. According to calculations productivity of the new test bench will be 3 magnets weekly.

5. Conclusions

The full-scale Nuclotron–type superconducting model dipole and quadrupole magnets for the NICA booster and collider were manufactured at LHEP JINR. In the near future cryogenic testing of the all four types of the superconducting magnets prototypes will be performed. The use of mathematical models were projected pace and timing of the production of superconducting magnets for the accelerator complex NICA. The manufacturing and testing of 266 superconducting magnets on the new test bench will take about two years. The commissioning of new test facility is scheduled at end of 2013.

6. References

[1] Nuclotron -based ion collider facility. Available: http://nica.jinr.ru/
[2] Khodzhibagiyan H, Akishin P, Bychkov A, K et al. 2011, Superconducting Magnets for the NICA Accelerator Complex in Dubna, *IEEE Trans. Appl. Supercond.*, **3**:1795-98.
[3] Agapov N N et al. 1998 "Thermodynamic analysis and operating experience of the nuclotron helium refrigerators in the "Satellite" mode", *Advances in Cryogenic Engineering., Vol.* ***43*** *edited by P. Kittel, Plenum Press, New York,* 557-564.

Characteristic Analysis of DC Electric Railway Systems with Superconducting Power Cables Connecting Power Substations

H Ohsaki[1,2], N Matsushita[2], T Koseki[2] and M Tomita[3]

[1] Department of Advanced Energy, Graduate School of Frontier Sciences, The University of Tokyo, Kashiwa 277-8561, Japan
[2] Department of Electrical Engineering and Information Systems, School of Engineering, The University of Tokyo, Hongo, Tokyo 113-8656, Japan
[3] Railway Technical Research Institite, Kokubunji, Tokyo 185-8540, Japan

E-mail: ohsaki@k.u-tokyo.ac.jp

Abstract. The application of superconducting power cables to DC electric railway systems has been studied. It could leads to an effective use of regenerative brake, improved energy efficiency, effective load sharing among the substations, etc. In this study, an electric circuit model of a DC feeding system is built and numerical simulation is carried out using MATLAB-Simulink software. A modified electric circuit model with an AC power grid connection taken into account is also created to simulate the influence of the grid connection. The analyses have proved that a certain amount of energy can be conserved by introducing superconducting cables, and that electric load distribution and concentration among the substations depend on the substation output voltage distribution.

1. Introduction

In conventional electric railways, a DC feeding system is widely used, especially in railway lines with dense train traffic. However, the DC feeding system has some essential problems due to its low feeding voltage. The problems are a relatively lower efficiency of power transmission, a larger voltage drop along the feeder, the regenerative braking cancellation, a large number of substations along the line, etc. We consider that the introduction of superconducting cables could be a solution to them [1-3].

By laying superconducting cables along the feeder in parallel and connecting them to the railway substations and, in some cases, to several points of the feeder, a voltage change along the feeder can be significantly reduced. This allows the trains to use more a regenerative braking system, thus improving energy efficiency of the whole system. Furthermore, the use of superconducting cables would enable effective load sharing among the substations. The substations have less peak power and can be designed to have a smaller capacity. It leads to a less expensive design of the substations and, as a result, a less expensive DC electric feeding system.

This paper focuses on analysis of the energy conservation and the levelling of load distribution among the substations by introducing superconducting cables.

2. Analysis method

A model railway line was first set up, on which the feasibility of introducing superconducting cables are to be evaluated. Then, an electric circuit model of the railway line was built based. The circuit model includes trains, substations, feeders, superconducting cables, and, if necessary, transmission lines on the grid side. Using the electric circuit model and MATLAB-Simulink software, the current and voltage distributions in the circuit, and electric energy flows in the system were analysed [2].

2.1. A model railway line

Figure 1 shows a schematic view of the model railway line used in this study. It is a double-tracked line of 26.5 km and with 24 stations. On each of its inbound and outbound lines, trains are operated so as to depart at each station in a five-minute interval, which means that the same train operation pattern is repeated every five minutes. Consequently, simulations are conducted for a certain five-minute period.

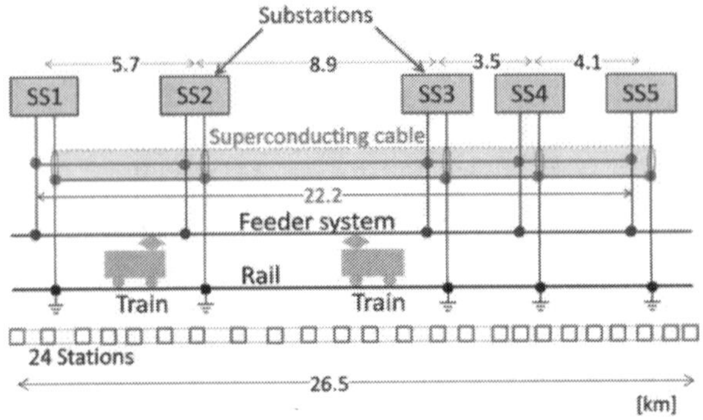

Figure 1. Schematic view of a model railway line.

In a conventional double-tracked feeder system, even if trains on both inbound and outbound lines are operated in the same interval, the amount of energy consumption may be changed just by changing time shift between the operation plans of these two lines. In order to clarify the effect of introducing superconducting cables alone, the simulation was conducted for 300 patterns of the "time shift". Superconducting cables are introduced so as to interconnect the output buses of adjacent substations.

2.2. An electric circuit model

Figure 2 shows schematic views of two electric circuit models used in the analysis. They consist of four parts: substations, feeders, trains, and superconducting cables. The substation model includes a diode. This represents a silicon diode rectifier commonly used in railway substations. In Model B shown in figure 2 (b), all substations are connected to a circuit equivalent to an AC power grid that supplies power to the substations. If the electric utility substation is located farther or a power distribution line has a larger electric resistance, the voltage drops becomes larger at the input and also the output of the railway substations. In addition, the voltages of the railway substations depend on the locations.

The train models are set to consume or regenerate certain amount of electric power, depending of its velocity and acceleration conditions. Electric resistance R_f on the feeder side changes with a train running.

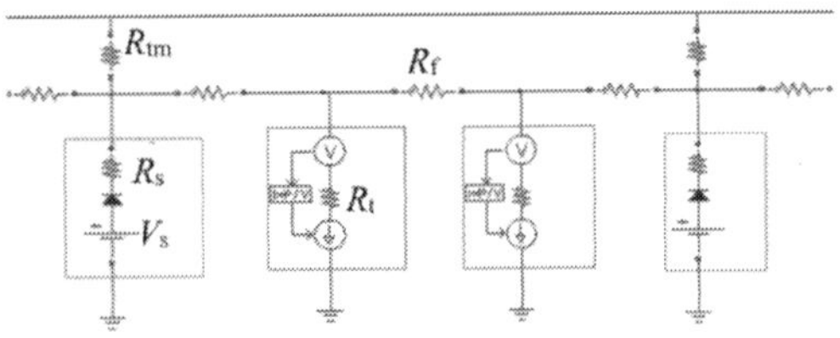

(a) Model A with a simple voltage source substation model

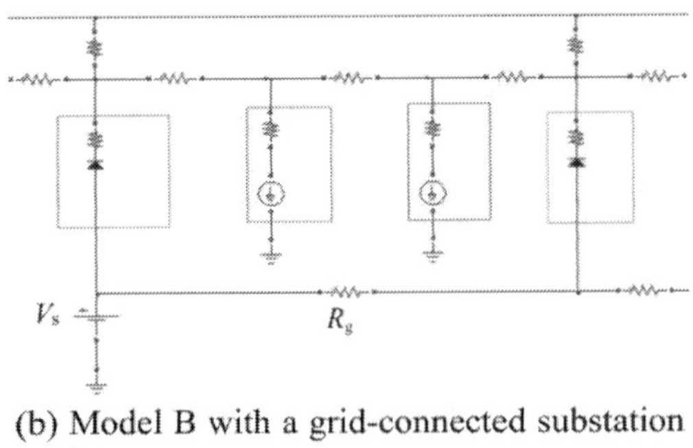

(b) Model B with a grid-connected substation model

Figure 2. Schematic view of the DC electric circuit model.

The superconducting cables are modelled as zero-resistance conductor connecting the adjacent substation outputs. Electric resistance of a connecting line between a superconducting cable and a conventional feeder cannot be ignored and therefore modelled as a single resistor [4].

The specifications of the DC electric railway system mode are summarized in table 1.

Table 1. Specifications of the DC electric railway system model.

Substation	Internal resistance (R_s)	0.0225	[Ω]
	Open-circuit voltage (V_s)	1590	[V]
Feeder	Resistance (R_f)	0.0333	[Ω/km]
Transmission line	Equivalent resistance (R_g)	0.1	[Ω /km]
Train	Internal resistance (R_t)	0.015	[Ω]
	Weight per train	186	[tons]
Superconducting cable	Terminal resistance (R_{tm})	0.01	[Ω]
	Cable length	22.2	[km]
	Heat penetration per length	1000	[W/km]
	Heat penetration per connector	500	[W]
	Number of connectors	2 × 5	
	COP[a] of a cooling system	0.1	
	Total electric power for cooling the superconducting cable system	27.2	[kW]

[a] COP: Coefficient of performance

3. Results and discussions

3.1. Analysis on energy conservation

Energy consumption and conservation were analysed using the model shown in figure 2 (a). Figure 3 shows total energy consumed by all trains for five minutes in a model with or without superconducting cables. The analyses were carried out using 300 different patterns of train operation mentioned in 2.1.

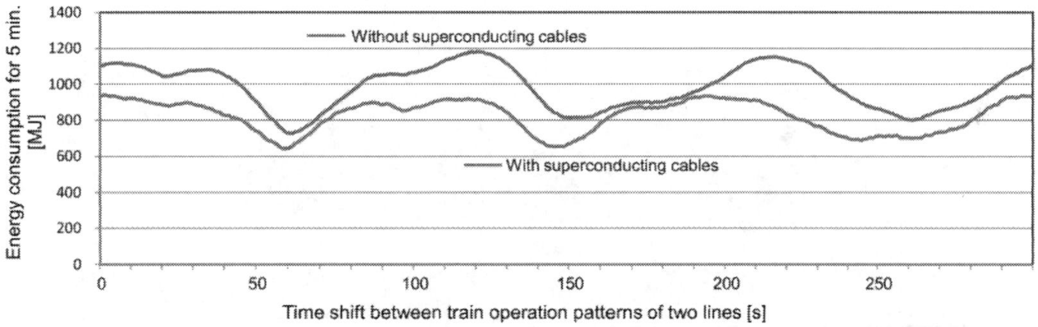

Figure 3. Energy consumption for five minutes in a model with or without superconducting cables.

It has been confirmed that the introduction of superconducting cables can conserve energy more than the cooling energy needed to keep the cable in a superconducting state. Although the amount of energy conserved differs according to the configured operation plan, or "time shift", the average energy conservation rate was estimated to be approximately 16% from figure 3.

3.2. Analysis of load distribution and concentration among substations

Figure 4 shows the time variations of substation output currents for a five-minute train-operation period. In the case without superconducting cables, shown in figure 4 (a), there are current peaks over 2 kA at some of the substations. On the other hand, in the case with superconducting cables, shown in figure 4 (b), these peaks are suppressed by supplying power from other substations. This clearly shows that the introduction of superconducting cables greatly contributes to levelling electric loads among the substations, thus decreasing the current capacity of substations.

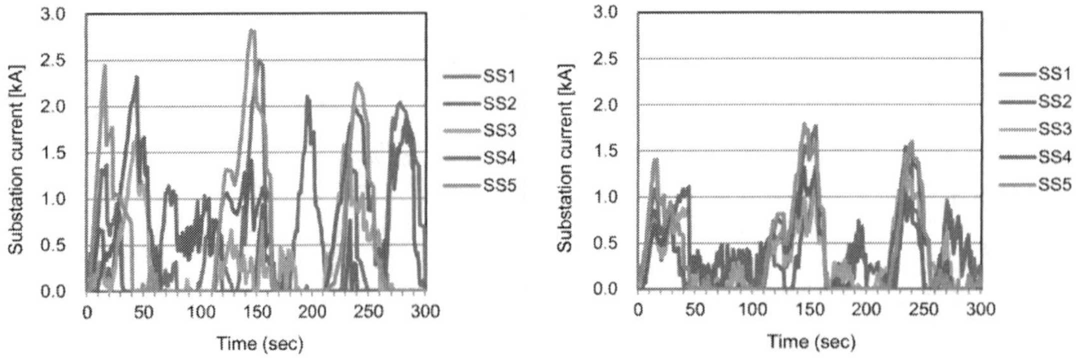

(a) Without superconducting cables (b) With superconducting cables

Figure 4. Substation output currents.

Figure 5 shows the dependence of substation peak current on the substations, with or without superconducting cables, and the grid-connected model shown in figure 2 (b) or not (figure 2 (a)). The left two groups of results shown in figure 5 were obtained using Model A shown in figure 2 (a) and the right two groups were obtained using Model B shown in figure 2 (b), where the railway substations receive power from the equivalent AC power grid model through a resistance of $R_g = 0.1$ [Ω/km]. As a result, the voltages of the substations are different from one another. The output voltage is likely to be the highest at SS1, and the lowest at SS5.

In the cases without superconducting cables, there is only a small difference in the peak currents between the results obtained using Model A and Model B. The case with superconducting cables and a 66 kV class AC power grid, the load current is still well shared among the substations. However, when a 22 kV class grid is assumed, the peak current at SS1 is as high as that before introducing superconducting cables. This suggests that, if unbalance of the substation output voltages is larger than a certain level, superconducting cables may cause a load current concentration to a certain substation.

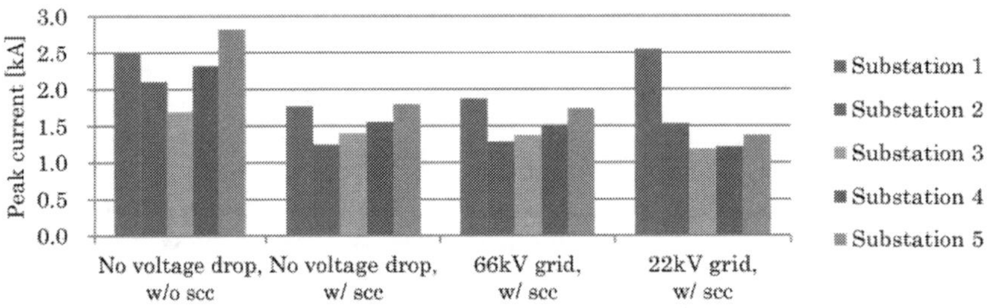

Figure 5. Peak current at each substation.

4. Conclusions

The effects of introducing superconducting cables to DC electric railway were analysed. When trains are operated in a five-minute interval on the model line, 16% conservation of electric energy is possible. When the output voltages of the substations are almost the same, electrical load can be well distributed among the substations through superconducting cables. This leads to a substation design of a reduced current capacity. On the other hand, the electrical load may be concentrated on a certain substation, if the output voltages of the substations are not even enough.

References

[1] Tomita M *et al* 2013 *IEEE Trans. Appl. Supercond.* **23** 3601504
[2] Ohsaki H *et al* 2013 *IEEE Trans. Appl. Supercond.* **23** 3600705
[3] Takagi R 2012 *IET Electr. Syst. Transp.* **2** 103
[4] A Superconducting DC Cable. EPRI, Palo Alto, CA: 2009. 1020458.

Journal of Physics: Conference Series **507** (2014) 032036 doi:10.1088/1742-6596/507/3/032036

Analysis of the effects of asymmetric faults in three-phase superconducting inductive fault current limiters

R Ferreira[1], J M Pina[1,2], N Vilhena[1], P Arsénio[1], A G Pronto[1,2], J Martins[1,2]

[1]Faculdade de Ciências e Tecnologia, FCT, Universidade Nova de Lisboa, 2829-516 Caparica, Portugal
[2]Centre of Technology and Systems, CTS, Uninova, Faculdade de Ciências e Tecnologia, FCT, Universidade Nova de Lisboa, 2829-516 Caparica, Portugal

E-mail: jmmp@fct.unl.pt

Abstract. Inductive fault current limiters of magnetic shielding type can be described in terms of the excursion in the plane defined by flux linked with primary and line current, and this methodology has been previously applied to single-phase devices. Practical applications, however, require three-phase limiters, which, for the sake of compactness, may be built by three legged cores, instead of three single phase units. This has the advantage of using well established methods of power transformers industry, but the performance of the devices depends on the type of fault, e.g. phase to ground or phase to phase. For instance, in a three legged core, a phase to ground fault affects healthy phases, and these are the most frequent faults in distribution grids, where such systems are envisaged. The effects of asymmetric faults are analysed in this paper, by means of measured excursions in the linked flux-current plane.

1. Introduction

Most developed topology of superconducting fault current limiters (SFCL) is the resistive one, due to its compactness. Nevertheless, air core reactors are usually added in order to decrease the amount of costly high temperature superconducting (HTS) tape material required to obtain the desired limiting impedance, and thus relatively large installations may result, see e.g. ECCOFLOW project [1]. Inductive SFCLs of the magnetic shielding type are intrinsically less compact, due to the need of a ferromagnetic circuit. Yet, they require only moderate amount of superconducting tape and they are robust against hot-spot formation, which makes them attractive in medium voltage grids with e.g. dispersed generation. After the successful demonstration of a magnetic shielding SFCL by ABB [2], nearly no projects have been developed, due to the need to build relatively large HTS bulk cylinders for secondaries. Recently, some projects appeared where bulk tubes are replaced by BSCCO [3] or coated conductor [4] tapes, which obviate that problem.

Previous work included developing a modeling methodology based on predicting the excursion of the operating point of the SFCL in the linked flux – current ($\psi - i$) plane [3]. This was first applied to single-phase devices, and to a three-phase limiter in the present work. These are preliminary studies for the development of a medium voltage device for a live grid in Portugal. One concern for utilities regarding the acceptance of SFCLs is their performance under asymmetric faults. Taking this into account, the limiter is tested under several types of short-circuits.

2. Prototype, experimental apparatus and methodology

The SFCL is built by a three legged iron core whose dimensions are shown in figure 1. The device, shown in figure 2, allows choosing 20, 60 or 80 turns in the primaries. These are placed on the bottom of each leg, while dewars with HTS rings made of single turns of Bi-2223 tape are placed on top. Different number of rings may be used (1, 2 or 3). Bi-2223 tape was supplied by Innost, and its critical current is 90 A. The test grid is shown in figure 3 where switches simulate faults, namely phase to ground, phase to phase and phase-phase to ground faults. Line impedances are built by 400 W/1 Ω resistors. Phase to ground voltage in the experiments was 80 V peak.

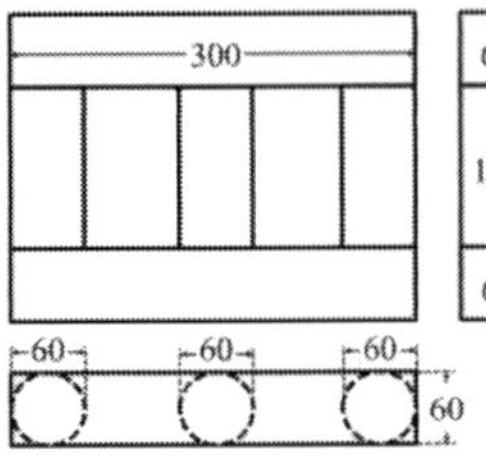

Figure 1. Dimensions (in millimetres) of the SFCL magnetic core.

Figure 2. SFCL with primaries (bottom) and dewars for secondaries (top).

3. Experimental results

In each experiment, prospective currents were measured, in order to evaluate limitation. The excursion in $\psi - i$ plane was also plotted. Different numbers of rings were used in the secondaries. The number of turns in the primaries, in the following experiments, was always 20.

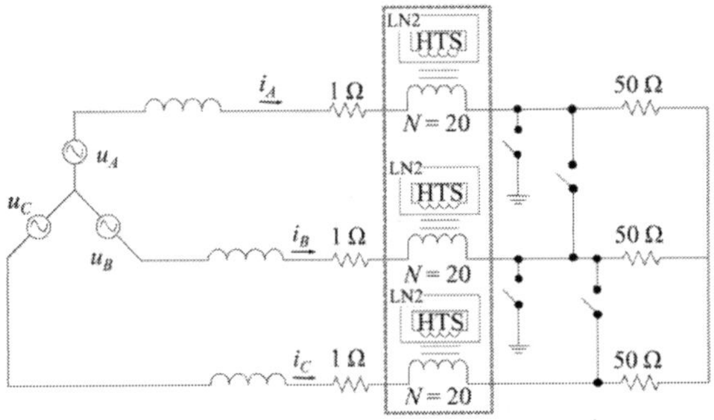

Figure 3. Diagram of the test circuit. Faults are simulated by means of different configuration of switches, implemented with circuit-breakers. Three-phase load (50 Ω resistors) is ungrounded wye connected.

3.1. Phase to ground fault

Phase B was short-circuited to ground. Prospective and limited currents are shown in figure 4, and $\psi - i$ excursion in figure 5. "Healthy" phase A is shown in figure 6. Each secondary had two turns.

3.2. Phase to phase fault

Phase B and phase C were short-circuited in this test. Prospective and limited currents are shown in figure 7, while figure 8 shows $\psi - i$ excursion. One turn secondaries were used.

3.3. Phase-phase to ground fault

Phases A and B were short-circuited to ground in this test. Currents are shown in figure 9, and $\psi - i$ excursion in figure 10. "Healthy" phase C is shown in figure 11. Three turns secondaries were used.

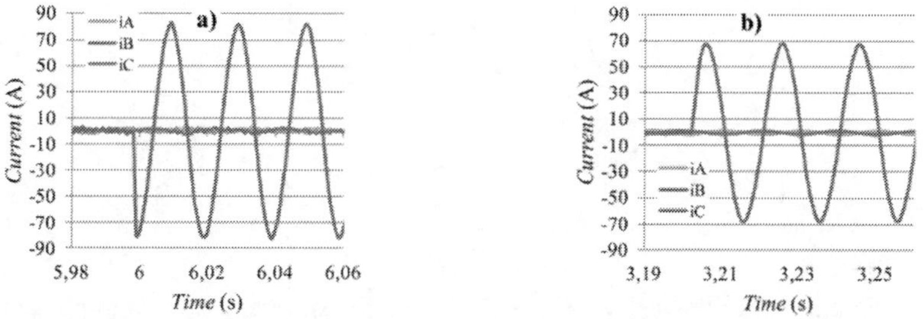

Figure 4. Prospective currents (a) and limited currents (b) on a phase to ground fault.

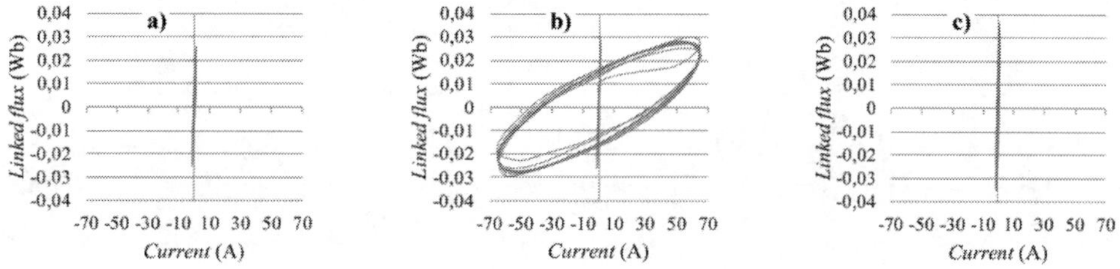

Figure 5. Excursion in the $\psi - i$ planes for phase A (a), phase B (b) and phase C (c).

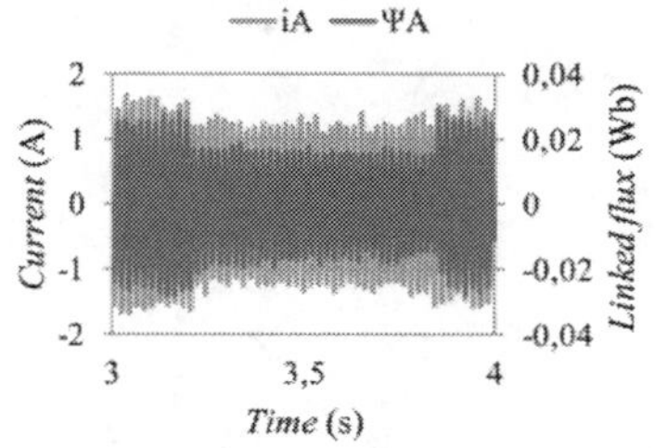

Figure 6. Evolution of linked flux and current in phase A, namely ψ_A and i_A, during the fault.

4. Conclusions

Taking into account the type of load connection (ungrounded wye) asymmetric faults will obviously always affect current in healthy ones. Nevertheless, when magnetic flux penetrates the core, it links also with the latter, and this has to be taken into account in the models, as different types of faults represent distinct magnetic paths in the core. This can be eliminated by using magnetic circuits of shell type. Limitation efficiency depends on the type of fault, as different electric and magnetic circuits are established (most effective limitation happens with only one turn in secondary). Even slight unbalances in the limiter are reflected in the hysteresis loops, see e.g. last fault. Future work includes designing an SFCL with shell type core as well as analysing grounded unbalanced loads.

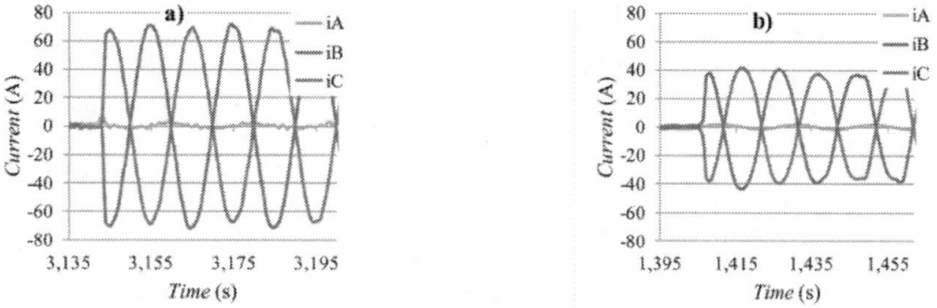

Figure 7. Prospective currents (a) and limited currents (b) on a phase to phase fault.

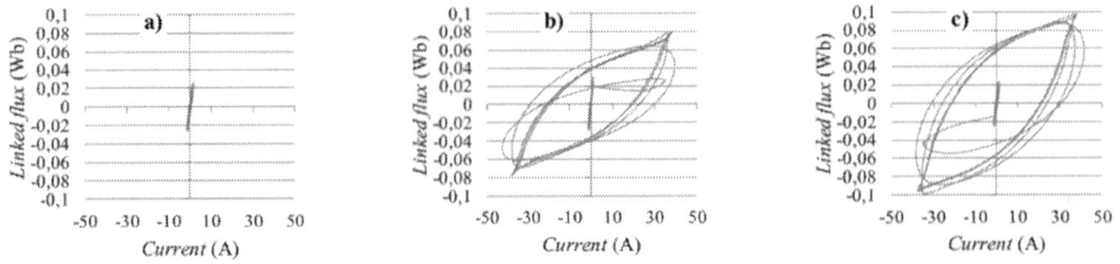

Figure 8. Excursion in the $\psi - i$ planes for phase A (a), phase B (b) and phase C (c).

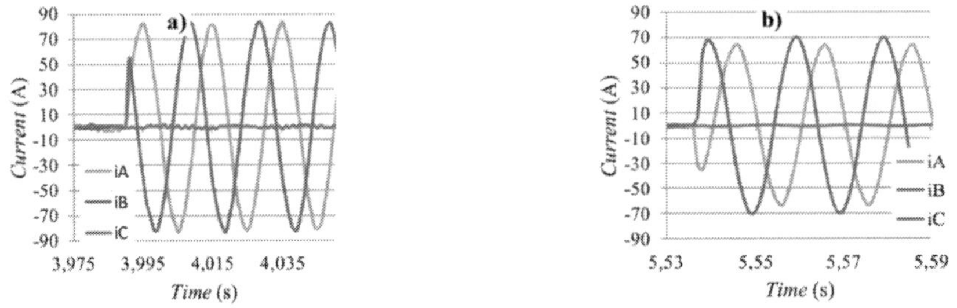

Figure 9. Prospective currents (a) and limited currents (b) on a phase-phase to ground fault.

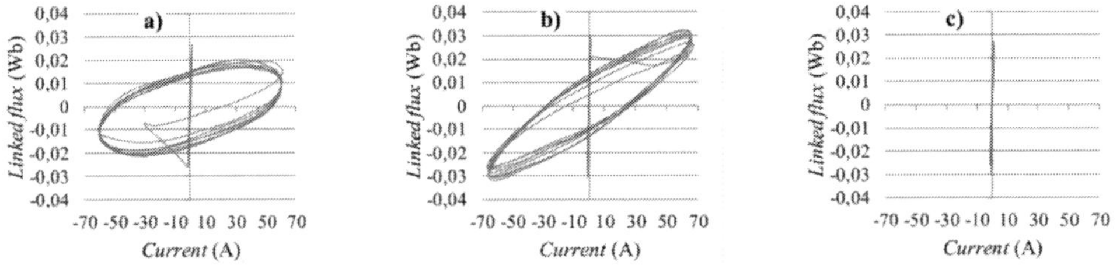

Figure 10. Excursion in the $\psi - i$ planes for phase A (a), phase B (b) and phase C (c).

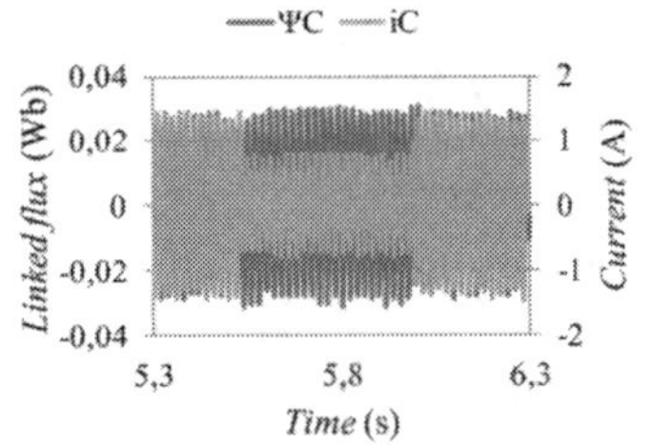

Figure 11. Evolution of linked flux and current in phase C, namely ψ_C and i_C, during the fault.

Acknowledgment
This work was supported by Fundação para a Ciência e a Tecnologia (Centre of Technology and Systems multiannual funding) under the framework of project PEst-OE/EEI/UI0066/2011.

References
[1] Hobl A, Goldacker W, Dutoit B, Martini L, Petermann A and Tixador P 2013 *IEEE Trans. App. Supercond.* **23** 5601804
[2] Paul W, Lakner M, Rhyner J, Unternährer P, Baumann T, Chen M, Widenhorn L and Guérig A 1997 Supercond. Sci. Tech. **10** 914-8
[3] Arsénio P, Silva T, Vilhena N, Pina J M and Pronto A 2013 *IEEE Trans. App. Supercond.* **23** 5601004
[4] Usoskin A, Mumford F, Dietrich R, Handaze A, Prause B, Rutt A and Schlenga K 2009 *IEEE Trans. App. Supercond.* **19** 1859-62

Prospects of long-distance HTS DC power transmission systems

M A Romashov[1], **V E Sytnikov, Y G Shakarian and Y V Ivanov**
R&D Center @ FGC UES, JSC, Kashirskoe shosse, 22/3, Moscow, 115201, Russia

E-mail: romashov@ntc-power.ru

Abstract. Continual improvement of technologies for the safe use of power resources is a key to sustainable development of a human society. In particular, high-temperature superconductivity (HTS) should be used to meet the growing needs of the electric-power industry. It is known that HTS power cables allow us to increase the level of transmitted energy to several GW at voltage of 66-110 kV. HTS power cables of a coaxial design are almost ideal non-polluting system shielding electromagnetic field. In the present work we have tried to analyze various configurations of HTS power transmission systems, estimate the cable transmission capacitance depending on distance, and characterize reliability and efficiency of the systems.

1. Introduction

Efficiency of power production, transmission, and distribution, improvement of energy quality become priorities in the field of electric power industry in the 21th century. Requirements to ecological and resource saving aspects at all phases of power production and distribution are simultaneously raise. Advanced technologies including superconducting one should be used to satisfy the growing needs of the people. Intensive research and development programs are supported to create different kinds of HTS electrical devices such as transformers, current limiters, engines and generators, and power cables. In fact, HTS power cables now are the most developed application of superconductivity in electric power industry. Using power cables made of the traditional materials (copper, aluminium), high voltage has to be applied in order to increase the transmitted power and reduce power dissipation. The maximum achieved voltage is at the level of 500 kV and does not exceed level of 1150 kV for systems under development. Therefore, transmitted power is limited at the level of 0.5-1.5 GW. A number of environmental problems such as ground currents, ground heating up, electromagnetic pollution, dielectric oil pollution, etc. raise, even if special conditions to be observed during cable installation. In this context, HTS power cables allow increasing the level of transmitted energy to several GW at voltage of 66-110 kV. The power transfer efficiency of the cable can be increased to 95-98 %.

In the world, there are a large number of experimental cable facilities constructed to explore the efficiency of HTS electric power transmission [1]. Two DC HTS cable lines being under construction in Russia [2] and Japan [3] have record length of 2.5 km. The construction will be completed in the next 2-3 years. Therefore, the problems of short-distance superconducting power transmission are close to be solved. However, estimates show that the benefits of HTS power cables will fully manifest

[1] To whom any correspondence should be addressed.

in long-distance transmission lines. The development of intermediate-distance (10 km class) and long-distance (100-1000 km class) power transmission technologies demand high investments and intensive researches in order to solve technological problems, generally related to HTS cable cooling system. Long-distance power transmission is especially important owing to non-uniformity of distribution of power resources on a planet. Megacities accumulate powerful energy consumers, but territory where electric power stations can be built are usually at considerable distances from consumers. It is possible to remember as an illustrative example the project of construction of solar batteries in the Sahara Desert and transmission of energy to Europe. The embodiment of the above project entirely depends on feasibility of long-distance HTS power transmission lines. Furthermore, DC HTS lines are well suited for collection of large amounts of electricity from renewable sources of different types, such as solar, wind and geothermal power plants, and so on [4].

2. DC HTS cable line length limitations

Let us estimate the possible level of transmitted power at different voltages, based on the achieved performance of superconducting materials. When the value of the working critical current density is 200 A/mm^2, the creation of DC cables with a rated current of 10-20 kA is real. Experimental cables with the operating current of 10 kA are already constructed. We assume working current of 15 kA to achieve 6 GW transmitting capacitance of two DC HTS cables at 100 kV bipolar voltage.

Let us determine the maximum theoretical length of the DC HTS line in terms of power losses. We can assume for estimations that the energy loss in the line depends on the following factors:

- Energy losses in the converters;
- Heat leakage through the current leads;
- Heat leakage through the cryostat;
- Heating due to hydraulic friction in the cryostat.

The first two quantities are independent of the cable length. The heat leakage through the optimized current leads in the case of long line will be comparatively small, which makes it possible to ignore this value. Losses of energy in the converting equipment may be assumed to be 2 % of the transmitting power. Heat leakage through the thermal insulation of the modern cryostats is at the level of 1.0-1.5 W/m (if the inner diameter is 50-75 mm), and the frictional heating is estimated to be 0.1 W/m. The sum of the last two values should be multiplied by the refrigerating factor, which is equal to 12-18 at the given temperature range. Therefore, the power loss in the range of 13-29 W per one meter should be considered. We assume an average value of 20 W/m. For a more efficient energy transfer let us limit the total loss at the level of 3 %. In this situation the loss in the cable line should not exceed 1 %, and then the total amount of specific losses is about 30 W/m. This value is comparable to the average annual loss due to corona discharge at 750 kV [5]. Results of calculations are summarized in the table 1.

Table 1. Length of DC HTS transmission line with 3 % losses.

Transmitting capacitance, MW	100	300	500	1000	3000	6000
Length, km	50	150	250	500	1500	3000

Let us now consider the limitation on the length of the line originates from the cooling system. Typically, liquid nitrogen (LN_2) pumping in the space between the cable jacket and the inner surface of the cryostat is used to cool HTS cable. It should make maximum use of the working temperature range of LN_2. Because temperature range is limited from below by the freezing point and from above by the boiling temperature and the critical temperature of HTS used, the temperature difference between the inlet and the outlet, ΔT, in practice will be equal to approximately 10-15 K. In this case it is easy to show [6] that under the condition of $\Delta T = const$ the dependence of the pressure drop Δp on the length L is close to a cubic, and on the specific heat load q is close to a quadratic:

$$\Delta p \propto L^{n+1} q^{n}, \text{ where } n = 1.75\text{-}2 \tag{1}$$

or, more conveniently,

$$\frac{\Delta p}{L^{n+1} q^{n}} = const \tag{2}$$

The same relationship holds also for corrugated cryostats. In principle, there is no limitation on the transmission range originated from the cryogenic system, but long line will have to include a large number of intermediate pumping/cooling stations. This can raise the cost to an unacceptable level. Therefore, it is basically need to achieve greater length of each segment of the line, not only in terms of the economy, but also for the laying out in inaccessible locations, such as the sea bed. The pressure drop of the cryogen is the limiting factor if the thermal insulation quality is unchanged. On the one hand, the flow area should be increased in order to facilitate circulation. On the other hand, the diameter of the cryostat must be as small as possible to minimize the heat leakage. Increasing the diameter will also results the loss of benefits of the flexible cryostat because of it can not be wound on a reel and delivered to the construction site. These considerations lead to the conclusion of the need to carefully examine the possibility of using cryostat composed of straight pipes for the construction of HTS transmission lines.

To define the maximum length that can be achieved between two intermediate pumping/cooling stations using both flexible cryostat and straight cryopipes of the same diameter we can take as a basis the geometry of the real project to be embodied in Russia [2]. Namely, the diameter of the corrugated cryostat is 60/66 mm, and the diameter of the HTS cable is 40 mm. The method of calculating the pressure drop in the straight pipe with the cable placed inside is described in [7], and in the corrugated pipe with cable inside – in [8]. The cryogen flow rate increases with the cable length under the fixed temperature difference. It will cause additional frictional heating, which should be compensated by additional amount of the circulating cryogen. A kind of positive feedback forms and at certain length the abrupt unlimited growth of Δp will be observed. The calculated dependences are given in figure 1.

To determine the maximum possible length of the segment it is necessary to define the maximum allowable pressure drop, which to a large extent be determined by the mechanical performance of the cryostat. Table 2 lists the maximum length of the corrugated and straight cryostats at $\Delta p = 1.5$ MPa.

Table 2. Maximum length of DC HTS transmission line segment.

Temperature difference, K	4	6	8	10	12	15
Length, km (straight)	5.59	7.67	9.47	11.1	12.6	14.7
Length, km (corrugated)	3.46	4.83	6.03	7.12	8.13	9.54
Frictional heating, W/m	0.35	0.21	0.15	0.12	0.10	0.08

The data show that the conventional corrugated cryostat is more convenient in the case of several kilometers line, since the cryogenic system can eliminate the intermediate pumping/cooling stations. However, if the length of the line will be a few tens of kilometers, it is preferable to use straight cryostat in order to significantly reduce the number of expansive intermediate pumping/cooling stations.

As may be seen from table 2, various practical problems of the energy transfer can be solved using two or three intermediate stations. These problems include the transfer of electricity from the nuclear power plants located at large distances from cities and consumers, resulting in the need for high voltage converters leading to the energy losses of 6-8 %. Capacitance of the any existing nuclear power plant can be transferred to the customers at the voltage of 50-100 kV using DC HTS power cable, and from one nuclear unit at generator voltage without step-up substations.

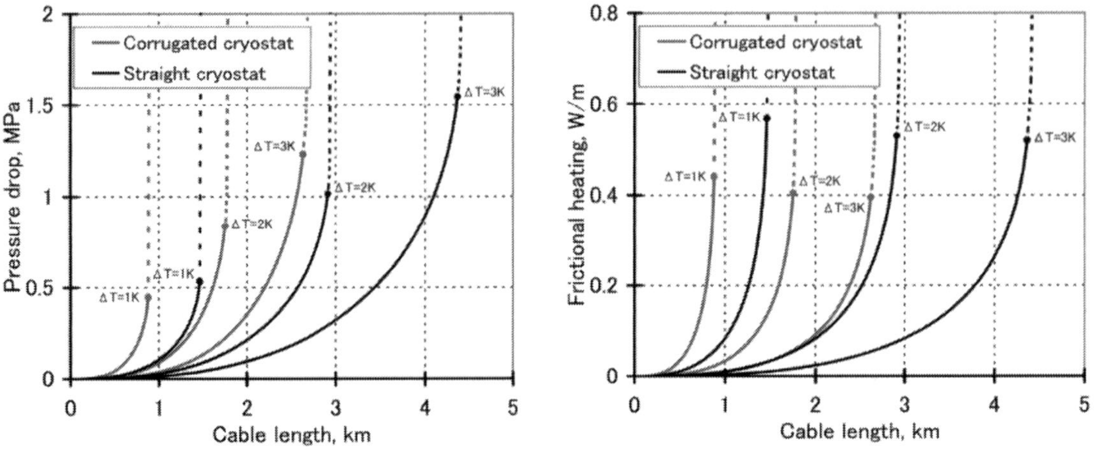

Figure 1. Behavior of pressure drop (left) and frictional heating (right) as a function of cryostat length in case of $\Delta T = const$, namely 1, 2, and 3 K. Straight cryostat I.D. is 60 mm, corrugated one is 60/66 mm. Cable O.D. is 40 mm in both cases. Heat leakage through thermal insulation is 1.2 W/m.

One more of the advantages of the DC HTS power cables is the possibility of doubling the existing lines capacitance without making capital expenditures in transmission infrastructure. This is due to the fact that the cooling system requires a closed loop of cryogen circulation consisting of two cryostats for forward and return flows. Since DC HTS cable does not practically produce heat, the additional cable can be laid in return cryostat as may be necessary without significant impact on the system. It is only required to foresee this possibility and choose the appropriate flow area of the cryostat.

3. Conclusion

It was shown that several kilometers' distance DC HTS power transfer can be embodied using conventional corrugated cryostat. However, it is necessary to apply straight cryostat in order to achieve best hydraulics of the long transmission system. Increasing diameter of the cryostat bring to nothing all advantages basically because of problems with transportation. Losses in HTS lines, in general, do not depend on the length and the transmitted power. The main limiting factor that requires attention in the construction of HTS lines is the design of cryosystem. The use of corrugated cryostat requires 50 % more intermediate stations, which significantly raise the price of the system.

4. References

[1] Sato K 2008 *SEI Tech. Rev.* **66** 55
[2] Sytnikov V E et al 2013 *IEEE Trans. Appl. Supercond.* **23** 5401904
[3] *Nihon keizai shinbun.* 01.01.2013 (in Japanese)
[4] Superconductor electricity pipelines A White Paper 2009 (American Superconductor)
[5] Faibisovich D L 2005 Electrical network design handbook (Moscow: NC ENAS) pp 76-77 (in Russian)
[6] Ivanov Y, Watanabe H, Hamabe M, Kawahara T, Sun J and Yamaguchi S 2012 *Phys. Procedia* **36** 1372
[7] Ivanov Y, Watanabe H, Hamabe M, Sun J, Kawahara T and Yamaguchi S 2011 *Physica C* **471** 1308
[8] Ivanov Y V, Romashov M A, Bemert S E and Sytnikov V E 2013 *Adv. Cryo. Eng.* (submitted)

Feasibility study of a superconducting motor for electrical helicopter propulsion

C A B A E Simons[1], C Sanabria-Walter[2], H Polinder[3]

[1,3] Delft University of technology, 2628 CD Delft, the Netherlands
[2] EADS Innovation Works, 81663 Munich, Germany

E-mail: [1] c.a.b.a.e.simons@student.tudelft.nl
[2] christian.sanabria-walter@eads.net
[3] H.Polinder@tudelft.nl

Abstract. During the past decades, superconducting electrical machines have become more suitable to replace conventional iron based designs, because of their lower weight and higher torque density. These properties make them good candidates for use in More Electric Aircraft (MEA). Especially helicopter propulsion systems could benefit from the increased performance. This paper describes the feasibility study of a superconducting motor to be used for helicopter propulsion as part of a More Electric Aircraft (MEA). For this, the armature, field windings and cryostat are designed, aiming at meeting the difficult specifications. Since superconductors have virtually no electrical resistance when cooled down below a certain critical temperature, they can be used to build high field and low weight coils for electrical machines. Especially the possibility to not use iron can make the superconducting motor lighter with a higher power density compared with conventional Permanent Magnet (PM) motors.

1. Introduction

Increasing the efficiency and therefore lowering the environmental impact of aircraft is becoming an increasingly important goal of the aerospace industry. One effective way to reduce the environmental impact of aircraft is to utilize a hybrid drivetrain, more efficient and reliable when compared to conventional solutions. By combining the advantages of electrical motors and high efficient diesel engines, a hybrid solution is currently considered [1]. An essential part of this solution is the main propulsion direct-drive electrical motor, which greatly determines the feasibility of the concept itself. Such a motor has to provide a very high torque and power density, and a high reliability.

In aerospace it is common to use Permanent Magnet (PM) machines, operating at high rotational speeds [1]. When considering direct drive propulsion for helicopters, no conventional machine can suffice and new topologies have to be considered, an example being a new concept machine in [1].

In this paper a superconducting machine is proposed that could offer the high torque densities required. This first study is to research the possibility of such a machine and therefore only the technical feasibility is considered. Up to now not much work has been done on superconducting propulsion motors for aerospace, with the exception of [4] and [5]. These motors are designed for high speed operation and much less on high torque properties.

In the following sections a machine design for this purpose is developed. First the machine setup is described. Then, the armature is designed, together with superconducting field windings and cryostat. Finally the machine performance is analyzed and a conclusion with recommendations are given.

2. Machine Setup

In this section the machine parameters are given and the major design choices are described. The specifications for the propulsion motor are based on an average civilian helicopter with a payload of 7 to 8 persons, 1.5 tons of empty weight and 3 tons of maximum take-off weight [1]. Technical and operational specifications and machine dimensions are presented in table 1. Based on them; the following design choices were made:

- Partially superconducting: Conventional aluminium armature coils are used together with superconducting field windings placed in a cryostat to avoid high AC losses.
- Rotating armature: A rotating armature using slip rings is adopted while the field windings remain stationary, avoiding a heavy failure-prone cryogenic rotating coupling [6].
- Air core: Since HTS coils produce a larger MMF without active iron, a lighter air core machine can be made [7], as low weight is essential to the design.

Table 1. Machine Specifications.

Property	Value	Property	Value
Nominal power (kW)	600	Airgap radius (cm)	51.25
Nominal rotation speed (rpm)	350	Mechanical airgap (cm)	0.5
Nominal torque (kNm)	16.4	Electromagnetic airgap (cm)	2.5
Axial length (cm)	25	Rotor radius (cm)	50
Expected weight (kg)	150	Machine radius (cm)	70
Operating temperature HTS (K)	65	Inner rotor radius (cm)	45

3. Armature Design

In this section the non-superconducting aluminum armature coils are proposed. The choice of the armature configuration for superconducting machines is usually distributed because of the significantly lower space harmonics. However, in aerospace it is preferred to choose Fractional Slot Concentrated Windings (FSCW) [9] for certain performance advantages as mentioned in [10].

A full pitch distributed winding, the "least distributed", is then compared to a 3/2 FSCW configuration, "the least concentrated", for use in an HTS machine (see figure 1). The decision criteria for the armature configuration choice are the severity of the space harmonics seen at the superconducting side and the losses induced in a 1cm thick aluminum EM shield. The rotating armature uses the mechanical parameters given in the previous section with 72 slots, 60% slot space and a fill factor of 0.4 for the distributed configuration and 0.7 for the concentrated one (5A/mm^2).

A Comsol FEM simulation of the space harmonics, normalized to the fundamental with and without the EM shield is shown in figure 3. The effectiveness of the shield in reducing harmonics is proven, and the eddy current loss is of 120W and 2.9kW for the distributed and concentrated windings respectively. The increased cooling needed for the concentrated windings means that for any given cooling capacity, the distributed configuration can operate at a higher current density compared to the concentrated configuration. Because of this advantage the distributed winding configuration is used.

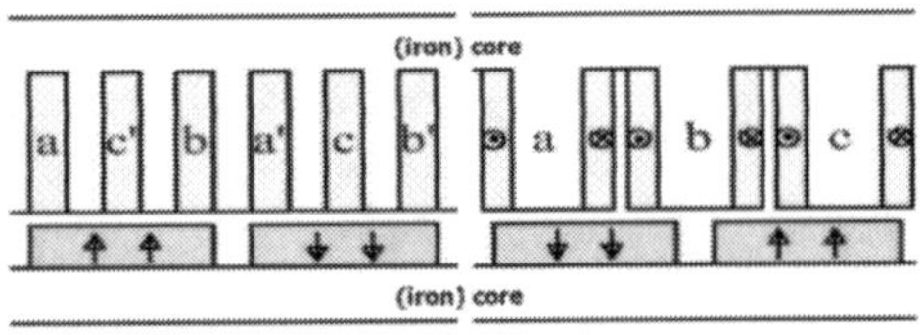

Figure 1. Full pitch distributed (left) and 3/2 fractional slot concentrated (right) windings [10]

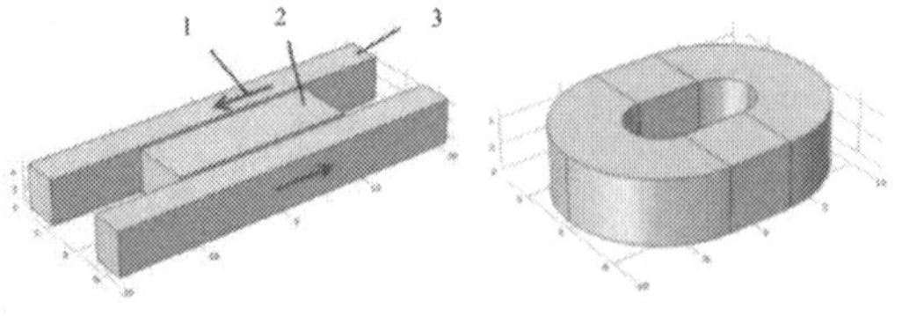

Figure 2. Model of bulk concentrating flux setup with current direction (1), HTS bulk (2) and HTS windings (3) (left), and model of the racetrack coil setup (right)

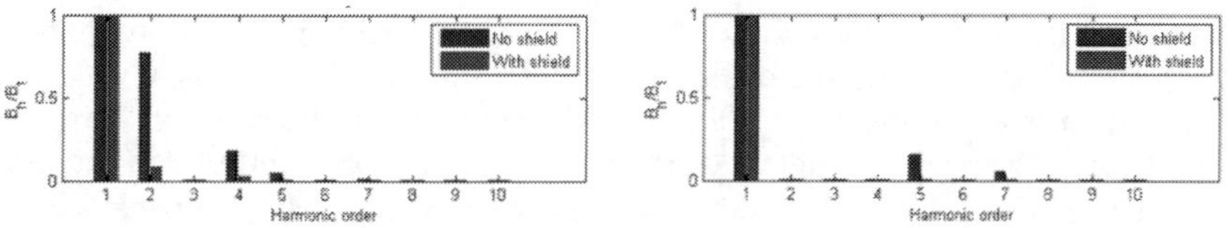

Figure 3. Space harmonics with and without EM shield for 3/2 concentrated (left) and full pitch distributed (right) winding configurations.

4. Field Winding Design

Using the methodology below, the superconducting field windings are designed in this section.

4.1. Superconducting topology selection

Two inductor topologies were considered for the field windings: the synchronous machine with superconducting racetrack coils shown in figure 2, the most popular and with very good performance, and the bulk concentrating flux topology shown also in figure 2, proposed in [4] and [5] and specifically developed for aerospace applications.

A comparison of the airgap field using Comsol for both designs using models of comparable size and weight is shown in figure 4. The resulting waveforms have peak values of 0.25T and 0.75T for the bulk concentrating flux and racetrack coil configuration respectively. It is evident that with three times better performance, race track coils are more appropriate for the ongoing design.

4.2. Superconducting topology selection

The racetrack coils consist of 12 stacked pancakes, each with 375 windings of 4mm YBCO tape from Superpower [12] and aluminum support material. Of interest are the parameters examined below.

4.2.1. Critical values. The critical current according to the *maximum normal and parallel magnetic field* are calculated using Comsol FEM simulations. The resulting critical current density is $144 A/mm^2$ and with a fill factor of 0.7, the critical current becomes 58A, incorporating a safety factor of 0.8. The corresponding maxima are 2.3T and 1.1T for the parallel and normal fields respectively.

4.2.2. Effective airgap field. Although the field inside the coil is as high as 2.3T, it decays rather fast further away from the coils due to the lack of iron to contain the flux, and due to the large airgap. Because of this, the average airgap field seen by the armature is taken: 0.8T.

4.2.3. Effective stack length. Also because of the absence of iron, air core machines suffer from large end-effects. These end-effects are taken into account by defining the effective stack length, which amounts to only 65% of the actual stack length.

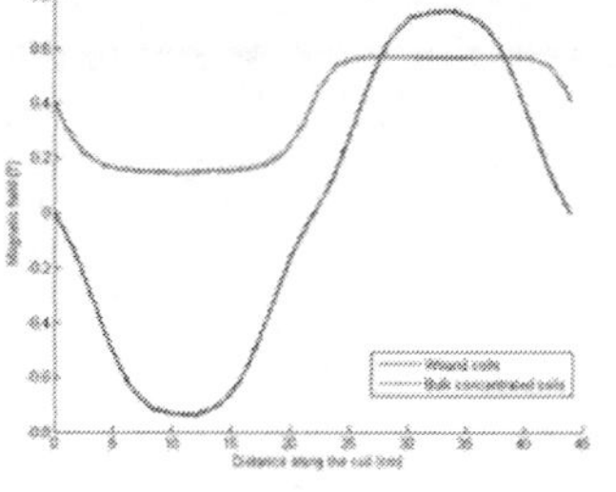

Figure 4. Magnetic field along the coil for bulk concentrating flux setup and race track coils.

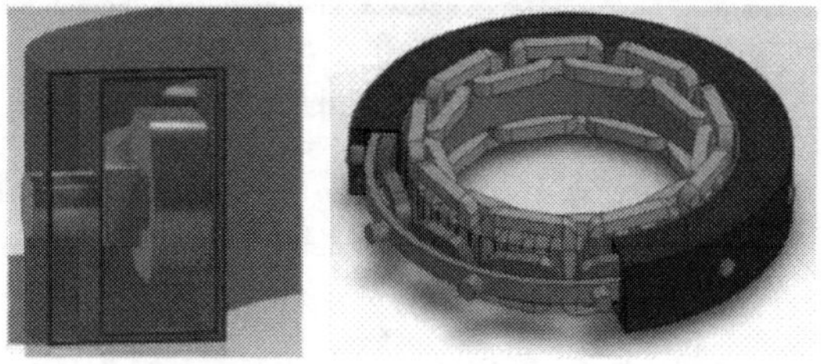

Figure 5. SolidWorks model of the cryostat (left) and of complete machine (right)

4.3. Cryostat design

The field coils are placed in a cryostat to keep them at 65K. Important design parts of a cryostat are the vacuum insulation and the cryostat mounting. Vacuum insulation is part of the airgap, therefore as thin as possible, and consists of two glass fiber reinforced plastic (GFRP) walls 3mm thick. The Multi-Layer Insulation (MLI) space between the walls is 4mm long for a total thickness of 1cm.

By using 10 GFRP rods for mounting, a thermally insulating and mechanically strong support is proposed. A thorough mechanical analysis, beyond the scope of this paper, must still be performed.

With the cryostat structure defined (see figure 5), the main heat losses inside the cryostat are calculated (see table 2): HTS conduction losses, mounting and support conduction losses, radiation losses from the outer to the inner cryostat wall, and current lead losses. For a thorough explanation on these calculations, see [13]. For cooling the cryostat the use of the cryo-coolers is preferred, since it will result in a much lighter cooling system. Using pulse tube cryo-coolers designed for aerospace made by Lockheed Martin [14], 15W can be removed at 65K, with 7kg of weight and consuming 600W. Using 5 cryo-coolers for redundancy the total weight amounts 35kg. The required input power is 3kW, which is only 0.5 percent of the target power.

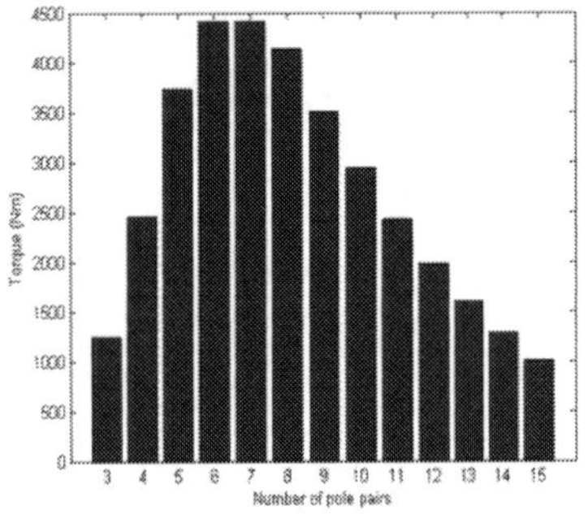

Figure 6. Torque as function of the number of pole pairs

Table 2. Cryostat heat sources.

Heat source	Losses [W]
Electrical conduction	0.07
Thermal conduction	39
Thermal radiation	0.7
Current leads	3.7
Total	**43.07**

Table 3. Specification comparison.

Property	Target	Acquired
Nominal power (kW)	600	251
Normal torque (kNm)	16.4	7.1
Power-to-weight ratio (kW/kg)	4-5	0.65
Torque-to-weight ratio (N/kg)	110-130	18.4
Torque volumetric density (kNm/m^3)	43-51	17.6
Expected weight (kg)	150	386

5. Machine Performance

In this section the armature, field windings and cryostat designs are combined to calculate the torque, torque density and efficiency of the final design.

5.1. Torque Calculation

The torque is calculated using FEM simulations on Comsol as function of the number of pole pairs and the results presented in figure 6. From the results it is concluded that the optimum torque of 4.4kNm occurs with 6 pole pairs, using an armature current density of 5A/mm^2.

The optimum is explained as follows: with a pole pair number lower than six, the field coils are too large compared to the machine and the excitation flux becomes less effectively linked to the armature coils. On the other hand, with a pole pair number higher than six, the field coils become smaller. Due to the lack of iron and the large airgap, the flux leakage is dominant and the torque decreases.

5.2. Weight Estimation

A very important aspect of the design is the total weight of the machine. Summing up the weight of the different components, the total weight is 386kg, including cooling system.

5.3. Performance of Final Design

In table 3 the results obtained are presented together with the target values (for an armature current density of 8A/mm^2). The main power losses are 3kW from the cryo-coolers and 12.3kW of resistive losses in the aluminum windings. With a machine power of 251kW, the efficiency becomes 94%.

From table 4, 42% of the targets are met with a weight 2.5 times higher. The main reasons behind these low values are explained below.

5.3.1. Operating temperature of superconductors. The operation at 65K limits the performance of the superconductors resulting in a low critical current of 60A, and in a low airgap field of 0.8T.

5.3.2. No iron use. Without the use of iron the machine weight was kept as low as possible, but at the same time no flux path could be designed, resulting in dominance of airgap leakage flux. The weight decrease is partly counteracted by the higher amount of superconductor material needed.

5.3.3. Large airgap. The airgap length, characteristic of a superconducting design, is 5% of the airgap radius and 17% of the active axial length, making it more advantageous to design larger machines.

6. Conclusions

A first technical feasibility study on a superconducting motor for helicopter propulsion has been made. The armature, superconducting field windings and cryostat were designed adopting a radial flux topology with an inner rotating armature removing the need for a cryogenic rotating coupling.

For the armature design, a comparison between a distributed winding configuration and a FSCW configuration yields unacceptable losses for the concentrated configuration, and favoring the choice of the distributed configuration. Regarding the superconducting field windings, a comparison between the bulk flux concentrating and the wound racetrack coil topology showed that the latter has much higher torque capabilities, favoring its choice. A tentative cryostat design was given. The heat load, estimated at 43W, is removed most effectively with high-power-density pulse tube cryo-coolers.

Finally the machine parameters were calculated. The result was that only 42% of the targets were met with a weight 2.5 times higher. According to these figures is estimated that a superconducting electrical propulsion motor may not be feasible with current technology and specifications.

7. Recommendations

Although the superconducting electrical propulsion motor may not be feasible under the current set of constraints, it is expected that the machine performance can be improved with future work.

The most important recommendation is to consider liquid hydrogen (LH2) for cooling, which is not unrealistic in aerospace, since it is also considered as a potential replacement for kerosene in the near future. This would bring the operating temperature from 65K down to 20K, resulting in significantly higher operating currents and a potential performance increase of 3 to 4 times, but also in a need to redesign the cooling scheme. This would bring the machine performance closer to the target values.

References

[1] P Jänker, J Stuhlberger, F Hoffmann, G Niesl and V Kloeppel 2010 The Hybrid Helicopter Drive – A Step to New Horizons of Efficiency and Flexibility, *36th European Rotorcraft Forum*

[2] E D Ganev 2007 *Power Engineering Society General Meeting* **1** 771-778

[3] C Sanabria-Walter, H Polinder, J A Ferreira, P Jänker and M Hofmann 2012 *Proc. of the XX conference on electrical machines (ICEM)* 2583-2590

[4] P Masson and C Luongo 2005 *IEEE Trans. App. Superconductivity* **15** 2226-2229

[5] C Luongo, P Masson and T Nam 2009 *IEEE Trans. App. Superconductivity* **19** 1055-1068

[6] R Flitney 2007 *Seals and Sealing Handbook* (Elsevier)

[7] J Frauenhofer et all 2008 *Journal of Physics: Conference Series* **97** 012189

[8] A Parviainen 2005 *Design of axial-flux permanent-magnet low-speed machines and performance comparison between radial-flux and axial-flux machines* PhD Thesis, Lappeenranta University of Technology, Finland

[9] E Tsampouris, M Beniakar and A Kladas 2012 *IEEE Trans. Magnetics* **48** 943-946

[10] A EL-Refaie 2010 *IEEE Trans. Ind. Electronics* **1** 107-121

[11] H Polinder, M J Hoeijmakers and M Scuotto 2007 *Electric Machines & Drives Conf.* **1** 652-657

[12] SuperPower 2012 [Online]. Available: http://www.superpowerinc. com/content/products.

[13] S S Kalsi 2011 Applications of High Temperature Superconductors to Electric Power Equipment (John Wiley & Sons, IEEE Press)

[14] J R Olson, P Champagne, E Roth and T Nast 2012 AIP Conf. Proc. **1434** 161-167

LIQHYSMES - spectral power distributions of imbalances and implications for the SMES

M Sander, R Gehring and H Neumann
Karlsruhe Institute of Technology, ITEP
Eggenstein, Germany, Hermann-von-Helmholtz-Platz 1
76344 Eggenstein-Leopoldshafen, Germany

E-mail: michael.sander@kit.edu

Abstract. LIQHYSMES, the recently proposed hybrid energy storage concept for variable renewable energies, combines the storage of LIQuid HYdrogen (LH2) with Superconducting Magnetic Energy Storage (SMES). LH2 as the bulk energy carrier is used for the large scale stationary longer-term energy storage, and the SMES cooled by the LH2 bath, provides highest power over shorter periods and at superior efficiencies. Both together contribute to the balancing of electric load or supply fluctuations from seconds to several hours, days or even weeks. Here different spectral power distributions of such imbalances between electricity supply and load reflecting different sources of fluctuations in the range between 1 sec and 15 minutes are considered. Some related implications for MgB2-based 100 MW-SMES operated at maximum fields of 2 T and 4 T, are considered for these buffering scenarios. Requirements as regards the storage capacity and correspondingly the minimum size of the LH2 storage tank are derived. The related loss contributions with a particular focus on the ramping losses are analysed.

1. Introduction

Energy storage systems providing tens to hundreds of MW and GWh and storing large amounts of excess energy from renewable energy sources could contribute to balancing electricity supply and demand from seconds to several days or weeks. For this purpose recently the LIQHYSMES approach has been proposed [1-4]. The imbalance (or difference) between (e.g. wind power) supply and (consumers') demand can vary significantly. The required power level and storage capacity of the SMES for buffering the momentary imbalance will then directly depend on the characteristics of this difference. In the following it will be assumed that the electrochemical energy (re-) conversion of H2 fully compensates the variations of the imbalance on a time scale above 15 minutes which is typical for the (primary) control and trading of electric power. In the following different remaining imbalances on time scales below 15 minutes will then be used to investigate the corresponding requirements for the SMES.

2. Simulation of Balancing Cases #1 to #3

Three different spectral power distributions are used to simulate the remaining imbalances between supply and load on time scales below 15 minutes. The power spectrum contains a geometric sequence made of 100 discrete frequencies f between 1/(15 min) and 1/(1 sec), and the relative phases have been created by generating random numbers. The relative peak intensities are chosen so that white, 1/f and

1/f2 spectral power densities are approximated. Widely Gaussian distributions are obtained for all three cases when averaging over periods of 1 sec, 10 sec, 1 minute and 5 minutes. All distributions are adapted so that the standard deviation for the power averaged over 1 second is in all three cases 25 MW (values for longer periods differ due to their specific spectral power distribution). Therefore also the mean absolute power is with 20 MW identical in all cases. Whereas the "white" case #1 has strong fluctuations on a second time scale, the "1/f2" case #3 shows only slow variations over several minutes, and the "1/f" case #1 lies in between. The maximum SMES power is then assumed to be four standard deviations or ± 100 MW, but the requirements as regards the stored SMES energy differ for the three distributions. For all SMES systems toroidal configurations having 20 identical solenoidal coils of Magnesium Diboride (MgB2) round wires with a diameter of 100 μm and maximum magnetic fields of 4 T and 2 T have been assumed. The data are summarized in Table 1.

Table 1:Characteristics of Test Cases #1 to #3 and 100MW-10kA-SMES for 4T & 2T

Spectral Variation of Power Density of Imbalance	"white"		"1/f"	"1/f2"
Standard Deviation in MW for Averaging over @ 1sec / 10 sec / 1 min / 5 min (20MW Mean Power)	25 / ~ 4.7 / 0.56 / 0.15		25 / 18 / 13 / 6.0	25 / 25 / 25 / 16
Test Case / Rated SMES Energy in GJ (discharged @ 50 % of max. Coil Current 10kA)	#1A / 0.25	#1B / 2.5	#2 / 10.0	#3 / 22.5
Maximum Magnetic Field at Coil in T	4 (top data in each line below) / 2 (bottom data in each line below)			
Outer Coil Radius R in m (Height = R / 5; Total SMES System: Radius / Height ~ 2.75 R / 2 R)	1.92 / 3.04	4.13 / 6.55	6.55 / 10.4	8.59 / 13.6
Required Conductor in GAm (on the Basis of the Critical Current Ic@4T,20K)	0.71 / 0.16	3.29 / 0.76	8.29 / 1.91	14.2 / 3.27
Max. Mean Operating Current Density in Coil Winding in kA/cm2	1.85 / 0.58	0.86 / 0.27	0.54 / 0.17	0.41 / 0.13
Full-Cycle Ramping Loss of SMES (round MgB2 wire, diameter 100 μm) in MJ	0.18 / 0.05	0.83 / 0.21	2.09 / 0.53	3.59 / 0.91
Electric Power for Removing Thermal Losses of Cryostat & Current Leads in MW (~ 93 W/W)	0.104 / 0.136	0.178 / 0.306	0.306 / 0.607	0.451 / 0.950
Mean Electric Power for Removing the SMES Ramping Loss in MW (~ 93 W/W)	0.964 / 0.225	0.426 / 0.101	0.247 / 0.060	0.214 / 0.050
Total Mean Loss of SMES in MW (w/o ~ 1.6 MW Mean PCC Loss)	1.07 / 0.36	0.60 / 0.41	0.56 / 0.67	0.67 / 1.00

For case #1 with the predominantly higher frequency components a discharge duration of 2.5 seconds at the maximum power of 100 MW would be sufficient (case #1A) for providing a reliable primary control. Due to the buffering requirements coming from the H2 components of a LIQHYSMES plant a ten-fold oversized SMES for 25 seconds of discharge duration is also included (case #1B). For cases #2 and #3 the much stronger low frequency contributions require higher storable energies for the SMES. To reduce the losses and to have sufficient safety margins as regards a quench, the maximum operating current is set to be only 50 % of the critical current. The assumed Ic(B) characteristic has been shown in [2]. Full-cycle ramping losses are calculated for the SMES systems when ramped between 50 % and 100 % of the operating current or magnetic field. The calculation method has already been described in [1]. In brief, for sufficiently slow ramping processes and high fields, the hysteretic magnetization losses strongly dominate and can be well approximated by the critical state model. Under these assumptions the ramping losses become independent of the ramping rate. The introduction of a differential ramping loss giving the loss per field change allows calculating the local conductor losses for any local magnetic field B in the coil winding and taking into account the Ic(B) characteristic. For the fields and the discharge times under consideration the ramping rates are in the range of about 4 mT/s (case #3 @ 2 T) to 800 mT/s (case #1A @ 4 T). Thus for discharge times of only 2.5 seconds the assumption for the calculation method may not be fully valid and loss data provide only a crude estimate for the lower bounds.

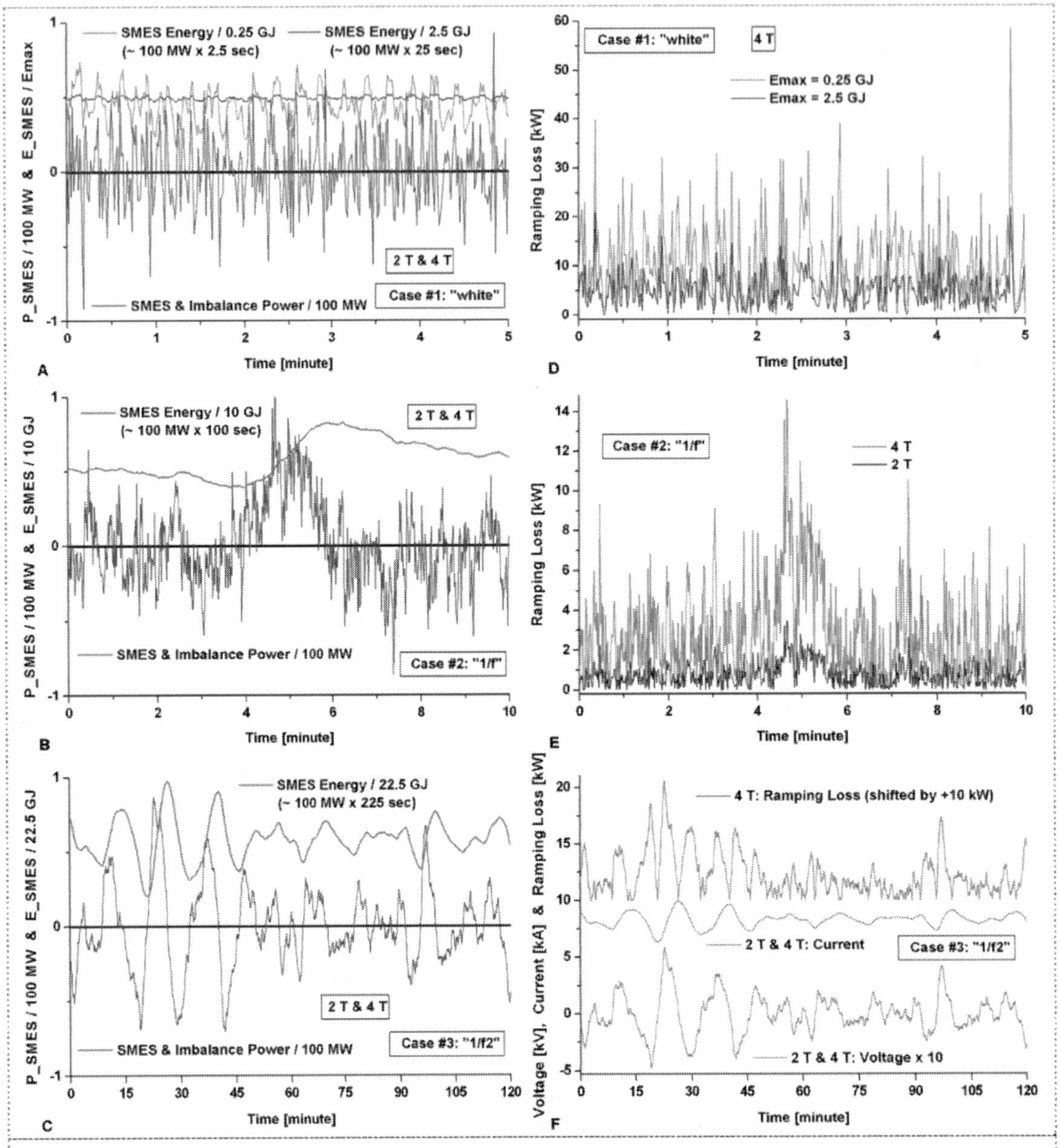

Figure 1. A-C) Momentary power level for the cases #1 to #3 and charging status for both magnet designs of 2 T and 4 T (in case #1 also for the two different storage capacities Emax); D) Ramping loss of one coil for both storage capacities and the 4 T magnet design (case #1); E) Ramping loss of one coil for both magnet designs (case #2); F) Ramping loss for the 4 T magnet design and momentary current & voltage levels (identical for both magnetic field designs) of one coil.

3. Simulation Results for Balancing Cases #1 to #3

The balancing processes for cases #1 to #3 have been calculated for a total period of four days which was found to give a sufficient statistics. These averaged data were then used to calculate the mean loss components which are included in Table 1. The heat created by the ramping losses has to be removed by a cooling system and the related electric power varies from 0.05 MW for the 2 T design of case #3 to almost 1 MW for the 4 T design of case #1A. The thermal load on the cryostat necessarily increases with its size, and, therefore, also the electric cooling power for cryostat and current leads goes up with the SMES size. For all cases the estimated losses of about 1.6 MW for the power conversion and control are the strongest individual loss contribution reflecting the assumed operational and standby losses.

Typical examples for the simulated balancing processes are given in Figure 1. The momentary power levels and the charging status of the SMES systems used in the cases #1 to #3 are given in the graphs A-C. For case #1 the two different storage capacities Emax corresponding to 2.5 and to 25 seconds are shown in Figure 1A. The corresponding ramping losses occurring in one coil of the SMES are provided for the magnetic field design for 4 T in Figure 1D. For a discharge duration of 2.5 seconds the strongly fluctuating power level of the imbalance results in quite significant changes of the charging status and, consequently, also in significant ramping losses. The tenfold larger storage capacity of 25 seconds reduces the fluctuations of the charging status and, in particular, the losses by roughly a factor of 2. For case #2 the ramping losses of one coil for both magnetic field designs are provided in Figure 1E. The lower field of 2 T shows only roughly a fourth of the losses of the 4 T design, and this is actually observed for all cases which reflects the general trend for field dependencies of ramping losses [2]. The storage capacity of 10 GJ is four times the one for case #1B, and the losses decrease almost by a factor of 2 when comparing the 4 T designs. Going to case #3 with 225 seconds of specified discharge duration gives only a modest additional reduction of losses (Figure 1F). Figure 1F also shows an example for the relations between the momentary power and charging status of the whole SMES system and the current and voltage levels seen by a single solenoidal coil. The latter applies to both magnetic field designs. The rated power of the SMES and the operating currents were assumed to be same for all cases (changing between 5 kA and 10 kA). Consequently also the voltage levels are about the same for case #1 to case #3 and stay well below 1 kV.

4. Conclusion

A simple calculation method is used to simulate imbalances between supply and load for different spectral power distributions and for a time scale below 15 minutes which is typical for the (primary) control and trading of electric power. In LIQHYSMES plants these contributions mainly have to be buffered by a SMES of adequate storage capacity. SMES operated at smaller magnetic field require more space, but show lower losses, and may be particularly attractive when very large amounts of excess energy i.e. tens to hundreds of GWh from variable renewable energy sources have to be stored as LH2 anyway.

5. References

[1] Sander M and Gehring R 2011 *IEEE Trans. on Appl. Supercond.* **21(3)** 1362
[2] Sander M and Neumann H 2011 *Supercond. Sci. Technol.* **24** 105008
[3] Sander M, Gehring R, Neumann H and Jordan T 2012 *Int. J. of Hydrogen Energy* **37** 14300
[4] Sander M, Gehring R and Neumann H 2013 *IEEE Trans. on Appl. Supercond.* **23** 5700505

Superconducting light generator for large offshore wind turbines

S Sanz[1], T Arlaban[2], R Manzanas[3], M Tropeano[4], R Funke[5], P Kováč[6], Y Yang[7], H Neumann[8], B Mondesert[9]

[1]TECNALIA, Parque Tecnológico de Bizkaia, C/Geldo, Edif. 700, 48160 Derio, Spain
[2]Acciona Windpower S.A., Pol. Ind. Ciudad del Transporte, 31119 Noain, Spain
[3]Acciona Energía, S.A., Av. de la Ciudad de la Innovación, 531621, Sarriguren, Spain
[4]Columbus Superconductors Spa, Via delle Terre Rosse, 30, 16133 Genova, Italy
[5]Oerlikon Leybold Vacuum GmbH, Bonner Str. 498, 50968 Köln, Germany
[6]Institute of Electrical Engineering of SAS, Dúbravska cesta 9, 841 04 Bratislava, Slovakia
[7]The University of Southampton, Highfield, SO17 1BJ, Southampton, UK
[8] Institute for Technical Physics Karlsruhe Institute of Technology, 76344 Eggenstein-Leopoldshafen, Germany
[9]d2m Engineering, 663 Av. de la 1ère Armée Française, 83500 La Seyne sur Mer, France

E-mail: santiago.sanz@tecnalia.com

Abstract. Offshore wind market demands higher power rate and reliable turbines in order to optimize capital and operational cost. These requests are difficult to overcome with conventional generator technologies due to a significant weight and cost increase with the scaling up. Thus superconducting materials appears as a prominent solution for wind generators, based on their capacity to held high current densities with very small losses, which permits to efficiently replace copper conductors mainly in the rotor field coils. However the state-of-the-art superconducting generator concepts still seem to be expensive and technically challenging for the marine environment. This paper describes a 10 MW class novel direct drive superconducting generator, based on MgB_2 wires and a modular cryogen free cooling system, which has been specifically designed for the offshore wind industry needs.

1. Introduction

Wind power is the renewable energy source that has seen the most successful deployment over the last two decades, from less than 3 GW to 282 GW of global cumulative capacity [1]. In the EU, five countries source more than 10 % of their electricity from wind [2] and wind energy will provide at least 12 % of EU electricity by 2020, significantly contributing to the 20/20/20 goals of European energy and climate policy [3].

Offshore wind energy is rapidly developing motivated by the stronger, more regular winds at sea and more limiting sitting restrictions on land. According to the International Energy Agency, global offshore wind cumulative capacity could reach 100 GW in 2020 and 375 GW in 2030 [4].

Table 1. Cumulative offshore wind capacity prediction [4][5]

	2020 (GW)	2030 (GW)	2050 (GW)
Europe	40	150	460
World	100	375	1150

The offshore wind industry, in order to be more competitive and be able to meet the market predictions (see table 1), demands important capital and operation costs reduction. Several cost reduction opportunities have been identified, which are mainly focused on technology development, improvement of the supply chain efficiency and finance. The increase of turbines' power rating has been pointed out as one of the issues with higher LCOE (Levelized Cost of Energy) reduction potential [6]. Indeed, there is already a trend toward larger turbines, in 2012 the average size of installed offshore wind turbines has been 4 MW, while ten years before it was 2 MW [7].

Nowadays, the dominant technology is the geared drive train induction generator, while direct drive permanent magnet synchronous generator still represents a small market share of 3.16 % [8]. Direct drive solutions are expected to dominate the offshore wind market in the coming years based on its higher reliability, efficiency and better scalability in comparison to geared solutions.

2. Superconducting generators for offshore wind

Wind turbines state-of-the-art shows that both geared and direct-drive synchronous and permanent magnet generators are difficult to scale up to 10 MW and beyond. Their huge size and weight drives up the cost of both fixed and floating foundations as well as installation and operation and maintenance costs. New solutions to provide better power scalability, topside weight reduction and reliability are needed. Superconductivity combines such features and allows scaling to 10 MW and beyond by radical reduction of the head mass.

In recent years, superconducting materials have emerged as real alternatives to conventional resistive materials such as copper. These materials have the potential to construct higher power rate generators with lower volumes and weights driven by the following advantages. They can achieve stronger magnetic fields by means of DC field superconducting windings. They require less iron in the magnetic circuit, which allows more space for AC stator winding. And they permit higher air-gap shear stress, which allows smaller and lighter direct-drive machine design.

Today superconductivity in wind turbines is an active R&D trend and there are several superconducting generator concepts, as those proposed by General Electric, AMSC, RISO-DTU or AML [9]. However these concepts face certain technical barriers that complicate their industrial feasibility for the very challenging offshore wind sector. Cryogenic fluids are used to cool down the rotor, and compatibility of the use of cryogens with the marine environment may not have been taken into account. Reliability of the cooling system has to be carefully considered with handling and availability of cryogenic fluids in offshore locations. The huge torque to transmit from the room temperature environment to the cryogenic temperature is also a challenging mechanical problem. Finally, most of the superconducting generator concepts are based on still expensive and not always commercially available materials, such as second generation high-temperature superconductors (2G HTS), or materials without attractive cost reduction perspectives in the future, such as 1G HTS.

3. Suprapower: Superconducting light weight generator

The present work describes SUPRAPOWER EU FP7, a cofounded research project that started in December 2012 and which is expected to finish at the end of 2016. It aims to provide an important breakthrough in offshore wind industrial solutions by designing and validating an innovative, lightweight, robust and reliable 10 MW class MgB_2 superconducting generator for offshore wind turbines, on the basis of the TECNALIA's patented generator concept [10].

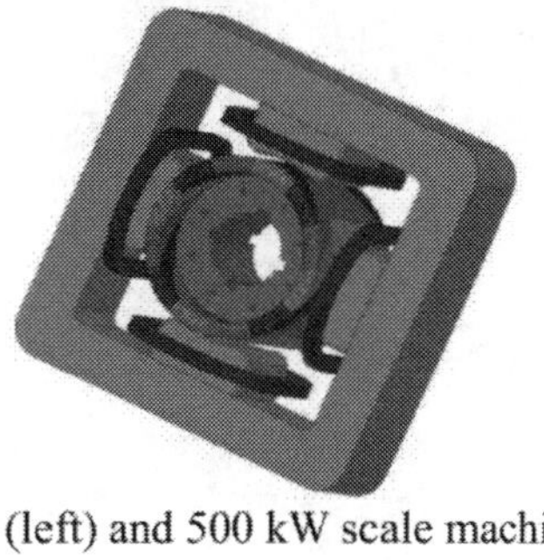

Figure 1. 10 MW superconducting generator concept (left) and 500 kW scale machine (right)

This generator overcomes the presented barriers for other superconducting concepts while the design is oriented to the offshore wind industry demands. The main advantages of this solution are the reduction of turbine head mass to about 30 % with respect to conventional generators, a modular cryogen free cooling system that requires a reduced maintenance and the no need of rare earth materials which have shown high price volatility and supply problems.

The generator concept will be validated through a scale machine in the range of 500 kW (figure 1). To keep the maximum similitude between the model and the full scale generator (figure 2), the power reduction will be obtained by reducing the number of poles from 60 to 4, maintaining the size of the SC rotor coils identical both in full and small scale generator. Basic full scale generator features as specific shear stress, superconducting and cryogenic implementation, modularity, quench detection or torque transmission will be similar, too.

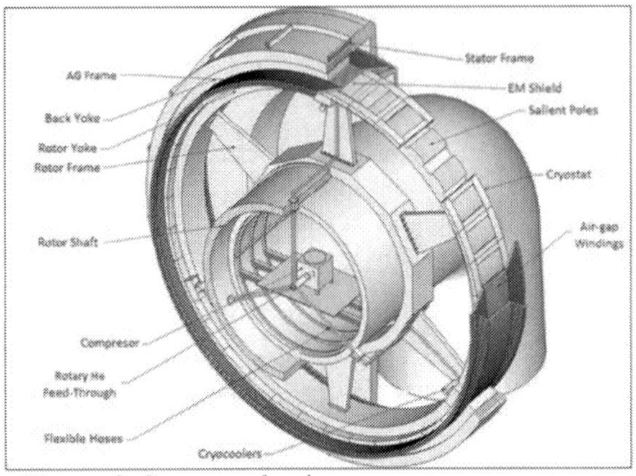

Summary of characteristics:
- 10 MW
- 8.1 rpm and 11.8 MN·m
- MgB_2 superconducting field coils
- Air-gap armature winding
- 60 warm iron poles, 230 kAm
- 11.9 m air-gap diameter
- 0.52 m stack length
- Overall weight including structural mass ~ 200 t
- Efficiency at full load over 95% (including cryogenic system)
- One year maintenance interval

Figure 2. Superconducting generator concept patented by TECNALIA (left). Main features of the 10 MW generator design (right)

The superconducting generator is a direct drive salient poles synchronous machine, based on the use of MgB_2 wire in the superconducting field coils, which is an industrial solution with a very competitive cost, near to copper, and several times lower than other HTS wires. MgB_2 wires in the form of sandwich tape with outer Cu stabilization layer, will be specifically designed, manufactured and characterized for this generator.

The field coils will be constituted by a stack of several double pancakes of MgB_2 racetrack coils, placed between 2 thick copper plates for the refrigeration of the assembly (see figure 3 right).

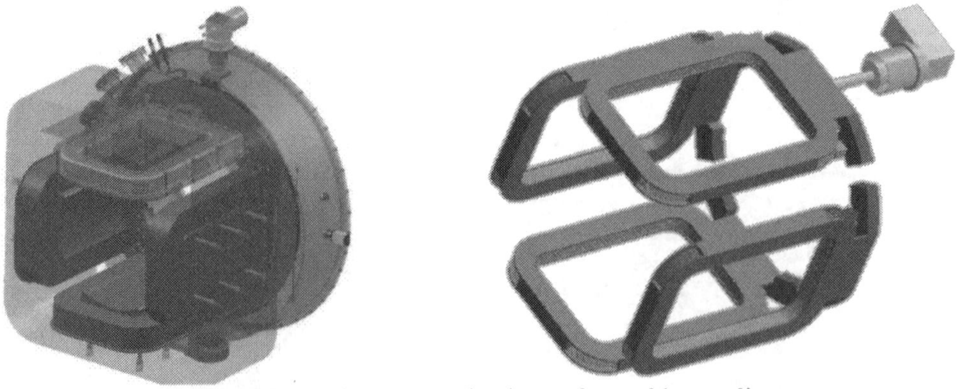

Figure 3. SUPRAPOWER's superconducting scale machine cooling system

The cooling system is an essential part of the superconducting rotating machines, up to now circulating cryogenic liquids solutions have been used, but this is a low reliable option for offshore applications. In this case a cryogen free cooling system in rotating configuration has been designed. As it is shown in figure 3, it basically consists of two parts, one modular cryostat able to accommodate one coil and a thermal collector which links all the modules, while the heat is extracted by two stage G-M cryocoolers, which rotate jointly with the rotor. Heat will be extracted by conduction through two high conductivity thermal circuits, enclosed by the thermal collector. One of them will be connected to the cryocoolers first stages and to the thermal shield (T~80K), and the other one will be thermally linked to the second stage of the cryocoolers and to the superconducting coils, maintaining them at operation temperature (T~20K).

The rotor has a warm iron poles configuration, which means that each modular cryostat only encloses the superconducting coils, thus the mass to be cooled is reduced and the high torque (11.8 MN·m) transmission will be shared between the cold part (superconducting coils) and the warm one (Fe poles). In addition, the modular cryostat solution makes the construction, assembly and maintenance operations affordable in offshore conditions.

4. Conclusions

A novel concept of a lightweight, reliable and robust 10 MW class superconducting generator has been presented. It aims to fulfil the offshore industry demands on LCOE reduction by means of higher power rates and more reliable turbines.

Starting from an already patent-applied concept by TECNALIA, this generator is based on the use of MgB_2 wires for the rotor field coils; cryogen free modular cooling system; warm iron poles configuration and modular cryostat.

The generator highlights are, a 30 % weight reduction with respect to conventional generator technologies, reduced maintenance requirements, design fully oriented to the offshore environment and very competitive life cycle costs.

Acknowledgments

The research leading to these results has received funding from the European Union Seventh Framework Programme (FP7/2007-2013) under grant agreement n° 308793.

References

[1] Global Wind Energy Counsil (GWEC) Global statistics.

[2] European Wind Energy Association (EWEA) report February 2013 *Wind in power 2012 European statistics*.

[3] EU agreement for 20% carbon emission reduction by 2020. February 2007.

[4] International Energy Agency 2010 *World Energy Outlook 2010* ISBN 978-92-64-08624-1.

[5] European Wind Energy Association report November 2011 *Wind in our Sails The coming of Europe's offshore wind energy industry*.

[6] The Crown Estate report May 2012 *Offshore Wind Cost Reduction Pathways Study*.

[7] European Wind Energy Association report January 2013 *The European offshore wind industry key trends and statistics 2012*.

[8] Zhang Z, Chen A, Matveev A, Nilssen A, Nysveen A 2013 High-power Generators for Offshore Wind Turbines Original Research Article. *Energy Procedia* **Volume 35** 52-61.

[9] Jensen BB, Mijatovic N, Abrahamsen AB. 2012 *Development of Superconducting Wind Turbine Generators*. In Scientific Proceedings of EWEA 2012 - European Wind Energy Conference & Exhibition. EWEA - The European Wind Energy Association.

[10] European Patent Application EP 2 5251 252 A1: "Direct-action superconducting synchronous generator for a wind turbine"; 7.12.2012; PCT/ES2009/070639.

Design, manufacturing and tests of first cryogen-free MgB_2 prototype coils for offshore wind generators

G Sarmiento[1], S Sanz[1], A Pujana[1], J M Merino[1], R Iturbe[2], S Apiñaniz[1], D Nardelli[3], I Marino[1]

[1]TECNALIA. Parque Tecnológico, Edificio 700. 48160 Derio. Spain
[2]ANTEC S.A. Ramón y Cajal, 74 48920 Portugalete. Spain
[3]COLUMBUS SUPERCONDUCTORS SpA, Via delle Terre Rosse, 30 16133 Genova. Italy

E-mail: gustavo.sarmiento@tecnalia.com

Abstract. Although renewable sector has started to take advantage of the offshore wind energy recently, the development is very intense. Turbines reliability, size, and cost are key aspects for the wind industry, especially in marine locations. A superconducting generator will allow a significant reduction in terms of weight and size, but cost and reliability are two aspects to deal with. MgB_2 wire is presented as one promising option to be used in superconducting coils for wind generators. This work shows the experimental results in first cryogen-free MgB_2 prototype coils, designed according to specific requirements of TECNALIA's wind generator concept.

1. Introduction

Offshore wind sector is being developed in recent times, choosing direct drive machines as one of the possible generator configuration to exploit the renewable resource. Such machines present larger size as generator power is increased, which impact directly in the turbines economy. Superconductivity has the potential to allow drastic reductions in terms of size and weight [1, 2], but cost and reliability are two key aspects required by industry in marine locations. So, intense efforts developing new concepts of superconducting (SC) wind generator are being done [3-6]; mostly based on high temperature superconductors (HTS). Among them, MgB_2 has an interesting position because it accomplishes a good compromise between performance and price, and additionally, it is available in km lengths. SC coils main parameters and cooling system have been selected according to TECNALIA's concept [7]. MgB_2 wire will be used for the field coils but its reliability must be validated. In the following sections, some results of testing first cryogen free cooled prototype coils are presented.

2. Experimental

Commercial MgB_2 wire from Columbus SpA has been selected for the field coils in the wind generator. Chosen wire is a sandwich class with 19 MgB_2 filaments and a Ni external sheath. SC filling factor is 24.1% and its size is 3 x 0.5 mm². An additional 0.2 mm thickness Cu strip soldered to one side of the Ni sheath provides stabilization. Maximum bending diameter is 150 mm. Critical temperature (T_c) at self-field is around 35.5 K. Important industrial issues are the length in a batch, more than 1.5 km and the performance vs. cost, with a reasonable price of 30 €/(kA m) at 1.8 T and 20 K. According manufacturer, the cost is foreseen to be reduced due to both performance improvement and cost reduction by a factor of 5 in a short-medium time. Figure 1(a) presents some wire pictures and its general performance under magnetic field. Generator field coils will be constituted of a stack of several racetrack shape MgB_2 double pancakes, with an overall size of about 914 mm x 662 mm x 90 mm. Before building such as field coils, it is compulsory to validate the wire and the coil manufacturing in order to demonstrate the coil performance under cryogen-free cooling and to obtain experimental results for the simulations models. After having tested straight samples, some small-scale coils have been built: a three-turn solenoid with one layer called SC1 and a solenoid of 7 turns with one layer called SC2.1. Similar coil architecture has been used by others [8]. The procedure for the SC2.2 is presented in this paper. It follows the same SC2.1 parameters. Seven turns of MgB_2 wire were wounded in one layer over a copper disk (110 mm radius, 10 mm thick). Coil cross section is around 3.13 x 6.39 mm^2 and external radius is 116.4 mm. About 5 wire meters were used. Inside coil terminal was soldered to the plate with SnPb in the Ni side. Outside one was soldered to a copper sector (Cu strip side). Disk and sector purpose is twofold: to energize the coil and to refrigerate the terminals. Soldering length is around 165 mm (outside) and 177 mm (inside) to ensure complete current transfer and refrigeration. Sector, coil and disk were vacuum impregnated using Araldit F. Winding, moulding, impregnation, curing and mould releasing were carried out by Antec S.A. Figure 1 ((b) to (e)) shows the manufacturing process.

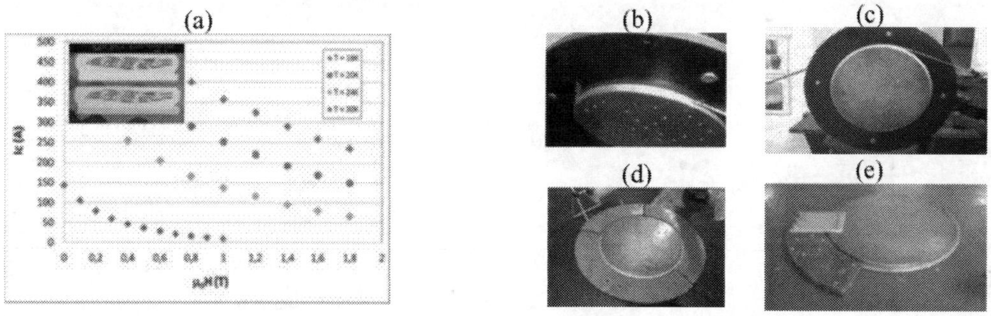

Figure 1. (a) MgB_2 sandwich class wire and current vs. magnetic field characteristics. Manufacturing process: soldering internal terminal (b), winding (c), impregnation tooling (d) and end coil (e)

Figure 2 shows the coil sketched; copper plate and sector are refrigerated by oxygen free copper braids, electrically insulated and connected to the cryocooler 2nd stage. Electric feeding is done through six DI-BSCCO tapes (manufactured by Sumitomo), put in contact by pressure using In interface to the plate and sector. Cooling is allowed by extracting heat through the Cu plate and sector. The most relevant voltage taps and thermometers are also defined in figure 2. Wiring is thermally anchored to the disk, to minimize heat entrances. Hot spot heater is also located at point (3), though induced quench results are not showed.

3. Results and discussion

Coil was tested under three thermal ranges: ~11 K, ~25 K, ~30 K. Cooling configuration was designed to have temperature gradients between disk and sector and, therefore, inside the coil. In figure 3, a current ramp in the ~11 K test is shown; similar trends were followed in higher temperature tests. As the coil aim is to work under DC conditions, current steps were kept more than one minute.

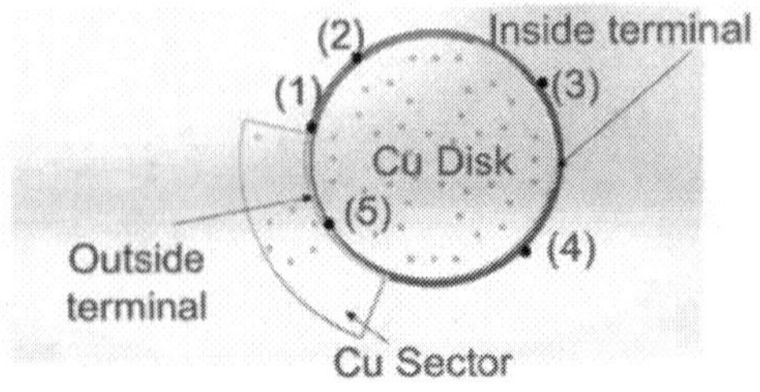

Figure 2. Coil SC2.2 sketched, showing voltage taps and thermometers distribution (1) to (5)

In order to extract current transfer effects, the inner and outer terminal performances are also presented. Inside one presents a mix between resistance and superconductivity behavior. For the 11 and 25 K tests, the electric field vs. temperature follows the same trend; with a resistance of about 10^{-8} Ω. It could be due to the solder SnPb and the Ni length that the current crosses during the current transfer process. For the 30 K test, the SC behavior is denoted with an I_c of 100 A and "n" value around 6. Outside terminal presents a thermal dependence and a resistance of ~1 μΩ, in which SC part has almost disappeared, probably because of manipulation during soldering. Estimated values for the SC2.2 load line present critical currents (I_c) higher than 500 A, 400 A and 80 A for temperatures of 15 K, 20 K, and 30 K respectively. Inside terminal I_c fits well with the expected critical current.

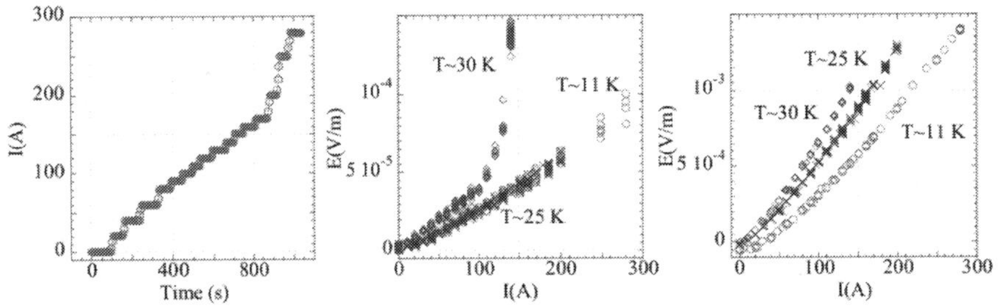

Figure 3. Current ramp for the test at ~11K (left). Electric field for the terminals, inside one (middle) and outside one (left), at different test temperatures.

Coil performance under current injection is presented in figure 4. A set of voltage taps and thermometers is distributed in the coil according to the sketch in figure 2, located in the most external turn. Temperature behavior is presented for the 11 K, 25 K and 30 K range, showing a thermal gradient through the wire in the coil. However, even with zero current, temperatures are not in the gap between disk and sector. That is explained as a heat entrance through wiring. In the lower part of figure 4, voltage drops in the whole coil (V_{coil}) and in selected voltage taps are presented. They are located at the end of wire length, close to the outside terminal (V2-V1) and in the last turn (V5-V4). Degradation due to the manipulation could be the reason of (V2-V1) voltage drop. An artificial defect generated in the wire during winding by accidental bending is probably the responsible of the (V5-V4) voltage drop. Subtraction of voltage drops (V2-V1) and (V5-V4) to the total V_{coil} (named "$V_{residual}$") shows the current injection evolution in the non-degraded part of the coil. Degraded points are the responsible of the thermal increase detected in all range tests, due to internal heat generation. Special interest has the 30 K range test, in which, above 100 A, $V_{residual}$ is increased. That fact is explained taking into account that inside terminal has quenched, and it is propagating outwards. Despite the heat generation, coil is rather stable below DC current of 100 A. In the 11 K range test applied current reaches 280A, the maximum allowed with the used current leads. Note that sector temperature is fairly constant and that the disk follows the same trend that temperature in selected voltage taps. That may be explained as the heat extraction is more effective through thin insulation to the disk than longitudinal through Cu strip to the sector.

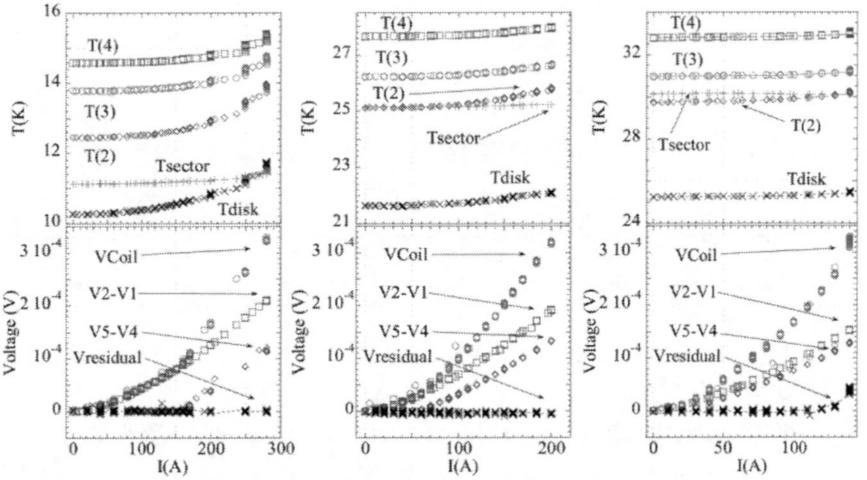

Figure 4. Temperature evolution in selected voltage taps and voltage drops, corresponding to 11 K (left), 25 K (middle) and 30 K (right) range tests.

4. Conclusions

Prototype MgB_2 coils have been designed, manufactured and tested, according TECNALIA's requirements for wind generators SC field coils. SC2.2 coil has been found superconductor and thermally stable for DC current up to 100 A in three temperature ranges of 11 K, 25 K and 30 K. Higher currents developed heating in the coil, though at 11 K up to DC 280 A were applied without quench. Degradation is found in outside terminal end wire part and in an internal defect. Two last ones are responsible of internal heat generation. Quench started in the inside terminal at its expected I_c. Further works are necessary to analyse the thermal process, quench propagation and the overall coil I_c.

Acknowledgment

The Spanish Ministry of Industry, Energy and Tourism support under the project SUPERTURBINES (IPT-2011-1664-920000).

References

[1] Kalsi S, 2011 Applications of High Temperature Superconductors to Electric Power Equipment (Wiley-IEEE Press).

[2] Nick W, Frank M, Klaus G, Frauenhofer J. and Neumuller H 2007 IEEE Trans. Appl. Supercond. 17 2030–2033

[3] Abrahamsen A B, Jensen B B, Seiler E, Mijatovic N, Manuel R C V, Andersen N H and Østergaard J 2011 Physica C 471 1464–1469

[4] Lewis C 2010 Direct drive superconducting wind generators. Wind Power Generation and Wind Turbine Design (Southampton, UK: Witpress)

[5] Snitchler G, Gamble B, King C and Winn P, 2011 IEEE Trans.Appl. Supercond. 21 1089–1092

[6] Fair R, 2012 Superconductivity for large scale wind turbines Final Scientific Report - DE-EE0005143, General Electric –Global Research, Niskayuna/DUNS 08618840

[7] Scuotto M, Sarmiento G, Merino Azcárraga J M, García-Tejedor J, Ibañez P, Apiñániz S 2009 Patent Application WO2011080357 A1, PCT/ES2009/070639

[8] Pelegrín J, Romano G, Martínez E, Angurel L A, Navarro R, Ferdeghini C, Brisigotti S, Grasso G and Nardelli D 2013 Supercond. Sci. Technol 26 045002

Design and test of a NbTi prototype coil for a low beta section

M Statera[1], G Ciullo[1], K Grigoryev[2], H Jagdfeld[3], L Kotchenda[2], P Kravtsov[2], P Lenisa[1], F Rathmann[3], H Soeltner[4], J Schillings[3], H Straatmann[3], V Trofimov[2], A Vasilyev[2]

1 Università di Ferrara and INFN Sezione di Ferrara - V. Saragat 1/C 44122 Ferrara - Italy

2 PNPI, Orlova Roscha, Gatchina, Leningrad district, 188300, Russia

3 Forschungszentrum Juelich GmbH - 52425 Jülich – Germany

Abstract. The design of superconducting quadrupoles for a proton and antiproton low beta section and the test of a prototype coil are presented. Previous studies [1,2] show that high gradient and short quadrupole magnets are required for a compact low beta section in order to allow the insertion of such a magnetic system with minor changes of the lattice [3]; each quadrupole is 400 mm long and has to provide a magnetic induction gradient of 60 T/m. A beam pipe of at least 120 mm diameter is required to avoid beam loss during injection and before the beam cooling. The magnetic design of the superconducting magnets for the low beta section is presented, together with a detailed discussion of the quench protection design. Two prototype coils were produced and one of them was tested. A detailed description of the test setup and a full discussion of the results will be presented.

1. Introduction

A low beta section to inject a proton or antiproton beam into an accumulation cell [4] having a low impact on the lattice has been investigated [1]. A proton energy range up to 3.7 GeV has been considered [2]. In order to reduce the impact of the insertion of the low beta section, high gradient and short superconducting quadrupole can be used to design a compact assembly of two magnets each side of the target region. Each quadrupole has a magnetic length of 400 mm and a gradient of 60 T/m, giving an integrated gradient of 24 T The maximum beam envelope diameter is 114 mm so that the reference radius is 57 mm.

2. Superconducting magnets

The superconducting magnets for the low beta section are an assembly of four NbTi racetrack coils (figure 1). Each coil is 443 mm long and has a cross section of 39 mm x 65 mm; the distance between facing coils is 140 mm. The quadrupole has an iron shield 30 mm thick with an inner radius of 180 mm, the shielding reduces the fringe filed in the radial direction and does not affect the field in the quadrupole. The NbTi wire by SUPERCON Inc. has an (insulated) outer diameter of 1.17 mm and a copper over superconductor ratio of 1.3. The loadline of the magnet together with the engineering critical line of the NbTi wire is plotted in figure 2; the engineering critical line is calculated reducing the wire performance of 20% assuming a 0.8 filling factor in coil windings [5]. The working point is

[1] To whom any correspondence should be addressed. E-mail statera@fe.infn.it.

set at 80% of the critical point on the loadline. Maximum field on coils at the working point is 6.88 T and the maximum radial magnetic field at the reference radius is 3.37 T (see also figure 1). The radial gradient at the center of the magnet is then 59.2 T/m, whereas the integrated gradient over a line parallel to z axis at the reference radius (57 mm) is 23.4 T; the latter includes the fringe field.

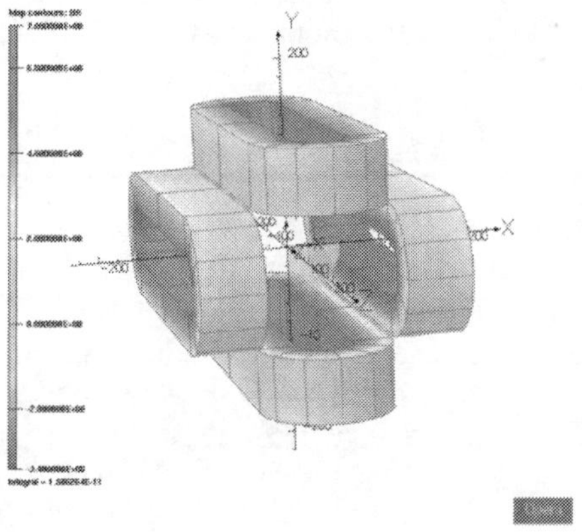

Figure 1. Field on coils (Bmod) and radial field (Br) at the center of the magnet (z=0); the radial field is plotted up to 57 mm radius.

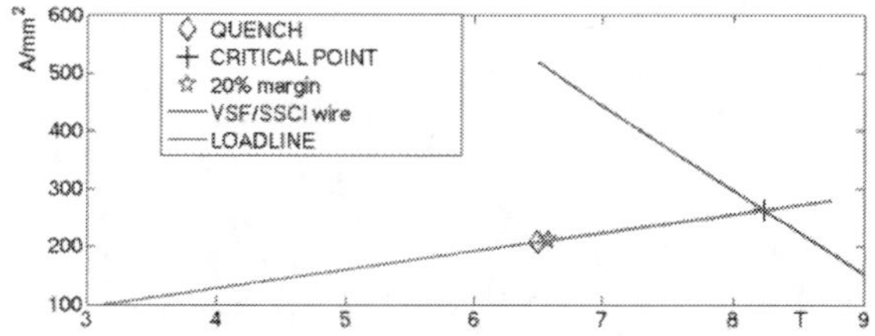

Figure 2. Load line of the superconducting coil and critical line of the VSF/SSCI NbTi wire. The latter is calculated using data by SUPERCON Inc. considering an 80% filling factor. The critical point, the calculated working point for 20% safety margin and the measured quench (QUENCH) point are plotted on the load line.

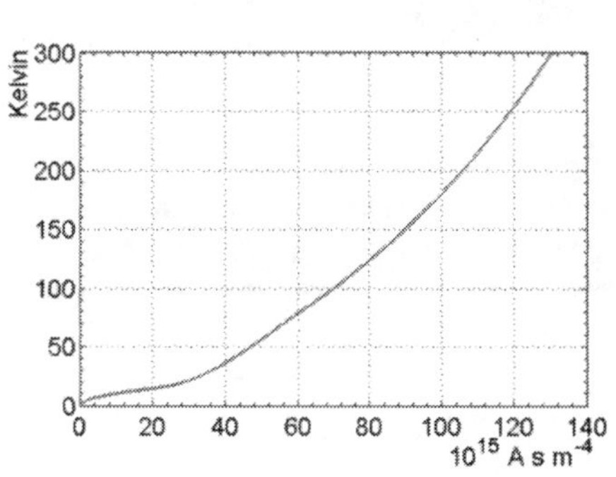

Figure 3. Adiabatic maximum temperature versus the ICD ($\int J_c^2 dt$) for the VSF/SSCI wire.

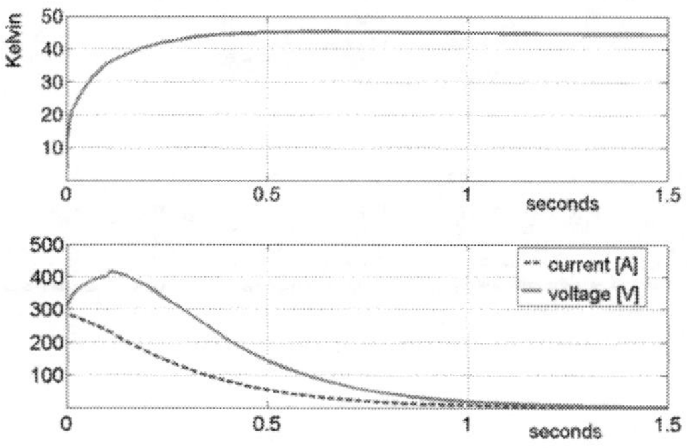

Figure 4. Maximum temperature (top) and voltage and current (bottom) versus time calculated by quench program by Cobham plc.

2.1. Quench protection

The engineering current density at the working point is 210 A/mm^2, the wire current is 285- A and the energy stored for a single coil is 30 kJ. The inductance is thus 0.75 H. Two quench calculations has been performed: the first in the adiabatic case [6] and a full finite elements model (FEM) quench simulation. The adiabatic maximum temperature as a function of the integrated current density (ICD) in the coil for the VSF/SSCI wire is plotted in figure 3; the calculated maximum temperature is about 130 K since the ICD is $8.3 \cdot 10^{16}$ $A \cdot s \cdot m^{-4}$ considering a shunt resistance of 7 Ω and a maximum voltage of 1 kV. The coil temperature for the FEM simulation, together with the voltage over the coil and the current versus time are plotted in figure 4. Maximum temperature is about 45 K and maximum voltage is below 450 V. In the simulation, a reaction time of 100 ms and a shunt resistance of 0.5 Ω are considered.

3. Cold mass design of the Prototype coil test

Two prototype coils were produced by ALSTOM; the G10 pole and case of the coil was designed and machined at Forshungzentrum Juelich (FZJ). One of the coils encased in the G10 support is visible in figure 5: the two NbTi wires to connect the coil are also visible. The cold mass assembly is plotted in figure 6. The design of the cold mass assembly considered two load cases: the cooldown from room temperature to 4 K and the Lorentz force. The latter were calculated up to a current density of 225 A/mm^2, 7% higher than the nominal working point, splitting the coil in 12 sections. The calculation by ANSYS showed that the cold mass assembly can work with a safety margin higher than 1.4 and requires the use of AL AW5083-H321. Both the G10 support and the aluminium case have to be machined with a 50 μm tolerance. The G10 support guarantees high electrical insulation, high mechanical resistance, no eddy currents and low thermal expansion coefficient.

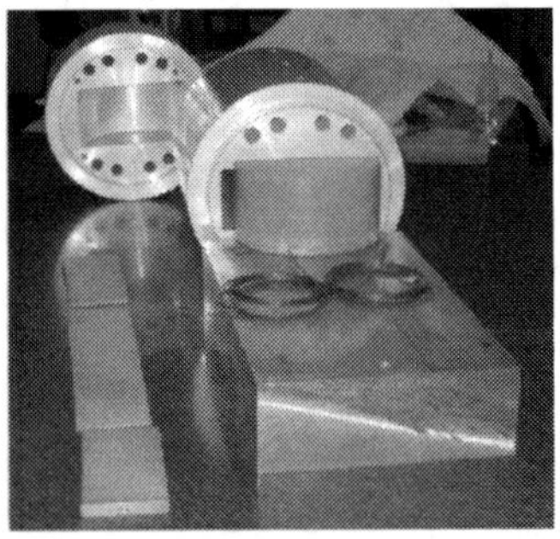

Figure 5. Cold mass of the single coil test during assembly. The G10 support of the coil, the aluminium case body, its end cap and the Al wedges are visible.

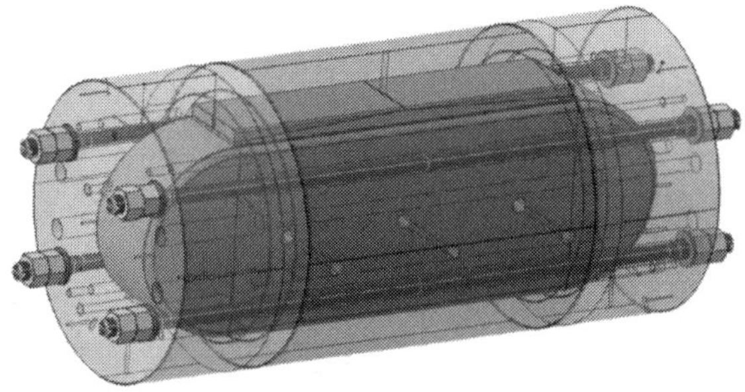

Figure 6. Drawing of the cold mass assembly: all the parts in Figure 5 and the aluminium tie rods which provide the required pre-stress are plotted.

4. Results

The cold assembly during the commissioning of the cold test is visible in figure 7. One prototype coil has been housed in a standard top access cryostat and cooled down to 4 K; the cooldown lasted about two days in order to avoid damages of the coil due to cooling speed. A quench protection system having a 10 ms reaction time and a 0.5 Ω shunt resistance together with a slow control system to control the current supplied (up to 400 A) and the current leads temperature were used. The coil was successfully loaded up to 200 A with a ramping speed up to 30 A/s and then up to 282 A with a ramping speed of 16 A/s (see figure 8). The limits of current in the first case and ramping speed in the latter are due to power supply system limitation. The coil survived different quenches due to heat leak

by current leads and it did not suffer any damage after the quench during the higher current test at 282 A. A full training of the coil could not be performed due to the amount of liquid helium available for the tests; the first current induced quench is 99% of the nominal current.
Both the maximum current simulations and mechanical design are validated since the full assembly and in particular the superconducting have not suffered any damage. The presented NbTi racetrack coils can be used to assemble superconducting quadrupoles for a compact low beta section.

Figure 7. Cold mass (see figures 5, 6) installed on the top flange of the cryostat

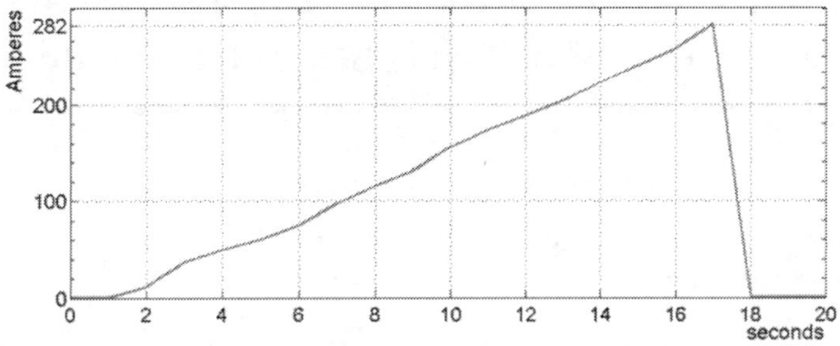

Figure 8. Measured Current ramp up to 282 A. The nominal working point current is 285 A.

References

[1] Statera M, Garishvili A, Lorentz B, Martin S A, Rathmann F, Lenisa P and Stancari G 2007 A low beta section for polarization studies of antiprotons by spin filtering, *22nd Particle Accelerator Conference Proceedings PAC'07, Albuquerque, USA*, Conf.Proc. C070625 pp 1451-1453

[2] Garishvili A, Nass A, Steffens E, Lehrach A, Lorentz B, Maier R, Rathmann F, Schleichert R, Stroher H, Lenisa P, Statera M and Martin S 2008 A Lattice Studies for Spin-filtering Experiments at COSY and AD *11th European Particle Accelerator Conference Proceedings, EPAC 2008, Conf.Proc. C0806233* pp 322-324

[3] Statera M, Martin S A, Garishvili A, Lorentz B, Lenisa P and Rathmann F 2008 A Superconducting Magnets System for a Low Beta Section at COSY and AD , *IEEE Transactions on Applied Superconductivity Vol. 18 Issue 2* pp 1463-1466

[4] Ciullo G, Barion L, Barschel C, Grigoriev K, Lenisa P, Nass A, Sarkadi J, Statera M, Steffens E and Tagliente G 2011 The polarised internal target for the PAX experiment, *J. Phys.: Conf. Ser. 295 012150* doi:10.1088/1742-6596/295/1/012150

[5] Statera M, Contalbrigo M, Pappalardo L, Barion L, Bertelli S, Ciullo G and Lenisa P 2013 Magnetic System for the CLAS12 Proposal *IEEE Transactions on Applied Superconductivity Vol. 23 Issue 3* pp 1463-1466

[6] White G K, White G and Meeson P J 2002 *Experimental Techniques in Low Temperature Physics* (Oxford) ISBN: 019851428X

Toroidal superconducting transformer with cold magnetic core - results of analysis and measurements

B Grzesik and M Stepien

The Silesian University of Technology, Faculty of Electrical Engineering
B. Krzywoustego 2, 44-100 Gliwice, Poland

mariusz.stepien@polsl.pl

Abstract. The paper is focused on a toroidal superconducting transformer with cold magnetic core. The transformer was developed aiming at the solution where magnetic core could operate immersed in LN2, not having much higher power losses than a core operating in room temperature. The second aim is concerned to the perpendicular component of the magnetic flux. It was assumed that it should be minimized. The third aim is required maximal coupling coefficient between windings. Magnetic material METGLAS 2605 SA1 (Metglas Ltd) was selected as the best taking electrical and mechanical properties into account. Numerical analysis of the transformer was carried using ANSYS software. It was assumed that magnetic core could be put inside windings and/or outside of it. The analysis and measurements yield that the best solution is the transformer with magnetic core put inside the windings. Such a construction reduces significantly perpendicular component of the magnetic field in HTS wire. Because outer winding is put on the inner one it gives maximal possible coupling coefficient. The primary and the secondary have the same number of turns. It means bifilar way of winding. The described transformer has the following constructional data: i) diameter of the main circle: 144 mm, ii) maximal diameter: 208 mm, iii) diameter of the small cross-section of the transformer (without external magnetic core): 50 mm and it results with the following parameters: output power 1.59 kVA, power density 700 VA/kg, efficiency: 99.5, coupling coefficient 0.99 at 100 A_{RMS} of input current and maximum flux density 1.2 T.

1. Introduction

A transformer is a one of the most important elements in the electric power system. The efficiency of the transformer influences on the efficiency of the whole power system. Application of new materials (HTS conductors, amorphous magnetic materials) increases the efficiency of the transformer. The paper contains the results of modeling and laboratory tests of HTS transformer with amorphous magnetic core that is kept in LN2 [1, 2, 3].

The idea of analyzed toroidal HTS transformer is presented in Fig.1. The windings of the transformer are uniformly distributed along whole torus circumference with 1:1 turn-to-turn ratio. The windings are made of HTS tape Bi-2223. The cold magnetic core (in LN2) is made of amorphous magnetic material METGLAS 2605SA1. The windings have the form of helix with the angle that can vary from near zero to right one. Magnetic core is put inside (not shown in Fig.1) as well as outside of the windings. The presented HTS transformer was theoretically analyzed by FEM for different variants of construction (only internal magnetic core, only external magnetic core and both types of magnetic core for various angles of windings). A transformer with internal core, similar to the classical toroidal transformer, was examined in laboratory. The results are presented in this paper.

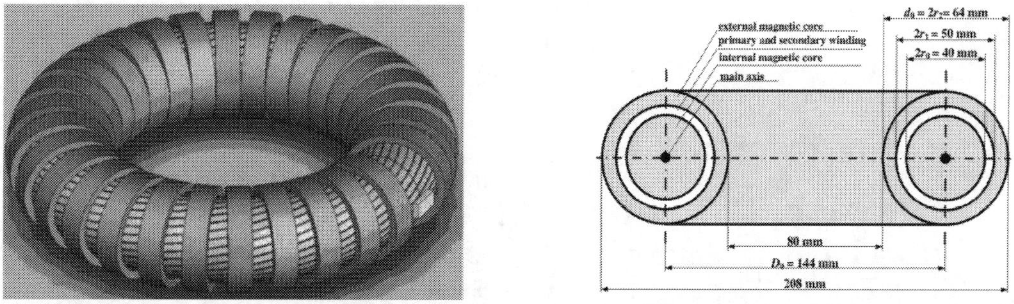

Figure 1. Idea of a toroidal transformer (left) and its main dimensions used in analysis (right).

2. An experimental toroidal transformer

The described toroidal HTS transformer was one variant of mentioned (see Fig. 1) where magnetic core is put inside windings. The windings are perpendicular to the main axis of the transformer. This is a typical construction of the transformer with toroidal shape of the magnetic core and copper windings.

The construction of the experimental HTS transformer is presented in Fig. 2. The internal magnetic core is made of amorphous tape of METGLAS 2605SA1 and put into the toroidal former. It is made of epoxy resin reinforced by glass fiber. The HTS windings are uniformly distributed in one layer on the spool of the transformer. The primary winding containing 65 turns is forms the inner layer while the secondary having the same number of turns - 65 forms outer layers. The distance between them equals twice of Kapton's thickness. The isolation (1 000 V) between primary and secondary winding is made of Kapton tape.

Figure 2. Components of the experimental transformer, magnetic core (left), non-magnetic former (middle) and final transformer wounded with HTS tape.

3. A stand for laboratory tests

The scheme of the laboratory stand is presented in Fig.3. The experimental HTS transformer was immersed into LN2 in the open cryostat. The transformer was supplied by the current source with adjustable range of amplitude and frequency of the output current (10 - 100 A; 40 - 71 Hz). The transformer was measured by digital oscilloscope Tektronix TDS 3032 with current probes (Chauvin-Arnoux) and voltage probes (Tektronix P5205)

The stand was used to carry out measurements of the transformer. A results of the laboratory tests embrace the main parameters of the transformer and parameters of the equivalent circuit. During the tests the input and output voltages u1(t), u2(t) and the input and output currents i1(t), i2(t) were measured in the wide range of current value (10A - 100A) and current frequency (40 - 71 Hz).

Journal of Physics: Conference Series **507** (2014) 032043 doi:10.1088/1742-6596/507/3/032043

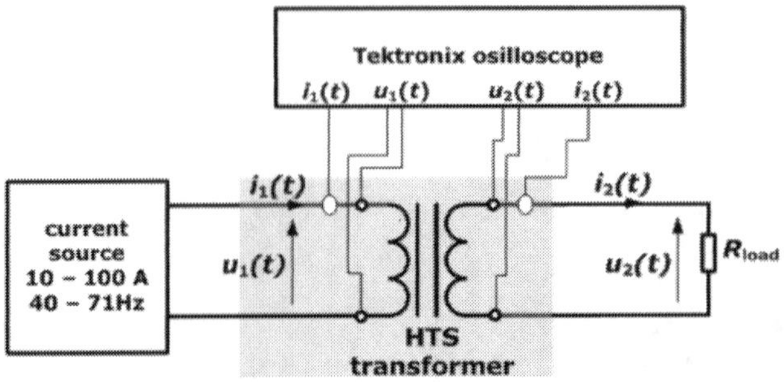

Figure 3. Scheme of the laboratory stand for HTS transformer.

4. Results of numerical modelling of toroidal transformer

A detailed numerical analysis was carried out using 2D and 3D ANSYS modeling. Assuming linear parameters of model (harmonic analysis) and HTS wire operated with rated current power properties were determined. The results were calculated for different shift *s* angle, which reflects twist of HTS wire (shift s is 0 at windings parallel to core axis – normal to core cross section, and it is 180 when windings are perpendicular) – cf. Fig. 5. In Fig. 5 output power is shown as a function of shift *s*.

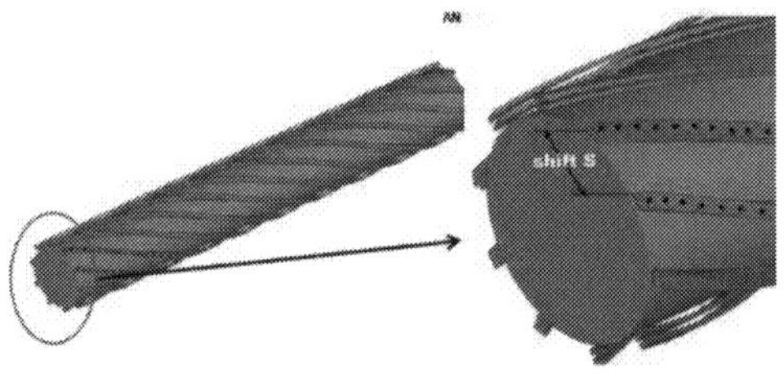

Figure 4. Geometry of FEM model with windings twisted with shift *s*.

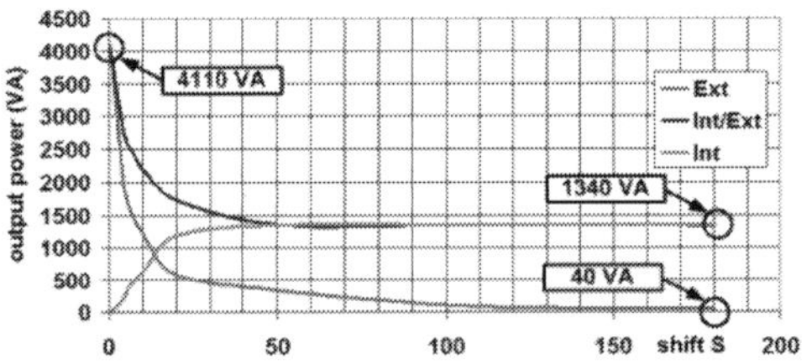

Figure 5. Output power of perpendicular transformer obtained by FEM

The perpendicular transformer (with shift *s* 180) is corresponding to the experimental one. The results obtained for this type are input voltage 13.5 V, output power 1340 VA (c. f. Fig. 5), efficiency 99,5% and power density 194 VA/kg.

5. Results of laboratory tests

The results of the laboratory tests embrace the main parameters of the transformer and parameters of the equivalent circuit. During the tests the input and output voltages u_1(t), u_2(t) and the input and output currents i_1(t), i_2(t) were measured in the wide range of current value (10A - 100A) and current frequency (40 - 71 Hz).

The results of the tests: no-load and short circuit, were used to determine the parameters of the equivalent circuit of the HTS transformer. The scheme of the equivalent circuit is presented in Fig. 6. The parameters R_1, R_2 (L_1, L_2) represents correspondingly: the resistance (leakage inductance) of primary and secondary HTS winding. The parameters R_{Fe} and L_μ represent the power losses and magnetizing reactance of the magnetic core. The parameters C_{11} and C_{22} represents the stray capacitance of the primary and secondary winding while the parameter C_{12} represents the stray capacitance between primary and secondary layer of winding. A values of the parameters of equivalent circuit are shown in Fig. 6.

The load tests were used to determine the efficiency η, power-to-mass ratio P-t-m and power losses ΔP of HTS transformer for given input parameters as voltage U_1, current I_1, power S_1 and frequency *f*. The results are collected in Tab. 1. Because of low power losses the transformer has high efficiency and high power-to-mass ratio for different frequencies of the input current.

Table 1. Power properties of the transformer (measured).

frequency f	Hz	40	45	50	57	61	66	71
input voltage U_1	V	12,5	14,1	15,4	17,8	19,1	20,5	22,2
input power S_1	kVA	1248	1411	1540	1776	1914	2052	2224
efficiency η	%	99,46	99,48	99,49	99,53	99,56	99,57	99,60
power-to-mass ratio *P-t-m*	VA/kg	574	648	709	815	880	940	1024

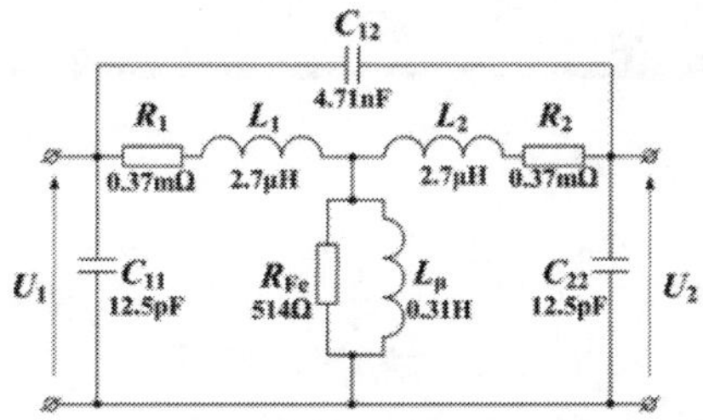

Figure 6. Equivalent circuit of the transformer and its parameters.

The transformer described in the paper having only one layer primary and one layer of secondary has power-to-mass ratio (P-t-m) of ~700 VA/kg that is higher value in comparison with reported one. For the transformer with higher number of layers one can obtain much higher than 700 VA/kg what makes it candidate for higher power and voltage construction.

6. Conclusions

The presented toroidal HTS transformer with cold magnetic core is made of new materials: HTS tape Bi-2223 and amorphous magnetic core METGLAS 2605SA1. It exhibit low power losses, high efficiency and high power-to-mass ratio (P-t-m) of ~700 VA/kg. The parameters of HTS transformer compared with other constructions of HTS transformer described in the literature [4, 5, 6].

The future work will be focused on the increasing the output power of the transformer and increasing its voltage. It can be achieved by changing the turn-to-turn ratio and operation in lower temperatures than 77K (increasing the critical current of HTS tape). The future works will embrace also voltage brake problems.

Acknowledgement

Presented work is supported by National Science Centre (Poland) under research project number DEC-2011/03/B/ST8/04908.

References

[1] Grzesik B and Jez R 2007 Highly efficient HTS-transformer requires low power losses *14th International Symposium on Power Electronics, Ee2007, Novi Sad, Serbia*, pp. 1 – 5
[2] Grzesik B, Janowski T and Stepien M 2008 HTS Toroidal Helical Transformer *J. Physics: Conf Series*, 012311 Vol. 97
[3] Grzesik B, Stepien M and Jez R 2010 Toroidal HTS transformer with cold magnetic core - analysis with FEM software *J. Physics: Conf. Series*, Vol. 234, Part 3
[4] Bohno T et al 2005 Development of 66 kV/6.9 kV 2 MVA prototype HTS power transformer *Physica C* 426-431
[5] Jelinek Z et al 2003 Test Results of 14 kVA Superconducting Transformer with Bi-2223/Ag Windings *IEEE Trans. on Appl. Supercond.* Vol. 13, no. 2, pp. 2310-2312
[6] Schlosser R et al 2003 Development of High-Temperature Superconducting Transformers for Railway Applications *IEEE Trans. on Appl. Supercond.* Vol. 13, no. 2, pp. 2325 – 2330

Influence of magnetic non-uniformity existing in a rigid rotor supported by a superconducting magnetic bearing on its whirling

Soichiro Kamada, Ryosuke Amano and Toshihiko Sugiura
Keio University, 3-14-1 Hiyoshi, Kohoku-ku, Yokohama-shi, Kanagawa, JAPAN
E-mail: sugiura@mech.keio.ac.jp

Abstract. Superconducting magnetic bearings (SMBs) have a significant feature over conventional bearings in terms of supporting a shaft without physical contact while attaining its stability without control. In their large-scale rotary applications, magnetization distribution of a rotor in the circumferential direction can be non-uniform and it would be better to know influence of such circumferential magnetic non-uniformity existing in a rotor on its dynamics, especially on its behaviors in the vicinity of the critical speed. In this study, further developing our previous research, we improved our analytical model so that we can adjust several different degrees of magnetic non-uniformity by arranging multiple magnetization vectors and investigated its influence. First, we simulated dynamical behavior of the system by numerical calculations and their results show that, with increasing the degree of magnetic non-uniformity, the whirling amplitude of the system, together with the difference of the amplitudes in the orthogonal directions in the whirling plane, get larger. Further, the rotational frequency at which the whirling amplitude takes its peak gets lower, which is caused by nonlinearity of the electromagnetic force. We carried out experiments and verified our numerical predicions.

1. Introduction

A superconducting magnetic bearing (SMB) has various merits such as low damping, supporting high rotational speed, high energy efficiency and utilization under special environment due to maintenance-free property, because they are able to support a rotor without physical contact. However, it shows complicated nonlinear vibration in the vicinity of the critical speed due to nonlinearity of electromagnetic force [1]. Nevertheless, it is often required to spin at high rotational speed over the critical speed. For these reasons, we need to evaluate behaviors near the critical speed of a system using a SMB. Magnetization distribution in the circumferential direction can be non-uniform, because SMBs, if applied to large systems including flywheel energy storage systems, will support a rotor with a single large magnet or multiple magnets connected with each other [2]. For mechanical design of such a rotary system, it would be better to know influence of such magnetic non-uniformity on rotordynamics, because it is conceivable that this magnetic non-uniformity can produce periodic change of electromagnetic force supporting a rotor, which may cause instability of rotor whirling. We previously investigated vibrations of the elastic shaft with and without circumferentially non-uniform magnetization distribution supported by a SMB and theoretically pointed out that anisotropy of its horizontal stiffness and periodic change of that stiffness leading to parametric excitation newly arose from magnetic non-uniformity [3]. We also found that the equations of motion of a rotor with magnetic non-uniformity supported by a SMB are equivalent to those of an asymmetrical shaft supported by a mechanical bearing [4]. We further numerically confirmed that an unstable frequency region, predicted in the linear theory, does not exist because of nonlinearity and that the amplitudes in the horizontal directions, where there is anisotropy of stiffness, take their peaks simultaneously, which is also different from linear prediction. In this study, we investigate influence of magnetic non-uniformity existing in a rigid rotor supported by a SMB on its whirling in more detail by numerical calculations and experiments. First, we improve our analytical model and perform numerical calculations with several different degrees of magnetic non-uniformity of a rotor. Further, by carrying out corresponding experiments, we verify our numerical predictions.

2. Theory

As shown in Fig. 1, our analytical model consists of a rigid shaft and a magnet attached at the bottom of the rotor shaft and a superconducting bulk fixed under the bottom end of the shaft. In this study, as shown in Fig. 2, we assume magnetic non-uniformity of the magnet in the circumferential direction, so that the magnetic flux density due to the magnet changes sinusoidally in a circumferential direction, as given by $B = B_0 + \Delta B \cos 2\theta$. Here θ is a circumferential angle around the central axis normal to the $x-y$ plane. The nondimensional equation of motion with and without magnetic non-uniformity are shown in (1), (2). The electromagnetic force due to the superconducting bulk is evaluated by the advanced mirror image method [5], and is expanded into Taylor series around the equilibrium position up to the 3rd order terms. We assume one permanent magnet as an individual magnetic dipole to simplify evaluation of the electromagnetic force on the basis of the advanced mirror image method. To satisfy above assumption of magnetic non-uniformity, we arrange additional magnetization vectors m_1 on both sides of a main central magnetization vector m_0 , as shown in Fig. 3 , and adjust the degree of magnetic non-uniformity by changing the value of m_1.

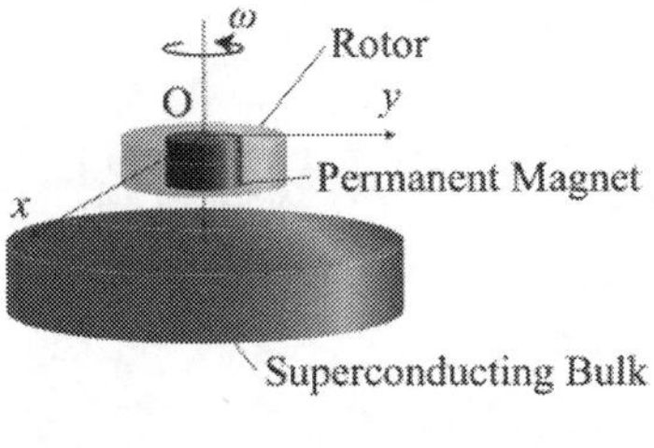

Figure 1. Analytical model.

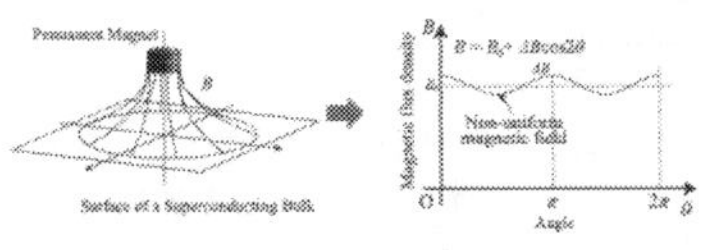

Figure 2. Assumption of magnetic non-uniformity in the circumferential direction.

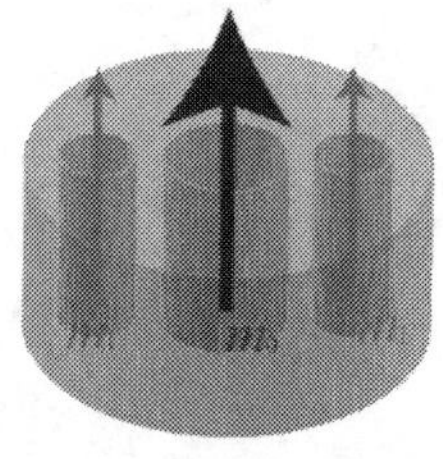

Figure 3. Arrangement of additional magnetization vectors.

$$\left\{ \begin{array}{l} \ddot{x} + 2\gamma\dot{x} + px = \epsilon\nu^2 \cos \nu t + \Delta_{xyy} xy^2 + \Delta_{xxx} x^3 \\ \ddot{y} + 2\gamma\dot{y} + py = \epsilon\nu^2 \sin \nu t + \Delta_{yxx} x^2 y + \Delta_{yyy} y^3 \end{array} \right. \quad (1)$$

$$\left\{ \begin{array}{l} \ddot{x} + 2\gamma\dot{x} + p_x x = \epsilon\nu^2 \cos \nu t + \Delta_{11} x \cos 2\nu t + \Delta_{12} y \sin 2\nu t + \Delta_{xyy} xy^2 + \Delta_{xxx} x^3 \\ \ddot{y} + 2\gamma\dot{y} + p_y y = \epsilon\nu^2 \sin \nu t + \Delta_{21} x \sin 2\nu t - \Delta_{22} y \cos 2\nu t + \Delta_{yxx} x^2 y + \Delta_{yyy} y^3 \end{array} \right. \quad (2)$$

Here γ is the nondimensional damping coefficient and ϵ is the nondimensional eccentricity. p_x and p_y are the nondimensional stiffnesses in the x and y directions, respectively . ν is the nondimensional excitation frequency. Δ_{11}, Δ_{12}, Δ_{21} and Δ_{22} are nondimensional coefficients of parametric exicitation terms and Δ_{xyy}, Δ_{xxx}, Δ_{yxx} and Δ_{yyy} are nondimensional coefficients of nonlinear terms. In our previous study [3], we discussed the force terms in the equations of motion and theoretically pointed out following significant characteristics inherent to SMB systems appearing under the condition of magnetic non-uniformity: anisotropy of its horizontal stiffnesses and periodic change of that stiffness at integral multiplication of rotational frequency which may lead to parametric excitation. We theoretically predicted that there might be some difference between the resonant frequencies in two orthogonal horizontal directions at which respectful horizontal stiffnesses differ, and that there may be also some regions of the rotational frequency at which the whirl amplitude can be unstable, and that the system could show the soft-spring nonlinearity so that the resonant peak might bend to lower frequencies in the frequency response curve and thus a phenomenon of jump in vibration to higher amplitudes may happen near the resonance. However, from our previous numerical calculations [3], we also confirmed that an unstable region does not exist and the amplitudes in the x and y directions take their peaks simultaneously.

3. Numerical calculation

Equations (1), (2) were solved numerically by the Runge-Kutta method. To adjust the degree of non-uniformity of magnetization distribution in the circumferential direction of a rotor, we changed the ratio of m_1 to m_0 to be 0, 1/36, 1/9 and 1/4. The degree of magnetic non-uniformity gets larger as the ratio of m_1 to m_0 increases. Fig. 4 (a) ~ (d) show calculated frequency response curves indicating relations between the rotational frequency and the whirling amplitude. The whirling amplitude and the rotational frequency are respectively nondimensionalized by the distance H between the superconducting bulk and the permanent magnet and the mean-square value of eigenfrequencies in the x and y directions. According to Fig. 4 (a) ~ (d), the resonant amplitudes in the x and y directions get larger as the degree of magnetic non-uniformity increases. Difference of amplitudes in the x and y directions similarly get larger. Moreover, the rotational frequency at which the whirling amplitudes in the x and y directions take their peaks gets lower.

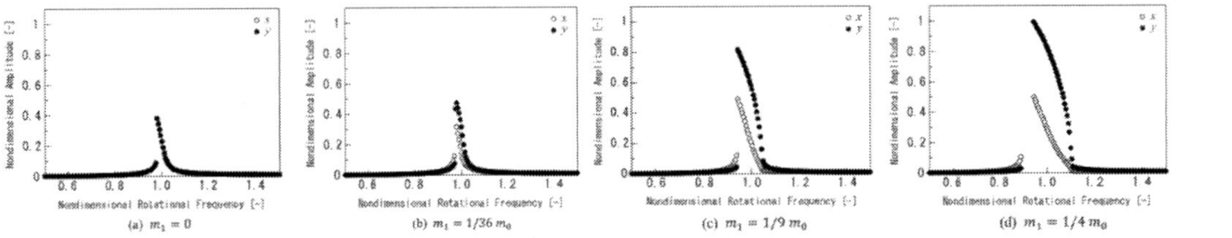

Figure 4. Numerically obtained frequency response curves.

According to Fig. 4 (b) ~ (d) , it seems that effect of soft spring characteristics due to nonlinearity increases as the degree of magnetic non-uniformity get larger. Therefore, the slope of the frequency response curves also gets larger and the amplitude in each direction has a stable solution of higher amplitude at lower rotational frequency. Thereby, the rotational frequency at which amplitudes in the x and y directions take their peaks gets lower and the peaks in respective directions grow as the degree of magnetic non-uniformity increases.

4. Experiment

We carried out experiments to verify our numerical predictions above. Our experimental model is shown in Fig. 5. As shown in Fig. 6, to adjust degrees of non-uniformity of magnetization distribution in the circumferential direction of a rotor, we additionally located two smaller permanent magnets with symmetry against a main central permanent magnet in the rotor and changed the radii of the additional magnets r_1 to be 0 mm, 1.5 mm, 2 mm and 3 mm. The radius of the main central magnet was 6 mm in all cases. Magnetic non-uniformity increases as the radius of r_1. By measuring displacements of the rotor, we obtained their time histories with their FFT spectra. Fig. 7 (a) ~ (d) show frequency response curves, nondimensionalized as in Fig. 4. Here $H = 6.6$mm and the mean-square value of eigenfrequencies are 7.69Hz in (a), 7.32Hz in (b), 7.33Hz in (c), 8.09Hz in (d). As the degree of magnetic non-uniformity increases, the resonant amplitudes in the x and y directions and their difference get larger. Furthermore, the rotational frequency at which the whirling amplitudes in the x and y directions take their peaks gets lower. Our experimental results were in good qualitative agreement with our numerical results in most cases. One of possible reasons for quantitative differences may be dipole approximation of permanent magnets.

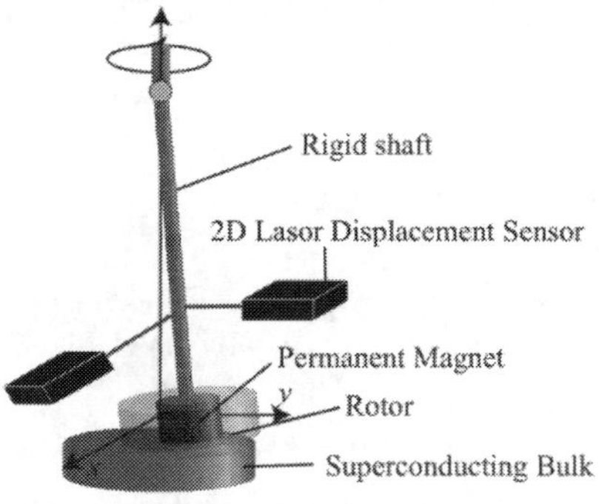

Figure 5. Schematic diagram of our experimental setup.

Figure 6. Arrangement of the additional magnets.

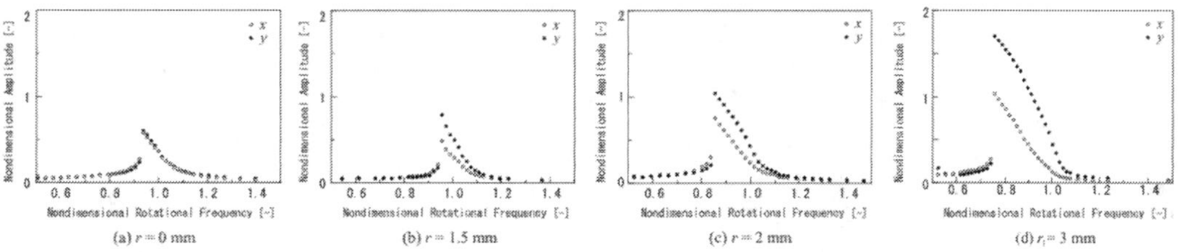

Figure 7. Experimentally obtained frequency response curves.

5. Conclusion

We numerically and experimentally investigated dynamical behaviors of a rotor with circumferentially non-uniform magnetization distribution supported by a SMB. Obtained results of numerical calculations with increasing the degree of magnetic non-uniformity showed that the whirling amplitude of the system gets larger and that difference of the amplitudes in the x and y directions also gets larger. Further, the rotational frequency at which the amplitudes in the x and y directions take their peaks gets lower. These numerical predictions were verified by experiments.

References

[1] Hull R J 2000 *Supercond. Sci. Technol.* **13** R1-R15.
[2] Frank. N. Werfel, et al. 2010 *IEEE Transactions on Applied Superconductivity* **20** 2272-2275
[3] Amano R, Kamada S and Sugiura T 2013 *IEEE Transactions on Applied Superconductivity* **23** 5202104
[4] Ishida Y and Liu J 2010 *Journal of vibration and acoustics* **132** 011011.1-011011.8
[5] Kordyuk A A *J. Appl. Phys.* **83** 610-2.

Vibration reduction using autoparametric resonance in a high-T_c superconducting levitation system

Hiroshi Yamasaki, Toyoki Takazakura, Ryunosuke Sakaguchi and Toshihiko Sugiura

Keio University, 3-14-1 Hiyoshi, Kohoku-ku, Yokohama-shi, Kanagawa, Japan

E-mail: sugiura@mech.keio.ac.jp

Abstract. High-T_c superconducting levitation systems have very small damping and enable stable levitation without control. Therefore, they can be applied to various kinds of application. However, there are some problems that small damping produces large vibration and nonlinearity of magnetic force can generate complicated phenomena. Accordingly, analysis of these phenomena and reduction of vibration occurring in the system are important. In this study, we examined reduction of vibration without using any absorbers, but utilizing autoparametric resonance caused by nonlinear coupling between vertical oscillation and horizontal oscillation. We conducted numerical analysis and experiments in order to investigate motions of a rigid bar levitated by the electromagnetic force from high-T_c superconductors. As a result, if the ratio of the natural frequency of vertical oscillation and that of horizontal oscillation is two to one, the vertical oscillation decreases while the horizontal oscillation is excited. Thus, it was confirmed that the amplitude of a primary resonance can be reduced by occurrence of autoparametric resonance without using any absorbers.

1. Introduction

High-T_c (High-Critical-Temperature) superconducting levitation systems have very small damping and stable levitation without control. Therefore, they can be applied to flywheels for energy storage, conveyors in clean room and so on [1]. However, in such small damping systems, there is a problem that vibration amplitude may increase easily. Too large vibration may cause destruction of the system, hence it is important to suppress vibration in a high-T_c superconducting levitation system.

In previous study, Takazakura et al. confirmed that parametrically excited oscillation in a high-T_c superconducting levitation system can be reduced by using a vibration absorber [2]. Besides, H. Yabuno et al. confirmed the validity of an autoparametric absorber to stabilize 1/3-order subharmonic resonance [3]. However, there are few studies of vibration reduction by using nonlinearity of a system itself.

The purpose of this study is reduction of the amplitude of the primary resonance of a levitated body by utilizing autoparametric resonance, one of nonlinear interaction between plural vibration modes, without using any absorbers. In general, if their natural frequency ratio is 2 to 1, the autoparametric resonance can occur in a multiple degree-of-freedom coupled nonlinear system [4]. Here we deal with a system which consists of a rigid body and two permanent magnets levitated above two high-T_c superconducting bulks. In this system, vertical motion and horizontal motion are coupled nonlinearly by magnetic force. We aim at transferring kinetic energy from vertical motion to horizontal motion through autoparametric resonance. In order to utilize the autoparametric resonance, we examined numerically and experimentally how it affects the vibration amplitude of the levitated body in the vertical direction when the primary resonance occurs.

2. Analytical model and equations

An analytical model is shown in Figure 1. The main system of the model is a symmetric rigid bar supported at both its ends by electromagnetic forces between permanent magnets and superconductors. Using this model, we considered the two degrees of freedom defined by the motion of the levitated bar. The coordinates are shown in Figure 1. The initial position of the center of the levitated body before field cooling is $(x, Z) = (0, z_0)$. After field cooling, the levitated body balances with the gravity at the equilibrium position $(0, z_0 - z_s)$. We further define the relative coordinate z in the vertical direction from the equilibrium position. The superconductors are excited in the vertical direction on the amplitude a and the frequency N. The magnetic forces exerted by superconductors to magnets, F_z and F_x, are evaluated here by using the frozen mirror image method [5]. Further, the Taylor expansion is applied them up to the third order terms. The equations of motion of the levitated body, nondimensionalized by the natural period in the vertical direction Ω_z and z_s, are derived as follows,

$$\ddot{z} + \mu_z \dot{z} + z - \alpha_{zzz} z^2 - \alpha_{xxz} x^2 + \alpha_{zzzz} z^3 + \alpha_{xxzz} x^2 z = b\nu^2 \cos \nu t \quad (1)$$

$$\ddot{x} + \mu_x \dot{x} + {\omega_x}^2 x - \alpha_{xzx} xz - \alpha_{xxxx} x^3 + \alpha_{xzzz} xz^2 = 0 \quad (2)$$

where μ_- are damping coefficients, α_- are coefficients of the Taylor series of the magnetic forces (F_z and F_x) up to the third order terms, $b \ (= \frac{a}{|z_s|})$ is the excitation amplitude, $\nu \ (= \frac{N}{\Omega_z})$ is the excitation frequency and ω_x is the natural frequency in the x direction.

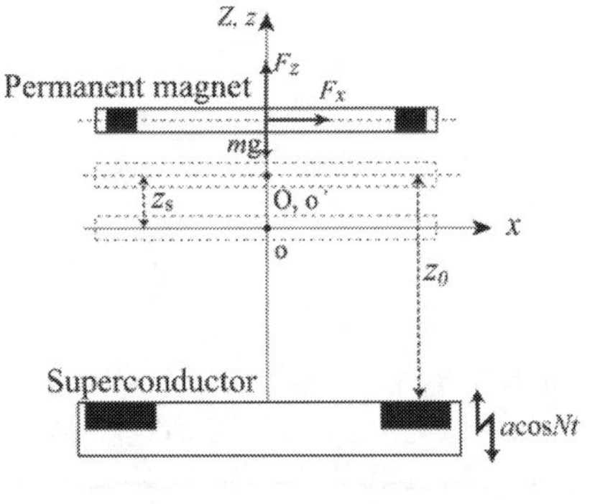

Figure 1. Analytical model.

3. Numerical calculation

We carried out numerical calculation of (1) and (2) by using the Runge-Kutta 4th order method (RK4). The length of a time step of calculation was one-128th of the period of excitation, and the number of steps was 2^{16}. We calculated on condition that the natural frequency of the levitated body in the vertical direction is twice as much as that in the horizontal direction, and also on condition that it is not twice. The parameters of the two cases are shown in Table 1. We increased the excitation frequency ν around $\nu = 1.0$. Figure 2 and Figure 3 show the numerical results of frequency responses of z and x, respectively, where solid circles show the results under $\omega_z : \omega_x = 2 : 1$ and open circles show those under $\omega_z : \omega_x \neq 2 : 1$. It can be found that if $\omega_z : \omega_x = 2 : 1$, vibration whose frequency is $\nu/2(\simeq \omega_x)$ occurs in the horizontal direction under the excitation frequency ν in the neighborhood of $\omega_z(= 1)$. Correspondingly, the amplitude of vibration in the vertical direction is lowered compared to that under $\omega_z : \omega_x \neq 2 : 1$. These numerical results show that the vertical resonance can be effectively reduced by occurrence of autoparametric resonance.

Table 1. Values of dimensionless parameters used in numerical calculation.

$\omega_z : \omega_x$	α_{zz}	α_{xx}	α_{zzz}	α_{xxz}	α_{xz}	α_{xxx}
$= 2 : 1$	5.24×10^{-1}	4.83×10^{-2}	1.90×10^{-1}	2.20×10^{-2}	9.65×10^{-2}	5.50×10^{-3}
$\neq 2 : 1$	9.30×10^{-2}	4.90×10^{-3}	4.83×10^{-1}	4.67×10^{-2}	1.60×10^{-1}	1.96×10^{-2}

$\omega_z : \omega_x$	α_{xzz}	ω_x	b	μ_z	μ_x
$= 2 : 1$	2.20×10^{-2}	5.00×10^{-1}	1.50×10^{-2}	2.81×10^{-2}	1.34×10^{-2}
$\neq 2 : 1$	1.96×10^{-2}	5.10×10^{-1}	1.50×10^{-2}	2.81×10^{-2}	1.34×10^{-2}

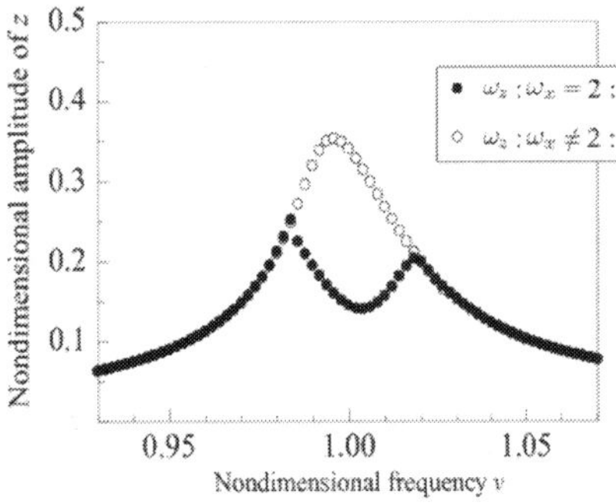

Figure 2. Numerically obtained frequency responses of z.

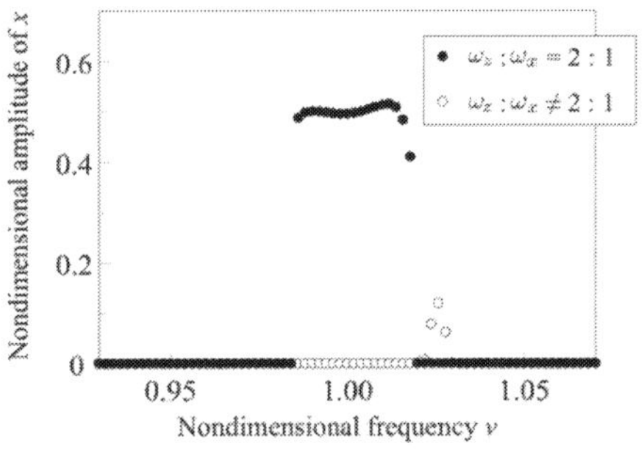

Figure 3. Numerically obtained frequency responses of x.

4. Experiments

We carried out experiments using an experimental setup shown in Figures 4 and 5. Two cylindrical shaped GdBCO high-T_c superconducting bulks of melt-quench type (ϕ45, 20H) were used here. They were cooled by LN_2 (77 K). Two cylindrical permanent magnets (ϕ20, 10H), consisting of Nd, Fe and B, were used. Residual magnetization of each magnet was 1.45 T. Each magnet, placed above the center of each superconducting bulk, was embedded in a hole made at each end of an aluminum bar shaped a rectangular parallelepiped (300W, 40D, 10H). The distance between the magnets was 230 mm. Two different initial cooling positions of the bar supported by acrylic spacers, were tried: $z_0 = 16.15$ mm and $z_0 = 16.10$ mm. They correspond to the case of $\omega_z : \omega_x = 2 : 1$ and the case of $\omega_z : \omega_x \neq 2 : 1$, respectively. By using a vibration generator, periodic vertical excitation was given to the superconductors. We measured the displacement of excitation together with the displacements in the vertical and horizontal directions by laser displacement sensors, while we changed the excitation frequency N around the neighborhood of the natural frequency of z. Figures 6 and 7 show experimental results of frequency responses in the vertical and horizontal directions, respectively, where solid circles show results under $\omega_z : \omega_x = 2 : 1$ and open circles show those under $\omega_z : \omega_x \neq 2 : 1$. The experimental results were qualitatively in good agreement with the numerical ones.

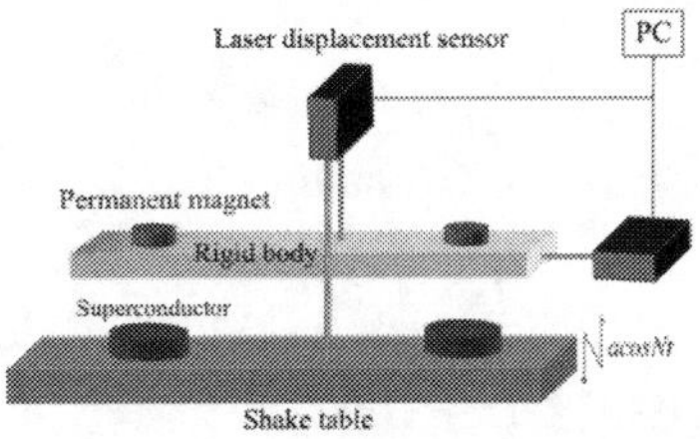

Figure 4. Sketch of our experimental setup.

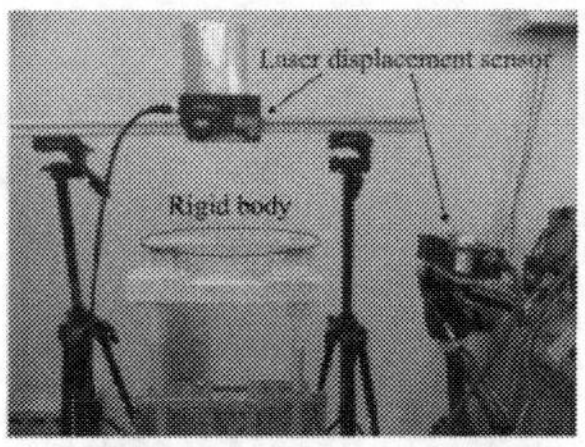

Figure 5. Photograph of our experimental setup.

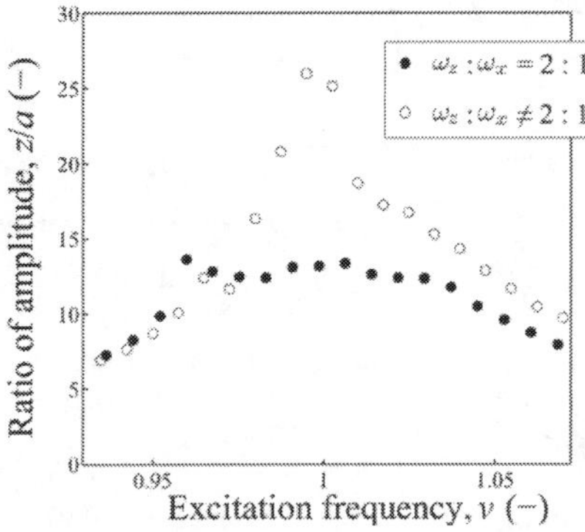

Figure 6. Experimentally obtained frequency responses of z.

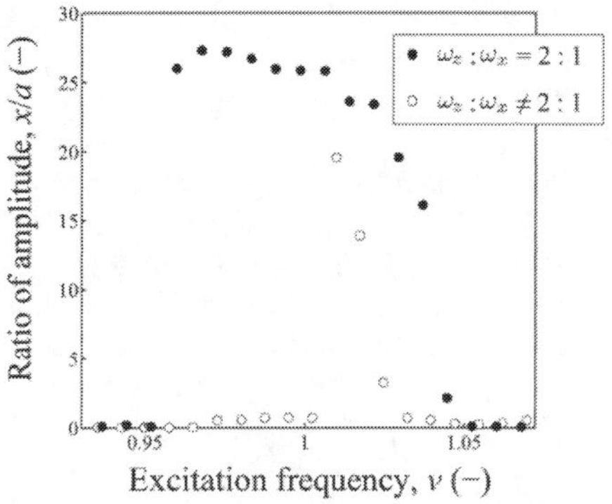

Figure 7. Experimentally obtained frequency responses of x.

5. Conclusion

We investigated reduction of vibration amplitude in a high-T_c superconducting levitation system by utilizing autoparametric resonance without using any absorbers. We conducted numerical calculations and experiments on a high-T_c superconducting levitation system. Both results show that the vibration amplitude of the primary resonance reduced with horizontal vibration resonating. The horizontal vibration resonated on condition that the natural frequency of the levitated body in the vertical direction is twice as much as that in the horizontal direction. Therefore we conclude that autoparametric resonance, caused by coupling between the vertical and horizontal motions due to nonlinearity of electromagnetic force, can occur if the natural frequency in the vertical direction is twice as much as that in the horizontal direction. Occurrence of autoparametric resonance can reduce the amplitude of the primary resonance in the vertical direction in exchange for occurrence of horizontal vibration.

References

[1] Hull J R and Murakami M 2004 *Prc. of the IEEE* **92** 1705-18
[2] Takazakura T, Sakaguchi R and Sugiura T 2013 *IEEE Transactions on Appl. Superconductivity* **23** 3600204
[3] Yabuno H, Endo Y and Aoshima N 1999 *J. Vibration and Acoustics* **121** 309-15
[4] Tondl A, Ruijgrok T, Verhulst F and Nabergoj R 2000 *Autoparametric Resonance in Mechanical Systems* (Cambridge: Cambridge University Press) pp 44-63
[5] Kordyuk A A 1998 *J. Appl. Phys.* **83** 610-2

Journal of Physics: Conference Series **507** (2014) 032046 doi:10.1088/1742-6596/507/3/032046

Testing of machine wound second generation HTS tape Vacuum Pressure Impregnated coils

D Swaffield, C Lewis, J Eugene, M Ingles, D Peach
General Electric
Power Conversion, Boughton Road, Rugby, CV21 1BU, UK

E-mail: david.swaffield@ge.com

Abstract. Delamination of second generation (2G) High Temperature Superconducting (HTS) tapes has previously been reported when using resin based insulation systems for wound coils. One proposed root cause is the differential thermal contraction between the coil former and the resin encapsulated coil turns resulting in the tape c-axis tensile stress being exceeded. Importantly, delamination results in unacceptable degradation of the superconductor critical current level. To mitigate the delamination risk and prove winding, jointing and Vacuum Pressure Impregnation (VPI) processes in the production of coils for superconducting rotating machines at GE Power Conversion two scaled trial coils have been wound and extensively tested. The coils are wound from 12mm wide 2G HTS tape supplied by AMSC onto stainless steel 'racetrack' coil formers. The coils are wound in two layers which include both in-line and layer-layer joints subject to in-process test. The resin insulation system chosen is VPI and oven cured. Tests included; insulation resistance, repeat quench and recovery of the superconductor, heat runs and measurement of n-value, before and after multiple thermal cycling between ambient and 35 Kelvin. No degradation of coil performance is evidenced.

1. Introduction

The exciting prospect of superconducting rotating machines with a vastly reduced footprint, mass and greater efficiency has tremendous promise for a range of applications, from high power wind generators (> 10MW) to Ship propulsion drivetrains [1-3]. It is these benefits that have motivated academic institutions and machine manufacturers alike to pursue High Temperature Superconducting (HTS) rotating machine design. Several promising prototype machine designs have now been realised, however the price of the HTS material, usually used in the form of tape, remains key to any commercial proposition. In the last decade this has prompted development of second generation (2G) tape which promised reduced manufacturing costs and tape prices.

It has been reported by some workers that testing of second generation tape sometimes resulted in delamination of the tapes when using resin based insulation systems for wound coils [4-7]. One proposed root cause is the differential thermal contraction between the coil former (the mechanical structure upon which the coil is wound) and the resin encapsulated coil turns resulting in the tape c-axis (direction normal to tape length and widest tape face) tensile stress being exceeded. This represents an engineering challenge as delamination of the tape results in an unacceptable degradation of the superconductor critical current level.

As part of a series of tests at GE Power Conversion to mitigate against this risk in HTS machine manufacture, and to prove winding, jointing and VPI processes in the production of coils for superconducting rotating machines, two scaled trial coils have been wound and extensively tested. This paper reports the results of these process validation tests.

2. Coil design

The coils are pre-production down-scaled coils. Winding processes including: winder tension, jointing, in-process test and VPI processes are as for full scale superconducting field coil production.

Two coils have been manufactured and tested as detailed and results are reported below. The first of these was manufactured by a third tier supplier and the second was fully manufactured at GE Power Conversion's Rotating machines factory in Rugby, UK (figure 1).

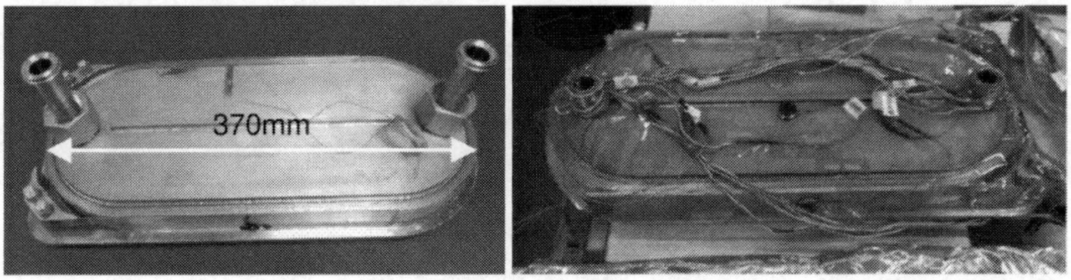

(a) Coil #1 (b) Coil #2

Figure 1. (a) Coil #1 (manufactured by Zenergy GmbH), (b) Coil#2 (manufactured by GE, Rugby), shown instrumented for testing

The coils are wound from 12mm wide 2G HTS tape Copper Laminated 'Amperium' (Specification: 225A Ic (77K, 0T), AMSC measured: 281A) supplied by AMSC. This product has a wide solder fillet promoted as increasing the c-axis strength [8].

The coils are a simple 'racetrack' with straights of 220mm and end radius of 75mm. Coils are formed in 2 layers with 30 turns each, 60 turns total, wound onto stainless steel coil formers. To prove jointing processes the coils are wound in two layers which include both in-line and layer-layer joints subject to in-process test. The chosen resin insulation system is vacuum pressure impregnated (VPI) and oven cured.

3. Experimental design

3.1. Objectives

The objectives of the testing are:

- To characterise the quench and recovery of 2G HTS coils.
- To measure critical current and qualify the winding process (i.e. test what if any impact is evident on the critical current)
- To check for degradation or failure due to delamination within HTS tape layers or between HTS tape and former/insulation.
- To qualify both: in-line and layer-layer jointing processes

3.2. Test arrangement

Cernox sensors have been connected to the test coil arrangement to monitor the temperatures at the following critical locations: coil former top and bottom plate, coil terminals, each side of the coil (outside faces), cryostat current lead feed-through and current lead thermal dump. A Lakeshore supplied cryogenic Hall effect probe is positioned in the centre of the top plate of the coil former to measure magnetic flux.

The whole arrangement, coil, current leads and mechanical support (also acting as thermal break to room temperature) is wrapped in Multi-Layer Insulation to prevent radiation heat transfer and placed in an vacuum chamber evacuated to better than $2x10^{-5}$ mbar to prevent convection heat transfer. The arrangement is cooled by copper buswork between a Cryomech AL330 G-M coldhead and the coil former. The temperature of test coil can be cooled to a minimum of 35K. Heaters mounted on the coldhead can be used to control the temperature of the thermal buswork and test coil using a Lakeshore temperature controller.

The coil power is supplied by an Agilent 6690A DC power supply (0-15 V, 0-440 A and can be controlled by setting either the voltage or current) via a 0.01 Ω ballast resistor. A 600 A diode is connected directly across the feed-though terminals to provide a 'freewheel' current path and prevent overvoltage from the inductive coil in the event of accidental disconnection of the cables from the power supply (high di/dt condition).

4. Results & Discussion

4.1. Coil performance – baseline characterization

At room temperature the insulation was verified post VPI, i.e. >10GΩ at 500V between coil and former.
With the coil cooled to 35K the coil current was raised in steps to 440A without allowing temperatures to stabilise. This resulted in flux density of approximately 136mT, see Table 1 for full results. Resistance of the coil was monitored and as predicted was not observed to reach the critical current limit at this operating point.

The measured flux density was 31.5mT at 100A compared with 30mT predicted by Finite Element modelling at the Hall probe position.

4.2. Coil critical current, pre thermal cycling

The purpose of this test is determine the critical current for the coil at a known temperature, to compare with that after thermal cycling. The temperature controller was set to hold the coil temperature stable at 59K (62-63K on the current leads) when not energised. The coil current was gradually increased until the coil voltage measurements achieve the equivalent of 1 μV/cm, an industry accepted level at which to record critical current. Critical current was recorded at 370A. This represents 95% of the calculated critical current assuming no degradation from that measured by the tape manufacturer. 5% degradation is as expected from coil winding and is within design margins. The n-value was extracted by Least Squares Regression curve fitting and recorded as 27.

4.3. Thermal cycling

Both test coils have been thermally cycled to 35K and back to greater than 0 °C eight times. This temperature range ensures that the coil has been through the superconducting transition every cycle and differential thermal contraction stresses within the coil assembly have been applied.

In addition critical current was measured before and after the thermal cycles on both coils. Table 1, outlines the main test results for the test coils chronologically as performed. Visual inspection of the coils after completion of the tests in Table 1 revealed no signs of de-bonding of the tape or insulation system selected.

4.4. Coil critical current, post thermal cycling

Critical current is recorded before, after 5 and after 8 thermal cycles. Neither coil showed degradation after thermal cycling. For example Coil 2 critical current values, based on an equivalent of 1μV/cm, are: 370A, 370A and 373A respectively. The variability is likely due to measurement error and not considered significant.

Table 1. Test results of the 2 test coils

Test	**Coil #1**	**Coil # 2**
Post wind IR*	PASS	PASS
Post VPI IR*	PASS	PASS
Number of thermal cycles (440A applied in superconducting state)	5	8
Final Ic	370A @60K	373A @59.0K
Inductance (FE model, 10Hz)	1.76mH	1.76mH
Inductance (measured, 440A, 35K)	1.75mH	1.73mH
Flux density (FE model, 440A)	136mT	136mT
Flux density (measured, 440A, 35K, at coil centre position)	Not measured	141mT
Post cycling IR*	PASS	PASS

* IR stands for Insulation Resistance, a PASS is recorded if >10GΩ at 500V between coil and former.

5. Conclusion

Further confidence in the insulation system, winding and jointing process is provided by verifying that the critical current has not degraded after thermal cycling and the resulting mechanical stresses due to the differential thermal contraction within the tested 2G HTS coils. This added to the evidence of practical testing of larger coils and the Hydrogenie machine effectively demonstrates the viability of the insulation system and coil design employed at GE Power Conversion for HTS rotating machine field coils using 2G HTS tapes.

6. Acknowledgments

The authors would like to acknowledge financial support for this work by the European Commission through its FP7 programme call FP7-SST-2007-RTD-1 project 218599 'POSEEIDON'.

7. References

[1] *Wind Power Generation and Wind Turbine Design*, Ed. Wei Tong, Chapter 9: Direct Drive Superconducting Generators, Lewis C, WIT Press: Southampton, 2010

[2] Eugene J, Lewis C, Cornhill P, "HTS Machines for Ship Propulsion Motors", *Proc. Conf. Schiffbautechnische Gesellschaft* 2011, Nov 2011, Rostock

[3] Frank M, van Hasselt P, Kummeth P, Massek P, Nick W, Rothfischer H, Schmidt H, Wacker B, Neumuller H.-W, Nerowski G, Frauenhofer J, Hartig R, Rzadki W, 2006 *IEEE Trans. App. Super.* **16** n2 1465-68

[4] Oomen M, Herkert W, Bayer D, Kummeth P, Nick W, Arndt T, 2012, *Physica C* **482** 111-18

[5] Takematsu T, Hu R, Takao T, Yanagisawa Y, Nakagome H,Uglietti D, Kiyoshi T, Takahashi M, Maeda H, 2010, *Physica C* **410** 674-7

[6] van der Laan D C, Ekin J W, Clickner CC and Stauffer T C, 2007, Supercon. Sci. Technol. **20** 765-70

[7] Trociewitz U P, Dalban-Canassy M, Hannion M, Hilton D K, Jaroszynski J, Noyes P, Viouchkov Y, Weijers H W, and Larbalestier D C, 2011, *App. Phys. Lett.* **99**

[8] AMSC 'Amperium' product datasheet

HTS DC Transmission Line for Megalopolis Grid Development

S Kopylov[1], V Sytnikov[1], S Bemert[1], Yu Ivanov[1], I Krivetskiy[1], M Romashov[1], Yu Shakaryan[1], V Keilin[2], A Shikov[2], V Patrikeev[2], V Lobyntsev[2] and V Shcherbakov [2]
[1]R&D Center @ FGC UES, Moscow 115201, Russia.
[2]NRC Kurchatov Institute, Moscow, 123185, Russia.

E-mail: vsytnikov@gmail.com

Abstract. Using of HTS AC and DC cables in electric power grids allows increasing of the transferred power, losses diminishing, decreasing of exclusion zone areas, the enhancement of the environmental conditions and fire/explosion safety of electric power systems. However, the use of DC superconducting cable lines together with converters brings additional advantages as reduction of losses in cables and suitable lowering of refrigerating plant capacity, as well as the realization of the function of short-circuit currents limitation by means of the appropriate setting of converter equipment. Russian Federal Grid Company and its R&D Center started the construction of the DC HTS power transmission line which includes the cable itself, cryogenic equipment, AC/DC converters, terminals and cable coupling boxes. This line will connect two substations in Saint-Petersburg - 330 kV "Centralnaya" and 220 kV "RP-9". The length of this HTS transmission line will be about 2500 meters. Nowadays are developed all the elements of the line and technologies of the cable manufacturing. Two HTS cable samples, each 30 m length, have been made. This paper describes the results of cables tests.

1. Introduction

The efficiency of electricity generation and its transportation to the consumer are becoming the priorities in the development of power industry in the XXI century. Simultaneously, the demands for environmental and resource-saving parameters in all phases of production and distribution of electricity are increasing. Meeting the growing demands is only possible by using the most progressive and advanced technologies, including superconductivity. AC and DC superconducting cables allow the increase of transmission capacity, the decrease of power losses, reducing of allotment areas and environmental conditions improvement. HTS DC cables give additional advantages including increased capacity without increasing of fault current levels, active and reactive power regulation, lower power losses and bigger crucial length of transmission as compared with HTS AC cables. These advantages generate additional interest to the HTS DC cable lines both from high voltage (HV) transmission [1] and medium voltage (MV) distribution [2-4]. The load growth in metropolitan power systems makes it necessary to put into operation new generating and network equipment. In order to improve the reliability of power supply the reservation of substations is implemented (new transmission lines are built and brought into service) that occurs in multilinked connection network creation. These trends contribute to higher levels of short circuit currents in power systems. To reduce the levels of short circuit currents to safe values current-limiting reactors and network sectioning are used. However current-limiting reactor application leads to additional power losses and network sectioning results in the reduction of consumer power supply reliability. MV DC transmission lines with HTS cable commissioning (instead of AC cross-linked polyethylene (XLPE) insulated cables with current limiting-reactors) are regulating element of the grid capable of power flow regulations and fault current limitation. The Federal Grid Company of Unified Energy System (FGC UES) started the activity aimed to the building of a HTS DC cable link in St. Petersburg by connection substation 330 kV "Centralnaya" and substation 220 kV "RP-9". The design, production technology and first tests of HTS cable for the project are describe below.

2. Choice of the pilot object for HTS DC Cable line.

An analysis of the current state of the power supply in large cities has been conducted in joint-stock company (JSC) "R&D Center of FGC" and JSC «R&D Center of UES». Both, the current state of the power grids of megalopolises, as well as the prospect development up to the year of 2020 have been taken into account. The steady state, overload regimes and short-circuit current value estimation were analyzed. As a result, a pilot research facility for the HTS DC cable line has been selected in St. Petersburg as a link between two substation 330 kV "Centralnaya" and 220 kV "RP-9" (figure 1). As seen from figure 1, HTS DC cable line provides mutual redundancy of two sections of power systems.

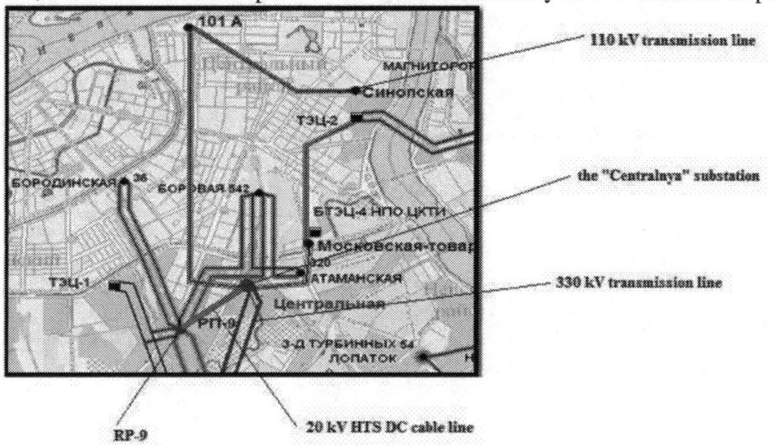

Figure 1. HTS DC cable line installation in the St. Petersburg's electrical grid (project).

As shown in Table 1, HTS DC cable line installation does not increase fault current, unlike the case of the AC cable line, therefore this installation obviates the necessity of circuit breaker replacement.

Table 1. Short-circuit current levels in the proposed facility installation site.

Object	Short-circuit current, kA			
	3-phase AC cable line	HTS DC cable line	1-phase AC cable line	HTS DC cable line
«Centralnaya» subst.	40	18	41	22
«K3-9» substation	40	26	46	27

Increased load capacity of HTS cables can reduce the number of parallel lines, as well as decrease the load on the existing lines. HTS DC cable line parameters are shown in Table 2.

Table 2. HTS DC cable line parameters, data are taken from [4].

HTS DC line length, km	Transmission power, MW	Voltage, kV	Operating current, A	Overload current, A	Power reverse
2.5	50.0	20.0	2500	2750	yes

3. HTS Cable design

As the basic design was chosen unipolar cable with the reverse conductor (figure 2).The cable consists of concentric layers containing the following elements:

- former and stabilizing element;
- superconducting central wire playing a role of the forward conductor;
- high voltage insulation;
- superconducting reverse conductor;
- external stabilizer;

- external (screening) insulation;
- electric (non- superconducting) screen;
- cryostat consisting of two corrugated tubes with the vacuum thermal isolation and protecting layer.

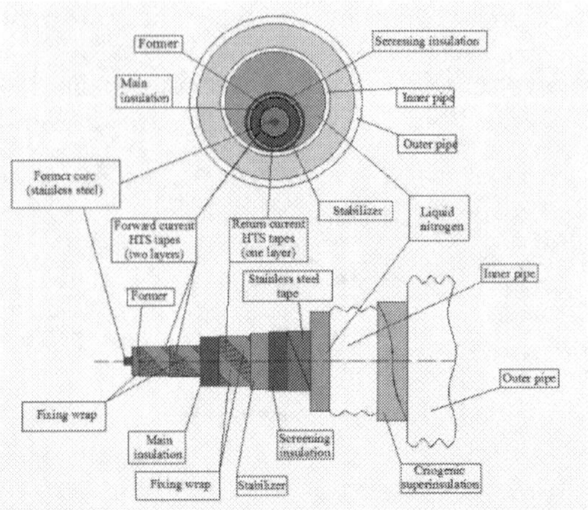

Figure 2. Schematic diagram HTS DC cable.

The external cable diameter is 39 mm. As a basic superconducting material for the cable the 1G HTS tape is used produced by SEI (Japan), type HT-CA. Forward conductor include 22 tapes with I_c=160A, placed in two layer. Reverse conductor include 19 tapes with I_c=180 A, placed in one layer.

In this cable the forward and reverse currents flowing only through superconducting elements that causes the absence of Joule losses and electromagnetic field location is only between two superconducting poles. The absence of stray fields and using of the liquid nitrogen as a coolant makes these cables environmentally friendly and fire-safe. The requirements to the cable placement are essentially simpler than other designs since there are neither electromagnetic nor thermal stray fields.

Two 30-meters cable samples were made in serial cable plant "Irkutskkable". Current leads and coupling box were designed and produced in NRC Kurchatov Institute [5].

4. Preliminary Cable test results.

Preliminary test was carried out by temporary scheme in NRC Kurchatov Institute. The main goals of the first test were: check of cabling technology and measurement of electrical resistance and vacuum tightness of current leads and coupling box.

Current source "plus" pole was connected to the direct conductor and "minus" pole to the reverse conductor on one side of the test line. Current leads on another side of the line were shorted. Consequently the current flows through four 30 meters HTS cable conductors, two joints and four current leads. Voltage taps were placed on each of the conductors and joints in cold area. Voltage taps on the reverse conductors were attached to the ends of it, and that's why there is a copper-to-superconductor connection between taps.

(a) (b)

Figure 3. Cable current leads (a) and coupling box (b) during testing.

As a result a small resistive component is present in the measured signal. Cooling cable line was made by the time scheme by pumping liquid nitrogen through the line between the two cryogenic tanks and temperature in the line was 79 K. The results of three conductors (one voltage tap was tear during assembling) critical current measurement are shown on figure 4.
The critical current of HTS conductors was within 3420 - 3550 A. The critical current of the cable is practically equal to the sum of critical currents used superconducting tapes that indicates the development of a reliable cabling technology. Current leads and coupling box (figure 3) has successfully passed the tests on vacuum tightness and overpressure. All fittings demonstrated reliable operation without the appearance of leaks or intensive freezing during two cooling sessions.

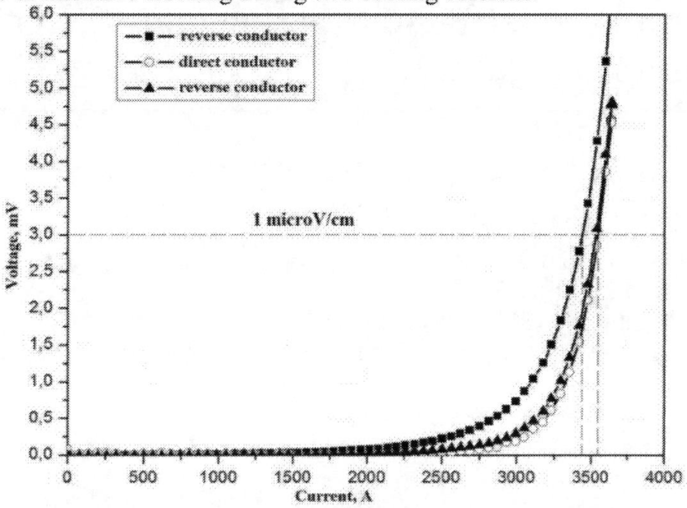

Figure 4. V-I curve of 30 meters pieces of HTS cable.

Electrical resistance of the current leads and joints remained stable in all range of currents. Resistance of two conductors in the coupling box were less than 1 μΩ. Resistance of the four current leads was within 21-23 mcOhm that corresponds to the total Joule heat generation ca. 140 Wt at the rated current.

5. Conclusions

Overall, the HTS DC cable line exerts numerous positive effects concerning power systems' regimes of megalopolises due to the ability to transmit high power, decrease of the load of parallel lines and due to the high level of controllability of these systems. The HTS DC cable line installation improves the reliability of energy supply to the consumers at the expense of mutual redundancy. Along with this, it does not increase short-circuit currents in the place of engagement and, consequently, it obviates the necessity of replacing the switching equipment.
Sant-Petersburg Project is the first network HTS DC cable demonstration in Russia. Cable manufacture technology was developed and two 30 meters samples was made. First tests confirmed the compliance of the measured parameters of HTS cable, current leads and coupling box to design value. All equipment for HTS DC line including cryogenic system and convertors will be made to the end of 2014 year.

6. References

[1] Hassenzahl W, Eckroad S, Grant P, Gregory B, and Nilsson S 2009 A high-power superconducting DC cable *IEEE Trans. Appl. Supercond.*, **19(3)** 1756
[2] Sytnikov V 2011 *Superconductivity: Reseach and Development* **15** 65-74.
[3] Sytnikov V E, Bemert S E, Ivanov Yu V, Kopylov S I, Krivetskiy I V, Rimorov D S, Romashov M S, Shakaryan Yu G, Berdnikov R N, 2013 *IEEE Trans. on Appl. Supercond.* **23(3)** 5401904.
[4] Yamaguchi S, 2011 *1st Asia-Arab Sustainable Energy Forum* , Nagoya, Japan, Aug. 23-26,.
[5] Naumov A, Diev D, Denisov A, Anashkin O, Keilin V, Kovalev V, Lobyntsev V, Shcherbakov V, Sytnikov V, 2013 *Report 3EPoB2-04 on CEC/ICMC*, Anchorage, June 17-21.

Superconducting properties of experimental YBCO coils for FFAG accelerator magnets

S Takayama[1], K Koyanagi[1], T Tosaka[1], K Tasaki[1], T Kurusu[1], Y Ishii[1], N Amemiya[2] and T Ogitsu[3]

[1] Toshiba Corporation Power System Company, 2-4 Suehiro-cho, Tsurumi-ku, Yokohama 230-0045, Japan
[2] Kyoto University, Kyoto-Daigaku-Katsura, Nishikyo, Kyoto 615-8510, Japan
[3] High Energy Accelerator Research Organization, 2-4 Shirane, Shirakata, Tokai 319-1195, Japan

E-mail: shigeki2.takayama@toshiba.co.jp

Abstract. A project to develop fundamental technologies for accelerator magnets using high-T_c coated conductors is currently in progress. The primary applications of this project are fixed field alternating gradient (FFAG) accelerators for carbon cancer therapy systems and accelerator-driven subcritical reactors. Several types of superconducting coils for FFAG accelerators have been conceptually designed. These coils have complicated shapes, including a negative-bend part or a three-dimensional bent part. One of the aims of the project is to establish winding technologies for complicated coil shapes using coated conductors. To demonstrate winding technologies for $YBa_2Cu_3O_{7-x}$ (YBCO) coils, small test coils having a negative-bend part or a three-dimensional bent part were designed and fabricated according to the present design of the FFAG magnet. The outside dimensions of the negative-bend test coil were 460 mm long and 190 mm wide, and the radius of curvature of the negative-bend part was 442 mm. The outside dimensions of the three-dimensional test coil were 380 mm long and 280 mm wide, and the radius of curvature of the mandrel of the three-dimensional coil was 700 mm. The test coils were wound using YBCO coated conductors with a length of about 100 m and were then impregnated with epoxy resin. The coils were placed in liquid nitrogen and excited to measure their *V–I* characteristics. From the *V–I* characteristics throughout a voltage range down to 10^{-9} V/cm, the *V–I* characteristics before and after impregnation were approximately the same, demonstrating that the superconducting properties were not degraded.

1. Introduction

A project to develop fundamental technologies for accelerator magnets using high-T_c coated conductors is currently in progress [1]. The target applications of this project are fixed field alternating gradient (FFAG) accelerators for carbon cancer therapy and accelerator-driven sub-critical reactors (ADSRs). Several types of superconducting coils for FFAG accelerators have been conceptually designed in this project, such as a spiral sector superferric type and a radial sector coil-dominated type [2], [3]. The superconducting coils for these types have complicated shapes including negative-bend parts or three-dimensional bent parts. One of the aims of the project is to establish winding technologies for such complicated coil shapes using coated conductors. This paper describes the fabrication and the test results of two coils, one including a negative-bend part and one including a three-dimensional bent part.

2. Coil designs for FFAG accelerators

The designed structures of the main rings of the spiral sector superferric type and the radial sector coil-dominated type FFAG accelerators for carbon cancer therapy systems are schematically shown in figure 1 and figure 2. The main rings are composed of a number of magnets and the outer diameters of the rings are approximately 10 m, which is about half the size of present ones. The spiral sector superferric type accelerator achieves strong focus by using edge-focusing [4]. For this reason, the coils have an inward winding part called a negative bend. The radius of curvature of the negative-bend portion is 2340 mm. In this type of accelerator, the magnetic field distribution required in the beam orbit can be optimized through appropriate design of the pole gap shape, and therefore, the superconducting coils can be built simply by winding the tape conductor flatwise into a pancake shape. In contrast, with the radial sector coil-dominated type of magnets, the magnetic field distribution is ensured by a precise suitable design of the coil shape, and a three-dimensional winding structure has been required. A schematic view of the designed coils is shown in figure 3.

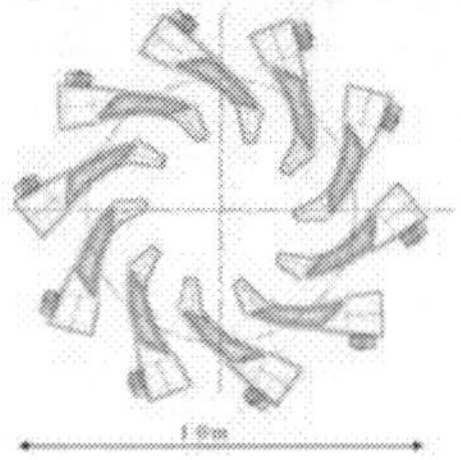

Figure 1. Schematic view of the spiral FFAG main ring composed of ten superferric magnets [2].

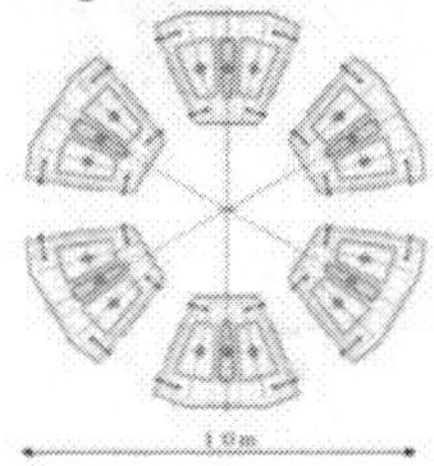

Figure 2. Schematic view of the radial sector FFAG main ring composed of FDF triplet magnets.

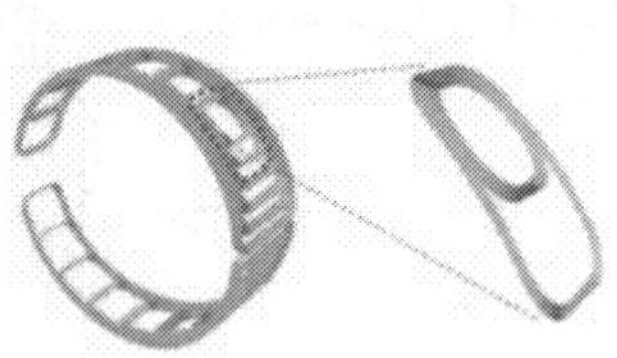

Figure 3. Schematic view of the three-dimensional coil (left) [3] and magnified view of three-dimensional two-block coil (right).

3. Test coil fabrication

A coil having a negative-bend part suffers from the problem that, when a winding tension is applied, the coated conductor tends to come off the bobbin. The coated conductor needs to be fixed at the negative-bend part of the coil. [5]. To demonstrate winding technologies for YBCO coils with a negative-bend shape, a small test coil, of about one-eighth scale, was designed and fabricated according to the present magnet design. The specifications of the test coil are listed in table 1. To construct the test coil, we used a YBCO coated conductor with a width of 5 mm and an I_c of 184 A (77 K, self field). The winding method is schematically illustrated in figure 4. Holding jigs were set to completely surround the bobbin. The tape was wound onto the bobbin by repeatedly removing and reapplying the holding jigs. Therefore, even if holding jigs in negative-bend part are removed, the coated conductor is restrained from coming off the bobbin because other jigs keep holding it in other area. The force of coming off the bobbin at the negative-bend part is increased in every turn. So, the force of applying the jigs was increased by using springs when thickness of coil increased. The coil was co-wound with a conductor composed of YBCO tape and polyimide tape for insulation and was then impregnated with epoxy resin. A photograph of the test coil is shown in figure 5. The mean thickness of the test coil is 31.5 mm, and the deviation of thickness of coil held down with 0.35 mm.

As shown in figure 3, the designed coils require a three-dimensional structure at the coil ends. On the other hand, YBCO conductor usually has a flat tape shape, and it is difficult to plastically deform the conductor in the edgewise direction. We designed the bobbin shape in such a way that the two edges of the tape wound along the bobbin follow curves of equal length, forming a constant-perimeter condition, without applying edgewise bending [6]. So, the innermost turn was the constant-perimeter condition. On the other hand, the outer turns were wound in accordance with the inner turn, and were diverged from the constant-perimeter condition. Therefore, stress occurred by the outer turns, and there was a possibility that the external coil shape was distorted from the shape of the outspread of the innermost turn. The coil end was separated into two blocks by a spacer as shown in figure 3. The second block coil was wound on the first block coil, and it had discontinuous connections in the coil end. On the premise of no deformation of the external shape of first block coil, the spacer, made of FRP, was manufactured in a curved surface calculated with the constant-perimeter condition. So, deformation of the external coil shape leads to a stress concentration or kinks around the discontinuous connections. In order to evaluate winding technologies for three-dimensional coils, we tried to fabricate a one-fourth scale test coil having a discontinuous coil end. Because of avoiding deformation, the conductors were fixed in places not to slide to the width direction, and the test coil was wound by the winding machine synchronizing the rotating and yawing motions. A 90-turn coil was wound on the bobbin with a 90 m length of 3 mm-wide YBCO coated conductor (I_c: 87 A at 77 K, self field), and an additional 40-turn coil was wound after setting the spacer. The specifications of the three-dimensional test coil are listed in table 2. The coil was co-wound with a conductor composed of YBCO tape and polyimide tape for insulation and was then impregnated with epoxy resin. A photograph of the coil is shown in figure 6.

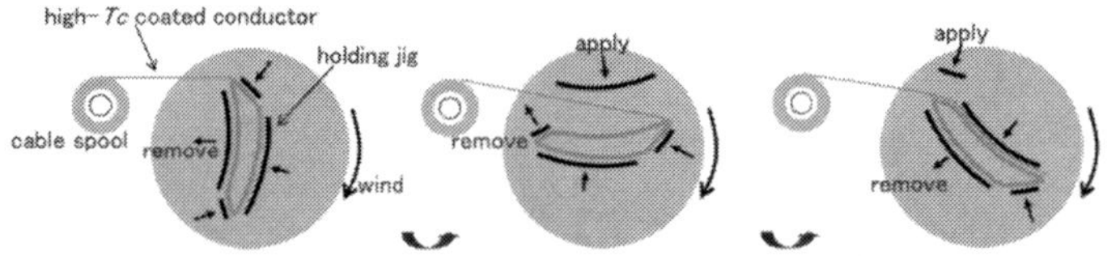

Figure 4. Schematic view of winding process.

Table 1. Specifications of the negative-bend test coil.

Length of the test coil	460 mm
Width of the test coil	190 mm
Radius of curvature of negative-bend part	442 mm
Number of turns	100
Length of conductor	100 m

Table 2. Specifications of the three-dimensional test coil.

Length of the test coil	380 mm
Width of the test coil	280 mm
Radius of the mandrel	700 mm
Number of turns	90+40
Length of conductor	140 m

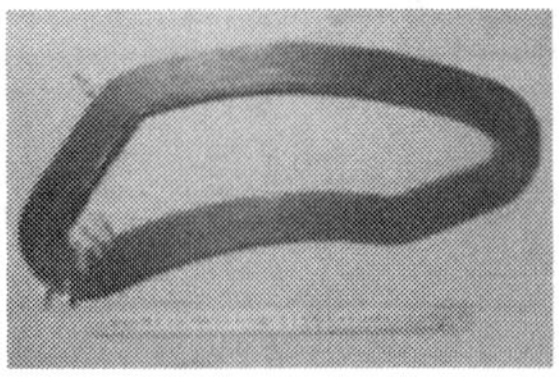

Figure 5. Photograph of the negative-bend test coil.

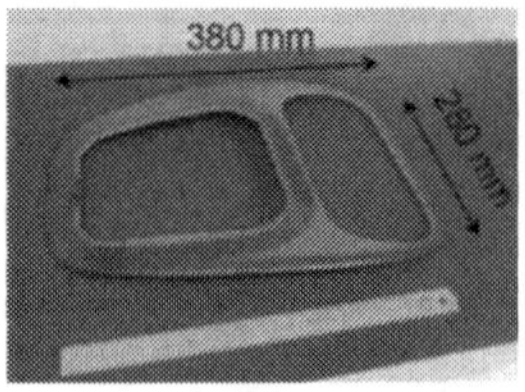

Figure 6. Photograph of the three-dimensional test coil.

4. Test results

In a previous study, degradation of impregnated YBCO coils was observed, and we have previously reported that the performance of impregnated circular coils was improved by dividing the coils into a number of parts in the radial direction [7], [8]. To verify that this method can be applied to complicated shaped coils, we measured the test coil's *V–I* characteristics before and after it was impregnated. The coils were cooled down carefully with liquid nitrogen for a few hours, and the electrical current was increased to observe the behavior of the voltage generated in the coil. The measured *V–I* characteristics of the negative-bend test coil and the three-dimensional test coil are shown in figure 7 and figure 8. The *V–I* characteristics before and after impregnation were approximately the same. Furthermore, an index values of as high as 30 were demonstrated all the way down to the low voltage region of 10^{-9} V/cm. The coated conductor wound into the shape of a coil having a negative-bend part or three-dimensional bent part exhibited superior superconducting properties, and the method of improvement reported in [7], [8] was suitable for the complicated shaped coils, too. These test results show that degradation-free superconducting properties were obtained and demonstrated the possibility that a coated conductor can be wound into a complicated shape for the magnet of a FFAG accelerator.

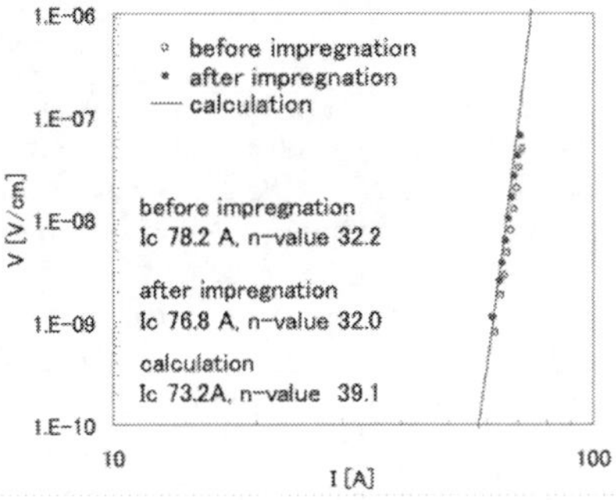

Figure 7. *V–I* characteristics of the negative-bend test coil.

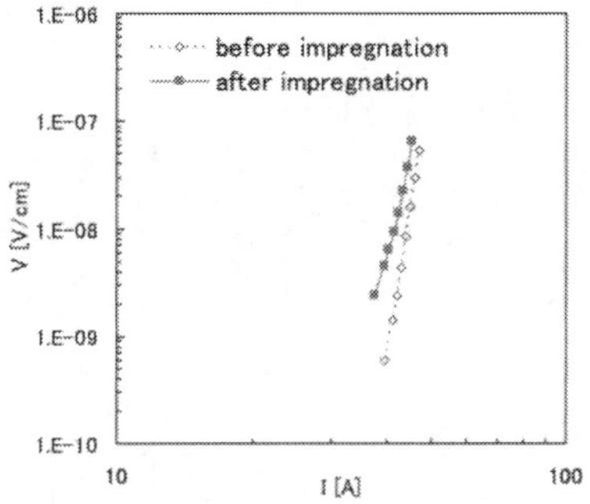

Figure 8. *V–I* characteristics of the three-dimensional test coil.

5. Conclusion

Complicated coil shapes including a negative-bend part or a three-dimensional bent part were experimentally fabricated to evaluate the superconducting properties. The results of experiments showed that impregnated test coils were successfully fabricated and did not show any degradation in the superconducting properties. The results of the present study will promote the adoption of the winding technology described here, but further investigation will be required in order to consider the required field distribution in accelerators. We will continue our R&D into winding technologies for coated conductors in order to realize a larger prototype coil for further experiments in the future.

References

[1] Amemiya N *et al.* 2012 *Phys. C, Supercond.* **482** 74.
[2] Qin B and Mori Y 2012 *Nucle. Instr. and Meth. A* **648** 28.
[3] Takahashi K *et al.* 2012 *IEEE Trans. Appl. Supercond.* **22** 3 4901705.
[4] Kerst D W, Terwilliger K M, Symon K R and Jones L W 1955 *Phys. Rev.* **98** 1153.
[5] Koyanagi K *et al.* 2012 *IEEE Trans. Appl. Supercond.* **22** 3 4101904.
[6] Koyanagi K *et al.* 2012 *IEEE Trans. Appl. Supercond.* **23** 3 4100404.
[7] Miyazaki H *et al.* 2014 *IEEE Trans. Appl. Supercond.* **24** 3 4600905.
[8] Iwai S *et al.* 2014 *IEEE Trans. Appl. Supercond.* **24** 3 4600705.
[9] Iwai S, Miyazaki H, Tosaka T, Tasaki K and Ishii Y 2013 *TEION KOGAKU* **48** 187.

A superconducting magnet feeder system that avoids the risk of Paschen discharge

T Taylor
AT Scientific LLC, Bernex, 1233 Switzerland

E-mail: tomtaylor@atscientific.com

Abstract. Current is traditionally fed to magnets in fusion devices via a system of superconducting busbars in a thermally insulating vacuum enclosure. Some of the magnets are pulsed, taking the busbars to a potential of several kilovolts, so these must be well insulated electrically. There is a risk of helium leaking into the insulation vacuum, so the electrical insulation of the busbars must be permanently hermetic in order to avoid the risk of Paschen discharge, subsequent arcing and concomitant heavy damage. Such so-called Paschen-tight insulation is difficult to achieve, especially considering that the system is pulsed and the required lifetime may be of the order of 20 years. This report describes a conceptual study of such a feeder system that avoids the risk of Paschen discharge by routing the superconducting busbars via a pressurized helium environment, thus eliminating the necessity of thorough hermetic sealing of the insulating envelope. In the concept that is presented several other features address design issues that have occurred in recent feeder systems for fusion magnets.

1. Introduction

The conductors related to the large pulsed magnets used for confining the plasma in experimental fusion reactors must be electrically insulated. It is customary to enclose the busbars and current leads in the associated feeders in a vacuum enclosure to provide thermal insulation. At each pulse or rapid discharge of a magnet large voltages are present on these busbars. A great deal of effort is put into providing good quality insulation [1]. If there is a crack it is imperative that the vacuum be good; should there be a leak of helium into the vacuum an electric discharge with destructive arc may occur. For this reason great care is devoted to making the insulation hermetic and to proving that it is so, which is both expensive and time-consuming. There have been instances of spectacular failures in the past, leading some experts to propose enclosing the whole system in earthed cans and filling the spaces with vacuum/pressure impregnated resin. It has also been suggested that by carefully monitoring the vacuum pressure and interlocking the power converters disaster could be avoided, but this is considered to be a risky alternative because leaks of helium are most likely to occur from the coolant in the bus and if there is a crack in the insulation then the local vacuum will be degraded before the degradation is detected by the monitoring system. This problem does not appear in superconducting magnet systems for accelerators: such magnets are based on the use of Rutherford cable purposely insulated electrically with permeable material that allows good thermal exchange with the liquid helium coolant. In this paper it is suggested that a viable alternative for fusion magnets using cable-in-conduit conductors (CICC) may be to avoid the risk of Paschen discharge by enclosing the entire bus system in a pressurized helium environment. It is further argued that CICC may not be the most appropriate topology for the feeder busbars.

Journal of Physics: Conference Series **507** (2014) 032049 doi:10.1088/1742-6596/507/3/032049

2. Feeder insulation

In this section we discuss alternative approaches to insulating the feeder busbars.

2.1. Traditional layout

The traditional layout of a tokamak busbar system is shown in figure 1. The superconducting coils are wound using cable in conduit conductor, the large Lorentz forces being withstood by the stainless steel conductor casing and/or by additional thick structural casing. At the exit the conductors are connected electrically to the busbars supplying the excitation current; coolant helium gas for each busbar is connected via an insulation break. A different cryogenic circuit supplies the coolant via another break for the coil. The busbars and cryogenic pipework are surrounded by a vacuum enclosure that provide heat insulation and leads through the shielding wall to the valve and connection cryostat (feedbox), where the busbars connect to the current leads [2]. As the feedbox is fixed, the busbars incorporate an S-bend to allow for thermal contraction. The magnets are large and produce high fields: the stored energy is large (some GJ), and the excitation current is of the order of 60 kA. The power source must provide high voltage (of the order of 10 kV) to enable fast ramping of the magnets, as well as fast de-excitation to extract the energy in case of problems. This means that the whole busbar and current lead system must be well insulated between conductors and to ground. This insulation must be hermetic in order to avoid Paschen discharge in case of helium leakage into the technical vacuum. This is achieved by wrapping the components in a combination of polyimide film and glass-fibre tape followed by epoxy impregnation. Experience has shown that this insulation must be strictly controlled for cracks, and subjected to extensive "Paschen testing" to verify its integrity before commissioning of the system. The insulation is required to remain hermetic for many years.

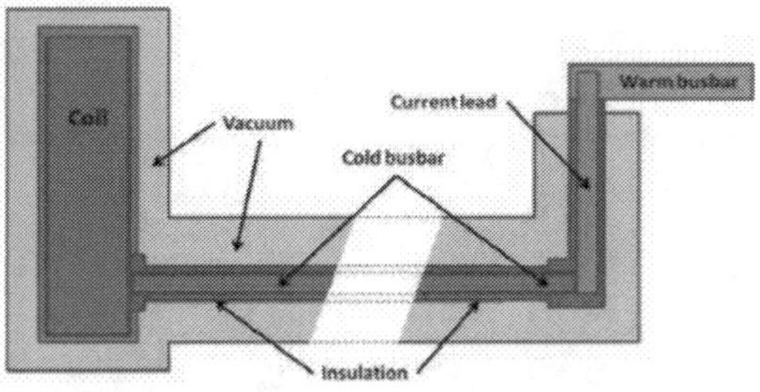

Figure 1. Schematic view of a feeder for which the insulation should be Paschen-tight.

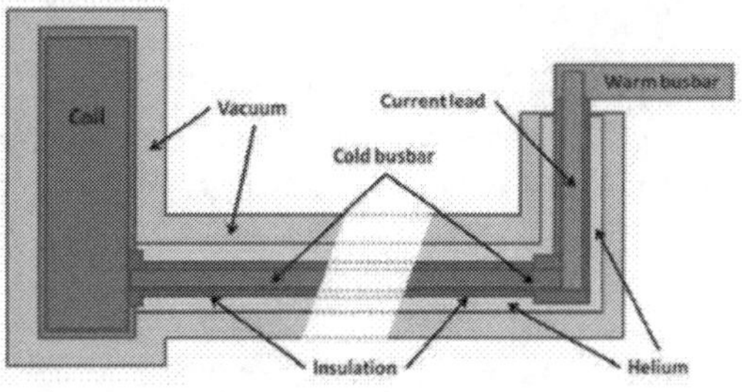

Figure 2. Modified general feeder layout that avoids the risk of Paschen discharge and arcing.

2.2. Alternative proposal

The required long-term hermetic nature of the feeder insulation is a concern. The epoxy resin impregnation must be executed with extreme care to avoid the formation of resin-rich pockets that are subjected to internal stresses and possible cracking. Ivanov [3] has proposed encapsulating the whole busbar system, wrapped in fibre-glass tape, in a stainless steel jacket and vacuum impregnating the system with epoxy resin. This should improve the long term viability of the insulation of the feeder system, but the risk of resin-rich pockets still exists, and while the system would be protected in such a case against degraded vacuum conditions outside the jacket it would not be protected against helium leaks occurring within (where they may be more likely due to the connections of the CICC busbars). Eventual maintenance would also be hampered by the presence of the impregnation.

Superconducting windings of magnets and busbars for accelerators are usually bath-cooled using liquid or supercritical helium; current leads operate in chimneys filled with stratified gaseous helium. High voltage never appears in the technical (insulating) vacuum: should this become degraded by a helium leak there will be no voltage breakdown. Large magnets for experiments are usually indirectly cooled: the busbars are in vacuum, but the magnets are not pulsed and the maximum voltage seen in the vacuum is of the order of 50 V. In the event of degraded vacuum there will not be a discharge.

In the light of these observations it is proposed to enclose the insulated busbar system in a pressurized helium environment as depicted in figure 2, such that should a helium leak occur in the busbar (and cryogenic) pipework this would not affect the environment of the insulation: components at high voltage would not be surrounded by low pressure helium associated with Paschen discharge. And should a leak occur in the technical vacuum ensuring thermal insulation, there would not be electrical repercussions as there would not be any component at high voltage within that enclosure.

The current leads are shown to be vertical in this arrangement: this is preferable because the helium in the annular chimney must be stratified by use of close-fitting baffles to avoid convective heat leak into the helium enclosure. The temperature of the enclosure could be maintained close to that of liquid helium by controlling the liquid feed via a level gauge in a local pot. The gas recovered from evaporating helium could be used to cool the heat shield (not shown in the figures) placed between the vacuum and helium enclosures.

2.2.1 Insulation breaks

The cryogenic pipes for cooling the coils should also be routed through the helium enclosure. If these are insulated the ensemble of insulation breaks associated with these pipes, and the current leads, could be grouped in a cryostat close to the feedbox, thus addressing another concern, i.e. the long-term reliability of these components, by providing easy access to eventually defective insulation breaks without having to enter the vacuum vessel of the tokamak. Other specific concerns related to the high current busbars feeding fusion magnets.

3. Other features

3.1. Torque

There is a considerable stray magnetic field surrounding a tokamak. This can be locally up to 1 T, and in conjunction with a pair of busbars carrying up to 60 kA the $J \times B$ forces on the conductors give rise to a large torque requiring a sophisticated mechanical support structure. For currents up to 10 kA this is less of a problem (and could in any case be solved by using concentric cables). For large currents it is suggested that an interesting solution would be to divide the cables as illustrated in figures 3 and 4. The four insulated cables are connected mechanically and there is no net torque on the assembly. For a 60 kA system the cable dimension would reduces from a diameter of about 40 mm to about 30 mm for 30 kA [4] and the required transverse space would be unchanged. An insulation thickness of < 2 mm (combined polyimide film layers and GFRP) should suffice. The additional complexity of having 4 conductors instead of 2 is offset by the advantages of increased flexibility (making joints easier), and less massive current leads. And eliminating gimbals from the support structure reduces complexity. Equalization of the current in the parallel conductors would be ensured by the resistance of the warm section of the busbars (connecting warm ends of current leads to the power converter) being far greater than that of the joints in the superconducting section.

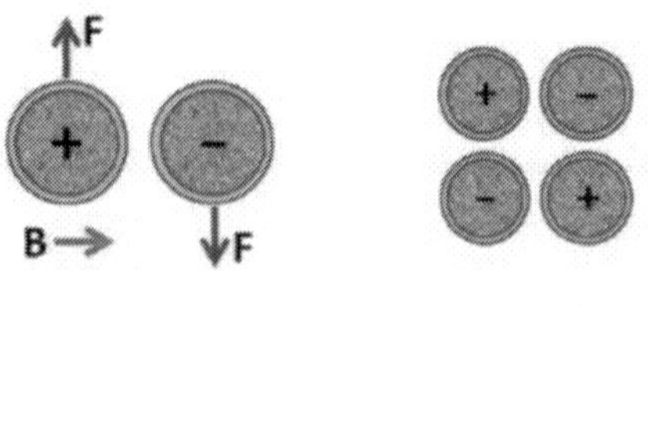

Figure 3. Left: standard arrangement. Right: arrangement eliminating torque.

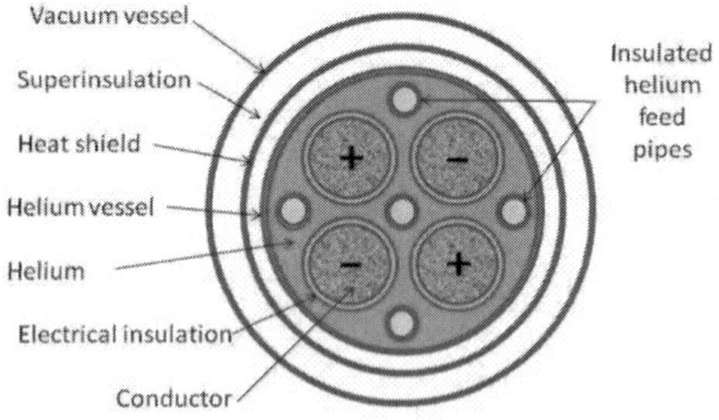

Figure 4. Schematic cross-section of the horizontal long straight section of a 60 kA feeder bus.

3.2. Cable variants
Because a magnet for a fusion device may be based on the use of cable-in-conduit conductor (CICC) does not imply that the feeder busbar should also be made of CICC. In particular if the insulation is allowed to be permeable (as can be the case in this proposal) other types of cable may be preferred, e.g. as shown in figure 5. Studies have been made [5] of using HTS material in such feeders, but not (yet) found to be economically viable. However, cables based on the use of MgB_2 could be interesting [6], if a conductor is developed that does not contain ferromagnetic material, and hybrid LTS/HTS conductors could also be considered [7].

3.3. Thermal contraction
An alternative to the classical S-bend is shown in figure 6, where the radius of curvature is about 1.5 m – typically that with which it is wrapped on the drum for transport. It should also be noted that for some types of cabled conductor the longitudinal modulus is low (for small extensions) and that the thermal contraction can be absorbed (at least partially) by allowing it to go into tension. The helium vessel can be of the corrugated (Nexans) type as used for the LHD feeder system [8], or made from Invar, so thermal contraction is not an issue. If the straight section is made long, to bring the feedboxes close to the power converters, thermal contraction can be conveniently taken into account by "snaking" the conductor, e.g. by supporting it periodically and allowing (encouraging) it to sag between supports using springs. One could consider housing these links out of the way in trenches as proposed in the original report [9] on superconducting power transmission.

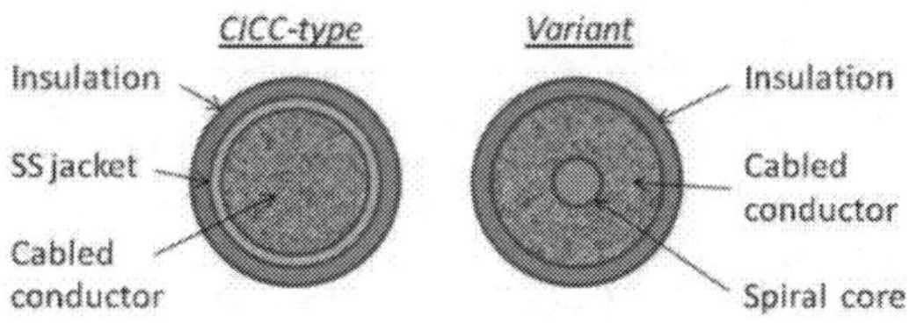

Figure 5. Example of alternative feeder cable.

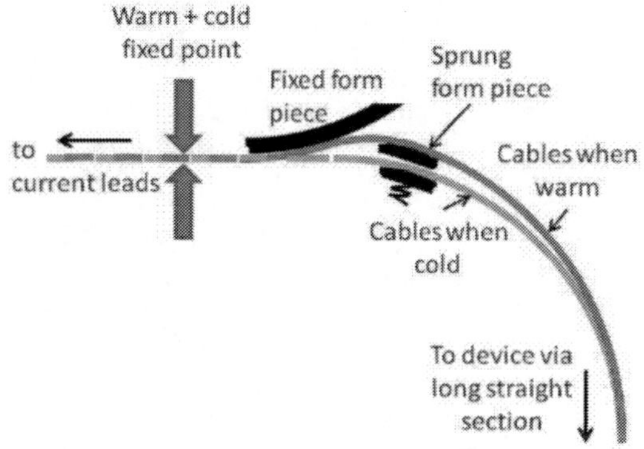

Figure 6. Compensation of contraction.

4. Conclusion
It should be possible to improve on the feeder systems being deployed for the present generation of fusion magnet systems. Integration of the design of the feeders should be given the same level of attention as that devoted to the magnets. Use of cables other than CICC should be considered.

5. References
[1] Rajainmaki H, Evans D, Knaster J, and Losasso M 2012 *IEEE Trans. Appl. Supercond.* **22** 3800505
[2] Bauer P *et al.* 2012 *IEEE Trans. Appl. Supercond.* **22** 4800504
[3] Ivanov D *et al.* 2012 *IEEE Trans. Appl. Supercond.* **22** 4200604
[4] Calvi M, Bauer P, Bessette D, Cau F, Marinucci C and Bruzzone P 2010 *IEEE Trans. Appl. Supercond.* **20** 407
[5] Tanna V L, Fietz W H, Heller R, Vostner A, Wesche R and Zahn G R 2006 *J. Phys.: Conf. Ser.* **43** 1023
[6] Ballarino A 2010 *J. Phys.: Conf. Ser.* **234** 032003
[7] Liu H, Wang Y, Zhang H, Wang Y and Qu Y 2010 *IEEE Trans. Appl. Supercond.* **20** 2176
[8] Uede T, Yamada S, Mito T, Hiue H, Itoh I and Motojima O 2001 *IEEE Trans. Appl. Supercond.* **11** 2563
[9] Garwin R L and Matisoo J 1967 *Proc. IEEE* **20** 538

Status of the Consolidation of the LHC Superconducting Magnets and Circuits

J Ph Tock, S Atieh, D Bodart, F Bordry, N Bourcey, P Cruikshank, K Dahlerup-Petersen, J M Dalin, C Garion, A Musso, R Ostojic, A Perin, M Pojer, F Savary and C Scheuerlein

CERN – European Organization for Nuclear Research
1211 Geneva 23 Switzerland
E-mail: Jean-Philippe.Tock@cern.ch

Abstract. The first LHC long shutdown (LS1) started in February 2013. It was triggered by the need to consolidate the 13 kA splices between the superconducting magnets to allow the LHC to reach safely its design energy of 14 TeV center of mass. The final design of the consolidated splices is recalled. 1695 interconnections containing 10 170 splices have to be opened. In addition to the work on the 13 kA splices, the other interventions performed during the first long shut-down on all the superconducting circuits are described. All this work has been structured in a project, gathering about 280 persons. The opening of the interconnections started in April 2013 and consolidation works are planned to be completed by August 2014. This paper describes first the preparation phase with the building of the teams and the detailed planning of the operation. Then, it gives feedback from the worksite, namely lessons learnt and adaptations that were implemented, both from the technical and organizational points of view. Finally, perspectives for the completion of this consolidation campaign are given.

1. Introduction

The LHC started regular operations in 2009 at an energy of 3.5 TeV per beam. With the experience acquired over two years of operation, the energy was increased to 4 TeV at the beginning of 2012. These beam energies were chosen on the basis of careful analysis of the machine safety following the incident in 2008 that occurred in the main dipole circuit of sector 3-4 [1]. Its origin is basically hih resistance in the superconducting circuits that initiated a quench associated with a poor connection of the copper stabilizer that should have carried the current during the current decay. The goal of the first long shutdown (LS1) of the LHC is to consolidate the 13 kA splices of the main superconducting circuits to allow safe and reliable operation of the LHC up to its design energy of 14 TeV center of mass. The CERN's plan and consolidation actions were successfully reviewed by an international committee of experts. It is also an opportunity to carry out repairs, maintenance and consolidation in the whole accelerators complex [2]. The LS1 started on 16th February 2013 with powering tests aiming to identify other possible limitations of the superconducting magnet circuits that could be repaired during LS1. About 540 circuits were tested in 10 days, with a total of 1300 powering cycles; only a few minor issues, which neither jeopardize the LS1 schedule nor the capability of LHC to operate at the design energy, were discovered.

Before and after warming-up (figure 1) of the LHC from 1.9 K to room temperature, electrical quality assurance tests were carried out to check the integrity of the superconducting circuits. They revealed a few weaknesses and defects that will be corrected during LS1. Leak tests were also carried out at different temperature levels to identify, measure and localise existing air and helium leaks prior to the consolidation interventions. Operation had coped with all of these leaks, but LS1 provides an opportunity to understand their origins and eliminate them. So far, 14 of the 20 known helium-to-insulation vacuum leaks have been localized to the component level.

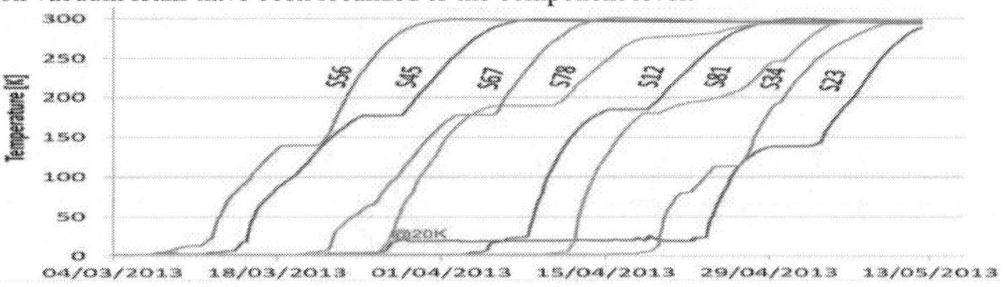

Figure 1. Warm-up of the 8 LHC sectors over 10 weeks

2. Consolidation of the 13 kA main splices

2.1. Design of the consolidated splices

The consolidation of the 13 kA splices [3] consists mainly of:

- Soldering a shunt on the existing splices to ensure the electrical continuity of the copper stabiliser of the busbars in all conditions as shown in figure 2. Tin/lead alloy is used for soldering to avoid unsoldering the underlying splices that were carried out with tin/silver-
- Installing an improved electrical insulation system that, in addition to providing dielectric strength, is also a mechanical restraint that limits stresses and deformations of the busbars and splices under the Lorentz forces as shown in figure 3.

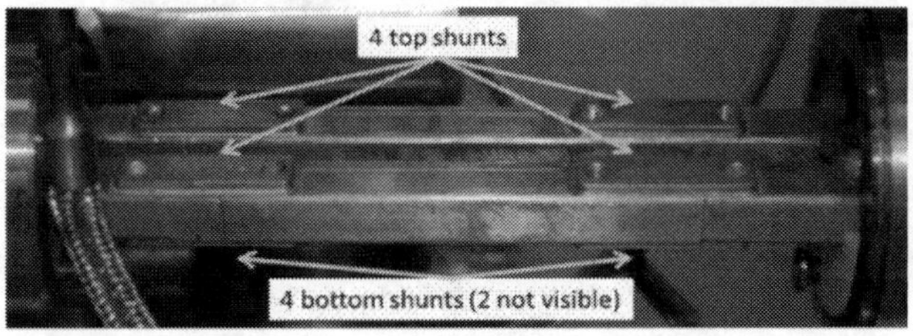

Figure 2. Shunts installed on the 13 kA splices.

Figure 3. Electrical insulation system

Strict quality controls (electrical and geometrical) are applied to existing splices. The on-going verification of more than 40 % of the 10 000 existing splices shows that about 30 % of splices have to be redone. This proportion, based on the repairs and measurements carried out in 2009, had been estimated at about 15 % [3]. This increase is mainly due to the need for appropriate geometry to allow installation of the shunts in optimum conditions. No compromise is made on the quality: the defined criteria in terms of resistance of the copper stabilizer and of geometry are not relaxed. To cope with the resulting extra work, the team has been reinforced with extra technicians, now trained and fully operational. Extra tooling and components were also procured.

2.2. Sequence of operations

Although the delicate operations on the splices are at the heart of the superconducting circuits consolidation, they occupy only about 15 % of the entire project team of 280 persons. Most of the resources are used to provide access to the splices and then reclosure after the consolidation interventions: each envelope of the interconnections (insulation vacuum external bellows, thermal shielding, superfluid helium enclosure, electrical insulation system) have to be opened and finally reclosed. A simplified sequence is shown in figure 4. These operations, together with other specific interventions, are part of the Superconducting Magnets And Circuits Consolidation (SMACC) project.

The external bellows and thermal shields of the first interconnection were opened on the 8th of April 2013 and about 75 % were open by the end of August 2013. This task is entrusted to two teams of collaborators from the National Technical University of Athens (Greece) and from the Wroclaw University of Technology (Poland).

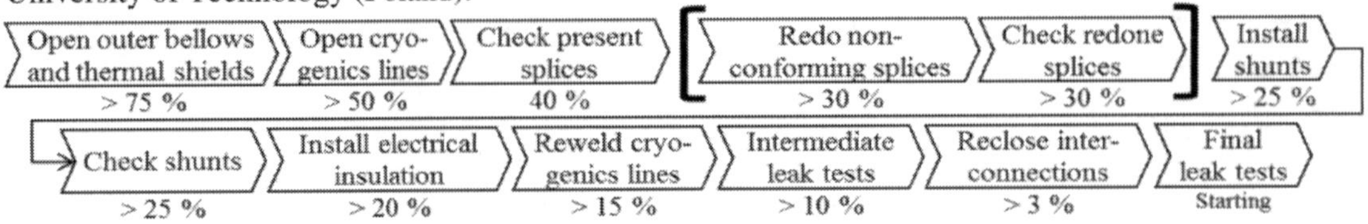

Figure 4. Simplified workflow and status updated on 9.09.2013.

Opening of the cryogenics lines enclosing the busbars gives access to the splices. During LS1, 5 000 welded stainless steel sleeves will be cut using orbital cutting machines designed and produced at CERN. This operation is particularly important because the quality of the orbital TIG weld used for re-closing the lines strongly depends on the geometry. This task is carried out by a team of collaborators from the National Center for Physics (Pakistan Atomic Energy Commission) [NCP(PAEC)].

After removal of the electrical insulation, the quality of the existing splices, mainly electrical and geometrical, is systematically checked. So far, about 30 % of the splices are not meeting the criteria and must be completely remade as shown in table 1.

Table 1. Preliminary results of existing splices quality control (9.09.2013).

Sector	Sector 56	Sector 67	Sector 78	Sector 81	LHC
Measured [%]	100	100	100	25	40
To redo [%]	28	32	29	31	30

Splices not satisfying the acceptance criteria are first de-soldered. The superconducting cables are then carefully inspected to identify all defects that could jeopardize the quality of the future splices. The splices are then re-soldered with improved procedures and induction soldering machines.

The installation of the shunt is the main activity: more than 27 000 shunts have to be installed. To meet the schedule, 20 shunts should be installed every working hour. The shunts are soldered across the splices with resistive ovens after preparing the interface by a careful milling operation. This surface quality and the integrity of the splice are validated by mechanical and electrical checks.

The next step is the installation of the electrical insulation systems: 5 000 units have to be assembled and validated by the QC team during LS1. Most of the QC steps are carried out by a team of engineers and technicians normally in charge of the operation of the CERN accelerators. They have been trained on mock-ups and are overseen by system experts for deeper analysis and handling of problematic cases.

The previously cut sleeves are re-welded. This task is carried out by a team constituted of collaborators from NCP(PAEC) and of experienced CERN welders (staff and industrial support). Rigorous quality assurance is also applied to this activity: pre-qualification of the welding procedures, equipment and personnel associated with recording of the actual welding parameters, visual endoscopic inspection by certified inspectors and local leak tests. Specific tooling was developed to allow testing of 2 welds simultaneously. On the first 2 000 welds, 2 leaks were discovered. Careful off-line analysis localised them in the vicinity of the welding area but not in the welds themselves.

After verification that all non-conformities have been closed, the interconnections are finally reclosed. At the end of August, 45 interconnections have been reclosed. This is according to the baseline schedule defined in March 2013. The opportunity is also taken to improve the thermal screens design [4]. The last operation, the final leak testing of the insulation vacuum, has just started.

3. Other consolidations of the superconducting magnets and circuits

Eighteen main superconducting magnets with different non-conformities will be replaced during LS1: 8 presenting a high internal resistance that could be the sign of a lack of mechanical strength, 5 being affected by minor electric integrity issues and 5 for other reasons. They have all been disconnected and replaced by spare ones. This operation is part of the mandate of the Special Intervention Team, composed of polyvalent and experienced technicians who also intervene on major non-conformities.

Following the incident of 19th of September 2008 [1], pressure relief devices were installed on the LHC cryostats. Their aim is to limit the extent of damages should the helium enclosure be punctured. At that time, for schedule reasons, only part of the machine was equipped. During LS1, 612 new pressure relief devices will be installed, bringing the total to 1344 units.

Sixteen electrical feed boxes power the main LHC superconducting circuits. As the 135 main circuit splices in the electrical feedboxes need to fulfill the same functionality as the ones between the magnets, they will be consolidated to the same standards [5].

4. Quality Assurance (QA)

The quality of the work has priority over schedule. To achieve the necessary high level of quality, about one third of the project team is assigned to QA tasks. Meetings are held several times a week to systematically review arising non-conformities, to agree on the action plan, to decide on the updates of the procedures and hence rapidly deal with open issues. An independent QA manager is nominated, reporting directly to the LS1 coordinator.

5. Conclusions

Starting new activities as early as possible to allow time to debug them has proven to be a profitable strategy. The CERN priorities are strictly followed: safety, quality, schedule, budget. Only one minor accident related to the work in the LHC tunnel occurred. The motivation of all the participants and the control of the routine risk are the present focus of the management team.

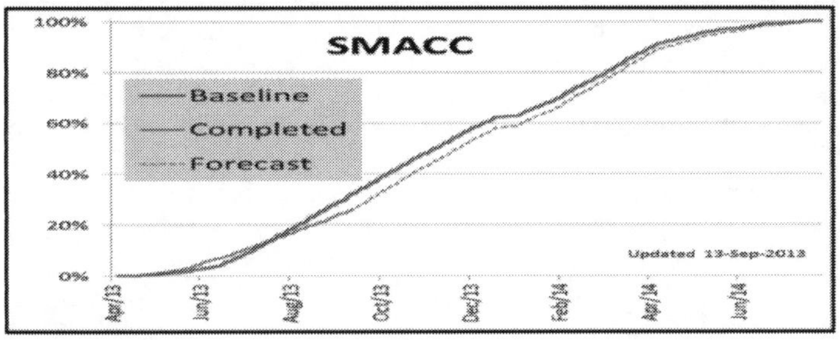

Figure 5. .SMACC main progress dashboard

Thanks to a good preparation, the end date of the SMACC project is still summer 2014, according to the baseline. Regular updates are available at: http://lhcdashboard.web.cern.ch/lhcdashboard/ls1/

Acknowledgments

The authors wish to thank colleagues throughout the whole organisation for their continuous support in the course of this project: logistics, transport, infrastructure, cryogenics,...

References

[1] Lebrun P et al., Report of the Task Force on the incident of 19 September 2008 at the LHC, CERN LHC Project Report 1168, 31st March, 2009
[2] Bordry F, Baird S, Foraz K, Perrot AL, Saban R, Tock J Ph, The first long shutdonwn (LS1) for the LHC, IPAC 13, Shanghai, May 2013
[3] Tock J Ph et al, Consolidation of the LHC superconducting circuits: a major step towards 14 TeV collisions, IPAC 12, New Orleans, May 2012
[4] Musso A et al, The Upgrade of the LHC Magnet Interconnections Thermal Shielding, CEC-ICMC 2013, Anchorage, June 2013
[5] Perin A et al, Consolidation of the 13 kA splices in the electrical feedboxes of the LHC, IPAC 12, New Orleans, May 2012

Investigation of DC flow effects on a 4K two-stage pulse tube cryocooler

A Tsuchiya[1] **and M Y Xu**[2]

[1]*Precision Equipment Group, Sumitomo Heavy Industries, Ltd.*
2-1-1, Yato-cho, Nishitokyo-city, Tokyo 188-8585 Japan
[2]*Technology Research Center, Sumitomo Heavy Industries, Ltd.*
2-1-1, Yato-cho, Nishitokyo-city, Tokyo 188-8585 Japan

Akh_Tsuchiya@shi.co.jp

Abstract. Sumitomo Heavy Industries, Ltd. (SHI) has been continuously developing 4K GM-type pulse tube cryocoolers for cooling superconducting magnets and pre-cooling ultra-low temperature refrigerators, such as dilution refrigerators, etc. In a double-inlet or a 4-valve GM-type pulse tube cryocooler, there is a gas circulation, called DC flow, generated by the pulse tube, the regenerator and the orifice. The performance of a pulse tube cryocooler is strongly dependent on the rate of such kind of flow. It is possible to improve the performance by optimizing the rate. In order to optimize the rate, a bypass line with a small valve is installed between the warm end of the pulse tube and the low pressure side of the compressor. The second stage cooling capacity at 4.2 K was improved by about 0.25 W after optimization of the rate by adjusting the opening of the valve. The details of the experiment results will be reported in this paper.

KEYWORDS: cryogenics, pulse tube cryocooler, 4K cryocooler, DC flow.

1. Introduction

Owing to its low vibration and high reliability, pulse tube cryocooler has been widely used for cooling superconducting magnets and pre-cooling ultra-low temperature refrigerators. SHI has been developing pulse tube cryocoolers for such applications.[1-4]

In a double-inlet or a 4-valve pulse tube cryocooler, there is a loop configured by the regenerator, the pulse tube and the double inlet orifice or the on-off valves at the pulse tube warm end. Through this loop, a gas circulation, called DC flow, is generated. A DC flow greatly affects the cooling performance of a pulse tube cryocooler, so optimization of its rate is required to achieve the best performance.[5-7]

SHI has been investigating the effect of a DC flow in a pulse tube cryocooler for the purpose of developing a higher efficiency pulse tube cryocooler.

The experimental evaluation results will be reported in this paper.

2. Experiment apparatus

The configuration of a pulse tube cryocooler used for this experiment is shown in figure 1. The effect of a DC flow was investigated in a two-stage 4K pulse tube cryocooler. In order to easily install a DC flow valve, a pulse tube cryocooler with a remote valve unit was selected for the test. (figure 2) As shown in figure 1, the warm end of the second stage pulse tube is connected to the low pressure line of the compressor using a bypass line with a small valve, called DC flow valve.

The frequency of the gas oscillation which is generated by the compressor is 1.7 Hz. The high and low pressures of the compressor are about 2.0 MPa and 0.9 MPa, respectively. The gas flow lines between the compressor and the pulse tubes are connected through needle valves to adjust the flow rate. 30 W and 1.3 W heat load are added to the first and second stages.

The second stage temperature is measured using a RuO sensor, and the temperature of the first stage and the temperature profile of the pulse tubes are measured with PtCo thermometers. All sensors are connected to a current supply at the room temperature with CuNi wires, and the heat conduction through the wires is negligible.

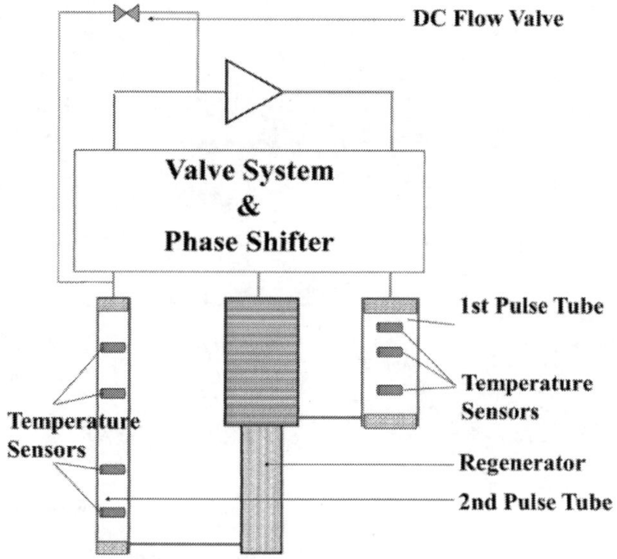

Figure 1. Schematic of a two-stage pulse tube cryocooler with temperature sensors for measuring temperature profile.

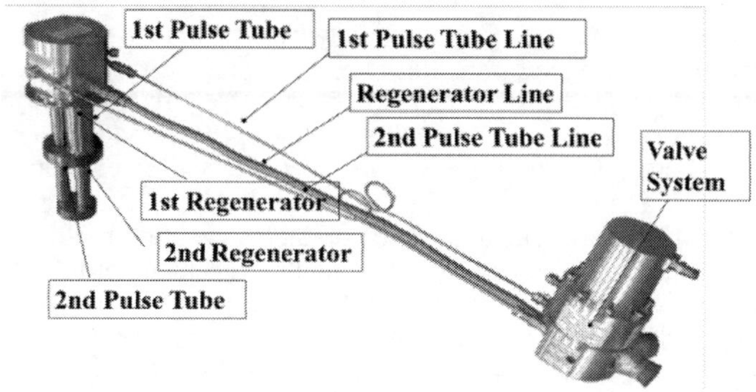

Figure 2. Picture of a two-stage pulse tube cryocooler with a remote valve unit.

3. Experiment results

The DC flow rate was adjusted by changing the DC flow valve opening. The temperature at each point where a sensor is mounted is measured at each valve opening. Thus the relation between the DC flow rate and the cooling capacity of the pulse tube cryocooler as well as the temperature profile of the pulse tubes was investigated. The results are summarized in table 1 and figure 3, 4.

Table 1. Temperature of the first and second stages at each DC flow valve opening.

T_1st stage (K)	T_2nd stage (K)	DC flow valve opening (turns)
47.23	4.53	0
55.31	4.28	0.1
55.83	4.29	0.5
63.32	5.26	1.0

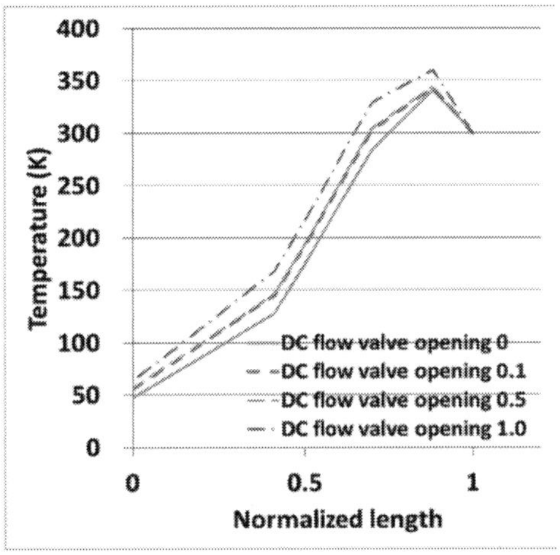

Figure 3. Temperature profile of the first stage pulse tube.

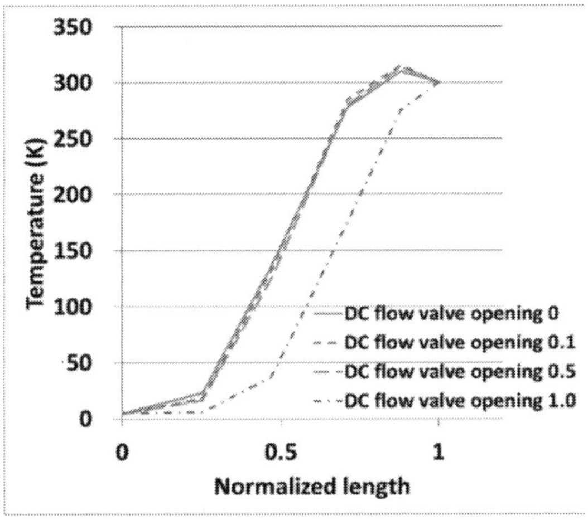

Figure 4. Temperature profile of the second stage pulse tube.

The experiment results show that the cooling performance is strongly dependent on the opening of the small valve.

Figure 4 shows the cold temperature front advances when the DC flow valve opening is set to an optimum position and the cooling capacity at the second stage increases. This tendency is similar to the results in reference 8. The second stage temperature decreased from 4.53 K to 4.28 K with 1.3W heat load. From our experiment results, it is known that the specific cooling capacity is about 0.1 W per 0.1 K at the second stage for this system. Therefore, the cooling capacity at the second stage increased about 0.25 W.

It also shows that if the valve opening is too large, the temperature profile significantly changes and the cooling capacity gets worse. Figure 3 shows that the temperature profile of the first stage pulse tube also changes as the valve opening varies and the cooling capacity at the first stage decreases. When the valve is opened, the temperature of the first pulse tube is increased. When the valve is opened, the gas flow from the cold end to the warm end is generated in the second stage pulse tube. With an optimized valve opening, reasonable DC flow is generated, which results in a cooling capacity improvement at the second stage. However, in the first stage pulse tube, a gas flow from the warm end to the cold end is generated, which results in an extra heat conduction loss from room temperature. Accordingly, the cooling capacity at the first stage decreases.

4. Conclusion

The dependence of the cooling performance on the rate of a DC flow was investigated. By installing a bypass line between the warm end of the pulse tube and the low pressure side of the compressor, the cooling capacity at the second stage has been improved by about 0.25 W at 4.2 K. Meanwhile, the temperature profile along the pulse tubes was measured. The results show that, a DC flow, of which direction is from the cold end of the pulse tube to the warm end, was generated. As expected, a better cooling performance at the second stage has been achieved with such kind of DC flow.

References

[1] Xu M Y, Yan P D, Koyama T and Li R 2003 *Proc. Intl. Cryocooler Conf. (Cambridge)* vol 12 (New York: Kluwer Academic/Plenum Publishers) p 301
[2] Xu M Y, Gao J L, Seitz E and Longsworth R C et al 2006 *Advances in Cryogenic Engineering* **49A** 837
[3] Xu M Y, Takayama H, Yamanouchi H and Saito M 2008 *Proc. Intl. Conf. on Cryogenics and Refrigeration (Shanghai)* p 120
[4] Xu M Y, Takayama H and Nakano K 2011 *Proc. Intl. Cryocooler Conf. (Atlanta)* vol 16 (Boulder: ICC Press) p 57
[5] Gedeon D 1997 *Proc. Intl. Cryocooler Conf. (Cambridge)* vol 9 (New York: Plenum) p 385
[6] Chen G, Qiu L, Zheng J, Yan P, Gan Z, Bai X, and Huang Z 1997 *Cryogenics* **37** 271
[7] Jiang Y L, Chen G B and Thummes G 2003 *Proc. Intl. Conf. on Cryogenics and Refrigeration-proceeding of ICCR* 77
[8] Wang C, Thummes G, and Heiden C 1998 *Cryogenics* **38** 689

Construction and test of a non-insulated insert coil using coated conductor tape

D Uglietti, R Wesche, P Bruzzone

EPFL-CRPP Fusion Technology, CH-5232 Villigen PSI, Switzerland

E-mail: davide.uglietti@psi.ch

Abstract. A small coated conductor coil (central field 1 T at 200 A) was constructed and tested in order to evaluate the benefit and the possible disadvantages of non insulated, double pancake construction, in particular regarding quench protection. The insert coil consists of three double pancakes without turn to turn insulation, and it was impregnated with beeswax. The coil was equipped with voltage taps on each pancake, thermocouples and a quench heater. Quench tests were carried out because quench protection is one of the main issues in developing coil technology for coated conductors. Quenches were induced either by a heater or by exceeding the coil critical current; in all cases the quench protection system was able to detect the quench.

1. Introduction

In the last few years remarkable results have been achieved in the construction of high field insert coils using coated conductors [1]. In the Superconductivity Group at CRPP a project aiming at the construction of 4 T insert coil is running [2]: it was planned to build a coil composed of double pancakes using a coated conductor tape without electrical insulation. There are just a couple of examples of non insulated magnets [3] and one of them is the CMD-2 solenoid [4]: it was constructed by winding the NbTi cable without any insulation on the steel former, which act as current shunt during a quench. This concept has been recently applied to coated conductors coils [5], the main advantage being the remarkable tolerance to over current compared with coils wound with insulated tapes.

If the coil under development at CRPP [2] was constructed with insulated tape, it could be easily protected with the same technique used for standard LTS magnet. Nevertheless, the use of a non insulated tape adds a generous margin to the quench protection, without any major drawback. The only disadvantage is the time constant that appears during transient (charging and discharging), but in this case it is limited to less than a second, and it does not perturb the normal operation.

In this work a test coil, composed of three double pancakes, is constructed and tested at 4.2 K in background field. Quench detection and protection have been also implemented and tested.

2. Coil construction

The coil consists of three double pancakes each wound on a common long former made of fibreglass. Each double pancake was wound with about 6.5 m of non insulated tape (Superpower SCS4050); inner diameter is 43 mm and outer diameter is 52 mm. Each pancake is insulated from the other with a cardboard ring, which has been previously impregnated with bee wax. In coils made of double pancakes wound on separate formers, an axial pressure can be applied to reduce the gaps in between the pancakes during the assembling of the double pancake modules. If the double pancakes are wound one after the other on the same common former it is inevitable that a gap is present in between each pancake. During operation the winding pack can move and hence the strain could locally exceed the irreversible limit. Therefore the coil should be impregnated, in order to fill up the voids in between the pancakes. In order to achieve this objective, the presence of a cardboard disk impregnated with wax will help the wax to diffuse, even if the impregnation is not carried out in vacuum. The bee wax was mixed with quartz powder (Quarzmehl K8 from ASTORIT) at about 20% in weight; the reason is to reduce the thermal contraction and improve the strength during thermal cycling.

The two joints were prepared using three short pieces of tape, cut diagonally and placed one besides the other (see figure 1) with the superconducting layer side facing upward (outward). The tapes from the two double pancakes to be jointed were overlapped upon the three tapes in diagonal, so that the sides with the superconducting layer are in contact. On the outer side of the joint a brass foil (8 mm wide, 40 mm long, 0.2 mm thick), acting as mechanical reinforcement, was soldered.

Voltage taps were attached to each double pancake, so that they could also be used to measure the voltage drop across the two joints. A heater (strain gauge) was placed between the uppermost pancake and the flange. One thermocouple was placed besides the heater, and two more were located close to the lowermost joint and between the lowermost pancake and the flange.

Finally the coil was wrapped with glass band and impregnated with wax. The function of the wax/glass over-banding is to insulate thermally the coil from the bath, so that the heater can efficiently heat the coil, without dispersing too much heat into the bath.

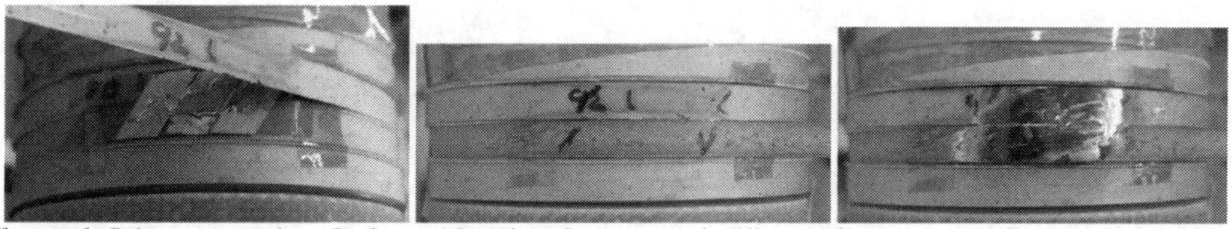

Figure 1. Joint preparation. Left: overlapping the tape to the diagonally cut tapes. Center: Overlapping completed; then it will be soldered. Right: A thin brass foil is soldered over the joint as mechanical reinforcement.

3. Coil test at 77 K

The coil was first tested in liquid nitrogen. The critical current was about 35 A at a total voltage drop of 0.1 mV; the exponential *n* value was about 18; these values are in line with the expectations.
Few quench experiments were carried out, either using the heater to initiate the quench, or by running the coil above the critical current, till the thermal runaway. During these tests no quench detection/protection was used, and for several seconds the power supply delivered currents of over 30 A with total voltage drop on the coil higher than 10 V; therefore a power between 400 W and 550 W was dissipated into the coil, with a large boil-off of liquid nitrogen. Such operation is possible only with non insulated coils, where the current can be shunted through the copper and the dissipation distributed eventually over the whole coil, instead of being localized into a small hot spot region, which would quickly burn out.

It was observed that the time constant was gradually reduced during the quench tests (see figure 2). The temperature measured in these tests was of the order of 250 K, but it can not excluded that the actual temperature in the winding pack was higher thus exceeding the melting temperature of beewax (about 350 K). In addition, the increase of temperature may have led to a reduction of tension in the winding because of the thermal expansion of the winding pack, while the former being in contact with liquid nitrogen was still cold; it can be speculated that the liquid wax may have infiltrated the turns, increasing the turn to turn resistance and thus reducing the time constant; the corresponding resistance was estimated to be increased from 5.4 to 13.5 mΩ. Such situation should not occur in a real operation of a wax impregnated coil, because the quench detection and protection is designed so that the temperature never exceeds 150 K.

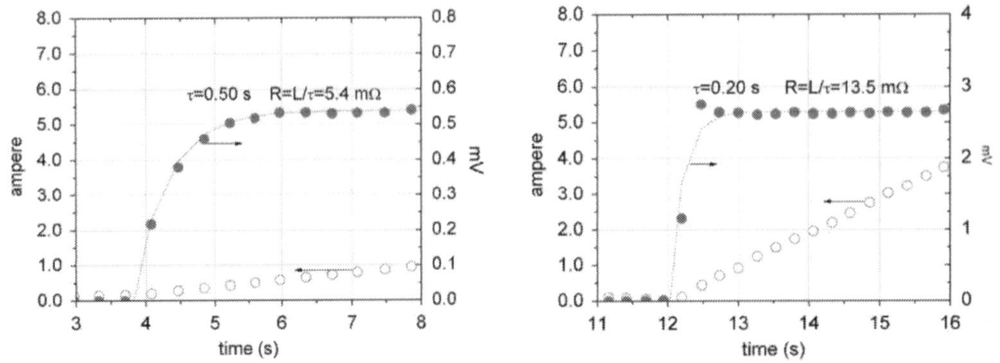

Figure 2. Time evolution of the voltage drop during the current ramp. The time constant is clearly reduced from the first test (left) and after quenches (center and right).

4. Coil test at 4 K

Despite the change in time constant, the coil was tested also in liquid He. The V-I characteristics are shown in figure 3. Measurements were carried out in self field and in background fields of 5 T and 9 T, generated by an Oxford 15 T magnet. As expected, the central double pancake (B) is the last to start the transition, because the radial component of the magnetic field is the lowest and thus the critical current is the highest. The critical currents have been determined using the resistivity criterion at 10^{-14} Ωm: the values are 400 A in self field, 378 A in background field of 5 T and 353 A in background field of 9 T. The little variation with the background field is due to the fact that the field is almost only parallel to the wide face of the tape, and the critical current is weakly affected by parallel fields.

The V-I characteristics of the joints are plotted in figure 3. The total resistance is acceptable for an insert coil, because the power dissipated at each joint would be less than 5 mW at 250 A.

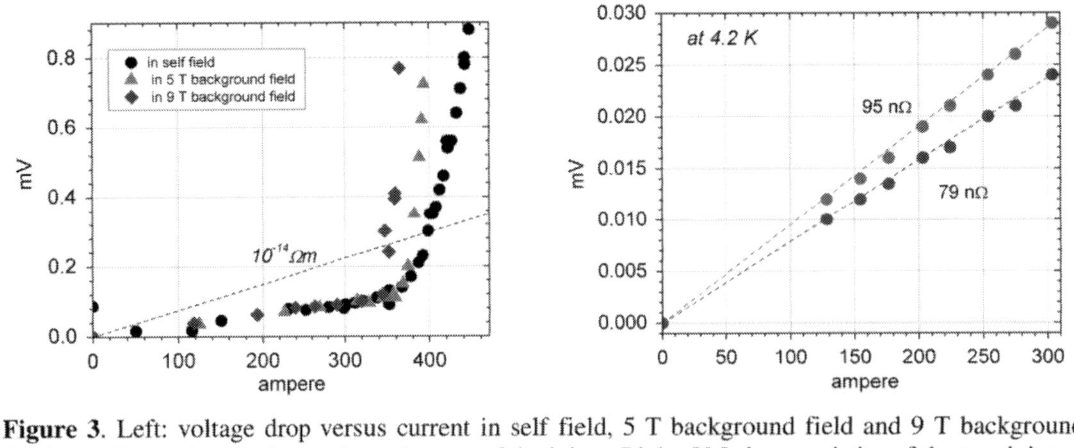

Figure 3. Left: voltage drop versus current in self field, 5 T background field and 9 T background field. The linear slope is due to the resistance of the joints. Right: V-I characteristics of the two joints.

For the tests at 4.2 K a simple quench detection system was installed: when the total voltage drop across the coil is larger than 15 mV a 5 V signal is sent to the power supply, which then switches off. One example of the system operation is shown in figure 4 (left): at 388 A the heater is first turned on at t=480 s, but the power is not sufficient to induce the quench; only a stable increase of voltage is observed. The current is increased to 402 A, and the heater is switched on at t=530 s; this time the self heating effect is sufficient to put the pancake in thermal run away. As soon as the voltage exceeds 15 mV the current drops to zero.

In figure 4 (right) the coil current is increased till the thermal run away appears. Again the quench protection switches the power supply off as soon as the voltage exceeds 15 mV.

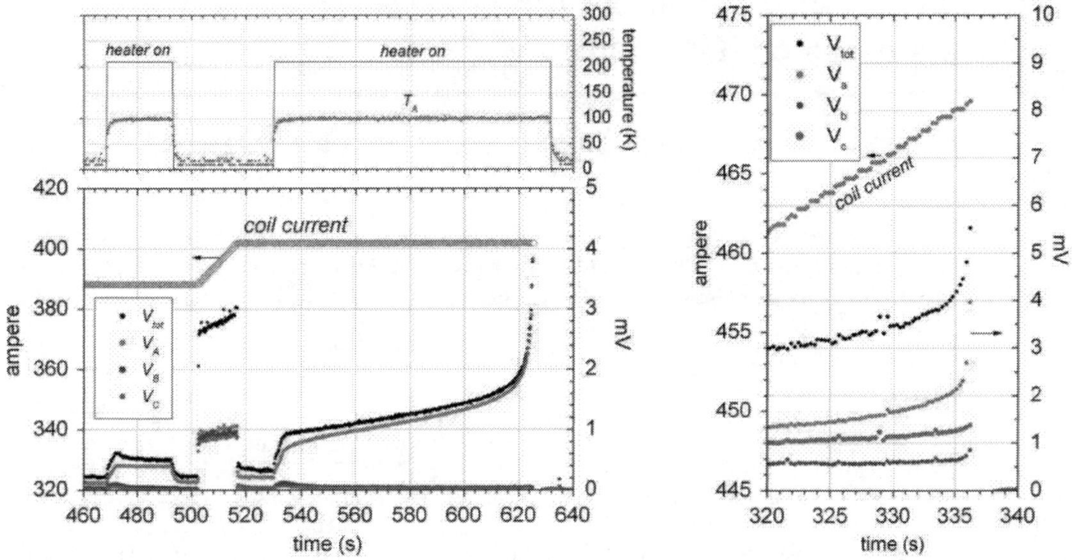

Figure 4. Example of quench detection. Left: after the second heat pulse the coil is driven in thermal runaway; when the total voltage drop (main contribution is from double pancake A) exceed 15 mV the current is set to zero. Right: the current is increase (470 A) till runaway.

5. Summary

A non insulated coil composed of three double pancakes has been prepared. The pancakes were wound on the same common former and insulated one from the other with cardboard disks; the coil was impregnated with wax at ambient pressure. During the over-current operation at 77 K the temperature exceeded 250 K. The coil was not damaged, but an increase in the turn to turn resistance was observed, probably because melted wax infiltrated in between the turns during the quenches.

The test at 4.2 K was carried out up to 9 T background field. Critical current in self field was 400 A and in 9 T background field about 353 A. During the tests at 4.2 K a quench detection circuit was used, which switches the power supply off when the total voltage drop exceeded 15 mV.

References

[1] Matsumoto S 2012 *Superconductor Science and Technology* vol 25 p 025017
[2] Wesche R 2012 *Physics Procedia* vol 36 p 1165
[3] Gomory F 1986 *Proc. IIR Conf., Recent achievements in cryoengineering Cryoprague* p 197
[4] Barkov L M 1999 *IEEE Transactions on Applied Superconductivity* vol 9 p 4644
[5] Hahn S 2011 *IEEE Transactions on Applied Superconductivity* vol 21 p 1592

New evidence for thermal boundary resistance effects in superconducting 6 GHz cavities

M Checchin [1,2], M Martinello [1,2], V Palmieri [1,2], AA Rossi [1], S Stark [1], F Stivanello [1], RK Thakur [1,2], G Yu [1,2,*], R Vaglio [3,°]

[1] Laboratori Nazionali di Legnaro, Istituto Nazionale di Fisica Nucleare, Legnaro Italy
[2] Università degli Studi di Padova, Dipartimento di Fisica ed Astronomia, Padova Italy
[3] CNR-SPIN and University of Napoli Federico II, Piazzale Tecchio 80, 80125, Italy

E-mail: vaglio@na.infn.it

Abstract. Thermal boundary resistance and, more specifically, Kapitza resistance effects have been often considered as a possible source of "non ideal" superconducting accelerating cavity behavior , through the formation of a temperature difference between the inner cavity superconducting surface and the helium bath. However, in the present literature the general reported assessment is that such effects could be neglected, at least at low or moderate input power. In this communication we present new data on small test bulk Nb 6Ghz cavities, showing that when the cavity surface resistance (or the Q) is plotted as a function of the temperature at constant input power, a clear anomaly occurs at the Helium superfluid transition point T_λ reflecting the abrupt change of the thermal boundary resistance at that temperature. The data analysis shows that this anomaly is consistent with the typically measured values of the thermal boundary (Kapitza) resistance. Implications on the cavity optimization strategy are finally discussed.

1. Introduction

The performance improvement of superconducting rf cavities is a relevant issue for future particle accelerators development. The parameters to be optimized are the absolute value of the cavity quality factor Q, the maximum value of the accelerating electric field (breakdown field) and the "Q-slope" i.e. the level of degradation of the Q for increasing values of the accelerating field.
The quality factor Q is directly related to the superconductor surface resistance R_s by the relation $Q = \Gamma/R_s$, where Γ is a constant (temperature and field independent) related to the cavity geometry.
The surface resistance is usually written as a sum of two terms [1]:

$$R_s(T) = R_{BCS}(T) + R_0 \qquad (1)$$

At $T << T_c$ and in the dirty-local limit we can assume $R_{BCS}(T) = \dfrac{A(\omega)}{T}\exp\left(-\dfrac{\Delta_0}{K_B T}\right)$.

* On leave from China Institute of Atomic Energy
° Author to whom any correspondence should be addressed.

R_0 is a temperature independent residual term that accounts for a large set of possible "spurious" effects . R_{BCS} has also a weak dependence on the rf field H_{rf} (through the gap energy Δ_0 field

dependence), directly proportional to the amplitude of the accelerating field E_{acc}. The field dependence of R_o depends on the specific mechanism that is supposed being active.
Independent of the possible intrinsic field dependence of the surface resistance, another effect has to be considered when analyzing the Q vs E_{acc} cavity data. In fact the heat dissipated by the rf field per unit area, $P_d = \frac{1}{2} R_s(T) H_{rf}^2$, can produce a temperature increase of the inner cavity superconducting layer in respect to the liquid helium bath temperature. Using a simple "one-dimensional" model, the temperature difference ΔT between the inner cavity surface T and the bath temperature T_0 can be written as :

$$\Delta T = T - T_0 = \left(\frac{d}{k_m} + R_B \right) P_d \qquad (2)$$

Here d is the thickness of the cavity wall, k_m is the thermal conductivity of the superconducting cavity material and R_B represents the thermal resistance present at the interface between the cavity outer surface and the He bath. The thermal boundary resistance R_B has a fully different nature above and below the superfluid He transition temperature T_λ. Above T_λ the heath transfer from a metal surface to a pool boiling liquid (HeI) occurs via a strong and complex convective process involving the formation of bubbles [2]. In the superfluid regime, below T_λ (He II), boiling is suppressed and the thermal boundary resistance can be identified with the Kapitza resistance R_k [3], much lower than the value of R_B above the transition.

In the presence of a thermal gradient between the inner cavity surface and the He bath, the BCS surface resistance can be rewritten as :

$$R_{BCS}(T_0) = \frac{A(\omega)}{T_0 + \Delta T} \exp\left[-\frac{\Delta_0}{K_B (T_0 + \Delta T)} \right] \cong R_{BCSO}(T_o) \left(1 + \frac{\Delta_o \cdot \Delta T}{K_B T_o^2} \right) \qquad (3)$$

where R_{BCS0} is the BCS surface resistance in the absence of a thermal gradient and $\Delta T \propto P_d \propto H_{rf}^2$; the linear approximation is valid for $\Delta T << T$, that always holds in our experiments. Eqs. 2-3 imply a dependence of the surface resistance on the rf power and therefore on the rf field, and a corresponding Q-slope effect.

The relevance of thermal effects on the accelerating cavity behavior has been often discussed in the literature [4],[5], however, the current assessment is that thermal effects can be essentially neglected at low or medium input power, and that the main mechanism causing the Q-slope is the intrinsic dependence of the surface resistance on the rf field.
In the present communication we present new surface resistance data on small test bulk Nb 6Ghz cavities, in a wide range of temperatures and input power. The body of our data suggest that thermal effects are indeed one of the main limiting factor on our accelerating cavity performances.

2.Experimental Techniques, Results and Discussion

Our monocell 6GHz cavities were realized by the spinning process developed at INFN-LNL. [6].
6GHz cavities can be produced in large number at low cost from a relatively small amount of Niobium and allow to make a large experimental statistics [7]. The Q-factor of the cavity is determined by measuring the decay time, while the accelerating field value is extracted by the measurement of the corresponding power injected into the resonator at the critical coupling. The rf coupler adjustment is made by a stepping motor that is remote controlled. Temperature, vacuum and cryogenic parameters are visualized and automatically acquired and stored in a database [8].
In Fig. 1 the temperature dependence ($1/T_o$) of the residual resistance R_s as extracted by Q measurements on one of our 6GHz cavities is reported, keeping constant the input power P_d.

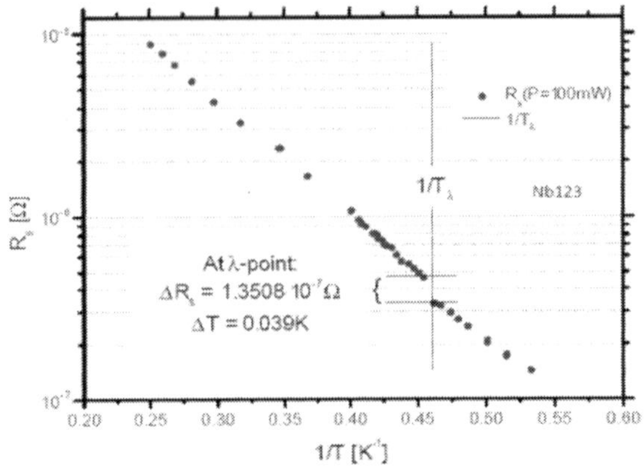

Fig. 1. Surface resistance as a function of the inverse bath temperature at constant input power.

The data clearly show an abrupt "jump" at the He_I-He_{II} transition ($T_\lambda = 2.18K$, $1/T_\lambda = 0.46K^{-1}$) . This behavior is always found, with the same general characteristics, in all our measured samples. We also found that the amplitude of the jump ΔR_s is a linear function of the input power P_d . This jump has to be attributed to the change of ΔT across the λ transition, in turn determined by the abrupt change in the thermal boundary resistance R_B. The linear P_d dependence of ΔR_s is a natural consequence of Eqs. 2-3. It is worth observing that in many similar experiments reported in the literature, the Q measurements are performed at constant rf field, and not at constant rf power. In that case the anomaly at T_λ is still seen, but appears as a smooth "bump" and has never been considered as a relevant effect (as an example, see ref. [9]). Inverting Eqs. 1,3 we can extract the overall thermal resistance :

$$\frac{d}{k_m} + R_B = \frac{K_B T_0^2}{\Delta_0 P_d}\left[\frac{R_s(T_0) - R_0}{R_{sBCS0}(T_0)} - 1\right] \qquad (4)$$

In Fig. 2, curve 1, the temperature dependence of the overall thermal resistance deduced by Eq. (4) is reported, using for $R_s(T_0)$ the experimental data reported in Fig. 1. The values of the parameters set to invert the data are reported in Tab. I.

Tab. 1. Parameters set to invert the data		
Δ_o	$A(\omega)$	R_o
1.40meV	2.18 10^{-3} Ω*K	20 nΩ

Δ_o depends only by the slope of the lnR_s vs $1/T$ curve at high temperatures and is independently determined. The reported value implies $\Delta_o/K_BT_c = 1.76$, very close to the theoretical BCS value.
The $A\omega^2$ value has been chosen to guarantee that all the reported values of the overall thermal resistance (including those of curve 2 that will be discussed later) were positive and within the limits of the typically literature reported values. Finally the value of the residual resistance has been determined to ensure that the low temperature data (below T_λ) were following a T^{-3} dependence as theoretically predicted for the Kapitza resistance (continuous fitting curve in the figure). In fact, due to the already discussed small value of d/k_m for our 6GHz cavities and the nature of the external surface our cavities , the Kapitza resistance should be the dominant term in the overall thermal resistance , as discussed in ref . [4] It is worth underlining that, though the described fitting procedure has some degree of arbitrariness, the deduced temperature dependence of the thermal resistance is relatively sensitive to the parameters choice (as an example the assumption of a $T^{-4.5}$ temperature dependence of the Kapitza resistance, as assumed, as an example, in ref. [5] , would not change much the choice of R_o nor the overall shape and values of the deduced thermal resistance curve.
It is important to observe that the absolute values of the thermal resistance both above and below the

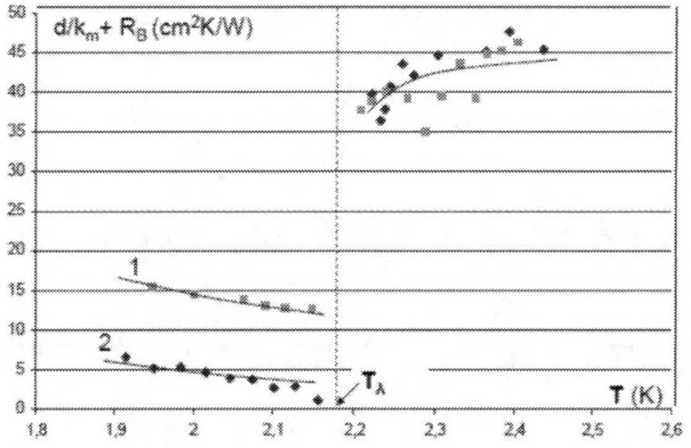

Fig. 2. Thermal boundary resistance for a 6GHz Nb cavity before (1) and after (2) external surface anodization treatment

lambda point are compatible with the data reported in the literature for heath transfer to pool boiling He_I [2]and for the Kapitza interface resistance between polished Nb and superfluid He_{II} [4].
Literature data clearly indicate that it is possible to reduce the Kapitza boundary resistance through appropriate external surface treatments that could optimise the acoustic phonon transmission to the He bath. To achieve this goal we used an Electrolytic Pad Anodization process to grow ,externally to the cavity surface, a bleu-violet porous Nb oxide a few tenth of nm thick. The process did not influence in any way the inner cavity surface.
The data points reported in Fig. 2, curve 2, refer to the same 6GHz cavity whose behavior was described above (Fig.1), and have been obtained by Eq. 4 using the surface resistance data measured after the external surface anodization process and the same set of parameters reported in Tab.1 .
Fig. 2 clearly shows that the thermal boundary resistance is not affected by our external surface anodization above T_λ , but it is reduced below T_λ (Kapitza regime), determining a reduction of the surface resistance and, in turn, an increase of the Q and an improvement of the cavity performances.

3. Conclusions

Our data on 6GHz Nb accelerating cavities clearly indicate the relevance of Nb-He thermal boundary resistance and allowed to deduce both its the magnitude and temperature dependence both above and below the He superfluid transition (Kapitza resistance). External surface treatments (anodization) have been proven to reduce the Kapitza resistance, improving the overall cavity performances.

References

[1] Halbritter, Z . Phys, 266, 209 (1974)
[2] Van Schiver S."Helium Cryogenics" Plenum Press (Mar 10, 2011)
[3] Khalatnikov I. M., Zh. Eksp. Teor. Fiz., 22, (1952) p.687
[4] Amrit J. and Francois M.X., J. Low Temp. Phys, 119 (2000), p.27
[5] Bauer P, Solyak N, Ciovati G L, Eremeev G, Gurevich A, Lilije L, Visentin B, PHYSICA C C441 (2006), p.51
[6] Palmieri V, Preciso R, Ruzinov V L, Stark S, Gambalonga S, Nuclear Instruments and methods in physics research A 342, issue 2/3 (1994) pp.353
[7] Badan L, Durand C, Palmieri V, Preciso R, Stark S, Stivanello F, Venturini W, Particle Accelerators, 62, (1998), p 637.
[8] Stark S, Bisoffi G, Boscagli L, Palmieri V. Porcellato A.M, Proc. EPAC 2004, p.692
[9] P. Dhakal; G. Ciovati, G. R. Myneni, K. E. Gray, N. Groll, P. Maheshwari, D. M. McRae, R. Pike, T. Proslier, F. Stevie, R. P. Walsh, Q. Yang, J. Zasadzinzki Phys Rev ST Accel. Beams, 16 (2013), p. 041001.

Quench Simulation for 9.4 T MRI Superconducting Magnet

Yi Li, Qiuliang Wang, Shunzhong Chen

Institute of Electrical Engineering, Chinese Academy of Science, P. O. Box 2703, Beijing 100190, P. R. China

Corresponding author, Q. Wang: qiuliang@mail.iee.ac.cn, Fax: +86-10-8254-7161

Abstract. The 9.4 T MRI superconducting magnet is being fabricated. A quench simulation code based on anisotropic continuum model has been developed for the quench protection design. The quench behavior of the magnet is studied by means of the code. A commercial software based on finite element method is also applied to the quench simulation of the magnet. The calculation results of the two numerical methods have been compared and analyzed.

1. Introduction to the 9.4 T MRI superconducting magnet

A 9.4 T whole-body MRI superconducting magnet composed of 5 main coils (named as A, B, C, D, and E), and 4 compensating coils (named as F1, F2, G1, and G2) is being fabricated, shown in figure 1. The coils are wound with Wire-In-Channel NbTi/Cu conductors with different cross-section sizes, Cu/SC ratios and critical currents.

All of the coils are powered in series with the operating current of 212.5 A. The inductance of the magnet is about 6000 H, and the stored energy is about 134 MJ. The magnet produces a central magnetic field of 9.4 T. The highest magnetic field of about 9.5 T locates inside the innermost coil A. Because of the high magnetic field inside the coils, the temperature margins ($\Delta T = T_{cs} - T_0$) of the coils are very low. Some design parameters and operating conditions of the coils are listed in reference [1].

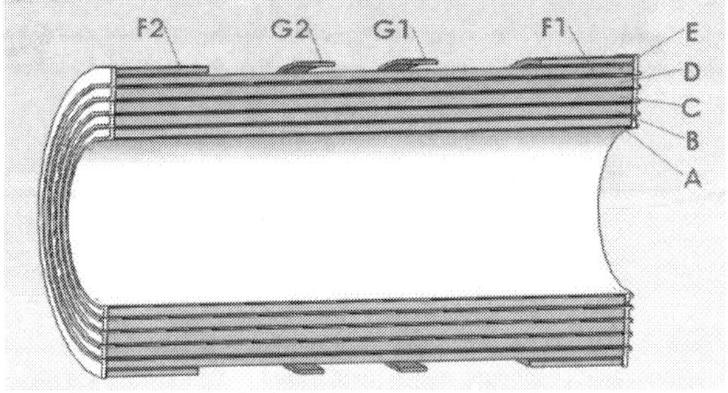

Figure 1. The 9.4 T MRI superconducting magnet

2. Quench protection scheme of the 9.4 T superconducting magnet

The passive quench protection method is adopted. The circuit subdivision and heater network are employed in the quench protection scheme. In order to restrict the coil voltages during the quench event, each main coil is subdivided into 2 sections and the 4 compensating coils are combined into one section. Each section is in parallel with a back-to-back diode and a shunt resistor. In order to accelerate the quench propagation and lower the temperature rise of the coil, the quench heaters are used, which are attached on the coil outer surfaces. The heaters are also used as the shunt resistors of the circuit sections and are powered by the currents split from the circuit sections during quench.

The circuit configuration and the heater network are illustrated in figure 2. The working process of the heater network is that if a coil initiates to quench, the current split from the section of the coil will power the heaters of H11-14, trigger the coils F1, F2, G1, and G2 to quench, and then the current split from the section of the 4 coils will power all other heaters, triggering the rest of coils to quench. The detailed quench protection design is introduced in [2].

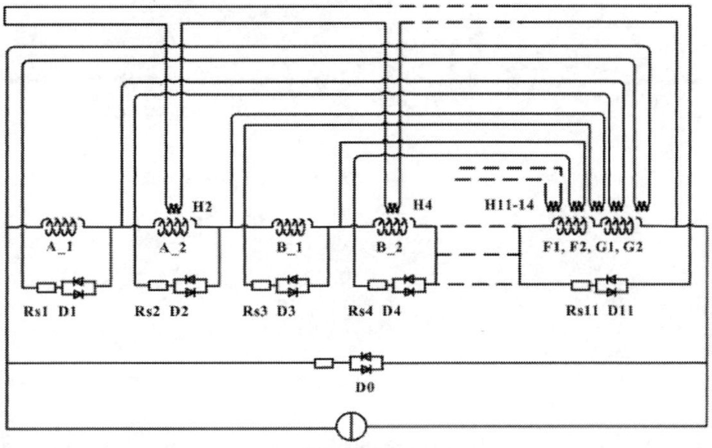

Figure 2. Configuration of the protection circuit and the heater network of the 9.4 T magnet

3. A quench simulation code based on the anisotropic continuum model

A quench simulation code has been developed in our laboratory for the quench protection design [3][4]. The anisotropic continuum model (ACM) is adopted for the prediction of the quench resistances, currents, and voltages of the solenoid superconducting coils. When a quench initiates from a spot in a coil, the normal zone propagates along longitudinal, radial and axial directions. Because the quench propagation velocities along the three directions are not the same, the quench evolution front is assumed to be an ellipsoidal envelope [3]. The ellipsoid is described in cylindrical coordinate as:

$$\frac{(r-r_0)^2}{\left[R_0+\sum v_r(t)\delta t\right]^2}+\frac{(z-z_0)^2}{\left[Z_0+\sum v_z(t)\delta t\right]^2}+\frac{\phi^2}{\left[\Phi_0+\sum \frac{v_l(t)}{r_0}\delta t\right]^2}=1 \qquad (1)$$

where r_0, z_0 are the radial and axial coordinates of the spot where the quench initiates. v_r, v_z, and v_l are the radial, axial and longitudinal quench propagation velocities. R_0, Z_0 and Φ_0 are the radial, axial and longitudinal dimensions of the original ellipse, which are usually triggered by heaters.

The volume of the normal zone is the volume of the ellipse that is confined inside the coil. With the ellipsoidal normal zone volume, the quench resistance of the coil is calculated, and then the coil current can be obtained at each time step. The wire material properties and critical properties dependent on the local time-varying magnetic field and temperature are considered.

The 9.4 T magnet comprises multiple coils, which are electromagnetic coupled. The quench behavior with respects to the coil currents and voltages is closely related to the quench sequences of the coils. The quench is started by the heater triggering (except the coil that initiates to quench). Therefore, the calculation of the times that the heaters trigger the coils to quench is required.

Since the ACM could not predict the quench starting times of the coils, the 2D control volume method (CVM) is adopted. It models the heat conduction between the coils and the heaters [4], shown in figure 3. The heat generation of the problem is the ohmic heating in the heater. It is assumed that when the temperature of a control volume in a coil rises above its current sharing temperature T_{cs}, the quench of the coil will be triggered by the heater.

In order to increase the computational speed of the simulation for many attempts of quench protection designs, the CVM will be altered to the ACM when a coil is triggered to quench by heater.

By means of the code, the quench simulation for the 9.4 T magnet with the above quench protection scheme is conducted. It is assumed that the quench event initiate in coil A. The calculation result shows that the highest temperature rise is below 100 K, the highest coil voltage is restricted below 400 V and the strongest hoop stress is limited below 150 MPa among all coils. It is concluded that the safety of the magnet is guaranteed during quench. The details are shown below.

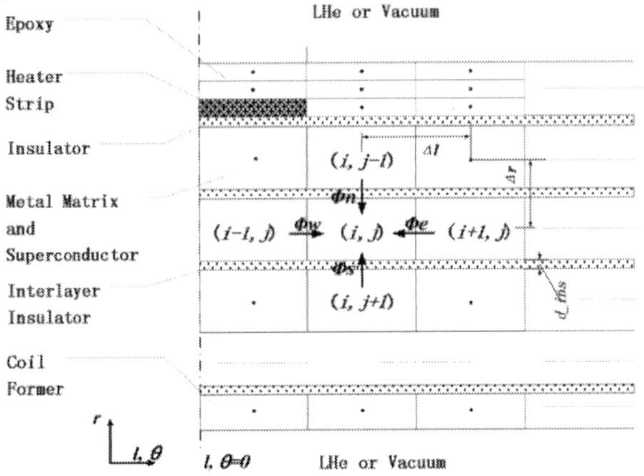

Figure 3. The 2D grid of the CVM for the problem of heat conduction between heater and coil

4. Quench simulation comparison between a commercial software and the ACM code

A commercial software 'Vector Fields Opera-Quench' is applied to the quench simulation of the 9.4 T superconducting magnet. The software is based on the finite element method (FEM), and combines the 3D magnetic field calculation into the 3D thermal calculation [5]. The software employs the same geometric parameters, circuit configuration, and material properties as those inputted into the code discussed above.

The simulation results of coil currents, voltages and temperature rises by both the FEM and ACM are shown in figure 4 to figure 6. Results by the two models show good agreement with respect to the shape of curve, the peak value, and the time of decay to a low level (figure 4, 5). In figure 6 the coil temperature rises by the ACM are higher than those by the FEM. This may be due to the adiabatic assumption in the ACM. Besides, the computation time for the FEM is near 16 hours, while the time for the code developed by us is about half an hour, with the same computer. The ACM code is preferable for the parametric optimization (such as circuit shunt resistance, heater width, etc.) of the quench protection design, and the FEM could be supplementary verification of the optimized design.

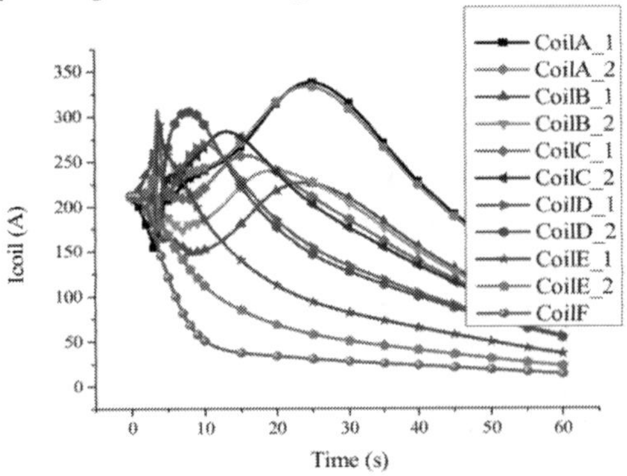

Figure 4 (a). Coil currents by the ACM model.

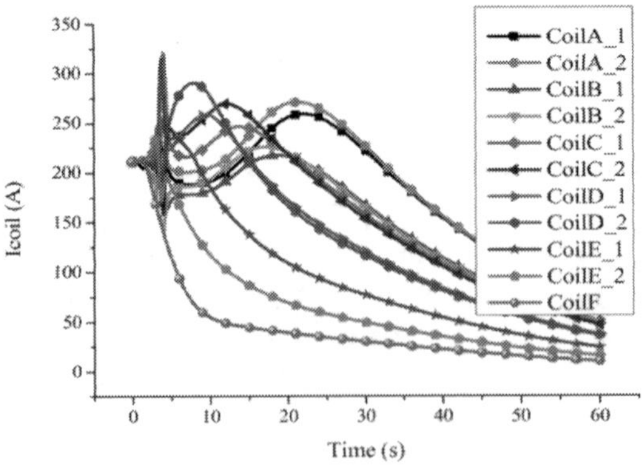

Figure 4 (b). Coil currents by the FEM model.

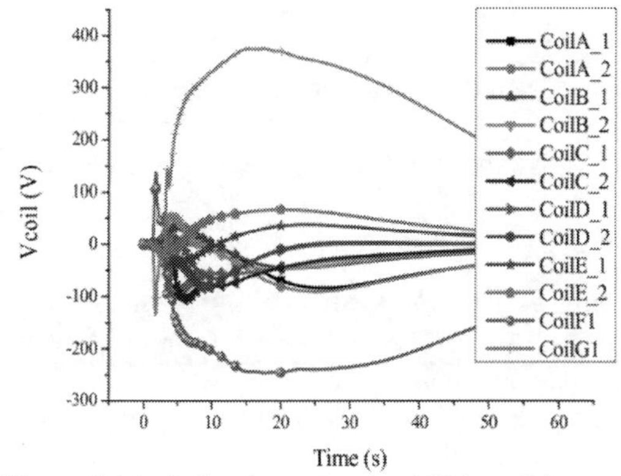

Figure 5 (a). Coil voltages by the ACM model.

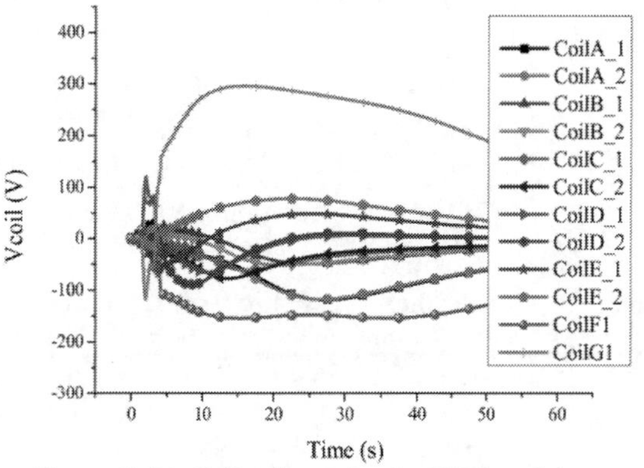

Figure 5 (b). Coil voltages by the FEM model.

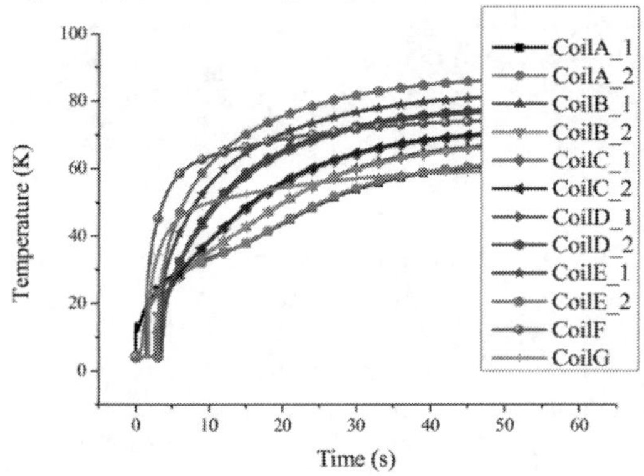

Figure 6 (a). Coil temperatures by the ACM model.

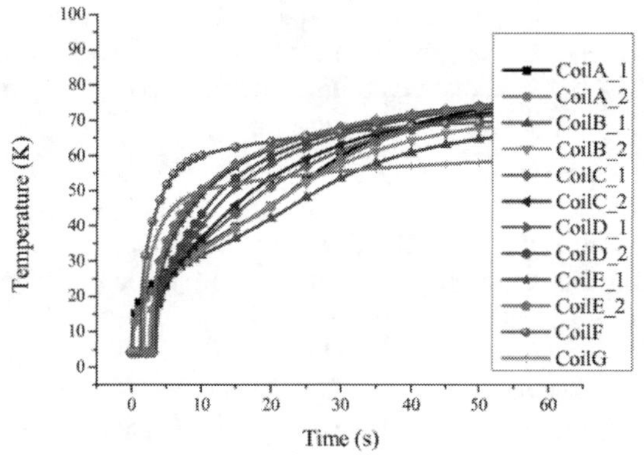

Figure 6 (b). Coil temperatures by the FEM model.

5. Conclusion

A quench simulation code based on the ACM has been developed by us and applied to the 9.4 T MRI superconducting magnet. The simulation results show that the temperature rise, the voltage, and the hoop stress of each coil are within the safety level during quench. The FEM is also employed to simulate the quench behaviour of the magnet. The calculation results of the FEM shows good agreement with the ACM, but the computational speed of the FEM is much slower than the ACM. Besides, the ACM code has been verified by experiments. The details are presented in [6].

References

[1] Wang Q *et al* 2012 A superconducting magnet system for whole-body metabolism imaging *IEEE Trans. Appl. Supercond.* **22** 4400905

[2] Li Y *et al* 2014 A Passive Quench Protection Design for the 9.4 T MRI Superconducting Magnet *IEEE Trans. Appl. Supercond.* **24** 4401605

[3] Wilson M 1983 *Superconducting Magnet* (OXFORD: Oxford University Press)

[4] Li Y *et al* 2012 Quench protection design of a 1.5 T superconducting MRI magnet *IEEE Trans. Appl. Supercond.* **22** 4703604

[5] Aird G J *et al* 2006 Coupled transient thermal and electromagnetic finite element simulation of quench in superconducting magnets *Proceedings of ICAP (Chamonix France)* pp 70–73

[6] Li Y *et al* 2013 Experimental study for the quench protection system of the 9.4-T whole-body MRI superconducting magnet *IEEE Trans. Appl. Supercond.* **23** 4401309

Next generation of HTS magnetic application: HTS bulk and coil interaction

F N Werfel, U Floegel - Delor, R Rothfeld, T Riedel, B Goebel, D Wippich, P Schirrmeister, R Koenig
Adelwitz Technologiezentrum GmbH (ATZ), Naundorfer Str. 29, 04860 Torgau, Germany
E-Mail: werfel@t-online.de

Abstract
The next generation of HTS magnetic bearings (SMB) will operate at magnetic field excitation higher than permanent magnets (> 1.0 T). The new bearing type is capable to support heavy -load rotors of more than one ton mass. Using FEM we calculate the interaction of HTS bulk and 2G coil to achieve higher magnetic flux excitation and flux gradients of HTS bearings at temperatures of 50 – 60 K. The new total HTS bearing type multiplies the present 10 -15 N/cm² force density obtained with PM's by a factor of 5 and passes the force properties of active magnetic bearings (AMB). HTS coil excitation is capable to increases the levitation forces to more than the present 10 kN level and reduces relative cooling and material effort per load. We design a magnetic bearing for HTS bulk - coil excitation.

1. Introduction

What is the future SMB development? Physically it is clear that the magnetic excitation due to PM's is basically limited by the available coercitive force of the PM material. This limitation is tried to overcome by the use LTS/HTS coils for excitation in magnetic bearing. Several groups in advanced laboratories are concentrated on HTS coil interaction with increased magnetic flux excitation. First success is reported by RTRI in Tokyo e. g. for a 10 kWh railway flywheel [1]. A promising load capacity of 20 kN has been demonstrated at cusp fields of 5 T (NbTi coil) and 2.8 T (Bi2223 coil) against 60 mm GdBCO bulks, respectively. The disadvantage of the NbTi coil excitation is necessity of Helium temperature and about 15 K for BSCCO coil, here provided by cryo-cooling.
For rotating application we favored in the past the journal–type superconducting magnetic bearing (SMB) interaction by using permanent magnet (PM) rings and bulk REBCO superconductors [2]. Subdivision of the PM rings in multi-pole arrangement increases the bearing stiffness. Because of the limited magnetization of PM material NdFeB and the strong flux decay in air the achievable magnetic force density is limited to 5-10 N/cm². An increase of the magnetic excitation flux B_{exc} in the air gap would enhance the forces by a square factor B^2_{exc}. Therefore, with B_{exc} = 1T instead of now 0.5 T e. g., the force density would approach the 30 – 40 N/cm² level of active magnetic bearings (AMB).
We investigate magnet field generators, calculate the optimum coil design und propose the achievable force density under coil bulk excitation.

2. Magnetic field generators

For preparing a magnetic field of a certain level one has in general some technical options: Conventional or superconducting wired coil, permanent magnet PM, or superconducting bulk magnet. The physical basis for generating magnetic forces is the ability to produce a magnetic field of a certain value and distribution. Magnetic fields may have different physical and technical origins. The conditions and constraints range from atomic scale collective electron spin behavior (PM), magnetic flux trapping in bulk superconductors, and magnetic coil circuits in solenoids. Between

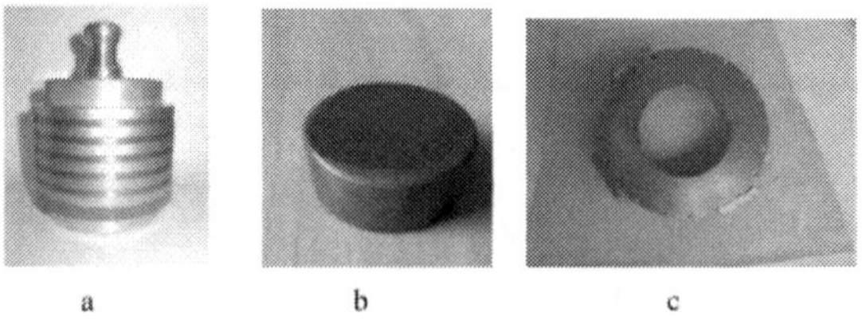

Figure 1 . Magnetic flux generators: 120 mm PM rotor , 4 kg, 30 mm YBCO bulk, T =77 K, 120 g, and 140 mm YBCO tape double pancake coil; T= 45 K, 2.5 kg.

Journal of Physics: Conference Series **507** (2014) 032055 doi:10.1088/1742-6596/507/3/032055

technical and economical constraints the question arises: *How we can provide a magnetic field as a universal tool at a minimum effort and cost?* In Fig. 1 we exhibit and compare three field generators providing similar magnetic flux. A 120 mm PM rotor consisting of 6 NdFeB PM rings separated by 5 Fe flux collectors (a). A melt-textured YBCO bulk superconductor of 30 mm diameter, 15 mm thick (b) , and a 140 mm experimental 2G pancake coil delivering homogeneous central field of 0.5 T at 77 K and 48 A (c) . For comparison the magnetic devices one has to consider the different technical environment and conditions, like the cryogenics in b and c, the power supply (c) or the mechanical structure for magnetic biasing (a).

Next, we will analyze and compare the properties of the field generating device and discuss the possibility of magnetic bearing improvement.

3. Magnetic design and performance

In Fig. 2 the magnetic circuits obey the conservation equation of magneto-statics. The provided magnetic flux and magnetomotive force of coils is considered rigorously in B and H units per volume.

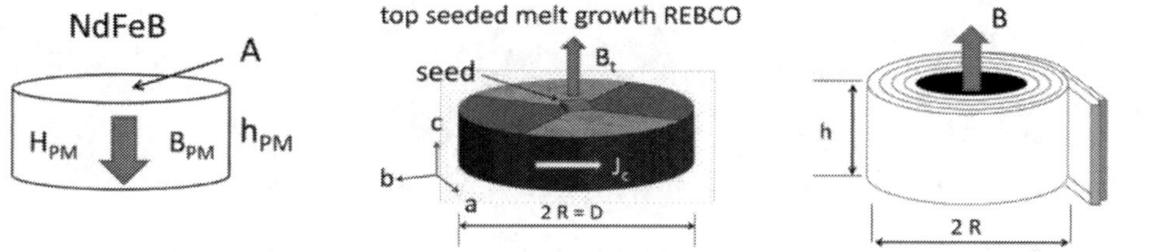

Permanent Magnet PM (a) YBCO bulk superconductor (b) REBCO coil (c)

Figure 2. Geometric and magnetic parameters of PM, HTS bulk, and HTS solenoid.

Permanent Magnet PM (Fig. 2a)

The magnetostatic field concepts are based on the existence of magnetic dipole, the equivalent torque, and on the phenomenon of magnetization. For a PM with the area A and the height h_{PM} under the condition of a constant flux density we can assume a simple expression with the magnet height and area:

Magnetomotive force mmf follows:

$$mmf = \int \mathbf{H}\, \mathbf{dl} \qquad h\, B/\mu_0 \rightarrow H_{PM}\, h_{PM}$$

The H- field arises from the free ends of the dipole chains and is conservative. In contrast the B -field has no sources, but consists of closed induction lines. $\nabla\mathbf{B}=0$ expresses the observation that no true magnetic mono-poles can exist in which the B- field could terminate. Magnetic flux Φ is,

$$\Phi \quad = \int \mathbf{B}\, dA \qquad \mu_0\, A\, n\, I\, /h \qquad B_{PM}\, A_{PM}$$

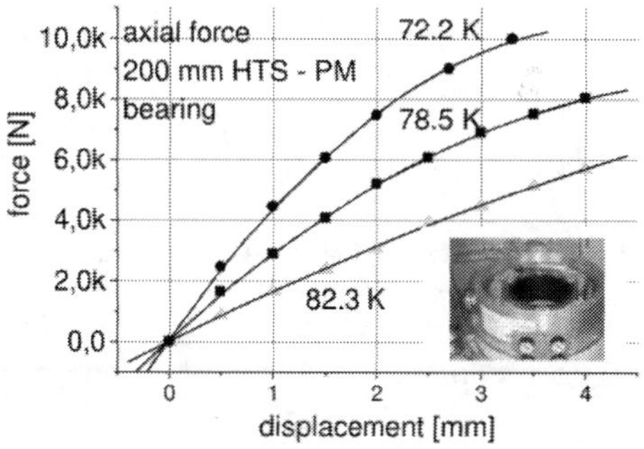

Figure 3. Temperature-dependent magnetic forces of a 200 mm superconducting bearing

The equation express the experience in the work with permanent magnets. The magnetic performance is simply correlated with the cross section area A_{PM} and the height H_{PM}. Both geometrical parameters give the magnetic volume, and together with the coercitivity force parameter (kA/m) the magnetic performance of a PM.

The maximum energy product of a PM $(B\bullet H)_{max}$ is about half of the coercitivity force H_c, giving the parameters $H_{PM} \approx H_c/2$, $B_{PM} \approx B_r/2$. Practically, the best PM material (NdFeB) shows flux densities on the surface between 0.4 – 0.6 T. According to the above formulas higher flux density is directly coupled with an increase of the volume respective the mass of PM.

Using available 200 mm PM ring biasing in a high- gradient magnet design at sub-cooled LN_2 (72 K) axial forces up to 10 kN are achieved (Fig. 3). The performance of magnetic bearings with PM's can be improved by lowering the HTS temperature, as it is demonstrated in Fig. 3. The maximum obtainable force is always limited by the excitation field.

Bulk superconductor (Fig. 2b)
Bulk superconductors can be considered in a rough scale as analogous magnets to PM's. Following the Maxwell equation, the rotation of the internal magnetic flux density determines the maximum trapped field flux density in z direction $B_{z,}^{max}$. For an infinite long cylindrical sample with a diameter of 2R the flux is given by the relation,

$$B_z = \mu_0 \int_0^R J\,dr \quad \rightarrow \quad B_z^{max} = \mu_0 J_c^{\theta} R$$

After this equation, the maximum trapped field depends on the critical current density J_c and the diameter D = 2R of the superconducting domain. In practice, the value is reduced by geometrical and demagnetization effects by about 20% relative to applied magnetic flux density B_{exc}. Bulk superconductors can trap routinely 1.2 – 1.5 T at 77 K, and may provide at least a two times higher B values compared to PM's. At reduced temperatures much higher trapped magnetic flux can be deposited into the HTS bulk material. In Fig. 4 the increased trapped flux maximum in large multi-seed YBCO bulks is demonstrated. At 50 K and 5 T excitation the material is capable to trap almost 3.5 T which is about four times higher compared to the 77 K value.

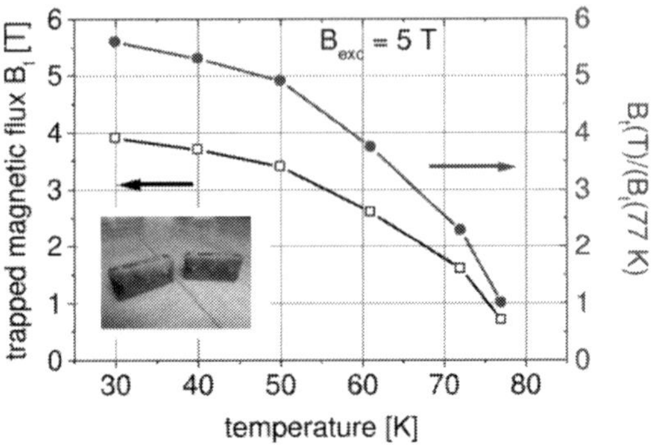

Figure 4. Trapped flux increase in 3-seed YBCO blocks at 5 T excitation with lowering the operation temperature.

Solenoid - circular coil (Fig. 2c)
A circular (thin) coil with a diameter D = 2R, height h, winded with n turns in a length l= n π *D* follows,

$$B = \mu_0\, n\, I$$

As a magnetic performance factor the flux density B and the volume V needed for a corresponding force interaction, can hold for a normalization and comparison,

$$B\,V = \mu_0\, D\, h\,/4 * (I \bullet l)$$

The coil performance is proportional to the geometrical parameters (D • h) and to electrical Ampere-meters (I • l). The above equation gives advices to the necessary material and power costs of a coil circuit producing a improved flux density B.

4. FEM calculation of optimum magnet configuration

Finite element model calculations are performed in an axi-symmetric model. Table 1 shows the chosen parameters for the FEM model. More than 60 different electric and magnetic configurations have been calculated. For a given maximum conductor length of 1000 m the number of pancakes and double pancakes (DP) are optimized. As parameters the DP iron collector stacking distance, the driving pancake current 40, 60 and 80 Amperes, the Fe collector shape, and Fe magnetic back circuit are varied.

Table 1 gives the FEM parameters used in the calculation by the geometrical position of rotor and stator. Fig. 5 displays a FEM result of an optimum gradient coil configuration consisting of 3 pairs of DP with 2 pairs of Fe collectors sandwiched by 2 mm thick Cu plates for efficient cooling.

An optimum set of parameters could be obtained according to the performed calculations:

- Double pancakes should be stacked in pairs.
- At 80 A the magnetic flux is 1.3 T within 2 mm of the rotor surface.
- At the inner part of the YBCO bulk rotor the flux is still 0.3 - 0.7 T.
- Magnetic flux gradients are dB_r/dr = 83 T/m, dB_z/dz = 33 T/m.
- The optimum distance DP - Fe collector is 20 mm.
- At 80 A the D-pancakes see a background flux density of 2.0 T; 50 K cooling is required.

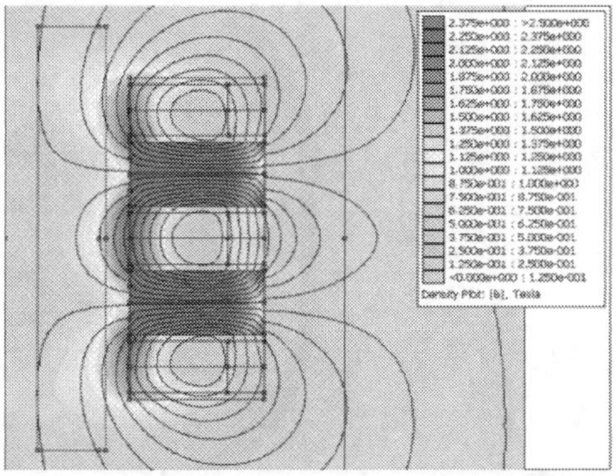

Figure. 5 FEM calculation of coil magnetic flux distribution (3x DP, 2x 20 mm Fe, 6 x Cu plate, 80 A)

TABLE 1: FEM MODEL PARAMETER

Rotor, radius	position [mm]
YBCO bulk	15 – 30
Air gap, insulation, bandage	30 – 37
Stator, radius	
Pancake winding	37 - 89
Fe collector	37 - 89
Double pancake, axial	2 x 4.2
Fe collector, axial	2 x 10
Cu plate, cooling, axial	2
Material parameter	**density [g/cm³]**
YBCO, melt textured	6.0
2G coated conductor + polyimide	7.25
Fe, collector + magnetic circuit	7.8
Glass fiber	2.05
Insulation, resin, epoxy	
	magnetic property
Fe, max. magn. flux density	2 T
Other materials	$\mu_{rel} = 1$

5. High -gradient magnetic flux concept

Due to the availability of 2G wires a promising possibility to increase the force density of SMB is given. The electromagnetic force is now generated by a magnetic field of stacked coils (Fig. 6). In contrast to most of the magnetic coil applications, where a great homogeneity of the magnetic field is gained, we follow our concept of a high-magnetic flux gradient by assembling Fe collectors between the coils after Fig. 5. The FEM calculations in Fig. 5 give the optimum coil design together with the Fe collectors. The essential features are taken from Fig. 5; at 80 A coil current about 1.3 T is the generating flux in the gap between bearing stator and cryostat rotor. The 1.3 T flux density is about 3 times higher than equivalent parameters with PM's. From this and the experiments in Figs. 3 and 4 we calculate the expected force density in Fig. 7. The Fe collectors are almost field saturated by 1.9 T which influences the coil J_c operation. To maintain the high flux excitation the coils have to operate at about 50 K because of the strong J_c (B) dependency. With a force density of 40 N/cm² at 50 K and 1.3 T gap excitation the HTS coil and bulk interaction opens a substantial improvement of HTS magnetic bearing. The performance approach or pass the values of active magnetic bearings AMB.

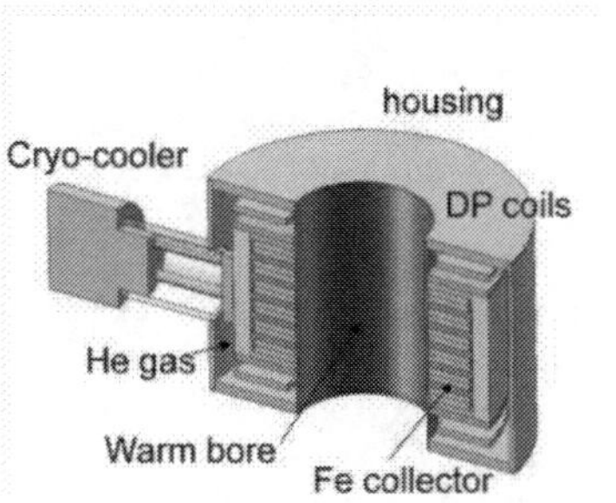

Figure 6. HTS coil bearing concept

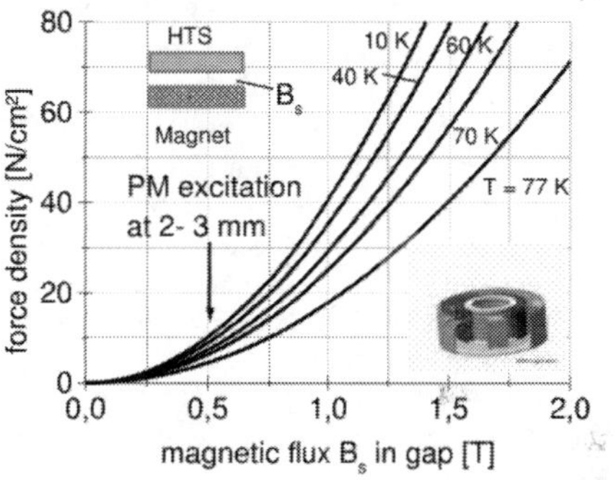

Figure 7. Expected force density of HTS coil bulk interaction

Conclusions

HTS coil bulk magnetic interaction has been analyzed to get a proof for the next generation of total HTS magnetic bearings. Magnetic flux generators are compared in the parameters and conditions. Higher flux is achieved by more PM volume only. FEM calculations give the optimum Double Pancake coil configuration assembled in pairs with Fe collectors. For an excitation of 1.3 T in the gap the estimated force density at 50 K is more than 40 N/cm² and passes the active magnetic bearing AMB parameters.

References

[1] H Seino, K Nagashima, Y Tanaka, M Nakauchi, Study of Superconducting Magnetic Bearing Application to the Flywheel Energy Storage System that consist of HTS bulk and Superconducting coils, J. Phys. C Conf. Series, p. 1076

[2] F N Werfel, U Floegel – Delor, R Rothfeld, T Riedel, G Goebel, D Wippich, and P Schirrmeister, Superconducting bearings, Flywheels and Transportation, Supercond. Sc. Technol. 25(2012)014007.

Results of the test of a pair of 20 kA HTS currents leads

R Wesche[1], P Bruzzone[1], C Fiamozzi Zignani[2], L Affinito[2], S Chiarelli[2], R Freda[2], A. Formichetti[2], M Marchetti[2], H Ehmler[3], J. Heinrich[3] and P. Smeibidl[3]

[1]EPFL-CRPP Fusion Technology, CH-5232 Villigen PSI, Switzerland
[2]ENEA C.R. Frascati, I-00044 Frascati, Italy
[3]Helmholtz Zentrum Berlin, D-14109 Berlin, Germany

E-mail: Rainer.Wesche@psi.ch

Abstract. A new series connected 25 T hybrid magnet system is being set up by the Helmholtz Zentrum Berlin (HZB) for neutron scattering experiments. CRPP has designed and manufactured a pair of 20 kA current leads for the powering of the outer superconducting coils of the hybrid magnet system. In connection with the test of joints for JT60SA, the current leads were tested at ENEA at low voltage up to a current of 18 kA. The mass flow rates required to cool the current leads at different currents measured in the test are in line with the design calculations. For the sum of the resistances of the warm and cold end copper contacts of the HTS module values of 13 (Lead A) and 11 nΩ (Lead B) were measured. In addition, the helium flow through the heat exchanger part was stopped at 10 and 12 kA to study the behaviour of the current leads in case of a loss of flow. The time elapsed between stopping of the helium mass flow and the initiation of a quench was found to be 117 s (Lead A) and 125 s (Lead B) compared to a calculated value of 86 s. The lower value obtained by the calculation can be attributed to the lower initial temperatures in the experiment.

1. Introduction

In the framework of a collaboration of CRPP, ENEA and HZB, CRPP designed and manufactured a pair of 20 kA high temperature superconductor (HTS) current leads [1,2], which are foreseen to power a 25 T hybrid magnet system at HZB. The current leads were used for the test of a JT60SA joint sample at ENEA. The test set-up at ENEA, dedicated for joint testing, was used also to test the 20 kA HTS current leads. The cold ends of the two current leads and the JT60SA joint sample connecting them were immersed in liquid helium. The HTS module composed of 28 AgAuMg/Bi-2223 stacks (8 Bi-2223 tapes per stack) soldered into a grooved stainless steel support tube with bulk copper end pieces is cooled only by heat conduction from the cold end. The copper part is actively cooled by helium gas of ≈44 K inlet temperature. Steady state operation of the HTS current leads at 20 kA would require a mass flow rate of 1.37 g/s. Due to the fact that the maximum achievable 44 K helium mass flow rate was around 1.2 g/s the current leads could be tested only up to 18 kA. In the present article, only the results of the current lead test are presented.

2. Experimental details

In later operation, the cold ends of the current leads will be forced flow cooled, while during the test they were immersed in liquid helium together with the joint sample. The intermediate temperature helium, required to cool the copper parts of the HTS current leads, was supplied from a cryogenic storage dewar, filled with liquid helium. Pressure was built up in the dewar by means of a heater.

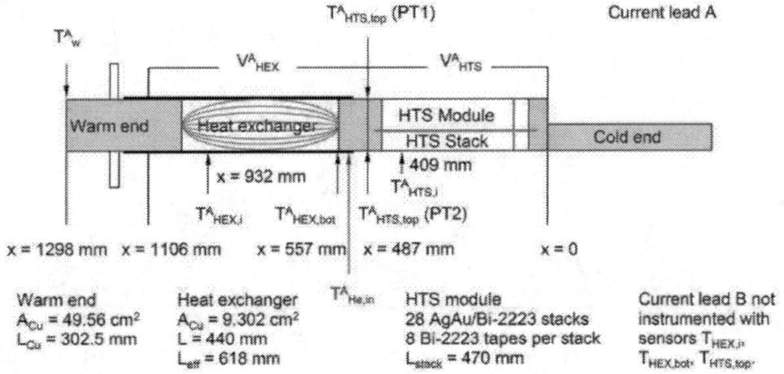

Figure 1. Sketch of the instrumentation of current lead A (x: length from cold end).

The cold helium gas is transferred via a short transfer line to the HTS current leads. A second heater was used to adjust the helium inlet temperature. During the current lead test it proved to be difficult to reach a stable mass flow rate and a constant helium inlet temperature for a time long enough to establish steady state operation of the current leads. In general, it was not possible to fully establish a steady state but for some runs the conditions were close to thermal equilibrium. This situation affects the comparison of the measured results with the design calculations based on operation in fully steady state.

A sketch of the instrumentation of current lead A is presented in figure 1. The helium temperature was measured at the inlet ($T^A_{He,in}$) and the outlet ($T^A_{He,out}$) of the current lead. One temperature sensor was placed at the warm end of the whole current lead (T^A_w) and the two redundant sensors PT1 and PT2 at the warm end of the HTS module ($T^A_{HTS,top}$). Additional sensors $T^A_{HEX,i}$, $T^A_{HEX,bot}$ and $T^A_{HTS,i}$ were applied to current lead A only for the test. The voltages across the heat exchanger (V^A_{HEX}) and the HTS module (V^A_{HTS}) including warm and cold end copper contacts were used to protect the current lead.

3. Results and discussion

The results of a measurement performed at 18 kA, a helium mass flow rate of ≈1.19 g/s and $T^A_{He,in}$ ≈28 K are presented in figure 2. At t = 4000 s all temperatures are almost constant and the temperature at the warm end of the current lead reached ≈316 K (figure 2, left). The average of the readings of PT1 and PT2 provides $T^A_{HTS,top} \approx 31.3$ K, while $T^A_{HEX,bot} \approx 47$ K and $T^A_{HTS,i} \approx 35$ K. The expected sequence would be $T^A_{HEX,bot} > T^A_{HTS,top} > T^A_{HTS,i}$. The readings of the sensors embedded in or attached on the electrical insulation may be affected by cold helium gas in the vicinity of the current leads. The voltage of ≈52.3 mV measured across the heat exchanger seems to be reasonable (figure 2, right).

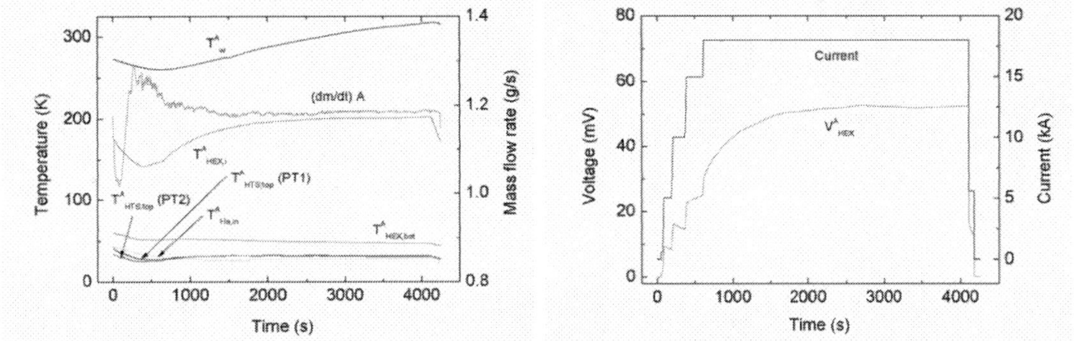

Figure 2. Evolution of temperatures (left) and V^A_{HEX} (right) during longer operation at 18 kA.

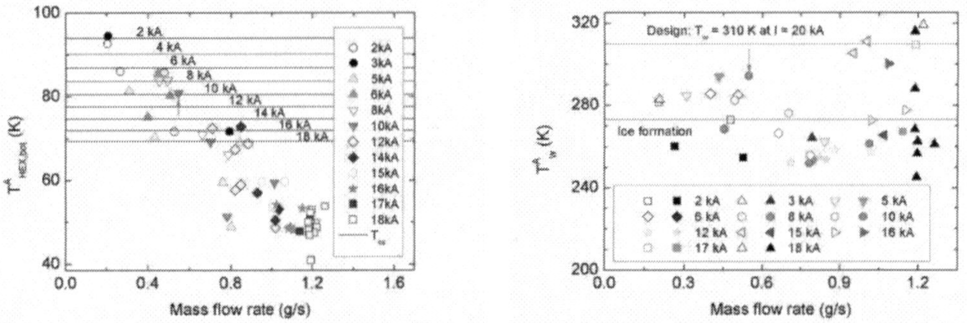

Figure 3. Temperatures at the bottom of the heat exchanger (left) and the warm end of the current lead (right) versus mass flow rate at various currents. Horizontal lines (left) indicate calculated T_{cs}.

In figure 3 (left), $T^A{}_{HEX,bot}$ versus the mass flow rate is shown for currents in the range of 2 to 18 kA. The temperature $T^A{}_{HEX,bot}$ is expected to provide an upper limit of the temperature at the warm end of the HTS module. The horizontal lines provide the calculated current sharing temperature T_{cs} based on a criterion of 1 µV/cm. With the exception of one data point for 3 kA and a mass flow rate of 0.207 g/s, the measured temperatures are lower than the corresponding T_{cs}. In general, the temperature at the bottom of the heat exchanger decreases with increasing mass flow rate at constant current. In figure 3 (right), the temperature $T^A{}_w$ at the warm end of the whole current lead is shown as a function of mass flow rate. The open symbols represent measurements performed in May 2013, while the solid symbols indicate the results of a second test campaign starting in June 2013. For the second test campaign the total copper cross-section of the cables between the warm end of the current leads and the permanently installed bus bars of the power supplies was enlarged. In the first campaign at a current of 15 kA a warm end temperature of 310 K was reached. In the second test campaign, $T^A{}_w$ stayed well below 310 K even at a current of 16 kA. The variation of $T^A{}_w$ for nearly equal mass flow rates and currents can be attributed to the fact that in some of the measurements no steady state was established. The results indicate that the current leads can be operated at 18 kA with a mass flow rate of 1.19 g/s and $T^A{}_{HE,in} \approx 28$ K. However, the comparison with the design is not straight forward because of the higher helium inlet temperature of 44 K. For the 10 kA run indicated by the arrow (see figure 3), the helium mass flow rate and $T_{He,in}$ were ≈0.55 g/s and ≈45 K, respectively. The measured values of $T^A{}_W = 294.2$ K, $T^A{}_{HEX,bot} = 80.8$ K and $V^A{}_{HEX} = 55.86$ mV are in good agreement with the corresponding design values of 293 K, 74.84 K and 58.42 mV.

The voltages across the HTS module (including HTS-Cu contacts) and the heat exchanger are presented in figure 4. Supposing a uniform current distribution among stacks the data in figure 4 (left) provide 13 nΩ (lead A) and 11 nΩ (lead B) for the sum of the warm and cold end contact resistances.

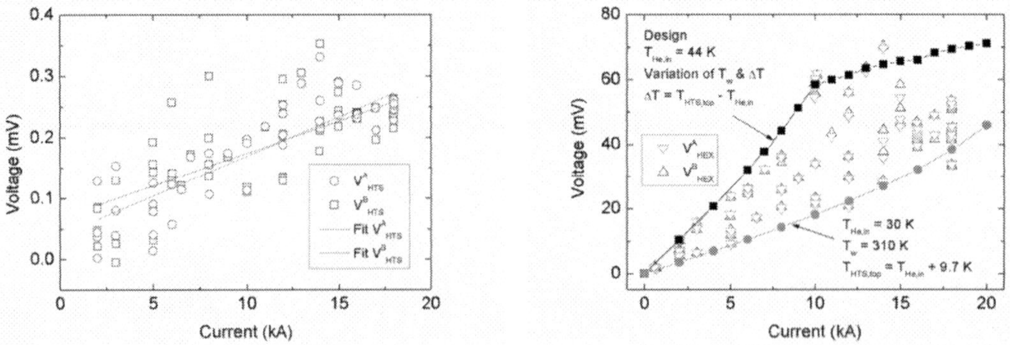

Figure 4. Voltages across the HTS module (left) and the heat exchanger (right) versus current.

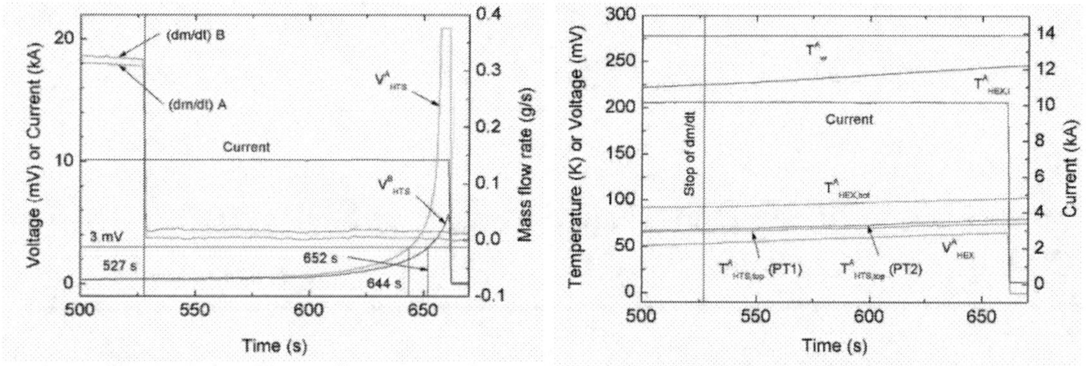

Figure 5. Evolution of V^A_{HTS}, V^B_{HTS} (left), V^A_{HEX} and various temperatures (right) after a stop of mass flow at a current of 10 kA. The time to reach V_{HTS} = 3mV is 117 s (lead A) and 125 (lead B).

The sum of the warm and cold end contact resistances of 13.8 nΩ found for the 18 kA EDIPO current leads [3] is comparable to the values obtained for the present current leads. In figure 4 (right) the measured values of V_{HEX} are presented. Most of the measured voltages are between the design values ($T_{He,in}$ = 44 K) and that obtained for $T_{He,in}$ = 30 K and a constant $\Delta T = T_{HTS,top} - T_{He,in}$ = 9.7 K.

The last aspect to be considered is the behavior of the current leads in case of a loss of flow. The evolution of V^A_{HTS}, V^B_{HTS} (left), V^A_{HEX} and the temperatures in lead A (right) after a stop of mass flow at t = 527 s are presented in figure 5. The initial temperature of the warm end of the HTS module was 66.4 K (lead A, average of PT1 and PT2). Times elapsed to reach a voltage of 3 mV across the HTS modules at I = 10 kA were 117 s for lead A and 125 s for lead B. In parallel, the temperature $T^A_{HEX,i}$ increased from 225 to 242 K leading to an increase of V^A_{HEX} from 52.9 mV to 63.7 mV. A simulation of the loss of flow behavior for a current of 10 kA provided a time of ≈86 s to reach a voltage of 3 mV. The shorter time obtained from the calculation may be attributed to a higher initial temperature of 75 K at the warm end of the HTS module in the calculation.

4. Conclusion

A pair of 20 kA HTS current leads, manufactured at CRPP, to be used to power a 25 T hybrid magnet system at HZB were tested at ENEA at low voltage and currents up to 18 kA. At a helium inlet temperature of ≈28 K a mass flow rate of 1.19 g/s was found to be sufficient to operate the current leads in steady state at 18 kA. The design value of the helium inlet temperature is 44 K and therefore a comparison with the design is not straightforward. Measurements at 10 kA with a helium inlet temperature of ≈45 K were found to be in reasonable agreement with the design values. The measured sums of cold and warm end Cu-HTS contact resistances were found to be 13 nΩ (A) and 11 nΩ (B). The measured time for quench to initiate, following a loss of flow at 10 kA, is larger than the calculated value. Therefore, it can be expected that the design calculations provide a roughly correct value of this time also for the rated current of 20 kA. The calculated design value of 94 s at 20 kA is acceptable and would allow the rapid ramp-down of the magnet system after detection of a loss of flow instead of a safety discharge with a time constant of 2 s starting at V_{HTS} = 10 mV.

References

[1] Wesche R et al 2012 Design and manufacture of 20 kA HTS current leads for a hybrid magnet system *Physics Procedia* **36** 927-930 doi:10.1016/j.phpro.2012.06.231

[2] Lewandowska M and Wesche R 2013 Parametric study for the cooling of high temperasture superconductor (HTS) current leads *Cryogenics* **53** 31-36 doi:10.1016/j.cryogenics.2012.06.014

[3] Wesche R et al 2011 Test results of the 18 kA EDIPO HTS current leads *Fusion Engineering and Design* **86** 1414-1417 doi:10.1016/j.fusengdes.2011.01.026

Journal of Physics: Conference Series **507** (2014) 032057 doi:10.1088/1742-6596/507/3/032057

Investigation of the Dielectric Strength of Syntactic Foam at 77 K under DC Stress

D. Winkel, R. Puffer, A. Schnettler

Institute for High Voltage Technology, RWTH Aachen University, Schinkelstr. 2, 52062 Aachen, Germany.

E-mail: winkel@ifht.rwth-aachen.de

Abstract. Liquid nitrogen (LN_2) based electrical insulation systems for superconducting equipment of electrical power distribution networks are state of the art. Since LN_2 is a cryogenic liquid it has some disadvantages when used as insulation. This paper deals with syntactic foam as an alternative insulation system for superconducting apparatus. Syntactic foam is a composite material consisting of a polymeric matrix and embedded hollow microspheres with diameters of several 10 µm. As hollow microspheres are gas-filled, using those as filling material features significant reductions of the relative permittivity and of the thermal contraction due to cooling the material to liquid nitrogen temperature (LNT, T = 77 K). In this study both an epoxy resin (ER) and an unsaturated polyester resin (UPR) serve as matrix material. The hollow microspheres used in this investigation are made of untreated and silanized glass. The results of measurements of the dielectric DC strength show, that the dielectric strength of all investigated syntactic foam compositions are significantly higher at LNT compared to ambient temperature (AT). Furthermore, the effect of a higher dielectric strength of syntactic foam with silanized glass spheres at ambient temperature vanishes at LNT. Hence, the dielectric strength at LNT is unaffected by silanization of glass microspheres.

1. Introduction

LN_2-based insulation systems are commonly used in superconducting equipment of electrical power distribution networks. In this case, LN_2 takes cooling and insulating function simultaneously. Disadvantageously, nitrogen bubble formation due to heat losses leads to a reduction of the dielectric strength [1]. Furthermore, the routine tests after manufacturing of those utilities lose their significance, as the liquid nitrogen has to be emptied for delivery. An alternative to LN_2-based insulation systems are solid insulations. Using solids means that LN_2 henceforth has only cooling function and the disadvantages of LN_2 as electrical insulation are eliminated. To apply polymeric insulation systems for superconducting power apparatus it is necessary to use fillers to reduce the thermal contraction due to cooling the insulation down to liquid nitrogen temperature (LNT). Otherwise the insulation system will delaminate from metallic or superconducting electrodes, whose thermal contractions are much lower than those of pure polymers [2]. This paper deals with syntactic foam as an alternative solid insulation system. Syntactic foam consists of a polymeric matrix and embedded hollow microspheres (HMS) which serve as filler and providing a foam-like structure. Using a half-wave rectification different syntactic foam constellations are investigated concerning their dielectric DC strength at ambient and liquid nitrogen temperature. A scanning electron micrograph of syntactic foam is shown in figure 1.

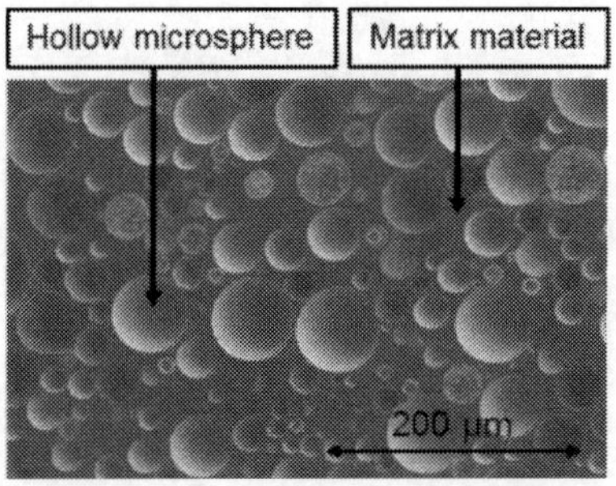

Figure 1. Scanning electron micrograph of syntactic foam.

2. Investigated Materials

2.1. Syntactic foam

To determine the influence of adhesion between the HMS and the matrix material on the dielectric DC strength, different constellations of syntactic foams are investigated. The matrix material is varied between epoxy resin (ER) and unsaturated polyester resin (UPR). Both matrix materials are hot cured polymers. The quality of adhesion is varied by using glass HMS which are on the one hand surface-modified by silanization and on the other hand untreated. The silanization results in an increase of bonding forces between the inorganic glass spheres and the organic matrix due to chemical bonding instead of physical bonding, as it occurs for untreated glass microspheres. The glass HMS have a mean diameter of 40 µm and a wall thickness of about 1 µm. Within these investigations filling degrees of 30 and 50 percentage of volume (vol. %) are examined to change the amount of boundary surfaces.

2.2. Test sample geometry

The dielectric DC strength of syntactic foam is determined with the aid of test samples with embedded spherical electrodes. The electrodes have diameters of 12 mm and the gap distance is 2 mm. This electrode configuration results in a field efficiency factor η of 0.865 which features a quasi homogeneous electrical field between the electrodes after voltage applying.

3. Experiments and Simulations

The DC test voltage is induced by a half-wave rectification. The rectified AC voltage is generated by a transformer cascade with a maximum output of 200 kV rms. To avoid surface discharges during the measurements at ambient temperature the test samples are placed in an oil vessel. The measurements at LNT are performed in LN_2, which is filled in a basin made of expanded polypropylene (EPP) for thermal insulation. To ensure a test sample temperature of 77 K and to avoid material cracking due to high temperature gradients the samples are cooled stepwise over a time period of one hour.

According to [3] the test voltage is applied stepwise. Starting with a voltage U_s, which is 40 % of the prospective breakdown voltage U_b, the voltage is increased every $t_h = 20$ s by $\Delta U = 2$ kV until breakdown occurs. By means of the breakdown voltage U_b, the dielectric strength E_b is given by

$$E_b = \frac{U_b}{\eta s} \quad (1)$$

with the field efficiency factor η and the gap distance between the sample electrodes s [4]. For each syntactic foam composition five samples are investigated.

Furthermore, simulations are carried out to obtain the electrical DC field distribution inside syntactic foam. Therefore, three HMS are arranged in parallel to a background field strength $E_0 = 45$ kV/mm, which is determined by the dielectric strength of ER respectively UPR filled with 50 vol. % silanized glass HMS at AT. The conductivities of the syntactic foams' components used in this simulations are set to $\lambda_{ER} = 10^{-15}$ S/m, $\lambda_{UPR} = 10^{-15}$ S/m, $\lambda_{glass} = 10^{-12}$ S/m and $\lambda_{gas} = 10^{-18}$ S/m.

4. Results

The results of the dielectric DC strength measurements are presented in figure 2 including the 95% confidence interval. The testing temperature differs between ambient temperature and liquid nitrogen temperature (LNT). The matrix materials are ER and UPR filled with untreated or silanized glass HMS. The filling degree varies between 30 vol. % and 50 vol. %.

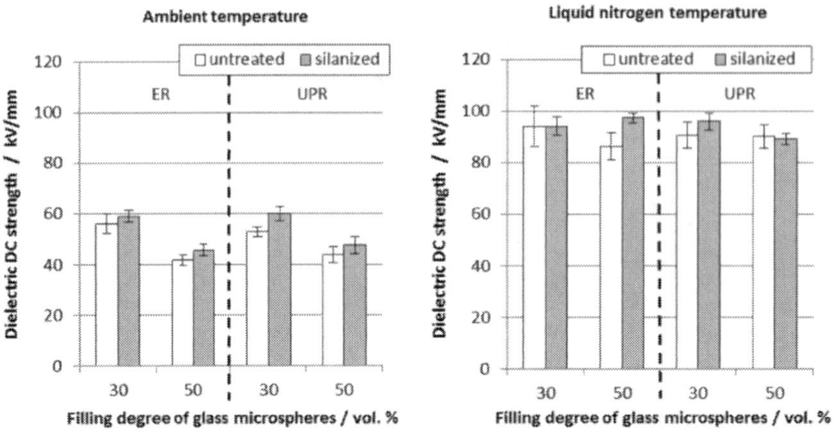

Figure 2. Dielectric DC strengths of different syntactic foam compositions with untreated and silanized HMS at ambient and liquid nitrogen temperature.

It can be observed that the choice of the matrix material between ER and UPR has no significant impact on the dielectric DC strength of syntactic foams. Furthermore, the dielectric strengths of all investigated syntactic foams increase significantly by decreasing the test temperature to LNT. Focussing on the impact of the filler material of syntactic foams, the dielectric strength at ambient temperature decreases significantly with increasing filling degree and increases with silanization of the glass HMS by trend for both matrix materials. Comparing the results at LNT, only a slight trend of decreasing dielectric strength with increasing filling degree is recognized. However, the silanization of the HMS does not show a significant impact on the dielectric strength at LNT.

Figure 3 shows the simulation results. It is found that the field distribution depends on the matrix material. The electrical field inside the HMS embedded in UPR is enhanced (Figure 3 left). It is assumed that discharges within the spheres will occur by increasing background field strength and that will lead to an enhancement of the field between the spheres like it is the case of HMS embedded in ER. There the field inside the HMS is reduced and the field between spheres is enhanced (Figure 3 right).

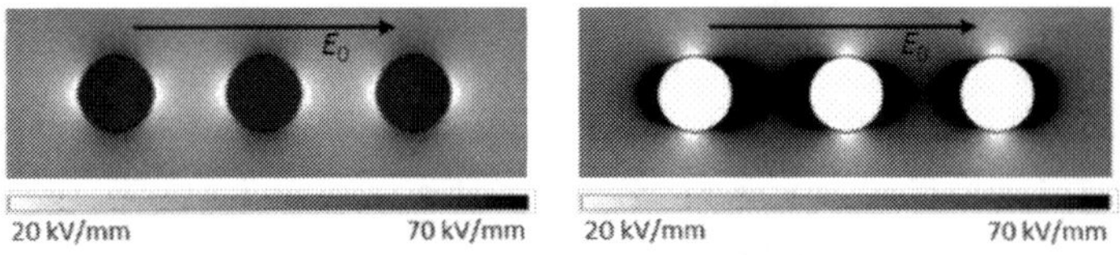

Figure 3. Electrical DC field distribution of three glass HMS in UPR (left) and ER (right) at ambient temperature.

5. Discussion

Due to higher adhesion of silanized HMS to the matrix materials there are no gaps between spheres and matrix. Those gaps could reduce the dielectric DC strength of syntactic foam at ambient temperature and could occur within the curing process of syntactic foam. Because of the much higher conductivity of glass compared to the conductivities of the filling gas of the HMS and ER, the field stress on the parts between two spheres is enhanced for higher filling degrees. It is assumed that the same field distribution would be found, if spheres embedded in UPR discharge due to enhanced field stress of the spheres interior. Thus, independent of the matrix material the breakdown starts at lower background field stresses for higher filling degrees [5]. When lowering the test temperature to LNT the matrix material begins to shrink so that the gaps between untreated HMS and the matrix seam to vanish and the dielectric DC strength of syntactic foam becomes unaffected of the silanization process. The decrease of dielectric DC strength with increasing filling degree is damped and can only be recognized by trend. In general, the dielectric DC strength increases significantly at LNT. Assumed that the electrical field distribution within syntactic foam at LNT is similar to that at ambient temperature the electrical field displacement leads to a similar enhancement of the field stress in the matrix material between the HMS. Thus, the higher dielectric strength of syntactic foam at LNT must arise from an increase in dielectric strength of the matrix material at LNT as it is found for some epoxy resins in [6].

6. Conclusion

In this paper syntactic foam is investigated regarding its dielectric DC strength under ambient and liquid nitrogen temperature. It was observed that the dielectric DC strength decreases with increasing filling degree of HMS. This results from a higher field stress of the matrix material between HMS since the electrical field is displaced out of the hollow spheres at DC stress. A silanization of HMS leads to higher dielectric strengths at ambient temperature but not at LNT. However, thermal contraction of syntactic foam leads to vanishing gaps at LNT between untreated HMS and the matrix which can occur during the material curing. Hence, the impact of silanization becomes negligible. At last, the dielectric strengths of syntactic foam increases significantly by lowering the test temperature to LNT. This effect seems to be affected by increasing dielectric strength of the matrix materials providing an electrical field displacement similar to that at ambient temperature.

7. Acknowledgment

The authors would like to thank Mr. Sebastian Blasche for his assistance in carrying out the experiments.

This project is funded by Deutsche Forschungsgemeinschaft (DFG) under the reference number SCHN 728/10-1.

References

[1] Sauers I, James DR, Ellis AR, Tuncer E, Polizos G, Pace MO. Breakdown in liquid nitrogen in the presence of thermally generated bubbles for different electrode geometries. Annual Report Conference on Electrical Insulation and Dielectric Phenomena, October 2009, Virginia Beach, USA, 2009.

[2] Ekin JW. Experimental techniques for low-temperature measurements. Boulder, CO, Oxford University Press, 2006.

[3] IEC 60243-2, "Electric strength of insulating materials (IEC 60243-2:2001); German version", VDE-Verlag, 2001.

[4] Kuffel E, Zaengel WS, Kuffel J. High voltage engineering: Fundamentals. Butterworth-Heinemann, Oxford, 2000.

[5] Strauchs A, Schnettler A. Enhancements of epoxy resin based syntactic foam by inner interface and matrix modifications. Annual Report Conference on Electrical Insulation and Dielectric Phenomena, October 2011, Cancun, Mexico, 2011.

[6] James DR; Sauers I; Pace MO; Ellis AR. High voltage breakdown of solid epoxies at room temperature and in liquid nitrogen. Annual Report Conference on Electrical Insulation and Dielectric Phenomena, October 2002, Cancun, Mexico, 2002.

Double storey three phase saturated cores fault current limiter

Y Wolfus, Y Nikulshin, A Friedman and Y Yeshurun
Institute of Superconductivity, Department of Physics, Bar-Ilan University, Ramat-Gan, Israel

E-mail: shuki.wolfus@biu.ac.il

Abstract. A novel saturated-cores Fault-Current-Limiter (FCL) configuration is described. This FCL is based on two parallel planes of iron rectangular cores, on which three-phase coils are mounted and connected in series to the grid. Two DC coils are mounted in between the planes on perpendicular core limbs connecting the two AC planes. The DC coils are set to magnetically saturate the AC cores. The transition to three-dimensional, double-storey design enables handling three-phase symmetrical faults while offering better decoupling between the AC and DC circuits. At the same time, it shortens the AC limb lengths and enables deeper magnetic saturation levels in comparison to other saturated cores FCL designs. Hence, this FCL configuration exhibits lower insertion impedance and higher ratio of fault to nominal state impedance in comparison with other designs.

1. Introduction

Recent installations of saturated cores fault current limiters (SCFCL) by Zenergy Power [1], InnoPower [2], Applied Superconductivity Ltd. [3] and GridOn [4] in live networks have made the SCFCL a leading candidate for limiting fault currents in medium to high-voltage networks. While the above installations are expected to provide data about the operation of the SCFCLs in various grid conditions, a continuous search for improved SCFCL designs and for better understanding of the underlying processes of the SCFCL mechanisms takes place. In a recent work, Aracil *et. al.* analyzed the electromagnetic forces of an SFCL design and pointed to the differences between such forces in transformers and SCFCL [5]. Nikulshin *et. al.* analyzed the dynamics of the transition between the saturated and non-saturated core states and suggested the "effective core length" concept as means for comparing different SCFCL designs and performances [6].

In this work we describe a novel, three-phase SCFCL design first suggested in a patent application [7]. This SCFCL "double-storey" concept is described here and its performances are studied for a bench model design. Insertion impedance lower than 1% is obtained for the nominal state and fault current clipping over 50% can be obtained for single-phase faults and over 30% for three-phase symmetrical faults. The advantages of this SCFCL concept are described below.

2. "Double-storey" SCFCL design

Figure 1a. exhibits a 3D CAD model of the "double-story" design. Two rectangular electric steel core frames are placed to form two parallel planes. Each frame contains two parallel long limbs connected by two short limbs with a 50% larger cross section. The AC coils of the three phases are mounted on the long limbs. Two further limbs are placed vertically between the horizontal planes connecting them magnetically and supporting the two planes. DC bias coils are mounted on these vertical limbs in opposing directions namely, the current circulates in the coils in reversed directions.

The flow of the DC flux lines generated by the DC coils is displayed in figure 1b. In the absence of current in the AC coils, flux lines generated in the DC coils point upwards in one DC limb, enter the upper frame and then split and flow along both AC limbs to the opposing DC limb. DC magnetic flux then flows downwards and closes by splitting again in the lower AC limb. The DC flux path forms a closed magnetic circuit. The two steps variation of cross sections between core limbs is introduced to ensure maximal transfer of DC flux to the AC core limbs for achieving deep saturation levels.

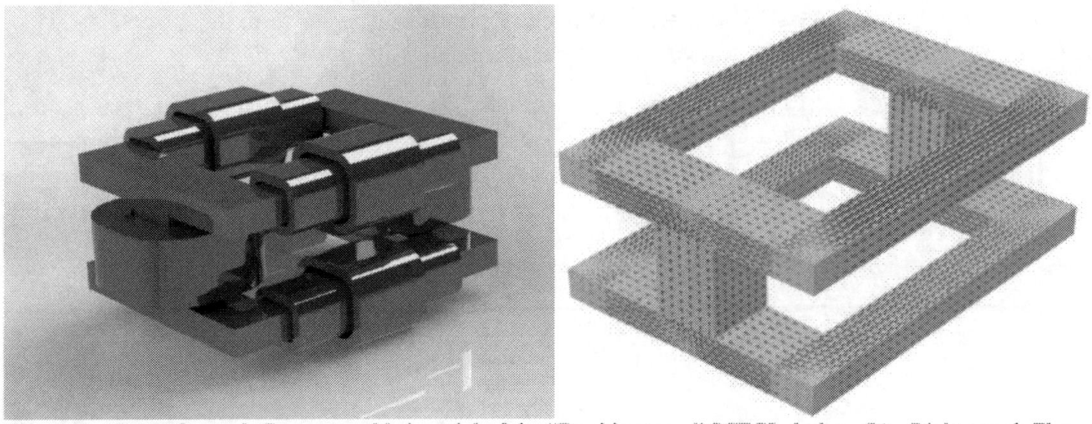

Figure 1. (a) - Left panel: Computer aided model of the "Double storey" SCFCL design. **(b)** - Right panel: Flow of DC magnetic flux lines in the core in the absence of AC currents.

The AC coils shown in figure 1a. are mounted on the AC limbs in a way that for each phase the coils are split and inversely connected in series across opposing limbs so that in the absence of DC bias their flux adds up. As described in ref [8], asymmetry between phase coils is a built-in feature of this design and achieved by varying the phase coil diameter, number of turns and/or the position along the AC limb. The amount of asymmetry is carefully selected to allow a net field vector sum high enough to drive the AC limb out of saturation in the event of a three-phase symmetrical fault, yet small enough to keep the negative sequence during normal grid state below allowed standards [7]. The model is designed for a 230 V/phase grid voltage and nominal current of 6.5 A namely, this model is rated at 4.5 kVA. The physical parameters of the core sections and coils are displayed in tables 1 and 2 correspondingly. The grid impedance is 1.5 ohm resulting in 154 A_{RMS} prospective current in the absence of the SCFCL.

Table 1. Dimensions of the core limbs.

	W [mm]	D [mm]	H [mm]
Vertical limb	60	62	90
Short horizontal limb	180	62	30
Long horizontal limb	354	40	30

Table 2. Coil parameters.

	Length [mm]	Turns	Inner cross-section [mm^2]	Coil thickness [mm]
DC coil	70	300	76x76	40
AC R-phase	85	95	46x36	10
AC S-phase	90	95	66x76	10
AC R-phase	85	95	46x36	10

3. Results

The performances of the "double-storey" model described above have been studied using static and transient COBHAM Vector Fields Opera simulation tools. All current and voltage waveforms obtained for nominal grid operation are sinusoidal and the current amplitude is practically identical for all three phases. The voltage amplitude is higher for the S phase in comparison to R and T phases. This is the result of the S coils having a larger inner cross-section. The normal state impedance obtained for the S phase is 0.59 Ω (1.66%) and 0.32 Ω (0.9%) for phase R. The relatively low impedance value obtained for phase R is the result of the "double-storey" design, which enables shortening the AC limbs length thus increasing saturation depth. The normal state impedances may be further reduced by increasing the DC bias field.

The most common fault scenario is a single-phase fault. Figure 2 displays the voltage and current curves for a fault in R-phase. The voltage curve is typical of SCFCL fault curve exhibiting the "split-peak" fingerprint of the rapidly varying coil impedance near zero AC current [7]. The voltage drop across the R coils is 133.05 Vrms namely, ~ 57% clipping. The calculated fault impedance is 1.95 Ω representing over 6-fold impedance increase over the normal state impedance.

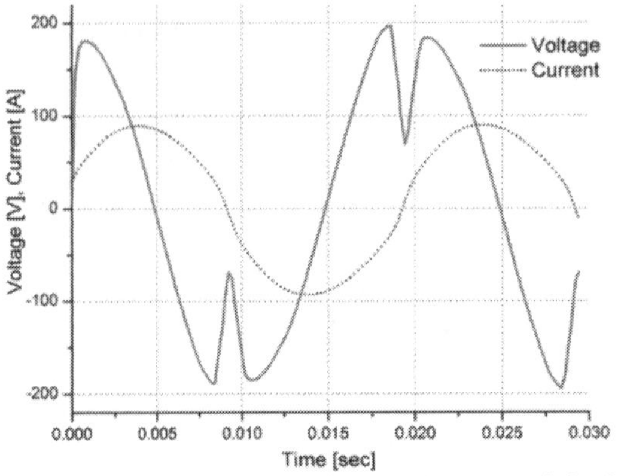

Figure 2. Voltage (red) and current (blue) curves for R-phase in the event of single-phase fault of phase R.

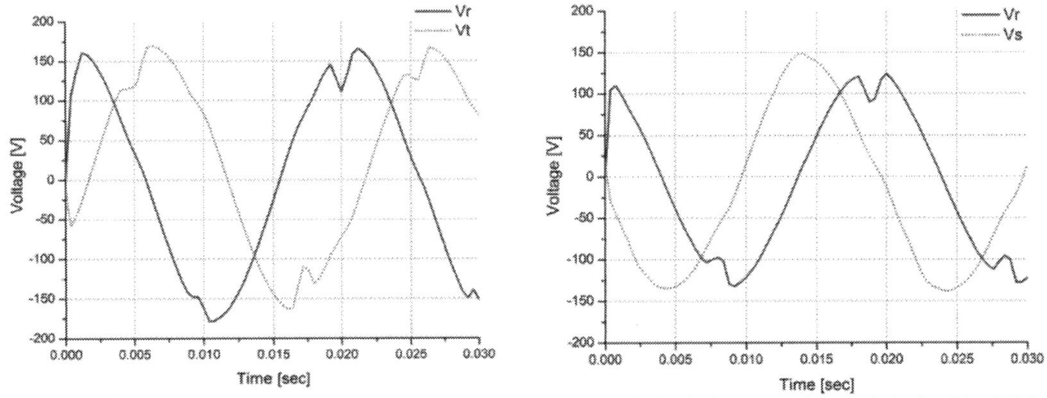

Figure 3. (a) - Left panel: Voltage waveforms obtained in the two faulted phases: R (blue), S (red). **(b)** - Right panel: Voltage waveforms obtained in the two faulted phases: R (blue), S (red).

A less common fault scenario involves a fault between two phases. We describe here two different cases of such faults: R-S and R-T faults. In both cases the faulted phases become directly connected to the ground. Figures 3a and 3b display the voltage curves correspondingly for the R-S and R-T fault scenario. The R-T fault scenario is described in figure 3a. Since the R and T phase coils are distanced from one another relative to their distance to the S coil, the obtained voltage curves are more "traditional" and both exhibit the "split-peak" feature. Voltage drop across each phase is about 113 V namely, about 49% clipping. It is interesting to note that the partial overlap between the R and S coils causes a strong coupling between them when the core is de-saturated resulting in the alteration of the "split-peak" feature in the voltage curve (figure 3b.). The voltage drop measured on each phase is 85.6 V and 98.3 V for phases R and S, respectively and their impedances are 0.83 Ω and 1.09 Ω namely, about 37% and 43% clipping in phases R and S, respectively.

The most challenging scenario is the three-phase symmetrical fault. For a single core SCFCL this is a serious challenge as the vector sum of the magnetic fields created by 3 coils wound on the same core shifted in phase by 120 degrees, cancels out when the vectors are identical. The built-in asymmetry of the phase coils in this design overcomes this challenge. In this example, the asymmetry introduced by the S coils is sufficient to drive the whole core out of saturation and provide current limiting protection. Figure 4 exhibits the voltage curves in this case.

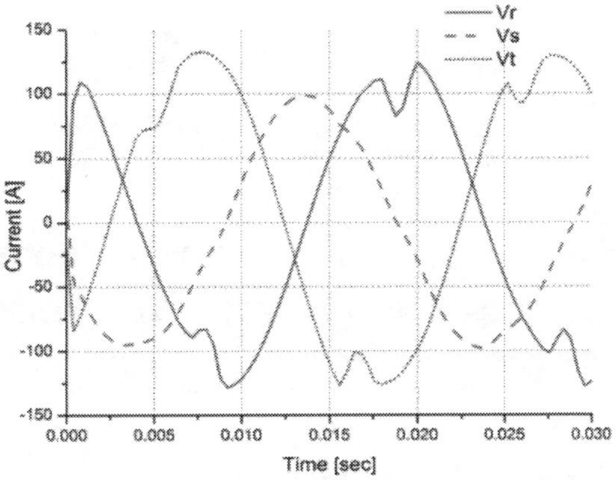

Figure 4. Voltage waveforms obtained in the event of a three phase symmetric fault: R (green), S (red), T (blue).

Clearly, a significant part of the grid voltage still falls on the SCFCL coils. The voltage across R, T phases is about 88 V (≈38% clipping) and about 69 V (≈30% clipping) for the S phase. In most cases such levels of clipping for a three-phase symmetrical fault are sufficient, therefore making the double-storey SCFCL design an interesting and promising candidate for future FCL implementation.

4. Summary

A new SCFCL design has been described. The model composed of two parallel core frames on which three AC phase coils are wound with a built-in impedance asymmetry. Two vertical DC core limbs connect between the AC core frames and generate the required saturation in the AC limbs. This out-of-plane placement of the DC and AC core sections results in a low coupling between the two closed AC and DC magnetic circuits. The results obtained for this design in various normal and fault state scenarios show strong clipping ability during a single-phase fault and fair clipping ratios for two and three-phase faults while maintaining low insertion impedance during normal grid operation. Further investigation of this design is taking place for comparing the dynamics of its de-saturation process with other SCFCL designs. The results will be published elsewhere.

Acknowledgments

The authors wish to thank Vladimir Rozenshtein for his contribution to discussions in the early days of this concept design. This research was funded by the Israeli Ministry of Energy and Water.

References:

[1] Moriconi F, De La Rosa F, Darmann F, Nelson A and Masur L 2011 *IEEE Trans. Appl. Supercon.* **21** 1288
[2] Yin X Gong W Z Niu X Y Gao Y Q Guo Q Q Xiao L X Cao Z J Hong H Wu A G Li Z H Hu X M Tian B Zhang J Y He Y Wang Y Cui J Ding S.Z Wang J.Z Ren A.L and Ye F 2009 *IEEE Trans. Appl. Supercon.* **19** 1934
[3] http://www.appliedsuperconductor.com/news/Project_Avanti_Circuit_of_.shtml
[4] http://www.eti.co.uk/news/article/gridons_fault_current_limiter_commissioned_into_service_
[5] Aracil J. C, Lopez-Roldan J Coetzee, J. C Darmann F and Tang T 2012 *J. Electrical Power and Energy Systems* **43** 1087
[6] Nikulshin Y, Wolfus Y, Friedman A, and Yeshurun Y 2013 *Supercond. Sci. Technol.* **26** 95013
[7] Wolfus Y, Yeshurun Y, Friedman A, Rozenshtein V and Bar-Haim Z 2010 *International Application:* PCT/IL2010/000714

Switching Properties of Liquid Nitrogen Cooled IGBTs and 24 kA Demonstration of Current Multiplier by Inductive Storage

S Yamada[§], H Nakayama[2], and Y Aso[2]
[1] National Institute for Fusion Science, Gifu 509-5292, Japan,
[2] The Japan Steel Works, Ltd., Hiroshima, 736-8602, Japan

E-mail: yamadas@LHD.nifs.ac.jp

Abstract. We had been developing a current multiplier by inductive storage (CMIS). The CMIS consists of 24 storage copper coils, which soaked into the liquid nitrogen, demonstrates a 24 kA of output current and the continuous current pulses of 3 pulses per second. Switching performance of the IGBTs and diode were tested in the liquid nitrogen bath. These experimental data were used to design the mega-ampere class CMIS. The system consists of the superconductive magnet section with a temperature of 20 K and the IGBT control switch section with a temperature of 77 K.

1. Introduction

The critical problem in a development of the inductive pulsed-power supply is in the opening switch to commutate a large current. This opening switch is required to commutate the current of several mega-amperes for the application to an electro-magnetic launcher. To reduce the duty of switch, the parallel operations of the inductive storages had been done [1, 2]. However, the current magnitude for each opening switch was still large because of the parallel operations of only several inductive storages.

For above solution, we have been developing the inductive pulsed power supply consisting of many coils. This inductive pulsed-power supply is excited in the series connection of each coil and discharged in the parallel connection to obtain the pulsed large electric current. The preliminary experiment showed that this process was correct [3-5].

In this paper, the preliminary experiments of CMIS with 24 inductors have been done. These inductors can be cooled down by liquid nitrogen ($L\text{-}N_2$) to decrease the resistance of the coils and bus-bar. Switching performance of the insulated-gate bipolar transistor (IGBT) and diode were tested in the liquid nitrogen bath. The conceptual design of mega-ampere CMIS and long pulse CMIS were also describes in this paper.

2. 24 kA CMIS performance via cupper coil at 77 K

2.1. Principle of CMIS circuit

Figure 1 illustrates the electrical circuit of the SC CMIS. Start-up and termination of the output current are formed by the on and off control of the series IGBTs and parallel IGBTs. Energizing of the SC magnet starts by turning on all series IGBTs. When the parallel IGBTs P_1 turns on and series IGBTs

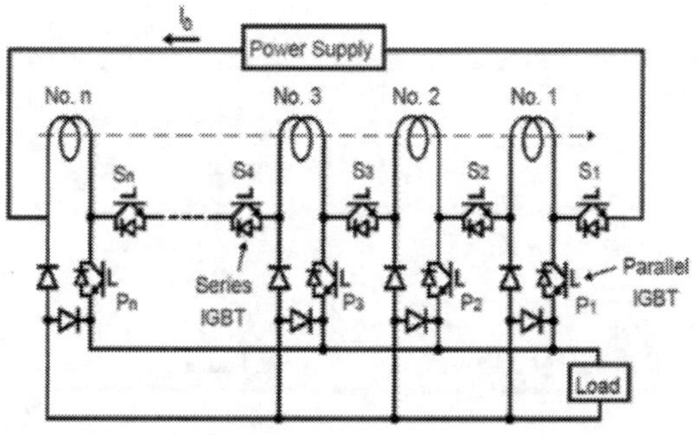

Figure 1.Circuit connection of the CMIS.

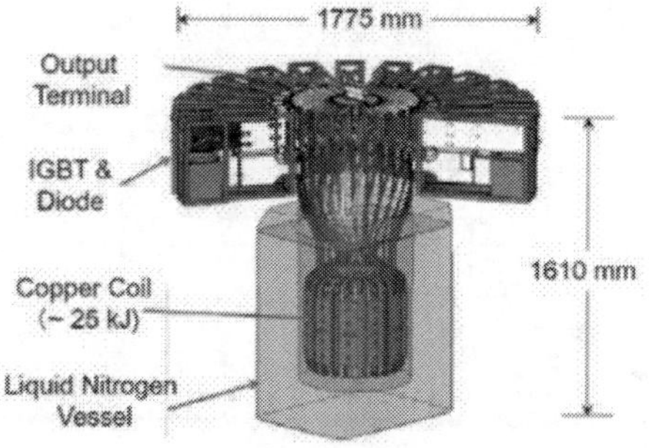

Figure 2 Schematic drawing of 24 kA CMIS .

S_1 turns off coil current I_0 is commutated to the Load. When P_2 is on and S_2 is off, current of No.1 coil and current of No.2 to No. n are commutated to the Load. In case of P_1 to P_n are off and S_1 to S_n are on, the all of module coil current flow into the load, and load current of nI_0 can be obtained. On the other hand, S_1 to S_n turn on and P_1 to P_n turn off, all module coil currents are commutated to the power supply.

2.2. Experimental apparatus 24 kA CMIS

The schematic of experimental apparatus is shown in Fig. 2. It has 24 inductors which form a toroidal coil. The inductance of the toroidal coil in parallel connection is 55 μH, and stored energy is about 25 kJ. An L-N_2 cooler was used to cool down the inductors until the temperature of L-N_2. The L-N_2 cooler cools only the toroidal coil, and the lead wire connecting to inductors cannot be cooled in whole. Thus the 24 lead wires were led out the cooler, and connected to IGBTs on the bench for the control devices. For the power supplies to energize the inductors, a 16 volts dc power supply with electric double layer capacitor is used.

2.3. Characteristics of the CMIS

As shown in Fig. 3, an output current of 24 kA was obtained, when the toroidal coil of serial connection was charged up to 1 kA. Decay time constant of the output current was about 70 msec. Equivalent resistance of the output circuit including switching device and coils was about 790 μΩ.

Continuous current pulse test was conducted with a time interval of 0.3 sec. Each pulse width was adjusted to 50 msec. Typical waveforms of continuous pulses are CMIS is shown in Fig. 4. Pulse heights of output current decreased about 10 % respectively. This was the cause of the large impedance of the electric double layer capacitor which was used in the power supply for charging the serial coils.

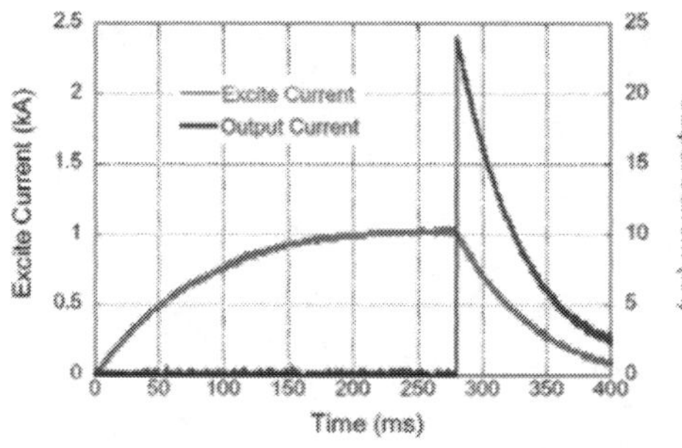

Figure 3. Current waveforms of CMIS.

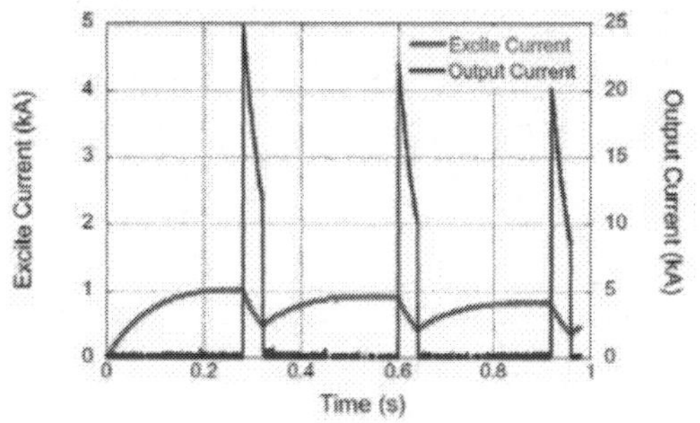

Figure 4 Continuous current pulses in CMIS.

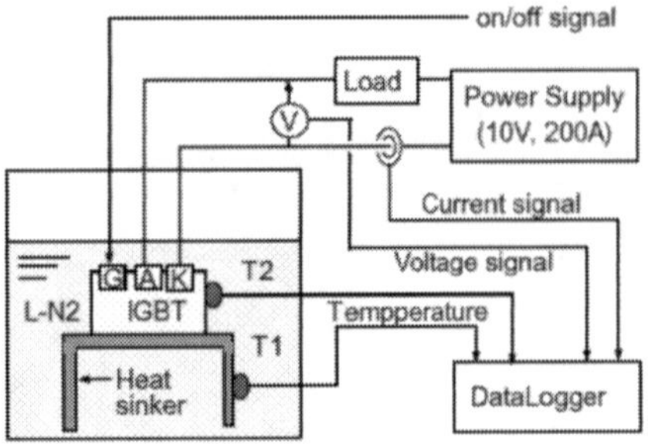

Figure 5. Test-bed for testing IGBTs.

Table 1. Parameters of tested switching device.

Type	Specification
SKM200GB176D (IGBT + IGBT)	V_{CE}: 1.7 kV, I_C: 200 A, P_C: 500 W T_J : -40 ~ 150 °C
PCHMB100B12A (IGBT + Diode)	V_{CE}: 1.2 kV, I_{op}: 100 A, P_C: 500 W T_J : -40 ~ 150 °C

3. Switching characteristics of IGBT and diode in the liquid nitrogen

3.1. Experimental set-up

The merit of using superconducting (SC) coil and SC bus-bar are to realize, 1) low energy consumption for long pulse operation, and 2) high current density and high magnetic field for downsizing the system. To decrease the number of the current leads and to suppress the heat load into the cryostat, the switching device should also be soaked into L-N_2. On the basis of this idea, they were tested in the L-N_2. Test circuit is shown in Fig. 5, and the parameters of test sample are listed in Table1. These samples were used in a prototype CMIS [4, 5].

3.2. Test results

The result of the IGBT cooled by L-N_2 is shown in Fig. 6. When temperature goes down to 77 K, Voltage drop of the IGBT increases. Voltage drop at 300 K and 77 K with zero current were 0.5 V and 1.2 V, respectively. Voltage waveform was non-linear in comparison with the current waveform, but maximum voltage drop did not exceed 2.6 V. When Joule heat occurs at the junction plane, temperature its will go up from 77 K. Increase of the temperature may leads to decrease of the voltage drop, even if the rated current is 100 A.

The result of diode test cooled by L-N_2 is shown in Fig. 6. Voltage drop at 300 K and 77 K with zero current were 0.5 V and 1.1 V, respectively. Non-linearity between voltage drop and current were smaller than that of IGBT. The maximum voltage drop was kept to around 1.5 V.

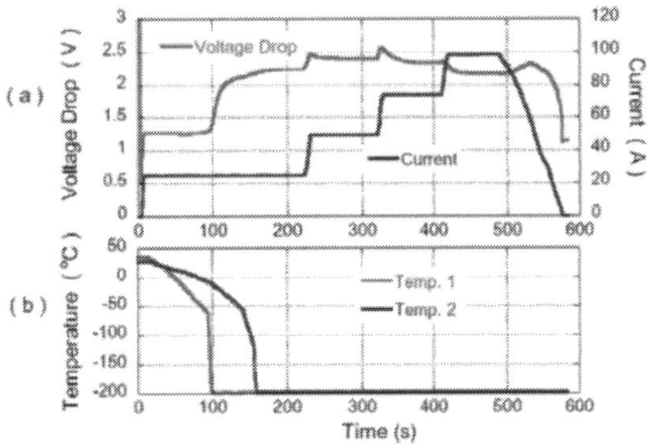

Figure 6. Time evolution of voltage, current (a), and temperature (b) of the IGBT.

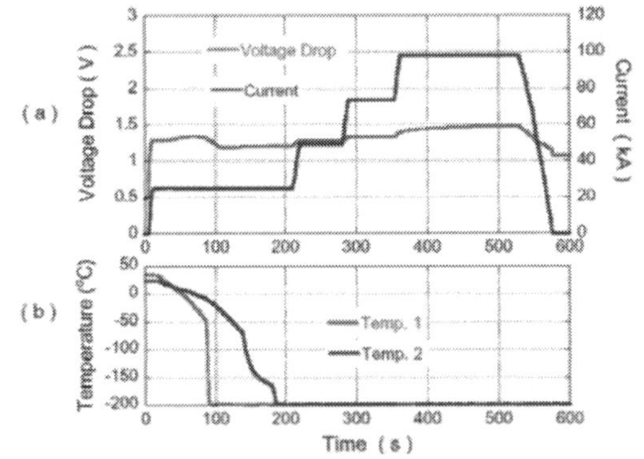

Figure 7. Time evolution of voltage, current (a), and temperature (b) of the diode.

Table 2. Parameters of SC CMIS and calculated results of output currents.

Objective	Mega Ampere CMIS	Long-pulse CMIS
Major radius / Minor radius	1.15 m / 0.45 m	1.15 m / 0.45 m
Number of module coils / turns of a coil	360 / 20	72 / 100
Inductance of Toroidal coil	4.99 H	4.99 H
Magnetic field / Stored energy	3.51 T / 19.6 MJ	3.51 T / 19.6 MJ
Output current / Exciting current	1 MA / 2.8 kA	100 kA / 2.8 kA

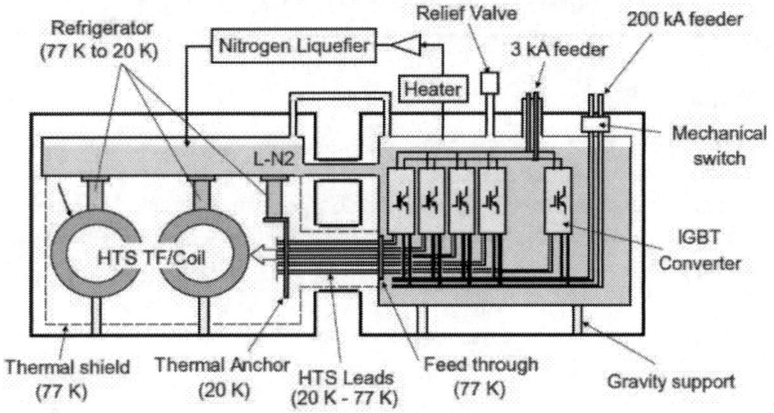

Figure 8. Schematic drawing of high temperature superconducting CMIS.

4. Conceptual design of MA CMIS by SC coils

Figure 8 shows a schematic drawing of the SC CMIS [5]. It consists of the SC module coils and the IGBT control switch. An SC coil is made of multi-filamentary BISCCO tape. The 360 module coils generate the current pulse of 1 MA. Since a decay time constant of the load current is 8.56 s, the pulse duration that the load current decayed to 80 % is 1.7 ms. To increase the pulse duration, repetition of current regulation is applied at the current decay phase. An energy stored coil consists of 72 module coils, and turn number of each coil is 100. Thirty-six modules were used for the current rise, and other 32 modules were used to compensate the current decay. In this way, duration of current flat-top increased to more than 0.1 s. Parameters of MA CMIS and long-pulse CMIS are listed in Table 2.

5. Conclusion

We proposed that new type of the current multiplier by the inductive storage (CMIS). The results are as follows: (1) A prototype CMIS consists of 24 copper coils. Output current of 24 kA was obtained, when the module coils charged up to 1 kA. (2) Current carrying performance of the IGBT and diode were tested in the liquid nitrogen. Voltage drop of the IGBT and diode were less than 3 volt and around 1.5 V, respectively. (3) Two types of large current multipliers by HTS SMES were designed conceptually. One is MA CMIS and the other is long-pulse CMIS. (4) It was confirmed that SC CMIS is effective to reduce a required space and to be flexible in operation.

References

[1] Ford RD and Hudson RD, IEEE Trans. on Magnetics: 1993; 29, No. 1, 949-953.
[2] Dedie Ph, Brommer V, and Scharnholz S, IEEE Trans. on Magnetics: 2009; 45, No. 1: 266-271.
[3] Aso Y, Hashimoto T, Abe T, Yamada S, IEEE Trans. on Magnetics: 2009; 45 No. 1: 237-240.
[4] Aso Y and Yamada S, IEEE Trans. on Magnetics: 2010, 47, No. 1, 237–240.
[5] Yamada S, Hishinuma Y, Aso Y, 2012, Physics Procedia, **36**: 741-746.

Effects of fluctuating magnetic fields on a superconducting bulk rotor shielded with superconducting rings

K Yamagishi[1], J. Ogawa[2] and O. Tsukamoto[3]

[1] Faculty of Engineering, Yokohama National University, 79-5 Tokiwadai, Hodogaya-ku, Yokohama 240-8501, JAPAN
[2] Department of Electrical and Electronic Engineering, Niigata University, 8050, Ninomachi, Ikarashi, Niigata 950-2181, JAPAN
[3] Faculty of Science and Technology, Sophia University, 7-1, Kioicho, Chiyoda-ku, Tokyo, 102-8554, JAPAN

yamagisi@ynu.ac.jp

Abstract. We study the effect of a fluctuating magnetic field, which is one of the technical problems for trapped magnetic fields in a bulk superconductor, to realize a practical bulk superconductor rotating machine. Previous research and other's research has shown that fluctuating magnetic fields reduce the strength of trapped magnetic fields in superconducting bulk modules [1, 2]. This deters development of applications of AC rotating machines because superconducting bulk modules are always exposed to a fluctuating magnetic field. Therefore, it is necessary to develop a method to control decrease of the trapped magnetic field. We propose a method to use the shielding ring of a superconducting wire to achieve this goal and the effects are confirmed experimentally [3]. We are now building test equipment for examining the performance of a shielding ring in a bulk rotating machine. This paper reports the test result for the shielding ring applied to the bulk superconducting rotor that is a part of the test equipment.

1. Design of test equipment

This section describes the design of rotating-type test equipment which we are now building. This test equipment is to be used to investigate how a fluctuating magnetic field, which is generated by a stator, affects a superconducting bulk rotor. The rotor of the test equipment and the structure of the stator are shown in Figure 1. Details of the rotor including the shielding ring are illustrated in Figure 2. The specifications of four High-Temperature Superconducting (HTS) bulk modules (A–D) attached to the rotor for use in this experiment are shown in Table 1. Specifications of the superconducting wire used for the shielding ring are shown in Table 2. The rotor of the test equipment is composed of four bulk modules. A heater is wound around the outside of each bulk module and heats the module when it is being magnetized. A superconducting shielding ring is attached just outside of each bulk module to protect it from fluctuating magnetic fields. The stator consists of eight coils and generates the magnetic field for rotation by an independent power supply. In this study, we investigate the effects of fluctuating magnetic fields on a part of the rotor module on which the shielding ring is attached. Based on a previous study [3], this test equipment is designed so that the bulk modules can be cooled more than was previously done by attaching a cooling channel to each bulk module holder. These channels were able to suppress temperature rises due to AC losses.

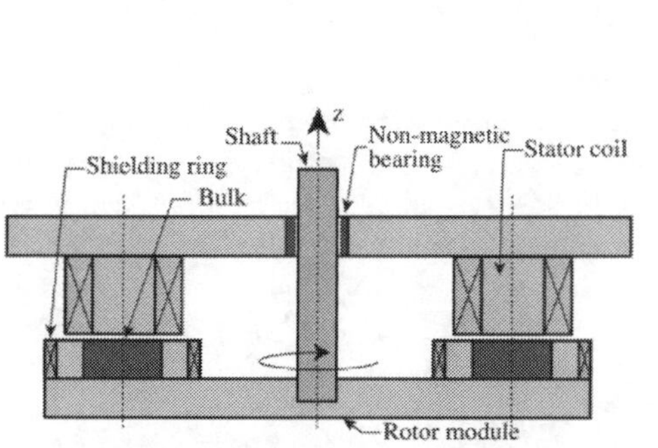

Figure 1. Illustration of the test equipment.

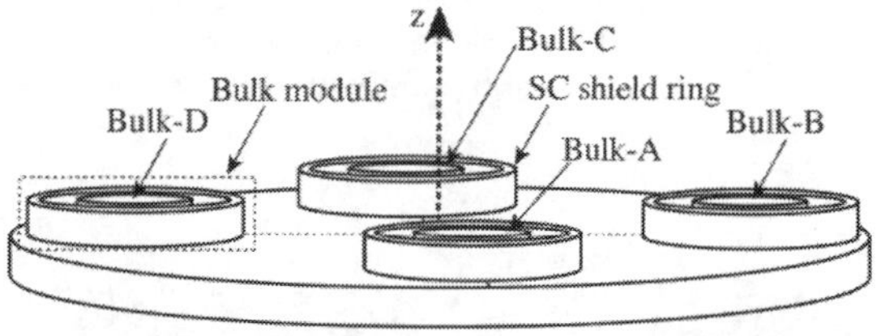

Figure 2. Illustration of the rotor module.

Table 1. Specification of HTS bulk modules.

	Material	D: Diameter (mm)	H: Height (mm)	Trapped magnetic field (T)	J_c (A/m^2)	n value
Bulk-**A**	Gd-Ba-Cu-O	34.55	10.46	1.164	6.84×10^7	20
Bulk-**B**	Gd-Ba-Cu-O	33.61	10.38	0.872	4.74×10^7	20
Bulk-**C**	Gd-Ba-Cu-O	32.52	10.36	0.904	5.13×10^7	20
Bulk-**D**	Gd-Ba-Cu-O	31.70	10.41	0.871	5.19×10^7	20

Table 2. Specification of wires used in HTS shielding ring.

Wire	Bi2223/Ag-sheathed
Thickness	0.32 mm
Tape wire width	2.8 mm
I_c	64 A at 77.3 K, 0 T
J_c	7.14×10^7 A/m^2
n value	20
Min. bending radius	25 mm

2. Effects of the shielding ring on the test equipment

2.1 Analysis

To confirm the design of the test equipment and the effects of the shielding ring of the test equipment, we first analysed the AC loss via magnetic-field analyses of a bulk module. To avoid reducing trapped magnetic fields of the bulk module, it is necessary to choose an arrangement of the shielding ring that minimizes AC losses. To design a shielding ring, we must consider the minimum bending radius of superconducting wire, the tape wire width and the dimensions of the rotor. We analysed the case in which no shielding ring is attached; then, we studied the arrangements of two kinds of superconducting shielding rings. These three cases are illustrated in Figures 3(a)–3(c). The particular specifications for the bulk module were those used for Bulk-A in Table 1. A fluctuating magnetic field was impressed along the z-axis of the bulk module; the fluctuating field had an amplitude (B_m) of 0.1 T. The resulting AC losses for both arrangements of the shielding rings, AR1 and AR2, are as shown in Figures 4(a) and 4(b). The plots in the figure show losses only the bulk module, only the shielding ring and both the bulk module and the ring. Also shown are losses in the bulk module without any ring attached. The losses in both shielding rings were very large and in comparison, while those in the bulk module were very small. Further, if the number of layers in the shielding ring increased, losses in the bulk module decreased. This shows that the shielding ring intercepted most of the fluctuating magnetic fields and the fluctuating magnetic field that reached the bulk module was very small. Although AC losses in the shielding ring were very large, if a sufficient cooling channel exists between the bulk module and the ring, the trapped magnetic field in the bulk module will hardly be affected. A previous study [3] showed that AC losses become large when the penetrating magnetic flux is concentrated on the edge of the bulk module; therefore, we expected that losses in AR2 would be low compared to those in AR1. However, we found that losses were almost the same; in fact, those in AR2 were only slightly smaller than those in AR1.

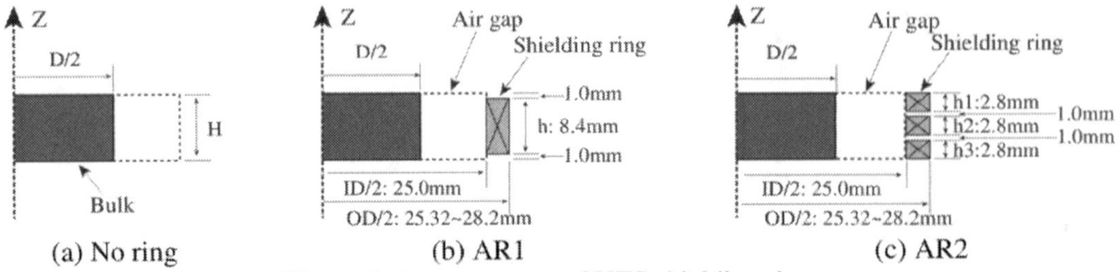

Figure 3. Arrangements of HTS shielding rings.

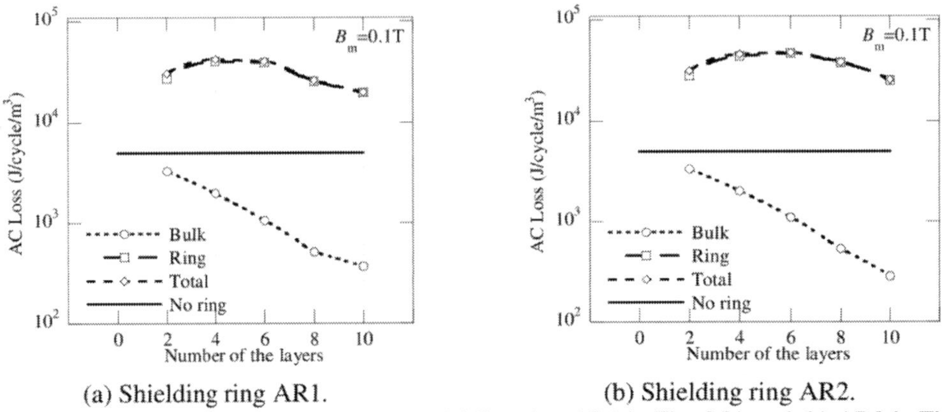

Figure 4. Analytic results of AC losses for (a) shielding ring AR1 in Fig. 3(b) and (b) AR2 in Fig. 3(c).

2.2 *Experiment*

Based on the results in Section 2.1, we built the shielding ring AR2 and performed experiments to verify the predicted performance of the ring. Therefore, we conducted this experiment for one bulk module using Bulk-A in Table 1 and this bulk module is fixed. The design of the experimental situation is shown in Figure 5. For the magnetizing method of the bulk, we used Zero Field Cooled Magnetization (ZFCM) by applying a DC magnetic field after cooling the bulk to 77 K. If enough heat is applied, while applying the DC magnetic field, with the heater wound around bulk, the trapped magnetic field in the bulk module is almost maximized. After magnetizing the bulk module by ZFCM, we applied a fluctuating magnetic field to the bulk module and measured variations in the trapped magnetic field and the surface temperature of the bulk. The temperature of the bulk module rises due to AC losses; those losses can be calculated from the measured temperature rise [4]. The measurement changed the amplitude of the fluctuating magnetic field, which was measured for 600 s. The amplitude of the fluctuating magnetic field applied to the bulk was 0.1 T at a frequency of 85.16 Hz. Figure 6 shows temperature rise of the bulk module when the shielding ring was attached, and when it was not attached. When no shielding was attached, the temperature of the bulk module increased from 77 K to 79 K due to the fluctuating magnetic field of amplitude 0.1 T. However, with the shielding ring attached, the temperature of the bulk increased to only about 77.5 K, i.e. the temperature hardly changed. Figure 7 shows the corresponding decrease in the factor (B_p/B_{p0}) of the trapped magnetic field. For the trapped magnetic field, B_{p0} = 1.164 T is the initial value of B_p. When no shielding ring was attached, we observed $B_p/ B_{p0} = 0.74$ after 600 s; this is a decrease of 26% from the initial value. However, when the shielding ring was attached, B_p/ B_{p0} hardly changed at all. This confirms that the shielding ring is very effective.

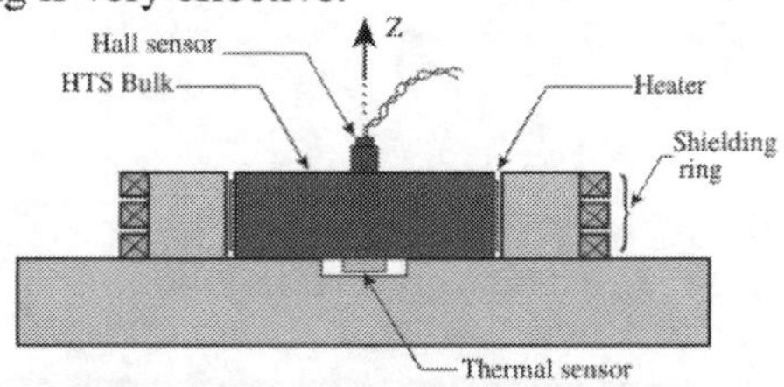

Figure 5. Illustration of the experimental set-up to test the performance of shielding ring AR2 in Fig. 3(c).

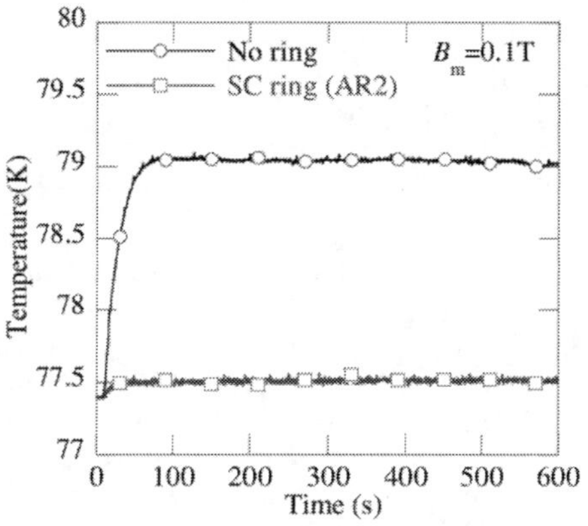

Figure 6. Time evolutions of surface temperature of bulk with and without the shielding ring AR2.

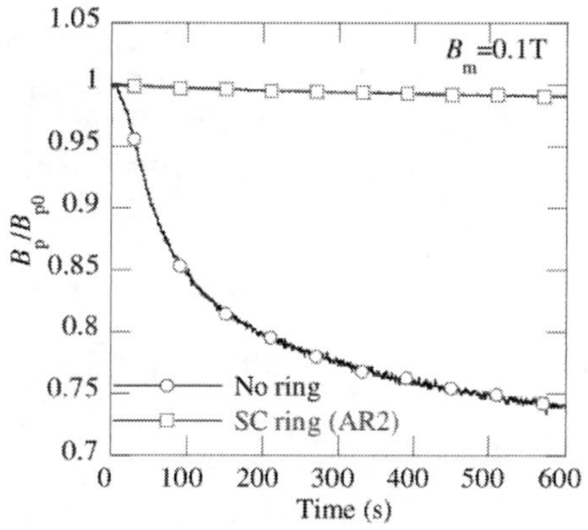

Figure 7. Time evolutions of B_p/B_{p0} of bulk with and without shielding ring AR2.

3. Conclusions

For rotating-type test equipment, we find that a shielding ring prevents a fluctuating magnetic field from affecting the rotor, and the analytic results show that AC losses in the bulk are very small. Further, we experimentally found that the shielding ring reduces temperature increases in the bulk module caused by AC losses, and the shielding ring prevents decreases in the magnetic field that is trapped in the bulk module. In the future, we plan to complete the design and construction of the test equipment, including a stator, and conduct rotating experiments. We intend to measure how fluctuating magnetic fields of realistic rotors affect the trapped magnetic field. These studies will be indispensable for developing practical superconducting bulk rotating machines.

Acknowledgements

This work was supported by JSPS KAKENHI Grant Number 24560325.

References

[1] Yamagishi K, Ogawa J, Tsukamoto O, Murakami M and Tomita M 2003 *Physica C*, **392–396** 659

[2] Vanderbemden P, Laurent P, Fagnard J-F, Ausloos M, Hari Babu N and Cardwe D A 2010 Supercond. Sci. Technol., 23 075006

[3] Yamagishi K, Tsukamoto O Ogawa J and Miyagi D 2011 *IEEE Trans. on Appl. Supercond.* **21**, 3320

[4] Yamagishi K, Sekizawa S, Tsukamoto O and Ogawa J 2006 *Journal of Physics: Conference Series* **43**, 295

Transport IV characterisation of MgB_2 conductor at a bend radius of 50mm

E A Young, I Falorio, C Beduz, W O S Bailey, Y Yang
Senior Lecturer
Institute of Cryogenics, Faculty of Engineering and the Environment, University of Southampton, SO17 1BJ UK

E-mail: e.a.young@soton.ac.uk

Abstract. Performance of state of the art MgB_2 multifilamentary conductor at a required bend radius is essential for many applications including but not limited to magnets and motors. The characterisation is generally done with benchmark transport Ic but further detail can be seen in IV characteristics which are undertaken in this paper. Two conductors with the same architecture but different diameters, 0.89 and 0.45 mm were measured from 32 K to 20 K in self-field in conditions of as received and deformed to a 50 mm bend diameter, corresponding to strains of 1.4 % and 0.7 % respectively. The qualifying 0.45mm conductor was further measured in background fields up to 3 T. The smaller diameter wire was found to have no signs of degradation of critical behaviour in I_c or IV characteristics.

1. Introduction

The performance of MgB_2 conductors in short lengths has improved dramatically since its discovery [1]. Research worldwide is being conducted into it suitability for winding in coils for applications such as magnets or motors.

The purpose of these measurements is to qualify a pre-selected conductor architecture for use in a magnet with a winding diameter of 100 mm. Distinguishing this work from others is qualifying measurements are not restricted to $I_c(B)$ but to a more complete IV characteristic, at different fields and temperatures and the combination of a state of the art architecture combined with an unusually small wire diameter.

Successful measurement of critical current and transport IV was previously demonstrated on various conductor architectures for currents up to 700 A and temperatures from 40 K down to 20 K [2,3].

2. Experimental

The conductors were supplied by Hyper Tech, USA, one with a diameter of 0.89 mm, the other 0.45 mm and both with the same architecture, a 6 filament, malic acid doped MgB_2 monel sheathed with copper stabilising core.

The samples were measured as received and after being deformed around a 50 mm bending radius and re-straightened using the same bend radius. Voltage current characteristics were measured in a conduction cooled cryostat with HTs current leads, and using a semi-dc method as described in previous publications [2,3]. In-field measurements applied a DC magnetic field in the radial direction.

The sample lengths were 15 cm zero field and 7 cm in field, with 1.5cm for current injection at each end and voltage taps separated by up to 7 cm and 3 cm respectively. I_c was determined using the 1 μV/cm criterion.

3. Results and Discussion

The IV data at 29.6K for the deformed 0.89mm conductor, and as received are shown in Figure 1. The straight sample has a clear power law behaviour with an I_c just over 90A. The bent sample has a lower I_c, about 87A, and a step like behaviour in IV at about 5microvolts, 90A. The step clearly shows mechanical damage in the superconductor filaments. There are two factors to consider when analysing the IV curves. Firstly the long length homogeneity of I_c has not been measured for MgB_2 conductors, presumably due to the cryogenic challenges of measurement in this temperature range and the associated cost. Secondly at these temperatures and high current densities care is taken to achieve an isothermal condition, (which is challenging with heat capacity rapidly dropping and power density increasing), yet inevitably at a certain power density heating will occur when currents are as high as 1000 A/mm^2 at 15 K. In Figure 1 the difference in I_c between the 2 conductors cannot simply be attributed to damage to the core as this may be due to long length inhomogeneity. Heating begins already at 10 μV/cm, so its behaviour above is dominated by the thermal condition of the sample, i.e. its dissipation and thermal contact with the surroundings.

The IV data in Figure 2 shows the smaller 0.45 diameter wire with the same deformation does not exhibit the step-like behaviour. The slightly higher I_c value again cannot be attributed to anything without either measuring the same sample, both deformed and as received or with a more significant change in I_c.

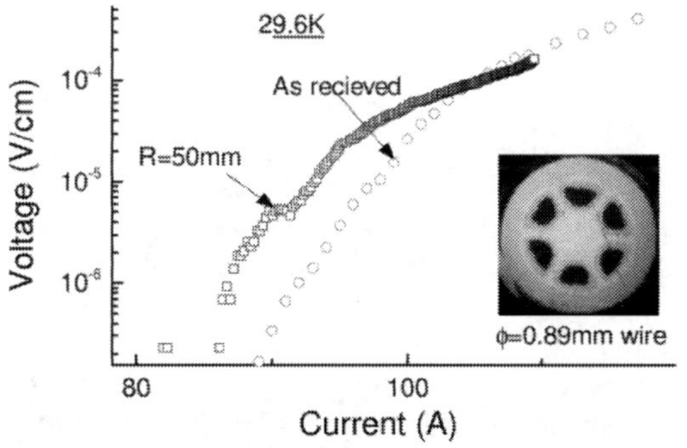

Figure 1. Current voltage characteristics of 0.89mm MgB_2 composite wire deformed and straight.

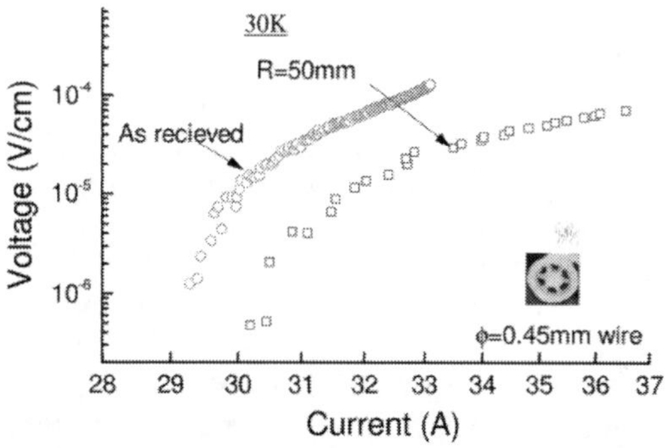

Figure 2. Current voltage characteristics of 0.45mm MgB_2 composite wire deformed and straight

The 0.45 mm diameter conductor was then measured in a range of fields and temperatures to explore if the power law behaviour was consistently present without any step-like characteristics. The data in Figure 3 at fields of 0.5, 1, 2 and 3 T show no signs of any step like behaviour in the IV curves. There is again some heating at higher voltages but this is to be expected above a certain dissipation level in a conduction cooled measurement.

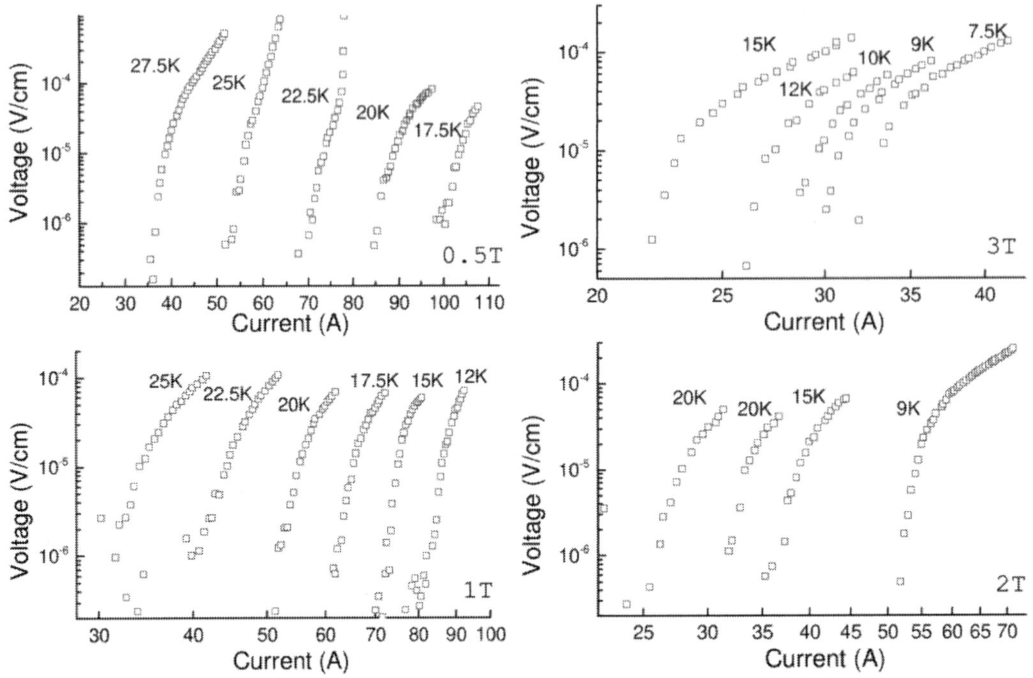

Figure 3. The voltage current characteristics of the deformed 0.45 mm conductor at a range of indicated temperatures and background magnetic fields 0.5 T, 1 T, 2 T and 3 T.

Is the strain consistent with other measurements? Measurements of strain are performed ether with a longitudinal tension or compression [4] or directly with bend radius tests [5, 6]. The tensile strain at degradation depends on the mechanical properties of the wire, the processing, and for the most recent publications varies from about 0.4 % [6] up to 0.9 % [5]. For the 0.89 mm wire the stain on the filaments using d/R where d is 0.7mm is 1.4 %, and correspondingly 0.7% for the smaller wire.

The same data for transport I_c of the 0.45mm diameter sample, bent to R=50 mm, is presented in Figure 4 at the different fields as a function of temperature and in Figure 5 as a function of field at different temperatures. The self-field values for the as received sample are plotted alongside in Figure 4 for comparison. The self-field values for Figure 5 were calculated using amperes law for an infinite wire and of the order of 0.01 T, so one order of magnitude off the plotted scale, however the interpolation can be seen.

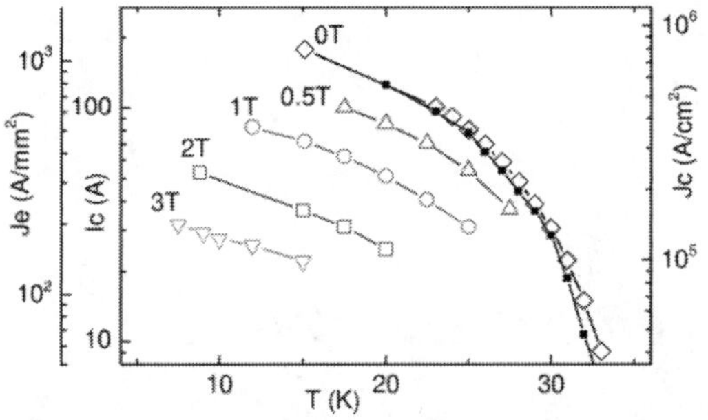

Figure 4. Log scale transport I_c , J_e, and J_c against temperature at different fields for the 0.45 mm diameter deformed sample, (open symbols). Plotted alongside is the self field for the as received sample (filled symbols).

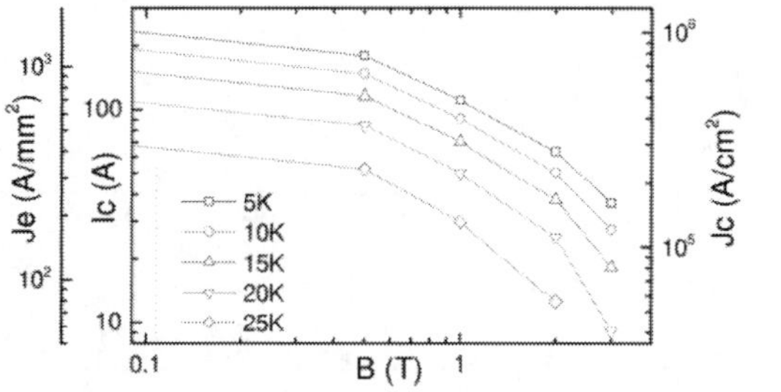

Figure 5. Log scale transport I_c, J_e, and J_c against temperature at different fields for the 0.45 mm diameter deformed sample. The lines are interpolated between data points. Self field is set artificially to 0.1 T.

4. Conclusion

Using transport IV as a characterisation tool two different diameters of state of the art conductor were tested at self field in a condition of as received and deformed to a 50 mm bend radius. The smaller diameter sample did not show any degradation, (0.7 % strain), and was further tested in fields up to 3 T and down to 7.5 K. The suitability of the 0.45 mm wire for winding at a bend radius of 50 mm is proven not just by I_c but by a more detailed picture that IV data provides.

5. References

[1] G Z Li, Y Yang, M A Susner, M D Sumption and E W Collings 2012 *Supercond Sci. Technol.* **25** 025001

[2] E. A. Young, W. O. S. Bailey, M. K. Al-Mosawi, C. Beduz, Y. Yang, S. Chappell, and A. Twin 2012 *Phys. Proc.* **36** 1343-1347

[3] E. A. Young, M. Paolella, and Y. Yang 2013 *IEEE Trans. Appl. Supercon.* **23** 8001304

[4] G Nishijima, S J Ye, A Matsumoto, K Togano, H Kumakura, H Kitaguchi and H Oguro 2012 *Supercond Sci. Technol.* **25** 054012

[5] P Kováč, I Hušek, T Melišek, L Kopera and M Reissner 2010 *Supercond Sci. Technol.* **23** 065010

[6] Kazumune Katagiri, Ryuya Takaya, KoichiKasaba, Kyoji Tachikawa, YutakaYamada, SatoshiShimura, Naoki Koshizuka and Kazuo Watanabe 2005 *Supercond Sci. Technol.* **18** S351-S355.

Updating and Test on 12kA HTSCL for Russia JINR

Tingzhi Zhou, K Ding, C Liu, C Xu, H Feng, H Wu, X Liu and Y Song
Institute of the Plasma Physics, CAS (ASIPP), Hefei, China,

E-mail: tingszhou@ipp.ac.cn

Abstract. The Joint Institute for Nuclear Research (JINR) started manufacture and test of superconducting magnets for the new accelerator complex Nuclotron-based Ion Collider fAcility (NICA). A pair of current leads rated 12 kA was developed for cold testing of the magnets in 2012. The operation of the facility needs ~350 leads rated from 100 A to 12 kA to convey current for the magnets. After the test of the pair of 12 kA leads in JINR a contract of supplying 3 pairs of 12 kA high temperature superconducting current leads (HTSCLs) were signed to ASIPP this year. In this paper the design updating and test of the current leads are introduced.

Index Terms—Current lead; Bi2223; NbTi; Insulation

1. Introduction

Current lead connects the room temperature busbar and cryogenic magnet to convey the large current. The high temperature superconducting (HTS) module is applied on the current lead that will minimize the heat load to the cold end and save 2/3 helium consuming in compare with the conventional copper lead. After the successful producing of Bi 2223 tapes many HTS current leads were developed in the world. The KIT and CRPP developed a 70 kA HTS current lead for ITER[1]. The current lead was designed with 50 K helium gas cooling of the fin type HEX and 4.5 K SHe cooing of the cold end. It was proved to be safety operation with 80 K helium. In CERN thousands of 6-13 kA leads were commercial produced and used on the Large Hadron Collider LHC[2].The HEX was cooled by 20K helium gas and the NbTi low temperature superconducting (LTS) strands were immerged in liquid helium. In ASIPP 15 kA HTS current leads were developed and assembled on EAST tokamak[3]. The HEX was designed with multi-helix. The cold end of the HEX was direct soldered to the HTS module and cold by LN_2. ASIPP also successfully developed 10 kA, 52 kA and 68 kA trial lead for ITER[4-6]. The foil type and fin types HEX were developed and compared. The foil type HEX was proved to be with much more cooling efficiency. The stacks soldered from Bi 2223 tapes and NbTi LTS strands were applied. The maximum current capacity of the 68kA trial lead reached 90 kA.

Currently in Russia the new accelerator complex Nuclotron-based Ion Collider fAcility (NICA) are in developing. ~350 leads rated from 100A to 12kA will be supplied for the facility operation. The Joint Institute for Nuclear Research (JINR) needs one pair of current leads to test its superconducting magnet. So in 2012 a pair of 12 kA current leads was developed and tested.

In 2013 ASIPP was awarded a contract of supplying 3 pairs of current leads rated 12 kA before June 2014. In this year the 12kA leads are design updated based on the fabrication experience and test results.

2. Design and test of 12 kA leads

The current leads developed in 2012 were optimized on 12 kA current and current capacity reached 21 kA. The current lead (see Figure 1) is 1.855m length and 40 kg net weight and mainly characterized by[7] 1) water cooling terminal; 2) foil type heat exchanger (HEX); 3) LN2 cooling transition; 4)Bi2223 high temperature superconductor (HTS); and 5) NbTi Nuclotron type low temperature superconductor and 6) high voltage (HV) insulation.

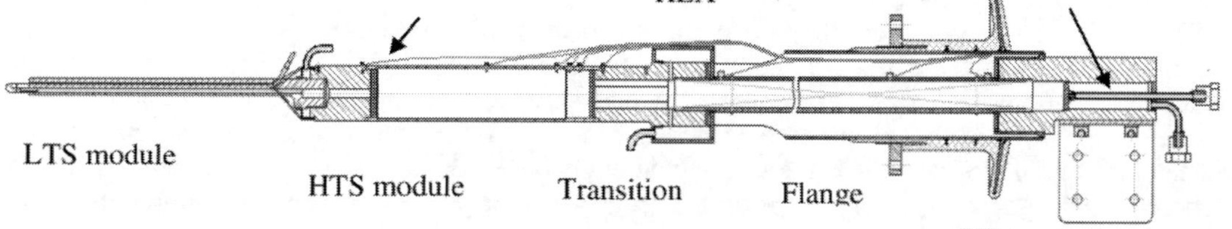

Figure 1. Drawing of the integrated 12kA HTSCL

The pair of 12kA leads was tested in ASIPP after the integration. Heat exchanger was cooled by liquid N_2 (LN_2) which was supplied from an external 500 L LN_2 tank. LTS section of the leads was cooled by liquid He. The pair of leads was soldered with a short length of an original Nuclotron cable as a superconducting jumper.

The current leads were assembled in a cryostat that was developed and fabricated for the test. A thermal shield in the cryostat was cooled with LN_2. The current from direct current (DC) generator was feed into the warm terminal. The current of power supply can be slowly ramped up to maximum 30 kA. The vacuum of the cryostat reached 1×10^{-5} Pa • m^3/s level in steady state operation.

Before cooling down of the current leads in the cryostat, the leakage was detected of all coolant circuit and current lead itself. The leakage rate of current lead was less than 10^{-8} Pa • m^3/s level. The maximum insulation voltage for flange was test to be more than 3 kV. After 1.5 hours cooling the leads reached a steady operation state. Then the LN_2 supply source was closed to measure the holding time in loss of flow accident until the quench protection was triggered. A maximum 21 kA over-load operation were performed. The test results are list in Table 2 and detail analyzed in [7].

Table 2 Main test results of a pair of 12kA current lead

Contents	1# lead	2# lead
High voltage test of insulation flange	3kV, leakage current < 1μ A	3kV, leakage current < 1μ A
5K heat load	3.68 W	3.67 W
Joint resistance of HTS/LTS	0.25 nΩ	1.01 nΩ
Temperature on LTS section	4.83±0.3K	6.5±0.4K
Pressure loss in LTS section	<0.02 bar	<0.02 bar
Holding time when loss of flow accident (hotspot < 180 K)	>5 min	>5 min
Over load current	21 kA	21kA

In February 2013 this pair of leads was transported to Russia JINR. However, there was leak between the insulation and the S.S tube. The reasons were analyzed that after the cooling down there were cracks between the tube and insulation because of different thermal shrinkage.

3. Updating on the 12kA leads

Even though the temporary solution stopped the leak to finish the test in Russia a lasting settlement of the troubles is also developing in parallel in ASIPP.

The interface of the current lead and the cryostat is a DN 200 O ring sealed flange (Figure 2). The original insulation flange is the structure of 10mm insulation cured from fiber glass and polymeric those are wrapped on the S.S tube. A S.S ring is welded to the flange and embedded into the insulation. The insulation with minimum 4mm thickness can separate the grounded flange and current lead with potential high voltage. The cap on the top of the insulation can increase the creepage distance.

The bonding between the insulation and S.S tube is the weakness of the insulation flange. The insulation layer is glued to the S.S tube with curing method. The epoxy is a weak contact. After the cooling the cracks happened and were extended.

Two possible solutions were developed and tested in ASIPP. In solution 1 (Figure 3) the S.S cap and reinforce rings welded to the S.S tube is embedded into the insulation. The reinforce rings anchored in the insulation can prevent the thermal shrinkage of the tube. The S.S cap on top of insulation can enlarge the curing surface between the insulation and flange. Because of the room temperature of the top of HTSCL the larger bonding between the insulation and S.S cap will be less thermal shrunk. Additionally the screw on the outer diameter of the tube can also enforce the curing between the insulation and the tube.

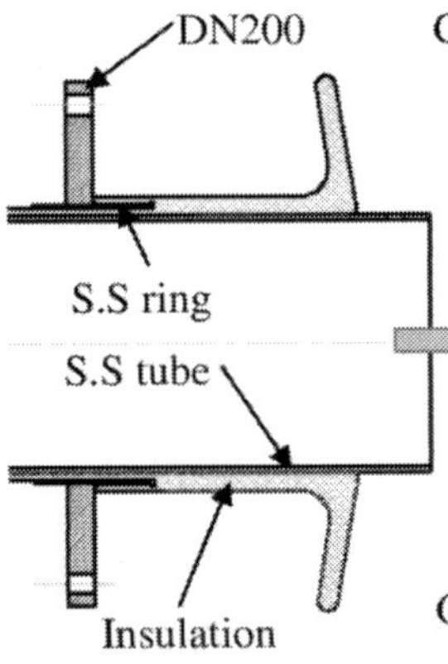

Figure 2. Original design

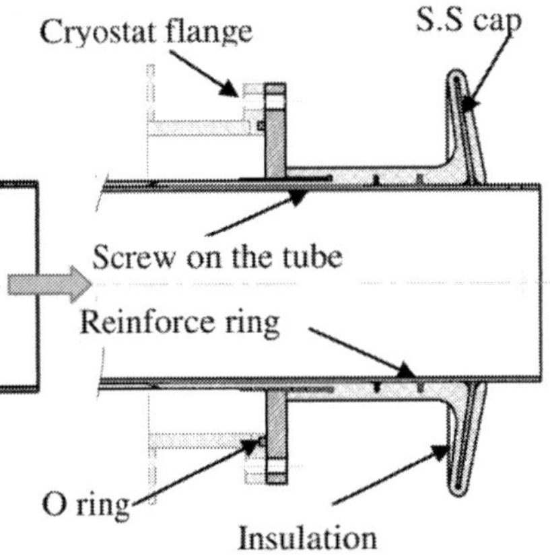

Figure 3. Solution 1

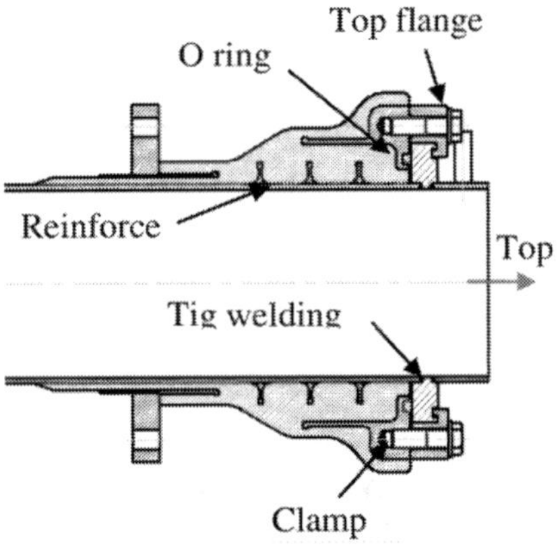

Figure 4. Solution 2

Another solution is more complicated as Figure 4 shows. On the top of the insulation a special flange is embedded. The O ring is sealed before the insulation interface. So the vacuum is totally sealed from the air even in the case of a crack between the insulation and tube.

Both solutions were tested under the real temperature and loads. The LN_2 was supplied to cool the flange down. 200 kg loads were applied by the 4 tighten springs. 3 PT100 sensors were attached on the top, middle and bottom of the insulation component (Figure 5).

The LN_2 was slowly feed into the cryostat. After the T3 temperature was less than -40℃ the status was kept for 5 minutes and then warmed it up to room temperature. 5 thermal cycles were tested. The temperatures are recorded in Figure 6. The T1(top) and T2(middle) keep the temperature larger than 0℃ during the thermal cycle that is well consistent with the design and the real lead test condition.

The vacuum leak rates were also tested. The leak rates on both solutions were less than 10^{-8} Pa • m^3/s level.

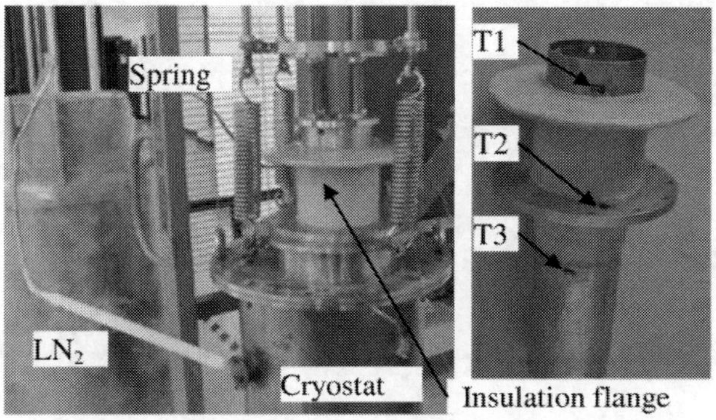

Figure 5. Test of the updated insulation flanges

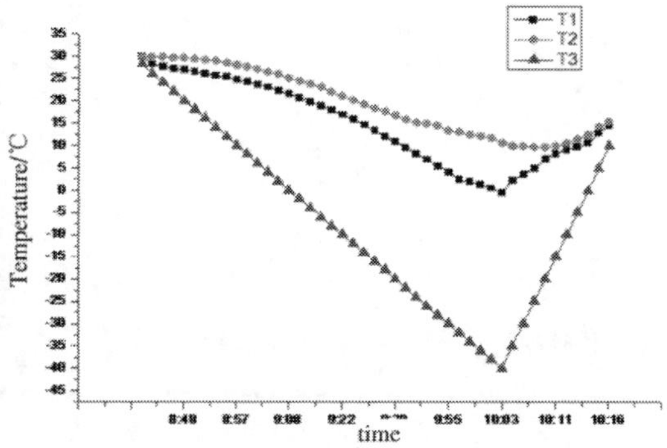

Figure 6. Temperature profiles during one thermal shock

After thermal shock the insulations were tested under high voltage to maximum 10kV. Under 2kV the leakage currents for both solutions were less than 8μ A (Figure 7 and Figure 8). On the 10kV voltage the leakage currents were less than 20μ A. Given 40μ A leakage current criteria the insulation of both solutions are safe enough and larger margins are given.

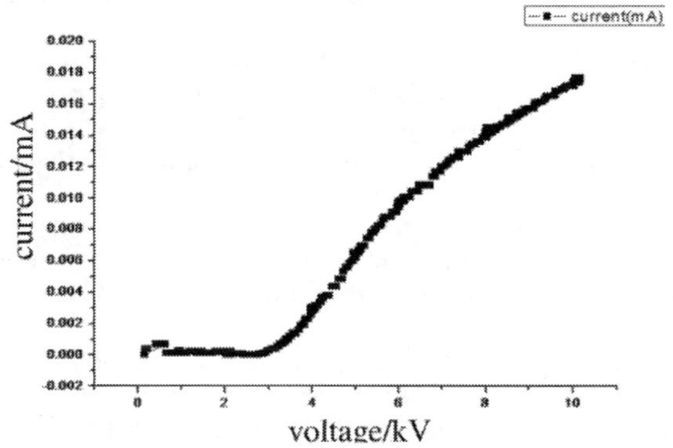

Figure 7. HV test on solution 1

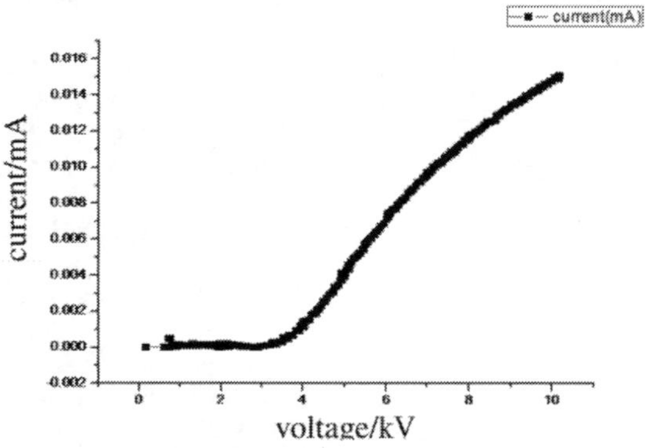

Figure 8. HV test on solution 2

The test results approve that both solutions can satisfy the design requirements. No obvious performance differences between two solutions. Solution 1 is chosen because of easy fabrication.

4. Conclusion

A pair of 12 kA HTS current leads was manufactured and tested in ASIPP. The insulation flange is updated after the cracks during the test and passes the vacuum leakage and HV test. Another 3 pairs of 12 kA leads will be prepared in ASIPP for JINR.

5. References

[1] R Heller 2005. IEEE Trans. Appl. SC. 15 1496.
[2] A Ballarino 2008 *J. Appl. Phys.* **468** 2142.
[3] Y Bi 2005 *J. Chinese Journal of Low Temperature Physics.***27** 1077.
[4] P Bauer, Y Bi and T Zhou 2009. *IEEE Trans. Appl. SC.* **19** 1500.
[5] Y.Bi, K.Ding and A.Cheng 2009 *J. Chinese Journal of Low Temperature Physics.* **1** 42.
[6] K.Ding and Y.Bi 2009 *J. Chinese Journal of Low Temperature Physics.* **31** 231.
[7] Y.Bi and K. Ding 2013. *IEEE Trans. Appl. SC.***23** 46.

First Model Power Cables Made of Russian 2G HTS Wires and their Test Results

S S Fetisov[1], A A Nosov [1], V V Zubko[1], N V Bykovsky[1], S Yu Zanegin[1], V S Vysotsky[1], G G Svalov[1], S Lee[2], V Petrykin[2], V Kalitka[3]

[1]Russian Scientific R&D Cable Institute (VNIIKP), Moscow, Russia, 111024
[2]SuperOx Japan LLC, Japan, 252-0243
[3]SuperOx, Moscow district, Russia, 143082

E-mail: vasily.zubko@gmail.com

Abstract. Recently launched production of 2G wires by SuperOx Company permitted us to start tests of representative models of power cable made of their wires. In Russian Scientific R&D Cable Institute we have the test facility for extensive testing of heavily instrumented HTS cable models with full size cross-section and length up to 5 m. Our standard test program includes critical current measurements, current distribution measurements among layers, AC loss measurements, etc. In this paper we present details of first power cable models made of 4 mm width 2G wires from SuperOx and their test results. Model cables have two counter-wound layers with 18 tapes in each layer. Critical currents of cables were from 3 kA to ~4.5 kA. We also measured AC losses and compared them with losses in model cables made of 2G wires of other producers. AC losses measured were compared with calculated ones and demonstrated the good coincidence between experiments and calculations. First model cables tests permitted us to conclude that SuperOx wires have good parameters for production of 2G HTS power cables with low AC losses.

1. Introduction

Russian SuperOx company along with its Japanese subsidiary SuperOx Japan LLC entered to worldwide contest in developing of second generation high temperature superconducting tapes (2G HTS) [1]. Recently first IBAD-PLD produced 2G wires with Hastelloy substrate become available in long enough lengths to make some HTS devices, particularly HTS cables [2].

In Russian Scientific R&D Cable Institute (known by Russian abbreviation: VNIIKP) we have the test facility for extensive testing of heavily instrumented HTS cable models with full size cross-section and length up to 5 m [3]. Our standard test program includes (but not exhausted) critical current measurements, current distribution measurements among layers at both: DC and AC conditions, AC loss measurements, etc. [3], [4].

The goal of this work is to make two first models of cable cores and to test them to validate properties of newly produced Russian 2G wires for using in HTS power cables.

2. Experiment and results

2.1. Cable core models design

Several HTS cable cores made of 2G wires where tested in VNIIKP [3]-[5]. Usually, we use a former made of stainless steel tube and two counter-wound layers of HTS tapes. Cable cores were equipped by voltage taps, Rogowski coils and other probes [3]-[5].

Similar two cable core models were made with newly produced SuperOx 2G HTS wires. These tapes were produced by IBAD-PLD technology with GdBCO superconductor. Parameters of wires used for models are listed in the Table 1.

Table 1. Parameters of 2G SuperOx wires used for models.

Marking	Width, mm	GdBCO layer thickness, μm	Minimal I_c guaranteed by supplier, A
First model	4	1.15	~80
Second model	4	1.6	~120

Both models have the same design but used different 2G tapes. SS former with 25 m diameter has been insulated by Kapton™ tape. Two counter-wound layers with 18 tapes per layer were separated by Kapton™ tape as well. On each layer Rogowski coil and two pairs of voltage taps (VT) were installed. Some details of cable models preparation are shown in figures 1 and 2.

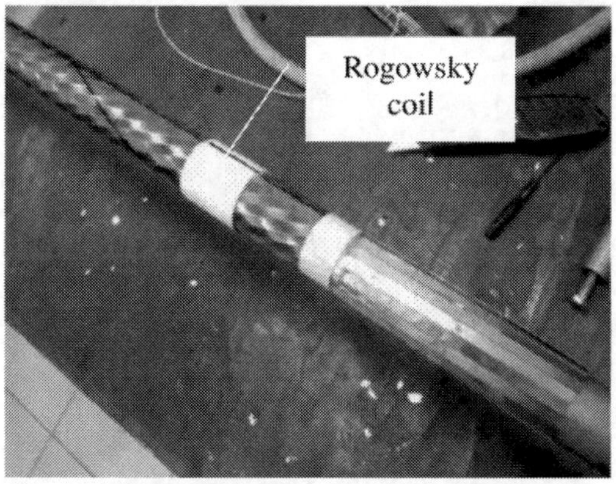

Figure 1. First layer of a cable core model with Rogovsky coil installed.

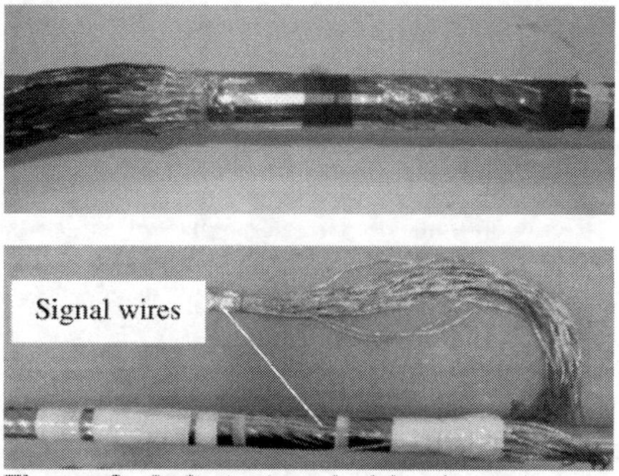

Figure 2. Left (top) and right (bottom) current terminations installed and signal wires.

2.2. DC test

The standard test program for cable core in VNIIKP includes (but not exhausted):

- DC test to determine critical currents and current distribution among layers caused by non-uniformity in termination resistance;
- AC test to determine current distribution among layers at AC conditions and AC loss measurements by electrical method [3], [5].

The test facility for testing of HTS cables models up to 5 m length has been described in details in [3]. The results of DC tests are shown in figures 3-5.

In figure 3 current distributions among layers are shown for two cases. In figure 3a terminations were soldered to 2G tapes via bunches of thin copper wires. One can see that due to asymmetry of 2G wires when superconducting layer is located from one side of a tape – current distribution is strongly non-uniform. We developed special connection method for 2G wires [6] that permits to achieve uniform current distribution at DC mode as it is shown in figure 3 b.

Voltage – current characteristics of models tested are shown in figure 4. In figure 4a VCC are shown for two layers of a cable model made of the first batch of 2G wires delivered.

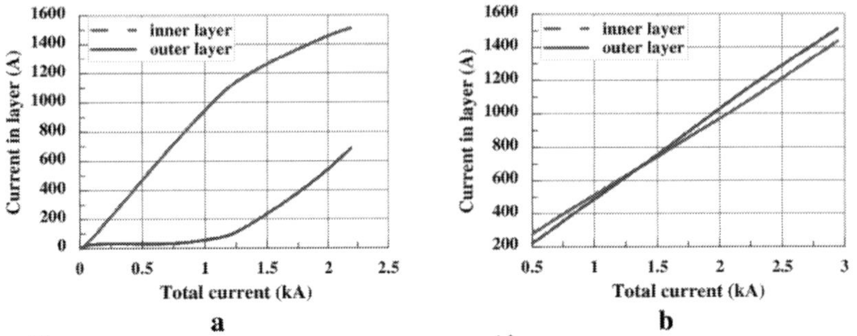

Figure 3. Current distribution among layers in DC mode for the first cable model. a – standard termination joint technique, b – using special joint uniforming technique [6].

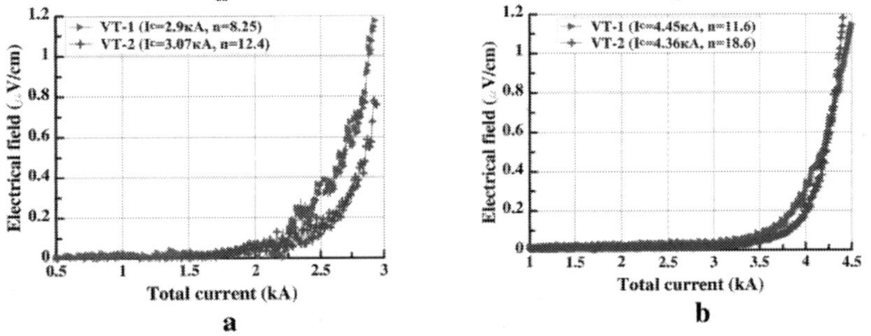

Figure 4. VCC of cable models tested, a – first cable model, b – second cable mode.

In figure 4a one can see the critical current measurements of the first cable model. I_c measured by voltage taps (VT) on inner layer is about 2903 A (or 83 A per tape) and 3065 A (85 A per tape) for VT on outer layer. Index value was ~8-12. This result is quite good for the very first production.

In figure 4b one can see that critical current of the second cable model is about 4360 A (or 121 A per tape) for inner layer and 4450 A (124 A per tape) for outer layer. Index value was ~11-18.5. Up to now it is the best result in critical current so far obtained in our measurements of 2G cable models.

2.3. AC test

During AC test current distribution among layers was quite uniform like in our previous tests [3]-[5]. In figure 5 AC losses measured are shown. In figure 5 a – total losses measured, in figure 5 b – normalized losses (I_c^2 of tape) compared with losses measured for 2G similar cable models made of Super Power 2G wires [3]. One can see that AC losses are typical for 2G HTS wires with non-magnetic substrate. The value of AC loss ~1W/m was achieved at current ~0.9I_c while for cables made of 1G HTS wires with same I_c such value of losses was achieved at ~0.3I_c [3].

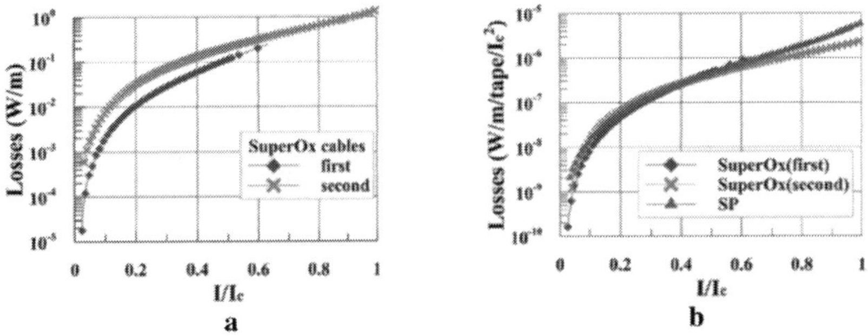

Figure 5. Measured AC losses in the cables, a – total, b – normalized.

3. Discussion

We analyzed AC losses in two-layers cable core with non-magnetic substrate. Calculations were done by Norris models (strip and ellipse) and developed 2D FEM model. In calculation with FEM model we took into account the non-uniformity density of critical current across of 2G tapes [7] in cable layers. Numerical calculation was made for two 2D cross-section in the cable: tape-on-tape cross-section and tape-on-gap cross-section [4].

Comparison calculated and measured AC losses (first model) are shown in figure 6. One can see good coincidence of numerical calculated and measured AC losses. That means that our model can be used for calculation of AC losses in thin superconducting layer.

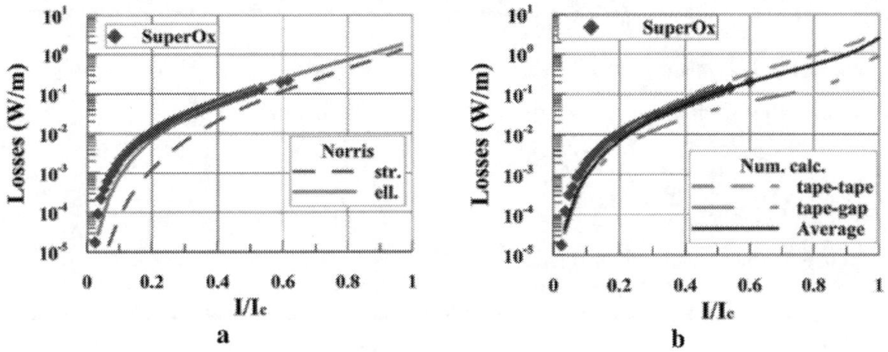

Figure 6. Measured AC losses, and calculated AC losses, a – Norris models, b – numerical model.

4. Conclusion

Two cable core models were made and tested made of 2G HTS IBAD-PLD wires produced by Russian SuperOx Company. Both models demonstrated good results. While the first cable demonstrated critical currents usual for other producers, the second model demonstrated quite high critical current along with good VCC index. Normalized AC losses are typical for cables made of 2G HTS wires with non-magnetic substrate and can be calculated by the numerical model developed. AC losses in our 2G cables are much less than those in cables made of 1G HTS wires. This allow for 2G cables to work close to I_c level with low losses.

IBAD-PLD wires produced by SuperOx can be used for HTS power cables with low AC losses and high current density.

5. References

[1] Available at http://www.superox.ru/en/

[2] Lee S, Petrykin V. Samoilenkov S, Kaul A, Vavilov A, Vysotsky V, Fetisov S 2013 Development and Production of Second Generation High T_c Superconducting Tapes in SuperOx and First Tests of Model Devices *this conference paper 2M-WT-06*

[3] Vysotsky V, Nosov A, Fetisov S, Shutov K 2011 AC Loss and Other Researches with 5 m HTS Model Cables *IEEE Transactions on Applied Superconductivity* Vol.21 N3 pp 1001-1004

[4] Zubko V, Nosov A, Polyakova N, Fetisov S, Vysotsky V, 2011 Hysteresis Loss in Power Cables Made of 2G HTS Wires with NiW Alloy Substrate *IEEE Transactions on Applied Superconductivity* Vol.21 N3 pp 988-990

[5] Sytnikov V, Vysotsky V et al 2010 AC Loss of a Model 5m 2G HTS Power Cable Using Wires with NiW Substrates *J. Phys.: Conf. Ser.* 234 032061 doi: 10.1088/1742-6596/234/3/032061

[6] Superconducting cable connection; Provisional patent # (RU) № 103986 МПК H01R 4/68 *(in Russian)*

[7] See for example: D.V.Sotnikov, I.P.Radchenko, V.V. Zubko, S.S. Fetisov, Influence of mutual directions of external magnetic field and transport current on critical currents of 2G HTS tapes. Available at: https://www.itep.kit.edu/cca2012/Downloads_CCA/Abstracts%20Extended/EA_PB-E11_Sotnikov.pdf

Journal of Physics: Conference Series **507** (2014) 032064 doi:10.1088/1742-6596/507/3/032064

Recent Development Status of Stirling Type Pulse Tube Cryocooler for HTS

Y. Hiratsuka, K. Nakano and T. Kato

Technology Research Center, Sumitomo Heavy Industries, Ltd.2-1-1, Yato-cho, Nishitokyo-city, Tokyo 188-8585 Japan

E-mail: Ysh_Hiratsuka@shi.co.jp

Abstract. Sumitomo Heavy Industries, Ltd. (SHI) has been developing a high power stirling type pulse tube cryocooler. For the purpose of cooling high-temperature superconductor (HTS) devices, such as superconductor motor, SMES and current fault limiter, requested specifications from the devices to a cryocooler are compact size, light weight, high efficiency and high reliability. Especially, the cryocooler must be demanded COP > 0.1 in the efficiency. The experimental results of prototype pulse tube cryocooler were reported in June 2012 [1]. For an In-line type expander, the cooling capacity was 210 W at 77 K and the minimum temperature was 37 K when the compressor input power was 3.8 kW and the operating frequency was 49 Hz. Accordingly, COP was about 0.055. Moreover, for miniaturization a U type expander was tested and the performance is about 10 % less than that of an In-line type expander. After that, we have estimated that the cooling performance is influenced by the environment such as the effect of the pulse-tube inclination, the temperature and the flowing quantity of cooling water. The detailed results are reported in this paper.

1. Introduction

Recently, technical innovation of HTS devices *at al.* has brought about the necessity of high efficiency cryocoolers, and pulse tube cryocoolers with 100~1000 W at 77 K cooling capacity have been developed for HTS applications. In 2004, J. H. Zia et al. reported a commercial prototype Stirling type pulse tube cryocooler (SPTC) with 200 W cooling capacity as civilian equipment for HTS electronics applications [2]. For the purpose of cooling HTS and semiconductor sensor, a small size Stirling cryocooler had been developed. But it became a problem of the reliability of cryocooler to be decreased a cooling capacity by a wear of the expander piston seal when it drove for a long time. Pulse tube cryocoolers are more attractive of high reliability than with other regenerative cryocoolers (e.g., GM and Stirling cryocoolers), because of having no moving parts in the expander. SHI has been developing a prototype high power SPTC to satisfy the demand specification from the device to the cryocooler, smaller lightweight, the high efficiency and high reliability and the experimental results were reported in June 2012[1]. The expander of cryocooler was an In-line type and the cooling capacity was 210 W at 77 K when the compressor input power was 3.8 kW at the operating frequency of 49 Hz, a compressor efficiency of 74 % and COP 0.055. In regard to the arrangement of the cryocooler system, a U type is more compact and is more convenient for this application and the performance was 191 W at 77 K, with the lowest temperature at 44 K. The performance decreased about 9.1 % compared to that of the In-line type. After that, we estimated that the cooling performance was influenced the change in the environment such as the effect of the pulse-tube inclination and changes in the temperature and flowing quantity of cooling water. The details of these experiment results are reported in this paper.

2. General design

Fig. 1 and Table 1 show a photograph of the prototype U type pulse tube cryocooler and its specifications, respectively. The cryocooler is an integral Stirling type pulse tube and the compressor that makes two pistons opposition is moved by the linear motor, with outer diameters of 330 mm and 670 mm in length. The compressor has a moving magnet type motor with high efficiency and the piston is guided by a flexure bearing. The piston position is monitored using a laser vibrometer. Pressure transducers are mounted near the compressor discharge head and a hot-end of the pulse-tube. The mass flow rate through the phase-shifter can be estimated by the pressure at the hot-end of the pulse-tube. These measurements are used to calculate both the pressure-volume P-V work of the compressor and an equivalent work (= work flow) of the pulse-tube. A PtCo sensor and a thermo-couple were set up to measure the cold-head temperature and the temperature distribution of the regenerator and the pulse-tube wall, respectively. The inlet-outlet cooling water temperature of the after-cooler and the hot-end of the pulse-tube ware measured with Pt100 sensors to calculate the heat capacity. We considered that the momentum equation of the inertance-tube was an equivalent LR circuit and the gas displacement in the inertance-tube was calculated by the measured pressure amplitude of the hot-end Ppt and the buffer-tank Pbf [3].

3. Results and Discussion

3.1. Prototype Cryocooler Performance

Fig. 2 shows the measured cooling performance of the prototype In-line and U type pulse tube cryocoolers. In regard to the arrangement of the cryocooler system, U type is smaller than an in-line type and more convenient for these applications. With the In-line type, the pulse-tube is located on the axis with the regenerator, but with the U type, the pulse-tube is parallel to the regenerator. The system was filled with helium gas up to 2.0 MPa and the vacuum chamber was evacuated to an order of <0.13 Pa (10^{-3} torr). A radiation shield was rolled in to surround the expander. With 3.8

Table. 1. Specifications of prototype U type STPC.

Maximum cooling power	: 190 W at 77 K
Maximum electric input power	: 3.8 kW (AC200 V)
Efficiency	:14.5 % Carnot (COP 0.053)
Cooling	: Water cooling
Operating frequency	: 49 Hz
Initial gas pressure	: He 2.3 MPa
Dimension	:H750 mm× L700 mm× W350 mm
Weight	: <150 Kg
Expander	: Integral U type with inertance-tube

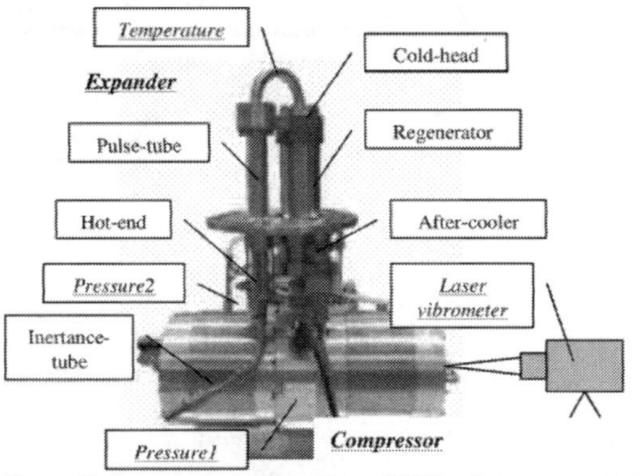

Fig. 1. Photograph of prototype U type STPC and measurement points of pressure, piston displacement and temperature.

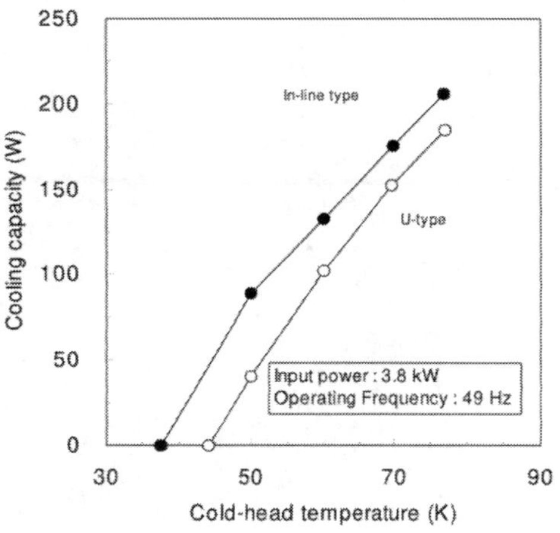

Fig. 2. Comparison of cooling capacity of the U type and the In-line type expander.

kW input power at the operating frequency of 49 Hz, the no-load temperature was 37 K, the cooling capacity was 210 W at 77 K and the compressor P-V work was about 2820 W. The compressor efficiency is 74 % at a motor efficiency 85 % at 77 K. The performance of the U type is 191 W at 77 K with the no-load temperature at 44 K. The performance decreased about 9.1 % compared to that of the In-line type, having a cooling capacity of 210 W at 77 K. It is suggested that a cold head's shape has a great effect on cooling performance. Because the volume of the cryocooler increases, the system was filled with helium gas up to 2.3 MPa. The pulse-tube work flow was calculated from the experimental results about 530 W at the cold-head temperature 77 K. These work flows go through the hot-end of the pulse-tube and are rejected as a heat in the phase controller. For improving efficiency of a pulse tube cryocooler, it's necessary to add a mechanism which can recover this work flow.

3.2. Influence of Pulse Tube Cryocooler Inclination

In 2010, G. W. Swift *et al.* reported the influence of a pulse-tube inclination by natural convection [4]. The performance of the pulse tube cryocooler depends on that inclination by influence of the natural convection in the pulse-tube. The operating conditions must be determined to minimize the effect of this inclination. In general, the convection loss can be reduced by high operating frequency and small pulse-tube size. We measured the effect of the U type pulse tube cyoclooer inclination on the performance at a temperature higher than 65 K, as used by HTS. The influence of inclination was a decreasing rate within only 3 %.

3.3. Influence of Expander and Compressor Configuration

It may be advantageous to use a split configuration of the compressor and the expander where only the expander is installed in a system, because the weight of the cryocooler increases when the cooling capacity increases. The cooling capacity of the split expander was measured and compared to the integral type. The connecting tube used between the expander and compressor was a stainless tube in a straight line, with a 25.4 mm outer diameter and 1m length. The results show that almost the same performance is achieved between the split and integral configuration.

3.4. Evaluation Cryocooler Performance

It is necessary to remake the structure of the cold-head to set up the cryocooler in HTS system. Because the prototype cold-head doesn't have the place for connecting the large size equipment, the

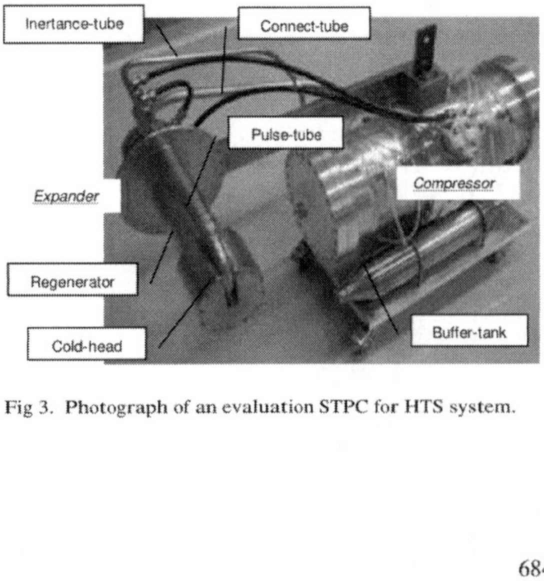

Fig 3. Photograph of an evaluation STPC for HTS system.

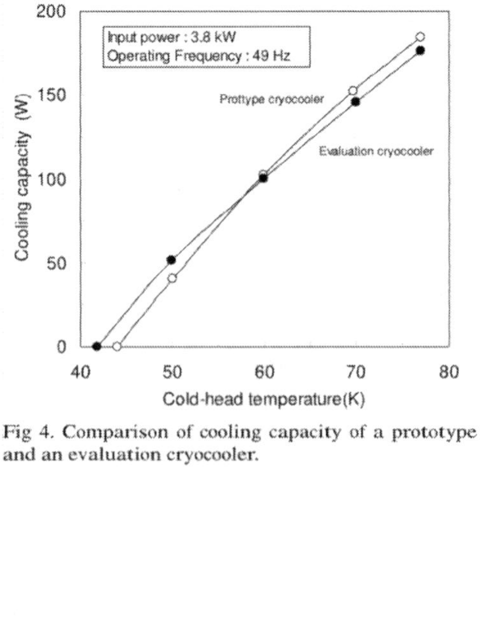

Fig 4. Comparison of cooling capacity of a prototype and an evaluation cryocooler.

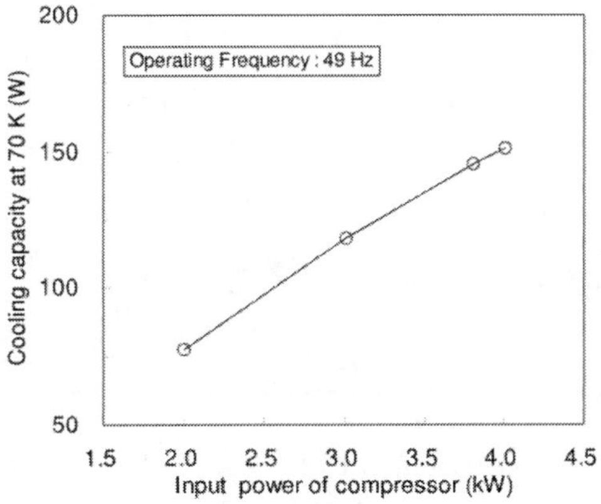

Fig 5. Measured cooling capacity at the cold-head temperature 70 K vs. the compressor input power for the evaluation SPTC.

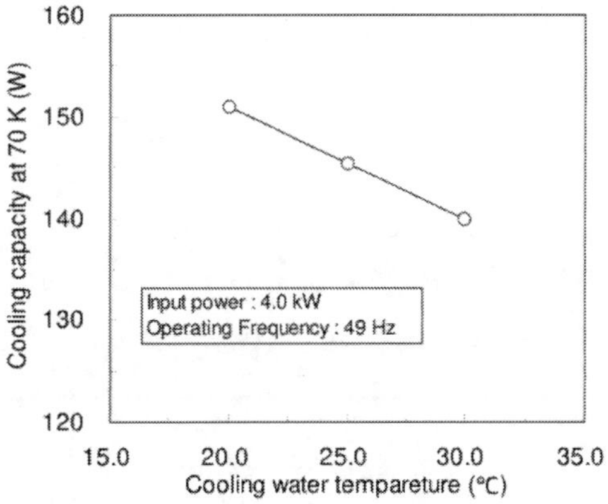

Fig 6. Measured cooling capacity at the cold-head temperature 70 K vs. the water cooling temperature for the evaluation SPTC.

low temperature part of the regenerator and the pulse-tube was connected in the tube as shown in Fig. 1. And an evaluation cryocooler re-designed an interface of the cold-head for HTS system and was examined. As for Fig. 3, 4, a photograph of the evaluation cryocoole and the comparison of the cooling performance examination results of the prototype and the evaluation cryocooler to estimate the HTS system are shown respectively. The performance of the evaluation cryocooler was 176 W at 77 K, with the no-load temperature at 42 K and decreased about 7.9 % compared to that of the prototype cryocooler. The heat conduction loss grows, because the distance of the place where the HTS device was set up from the low temperature heat exchanger in the end of the regenerator was so long and the cooling capacity decreased as a result. The cooling capacity that the HTS system needs has been over 150 W at 70 K and a necessary cooling capacity could be obtained 151 W at 70 K by the compressor input 4 kW as shown in Fig. 5. Moreover, the result of examining the influence on the cooling capacity by the difference of the inlet cooling water temperature is shown in Fig. 6. It is understood that the cooling capacity at 70 K decreases by 12 W when the inlet temperature of cooling water increases by about 10 °C.

4. Conclusion

We developed a prototype SPTC with cooling capacity of 191 W at 77 K and a COP of 0.05. However, it was necessary to remake the structure of the cold-head to set up the cryocooler in HTS system and the evaluation cryocoole was designed and estimated. The performance of the evaluation cryocooler was 151 W at 70 K when the compressor input power was 4 kW at the operating frequency of 49 Hz.

References

[1] Y.Hiratsuka and K.Nakano, "Development of Orientation-Free High Power Stirling-Type Pulse Tube Cryocooler" in Cryocoolers 17, edited by S.D.Miller and R.G.Ross Jr., Proceedings of the 17th International Cryocooler Conference, ICC Press, Boulder, CO, 2012, pp. 129-133.

[2] J.H.Zia., "A Commercial Pulse tube cryocooler with 200W Refrigeration at 80K" in *Cryocoolers 13*, edited by R. G. Ross Jr., Proceedings of the 13th International Cryocooler Conference, ICC Press, Boulder, CO, 2004, pp. 165-171.

[3] K.Nakano and Y.Hiratsuka "Loss Analysis of High Power Stirling-Type Pulse Tube Cryocooler", CEC-ICMC 2013, Washington to be published.

[4] G.W.Swift and S. Backhaus, "Why High-Frequency Pulse-tubes Can be Tipped", Cryocoolers 16, California(2010), pp. 183-192.

Simplified spinal cord phantom for evaluation of SQUID magnetospinography

Y Adachi[1], D Oyama[1], N Somchai[2], S Kawabata[3], and G Uehara[1]

[1] Applied Electronics Laboratory, Kanazawa Institute of Technology, Ishikawa, Japan
[2] Osaka University Graduate School, Osaka, Japan
[3] Tokyo Medical and Dental University, Tokyo Japan

E-mail: adachi.y+ieee@gmail.com

Abstract. Spinal cord functional imaging by magnetospinography (MSG) is a noninvasive diagnostic method for spinal cord diseases. However, the accuracy and spatial resolution of lesion localization by MSG have barely been evaluated in detail so far. We developed a simplified spinal cord phantom for MSG evaluation. The spinal cord phantom is composed of a cylindrical vessel filled with saline water, which acts as a model of a neck. A set of modeled vertebrae is arranged in the cylindrical vessel, which has a neural current model made from catheter electrodes. The neural current model emulates the current distribution around the activated site along the axon of the spinal cord nerve. Our MSG system was used to observe the magnetic field from the phantom; a quadrupole-like pattern of the magnetic field distribution, which is a typical distribution pattern for spinal cord magnetic fields, was successfully reproduced by the phantom. Hence, the developed spinal cord phantom can be used to evaluate MSG source analysis methods.

1. Introduction

Magnetospinography (MSG) is a newly developed noninvasive diagnostic method for the spinal cord diseases; MSG is based on superconducting quantum interference device (SQUID) biomagnetic measurements and magnetic source analysis [1]. Weak magnetic fields produced by the neural activity of a spinal cord are detected by an array of SQUID magnetic flux sensors. The transition of the neural current along the spinal cord is visualized by applying magnetic source analysis to the acquired magnetic field data [2]. This is vital information for the diagnoses of spinal cord diseases because the neural signal propagation along a spinal cord reconstructed by magnetic source analysis shows decreases and delays at spinal cord lesions. This is impossible to do when relying only on traditional anatomical information from X-rays computed tomography (CT), or magnetic resonance imaging (MRI). MSG is the only method that provides the functional information of the spinal cord.

To evaluate the accuracy and spatial resolution of lesion localization by MSG, a well-defined spinal cord phantom is necessary, just as in the cases of magnetoencephalography (MEG) [3] or magnetocardiography (MCG) [4]. Therefore, we are currently developing a spinal cord phantom to evaluate the accuracy and spatial resolution of MSG. The phantom simulates the structure of the conductivity distribution around the spinal cord by means of a plastic vessel filled with saline water and plastic vertebrae models. The distribution of the current produced by the neural activity of the spinal cord is artificially reproduced by electrodes dipped in the saline water. The magnetic field generated by this artificial neural current is recorded by the MSG system, and magnetic source analysis is applied to the magnetic field distribution data. The accuracy of the MSG measurement system and algorithm of the magnetic source analysis can be evaluated by comparing the location of the magnetic sources estimated from the magnetic field measurement with the actual location of the magnetic sources known in advance.

In this paper, we describe a simplified spinal cord phantom comprising a neural current model using catheter electrodes and vertebral bone model in detail.

2. Simplified spinal cord phantom

The source of the magnetic field produced by the spinal cord is the local current distributed in the vicinity of the neural signals propagating along the spinal cord. This local current distribution comprises an intracellular current, which is modeled by using two current dipoles oriented in opposite directions, and an extracellular volume current that compensates for the intracellular current, as shown in figure 1(a). Each vertebral bone has a circular structure and is arranged on a line to make a tubular structure surrounding the spinal cord. The vertebral bones act as insulators, but the extracellular volume current passes through the intervertebral space and is distributed outside the spine. This extracellular volume current out of the spine also contributes to the MSG signals along with the extracellular volume current in the vertebral canal.

The spinal cord phantom has the same topological structure as the spine and its surroundings to emulate the intracellular current and extracellular volume current distributed in both the inner and outer areas of the spine.

2.1. Neural current model using catheter electrodes

The neural current model using catheter electrodes, which we previously reported [5], was applied to emulate two oppositely directed current dipoles and the extracellular volume current distribution in the spinal cord phantom. Figure 1(b) shows the structure and dimensions of the catheter electrodes. Two sets of two platinum electrodes with thin platinum wire are embedded in a Teflon tube. Two small current elements are formed when the tube is dipped in saline water and a current is run through the electrodes, as shown in figure 1(b); between electrodes A and B and between electrodes C and D. These current elements were assumed to be the current dipoles for the intracellular current. The magnetic field from the currents flowing in reverse to each other can be negligible because the currents cancel each other out. The volume current in the saline water first and foremost closes the current loop, as shown in figure 1(c).

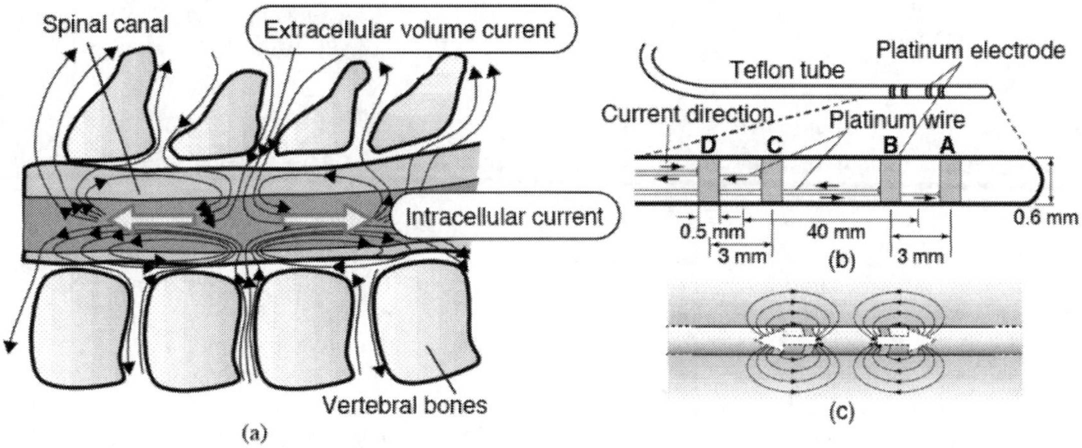

Figure 1. (a) Schematic diagram of neural current distribution accompanying spinal cord neural activity; (b) structure and dimensions of catheter electrodes for neural current model; (c) emulated current distribution by neural current model in saline water.

2.2. Vertebral bones model

For simplification, an acrylic plastic equilateral triangle prism is used to model a vertebral bone. The thickness and side length of each prism are 12 mm and 22.5 mm, respectively. One of the prism tips represents a spinous process. Each prism has a circular aperture of 20 mm in diameter at the center of the triangle. An array of 18 prisms arranged along a straight line is fixed in a cylindrical vessel positioned horizontally, as shown in figure 2. The gap between adjacent prisms is 1 mm. The center holes of each prism are aligned; this represents the spinal canal. The prism tips representing the spinous processes come close to the lower inner side of the cylindrical vessel to represent the spine of a subject lying in a supine position. There is a slit to embed the prism array on the upper side of the cylindrical vessel. The neural current model described in the previous section is inserted into the holes of the prism array in the cylindrical vessel. The position of the neural current model is shifted 5 mm below the center of the holes. When the cylindrical vessel is filled with saline water, the conductivity distribution along the spinal cord and the surrounding vertebrae bones is simplified but topologically reproduced. This is called a simplified spinal cord phantom.

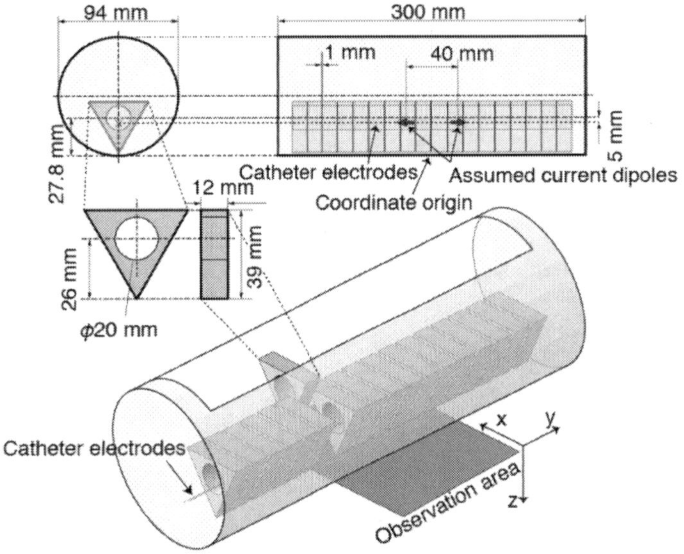

Figure 2. Structure and dimensions of spinal cord phantom with vertebral bones model.

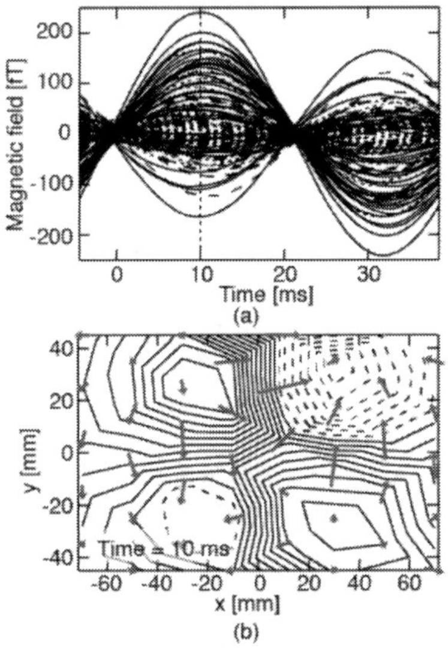

Figure 3. (a) Example of magnetic signal waveforms from phantom; (b) magnetic field distribution from phantom.

3. Evaluation of magnetic source analysis methods using the simplified spinal cord phantom

3.1. Measurement of the emulated spinal cord produced magnetic field

The distribution of the magnetic field produced from the simplified spinal cord phantom was recorded with the previously developed SQUID MSG system [6]. The SQUID MSG system was equipped with an array of 40 vector-type SQUID gradiometers covering the observation area of 90 mm × 140 mm.

The simplified spinal cord phantom was placed on the observation area of the SQUID MSG system. The relative position between the sensor array and phantom was given by localization of the marker coils attached to predetermined sites of the phantom [7]. The two current dipoles of the neural current model were activated by a 23-Hz sinusoidal electric current alternatively chopped at 1 kHz [5]. The signals from each SQUID sensor were filtered by a 100-Hz low-pass filter and digitized at 1 kHz. Figure 3 shows the acquired waveforms and magnetic field distribution at the peak latency. A quadrupole-like pattern was clearly observed along with the typical magnetic field distribution obtained by the actual MSG measurement.

3.2. Localization of equivalent current dipoles

To demonstrate the effectiveness of the phantom, equivalent current dipole (ECD) localization was applied to the magnetic field distribution shown in figure 3(b). Three methods were examined to solve the forward problem with searching for the appropriate ECDs; (a) Sarvas' equation, which assumes a horizontally layered conductor model [8]; (b) the boundary element method (BEM), which is based on Geselowitz' equations [9] assuming a cylinder conductor model composed of 20,000 triangle meshes and 10,002 vertices; and (c) the BEM based on Geselowitz' equations assuming a cylinder conductor and prism array insulator model composed of 39,610 triangle meshes and 19,807 vertices.

Table 1 summarizes the ECD localization results; displacement from the assumed ECDs, distance between two of the estimated ECDs, moment of the estimated ECDs, and goodness of fit. The assumed distance and moment were 40 mm and 48 nAm, respectively.

The results of the three methods were compared to each other. In terms of the distance between the two ECDs and their moments, method (c) provided the most reliable ECDs, although the computational cost of the boundary element method was the highest. Thus, the simplified spinal cord phantom was found to be useful for evaluating the performance of the MSG system and magnetic source analysis methods.

Table 1. Results of equivalent current dipoles localization.

		(a)	(b)	(c)	Assumed
Displacement	ECD 1	4.7 mm	4.9 mm	4.3 mm	
	ECD 2	3.1 mm	2.5 mm	2.6 mm	
	Average	3.9 mm	3.7 mm	3.5 mm	
Distance between ECDs		38.7 mm	45.0 mm	41.8 mm	40.0 mm
Moment	ECD 1	19.5 nAm	42.6 nAm	49.8 nAm	48.0 nAm
	ECD 2	16.9 nAm	27.9 nAm	32.9 nAm	48.0 nAm
	Average	18.2 nAm	35.3 nAm	41.4 nAm	48.0 nAm
Goodness of fit (GoF)*		83.5%	86.3%	86.6%	–

*GoF = 1 – Σ(calculated field – measured field)^2 / Σ(measured field)^2

4. Conclusion

We developed a simplified spinal cord phantom to evaluate the accuracy and resolution of the MSG system. The developed phantom was demonstrated to be useful for the evaluation and comparison of several magnetic source analysis methods. We will investigate the origin of the estimation errors for the ECDs in a future study.

References

[1] Curio G, Erné SN, Sandfort J, Sheer J, Stehr R, Trahms L 1991 *Electroenc. Clin. Neurophysiol.* **81** 450–453
[2] Adachi Y, Oyama D, Kawabata S, Sekihara K, Haruta Y, Uehara G 2013 *IEICE Trans. Electron.* **E96-C**, 326–333
[3] Yamamoto T, Williamson SJ, Kaufman L, Nicholson C, Llinás R 1988 *Proc. Nat. Aca. Sci. USA* **85** 8732–8736
[4] Tenner U, Haueisen J, Nowak H, Leder U, Brauer H 1999 *Phys. Med. Biol.* **44** 1969–1981
[5] Adachi Y, Oyama D, Kawabata S, Sato M, Uehara G 2011 *IEEE Trans. Mag.* **47** 3837–3840
[6] Adachi Y, Miyamoto M, Kawai J, Uehara G, Ogata H, Kawabata S, Sekihara K, Kado H 2011 *IEEE Trans. Appl. Supercond.* **21** 485–488
[7] Erné SN, Narici L, Pizzella V, Romani GL 1987 *IEEE Trans. Mag.* **MAG-23** 1319–1322
[8] J. Sarvas 1987 *Phys. Med. Biol.* **32** 11–22
[9] Geselowitz DB 1970 *IEEE Trans. Mag.* **MAG-6** 346–347

Acknowledgement

This study was partly supported by the SECOM Science and Technology Foundation.

Possibility to enhance teraherz emission from intrinsic Josephson junction by external local heating

Hidehiro Asai and Shiro Kawabata

Electronics and Photonics Research Institute, National Institute of Advanced Industrial Science and Technology (AIST), Tsukuba, Ibaraki 305-8568, Japan

E-mail: `hd-asai@aist.go.jp`

Abstract. We theoretically propose a practical method for realizing intense terahertz (THz) emission from intrinsic Josephson junctions (IJJs) using an external heat source. An arti cial inhomogeneous temperature distribution by the local heating strongly excites the Josephson plasma wave inside IJJs and enhances THz emission power. We show optimum heating conditions for achieving high power THz emission. Our result indicates that local heat control is a powerful method to realize practical solid-state THz-emitters utilizing IJJs.

1. Introduction

Recently, cuprate high-T_c superconductors have attracted much attention as promising candidates for compact solid-state terahertz (THz) sources. The layered structure in cuprate superconductor behaves as a natural stack of Josephson junctions referred to as intrinsic Josephson junctions (IJJs), and THz wave from cuprate superconductor is induced by the AC Josephson e ect in IJJs. As a mechanism that induces THz emission, the coherent ow of Josephson-vortex was proposed rst in Ref [1] in the presence of an applied magnetic eld. On the other hand, THz emission can be also realized by the excitation of electromagnetic cavity modes in the IJJs without an applied magnetic eld [2, 3, 4, 5, 6]. After an experimental realization of intense THz emission from $Bi_2Sr_2CaCu_2O_{8+\delta}$ (Bi2212) single crystal without an applied magnetic eld[7], THz emission from mesa-structured IJJ in zero magnetic eld condition has been intensively studied[7, 8, 9, 10]. However, the emission powers of these IJJ emitters are much lower than 1 mW which is required for practical applications. Thus, further investigation toward the high power emission is indispensable to realize IJJ THz emitter for practical use.

Recently, a *hot spot* in the mesa where the temperature is locally higher than the superconducting critical temperature T_c has been observed during THz emission processes[9, 10]. Hence, the temperature inhomogeneity such as the hot spot is considered to play a crucial role in the strong THz emission. Since the critical current density j_c depend on the temperature, the appearance of hot spots indicates the inhomogeneous j_c distribution. Previous theoretical studies show such an inhomogeneous j_c distribution strongly excites the Josephson plasma wave inside IJJs[3, 6]. Therefore, the arti cial control of the local temperature by some external heat source is expected to be a promising method to realize intense THz emission.

In this study, the THz emission from an IJJ mesa which is heated locally by an external heat source, e.g. laser irradiation, is theoretically investigated in the absence of an applied magnetic eld. We investigate the THz emission by solving the sine-Gordon, Maxwell, and thermal di usion equation simultaneously and found that the emission power is largely enhanced by the local heating. Based on the above analyses, we will clarify the optimum heating conditions for intense THz emission.

2. Calculation method

In Fig. 1, we show a 2D model of the IJJ emitter which is locally heated by laser irradiation. The IJJ mesa is fabricated on the IJJ base crystal, and the mesa is covered by the upper electrode. The upper electrode is locally heated by the focused laser beam, and the temperature of the IJJ mesa beneath the heating spot locally increases. In this calculation, we use the following dimensions : width of the electrode and the IJJ mesa $w_m = 60$ m, width of the IJJ base crystal $w_b = 220$ m, height of the electrode $h_e = 1$ m, height of the IJJ mesa $h_m = 1$ m and height of the IJJ base crystal $h_b = 20$ m[7, 8, 9, 10]. The DC voltage is applied to the IJJ mesa region, and the base crystal is attached to the in nite ground plane. In this study, we rstly solve the thermal di usion equation with Joule heating in the IJJ mesa. After that, we solve the sin-Gordon equation based on the obtained temperature distribution.

The thermal di usion equation in IJJs is given by,

$$0 = \frac{\partial}{\partial x}\left[\kappa_{ab}(T)\frac{\partial T}{\partial x}\right] + \frac{\partial}{\partial z}\left[\kappa_c(T)\frac{\partial T}{\partial z}\right] + \frac{j_{ex}^2}{\sigma_c(T)}. \quad (1)$$

Here, T is the temperature, σ_c is the c-axis conductivity in the IJJ mesa, and κ_{ab}, κ_c is the thermal conductivity along the ab plane and c axis respectively. j_{ex} is the external current density injected to the IJJ mesa region. We assume the local heating by the external energy source and add the constant heat source Q_L in the electrode region which is indicated by the red shaded area in Fig. 1. The position of the heating spot is given by x_L, and we assume that the spot size is 5 m. In this study, we take $x_L = 13.3$ m. The dependence of emission power on the x_L will be discussed in the future work. We use the temperature dependent κ_{ab}, κ_c and σ_c adopted in a previous theoretical study[11]. In the thermal di usion equation for upper electrode, we use an isotropic di usion constant and take $\kappa = 20$ W/m $\cdot$ K. We impose the boundary condition $T = T_{bath}$ at the boundary between IJJ base crystal and the ground plane. In this study, we take $T_{bath} = 0.3T_c$. The open boundary condition $\nabla T = 0$ is used for other boundaries.

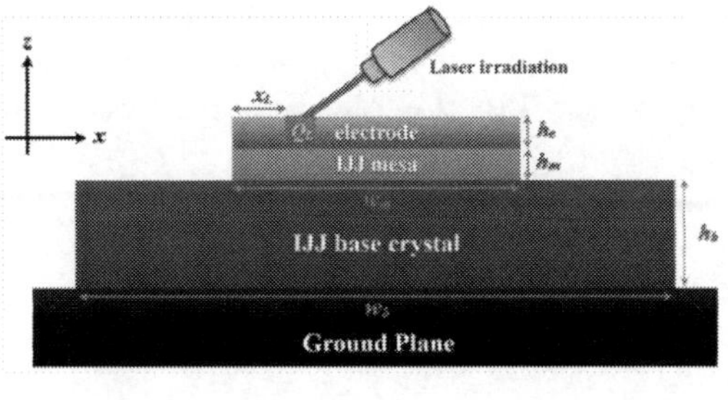

Figure 1. A two dimensional model of IJJ mesa which is locally heated by laser irradiation. The IJJ mesa is fabricated on the IJJ base crystal. The mesa is covered by a metallic electrode. The laser is irradiated on the upper electrode and locally increases the temperature of the mesa beneath the electrode. x_L indicates the position of the heating spot. A DC voltage is applied to the IJJ mesa region. $w_m = 60$ m and $w_b = 220$ m are width of the IJJ mesa and the IJJ base crystal. $h_e = 1$ m, $h_m = 1$ m and $h_b = 20$ m are height of the electrode, the IJJ mesa and the IJJ base crystal. We take bath temperature $T_{bath} = 0.3\ T_c$

The dynamics of the phase di erences in an IJJ mesa are described by the sine-Gordon equation within the in-phase approximation where all phase di erences between IJJ layers are equal to the common phase di erence [4, 6],

$$ {}_c\frac{\partial^2}{\partial t^2} = c^2\frac{\partial B_y}{\partial x} \quad \frac{1}{{}_0}\left[j_c(T)\sin \quad + \sigma_c(T)E_z \quad j_{ex}\right], \tag{2} $$

where, $j_c(T)$ is the critical current density, ${}_c$ is the dielectric constant of the junctions, d is the thickness of insulating layers of IJJs, c is the light velocity and Φ_0 is the ux quantum. The electromagnetic (EM) elds in the IJJs are given by $E_z(x,t) = \frac{\hbar}{2ed}\frac{\partial \phi}{\partial t}$, $B_y(x,t) = \frac{\hbar}{2ed}\frac{\partial \phi}{\partial x}$. We use the Ambegaokar-Barato relation for temperature dependence of the j_c[12]. In this study, we take ${}_c = 17.64$, $d = 1.2$ nm and $j_c(0) = 4 \cdot 10^2$ A/cm^2[8, 13].

3. Results and discussions

We calculate the THz emission power by changing Q_L for clarifying optimum "heating power" to achieve high power emission. Figure 2 (a) shows the current I and the emission power P as a function of the voltage V for $Q_L = 0.3$ W/cm. The voltage here is divided by the number of IJJ layer in order to show the voltage applied to each IJJ layer. At $V = 1.23$ mV, the I vs. V curves show a small kink, and the P vs. V curve show a sharp peak. Figure 2(b) shows the frequency spectrum of EM wave emitted by the mesa. The peak frequency 0.59THz is equal to the AC Josephson frequency $f_J = 2eV/h$ and the cavity resonance frequency given by $f_c = cn/(2\sqrt{{}_c}w_m)$, where $n = 1$. Moreover, we can also see a small peak around 1.18THz which corresponds to the higher harmonics. In this manner, strong THz wave is emitted from the mesa when the cavity resonant EM mode appears in the mesa.

In Fig 3(a), we shows the emission power as a function of Q_L to discuss the optimum heating power for high power emission. This gure shows a peak around $Q_L = 0.3$ W/cm. The spatial distribution of the temperature T in the mesa during the intense emission are shown in Fig. 3(b). The hot spot where T is locally higher than T_c appears for $Q_L \leq 0.2$ W/cm. Remarkably the emission power P is largely enhanced by external heat source Q_L in comparison to the case without the external local heating as shown in Fig 3 (a).

Our results indicate external local heating is powerful method to achieve high power THz emission. Of particular note is that the strongest emission is obtained around $Q_L = 0.3$ W/cm. In this case, the temperature of the heating spot is slightly lower than T_c, and thus the hot spot ($T > T_c$) is not formed. The change of j_c becomes remarkable when T is slightly below T_c as expected from the temperature dependence of j_c[12]. Therefore the drastic j_c modulation by the local heating strongly excites the THz Josephson plasma wave inside the IJJ mesa. On the

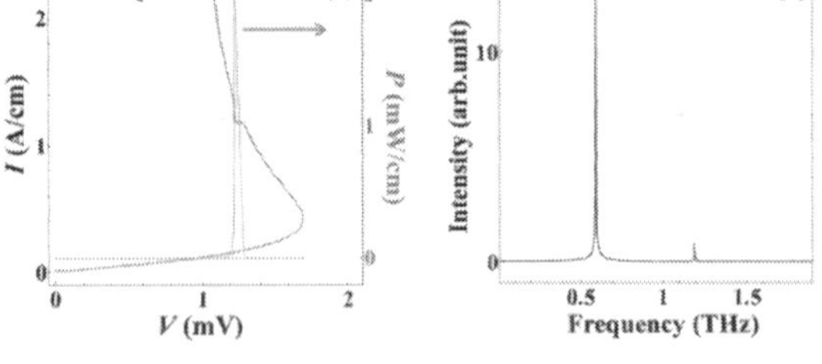

Figure 2. Calculation results for for $Q_L = 0.3$ W/cm and $x_L = 13.3$ m. (a) The current I and the emission power P as a function of the voltage V. (b) The frequency spectrum of the emission from the mesa.

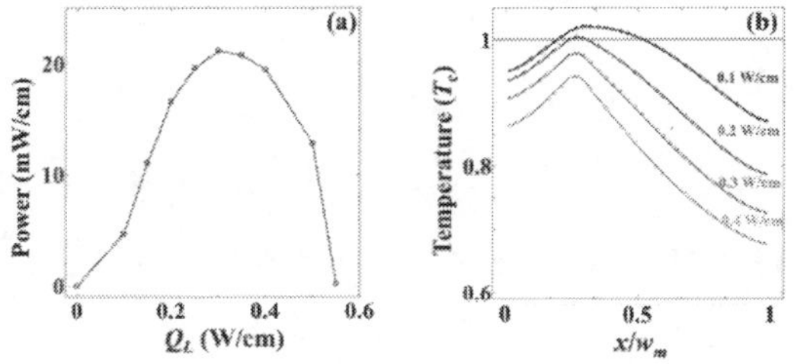

Figure 3. (a) The emission power as a function of Q_L for $x_L = 13.3$ m. (b)The distribution of T in the mesa for $Q_L = 0.1, 0.2, 0.3, 0.4$ W/cm.

other hand, for $Q_L < 0.3$ W/cm, a hot spot region is formed during the emission. Since a hot spot region does not contribute to THz emission because the $j_c = 0$ in this region, the formation of hot spot results in the reduction of the emission power. Conversely, for $Q_L > 0.3$ W/cm, the mesa temperature becomes much lower than T_c. In this case the modulation of j_c becomes small because the j_c has little temperature dependence unless T is close to T_c. Consequently, the excitation of the THz Josephson plasma wave becomes weak. Thus, we can conclude that *the local heating which keeps the mesa temperature "slightly lower" than T_c is preferable for high power emission.* Moreover, we would like to emphasize that our result is consistent with the recent experimental study which reported the strongest emission is realized when T is slightly lower than T_c[10]. In addition, we have compared the temperature distributions in our calculation with those in the experiments, and we found that our results are qualitatively similar to those reported in Ref. [10].

4. Conclusion

We have investigated THz emission from the IJJ mesa which is locally heated by an external heat source. We clari ed the optimum heating condition in order to achieve high power THz emission. The key point to design an intense THz emitter is a control of the heating power to keep mesa temperature "slightly lower" than T_c. We believe that the precise control of local temperature based on our guidelines will be a one of pragmatic way to realize the practical THz emitter using intrinsic Josephson junctions.

We wish to thank K. Kadowaki, I. Kakeya, T. Kashiwagi, F. Nori, H. Minami Y. Ota, M. Tachiki, M. Tsujimoto and H. B. Wang for fruitful discussions and comments.

[1] T. Koyama and M. Tachiki, Solid State Commun. **96**, 367 (1995).
[2] S. Lin and X. Hu, Phys. Rev. Lett. **100**, 247006 (2008).
[3] A. E. Koshelev and L. N. Bulaevskii, Phys. Rev. B **77**, 014530 (2008).
[4] T. Koyama, H. Matsumoto, M. Machida, and K. Kadowaki, Phys. Rev. B **79**, 104522 (2009).
[5] S. Savel'ev, V. A. Yampol'skii, A. L. Rakhmanov and F. Nori, Rep. Prog. Phys. **73**, (2010) 026501.
[6] H. Asai, M. Tachiki, and K. Kadowaki, Phys. Rev. B **85**, 064521 (2012).
[7] L. Ozyuzer *et al.*, Science **318**, 1291 (2007).
[8] K. Kadowaki, M. Tsujimoto, K. Yamaki, T. Yamamoto, T. Kashiwagi, H. Minami, M. Tachiki, and R. Klemm, J. Phys. Soc. Jpn. **79**, 023703 (2010).
[9] H. B. Wang, *et al.*, Phys. Rev. Lett. **105**, 057002 (2010).
[10] T. M. Benseman, A. E. Koshelev, W.-K. Kwok, U. Welp, V. K. Vlasko-Vlasov, K. Kadowaki, H. Minami, and C. Watanabe, J. Appl. Phys. **113**, 133902 (2013).
[11] B. Gross *et al.*, Phys. Rev. B **86**, 094524(2012) .
[12] S. Kashiwaya and Y. Tanaka, Rep. Prog. Phys. **63**, 1641 (2000).
[13] K. Inomata, T. Kawae, K. Nakajima, S. J. Kim and T. Yamashita, Appl. Phys. Lett. **82**, 769 (2003).

11th European Conference on Applied Superconductivity (EUCAS2013) IOP Publishing
Journal of Physics: Conference Series **507** (2014) 042003 doi:10.1088/1742-6596/507/4/042003

Cryogenic time-domain multiplexer based on SQUID arrays and superconducting/normal conducting switches

N Beev [1], M Kiviranta [1], J van der Kuur [2], M Bruijn [2], O Brandel [3], S Linzen [3], L Fritzsch [3], J Ahoranta [4], J Penttilä [5] and L Roschier [5]

[1] VTT Technical Research Centre of Finland. Tietotie 3, 02150 Espoo, Finland

[2] SRON, Netherlands

[3] IPHT Jena, Germany

[4] University of Helsinki, Finland

[5] Aivon OY, Finland

E-mail: nikolai.beev@vtt.fi

Abstract. We have demonstrated the operation of a 12-channel Beyer-style SQUID-based time domain multiplexer. It was manufactured using a fabrication process that is cross-compatible between VTT and IPHT-Jena. The multiplexer consists of twelve 12-SQUID series arrays, each shunted by a Zappe-style interferometer array acting as a flux-controlled superconducting/normal conducting switch. By keeping all switches but one in the superconducting state, it is possible to select one active readout channel at a time. A flux feedback coil common to all SQUID arrays allows realization of a flux-locked loop. We present characteristics of the multiplexer and measurement data from experiments with a 25-pixel X-ray calorimeter array operated at $T < 100$ mK in a dilution refrigerator.

1. Introduction

Transition-edge sensors (TES) designed for mK temperatures offer state of the art energy resolution in single-photon spectroscopy applications from soft X-rays to γ-rays. Under voltage bias condition, the TES operates with strong negative electro-thermal feedback, converting the energy of an absorbed incident photon to a decrease in Joule heating measured as an output current pulse.

Multiplexing is the only practical way to read out large arrays of TES pixels. TES calorimeter multiplexing has been demonstrated in the time [1], frequency [2] and code [3] domains (*TDM*, *FDM* and *CDM*). TDM is so far the most common and mature approach, however it suffers from a noise penalty [4] which limits the number of channels that could be multiplexed without degradation in energy resolution. FDM and CDM do not have the noise penalty but are more difficult to implement.

Thanks to their low noise, low power dissipation and versatility, SQUIDs have been widely used both as amplifying and modulating elements in the readout of TES detectors [1], [2], [3], [4]. The output signals of TES calorimeters optimized for X-rays are particularly difficult to multiplex due to their fast characteristic times, requiring a fast multiplexer and readout channel with high Shannon information capacity.

2. SQUID multiplexer

The SQUID multiplexer chip was designed at VTT and fabricated at IPHT-Jena using a process that is cross-compatible between the two institutes. It consists of 12 SQUID arrays each having 12 individual SQUID cells connected in series. The arrays have separate input coils with mutual inductance $M^{-1} = 10$ $\mu A / \Phi_0$. In addition, there is a flux feedback coil common to all arrays. The chip has a compact 3 x 3 mm form factor.

In a topology resembling Beyer's [5], each array is shunted by a flux-controlled on-off normal conducting/superconducting switch. However, in our design we have used 48-series arrays of Zappe-style three-junction interferometers [6] to realize the switches. These devices were initially proposed as building blocks for fast digital logic; in our work we take advantage of their square-shaped flux-voltage response, which alleviates the effect of noise in the address lines.

3. Experimental setup

Experiments were carried out in a Bluefors BF-LD400 cryogen-free dilution refrigerator. A schematic diagram of the setup is shown on Figure 1.

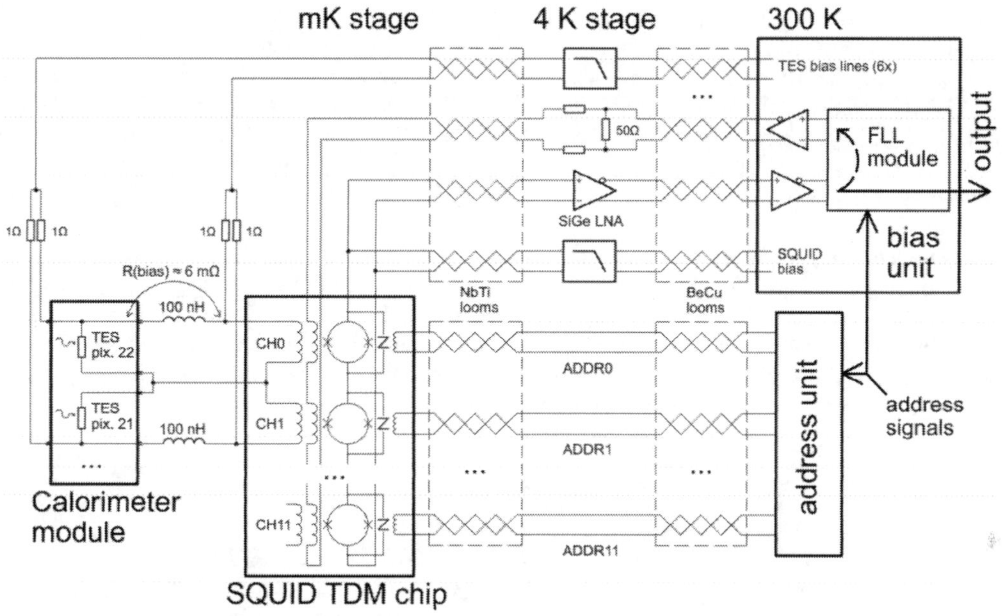

Figure 1. Schematic diagram of the experimental setup.

A. B.

Figure 2. Photographs of the **A.** SQUID multiplexer chip **B.** mK stage experimental box.

3.1. mK stage
A copper box containing the calorimeter module, anti-alias filters, TES bias-splitting resistors and the SQUID multiplexer (Figure 2 A) was attached to the mixing chamber plate (Figure 2 B). The 25-pixel X-ray calorimeter module from SRON [7] was mounted on a circuit board equipped with a 1.27 mm pitch double-row connector. Commercial 100 nH SMD inductors were used as Nyquist filters and bias resistors simultaneously. The series connection of one such inductor and two mated connector contacts provided the bias resistance of 6 mΩ, while all PCB traces and bondwires in the TES-SQUID input loop were superconducting. A ^{55}Fe radioactive source was used to irradiate the sensor array through a hole drilled in the copper box lid.

3.2. 4 K stage
The electronics at the quasi-4 K plate were contained within a steel shielding box. Apart from various passive filters and attenuators, this box housed the cryogenic low-noise SiGe transistor amplifier [8] used to read out the SQUID multiplexer output voltage.

3.3. Room-temperature electronics
The room-temperature electronics system was divided into two functional blocks: bias unit and address unit. The bias unit contained drivers for various bias lines, as well as the differential receiver for the cold SiGe amplifier and the flux-locked loop (FLL) module. The address unit contained a 16-channel CMOS multiplexer which translated binary codes into boxcar current signals driving the address lines. Binary address signals were generated by a CPLD and fed simultaneously to the address unit and the FLL module. The output signals were digitized by a commercial 14-bit acquisition unit with Ethernet interface [9].

3.4. Analog multiplexed flux-locked loop
In order to accommodate full X-ray pulses within the SQUID dynamic range and linearize the response of the arrays, a FLL was implemented. Given the high signal bandwidth of the pulses and the multiplexing factor, one needs sufficiently fast loop dynamics. Typically, in TDM systems the FLL has been implemented digitally using FPGA logic with memory for each channel [10]. In the present work we have developed a different approach, which is fully analog.

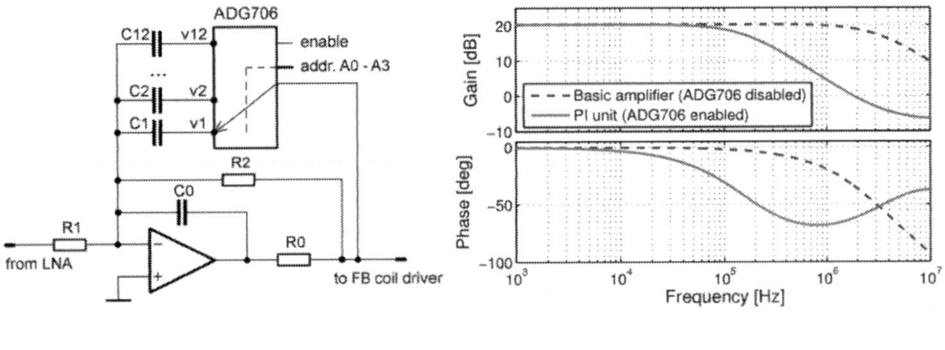

Figure 3. Multiplexed FLL unit: **A.** circuit diagram **B.** gain and phase vs frequency.

The circuit shown on Figure 3 A. acts as a proportional-integral (PI) unit (Figure 3 B.) within the FLL. At low frequencies the gain is constant and is set by the ratio R2/R1. At high frequencies, the C1 – C12 capacitor bank implements the integrating function and analog memory. The left-hand side terminals of the capacitors are all kept at the 0 V virtual ground, while the right-hand side is either within the feedback path (channel on) or effectively floating (channel off). This configuration resembles a multi-channel track-and-hold circuit. A fast operational amplifier is used to guarantee rapid switching between channels.

4. Results

Figure 5 A shows a typical X-ray pulse obtained in quasi-static (non-multiplexed) mode. The pulse has a very fast rising edge (Figure 5 B) due to the faster-than-optimal anti-alias filters. An energy resolution of ≈ 20 eV was measured at 5.9 keV. The raw output sampled at 2 MHz is shown on Figure 5 C. From the pixel dwell time of 2 µs, 4 samples (1 µs) on the transient were discarded and the remaining 4 within the settled period were averaged to produce one sample in the demultiplexed signal (decimation factor of 8). Figure 5 D shows demultiplexed X-ray pulses from 8 channels. Pulse acquisition was triggered from 6 of the channels, while events from the other two were captured randomly. All events are clearly resolved into the respective channels and no cross-talk is evident.

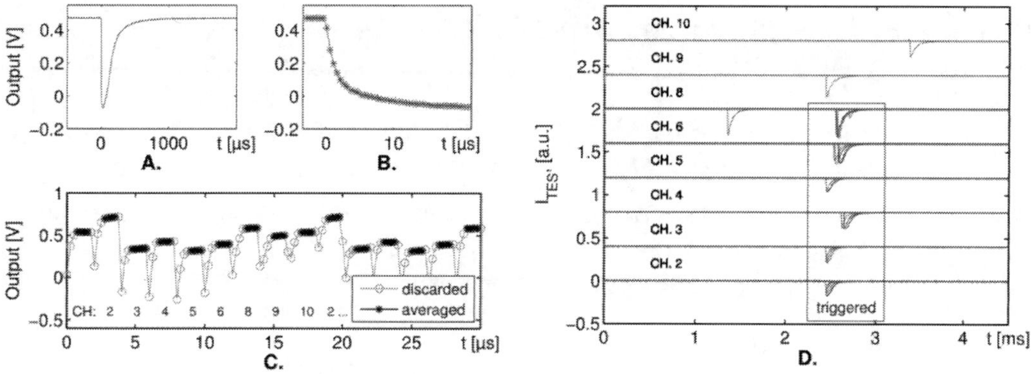

Figure 5. A. Typical X-ray pulse **B.** zoom-in of the falling edge of the pulse from A. **C.** raw multiplexed output **D.** demultiplexed pulses from 8 channels.

5. Conclusion

We have designed and fabricated a compact 12-channel SQUID multiplexer that is useful for quasi-static measurements and characterization. Dynamic operation in TDM is possible, but with added noise. The practical use of the component was demonstrated by multiplexing the signals of 8 TES X-ray calorimeters. Unlike more standard setups [4] that have two- or three-SQUID amplifier cascades, our system uses one SQUID stage and a cryogenic SiGe amplifier. An analog multiplexed FLL was successfully implemented and performed in a stable manner, despite the fast pulse rising edges. The ultimate energy resolution capability of the sensor array was not reached and we are working on improving the EMI and thermal performance of our experimental setup. As a parallel development, we are considering the use of three-junction Zappe interferometers as flux-controlled inductances for current-steering CDM [11].

Acknowledgment

This work has received funding from Grant No. 262947 of the European Community's seventh framework programme (FP7/2007–2013).

References

[1] Kilbourne C et al. *Proc. SPIE* **7011**, 701104 (2008)
[2] Ullom J et al. *IEEE Trans Appl Supercond* **13**, No. 2 (2003)
[3] Stiehl G et al. *Appl Phys Lett* **100**, 072601 (2012)
[4] Irwin K. *Physica C* **368**, 203–210 (2002)
[5] Beyer J and Drung D. *Supercond Sci Technol* **21**, 105022 (2008)
[6] Zappe H. *Magn* **MAG-13**, No. 1, 41–47 (1977)
[7] Gottardi L et al. *J Low Temp Phys* **151**, 106–111 (2008)
[8] Beev N and Kiviranta M. *Cryogenics* **57**, 129-133 (2013)
[9] Model CS328A-XSE, www.cleverscope.com
[10] Reintsema C et al. *Rev Sci Instrum* **74**, 4500 (2003)
[11] Kiviranta M and Beev N. *IEEE Trans Appl Supercond* **23**, No. 3 (2013)

Large Area Superconducting TES Spiderweb Bolometer for Multi-mode Cavity Microwave Detect

M Biasotti[1,2], D Bagliani[1,2], D Corsini[1], P De Bernardis[3], F Gatti[1,2], R Gualtieri[3], L Lamagna[3], S Masi[3], G Pizzigoni[1,2], and A Schillaci[3]

[1] University of Genoa, Dept. of Physics, Via Dodecaneso 33, 16147, Genova, Italy
[2] INFN of Genoa, Via Dodecaneso 33, 16147, Genova, Italy.
[3] University of Rome "La Sapienza" Roma, Italy

E-mail: michele.biasotti@ge.infn.it

Abstract. For the cosmic microwave background, the increase of the sensitivity of present superconducting TES Spiderweb Bolometers can be done coupling them to a large set of modes of the EM radiation inside the cavity. This will require a proper shaping of the horn-cavity assembly for the focal plane of the microwave telescope and the use of large area bolometers. Large area spiderweb bolometers of 8 mm diameter and a mesh size of 250 μm are fabricated in order to couple with approximately the first 20 modes of the cavity at about 140 GHz. These bolometers are fabricated with micro machining techniques from silicon wafer covered with SiO_2 - Si_3N_4 CVD thick films, 0.3 μm and 1 μm respectively. The sensor is a Ti/Au/Ti 3 layer TES sensor with Tc tuned in the 330-380 mK and 2 mK transition width. The TES is electronically coupled to the EM gold absorber that is grown on to the spiderweb mesh in order to sense the temperature of the electron gas heated by the EM radiation. The gold absorber mesh has 5 um beam size over a Si_3N_4 10 μm beam size supporting mesh. The Si_3N_4 mesh is then fully suspended by means of DRIE back etching of the Si substrate. Here we present the first results of these large area bolometers.

1. Introduction

Measuring the large scale polarization of the Cosmic Microwave Background (CMB) is one of the major challenges of modern observational cosmology. The inflationary scenario, proposed within the framework of the standard cosmological model, implies the existence of primordial gravitational waves which generate a tiny curl component ("B-mode") in the pattern of CMB polarization at large angular scales [1]. The amplitude of this signal, if any, is expected to be dependent on the energy scales of the inflationary phase transition, and on the nature of the small density perturbations in the primordial universe. Therefore, a direct measurement of large scale B-modes of the CMB would provide a strong confirmation of the inflationary theories [2]. This signal has been, so far, very elusive. Being at most 1% of the main CMB polarized component, its detection requires high sensitivity, state-of-the-art system stability, exquisite control and characterization of the instrumental systematics, and careful monitoring of polarized foregrounds through multifrequency observations across the mm-band. Modern bolometric balloon-borne polarimeters, like EBEX [3] and SPIDER [4], are designed to operate thousands of independent single-mode detectors to beat the single-detector photon noise limit and achieve sub-μK effective sensitivity. An alternative approach is to meet the sensitivity requirements by designing a multi-mode system. In principle, when the detectors efficiently coupled to a number N_{eff} of EM modes at a given frequency, the photon noise limit scales down as $1/\sqrt{N_{eff}}$. Such a detector suffers, of course, of degradation in angular resolution respect to a single-mode bolometer operating at the same wavelength from the same telescope. Anyway, this is an acceptable tradeoff for B-mode searches, since most of the signal is distributed at large angular scales (~10 deg) across the sky. Such a solution is adopted in the PIXIE spectro-polarimeter [5] and is currently planned as baseline field optics design for the SWIPE polarimeter [6], the high frequency (95-145-245 GHz) instrument aboard of the Large Scale Polarization Explorer balloon-borne experiment. The use of multi-mode field optics is not a novelty in mm astronomy; anyway, the planned implementation into a system with strict requirements on calibration accuracy and subsystem characterization has motivated the development of a modeling-prototyping-validation pipeline which complies with the standards of a CMB polarization experiment for precision cosmology.

In this work we present the early results on the development of a large area superconducting spiderweb bolometer. Spiderweb superconductive bolometer has been already presented in the past [8-9]. Anyway one of largest micromesh absorber was presented by Mauskopf et al. who shows an absorber with 5.6 mm in diameter, and use a neutron transmuted doped germanium thermistor [10]. Our device achieves 8 mm of diameter for a better multi-mode coupling in the cavity, and uses a Ti/Au TES.

In section 2 we describe the detector design and the fabrication process, in section 3 the preliminary low temperature measurements.

2. Detector design and fabrication process

The detector is designed to work at about 350 mK, with a thermal conductance of about 10^{-10} W/K that should give a NEP of 10^{-17} W/Hz$^{1/2}$, and couples with the first 20 modes at 140 GHz. The spiderweb design is required for minimizing the cross section to the cosmic rays and the heat capacity of the absorber. For these reasons we have chosen to produce a freestanding 1 μm thick silicon nitride spiderweb membrane with a mesh size of 250 μm onto which the gold absorber and the titanium/gold TES are grown. The silicon nitride spider-web with the gold circular absorber (8 mm in diameter) has been modeled together with the cylindrical resonant cavity where it is suspended. The cavity and detector are assembled within the feed-horn antenna and the EM coupling was studied using High Frequency Structure Simulator (HFSS). The simulation study was presented by Lamagna et al. to the International Conference on electromagnetics in advanced application [11]. This is a typical assembly for single-pixel microwave bolometers, and its single-mode performance has been already extensively modeled and optimized with respect to different technical needs [12].

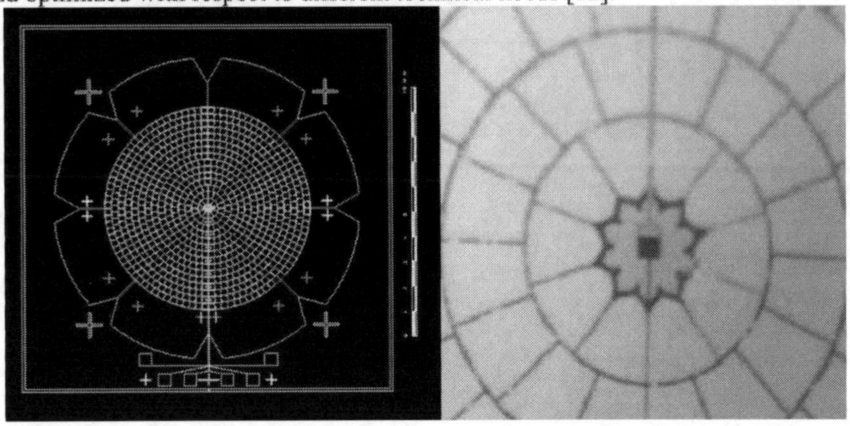

Figure 1. The full set of drawings of all photolithographic masks shown with different colours and a centre of a finished device.

In figure 1 the full set of drawings of all photolithographic masks are reported to give the overall view of the detector. The absorber has a diameter of 8 mm and 8 legs, 1 mm long, that connect the spiderweb absorber to the substrate frame.

The fabrication process make use of a commercially available silicon wafer 380 μm thick, with both sides covered by two layers of thermal silicon oxide, 300 nm thick, and CVD silicon nitride, 1 μm thick. The wafer is diced to square 15x15 mm chips and then rinsed in soap solution, deionized water and acetone.

The fabrication process is shown in figure 2: the first step is the TES film growth and patterning. This is obtained by lift-off using negative photolithography. The three-layer Ti/Au/Ti TES deposition is obtained by e-beam evaporation without breaking the vacuum. The base pressure is typically less than 10^{-6} Pa and the thickness are: 5 nm of Ti as sticking layer, 4 nm of Au and 38 nm of Ti (fig. 2a). The second step is the TES wires fabrication (fig. 2b) with e-beam evaporation of Ti/Al/Ti three-layers and lift-off lithography. Before the wire film deposition, the TES is sputtered with Ar ion beam to remove titanium oxide from its surface. The spiderweb gold absorber made by 65 nm thick Au patterned film, is grown on top of 5 nm of Ti sticking layer (fig. 2c).

A positive photolithography is then used to cut the silicon nitride and silicon oxide supporting layers. The parts uncovered by the photoresist are etched by Reactive Ion Etching (RIE) with a CF_4 + O_2 mixture for silicon nitride and CF_4 for silicon oxide (fig. 2d). The final step in which the spiderweb structure is suspended with deep RIE of Si chip from the back side. In this case an aluminum hard mask is deposited on the chip back side (fig. 2e). The deep RIE process is done with an SF_6 + O_2 mixture. The endpoint detection is performed with plasma light spectroscopy (fig. 2f). The final result is shown in figure 3 and in figure 1.

In addition to low temperature detector, a room temperature bolometer was produced in order to perform fast EM tests into the cavity. In this test version the Ti/Au multilayer of TES and the aluminum films of the wiring were substituted with platinum ones. The early results showing that we have successfully coupled the bolometer to several modes of the EM field in the cavity was reported by Lamagna together with the calculation [11].

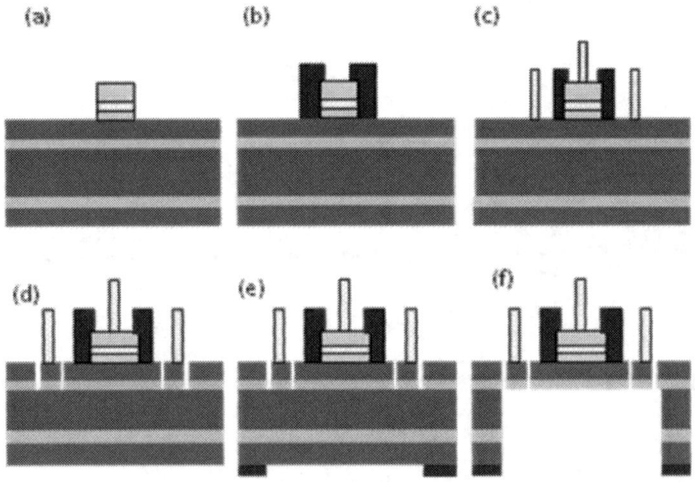

Figure 2. Scheme of spiderweb detector fabrication process. (a) TES growth; (b) wiring growth; (c) absorber growth; (d) spiderweb etch; (e) hard mask growth; (e) deep back etching.

Figure 3. Final spider web bolometer compared with a 1 Euro coin.

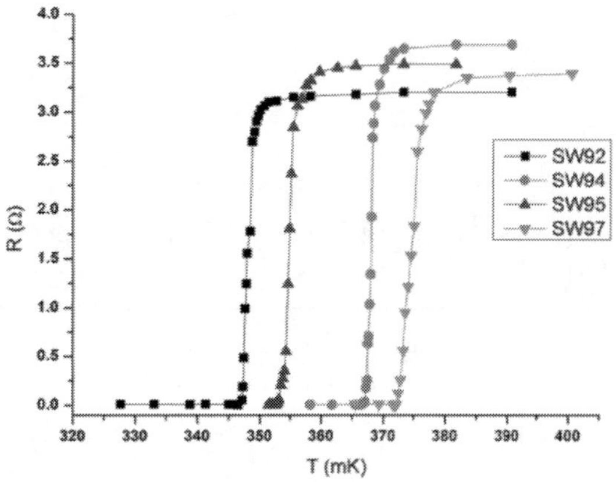

Figure 4. Suoerconducting to normal transitions of 4 devices reported as example.

3. Critical temperature measurements

The critical temperature of several devices were measured using a $^3He/^4He$ dilution cryostat. In figure 4 are reported the transition of few samples. All critical temperatures are grouped between 345 and 375 mK. All transitions are very narrow with a width of 2 mK or less. In tab.1 are reported the effective thicknesses of the TES layers. The correlation among thickness and critical temperature, due to the proximity effect, seems to be driven mainly by the topmost and thicker titanium layer and weakly by the gold thickness. Further measurements of TES parameters are on the way.

Table 1 Effective TES layers thickness of the devices considered in fig 4, the layers are reported as the deposition order: the last row is related to the topmost layer.

	SW92	SW94	SW95	SW97
Ti	5.0	5.0	5.0	5.0
Au	4.4	4.4	3.8	3.8
Ti	37.2	38.0	37.2	38.0

References

[1] Seljak U, and Zaldarriaga M 1997 *Phys. Rev. Lett.* **78**, 2054
[2] Boyle L A, Steinhardt P J and Turok N 2006 *Phys. Rev. Lett.* **96** 111301
[3] Reichborn-Kjennerud B, Aboobaker A M, and Ade P. 2010 *Proc. SPIE* **7741** 77411C
[4] Amiri M Burger B, Halpern M, and Hasselfield M 2010 *Proc. SPIE* **7741**, 77411N
[5] Kogut A et al. 2011 *Proc. SPIE* **8146** 81460T
[6] P. de Bernardis et al. 2012 *Proc. SPIE* **8452** 84523F
[7] The LSPE Collaboration 2012 *Proc. SPIE* **8446** 84467A
[8] Lee S, Gildemeister, J M, Holmes W, Lee, A T; and Richards P L 1998 *Appl. Opt.* **37** 3391
[9] Gildemeister J M, Lee A T, and Richards P. L. 1999 *Appl. Phys. Lett.* **74** 868
[10] Mauskopf P D, Bock J J, Del Castillo H, Holzapfel W L, Lange A E 1997 *Appl. Opt.* **36** 765
[11] Lamagna et al. 2013 *Proc. ICEAA* 6632435
[12] Glenn J, Chattopadhyay G, Edgington S F; Lange A E; Bock J J; Mauskopf P D, Lee A T 2002 *Appl. Opt.* **41** 136

Journal of Physics: Conference Series **507** (2014) 042005 doi:10.1088/1742-6596/507/4/042005

mm-wave harmonic generation in an array of SIS junctions

Bhushan Billade[1], Alexey Pavolotsky[2] and Victor Belitsky[2]

[1]Onsala space observatory.
[2]Group for Advanced Receiver Development.
[1,2]Chalmers University of Technology, Gothenburg, Sweden.

E-mail: [1]Bhushan.Billade@Chalmers.se

Abstract. We report the first experimental off-chip detection of frequency multiplication in a distributed array of Superconductor-Insulator-Superconductor (SIS) junctions. A test device consisting of series array of 68 Nb/Al-AlOx/Nb tunnel junctions was designed to study generation of the second harmonic. We measured extremely sharp spectral signals, associated with the $\times 2$ frequency multiplication. Distinct single and multi-photon processes were observed in the test device response operated in quantum mode. The mechanism of device saturation was experimentally studied. The test device when connected to the input of an SIS mixer, and pumped, showed $10 - 20\%$ increase in the SIS junction dark current.

1. Introduction

The ability of a Superconductor-Insulator-Superconductor (SIS) tunnel junction to respond to a single absorbed photon and extremely sharp nonlinearity in their current-voltage (I-V) characteristics, due to quasi-particle tunneling, has been utilized in ultra-low noise millimeter wave receivers [1]. The possibility of using this non-linear property of an SIS junction for frequency multiplication was suggested earlier[2], but there has not been any experimental effort to investigate this possibility further.

In this paper we present the first experimental demonstration of frequency multiplication using SIS junction-array operating in quasi-particle branch. We recognize that, in principle, the SIS junction as a frequency multiplier could operate in different modes. One can think of a purely classical scenario, when both input and output frequencies are less than the voltage nonlinearity ($\hbar\omega/e < \delta V_{gap}$). A classical-quantum mode, where the input pumping frequency does not produce quantum response ($\hbar\omega/e < \delta V_{gap}$) whereas output frequency is in the quantum regime. For a quantum-classical mode, one can think of a scenario, where the output frequency exceeds twice the gap frequency ($\hbar\omega/e > 2V_{gap}$). In this paper we investigate a quantum-quantum mode, where both the pumping and the output frequencies produce quantum response ($\hbar\omega/e > \delta V_{gap}$).

2. Design of test device

The quantum theory of mixing[3] describes in great details the mechanism of tunneling current at different harmonics. The tunneling current through the SIS junction as a response to the applied local oscillator signal (ω) is represented as [3], $I_{LO}(t) = a_0 + \sum_{m=1}^{\infty}(2a_m \cos m\omega t + 2b_m \sin m\omega t)$ where, the coefficients a_m and b_m are given by,

$$2a_m = \sum_{n=-\infty}^{\infty} J_n(\alpha)\left[J_{n+m}(\alpha) + J_{n-m}(\alpha)\right] \times I_{dc}\left(V_0 + n\hbar\omega/e\right) \quad (1)$$

$$2b_m = \sum_{n=-\infty}^{\infty} J_n(\alpha)\left[J_{n+m}(\alpha) - J_{n-m}(\alpha)\right] \times I_{KK}\left(V_0 + n\hbar\omega/e\right) \quad (2)$$

where, J is the Bessel function, a_0 is the DC component of the current, and I_{KK} Kramers-Kronig transform of the DC current-voltage characteristics of the tunnel junction.

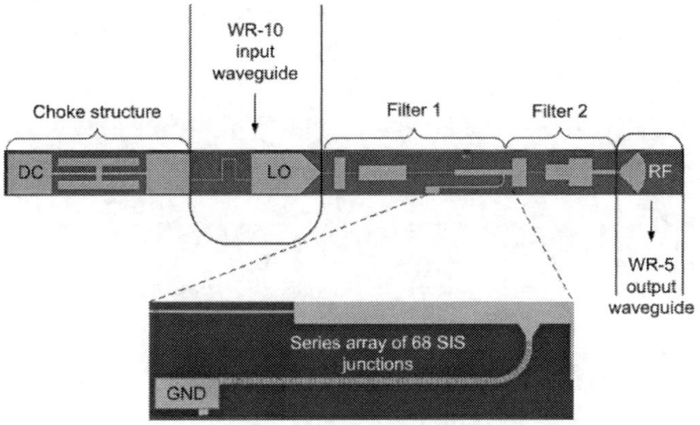

Figure 1. Photograph of the test chip showing the series array of 68 SIS junctions, two filter sections, RF and LO probes, position of the input and the output waveguides, and DC bias and ground (GND) ports. Each junction is 25 μm^2 in area with the overall chip dimensions of $415\,\mu m \times 6700\,\mu m$.

An array of SIS junction was used to boosting the total power in the multiplied signal. Figure 1 shows a test device consists of 68 identical $Nb/Al\text{-}AlO_x/Nb$ junctions connected in series, was fabricated on a 90 μm thick crystal Quartz substrate using in-house Nb thin-film technology[4]. Each junction has an area 25 μm^2 and a R_nA value of 50 $\Omega \cdot \mu m^2$. The test device was placed inside a split-block waveguide mount with possibility of applying magnetic field across the chip. The two filter structures (Filter 1 & Filter 2) provides isolation between the pumping frequency and the output ($\times 2$) frequency, and also matching between the SIS array and the embedding circuitry. In principle, different multiplication factors are possible, though in this experiment ($\times 2$) scheme was chosen because of the availability of measurement system at the output frequency.

For DC biasing, one end of the SIS array was connected directly to the waveguide block using bond wires, which provides the DC ground for the array. The bias current was supplied through a choke structure at the low frequency side.

3. Results and Discussion

The characterization of the test device was carried out at 4 K in a cryostat, Figure 2, with a closed cycle refrigerator. From the DC I-V characteristics, the gap voltage and the normal state resistance (R_n) of the whole array was found to be 195 mV and $R_n = 137\,\Omega$ respectively. Which corresponds to 2.86 mV gap voltage and $R_n = 2.01\,\Omega$ per junction, showing excellent agreement with the design value of $R_n = 2\,\Omega$.

To investigate the response from distributed SIS array, we use a double sideband SIS mixer [5] that covers the RF frequencies from 163 to 211 GHz with 4 to 8 GHz intermediate frequency (IF) bandwidth, connected at the output of the test device. The SIS array was pumped using a quasi-optically coupled W-band ($83 - 105$ GHz) Gunn oscillator and the output from the SIS array was coupled to the input of the mixer using a short piece of waveguide. In the experiment,

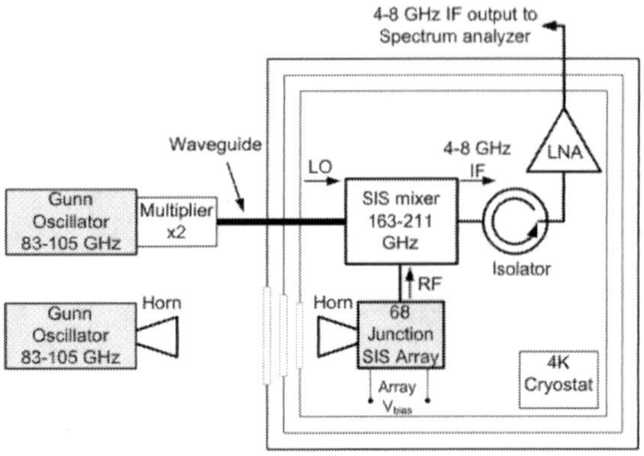

Figure 2. Schematic of the experimental setup used to study the test device. The cryostat contains the SIS junction array connected directly to a 163 – 211 GHz SIS mixer. Both the SIS mixer and the test array are pumped using two separate Gunn oscillators.

the Josephson effect was carefully suppressed using the magnetic field. The SIS mixer was pumped using a similar W-band Gunn oscillator and a frequency doubler [6], and the IF output was analyzed with a spectrum analyzer.

The LO for the SIS detector mixer $\left(f_{LO}^{SIS}\right)$ was swept between 171 – 203 GHz, with the array excitation in the frequency range of, $f_{array} = \frac{f_{LO}^{SIS} \pm (4 \rightarrow 8\,GHz)}{2}$ and the corresponding down converted IF response was studied.

Figure 3 shows the measured spectral response of the device, when the SIS detector mixer was pumped with LO frequency ≈ 190.3 and the SIS array excitation frequency was swept between 95 GHz to 100 GHz. We observed extremely sharp spectral signal between the 4 – 8 GHz IF band, corresponding to the second harmonic of the input pump frequency. The solid black curve at 5.7 GHz IF in Figure 3 corresponds to a 98 × 2 GHz output frequency from the SIS multiplier array. Similarly, the dashed gray curve at 7.7 GHz IF corresponds to output frequency of 99 × 2 GHz. Measurements with several combination of the LO and RF frequencies were performed, all confirming frequency multiplication at second harmonics by the SIS array.

The dependence of frequency multiplication on the array bias voltage was investigated by fixing both, the LO signal for the SIS mixer, and the excitation signal for the array, and measuring the IF power while sweeping the array bias voltage. Figure 4 shows the dependence of frequency multiplication on the array bias voltage, where peak power of the multiplied signal (after down conversion to IF) was plotted with respect to array bias voltage. In this case, the LO frequency for the SIS detector mixer was set to 190 GHz, and the input frequency for array was set to 98 GHz. We observed multi-photon process in the SIS array response to the applied microwave excitation. The first peak below the gap voltage is believed to be associated with absorption of one photon whereas the second peak below the gap voltage corresponds to absorption of two photons at the pumping frequency. We also observed significant conversion of the input pumping signal into the second harmonic for DC bias above the gap voltage. The plot shows the IF power hence includes, mixer conversion loss, gain of the LNA at 4 K, room temperature amplifier, and IF cables.

4. Conclusion

In summary, we present results of the first experimental study of frequency multiplication in a distributed SIS multiplier array consisting of 68 SIS junctions. We experimentally observed off-chip generation of the second harmonic signal in 180 – 200 GHz frequency band. The generated signal shows dependence on the DC bias voltage. The dependence exhibits features corresponding to single and multi-photon processes in the tunnel junction array. The device modeling shows good resemblance with the measurements. The device, although far from

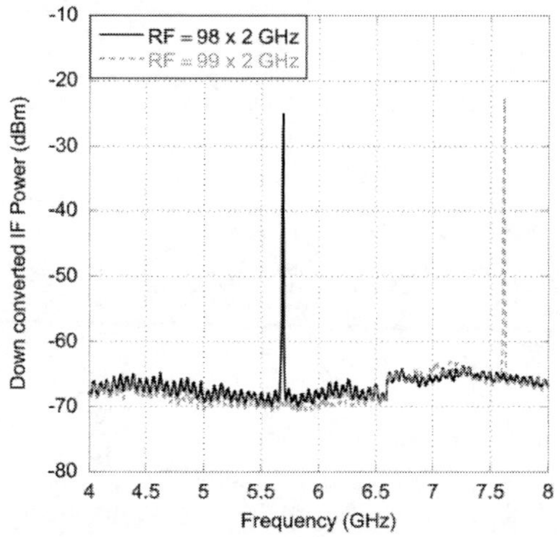

Figure 3. Measured down converted output response of the SIS array. The SIS detector was pumped with a LO frequency of ≈ 190.3 GHz, and the SIS array was exited with a signal at 98 GHz (solid black curve) and with 99 GHz (dashed gray curve).

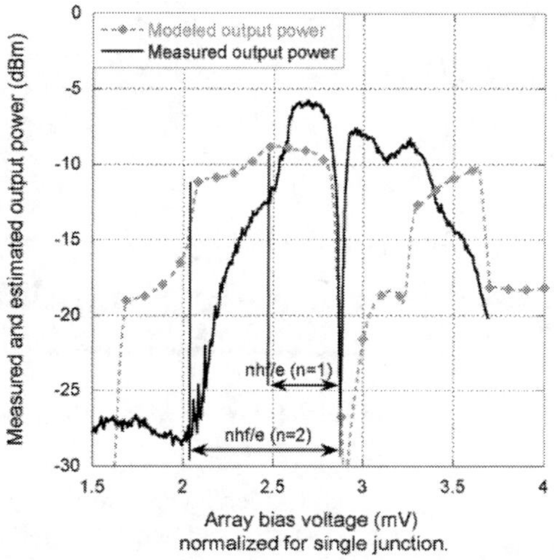

Figure 4. Bias voltage dependence of the multiplied signal. Down converted power of the 196 GHz signal was measured (solid black curve) at 6 GHz IF using spectrum analyzer. The voltage axis for the entire array is normalized for single junction. The dashed gray curve shows the estimated output power from the model.

providing enough power to pump a practical SIS mixer, may be considered as a first experimental step towards SIS frequency multipliers. Further development of the demonstrated principle of frequency multiplication may lead to a practical frequency multiplier device.

Acknowledgments

The authors would like to thank their colleagues at GARD Dr. Denis Meledin and Dr. Vincent Desmaris and Magnus Strandberg for their help. The authors would also like to thank Prof. Brian Ellison and Dr. Manju Henry, RAL, UK for providing the Schottky room temperature LO frequency doubler for the measurements.

References

[1] Kerr A and Pan S K 1990 *International Journal of Infrared and Millimeter Waves* **11**(10) 1169–1187 ISSN 0195-9271
[2] Foltz H and Davis J 1990 Calculated performance of SIS junctions as frequency multipliers *IEEE MTT-S International Microwave Symposium Digest, 1990.* vol 1 pp 307–310
[3] Tucker J R and Feldman M J 1985 *Rev. Mod. Phys.* **57**(4) 1055–1113
[4] Pavolotsky A, Dochev D and Belitsky V 2011 *Journal of Applied Physics* **109** 024502
[5] Billade B, Pavolotsky A and Belitsky V 2013 *IEEE transaction on, Terahertz science and technology* **3**(4) 416–421
[6] Henry M, Wang H, Ellison B, Alderman B, Huggard P, Billade B, Pavolotsky A and Belitsky V 2011 Oscillator system development for the ALMA band 5 receiver *Proceedings of the 4th UK/Europe-China Conference on Millimetre Waves and Terahertz Technologies* (Glasgow, UK)

Journal of Physics: Conference Series **507** (2014) 042006 doi:10.1088/1742-6596/507/4/042006

Observation of 65 GHz Coherent Oscillation in a Superconducting Flux Qubit Manipulated by Pulses

Chiarello F [1], Carelli P [2], Castellano M G [1], Cosmelli C [3], Torrioli G [1]

[1] IFN-CNR, via Cineto Romano 42, 00156 Rome, Italy.
[2] DSFC, Università dell'Aquila, via Vetoio 1, 67100 L'Aquila, Italy
[3] Dep. Physics, "Sapienza" Università di Roma, P.le A. Moro 5, 00185 Rome, Italy

fabio.chiarello@cnr.it

Abstract. A superconducting flux qubit can be manipulated by a fast modification of its potential, with a rapid transition from a double well to a single well shape, and coming back to the initial condition. This mechanism is based on a non-trivial quantum phenomenon, involving "partial" Landau-Zener transitions, coherent evolution in an harmonic potential and quantum interference. The study of this system enables a deep insight in decoherence mechanisms typical of superconducting qubits. Moreover, this procedure allows quantum operations with extreme high speeds, not possible with other standard manipulations. We present the experimental observation of coherent oscillations showing tunable frequencies with a 65 GHz top value.

1. Introduction

Superconducting devices based on Josephson effect are important candidates for the study of macroscopic quantum effects [1] and for the implementation of quantum computing [2]. They allow the realization of "artificial atoms" [3] that can be used as qubits, manipulated by microwave trains or by non-adiabatic pulses, and integrated in complex structures by using microelectronics techniques. The main problem in their use as qubits and quantum logic gates concerns the number of achievable coherent operations. This number must be well above $10^3 - 10^4$ in order to ensure the feasibility of real quantum algorithms. Great efforts are done in order to improve the coherence time by studying and removing the main causes of decoherence, and recent results show typical values of microseconds up to tens of microseconds [4–6]. On the other hand one can try to improve the operating rate. In previous works we showed the possibility to reach operation rates up to 21 GHz by using non adiabatic pulses in a double SQUID system [7–9]. In this paper we present an improved system that allowed the observation of operations up to 65 GHz. We present and discuss the system and the experimental results.

2. Double SQUID and pulse manipulation

The basic element for the manipulation with fast pulses is the double SQUID, a superconducting loop of inductance L interrupted by a second SQUID (a dc SQUID) consisting on a smaller loop of inductance l interrupted by a couple of nominally identical Josephson junctions of critical current i_0 and capacitance c_0 (Fig.1a).

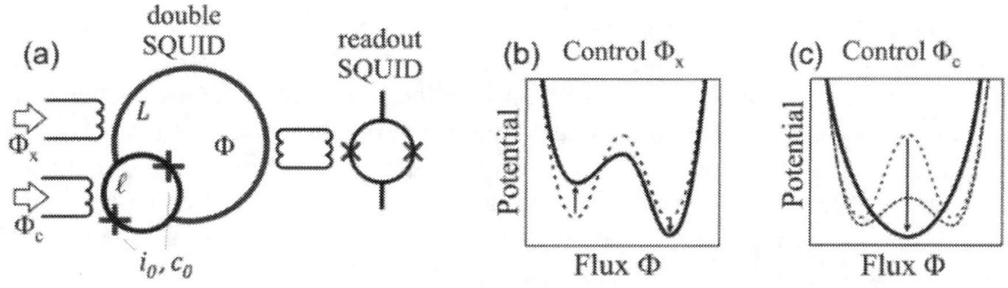

Figure 1. (a) Scheme of the double SQUID. (b) Modification of its potential by using magnetic fluxes applied respectively to the large (b) and to the small loops (c).

In the limit $l \ll \Phi_0/i_0$ (where $\Phi_0 = h/(2e)$ is the flux quantum) the dc SQUID behaves approximately as a single Josephson junction with total capacitance $C = 2c_0$ and modifiable critical current $I_0 = 2i_0 cos(\pi\,\Phi_c/\Phi_0)$, controlled by the magnetic flux applied to the small loop Φ_c. In the small loop limit the double SQUID can be described by a mechanical equivalent, with a particle of effective mass C moving along the "coordinate" Φ (the flux response in the large loop) in a potential $U = (\Phi - \Phi_x)^2/(2L) - I_0\Phi_b \cos(\Phi/\Phi_b)$ [10], where Φ_x is the magnetic flux applied to the large loop and $\Phi_b = \hbar/(2e)$ is the reduced flux quantum. In the π-junction case, for $2\pi L I_0/\Phi_0 < -1$, the potential can present a double well shape, with two minima separated by a central barrier.
For $\Phi_x = 0$ the potential is symmetric, and the flux states in the two wells are degenerate. Applied magnetic fluxes modify the potential shape and can be used to manipulate the SQUID state (Fig.1b and Fig.1c). In particular it is possible to prepare the flux state in one of the two wells by strongly unbalancing the potential (acting on Φ_x) and waiting for the complete relaxation in the minima. An oscillation between the flux states can be obtained by using an interferometric technique [8]. First of all the barrier is completely and abruptly removed by changing the flux Φ_c in order to arrive to a single well potential condition. This turns the system away from the degeneracy point with the effect of an half Landau Zener transition ("half" just because starting from the degeneracy point). Thanks to the particular energy levels structure, it is possible to have transitions that are completely non-adiabatic from the first two levels, but adiabatic for all the upper states. The effect of this fast transition is an equal population of the first two energy levels with a relative phase given by the initial flux state. Far from the degeneracy point (single well condition) there will be the accumulation of a phase difference between the two energy states. A fast return to the degenerate case (double well condition) stops the phase accumulation and causes a second half Landau Zener transition that projects energy states in flux states. The final state, a superposition of left and right flux states, will depend on the accumulated phase and on the pulse duration and depth too. The final flux state can be read out by means of an unshunted SQUID magnetometer, inductively coupled to the qubit large loop by means of a superconducting transformer (Fig.1a).
The oscillation frequency is given by the gap between energy levels and can be significantly higher than that obtainable with a manipulation based on Rabi oscillations with microwave pulses. This gap, expressed as a frequency, is approximately given by $f = f_0\sqrt{1 + \beta_0 cos(\pi\,\Phi_c/\Phi_0)}$, where $f_0 = 1/(2\pi\sqrt{LC})$ and $\beta_0 = 2i_0 L/\Phi_b$.

3. Experimental setup and results

In a previous work we obtained oscillation frequencies up to 21 GHz [9]. In this work we present a system that allows oscillation frequencies up to 65 GHz. This has been possible thanks to an advanced fabrication technology, the deep submicron (DSM) with fully-planarized Nb-(Al-AlOx)-Nb trilayer process available at the MIT Lincoln Laboratory facility (current density of 500 A/cm^2), which allowed the realization of high quality Josephson junctions with small dimensions (1 μm side) and reduced capacitance (ten times smaller than in the previous setup). In the new design the Josephson junctions have nominal critical currents $i_0 \cong 6\ \mu A$ and capacitances $c_0 \cong 30\ fF$, and the SQUID has large and small loops with nominal inductances $L = 120\ pH$ and $l = 10\ pH$ respectively, with gradiometric configurations ("8-shape" loops). The device has been first tested at 4.2 K, and then cooled down to 30 mK in a dilution refrigerator equipped for ultra low-noise qubits measurements, with concentric metal, μ-metal and superconducting shields. The dc lines are filtered by means of passive filters and thermocoax cables at different stages, while the fast lines (used for pulses) pass through attenuators at three different temperature stages. The fast pulses are generated by an Agilent 81130A pulse generator, characterized by a (measured) rise-time of 0.8 ns. For fixed amplitude and duration of the fast pulse, a sequence of operation is repeated: first of all a flux pulse (a Gaussian pulse of width 5 μs) is applied to Φ_x in order to unbalance the potential and prepare the initial state (the left state). Then (at 30 μs from the beginning of the sequence) a fast pulse is applied on Φ_c to perform the rotation between the flux states. After some time (at 40 μs) a current ramp is applied to the readout dc SQUID. The measured transition to the voltage state is used to trigger the acquisition of the critical current, and the double SQUID flux state is determined from this value . This sequence is repeated many times (from 1000 to 10 000, according to the required accuracy) with a repetition rate of 20 kHz in order to evaluate the mean flux state, and so the probability to find the left well occupied. By repeating this procedure for different pulse durations Δt it is possible to plot the oscillation curve, showing the rotation between the flux states. By using pulses with different heights $\Delta\Phi_c$ the frequency of oscillations changes. A series of curves corresponding to different oscillation frequencies are plotted in (Fig.2a) by using a waterfall graph.

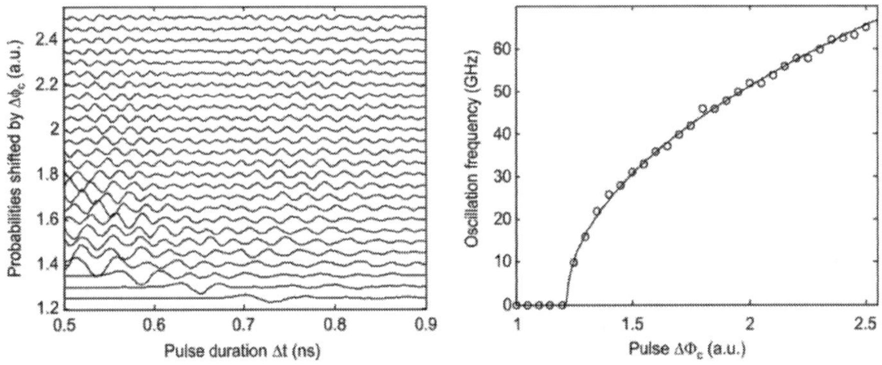

Figure 2. (a) Waterfall plot of the experimental oscillation curves obtained for different pulse heights $\Delta\Phi_c$ (the different curves are vertically shifted by the relative $\Delta\Phi_c$). (b) Measured oscillation frequencies (circles) compared with the expected values (line).

These curves presents an unwanted modulation in frequency and amplitude that can be explained by considering a small spurious fixed resonance at about 10 GHz, probably related to our setup. From these curves it is possible to extract a plot of the oscillation frequencies vs. pulse heights (Fig.2b), which is in good agreement with the theoretical expectation. The higher measured point in Fig.2a corresponds to a frequency of 65 GHz.

4. Discussions

This result is particularly interesting and promising, but emphasizes technological difficulties and limits reached with this kind of manipulation. First of all, in order to perform a single manipulation it is necessary to have pulses shorter than those of operation period. For a 65 GHz frequency this corresponds to use pulses shorter than 15 ps with rise times much shorter. Our setup allows only rise/fall times of 800 ps, so that we are no able to observe a single oscillation and also our shorter pulse contains a lot of rotations. Moreover, in order to have a slew-rate suitable for the Landau Zener transition and because of our "small" rise times, we have to use pulses with amplitude higher than necessary. Such large pulses have harmful effects on the oscillations stability and coherence because of overshoots, unwanted resonances and spurious couplings with other lines. The use of a faster pulse generator would mitigate this problem, but it would not be the final solution. In any case the control pulses must be large enough to ensure the modification of the potential from the double well to the single well shape (larger than in the case of manipulation with microwaves, where such a strong modification of the potential is not required). These pulses must be very well formed since small distortions strongly modify the dynamics. Therefore, well-shaped picoseconds large pulses that could be transmitted with small distortions and reflections from room temperature to 30 mK stage would be needed. This is an hard technological challenge that must be addressed in order to make really advantageous the use of SQUID qubits manipulated by fast pulses.

Acknowledgments

This work was supported by Italian MIUR under the PRIN2008 C3JE43 project. We thank W. Oliver and the MIT Lincoln Laboratory LTSE team for the device fabrication, and A. Marletta for support in manuscript revision.

References

[1] A. Wallraff, Schuster D I, Blais A, Frunzio L, Huang R-S, Majer J, Kumar S, Girvin S M and Schoelkopf R J 2004 Strong coupling of a single photon to a superconducting qubit using circuit quantum electrodynamics *Nature* **431** 162–7

[2] Reed M D, DiCarlo L, Nigg S E, Sun L, Frunzio L, Girvin S M and Schoelkopf R J 2012 Realization of three-qubit quantum error correction with superconducting circuits *Nature* **482** 382–5

[3] Cooper K B, Steffen M, McDermott R, Simmonds R W, Oh S, Hite D A, Pappas D P and Martinis J M 2004 Observation of Quantum Oscillations between a Josephson Phase Qubit and a Microscopic Resonator Using Fast Readout *Phys. Rev. Lett.* **93** 180401

[4] Paik H, Schuster D I, Bishop L S, Kirchmair G, Catelani G, Sears A P, Johnson B R, Reagor M J, Frunzio L, Glazman L I, Girvin S M, Devoret M H and Schoelkopf R J 2011 Observation of High Coherence in Josephson Junction Qubits Measured in a Three-Dimensional Circuit QED Architecture *Phys. Rev. Lett.* **107** 240501

[5] Rigetti C, Gambetta J M, Poletto S, Plourde B L T, Chow J M, Córcoles A D, Smolin J A, Merkel S T, Rozen J R, Keefe G A, Rothwell M B, Ketchen M B and Steffen M 2012 Superconducting qubit in a waveguide cavity with a coherence time approaching 0.1 ms *Phys. Rev. B* **86** 100506

[6] Barends R, Kelly J, Megrant A, Sank D, Jeffrey E, Chen Y, Yin Y, Chiaro B, Mutus J, Neill C, O'Malley P, Roushan P, Wenner J, White T C, Cleland A N and Martinis J M 2013 Coherent Josephson Qubit Suitable for Scalable Quantum Integrated Circuits *Phys. Rev. Lett.* **111** 080502

[7] Poletto S, Chiarello F, Castellano M G, Lisenfeld J, Lukashenko A, Cosmelli C, Torrioli G, Carelli P and Ustinov A V 2009 Coherent oscillations in a superconducting tunable flux qubit manipulated without microwaves *New J. Phys.* **11** 013009

[8] Castellano M, Chiarello F, Carelli P, Cosmelli C, Mattioli F and Torrioli G 2010 Deep-well ultrafast manipulation of a SQUID flux qubit *New Journal of Physics* **12** 043047

[9] Chiarello F, Paladino E, Castellano M G, Cosmelli C, D'Arrigo A, Torrioli G and Falci G 2012 Superconducting qubit manipulated by fast pulses: experimental observation of distinct decoherence regimes *New Journal of Physics* **14** 023031

[10] Castellano M G, Chiarello F, Leoni R, Mattioli F, Torrioli G, Carelli P, Cirillo M, Cosmelli C, de Waard A, Frossati G, Grønbech-Jensen N and Poletto S 2007 Catastrophe Observation in a Josephson-Junction System *Phys. Rev. Lett.* **98** 177002

Hybrid superconducting mesa-heterostructure with manganite-ruthenate interlayer

K Y Constantinian[1], G A Ovsyannikov[1,2], A E Sheyerman[1], A V Shadrin[1,2], Y V Kislinskii[1], A Kalabukhov[2] and D Winkler[2]

[1] Kotel'nikov Institute of Radio Engineering and Electronics RAS, Moscow, Russia
[2] Chalmers University of Technology, Gothenburg, Sweden

E-mail: karen@hitech.cplire.ru

Abstract. We present experimental data on Josephson effect in hybrid superconducting mesa-heterostructures (HSMH) with composite manganite-ruthenate interlayer. The HSMH base electrode consisted of the cuprate superconductor $YBa_2Cu_3O_7$ grown epitaxially on a NdGaO substrate using laser ablation. The interlayer was composed from in-situ deposited $SrRuO_3$ (F1) and $La_{0.7}Sr_{0.3}MnO_3$ (F2) thin films, each characterized by different directions of magnetization. The top electrode was Nb/Au thin film. A superconducting current was observed when the interlayer thickness was well above the correlation length, determined by the exchange field in F1 and F2. Obtained $I_C(H)$ dependences and non-sinusoidal current-phase relation evaluated from microwave measurements are discussed in terms of generation of long-range spin triplet superconducting current component in heterostructures with interfaces of singlet superconductors and bilayer ferromagnetic materials with different spatial directions of magnetization.

1. Introduction

In superconducting heterostructures with a non-uniform magnetized ferromagnetic layer (F) between two singlet superconductors (S) a long-range triplet superconducting correlations may occur [1]. In SFS structures with uniform magnetization the spin projection on direction of magnetization is conserved and the both singlet and triplet superconducting correlations with zero spin projection appear [1, 2], penetrating into F with a characteristic length $\xi_F=(\hbar D/E_{ex})^{1/2}$ determined by the magnetic exchange energy E_{ex}, where in the dirty limit $D=v_F l/3$ is the diffusion coefficient, v_F is the Fermi velocity, and l is the mean free path. The long-range triplet superconducting correlations with nonzero (±1) spin projection are not suppressed by the exchange interaction and is determined by temperature T: $\xi_N=(\hbar D/k_B T)^{1/2}$ as for SNS junctions with normal metal (N) interlayer. If $k_B T<<E_{ex}$ then a long-range triplet superconducting correlations leads to existence of superconducting current in SFS structures at sufficiently large distances between superconductors [3-10]. Here we report on experimental studies of $S/M/S_d$ hybrid superconducting mesa-heterostructures (HSMH), where S is the Nb/Au bilayer, S_d is the $YBa_2Cu_3O_{7-\delta}$ (YBCO) cuprate superconductor, and M is a composite magnetic oxide interlayer consisting of two thin F-films, $La_{0.7}Sr_{0.3}MnO_3$ and $SrRuO_3$ with noncollinear magnetization. SFS structures with spin triplet superconducting correlations were theoretically considered in [11–15].

2. Samples and measurements

Square shape HSMH with in-plane dimensions L = 10 – 50 μm were fabricated from S_d hybrid superconducting heterostructures deposited on (110) $NdGaO_3$ substrates using photolithography, plasma chemical, and ion etching [16, 17]. The base electrode was $YBa_2Cu_3O_{7-\delta}$, the interlayer M was consisted of two epitaxially grown ferromagnetic films: F1 – $SrRuO_3$ (SRO) and F2 – $La_{0.7}Sr_{0.3}MnO_3$ (LSMO) with thicknesses from 5 to 30 nm (see inset in figure 1). The upper electrode was Nb/Au bilayer as the LSMO/YBCO interface exhibits insufficient transparency [16]. The magnetization vector of the LSMO is lying in the plane of the substrate [17], whereas the magnetization vector of SRO is directed at ~ 23° from the normal to the substrate [18]. Ferromagnetic resonance in LSMO/SRO heterostructure was studied at 10 GHz and showed a uniaxial magnetic anisotropy that is characteristic of LSMO films deposited on (110)NGO substrate [17]. However, the ferromagnetism of the SRO film was not revealed from ferromagnetic resonance studies because of high, about 1 T, saturation field of SRO. Note, peaks in X-ray spectrum from YBCO, LSMO, SRO were observed indicating epitaxial growth of all oxide films and the absence of materials mixing at the interfaces.

3. Results and discussion

The superconducting current was observed in all HSMH with a total thickness d of the composite M-interlayer up to 53 nm, which is much larger than the coherence lengths in F1 and F2, determined by the exchange field. We estimated the coherence length ξ_{LSMO} = 7 nm in the LSMO film using the exchange energy E_{ex} =2.3 meV taken from electron specific heat [19], and $v_F = 2\cdot10^7$ cm/s in dirty limit for the mean free path $l_{LSMO} \approx 10$ nm. For SRO the E_{ex} = 13 meV was taken from the data on proximity effect at SRO/YBCO interface [20], and in the dirty limit $v_F = 10^7$ cm/s taking mean free path $l_{SRO} \approx 10$ nm we estimate ξ_{SRO} = 4 nm, close to the value obtained in [21].

Control measurements of the HSMH with only the LSMO [16] or the SRO interlayer, showed [22] that the critical current is absent if the SRO and LSMO films are thicker than d_1=14 nm and d_2=2 nm, respectively. The critical current density j_C of the HSMH decreases by an order of magnitude, when increasing the total thickness of $d = d_1 + d_2$ from 8.5 to 53 nm. The value j_C = 9 A/cm^2 was observed for a sample with d = 11.5 nm, L = 10 μm. Figure 1 shows that critical current density j_C depends non-monotonically on the thickness d_2 of the LSMO. A relatively low j_C at small d_2 (1.5 and 3 nm) could be explained by existence of a "dead" nonmagnetic layer at the interface. Note, a non-monotonic thickness dependence of the critical current was predicted [14, 15] for structures with long-range triplet superconducting correlations with a maximum of $j_C(d)$ at $d \approx \xi_F$ [14] In our HSMH further significant increase in d resulted in a decrease of j_C as expected from theoretical calculations [11–14].

The measurements of the critical current I_C as a function of magnetic field H (see figure 2) show that maxima of I_C are observed in the range 5 – 15 Oe. Note the HSMH and the H-field coil were screened by the cooled μ-metal shield. Similar $I_C(H)$ dependences were observed for SFS junctions where long-range triplet superconducting component was induced [5, 10]. In a wider variation of the external magnetic field, a hysteretic behaviour was observed, indicating the existence of ferromagnetism in the M-interlayer [23]. The $I_C(H)$ dependence shown in figure 2 does not exhibit hysteresis because the total range of H variation is much lower than the saturation magnetic field of the ferromagnetic layer [23].

The theoretical calculations reported in [12, 13, 15] predict a strong increase of the second harmonic in the current-phase relation of the superconducting current $I_S(\varphi) = I_{C1}\sin(\varphi) + I_{C2}\sin(2\varphi)$ for the asymmetric M-interlayer ($d_1 \neq d_2$) under the variation of the angle between the directions of the magnetizations of the F films, giving $I_{C2} >> I_{C1}$. The measurements of Shapiro steps of HSMH demonstrate deviation from the sinusoidal current–phase relation. The I-V characteristics of the HSMH with L = 10 μm, I_C = 88 μA, and R_N = 0.16 Ω under microwave radiation at f_e = 41 GHz exhibit both integer and fractional Shapiro steps (see figures 3 and 4). The maximum of the first Shapiro step was I_1 = 94 μA and, correspondingly, the ratio I_1/I_C = 1.1 is in well agreement with the resistively shunted junction (RSJ) model, as the critical frequency $f_C = (2e/h)I_CR_N$ = 6.8 GHz, which

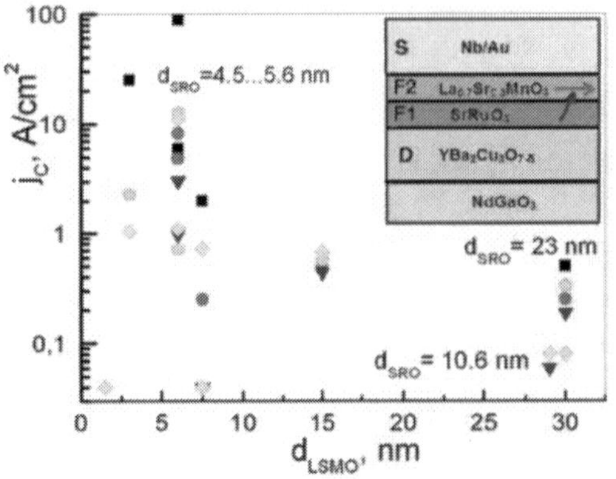

Figure 1. Superconducting current density in HSMH vs. thickness of the LSMO film at T = 4.2 K and H=0. The squares, circles, pentagons, triangles, and diamonds mark L equal to 10, 20, 30, 40, and 50 μm, respectively. The inset shows the cross section of the HSMH with different directions of the magnetization vectors.

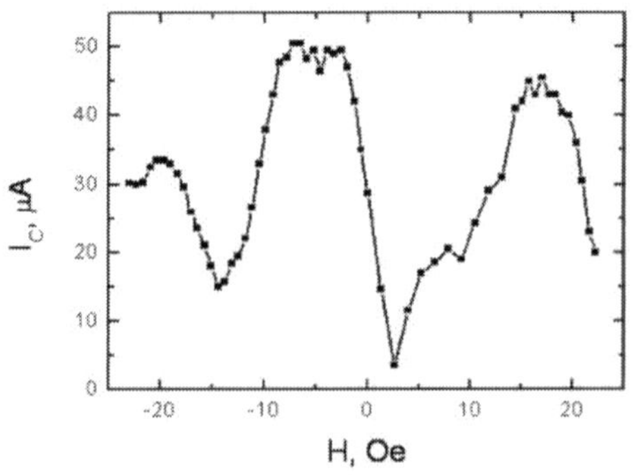

Figure 2. H-field dependence of the critical current for HSMH with L = 50 μm, d_1 = 4.5 nm, and d_2 = 3 nm, T = 4.2 K. The range of H-field is much below the level of saturation field of the ferromagnetic interlayer.

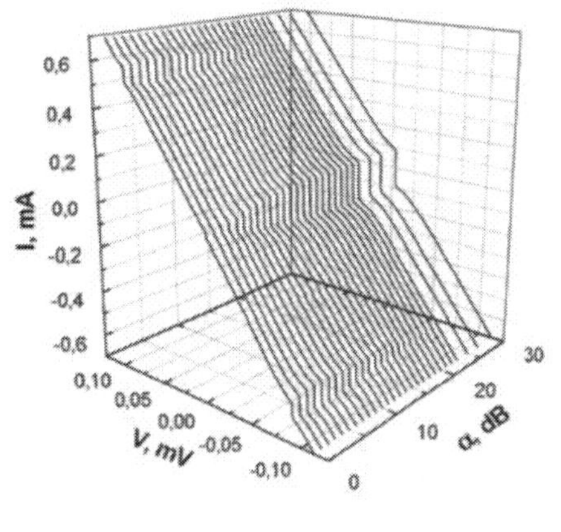

Figure 3. Family of the current–voltage characteristics of the HSMH under microwave irradiation at 41 GHz, where α is the inserted damping by external polarizing attenuator.

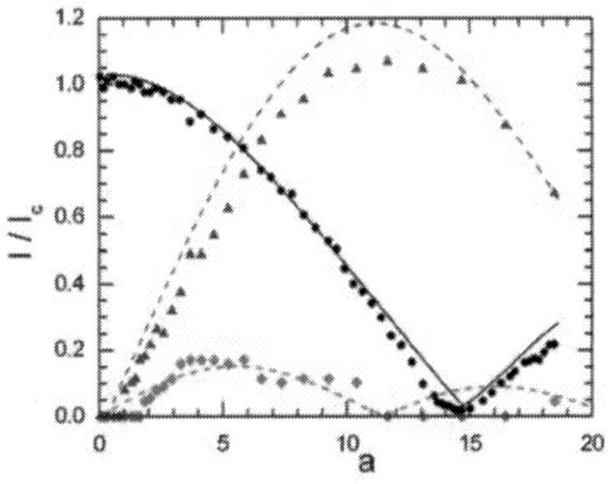

Figure 4. Shapiro steps for a HSMH with d_1 = 6 nm, d_2 = 5.5 nm, and L = 10 μm, T = 4.2 K. I_C (circles), I_1 (triangles), and $I_{1/2}$ (diamonds), are plotted vs. a = I_{RF}/I_C

gives the ratio f_e/f_C = 6, very well corresponds to the high frequency limit of the RSJ model. The maximum of the half-integer Shapiro step was $I_{1/2}$ = 15 μA and within the modified RSJ model [24], which takes into account a non-sinusoidal current–phase relation, the fraction of the second harmonic I_{C2}/I_{C1} is about 13%. It is worth noting that the direct comparison of experimental results with theory [11–15] is complicated because of the presence of a barrier between the manganite and superconducting electrodes. The authors of paper [14] stated that a long-range triplet superconducting correlations hardly may exist in a structure with only two ferromagnetic layers and the "third component" is required. In our case, likely, one of the S/M barriers is magnetically active and serves as the "third component." Also, it is not excluded that an antiferromagnetic barrier may appear at the SRO/LSMO interface [25] providing the function of the "third component" of a spin active interface.

3. Conclusions

We have experimentally observed the superconducting current in hybrid mesa-heterostructures with a composite oxide ferromagnetic bilayer with non-collinear directions of the magnetizations in the layers. It has been shown that the total thickness of the magnetic interlayer is much larger than the length of superconducting correlations in ferromagnetic layers, determined by the exchange field. The Josephson effect observed in these structures is explained by the penetration of the long-range triplet component of the superconducting order parameter into the magnetic interlayer. The deviation of the current–phase relation from a sinusoidal dependence has been measured having 13% fraction of the second harmonic that can also be explained by the triplet component of the superconducting order parameter.

Acknowledgment

The work was supported by Russian Academy of Sciences, Russian Ministry of Education and Science, Leading scientific schools grant NSh-2456.2012.2, RFBR projects 11-02-01234a, 12-07-31207mol_a, and VISBY Program of Russian-Swedish collaboration.

References

[1] Bergeret F S, Volkov A F, and. Efetov K B 2005 *Rev. Mod. Phys.* **77**, 1321
[2] Buzdin A I 2005 *Rev. Mod. Phys.* **77**, 935
[3] Sosnin I, Cho H, Petrashov V T, et al 2006 *Phys. Rev. Lett.* **96**, 157002
[4] Keizer R S, Goennenwein, S T B, Klapwijk T M, et al. 2006 *Nature* (London) **439**, 825
[5] Anwar M. S, Czeschka F, Hesselberth M, et al. 2010 *Phys. Rev.* **B82**, 100501
[6] Wang J, Singh M, Tian M, et al. 2010 *Nature Phys.* **6**, 389
[7] Sprungmann D, Westerholt K, Zabel H, et al. 2010 *Phys. Rev.* **B82**, 060505(R)
[8] Robinson J W A, Witt J D S, and Blamire M. G, 2010 *Science* **329**, 59
[9] Khaire T S, Khasawneh M A, Pratt W P, et al., 2010 *Phys. Rev. Lett.* **104,** 137002
[10] Leksin P V, Garif'yanov N N, Garifullin I A, et al. 2012 *Phys. Rev. Lett.* **109**, 057005
[11] Trifunovic L, Popovic Z, and Radovic Z, 2011 *Phys. Rev.* **B84**, 064511
[12] Mel'nikov A S, Samokhvalov A V, Kuznetsova S M, et al. 2012 *Phys. Rev Lett.* **109**, 237006
[13] Sperstad I B, Linder J, and Sudbo A, 2008 *Phys. Rev.* **B78**, 104509
[14] Volkov A F and Efetov K B, 2010 *Phys. Rev.* **B81**, 144522
[15] Richard C, Houzet M, and Meyer J S, 2013 *Phys. Rev. Lett.* **110**, 21704
[16] Petrzhik A M, Ovsyannikov G A, Shadrin A V, et al. 2011 *JETP* **112,** 1042
[17] Ovsyannikov G A, Petrzhik A M, Borisenko I V, et al. 2009 *JETP* **108**, 48
[18] Koster G, Klein L, Siemons W, et al. 2012 *Rev. Mod. Phys.* **84**, 253
[19] Woodfield B F., Wilson M L, and Byers J. M., 1997 *Phys. Rev. Lett.* **78**, 3201
[20] Asulin I, Yuli O, Koren G, and Millo O 2009 *Phys. Rev.* **B79**, 174524
[21] Mechin L, Flament S, Perry A, et al. 2005 *J. Appl. Phys.* **98**, 103902
[22] Ovsyannikov G A, Sheyerman A E, Kislinskii Y V, et al. in Hybrid Superconducting Heterostructures with Magnetic Interlayer, Proceedings of the 19th Workshop on Oxide Electronics (Apeldoorn, The Netherlands, 2012).
[23] Bol'ginov V V, Stolyarov V S, Sobanin D S, et al. 2012 *JETP Lett.* **95,** 366
[24] Komissinskiy P, Ovsyannikov G A, Constantinian K Y, et al. 2008 *Phys. Rev.* **B78**, 024501
[25] Ziese M, Vrejoiu I, Pippel E, et al. 2010 *Phys. Rev. Lett.* **104**, 167203

HTS ion damage Josephson junction technology for SQUID arrays

S Ouanani[1], J Kermorvant[2], D-G Crété[1], Y Lemaître[1], J-C Mage[1], B Marcilhac[1], N Bergeal[3], M Malnou[3], J Lesueur[3], D Mailly[4], C Ulysse[4]

[1]Unité Mixte de Physique CNRS/Thales and Université Paris-Sud, 1 Av. A Fresnel, FR-91767 Palaiseau, France
[2]THALES Communication and Security, 4 Av. des Louvresses, 92622 Gennevilliers CEDEx, France
[3]LPEM / ESPCI-CNRS-UPMC, 10 Rue Vauquelin, FR-75005 Paris, France
[4]Laboratoire de Photonique et de Nanostructures, Route de Nozay FR-91460 Marcoussis, France

denis.crete@thalesgroup.com

Abstract. The high temperature superconducting (HTS) Josephson Junction (JJ) ion damage technology we are developing is well suited for making large SQUID arrays. We have studied arrays of similar SQUIDs together with large SQIFs (Superconducting Quantum Interference Filter) with 2000 SQUIDs of different loop areas. Magnetic field sensitivity has been measured in both types of devices as a function of bias current and temperature. The effects of the barrier thickness (from 20 to 80 nm) and JJ length (2 or 5 μm) on characteristics have been investigated.

1. Introduction

SQUIDs (Superconducting Quantum Interference Device) made with low critical temperature (T_c) superconductors are known as the most sensitive magnetic sensors and have been used in a wide range of applications [1]. High critical temperature (HTc) devices, as they operate at higher temperature, can expand further the domain of application, provided they have comparable performance. One way to improve the sensitivity and the dynamic range is to create SQUID arrays, either with similar areas or with a random distribution of hole sizes. For such multiple-loop configuration named SQIFs (Superconducting Quantum Interference Filters), the magnetic flux to voltage transfer function shows a unique dip around zero magnetic flux. Therefore, SQIFs can be used to measure absolute magnetic field. Besides, their sensitivity and dynamic range are expected to increase as $N^{1/2}$ [2].

In this paper, we present experimental results on SQUID arrays produced by ion damage technology on $YBa_2Cu_3O_7$ (YBCO) thin films. Two kinds of arrays were investigated: arrays of N identical SQUIDs in series with N=1, 5, 12 and 22 and a series-array of 2000 SQUIDs with pseudo-random distribution of hole sizes from 6 to 60 μm^2.

2. Fabrication Method

The process, already described in [3, 4], is briefly be summarized here. A 150 nm thick c-axis oriented YBCO film was thermally coevaporated on a sapphire substrate [5] and then covered by an in situ 100 nm thick gold layer. After Ar-ion beam etching of the gold contact layer, a photoresist is patterned to protect superconducting strips from subsequent irradiation with 110 keV oxygen ions at large fluence to make the unprotected YBCO insulating. A PMMA (polymethylmethacrylate) resist is then deposited all over the sample, and a very narrow (20-nm wide) slit is opened across each arm of the superconducting loop by electron beam lithography. A second 110 keV oxygen ion irradiation performed at smaller fluence defines the junction barrier. This technique allows one to make controllable and reproducible Josephson junctions [6] and SQUIDs [7].

3. Experimental Results

3.1. Identical SQUID arrays

Arrays used in this study are made of 64 identical SQUIDs of 30 μm² square loop areas connected in series, with intermediate contacts to test several series of 1, 5, 12 and 22 SQUIDs. The JJ geometry has been varied from array to array: three barrier thickness (20, 40 and 80 nm) and two JJ lengths (2 and 5 μm) have been used. The arm width was maintained at 5 μm.
We define the parameter T_j as the highest temperature for Josephson coupling. All arrays have T_j ~74-76 K, depending on the barrier thickness (see below), with a slightly higher value for a single SQUID. The current-voltage characteristics are RSJ-like [1] at moderate bias currents.

Figure 1 shows the output voltage V across one single SQUID with JJ length of 2 μm and barrier thickness of 40 nm, versus applied magnetic field for several values of the dc bias current between 0.9 I_c and 1.15 I_c (fig. 1). For I_{bias}=340 μA (~1.04 I_c), the amplitude is maximum with a peak to peak output voltage of 13 μV. The period for one SQUID is 130 mG, corresponding to an effective area of ~130 μm² which has been evaluated around 85 μm²: this enhancement is presumably due to a flux-focusing effect caused by a nearby superconducting plane [7]. In addition to the SQUID modulations, the characteristic Fraunhofer pattern associated with the junctions themselves is observed (not shown here). This behaviour demonstrates the high quality of the junctions.

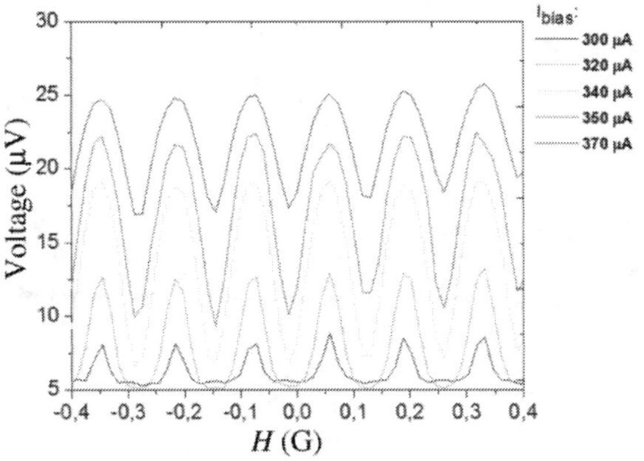

Figure 1: Voltage V vs. applied magnetic field H for a single SQUID at different bias currents.

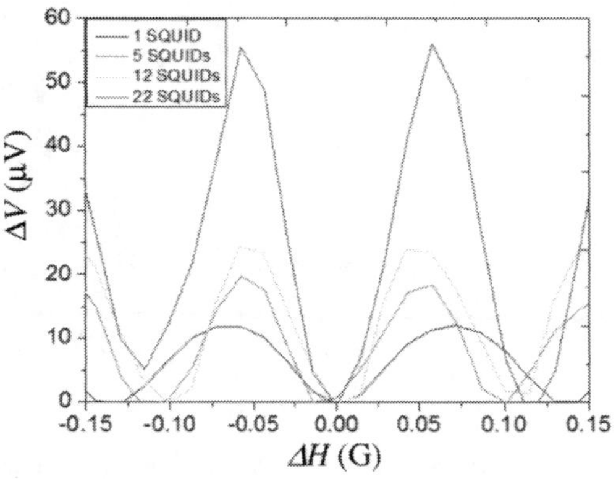

Figure 2: Voltage vs. magnetic field for 4 N-series arrays with N = 1, 5, 12 and 22 SQUIDs. Curves are shifted both horizontally and vertically for easier comparison.

The amplitude of the modulation increases with N, the number of SQUIDs in series, but not linearly as expected (see fig. 2) [8]. Non-uniformity in the magnetic environment may create destructive interferences and therefore, a decrease of the modulation amplitude. Also, as seen from the field period, the field focusing factor is not the same for all arrays, due to the different embedding superconducting circuit.

3.2. SQIF

Figure 3 presents the voltage response as a function of the applied magnetic field for a series array of 2000 SQUIDs with randomly distributed loop areas between 6 and 60 µm^2 and JJ geometry 40 nm × 2 µm (at T= 73 K and I_{bias}= 40 µA).

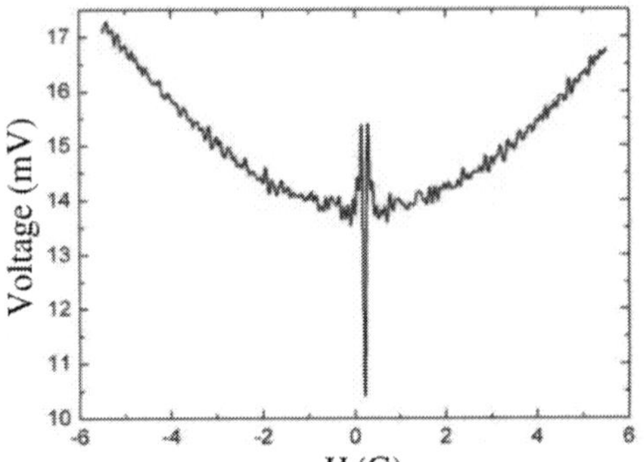

Figure 3: The voltage vs external magnetic field transfer function of a SQIF shows a unique delta-peak around $B = 0$.

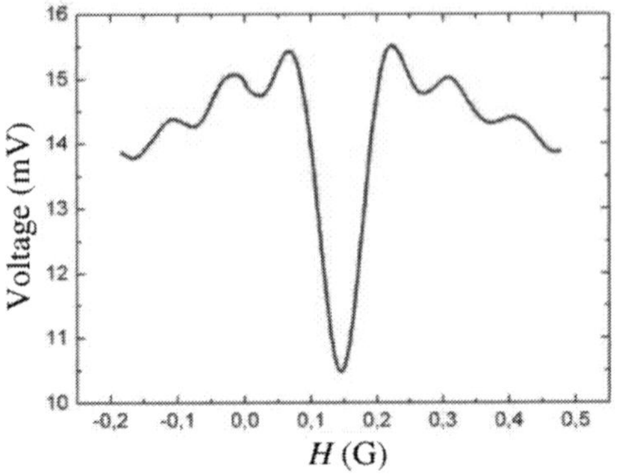

Figure 4: Enlarged central peak showing residual SQUID oscillations and offset of about 0.15 G.

As expected [2], an aperiodic $V(H)$ response is observed and characterized by a single sharp dip centred near zero magnetic field. In addition, the parabolic background seen in fig. 3 is due to the response of the JJs to the magnetic field and residual SQUID oscillations are also visible on both sides of the dip (cf. fig 4). One can note the slight shift in the peak position regarding to the zero field value due to the ambient magnetic flux trapping during the cooling down of the sample. The amplitude of the voltage swing and the slope dV/dH of the linear part of the SQIF dip strongly depend on both operating temperature and bias current (fig. 5 & 6). We find the optimal SQIF modulation conditions to be T = 73 K which is 0.8 K lower than T_j and I_{bias}= 40 µA which is 0.5 µA higher than the corresponding critical current of the whole structure. The maximum voltage is 5 mV and the transfer factor is equal to 105 mV.G^{-1}.

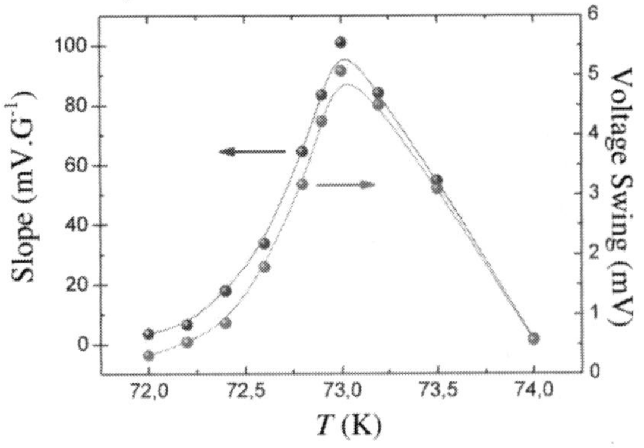

Figure 5: Evolution of the transfer factor and the voltage swing vs. T at I_{bias} = $35\ \mu A$. Lines are guides to the eye.

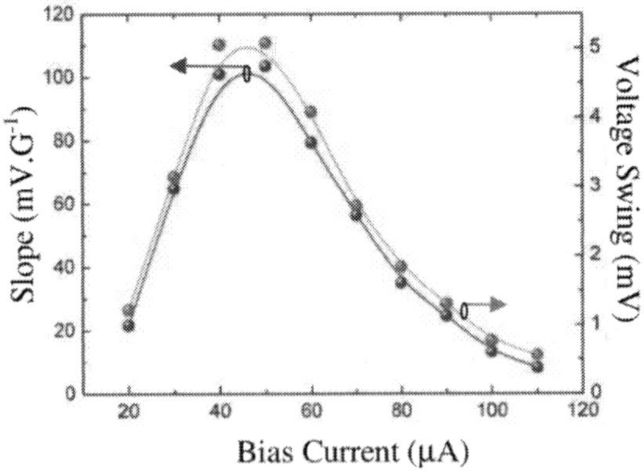

Figure 6: Evolution of the transfer factor and the voltage swing vs. I_{bias} at T = 73 K. Lines are guides to the eye.

3.3. Influence of the barrier thickness

To illustrate the scaling of barrier properties with thickness t_B, we report on the resistance R measured at a given temperature for a given probe current I_{probe}. Experimentally, R is measured versus T using I_{probe}= 10 µA, for series arrays on a 2 µm wide microbridge with t_B= 20, 40 and 80 nm.

We define T_i as the temperature at which the resistance vanishes, i.e. $I_c(T_i)$ ~I_{probe}, and $R_{Plateau}$, the resistance at midpoint of the plateau between T_i and the critical temperature T_c. The dependences of T_i

and $R_{Plateau}$ as a function of t_B are shown fig. 7: as expected, T_i decreases (solid line) and $R_{Plateau}$ increases (dashed line), with the barrier thickness. Linear extrapolation of $R_{Plateau}(t_B)$ at $R_{Plateau}= 0\ \Omega$ defines an "excess barrier thickness" t_{eb} of the order of 88 nm. Because the resistance depends only weakly on temperature in the range of the plateau, and because the $V(I)$ curve is linear in this range, this result very weakly depends on the arbitrary choice of I_{probe}. It indicates some straggling of the damaged area, as proposed by Wolf [9]: the irradiated material can be regarded as a superconductor whose local critical temperature and critical current density are reduced depending on the local defect density and induces a bias-current dependence of the normal resistance (visible at large bias currents). In order to investigate further the transport properties in the barrier area, we have also used $V(I)$ measurements at various temperatures, to apply the above procedure and produce the graph on fig. 8, t_{eb} versus T and I. Here, the ratio $V(I,T)/I$ has been substituted for $R_{Plateau}$. It shows that the excess barrier thickness increases with both current and temperature: again this is in qualitative agreement with the effect of local defect density which is the highest in the barrier region facing the aperture of the mask, and decreases as the distance to the barrier region increases.

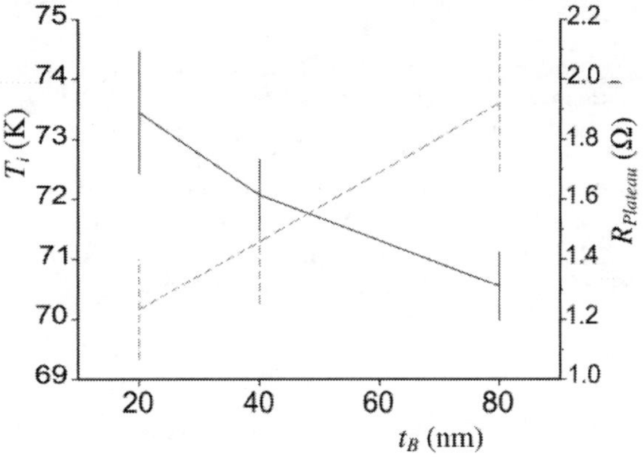

Figure 7 : Variations of T_i, the temperature at which I_c=10 µA (solid line), and $R_{Plateau}$, the device resistance at $(T_i+T_c)/2$ (dashed line), for 3 values of the nominal barrier thickness t_B. Vertical bars are 1-σ scattering over a dozen of devices.

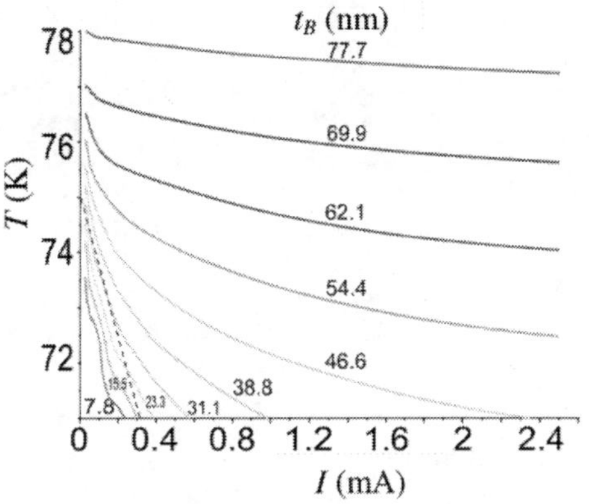

Figure 8 : Contour plot of the excess barrier thickness t_{eb} (labelled in nm) versus T and I. The lower left part of the diagram delimited by the dashed line is the region where I becomes comparable to $I_c(T)$.

4. Conclusion

Ion damage technology is a successful method to achieve large SQUID arrays. Results on SQIF are very promising and need to be investigated more deeply. Noise measurements to characterize the sensitivity of single SQUID and SQIFs have to be done in a close future.

References

[1] SQUID Handbook, J. Clarke and A.I. Braginski Eds., 2004 Wiley-VCH
[2] J. Oppenländer *et al.*, 2001 *IEEE Trans Appl. Supercond.* **Vol. 11**, pp.1271-1274
[3] S. Ouanani *et al.*, 2013 jul.7-12, *IEEE, Intern. Supercond. Electron. Conf.*, Cambridge, Ma. http://ieeexplore.ieee.org/stamp/stamp.jsp?tp=&arnumber=6604297
[4] N. Bergeal *et al.*, 2007 *J. Appl. Phys.* **102**, 083903
[5] http://www.ceraco.de/
[6] N. Bergeal *et al.*, 2005 *Appl. Phys. Lett.* **87**, 102502
[7] N. Bergeal *et al.*, 2006 *Appl. Phys. Lett.* **89**, 112515
[8] K. Li *et al.*, 1995 *IEEE, Trans. Appl. Supercond.*, **Vol 5**, pp.3255-3258
[9] T. Wolf, 2011 PhD dissertation : « Etude de nanojonctions Josephson à haute température critique en vue d'application terahertz » Université P& M. Curie, Paris

Graphoepitaxial high-T_c SQUIDs

M. I. Faley[1], D. Meertens, U. Poppe and R. E. Dunin-Borkowski
Peter Grünberg Institute (PGI-5: Microstructure Research)
Forschungszentrum Jülich GmbH, D-52425 Jülich, Germany

E-mail: m.faley@fz-juelich.de

Abstract. The fabrication process and physical properties of graphoepitaxially engineered high-T_c direct current superconducting quantum interferometer devices (DC SQUIDs) are studied. Double buffer layers, each comprising a graphoepitaxial seed layer of $YBa_2Cu_3O_{7-x}$ and an epitaxial blocking layer of $SrTiO_3$, were deposited over textured step edges on (001) surfaces of MgO substrates. Scanning electron microscopy and high-resolution transmission electron microscopy were used to investigate the microstructural properties of DC SQUIDs with graphoepitaxial Josephson junctions. Both direct coupled and inductively coupled high-T_c DC SQUIDs with graphoepitaxial step edge junctions and flux transformers were studied.

1. Introduction

High-T_c thin film superconducting devices utilize the unique properties of macroscopic quantum phenomena in superconductors and much lower cryogenic costs for their operation at 77 K, when compared to low-T_c devices, which are affected by the increasing cost of liquid helium. However, the low noise performance of high-T_c circuits requires sophisticated engineering of the microstructures of the films and Josephson junctions, taking into account the $d_{x^2-y^2}$ pairing symmetry of the superconducting gap function and the short coherence length in high-T_c superconductors. A single grain boundary in a high-T_c film that has been fabricated on a properly prepared substrate surface can serve as a high quality Josephson junction. Step edge Josephson junctions are cheaper in production when compared to ramp- and bicrystal Josephson junctions and demonstrate excellent superconducting parameters [1-4]. The reproducibility of step edge junctions can be improved using multilayer epitaxial buffering of step edges. Epitaxially buffered MgO substrates [5] are advantageous for the growth of $YBa_2Cu_3O_{7-x}$ (YBCO) films due to the similar thermal expansion coefficients of MgO and YBCO. Epitaxial buffers can also be used to improve the growth of YBCO films over the step edges: enhanced graphoepitaxial growth of YBCO films on a texturing layer with an artificially created micro-relief on the surface steps of MgO substrates has been proposed and verified [4, 6, 7]. These Josephson junctions were intended for integration in high-T_c direct current superconducting quantum interference devices (high-T_c DC SQUIDs). Such SQUIDs can be used, for example, in measurement systems for low field magnetic resonance imaging, geomagnetic surveys or magnetoencephalography (MEG). In the present paper, we describe high-T_c DC SQUIDs prepared with an additional epitaxial buffer layer of $SrTiO_3$ (STO), which serves as a blocking barrier against possible contamination of the superconducting YBCO film as a result of diffusion of Mg from the MgO substrate [8].

[1] Corresponding author. E-mail address: m.faley@fz-juelich.de.

2. Experimental details

Heterostructures of Josephson junctions and DC SQUID magnetometers were deposited by high-oxygen-pressure magnetron sputtering from stoichiometric polycrystalline targets [9, 10]. The formation of steps on MgO substrates was performed by ion beam etching (IBE) at an incident angle of 45° over the edge of an AZ TX1311 photoresist reflowed to achieve a 45° slope angle. The photoresist was removed using acetone and methanol. A second IBE step was used at an incident angle of 90° to clean the surface of fences of resputtered material [11] and to make an initial texture in the form of linear trenches along the [100] and [010] directions of the MgO substrate [4]. Enhancement of the texture was achieved by depositing a 10 nm thick homoepitaxial MgO film [12]. This texture was used to achieve graphoepitaxial growth of the YBCO films, resulting in an in-plane orientation of the YBCO films on the step edge and on the rest of the substrate surface. Contamination of the YBCO film and grain boundaries by Mg from the MgO substrate [8] was avoided by using an epitaxial blocking layer of STO of thickness about 30 nm deposited above a 10 nm thick YBCO seed layer. Atomic force microscopy (AFM), scanning electron microscopy (SEM) and high-resolution transmission electron microscopy (HRTEM) were used to characterise the textured surfaces of the step edges and the microstructure of the films. In-plane alignment of the grains in the YBCO films was confirmed by observing the orientations of growth spirals on the film surfaces using AFM and SEM.

The DC SQUIDs were made from the YBCO-STO-YBCO heterostructures and each consisted of two 2-μm-wide step-edge Josephson junctions. Multilayer superconducting flux transformers for inductively coupled SQUID magnetometers were prepared using YBCO, $PrBa_2Cu_3O_{7-x}$ and STO films on single crystal MgO (001) wafers buffered by epitaxial $BaZrO_3$ and STO films and assembled in a flip-chip geometry with the DC SQUIDs.

3. Results and discussion

HRTEM images of an YBCO-STO-YBCO heterostructure deposited on a textured step edge on an MgO substrate is shown in Fig. 1. The lower YBCO layer serves only as a seed layer and was not superconducting at 77 K due to contamination by Mg atoms sustained during deposition of the YBCO. Thanks to the use of the STO blocking layer, the superconducting transition temperature T_c of the top YBCO layer increased from approximately 89 K to above 91 K. Similar to the results obtained in Ref. [4], two [100]-tilted 45°-misoriented grain boundaries were observed at each step edge junction.

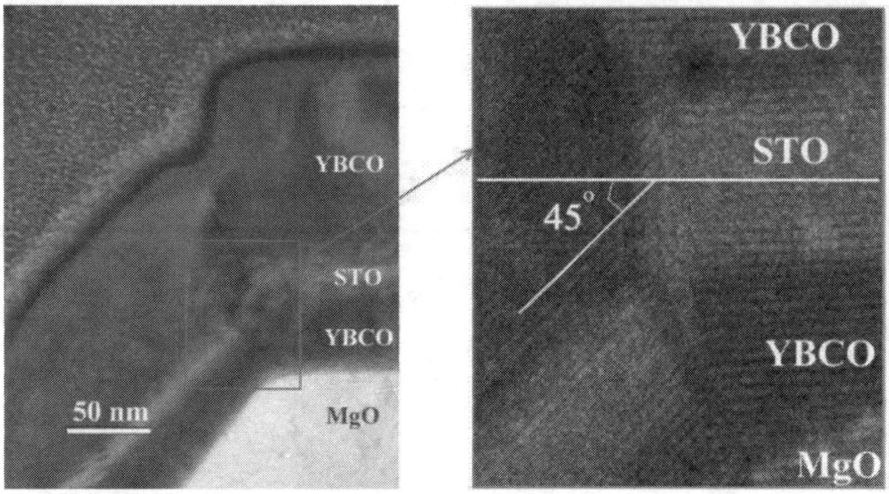

Figure 1. HRTEM images of an YBCO-STO-YBCO heterostructure deposited on the top corner of a textured step edge on an MgO substrate.

The implementation of the surface texture changed the growth mode of the YBCO film on the MgO surface from epitaxial to graphoepitaxial, resulting in alignment of the in-plane orientation of the YBCO film on the substrate steps. The orientation of the growth spirals showed that all of the grains in the YBCO film were aligned in-plane, with their *a*- or *b*- axes normal to the step. This property was reflected in improved reproducibility and improved parameters of the step-edge Josephson junctions made by patterning 2 μm wide bridges in the YBCO films across the steps. The I-V characteristics of the Josephson junctions showed RSJ-like behavior, with negligible excess supercurrent and values of I_cR_n product of approximately 0.6 mV at 77 K.

The high value of the I_cR_n product observed in the graphoepitaxial Josephson junctions can be explained by the fact that the grain boundaries in these junctions are straight, leading to a more homogeneous distribution of the Josephson current compared to that in [001]-tilted bicrystal junctions and to step edge junctions without in-plane alignment of the grains. In addition, Andreev bound quasiparticle states at the midgap energy (zero-energy states) probably do not appear in the case of a [100]-tilted grain boundary junctions [13]. These states could otherwise be responsible for partial electrical shunting of the resistance and for deviation of the critical current from the Sigrist–Rice phenomenological approach [14] in the case of [001]-tilted grain boundary junctions.

Step edge junctions provide much more freedom with regard to their positioning on the substrate, when compared to bicrystal junctions. Also, the steps on the substrate can be placed so that they influence the YBCO films only in the areas of the Josephson junctions, with the rest of the YBCO films remaining unperturbed and retaining the highest critical current and lowest noise values. These features are advantageous for the construction of both direct coupled DC SQUIDs (see Fig. 2) and also SQUIDs that are inductively coupled to superconducting flux transformers with multiturn input coils, due to the improved availability and lower costs of the substrates, improved reproducibility of the Josephson junctions due to a smaller concentration of voids, higher I_cR_n product, lower junction capacitance and easier alignment during photolithography, when compared with bicrystal junctions.

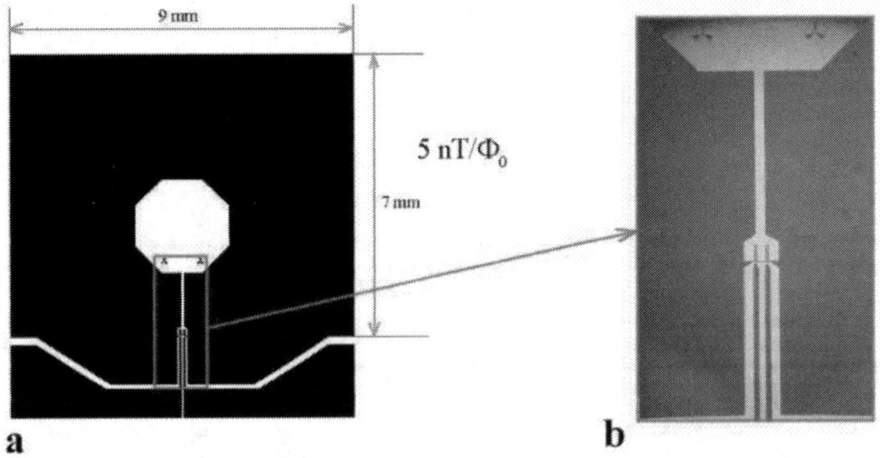

Figure 2. (a) Sketch of a direct coupled high-T_c DC SQUID magnetometer with an 8 mm pick-up loop and step edge Josephson junctions. (b) Photograph of the inner part of the magnetometer with two DC SQUIDs. Each SQUID has an inductance of approximately 100 pH.

Figure 2 shows a sketch and a photograph of a direct coupled DC SQUID magnetometer with step edge Josephson junctions prepared on a 10 mm x 10 mm MgO substrate. The SQUID magnetometer consists of a 9 mm x 7 mm pick-up loop and two DC SQUIDs. Each DC SQUID has a loop inductance of approximately 100 pH, according to estimates made with the help of the software package 3D-MLSI [15]. Such direct coupled DC SQUID magnetometers with graphoepitaxial step edge Josephson junctions demonstrated an ~ 5 nT/Φ_0 field-to-flux transformation coefficient and a magnetic field resolution of ~ 50 fT/√Hz at 77 K.

An increase in the size of the pick-up loop to about 25 mm leads to an improvement in the field-to-flux transformation coefficient to approximately 1.5 nT/Φ_0, and to a magnetic field resolution of ~ 15 fT/√Hz at 1 kHz and 77 K. This value is comparable to that reported in Ref. [16] for similar frequency and temperature conditions. A moderate improvement of sensitivities can be obtained by inductive coupling of the direct coupled SQUIDs to single layer flux transformers made from relatively thick superconducting films [5]. Such coupling leads also to a reduction in the pick-up loop inductance, which improves the sensitivity and operational stability of the sensor even if the flux concentrator has a similar area to that of the pick-up loop of the SQUID.

Much greater sensitivity improvement was achieved by the implementation of inductive coupling of the SQUIDs to a multiturn input coil of 8-mm or 16-mm superconducting flux transformers [17]. Magnetometers with 8-mm flux transformers demonstrated field-to-flux transformation coefficient of approximately 1.2 nT/Φ_0 and magnetic field resolution of ~ 12 fT/√Hz at 77 K. Magnetometers with 16-mm flux transformers observed field-to-flux transformation coefficients of ~ 0.45 nT/Φ_0 and magnetic field resolutions of ~ 5 fT/√Hz at 77 K. The noise spectrum of these magnetometers was white down to the frequencies of about 10 Hz and achieved of ~ 20 fT/√Hz at 1 Hz.

Acknowledgments

The authors gratefully acknowledge fruitful discussions with Yu. V. Maslennikov, A. S. Sobolev, G. A. Ovsyannikov and V. P. Koshelets and the technical assistance of V. Yu. Slobodchikov and R. Speen. The authors acknowledge IB-BMBF (project 01DJ13014) for partial financial support.

4. References

[1] Daly K P, Dozier W D, Burch J F, Coons S B, Hu F L, Platt C E, and Simon R. W 1991 *Appl. Phys. Lett.* **58**, 543

[2] Jia C L, Kabius B, Urban K, Herrmann K, Schubert J, Zander W, and Braginski A I 1992 *Physica C* **196** 211

[3] Foley C P, Lam S, Sankrithyan B, and Wilson Y 1997 *IEEE Trans. Appl. Superconductivity* **7** 3185

[4] Faley M I, Poppe U, Dunin-Borkowski R E, Schiek M, Boers F, Chocholacs H, Dammers J, Eich E, Shah N J, Ermakov A, Slobodchikov V Yu, Maslennikov Yu V, and Koshelets V P 2013 *IEEE Transactions on Appl. Supercond.* **23** 1600705

[5] Faley M I, Mi S B, Petraru A, Jia C L, Poppe U, and Urban K 2006 *Appl. Phys. Lett.* **89** 082507

[6] Faley M I 2012 Reproduzierbarer Stufen-Josephson-Kontakt *Patent pending DE102012006825*

[7] Faley M I, Meertens D, Poppe U, and Dunin-Borkowski R E 2013 *Extended abstracts of 14th International Superconductive Electronics Conference (ISEC 2013)* 247-249

[8] Hao Z, Wu Y, Enomoto Y, Tanabe K, and Koshizuka N 2002 *Journal of Appl. Phys.* **91** 9251

[9] Poppe U, Klein N, Dähne U, Soltner H, Jia C L, Kabius B, Urban K, Lubig A, Schmidt K, Hensen S, Orbach S, Müller G, and Piel H 1992 *J. Appl. Phys.* **71** 5572

[10] Faley M I, and Poppe U 2012 *Patent* WO2012051980

[11] Mitchell E E and Foley C P 2010 *Supercond. Sci. Technol.* **23** 065007

[12] Copetti C A, Schubert J, Klushin A M, Bauer S, Zander W, Buchal Ch, Seo J W, Sanchez F, Bauer M 1995 *J. Appl. Phys.* 78 5058

[13] Löfwander T, Shumeiko V S, and Wendin G 2001 *Supercond. Sci. Technol.* **14** R53

[14] Barone A, Lombardi F, Monaco A, Sarnelli E, Tafuri F, and Testa G 2004 *Phys. Stat. Sol. (b)* **241** 1192

[15] Khapaev M M, Kupriyanov M Yu, Goldobin E, Siegel M 2003 *Supercond. Sci. Technol.* **16** 24

[16] Lee L P, Teepe M, Vinetskiy V, Cantor R, and Colclough M S 1995 *Appl Phys Lett.* **66** 3058

[17] Faley M I, Poppe U, Urban K, Paulson D N, Starr T, and Fagaly R L 2001 *IEEE Transactions on Appl. Supercond.*, **11** 1383

Superconducting Microresonator Detectors for Neutrino Physics in Milano

E. Ferri[1], P. Day[2], P. Falferi[3], M. Faverzani[1], A. Giachero[1], C. Giordano[4], H.G. LeDuc[2], B. Marghesin[4], F. Mattedi[4], R. Mezzena[5], R. Nizzolo[1] and A. Nucciotti[1]

[1] Università di Milano-Bicocca and INFN Milano-Bicocca, Milano, Italy
[2] Jet Propulsion Laboratory, Pasadena, CA, U.S.A
[3] Istituto di Fotonica e Nanotecnologie, CNR-Fondazione Bruno Kessler, Trento, Italy
[4] Fondazione Bruno Kessler, Trento, Italy
[5] Dipartimento di Fisica, Università di Trento, Trento, Italy

E-mail: elena.ferri@mib.infn.it

Abstract.
Superconducting microwave microresonators are low temperature detectors compatible with large-scale multiplexed frequency domain readout. Our aim is to adapt and further advance the technology of microresonator detectors to develop new devices applied to the problem of measuring the neutrino mass. More specifically, we aim to develop detector arrays which can be applied to the calorimetric measurement of the energy spectra of ^{163}Ho EC decay (Q $\sim$ 2-3 keV) for a direct measurement of the neutrino mass. In order to achieve this goal, a study aimed to the selection of the best design and material for the detectors is required. A recent advance in microwave microresonator technology was the discovery that some metal nitrides, such as TiN, possess properties consistent with very high detector sensitivity. In this contribution, our progress on the design and test of Ti/TiN multilayer films is presented. We report measurements made on stoichiometric TiN, sub-stoichiometric TiN and multilayer Ti/TiN films including the critical temperature, the gap parameter and the quasi-particle recombination time extrapolated from $\sim$keV X-ray pulses.

1. Introduction

The determination of the neutrino mass is still an open question in particle physics. Direct neutrino mass measurements, performed by analyzing the kinematics of suitable weak decays, are able to tackle this problem. Exploiting only energy and momentum conservation, they are essentially free of theoretical assumptions about neutrino properties and are truly model-independent. These experiments look for a tiny deformation of the beta spectrum near the end point energy due to a non vanishing neutrino mass. The calorimetric technique, where the β source is embedded in the detectors, is a promising tool since the entire energy except the fraction carried away by neutrinos is measured. This approach eliminates the systematic effects connected with the use of an external source as in a β spectrometer. A disadvantage is that the isotope used in calorimetric measurements must be characterized by a low end point energy Q in order to allow a trade off between keeping the isotope activity low to minimize pile-up events and still achieving a high statistical sensitivity. An appealing isotope applicable to neutrino mass measurement could be the ^{163}Ho ($Q \sim$2.5 keV), which decays via electron capture EC to ^{163}Dy with a half-life of about 4570 y. The capture is only allowed from the M shell or higher. One of the methods to estimate the neutrino mass from the ^{163}Ho consists in studying the end-point of the total absorption spectrum as proposed by Rujula and Lusignoli in 1982 [1]. The absorption spectrum is made up of peaks with Breit-Wigner shapes and it ends at $Q - m_\nu$. In order to obtain a sub-eV sensitivity on neutrino mass, it would be necessary to have a large number of detectors multiplexed ($\sim 10^4$) with a good time response ($\sim \mu$s) and a good energy resolution ($\sim$ eV @ keV).

Our aim is to develop arrays of athermal micro-resonator detectors for the electron capture end point measurement of the neutrino mass using ^{163}Ho as source material [2, 3]. Currently, a study aimed to select the best material and the best design for the detectors is in progress. In this contribution the current progress on the design and test of Ti/TiN multilayer film is presented. Furthermore, a comparison between the characteristic parameters (i.e. critical temperature, gap parameter, quasiparticle recombination time) of films made of stoichiometric, sub-stoichiometric TiN and Ti/TiN multilayer films is reported. The sub-stoichiometric was provided by Jet Propulsion Laboratory while the stoichiometric and multilayer films by Fondazione Bruno Kessler in Trento.

2. Detector layout

Our array is composed by 16 individual detectors (lumped-element resonators) with resonance frequencies distributed from 3 to 6 GHz and with a variety of coupling quality factors. Their layout consists of two interdigital capacitors (IDC) connected with a coplanar strips (CPS) transmission line that works as an inductor. Each resonator is capacitively coupled on one side to a coplanar waveguide (CPW) line that is used for the readout. The strength of the coupling, which set the coupling quality factor Q_c, is determined by the width of the gap between the CPW line and the resonator. The spacing between the IDC finger is 10 or 30 μm. This large spacing is intended to reduce the two-level system (TLS) noise associated with amorphous dielectric layers at surfaces. For more details see [2].

3. Film production

Our goal is to produce films with a T_c in the range 1÷1.5 K in order to have a low energy gap and long quasi-particle recombination time (i.e. low generation-recombination noise).
Titanium Nitride (TiN) is a well suited for the production of MKIDs. Indeed TiN resonators show very high quality factors and large ratios of kinetic to magnetic inductance. Furthermore, by changing the amount of Nitrogen added to the Titanium film it is possible to tune the critical temperature T_c [4] in the range between 0 and 5 K. The T_c lowering occurs for a narrow range of nitrogen concentration and requires a very precise control of the nitrogen incorporation. An alternative that we preferred deals with Ti/TiN multilayer [5]. The basic idea is to fabricate films composed by a sequence of pure Titanium and stoichiometric TiN layers. Indeed, it is known that the T_c of a superconducting material can be reduced by the superposition of a normal metal or a metal with a superconductive transition at lower temperature [6, 7]. In this configuration the Cooper pairs leak into the normal metal, consequentially the pair density in the superconducting metal is reduced as well as the T_c. The magnitude of this effect depends on the thickness of the two layers. By adjusting these thickness one is able to tune the T_c between the transition temperature of Ti (0.4 K) and TiN (5 K).

The Ti/TiN multilayer have been produced by the superposition of layers Ti/TiN with Ti at the bottom interface and TiN on the top. The maximum number of stacked layers is 8. The films have been sputtered on high resistance Silicon wafers, which had been cleaned and etched with hydrofluoric acid (HF) to remove the native oxide. The DC 4-wire measurement of the electrical resistance as function of temperature was performed at the Cryogenic Laboratory of the University of Trento. The results have shown that it has been possible to tune the T_c

between 0.5 and 4.6 K by properly choosing the Ti thickness in the $(0 \div 15)$ nm range, and the TiN thickness in the $(7 \div 100)$ nm range. The T_c can be lowered by reducing the TiN thickness with fixed thickness of Ti or by keeping fixed the TiN thickness and increasing that of Ti. In this latter case the T_c is more sensitive to layer thickness variations in the $(5 \div 10)$ nm range and it is nearly stable for Ti thickness < 5 nm and > 10 nm. The produced Ti/TiN multilayers have shown a high uniformity and high quality factor [8].

4. Device characterization

Microwave measurements were performed to determine the characteristic parameters of the resonators made of stoichiometric, sub-stoichiometric TiN and Ti/TiN multilayer films. The critical temperature has been evaluated by measuring the amplitude of the RF signal transmitted through the resonator feed line as a function of the temperature. The energy gap parameter has been calculated by measuring the internal Qs of the resonators vs temperature (figure 1) and fitting to the relation:

$$\frac{1}{Q(T)} = \frac{1}{Q(0)} + \frac{\alpha\sigma_1(T)}{2\sigma_2(T)} \tag{1}$$

where the conductance depends (in the limit $k_BT \ll \Delta(0)$, $\hbar\nu \ll \Delta(0)$) on the gap parameter Δ as follow [9]:

$$\frac{\sigma_1}{\sigma_n} \approx \frac{4\Delta(T)}{\hbar\omega} exp(\frac{-\Delta(0)}{k_BT}) K_0(\frac{\hbar\omega}{2k_BT}) sinh(\frac{\hbar\omega}{2k_BT}) \tag{2}$$

$$\frac{\sigma_2}{\sigma_n} \approx \frac{\pi\Delta(T)}{\hbar\omega}[1 - 2exp(-\frac{\Delta(0)}{k_BT}) exp(-\frac{\hbar\omega}{2k_BT}) I_0(\frac{\hbar\omega}{2k_BT})], \tag{3}$$

where I_0 and K_0 are modified Bessel functions, σ_n is the normal state resistance just above T_c.

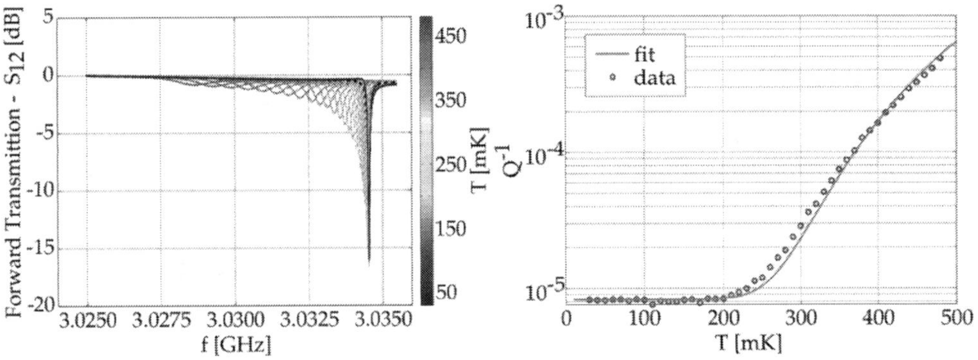

Figure 1. Example of the gap measurement for a multilayer film: it is possible to see how the resonant frequency and the Q of one single resonance decrease while the temperature increase.

The measured values for four different films are reported in the table 1.

5. Quasiparticle recombination time and energy spectra

Tests were performed radiating the detectors with two calibration sources (^{55}Fe, $\sim$ 6 keV; fluorescence of Al, $\sim$ 1.5 keV) to extrapolate the quasi-particle lifetime at different temperatures

Table 1. Critical temperature, gap parameter and Qs obtained for our films. The sub-stoichiometric was provided by Jet Propulsion Laboratory while the stoichiometric and multilayer films by Fondazione Bruno Kessler in Trento. In the table it is also listed the quasi-particle recombination time at 25 mK.

Film	T_c(K)	Δ(meV)	Q_i	$\tau_{fast}(\mu s)$	$\tau_{slow}(\mu s)$
Stoichiometric TiN	4.6	0.8	10^5	–	8
Sub-stoichiometric TiN	2.5	0.37	10^6	25	145
Ti(10nm)/TiN(15nm) 8 layers	1.6	0.26	10^5	–	–
Ti(10nm)/TiN(12nm) 8 layers	1.2	0.17	$5\cdot10^4$	31	105

and to build the first energy spectrum. The observed pulses showed several different behaviors: slower pulses, with a single exponential decay constant, very fast pulses decaying faster than exponential, and pulses with two exponential decay times. We interpret the slow component as the quasi-particle recombination time and the fast component as a result of a recombination of quasi-particles due to a local excess of quasi-particles. The results are listed in table 1.

Unfortunately, in the energy spectra acquired no peaks were observed. This feature could be explained by an escape of phonons from the detector to the substrate. Besides, since the large majority of X-rays are absorbed in the substrate ($\sim$ 90 %), a response to the phonons thereby produced could induce an energy degraded response of the detector. Finally, given the slow diffusivity of the TiN, a transition of the superconductor into normal metal state, caused by a local excess of quasi-particles, is also possible. To obtain an energy-resolving resonator, we are considering 3 different future upgrades: resonators with the inductor line suspended from the substrate to prevent the exchange of phonons between the film and the substrate; resonators with the inductor coupled to a higher-gap absorber with a faster diffusion to prevent the recombination of the quasi-particles before they diffuse; and ultimately, thermal devices.

6. Conclusion

Ti/TiN multilayer films suitable for a next generation of neutrino experiments are produced. The characteristic parameters, such as the T_c, the gap parameter and the quasi-particle lifetime of different TiN superconducting films have been measured. The energy spectra acquired have not yet shown well defined peaks, so new resonators are under investigation.

Acknowledgments

These activities are supported by Fondazione Cariplo through the project Development of Microresonator Detectors for Neutrino Physics (grant 2010-2351).

References

[1] A. De Rujula and M. Lusignoli 1982 *Phys. Lett. B* **118** 72
[2] M. Faverzani et al 2012 *J. Low Temp. Phys.* **167** 1041
[3] M. Faverzani et al 2013 *NIMA* **167** 492
[4] H. G. Leduc et al 2010 *Appl. Phys. Lett.* **97** 102509
[5] M. R. Vissers at al 2013 *Appl. Phys. Lett.* **102** 232603
[6] N. R. Werthamer 1963 *Physical Review* **132** 2440
[7] L. Yu, N. Newman, and J. M. Rowell 2002 *IEEE Trans. on App. Superconductivity* **12** 1795
[8] A. Giachero et al 2013 arXiv:1307.3781 [cond-mat.supr-con]
[9] D. C. Mattis and J. Bardeen 1958 *Phys. Rev* **111** 412
[10] B. Geilikman and V. Kresin 1974 Kinetic and Nonsteady-state Effects in Superconductors. J. Wiley and Sons

Niobium nano-SQUIDs based on sub-micron tunnel junction fabricated by three-dimensional Focused Ion Beam sculpting

M Fretto[1], E Enrico[1], N De Leo[1], L Boarino[1], V Lacquaniti[1], C Granata[2], R Russo[2] and A Vettoliere[2]

[1] INRIM, Istituto Nazionale di Ricerca Metrologica, Strada delle Cacce, 91 Torino 10135, Italy
[2] Istituto di Cibernetica "E. Caianiello" del Consiglio Nazionale delle Ricerche, Pozzuoli (Napoli), Italy

E-mail: m.fretto@inrim.it

Abstract. A three dimensional nano-SQUID (Superconducting Quantum Interference Device) has been realized in a vertical configuration (with the loop in the same plane of Josephson Tunneling Junctions, JTJs). The loop area is 0.25 μm^2 corresponding to a modulation period of about 5 mT, the square JTJs have a side length of 0.3 μm. Josephson junction's fabrication is carried out combining optical lithography to pattern trilayer and three dimensional (3D) Focused Ion Beam (FIB) sculpting technique to define the junctions' and the loop's areas. Two different ion etching processes were performed, perpendicular and parallel to the multilayer, resulting in a precise 3D structure. Finally, a standard anodization was performed to eliminate the unstructured surface material generated by the high energetic ion beam assuring high quality junctions. Electric transport characteristics of the nanodevice measured at T = 4.2 K are reported, in particular the current-voltage characteristics and critical current vs external magnetic field. The high modulation depth of the critical current (up to 70% of the I_c at zero magnetic flux) and the device reliability are very encouraging in view of nanoscience applications.

1. Introduction

In last years, there is a growing interest to investigate small cluster of magnetic nanoparticles, small spin population or magnetic nano-objects. The challenge is the measurement of the single atomic or molecule spin. In this framework nano superconducting quantum interference devices (nano-SQUIDs) having a flux capture area less than 1 μm^2 represent a powerful tool to study the magnetic properties of the nanoparticles at a microscopic level. Indeed, a small capture area improves the magnetic coupling between the SQUID and the nano-object under investigation. Due to the ultra low intrinsic magnetic flux noise (less than 1 $\mu\Phi_0/Hz^{1/2}$), it is able to detect few tens of elementary magnetic moment (Bohr magnetons) in the unit of bandwidth [1-5].
Beside the nanomagnetism applications, these nanodevices have been proposed for spintronics, single photon and macromolecule detection, quantum computing, nanoelectronics including memory, quantum metrology and scanning magnetic microscopy with a very high spatial resolution [6]. Typically, a dc nano-SQUID consists of a square loop having a side length less than 1 μm interrupted by two nanometric niobium constrictions (Dayem bridges). Contrary to the tunnel junctions, Dayem nanobridges can be easily fabricated by a single nanopatterning step, and they are resilient to the

magnetic field applied in the plane of the SQUID loop. However there are many disadvantages limiting the potentiality of nano-SQUIDs. Due to the difficulty to fabricate Dayem bridges having a non-hysteretic current-voltage characteristic in a reliable and reproducible way, in several applications, the dc nano-SQUIDs are employed as a magnetic flux-to-current transducer [7] preventing the low noise readout scheme employed in the standard configuration where the SQUID is used as a magnetic flux to voltage converter. Due to a non-trivial current-phase relationship, a SQUID based on Dayem bridges exhibits a quite different behaviour with respect to a standard one based on the Josephson tunnel junctions [8]. In particular the modulation critical current depths are small (maximum 20 - 30%) leading to an increase of magnetic flux noise. Moreover the nanobridge based SQUIDs show a limitation in the critical current reproducibility: since the critical current depends on the ratio between the geometrical dimension of the bridges (mainly the length and the width) and the superconducting coherence length, small variation of the latter can induce relative large variation in the critical current. In the present work we will report experimental results about the fabrication and the characterization of nano-SQUIDs in a vertical configuration based on sub-micron Josephson tunnel junctions (JTJs) fabricated by the Focused Ion Beam (FIB) sculpting [9, 10]. With respect to nanobridge based SQUID, JTJs nano-SQUIDs exhibit high modulation depths, ultra low noise and a better control of the critical current.

2. 3D Nano-SQUID fabrication process

The reported nano-SQUIDs are based on two overdamped Nb/Al-AlO_x/Nb SNIS (Superconductor-Normal metal-Insulator-Superconductor) Josephson junctions [11-13] coupled through a superconducting loop directly realized in the multilayered structure. In particular, the junctions' active area and the loop were patterned by an unconventional lithographic technique based on a FIB sculpting method arising in a 3D structure [9]. Compared to the mainstream optical and/or electronic lithography, this approach represents an improvement for the accurate control of the device dimensions by means of a Ga^+ energetic beam able to selectively and precisely remove portions in the multilayer. The overall fabrication process is based on different steps. Initially, the Nb/Al-AlO_x/Nb superconductive multilayer is grown by the sputtering technique through an RF deposition system, in a high vacuum atmosphere. It was previously patterned by optical lithography to draw either electrical contact pads or microstripes, and defined by lift-off in an acetone bath.

Detailing the SNISs fabrication parameters, their tunneling barrier is represented by a metallic aluminum (Al) layer 80 nm thick, along an aluminum oxide (AlO_x) grown by oxidizing Al in pure oxygen ambient, directly into the sputtering system maintaining a high degree of vacuum. On the other hand, two niobium thin films represent the superconducting electrodes, with a thickness of 400 nm.

Subsequently, the following step was the definition of the JTJs and consequently the loop by the 3D FIB sculpting method. A FEI Quanta 3D ESEM FEG equipped with a J. C. Nabity NPGS pattern generator was employed, carrying the sample on a customized sample holder pre-tilted at 30°. In this way it is possible to manage the micro movements of the substrate to orient it in parallel and vertically respect to the ion beam trajectory. In detail, positioning the sample perpendicular to the ion beam trajectory (Figure 1a), a multilayered lamella was roughly narrowed by a multi-step etching down to 0.6 μm at 30 pA of beam current. A further sidewalls refinement was performed at 10 pA until a lamella thickness of about 0.4 μm. This dimension will represent either the junctions or the loop width. Afterwards, the sample was tilted parallel to the beam and two (0.2 x 0.6) μm^2 rectangular holes were realized in the lamella at a distance of 400 nm and vertically respect to the substrate. In particular, the first incision is cutting the top Nb electrode, while the second one crosses the base Nb electrode. Thereafter, a horizontal hole is performed with a dimension of (1 x 0.25) μm^2 between the two previous apertures (Figure 1b). At this point, the junctions' and the loop's length of the nano-SQUID device are defined. The last fabrication step is represented by an anodization process in order to erase the amorphous layer arising from the Ga^+ ion implantation occurring at the sidewalls of Josephson junctions. This operating phase is necessary in order to avoid any contribution to the critical current due to spurious conducting paths, and coming from the redeposited metallic material during the FIB etching. This process was carried out with a constant current density of 0.5 mA/cm^2 up to a voltage of 50 V. As a consequence of the anodization process a reduction of the effective Josephson junctions' area down to about (0.3 x 0.3) μm^2 occurred. In figure 1c, a scanning electron micrograph (SEM) image of a 3D nano-SQUID is reported; the areas of the loop and the junctions are 0.25 μm^2 and 0.09 μm^2 respectively.

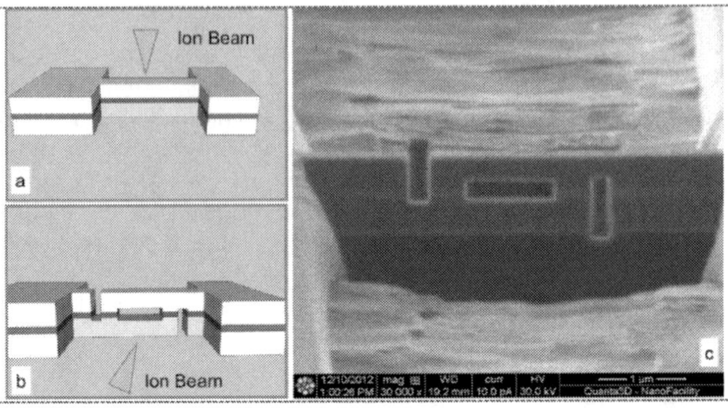

Figure 1. 3D Nano-SQUID fabrication steps: a) The sample surface is positioned perpendicular to the ion beam trajectory; b) The sample surface parallel to the ion beam trajectory; c) A scanning electron microscope (SEM) photograph of a 3D nano-SQUID fabricated by focused ion beam lithography.

3. Nanodevice characterization

The nanodevices have been characterized at T = 4.2 K by using an electromagnetically shielded Dewar. Current vs voltage (I-V) and critical current vs external magnetic field (I_c-Φ) have been measured (Figure 2). All the electrical connections to room temperature went through customized twin axial cables. The I-V characteristic has been obtained by biasing the SQUID with a current ramp and measuring the voltage across it. For this purpose, a low-noise 4-contact current-voltage technique has been used. The I-V shows a little hysteresis due to a critical current higher than expected, therefore at T = 4.2 K it can operate as a magnetic flux to current transducer. In this configuration, an external magnetic flux threading the SQUID loop causes a critical current modulation ΔI_c, which can be accurately measured. Knowing the transfer factor between the external magnetic flux and the critical current $I_* = \partial I_c/\partial\Phi$ (critical current responsivity) it is possible to obtain the external magnetic flux value $\Delta\Phi_{ext} = \Delta I_c/I_*$. By figure 2a, a normal resistance value R_N of 1 Ω can be evaluated, resulting in a value of resistance times junction area of $R_N \cdot A = 80$ m$\Omega\cdot$ μm^2.

Furthermore, the magnetic pattern was measured by applying an in plane magnetic field provided by a solenoid surrounding the sample. The I_c-Φ curve shows a sinusoidal-like behaviour and it is resonance-free allowing the sensor to work in a wide region of bias points, the field to flux transformation coefficient is 10 mT/Φ_0 corresponding to an effective area of 0.2 μm^2, in close agreement with the measured geometrical loop area. The maximum I_c was of 360 μA with a critical current modulation depth $\delta I_c = 180$ μA (50% of I_c). The critical current responsivity I_* is obtained by taking the maximum derivative of figure 2b. Due to high modulation depths, a I_* as high as 400 μA/Φ_0 has been evaluated. Considering the above responsivity value and a measurement error of the critical current $I_{c,n}$ of about 1 part 10^4 (0.05 μA), an overall magnetic flux sensitivity of $\Delta\Phi_n = I_{c,n}/I_* = 125$ $\mu\Phi_0$ can be estimated. It is worth to note that, in the case the sensor works as magnetic flux to voltage transducer, high values of both δI_c and $R_N \cdot A$ produce high values of voltage modulation and responsivity and a consequent low spectral density of magnetic flux noise.

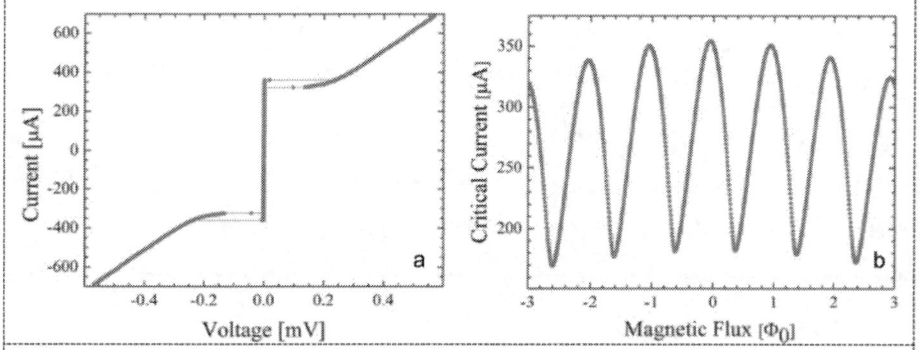

Figure 2. a) Current-Voltage characteristic of a vertical nano-SQUID measured at T = 4.2 K; b) Critical current vs external magnetic flux measured at T = 4.2 K.

4. Conclusion

A SQUID nanosensor based on submicron tunnel Josephson junctions has been fabricated and characterized. The high current responsivity allowed reaching an overall magnetic flux noise of 125 $\mu\Phi_0$. Due to the reliability and the performances of these nanosensors, we believe that the presented results are very encouraging in view of a wide employment of nano-SQUIDs in the nanomagnetism investigations. Work is in progress to reduce the critical current density in order to completely eliminate the hysteresis. In such a way, it will be possible to employ the nano-SQUID as a magnetic flux to voltage converter and use it to its fullest potential. For this new fully overdamped device both the spectral density of flux noise (in the maximum derivative point) and the 1/f noise will be also evaluated, as already carried out on other SNIS-based SQUID nanosensors [10].

5. References

[1] Lam S K H and Tilbrook D L 2003 *Appl. Phys. Lett.* **82** 1078
[2] Troeman A G P, Derking H, Boerger B, Pleikies J, Veldhuis D and Hilgenkamp H 2007 *Nano Lett.* **7** 2152
[3] Granata C, Esposito E, Vettoliere A, Petti L and Russo M 2008 *Nanotech.* **19** 275501
[4] Hao L, Aßmann C, Gallop J C et al. 2011 *Appl. Phys. Lett.* **98** 092504
[5] Granata C, Vettoliere A, Walke P, Nappi C and Russo M 2009 *J. Appl.Phys.* **106** 023925
[6] Foley C P and Hilgenkamp H 2009 *Supercond. Sci. Technol.* **22** 064001
[7] Russo R, Granata C, Esposito E, Peddis D, Cannas C and Vettoliere A 2012 *Appl. Phys. Lett.* **101** 122601
[8] Granata C, Vettoliere A, Russo M and Ruggiero B 2011 *Phys. Rev.* B **84** 224516
[9] Fretto M, Enrico E, De Leo N, Boarino L, Rocci R and Lacquaniti V 2013 *IEEE Trans. Appl.Supercond.* **23** 1101104
[10] Granata C, Vettoliere A, Russo R, Fretto M, De Leo N and Lacquaniti V 2013 *Appl. Phys. Lett.* **103** 102602
[11] Lacquaniti V, Cagliero C, Maggi S and Steni R 2005 *Appl. Phys. Lett.* **86** 042501
[12] Lacquaniti V, De Leo N, Fretto M, Maggi S and Sosso A 2007 *Appl. Phys. Lett.* **91** 252505
[13] Lacquaniti V, Andreone D, De Leo N, Fretto M, Sosso A, Khachaturova T and Belogolovskii M 2009 *IEEE Trans. Appl. Supercond.* **19** 3 234-237

Acknowledgments

The 3D FIB sculpting fabrication has been performed at NanoFacility Piemonte, INRIM, a laboratory supported by Compagnia di San Paolo.

Quantum non-locality vs. quasi-local measurements in the conditions of the Aharonov-Bohm effect

Armen M. Gulian

Institute for Quantum Studies, Chapman University, 15202 Dino Dr., Burtonsville, MD 20866, USA

E-mail: gulian@hotmail.com

Abstract. Theoretical explanation of the Meissner effect involves proportionality between current density and vector potential, which has many deep consequences. As noticed by de Gennes, superconductors in a magnetic field "find an equilibrium state where the sum of kinetic and magnetic energies is minimum" and this state "corresponds to the expulsion of the magnetic field". This statement still leaves an open question: from which source is the superconducting current acquiring its kinetic energy? A naïve answer, perhaps, is from the energy of the magnetic field. However, one can consider situations (Aharonov-Bohm effect), where the classical magnetic field is locally absent in the area occupied by the current. Experiments demonstrate that despite the local absence of the magnetic field, current is, nevertheless, building up. From what source is it acquiring its energy then? Locally, only a vector potential is present. How does the vector potential facilitate the formation of the current? Is the current formation a result of a truly non-local quantum action, or does the local action of the vector potential have experimental consequences? We discuss possible experiments with a hybrid normal-metal superconductor circuitry, which can clarify this puzzling situation. Experimental answers will be important for further developments.

1. Introduction

Superconductors occupy a peculiar position among solid-state objects: they demonstrate quantum properties not only at the microscopic and mesoscopic levels, but also – and most importantly – at the macroscopic level. One encounters this feature when considering superconductors main property: charge transport without resistance. It turns out [1] (F. London and H. London, 1935) that one should adopt a relation $j \propto -A$ (we drop unimportant factors in intermediate expressions) to avoid infinitely large values of conductivity and to describe the Meissner effect [2]. This immediately separates superconductors from classical objects: in classical physics observables cannot depend on A, since A is not gauge invariant. As soon as superconductors belong to the quantum world, they can "supply" a quantity, which will make the expression for the current gauge-invariant. Such a quantity is the phase θ of the quantum wave function $\psi = |\psi| \exp(i\theta)$. At the gauge transformation $A \to A + \nabla f$, with an arbitrary function $f(x,y,z,t)$, the phase transforms as $\theta \to \theta + f$, so that $j \propto -A + \nabla\theta$ is then gauge invariant. Does this indicate "sensitivity" of the superconductor to the vector potential? If so, then how could one measure the response of superconductors to the value of A? These questions are beyond pure theoretical analysis. In typical situations, when one uses superconductor-based electronics, such as SQUIDs, the responses are being expressed in terms of magnetic fluxes, ϕ. That is easily doable using Stokes' theorem, according to which $\oint_C A \cdot dl = \int_S curlA \cdot ds = \int_S H \cdot ds = \phi$, where S is a surface bounded by the contour C. However, the measurement of magnetic fields H and corresponding fluxes is always possible using closed superconducting circuits. Such measurements in our case would be categorized as "non-local" responses. *Non-locality is at the heart of quantum mechanics. Our task here is not in debating that principle. The intention is to find out whether or not superconductors react locally to the presence of the vector potential.*

Classical electromagnetic field theory operates with the local fields E and H. It is customary to express these fields via the 4-vector potential $A_i = \{\varphi, A\}$, $i=0,1,2,3$: $H = \nabla \times A$, $E = \nabla\varphi - \partial A/\partial t$. In classical electrodynamics, the value of A is defined up to the gradient of an arbitrary (gauge) function $f(x,y,z,t)$. Indeed, one can always perform a transformation $A = A' + \nabla f$, which will not change the magnetic, or the electric field provided the scalar potential φ is also transformed as $\varphi = \varphi' - \partial f / \partial t$.

In quantum theory, this function f couples with the phase θ of the wave function ψ. As was mentioned above, this opens opportunities to measure the value of the vector potential quasi-locally using specific devices with quantum elements. We will describe the details below.

2. The Aharonov-Bohm potential

In the Coulomb gauge $\nabla \cdot A = 0$, $A_{\text{int}}(x,y,z) = (\alpha / R^2)\{-y, x, 0\}$ inside the solenoid $x^2 + y^2 < R^2$, which corresponds to a uniform magnetic field $H_0 = \nabla \times A$ (see, *e.g.*, [3,4]). Of immediate interest is the external field described by $A_{ext}(x,y,z) = \alpha\{-y/(x^2+y^2), x/(x^2+y^2), 0\}$, $x^2+y^2 > R^2$. Direct calculation yields $H = \nabla \times A_{ext} = 0$ at any point outside the solenoid – no magnetic field associated locally with $A_{ext} \neq 0$. In cylindrical coordinates, $A_{ext}(z,\rho,\theta) = (0, 0, \alpha/\rho)$, which will be used in subsequent analysis. Here $\alpha = H_0 R^2 / 2$, where R is the radius of solenoid.

3. Basic idea of the measuring device

Let us first consider a loop in the plane orthogonal to the axis of the solenoid (figure 1).

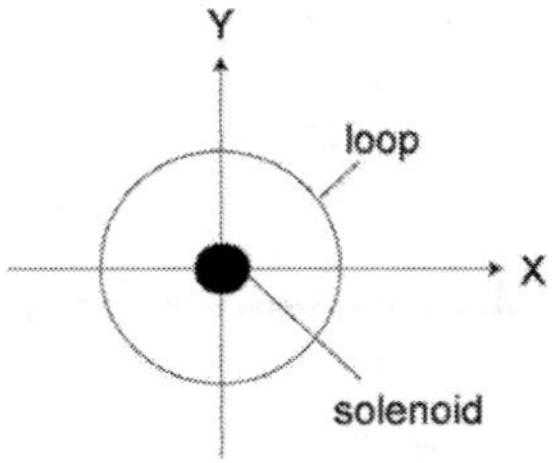

Figure 1. Circular loop in the plane orthogonal to the axis of infinitely long solenoid. $|A_\theta(\rho)|$ is constant along the loop.

Using this figure, one can describe a simple device for measuring the A-potential. Suppose that the loop is made of a superconducting wire. We consider a very thin wire so that the motion of the electrons is quasi-one dimensional and homogeneous throughout its cross section (*i.e.*, the diameter of the wire is much smaller than the London penetration depth λ_L). Then, if the loop is cooled down below the superconducting transition temperature T_c in a pre-existent solenoid field, a stationary supercurrent will settle in the wire. This effect has been demonstrated experimentally [5].

It is possible to detect the current in the looped wire by classical instruments of the magnetic field induced in the vicinity of the wire. The whole system represents then a primitive detector of the A-potential. At the same time, the topology of the detector allows us to integrate along the wire (the closed contour C in Fig. 1), and express the answer via the magnetic flux inside the solenoid. Then it may be regarded as a measuring device of a non-local influence of the magnetic field H_0 inside of the solenoid onto the superconducting loop.

To exclude such options, one should not work with a closed loop. Accordingly, let us consider a finite superconducting bar, such as the one shown between points *1* and *2* in figure 2.

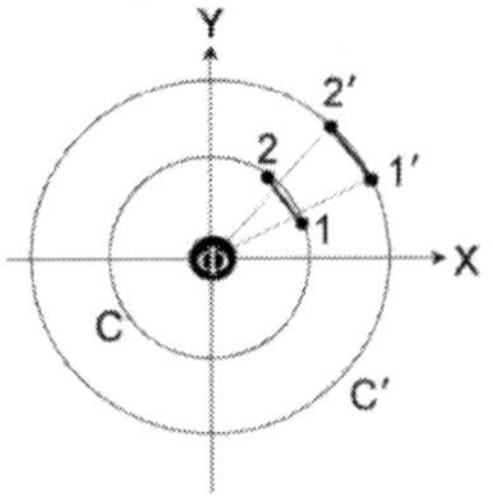

Figure 2. A superconducting bar in the field of the solenoid. The phase difference created by the *A*-field at the ends of the bar is proportional to the angle subtended by the chord at the axis of the solenoid: $|\delta\Theta_{12}| > |\delta\Theta_{1'2'}|$.

The current density in superconducting wire is described by the usual (Ginzburg-Landau-type) equation $j = j_s = (\hbar e^* / m^*)(\nabla\theta - A)$ (see, *e.g.*, [6]), which is a standard quantum-mechanical expression for the current density of charged particles (Cooper pairs) in the magnetic field [7] described by the wave-function ψ, normalized by the density n_S of superconducting electrons: $|\psi|^2 = n_S$. The velocity of superfluid motion is $v_s = (\hbar / m^*)[\nabla\theta - (e^* / \hbar c)A]$ and the effective charge and mass are $e^* = 2e$ and $m^* = 2m$. In the 1D-motion of electrons, assuming a constant cross section for the conductor, one can deal with the current density instead of the current (*i.e.*, $|\psi|$=*const*, and $|A|$ and $|j|$ are constant on the contour *C*). Since j is obviously zero, $A = (\hbar c / e^*)\nabla\theta$, or in other words, the bar possesses a phase difference at its ends $\delta\theta = \int_1^2 A \cdot dl$, which compensates the influence of the *A*-field. Is there any physical consequence of this phase difference $\delta\theta$, and if yes, how to measure it?

4. Two experimental opportunities: oscillating and pulsed fields

Let a superconducting bar (*1-2*), located at a distance *r* from the axis of solenoid on a circumference shown in Fig. 2, be a part of a normal circuit containing classical measurement devices. These external devices put current into the circuit, and are able to measure the energy required for the motion of charged carriers. If the superconducting bar has a thickness λ_L and a height *w*, then kinetic energy of the current is $\delta E = \varepsilon \lambda_0 w \int_1^2 A \cdot dl / |A|$, where $\varepsilon = n_s m v_s^2 / 2$ ε is related to the *A*-potential via $v_s = -(e / mc)A$ in the case of unrestricted motion of the supercurrent $(\nabla\theta = 0)$. Our suggestion is that the external current source mentioned above mimics the unrestricted motion by removing charges from the ends of the bar. If there is *no flux* in the solenoid, the current source provides the energy δE when switched on, and sets up a superfluid motion in the bar. Suppose now that there *is a flux* ϕ in the solenoid (and a corresponding *A*-field, figure 1). Because of the action of the *A*-field, at a fixed value of current *j* in our measuring device, the contribution of the external source should be less by an amount of δE, if the direction of the current it is initiating matches the direction of flow being initiated by the *A*-field. Reciprocally, the source should provide δE more energy if the directions are opposite. By the same logic, in the case of symmetric *AC* excitation of the device, the presence of the *A*-field will cause an asymmetric response in the circuit (figure 3a).

Our second approach is a pulse-type process. A strip of a superconductor with a cross-section *S* and length *L* is placed in the AB-field (figure 2), so that the *A*-vector is either parallel or anti-parallel to the length *L* (figure 3b). For a given orientation and value of the *A*-potential due to the solenoid,

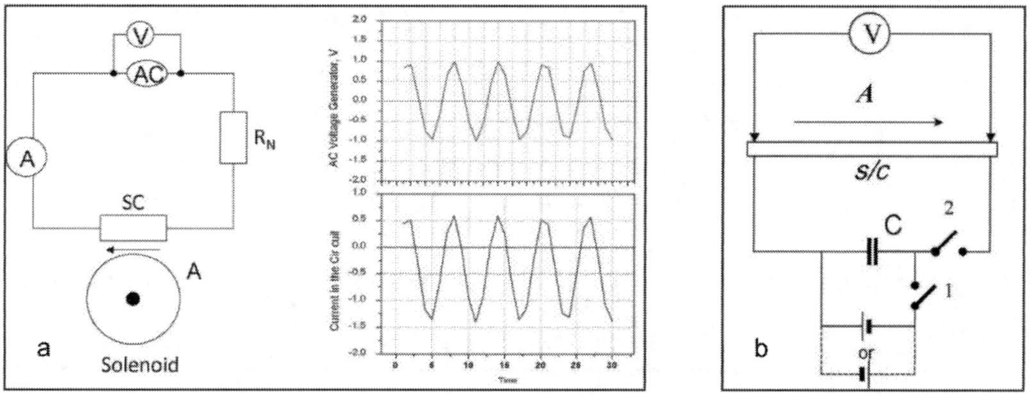

Figure 3. (a)-AC and (b)-pulse detection schemes of the *A*-potential.

one charges the capacitor *C* to a level *Q* (with a chosen voltage *V* and polarity), then opens switch 1 to disconnect the voltage source, and closes switch 2, thus putting current trough the superconducting strip. If the current is less than the critical value, then voltmeter will show no voltage. One can repeat the operations, and increase the voltage, thus increasing the energy $E_C=Q^2/2C=CU^2/2$ stored in the capacitor. When the energy reaches the amount required to destroy the superconducting state in the strip, one stops, and repeats the same procedure with the reversed polarity of the battery. Asymmetry between the threshold values will reveal the role of the vector potential, thus detecting it quasi-locally.

5. Discussion and conclusions

Let us now make some estimates. In the case (a), performing integration, one obtains $\delta E = (e^2 H_0^2 R^4 n_s \lambda L w \vartheta)/(8rmc^2)$, where $\vartheta = f_2 - f_1$. Let us estimate the value of this energy. We choose $\vartheta \sim 1$, $R\sim 0.1cm$, $r \sim 3R$, $\lambda_L\sim 10^{-5} cm$, $w \sim 10^{-5} cm$, $n_S\sim 10^{23} cm^{-3}$, and $H_0\sim 10\ Oe$. Then $\delta E\sim 1nJ$, which is a detectable value. For comparison, the energy of the so-called RSFQ pulses used in superconducting electronics is $\sim 0.2\ aJ$, *i.e.*, δE is equivalent to five billion RSFQ pulse energies.

In the case (b), suppose the bar has $L=1000\mu m$ and $S=0.01\mu m^2$, as in the previous case. Then the number of Cooper pairs is about 10^{12}. Each Cooper pair has $\sim 1\ meV$ energy, which means that the upper limit to the energy stored in the capacitor should be about $10^9 eV$, or about $100\ pJ$. In principle, the energy can be delivered using a more sophisticated device, such as a pulsing electronics circuit.

These conceptually simple experiments can be used to detect the local influence of the Aharonov-Bohm vector potential on superconductors. The outcome of these experiments can demonstrate whether or not quantum systems can sense the vector potential locally, thus serving the perplexed [9].

Acknowledgement – I am very grateful to Yakir Aharonov, Jeff Tollaksen, Alan Kadin, Lou Sica, Joe Foreman, Michael Steiner, and Mamikon Gulian for numerous discussions.

References

[1] London F and London H 1935 *Proc. Roy. Soc.* **A149** 71
[2] De Gennes P G 1966 *Superconductivity of metals and alloys* (W.A. Benjamin, Inc.) p.4
[3] Bateman H and Tonomura A 2009, *Phys. Today* 62 issue 9 38
[4] Landau L D and Lifshitz E 2000 *The classical theory of fields*, v.2 (Butterworth-Heinemann)
[5] Owen G E 1963 *Introduction to electromagnetic theory* (Allyn and Bacon)
[6] Gulian A M and Zharkov G F 1999 *Nonequilibrium electrons and phonons in superconductors* (New York: Kluwer-Plenum)
[7] Landau L D and Lifshitz E M 1981 *Quantum mechanics*, v. 3 (Butterworth-Heinemann)
[8] Kadin A 1999 *Introduction to superconducting circuits* (New York: John Willey and Sons, Inc.)
[9] Aharonov A and D. Rohrlich D 2005 *Quantum paradoxes* (Weinheim: Wiley-VCH)

Journal of Physics: Conference Series **507** (2014) 042013 doi:10.1088/1742-6596/507/4/042013

"Spaghetti" design for gravitational wave superconducting antenna

A Gulian[a1], J Foreman[b], V Nikoghosyan[a,c], S Nussinov[a,d], L Sica[a] and J Tollaksen[a]

[a]Chapman U., Institute for Quantum Studies, Orange, CA & Burtonsville, MD, USA
[b]Independent Researcher, Alexandria, VA, USA
[c]Physics Research Institute, National Academy of Sciences, Ashtarak, 0203, Armenia
[d]School of Physics&Astronomy, Tel-Aviv University, Ramat-Aviv, Tel-Aviv, Israel

E-mail: gulian@chapman.edu

Abstract. A new concept for detectors of gravitational wave radiation is discussed. Estimates suggest that strain sensitivity essentially better than that of the existing devices can be achieved in the wide frequency range. Such sensitivity could be obtained with devices about one meter long. Suggested device consists of multi-billion bimetallic superconducting wires ("spaghettis") and requires cryogenic operational temperatures (~0.3K in the case considered).

1. Introduction

To date, the most advanced approaches (LIGO, LISA) to gravitational wave (GW) detection measure change of the distances between macroscopic bodies. The longer the baseline the bigger the change. If the baseline distance is "2400 trillion miles, or one million times the distance to Neptune. ... then a *strong* GW will briefly change that distance by less than the thickness of a human hair. We have perhaps less than a few tenths of a second to perform this measurement. And we don't know if this infinitesimal measurement will come next month, next year, or perhaps in thirty years." [1]. For the known sources of GW radiation, like the Crab Pulsar, the change will be far smaller. The difficulty of the task justifies searching for novel highly sensitive GW antennas [2]. In this note, we will describe a device based on the Tolman-Stewart (TS) effect: when a rotating coil is decelerating, an electric field $E_{TS}=mg/e$ is generated in the conductor, which yields an electric current in the circuit containing the coil [3]. Here g is the deceleration (or, equivalently, acceleration) of the conductor, m is the rest mass of the electron, and e is its charge.

Gravitational waves accelerate charges by tidal gravitational forces, which are symmetric relative to the centre of mass of the charge carriers, generating a potential difference between the end of the conductor and its middle part. As early as 1976, Adler [4] noted that two different metallic conductors could be used to detect GWs by measuring the potential difference between the ends of the metals. Adler used the fact that acceleration in metallic wires acts not only on electrons but also on ions. By inhomogeneously deforming the lattice, it creates a gradient of chemical potential, which redistributes the electrons more strongly than the TS mechanism. This feature was first pointed out by Dessler *et al.* [5], confirmed experimentally by Beams [6], and recognized by many physicists, including Schiff [7]. The consensus [8] is that the electric field acting on electrons can be stronger than the TS field by a

[1] To whom any correspondence should be addressed.

factor of M/m, where M is the ionic mass. It also has an opposite sign, so that, say, in the gravity field of the earth the electrons will move opposite to the direction of g. The total electric field is therefore

$$E = -\alpha Mg/e + mg/e \approx -\alpha Mg/e. \tag{1}$$

The dimensionless coefficient α depends on the parameters of the material, and is estimated to be α~*0.1–1*. Then, two wires of materials with different αM will develop different electric potentials in the GW field. Adler did not consider the noise factors such as the Johnson-Nyquist noise related to his very long antenna. We show below how to modify this approach by taking advantage of superconductivity. In particular, this will allow making the antenna much shorter, say, *1m* long, so that it will be feasible to cool it to sub-Kelvin cryogenic temperatures. We will discuss the inherent limitations of the sensitivity due to thermal fluctuations. It will be shown by using a reasonable experimental design, that GWs from the Crab Pulsar may become detectable.

2. Basic principle

We start with the antenna geometry shown in figure 1.

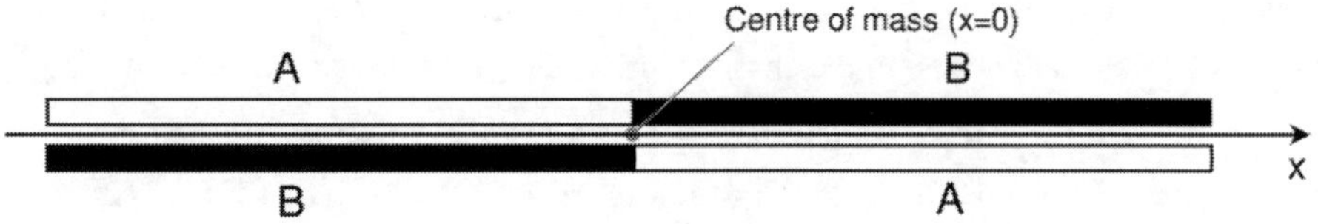

Figure 1. The GW $h=h_0 Sin\omega t$ is incident perpendicularly to the plane of the two bimetallic conductors. Each conductor, A or B, is of length L. The ends of the conductors are connected (not shown) to close the loop.

The force acting on a lattice ion at a distance x from the centre of mass is [4]

$$F(x) = M\ddot{h}x/2\,;\ \ddot{h} \equiv d^2h/dt^2, \tag{2}$$

where the dimensionless amplitude h of the GW is defined via induced small coordinate deviation $\Delta x = xh/2$. Then the analog of the Eq. (1) for the case of GW is

$$E(x) = \alpha M\ddot{h}x/2e. \tag{3}$$

The amplitude of the electromotive force $\mathscr{E}$ is the integral of the E-field over the closed loop [10,11]:

$$\mathscr{E} = \oint_{A-B-A-B} E(x)dx = -(\ddot{h}L^2/2e)(\alpha_A M_A - \alpha_B M_B). \tag{4}$$

In writing this we neglected the contribution of the connectors between two bimetallic wires in Fig.1. Time-periodic electric field $E = E_0 \exp(i\omega t)$ generates the following total current density in a superconductor [11, 12]:

$$j = j_n + j_s = (\sigma_1 - i\sigma_2)E, \tag{5}$$

where

$$\sigma_1 = (n_n e^2\tau)/[m(1+\omega^2\tau^2)],\quad \sigma_2 = n_s e^2/(m\omega) + n_n e^2(\omega\tau)^2/[m\omega(1+\omega^2\tau^2)]. \tag{6}$$

In Eq. (6), n_s and n_n are superconducting and normal electron densities, ω is the frequency, and τ is the relaxation time of the normal electrons (typically, τ~10^{-13} s). The smallness of $\omega\tau$ allows keeping only the first term in σ_2 of Eqs. (6) with the result

$$j \approx j_s = -i\frac{n_s e^2}{m\omega}E. \tag{7}$$

Notice that in the case of normal metals ($n_s \equiv 0$ in Eq. (6)), the dominant contribution to the current (5) would come from the σ_1-term, and the current would be smaller than (7) by a factor of $\omega\tau$, *i.e.*, ten orders of magnitude smaller than the superconductor response at 100 Hz frequency range. We assume for simplicity that the ratio n_s / m in (7), as well as the wire cross sections S are the same for both materials A and B, so that

$$E = \mathcal{E}/(4L) = (\ddot{h}L/8e)(\alpha_A M_A - \alpha_B M_B). \qquad (8)$$

The requirement of having the same current flow through all parts of our circuit implies via Eq. (7) that the electric field E in (8) is also spatially constant. For the harmonic GW mentioned in the caption to figure 1 with amplitude h_0 and frequency ω, the current amplitude is:

$$I = |jS| = [n_s e\omega h_0 LS/(8m)]|\alpha_A M_A - \alpha_B M_B| \qquad (9)$$

- this expression is immediately applicable to periodic sources, such as the Crab Pulsar.

3. Inductance and magnetic energy

In the above calculations we neglected the magnetic inductance effects. To justify such a procedure the magnetic energy: $E_{mag} = \mathcal{L}I^2/2$ (where $\mathcal{L}$ is the circuit inductance) should not exceed the kinetic energy associated with the current flow: $E_{kin} = LSn_s mv^2$. Otherwise, the energy would be transferred to the magnetic field rather than to the charge motion. For a single loop in figure 1 their ratio is

$$E_{mag}/E_{kin} \sim \mu_0 L e^2 n_s^2 v^2 S^2 / LSn_s mv^2 = \mu_0 e^2 n_s S/m \qquad (10)$$

(where $\mu_0 = 4\pi \cdot 10^{-7}$ *Henry/meter*). Substituting $n_s \sim 10^{22} cm^{-3}$ into Eq. (10), we find that $E_{mag}/E_{kin} \sim 1$ when $S \sim 10^{-14} m^2$, or the wire diameter is about 0.1μm. This choice has the extra advantage of using superconducting materials most effectively, since the diameter of the wire is close to the London penetration depth λ of typical superconductors. Substituting S and $[(\alpha M/m)_A - (\alpha M/m)_B] \sim 10^4$ (which one can expect, say, for vanadium-lead bimetallic pair) into Eq. (10), we find $I \approx 2 \cdot 10^{-2}(\omega/Hz)h_0$. Thus, for $h_0 \sim 10^{-26}$ and $\omega \sim 2\pi(60)Hz$ we have $I \approx 0.7 \cdot 10^{-25}$ $[Amp]$ in a single loop. In principle, one can imagine a design made of a big number N of bimetallic wires with currents in opposite direction in neighboring conductors. The wires can be merged into a single read-out circuit, as shown in figure. 2.

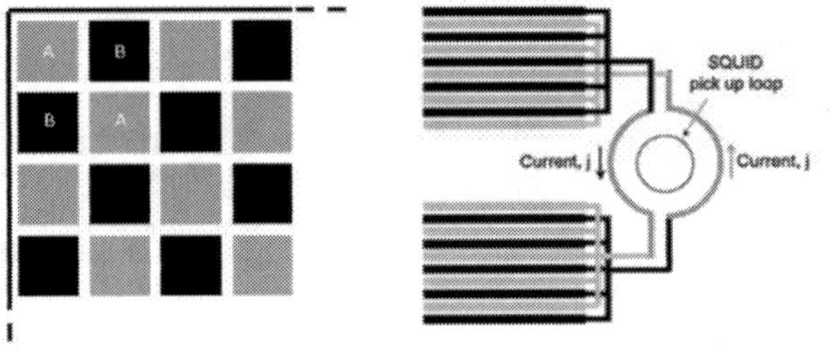

Figure 2. *Left* - Cross sectional view of wire arrangement. *Right* - Schematics of multi-wire interconnections for non-invasive current measurement via SQUID. Since the flow of current in materials A and B takes place in opposite directions, two semi-circles constitute a full current loop for inductive pick up.

One can, in principle, choose N as big as required to make the current detectable.

4. Sensitivity

The sensitivity of the antenna is determined by the signal-to-noise ratio: at unknown frequency I_{signal}/I_{noise} should be at least >1 for detection. Comprehensive consideration of all noise factors is beyond the scope of this paper. However, thermodynamic Johnson-Nyquist noise is unavoidable and we consider it as a first approximation. It is generated by unpaired electrons, and has magnitude $\langle I_{noise} \rangle = [4(k_B T/R_n)\delta\nu]^{1/2}$ [11,12], where k_B is the Boltzmann constant, R_n is the resistance to normal (unpaired) electrons in superconductors, and $\delta\nu$ is the measurement bandwidth:

$$I_{signal}/I_{noise} \sim en_s SL\omega(\alpha M/m)h_0/\{8[4(k_B T/R_n)\delta\nu]^{1/2}\} > 1 . \qquad (11)$$

This estimate is for a single loop. For the N loops signal will scale linearly with N, and noise will scale as $N^{1/2}$, thus signal to noise will scale as $N^{1/2}$. That means the amplitude h_0 can be no smaller than $h_0 \approx 16[(k_B T / R_n)\delta\nu]^{1/2} / N^{1/2} e n_s S L \omega(\alpha M / m)$. Since $R_n = (\rho L / S)\exp(\Delta / k_B T)$, where ρ is resistivity of the material above the transition temperature T_c, and $\Delta = \Delta(T)$ is the BCS gap in the spectrum of single-electron excitations [11,12], at $T << T_c$, $2\Delta(0) \approx 3.52 k_B T_c$:

$$h_0 \approx [16(k_B T\delta\nu)^{1/2} / e n_s S^{1/2} L^{3/2} \rho^{1/2} \omega(\alpha M / m) N^{1/2}]\exp(-0.88 T_c / T). \qquad (12)$$

At the parameters $L \sim 1m$, $S \sim 10^{-14} m^2$, $n_s \sim 10^{28} m^{-3}$, $\omega \sim 2\pi(60)Hz$, $\alpha M/m \sim 10^4$, $\rho \sim 0.1 m\Omega\text{-}cm$, $T \sim 0.3K$, and $N \sim 10^{12}$: $h_0 \approx 0.5 \cdot 10^{-22} (\delta\nu / Hz)^{1/2} \exp(-0.88 T_c / T)$. Since $T_c \geq 5K$ (for the V-Pb bimetallic pair), at $T_{op} \sim 0.3K$ the exponential factor is $\sim 4 \cdot 10^{-7}$, and thus $h_0 \approx 2 \cdot 10^{-29} (\delta\nu / Hz)^{1/2}$, making the Crab Pulsar GW radiation detectable. At the chosen number of wires N the value of the current signal to be detected is $I_{signal} \sim 10^{-13} A$, which, in principle, can be readily done.

5. Discussion and conclusions

To summarize, a concept for a superconducting GW is proposed. The antenna transforms part of the GW energy into the motion of superfluid electrons which is then detected electronically. The technical realization of the antenna may require some technological efforts, but there is no showstopper. For example, each "spaghetti" should have $\sim 0.1 \mu m$ diameter. With the chosen number $N \sim 10^{12}$, the total cross-section of the current loop is $\sim 10^{-2} m^2$. With $\sim 1Hz$ bandpass filter that will allow direct detection of Crab Pulsar, since signal-to noise will be >1. Moreover, because in this and similar cases the source frequency is known with high accuracy, one can use the lock-in amplification, and detect signals much smaller than the noise level. That brings in an opportunity to reduce the number of spaghettis and making the device easier to fabricate. The fabrication of the device will require thin film deposition and nanolithography methods. Another possibility is the eventual use of superconducting nanotubes as spaghettis. The estimates we provided above demonstrate that very high sensitivity could be obtained with these devices. They are non-resonant and may be used for a wide variety of sources. Hopefully, in parallel with other recognized efforts [2], the suggested superconducting device will become efficient for a most challenging physics measurement – the detection of gravitational waves.

Acknowledgements. One of us (AMG) would like to express his gratitude to L.P. Grishschuk, P.B. Abramian-Barco, and A.E. Sargsyan for encouraging discussions.

References

[1] Collins H 2004 *Gravity's Shadow* (Chicago: The University of Chicago Press) p 2
[2] Blair D G, Howell E J, Ju L, and Zhao C (Eds) 2012 *Advanced Gravitational Wave Detectors* (Cambridge: Cambridge University Press)
[3] Tolman R C and Stewart T D 1916 *Phys. Rev.* **8** 97
[4] Adler R J 1976 *Nature* **259** 296
[5] Dessler A J, Michel F C, Rorschach H E and G.T. Trammel G T 1968 *Phys. Rev.* **168** 737
[6] Beams J W 1968 *Phys. Rev. Lett.* **21** 1093
[7] Schiff L I 1970 *Phys. Rev. B* **1** 4649
[8] Kogan Sh M 1971 *Usp. Fiz. Nauk* **105** 157 [1972 *Sov. Phys. Uspekhi* **14**, 658]
[9] Landau L D and Lifshitz E M 1960 *Electrodynamics of continuous media* (Oxford: Pergamon Press) p 101
[10] Feynman R P, Leighton R B and Sands M 1964 *The Feynman lectures on physics* vol 2 (Reading, MA: Addison-Wesley) p 16-2
[11] Van Duzer T and Turner C W 1998 *Principles of superconducting devices and circuits* (Prentice Hall) p 5
[12] Tinkham M 1975 *Introduction to Superconductivity* (New York: McGrow-Hill) p 37

Optimization of electrical and structural parameters of $YBa_2Cu_3O_{7-x}$ thin-film bicrystal Josephson junctions with chemical and thermal treatments of substrates

I Gundareva[1,2] and Y Divin[1,2]
[1]Peter Grünberg Institute, Forschungszentrum Jülich, Jülich, Germany
[2]Kotel'nikov Institute of Radio Engineering and Electronics, Russian Academy of Sciences, Moscow, Russian Federation

E-mail: i.gundareva@fz-juelich.de

Abstract. The [100]-tilt high-T_c Josephson junctions are characterized by high characteristic voltages I_cR_n but their current range is limited by low critical currents I_{cf} of the thin-film electrodes with tilted c-axes. The impact of chemical and thermal treatments of $NdGaO_3$ substrates on morphology and IV curves of the $YBa_2Cu_3O_{7-x}$ thin films and the bicrystal thin-film junctions have been studied. It was found that IV curves of the films above a critical current I_{cf} are described by a power-law dependence $V = V_0(I/I_{cf} - 1)^n$, where I_{cf}, V_0 and n are dependent on the c-axis tilt of the film and treatment of the substrate. Critical current densities j_c(78K) up to $1.4 \cdot 10^7$ A/cm^2 have been reached in 1.7°-tilt $YBa_2Cu_3O_{7-x}$ films perpendicular to the tilt. Bicrystal junctions have been fabricated with optimized $YBa_2Cu_3O_{7-x}$ thin films, and RSJ-like IV-curves in the current ranges from I_c up to 5 I_c with I_cR_n (78K)-values of 1 mV have been achieved.

1. Introduction

High-T_c Josephson junctions are often fabricated by thin-film deposition on bicrystal substrates [1]. When compared with the c-axis $YBa_2Cu_3O_{7-x}$ thin-film bicrystal junctions, the bicrystal junctions made from the $YBa_2Cu_3O_{7-x}$ films with mutually tilted c-axes have demonstrated higher structural homogeneity, higher critical current density j_{cj}, and higher characteristic voltage I_cR_n up to 10 mV [2]. Such junctions can be used in THz detection and spectral analysis [3, 4] even at the voltages V above I_cR_n and corresponding Josephson frequencies exceeding 5 THz. However, due to the tilt of the c-axes the critical current densities j_{cf} of the tilted thin films along the tilt [5] are lower than that of the c-axis films. The motion of magnetic vortices in thin-film electrodes at $j > j_{cf}$ might contribute to the total voltage and this phenomenon limits the maximum current I where only the junction contributes to the measured voltage to the values below I_{cf}. Thus, optimization of substrate surface treatment, a crucial step for fabrication of high-quality epitaxial thin films, should be carried out.

Previous studies [6] on chemical and thermal treatments of substrates and epitaxial $YBa_2Cu_3O_{7-x}$ films did not report on electrical characterisation. Here, we present electrical characteristics of $YBa_2Cu_3O_{7-x}$ thin films, both c-axis and with tilted c-axes, fabricated on $NdGaO_3$ substrates after such treatments. The electrical characteristics of bicrystal junctions fabricated with optimized thin films are also presented.

2. Experimental details

The $NdGaO_3$ substrates, with the orientations of the main planes of (110), (110) with 1.7°-tilt around [001], (320) and (230), were used in fabrication of $YBa_2Cu_3O_{7-x}$ thin films. Bicrystal substrates made from (320) parts were used for junction fabrication. After chemical cleaning in acetone and ethanol, the substrates were rinsed with deionized water in an ultrasonic bath for 10 min, etched for 0.5 - 8 min in a buffered hydrofluoric (BHF) solution prepared with 10 ml commercial BHF (12.5 vol% HF + 87.5 vol% NH_4F) + 4 ml NH_4OH (37 vol%) + 90 ml H_20, after ref. [6]. The substrates were subsequently annealed in oxygen [6], [7], but the temperature of the heater was varied from 800°C to 1000°C. The annealing was made under constant oxygen flow during 20- 240 min and at pressure of 1 bar.

$YBa_2Cu_3O_{7-x}$ thin films were deposited on $NdGaO_3$ substrates at a heater temperature of 950°C by dc sputtering in pure oxygen at pressure of 3.4 mbar. These sputtering parameters were optimized to decrease *a*-axis inclusions in the predominantly *c*-axis thin films. The thickness t of the films was in the range of 60-200 nm. The films and junctions were patterned in the form of bridges by UV lithography with wet etching in 0.1% bromine-ethanol solution. The widths w of the bridges were measured by AFM in the range from 2 to 18 μm with a width error of ±0.5μm, which was due to inhomogeneous etching of the film edges.

3. Results and discussion

The effect of substrate treatment on the morphology of the *c*-axis thin films was studied by varying etching time in the BHF solution from 0.5 to 8 min. The smoother film surface was obtained with an etching time of 2 min (Figure 1a). The films sputtered on the substrates without chemical treatment as well as on the substrates subjected to longer BHF treatment (more than 2 min) had up to 5 times more *a*-axis grains. So, short-time chemical etching of $NdGaO_3$ substrates in the BHF solution improves the quality of the substrates as in [6]. The thermal treatment used in [6] did not result in improvement of the morphology of the films. After substrate annealing at temperatures T_a of 950-1000°C and at T_a = 910°C the concentrations of the *a*-axis grains in the *c*-axis thin films were 8 and 3 times larger, correspondingly, than those in films without thermal treatment of the substrates.

The $YBa_2Cu_3O_{7-x}$ thin films grown on (110) $NdGaO_3$ substrates with 1.7°-tilt demonstrated a transition from island growth to step-flow growth (Figure 1b). The morphology of an $YBa_2Cu_3O_{7-x}$ thin film deposited on (320) $NdGaO_3$ substrate is shown in Figure 1c. The (320) $NdGaO_3$ substrate was etched in BHF and, additionally, annealed in oxygen atmosphere. A set of terraces and steps between them was observed on the surface of the $YBa_2Cu_3O_{7-x}$ film (Figure 1c), with the step heights much larger than the $YBa_2Cu_3O_{7-x}$ cell size. This step-bunching on the surface of the $YBa_2Cu_3O_{7-x}$ films with tilted *c*-axes was found to depend on the duration and temperature of substrate annealing and could be varied from 10 nm for low-temperature (T = 800°C) or short-term (about 20 min) high-temperature (T = 950°C) annealing to 20-30 nm for high-temperature annealing (T = 950°C) for more

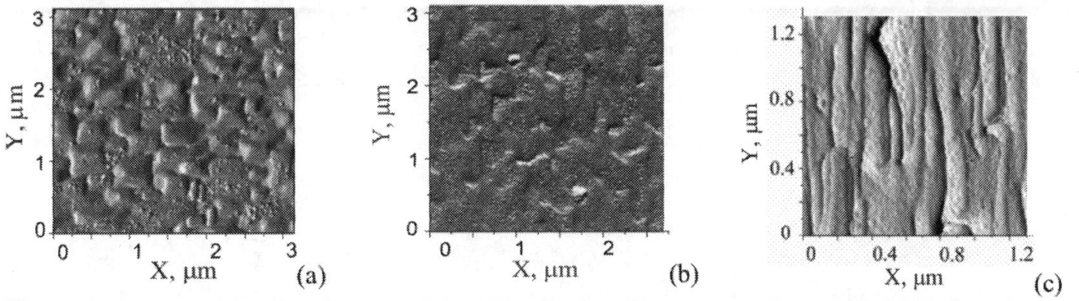

Figure 1. Topographic AFM images of the $YBa_2Cu_3O_{7-x}$ films sputtered on (110) $NdGaO_3$ substrate after 2′-BHF treatment (a); (110) $NdGaO_3$ with 1.7°-tilt after 5′-BHF treatment (b); (320) $NdGaO_3$, after 2′-BHF treatment and 20′-annealing at T=950°C (c). The tilt direction is to the left in (b) and (c).

than 0.5 hour, and this step-bunching is a consequence of the step-bunching on the surface of the substrate due to high-temperature annealing in oxygen atmosphere [7].

The treatment of the substrate surfaces had a large effect on the electrical characteristics of the $YBa_2Cu_3O_{7-x}$ thin films. The critical current densities j vs. voltages V for $YBa_2Cu_3O_{7-x}$ thin-film bridges deposited on $NdGaO_3$ substrates of various orientations and treatments are presented in Figure 2. Current densities $j = I/wt$ have been calculated for all fabricated bridges from the experimental current I values. The resulting $j(V)$-curves of fabricated $YBa_2Cu_3O_{7-x}$ thin films were fitted to the following expression:

$$V = V_0 \,\mathrm{sign}j\,(|j|/j_{cf} - 1)^n, \quad \text{if } |j| > j_{cf};$$
$$V = 0, \quad \text{if } |j| \le j_{cf}. \qquad (1)$$

Using (1) and a nonlinear-fit program [8], an accuracy of a few percent have been reached for j_{cf} and n. The c-axis $YBa_2Cu_3O_{7-x}$ films sputtered on the untreated (110) NGO substrates show j_{cf} values around $4{\cdot}10^6$ A/cм^2 (Figure 2, curve 1), and these values were increased up to $5{\cdot}10^6$ A/cм^2 for the c-axis $YBa_2Cu_3O_{7-x}$ films sputtered on the substrates, which were shortly (2 min) etched in BHF (Figure 2, curve 2). The index n also increased from 3.2 to 4.1. The $YBa_2Cu_3O_{7-x}$ films with 1.7°-tilt of the c-axes showed a considerable increase of the j_{cf} up to $1.4{\cdot}10^7$ A/cm^2 (Figure 2, curve 3) in the direction perpendicular to the tilt, with $n = 1.4$. The j_{cf} -values for $YBa_2Cu_3O_{7-x}$ films sputtered on unannealed (320) $NdGaO_3$ substrates were around $4.5{\cdot}10^5$ A/cm^2 along the tilt (curve 4), for the films on shortly annealed substrates these values dropped to $2{\cdot}10^5$ (curve 5) and the j_c-values increased to $4{\cdot}10^5$ A/cm^2, when (320) substrates were annealed in oxygen for 4 hours (curve 6). The n-indices were equal to 2.5; 1.9 and 2.9 for curves 4, 5 and 6, correspondingly.

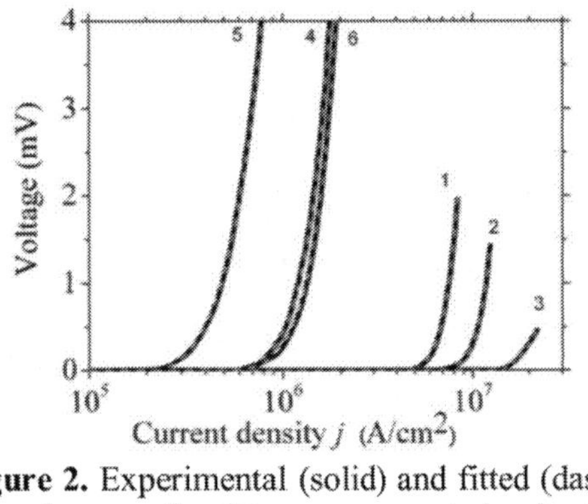

Figure 2. Experimental (solid) and fitted (dash) dependences of voltage vs. current density for $YBa_2Cu_3O_{7-x}$ thin films on the substrates: 1 - (110) NGO untreated, 2 - (110) NGO etched in BHF, 3 - 1.7°tilted (110) NGO etched in BHF, 4 - (320) NGO untreated, 5 -(320) NGO annealed 20 min., 6 - (320) NGO annealed 4 hours. T= 77.8 K

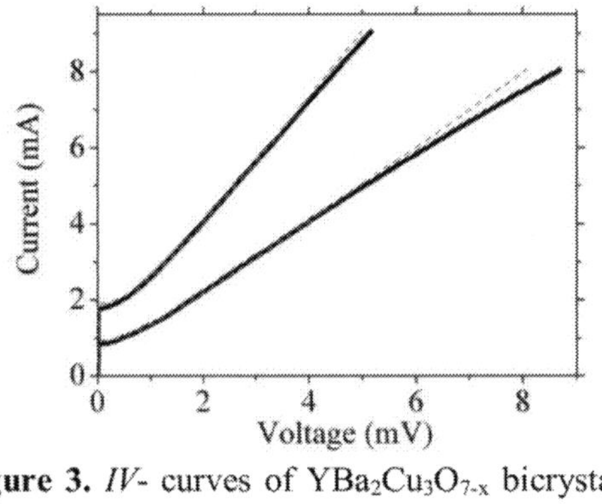

Figure 3. IV- curves of $YBa_2Cu_3O_{7-x}$ bicrystal junctions fabricated on (320) $NdGaO_3$ bicrystal substrates (solid lines) and corresponding IV-curves from the RSJ model (dash lines) with the following fitting parameters: 1 - I_cR_n = 1.08 mV, R_n = 0.56 Ω; 2 - I_cR_n = 0.93 mV, R_n = 1.0 Ω. T = 77.8 K

The resistive state of superconductors is explained by motion of the magnetic vortices. In the presence of static disorder in a superconductor, the IV-curve at $j \ge j_c$ has a non-ohmic form, which depends on the dimension of the superconducting material and on the topology of the induced flux motion [9]. The critical scaling applied to the IV-curves of the superconductors with pinning centers gives the power law $V \propto j^{\,n}$ [9], which is similar to the IV curves in Figure 2. The n-value of 3.2

obtained for our *c*-axis $YBa_2Cu_3O_{7-x}$ thin films is close to the corresponding value of $n = (2.9\pm0.2)$ obtained earlier for the *c*-axis $YBa_2Cu_3O_{7-x}$ thin films on $SrTiO_3$ substrates [10].

The [100]-tilt $YBa_2Cu_3O_{7-x}$ bicrystal junctions consisting of optimized $YBa_2Cu_3O_{7-x}$ thin-film electrodes with 2x11.3° mutually tilted *c*-axes were fabricated on (320) bicrystal $NdGaO_3$ substrates. The measured *IV* curves of the junctions at $T = 77.8$ K are presented in Figure 3 by solid lines, and corresponding RSJ-like *IV* curves - by dash lines. The optimized bicrystal junctions are characterized by high I_cR_n -values of around 1 mV and follow the RSJ-like *IV* curves in the current range from I_c to $4I_c$ for the junction 1 and from I_c to $5I_c$ for the junction 2. The deviations of the experimental curves from the RSJ curves at the currents *I*, which are more than 4-5 times larger than the critical currents I_c of the bicrystal junctions, originate from the vortex motion in the junction thin-film electrodes with tilted *c*-axes at $I > I_{cf}$. These optimized junctions at the temperature of 78 K can be used as Josephson detectors for electromagnetic radiation up to the voltages of 4 mV and Josephson frequencies of 2 THz.

4. Conclusions

In this paper, electrical and structural characteristics of $YBa_2Cu_3O_{7-x}$ thin films fabricated on the $NdGaO_3$ substrates subjected to thermal and chemical treatments are presented. Optimization of the sputter deposition parameters and substrate treatment yielded critical current density j_c up to $1.4\cdot10^7$ A/cm^2 for $YBa_2Cu_3O_{7-x}$ films with 1.7°-tilt of the *c*-axis. Critical current densities for [100]-tilt $YBa_2Cu_3O_{7-x}$ films sputtered on (320) $NdGaO_3$ substrates strongly depended on the time and temperature of preliminary annealing of the substrates in oxygen and on the size of the bridges. The shape of the *IV*-curves of $YBa_2Cu_3O_{7-x}$ films was found to be described by a power law $V = V_0 (I/I_c - 1)^n$, $I > I_c$, where the index *n* is in the range of $1.4 < n < 4$ and depends on the *c*-axis tilt and on the time and temperature of preliminary annealing of the substrate in oxygen atmosphere. Obtained dependencies and reproducibility of the electrical characteristics of $YBa_2Cu_3O_{7-x}$ thin films were used to fabricate high quality [100]-tilt $YBa_2Cu_3O_{7-x}$ bicrystal Josephson junctions with I_cR_n-values of 1 mV and RSJ-like *IV*-curves in the current range from I_c up to $5I_c$.

Acknowledgments
Authors are thankful to I. Kotelyanskii for fabrication of $NdGaO_3$ substrates, V. Gubankov and U. Poppe for interest to this work. This work is partially supported by the joint project HRJRG-207 of Helmholtz Association of German Research Centers and Russian Fund for Basic Research (2011-2014).

References

[1] Hilgenkamp H and Mannhart J 2002 *Rev. Mod. Phys.* **74** 485-549
[2] Divin Y and Poppe U 2012 *Physics Procedia* **36** 42-7
[3] Divin Y, Poppe U, Gubankov V N and Urban K 2008 *IEEE Sens. J.* **8** 750-7
[4] Volkov O Y, Divin Y Y, Gubankov V N, Gundareva I I and Pavlovskiy V V 2011 *IEEE Trans. Appl. Supercond.* **21** 306-10
[5] Divin Y Y, Poppe U, Seo J W, Kabius B and Urban K 1994 *Physica C* **235** 675-6
[6] Leca V 2003 *Heteroepitaxial growth of copper oxide superconductors by pulsed laser deposition.* (PhD Thesis, Enschede: Printpartners Ipskamp)
[7] Dirsyte R, Schwarzkopf J, Wagner G, Lienemann J, Busch M, Winter H and Fornari R 2009 *Appl Surf Sci* **255** 8685-7
[8] OriginPro 9.0, OriginLab Co.
[9] Cohen L F, Jensen H J 1997 *Rep Prog Phys* **60**, 1581-1672.
[10] Koch R H, Foglietti V, Gallagher W J, Koren G, Gupta A, and Fisher M P A 1989 *Phys Rev Lett* **63** 1511-14.

Design and Fabrication of Microwave Kinetic Inductance Detectors using NbN Symmetric Spiral Resonator Array

K Hayashi[1], A Saito[1], Y Ogawa[1], M Murata[1], T Sawada[1], K Nakajima[1], H Yamada[1], S Ariyoshi[2], T Taino[3], H Tanoue[3], C Otani[4], and S Ohshima[1]

[1] Graduate School of Science and Engineering, Yamagata University, 4-3-16 Jonan, Yonezawa, Japan
[2] Graduate School of Engineering, Nagoya Institute of Technology, Gokiso-cho, Showa-ku, Nagoya, Japan
[3] Graduate School of Science and Engineering, Saitama University, 255 Okubo, Sakura-ku, Saitama, Japan
[4] RIKEN Advanced Science Institute Terahertz Sensing and Imaging Team, 2-1 Hirosawa, Wako, Japan

E-mail: tse80358@st.yamagata-u.ac.jp

Abstract. We designed and fabricated a microwave kinetic inductance detector (MKID) using a niobium nitride (NbN) symmetric spiral resonator array. Previously we revealed that a rewound spiral structure works as not only a high-Q half-wavelength resonator but also as a broadband terahertz antenna. We conducted simulations for a 9 resonator array assuming NbN as the superconducting material and sapphire as the dielectric substrate, and obtained a maximum attenuation of over 30 dB and unloaded quality factors of over 2×10^5 for frequencies between 4.4 and 4.9 GHz. We fabricated the 9 resonator array MKID using NbN thin film deposited on an m-sapphire substrate by using dc magnetron sputtering. We observed half-wavelength resonances of around 4.5 GHz at 4 K. We measured the optical response of the MKID. The frequency shift was 0.5 MHz when illuminated with 650 nm photons.

1. Introduction

Microwave kinetic inductance detectors (MKID) [1] are a type of non-equilibrium superconducting photon detector made of high quality factor superconducting microwave resonators which are coupled to a suitable antenna or an absorber structure. MKIDs are powerful detectors for large arrays due to using only a pair of feed lines. Furthermore, they can be easily made with a single layer of superconducting film deposited on a substrate and patterned using conventional optical lithography. Research groups have been developing MKIDs for sub-millimeter astronomy [2], optical astronomy [3], and dark matter detection experiments [4] in recent years. These MKIDs require an operating temperature of 0.1 K, which is attained using expensive dilution refrigerators or adiabatic demagnetization refrigerators. Therefore, there are so far no applications in which MKIDs can be used to perform general-purpose measurements, particularly for terahertz spectrometers. To realize a larger array MKID, we need smaller resonators with a high quality factor, a large attenuation at the resonant frequency, and high efficiency coupling to THz waves. In past studies, Baryshev et al. proposed an antenna-coupled MKID using $\lambda_g/4$ resonators with twin arc slot antennas [5], and Doyle et al. proposed a lumped-element MKID (LEMKID) [6]. Previously we identified a similarity between the structures of a spiral terahertz antenna [7] and a rewound spiral microwave resonator [8] and proposed an MKID using a rewound spiral resonator (spiral-MKID) [9]. Recently, we proposed the use of symmetric spiral resonators for an MKID array, feasible for the realization of large arrays in electromagnetic simulations [10]. We foresaw the use of spiral-MKID operating at 4 K by using a niobium nitride (NbN) thin film. However, it has not been verified experimentally yet. In this experiment, we verified that a symmetric spiral resonator is feasible for large arrays. We fabricated a 9 resonator array MKID using an NbN spiral resonator array and measured the frequency and optical responses of the MKID at 4 K.

2. Resonator and array designs

We designed an MKID with rewound spiral resonators. Fig. 1 illustrates the design of a 9 resonator array spiral-MKID. Enlarged view shows one pixel of the spiral resonator. The frequency responses of the resonators were simulated using an electromagnetic simulator (Sonnet-EM software) on the basis of the finite integral method. We assumed the simulation parameters using an NbN thin film as the superconducting film and m-plane sapphire as the dielectric substrate. We defined a function for the RF resistivity (R_{rf}), which is directly proportional to the square of the frequency. Surface resistances (R_s) of the NbN films were measured using the dielectric resonator method at 21.8 GHz. We found that R_s = 0.19 mΩ at 21.8 GHz and 4.2 K. Then, an R_{rf} of 3.924 × 10^{-26} ΩHz^2/sq was used in simulations. A kinetic inductance value of 1.225 pH/sq was used for the simulation. Conventional coplanar waveguides (CPWs) without a ground plane were selected as the readout through the line. The width of the signal line and gap were 40 and 10 μm, respectively, corresponding to a specific impedance of 50 Ω. The spiral resonator was placed into the space near the signal line from which the ground plane of the CPW was removed. The size of the 12-turn spiral resonator was 480 μm square. Both the width of the line and space of the resonator were 10 μm. The total length of the resonator was adjusted to that of the effective guided half-wavelength. The distance between ground and resonator was 30 μm.

Fig. 2 shows the simulated transmission (S_{21}) response of a spiral resonator for the spiral-MKID. We observed a sharp resonance around 4.498 GHz and obtained maximum attenuation (A_{max}) of over 20 dB and an unloaded quality factor (Q_u) of over 5×10^4. Fig. 3 shows the current density distribution of a spiral resonator at 4.49824 GHz. The white and black areas represent higher and lower current density, respectively. The current is concentrated in the central part of the resonator, which means there is an effective guided half-wavelength resonance.

To realize large array spiral-MKIDs, frequency characteristics of all resonators with different resonant frequencies should be equal. We designed two models with different couplings between the signal line and the resonators, namely, an asymmetrical and symmetrical model shown in Figs. 4(a) and (b), respectively. The length of the resonators was varied for every 100 μm increment, which caused a frequency shift of around 30 MHz [10]. Fig. 5 shows the dependence of Q_u for the respective resonances. We obtained a Q_u of over 5×10^4 in both models. As resonant frequency of the resonator increased, the Q_u decreased in the asymmetric model shown in Fig. 4(a). On the other hand, we obtained a Q_u of approximately 4×10^4 for all resonators in the symmetric model shown in Fig. 4(b). Fig. 6 shows the simulated S_{21} responses of the 9 resonator array spiral-MKID. According to the preceding results, we found that the symmetric spiral resonator was superior to the asymmetric resonator concerning microwave response.

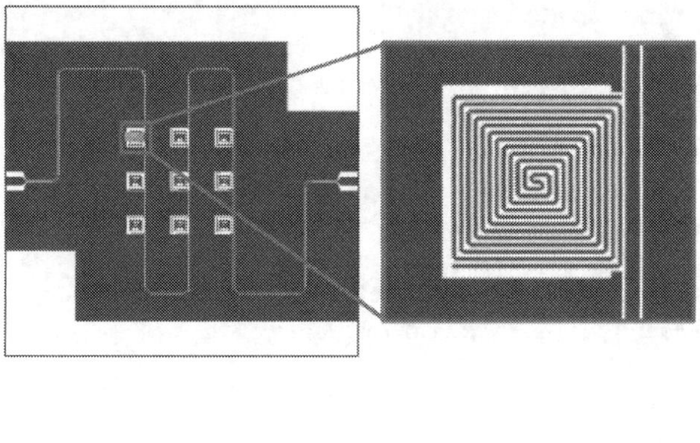

Figure 1. Typical patterns of 9 resonator array spiral-MKID. Enlarged view is the rewound spiral resonator.

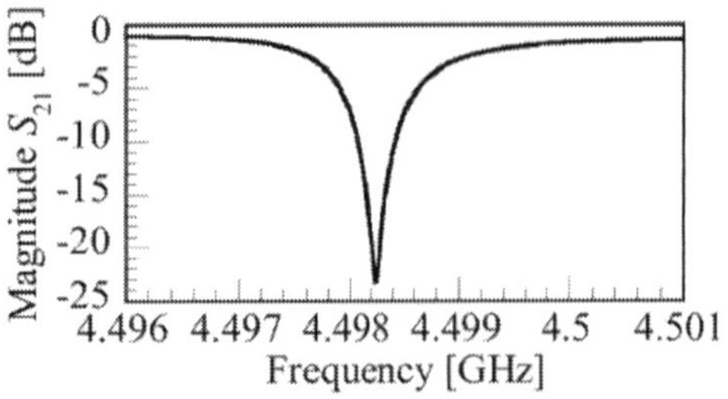

Figure 2. Simulated microwave response of rewound spiral resonator.

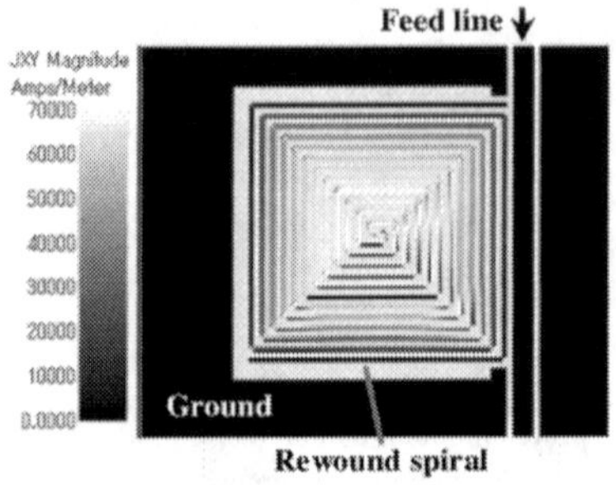

Figure 3. Current density distribution of spiral resonator at 4.49824 GHz.

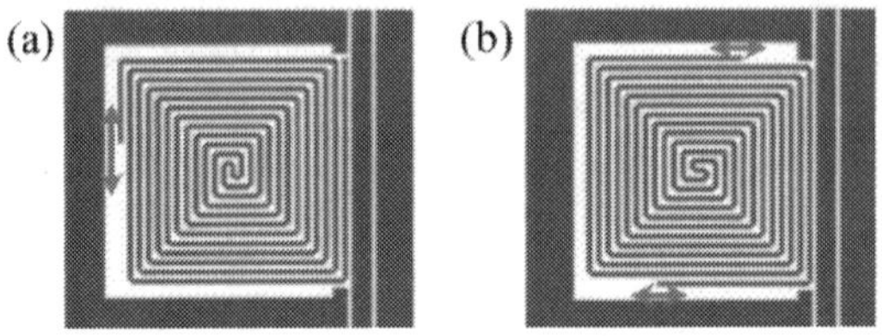

Figure 4. Schematic of spiral-MKIDs. (a) Asymmetric model. (b) Symmetric model.

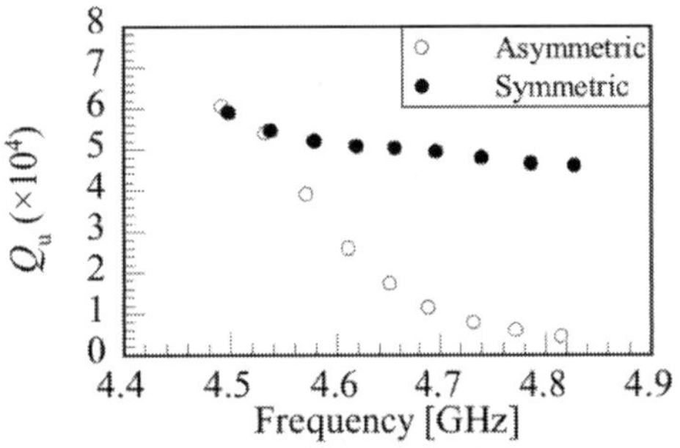

Figure 5. Dependence of unloaded quality factor for respective resonators.

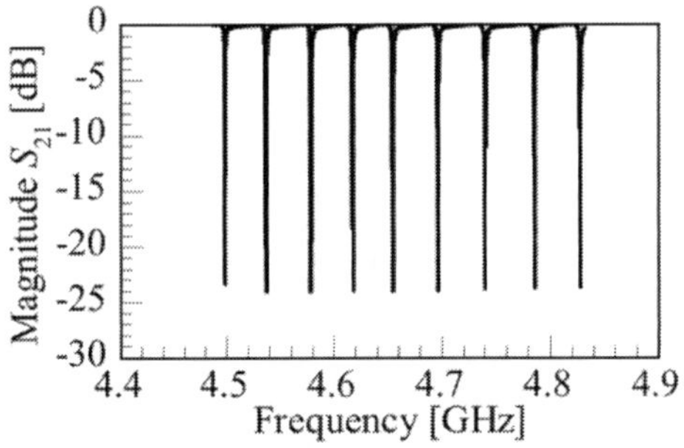

Figure 6. Simulated microwave response of 9 resonator array MKID using symmetric spiral resonator.

3. Experiments

3.1. Fabrication and measurement of MKID array

We selected NbN films for the fabrication of the spiral-MKID. The films were deposited on m-plane sapphire substrates using reactive DC magnetron sputtering with a discharge current source. The sputtering conditions of the films were investigated by varying the gas ratio of Ar and N_2 for the same current, and a suitable T_c of 16 K was obtained for an NbN film with a thickness of 150 nm on an m-plane sapphire substrate. The 9 resonator array spiral-MKID was patterned using standard photolithography and reactive ion etching, as shown in Fig. 4(b). The MKID chip was mounted to a polychlorinated biphenyl (PCB) chip carrier ($\varepsilon_r = 11.6$) coated on both sides with Au films and bonded using Al wires (ϕ 25 μm). The chip carrier was placed in a cryocooler and we measured the resonance frequency, the A_{max}, and the quality factor of the spiral-MKID. We measured at 4 K since we target to an application for general-purpose measurements. The microwave response, S_{21}, passing through a cryogenic low-noise amplifier (LNA) was measured using a vector network analyzer (VNA).

3.2. Frequency and optical responses

Fig. 7 shows the S_{21} response of the 9 resonator array spiral-MKID measured at 4 K without calibration. 9 resonances are clearly shown in the frequency range of 4.4 to 5.0 GHz at 4 K. There is a difference in the interval of resonant frequency. The reason for the different interval is still under investigation.

Fig. 8 shows the optical response of a single resonator measured at 4 K. The solid line indicates the S_{21} response of the spiral-MKID when illuminated with 650 nm photons, and the dotted line indicates the response without optical illumination. The frequency shift of 0.5 MHz is clearly observable. This result suggests that the spiral-MKID is capable of detecting terahertz waves.

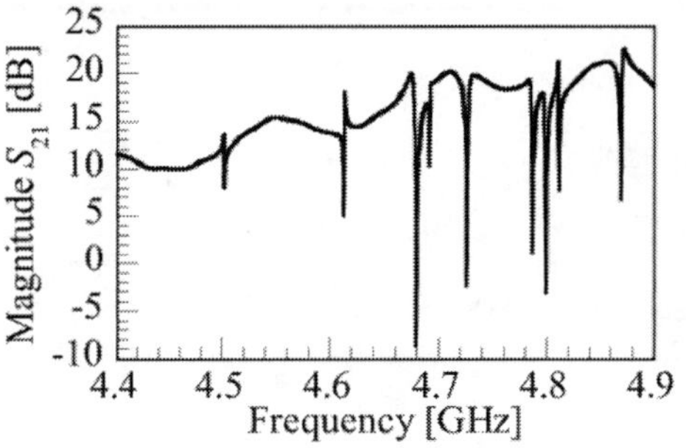

Figure 7. Microwave response of 9 resonator array MKID using a symmetric spiral resonator.

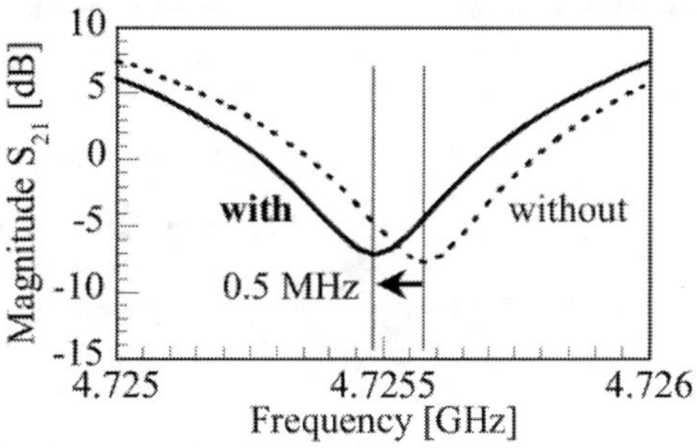

Figure 8. Optical response of single resonator.

4. Conclusions

We designed and fabricated a 9 resonator array MKID using an NbN symmetric spiral resonator, and then we measured the frequency response of the MKID. We observed 9 resonances of the S_{21} response. The frequency of 0.5 MHz was observed when illuminated with 650 nm photons. The preceding results suggest the spiral-MKID is capable of detecting terahertz waves.

Acknowledgements

This work was supported in part by JST (development of systems and technology for advanced measurement and analysis), JSPS KAKENHI Grant Numbers 22560317, 24560393, Nippon Sheet Glass Foundation for Materials Science and Engineering, Casio Science Promotion Foundation, The Mazda Foundation, and Kato Foundation for Promotion of Science. A part of this work was carried out in the clean room of Yamagata University.

References

[1] Day P K *et al* 2003 *Nature* **425** 817
[2] Monfardini A *et al* 2011 *Astrophys. J. Suppl. Series* **194** 24
[3] Mazin B A *et al* 2010 *Proc. SPIE* **7735** 773518
[4] Moore D C *et al* 2012 *Appl. Phy. Lett.* **100** 232601
[5] Baryshev A *et al* 2011 *IEEE Trans. Terahertz Sci. Technol.* **1** 112
[6] Doyle S *et al* 2008 *J. Low Temp. Phys* **151** 530
[7] Brown E R *et al* 2006 *Microwave and Opt. Tech. Lett.* **48** 524
[8] Ma Z and Kobayashi Y 2005 *IEICE Trans. Electron.* **E88-C** 1406
[9] Ariyoshi S *et al* 2013 *Appl. Phys. Express* **6** 064103
[10] Hayashi K *et al* 2013 *Physics Procedia* **45** 213

Journal of Physics: Conference Series **507** (2014) 042016 doi:10.1088/1742-6596/507/4/042016

Investigation of nonlinear superconducting microwave resonators including Nb Josephson junctions and SQUID arrays

M Khabipov, B Mackrodt, R Dolata, T Scheller and A Zorin

Physikalisch-Technische Bundesanstalt (PTB), Bundesallee 100, 38116 Braunschweig, Germany

E-mail: marat.khabipov@ptb.de

Abstract. Motivated by the realization of bifurcation regimes of operation, we report on experiments with superconducting microwave resonators comprising short serial arrays of Nb/AlO$_x$/Nb SIS Josephson junctions and low-inductance SQUIDs. These elements provide sufficiently large Josephson inductance $L_J \propto N$ (N is the number of elements in the chain) enabling nonlinear dynamics of the circuits and detection of low power signals.

1. Introduction

Detection of small electrical signals based on the bifurcation effect in nonlinear resonance circuits has been recently exploited in solid-state electronics and quantum information science. The bifurcation circuits are used, for example, in microwave kinetic inductance detectors, superconducting qubits and quantum electrodynamics circuits [1-3]. Due to their strong nonlinearity stemming from a periodic current-phase relation I (φ), the Josephson junctions and serial arrays of the Josephson junctions and SQUIDs perfectly play the role of nonlinear inductors. These nonlinear elements can be incorporated either into lumped element resonance circuits or included in a microwave cavity (for example, a coplanar waveguide (CPW) resonator) close to the current antinode [1]. Usually, the superconducting bifurcation circuits operate in the regime of zero current bias of the Josephson inductance, i.e. when the nonlinearity takes a simple symmetric form $L_J(\varphi) = L_J(-\varphi) \approx L_J(0)(1 + \varphi^2/2)$. In this work, we address the bifurcation regimes of both nonzero current bias and magnetic control of the inductance (in SQUID arrays) leading to asymmetric nonlinearities $L_J(\varphi) \neq L_J(-\varphi)$.

2. Sample Fabrication

Standard Nb/AlOx/Nb trilayers are deposited on oxidized three-inch silicon wafers (300 nm SiO_2). First, a Nb base layer, 160 nm in thickness, is deposited by dc-sputtering and covered by an Al layer of 10 nm thickness. The barrier is created by oxidation in pure O_2 at a pressure of 100 Pa for 20 minutes at a temperature of 20 °C. Thereupon the Nb counter electrode, also 160 nm in thickness, is deposited. The critical current density j_c of the trilayer was set to 1 kA/cm^2. On top of the trilayer, a 100 nm thick SiO_2 layer is deposited by plasma enhanced chemical vapour deposition (PECVD). This layer was found to improve the reproducibility of the first etching process and serves as a mask in a forthcoming anodization process. Using electron beam lithography (EBL) and a two-layer polymethylmethacrylate

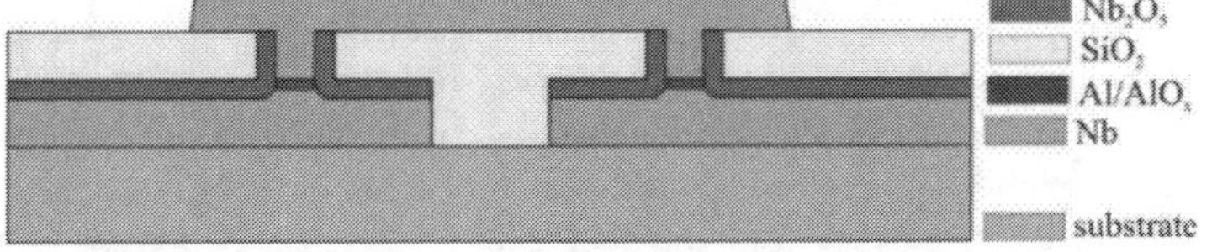

Figure 1. Cross-section (not to scale) of the $Nb/Al/AlO_x/Nb$ Josephson junctions.

(PMMA) resist the mask for the first etching process is created by the lift-off of a thermally evaporated Al layer, 25 nm in thickness. This Al mask defines the area of the Josephson junctions. The SiO_2 layer and the Nb counter electrode are patterned one after the other by reactive ion etching (RIE) in CHF_3 and CF_4, respectively. The etching parameters have been carefully optimized to ensure a good pattern transfer from the Al mask to the Nb counter electrode with minimal size deviations. Since the CF_4 RIE process does not etch Al, the AlO_x/Al barrier and the Al mask layers are removed by an Ar ion beam etching step. Then the wafer is anodized up to a voltage of 18 V, yielding an Nb_2O_5 layer thickness of about 46 nm. This anodization step was found to decrease subgap leakage currents, particularly in very small Josephson junctions [4]. By use of a second EBL process, now in combination with a negative tone resist, a mask for patterning the base electrode is defined. Then the Nb_2O_5/Nb layer is etched in an SF_6 RIE process. After removing the resist, a PECVD SiO_2 layer of 500 nm thickness is deposited. This layer is planarized by chemical mechanical polishing (CMP) until the Nb counter electrode is exposed at the surface. Since the optimum polishing time for each device on the wafer depends on the junction sizes and local pattern density, large dummy junctions are placed around the active junctions to ensure the same clearing time, independent of the size of the active junctions. Finally, after thorough Ar plasma cleaning of the samples, a 200 nm thick Nb wiring layer is deposited by dc-sputtering and patterned by RIE in SF_6 through a third, EBL-defined, negative tone resist mask (see figure 1). CPW resonators as well as interdigital capacitors and lumped inductors are realized in this layer.

3. Design of microwave CPW resonators

We developed and fabricated a variety of on-chip superconducting microwave CPW resonators operating as half-wave transmission line cavities with capacitive coupling. The resonance frequencies were designed to be around 2.4 GHz. The quality factor Q of the resonators measured at 4.2 K was

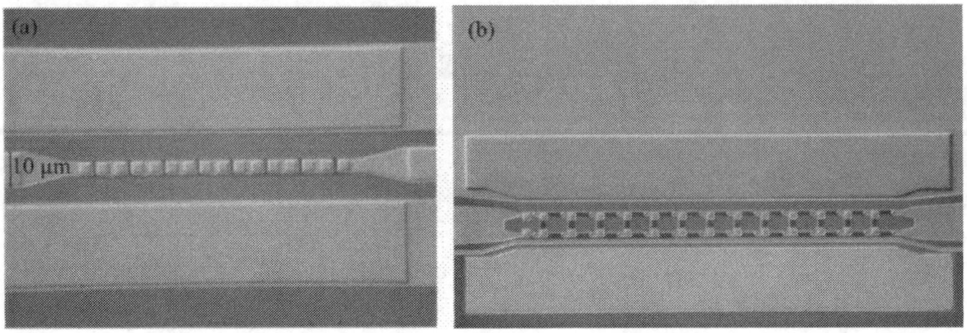

Figure 2. Microphotograph of the circuits: a) fragment of CPW resonator including array of N = 16 Josephson junctions of area 2.5×2.5 μm^2 and b) array of DC SQUIDs inserted in asymmetric CPW [5] resonator. This resonator has a ground plane on one side only and provides easy access to the array. The elongated metallic island seen below the array was patterned for the sake of protecting the junctions at the CMP step. The flux bias of the SQUIDs is done by current in the control wire located between this island and the array.

found to be about 3×10^4 for the smallest coupling capacitances (a few fF) to the feed line, indicating a sufficiently high quality of the deposited Nb films. Single Nb/AlO_x/Nb SIS Josephson junctions and low-inductance ($L \ll L_J$) SQUIDs, as well as short arrays (up to N = 16) of these elements, were integrated into these superconducting resonators with the goal of achieving sufficiently large Josephson inductance $L_J \propto N$ enabling nonlinear dynamics of the circuits. The control of this inductance was made by both magnetic flux applied to the SQUID loops and by biasing the junction/SQUID arrays by constant current, $|I_b| < I_c$, where $I_c \sim 50$–100 μA is the critical current of these serial Josephson circuits. This dc current bias $I = I_c \sin(\varphi + \varphi_0)$, where $\varphi_0 = \arcsin(I_b/I_c)$ is the constant phase bias of individual Josephson junctions and φ is the ac part of the phase, has allowed varying not only the conventional cubic term in the Josephson supercurrent, i.e. $-I_c \cos\varphi_0\, \varphi^3/6$ (ensuring conventional quadratic nonlinearity of the Josephson inductance L_J), but also adding an otherwise absent quadratic term, $-I_c \sin\varphi_0\, \varphi^2/2$ (admixing an antisymmetric component $\propto \varphi$ in nonlinear inductance L_J). The junction/SQUID arrays were located close to the antinodes corresponding to both $\lambda/2$ and λ resonances, i.e. somewhat shifted from the geometrical centre of the resonator. The later regime of driving at a double frequency should enable the period doubling bifurcations [6]. Figure 2 shows the fragments of circuit layouts in the vicinity of the arrays. To integrate arrays, the CPW structures were gradually downsized without changing the value of the impedance of 50 Ω. In the symmetrical CPW the current bias and flux bias control lines were fabricated in the base layer. The bias wire crosses the ground plane and the flux wire is put below the CPW ground plane.

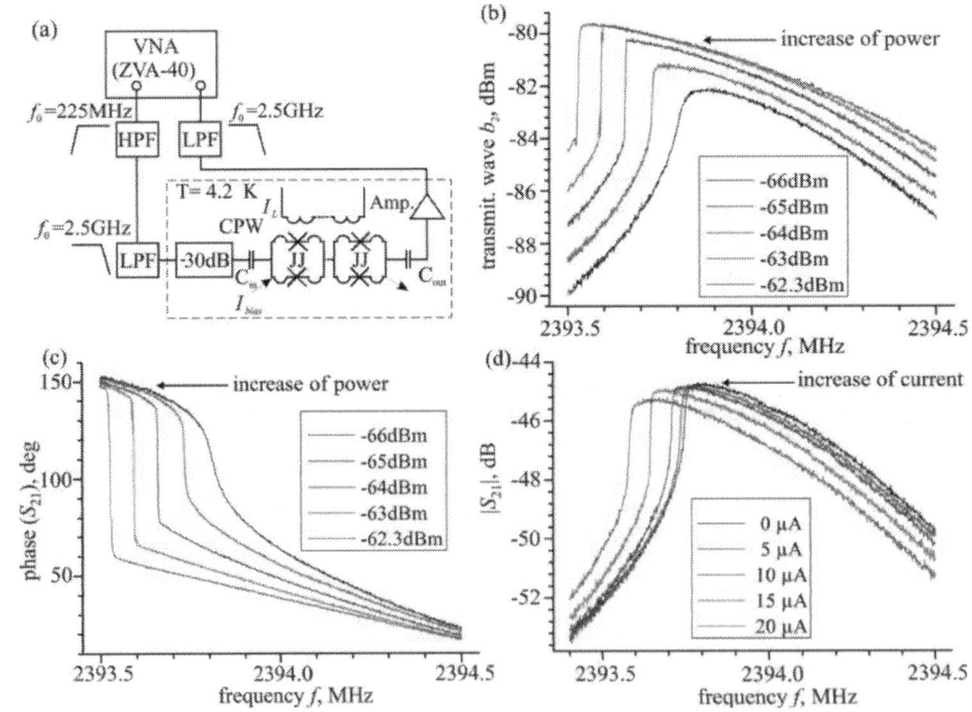

Figure 3. Experimental results: a) Measurement setup includes the vector network analyzer (VNA) with high pass filter (HPF), low pass filters (LPF) and cold attenuator and amplifier. b) Transmission parameter b_2 and c) phase of the transmission parameter S_{21} versus frequency for the different power levels P. At $P \approx -65$ dBm the circuit starts to bifurcate. d) Transmission parameter S_{21} for the fixed applied power level $P = -63$ dBm and different values of the dc current I_b bias of the array.

4. Experimental results

Figure 3a shows the configuration of the experimental setup. The chips consisting of microwave resonators with integrated Josephson junctions/SQUIDs were bonded on a special plate with SMA connectors and mounted in a metallic box. All low temperature parts were inserted into a cryoperm shield (screening factor better than 10^3). The measurements were carried out in liquid He at temperature T = 4.2 K. The resonance frequency was close to the designed value and the quality factor was close to 4000. We measured the transmission through the CPW (parameters S_{21} and b_2) as a function of frequency and applied microwave power. Figure 3b shows transmission parameter b_2 through the CPW with an integrated array of 8 Josephson junctions with critical current $I_c = 60$ μA for different power levels. With increasing applied power the resonance frequency is shifted to the lower values, whereas at the power level of −65 dBm the threshold of bifurcation regime is achieved. At larger power, one can see progressing hysteresis (for one-direction measurements, as here, a larger jump in S_{21}). Figure 3c shows the phase of parameter S_{21} for the same applied power levels as in figure 3b, also indicating the effect of bifurcation at $P > -65$ dBm. Figure 3d demonstrates the dependence of parameter S_{21} on dc current I_b biasing the array for the fixed power level $P = -63$ dBm. Due to suppression of the effective critical current I_c by factor $\cos\varphi_0$ and corresponding increase of L_J, the resonance frequency is reduced. Moreover, due to a decrease in the resonance frequency the normalized (effective) amplitude of ac drive increases and, therefore, causes bifurcation at lower power. This effect is seen in figure 3d for the bias current range up to 15 μA. At larger bias ($I_b \geq$ 20 μA), the amplitude of the drive is presumably comparable with the effective critical current and our assumption of small oscillations is no longer valid. Then the higher nonlinear terms lead to suppression of the bifurcation. Note that the shift of the transmission curves is, as expected, the same for the dc bias currents I_b of both polarities.

5. Conclusion

The bifurcation regimes were observed in the circuits with junctions/SQUIDs arrays (N = 16) inserted in both symmetric and asymmetric CPW resonators having a quality factor Q of 2000 or higher. The control of the Josephson inductance made by both magnetic flux (applied to the SQUID loops) and by biasing the arrays with a constant current was rather efficient. The integration of Josephson π-junctions (for example, of Superconductor-Ferromagnet-Superconductor type [7]) in these circuits should make it possible to control the effective Josephson inductance and, therefore, the bifurcation regime in a wider range of parameters.

Acknowledgements

The authors wish to thank Thomas Weimann, Rüdiger Wendisch and Hans-Peter Duda for technical support. This work is supported in part by the German DFG research project Zo124/2-2 and the European Union under the Joint Research Projects MICROPHOTON within the European Metrology Research Programme (EMRP). The EMRP is jointly funded by the EMRP participating countries within EURAMET and the European Union.

References

[1] Vijay R, Devoret M H and Siddiqi I 2009 *Review of Scientific Instruments* **80** 111101
[2] Manachuryan V E, Boaknin E, Metcalfe M, Siddiqi I and Devoret M 2007 *Physical Review B* **76** 014524
[3] Metcalfe M, Boaknin E, Manachuryan V E, Vijay R, Siddiqi I, Rigetti C., Frunzio L, Schoelkopf R J and Devoret M 2007 *Physical Review B* **76** 174516
[4] Dolata R, Scherer H, Zorin A B and Niemeyer J 2005 *Journal of Applied Physics* **97** 054501
[5] Lampe R W 1985 *IEEE Transactions on Antennas and Propagation* **AP-33** 1028
[6] Zorin A B and Makhlin Yu. 2011 *Physical Review B* **83** 224506
[7] Ryazanov V V, Oboznov V A, Rusanov A Y, Veretennikov A V, Golubov A A and Aarts J 2001 *Physical Review Letters* **86** 2427

11th European Conference on Applied Superconductivity (EUCAS2013) IOP Publishing
Journal of Physics: Conference Series **507** (2014) 042017 doi:10.1088/1742-6596/507/4/042017

Some phenomena due to SQUID input properties when local feedback is present

Mikko Kiviranta[1], Leif Grönberg[1], Nikolai Beev[1] and Jan van der Kuur[2]
[1]VTT, Tietotie 3, 02150 Espoo, Finland
[2]SRON, Sorbonnelaan 2, 3584CA Utrecht, the Netherlands

E-mail: Mikko.Kiviranta@vtt.fi

Abstract. We have constructed a two-stage SQUID amplifier, in which series-mixing local feedback has been used to screen the SQUID input inductances and hence to boost the power gain of the amplifier. 2.9 pA/Hz$^{1/2}$ current noise and 2.9 nH input inductance of the lower SQUID stage imply energy resolution of 18 times Planck constant at 4.2 K, which, arguably, can be further improved by input inductance screening. The upper stage consists of a 184-series 4-parallel SQUID array, which, when used alone, shows lower than 0.03 $\mu\Phi_0$/Hz$^{1/2}$ flux noise, but which as a part of the two-stage amplifier is operated at a higher flux noise level to provide robust, EMI-tolerant output signal. The series-mixing feedback also facilitates negative SQUID input impedance, which would allow self-oscillating SQUID-based frequency domain multiplexing.

1. Introduction

Signal chains for frequency-domain multiplexed arrays of transition edge sensors (TESes), such as the SPICA / SAFARI [1] and ATHENA+ / X-IFU [2] instruments require SQUID amplifiers with a large power gain. In order such instruments to be detector limited rather than readout limited, the noise equivalent signal power, roughly -160 dBm/Hz at the detector outputs, must be amplified within the cryogenic stage sufficiently to overcome the noise of the subsequent low noise amplifier (LNA). Even more important may be the electromagnetic interference (EMI), which couples to the cable between the deep cryogenic detector stage and the moderate-cryogenic stages capable of housing semiconductor LNAs. In the SPICA spacecraft that cable is estimated to be up to 10 meters in length. Additionally, amplification should be performed with as few cascaded SQUID stages as possible, because each bias and flux setpoint line in the wiring harness adds complexity and heat leakage. We have approached the power gain quest in two ways: (a) two-stage amplification chain using local feedback to boost the gain of individual stages, and (b) three-stage amplification chain where the two upper stages share the bias and flux setpoint lines and hence don't require additional lines in the wiring harness. The bias-reusing chip is a subdivided version of the 184x4 SQUID array chip described later, where a 4x4 intermediate array drives a 180x4 final array. In this paper we concentrate on the local feedback version, however.

The power gain of a single SQUID stage, or rather, its available output power to the (reactive) input power ratio is [3]

Figure 1. Microphotograph of the 6-subloop fractional-turn SQUID

Figure 2. Photograph of the 184x4 array with the screening transformer chip.

$$\frac{dP_O}{dP_I} = \frac{R_D \left({dI}/{d\Phi} \times M \right)^2}{\omega_C \, L_{IN}} \tag{1}$$

when expressed in terms of SQUID parameters: mutual- and self-inductances M and L_{IN} of the input coil, dynamic resistance R_D and flux-to-current response $dI/d\Phi$. One might consider using local positive feedback [4,5] to boost the $dI/d\Phi$. In the simple-minded approach R_D and L_{IN} degrade, however, leaving the dP_O/dP_I unchanged. We have implemented series-mixing feedback [6] rather than the traditionally used shunt-mixing. The technique can be seen as a virtual negative inductance in series with input, leading to improvement of L_{IN} rather than $dI/d\Phi$ in Eq. 1.

2. Devices and experiments

We used a 6-subloop fractional-turn SQUID as the lower stage in the experiments (Fig 1). The counterwound loop construction resembles [7], but the input coil only contains only one turn. We measure input sensitivity $M^{-1} = 5.8$ μA/Φ_0, and output swing $\Delta I = 12$ μA. The device is equipped with transformers for series-mixing feedback.

The upper stage is a 184-series 4-parallel array, constructed out of the same cascadeable cells as [3]. The purpose of the large form factor is to reach a large noise-equivalent output power, when driven by the lower SQUID with roughly 1:1 flux ratio between lower and upper stage. Although not taken advantage of here, the array shows routinely 0.06 μΦ_0/Hz$^{1/2}$ flux noise at 4.2 K when read out with a 20 MHz room-temperature SiGe LNA.We have measured 0.022 μΦ_0/Hz$^{1/2}$ flux noise at 2.8 K with a cryogenic LNA [8]. Series-mixing feedback for the array was accomplished by wire-bonding a separate thin-film transformer chip (Fig.2), with a choice of 25 nH or 120 nH mutual inductance.

The traditional shunt-mixing feedback [4,5] creates steep and shallow slopes in the flux response of the SQUID (Fig. 4, lower). The series-mixing feedback creates a fast slope where SQUID input

inductance is supressed and R/L cutoff frequency is higher; and slow slope where SQUID inductance is enhanced and R/L cutoff is lower. The bandwidth is visible as different RMS noise on different slopes in (Fig. 4, upper).
The natural L_{IN} = 160 nH of the 184x4 array is screened to L_{IN} = 60 nH on one slope and enhanced to L_{IN} = 260 nH on other slope. This is consistent with M = 25 nH of the transformer and roughly 4 μA/μA current gain provided by the SQUID array. When the M = 120 nH winding is used, one slope becomes unstable due to the negative total input impedance. We intend to use this configuration to implement the self-oscillating frequency domain multiplexer [9]. The array exhibits a clean 3 MHz oscillation when coupled to an external LC resonator in this mode.

The two-SQUID amplifier (Fig. 3) has shown the performance summarized in table 1. Input inductances were measured via the R/L cutoff, with a 5 mΩ resistive wire bonded across the input. The bandwidth limitation is roughly consistent with the R_D/L_{IN} cutoff between the first and second stage, including the interstage parasitic inductance, and R_D variation of the first stage as a result of applied feedback. Currently, no precautions against R_D variation has been taken.

Note that our series-mixing balance circuit against the L_{IN} enhancement on the steep slope (Fig. 4, lower) appears to be over-compensating, as the L_{IN} measures *lower* on the steep slope.

Finally note that the energy resolution,

$$\varepsilon = \frac{1}{2} L_{IN}\, i_N^2 \qquad (2)$$

when expressed in terms of input-referred current noise spectral density i_N and the SQUID input inductance L_{IN} appears to improve by the series-mixing feedback. In fact, eq. 2 is not valid for an inductive load, because it lacks the back-action noise terms. As a figure-of merit for resistive loads at low frequencies, such as TES arrays, eq. 2 is convenient, however.

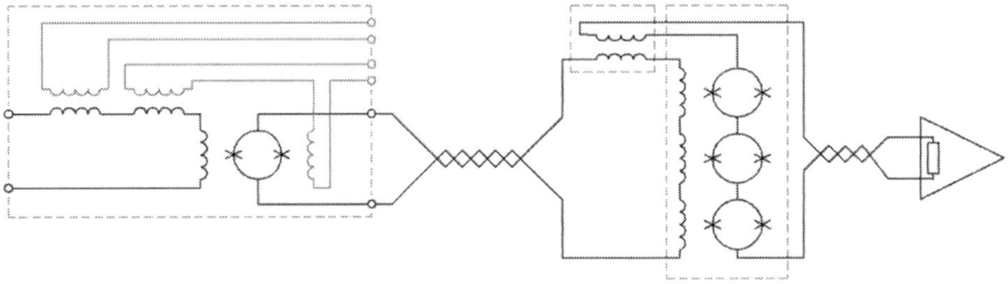

Figure 3. The two-stage amplifier schematic. Local feedback is not active in the lower stage, as drawn.

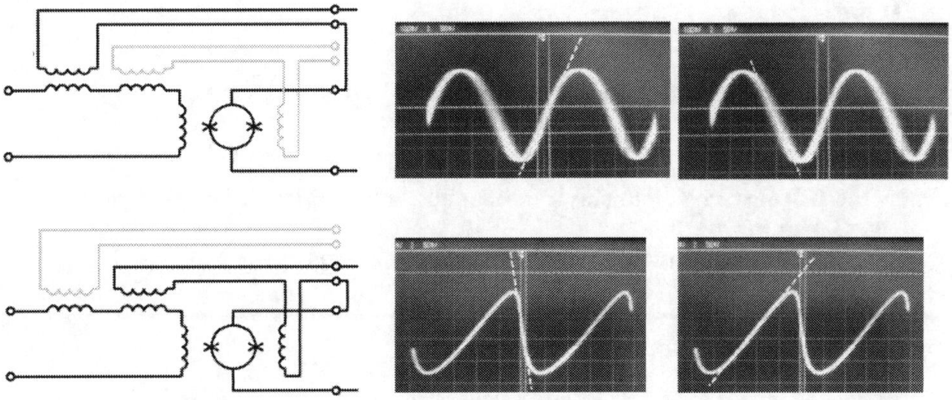

Figure 4. Local feedback configurations of the lower stage. (Upper) The plain series-mixing feedback and the associated flux-to-current characteristics. A 5 mΩ noise-providing resistor is connected at the SQUID input. (Lower) Shunt-mixing feedback produces steep and shallow slopes to the flux-to-current characteristics, but its L_{IN}-enhancing effect is counteracted by the additional series-mixing feedback.

Table 1. Bandwidth, flux noise and input inductance of the two-stage amplifier at various configurations.

	Lower SQUID configuration ⇒	Steep slope	Shallow slope	L-enhanced	L-suppressed
	L_{IN} [nH]	1.5	3.4	4.7	1.2
Upper array L-suppressed	Bandwidth [MHz]	2.1	7.1	6.8	
	Flux noise [$\mu\Phi_0/\mathrm{Hz}^{1/2}$]	0.45	0.6	1.3*	
Upper array L-enhanced	Bandwidth [MHz]	0.9	4.9	3.0	
	Flux noise [$\mu\Phi_0/\mathrm{Hz}^{1/2}$]	0.6	0.6	1.1*	
* Symptoms of instability at a few tens of MHz observed					

Acknowledgement.
Device fabrication was supported from GSTP program of the European Space Agency. Oscillator and feedback development has received support from the grant no. 262947 of the European Communitys seventh framework programme (FP7/2007-2013).

[1] Jackson B D et. al. 2012 *IEEE Tran. THz Sci. Tech.* **2** 12-21. Doi: 10.1109/TTHZ.2011.2177705
[2] Barret D et. al. 2013 *Preprint* astro-ph/1308.6784
[3] Kiviranta M, Grönberg L and Sipola H 2011 *SuST* **24** 045003. Doi: 10.1088/0953-2048/24/4/045003
[4] Drung D, Cantor R, Peters M and Scheer H J 1990 *Appl. Phys. Lett.* **57** 406. Doi: 10.1063/1.103650
[5] Kiviranta M 2008 *SuST* **21** 045009. Doi:10.1088/0953-2048/21/4/045009
[6] Sedra A S and Smith K C 1998 *Microelectronic circuits*, 4th edition, Oxford University Press
[7] Kiviranta M, Grönberg L and Hassel J 2012 *IEEE Tran. Appl. Supercond.* **22** 1600105. Doi: 10.1109/TASC.2012.2190286
[8] Beev N and Kiviranta M 2013 *Cryogenics* **57** 129-33. Doi:10.1016/j.cryogenics.2013.06.004
[9] Van der Kuur J and Kiviranta M 2013 *APL* **102** 023505. Doi: 10.1063/1.4788683

Journal of Physics: Conference Series **507** (2014) 042018 doi:10.1088/1742-6596/507/4/042018

High Linearity Voltage Response Parallel-Array Cell

V Kornev[1], N Kolotinskiy[1], V Skripka[1], A Sharafiev[1], I Soloviev[1] and O Mukhanov[2]
[1] Lomonosov Moscow State University, Leninskie Gory 1, Moscow, 119991 Russia
[2] Hypres, Inc., 175 Clearbrook Road, Elmsford, NY 10523 USA

E-mail: kornev@phys.msu.ru

Abstract. We studied in detail a cell consisting of two parallel SQUID arrays or two parallel superconducting interference filters (SQIFs) connected differentially with the goal of achieving highly linear voltage response to magnetic signal. In these different cell designs, we accounted for realistic values of coupling inductances in contrast to limiting case of vanishing inductances considered earlier. We found that a cell based on regular parallel SQUID arrays produces higher linearity as compared to the cell based on SQIFs. This high-linearity cell can be used for realizing Superconducting Quantum Arrays (SQA) capable of providing a broadband, highly-linear magnetic field-to-voltage transfer function and high dynamic range.

Introduction

A parallel array of Josephson junctions (a parallel SQUID array) is one of the frequently used elements of superconductive electronics [1-6]. Among many applications, the parallel array was suggested to design quantum cells for Superconducting Quantum Arrays (SQA) capable of providing both highly linear magnetic signal to voltage transformation and high dynamic range [7-9]. Such an implementation exploits the close-to-parabolic form of the main peak sides of the parallel array voltage response. As long as subtraction of two identical parabolic functions with the mutually shifted vertex positions results in linear dependence, differential connection of two parallel arrays oppositely biased by some magnetic flux $\delta\Phi$ results in the cell providing a linear voltage response (quantum cell) [9]. Figure 1 shows schematically the differential cell and a realistic schematic of the parallel array.

An accurate analysis of the parallel array is a complex numerical problem. Therefore in many cases, one has to consider the array in the limit of zero coupling inductances between Josephson junctions in order to use an analytic formula to calculate a voltage response to magnetic flux. However, the realizable in practice arrays have inevitably non-zero coupling inductances which lead to characteristics quite different from the ones obtained for the idealized array with zero coupling inductances.

This paper presents an analysis of achievable linearity of the differential cell voltage response. The

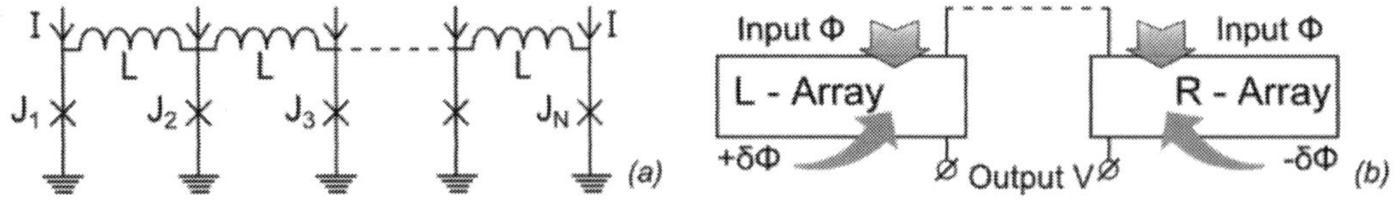

Figure 1. Schematics of a practicable parallel array of Josephson junctions (*a*) and a differential cell (*b*). Here Φ is an input signal, $\delta\Phi$ is a frustration magnetic flux, $I = I_b/N$ is a Josephson-junction bias current, where I_b is a total bias current of the array, and N is total number of Josephson junctions.

study is based on the detailed numerical simulation of the parallel array with realistic parameters and an accurate analysis of the array characteristics.

Approximation Approach

To analyze a linearity of the differential cell response, the terms of the next order of infinitesimality in the parallel array voltage response approximation should be taken into account. The terms can be expressed as a sum of the 4-th and 6-th order parabolas, and therefore the shape of the array peak sides can be described by the following relation:

$$V_{R,L}(\Phi) = C_0 - k\cdot(\Phi - \Phi^*)^2 + a_4\cdot(\Phi - \Phi_4)^4 + a_6\cdot(\Phi - \Phi_6)^6\,, \qquad (1)$$

where C_0, k, a_4, a_6 are constants, Φ^*, Φ_4, Φ_6 - the vertex coordinates, which are positive for the right side relation $V_R(\Phi)$ and negative for the left side relation $V_L(\Phi)$. This allows us to write the differential voltage response as follows:

$$V(\Phi) \equiv V_R(\Phi + \delta\Phi) - V_L(\Phi - \delta\Phi) =$$
$$= [4k(\Phi^* - \delta\Phi) - 2a_4(\Phi_4 - \delta\Phi)^3 - 3a_6(\Phi_6 - \delta\Phi)^5]\cdot\Phi - 8[a_4(\Phi_4 - \delta\Phi) + 5\,a_6(\Phi_6 - \delta\Phi)^3]\cdot\Phi^3 -$$
$$-12a_6(\Phi_6 - \delta\Phi)\cdot\Phi^5. \qquad (2)$$

At $\delta\Phi = \Phi_4$, the 4-th order parabola does not generate nonlinear terms in (4). Similarly, at $\delta\Phi = \Phi_6$, the 6-th order parabola does not generate any nonlinear terms. If one applies sinusoidal signal $\Phi = A\cdot\sin(\omega t)$ to the differential cell, the output signal $V(\Phi)$ consists of the fundamental tone with amplitude $B_1 \approx 4k(\Phi - \Phi^*)\cdot A$ and two harmonics with the following amplitudes:

$$B_3 = [10a_6(\Phi_6 - \delta\Phi)^3 + 2a_4(\Phi_4 - \delta\Phi)]\cdot A^3 + (15/4)a_6(\Phi_6 - \delta\Phi)\cdot A^5, \qquad (3)$$

$$B_5 = -(3/4)a_6(\Phi_6 - \delta\Phi)\cdot A^5, \qquad (4)$$

In accordance with the one-tone analysis technique, the linearity of the magnetic flux to voltage transformation can be expressed by formula

$$Lin = B_1/\max\{B_3, B_5\}. \qquad (5)$$

To realize the highest linearity, one has to apply and hold the needed magnetic frustration $\delta\Phi$ with the utmost precision.

Array Response Fitting

As it was found earlier [10,11], at high enough number N of Josephson junctions ($N > 10$), voltage response of the idealized parallel array with critical current biasing $I_b = I_c$ approaches the ultimate shape of its main peak and the side peak shape becomes close to parabolic starting right from the extreme point at $\Phi = 0$. Moreover, the range of the magnetic frustration $\delta\Phi$ needed to realize the high linearity of the response can be increased by implementation of a parallel SQIF to provide an inhomogeneous flux distribution along the array with minimum in the array center [10]:

$$\Phi_m \propto \{1 - \chi\cdot\sin^3[\pi(m-1)/(N-1)]\}. \qquad (6)$$

As the factor χ increases up to 1, the range of $\delta\Phi$ becomes 1.5 to 2 times larger.

It seems to be impractical to realize a parallel array of Josephson junctions with normalized coupling inductances less $l = 2\pi I_c L/\Phi_0$ than about 0.3, where I_c is Josephson-junction critical current (~ 100 μA for niobium process with critical current density 4.5 kA/cm^2). Our study shows that at low (but realizable) normalized coupling inductances l ~ 0.3...0.7, the use of the aforementioned SQIF structure instead of regular array does not give any pronounced change in the array voltage response and therefore gives no improvement in linearity of the differential cell response. The insensitivity to the flux distribution along the array results from the weakening of the collective interaction between the junctions at Josephson oscillation frequency. Further increase in coupling inductances leads to a gradual degradation of the response shape due to a progressive reduction in the number of junctions which can interact with each other and, therefore, contribute to the array response.

Voltage response of the realistic parallel array has significantly wider main peak than the one at zero coupling inductances and its side has two domains which can be well approximated by parabolas (1) with different parameters. Figures 2(*a*) and 2(*b*) show two parabolic approximations of the main peak side of the voltage response of the array of 10 Josephson junctions coupled by inductances of value $l = 0.5$. Some increase in bias current I_b above I_c (up to ~ 1.06 I_c) allows widening this domain to $\Phi = 0$. The main fitting parabola has vertex position $\Phi^* \approx 2.47\Phi_0$ and factor $k \approx 0.132\cdot V_c/(\Phi_0)^2$.

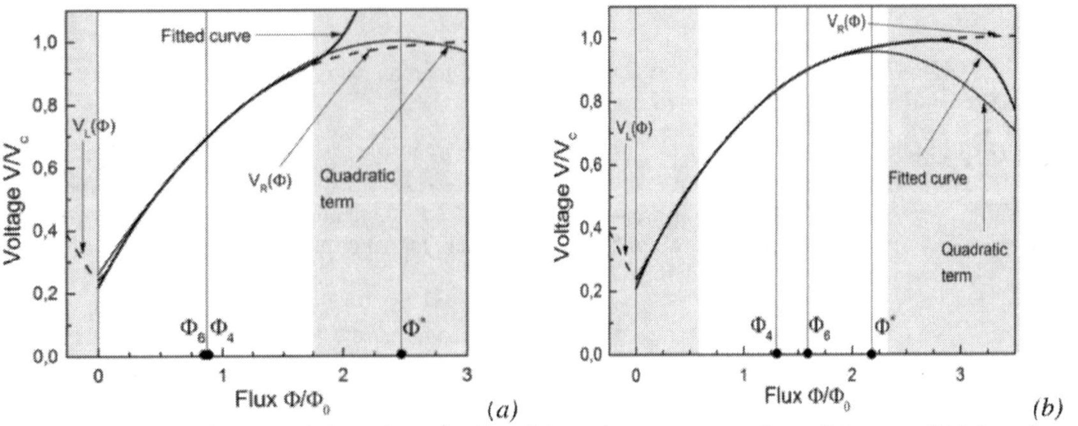

Figure 2. *(a)* Approximation of the main peak side of the voltage response of parallel array of 10 Josephson junctions by parabolas (1) in case of $l = 0.5$ for *(a)* the first domain (parabola vertex positions $\Phi^* \approx 2.47\Phi_0$, $\Phi_4 \approx 0.86\Phi_0$, $\Phi_6 \approx 0.85\Phi_0$); and *(b)* for the second domain ($\Phi^* \approx 2.2\Phi_0$, $\Phi_4 \approx 1.36\Phi_0$, $\Phi_6 \approx 1.57\Phi_0$). The exact voltage response is shown by dash line, bold solid line shows the fitted curve, and thin solid line shows the only quadratic parabola.

Positions of the 4-th and 6-th parabola vertexes are localized within a range from 0.79 Φ_0 to Φ_0. The vertex positions are very sensitive to the current biasing and at $I_b = 1.06\ I_c$ they well approach each other.

For the second domain, the fit parameters are practically insensitive to a change in the bias current. The main fitting parabola has the practically fixed factor $k \approx 0.15 \cdot V_c/(\Phi_0)^2$ and vertex position $\Phi^* \approx 2.2\Phi_0$. Vertex positions for the 4-th and 6-th parabolas are about fixed and distinct: $\Phi_4 \approx 1.36\ \Phi_0$ and $\Phi_6 \approx 1.57\ \Phi_0$.

Response Linearity

Figure 3 shows the differential cell response linearity vs magnetic frustration $\delta\Phi$ at bias current $I_b = 1.06\ I_c$ and different amplitudes of the applied magnetic signal. The data reveal three domains of parabolic approximation on the voltage responses of the parallel arrays forming the differential cell. The additional domain with center at $\Phi_{add} \approx 0.4$ is superimposed on some part of the first domain shown in figure 2*(a)*. When magnetic frustration $\delta\Phi$ coincides with the centres of the additional domain (the first one and the second one), the response linearity reaches the maxima.

Similarly to the second domain discussed above, the additional domain demonstrates the lack of convergence of the high-order parabola vertex positions and therefore both the side maxima in figure 3 show a shift with the input signal amplitude in accordance with the formulae (3), (4). In contrast to these peaks, the middle maximum does not shift with the signal amplitude due to the fact that the 4-th and 6-th parabola vertexes approach each other at $I_b = 1.06\ I_c$. Thus regardless of the signal amplitude, the highest linearity is achievable at the same frustration $\delta\Phi \approx 0.71\ \Phi_0$ and can be as high as ~95 dB and ~80 dB for the signals over 30% and 60% of the total response swing, correspondingly. Moreover, the requirement for an accuracy of the magnetic frustration $\delta\Phi$ seems not too stringent.

Load Impact Balancing

If a load R_e is connected to the differential cell, some part of the bias currents I_b applied to the cell shoulders (parallel arrays) flow through the load and this causes essential changes in the cell response. The effective currents I_R, I_L flowing through the cell shoulders depend on output voltage $V = V_R - V_L$ and change in antiphase as follows:

$$I_R = I_b - V/R_e, \qquad I_L = I_b + V/R_e. \tag{7}$$

Maximum distortion of the response takes place when signal Φ comes close to $-\delta\Phi$ or $+\delta\Phi$. In fact, at $\Phi \approx -\delta\Phi$ (output voltage $V < 0$) total magnetic fluxes Φ_R and Φ_L applied to the right and left shoulders approach correspondingly 0 (peak point of the shoulder response) and $2\delta\Phi$ (far from peak point), and vice versa at $\Phi \approx +\delta\Phi$. Reduction in the currents I_R and I_L occurs at total fluxes Φ_R and Φ_L close to $2\ \delta\Phi$ where the shoulder responses are not sensitive to the current reduction. At the same time, the increase in these currents takes place when the total fluxes Φ_R and Φ_L approach 0 (peak point) and the

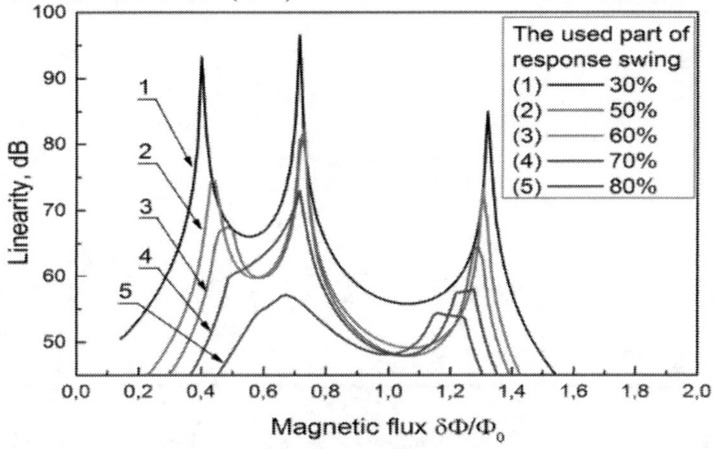

Figure 3. Linearity of the output signal of the practicable differential cell at optimal current biasing $I_b = 1.06$ versus magnetic frustration $\delta\Phi$ of the cell shoulders for different amplitudes of the input sinusoidal signal exploiting 30% (1), 50% (2), 60% (3), 70% (4), and 80% (5) of the total swing of the cell response. Normalized coupling inductance $l = 0.5$, each parallel array contains N = 10 Josephson junctions.

shoulder responses are extremely sensitive to the current change. This fact points at the possible solution of the problem through the proper decrease in the bias currents.

Our numerical simulations show that the increase in the current biasing of the cell shoulders allows us to balance the load impact within the load impedance R_e down to about $10R_N$, where R_N is the shoulder normal resistance. This means that output current can reach one tenth of the bias current.

Conclusion

Our detailed numerical analysis of the realistic differential cell based on two parallel arrays with non-zero coupling inductances confirms its capability of providing highly linear voltage response at low, but practical normalized coupling inductances between Josephson junctions l ~ 0.3...0.7. The high linearity is achievable within significantly wider range of magnetic frustrations then the one following from analysis of the idealized cell with zero inductances. Moreover, the cell characteristics including response linearity show good tolerance to output load. The results have been used in designing of SQA reported in [8,9].

Acknowledgment

The work was supported by ONR via CRDF GAP grant and the grant of Russian Ministry of education and science (GK 14.513.11.0128).

References

[1] Oppenländer J, Häussler C and Schopohl N 2001 *Phys. Rev. B* **63**(2) 024511-1–9
[2] Kornev V K, Soloviev I I, Klenov N V and O. A. Mukhanov O A 2008 *Phys. C* **468** 813–16
[3] Cybart S A, Anton S M, Wu S M, Clarke J, Dynes R C 2009 *Nano Letters* **9**(10) 3581-85
[4] Kornev V K, Soloviev I I, Klenov N V and Mukhanov O A 2009 *IEEE Trans. Appl. Supercond.* **19**(3) 741–44
[5] Berggren S *et al. IEEE Trans. Appl. Supercond.* **23**(3) 1400208
[6] Kornev V K, Soloviev I I, Klenov N V, Filippov T V, Engseth H, and Mukhanov O A 2009 *IEEE Trans. Appl. Supercond.* **19** 916-19
[7] Kornev V K, Soloviev I I, Sharafiev A V, Klenov N V and Mukhanov O A 2013 *IEEE Trans. Appl. Supercond.* **23**(3) 1800405
[8] Kolotinskiy N V, Kornev V K, Sharafiev A V, Soloviev I I and Mukhanov O A 2013 *Phys. Wave. Phenom.* **21**(4) 294-99
[9] Kornev V K, Sharafiev A V, Soloviev I I and Mukhanov O A 2013 Superconducting Quantum Arrays for broadband RF systems, *IEEE 14th Int. Supercond. Electron. Conf. (ISEC)*, PF10.
[10] Kornev V K, Soloviev I I, Klenov N V and Mukhanov O A 2011 *IEEE Trans. Appl. Supercond.* **21**(3) 394–98
[11] Kornev V K, Kolotinsky N V, Skripka V A and Mukhanov O A 2013 High linearity voltage response differential cell *IEEE 14th Int. Supercond. Electron. Conf. (ISEC)*, PG11.

11th European Conference on Applied Superconductivity (EUCAS2013) IOP Publishing
Journal of Physics: Conference Series **507** (2014) 042019 doi:10.1088/1742-6596/507/4/042019

Superconducting Quantum Arrays for Broadband RF Systems

V Kornev[1], A Sharafiev[1], I Soloviev[1], N Kolotinskiy[1] and O Mukhanov[2]

[1] Lomonosov Moscow State University, Leninsky Gory 1, Moscow, 119991 Russia
[2] Hypres, Inc., 175 Clearbrook Road, Elmsford, NY 10523 USA

E-mail: kornev@phys.msu.ru

Abstract. Superconducting Quantum Arrays (SQAs), homogenous arrays of Superconducting Quantum Cells, are developed for implementation of broadband radio frequency (RF) systems capable of providing highly linear magnetic signal to voltage transfer with high dynamic range, including active electrically small antennas (ESAs). Among the proposed quantum cells which are bi-SQUID and Differential Quantum Cell (DQC), the latter delivered better performance for SQAs. A prototype of the transformer-less active ESA based on a 2D SQA with nonsuperconducting electric connection of the DQCs was fabricated using HYPRES niobium process with critical current density 4.5 kA/cm^2. The measured voltage response is characterized by a peak-to-peak swing of ~100 mV and steepness of ~6500 μV/μT.

1. Introduction

A search for multi-element Josephson structures capable of providing a highly linear magnetic signal to voltage transfer with high dynamic range lead to development of Superconducting Quantum Arrays (SQAs). SQA is a homogeneous array of unique superconducting cells with a highly linear voltage response to magnetic field - Superconducting Quantum Cells (SQCs). Such an array is characterized by independent operation of the cells collectively forming a high linearity output signal. By increasing the number of the cells, N, one achieves the increase in dynamic range as a square root of N. This is true for both serial and parallel connections of the cells. In fact, spectral density of the low-frequency (at signal frequency) noise voltage $S_V(0)$ across the N serially connected SQCs is $N{\cdot}S_{V0}(0)$, where $S_{V0}(0)$ is the one across single SQC, and therefore the mean-square-root noise voltage across the array $V_F = [N{\cdot}S_{V0}(0)\Delta F]^{1/2}$ (ΔF is an output frequency band) increases with N as $N^{1/2}$, while transfer factor $K = dV/d\Phi$ and output signal V both increase as N. At parallel connection, the transfer factor does not change, and spectral density of the low-frequency noise current $S_I(0)$ through the parallel array is $N{\cdot}S_{AI}(0)$, where $S_{AI}(0)$ is the one for fluctuation-current source connected to each SQC in accordance with Langevin method [1]-[3], and therefore the mean-square-root noise voltage across the parallel array $V_F = [N{\cdot}S_{V0}(0)\Delta F]^{1/2}R_d/N$ decreases with N as $1/N^{1/2}$ (here R_d is differential resistance of a single SQC). In both cases, the reduced-to-input mean-square-root noise V_F/K determining minimum input signal decreases with N as $N^{1/2}$. Since maximum input signal does not depend on N, dynamic range increases with N as $N^{1/2}$.

One can design a high-performance broadband amplifier by integrating SQA with a broadband input line to apply an input magnetic signal to all array cells (see, for example, the designs proposed in [4-6]). Moreover, active electrically small antennas (ESAs) can be developed on the basis of SQAs to

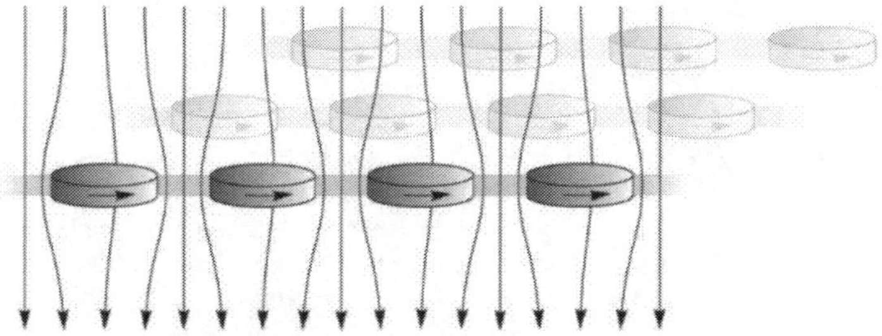

Figure 1. Superconducting Quantum Array (SQA) in an external magnetic field. Meissner effect is shown schematically for only one row of superconducting quantum cells (SQCs).

implement simultaneously broadband reception and amplification of electromagnetic signals. Practical implementation of these active ESAs can be significantly simplified with absence of input RF line.

2. SQA as an active electrically small antenna

There are two possible approaches to design the SQA-based active electrically small antenna (ESA). The first one is based on integration of SQA with a superconducting flux transformer (concentrator) [7, 8]. Such antenna prototypes based on a 1D array of about 80 cells were designed and experimentally evaluated [8]. With circuit area of 3.3 x 3.3 mm^2, the devices were characterized by transfer factor of ~750 μV/μT and peak-to-peak linear response of ~8mV.

According to the second approach, the SQAs in form of 2D arrays of superconducting cells with nonsuperconducting electric connection of the cells can be used directly as active ESAs of a transformer-less type. Figure 1 shows schematically such a 2D SQA in an external magnetic field. Meissner effect is shown for only one row of the superconducting quantum cells (SQCs). Two types of quantum cells with linear voltage response to magnetic signal were proposed as a basic block for SQAs - SQCs: (i) bi-SQUID [9-11] and (ii) the so-called differential cell [12-14]. The latter delivered better performance for SQAs.

Differential Quantum Cell (DQC) consists of two elementary parallel arrays of K = 10 *to* 15 Josephson junctions, differentially connected and oppositely biased by some magnetic flux $\delta\Phi$. Such an implementation exploits a close-to-parabolic form of the main peak sides $V_R(\Phi)$ and $V_L(\Phi)$ of the parallel array voltage response. As long as subtraction of two identical parabolic functions with the mutually shifted vertex positions results in a linear dependence, the differential connection of two parallel arrays oppositely biased by a certain magnetic flux δΦ results in a circuit linear voltage response. An accurate analysis of the parallel array is a complex numerical problem. Therefore in many cases, one has to consider the array in the limit of zero coupling inductances between Josephson junctions in order to use an analytic formula to calculate a voltage response to magnetic flux.

Our detailed numerical analysis of the realistic DQC based on two parallel arrays with non-zero coupling inductances confirms its capability of providing highly linear voltage response at low, but practical normalized coupling inductances between Josephson junctions l ~ 0.3 to 0.7 (here $l = 2\pi I_c L/\Phi_0$, and I_c is Josephson-junction critical current) [13, 14].

The array uniformity ensures a uniform distribution of the magnetic component of electromagnetic wave in the array and, therefore, provides an identical magnetic signal at the input of each SQC (maybe, except for the periphery rows, which can be excluded from the output signal formation). Figure 2 schematically shows a mechanism of conversion of the perpendicular magnetic flux applied to the entire area of the superconducting shoulder of DQC into a parallel magnetic flux applied to an elementary parallel circuit. Screening Meissner currents flow through the edges of the superconducting films $S1$ and $S2$, which form the cell, and connect over the inner surfaces of these films in the region of their overlap (where Josephson junctions are formed), and thus induce a magnetic flux to be applied to a parallel circuit of Josephson junctions. In fact, magnetic field B in the gap between the superconductor films satisfies Maxwell's equation rot($\boldsymbol{B}$) = $\mu_0\mathbf{j}$, and hence $B = \mu_0 j_0 \lambda$ where j_0 is the current density at the film surface and λ is the magnetic penetration depth. Therefore, magnetic flux applied to every section of the Josephson-junction array equals to $B{\cdot}(d+2\lambda){\cdot}dx$ =

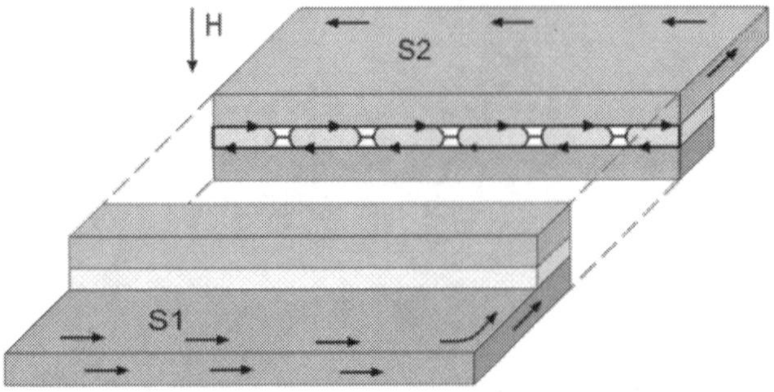

Figure 2. Schematic explanation of a mechanism of magnetic flux application to the SQIF-based cell when magnetic vector H of the incident electromagnetic wave is perpendicular to the cell plane. Meissner currents flow on borders of the S1 and S2 superconductor layers and connect on the interior surfaces in the layer intersection.

$\mu_0 j_0 \lambda(d+2\lambda)dx$, where d is thickness of an isolator film, and dx is a distance between Josephson junctions.

Figure 3 presents the measured voltage response of the transformer-less antenna prototype based on SQA consisting of 560 DQC and occupying area of 3.3 x 3.3 mm^2. The device was fabricated using HYPRES niobium process with critical current density 4.5 kA/cm^2 and Josephson-junction critical current I_c = 125 μA [15]. All cells of the SQA are sectioned into two serial arrays of cell shoulders (twelve-junction parallel arrays). Insets in figure 3 show typical traces of the voltage responses of the two arrays. The arrays are differentially connected to the antenna output. To realize an opposite dc magnetic flux biasing of the arrays, each array is equipped with a control line with normal-metal sections between the array elements. An external multi-turn coil was used to apply magnetic field to the antenna prototype. The total peak-to-peak voltage response of this antenna reaches almost 100 mV, and the steepness of the magnetic signal conversion into output voltage is $dV/dB \approx 6500\ \mu V/\mu T$. Our experimental setup was not designed for a two-tone technique to measure linearity of the antenna voltage response. However, our prior testing of the transformer-based antenna prototype based on DQC type cells showed linearity of about 70 dB [8]. This was measured at signal frequency 300 kHz within ~30% to ~80% of the linear region of the voltage response.

3. Conclusion

Superconducting Quantum Arrays (SQAs) can be used as front-end circuits for broadband radio frequency systems capable of providing highly linear magnetic signal to voltage transformation with high dynamic range, e.g., as active electrically small antennas. We believe that the design of an active antenna using the transformer-less antenna based on 2D SQA with nonsuperconducting electric connection of the quantum cells can lead to the most efficient design. This design allows a significant increase in a number of integrated quantum cells in fixed antenna area. Among the proposed quantum cells such as bi-SQUID and Differential Quantum Cell (DQC), the latter delivered the better performance for SQAs. The transformer-less antenna prototype containing 560 DQCs with nonsuperconducting cell-to-cell connections was realized using HYPRES niobium process with

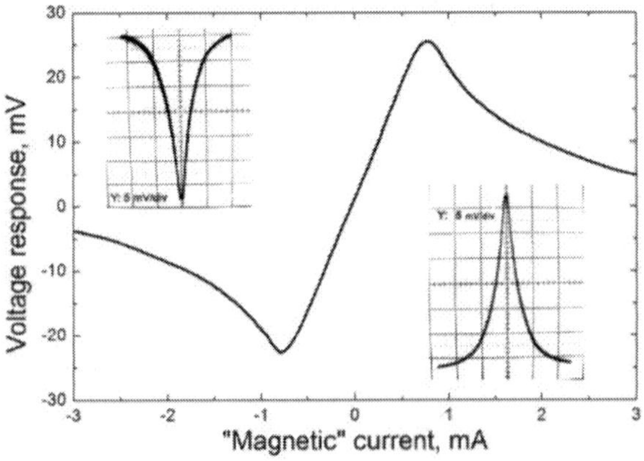

Figure 3. Antenna prototype voltage response. "Magnetic" current I_M is the current flowing through external multi-turn coil. Insets show typical traces of the voltage responses of the two shoulders of the antenna (the second one is taken with negative sign).

critical current density 4.5 kA/cm^2. The peak-to-peak voltage response of this antenna reaches almost 100 mV, and the steepness of the magnetic signal conversion into output voltage is $dV/dB \approx 6500$ μV/μT. To meet matching requirement and increase load capability, a parallel connection of the 2D SQAs can be used. Such superconducting devices can be used in broadband receiving systems with direct digitization of input signal [16-24] for communications, radar, and signal intelligence. The SQAs are also beneficial for many applications, in which SQUIDs and SQUID arrays are used [25-27].

Acknowledgment
Authors thank D. Van Vechten for attention to this work, Hypres fabrication team of D. Yohannes, J. Vivalda, R. Hunt, D. Donnelly, D. Amparo for manufacturing the integrated circuits, A. Kirichenko, T. Filippov and D. Kirichenko for help and valuable discussions.

The work was supported by ONR Program 6.1 under grant RUP1-1493-05-MO via CRDF GAP and in part by grants PGSS 4871.2014.2 and GK 14.513.11.0128. One of the authors, Igor Soloviev acknowledges support from Russian President Grant MK-1841.2014.2.

References

[1] Stratonovich R L 1967 *Selected Topics in the Theory of Random noise* (New York: Gordon and Breach)
[2] Whalen A D 1971 *Detection of Signals in Noise* (New York: Academic Press)
[3] Likharev K K 1985 *Dynamics of Josephson Junctions and Circuits* (New York: Gordon and Breach)
[4] Prokopenko G V *et al* 2013 *IEEE Trans. Appl. Supercond.* **23**(3) 1400607
[5] Berggren S *et al* 2013 *IEEE Trans. Appl. Supercond.* **23**(3) 1400208
[6] Luine J *et al* 1999 *IEEE Trans. on Appl. Supercond.* **9** 4141–44
[7] Kornev V, Soloviev I, Klenov N, Sharafiev A and Mukhanov O 2012 *Physica C* **479** 119–22
[8] Kornev V K, Soloviev I I, Sharafiev A V, Klenov N V and Mukhanov O A 2013 *IEEE Trans. Appl. Supercond.* **23**(3), 1800405
[9] Kornev V K *et al* 2009 *Supercond. Sci. Technol.* **22** 114011
[10] Kornev V K *et al* 2010 *Physica C* **470** 886–89
[11] Kornev V K, Soloviev I I, Klenov N V, Sharafiev A V and Mukhanov O A 2011 *IEEE Trans. Appl. Supercond.* **21** 713–16
[12] Kornev V K, Soloviev I I, Klenov N V, and Mukhanov O A 2011 *IEEE Trans. Appl. Supercond.* **21** 394- 98
[13] Kornev V K, Kolotinsky N V, Skripka V A and Mukhanov O A 2013 High linearity voltage response differential cell *IEEE 14th Int. Supercond. Electron. Conf. (ISEC)* PG11
[14] Kornev V K *et al* 2014 High Linearity Voltage Response Parallel-Array Cells *J. Physics: Conf. Ser. (EUCAS 2013)* to be published
[15] HYPRES Nb Process Design Rules. Revision #24 Jan 11, 2008. http://www.hypres.com
[16] Gupta D, Kirichenko D E *et al* 2011 *IEEE Trans. Appl. Supercond.* **21** 883-890
[17] Mukhanov O A, Kirichenko D, Vernik I V *et al* 2008 *IEICE Trans. Electron.* **E91-C** 306–17
[18] Vernik I V, Kirichenko D E, Dotsenko V V *et al* 2007 *Supercond. Sci. Technol.* **20** 323-327
[19] Wong J, Dunnegan R *et al* 2007 High performance, all digital RF receiver tested at 7.5 GigaHertz *IEEE Military Comms Conf* (*MILCOM'07*)
[20] Leese de Escobar A *et al* 2006 High performance HF-UHF all digital RF receiver tested at 20 GHz clock frequencies *IEEE Military Comms Conf. (MILCOM'06)*
[21] Rylov S V, Bunz L A, Gaidarenko D V *et al* 1997 *IEEE Trans. Appl. Supercond.* **7** 2649-2652
[22] Vernik I V, Kirichenko D E *et al* 2007 *IEEE Trans. Appl. Supercond.* **17**(2) 442–445
[23] Inamdar A, Rylov S et al. 2009 *IEEE Trans. Appl. Supercond.* **19** 670-675
[24] Gupta D, Inamdar A, Kirichenko D E, Kadin A M and Mukhanov O A 2011 Superconductor analog-to-digital converters and their applications *Microwave Symposium Digest* (MTT)
[25] Nishijima S, Eckroad S, Marian A, Choi K *et al* 2013 *Supercond. Sci. Technol.* **26** 113001
[26] Kleiner R, Koelle D, Ludwig F, Clarke J 2004 *Proc. IEEE* **92** 1534-1548
[27] Fagaly R L 2006 *Rev. Sci. Instrum.* **77** 101101.

High-T_C micro and nano-constrictions modeling: hot spot approach for DC characteristics and HEB THz mixer performance

Romain G Ladret, Annick F Dégardin and Alain J Kreisler[1]
SUPÉLEC LGEP; CNRS UMR 8507; UPMC Univ Paris 06; Univ Paris-Sud 11;
11 rue Joliot-Curie, 91192 GIF SUR YVETTE Cedex, France

E-mail: alain.kreisler@supelec.fr

Abstract. High-T_C hot electron bolometers (HEB) are promising THz mixers due to their expected wide bandwidth, large mixing gain, and low intrinsic noise. We have simulated the characteristics of YBaCuO HEBs with a hot spot model usually dedicated to low-T_C phonon cooled devices. With respect to e.g., NbN, the high-T_C specificities are mainly arising from the large values of YBaCuO phonon thermal conductivity, and the film to MgO substrate phonon escape time. The consequent effects on the electron temperature profiles along the YBaCuO constriction and the current voltage DC characteristics are considered. The conversion gain G and noise temperature T_N were computed for two constriction dimensions. For a 100 nm long × 100 nm wide × 10 nm thick constriction at 9 microwatt local oscillator (LO) power, the expected double sideband (DSB) T_N = 1520 K (G = −13.7 dB). For a larger (but more realistic according to YBaCuO aging effects) 400 nm long × 400 nm wide × 35 nm thick constriction, minimum T_N = 1210 K DSB at 35 microwatt LO power (G = −13.1 dB).

1. Introduction

During the past decade, superconducting hot electron bolometers (HEB) have been the object of unceasing progress. The performances achieved by low-T_C devices are at a very high level in terms of heterodyne mixing in the THz. For example, double sideband (DSB) noise temperatures T_N = 950 K [1] and 600 K [2] at 2.5 THz were reported for NbN HEBs. Efforts on high-T_C HEBs ($YBa_2Cu_3O_{7-\delta}$ - YBCO) have been, however, more limited, with few published results from microwave to THz frequencies [3-6]. Indeed, YBCO HEB technology has proven difficult due to the chemical reactivity and the aging effects of YBCO ultrathin films. Such films are required to achieve sensitive HEBs, but a sharp increase in conversion losses results from aging [7,8]. However, predicted performance was particularly encouraging, with T_N = 2000 K (single sideband – SSB) at P_{LO} = 11 μW local oscillator (LO) power for an YBCO constriction of 10 nm thickness and 100×100 nm^2 transverse dimensions [9]. This first model was based on the "0-D" point bolometer approach, which describes the system in terms of thermal reservoirs only, namely the electrons of the superconducting film at temperature T_e and the phonons of the film at temperature $T_p < T_e$. The electron - phonon interaction time τ_{ep} drives the HEB transient response or its instantaneous bandwidth as a mixer. With τ_{ep} = 1-2 ps, the YBCO HEB gives the hope of a bandwidth close to 100 GHz (but so far unobserved), whereas NbN exhibits

[1] To whom any correspondence should be addressed.

$\tau_{ep} \approx 20$ ps. Moreover, the "1-D" hot spot model, which takes into account the spatial dependence of T_e along the superconducting constriction, was implemented for low-T_C HEBs [10]. Our motivation was to apply this model to YBCO, while taking the high-T_C specificities into account, which are mainly due to the large phonon thermal conductivity and the long film to substrate phonon escape time τ_{esc}.

2. HEB geometry and thermal equations

The main constituents of an YBCO-based HEB are shown in figure 1. The superconducting ultrathin film is deposited on an MgO substrate (with lattice matching buffer layer). The YBCO constriction is coupled to the rf radiation by means of a planar metal antenna.

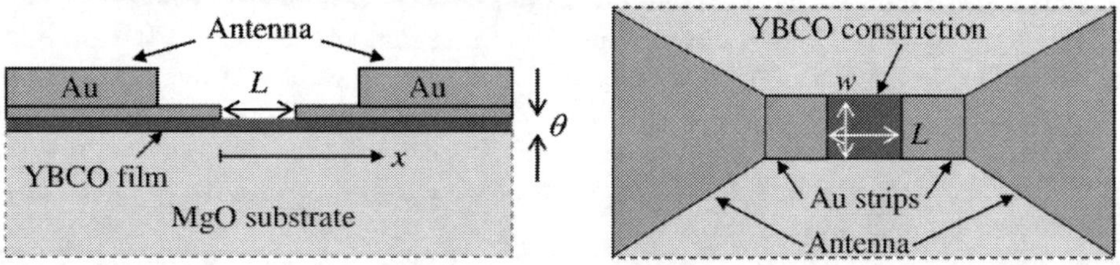

Figure 1. HEB schematics (not to scale) depicting the YBCO constriction connected to the arms of a planar antenna. Left: cross-section; right: top view. The constriction dimensions are: length L, width w, and thickness θ. The hot-spot model coordinate x is parallel to L.

The thermal exchange in the device is governed by the heat equations pertaining to the electrons (1) and the phonons (2) of the YBCO film, bearing indices e and p, respectively:

$$V_c\kappa_e(T_{ec})\frac{\delta^2 T_e}{\delta x^2} - V_c C_e(T_{ec})\frac{\delta T_e}{\delta t} = \frac{V_c C_e}{n\tau_{ep}T_{ec}^{n-1}}\left(T_e^n - T_p^n\right) - P_{LO} - P_J , \quad (1)$$

$$V_c\kappa_p(T_{pc})\frac{\delta^2 T_p}{\delta x^2} - V_c C_p(T_{pc})\frac{\delta T_p}{\delta t} = \frac{V_c C_p}{m\tau_{esc}T_{pc}^{m-1}}\left(T_p^m - T_0^m\right) - \frac{V_c C_e}{n\tau_{ep}T_{ec}^{n-1}}\left(T_e^n - T_p^n\right) , \quad (2)$$

where the only spatial coordinate x is along the length of the constriction (figure 1), which is anchored to its ends at the thermostat temperature T_0; i.e., $x \in [0,L]$, with $T_e(0) = T_e(L) = T_0$. V_c is the constriction volume; $\kappa_{e,p}$ and $C_{e,p}$ are the thermal conductivities and volumic specific heats, respectively (evaluated at the assumed constant temperatures T_{ec}, T_{pc}). The power source terms are P_{LO} and the DC bias Joule power P_J. Equations (1) and (2) are depicting a phonon cooled device, because cooling by the electron diffusion to the metal contacts can be neglected in YBCO [9].

3. Solving hypotheses and DC characteristics

Apart from general simplifying assumptions (YBCO material homogeneity, uniform current density in the constriction section, temperature independent thermal conductivities and specific heats, no LO losses), other hypotheses are more specific.

i) As the phonon thermal conductivity is one order of magnitude larger for YBCO at ~ 80 K than for NbN at ~ 10 K, the second order derivative in (2) cannot be neglected as in [10].

ii) Due to film/substrate interface nature (lattice mismatch and/or buffer layer), the phonon escape time is much longer in YBCO/MgO; i.e., τ_{esc} (ps) $\approx 75\times\theta$(nm) [11] ($\approx 13\times\theta$ in NbN/Si [10]).

iii) The thermal exchange exponents are $m = n = 3$ for YBCO ($m = 4$ and $n = 3.6$ for NbN) [9,10].

The numerical values for the various parameters are gathered in table 1. Two devices have been simulated; Device A represents a constriction close to the limits of achievable e-beam lithography (same dimensions as in [9]) and Device B represents our currently achievable dimensions [8].

Table 1. Simulated device parameter values.

	L nm	W nm	θ nm	τ_{ep} ps	τ_{esc} ps	κ_e $Wm^{-1}K^{-1}$	κ_p $Wm^{-1}K^{-1}$	T_C[a] K	ΔT[b] K	J_C[c] Acm^{-2}	ρ[d] Ωcm	T_0 K
Dev. A	100	100	10	1.0	0.75	1	10	85	1.2	2×10^6	3×10^{-4}	60
Dev. B	400	400	35	1.7	2.6	1	10	89	1.2	2×10^6	3×10^{-4}	70

[a]At mid-transition [b]Transition width [c]At 77 K [d]Normal state resistivity at ≈ 100 K

Solving was performed in steady-state conditions (time derivatives omitted) using the MATLAB® ordinary differential equations toolbox. Temperature profiles for Device A are shown in figure 2 and DC current-voltage plots are shown in figure 3 for both devices. The hot electron effect ($T_e > T_p$) is clearly visible in figure 2a, as well as the hot spot central region ($T_e > T_C$ at $I = 188$ μA). Negative resistance is observed for device A at low LO levels (figure 3a), whereas it is not visible for device B (figure 3b) seemingly due to the damping effect of the larger thermal resistance associated to the longer τ_{esc}. The effect of neglecting the phonon diffusion along the constriction (second order derivative term in (2) suppressed) shows a much larger electron heating effect (figure 2b) and no negative resistance (figure 2b, inset), as discussed in [12].

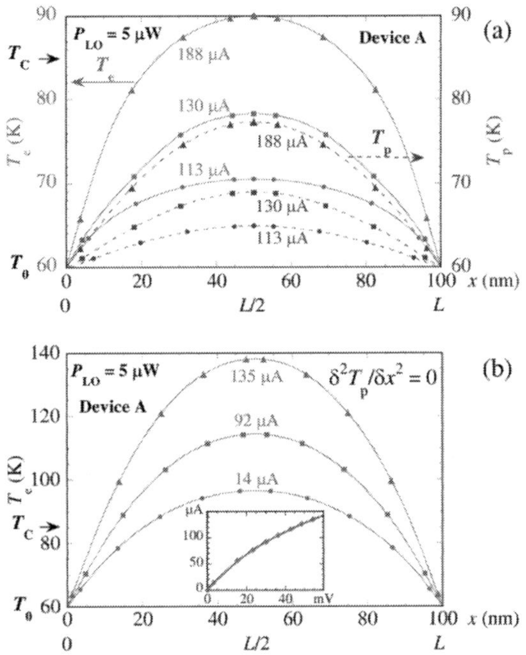

Figure 2. Temperature profiles for device A at various DC bias current values. (a): T_e and T_p plots. (b): T_e plots with second order T_p term neglected (inset: corresponding DC I-V plot).

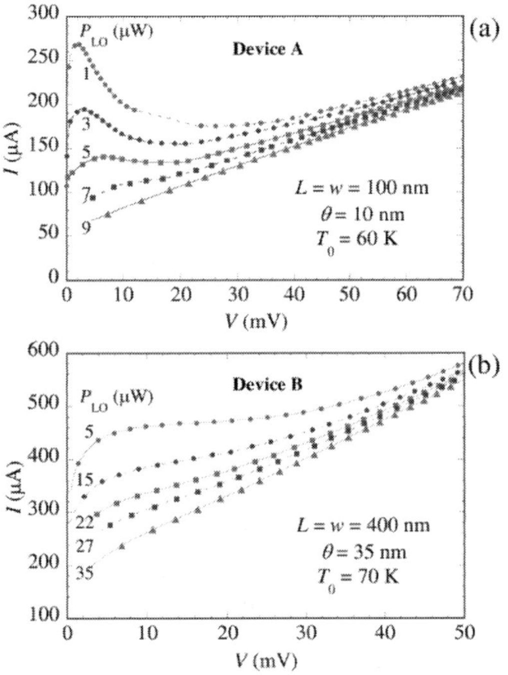

Figure 3. DC I-V plots at various LO power values, for device A (a) and device B (b). The sign of dI/dV is sensitive to the second order T_p term ($P_{LO} = 5$ μW in (a) and inset in figure 2b).

4. HEB mixer performances

The conversion gain G and noise temperature T_N were extracted from the large signal analysis, as performed in [10] for NbN devices. G is defined as the ratio between the power dissipated into the device loading resistance (50 Ω) and the signal power absorbed by the device. It contains both the DC and LO power efficiencies. T_N is the sum of the DSB temperatures due to the thermal fluctuations and Johnson noise. The most significant performances are gathered in table 2.

Table 2. Minimum DSB noise temperature and maximum conversion gain for both YBCO devices.

	Device A			Device B		
P_{LO} (μW)	1	5	9	5	15	35
T_{Nmin} (K)	6200	2000	1520	4000	1730	1210
V (mV) / I (μA)[a]	24 / 175	19.5 / 135	10.5 / 83	19 / 470	16.5 / 400	9.5 / 260
G (dB)[a]	−19	−14.9	−13.7	−16.6	−13.9	−13.1
G_{max} (dB)	−2	−6	−13.4	−15.6	−12.2	−12.6
V (mV) / I (μA)[b]	3.6 / 255	2.8 / 130	6.5 / 74	12.5 / 465	8 / 375	7 / 235
T_N (K)[b]	19800	5700	1620	5150	2250	1350

[a]For T_{Nmin}: DC bias point and associated gain [b]For G_{max}: DC bias point and associated noise temperature

For device A, P_{LO} = 9 μW allows to achieve T_{Nmin} = 1520 K DSB with $G = -13.7$ dB (close to $G_{max} = -13.4$ dB), whereas the point bolometer model provides $T_N \approx 1000$ K (2000 K SSB) at P_{LO} = 11 μW [9]. High G values could be achieved at low P_{LO} levels, due to the negative resistance effect (figure 3a), but the noise dramatically increases. For device B, optimal performance is achieved at P_{LO} = 35 μW, with $T_N \approx 1200$ to 1350 K and $G \approx -13$ to −12.5 dB, respectively.

5. Conclusions

Starting from a hot spot model developed for NbN HEBs, we have revised the physical hypotheses to fully represent the YBCO specificities. Good mixer performance could be obtained with a device of reasonable dimensions, according to the YBCO sensitivity to technological process. Further improvements will concern the behavior at THz frequencies and the mixer bandwidth.

Acknowledgment

This work has been supported by the French National Research Agency (ANR) under grant # 2011-BS03-008-01.

References

[1] Baselmans J J A, Hajenius M, Gao J R, Klapwijk T M, de Korte P A J, Voronov B and Gol'tsman G N 2004 *Appl. Phys. Lett.* **84** 1958
[2] Tretyakov I, Ryabchun S, Finkel M, Maslennikova A, Kaurova N, Lobastova A, Voronov B and Gol'tsman G 2011 *Appl. Phys. Lett.* **98** 033507
[3] Li C-T, Deaver B S, Lee M, Weikle II R M, Rao R A and Eom C-B 1998 *Appl. Phys. Lett.* **73** 1727
[4] Harnack O, Il'in K S, Siegel M, Karasik B S, McGrath W R and De Lange G 2001 *Appl. Phys. Lett.* **79** 1906
[5] Gousev Y P, Semenov A D, Nebosis R S, Pechen E V, Varlashkin A V and Renk K F 1996 *Supercond. Sci. Technol.* **9** 779
[6] Cherednichenko S, Rönnung F, Gol'tsman G, Kollberg E and Winkler D 2000 *Proc. 11th Int. Symp. Space Terahertz Technology* www.nrao.edu/isstt/papers/200517528.pdf
[7] Péroz Ch, Dégardin A F, Villégier J-C and Kreisler A J 2007 *IEEE Trans. Appl. Supercond.* **17** 637
[8] Aurino M, Kreisler A J, Türer I, Martinez A, Gensbittel A and Dégardin A F 2010 *J. Phys.: Conf. Ser.* **234** 042002
[9] Karasik B S, McGrath W R and Gaidis M C 1997 *J. Appl. Phys.* **81** 1581
[10] Khosropanah P, Merkel H, Yngvesson S, Adam A, Cherednichenko S, and Kollberg E 2000 *Proc. 11th Int. Symp. Space Terahertz Technology* 474-488
[11] Harrabi K, Ladan F-R, Vu Dinh Lam, Maneval J-P, Hamet J-F, Villégier J-C and Bland R W 2009 J. Low Temp. Phys. **157** 36
[12] Ladret R G, Dégardin A F and Kreisler A J 2013 *IEEE Trans. Appl. Supercond.* **23** 23003305

Detection of X-ray photons by niobium Josephson tunnel junctions with trapped Abrikosov vortices

M P Lisitskiy[1,2], **C Camerlingo**[1]

[1]C.N.R.- Istituto di Cibernetica "Eduardo Caianiello", Via Campi Flegrei 34, I-80078, Pozzuoli, Napoli, Italy
[2]Istituto Nazionale di Fisica Nucleare, Sezione di Napoli, Via Cinthia, I-80126 Napoli, Italy

E-mail: m.lisitskiy@cib.na.cnr.it

Abstract. The high atomic number of niobium (Z=41) can be exploited to develop a high efficiency superconducting gamma-ray detector based on a novel detection principle, namely, the interaction of a single gamma-ray photon with Abrikosov vortices trapped inside a niobium bulk absorber. To study the feasibility of this principle, niobium type Josephson tunnel junctions with the aluminium oxide as a tunnel barrier and with a thick (0.3 mm) niobium base electrode have been fabricated. The devices have been tested at T = 4.2 K in terms of the current-voltage characteristic and of the magnetic field dependence of the Josephson critical current. The feasibility of the detection principle has been tested under X-ray irradiation from the ^{55}Fe source. The time dependence of the Josephson critical current of the junction with trapped Abrikosov vortices has been recorded without and with X-ray irradiation. The data analysis of obtained experimental curves has confirmed the effect of the X-ray photon absorption on the Josephson critical current caused by jumping of Abrikosov vortices.

1. Introduction

An innovative detection principle - based on the interaction of a single gamma-ray photon with the Abrikosov vortices trapped inside a type II superconducting absorber with high atomic number Z - has been proposed and theoretically explained in [1], with the aim of developing a gamma-ray superconductor single photon counter with high intrinsic detection efficiency in the energy range up to 100 keV. The gamma-ray absorption results in a photoelectronic hot-spot formation and produces a variation of the magnetic inductance around the absorber surface. The high sensitivity of the Josephson tunneling to weak magnetic field can be exploited to readout this variation. Two device designs were proposed in [1] for readout the detection events. The first one consists of a single Josephson tunnel junction device. In this case the superconducting absorber is one of two electrodes of the junction and the Abrikosov vortex jumps are revealed by step-like variations of the Josephson critical current (I_c). The second design is based on a superconducting quantum interference device (SQUID) magnetometer with the pickup loop placed in proximity of the absorber site where Abrikosov vortices are trapped. The motion of Abrikosov vortices, caused by gamma-ray absorption, results in the variation of the magnetic inductance around the absorber surface which, in turn, can be revealed by the SQUID.

As regards the absorber material, superconducting Nb (Z=41) is a good material for practical realization of a detector with high detection efficiency based on Abrikosov vortices [1]. First, the gamma-ray detection efficency of Nb with thickness as great as 0.3 mm is of 65 % for gamma-ray energy of 60 keV while the gamma-ray absorption of Si (Z=14) detector with the same thickness and for the same gamma-ray energy is only of 2 % [1]. Second, Nb is a type II superconductor and the magnetic field penetrates inside it spontaneously in form of Abrikosov vortices which occupy the whole volume. In addition, Nb absorber permits the detector to operate at liquid Helium temperature at normal pressure (T=4.2 K), facilitating the practical use of the device.

The operation principle requires that Abrikosov vortices have to be trapped on pinning centers in the Nb absorber by field-cooling process. For this reason it is preferable to choose a polycrystalline bulk Nb as device absorber because the grain boundaries between grains are itself effective pinning centers for vortices [2]. It is worth noting that spurious signals in absence of radiation might occur in a detector based on Abrikosov vortices (dark counts) due to thermally activated flux creep [3]. In [4], a low temperature magnetic force microscopy was used for experimental evaluation of the vortex depinning threshold temperature for a 0.3 mm thick Nb polycrystalline foil. This foil is analogous to the absorber chosen for fabrication of the Josephson tunnel junctions as vortex sensor element in the present research. As demonstrated in [4], the thermal activated dark counts of a thick (0.3 mm) Nb polycrystalline bulk absorber at T=4.2 K is negligible.

The present paper is devoted to the experimental test of the feasibility of the single Josephson tunnel junction device configuration based on the detection principle based on the Abrikosov vortices. X-ray irradiation from the ^{55}Fe source with energy of 5.9 keV was used in the present experiment in order to test the working principle in the case of absorption of low energy gamma-ray photons.

2. Experimental details

2.1. Sample description
A Nb foil with thickness of 0.3 mm and with 5x5 mm^2 size, identical to that investigated in [4], was used as a base electrode of the Josephson tunnel junctions with the following layer configuration: Nb(0.3 mm foil-base electrode)/Al(20 nm)/Al_2O_3/Al(20 nm)/Nb(100 nm-top electrode). Before the deposition of the first 20-nm thick Al layer, the native oxide of the Nb foil was removed by Ar-ion beam cleaning. A secure electric disconnection between the Nb wiring and the Nb foil was reached by growing of the Nb oxide by anodization process with consequent deposition of two SiO insulating layers. The sample contains Josephson tunnel junctions with circular geometry with diameters of 20 μm, 60 μm, 80 μm and 100 μm. The Josephson critical current density was of 3 A/cm^2.

2.2. Experimental cryogenic setup and measurement procedure
The special cryogenic setup was used for measurements of the Josephson critical current of the Nb Josephson tunnel junction with trapped Abrikosov vortices exposed to X-ray irradiation from a ^{55}Fe radioactive source (5.9 keV X-rays). The sample was mounted inside a vacuum cylindrical chamber of a cryogenic insert fabricated from a thin wall stainless steel tube. The cooper sample holder was attached to the thin wall stainless steel tube. The chamber was immersed in a liquid Helium bath. Helium exchange gas was introduced into the chamber. Besides the exchange helium gas, the sample holder stayed in a good thermal contact with the copper bottom of the cylindrical chamber in order to keep the stable temperature T=4.2 K. A 100 Ohm micro resistor was used as a heater in contact with the sample holder, in order to control the temperature of the sample with a high accuracy at values above the liquid Helium temperature. The temperature of the sample was measured by a Lake Shore Germanium thermometer. The junction plane was oriented perpendicular to the axis of the cylindrical chamber. Under the sample, the encapsulated ^{55}Fe disc source was mounted inside a copper box with an externally controllable mechanical copper shutter to shield the sample from X-ray irradiation. A special magnetic coil system was used to generate both the magnetic field perpendicular (B_{per}) and parallel (B_{par}) to the junction plane. The shielding of the earth magnetic field and of the electromagnetic noise was obtained by a superconducting lead shield and three coaxial μ-metal shields.

Abrikosov vortices were trapped by cooling the sample from a temperature $T>T_c$ down to $T<T_c$ in the presence of a perpendicular magnetic field (T_c is the superconducting transition temperature). The perpendicular magnetic field was turned off at T=4.2 K. The Josephson critical current was recorded in time by a data acquisition electronic system. The current-voltage (I-V) characteristic of the Josephson junction was measured by application of the triangle shape current through the junction with the frequency of 10 Hz and with the amplitude greater than I_c. The junction voltage was applied to a precise comparator which produced the digital signal when the non-zero voltage appeared on the junction. By acquiring and analyzing the output signal from the comparator, the acquisition system stored the current values when the non-zero voltage occurred on the junction. These current values represented the Josephson critical current I_c values. Both the parallel magnetic field dependences of the I_c and the dependences of the I_c on the time (t) were recorded by suitable software.

3. Experimental results and discussion

As predicted in [1], the absorption of 5.9 keV X-ray photons in Nb results in the formation of a hot spot with dimension of order of 1 μm. In order to ensure a high probability of the interaction between a hot spot and an Abrikosov vortex, a perpendicular magnetic field B_{per} of 20 G was chosen for trapping the magnetic field. Indeed, the field cooling process in this magnetic field creates a flux structure with an average distance between vortices of order of 1 μm, which is close to the expected dimension of the hot spot. The Josephson tunnel junction with the diameter of 20 μm was employed in the experiment described in the present paper. After Abrikosov vortex trapping, the $I_c(B_{par})$ dependence was recorded. As expected from a previous research [5], the maximum value of the Josephson critical current was suppressed with respect to the vortex free state and oscillation in I_c occured. The Josephson critical current was tuned by applying a parallel magnetic field in order to achieve the maximum value of the dI_c/dB_{par} derivative of the $I_c(B_{par})$ dependence. At this point the measurement of the Josephson critical current in time was started and the data were recorded both with and without X-ray irradiation. A typical experimental time dependence of the Josephson critical current I_c without X-ray irradiation (first part of the time track) and then after X-ray irradiation is shown in Figure 1 by black points (raw data). Full color line in Figure 1 reports the signal after a denoising procedure made by a wavelet transform method (OriginPro 8 SR3 Software, wavelet type - Haar, thresholding level - 6, threshold of every level - 100%). The derivative of the denoised signal with respect to time was calculated. The threshold for event detection was considered as the mean value of the non-zero spikes of the derivative for the time interval without X-ray irradiation, plus its standard deviation multiplied by three. Open circles in Figure 1 are the denoised signal values at time moments where the derivative spikes exceed the threshold. As can be seen from Figure 1, the derivative spikes exceed the threshold mainly in case of the signal recorded under X-ray irradiation. Therefore for the part of signal stored under X-ray irradiation, the open circles indicate the variations of the magnitude of I_c which can be attributed to the jumps of trapped Abrikosov vortex/vortices caused by the absorption of single 5.9-keV X-ray photons. The measured average time interval between the events is 448 s while the expected time interval between absorption events of 5.9 keV X-photons inside 0.3 mm thick Nb foil is of 0.2 s (for the given source emission rate and measurement geometry). This discrepancy is taken as an indication of low observed detection efficiency, with respect to calculations. Based on these measurements we speculate that, for reasons still unclear, only thin part of Nb absorber just near the junction barrier are effectively producing a vortex jump to be recorded by the Josephson junction in critical current operation mode. The reason of the low detection efficiency will be studied in future experiments.

4. Conclusions

Josephson tunnel junctions with the 0.3-mm thick Nb polycrystalline foil as a base electrode were fabricated. Preliminary experiment was carried out under 5.9 keV X-ray irradiation from the ^{55}Fe source, namely, under absorption of single low energy gamma-ray photons. After Abrikosov vortex trapping, single X-ray photons were detected by registration of step-like changes of the Josephson critical current recorded vs. time.

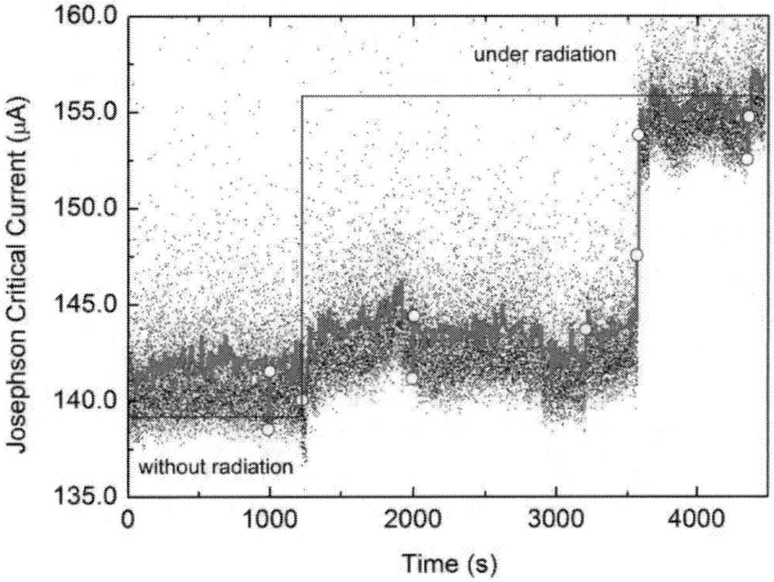

Figure 1. The $I_c(t)$ dependence for a 20 μm circular Josephson tunnel junction after field cooling process in B_{per}=20 G, without and with X-ray irradiation (black point). Red line shows the signal after a denoising procedure. Open circles are I_c changes which are attributed to the absorption events of single X-ray photons, by selection procedure described in the text.

Acknowledgements

The work was supported by the Italian Institute of Nuclear Physics (INFN, Experiment SUPERGAMMA) and by the Department of Materials and Devices of the National Research Council of Italy (CNR) (at present time - Department of Physical Sciences and Technologies of Matter). The authors are grateful to Dr. R. Cristiano for help in the experiment with X-ray radiation source and to Prof. P. Russo for scientific discussions and suggestions in the preparation of the present paper.

References

[1] Lisitskiy M P 2009 *J. Appl.Phys.* **106** 103927
[2] Mezzetti E, Chiodoni A, Gerbaldo R, Ghigo G, Gozzelino L, Minetti B, Camerlingo C, Giannini C 2001 *Eur. Phys. J.* **B19** 357
[3] Anderson P W and Kim Y B 1964 *Rev. Mod. Phys.* **36** 39
[4] Lisitskiy M P, Volodin A P, Bruno A and Van Haesendonck C 2012 *Physics Proceda* **36** 405
[5] Gubankov V N, Lisitskii M P, Serpuchenko I L, and Fistul' M V 1991 *Sov. Phys. JETP* **73** 734

Testing of Josephson Spectrometer with Waveguide Coupling

M Lyatti[1,2], I Gundareva[1,2], V Pavlovskii[2], U Poppe[1], Y Divin[1]

[1] Peter Grünberg Institute, Forschungszentrum Jülich, D-52425 Jülich, Germany
[2] Kotel'nikov Institute of Radio Engineering and Electronics, Russian Academy of Sciences, 125009, Moscow, Russia

E-mail: m.lyatti@fz-juelich.de

Abstract. One of the challenges in public security is the quick and reliable identification of threat liquids in bottles, when vapour analysis is not possible. Recently, we demonstrated that it is possible to rapidly identify liquids by EM measurements of their dielectric functions in the sub-THz range with a high-T_c Josephson spectrometer. Following this approach, we have developed a Josephson spectrometer with a new radiation coupling system, based on dielectric waveguides. In this paper, we present the results of spectroscopic measurements on liquid samples of various purities including 30% H_2O_2/H_2O, performed using our Josephson spectrometer with waveguide coupling. Also, the signal and noise characteristics of a classical Josephson detector used in our liquid identifier were numerically simulated and the power dynamic range was estimated for a wide spread of junction parameters.

1. Introduction

Among various discussed ways, the techniques using electromagnetic radiation for liquid explosive detection are considered as one of the most promising [1]. It has been shown that the fast identification of harmless and threat liquids, including 30% H_2O_2/H_2O solution, can be performed with a Josephson spectrometer in the sub-THz frequency range [2]. Recently, we demonstrated a new type of liquid identifier based on the Josephson spectrometer with waveguide coupling [3]. Both classical and frequency-selective modes of a Josephson detector were used. The frequency bandwidth of the liquid identifier was extended to include lower frequencies down to 1 GHz. Due to two different detection techniques at low and high frequencies, the response of the liquid identifier with waveguide coupling was very specific, which was favourable for reliable identification. In this paper, we present a theoretical evaluation of the power dynamic range of the classical Josephson detector and results on the identification of liquids with various purities by the liquid identifier with waveguide coupling.

2. Power dynamic range of classical Josephson detector

At low frequencies, a classical mode of the Josephson detector is used in our liquid identifier. In this mode, a nonlinearity of the dc *IV*-curve of a Josephson junction is used to detect radiation with a frequency f, which is low enough compared with the characteristic Josephson frequency $f_c = (2e/h)\cdot I_cR_n$, where I_c is the critical current and R_n is the normal-state resistance. In the first approximation, a response of the classical Josephson detector is proportional to the second derivative of the *IV*-curve of the Josephson junction [4].

In contrast to the well-studied power dynamic range of the frequency-selective Josephson detector

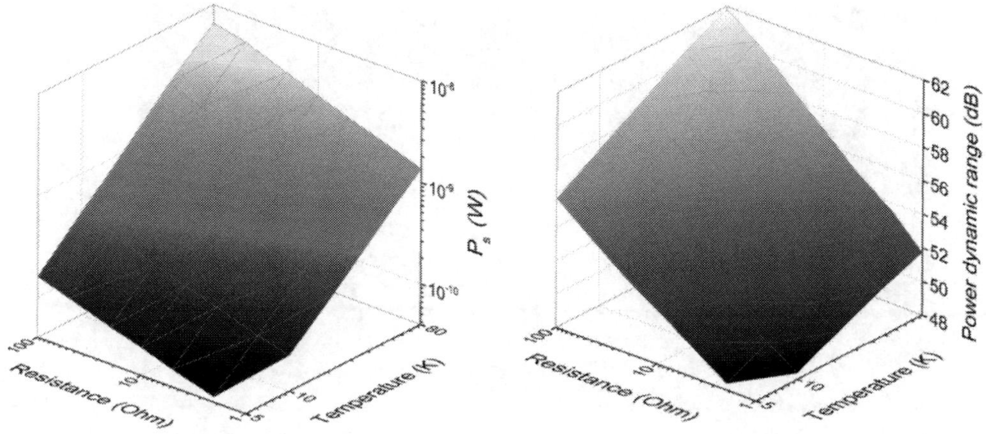

Figure 1. High-power limit P_s of the power dynamic range (left) and the power dynamic range (right) of the classical Josephson detector with I_cR_n = 1 mV vs. normal-state resistance and temperature.

[5]-[7], there is a lack of data on the power dynamic range of the classical Josephson detector. Numerical simulation of signal and noise characteristics of the classical Josephson detector based on the solution of the Fokker-Planck equation was carried out for the resistively shunted junction model with thermal fluctuations. The simulations were carried out over the range of normal-state resistances of the Josephson junction R_n = 1 - 100 ohm, critical currents I_c = 0.01 – 1 mA, and temperatures T = 5 - 80 K. The frequency f of incident radiation was set at $0.05f_c$.

The power dynamic range D of the detector is by definition $D = P_s/NEP$, where P_s is the high-power limit of the power dynamic range, and NEP is the noise-equivalent power of the detector. The high-power limit value P_s was taken as the level of the input power P, at which the detector responsivity $R = \Delta V/P$ decreases by a factor of two. The P_s-value was determined from the numerically simulated dependence of the classical response ΔV of a current-biased junction on input radiation. The ΔV-values were taken at the bias, corresponding to the response maximum at a low power of incident radiation. An example of dependence of P_s of the Josephson detector with I_cR_n = 1 mV on the normal-state resistance and temperature is presented in figure 1 (left). Analysis of the numerical simulation data revealed that the high-power limit of the power dynamic range of the classical Josephson detector can be described by equation $P_s[\text{watt}]=4.07\cdot10^{-10}\ R_n[\text{ohm}]\cdot I_c^{2/3}[\text{A}]\cdot T^{4/3}[\text{K}]$ with 3% accuracy.

The values of NEP were numerically simulated at the same points as the values of P_s. Finally, the power dynamic range of the classical Josephson detector was obtained from the P_s and NEP values. An example of the dependence of the power dynamic range of the classical broadband Josephson detector on normal-state resistance and temperature is shown in figure 1 (right). It is clear from the numerical simulation that the power dynamic range of the classical Josephson detector can exceed 60 dB and both the normal-state resistance R_n and the I_cR_n-product of the Josephson junction should be increased to obtain an extended power dynamic range.

3. Experimental setup

A photograph of the liquid identifier with waveguide coupling of radiation is shown in figure 2. A description of the experimental setup can be found in [3]. The liquid to be tested was poured into a cell with a thin high-density polyethylene (PE) window. The prism (see figure 2) and the cell window were made of the same material to eliminate any reflection from the cell window. At low frequencies, the radiation was coupled to the liquid by an evanescent field at the waveguide edge. At high frequencies, the radiation was attenuated by reflection from a PE-liquid interface. To reduce possible errors due to the procedure of inserting and removing the cell, the liquids were changed without moving the cell.

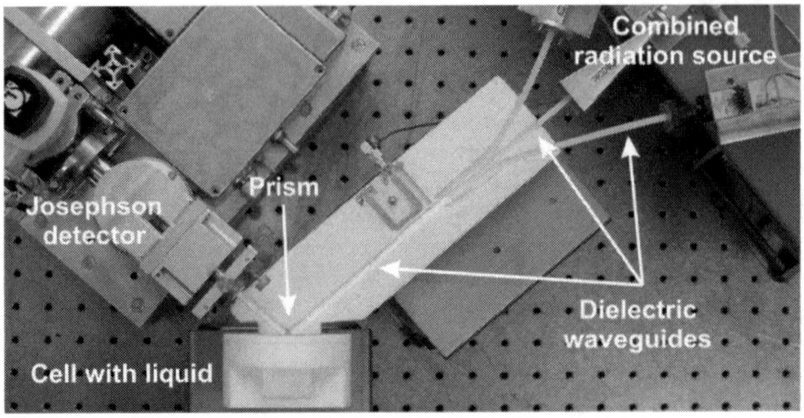

Figure 2. Photograph of liquid identifier with waveguide coupling

The combined radiation source consisted of 3 multipliers with output frequencies of 93 GHz (6×15.5 GHz), 186 GHz (12×15.5 GHz), and 282 GHz (3×94 GHz), and a frequency synthesizer HP83630A.
The Josephson detector was based on a [001]-tilt $YBa_2Cu_3O_{7-x}$ bicrystal junction with the normal-state resistance of 1.9 ohm and I_cR_n-product of 0.23 mV at 69 K. The junction was cooled by a Stirling cryocooler (LSF9589, Thales Cryogenics B.V.). The operating temperature was maintained with an accuracy of 10 mK by a cryocooler controller (HPCDE 2465, Thales Cryogenics B.V.)
Both classical and frequency-selective modes of the Josephson detector were used to provide an extended frequency range. The classical mode was used for low-frequency measurements and the frequency-selective mode was used for high-frequency measurements. During low-frequency measurements, the output frequency of the frequency synthesizer was scanned in 1 MHz steps close to the resonance frequencies of the PE waveguide and the response of the Josephson detector was measured at each frequency. During high-frequency measurements, all multipliers were switched on simultaneously and the frequency-selective response of the Josephson detector was recorded. The line intensities of polychromatic radiation were then determined from the recorded frequency-selective response. All responses were measured by a lock-in amplifier at a modulation frequency of 100 kHz.

4. Identification of liquids

The liquid identifier was tested using different liquids of various purities including distilled deionized water (18.2 Mohm·cm), distilled water, tap water, acetone, methanol, ethanol, propanol, 30% H_2O_2/H_2O suprapur©, and stabilized 30% H_2O_2/H_2O. The distilled deionized water was used as a reference liquid.
At low frequencies (< 20 GHz), the PE waveguide with metal plates above and below can be considered as a resonator. Multiple resonances with quality factors up to 400 were observed due to standing waves in the waveguide. Resonance frequencies and quality factors of the resonances were changed when the cell with liquid was attached to the prism. The response values at frequencies of 0.95, 1.7, 3.4, 9.5, 18.7 GHz were taken at the maxima of the resonances. Results of low-frequency measurements are presented in figure 3. Different measurement frequencies are colour-coded. The reference level of the responses for deionized water is shown by a dashed line.
The experimental results at low frequencies are correlated, as expected, with the imaginary part of the dielectric constants of the liquids. A higher response corresponds to lower losses. For example, the losses in ethanol are higher than the losses in water at about 1 GHz. The result is similar at a frequency of 2 GHz while losses are much lower at 20 GHz [8]. It can be seen that the response of a 30% solution of hydrogen peroxide in water is higher than the response of water of various purities at frequencies of 0.95 GHz and 1.7 GHz. These frequencies can be used as low-frequency markers for hydrogen peroxide.

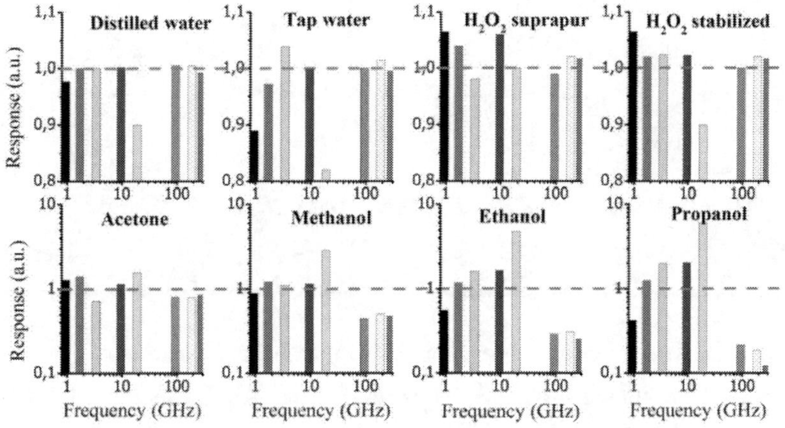

Figure 3. Responses of liquid identifier to different liquids of various purities. The 30% solutions of H_2O_2 in water are designated H_2O_2.

At higher frequencies, the response is proportional to the reflectance of the liquid. The higher response corresponds to the higher reflection coefficient. The results of high-frequency measurements are also plotted in figure 3. There is a large difference between water-based liquids and the other liquids tested. No difference was found between water and both types of 30% H_2O_2/H_2O at 93 GHz, and the response of 30% hydrogen peroxide was 2 % higher at 186 and 282 GHz. These results are consistent with our previous work in which quasioptical radiation coupling was used [2]. The difference between water of various purities and 30% hydrogen peroxide is not so pronounced at high frequencies as at low frequencies, but still large enough for a reliable distinction to be made.

Conclusions

The signal and noise characteristics of the classical Josephson detector were numerically simulated and the power dynamic range was estimated. Values of the power dynamic range above 60 dB are expected for high-resistance Josephson junctions.

A liquid identifier with waveguide coupling was tested using liquids of various purities. The ability to distinguish different liquids including water and diluted hydrogen peroxide of various purities was demonstrated both at low and high frequencies.

Acknowledgements

This work is supported by the joint project HRJRG-207 of Helmholtz Association of German Research Centers and Russian Fund for Basic Research (project 11-02-91321-SIG_a).

References

[1] Existing and potential standoff explosive detection techniques 2004 (Washington, D.C.: Nat. Academies Press)
[2] Divin Y, Lyatti M, Poppe U, Urban K and Pavlovskiy V 2011 *IEEE Trans. Appl. Supercond.* vol. 21(3), pp. 311-314 (2011)
[3] Lyatti M, Poppe U and Divin Y 2012 *Physics Procedia* **36** 193
[4] Kanter H and Vernon F L 1972 *J Appl Phys* **43** 3174
[5] Shirotov V V and Divin Y Y 2004 *Techn. Phys. Lett.* **30** 522
[6] Chen J, Wu P H, Nakajima K and Yamashita T 2005 *IEEE Trans. Appl. Supercond.* **15(2)** 530
[7] Lyatti M, Divin Y, Volkov O, Pavlovskii V, Gubankov V and Urban K 2007 *IEEE Trans. Appl. Supercond.* **17(2)** 332
[8] Barthel J and Buchner R 1991 *Pure Appl. Chem.* **63** 1473

Integrated quantum voltage noise source for Johnson noise thermometry

Masaaki Maezawa, Takahiro Yamada and Chiharu Urano

National Institute of Advanced Industrial Science and Technology, 1-1-1 Umezono, Tsukuba, Ibaraki 305-8568, Japan

E-mail: masaaki.maezawa@aist.go.jp

Abstract. We have been developing a new implementation of quantum voltage noise source (QVNS), an integrated QVNS (IQVNS), for Johnson noise thermometry (JNT) in thermal metrology. We designed an IQVNS chip based on rapid single flux quantum (RSFQ) technology. A general formula for the output power spectrum has been derived by using a Fourier analysis method, which shows that the spectrum is calculable with sufficiently small errors for practical JNT measurements.

1. Introduction

Redefinition of Boltzmann's constant is an on-going hot topic in thermal metrology [1]. In other words, a precise value of Boltzmann's constant is being determined by a number of experimental data of thermodynamic temperature measurements at the triple point of water. Johnson noise thermometry (JNT) is a type of primary thermometry capable of precise measurements of thermodynamic temperatures, which is based on the Johnson-Nyquist formula for the thermal noise in resistance. An advantage of JNT is a wide temperature range of 50 K to 1400 K, enabling systematic measurements of the temperature fixed points [2].

In the past decade, the measurement accuracy of JNT was greatly improved by use of quantum voltage noise sources (QVNSs) which calculably generate reference noise voltages based on the Josephson effect [3]. The uncertainty of the state-of-the-art QVNS-JNT has reached as small as 10 ppm at the triple point of water, 273.16 K [4]. On the other hand, the conventional QVNS, which consists of discrete components at room- and low-temperature stages, has several drawbacks such as crosstalk contamination, high cost, and inconvenience. To overcome the problems, we have proposed an integrated QVNS (IQVNS) in which all the QVNS components are integrated on a single chip by using a superconducting integrated circuit technology [5]. Interconnection with superconducting microstrip lines effectively suppresses crosstalk among the components, further improving the accuracy of the JNT measurements. Disuse of expensive high-speed semiconductor electronics reduces the system cost. Simplification of the system improves the usability. In this paper we present design of IQVNS and results of calculations of the output power spectrum.

2. Design

Figure 1 shows a block diagram of an IQVNS chip consisting of three main subsystems, a pseudorandom number generator (PRNG), a variable pulse number multiplier (VPNM) and a voltage multiplier (VM). All the circuits are implemented with rapid single flux quantum (RSFQ) circuitry.

The operation is shown in figure 2. A reference clock at f_{clk} is fed from a stabilized oscillator at room temperature and is then converted into a SFQ pulse train Clk with a DC-to-SFQ converter (DC/SFQ). The PRNG is triggered with Clk and generates a SFQ pulse pattern PRN corresponding to a sequence of pseudorandom numbers based on the maximum-length sequence (MLS) algorithm. The PRN pulse pattern is multiplied by a factor N_1 which determined by a spectrum density control signal SD, and is then converted into complementary signals PRN_P and PRN_N with the VPNM. The PRN_P and PRN_N are multiplied by a factor N_2 and combined into a bipolar waveform with the VM. Finally we obtain a reference noise voltage v_{IQ}.

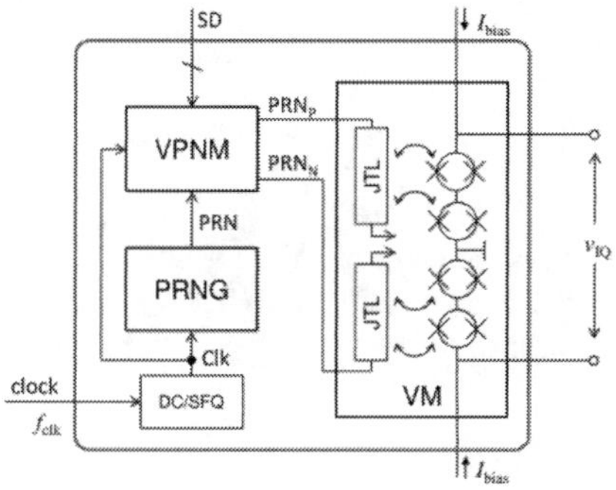

Figure 1. Block diagram of IQVNS.

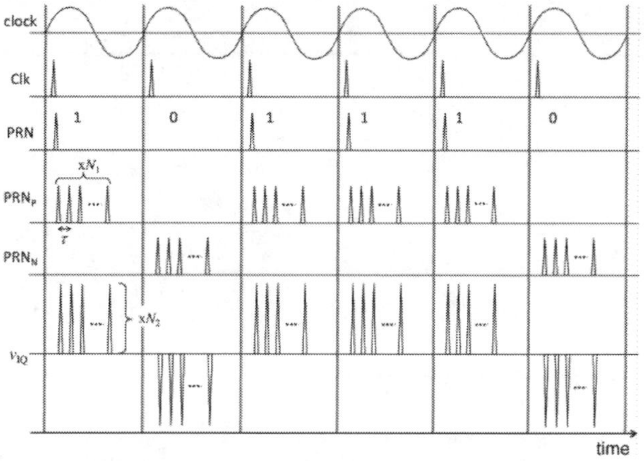

Figure 2. Time chart of IQVNS operation.

Figure 3 shows a schematic of a PRNG which is simply implemented with a shift register and an exclusive OR (XOR) gate. A PRNG with m-stage shift register produces a MLS of a length $2^m - 1$. For instance, a 21-stage PRNG, which consists of less than 500 Josephson junctions, generates a sufficiently long MLS with a length of $2^{21} - 1 \sim 2 \times 10^6$ for the JNT applications.

Figure 4 shows a schematic of a VPNM, which is based on a counter-forwarding type VPNM developed for RSFQ digital-to-analog converters [6]. Main differences from the original circuit are a separate ring oscillator and complementary outputs. The new VPNM for IQVNS is operated with a high-frequency SFQ clock at $1/\tau$ supplied with either the separate RSFQ ring oscillator or an external oscillator in order to stabilize frequencies of the output SFQ pulses. The other modification is generation of the complementary outputs, PRN_P and PRN_N, so as to synthesize bipolar waveforms in the next stage. For a prototype IQVNS, we use an 8-bit version of the modified VPNM which is estimated to consist of about 500 junctions.

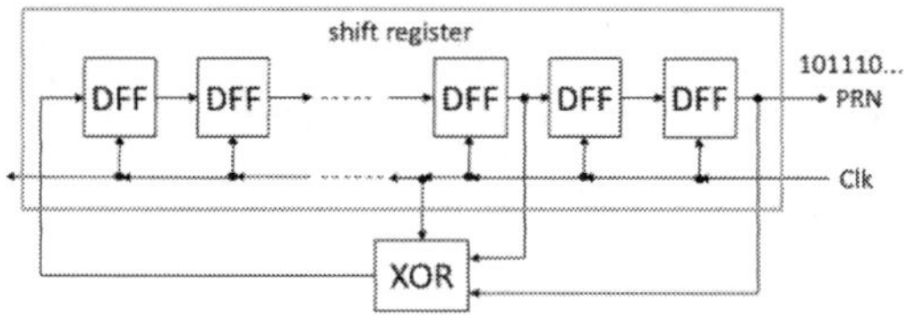

Figure 3. Pseudorandom number generator (PRNG) based on maximum-length sequence (MLS) algorithm.

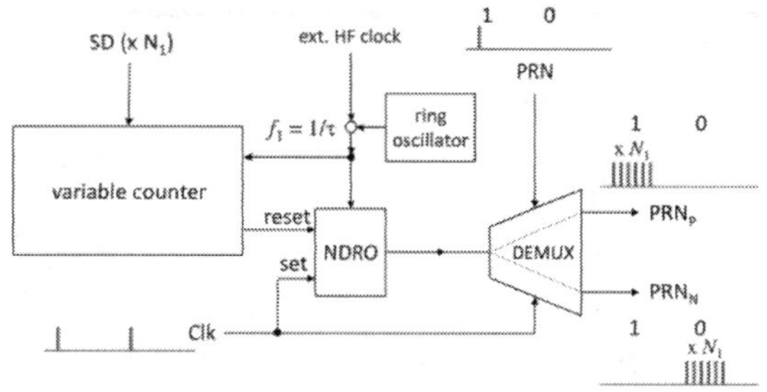

Figure 4. Variable pulse number multiplier (VPNM) with complementary outputs.

As shown in figure 5, a VM consists of input Josephson transmission lines (JTLs) and an output SQUID array which are magnetically-coupled via transformers, originally developed for RSFQ digital-to-analog converters [7]. A $2N_2$-stage VM include an array of $2N_2$ SQUIDs and converts complementary SFQ pulse patterns into a N_2-fold bipolar waveform. A big advantage of the magnetically coupled VM is good input/output isolation at DC and low frequencies, which effectively reduces common-mode noise in the output voltage waveforms.

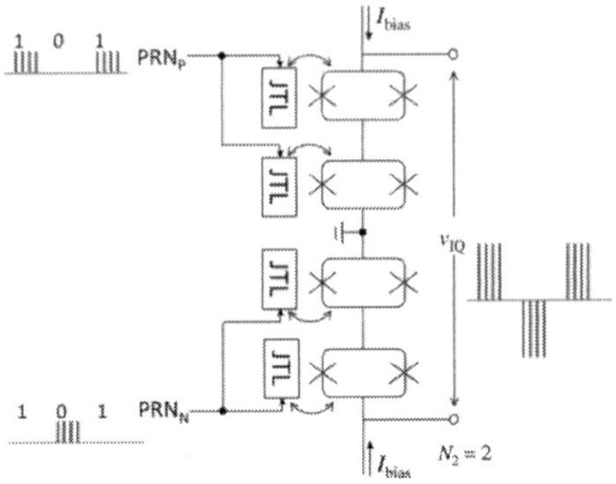

Figure 5. Voltage multiplier (VM) with complimentary inputs and a bipolar output. An example for $N_2 = 2$.

3. Output power spectrum

3.1. Calculability

By using a Fourier analysis method, we obtain the following formula for power spectrum densities $PSD_{IQ}(f)$ of the output voltage waveform $v_{IQ}(t)$ shown in the bottom of figure 2:

$$PSD_{IQ}(f) = \frac{f_{clk}}{2^m - 1} N_2^{\ 2} \left[\frac{\sin(\pi N_1 f\tau)}{\sin(\pi f\tau)}\right]^2 \Gamma(f), \tag{1}$$

where $\Gamma(f)$ is a frequency domain function determined by a SFQ pulse waveform.

For quantitative discussions on $PSD_{IQ}(f)$, we have to determine the SFQ waveform function to derive $\Gamma(f)$ in equation (1). In general, it is difficult to know a SFQ-pulse waveform exactly, so that the SFQ pulse shape should be approximated by simpler functions. However, such approximations might degrade the calculability of $PSD_{IQ}(f)$ and thus the accuracy of the JNT measurements. To evaluate the degradation quantitatively, we have made a comparison for two typical approximations, rectangular and triangular pulses, as follows. For a rectangular SFQ pulse with a width τ_R and a height Φ_0/τ_R, $\Gamma(f)$ is calculated as

$$\Gamma_R(f) \equiv \Gamma(f) = 2^{m+1}\Phi_0^{\ 2}\mathrm{sinc}^2(\pi f\tau_R). \tag{2}$$

For an isosceles triangular SFQ pulse with a base $2\tau_T$ and a height Φ_0/τ_T, $\Gamma(f)$ is calculated as

$$\Gamma_T(f) \equiv \Gamma(f) = 2^{m+1}\Phi_0^{\ 2}\mathrm{sinc}^4(\pi f\tau_T). \tag{3}$$

Here, Φ_0 is the single flux quantum and $\mathrm{sinc}(x) = \sin(x)/x$ is a sampling function. We defined a relative error as

$$\xi \equiv 1 - \frac{PSD_{IQ\_R}(f,\tau_R)}{PSD_{IQ\_T}(f,\tau_T)} = 1 - \frac{\Gamma_R(f,\tau_R)}{\Gamma_T(f,\tau_T)}. \tag{4}$$

Journal of Physics: Conference Series **507** (2014) 042023 doi:10.1088/1742-6596/507/4/042023

By introducing a parameter $\gamma = \tau_T/\tau_R$ and substituting τ_{SFQ} for τ_R and τ_T/γ, equation (4) reduces to

$$\xi = 1 - \frac{\mathrm{sinc}^2(\pi f \tau_{SFQ})}{\mathrm{sinc}^4(\gamma \pi f \tau_{SFQ})}. \tag{5}$$

In figure 6, the calculated relative error ξ is plotted as a function of a non-dimensional parameter $f\tau_{SFQ}$. The relative error ξ is smaller than 10^{-7} for practical JNT measurements with our system in which $f\tau_{SFQ}$ is less than 7×10^{-5} (f < 0.65 MHz, τ_{SFQ} < 100 ps). The result means that $PSD_{IQ}(f)$ is almost independent of the SFQ pulse shape approximation when $f\tau_{SFQ}$ is sufficiently small, implying that $PSD_{IQ}(f)$ is practically calculable with errors less than 10^{-7}.

3.2. *Measurements of the triple point of water*

Figure 7 shows an example $PSD_{IQ}(f)$ calculated with parameters for measurements of the triple point of water, 273.16 K, using a 100-ohm resistor sensor: $m = 21$, $f_{clk} = 2.013$ MHz, $\tau = 200$ ps, $N_1 = 74$, and $N_2 = 4$. The SFQ pulse was approximated by a triangular pulse with a base of $2\tau_{SFQ} = 40$ ps and a height of Φ_0/τ_{SFQ}. Although slight dependence on frequency is shown, the spectrum is precisely calculable with a sufficiently small error, < 10^{-7}, in a frequency range of interest. These results show that IQVNS is useful for the JNT applications.

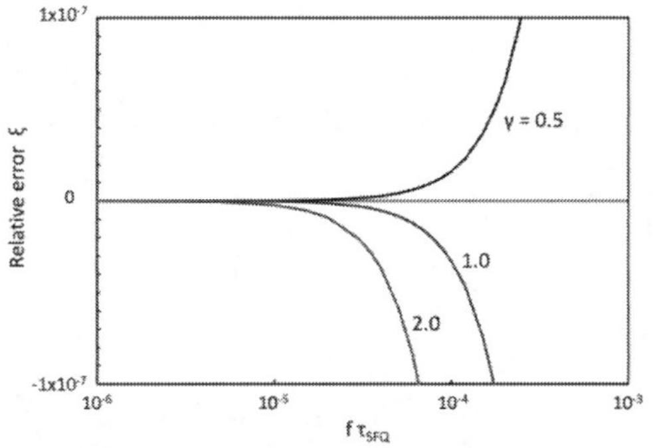

Figure 6. Dependence on the SFQ pulse waveform approximation.

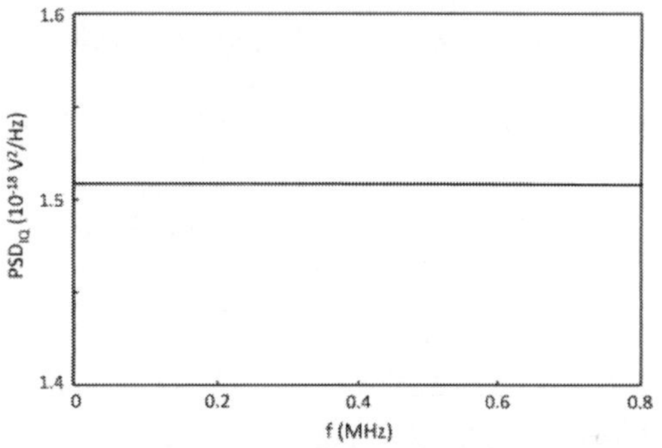

Figure 7. Example power spectrum density for measurements of the triple point of water.

4. Summary

We have designed an integrated quantum voltage noise source (IQVNS) for Johnson noise thermometry (JNT). The IQVNS consists of RSFQ subsystems, all of which are integrated on a single chip. A Fourier analysis has proven that the IQVNS power spectrum density is highly calculable with sufficiently small errors, which implies that the IQVNS is applicable to JNT for thermal metrology.

Acknowledgements
This work was supported by JSPS KAKENHI Grant Number 25289126.

References

[1] Jones N 2009 *Nature* **459** 902
[2] Fischer J *et al* 2011 *Int. J. Thermophys.* **32** 12
[3] Benz S P *et al* 2003 *IEEE Trans. Instrum. Meas.* **52** 545
[4] Benz S P *et al* 2011 *Metrologia* **48** 142
[5] Maezawa M, Yamada T and Urano C 2012 *SSV2012 Prog. and Proc. (Nagoya).* p 67
[6] Mizugaki Y *et al* 2011 *IEEE Trans. Appl. Supercond.* **21** 43
[7] Maezawa M *et al* 2006 *J. Phys.: Conf. Series* **43** 1187

Superconducting nanowires connected in series for photon number resolving functionality.

F. Mattioli[1], S. Jahanmirinejad[2], Z. Zhou[2], A. Gaggero[1], G. Frucci[2], D. Sahin[2], R. Leoni[1], and A. Fiore[2]

[1] Istituto di Fotonica e Nanotecnologie, CNR, Via Cineto Romano 42, 00156 Roma, Italy

[2] COBRA Research Institute, Eindhoven University of Technology, PO Box 513, 5600 MB Eindhoven, The Netherlands

Francesco.mattioli@ifn.cnr.it

Abstract. The experimental demonstration of a superconducting photon-number-resolving detector, based on the series connection of N superconducting nanowires, is presented. An integrated resistor is connected in parallel to each section of the device that provides in this way a single voltage-readout, proportional to the number of photons detected in distinct nanowires. As a proof of principle a four element detector has been fabricated from an NbN film on a GaAs substrate and fully characterized. Clearly separated output levels corresponding to the detection of n = 1 - 4 photons are observed achieving a single-photon system quantum efficiency of 2.6% at λ=1.3 μm. In order to demonstrate the potential scalability of the series-nanowire detector to a larger number of photons, we report our preliminary results in the characterization of detectors fabricated with 8 and 12 pixels. Clear evidence of n= 1-8 photon absorption in the 8-pixel detector has been achieved.

1. Introduction

Single-photon detection needs highly sensitive devices, which, for this reason, operate in a strongly nonlinear mode. Their response in this regime is independent from the number of detected photons. On the contrary, conventional optical detectors generate an electrical signal that is proportional to the intensity of the incident light, but their sensitivity is strongly limited by the electrical noise in the circuit needed for the signal amplification. In the last years, great efforts have been devoted to fill the gap between those two distinct regimes, creating a device capable of determining the number of incident photons: a photon-number-resolving (PNR) detector. Even though several detector technologies combining linear operation and single-photon sensitivity have been proposed, in the telecommunication wavelength region, which is interesting for many applications, their performance has remained limited in terms of speed and dynamic range. Transition-edge-sensors operating at sub-Kelvin temperatures show relatively slow response times [1]. Self-differencing InGaAs avalanche photodiodes (APDs) offer limited photon number discrimination ability [2]. APDs have been time-multiplexed to achieve photon number resolving capability, but with a necessarily reduced count rate due to the delay loops [3, 4]. The silicon photomultiplier (SiPM) combined with frequency up-conversion provides PNR functionality with large dynamic range, but introduces added noise due to the up-conversion process and cross-talk between the pixels [5]. In this framework it is natural to take advantage of the very high sensitivity, short dead time, and low timing jitter of SSPDs in the near-infrared range [6]. While multi-element SSPD arrays with separate readout for each element have been proposed [7], a configuration with a single readout is preferable in terms of simplicity and scalability to large photon numbers. Indeed, exploiting new geometries and using the spatial multiplexing it is possible to achieve a photon number resolving capability with SSPDs. For this reason in the past a parallel configuration with single output (parallel nanowire photon-number-resolving detector, PND) [8] has been proposed. However, this configuration has a limited dynamic range and efficiency, due to the current redistribution issue that increases the bias current in the unfiring sections, potentially leading to spurious switching [9].

In this article, we describe recent results on the photon-number-resolving detector based on the series connection of nanowires (series-nanowire detector, SND), theoretically proposed in Ref. [10]. A 4-pixel proof-of-principle detector was already characterized in Ref. [11]. Here, to demonstrate the scalability to large photon numbers of the series-nanowire structure, we present results on 8- and 12-pixels devices. The results on an 8-pixel SND show clear evidence of detection of n-photons with n ranging from 1 to 8.

2. Design and sample fabrication

2.1. Design

The electrical structure of the SND is based on the series connection of several wires, each shunted by a resistance (see figure 1) [10]. As in standard SSPDs, the nanowire is biased with a current close to its critical current (I_c), and after a single photon is absorbed, a resistive region is formed across the nanowire [6]. Due to the presence of the resistances integrated in parallel to each meander section (figure 1), the bias current (I_b) flows into these resistances rather than into the external load, producing a voltage pulse. If several sections switch simultaneously, the bias current flows into several parallel resistances, giving rise to a voltage across the whole device proportional to the number of switching sections [10].

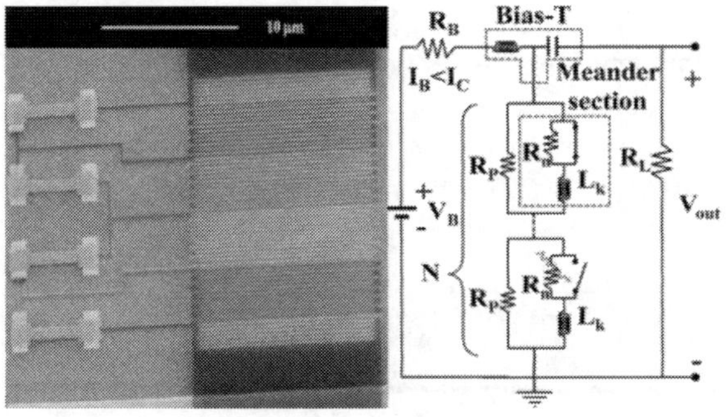

(a) (b)

Figure 1. (a) SEM image of the 4-pixel SND where the active wires, the parallel resistors, and their contact pads have been colored for clarity. (b) Equivalent electrical circuit of an N-element SND structure in which the last section has absorbed a photon. Each meander section is represented as a switch in parallel with the nanowire normal resistance and then in series with its kinetic inductance (the switch is closed in the superconducting state and opens when a photon is absorbed).

This device is designed as the electrical dual of the parallel nanowire photon-number-resolving detector in order to solve the current redistribution problem. All the detecting sections are biased with the same bias current (I_B) close to the critical one (I_C) (figure 1(b)). Switching of a wire in the SND reduces the bias current in the other unfiring wires (because a part is also diverted into the external load), as compared to the PND where the current in the unfiring wires is increased [9]. In this way, the SND can be biased closer to the critical current reaching higher quantum efficiency and more sections can be added in series allowing scaling to large photon numbers. In terms of response amplitude and speed, the device performance can be further optimized using a cryogen high-impedance preamplifier stage (high electron mobility transistor mounted close to the SND).

2.2. Fabrication process for 4-, 8- and 12-pixel SNDs

The SNDs are fabricated starting with 4.5nm, 4.8nm and 4.8nm thick NbN film (for the 4-SND, 8-SND and 12-SND, respectively) grown on GaAs substrates by reactive magnetron sputtering, at 410°C. The film parameters are a superconducting transition temperature of 9.5K and a transition width of 0.7K. Four steps of direct-writing electron beam lithography (EBL) are used to define the geometry of the SPD. For electron beam lithography a high resolution Vistec EBPG 5HR is used: this system is equipped with a field emission gun with 100 kV acceleration voltage. In the first step, Ti(10nm)/Au(60nm) electrical contact pads are patterned as a 50 Ω coplanar transmission line, together with alignment markers, using a positive tone PMMA electronic resist, followed by metal evaporation and lift-off. To obtain the electrical contact between the nanowires and the parallel resistances, additional Ti(5nm)/Au(200nm) pads are needed and are defined in a second step (light yellow-coloured pads in figure 1(a)). In the third step a 140 nm thick hydrogen silsesquioxane (HSQ) mask is used to define the 100 nm wide meandered nanowires, covering a total active area of 12 x 12 μm^2 with a 40% filling factor. Great care has been devoted to this lithographic step: in order to obtain a uniform width of the nanowire, during the e-beam exposure of the meander we varied locally the electronic dose reducing it towards the centre of the pattern. The pattern obtained in this way on the HSQ mask, is then transferred onto the NbN film underneath with a reactive ion etching (RIE) process based on a CHF_3, SF_6 and Ar mixture. The fourth step is used for the Ti (10 nm)/AuPd (50 nm) resistances definition. Each of them is 500 nm wide and 3.5 μm long corresponding to a design value of R_p= 50Ω. The fabricated structures are shown in the scanning electron microscope (SEM) image in figure 1 (a) and figure 2.

3. Optical characterization

The electro-optical characterization of the SND has been obtained in a closed-cycle cryocooler with a base temperature of 1.2K. The bias current is supplied through the DC port of a bias-T by a voltage source in series with a 10Ω bias resistor: the IV curve of this device is different from the standard SSPD IV curves due to the presence of the parallel resistance.

A pulsed laser-diode (50 ps pulse width, 1.3 μm wavelength, and 10 MHz repetition rate) has been used for the optical characterization. The coupling to the device has been achieved by means of a single-mode polarization maintaining lensed fiber, mounted on a XYZ piezo stage for the alignment. The beam has been defocused to a spot diameter of about 20 μm in order to illuminate equally all sections of the SND (12 x 12 μm^2 active area). A 40-GHz bandwidth sampling oscilloscope or a 350 MHz counter have been used for optical characterization to analyze the output signal collected at room temperature through the RF arm of the bias-T and amplified using a chain of wideband amplifiers.

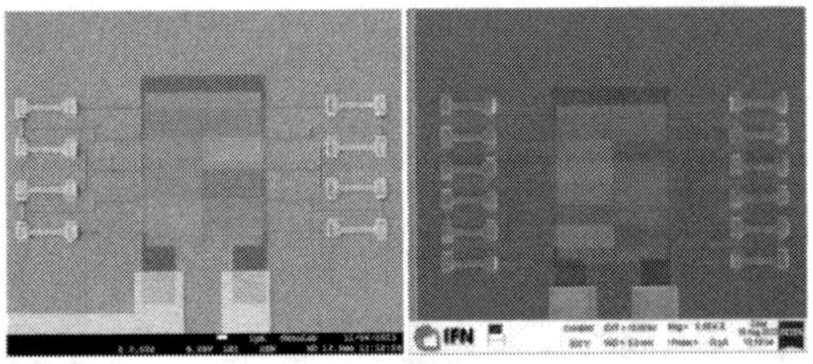

(a) (b)

Figure 2. (a) SEM image of the fabricated 8-SND with pixel colored with different colors. (b) SEM image of the 12-SND with pixel colored with different colors.

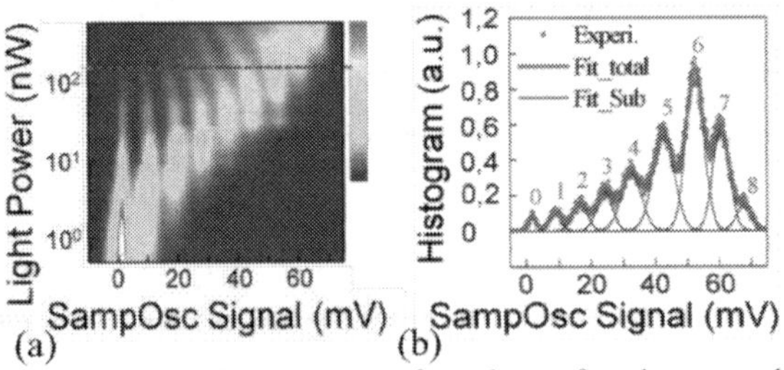

(a) (b)

Figure 3. (a) Counts as a function of voltage and light power taken with λ=1.3 μm and 14 MHz repetition rate (b) Histogram (gray line) at the light power indicated as a red dashed line in figure 3 (a), showing 8 clearly spaced peaks, and fit (red line), assuming each photon level contributes a Gaussian voltage distribution (blue lines)

3.1. 4-pixel SND

Results of 4-pixel SNDs were already reported in ref. [11]. Four clear peaks have been observed and it has been proved that they correspond to the detection of 1–4 photons. The highest value of the single-photon system quantum efficiency (SQE) was 2.6% (with the spot focused on the device) at a bias current of I_B=12.4 μA (0.99 I_C), with the light polarized parallel to the wires. The capability of biasing the device at 99% of the I_C demonstrates the correct functionality of the series nanowire configuration. At low bias currents (I_b<10 μA) a dark count rate (DCR) as low as 1 Hz is measured. At I_B=10 μA, a SQE of 1% and a DCR of 5Hz are obtained. The approaches used to increase the quantum efficiency of SSPDs can also be applied to SNDs: Improvement of the superconducting properties of the thin film [12, 13], reduction of the width of the nanowires [14] and implementation of advanced optical cavities [15-17] or waveguides [18].

3.2. 8-pixel SND

The fabrication process already described and a similar optical characterization has been applied to SNDs with 8 and 12 pixels, shown in figure 2 (a) and (b), respectively. Figure 3 shows clear evidence that n= 1-8 detection events have been observed in the 8-pixel SND. Figure 3 (a) shows the color plot of the counts as a function of the pulse height voltage taken for different attenuations of the laser pulse: the pulse height distribution taken for a particular light attenuation (red dashed line in figure 3 (a)) is showed in figure 3(b). All the eight peaks are clearly visible (every peak is separated from the others) and evenly spaced. A full optical characterization of this 8-SND together with the first tests on the 12-SND are in progress.

4. Conclusions

In conclusion, a design for PNR detectors based on the series connection of N superconducting nanowires sections, each shunted with a resistance, has been described and implemented. The detection of n=1–4 photons in the telecom wavelength range has been demonstrated in a 4-pixel SND with maximum system quantum efficiency of 2.6% and recovery time constant in the 10 ns range. The integration of the SND with an optical cavity will lead to the improvement of the SQE. Scaling to large number of sections has been realized up to 12 element detector, keeping the same active area of the device. Clear evidence of the detection of n= 1-8 in the 8-pixel SND has also been reported.

References

[1] Lita A E, Miller A J and Nam S W 2008 *Opt. Express* **16**(5) 303.
[2] Kardynal B E, Yuan Z L and Shields A J 2008 *Nat. Photonics* **2** 425.
[3] Fitch M J, Jacobs B C, Pittman T B and Franson J D 2003 *Phys. Rev. A* **68** 043814.
[4] Eraerds P et all.2010 *Rev. Sci. Instrum.* **81** 103105.
[5] Pomarico E, Sanguinetti B, Thew R and Zbinden H 2010 *Opt. Express* **18**(10) 10750.
[6] Gol'tsman G et all. 2001 *Appl. Phys. Lett.* **79** 705.
[7] Dauler E A et all. 2009 *J. Mod. Opt.* **56** 364.
[8] Divochiy A et all. 2008 *Nat. Photonics* **2** 302– 306.
[9] Marsili F, et all. 2009 *New J. Phys.* **11** 045022.
[10] Jahanmirinejad S and Fiore A 2012 *Opt. Express* **20** 5017.
[11] Jahanmirinejad S et all. 2012 *Appl. Phys. Lett.* **101** 072602.
[12] Tanner M G et all. 2010 *Appl. Phys. Lett.* **96** 221109.
[13] Baek B, Lita A E, Verma V, and Nam S W 2011 *Appl. Phys. Lett.* **98** 251105.
[14] Marsili Fet all. 2011 *Nano Lett.* **11**(5) 2048–2053.
[15] Gaggero A et all. 2010 *Appl. Phys. Lett.* **97** 151108.
[16] Rosfjord K M et all. 2006 *Opt. Express* **14** (2) 527–534.
[17] Baek B, Stern J A, and Nam S W 2009 *Appl. Phys. Lett.* **95** 191110.
[18] Sprengers J Pet all. 2011 *Appl. Phys. Lett.* **99** 181110.

Structuring metallic coatings to reduce eddy currents and thermal noise in super insulation

R.Mayrhofer[1], J.Stipsitz[1], R.Koerber[2], J.Stampfl[3], P.Schoenhuber[4]

[1] RUAG Space GmbH, Stachegasse 16, 1120 Vienna, Austria
[2] Physikalisch-Technische Bundesanstalt, Abbestrasse 2-12, 10587 Berlin, Germany
[3] Institute of Materials Science and Technology, Vienna University of Technology, 1040 Vienna, Austria
[4] Institute of Electrodynamics, Microwave and Circuit Engineering, Vienna University of Technology, 1040 Vienna, Austria

Email: Robert.Mayrhofer@ruag.com

Abstract. The performance of multi-layered super insulation in cryogenic systems is highly dependent on the radiative properties of reflector foil surfaces. Aluminum coated polyester foils are utilized for their high reflectivity in the infrared spectrum. This coating is an electrical conductor and in applications with variable magnetic fields, eddy currents are induced in the coating. These eddy currents can have two effects. In high energy applications like for example superconducting fault current limiters, heat is dissipated in the super insulation, degrading its performance and potentially even damaging the super insulation. For high precision magnetic measurements using SQUIDs, switching ambient fields induce eddy currents in the super insulation which result in a transient response superposing the signal to be measured. This, in addition to a higher thermal noise background reduces the quality of the magnetic measurements. Our study developed Coolcat 2 NI with a 10x10 mm grid in the metallic coating. In this way the high reflectivity provided by the aluminum has been retained and the eddy currents were reduced to a level acceptable for different applications. Calorimetric and magnetic experiments are presented to show the merits of this structuring.

1. Introduction

Polyester foils with vacuum deposited aluminum (VDA) are widely used in cryogenics to decrease the heat load on superconductors and cold masses. For some applications the continuous coating of the non conducting polyester with a conductor like aluminum causes problems. One can define two different applications with slightly different requirements.

Changing magnetic fields or rotating cold masses induce eddy currents in the aluminum and heat dissipated by these eddy currents may damage the super insulation like in superconducting Fault Current Limiter or other energy applications. Another application are high precision measurements utilizing SQUIDS, as ambient magnetic fields interact with the super insulation and lead to high background levels for the measurements. Our study therefore investigated various processes to structure the metallic coatings to address these issues. The aim was to retain the high reflectivity provided by the aluminum and to reduce the eddy currents to a level acceptable for different applications.

The design of the foil parameter has to be a compromise between thermal and electrical properties. To test und evaluate these properties several calorimetric and magnetic experiments were conducted. Based on these the most effective structured coating was chosen.

2. Materials

A simulation with Comsol Multiphysics 4.3 shows that eddy currents get smaller with each division of the coating. The reduction of the eddy currents gets smaller with each additional division, so it was determined that at least 5-10 divisions per area would be best. Typical super insulation designs feature areas larger than 100 mm, so a pattern size of 10 mm x10 mm was utilized.

Coolcat 2 NI (Non Inductive) is a spacered superinsulation composed of 10 layers of 12μm polyester foil, single-side aluminized in squares of 10 mm x 10 mm, interleaved with 10 layers of non-woven polyester spacer material. It is available with two different surface resistances, 0.8 Ohms per square and 1.6 Ohms per square. A lower surface resistance means that the aluminum coating is thicker. The distance between the squares is approximately 100 μm.

The foils used in Coolcat 2 NI were developed in a cooperation of the Vienna University of Technology and the Thermal Systems department of RUAG Space GmbH.

Details of the production technique are still confidential at this time, but the gaps in the aluminum were produced using a chemical release method. Before depositing the aluminum on the polyester by vacuum deposition, a chemical gets printed onto the foil in the desired pattern. This chemical is then later activated and the aluminum can be removed locally.

3. Magnetic field measurements

Magnetic thermal noise measurements were performed with 20 layers of structured foil interleaved with 20 layers of polyester spacer material, on a sample size of 100 mm × 100 mm. The measurements were conducted with a SQUID system inside a magnetically shielded room of the Physikalisch-Technische Bundesanstalt [1]. The SQUID current sensor, coupled to a first order axial gradiometer, baseline 120 mm and radius 20 mm, was operated in a low-noise dewar with a warm-cold distance of 8.8 mm (measured at room temperature) [2]. The stack was positioned centrally under the dewar bottom.

Table 1. Spectral noise density - Values in brackets give the calculated noise contribution.

Frequency [Hz]	**Noise spectral density for SQUID system [fT/√Hz]**	**Noise spectral density for 0.8 Ohms per square [fT/√Hz]**	**Noise spectral density for 1.6 Ohms per square [fT/√Hz]**
40	2.30	2.68 (1.37)	2.60 (1.21)
100	2.06	2.39 (1.21)	2.40 (1.23)
500	1.65	2.03 (1.18)	2.06 (1.23)
1000	1.66	2.14 (1.35)	2.12 (1.32)
10000	1.48	1.89 (1.18)	1.80 (1.02)

The calculated noise contributions shown in **Table 1** are very similar for both coating thicknesses. For SQUID Applications the noise contribution shall be as low as possible. A value in the range of 1-2 fT/√Hz is considered acceptable. **Figure 1** shows the spectral noise density for both materials. The white noise contribution for the 0.8 Ohm/square and 1.6 Ohms/square samples for this geometry at room temperature are measured as 1.0 fT/√Hz and 1.1 fT/√Hz, respectively (from the noise spectra).

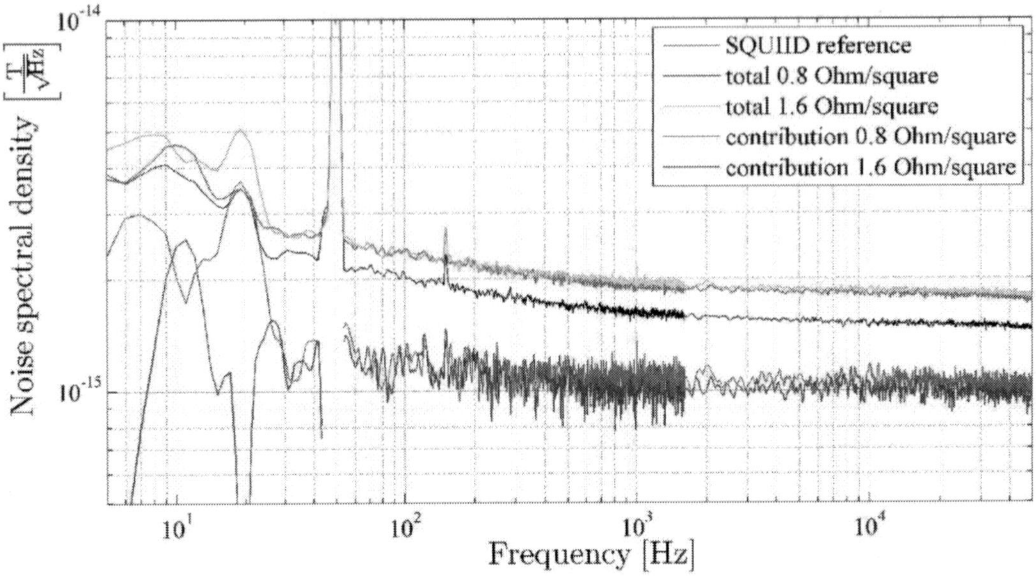

Figure 1. Noise spectral density spectra for Coolcat 2 NI, with 0.8 and 1.6 Ohm/square coating; the contribution without SQUID background levels is shown individually

4. Calorimetric measurements

The emissivities of the different foils were measured and compared to a former material type, which is no longer available. In Figure 2 one can compare the emissivities of a continuously VDA coated foil, of the former patterned material and of the new 0.8 Ohms per square surface resistance patterned material utilized in Coolcat 2 NI. The emissivity of the new material is significantly better than the values of the former material. The measurements were performed by using a calorimeter as described in ref. [3].

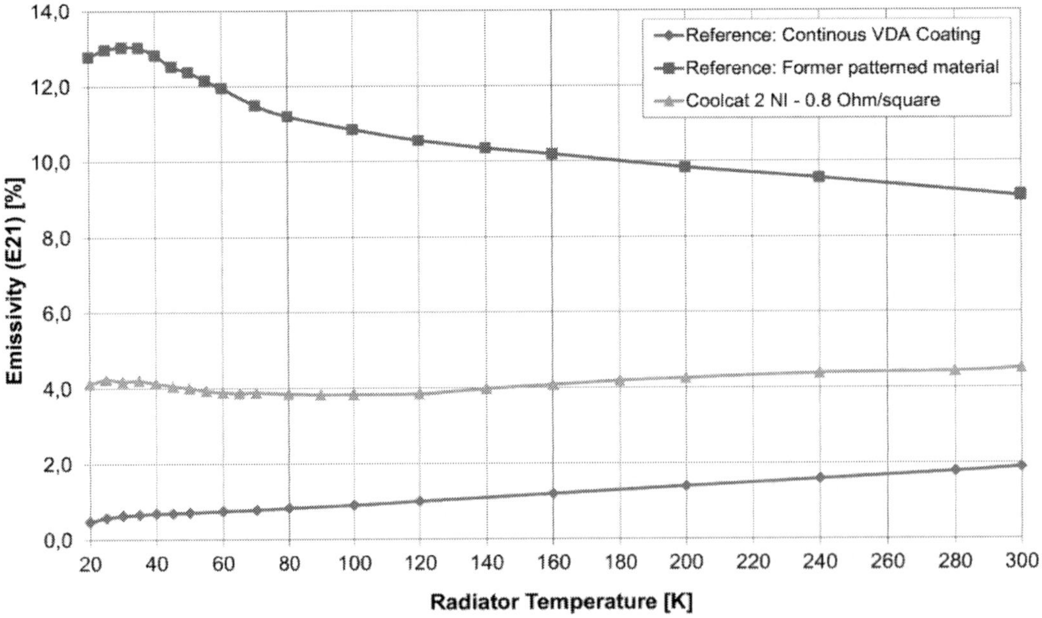

Figure 2. Emissivity single foil layer

Journal of Physics: Conference Series **507** (2014) 042025 doi:10.1088/1742-6596/507/4/042025

In **Table 2** we summarize the results of calorimetric measurements on Coolcat 2 NI MLI blankets performed at THISTA, Karlsruhe Institute of Technology (KIT) [4].

Table 2. Calorimetric performance of Coolcat 2 NI with the 0.8 Ohms per square variant

Temperature range [K]	**Number of layers**	**Heat flux [W/m²]**
300 to 77	10 foils + 10 spacers	> 4.60
	20 foils + 20 spacers	> 3.40

5. Conclusion

The work provided a new material for application in SQUIDs and high energy superconducting applications to reduce eddy currents and magnetic background noise in super insulation. The thermal performance was improved compared to state of the art materials and important properties of the materials were tested with application in superconducting fault current limiters, superconducting generators, superconducting motors and SQUIDs in mind. Coolcat 2 NI has a white noise contribution of 1.0 $fT/\sqrt{Hz}$ and reduces the heat flux from 300 to 77 K to approximately 4 W/m^2.

6. Acknowledgements

Coolcat 2 NI was developed in a cooperation of the Vienna University of Technology and the Thermal Systems department of RUAG Space GmbH. The authors thank RUAG Space GmbH to let them use available state-of-the-art equipment and analyses tools, clean rooms and automated cutting machines.

The authors acknowledge support from RUAG Space GmbH and the Austrian Federal Ministry of Transport, Innovation and Technology (Austrian Research Promotion Agency).

7. References

[1] J.Voigt, S.Knappe-Grüneberg, A.Schnabel, R.Körber, M.Burghoff, "Measures to reduce the residual field and field gradient inside a magnetically shielded room by a factor of more than 10", Metrology and Measurement Systems, 20, 239-248, (2013)

[2] I.Hilschenz, R.Körber, HJ.Scheer, T.Fedele, HH.Albrecht, A.Cassara, S.Hartwig, L.Trahms, M.Burghoff, „Magnetic Resonance Imaging at frequencies below 1 kHz", Mag. Res. Imag. 31, 171-177; (2013)

[3] T.Kralik, et al., "Device for measurement of thermal emissivity at cryogenic temperatures", Proceedings of the 8th IIR Int. Conf. CRYOGENICS, Prague (2004), 23-29

[4] H.Neumann: "Thermal performance of different glass microspheres in comparison to perlite between 77 K and 300 K"; CEC-ICMC; 28.06. – 02.07.2009; Tucson, Arizona, US

DC current distribution mapping system of the solar panels using a HTS-SQUID gradiometer

Shingo Miyazaki[1], Syohei Kasuya[1], Mohd Mawardi Saari[1], Kenji Sakai[1], Toshihiko Kiwa[1], Akira Tsukamoto[2], Seiji Adachi[2], Keiichi Tanabe[2] and Keiji Tsukada[1]

[1]Graduate School of Natural Science and Technology, Okayama University, 3-1-1 Tsushimanaka, Kitaku, Okayama 700-8530, JAPAN
[2]Superconductivity Research Laboratory – ISTEC, 1-10-13 Shinonome, Koto-ku, Tokyo 135-0062, JAPAN

Email: tsukada@cc.okayama-u.ac.jp

Abstract. Solar panels are expected to play a major role as a source of sustainable energy. In order to evaluate solar panels, non-destructive tests, such as defect inspections and response property evaluations, are necessary. We developed a DC current distribution mapping system of the solar panels using a High Critical Temperature Superconductor Superconducting Quantum Interference Device (HTS-SQUID) gradiometer with ramp edge type Josephson junctions. Two independent components of the magnetic fields perpendicular to the panel surface ($\partial Bz/\partial x$, $\partial Bz/\partial y$) were detected. The direct current of the solar panel is visualized by calculating the composition of the two signal components, the phase angle, and mapping the DC current vector. The developed system can evaluate the uniformity of DC current distributions precisely and may be applicable for defect detection of solar panels.

1. Introduction

Effective means of using sustainable energy such as solar panels are required due to the depletion of fossil fuels, global warming and energy problems. Non-destructive evaluation of solar panels is important to develop solar panels with high efficiency and high durability. For example, the photoexcitation method using laser and microwave technologies has been developed to assess the concentration, lifetime, and mobility of carriers in semiconductors. The thermography method [1], laser Superconducting Quantum Interference Device (laser–SQUID) method [2] and electro luminescence method [3] have been used to evaluate these technologies. The above evaluation methods are useful techniques in evaluating the solar panels because they are capable of measuring the excitation current and detecting micro-cracks.

We developed a system of evaluating solar panels using a SQUID, which is an ultra-high sensitivity magnetic sensor. Although the sensitivity of LTS (Low Critical Temperature)-SQUIDs made of a low-temperature superconductor is higher compared to HTS- (High Critical Temperature) SQUIDs, there is the problem that LTS-SQUIDs require a large operating system due to coolant recovery apparatus and the high running costs of liquid helium. Therefore, we developed a system using a HTS-SQUID made of a high temperature superconductor. The HTS-SQUID is inferior to LTS-SQUID in sensitivity; however, it has reasonable running costs and requires a smaller system, because it works at liquid nitrogen temperature. The developed system allows to us visualize the electrical characteristics of the solar panels using HTS-SQUID and an external normal conducting pickup coil [4]. In addition, this system was able to evaluate the solar panels depending on the differential conductance characteristics in the solar panels. Furthermore, the solar panels could be separated from the HTS-SQUID using the pick-up coil to reduce interference. However, the intensity of the current mapping flowing through the solar panel was difficult to detect because the modulation voltage should be applied to the solar panels. Therefore, in this study, we developed a direct detection system for detecting the magnetic field from the sample using the HTS-SQUID gradiometer. We report the current distribution measurement of the solar panels by using this system.

2. Experimental

2.1. *HTS-SQUID gradiometer*

Cross-sectional view of the HTS-SQUID and the HTS-SQUID chip are shown in Fig. 1. The HTS-SQUID has ramp-edge type Josephson junctions[5].The base and counter superconducting electrodes were $La_{0.1}Er_{0.95}Ba_{1.95}Cu_3O_y$ and $SmBa_2Cu_3O_y$, respectively. $SrSnO_3$ (SSO) was used in the insulating layer. The size of the SQUID chip was 15 mm × 7.5 mm with an on-chip 3 turns feedback coil provided in each SQUID, and four SQUID elements were fabricated on the same substrate in the chip. Also, a gradiometer connected to these SQUIDs was fabricated on the same substrate and the baseline is about 7.5 mm.

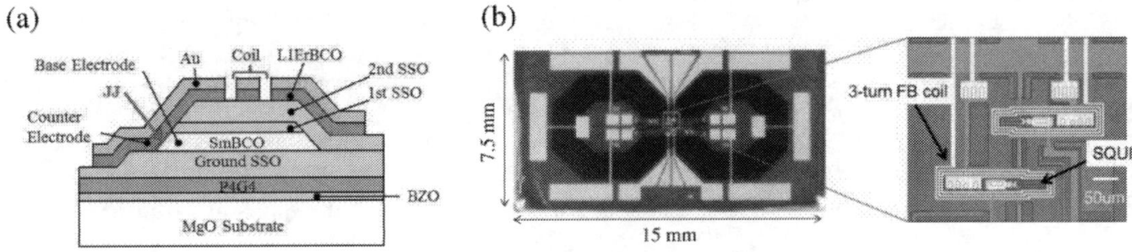

Figure 1. (a) Cross-sectional view of a SQUID, (b) HTS-SQUID gradiometer.

2.2. *System configuration*

The measurement system developed in this study is shown in Fig 2. Two independent components of the magnetic fields perpendicular to the panel surface ($\partial Bz/\partial x$, $\partial Bz/\partial y$) of the magnetic field were detected by the HTS-SQUID gradiometer. The HTS-SQUID was placed on a sapphire rod and cooled by liquid nitrogen through heat conduction. The measurement sample and the HTS-SQUID were placed in a two-layer permalloy magnetic shield. The size of the magnetic shield is 912 mm × 700 mm × 942 mm. The SQUID is operated by a Flux Locked Loop (FLL) circuit (SEL-1, MAGNICON). The obtained signals detected by the HTS-SQUID gradiometer were recorded through an analog-to-digital converter (NI USB-DAQ, National Instruments). Furthermore, in order to enable highly reproducible, high spatial resolution and automatic measurement, a measurement sample was placed on a PC-controlled XY stage. The magnetic flux noise level of the developed system did not show significant changes at the time of excitation and operation of the XY stage.

2.3. *Examined sample and procedure*

A commercially available amorphous silicon solar cell was measured as the test sample. The effective area of the panel is 141.9 mm × 107.0 mm, the thickness was 2.3 mm. First, the characteristics of the magnetic field to the applied voltage were measured. This measurement was carried out in 2 places while being swept by 0.1 V intervals from 0 to 10 V of DC voltage. Then, to evaluate the DC current distribution of this sample, 9 V of the DC voltage was applied to the test sample. In this case, the scanning area was 80 mm × 60 mm, the scanning interval was 2 mm, and measurements were performed at 1200 points on the solar panel.

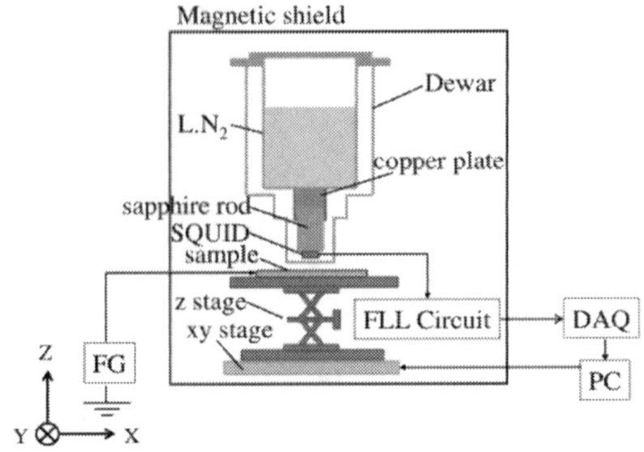

Figure 2. Schematic diagram of the developed system.

3. Results and discussion

The magnetic field characteristic with respect to the voltage of the solar panel is shown in Fig. 3 (left) and (right). The output of the SQUID changed with the increase in voltage. With the use of a gradiometer in this system, when the current is uniformly distributed, the output of the SQUID does not change. Therefore, the results show that the current flowing through the solar panel was not uniform. The value of $\partial B_z/\partial x$ showed that the current flowing was not uniform between the positive and negative electrodes, and the direction of the current flowing perpendicular to the electrodes was also different in the two measurement points. The variations in the initial value of the signal were due to the DC offset of the SQUID by the FLL circuit. In addition, the change of $\partial B_z/\partial y$ component was similar because current flow horizontally between the electrodes was in the same direction.

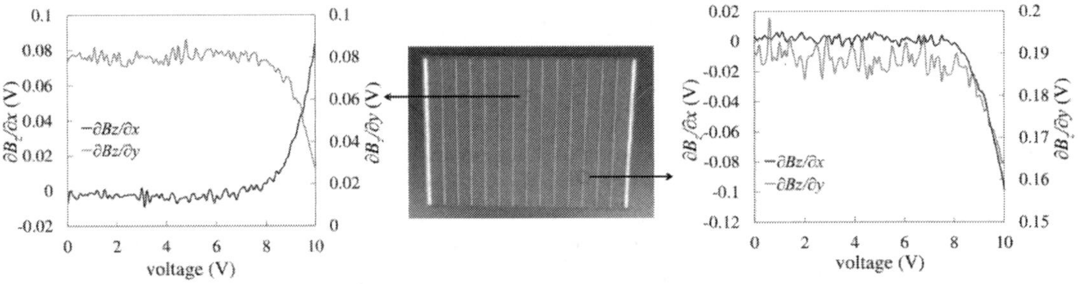

Figure 3 (left) and (right). Voltage-Magnetic field characteristic of solar panel.

The signals obtained in this system were two independent components of the magnetic fields perpendicular to the panel surface below the HTS-SQUID. As the gradiometer was used in this system, the two signals of the magnetic field gradient generated by the sample were obtained. Therefore, it was possible to calculate the intensity of the current and phase angle of the current flowing in the samples by using two components of the obtained signals. The intensity of the current ($I_x=\partial B_z/\partial x,\ I_y=-\partial B_z/\partial y$) is expressed by the following equation (1).

$$I=\left(I_x^{\,2}+I_y^{\,2}\right)^{1/2}=\left\{(\partial Bz/\partial x)^2+(-\partial Bz/\partial y)^2\right\}^{1/2} \tag{1}$$

The current phase angle θ of the current is expressed by the following equation (2).

$$\theta = a\tan\left(\frac{Ix}{Iy}\right) - \frac{\pi}{2} = a\tan\left(\frac{-\partial Bz/\partial y}{\partial Bz/\partial x}\right) - \frac{\pi}{2} \tag{2}$$

The vector using arrows displays the current phase obtained by the calculation described above, depicting the current arrow map that shows the contour lines of the current intensity.

The result of DC current distribution measurement of the solar panel is shown in Fig.4 left. The measurement was performed from the rear surface of the solar panel without a light source. Around the external positive electrode, it is found from Fig. 4 that the direct current that flows in a direction perpendicular to the inter-electrode is not flowing uniformly. Thus, we succeeded in identifying non-uniform current distribution points by using the developed system.

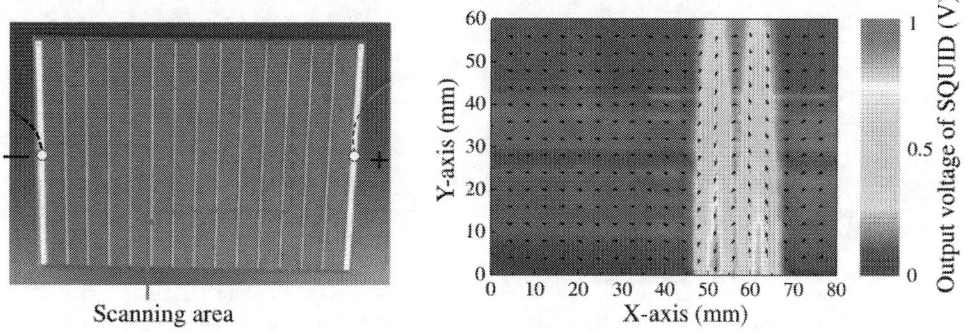

Figure 4 (right). DC current distribution of a solar panel.

4. Conclusion

We developed a system for visualizing the DC current distribution of a solar panel using an HTS-SQUID gradiometer. The current vector was detected by using a gradiometer. Therefore, it is possible to visualize the current distribution by the composition of two independent components of the magnetic fields perpendicular to the solar panel. The developed system was able to visualize a non-uniform current phase and the place of the DC current distribution in an amorphous silicon solar cell. Therefore, the developed system is useful as a non-destructive testing method in the process of solar cell development. The developed system can be applicable to the evaluation of a defective portion and the uniformity of the current in various types of solar panel.

Acknowledgments
This work is supported by the "Strategic Promotion of Innovative R&D" of the Japan Science and Technology Agency (JST).

References

[1] Breitenstein O, Iwig K and Konovalov I 1997 *phys. stat. sol. (a)* **160** 271–82
[2] Daibo M, Shikoda A and Yoshizawa M 2002 *Physica C: Superconductivity* **372-376** 263–6
[3] Pankove J 1980 *Solar Cells* **2** 443–9
[4] Kiwa T, Maedaa S, Miyake K, Kataoka N, Tsukamoto A, Adachi S, Tanabe K, Kandori A and Tsukada K 2011 *Physica C: Superconductivity* **471** 1238–41
[5] Adachi S, Hata K, Sugano T, Wakana H, Hato T, Tarutani Y and Tanabe K 2008 *Physica C: Superconductivity* **468** 1936–41

Improvement of power-handling capability of superconducting filters using 3D-matrix microstrip lines

S. Ohshima, S. Takahashi, M. Endo, and A. Saito
Graduate School of Science and Engineering, Yamagata University
4-3-16, Jonan Yonezawa, 992-8510, Japan
E-mail: ohshima@yz.yamagata-u.ac.jp

Abstract. We examined power-handling capability of superconducting filters using three different types of resonator: type one, a conventional microstrip line, type two, a layered-film microstrip line, and type three, a 3D-matrix microstrip line. The 3D-matrix microstrip line was made by dividing a type two microstrip line into narrow lines. We used NbN thin films as superconducting microstrip lines. The optimal shape of the filters was designed by using a Sonnet electromagnetic simulator, and the power-handling capability of these filters was measured at 6.5 K. The values of a conventional microstrip line, a layered-film microstrip line, and a 3D-matrix microstrip line filters were 21.2, 22.9, and 24.6 dBm, respectively.

1. Introduction

Superconducting bandpass filters (SBFs) have advantages such as a high frequency selection, small insertion loss, and a large out-of-band rejection. Consequently, such filters have been put to practical use in receiving systems of base stations for wireless communications in the U.S.A and China. SBFs have only been used in receiving systems in which small amounts of electric power flow. If the power-handling capability of the SBFs can be improved, it could be used as a transmitting filter, and hence its range of use could be expanded. Wide microstrip line resonator filters [1-3] and the patch resonator filters [4, 5] have been reported to increase power-handling capability. Filters with power-handling capability over one watt can be obtained; however, they are not useful for designing multi-resonator filters. This problem may be overcome if we can design filters by using microstrip lines with high power-handling capability. We think that the limit of power-handling capability of filters is caused by the current concentration at the outer edge of a microstrip line. Therefore, it is very important to reduce the current concentration to increase power-handling capability of SBFs.

With this goal in mind, we have proposed some resonator shapes to reduce the current concentration at the outer edge of a microstrip line in previous papers [6-11]. The sliced microstrip line filters and layered-film microstrip line filters could increase the power-handling capability of the filter compared with a conventional microstrip line filter.

In this paper, we present a new microstrip line shape (3D matrix microstrip line) to reduce the current concentration at the outer edge of a microstrip line. The 3D microstrip line was constructed by dividing it into a narrow line of a layered film microstrip line. The narrow line consists of alternately stacked superconducting thin film (NbN) and dielectric thin film (AlN) layers on an MgO substrate. We describe how the power-handling capability of SBFs was improved experimentally using the 3D matrix resonators.

2. Experimental Procedure

2.1. Filter design

We designed a three-pole Chebyshev type bandpass filter based on the following specifications; a center frequency of 5.0 GHz, band width of 100 MHz, and pass-band ripple of 0.01 dB. The frequency response of the SBFs was simulated using an electric magnetic simulator (Sonnet EM). It is difficult to simulate the shape of 3D-matrix microstrip line SBFs using Sonnet EM, and therefore we used the fact that the shape of the 3D-matrix microstrip lines was the same as that of the layered-film microstrip line. The sliced number for 3D-matrix microstrip line was thirteen. Figure 1 shows schematic drawings of the conventional microstrip line filter (Fig.1(a)) and 3D-matrix microstrip line filter (Fig.1(b)) and the simulation results of frequency response. S_{11} of these filters was less than -30 dB, and the in-band ripple of these filters was less than 0.02 dB.

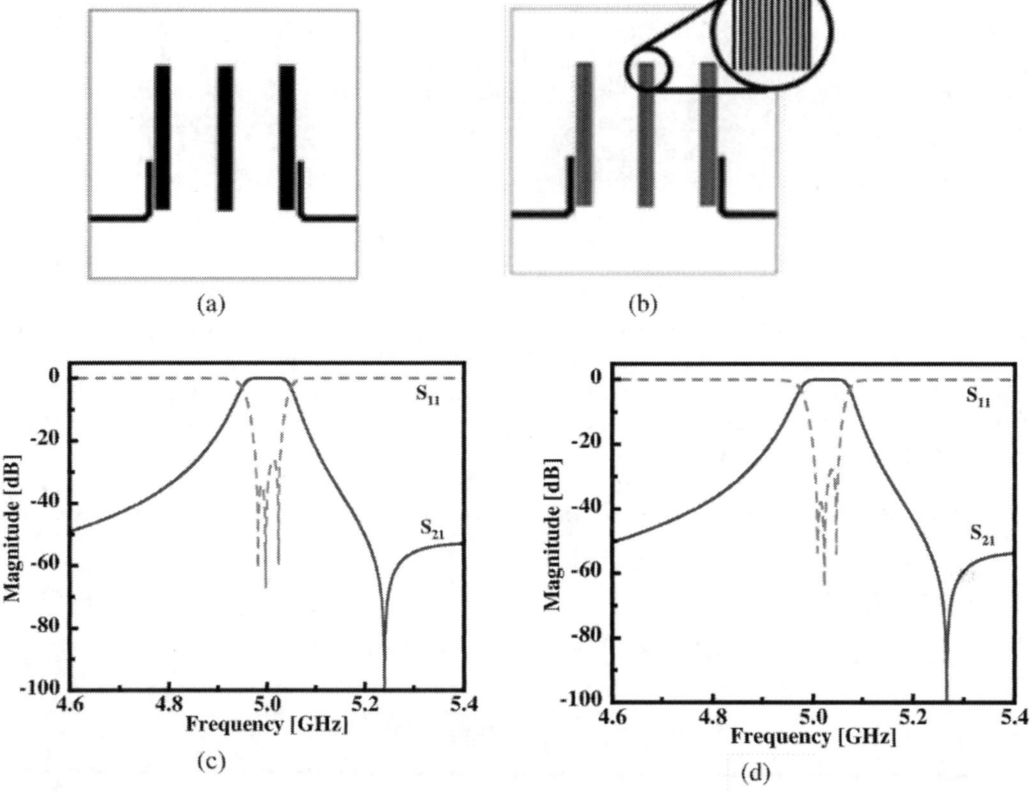

Fig.1. Schematic drawing of filter shapes and simulation results of frequency responses of the filters. (a) Shape of a three-pole conventional filter; (b) Shape of a three-pole 3D-matrix filter; (c) simulation results for a conventional filter; (d) simulation results for 3D-matrix filter.

2.2 Fabrication of the 3D-matrix microstrip line filter

We prepared NbN single-sided 600-nm thick film and NbN/AlN/NbN single-sided 300/20/300 nm multi-layered film on MgO substrate by dc magnetron sputtering. The schematic drawings of the films are shown in Fig.2. The Tc of these films was about 14 K, and their surface resistance was about 0.2 mΩ at 21.8 GHz below 7 K, which was the same as that of high quality YBCO thin films at 7 K. These characteristics were reported in a previous paper [6]. The grain orientation of NbN was

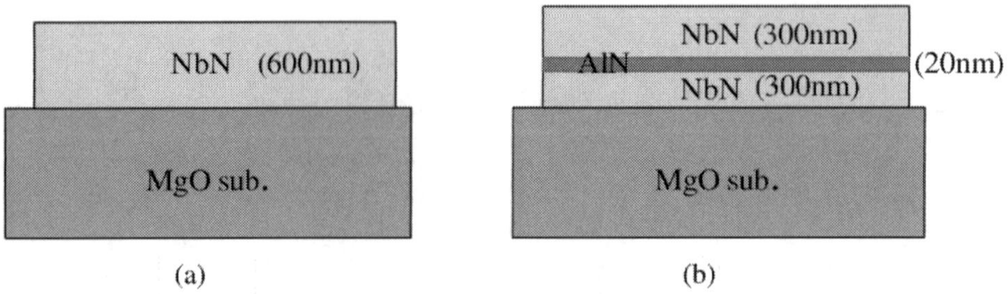

Fig.2. Schematic drawing of NbN mono-layered film and NbN/AlN/NbN multi-layered film on MgO substrate.

measured by X-ray ϕ-scan diffraction. We found that NbN and NbN/AlN/NbN showed hetero-epitaxial growth on the MgO substrate. We fabricated a three-pole filter using NbN and NbN/AlN/NbN thin films by conventional photolithography and ECR dry-etching techniques. We examined three differently shaped resonator filters: type one, a conventional microstrip line, type two, a layered-film microstrip line, and type three, a 3D-matrix microstrip line. Type one was made of mono-layered NbN thin film (shown in Fig.2 (a)), and types two and three were made of multi-layered thin film (shown in Fig.2(b)). The 3D-matrix microstrip line was obtained by dividing the layered-film microstrip line into narrow lines. The number of slits was thirteen.

3. Results and Discussion

Figure 3 shows the effective input power dependence on the output power in the filters made of conventional, layered-film, and 3D-matrix microstrip lines. The effective input power means the input power minus the reflected power. The vertical axis shows P_{in}- P_{out}, which is the calculated value of the input power minus the output power. If the filter has thermal loss, P_{in}- P_{out} becomes negative. We define the power proof of the filters as the value at which P_{in}- P_{out} is -0.5 dBm (dashed line in the figure). As shown in Fig.3, the electric power proofs of the conventional microstrip line filter, the layered-film microstrip line filter, and 3D-matrix microstrip line filter were 21.2, 22.9, and 24.6 dBm,

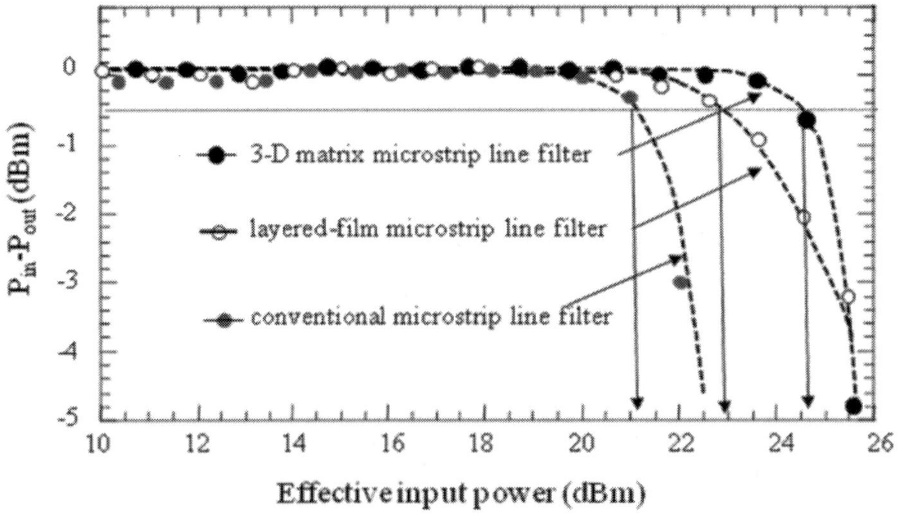

Fig.3. Effective input power dependence on P_{in}-P_{out} of the filters.

respectively. The electric power proof of the 3D-matrix microstrip line filter was larger than those of the conventional filter and layered-film filter by about 3.4 and 1.7dBm, respectively. From the simulation of the current concentration of the microstrip line, we can estimate the increase in the power proof of the filter. We reported the concentrated current of the conventional microstrip line, a sliced microstrip line, and a layered-film microstrip line [5]. The concentrated current of the conventional microstrip line is about two times larger than that of the layered-film microstrip line and 1.33 times larger than that of the sliced microstrip line. Therefore, we can estimate the difference in the power proof among these filters. The power proof of the 3D-matrix microstrip line filter may be larger than that of the conventional microstrip line filter and layered-film microstrip line filter by about 5.7 dB and about 2.9dB, respectively. The experimental results for the power proof were about 60 % of the simulation results. The estimated power proof can be realized by the optimal thickness of the microstrip line, and therefore, we must determine the optimal thickness of the superconducting and insulator films. We will be able to obtain a larger power proof with the 3D-matrix microstrip line filter.

4. Conclusion

Superconducting bandpass filters using a 3D-matrix microstrip line were designed and fabricated. The power- handling capability of the 3D-matrix microstrip line filters was 1.7 dB larger than that of a layered-film microstrip line filter and 3.4 dB larger than that of a conventional microstrip line filter. This different in the power proof occurred due to the current concentration of a microstrip line at the outer edge of a microstrip line. We clarified that 3D-matrix microstrip line was useful to increase the power proof of the microstrip line.

Acknowledgements

This research was partially supported by the Ministry of Education, Science, Sports and Culture, Grant-in-Aid for Science Research (C) (22560317)

References

[1] G.C Liang, D. Zhang, C.F.Shih, A.E.Johansson, R.S.Withers, D.E.Oates, A.C.Anderson, P.Polakos, P.Mankiewich, E.de Obaldia, R.E.Miller, 1995, IEEE. Trans. M.T.T, **43**, 3020.
[2] RR Mansour, S.Ye, U.Dokas, B.Jolley, WC Tang, CM Kudsia, 2000, IEEE Trans. M.T.T, **48**, 1199.
[3] N. Sekiya, K. Yamamoto, S. Kakio, A.Saito, S. Ohshima, 2012, Pysics Procedia, **27**, 328.
[4] K. Yamanaka, M. Ishii, A. Akasegawa, T. Nakanishi, J. D.Banieecki, K. Kurihara, 2008, Pysica C **468**, 1950.
[5] N. Imai, N. Sekiya, S. Kakio, S. Ohshima, 2012, Pysics Procedia, **27**, 332.
[6] Y. Endo, S. Ono, M. Uno, T. Saito, A. Saito. K. Nakajima, S. Ohshima, 2011, IEEE.Trans. Applied Superconductivity **21**, 559.
[7] S. Ohshima, M. Endo, K. Takeda, K. Nakagawa, T. Honma, S. Sato, S. Takahashi, Y. Tanaka, A. Saito, 2012, Pysica C **36**, 429.
[8] S. Ohshima, 2000, Supercond. Sci. Technol. **103-108**, 61.
[9] S. Ohshima, M. Endo, K. Takeda, K. Nakagawa, T. Honma, S. Sato, S. Takahasi, Y. Tanaka, A.Saito, 2012, Pysics Procedia **36**, 429.
[10] S. Ohshima, M. Uno, Y. Endo, S. Tanaka, S. Ono, A. Saito, N. Sekiya, 2010, J. Phys. Conference Series **234**, 042025.
[11] N. Sekiya, Y. Nakagawa, S. Ohshima, 2010, Pysica C **470**, 1503.

Evaluation of superconducting pickup coils with high Q for 700 MHz NMR

Satoshi Oikawa[1], Yuto Tanaka[1], Takahiro Yamada[1], Akira Kanamaru[2], Masato Takahashi[3], Atsushi Saito[1] and Shigetoshi Ohshima[1]

[1]Graduate School of Science and Engineering, Yamagata University, 4-3-16 Jonan, Yonezawa, Yamagata 992-8510, Japan

[2]Graduate School of Life Science and Nano Systems, Yokohama City University, Kanagawa 1-7-29 Suehiro-cho, Tsurumi-ku, Yokohama, 230-0045, Japan

[3]RIKEN, 1-7-22 Suehiro-cho, Tsurumi-ku, Yokohama, Kanagawa 230-0045, Japan

E-mail : tme59702@yz.yamagata-u.ac.jp

Abstract. In this study, we propose a helmholtz-type coil using a high-temperature superconductor (HTS) for 700 MHz NMR pickup coils with a high quality factor (Q). We analyzed the coil microwave resonant properties of the helmholtz-type one turn coil made using an HTS ($DyBa_2Cu_3O_{7-\delta}$: DyBCO coil) and the same type of coil made from normal metal (Cu coil) through electromagnetic simulations. The Q of the DyBCO coil at 77 K and the Cu coil at 77 K were 3295 and 482 in the simulations. The one turn coils were fabricated using 200-nm-thick DyBCO thin film deposited on an r-sapphire substrate and 3-μm-thick Cu thin film on an r-sapphire substrate. The Q of the DyBCO coil was 2920 at 77 K, and that of the Cu coil was 423 at 77 K. These experimental results indicate that the DyBCO coils are more advantageous than the Cu coils for high-sensitivity NMR measurements.

1. Introduction

Nuclear Magnetic Resonance (NMR) spectroscopy is useful for studies in organic chemistry and biology, for example protein structure analysis and new drug development. However, NMR has its drawbacks: it has low sensitivity and requires a long time for conducting measurements. The signal to noise ratio (S/N) of NMR depends on the filling factor of the pickup coil system, the quality factor of the pickup coil, and the noise temperature of the pickup coil and pre-amplifier as shown in the following equation: $S/N \propto \sqrt{\eta Q/(T_c+T_a)}$. η is the filling factor of the pickup coil system, Q is the quality factor of the pickup coil, T_c is the noise temperature of the pickup coil and T_a is the noise temperature of the pre-amplifier. To increase the sensitivity of NMR spectroscopy it is necessary to increase Q. Thus, superconductors are a promising material for NMR pickup coils. The low surface resistance (R_s) can significantly improve the Q and contribute to the enhancement of the S/N. Recently, high sensitivity pickup coils for NMR have been investigated by using superconductors [1-4]. However, NMR with superconducting NMR pickup coils is still not widely utilized. This is because superconducting NMR pickup coils with a Q higher than 10,000 are rarely reported, tuning of higher Q pickup coils is difficult, and a high RF field homogeneity is difficult to obtain.

The final goal of this study is to demonstrate higher sensitivity in NMR spectrometers by using high-temperature superconductor (HTS) pickup coils. The immediate objective is to realize the advantages of HTS pickup coils for the enhancement of Q and establish high precision simulations. In this study, we investigated the Q of NMR pickup coils by applying HTS thin film (DyBCO) to the coils. HTS thin films have low R_s in a dc magnetic field [5, 6]. In this paper, the simulation, fabrication, and characterization of one turn HTS (DyBCO) coils are described. The DyBCO coil fabricated for this study was evaluated at 77 K, and it was compared with that of a normal metal coil (Cu coil).

2. Simulation

We analyzed the microwave resonant properties of the DyBCO and Cu helmholtz-type one turn coils through electromagnetic simulations using Sonnet-EM software. Fig.1 shows the simulation and three dimensional models of the one turn coils. For the substrate, we used r-sapphire (25 mm×25 mm) with a dielectric constant (ε_r) of 10.9 and a dielectric loss tangent of 10^{-7}. For the one turn coils, we used half wavelength resonators with a line width of 0.5 mm. We used the dielectric resonator method to obtain the R_s value of the DyBCO thin film for the initial simulation [5]. We adopted a contactless power feeding method by using a feeding loop antenna made of Cu. Matching was obtained by adjusting the distance between the one turn coils and the feeding loop antenna from 0.01 mm to 9.5 mm. Self resonant frequency of the one turn coils was set to 750-850 MHz. This is because the resonant frequency of the one turn coils shifts, when using a loaded sample tube and tuning with a dielectric plate. For the simulated results, we obtained an unloaded quality factor (Q_u) of 102 for the Cu coil at room temperature, 482 for Cu coil at 77 K and 3295 for the DyBCO coil at 77 K.

3. Experiment

In this study, two kinds of one turn coils were prepared using normal metal (Cu coil) and HTS (DyBCO coil). For the Cu coil, Cu thin film (3 μm) was deposited on the r-sapphire substrate by using a DC-sputtering system. Then, the film was patterned in a one turn shape by photolithography and wet etching by using aqueous ferric chloride. For the DyBCO coil, DyBCO (200 nm) thin film was deposited on the r-sapphire substrate by the thermal co-evaporation method (produced by THEVA GmbH). Then, the film was patterned in a one turn shape by photolithography and Ar ion-beam etching. Then, the fabricated one turn coils were mounted on a jig made of Cu. Fig. 2 shows a photograph of the fabricated DyBCO coil mounted on the jig. Under this setup, the frequency responses were measured by a network analyzer at 77 K. Matching for the 1 turn coils was achieved through adjusting the distance between the one turn coils and the feeding loop antenna. Based on the measurement, we estimated the Q_u of the one turn coils.

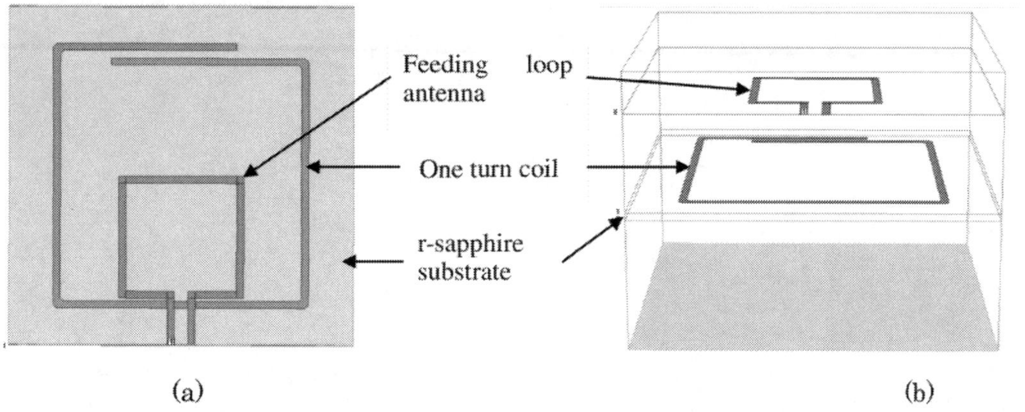

Fig.1. Simulation model for one turn coil. (a) Top view. (b) Three dimensional model.

4. Results and Discussion

4. 1. Actual value of the Q_u

In this study, one turn coils were prepared using normal metal (Cu coil) and HTS (DyBCO coil). Fig. 3 shows the frequency responses of the one turn coils for the simulation results and the experiment results. Solid lines represent the frequency characteristics obtained from the experiment and broken lines represent the simulated frequency characteristics for the one turn coils. In each resonance curve, matching for the one turn coils was achieved by adjusting the distance between a movable feeding loop antenna and the one turn coils. From the experiment, we obtained a quality factor (Q_u) of 102 for the Cu coil at room temperature, 423 for the Cu coil at 77 K and 2920 for the DyBCO coil at 77 K. The actual and simulated self resonant frequency of the one turn coils displayed approximately the same value; however, the actual Q_u was lower than the simulated Q_u. The cause of this could be that the R_s of HTS and Cu thin films deteriorated, which resulted from the thin films being patterned in a one turn shape. The actual Q_u of the DyBCO coil was 6.9 times higher than that of the Cu coil. This result indicates that DyBCO coil is more advantageous than Cu coil for high-sensitivity NMR measurements. However, since the Q_u of DyBCO coil is low, it is difficult to utilize for NMR. Q_u is defined using following equation: $1/Q_u = 1/Q_r + 1/Q_d + 1/Q_c$. Q_r (radiative Q) is the Q resulting from radiation loss, Q_d (dielectric Q) is the Q resulting from dielectric loss by using a substrate and Q_c (conductive Q) is the Q resulting from conductor loss. DyBCO coils fabricated by using r-sapphire substrates have low dielectric loss and DyBCO thin films have low conductor loss. Therefore, Q_r is thought to be a dominant factor for determining Q_u.

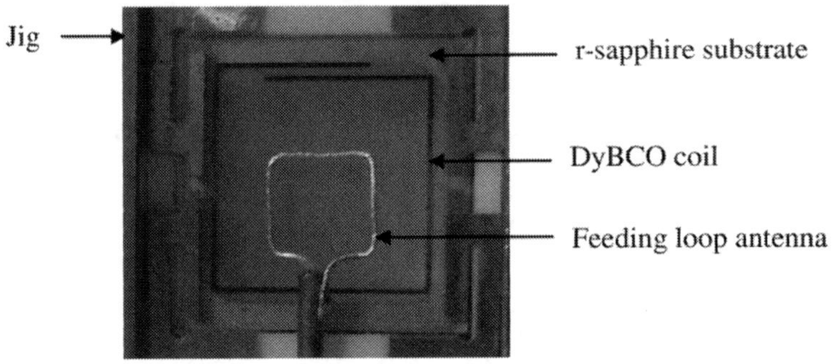

Fig.2. Photograph of fabricated DyBCO coil mounted on the jig.

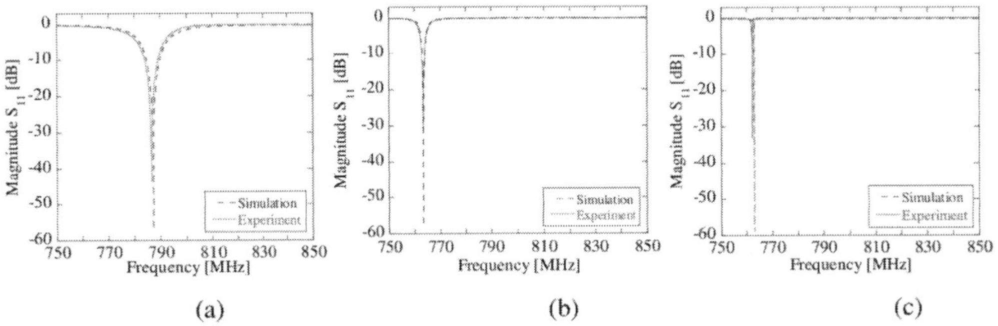

Fig.3. Frequency characteristics of one turn coils for simulated results and actual results. (a) Cu coil evaluated at room temperature. (b) Cu coil evaluated at 77 K. (c) DyBCO coil evaluated at 77 K.

4. 2. Evaluation of R_s of patterned thin films

The simulated R_s value was altered in order to obtain the same Q_u as observed in the experiment. Consequently, the R_s of the Cu thin film at resonant frequency and 77 K increased from 1.9 mΩ to 2.3 mΩ. Similarly, the R_s of the DyBCO thin film at resonant frequency and 77 K increased from 2.5 μΩ to 37 μΩ. Additionally, we established a high precision simulation for measuring the R_s of patterned thin films.

5. Conclusions

In this study, one turn coils for 700 MHz NMR were designed and fabricated using DyBCO and Cu thin films and their Q_u were evaluated at 77 K. In the actual experiment, the Q_u of the DyBCO coil at 77 K was 2920, and the Q_u of the Cu coil at 77 K was 423. Therefore, the actual Q_u of DyBCO coil was 6.9 times higher than that of Cu coil. This experimental result indicates that DyBCO coil is more advantageous than Cu coil for high-sensitivity NMR measurements. For the simulation, the simulated R_s value was altered so as to obtain the same value for Q_u as in the experiment. Consequently, the R_s of the Cu thin film at resonant frequency and 77 K increased from 1.9 mΩ to 2.3 mΩ. Similarly, the R_s of the DyBCO thin film at resonant frequency and 77 K increased from 2.5 μΩ to 37 μΩ. Additionally, we established a high precision simulation for measuring the R_s of patterned thin films.

Acknowledgments

This work was supported by the Strategic Promotion of Innovative Research and Development of JST (Development of the next-generation NMR technology using HTS materials). A part of this work was carried out in the clean room of Yamagata University Faculty of Engineering.

References

[1] Hill H D W 1997 IEEE Trans. Appl. Supercond. **7** 3750

[2] Brey W W, Edison A S, Nast R E, Rocca J R, Saha S, and Withers R S 2006 J. Magn. Reson. **179** 290

[3] Hiroyuki Yamamoto, Kazuo Saitoh, Michiya and Hitoshi Kitaguti 2008 Journal of Physics **97** 012137

[4] H. Yamamoto, A Tsukamoto, H. Hasegawa, K. Saitoh, Y. Fukuda and M. Okada 2007 Physica C **463-465** 948-951

[5] S. Nakayama, T. Honma, K. Nakagawa, A. Saito and S. Ohshima 2010 Physica C **470** 1042-1044

[6] T. Honma, S. Sato, M. Watanabe, K. Koike, H. Kato and S. Ohshima 2013 Physica C **484** 46-48

An improved superconducting neural circuit and its application for a neural network solving a combinatorial optimization problem

T Onomi and K Nakajima

Research Institute of Electrical Communication, Tohoku University,
2-1-1 Katahira, Aoba-ku, Sendai, Miyagi, 980-8577, Japan

E-mail: onomi@riec.tohoku.ac.jp

Abstract. We have proposed a superconducting Hopfield-type neural network for solving the N-Queens problem which is one of combinatorial optimization problems. The sigmoid-shape function of a neuron output is represented by the output of coupled SQUIDs gate consisting of a single-junction and a double-junction SQUIDs. One of the important factors for an improvement of the network performance is an improvement of a threshold characteristic of a neuron circuit. In this paper, we report an improved design of coupled SQUID gates for a superconducting neural network. A step-like function with a steep threshold at a rising edge is desirable for a neuron circuit to solve a combinatorial optimization problem. A neuron circuit is composed of two coupled SQUIDs gates with a cascade connection in order to obtain such characteristics. The designed neuron circuit is fabricated by a 2.5 kA/cm^2 Nb/AlOx/Nb process. The operation of a fabricated neuron circuit is experimentally demonstrated. Moreover, we discuss about the performance of the neural network using the improved neuron circuits and delayed negative self-connections.

1. Introduction

We have proposed a superconducting Hopfield-type[1] neural network for solving the N-Queens problem which is one of combinatorial optimization problems[2]. The N-Queens problem is a famous combinatorial optimization problem. Its solution involves finding a placement of N queens on an N x N chess board such that none of the queens are able to capture any other using standard chess moves for a queen. Figure 1 shows the circuit diagram of a neural network for solving the 4-Queens problem; it consists of 16 neurons representing a 4 x 4 chess board. The neuron circuits are composed of coupled SQUIDs(c-SQUIDs) gates[3] with sigmoid-shape functions. The output of a neuron, representing a queen, is connected to the inputs of other neurons restricted by the queen. These synaptic connections are represented by electrical resistances. In Fig. 1, the output of neuron(1,1) is connected to inputs of neurons placed to vertical, horizontal, and diagonal directions by cascade connections of inductances and resistances. Figure 2 shows an example of a dynamic simulation for solving the 4-Queens problem. The neuron outputs have negative polarity, which represents the constraint signals. We can confirm that the state shown in Fig. 2 fall into a correct pattern.

Unfortunately, not all network states can fall into correct patterns because of the local minima of the energy function of a neural network. We are interested in some characteristics of networks using analog Josephson devices; For example, network fluctuations caused by intrinsic Josephson voltage oscillations, network dynamics when a threshold of a neuron device has a hysteretic function, and so on. These influences can be changed by changing device parameters, and the network can be effectively operated to avoid local minima of an energy function. One of the important factors for an improvement of the network performance is an improvement of a threshold characteristic of a neuron circuit. In this paper, we report an improved design of c-SQUIDs gates for a superconducting neural network. Moreover, we discuss about the performance of the neural network using the improved neuron circuits and delayed negative self-connections.

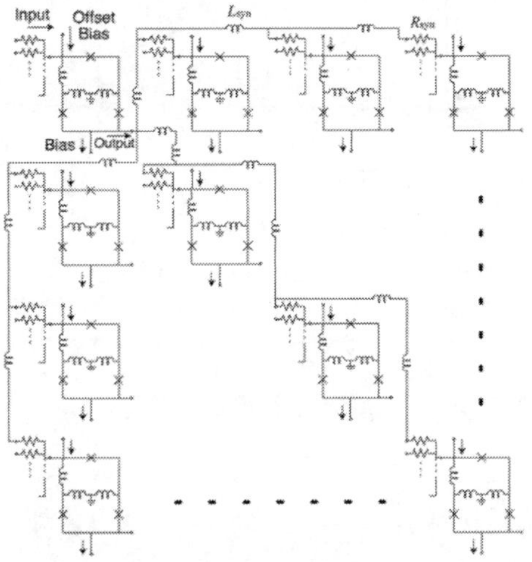

Figure 1. Circuit diagram of a neural network for solving the 4-Queens problem[1]. Synaptic connections from neuron(1,1) only are shown in this figure.

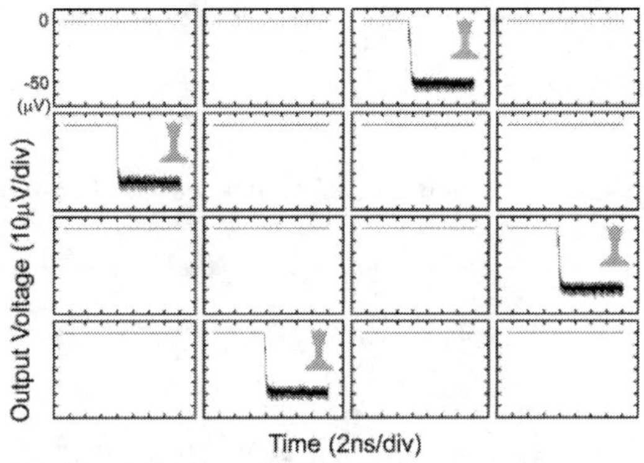

Figure 2. Dynamic simulation of a neural network for solving the 4-Queens problem. The polarity of the output voltages is negative to represent the constraining signals of neurons.

2. Improvement of the threshold characteristic of a neuron circuit

2.1. Design of an improved neuron circuit
The input-output characteristic of a conventional c-SQUIDs gate has a round shape at a vicinity of the rising edge of the threshold. This round shape causes a decrease of an output voltage, namely, a constraint signal to other neurons. A step-like function with a steep threshold at a rising edge is desirable for our neural network, because neurons generating intermediate states except 0 or 1 cause the network to tend to be trapped in local minima on the energy function. We propose two-stage c-SQUIDs with a cascade connection for obtaining such an activation function. Figure 3 shows the two-stage c-SQUIDs gate which is the improved neuron circuit. The first stage c-SQUIDs gate has the same parameters of the original one[2] therefore has the round characteristic(Output1) at the rising edge of the threshold as shown in Fig. 3(b). To obtain a flat output characteristic, LI_c product of rf-SQUID(J4,L4-6) in the second stage is larger than that of the first stage. Although a hysteretic characteristic comes to appear on the single input-output characteristic because of the increase of LI_c product, the output of the second stage(Output2) is not hysteretic because a discrete input current from Output1 passes through the hysteretic region on the second stage.

2.2. Fabrication and measurement of an improved neuron circuit
A neuron circuit consisting of two-stage c-SQUIDs was fabricated using the AIST standard process(STP2) in order to confirm the characteristic. Figure 4 shows a measurement result of a fabricated circuit. Figures 4 (a) and (b) denote the first stage and the second stage output voltages, respectively. An offset current I_{offset} was injected from the offset bias point shown in Fig. 3(a). The output characteristic of the first stage c-SQUIDs shown in Fig. 4(a) has the similar round shape to

conventional c-SQUIDs. On the other hand, that of the second stage has a step-like function with a steep threshold at a rising edge as shown in Fig. 3(b).

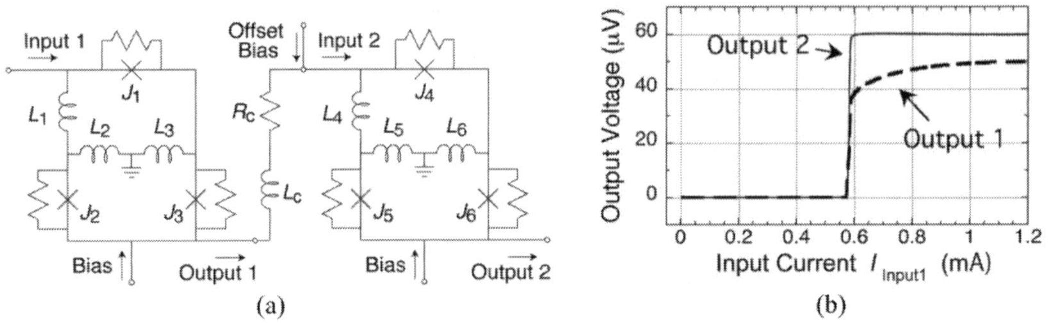

Figure 3. Improved design of a neuron circuit. (a) Circuit diagram. (b) Numerical simulation. The device parameters are designed as $I_{c1} = I_{c4} = 0.188$ mA, $I_{c2} = I_{c3} = I_{c5} = I_{c6} = 1.50$ mA, $L_1 = 1.5$ pH, $L_2 = L_3 = 0.75$ pH, $L_4 = L_5 = L_6 = 1.35$ pH, $L_c = 10.0$ pH, $R_c = 0.10\ \Omega$, $I_{\text{offset}} = 0.32$ mA, $I_{\text{bias}} = 2.85$ mA.

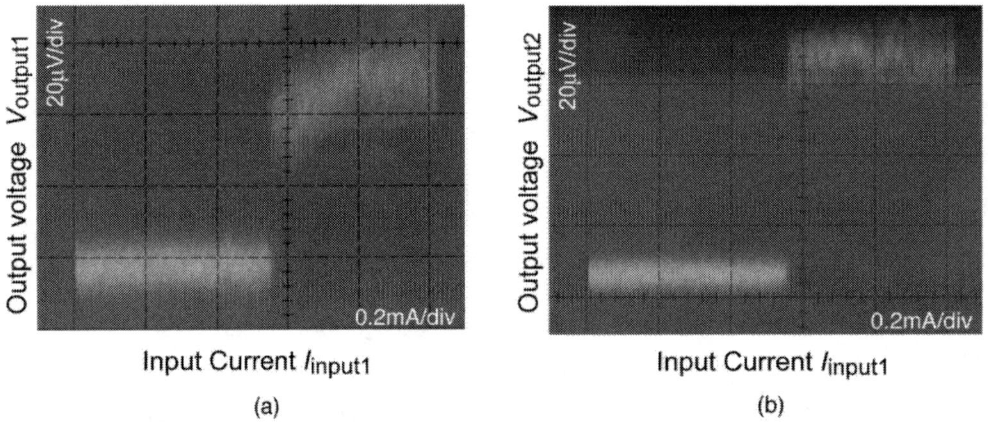

Figure 4. Measurement result of a fabricated neuron circuit. (a) Output voltage V_{output1} of 1st stage c-SQUIDs. (b) Output voltage V_{output2} of 2nd stage c-SQUIDs.

3. Numerical estimation of the network performance

To improve the network performance, we introduce the improved neuron circuits described in the previous section and delayed negative self-connections. The self-connection consists of a feedback loop of an inductance (L = 500 pH) and a synaptic weight (R = 1.0 Ω). The self-connection is added to each of 16 neuron circuits. The delayed negative self-connections cause oscillations when the neuron output is close to the threshold value. Figure 5 shows an example of a dynamic simulation for solving the 4-Queens problem. We can confirm that an incorrect pattern changes into a correct one.

We investigate the network performance using original c-SQUIDs gates and that using the proposed method by numerical simulations. A performance of one network is evaluated using 1000 individual simulation results including unavoidable fluctuations, i.e. thermal noise and circuit parameter variations[2]. Figure 6 shows the dependence of the correct pattern ratio of the 4-Queens problem on the MaCumber parameter β_c ($= 2\pi I_c C R^2/\Phi_0$). β_c is changed by adjusting the two shunt resistors in the double-junction SQUID. Changing β_c causes changing the fluctuation of the network due to Josephson oscillation[2]. Open squares denote simulation results of the network using original c-SQUIDs gates. Black circles denote those using two-stage c-SQUIDs gates. We assume that the networks start to operate from two initial conditions[2] when the offset biases were decreased from an over-biased state (I), and when those were increased from zero (II), respectively. An increase of the ratios of achieving a correct answer is confirmed from Fig. 6. Increasing a fluctuation of the network by increasing β_c induces an increase in escape probability from local minima. So, these results suggest that flat output characteristics generated by two-stage c-SQUIDs gates contribute an increase in escape probability from local minima when a fluctuation arising from Josephson oscillation is relatively large. We confirm that the improved design of neuron circuits and delayed negative self-connections contribute better network performance.

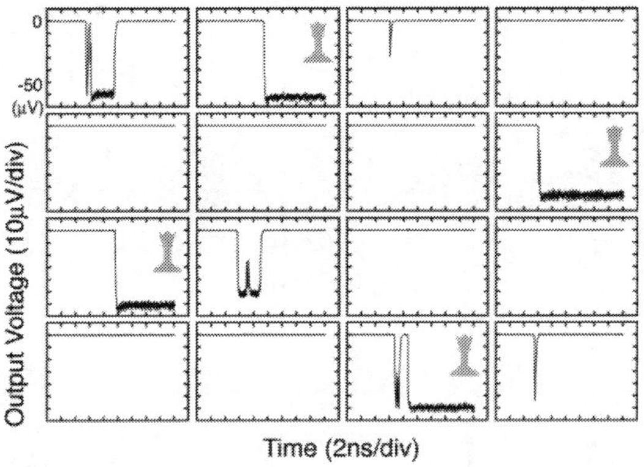

Figure 5. Dynamic simulation of a neural network two-stage c-SQUID gates and delayed negative self-connections for solving the 4-Queens problem.

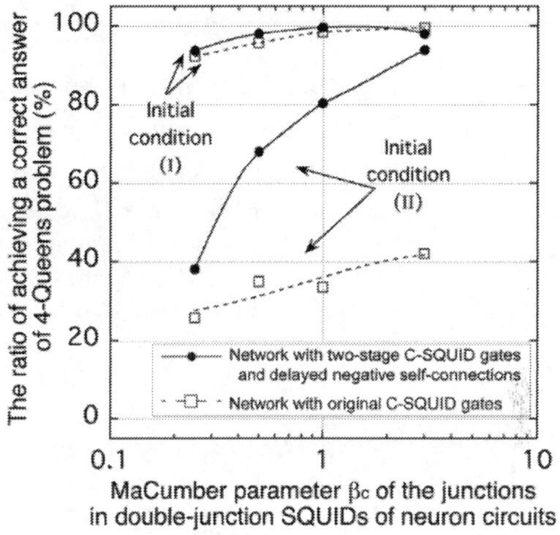

Figure 6. MaCumber parameter dependence of the correct pattern ratio of the 4-Queens problem. (I) and (II) denote the two cases where network offset biases were decreased from an over biased state and increased from zero, respectively.

4. Conclusion

We proposed an improved design of coupled SQUIDs gates for a superconducting neural network. A neuron circuit was composed of cascade connected two coupled SQUIDs gates to obtain the flat output characteristics. The designed neuron circuit was fabricated by AIST standard process(STP2). The operation of the neuron circuit was experimentally confirmed. The improved neuron circuits and delayed negative self-connections were introduced to the network for solving 4-Queens problem in order to improve the network performance. An increase of the ratios of achieving a correct answer was confirmed by numerical simulations of the improved network.

Acknowledgments

This research was supported by JSPS KAKENHI Grant Number 24500271. The circuits were fabricated in the clean room for analog-digital superconductivity (CRAVITY) of National Institute of Advanced Industrial Science and Technology (AIST) with the standard process 2 (STP2).

References

[1] Hopfield J J and Tank D W 1985 *Biol. Cybern.* **52** 141
[2] Onomi T, Maenami M and Nakajima K 2011 *IEEE Trans. Appl. Superconduct.* **21** 701
[3] Mizugaki Y, Nakajima K, Sawada Y and Yamashita T 1994 *IEEE Trans. Appl. Superconduct.* **4** 1

Investigation of Magnetic Interference Induced via Gradient Field Coils for Ultra-Low-Field MRI Systems

D Oyama[1], J Hatta[2], M Miyamoto[1], Y Adachi[1], M Higuchi[1], J Kawai[1], J Fujihira[3], N Tsuyuguchi[4], and G Uehara[1]

[1] Applied Electronics Laboratory, Kanazawa Institute of Technology, 3 Amaike, Kanazawa, Ishikawa, 920-1331, Japan
[2] Department of Environmental and Life Sciences, Toyohashi University of Technology, 1-1 Hibarigaoka, Tenpaku-cho, Toyohashi, Aichi, 441-8580, Japan
[3] FEDLIC Co. Ltd., 2-10-14, Ninomiya, Tsukuba, Ibaraki, 305-0051, Japan
[4] Department of Neurosurgery , Osaka City University Graduate School of Medecine, 1-4-3 Asahimachi, Abenoku, Osaka, 545-8586, Japan

E-mail: oyama@neptune.kanazawa-it.ac.jp

Abstract. We are developing a compact ultra-low-field MRI system that is composed of a SQUID gradiometer and a coil set that generates magnetic fields for capturing MR images. The magnetic interference induced from a power amplifier potentially disturbs MRI measurements. We investigated the path of the interference by experimental measurements and calculation of the magnetic field generated by the coil set. We found that the magnetic field generated from a particular gradient coil affected the SQUID gradiometer and that the level of the interference was strongly dependent on the shape of the gradient coils. When the coils' shapes are designed, minimizing the noise introduced from the power amplifier is crucial, in addition to consideration of the homogeneities of the magnetic field.

1. Introduction

Ultra-low-field nuclear magnetic resonance (NMR) and magnetic resonance imaging (MRI) have been widely studied in the past 10 years [1-2]. Our group is now developing a compact ultra-low-field MRI system that can be integrated with a magnetoencephalography (MEG) system for small animals based on SQUID techniques [3-4]. This integration will greatly benefit the development of new medicines and pathological studies because it enables concurrent capturing of the biomagnetic signals and anatomical images of small animal.

The ultra-low-field MRI system was equipped with a set of coils to generate magnetic fields oriented in the direction of the *y*-axis. The coil set was composed of coils for so-called the measurement field (B_m) and coils to produce gradient fields along the *x*-, *y*-, and *z*-axes (G_x, G_y, and G_z, respectively). The electric noise induced from a power amplifier via such coils sometimes affects the magnetic sensors and eventually disturbs MRI measurements even if the power amplifiers have been carefully designed to have low electric noise. We found that the magnetic noise induced from the G_y coil was significantly greater than the noise induced from the other coils.

Therefore, we investigated the magnetic interference induced via the coils through the experimental measurement and numerical calculation of the magnetic field distribution.

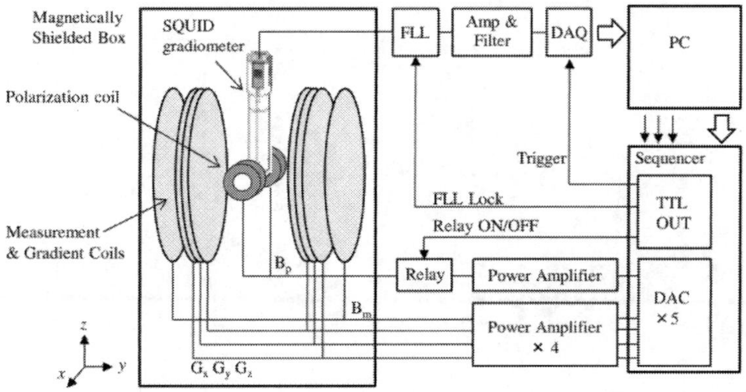

Figure 1. Block diagram of the ultra-low-field MRI System

2. Configuration of the System

Figure 1 shows a block diagram of our ultra-low field MRI system. The coil set was fabricated to be desktop-sized for installation inside a magnetically shielded box.

The four pairs of shielded planar coils for the measurement field and the 3D gradient fields were optimized through a target field method and sufficient homogeneity for ultra-low-field MRI measurements was obtained [3]. In this paper, the noise induced via the polarization coil (B_p coil) was not considered because the B_p coil was disconnected from the power amplifier using a mechanical relay during the acquisition of the MR signal.

A SQUID sensor with a first-order axial gradiometer pick up coil was employed to detect the MR signal. The pickup coil was perpendicular to the *z*-axis in Figure 1, and its diameter was 15 mm. The baseline of the gradiometer was set along the *z*-axis and its length was 50 mm. The noise level of the SQUID gradiometer was 5.3 fT/Hz$^{1/2}$.

3. Magnetic Field Measurements

Experimental magnetic field measurements were performed in order to examine the path of the magnetic noise, which we discuss in this study.

First, we measured the free-induction decay (FID) signal from 100 ml of tap water; this signal is shown in Figure 2. The intensity of the measurement field was 33 μT; the corresponding Larmor frequency of protons was 1.4 kHz. The FID signal was clearly observed when only the G_x and G_z coils were connected to the current driver unit, whereas the noise level of the measured waveform considerably increased to 520 fT/Hz$^{1/2}$ and the FID signal could not be seen when the G_y coil was connected to the power amplifier. This result suggests that the G_y coil was the main noise introduction path from the power amplifiers.

Second, a sinusoidal electric current was individually applied to each coil from the power amplifier which was connected to a function generator. The frequency and the amplitude of the electric current were 5 Hz and 10 mA, respectively. A fluxgate magnetometer, instead of the SQUID gradiometer, was employed to measure the magnetic signal, in order to clarify magnetic field distribution.

Figure 3 shows the results of the measurements. The horizontal axis shows the position of the fluxgate magnetometer along the *z*-axis. Here, $z = 0$ was defined to be equal to the center of the planar coils. The measured range corresponded to the position of the SQUID gradiometer during MRI measurement. The vertical axis shows the *z*-component of the measured magnetic signals (B_z). The intensity of the measured signal generated by the G_y coil was 25 to 400 times larger than those generated by the other coils. This means that the SQUID sensor's sensitivity to the magnetic field induced from the G_y coil was 25 to 400 times greater than its sensitivity to the magnetic fields induced from the other coils.

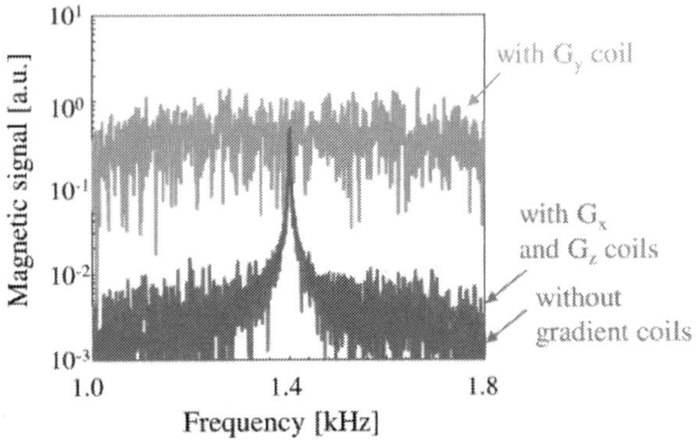

Figure 2. The FID signal from 100 ml of tap water in the measurement field of 33 μT. The noise level of the measured waveform increased when the G_y coil was connected to the power amplifier.

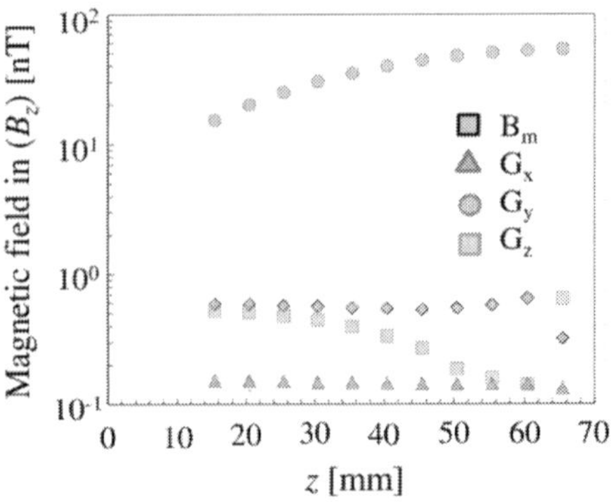

Figure 3. Intensity of B_z generated from the coil set.

4. Analysis of Magnetic Field Distribution

In order to investigate why the G_y coil induced the most magnetic noise, the distributions of the magnetic fields generated by each coil were calculated around the field-of-view (FOV) area and the SQUID sensor. The coils were assumed to be composed of small segments, and the magnetic field generated by each segment was calculated using the Biot-Savart equation. Thus, the magnetic fields generated by the coils were sums over the calculated magnetic fields corresponding to the coil segments.

The directions of the calculated magnetic fields are indicated by the arrows in Figure 4. The arrows have a fixed length in this figure and hence do not represent the field magnitudes. The circles drawn with dashed lines indicate the FOV area; its diameter was 40 mm. The position of the pickup coil of the SQUID gradiometer for the MRI measurements is also shown in the Figure 4 diagrams.

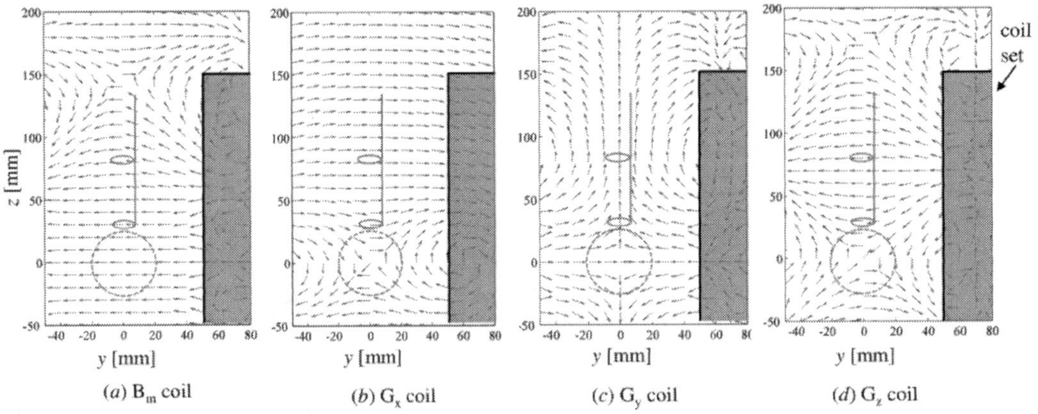

Figure 4. Directions of the calculated magnetic field corresponding to each coil: (*a*) B_m coil, (*b*) G_x coil, (*c*) G_y coil, and (*d*) G_z coil. The length of each arrow is fixed and do not represent the intensity of the magnetic field.

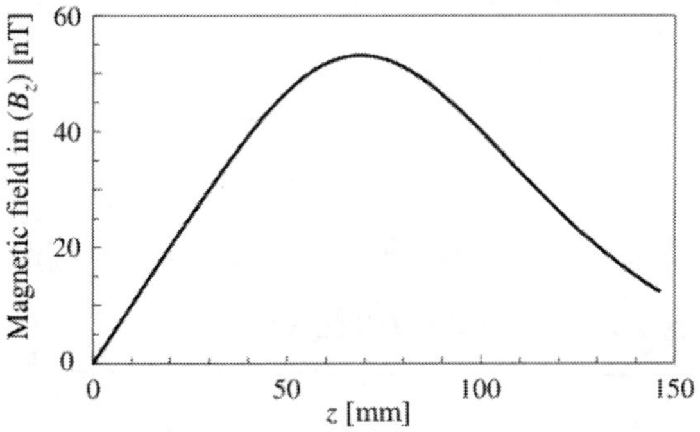

Figure 5. Calculated intensity of B_z generated from the G_y coil. The amplitude of assumed current to be applied was 10 mA.

Only the magnetic field generated by the G_y coil pointed toward the *z* direction around the SQUID gradiometer. Therefore, the intensity of the magnetic field generated by the G_y coil in *z* direction was much larger than that generated by the other coils.

Figure 5 shows the calculated intensity of B_z generated by the Gy coil along the *z*-axis. The B_z generated from the G_y coil had a gradient parallel to the baseline of the gradiometer. This calculation agreed well with the result of the magnetic measurement shown in Figure 3. The gradient of B_z functioned to deteriorate the gradiometric cancellation effect of the gradiometer.

5. Discussion

It is possible to deduce from the section 3 and 4 that the level of magnetic interference to the SQUID sensor strongly depends on the shape of the coils. In our system, large magnetic noise induction resulted from the shape of the G_y coil, because the magnetic field generated by the G_y coil was dominated over B_z and had a gradient along the *z*-axis.

Previously, only the direction parallel to the measurement field has been considered in designing coil patterns. In addition to the consideration of the homogeneities of the magnetic field, minimizing the noise introduced from the power amplifier to SQUID sensors is also crucial.

With regard to the SQUID sensor design, the second-order axial gradiometer (d^2Bz/dz^2) or the planar gradiometer (dB_z/dx of dB_z/dy) might be useful to reduce the magnetic interference induced via the G_y coil, which we plan to introduce in our ultra-low-field MRI system.

6. Conclusion

The level of magnetic interference strongly depends on the shape of the gradient coils. Reducing the magnetic interference induced via gradient coils requires optimization of coils' shape or appropriate design of the SQUID gradiometer.

References

[1] Espy M, Matrashov A, and Volegov P 2013 SQUID-detected ultra-low field MRI *Journal of Magnetic Resonance* **229** 127–141

[2] Hartwig S, Albrecht H H, Scheer H J, Burghoff M, and Trahms L 2013 A Superconducting Quantum Interference Devices Measurement System for Ultra Low-Field Nuclear *Magnetic Resonance Applied Magnetic Resonance* **44** 9–22

[3] Hatta J, Miyamoto M, Adachi Y, Kawai J, and Uehara G 2011 SQUID-based low field MRI system for small animals *IEEE Transactions on Applied Supercondcuctivity* **21** 526–529

[4] Oyama D, Hatta J, Miyamoto M, Adachi Y, Kawai J, Higuchi M, and Uehara G 2013 Development of SQUID-Based Compact Low-Field MRI Systems *IEEE Transactions on Applied Supercondcuctivity* **23** 1601604

Acknowledgments

This research was supported by SECOM Science and Technology Foundation.

Large Area A-thermal Phonon TES Detector Mediated by the quasi-particle Diffusion Signal for Space Application

G. Pizzigoni[1,2], D. Bagliani[1,2], M. Biasotti[1,2], D. Corsini[1], M. De Gerone[2], F. Gatti[1,2], C. Macculi[3], L. Piro[3]

[1]University of Genova, Dept of Physics, Via Dodecaneso 33, 16146 Genova.
[2]INFN of Genova, Via Dodecaneso 33, 16146 Genova.
[3]INAF-IASP Roma, Via Fosso del Cavaliere 100, 00133 Roma.

E-mail: giulio.pizzigoni@ge.infn.it

Abstract. Low temperature detectors operated at about 0.1K have achieved excellent spectral performances in the soft X-rays, becoming appealing for new challenging measurements with space missions in Astrophysics. In order to exploit their full sensitivity, it is necessary to minimize the background signals generated by the cosmic rays, i.e., high energy protons and light nuclei, that leave sizable amounts of energy in the same spectral window of the astrophysics signals. Detectors for GeV protons and nuclei operating few millimeters from the X-ray detector at 0.1K can act as anti-coincidence to disentangle the fake signal of cosmics. Fast and large detectors are designed and fabricated. These operate by mixing the fast a-thermal phonon signal with the slow diffusive thermal ones. A greater uniformity in the response should be obtained using large shaped superconducting aluminium films that acts as phonon collectors: the quasi-particles created by high energy phonons diffuse along the film toward a small Ir TES sensor giving out to a fast rise time. Here we present the measurement of an operating prototype of a superconducting anticoincidence detector for the proposed space mission ATHENA+.

1. Introduction

To the present state-of-the-art, superconducting Transition Edge Sensors (TES) allow very accurate measurements of X-rays from several kind of sources. The proposed next x-ray observatory, ATHENA+, includes a high resolution X-ray micro-calorimeter spectrometer, based on the TES-technology [1], with a working temperature around 50mK. In order to achieve the science requirements it is necessary to reduce the cosmic background. High energy protons passing through the microcalorimeters release an energy comparable to the one of an X-ray. The rejection of these fake events is possible with a second detector, the so-called Cryogenic Anti-Coincidence (Cryo-AC), just few millimeters beneath the former; in this way the cosmic ray hits both detectors and a signature to reject this event is produced. It has been demonstrated that the Cryo-AC can reduce the background up to a factor of 20 [2]. M. Loidl et al investigated the properties of quasi-particles diffusion over several mm in cryogenic detectors finding good performances using Ir [3], we therefore decided to use this element, as its critical temperature is near that of the main micro-calorimeter.

2. A-thermal phonon physics

The Cryo-AC must fulfil two main requirements: fast response, for the prompt time coincidence of false events, and energy linearity to fix the threshold and optimize the efficiency. In order to get fast signals with the Cryo-AC we need to operate out of the equilibrium, then it is necessary to absorb the high energy phonons before they thermalize. In the most general case, the primary radiation will produce a highly ionized track in the detector, actually a silicon chip. The process of the primary energy thermalization goes as described in the following [4]. In silicon the primary energy produces energetic electron-hole pairs and phonons. Phonons anharmonicaly decay in lower energy pairs of acoustic ones. These propagate balistically reaching the surface of the absorbing crystal being the life time of few microseconds [5]. If an event in the absorber occurs near a TES we have a fast rise time pulse due to the shorter travelling path of the phonons. The low energy diffusive phonon gives out to the slow thermal signal in the whole crystal. The Cryo-AC returns to the equilibrium temperature via the thermal link to the heat sink. Due to the radiative transport of the first burst of fast phonons it is necessary to increase the collecting area as much as possible, without increase of heat capacity. If we use a low specific heat superconducting film coupled with the TES, the phonons break the Cooper pairs into quasi-particles that diffuse toward the lower gap TES film. Thus the critical parameter is the whole collection efficiency which includes also the quasiparticle transport to the TES.

3. First detectors

Several samples of Cryo-AC have been produced and tested, which were based on Silicon absorbers and sensed by Ir TES [6][7] (Figure 1). Further, to enhance the a-thermal phonon collection, aluminium-pads were grown for increasing the detector efficiency. Roughly, the key element in the design and optimization of the Cryo-AC is the understanding of the physics processes underlying the quasi-particle and the a-thermal phonons phenomenology at short time. Figure 1 shows some preliminary results of the samples S1, S3 and S4. The so called Cryo-AC-S1 was built with a large Ir TES (with a surface of 16.5 mm^2) covering 22% of the silicon chip absorber. It showed a fast rise time of about 1 μs. The pulse shape is well fitted with a model that mix 10% of fast a-thermal response and 90% slow thermal pulse. The main disadvantage of this sample was the insufficient spectral linearity. The following others Cryo-AC-S3,S4 have similar full collecting area but made partially with 4 smaller Ir TESs (area 6 mm^2) and large Al films (area 20 mm^2). This design allows lower heat capacity from TES while keeping the same coverage. The aluminum fingers act as phonon collectors, giving rise to quasiparticles that diffuse to the TES, in order to recover additional energy. In Cryo-AC-S3 the pulse shapes split in two sets: fast energetic pulses and slower weak pulses. This behavior of the response is not well understood. Our hypothesis is that a considerable amount of quasiparticles was trapped and condensed to Cooper pairs before entering the TES. Thus fast pulses are associated to direct TES excitation, while slow pulses to indirect - via quasiparticle re-condensation [6][7]. In order to investigate this phenomena we decided to study a new geometry: 65 TES parallel connected with and without a tree shaped Al collectors; in the first configuration the total collecting area is 50%. In both chips we are interested in the global signal coming from all the sensors. In order to measure the quasi-particle diffusivity of the collectors we built, in the same fabrication process, a one millimeter long Al-strip readout by one TES at one side.

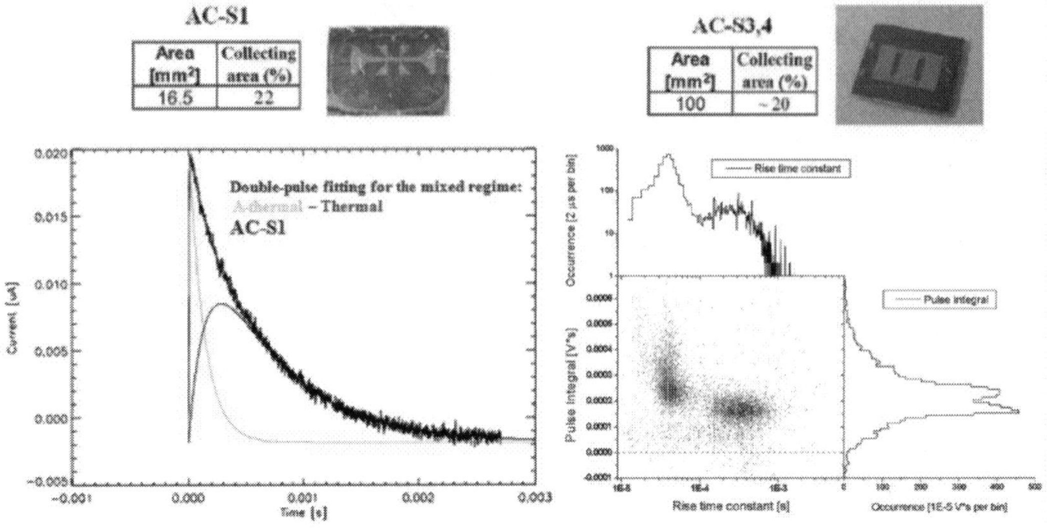

Figure 1. sketch of the preliminary results on AC-S1 and AC-S3,4 analysis; AC-S1 pulse shape with 1 µs rise time constant (left) and pulse integral vs PH analysis for AC-S3 (right)

Figure 2. (left) AC-S6 detector and (right) a detail of one of the 65 Ir TES surrounded by his own Al collecting tree fins, up and down, and the wiring line.

4. Cryo-AC-S6, analysis and results

This device has the following features: on a silicon chip of 1 cm^2, 300 µm thick, we have grown 65 Ir TES (100x100 μm^2) and 2 tree-shaped Al phonon collectors for each TES. The TES area coverage was 0.65%, while the Al collector coverage was about 50%. The AC-S6 has been tested in a dilution fridge with 60 keV gamma ray from ^{241}Am. The RvsT plot shows a transition width 10%-90% of 5 mK at 114 mK. The working point was 18 mOhm with a bias voltage of 0.94 µV. The pulses corresponding to 60 keV absorption were digitized and analyzed off line. Rise times and integrals of pulses are shown in figure 3. The pulse integral histogram, which is proportional to the collected energy spectrum, shows a peak nearby zero, that is filled by very small pulses at the threshold. At higher energies there is a structure with two peaks with slight different rise times. The high energy peak can be assigned to 60 keV absorption in the whole volume, being the gamma ray attenuation length about 1.4 cm in silicon. The 10-20 keV lines of ^{237}Np should stay in the lowest peak if we assume linear the energy scale. The intermediate peak is due to events with shorter rise time that should be caused by hits in the TES neighboring volumes.

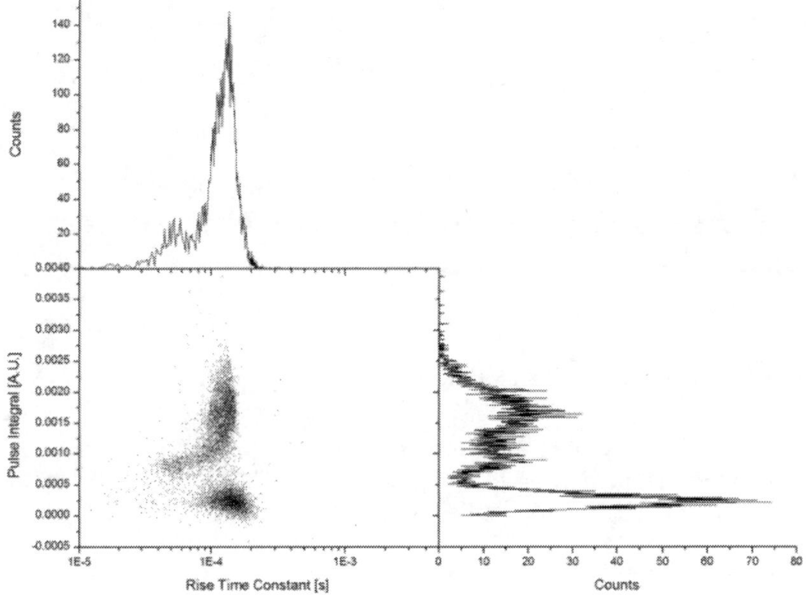

Figure 3. AC-S6 pulse integral vs pulse height analysis

5. Conclusions

With AC-S6 we fabricated a device with a larger collecting area then AC-S3, 50% in respect to the 20%. In this way we uniformed the rise time, around 100μs, obtaining only one family of pulses, unlike AC-S3. To better understand the behaviour of our new devices, in order to achieve the goal to lower the rise time down to 30μs, two other detectors have been fabricated: a version of AC-S6 without collector fins and a device to measure the quasiparticle diffusion in aluminium. These last two are currently under study.

References

[1] "The X-ray Integral Field Unit (X-IFU) for Athena+", D. Barret, J. W. den Herder, L. Piro, 2013

[2] "The Cryogenic Anticoincidence Detector for ATHENA+", C. Macculi et al., 2012 *J Low Temp Phys* template

[3] "Quasiparticle diffusion over several mm in cryogenic detectors", M. Loidl et al., 2001 *Nuclear Instruments and Methods in Physics Research A* **465**: 440-446

[4] "Physics of electrons and phonons in low-temperature detectors", M. Chapellier, 2004 *Nuclear Instruments and Methods in Physics Research A* **520**: 1-3, 21-26

[5] "Model for Cryogenic Particle Detectors with Superconducting Phase Transition Thermometers", F. Probst et al., 1995 *J Low Temp Phys* **100**: 69-104

[6] "The Cryogenic Anticoincidence Detector for ATHENA-XMS", C. Macculi et al.,2012 *J Low Temp Phys* **167**: 783-794

[7] "The TES-based Cryogenic Anticoincidence detector for IXO: First results from large area prototypes", C. Macculi et al.

Measurement of quality factor and losses in superconducting microwave resonator integrated with NbN/AlN/NbN qubit circuit

W Qiu[1,*], K Makise[1], H Terai[1], Y Nakamura[2], Z Wang[1,3]

[1]Advanced ICT Research Institute, National Institute of Information and Communications Technology (NiCT), Iwaoka-cho, Nishi-ku, Kobe 651-2492, Japan
[2]Research Center for Advanced Science and Technology (RCAST), The University of Tokyo, Meguro-ku, Tokyo 153-8904, Japan
[3]Shanghai Institute of Microsystem and Information Technology, Chinese Academy of Sciences, 865 Changning Road, Shanghai 200050, China

[*]E-mail: qiuwei@nict.go.jp

Abstract. Dielectric loss from two-level systems (TLSs) formed by local defects have shown a significant impact on the qubit coherence time. These defects can originate in the insulation material for superconducting wires isolation or in the Josephson junction tunnel barrier material. Due to the complexity of the qubit circuit fabrication process, identifying the contribution from each decoherence source is challenging. In an effort to address this issue, we have developed superconducting qubit that consists of full epitaxially-grown NbN/AlN/NbN Josephson junctions in NbN coplanar waveguide (CPW) resonator circuit. The dielectric loss introduced from TLFs in tunnel junction barrier has been largely reduced due to the unique epitaxial feature of the tunnel junction. The quality factor Q_i of the CPW resonator was measured and the dielectric loss tanδ is 3×10^{-4}. The relaxation time inferred from the measured resonator quality factor was comparable to the qubit relaxation time.

1. Introduction

Measure of the material loss at microwave frequency in superconducting resonator is one the most effective methods to study the TLS related decoherence from different materials in superconducting quantum bit (qubit) circuits. Fabricating epitaxially-grown Josephson junctions for qubit circuit, one can expect the decoherence mechanisms within the Josephson devices would be largely removed. Motived by this purpose, we fabricated superconducting qubit that embedded in a NbN thin film CPW resonator. In this letter, we focus on the experimental results from the rest part of the circuit, the NbN CPW resonator.

2. Experiment and results

The circuit is fabricated by photolithography and dry-etching process. Detailed fabrication process can be found in our recent report [1]. The NbN/AlN/NbN tri-layers films were continuously deposited on a single-crystal MgO (100) substrate by dc-magnetron sputtering in a load-locked sputtering system with a base pressure lower than 2×10^{-5} Pa. Photo masks are used to define base, junctions, and counter electrodes in an i-line stepper followed by CF_4 plasma for reactive ion etching (RIE); as well as remaining NbN thin film CPW resonator. Amorphous SiO_2 film with thickness of 250~300 nm is sputtered as an insulation layer. Amorphous SiO_2 is generally very lossy. Therefore, most of them are etched away by CHF_3 and only less than 3% remained in the circuit, mostly near the junctions. As a result, the thickness of NbN film for CPW resonator is about 200 nm.

The circuit consists a dc SQUID with two small Josephson junctions in parallel enclosed in a small superconducting loop. One end of the SQUID is connected to a superconducting island that is capacitively coupled to the center conductor of a half-wavelength CPW resonator. The other end of the SQUID is connected to the ground plane of the CPW. The width of the center conductor is s = 23 μm with a gap w = 10 μm between center conductor and the ground plane. This configuration gives an impedance of Z_0 = 50 Ω for MgO substrate. The resonant frequency of the CPW is determined by

$$f_r = \left((L_m + L_k)C\right)^{-1/2}, \qquad (1)$$

, where L_m is the geometric inductance, L_k is the kinetic inductance, C is the total capacitance. Figure 1(a) shows a schematic of our measurement set-up. The circuit is located at the base temperature of 20 mK in a dilution refrigerator. The resonator transmission S_{21} is amplified by a low noise high electron mobility transistor (HEMT) located at 1.5 K and microwave amplifiers at room temperature before recorded by a vector network analyzer. The loaded resonator quality factor $Q_m = f_r/\delta f_r$ is obtained from the Lorentzian fit of resonant peak of S_{21} figure 1(c).

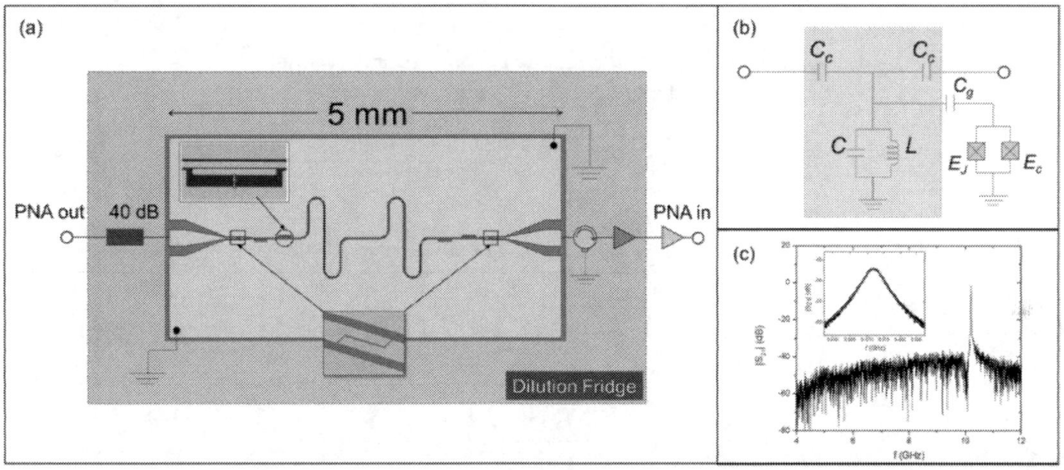

Figure 1. Determination Q factor of NbN CPW resonator and system set-up. (a) Schematic of the measurement set-up at low temperature with chip layout. The CPW resonator has two interdigital coupling capacitances C_c (zoom-in from the boxed regions) at the input and output of the CPW center conductor. The top inserted shows the SEM image of the qubit. (b) Equivalent circuit of the resonator is highlighted in the red region. (c) Transmitted signal of the CPW resonator with resonator frequency near 10 GHz. The inserted graph shows the typical data with a fitted Lorentzian peak.

As the kinetic inductance L_k is temperature dependent below critical temperature T_c [2], one can define the kinetic inductance factor $\alpha = L_k/L$, L is the total inductance. Equation (1) can be re-written as,

$$f_r = \left(((1-\alpha)L + \alpha L)C\right)^{-1/2} \qquad (2)$$

It is clear that the resonant frequency is also temperature dependent. A measured resonance frequency as a function of temperature is depicted in figure 2. The resonator parameters obtained by fitting the data with Equation (2) are listed in table 1. From the best fit, kinetic inductance factor $\alpha = 6.1\%$, which is in the range of typical superconducting thin film resonator [2]. Critical temperature of $T_c = 15.8$ K is in good agreement with the result of independently measured Tc obtained from R-T measurement of NbN thin film [3, 4].

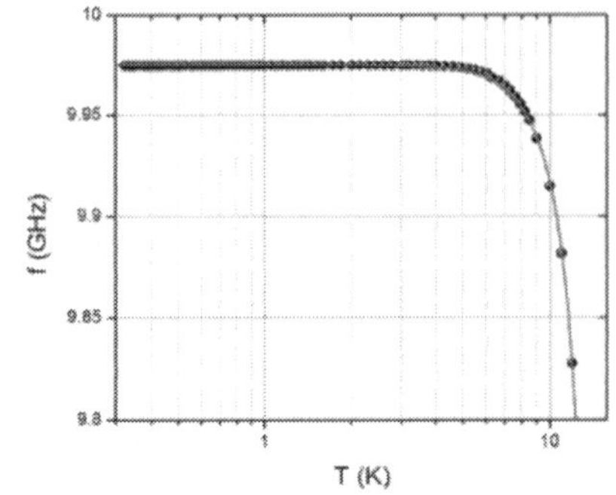

Figure 2. Measured temperature dependent of resonant frequency shifting (black dots) fitted by equation (2) (red solid curve).

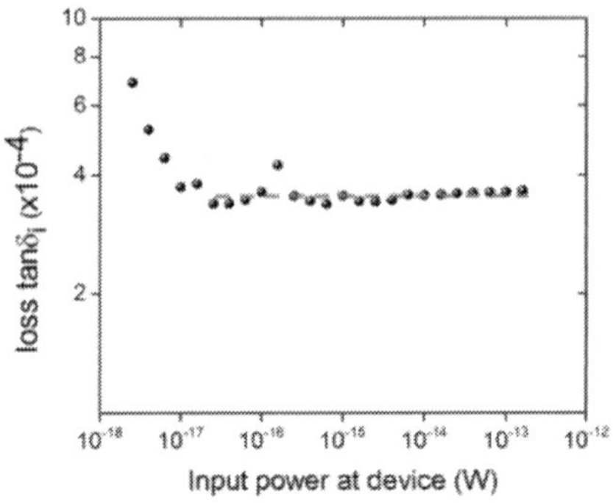

Figure 3. Dielectric loss $\tan\delta$ at various input power to the device (black dots). Red dashed line gives the eye-guide line for loss level at 3.3×10^{-4}. (Note: data below 10^{-17}W that show higher loss are probably due to the poor SNR of amplifiers)

Table 1. Values for resonator design and fitting temperature dependent $f_r(T)$ by Equation (2).

Parameter	Value	Unit
ε_{eff}	9.8	
s	23	μm
w	10	μm
C	1×10^{-12}	F
L	2.51×10^{-14}	H
α	6.1%	
T_c	15.8	K

The microwave power applied to the resonator is heavily attenuated at both room temperature and cryogenic temperature. Single-photon level microwave power applies to the input port of the NbN CPW resonator. The internal loss due to the conductor and substrate is $\tan\delta = Q_i^{-1} = Q_m^{-1} - Q_c^{-1}$, where Q_i is the internal quality factor, and Q_c is the external quality factor. The loss $\tan\delta$ at various input power is plotted in figure 3. At low input power (*e.g.* close to single-photon level) and temperature (*e.g.* $T \ll T_c$), the internal loss appears to saturate at the level $\tan\delta = 3.3 \times 10^{-4}$ (red dashed line in figure 3).

3. Discussion

We measured the dielectric loss of NbN resonator embedded with qubit circuit on a single crystal MgO (100) at close to single photon power and at low temperature of 20 mK. We also noted a recent report of the dielectric loss on MgO substrate that only showed the loss is at least one order of magnitude higher than our result [5]. Although the superconducting materials (Al and NbN) used for both studies were different, the lattice orientation of the substrate MgO leads to significant different results. When taking into account the geometric participation of ~10% across the whole chip for MgO substrate, we estimate the relaxation time T_1~ 1 μs. This is indeed within the range of experimentally measured relaxation time [6]. The reason that the loss of our epitaxial NbN film on MgO (100) is higher than the losses of nitride and single-crystal silicon substrates is probably due to the oxide nature of the MgO substrate [5, 7-9]. Qubit circuit fabrication consists of epitaxial Josephson junctions and low dielectric loss substrate would be an ideal case but technical challenges remained to be solved in future studies.

References

[1] Z Wang, H Terai, W Qiu, K Makise, Y Uzawa, K Kimoto and Y Nakamura 2013 Appl. Phys. Lett. **102** 14 142604
[2] J Gao 2008 Ph.D. thesis, California Institute of Technology
[3] Z Wang, A Kawakami, Y Uzawa, and B Komiyama 1994 Appl. Phys. Lett. **64** 15 2034
[4] Z Wang, A Kawakami, Y Uzawa, and B Komiyama 1996 J. Appl. Phys. **79** 10 7837
[5] A D O'Connell, M Ansmann, R C Bialczak, M Hofheinz, N Katz, E Lucero, C McKenney, M Neeley, H Wang, E M Weig, A N Cleland, and J M Martinis 2008 Appl. Phys. Lett. **92** 112903
[6] Unpublished result 2013 NiCT & The University of Tokyo
[7] P Macha, S H van der Ploeg, G Oelsner, E Il'ichev, H –G Meyer, S Wunsch, and M Seigel 2010 Appl. Phys. Lett. **96** 062503
[8] R Barends, N Vercruyssen, A Endo, P J de Visser, T Zijlstra, T M Klapwijk, P Diener, S J C Yates, and J J A Baselmans 2010 Appl. Phys. Lett. **97** 023508
[9] M R Vissers, J Gao, D S Wisbey, D A Hite, C C Tsuei, A D Corcoles, M Steffen, and D P Pappas 2010 Appl. Phys. Lett. **97** 232509

Mitigation of position dependence in x-rays absorbers for Thermal Kinetic Inductance Detectors

O Quaranta[1], T Cecil[1], L Gades[1] and A Miceli[1]

[1]Argonne National Laboratory, 9700 S. Cass Avenue, Argonne, IL 60439 - USA

oquaranta@ anl.gov

Abstract. The currently available microcalorimeter technologies are all affected by the difficulty of multiplexing large numbers of pixels (>1000). Microwave kinetic inductance detectors, on the other hand, are intrinsically highly multiplexible. We have recently shown that it is possible to use a superconducting resonator as a thermometer for a microcalorimeter (TKID). The moderate sensitivity of the resonator of a TKID requires a low heat capacity absorber (e.g. like Tantalum, Tin) in order to achieve sufficient signal to noise and absorption efficiency. These materials are often characterized by poor thermalization properties, which give rise to a position dependent response. In this work we introduce a new design solution based on a bilayer of Tantalum and Gold grown on a perforated SiN island that provides better thermalization properties and reduces the position dependence, while maintaining a small total heat capacity.

1. Introduction

Current microcalorimeter detectors, independent of their detection mechanism, require a fairly complex read-out system, which limits the scalability to big arrays. Microwave Kinetic Inductance Detectors (MKIDs) are attracting an increasing interest, due in part to the relative ease of multiplexing MKIDs. An array of MKIDs is readout by sending a comb of microwave frequencies matched to the resonators frequency past the resonators. Each resonator interacts with the comb signal at its designed frequency, changing the phase and amplitude of the input signal. The responsivity of MKIDs using phase readout is governed by several parameters according to $d\theta/dN_{qp} \propto \alpha Q/V$ where $d\theta/dN_{qp}$ is the phase change per change in quasiparticle density, α is the kinetic inductance fraction (i.e. the ratio between the kinetic inductance and the total inductance, kinetic plus magnetic inductance), V is the volume and Q is loaded quality factor [1]. This formula shows that in order to increase the responsivity, resonators with high Q, high α and small volumes are needed. These conditions put strong constraints on the material selection used for the resonator, especially in the case of detectors for x-rays where high-Z materials and big volumes are needed to achieve reasonable stopping powers. In our previous work [2] we introduced a new MKID concept, called a Thermal Kinetic Inductance Detector (TKID), where a superconducting resonator is used as a microcalorimeter, rather than as a non-equilibrium detector. The device consists of an absorber and the inductor of a superconducting resonator deposited on a SiN suspended membrane, which provides a weak thermal link to the bath. The thermalization of the photon in the absorber causes an increase in the temperature of the absorber itself and of the SiN membrane and inductor. As shown in Gao et al [3], the change in surface impedance of a superconductor due to an increase in temperature is nearly identical to the change due to an excess of quasiparticles. Consequently, by fabricating a resonator on the same SiN we were able to measure the temperature rise by examining the induced frequency shift and reduction in amplitude of a microwave signal transmitted past the resonator. In this design there is no electrical contact between the absorber and the resonator; the response of the resonator is purely due to the temperature evolution of the island and not to the injection of non-equilibrium quasiparticles.

Figure 1. (a) SEM image of a representative device (Dev-H) in false colors; the resonator is green and the absorber is blue. **(b)** Detail of a corner of the absorber where showing the perforation in the SiN membrane.
(c) Schematic of the cross section of the device (not to scale).

2. Device and Measurements

The TKID device consists of three main components: a superconducting resonator, an x-ray absorber, and a SiN membrane. A false color SEM image of a finished device is shown in Figure 1(a) with the absorber blue and the resonator green. All of the devices described here use a 100 nm thick WSi_2 film patterned into a lumped element resonator as the thermometer. The inductor portion of the resonator extends onto the SiN membrane while the capacitor remains on the bulk substrate. The SiN membrane that serves as the weak thermal link is 500 nm thick and patterned into an island to further decrease thermal conductance between the absorber and substrate. Given the sensitivity of our resonators ($\mathrm{Im}[\alpha] = -Q_i \times (T/L) \times (dL/dT) \sim 50$), a low heat capacity absorber is needed. Consequently, we used a superconducting material (Tantalum) as the basis for our tests of the absorber. We have looked at two design solutions: pure superconducting absorber (500 nm of Tantalum, 300 × 300 μm lateral dimensions, from this moment Dev-S) and a hybrid bilayer of a superconductor over a normal metal (500 nm of Ta over 25 nm of Au, 200 × 200 μm lateral, from this moment Dev-H). In addition, the SiN between the absorber and the resonator has been perforated in Dev-H to further decouple the absorber from the resonator. All the devices have been characterized in an ADR cryostat with 6 keV photons from an uncollimated Fe-55 radioactive source, following the procedure described in [2].

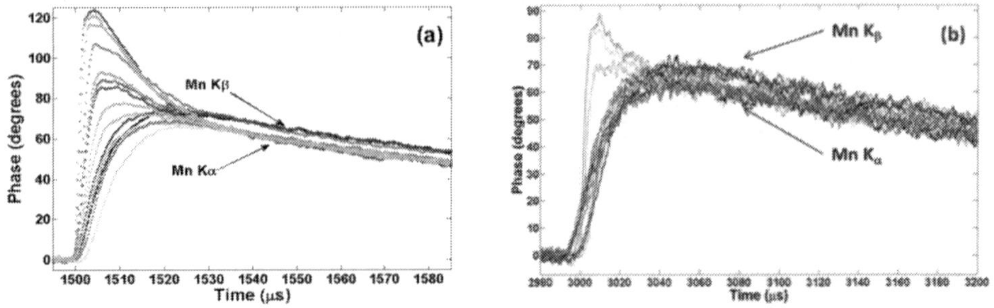

Figure 2. (a) Pulses from Dev-S under illumination with a Fe-55 source. (b) Pulses from Dev-H under illumination with a Fe-55 source.

Figure 2(a) shows a series of phase pulses from Dev-S when illuminated with photons. Clearly the rise time of the pulses is extremely variable, ranging from a few μs up to ~ 10-15 μs, giving rise to very different pulse heights and pulse shapes. After ~30 μs all the pulses collapse on two lines, corresponding to the K_α and K_β emission lines of the source. Applying a matched filter to the data, based on a template generated by averaging tens of pulses we have obtained a distribution of pulse heights. Fitting the rise time of each pulse and plotting the distribution of the rise times as a function of the pulse height, as in figure 3(a), a clear correlation can be seen. This is a strong indication of a position dependent response, where events closer to the edge of the absorber generate bigger pulses with a faster rise time. Due to poor thermalization in the absorber, the photon energy is initially confined to a limited area of the absorber, resulting in a larger pulse. Because this pulse is closer to the resonator, the heat arrives at the resonator sooner than for a photon absorbed in the middle of the absorber. Conversely, for events toward the center of the absorber the heat is forced to flow through the entire volume of the absorber; consequently the effective heat capacity will be higher, giving rise to smaller pulses and slower rise times. A similar effect has been seen in other microcalorimeters that used superconducting absorbers, with the effect particularly evident in Ta [4]. This dependence can strongly affect the energy resolution of the detector: collapsing the distribution in figure 3(a) on the x-axes the spread in phase is wide, with overlapping between the two emission lines. One way to improve the resolution could be by rejecting all the pulses with a rise time faster than 10 μs, at the cost of strongly reduced detection efficiency. The application of a low pass filter to the data could mitigate the problem, but this will reduce the usable bandwidth of the pulse.

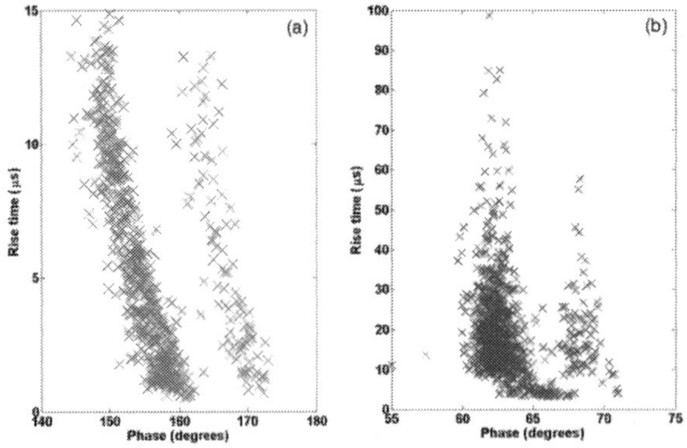

Figure 3. Pulse rise time as a function of pulse height for (a) Dev-S and (b) Dev-H.

We have begun exploring ways to mitigate this position dependence. In this paper, we present a hybrid device (Dev-H) that incorporates a thin layer of Au underneath a thick Ta absorber. Additionally, this structure is grown on a SiN island with perforations between the absorber and the resonator. A SEM picture of such device is shown in figure 1(a). In figure 1(b) is the detail of the device, where the perforation is clearly visible. The detail is schematically depicted in figure 1(c). An example of the pulses from this device is shown in figure 2(b). The vast majority of the pulses are characterized by a longer rise time (~ 10-30 μs) with respect to Dev-S. This is due to the reduced thermal conductance between the absorber and the resonator due to the perforation of the SiN membrane. The pulses have nearly identical rise time and shape (in the limit of the noise), indicating much less position dependence. In this device the gold underlayer helps with the thermalization of the Ta absorber. A role is also played by the perforation of the membrane, which, by reducing the thermal conductance between the absorber and the resonator, confines the heat in the absorber longer, allowing its uniform thermalization before the temperature rise is sensed by the resonator. To distinguish between the two effects a device with an absorber of Ta only on a perforated island is needed; this will be subject of future studies. A few fast rising pulses are still present, even if in very low number. Those are probably due to photon absorptions in the SiN membrane, close to the resonator. Even in this case two emission lines of the source are visible.

Plotting the rise time as function of the pulse height (figure 3(b)) the points align on two vertical straight lines, showing no relationship between the rise time and pulse height. The different rise times are actually an artifact of the poor quality of the fit, due to the relatively high level of noise in this device. The distribution in figure 3(b) indicates that the response of the device is independent of the absorption position in the Ta. Collapsing the data on the x-axes, it is possible to generate a histogram of the pulse height distribution. Fitting two Gaussian functions on the histogram a resolution of $\Delta E \sim$ 130 eV has been obtained, without having to reject the vast majority of the pulses (i.e., limited to rise times greater than 8 μs).

3. Conclusions

In this work we have analyzed two different kinds of absorbers for TKIDs. A low heat capacity absorber is needed, given the sensitivity of the resonator. Although a superconducting absorber (Ta) has a low enough heat capacity, the poor thermalization properties generate a position dependent response, which ruins the resolution of the detector. Consequently, we have introduced a new design of our TKID for x-rays based on a hybrid Ta/Au absorber, deposited on a perforated SiN island. This solution kept the total heat capacity low but improved the thermalization properties, reducing the position dependence.

References

[1] B Mazin, (2004) *Microwave Kinetic Inductance Detectors*, Dissertation (Ph.D.), California Institute of Technology

[2] O Quaranta, T W Cecil, L Gades, B Mazin and A Miceli 2013 Supercond. Sci. Technol. 26 105021

[3] Gao J, Zmuidzinas J, Vayonakis A, Day P, Mazin B and Leduc H 2008 J. Low Temp. Phys. 151 557

[4] Perinati E, Barbera M, Varisco S, Silver E, Beeman J and Pigot C 2008 Rev. Sci. Instrum. 79 053905

Numerical evaluation of a novel high-temperature superconductor-based quasi-diamagnetic motor

Arpad Racz, Istvan Vajda

Budapest University of Technology and Economics, Department of Electric Power Engineering, Egry J. u. 18., H-1111 Budapest, Hungary
racz.arpad@vet.bme.hu, vajda.istvan@vet.bme.hu

Abstract. An investigation is being pursued at the Budapest University of Technology and Economics, Department of Electric Power Engineering for the application of high-temperature superconductors (HTS) in electrical power systems. In this paper we are going to propose a novel electrical machine construction based on the quasi-diamagnetic behaviour of the HTS materials. The basic operation principle of this machine will be introduced with detailed numerical simulations. Also a possible geometric outline will be presented.

1. Introduction

Most of the electrical machines are using the magnetic field to convert the electrical energy into mechanic energy. This conversion basically relies on the attraction of the opposite magnetic poles. The discovery of high-temperature Type II superconductor opened the possibility of using the superconducting materials in electrical power systems. These materials maintain their superconducting state in case of high current densities and high external fields. Also, if we consider the first magnetization curve of these materials with lower external fields, such as 1...2 T, we can see that the penetration of the field is very low. (Fig. 1.) This magnitude of field is already high enough for technical applications. At this interval we are going to consider the material as an almost ideal diamagnetic (quasi-diamagnetic) material. Based on this behavior it is possible to design a novel type electrical machine where we can utilize this quasi-diamagnetic effect. The operation principle of this new machine is different from any currently used or proposed machine.

2. Application of superconductors in electrical machines

The most common application of HTS material in electrical machines is the use of HTS wires for winding. This solution also gives the possibility to create homopolar machine [1][2]. These solutions use the HTS in the stator. There are also several applications where they used it in the rotor. With the application in the rotor it is possible to create hysteresis motor or it can be used to replace permanent magnet in synchronous machines (with Field Cooling). [3] Special application of HTS is to concentrate the magnetic flux in the rotor of synchronous reluctance machines or switched reluctance machines. [1][3]

The only application of pure HTS rotor is the Meissner-motor. This motor utilizes the Meissner-effect and the transition between superconducting and normal state. This solution is basically a thermodynamic motor. [4] The stator consists of a single permanent magnet. The torque and the possible application of this motor are very limited. The construction is known for more 20 years and no practical application came forward so far.

3. Operation principle of the HTS quasi-diamagnetic motor (QDM)

For the proposed quasi-diamagnetic motor we are going to use the following properties of the high-temperature superconducting materials. The typical level of magnetic induction (the magnitude of 1T) used in an electrical machine is below the total penetration field of an advanced HTS Type II material. Based on this we expect a quasi-diamagnetic behavior of these material as long as we remain below this total penetration field level. Our proposal is to utilize this quasi-diamagnetic effect to create a repulsive force between the stator and the rotor. This effect was already used in the Meissner-motor. In our case it is not necessary to change between the superconducting and non-superconducting state in order to create rotation. Instead of that one should use a suitable geometric configuration and switching of the poles' current to create rotation without state-change.

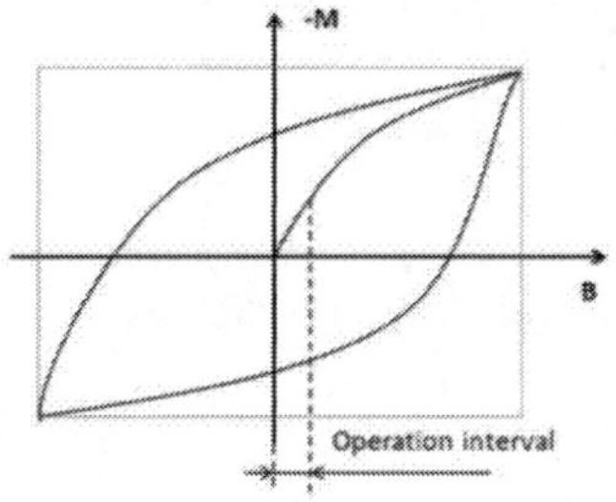

Figure 1: The operation interval for the proposed motor

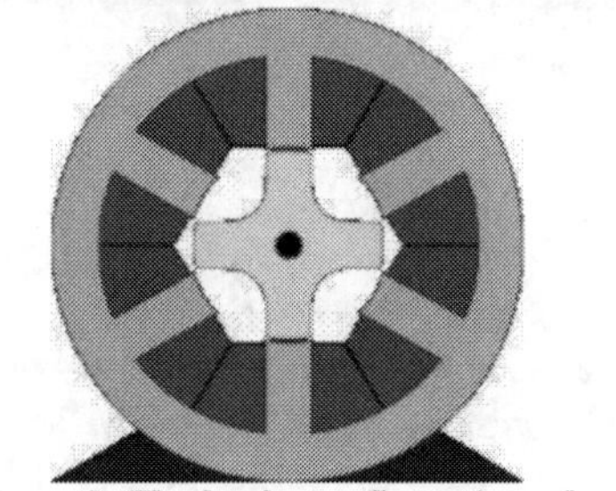

Figure 2: The basic configuration of a 6/4 switched reluctance machine

The idea of the geometric setup comes from the switched reluctance machines (SRM). The machines have salient poles on the stator as well as on the rotor. (Fig. 2.) The rotor consists of ferromagnetic material only. To create the rotation the poles are energized one after another with pulses. One pair of the salient poles of the rotor align with the nearest energized poles. As the next pole is being energized the rotor again aligns and the rotation is created. This is basically an attractive force. The SRM has many configuration usually indicated by the following convention: number of poles on the stator slash number of poles on the rotor. We are going to use this convention for our setups as well.

The idea is to change the rotor material of an SRM with a diamagnetic material. In this way the attractive force acting on the rotor becomes a repulsive force, which is quite unique among the electrical motors. The only occurrence is the Meissner-motor mentioned earlier. The operation Meissner-motor requires state-change in the superconducting material. For this reason it is not so interesting for practical application. In the following sections we are going to show that it is possible to create a motor without state-change and the expected torque of this machine is at the same magnitude as a similar SRM.

For the analysis of the proposed machine we used a software called Finite Element Method Magnetics. The simulation was performed in 2D. The parameters of the machine are: the length of active part 150 mm, outer diameter of the stator is 300 mm, outer diameter of the rotor is 200 mm, airgap between the stator and rotor is minimum 1 mm. The excitation current for poles is 20 A/mm^2 which is higher than normally, but it makes easier to see the effect. For modeling the superconducting material, relative permeability value of 10^{-6} was set.

For the checking of the principle we choose a 6/2 configuration as shown in Fig. 3. This is a very simple setup just to demonstrate the base effect. One can see that the electro-magnetic force is created as the field "pushes" away the quasi-diamagnetic material. During the simulations a torque-angular position was calculated in order to check the behavior and to prove the appropriate level of torque for technical application. For reference an SRM motor with the same parameters was calculated. The peak torque of QDM is around 4 Nm compared to the SRM which has around 60 Nm. The difference is around one order of magnitude, which indicates that with further improvements the QDM can be a valid alternative to other motors.

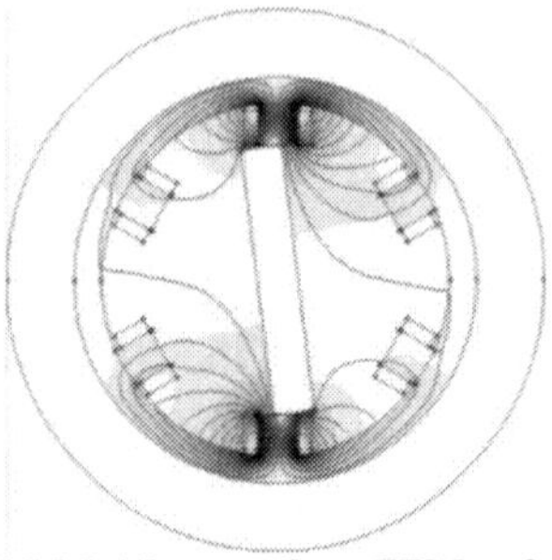

Figure 3: QDM 6/2 setup from FEM software with magnetic flux density plot

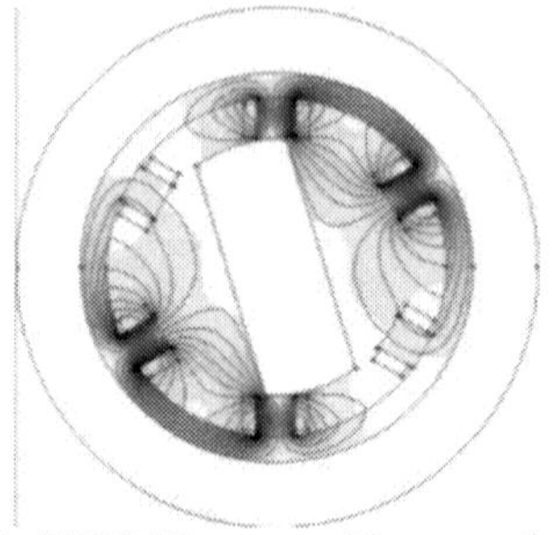

Figure 4: QDM 6/2 setup wider rotor from FEM software with magnetic flux density plot

4. Geometric optimization

During our examination we found two ways to improve the torque of this QDM. First solution is to make the rotor wider than the original construction, and the second is to energize two pole pairs at the same time. Figure 4. shows the solutions applied together.

The peak torque of the reference SRM doesn't change significantly (from 59.15 Nm to 63.47 Nm). Meanwhile the peak torque of the QDM increased five times (from 4.16 to 20.93 Nm). This means that with appropriate changes in the geometric outline and in the control strategy it is possible optimize the QDM.

5. Geometric outline

We already showed that it is possible to create electromagnetic torque based on the quasi-diamagnetic behavior of the Type II superconductors. In this section we are going to propose a possible geometric outline of the machine. Based on torque curve of the 6/2 QDM one see that the peak torque is created when rotor pole is slightly misaligned with stator pole. One effective way to decrease the torque ripple is to increase the number of poles. This will result more peaks in the torque during one full turn.

When we make the concept of rotor we have to take into account the followings:

- increasing the number of the poles it is not necessary for them to be wide,
- the length of the salient poles on rotor affects the torque of the machine,
- the distance between two poles should be wide enough in order to allow the magnetic flux to act on the side of the pole.

Based on these requirements one possible geometric outline is shown in Figure 5. This rotor is like a wheel of a water mill (part (a) in Fig. 5.). One can imagine that the flow of the magnetic flux is pushing this wheel. Figure 6. shows the torque characteristics of this machine. The figure shows the electromagnetic torque dependence on the angular position when different poles are energized. One can notice that only energizing one pair of stator poles we can produce almost the same torque as the best result with the 6/2 setup. With this solution the control strategy will be simpler as well.

6. Design considerations

One can notice that for the ease of the simulation we used pure HTS material for the rotor. When we are going to realize this machine it is not necessary to do it this way. It is enough to have thin superconducting layer over inner core of the rotor. The thickness of the layer is depending on the critical current/field of the HTS material.

One can also notice that the current density at the stator coil is higher than the normally allowed current density of copper conductors. Partly this is not a problem since the stator pole is only operating with short pulses. Other solution to this issue can be the application of HTS wires at the stator coils. Since it is necessary to cool the rotor we will have cryogenic system anyway.

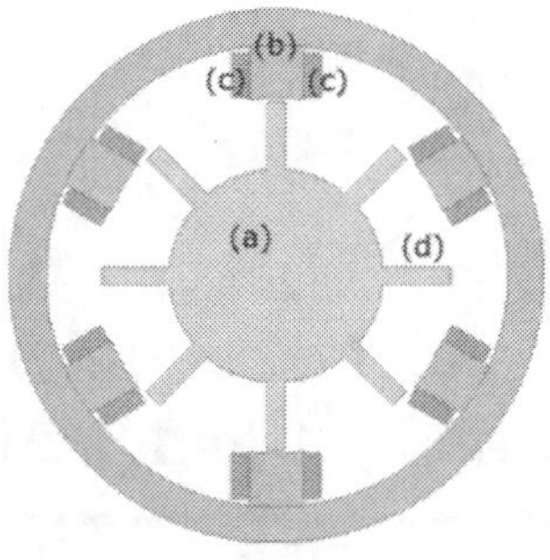

Figure 5: The schematic of the proposed 6/8 QDM geometric layout.

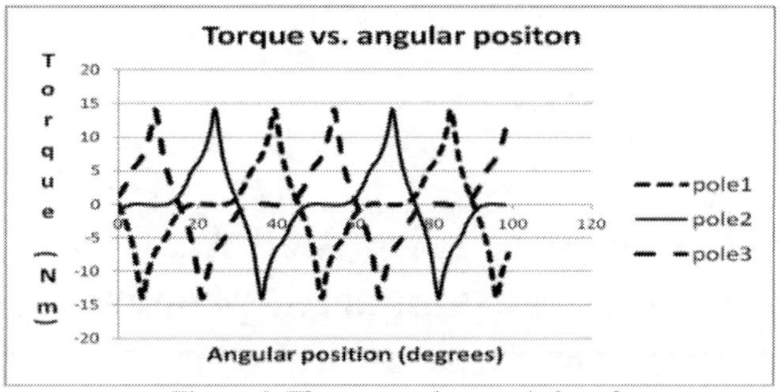

Figure 6: The torque characteristics of the 6/8 QDM

Also the question of cooling can arise. For cooling the rotor we can think about two possible solutions. One is the encapsulation of the rotor and the other is to flood the airgap with liquid nitrogen. The second solution certainly has the advantage of smaller airgap which increases the peak torque of the machine. With this solution it is also quite evident to use HTS coil windings at the stator. This gives the possibility to increase current of the stator coils. As a first approximation we can say that the torque has quadratic dependence on the current. Even if we increase the current of one order of magnitude the steel components will not saturate. This can compensate the difference between this concept and the currently used machines.

7. Summary

In this paper we presented a novel electrical machine concept based on the quasi-diamagnetic behavior of the Type II high-temperature superconductors. Our suggestion is to name it as "quasi-diamagnetic machine", and QDM in a short form. We showed the results of detailed finite-element method simulations of different configurations of this machine. The results indicate that this machine can be real alternative to the currently used electrical machines.

We also presented a possible geometric outline of the new machine which can enhance the dynamic behavior of the machine. It also gives relatively high torque and simpler control strategy.

Finally we presented design considerations which should be taken into account during further design of this new QDM.

We would like to emphasize that this paper is the first publishing of this idea. Everything in it should be considered as a preliminary conceptual design.

References

[1] Kovalev K L, Dezhin D S, Kovalev L K, Poltavets V N, Ilyasov R I, Golovanov D S, Oswald B, Best K J, and Gawalek W 2010 HTS High-dynamic Electrical Motors *IEEE/CSC & ESAS European Superconductivity News Forum (ESNF)* **11**

[2] Hazelton D W, Gardner M T, Rice J A, Walker M S, Trautwein C M, Haldar P, Gubser D U, Superczynski M, and Waltman 1997 D HTS coils for the Navy's superconducting homopolar motor/generator *IEEE Trans Appl Supercond* **7** pp 664–667

[3] Barnes G and Barnes C 2000 Electric Motors with High Temperature Cuprate Superconductors in the Rotor *Copper Development Association Inc.*

[4] Takeoka A, Ishikawa A, Suzuki M, Niki K, and Kuwano Y 1989 Meissner motor using high-Tc ceramic superconductors *IEEE Trans. Magn.* **25** pp 2511–2514

[5] Frauenhofer J, Kaufhold M, Kummeth P, Nerowski G, and Nick W 2006 High-temperature-superconducting machines- a high-technology step for large rotating electric machines *Proc. SPEEDAM 2006* pp 379–383

Journal of Physics: Conference Series **507** (2014) 042035 doi:10.1088/1742-6596/507/4/042035

Development of integrated AC-DC magnetometer using high-T_c SQUID for magnetic properties evaluation of magnetic nanoparticles in solution

Mohd Mawardi Saari [1], Ryuki Takagi, Toki Kusaka, Yuichi Ishihara, Yuya Tsukamoto, Kenji Sakai, Toshihiko Kiwa and Keiji Tsukada

Graduate School of Natural Science and Technology, Okayama University, 3-1-1Tsushima Naka, Kita Ku, Okayama 700-8530, JAPAN

E-mail: en19463@s.okayama-u.ac.jp

Abstract. We developed an integrated AC-DC magnetometer using a high critical temperature superconducting quantum interference device (high-T_c SQUID) to evaluate the static and dynamic magnetic properties of magnetic nanoparticles (MNPs) in solution. The flux-transformer method consisted of first-order planar and axial differential coils that were constructed for static and dynamic magnetization measurements, respectively. Vibrating-sample and harmonic detection techniques were used to reduce interference from excitation magnetic fields in the static and dynamic magnetization measurements, respectively. Static and dynamic magnetization measurements were performed on commercially available iron oxide nanoparticles in diluted solutions. The magnetic responses increased with the increase in concentration of the solutions in both measurement results. The magnetization curves showed that the diamagnetic signal due to the carrier liquid of the iron oxide nanoparticles existed in a dilute solution. Biasing with a proper DC magnetic field in the dynamic magnetization measurement resulted in improved signals of the second and third harmonics. Therefore, highly sensitive magnetic characterizations of MNPs utilizing the static and dynamic magnetization measurement are possible via the developed system.

1. Introduction

Magnetic nanoparticles (MNPs) are attracting interest in fields such as bio-immunoassay, magnetic nanoparticle imaging, and magnetic drug targeting due to their promising results on a nanometer scale. In order to detect MNPs, their inherent magnetic properties are measured by magnetic remanence [1], relaxation [2] and susceptibility methods [3]. Measurement systems that utilize the former two methods can achieve high sensitivity due to less/no interference from excitation magnetic fields. However, magnetic susceptibility requires measurement in the presence of excitation magnetic fields, thus, the interference from these fields will limit the sensitivity, particularly in AC systems. Harmonics generated from the nonlinear magnetization characteristics of MNPs have been utilized to reduce the interference of excitation magnetic fields in the dynamic magnetization measurement [4]. These harmonics are used to quantify MNPs with fast measurements and improve sensitivity by isolating the frequency component of excitation magnetic fields. Large amplitudes of excitation magnetic fields, which can cover wide regions of the nonlinear characteristics, are necessary to

[1] To whom any correspondence should be addressed.

increase the generation of harmonics. In the static magnetization measurement, high sensitivity is achieved using the vibrating-sample technique and a superconducting quantum interference device (SQUID) with a flux-transformer [5].

The combination of the static and dynamic magnetization measurement techniques in a magnetometer system can reveal further information for analysis of MNPs in solutions. With the aim of achieving a compact, low-running cost and highly sensitive magnetometer to evaluate the magnetic properties of magnetic nanoparticles in solutions, we developed an integrated AC-DC magnetometer using a high critical temperature superconductor (high-T_c) SQUID with a flux transformer method based on our previously developed system [6]. For measurements of static and dynamic magnetizations, different coil arrangements were adopted to increase the sensitivity. To demonstrate the sensitivity of the system, we measured the magnetic properties of iron oxide nanoparticles in solutions with respect to different concentrations.

2. Experimental

2.1. Measurement system configuration

An overview of the developed system and coil arrangements for static and dynamic magnetization measurements is shown in figure 1. We employed first-order planar and axial differential coils for static and dynamic magnetization measurements, respectively, as the pickup coils in order to reduce the environmental noise being transferred to the SQUID. Both types are constructed from identical elliptical coils with two coils connected in a series-opposing configuration. The external major and minor axes, internal major and minor axes, length, resistance, inductance and Cu wire diameter of one elliptical coil were 26 mm, 13.5 mm, 15 mm, 1.5 mm, 3.9 mm, 2.99 Ω, 451.6 μH, and 0.32 mm, respectively. The baselines of the planar and axial differential coils were 13.5 mm and 18 mm, respectively.

In the static magnetization measurement, a sample was exposed in a DC magnetic field generated by an electromagnet and vibrated perpendicularly to the axis of the DC magnetic field using an actuator. The sensitive axis of the planar differential coil was parallel to the axis of the DC magnetic field as shown in Fig. 1 (a). The induced signals were transferred to the inductively coupled SQUID and the SQUID output was lock-in detected. The motion of the actuator detected by a laser positioning sensor was used as a reference signal at a lock-in amplifier.

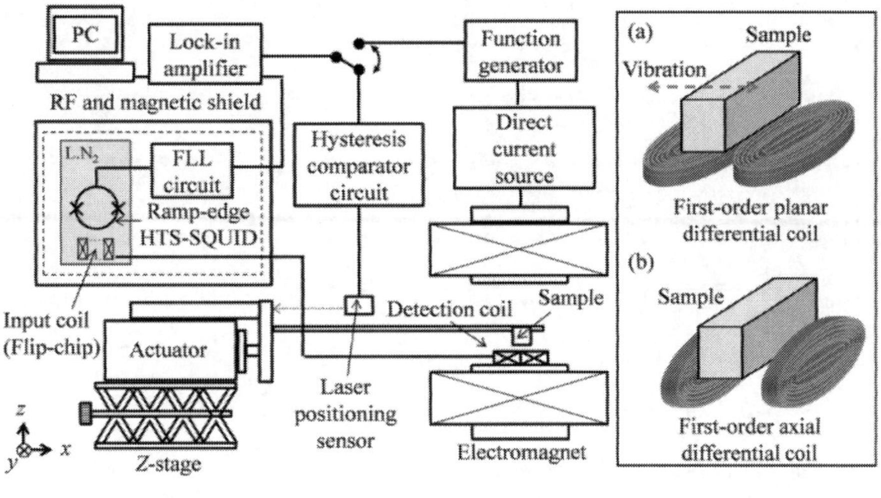

Figure 1. Schematic diagram of the developed system with coil arrangements of (a) first-order planar differential coil and (b) first-order axial differential coil for static and dynamic magnetization measurements, respectively.

In the dynamic magnetization measurement, the sensitive axis of the axial differential coil was placed perpendicular to the excitation magnetic field so that interference of the excitation magnetic field could be reduced. The sample was placed at the center of the axial differential coil as shown in figure 1 (b). An AC magnetic field superimposed with a DC magnetic field was applied to the sample and the induced signal, which was amplified at the SQUID, was lock-in detected using a reference signal at a function generator. The excitation magnetic fields were measured using a Hall sensor.

In this study, we used a ramp-edge type SQUID. The SQUID was fabricated on an MgO substrate using a multilayer fabrication technique. $SmBa_2Cu_3O_y$ (SmBCO) and $Er_{0.95}La_{0.1}Ba_{1.95}Cu_3O_y$ (L1ErBCO) were used as the base and counter electrodes, respectively. The input coil substrate was placed over the high-T_c SQUID gradiometer substrate with an overall size of 7.5 × 15 mm^2, and their mutual inductance was calculated as being 1.88 nH. Details of the SQUID can be found in [6].

2.2. *Solutions of iron oxide nanoparticles*

The dilute solutions of iron oxide nanoparticles (nanomag®-D-spio) were prepared by suspending iron oxide nanoparticles in purified water. The diameter of iron oxide nanoparticles amounts to 100 nm and consists of dextran iron oxide composites. Iron oxide concentrations of the prepared solutions were 24 μg/ml, 48 μg/ml, 72 μg/ml and 96 μg/ml, and the solutions were encased in 10 × 30 × 10 mm^3 of acrylic cases, respectively.

3. Results and discussion

3.1. *Static magnetization measurement*

The measurement results of the static magnetization of iron oxide nanoparticle solutions in the range of −260 to 260 mT with 20-mT intervals are shown in figure 2. The magnetization curves showed the superparamagnetic characteristic with the saturation magnetic field H_s of approximately $\mu_0 H$ = 40 mT, where μ_0 is the vacuum permeability and H is the magnetic field. The curves responded almost linearly with the change of concentrations. The diamagnetic susceptibility was observed in the magnetic field region higher than 40 mT, indicating the diamagnetism of water as the carrier liquid. This result suggests that the diamagnetic signal from the carrier liquid should be taken into account when measuring dilute solutions of MNPs. The high sensitivity of the developed system showed that it was able to measure the iron oxide nanoparticles in low-concentrated solutions.

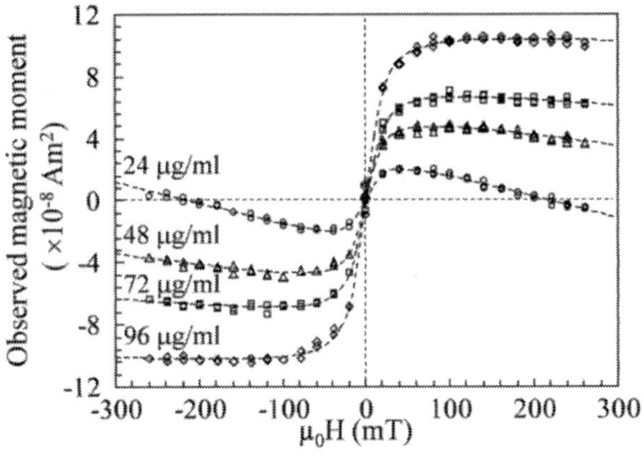

Figure 2. Magnetization curves of iron oxide nanoparticles solutions in different concentrations.

3.2. *Dynamic magnetization measurement*

To cancel the circumference noise, subtractions between lock-in detected signals of with-sample and without-sample measurements were performed in the dynamic measurement technique. Figure 3 shows the characteristics of second and third harmonics of the 24-μg/ml iron oxide nanoparticles solution with respect to the different AC magnetic field amplitudes biased with the DC magnetic fields from 0 to 100 mT. The frequency of the AC magnetic field was 5 Hz. The signals of second and third harmonics increased with larger amplitudes and a proper bias of the DC magnetic field. Figure 4 shows the second and third harmonics at 20 and 0 mT of the DC magnetic field, respectively, during the excitation of 30-mT amplitude of the AC magnetic field. The second and third harmonics responded linearly with the increasing concentrations. The results showed that highly sensitive dynamic magnetization measurement was possible using the developed system.

4. Conclusion

We developed an integrated AC-DC magnetometer using a high critical temperature superconductor (high-T_c) SQUID with a flux transformer method for the magnetic properties evaluation of MNPs in solutions. The coil arrangements of the planar and axial differential coils were used for the static and dynamic magnetization measurement techniques, respectively. Using commercially available iron oxide nanoparticles to demonstrate the developed system, high sensitivity was shown in both techniques. Highly sensitive magnetic characterizations of MNPs utilizing the static and dynamic magnetization measurement were possible via the developed system.

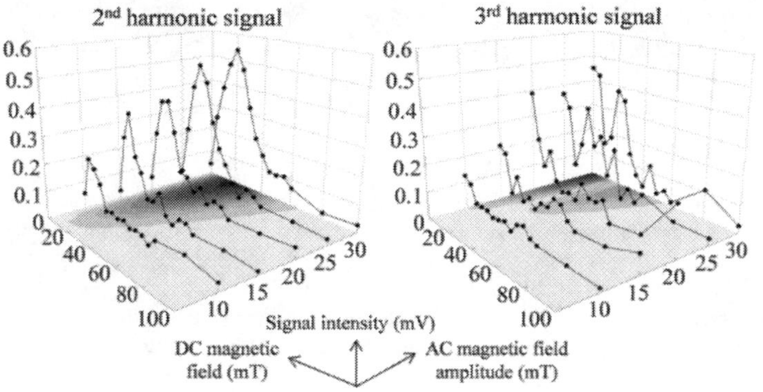

Figure 3. Second and third harmonic signals of the 24-μg/ml iron oxide nanoparticles solution with respect to the different DC and AC magnetic fields.

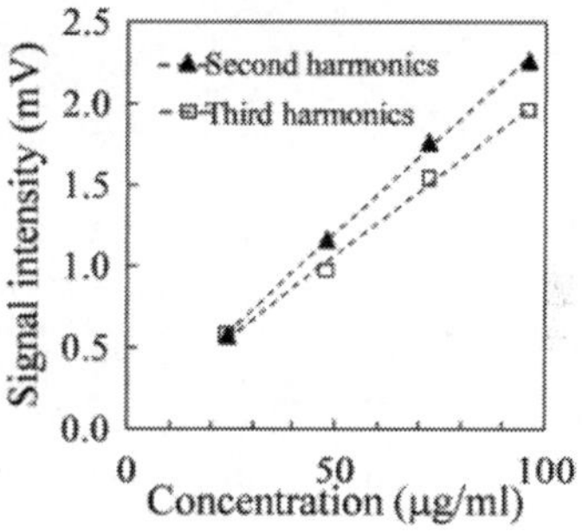

Figure 4. Second and third harmonic signals upon excitation of 30-mT amplitude of the AC magnetic field with the DC biases of 20 and 0 mT, respectively.

Acknowledgements

The high-T_c SQUID was developed and provided by the Superconductivity Research Laboratory, International Superconductivity Technology Center (SRL-ISTEC) in Japan. This work is supported by the "Strategic Promotion of Innovative R&D" of the Japan Science and Technology Agency (JST).

References

[1] Tsukamoto A, Saitoh K, Suzuki D, Sugita N, Seki Y, Kandori A, Tsukada K, Sugiura Y, Hamaoka S, Kuma H, Hamasaki N and Enpuku K 2005 *IEEE Trans. Appl. Supercond.* **15** 656–9
[2] Enpuku K, Tanaka T, Matsuda T, Dang F, Enomoto N, Hojo J, Yoshinaga K, Ludwig F, Ghaffari F, Heim E and Schilling M 2007 *J. Appl. Phys.* **102** 054901
[3] Chung S-H, Hoffmann A, Guslienko K, Bader S D, Liu C, Kay B, Makowski L and Chen L 2005 *J. Appl. Phys.* **97** 10R101
[4] Gleich B and Weizenecker J 2005 *Nature* **435** 1214–7
[5] Pelizzone M and Treyvaud A 1981 *Appl. Phys.* **24** 375–9
[6] Saari M M, Sakai K, Kiwa T, Tsukamoto A, Adachi S, Tanabe K, Kandori A and Tsukada K 2012 *Jpn. J. Appl. Phys.* **51** 046601

Influence of d-wave pairing on electrical properties of d_0-d_0 submicron $YBa_2Cu_3O_{7-x}$ bicrystal grain boundary junctions

E Sarnelli[1,2], M Adamo[1,2], S De Nicola[2], S Cibella[3], R Leoni[3], and C Nappi[1,2]

[1]Istituto di Cibernetica "E. Caianiello" CNR, Pozzuoli, Italy
[2]Istituto "SPIN" CNR, Napoli, Italy
[3]Istituto di Fotonica e Nanotecnologie CNR, Roma, Italy

ettore.sarnelli@cnr.it

Abstract. We have fabricated submicron $YBa_2Cu_3O_{7-x}$ grain boundary junctions using [100] tilt bicrystal substrates. In these substrates, crystal planes are tilted around the b-axis, parallel to the bicrystal line. We have obtained high-quality junctions and fitted experimental data using a recently theory accounting for d-wave symmetry of the order parameter in layered superconductors with arbitrary orientation. In our experiment, sensitive to the amplitude of the pairing state, lobes of different signs do not influence the electrical transport in the junction. Nevertheless, the effects of a non-conventional symmetry are clearly observed on the temperature dependence of the critical current.

Since their discovery, an impressive number of papers dedicated to the understanding of physical mechanisms in high-critical temperature superconductors has been produced. Among these materials, intensive studies have been pursued on $YBa_2Cu_3O_{7-x}$ (YBCO) superconductor, that has generated large curiosity in the researchers, because of its complicated phenomenology. One of the methods largely used for analyzing properties of superconductive materials is the Josephson effect [1]. The study of the Josephson effect allows the analysis of several electrical and magnetic phenomena, mirroring intrinsic properties of the junction. Typically planar technologies, used for the fabrication of tunnel junctions with conventional superconductors, generally failed. Indeed, the coherence lengths extremely short in YBCO, of the order of 2 nm in the plane and even shorter along the c-axis, forced the scientific community to look for alternative techniques for the fabrication of Josephson devices. Since then, several technologies have been introduced, as for instance junctions known as step-edge [2], ramp-edge [3], or grain boundary junctions [4]. The simplest and most used method was the fabrication of grain boundary junctions (GBJs) through the employment of bicrystal [4], tricrystal [5], or tetracrystal [6] substrates. This was also the fastest method for fabricating YBCO junctions and starting the analysis of electrical properties of such fascinating superconductors. Thanks to such technologies, most of exotic properties have been determined. In particular, the most important difference with respect conventional superconductors is the d-wave symmetry of the order parameter, in the form d_{x2-y2} [7]. This important result has been determined through phase-sensitive experiments using peculiar properties of dc-SQUIDs [5,7,8] (Superconducting Quantum Interference Devices) or rf-SQUIDs [9]. About this point, more than one convincing experiment has been carried out. However, to date, whether the symmetry of the order parameter in YBCO is fully represented by a d_{x2-y2} symmetry [5, 10] or small components of other symmetries [11] are present, is still matter of debate. A step forward the understanding of physics of YBCO junctions came from the introduction of nanotechnologies. Indeed, as long as qualitative phase-sensitive tests are the core of the experiment, the use of junctions with micrometric dimensions is adequate. However, once experiments involve quantitative measurements, micrometric grain boundary junctions do not guarantee reliable devices with repeatable properties. Considering the difficulties for fabricating sub-micron YBCO devices, only few experiments involving these type of junctions are available in the literature, see for instance [9, 12, 13]. Moreover, as far as we know, the literature is lacking of experiments testing the amplitude of the order parameter, which is a quantitative test. In the present paper we show experiments on the temperature dependence of the critical current ($I_c(T)$) for submicron YBCO GBJs, in which the transport of the Josephson current is regulated only by d-wave lobes with the same sign (d_0-d_0 junctions). In such structures, the relevant phases never change sign and we show that different electrical behaviors, with respect conventional s-wave symmetries, can be ascribed exclusively to the different symmetry itself and, more specifically, to the presence of nodes in the pairing state function.

The behaviour of the temperature dependence of the critical current strongly depends of interface properties. The shape, as well as the slope of the $I_c(T)$ curves and its anomalies are direct expression of both the superconducting gap at the interface and the current-phase relation in the junction. Essentially, $I_c(T)$ behaviors can be fitted by theoretical expressions only if the interface properties for that junction are well known (for an extended review, see [14]). In particular, relevant parameters at the interface influencing the current-phase relation of the Josephson current are the pairing state, the barrier transparency, the dispersion relation of quasiparticles crossing the barrier, etc. On the contrary, for phase-sensitive experiments, only considerations based on general aspects of the Josephson effect are necessary [7].

We have fabricated [100] tilt submicron YBCO bicrystal grain boundary junctions, as shown in Fig. 1. We would like to outline here that, in such a junction configuration, the positive lobes of the two electrodes are aligned along a direction perpendicular to the junction interface (d_0-d_0 junctions). In the particular case of YBCO junctions, the effect on the transport critical current coming from the nodes of the order parameter should be manifested. Based on previous experiments carried out on bicrystal YBCO GBJs, in order to have a deterministic behaviours of the

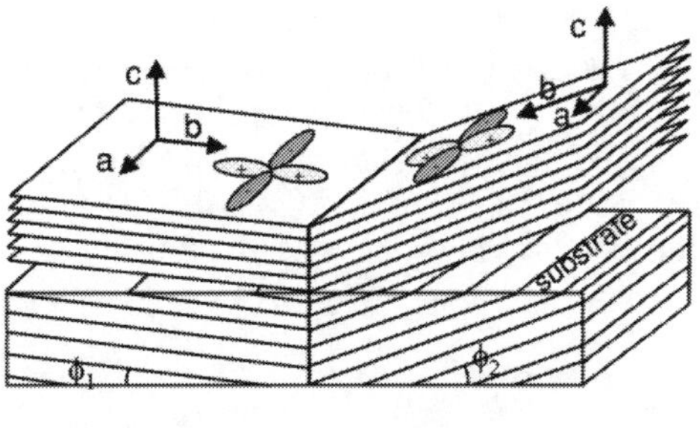

Fig. 1 Scheme of [100] tilt bicrystal geometry. For the experiment we used $\phi_1 = 0$ and $\phi_2 = 45°$

transport current, we developed a suitable technology for fabricating submicron junctions [15]. We have fabricated [100] tilt submicron YBCO GBJs starting from high-quality YBCO films deposited by DC diode on [100] 45° tilt bicrystal substrates. Films showed zero-resistance critical temperatures in the range 89-91 K. Electrodes 100 nm thick are covered by a Au cap layer 20 nm thick. Nanolithography is obtained by lift-off technique of Ti/Au bilayer (100 nm/10 nm in thickness) deposited on PMMA resist, previously exposed through an electron beam lithography [15].

Another important aspect for setting up our experiment was certainly the theoretical framework. Indeed, many theoretical models have been developed for describing the phenomenology of high-temperature superconductive junctions. In particular, it is worth mentioning the work by Tanaka and Kashiwaya [16] describing the electrical transport in d-wave junctions constituted by YBCO electrodes with their c-axis perpendicular to the substrate plane. Starting from the Bogoliubov-de

Journal of Physics: Conference Series **507** (2014) 042036 doi:10.1088/1742-6596/507/4/042036

Gennes equations, the authors describe the transport of quasiparticles across a barrier with arbitrary transparency. Although very complete, using their theory for describing [100] tilt YBCO junctions was not an immediate task, because in our case the c-axis (only one or both of them) are tilted with respect the substrate surface. Similar considerations apply in the case of other theoretical approaches as well, see for instance [17]. Then we decided to develop an original theoretical model for describing the electrical transport in this specific case [18]. Although some differences exist, we can consider our model as an extension of the theory by Tanaka and Kashiwaya [16]. Indeed, we have solved the Bogoliubov-de Gennes equations in the case of twofold-tilted Josephson junctions made by pure d-wave anisotropic layered superconductors, i.e. YBCO. The misalignment of the superconductive YBCO planes in the two electrodes (angles ϕ_1 and ϕ_2 in Fig. 1) is responsible for the appearance of an intrinsic electrical resistance in the junction, producing an effective barrier given by:

$$Z_{eff} = \sqrt{\left[4Z(\theta)^2 + (\cos\phi_1 - \cos\phi_2)^2\right]/4\cos\phi_1\cos\phi_2} \tag{1}$$

where $Z(\theta) = Z_0\cos\theta$, and Z_0 is the barrier amplitude parameter [18]. With this approach we find the Andreev bound states at the junction interface [18]. Then, the Josephson current is derived using the following expression:

$$I_x(\varphi,\phi_1,\phi_2) = \frac{2e}{\hbar}\frac{k_F L_y}{2\pi}\sum_n \int_{-\pi/2}^{\pi/2} d\theta\cos(\theta)\frac{dE_n(\theta,\varphi)}{d\varphi} f\left[E_n(\theta,\varphi)\right] \tag{2}$$

where the index n labels the bound Andreev energy levels, θ is the quasiparticle incidence angle, e is the electron charge, $\hbar$ is the Plank constant divide by 2π, k_F is the Fermi momentum, L_y is the GB length, and $f[E_n(\theta)]$ is the Fermi function. Using Eq. 2 we can derive the temperature dependence of the critical current for d-wave junctions with arbitrary orientations. Similarly, in order to compare our results with conventional s-wave junctions, we have found the Andreev energy states for layered s-wave superconductors. With this tool in our hands we can describe the $I_c(T)$ behaviors for any d-wave or s-wave junctions, with electrodes oriented arbitrarily, and arbitrary barrier transparency.

We have measured the $I_c(T)$ behaviors of several [100] tilt YBCO GBJs with subicrometer dimensions. In Fig. 2 we show the current voltage characteristics (IV) of a junction 0.3 μm wide at several temperatures. In the inset, the $I_c(T)$ behavior (open dots) and the fit (solid line) obtained from Eq. (2) are shown respectively. The only free parameter is the barrier transparency Z_0. The agreement with experimental data is obtained in the clean limit ($Z_0 = 0$), that implies that the junction behaves as a SNS structure, where N indicates a vanishing size normal layer. No fit was possible with low-transparency barriers or using conventional s-wave symmetries. We also have measured the $I_c(T)$

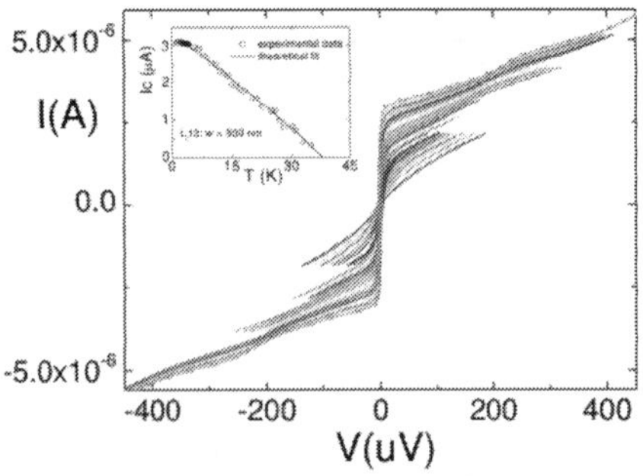

Fig. 2 Current-Voltage characteristics at several temperatures for a junction 0.3 μm wide. In the inset the $I_c(T)$ behavior (open dots) and the fit (solid line) using Eq. 2 are shown.

dependence of another junction 0.7 μm wide on the same chip, that showed similar behavior than reported in Fig. 2 [15]. It is important to outline that all the junctions on the chip showed a linear dependence of the critical current vs the junction dimension, giving a best fit value for the critical current density $J_c = 7.62x10^3$ A/cm^2. This aspect confirms the high-transparency of barriers in our junctions and the uniformity of our devices along the substrate.

In conclusions, we have developed a suitable technique for fabricating high-quality d_0-d_0 submicron YBCO bicrystal junctions, showing high-transparency barriers. In this geometry, no effect due to the different signs of the d-wave lobes is expected. Hence, the electrical transport is sensitive only to the amplitude of the order parameter. The $I_c(T)$ behaviors have been described starting from Andreev energy states at the interface. A very good agreement between experiment and theory has been obtained, confirming the influence of nodes in the pairing state on the electrical transport in the junctions.

Acknowledgements
The financial contribution of EU NMP.2011.2.2-6 IRONSEA project no. 283141 is gratefully acknowledged.

References

[1] Barone A and Paternò G 1982, *Physics and applications of the Josephson effect* (Wiley, New York)
[2] Daly K P, Dozier W D, Burch J F, Coons S B, Hu R, Platt C E, and Simon R W 1991, Appl. Phys. Lett. **58**, 543–545.
[3] Ramp Edge
[4] Chaudhari P, Dimos D and Mannhart J 1989, IBM J. Res. Dev. **33**, 299–306.
[5] Tsuei C C, Kirtley J R, Chi C C, Yu-Jahnes L S, Gupta A, Shaw T, Sun J Z, and Ketchen M B 1994, Phys. Rev. Lett. **73**, 593–596.
[6] Tsuei C C, Kirtley J R, Ren Z F, Wang J H, Raffy H, and Li Z Z, 1997, Nature **387**, 481–483.
[7] Van Harlinghen D J 1995, Rev. Mod. Phys. **67**, 515-535
[8] Testa G, Monaco A, Esposito E, Sarnelli E, Kang D-J, Mennema S H, Tarte E J and Blamire M G 2004, Appl. Phys. Lett. **85,** 1202; Testa G, Sarnelli E, Monaco A, Esposito E, Ejrnaes M, Kang D-J, Mennema E H, Tarte E J and Blamire M G 2005, Phys. Rev. B **71** 134520
[9] Il'ichev E et al 2001, Phys. Rev. Lett. **86**, 5369
[10] Lombardi F, Tafuri F, Ricci F, Miletto Granozio F, Barone A, Testa G, Sarnelli E, Kirtley J R, and Tsuei C C 2002, Phys. Rev. Lett. **89**, 207001
[11] Komissinski P V, Il'ichev E, Ovsyannikov G A, Kovtonyuk S A, Grajcar M, Hlubina R, Ivanov Z, Tanaka Y, Yoshida N and Kashiwaya S 2002, Europhys. Lett. **57**, 585
[12] Testa G, Monaco A, Sarnelli E, D'Agostino A, Kang D J, Tarte E J, Mennema S H, Bell C and Blamire M G 2004, Supercond. Sci. and Technol. **17**, 287-290; Papari G, Carillo F, Born D, Bartoloni L, Gambale E, Stornaiuolo D, Pingue P, Beltram F, Tafuri F 2009, IEEE Trans. On Appl. Supercond. **19**, 183-186.
[13] Schwarz T, Nagel J, Wlbing R, Kemmler M, Kleiner R, Koelle D 2013, ACSNano **7**, 844-850
[14] Delin K A and Kleinsasser A W 1996, Supercond. Sci. and Technol. **9**, 227-269
[15] Sarnelli E, Adamo M, De Nicola S, Cibella S, Leoni R, Nappi C 2013, Supercond. Sci. And Technol. **26**, 105013.
[16] Tanaka Y and Kashiwaya S 1997, Phys. Rev. B **56**, 892-912
[17] Komissinski P V, Il'ichev E, Ovsyannikov G A, Kovtonyuk S A, Grajcar M, Hlubina R, Ivanov Z, Tanaka Y, Yoshida N and Kashiwaya S 2002, Europhys. Lett. **57,** 585-591
[18] Nappi C, De Nicola S, Adamo M and Sarnelli E 2012, Phys. Rev. B **86**, 024517

Grain boundary engineering with gold nanoparticles

F Schmidl, C Katzer, P Michalowski, S Koch and V Tympel
Institute of Solid State Physics, Department of Low Temperature Physics, Friedrich-Schiller University Jena, Helmholtzweg 5, D-07745 Jena, Germany

E-mail: frank.schmidl@uni-jena.de

Abstract. We investigated high-T_C grain boundary Josephson junctions with and without incorporated gold nanoparticles. Pulsed laser deposition was used for the deposition of $YBa_2Cu_3O_{7-\delta}$ thin films on $SrTiO_3$ bicrystal substrates with different grain boundary angles. During the deposition process, single-crystalline nanoparticles self-assembled from a thin gold layer which was sputtered on the substrate before the YBCO deposition. The interaction between nanoparticles and thin film growth significantly influences the quality of the YBCO films [1]. The critical current density and the critical temperature of the superconducting films can be increased in a defined manner. Furthermore, the nanoparticles influence the growth conditions in the region of the grain boundary and thus the properties of the later patterned Josephson junctions. The comparison between Josephson junctions with and without nanoparticles on the same substrate shows a reduction of the critical current I_C and an increase of the normal state resistance R_N for all investigated types of grain boundaries in the areas with gold nanoparticles. In some cases we even found an increase of the resulting I_CR_N product. We present the influence of light irradiation on the properties of the Josephson junctions.

1. Introduction

Josephson junctions are the basic elements of superconducting analogue and digital devices, e.g. SQUIDs and voltage standards. For practical applications it is necessary to develop techniques which allow the preparation of Josephson junctions with specific superconducting properties. Critical current density J_C and specific normal resistance ρ_N as well as the capacitance C of Josephson junction should be optimized independent from each other. In this paper we describe the possibilities and limitations of a new method for tuning the superconducting properties of bicrystal Josephson junctions based on $YBa_2Cu_3O_{7-X}$ (YBCO) thin films. The incorporation of Au nanoparticles changes the growth conditions of the YBCO film in the grain boundary region so that critical current density and normal resistance of such Josephson junctions can be influenced on the same substrate in a different way.

2. Sample preparation

For our investigation we used $SrTiO_3$ bicrystal substrates (symmetric and asymmetric grain boundaries) with grain boundary angles of 24°, 30° and 36.8°. YBCO films with a thickness between 50 nm and 150 nm were epitaxialy grown on these substrates by pulsed laser deposition under standard deposition conditions [2]. The superconducting YBCO films in these investigations had a critical temperature of about 90 K, the critical current density without grain boundaries was better than $3 \cdot 10^6$ A/cm^2 at 77 K. The films were patterned by photolithography and Ar-ion beam etching on a cooled sample holder to avoid degradation in a-b-direction. Using this technique, Josephson junctions and dc-SQUIDs with a junction width between 3 μm and 10 μm across the grain boundaries were prepared.

In our investigations we use a very thin sputtered Au seed layer (between 1 nm and 10 nm thickness) which only covers one side of the substrate to ensure a direct comparison between both types of junctions on the same substrate. We found the dewetting of these gold layers to be depending on the seed layer thickness, substrate material, temperature and the vacuum conditions (partial pressure) in the preparation chamber. Thus, by dewetting the seed layer a defined number of Au nanoparticles with a special distribution (size and distance) were formed. During the deposition of the YBCO layer this particle ensemble is modified. Particle diameter, volume distribution and particle density were changed depending on the seed layer thickness as well as on the film thickness of the YBCO layer itself. Meanwhile, the incorporation of self-organized Au nanoparticles also changes the growth conditions of the YBCO thin films [2]. In REM and TEM investigations we found a smoother surface, the suppression of a-axis growth and bending of the YBCO layers in vicinity of the nanoparticles. The result is a small increase of the critical temperature T_C and a significant increase of the critical current density J_C at 77 K of the investigated samples with a seed layer thickness between 1 nm and 5 nm. Based on these results we use the same preparation conditions for our bicrystal experiments.

3. Experimental results

Near T_C of the superconducting film we observe a typical foot structure in the resistance-temperature dependence. Depending on the grain boundary angle and the junction width this behaviour can reduce the operation temperature of the Josephson junctions even below the boiling temperature of liquid nitrogen. The current-voltage characteristics of our prepared Josephson junctions with Au nano-particles show a RSJ like behaviour in a wide temperature range (figure1). The slope of the extracted temperature dependence of the critical current of Josephson junctions with nanoparticles follows the dependence of junctions without particles [3]. Changes in the seed layer thickness did not affect this behaviour (figure 2).

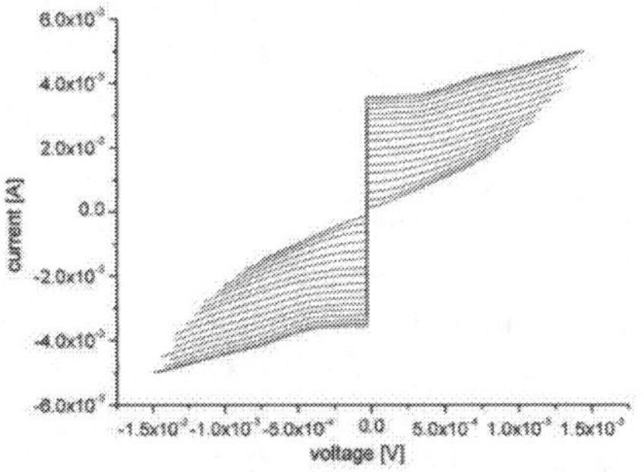

Figure 1. Current – voltage characteristics of a Josephson junction (30° grain boundary) at different temperatures (5 K to 85 K, step width 5 K). The Au seed layer thickness was 3 nm.

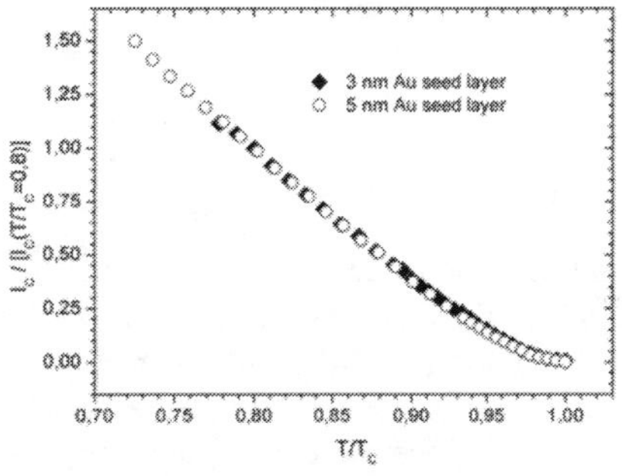

Figure 2. Normalized critical current density as a function of normalized temperature for Josephson junctions (30° grain boundary) with different Au seed layer thicknesses (open symbol 3 nm, filled symbol 5 nm)

In contrast to the experiments with simple microbridges on single crystal substrates, we found a clear reduction of the critical current density [2] of the Josephson junctions in all of the investigated samples for all grain boundary angles. This cannot be explained by a simple reduction of the critical current density by the area or volume of the incorporated normal conducting nanoparticles itself. Furthermore, from REM and TEM investigations we have no indication for any additional agglomeration of nanoparticles along the grain boundary region. The reduction of the critical current density is correlated with an increase of the specific normal resistance of the Josephson junctions. Both dependencies influence the I_CR_N product of the Josephson junctions. Figure 3 shows an overview over different Josephson junctions and dc-SQUIDs with Au nanoparticles (filled symbols). In the background we marked former results without nanoparticles (open symbols). In this more or less statistic overview we see that the $I_C R_N$ product follows the well-known scaling law $I_C R_N \sim J_C^n$ for bicrystal Josephson junctions with n close to 0.5 [4]. A more detailed description is only possible if we focus our interest on junctions prepared on the same substrate but with different Au seed layer thickness. This is necessary to minimize the influence of changed preparation conditions during the deposition of different YBCO films on different substrates. Figure 4 shows the same relation for junctions on two different 36.8° bicrystals. The incorporation of nanoparticles from a 3 nm seed layer (filled symbols) creates a reduction of the critical current density but a small increase of the related $I_C R_N$ product of the Au modified on the same substrate [5]. If we increase the seed layer thickness from 3 nm (filled symbols) to 5 nm (half filled symbols), the reduction of J_C is stronger in the case of a 30° grain boundary (figure 4). In this case we also found a reduction of the $I_C R_N$ product on the same substrate.

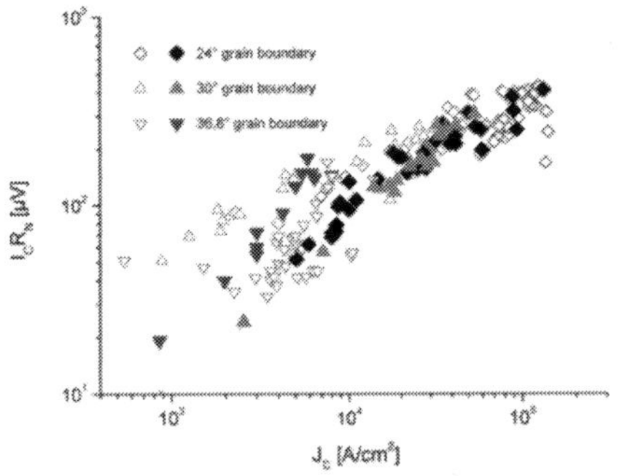

Figure 3. $I_C R_N$ product as a function of J_C for Josephson junctions and dc SQUIDs with (filled symbols) and without Au nanoparticles (open symbols)

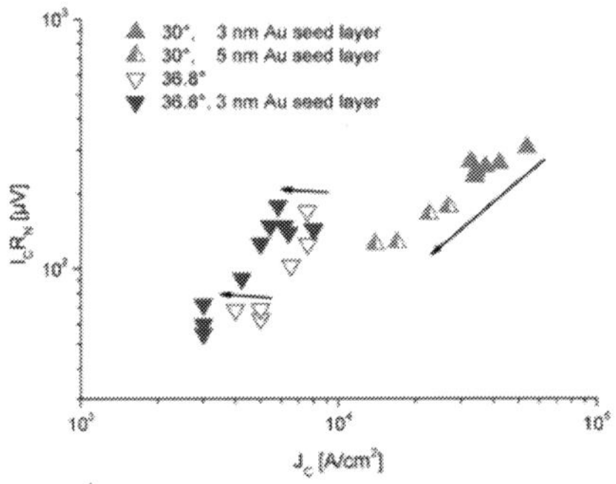

Figure 4. $I_C R_N$ product as a function of J_C for dc SQUIDs on two 36.8° and one 30° bicrystal substrate. Au seed layer thickness 3 nm (filled symbols) and 5 nm (half-filled symbols). The change in the junction properties on the same substrate is marked by arrows.

The magnetic field dependence of the critical current (77K) for all grain boundary angles shows a more regular, Fraunhofer-like pattern compared to junctions without nanoparticles on the same substrate. A possible leakage current is significantly supressed. Furthermore, these investigations give no evidence for a more inhomogeneous current distribution inside the Josephson junctions with Au nanoparticles. Another interesting possibility to tune the junction properties is the irradiation with light. Persistent superconductivity can induce additional carriers in oxygen deficient film or grain boundaries, thus increasing the critical current of the Josephson junctions, in the most cases in the order of 10 % or 20 % [7]. The main problem of all these former experiments was the reduction of the normal resistance of the grain boundary junctions during the irradiation. We have done the same experiments for our junctions. A blue light emitting diode (LED, wavelength between 400 nm and 500 nm) was placed near by the junctions inside the magnetically shielded measurement stick. Figure 5 shows the current - voltage characteristics of two junctions with the same junction width on the same bicystal substrate again. The characteristics were measured before and after the irradiation (30 min) with the LED at 77 K. After the irradiation the sample was warmed up above T_C and cooled down again to 77 K. In both cases we found an increase of the critical current. But only the Josephson junction with Au nanoparticles also showed a minor increase of the normal resistance and not the reduction measured in all former experiments. In our investigations we found a maximum enhancement of the values of critical current and normal resistance of about 250% and 5%, respectively. Using nanoparticle modified Josephson junctions it was even possible reaching higher critical currents and normal resistances than in non–modified junctions on the same substrate. Correspondingly, a real increase of the $I_C R_N$ product after irradiation was observed (figure 5). The physics behind this behaviour is still unclear. The changing growth conditions of the YBCO can create a grain boundary with a lower electron transparency (higher R_N) compared to the part without Au nanoparticles. The explanation of the strong increase of the critical current density after irradiation is more complicated. An interaction between radiation and our plasmonically active nanoparticles might be a possible candidate [7]. Additional experiments with different wavelengths should help to solve this question.

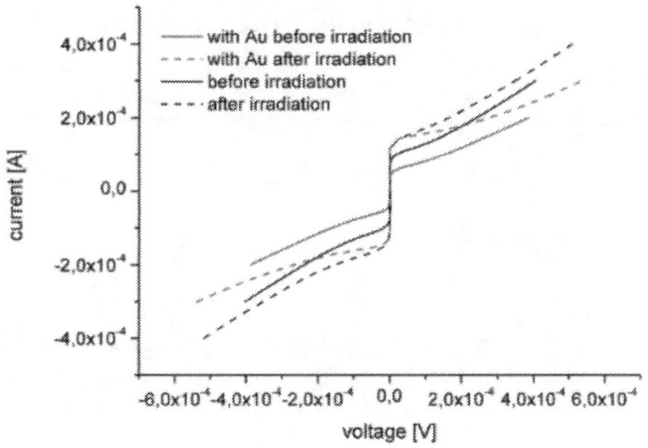

Figure 5. Current-voltage characteristics for Josephson junctions on the same 24° bicrystal substrate (junction width 7 μm, film thickness 150 nm, red curves with Au nanoparticles) before and after irradiation with blue LED.

4. Summary

The incorporation of Au nanoparticles offers the possibility to tune the superconducting properties of Josephson junctions. Critical current density and specific normal resistance can be changed on the same bicrystal substrate depending on the Au seed layer thickness with a local resolution depending on the resolution of the used lithography technology. Irradiation with light can change the junction properties in a more significant way as in the case of unmodified Josephson junctions. Both parameters I_C and R_N could be increased in such a manner.

5. Acknowledgment
C. Katzer would like to thank the Landesgraduiertenförderung Thüringen for financial support.

6. References

[1] Katzer C, Stahl C, Michalowski P, Treiber S, Schmidl F, Seidel P, Albrecht J, Schütz G 2013 *New Journal of Physics* **15**, 113029

[2] Katzer C, Westerhausen M, Uschmann I, Schmidl F, Hübner U, Seidel P 2013 *Supercond. Sci. Technol.* **26**, 125008

[3] Michalowski P, Schmidt M, Schmidl F, Grosse V, Kuhwald D, Katzer C, Hübner U, Seidel P 2011 *Physica Status Solidi RRL* **5** (8) 268-70

[4] Gross R, Chaudhari P, Kawasaki M and Gupta A 1990 *Phys. Rev. B, Condensed matter* **42** 10735-37

[5] Michalowski P, Katzer C, Schmidl F, Seidel P 2012 *J. Phys.: Conf. Ser.* **393** 012003

[6] Elly J, Medici M G, Gilabert A, Schmidl F, Seidel P, Hoffmann A, Schuller I K 1997 *Phys. Rev. B, Condensed matter* **56** R8507-10

[7] Katzer C, Grosse V, Schmidl F, Michalowski P, Schmidl G, Mueller R, Dellith J, Schmidt C, Jatschka J, Fritzsche W 2012 *J Nanopart Res* **14**, 1285

Coherent emission of intrinsic Josephson junctions

Alexander Grib[1, 2] **and Paul Seidel**[1]*
[1]Institut für Festkörperphysik, Friedrich-Schiller Universität Jena, D-07743 Jena, Germany
[2]Physics Department, Kharkiv V. N. Karazin National University, Svobody sq. 4, 61022 Kharkiv, Ukraine

The model of synchronization of intrinsic Josephson junctions in a mesa structure of a high-temperature superconductor is developed. According to our model, the mesa structure does not consist of continuous layers but is divided into small stacks of layers. These stacks form the continuous environment which is considered as a transmission line. We showed that the conventional s-wave dynamic model for the Josephson junction and the d-wave dynamic model give nearly the same particularities in IV-characteristics. We showed also that the model describes numerically the dependence of the frequency of the coherent emission on the width of samples obtained in experiments.

1. Introduction

It was experimentally shown that systems of intrinsic Josephson junctions of high temperature superconductors radiate coherently on frequencies that coincide with frequencies of geometrical resonances of the system [1]. For the explanation of this effect the hypothesis about inhomogeneous distributions of critical currents in samples was made [1]. Structures of intrinsic Josephson junctions usually have defects like boundaries of grains and dislocations, and it is naturally to suppose that the system does not consist of continuous layers but is divided into 'grains'. At the same time, the set of grains forms a continuous environment, which has a definite refraction index for waves with the length which exceeds the size of the 'grain'. Each of the 'grains' we consider as the stack of n junctions with the linear size κ_0. Earlier we showed that our approach explains the appearance of self-induced steps in IV-characteristics of junctions, the coherent radiation which appears in vicinity of these steps and the strongly inhomogeneous distribution of the temperature along the sample [2]. In the present paper we compare calculations of self-induced steps obtained with the use of the d-wave and the s-wave dynamic model of the Josephson junction. Then we model the mesa structure with the width of 80 micrometers and the height of 1 micrometer and calculate the dependence of the frequency at which the first maximum of emitted power appears as a function of the reciprocal width of the system and compare it with the experimental data [1].

2. The model

The high-frequency scheme of the system and the meaning of indices for layers, cells and stacks in our model is shown in figure 1. Each of the stacks has the source of the direct current I_b. In the conventional s-wave resistive model of the Josephson junction [3], the junction with the indices (i, l) is

* Corresponding author; E-mail: Paul.Seidel@uni-jena.de

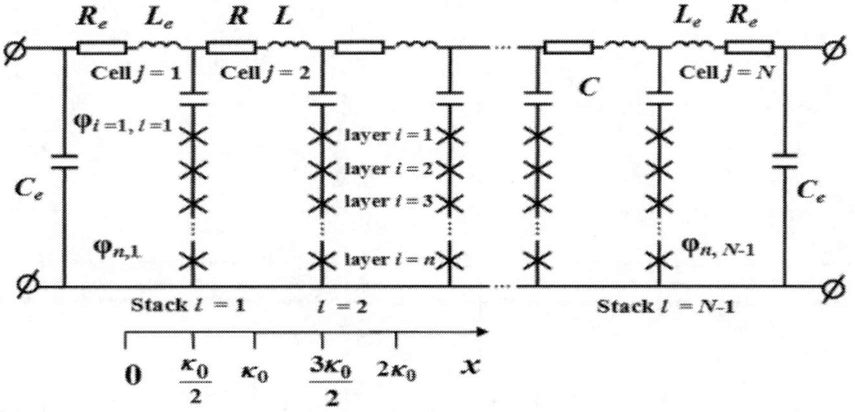

Figure 1. The high-frequency scheme of the system. Josephson junctions are marked by crosses.

represented as a parallel connection of the capacitance $C_{0i,l}$, the electric resistance $R_{0i,l}$ and the source of the Josephson current $I_{i,l} = I_{cj,l}\sin\varphi_{i,j}$, where $\varphi_{i,l}$ is the difference of phases of the order parameter of superconductor across the junction and $I_{ci,l}$ is the critical current of the junction. Dynamics of phases $\varphi_{i,l}$ of junctions in the system is described by a system of equations [2,3]:

$$\alpha C_{0i,l}\ddot{\varphi}_{i,l} + (\alpha/R_{i,l})\dot{\varphi}_{i,l} + I_{ci,l}\sin\varphi_{i,l} = I_b - \dot{q}_j + \dot{q}_{j+1}, \quad l = j = 1...N-1, \tag{1}$$

$$L\ddot{q}_j + R\dot{q}_j + (2q_j/C) = \alpha\left(\sum_{i=1}^{n}\dot{\varphi}_{i,l-1} - \sum_{i=1}^{n}\dot{\varphi}_{i,l}\right), \quad l = j = 2...N-1, \tag{2}$$

$$L_e\ddot{q}_1 + R_e\dot{q}_1 + (2q_1/C_e) = -\alpha\sum_{i=1}^{n}\dot{\varphi}_{i,1}, \quad L_e\ddot{q}_N + R_e\dot{q}_N + (2q_N/C_e) = \alpha\sum_{i=1}^{n}\dot{\varphi}_{i,N-1}, \tag{3}$$

where one or two dots over symbols mean the first or the second derivatives with respect to time, correspondingly, $\alpha = \Phi_0/2\pi$ with Φ_0 is the quantum of magnetic flux, q_j is the electric charge that passes through the inductance L of the j-th cell of the transmission line. It is supposed that $I_{ci,j}R_{0i,j} = V_c = const$ for all junctions. The velocity of the propagation of waves in a transmission line is equal to $c' = c_0 n'^{-1} = \kappa_0(LC)^{-1/2}$, where c_0 is the velocity of light in vacuum, $n' \approx 4.47$ is the refraction index. Voltages across junctions are equal to $V_{i,l} = \alpha\dot{\varphi}_{i,l}$. The averaged frequency and the averaged voltage are equal to $\langle\nu\rangle = [2\pi n(N-1)]^{-1}\left\langle\sum_l\sum_i\dot{\varphi}_{i,l}\right\rangle$ and $\langle V\rangle = \Phi_0\langle\nu\rangle$, where $\langle...\rangle$ means averaging over time. The mean square of the ac voltage is $S = \left\langle\left(\sum_{l=1}^{N-1}\sum_{i=1}^{n}(\alpha\dot{\varphi}_{i,l} - \langle V_{i,l}\rangle)\right)^2\right\rangle$. The value of S is proportional to the power of ac radiation emitted by the system. The relation of the ac power emitted in the l-th row to the averaged ac power emitted by one junction in this row is equal to $p_l = \left\langle (1/n)\left[\sum_{i=1}^{n}(\alpha\dot{\varphi}_{i,l} - \langle V_{i,l}\rangle)\right]^2\right\rangle\left[(1/n)\left\langle\sum_{i=1}^{n}(\alpha\dot{\varphi}_{i,l} - \langle V_{i,l}\rangle)^2\right\rangle\right]^{-1}$. If all junctions oscillate in-phase,

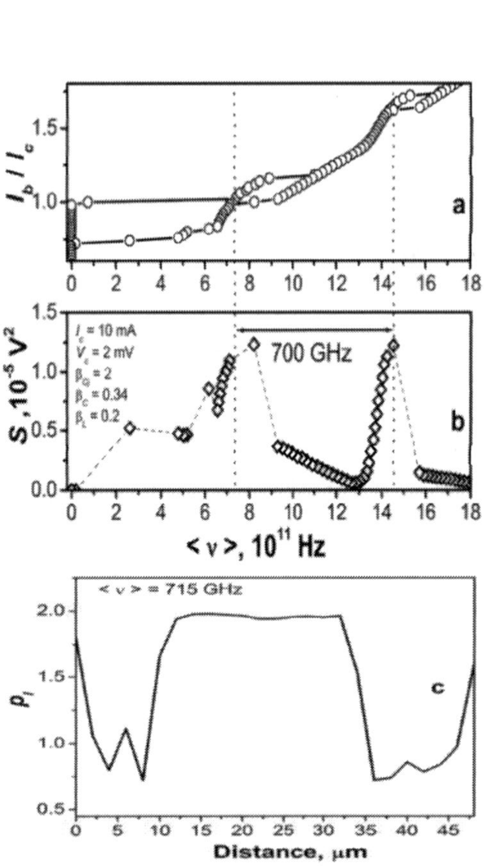

Figure 2. The s-wave model for the system of 2 layers. Dependences of $\langle \nu \rangle = f(I_b/I_c)$ (a), $S = f(\langle \nu \rangle)$ (b) and $p_l = f(x)$ (c). The width of the system is 48 μm.

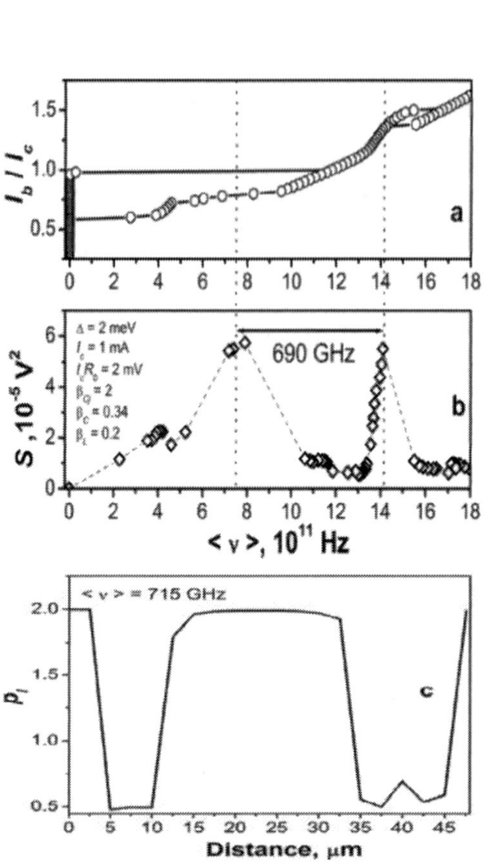

Figure 3. The d-wave model for the system of 2 layers. Dependences of $\langle \nu \rangle = f(I_b/I_c)$ (a), $S = f(\langle \nu \rangle)$ (b) and $p_l = f(x)$ (c). The width of the system is 48 μm.

then $p_l = n$. We developed also the dynamic model of $d_{x^2-y^2}$-wave symmetry of pairing [4]. In this case the second terms in equations (1) are changed to quasiparticle tunnel currents $I_q(\dot{\varphi}_{i,l})$ calculated according to standard expressions for the superconductor with $d_{x^2-y^2}$-wave symmetry of pairing [4].

3. Results of calculations

Dependences of the averaged frequency of Josephson radiation on the normalized bias current calculated with the use of the s-wave model are shown in figure 2a for the system with two layers. Valued of parameters were as follows: $I_{ci,1} = I_c(1-\delta)$, $I_{ci,2} = I_c(1+\delta)$, with I_c = 10 mA and δ = 0.02, V_c = 2 mV, C_{0il} = const = 25 pF, C = 0.28 pF, C_e = 0.22 pF, L_e = 30 pH, L = 3 fH, $R = 10^{-4}$ Ohm, R_e =10 Ohm, $\kappa = 2\,\mu$m. Both self-induced steps of IV-curves and maxima of S at frequencies of these steps are clearly seen in figure 2a, b. The distribution of values of p_l along the system shows that junctions are synchronized in the center of the transmission line (figure 2c). The value of the energy gap in the d-wave model was Δ = 2 meV. Values of resonant frequencies and the distribution

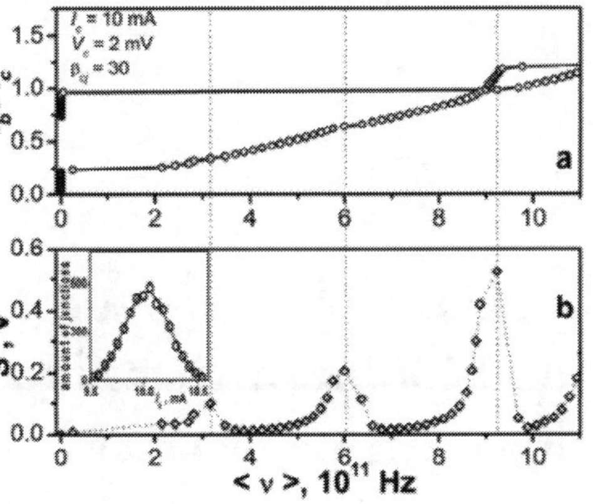

Figure 4. The s-wave model. Dependences of the averaged frequency of Josephson radiation on the normalized bias current (a) and $S = f(\langle\nu\rangle)$ (b) for the system of 80 μm width and with 667 layers. In inset the distribution of critical currents is shown.

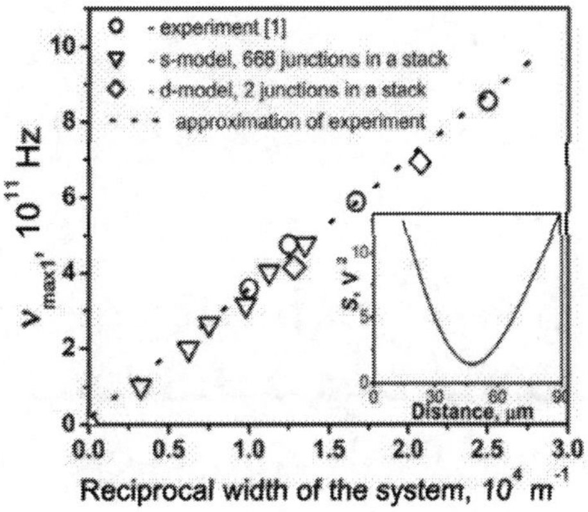

Figure 5. The dependence of $\nu_{\max 1}$ on the reciprocal width of the system: experimental data [1] (circles), the approximation of experimental data (dot line), our calculations (triangles and diamonds). In inset the distribution of S along the length for the 90 μm system and $\langle\nu\rangle$=300 GHz is shown.

of p_l (figure 3) calculated with the use of the d-wave model are nearly the same as for the s-wave model, though they obtained at different values of critical currents (see figure 2a and 3a). In figure 4 dependences of $\langle\nu\rangle = f(I_b/I_c)$ and $S = f(\langle\nu\rangle)$ are shown for 667 intrinsic junctions that corresponds to the thickness of the mesa structure of about 1 μm [1]. Parameters of calculations were as follows: $C_{0i,l}$=20 pF, $L = 0.5$ pF, R_e=10^4Ohm, L_e=10L, C_e=C, $\kappa = 5\,\mu$m. We set the Gaussian random distribution of critical currents (see inset in figure 4b). We plotted frequencies of the first maximum of emitted radiation $\nu_{\max 1}$ on the reciprocal width of the system with 668 intrinsic junctions (figure 5). Parameters of calculations were the same as in figure 4. It is seen that our calculations are in good agreement with the experimental data [1]. The distribution of S along the system (see inset in figure 5) shows that in this case junctions are synchronized at ends of the system.

4. Conclusions

We developed the model of coherent emission from the mesa structure of intrinsic Josephson junctions. We showed that different models of junctions give the same particularities of IV-characteristics, i.e. that the effect does not depend on the model for the individual junction and that self-induced steps are produced under the influence of the transmission line. The calculated resonant frequencies of coherent emission are in good agreement with the experiment [1].

References

[1] Ozyuzer L, Koshelev A E, Kurter C, Gopalsami N, Li Q, Tachiki M, Kadowaki K, Yamamoto T, Minami H, Yamaguchi H, Tachiki T, Gray K E, Kwok W-K, Welp U 2007 *Science* **318** 1291

[2] Grib A and Seidel P 2012 *Low Temp. Phys.* (*Fiz. Nizk. Temp.*) **38** 321

[3] Likharev K K 1991 *Dynamics of Josephson junctions and circuits* (Philadelphia, Gordon and Breach) p 750

[4] Won H, Maki K 1994 *Phys. Rev.* B **49** 1397

Different shape normal metal interlayers between niobium based SIS junctions and niobium titanium nitride leads and their influence on the electron temperature

S Selig, M P Westig, K Jacobs and C E Honingh

Kölner Observatorium für Submillimeter Astronomie (KOSMA), I. Physikalisches Institut, Universität zu Köln, Germany

E-mail: selig@ph1.uni-koeln.de

Abstract. In this paper we demonstrate the reduction of heating in a niobium superconductor-insulator-superconductor (SIS) junction with aluminum-oxide tunnel barrier embedded in a niobium-titanium-nitride circuit. Nonequilibrium quasiparticles which are created due to the Andreev trap at the interface between the niobium and the niobium-titanium-nitride layers are relaxed by inserting a normal-metal conductor of gold between these two layers. In an earlier work we explained the observed relaxation of nonequilibrium quasiparticles due to the geometrically assisted cooling effect. In this paper we investigate this cooling effect in dependence of the normal-metal layer shape and size. We expect that an adapted normal-metal layer is necessary for implementation in practical terahertz SIS heterodyne mixer circuits. We observe in DC-measurements of a large number of devices a clear relation between the volume of the gold layer and the effective electron temperature in the device. Our central finding is that the shape of the gold layer does not influence the cooling provided that the volume is sufficient.

1. Introduction

Heterodyne receivers are a powerful tool for astronomy at terahertz (THz) frequencies. With their high sensitivity and spectral resolution of the order of 10^5-10^6 they enable the detection of complicated emission-line spectra of astronomical objects like cold molecular clouds. These spectra are used to study the chemical and physical conditions for star formation which takes place in these clouds. In the range of 0.3 to 1.3 THz, heterodyne receivers equipped with SIS devices as frequency mixer offer the highest sensitivity. Due to their superior tunnel barrier quality, mainly Nb-AlO_x-Nb junctions are used. These are suitable for mixing up to 1.4 THz which is twice the gap frequency of Nb. Up to its gap frequency of 700 GHz, Nb is also used for the embedding HF circuit of the junction. Above this frequency the loss in Nb increases due to Cooper-pair breaking. In this case a superconductor with a higher gap like NbTiN would be the material of choice. Devices with a NbTiN embedding circuit have been fabricated before and resulted in heating caused by nonequilibrium quasiparticles trapped in the Nb junction [1]. Therefore, the presently best performing mixer devices are fabricated with one layer of the circuit made of a normal-metal conductor where quasiparticles are not trapped [2, 3, 4]. However, this decreases the sensitivity in comparison to the use of a superconductor due to signal losses in the normal conducting embedding circuit.

2. Heating and device considerations

We want to solve these heating issues by inserting a normal-metal cap of gold between the Nb-junction top electrode and the NbTiN top layer of the embedding HF-circuit. This layer provides extraction of nonequilibrium quasiparticles out of the Nb-junction and transfers their energy to the bath. We have already shown in DC measurements that heating can be significantly reduced with this approach because of the geomatrically assisted cooling effect [5]. Here we have shown that heating is substantially weaker in a device with a rather large gold cap. On the other hand, for the development of a mixer device the cap has to be as small as possible to minimize ohmic losses in the HF embedding circuit. Hence, the optimum dimensions have to be determined. Most importantly the lateral shape of the cap will have an impact on the performance of the HF circuit. Fig. 1(b) shows two examples to illustrate this. First, we choose a circular gold cap centered on the junction. In this case a substantial part of the inductive line contacting the junction has gold underneath. Secondly, we choose a half-circular shape and match the edges of gold cap and junction area. Here, the overlap of the gold cap and the matching circuit is minimized, although the penalty is a larger gold cap extending at the other side of the SIS device that will have some influence on the HF circuit as well. For the calculation of the required HF-circuit dimensions the surface impedance of the circuit material is needed. For a conventional homogeneous superconductor it can be calculated with the help of the Mattis Bardeen theory [6]. The circular gold cap together with the NbTiN embedding circuit is effectively a normal/superconducting (NS) bilayer system in which an inhomogeneous superconducting state is created due to the Andreev reflection process (proximity effect) [7]. In this case a more general theory [8] has to be applied to evaluate the surface impedance. The ohmic losses in the gold will have the largest impact if the bilayer covers the part of the circuit with the highest HF current density. For the half-circular cap, compared to the full-circular cap, a smaller part of the circuit consists of a NS bilayer and in the part behind the junction the current density is lower so its contribution in terms of ohmic losses will be smaller. To provide a foundation for future HF circuit designs we quantify in this paper the correlation between electron temperature elevation in the Nb SIS junction and the volume and shape of the gold layer in DC measurements.

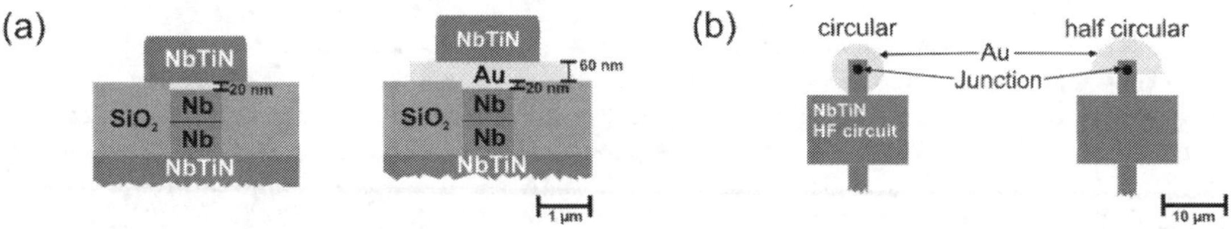

Figure 1. (a) Layer schemes of devices with a 20 nm thick layer of gold and an added 60 nm thick gold cap. (b) Top-view schematics of a NbTiN matching circuit and junction with circular or half-circular gold cap.

3. Device variations and fabrication

We fabricated devices with a variety of gold layers between the Nb top electrode and the NbTiN top wiring layer. Fig. 1(a) shows layer schemes of devices with and without such a gold cap. The devices are fabricated on a 525 μm thick low resistivity silicon substrate. The NbTiN-Nb-AlO$_x$-Nb-Au layers are patterned by optical UV-lithography and are deposited by DC magnetron sputtering. The NbTiN groundplane has a thickness of 350 nm and the Nb junction electrodes

Table 1. Selection of devices with parameters for the gold cap.

#	cap type	t_{Au} [nm]	r_{Au} [μm]	V_{Au} [μm^3]
1	-	-	-	0.02
2	half-circular	60	1.07	0.13
3	half-circular	60	1.57	0.25
4	circular	60	1.57	0.48
5	half-circular	60	2.57	0.64
6	circular	120	3.59	4.86

are both 100 nm thick. The barrier is formed by thermal oxidation of a 8 nm thick Al layer. The 20 nm thick gold layer on top provides a clean interface to the thicker gold cap. To pattern the junction area we use electron beam lithography (EBL) and reactive ion etching. An AlN etch mask is patterned on top of a polyimide layer by EBL and liftoff using PMMA as EBL resist. After etching, the SiO_2 insulating layer is deposited by RF magnetron sputtering followed by the self-aligned liftoff of the junction etch mask supported by chemical mechanical polishing (CMP). After this step, gold caps with a thickness of 60 nm are patterned on top by EBL and liftoff. In the last step the 400 nm NbTiN top wiring layer is patterned by UV-lithography and liftoff. All measured devices have the same junction area size of 1 μm^2. We used circular and half-circular shaped gold caps like those shown in Fig. 1(b) and varied the cap diameters. In table 1 a selection of the devices and their gold layer dimensions are summarized.

4. Analysis of the measurement results

In this section we present measured DC I-V curves of the fabricated devices. Since the electron temperature in the Nb electrodes depends on the current through the device these have a negative slope in the quasiparticle current onset (backbending). Datapoints from this part of the I-V curve can be translated into effective electron temperatures (T_e) using the BCS theory [9]:

$$\frac{1}{N(E_F)\mathcal{V}} = \int_0^{k_B\theta_D} d\epsilon \frac{1 - 2f[(\epsilon^2 + \Delta(T_e)^2)^{1/2}]}{(\epsilon^2 + \Delta(T_e)^2)^{1/2}}. \tag{1}$$

Here $N(E_F)$ is the single-spin density of states at the Fermi energy, $\mathcal{V}$ is the average attractive electron-phonon interaction potential, k_B is the Boltzmann constant, θ_D is the Debye temperature of the material, ϵ is the independent quasiparticle energy relative to the Fermi energy and $\Delta(T_e)$ is the electron temperature dependent superconducting gap-energy of the Nb junction and f is the Fermi distribution function at the same electron temperature. For more details we refer the reader to [5]. Fig. 2(a) shows I-V curves of the devices specified in table 1 and Fig. 2(b) shows the corresponding electron temperature in the Nb junction as a function of the DC power P. The phonon bath temperature in all measurements was 4.2 K. For all devices except #6 we observe an elevation of T_e over the phonon bath temperature. The I-V curves of devices #1 to #5 in Fig. 2(a) show backbending which decreases with increasing gold layer volume. No signs of backbending are found for device #6 suggesting an electron temperature equal to the phonon bath temperature. An effective heat-transfer coefficient α can be extracted from the $T_e(P)$ plots in Fig. 2(b), which is equal to the inverse of the slope of the curves. The larger the gold volume, the higher is α and the more the heat transfer is enhanced by the gold cap. In fact there is an approximately linear relation between α and the gold cap volume as it is illustrated in Fig. 2(c). Additionally, we find that the shape of the gold cap has no impact on α. E.g. for devices with a 0,23 μm^3 circular cap we find 0.41 $\pm$ 0.09 μW/K in comparison to 0.44 $\pm$ 0.05 μW/K for a 0,25 μm^3 half cirular gold cap.

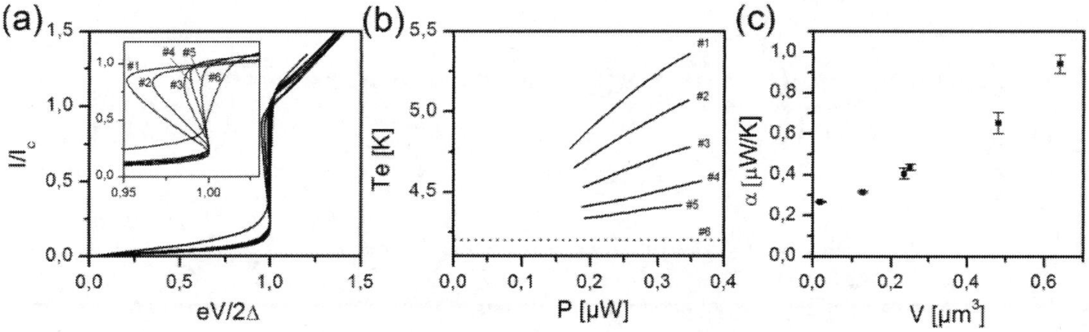

Figure 2. (a) I-V curves of the devices listed in table 1. (b) Electron temperature T_e as a function of DC power. The dotted line indicates the phonon bath temperature of 4.2 K which is equal to the electron temperature of device #6. (c) Heat transfer coefficient α as a function of gold volume. The data points are averaged over 30, 12, 11, 12, 11 and 8 (from left to right) values of individual devices. The errorbars represent the corresponding mean squared errors.

5. Conclusion and outlook

We have shown that the heat transfer out of the junction depends on the volume but not the shape of the added gold layer. This is an important finding because we can avoid the drawbacks of a circular shape like discussed in section 2 and don't lose cooling efficiency because of the different shape. HF measurements with well matched devices and an optimized gold cap geometry will have to show if the heating effects due to THz radiation are eliminated in the same way as DC heating. If this is successful this new device type will be the basis for future developments of very high sensitivity mixers at 800 GHz and 1.1 THz.

6. Acknowledgments

This work is carried out within the Collaborative Research Council 956, subproject D3, funded by the Deutsche Forschungsgemeinschaft (DFG) and by BMBF, Verbundforschung Astronomie under contractno. 05A08PK2. The devices were fabricated in the KOSMA microfabrication laboratory and measured at the I. Physikalisches Institut, Universität zu Köln.

References

[1] Leone B, Jackson B D, Gao J R and Klapwijk T M 2000 *Applied Physics Letters* **76** 780–782 ISSN 0003-6951

[2] Jackson B D, de Lange G, Zijlstra T, Kroug M, Kooi J W, Stern J A and Klapwijk T M 2006 *IEEE Transactions on Microwave Theory and Techniques* **54** 547–558 ISSN 0018-9480

[3] Karpov A, Miller D, Rice F, Stern J A, Bumble B, LeDuc H G and Zmuidzinas J 2007 *IEEE Transactions on Applied Superconductivity* **17** 343–346 ISSN 1051-8223

[4] Kroug M, Endo A, Tamura T, Noguchi T, Kojima T, Uzawa Y, Takeda M, Wang Z and Shan W 2009 *IEEE Transactions on Applied Superconductivity* **19** 171–173 ISSN 1051-8223

[5] Westig M P, Selig S, Jacobs K, Klapwijk T M and Honingh N 2013 *Journal of Applied Physics* **114** 124504–124504–11 ISSN 0021-8979

[6] Mattis D C and Bardeen J 1958 *Physical Review* **111** 412–417

[7] Andreev A F 1964 *Soviet Physics JETP* **19** 1228–1231

[8] Nam S B 1967 *Physical Review* **156**(2) 470–486

[9] Bardeen J, Cooper L N and Schrieffer J R 1957 *Physical Review* **108**(5) 1175–1204

Cryocooler operation of SNIS Josephson arrays for AC Voltage standards

A Sosso[1], N De Leo, M Fretto, E Monticone, L Roncaglione, R Rocci and V Lacquaniti
INRiM Istituto Nazionale di Ricerca Metrologica,
Strada delle Cacce 91 - 10135 Torino - Italy

E-mail: a.sosso@inrim.it

Abstract. Avoiding liquid helium is now a worldwide issue, thus cryocooler operation is becoming mandatory for a wider use of superconductive electronics. Josephson voltage standards hold a peculiar position among superconducting devices, as they are in use in high precision voltage metrology since decades. Higher temperature operation would reduce the refrigerator size and complexity, however, arrays of Josephson junctions made with high temperature superconductors for voltage standard applications are not to date available. The SNIS (Superconductor-Normal metal-Insulator-Superconductor) junction technology developed at INRIM, based on low temperature superconductors, but capable of operation well above liquid helium temperature, is interesting for application to a compact cryocooled standard, allowing to set a compromise between device and refrigerator requirements. In this work, the behavior of SNIS devices cooled with a closed-cycle refrigerator has been investigated, both in DC and under RF irradiation. Issues related to thermal design of the apparatus to solve specific problems not faced with liquid coolants, like reduced cooling power and minimization of thermal gradients for uniform operation of the chip are discussed in detail.

1. Introduction

Cryogen-free operation of superconducting devices is becoming more and more crucial to widespread the applications of superconductivity and is indeed a necessity in some cases, e.g. operation in remote locations. Considering the recent shortages, avoiding liquid helium has now become a worldwide issue, fundamental for the success of superconductive electronics [1]. Josephson voltage standards hold a peculiar position among superconducting devices, as they are used for high precision voltage metrology since the early years after the discovery of the physical phenomenon. Today, they still represent one of the most complex, yet successful, instances of a superconducting device. Cryocooler operation bears a special interest for Josephson standards, owing to the potentially tremendous impact of a new generation of digital voltmeters providing quantum accuracy through a Josephson standard operating into an integrated refrigerator. Due to the reduction in size and costs of coolers, the perspective of a quantum measurement provided by a commercial-grade instrument might be foreseeable in the future. Unfortunately, nowadays cryocoolers suitable for low temperature superconductors operation are still expensive, weighty and far from any possible integration inside an instrument case. Higher temperature operation is needed to dramatically reduce the refrigerator size and complexity [2]. At present, however, arrays of Josephson junctions made with high temperature superconductors for voltage standard applications are not yet available.

[1] To whom any correspondence should be addressed.

2. Experimental system

2.1.1. Cooling apparatus design

Reliable cryocooler operation necessitates a very specific thermal design to cope with problems not faced with liquid coolants, like minimization of thermal gradients to allow uniform operation of the chip. Moreover, the refrigerator has a reduced cooling power, thus the microwave guide to transmit the RF signal to the chip must be carefully designed in order to limit the heat load on the low temperature stage, while maintaining an effective the signal transmission.

Our system is based on a GM refrigerator with 1 W cooling power on the cold finger at 4.2 K and a minimum temperature achievable lower than 3 K, without thermal load. An additional aluminum disk was tightly connected at the top of the upwards-directed cold finger. A germanium thermometer fastened inside the disk along with a heater wire wound around it allow to monitor and control the temperature of the cooling surface. A first attempt to use a copper disk was unsuccessful because of the very long time needed to reach the operating temperature, that can be explained by a high heat capacity. The thermal contact between the disk and the cold finger was significantly improved by the insertion of an indium sheet at the interface. Eventually, the whole process to cool samples from laboratory temperature down to 4 K requires typically less than 3 hours.

The chip under test is fastened with a cryogenic glue to the copper sheet of a printed circuit board (PCB) with pads for soldering the cryocooler wires, and then bonded. Brass screws provide the thermal link between the cold finger and PCT copper sheet that holds the device. The PCB copper layer is also used as thermal conductive support for a thermometer monitoring the chip temperature. The operating temperature can be monitored and controlled with commercial instrumentation both with the disk and the chip support sensors.

Figure 1.- Overview of the cryocooler setup. On the right, detail of the cold head, including k-band waveguide, sample holder, connecting wires and holder thermometer.

In voltage standard applications, arrays of Josephson junctions are irradiated with high frequency (about 70 GHz) microwaves to generate steps with quantized voltages. To study the step properties of cryocooled Josephson devices a microwave guide was used for the transmission of the RF signal to the chip. Despite the good cooling power available, the thermal link to the outer environment introduced by a standard copper/brass waveguide would compromise the refrigerating effectiveness. A specifically designed stainless steel guide was adopted, with a significant reduction of the heat transmission to the cold area. Gold plating allowed to reduce the dramatic signal attenuation due to the low electrical conductivity of stainless steel (20 dB/m for the untreated surface). Furthermore, a beneficial side-effect of the low power dissipation along the guide is to eliminate the added thermal load inside the cryocooler due to Joule heating.

Several aspects have to be considered to minimize the heat transmission while keeping the electrical signal attenuation as low as possible. Owing to the many orders of magnitude higher thermal conductivity at cryogenic temperature of gold with respect to stainless steel, plating thermal conduction is non negligible within the overall balance, despite its extremely small cross-sectional area. We found a satisfactory compromise can be reached by plating only the inner guide surface, the outer being not relevant to signal transmission, and keeping gold thickness below 1 μm. The total attenuation obtained is below 7 dB. The semiconductor thermometer placed close to the chip to for precisely monitoring its operating temperature was used to study the guide loading and careful optimization of the thermal anchoring. The minimum temperature attainable was observed to be strongly dependent on anchor placement and effectiveness, together with a proper thermal insulation of guide sections, with variations in the Kelvin range. At present, the devices under test can be cooled with the guide in place down to temperatures near 3 K.

3. SNIS junctions for cryocooler operation

Nb/Al-AlOx/Nb SNIS junctions achieve some specific properties concerning the damping of the *IV* characteristic and the values of the critical current density and characteristic voltage through a trimming of some fabrication parameters [3,4,5].
A specific feature of SNIS is the temperature stability of its electrical parameters, measured as the temperature derivative of I_c (V_c) *vs.* T [6,7].
It is in fact possible to minimize this derivative as function of a parameter $\gamma_{eff} = \gamma_{S/N}\, d_{Al}/\xi_{Al}$ where $\gamma_{S/N} = R_{Nb/Al}/\rho_{Al}\ \xi^*_{Al}$ with $R_{Nb/Al}$ the product of the interface resistance and its area, and $\xi^*_{Al} = \xi_{Al}(T_{c,Al}/T_{c,Nb})^{1/2}$ [8].
In particular the devices measured in this experiment were series arrays for programmable voltage standard fabricated in cooperation with the Physikalisch Technische Bundesanstalt (PTB) [3] and we measured segments from 1 to 1024 junctions, to evaluate the effect of the RF and DC dissipated power by increasing the junction numbers.
The SNIS junctions tested at different temperatures from 3.5 to 6 K, had critical current density ranging from 5 to 20 kA/cm^2 and V_c from 250 to 500 μV.
These parameters were obtained using a thickness of aluminum from 80 to 100 nm, while Nb electrodes were 200 nm.
Correspondingly the parameter γ_{eff} varied between 10 and 20 .

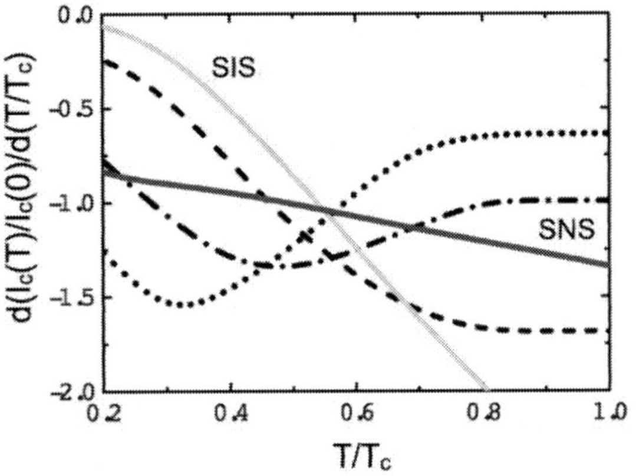

Figure 2.- Normalized temperature derivative of the critical supercurrent calculated for SNIS junctions with γ_{eff} =2, 10, and 20 (dashed, dashed-dotted, and dotted lines, respectively). In the figure are also reported the curves for conventional SIS and SNS junctions.

From figure 2, we see that this corresponds to a temperature derivative of I_c normalized to $I_c(0)$ from 0.5, optimal condition T/T_c> 0.7 K for γeff=20, which however becomes about 1.15 at 4.2 K and 1.5 at 3.5 K. On the other hand, an almost constant value of 1 of the derivative for γ_{eff}=10, in the experimental temperature range is possible.
First experiments have been carried on with both single junction and array devices, checking for the effect of DC current and microwave bias.
The experimentally estimated dissipated power ranged between fraction of mW (single junction, step n=1, up to 12 mW, 1024 junctions step n=2.
Quantized steps were observed in segments with a low number of junctions up to temperatures well above 4.2 K (Fig. 3), but were not attainable for the higher voltage sections, with many series-

connected junctions. At present we explain this behavior as a combined consequence of a localized temperature raise, due to an insufficient heat exchange between the chip and the cold finger, along with a limited strenght of electromagnetic shielding, as shown by the flux-trapping observed more and more frequently with the number of junctions in the section.

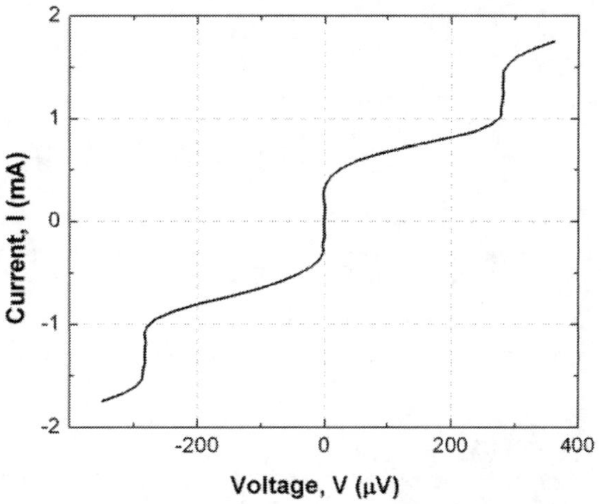

Figure 3. - Quantized steps of a low segment of a SNIS array (JOSY PTB 4-18) measured in cryocooler setup (estimated temperature: 6K).

4. Conclusions

Preliminary experiments of series arrays in a cryocooler set-up have been carried out, searching for optimal operating conditions.

Further experiments are needed for increasing operation to a high number of series connected junctions. This requires from one hand to use devices which show a minimal temperature dependence of the electrical parameters, and on the other hand an improved measurement environment, with a good thermal exchange between the junctions and the cold finger.

Acknowledgments

This work was partially supported by EMRP, a research program jointly funded by the EMRP participating countries within EURAMET and the European Union.

References

[1] Howe L, Burroughs C, Dresselhaus P, Benz S and Schwall R 2013 *IEEE Trans. Appl. Supercond.* **23** 1300605-1300605

[2] Yamamori H, Yamada T, Sasaki H and Kohjiro S 2012 *Trans. on Electronics* **95** 329-336

[3] Lacquaniti V, De Leo N, Fretto M, Sosso A Müller F and Kohlmann J 2011 *Supercond. Sci. Technol.* **24** 045004 doi:10.1088/0953-2048/24/4/045004

[4] Lacquaniti, V, Cagliero C, Maggi S, Sosso A and Steni R 2005 *IEEE Trans. Appl. Supercond.* **11** 1130-1133

[5] Lacquaniti V, Maggi S, Polcari, A Steni, R. and Andreone D *Appl. Phys. Lett.* 2005 **86** 04-2501

[6] Lacquaniti V, De Leo N, Fretto M, Maggi S. and Sosso A *Appl. Phys. Lett.* 2007 **91** 252505-252505-3

[7] Lacquaniti, V, Andreone D, De Leo N, Fretto M, Maggi, S, Sosso A and Belogolovskii M 2007 *IEEE Trans. Appl. Supercond.* **17** 609-612.
Lacquaniti V, De Leo N, Fretto M, Sosso A and Belogolovskii M *Journ. App. Phys.* 2010 **108** 093701-093701

[8] Lacquaniti V, Belogolovskii M, Cassiago C, De Leo, N, Fretto, M and Sosso A 2012 *New Journal of Physics* **14** 023025

Fabrication of Niobium Nanobridge Josephson Junctions

T Tachiki, K Horiguchi and T Uchida

Department of Electrical and Electronic Engineering, National Defense Academy, 1-10-20 Hashirimizu, Yokosuka, Kanagawa 239-8686, Japan

E-mail: tachiki@nda.ac.jp

Abstract. To realize antenna-coupled Josephson detectors for microwave and millimeter-wave radiation, planar-type Nb nanobridge Josephson junctions were fabricated. Nb thin films whose thickness, the root mean square roughness and the critical temperature were 20.0 nm, 0.109 nm and 8.4 K, respectively were deposited using a DC magnetron sputtering at a substrate temperature of 700°C. Nanobridges were obtained from the film using 80-kV electron beam lithography and reactive ion-beam etching in CF_4 (90%) + O_2 (10%) gases. The minimum bridge area was 65 nm wide and 60 nm long. For the nanobridge whose width and length were less than 110 nm, an *I-V* characteristic showed resistively-shunted-junction behaviour near the critical temperature. Moreover, Shapiro steps were observed in the nanobridge with microwave irradiation at a frequency of 6 - 30 GHz. The Nb nanobridges can be used as detectors in the antenna-coupled devices.

1. Introduction

It is well known that the Dayem bridge has one of the simple device structures which show the Josephson effect. The bridge width and length should be comparable to the coherence length ξ of a superconducting material. Since $\xi \approx 38$ nm for Nb, this bridge is called nanobridge [1]. Recently, nanobridges can be successfully obtained, because electron beam lithography (EBL) technology is highly developed, e.g. the minimum width of a line pattern < 10 nm [2].

The nanobridge Josephson junction is a planer junction. This junction is suitable to fabricate the antenna-coupled detector compared to a conventional vertical junction such as a superconductor-insulator-superconductor junction, because a high-frequency current induced in a thin-film planer antenna such as slot, bow-tie and spiral antennas is strongly coupled to the Josephson current flowing along the planer junction [3]. Using the antenna-coupled detector, it is expected to realize a harmonic mixer that detects microwaves and millimeter-waves with low local oscillator (LO) power and low LO frequency.

To fabricate nanobridges from Nb thin films, a thickness of the film should be less than 20 nm, because an aspect ratio of the thickness to the bridge width (or length) can hardly exceed 0.5 due to reactive ion etching. In this study, deposition conditions for 20-nm-thick Nb-sputtered thin films were evaluated. And then Nb nanobridges were fabricated from the thin films, and their electrical properties were evaluated for the evidence that the nanobridges acted as Josephson junctions.

2. Fabrication of 20-nm-thick Nb thin films

Superconducting Nb thin films less than 20 nm with a flat surface morphology is required to fabricate Nb nanobridges. The thin films were deposited onto $10\times10\text{-mm}^2$ SiO_2/Si substrates by a DC magnetron sputtering system with a DC power of 200 W and 100-mm target-substrate distance in the Ar atmosphere of 0.8 Pa. The base pressure of the system was 7.0×10^{-6} Pa. Under these conditions, a deposition rate was 0.2 nm/s, and 20 nm-thick Nb films having the critical temperature T_c of 7.3 K were obtained. In order to improve T_c, a substrate temperature T_s was increased from room temperature to 700°C [4]. Figure 1 shows dependencies of T_c and the residual resistance ratio (RRR) on T_s. RRR is defined as R(273 K) / R(10 K). T_c increased with increasing T_s and almost saturated around 700°C. The maximum T_c of 8.4 K was obtained at T_s = 700°C. Therefore, the films have a sufficient value of T_c even if the films are slightly degraded during the fabrication process of the nanobridge described below. Since T_s dependence of T_c is similar to that of RRR, the origin of T_c degradation is residual impurities such as Niobium oxide in the films.

Figure 2 shows an atomic force microscope (AFM) image of the Nb film deposited at T_s = 700°C. Dense Nb grains with diameters of 40 – 100 nm were formed, as shown in the figure. The root mean square (RMS) of surface roughness in the film was 0.109 nm. This value was half of the RMS in the film deposited at room temperature.

From these results, the obtained films are applicable to the nanobridge Josephson junctions.

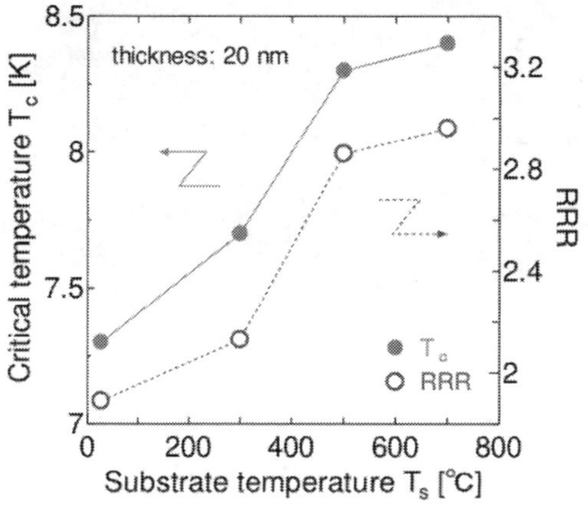

Figure 1. Dependencies of critical temperature and residual resistance ratio (RRR) on substrate temperature.

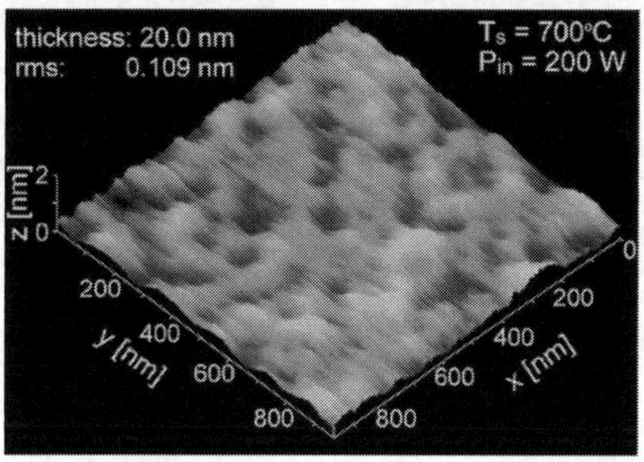

Figure 2. AFM image of 20-nm-thick Nb thin film deposited at a substrate temperature of 700°C.

3. Fabrication and electrical properties of Nb nanobridges

Nb nanobridges were fabricated from the 20-nm-thick Nb thin films mentioned above. The device structure of the nanobridge is shown in figure 3 (a). The bridge length ℓ and width w are defined in the magnified view in the figure. The taper angle of Nb banks, θ = 45°, was selected to avoid the microloading effect [5] in reactive ion-beam etching (RIBE), i.e. decrease in an etching rate of the Nb film around the nanobridge.

3.1. Device fabrication

Prior to the nanobridge fabrication, Nb contact pads and wires, as shown in figure 3 (a), were patterned on the Nb thin film by conventional photolithography. Then, the film was etched using RIBE by EIS-200ER (Elionix Co.) in a mixture of CF_4 (90%) and O_2 (10%) gases. The ion acceleration voltage and the ion current density were 400 V and 1.05 mA/cm^2, respectively. EBL process was performed by ELS-7800 (Elionix Co.) to form a nanobridge pattern onto an EB positive resist ZEP-520A (ZEON Co.) on the film. The electron acceleration voltage and the electron beam dose were 80 kV and 204 μC/cm^2, respectively. After the EBL process, the film was etched using RIBE under the same condition as the first RIBE process.

The minimum bridge area of 65 nm in width and 60 nm in length was obtained. These bridge sizes are comparable to the coherence length of Nb near T_c. This means that these nanobridges are regarded as weak links in the Nb films and act as Josephson junctions [6]. The nanobridges whose width and length were less than 110 nm actually showed the Josephson effect as described below. An SEM image of typical Nb nanobridge is shown in Figure 3 (b). The width and length of this nanobridge were 110 nm and 51 nm, respectively.

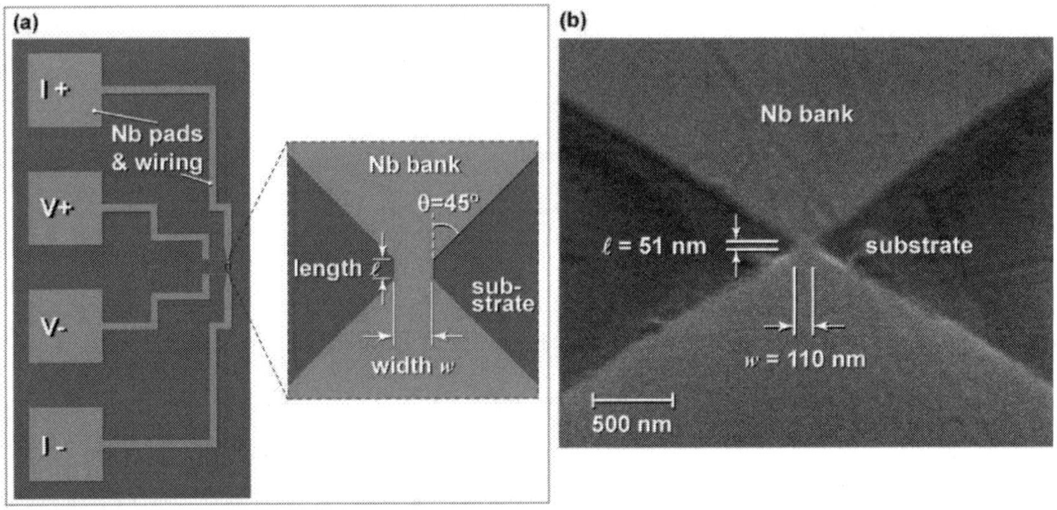

Figure 3. (a) Device structure of Nb nanobridge and (b) SEM image of the fabricated Nb nanobridge.

3.2. I-V and microwave response characteristics

Figure 4 graphs typical *I-V* characteristics of the fabricated nanobridge shown in figure 3 (b) at 7.1 K (a) without and (b) with microwave irradiation at 6.2 GHz. As shown in figure 4 (a), the *I-V* characteristic displays the resistively shunted junction (RSJ)-like behaviour. For this junction, the critical current I_c and the normal resistance R_n were 450 μA and 1.2 Ω, respectively. From these values, the I_cR_n product was estimated to be 540 μV.

With the microwave irradiation at a frequency in the range from 6 to 30 GHz, constant current steps were observed. The microwaves were directed from a semi-rigid cable with an open end onto the nanobridge by a signal generator (SG). Each irradiation frequency f and a voltage interval ΔV between the steps well satisfied the Josephson voltage-frequency relation. This implies that these steps are Shapiro steps and that the nanobridge acts as a Josephson junction. The highest order of the step was 7 at f = 6.2 GHz and a SG power of -0.3 dBm as shown in figure 4 (b). However, the order can increase and the maximum response frequency f_{max} can be much higher than 7 × 6.2 GHz = 43.4 GHz, if coupling efficiency between the junction and microwaves are improved by introducing the antenna-coupled structure into the device.

Although f_{max} can increase by improving the microwave coupling, f_{max} is probably lower than the characteristic frequency $f_c = (2e/h)\ I_cR_n$ = 260 GHz. This implies that the observed I_c includes the excess current which is a supercurrent except the Josephson current. An EB resist pattern of the nanobridge whose width and length are less than 40 nm can be formed by our EBL system, whereas it is difficult to fabricate the Nb nanobridge with no etching error and no film damage by the RIBE in the present stage. The nanobridge Josephson junctions can respond to higher frequency signals when the excess current component decreases by improving the RIBE process.

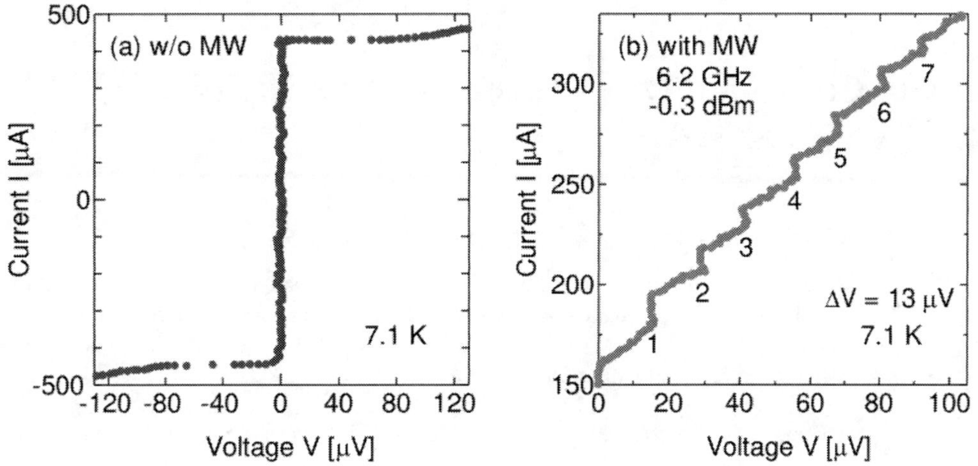

Figure 4. *I-V* characteristics of the fabricated nanobridge at 7.1K (a) without and (b) with 6.2-GHz microwave (MW) irradiation. The right figure graphs the *I-V* characteristic in the positive bias region.

4. Conclusion

Nb nanobridge Josephson junctions were fabricated from 20 nm-thick Nb thin films in order to realize antenna-coupled Josephson detectors for microwave and millimeter-wave radiation. The Nb-sputtered thin films whose thickness, the RMS and T_c were 20.0 nm, 0.109 nm and 8.4 K, respectively were obtained at T_s = 700°C. Nanobridges were obtained from the film using 80-kV EBL and RIBE in CF_4 + O_2 gases. The minimum bridge area was 65 nm wide and 60 nm long. For the nanobridge whose width and length were less than 110 nm, the *I-V* characteristic showed RSJ-like behaviour near T_c. Moreover, Shapiro steps were observed in the nanobridge with microwave irradiation at a frequency in the range from 6 to 30 GHz. These results are useful to realize Josephson detectors in the antenna-coupled devices. For the other application, the junctions are applicable to nano-SQUIDs [7].

Acknowledgment

This work was supported by KAKENHI (23560381).

References

[1] Harada Y, Hirose N, Uzawa Y, and Sekine M 1991 *Japan. J. Appl. Phys.* **30** 3933
[2] Bartolf H, Engel A, Schilling A, Il'in K and Siegel M 2008 *Physica* C **468** 793
[3] Nozue T, Hasegawa A, Yasuoka Y and Yamashita T 1996 *Japan. J. Appl. Phys.* **35** L1181
[4] Saito Y and Anayama T 1975 *J. Low Temperature Phys.* **21** 169
[5] Hedlund C, Blom H O and Berg S 1994 *J. Vac Sci. Technol* A **12** 1962
[6] Likharev K K 1979 *Rev. Mod. Phys.* **51** 124
[7] Vettoliere A, Granata C, Esposito E, Russo R, Petti L, Ruggiero B and Russo M 2009 *IEEE Trans. Appl. Supercond.* **19** 702

Journal of Physics: Conference Series **507** (2014) 042042 doi:10.1088/1742-6596/507/4/042042

Study of Cu-wound Flux transformer for High-Tc SQUID Ultra-Low Field MRI

S. Tanaka[1], H. Murata, K.Imamura and Y. Hatsukade

Department of Environmental & Life Sciences, Toyohashi University of Technology, 1-1 Hibarigaoka, Tenpaku-cho, Toyohashi, Aichi 441-8580, Japan

E-mail: tanakas@ens.tut.ac.jp

Abstract. We constructed a ultra-low field (ULF) nuclear magnetic resonance (NMR) / magnetic resonance imaging (MRI) system employing a high-temperature superconductor (High Tc) SQUID with a separated Cu-wound flux transformer. The pickup coil consisted of two single solenoid coils and each coil was differentially connected each other. The flux transformer consisted of a pickup coil at room temperature and an input coil, which was put in liquid nitrogen and was magnetically coupled with a high-Tc SQUID. The ratio of the transformer was considered and optimized. A water phantom of 10 mL was located in the one side of pickup coil. In the system, we applied polarizing field B_p perpendicular to the measurement field B_m before measurements. B_p was 0.8 T and permanent magnet was used. By using this system, free induction decay (FID) signals of ^{1}H were measured at B_m of 30 μT to evaluate the system. The longitudinal relaxation times T_1 of water were also estimated by changing the polarizing time of B_p.

1. Introduction

The ultra low field (ULF) magnetic resonance imaging (MRI) systems using high sensitive and frequency independent low-T_C SQUID sensors, which feature small-sized system, lower cost, higher frequency-resolution and shorter measurement time due to less averaging number with narrow bandwidth, have attracted attention in recent years [1-3]. The use of high-T_C (HTS) SQUIDs for ULF MRI may enhance these advantages further due to much easier handling with cryoliquid or cryocooler. It is not the easy way while in present technology of high-temperature superconductor (HTS) a possibility to fabricate and to use superconducting HTS transformer with the aim to obtain high magnetic field sensitivity comparable with one for low-Tc SQUID sensors is still the open question . Nevertheless a few attempts have been done to employ HTS SQUIDs for nuclear magnetic resonance (NMR) detection and approaching to MRI [4-7]. In this study we constructed and studied an ultra low-field (ULF) nuclear magnetic resonance (NMR) imaging system using a HTS rf SQUID with a room-temperature (RT) flux transformer. By using this system, free induction decay (FID) signals of ^{1}H were measured to evaluate the system.

2. ULF MRI System Using HTS rf SQUID

2.1. System Configuration

[1] To whom any correspondence should be addressed.

The ULF-SQUID-MRI system was constructed in an electromagnetic shield room of 2.2D×2.2W×2.2H m^3. The system consists of a polarizing magnet, HTS rf SQUID, SQUID driving electronics, input coil, Helmholtz-type measurement field coil, pick up coil, Maxwell gradient coil, two Golay gradient coils [6], delay pulse generator, function generators, power suppliers, mixer with filters, digital multi channel oscilloscope, and spectrum analyzer. The schematic diagram of the system is shown in figure 1. The rf SQUID and driving electronics used in the experiment are made by *Jülicher SQUID GmbH* in Germany. The rf SQUID is a washer-type $YBa_2Cu_3O_{7-x}$ (YBCO) SQUID ring with an outer diameter of 3.5 mm and with one step-edge Josephson junction deposited on a $SrTiO_3$ (STO) substrate [6]. The rf SQUID completed with a YBCO flux focuser of 10 mm x 10 mm size on a STO substrate, which works as a substrate resonator, and an LC tank circuit in liquid nitrogen in the cryostat, is operated by the SQUID electronics. The white noise level of the SQUID in the MSR is approximately 80 $fT/Hz^{1/2}$.

The SQUID is placed at the center of the copper wound input coil. The SQUID and input coil are then cooled by liquid nitrogen in an aluminum Dewar (30 *L*), which was surrounded by a three-layer mu-metal cylinder with a 2-mm thickness. The measured mutual inductance between the input coil and the SQUID is 16 nH. The pickup coil consists of two single solenoid coils and each coil is differentially connected each other. The pickup coil is connected to the input coil in the LN_2 in series so that these coils make a flux transformer as shown in figure 2. All of the coils used in the system except for the input coil are the room temperature coils.

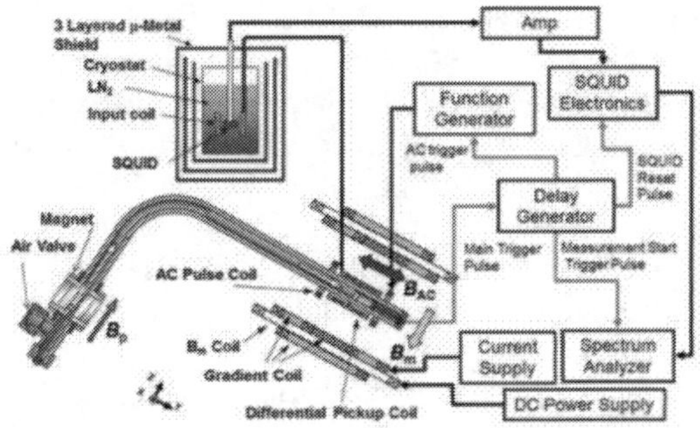

Figure 1. Schematic diagram of the system

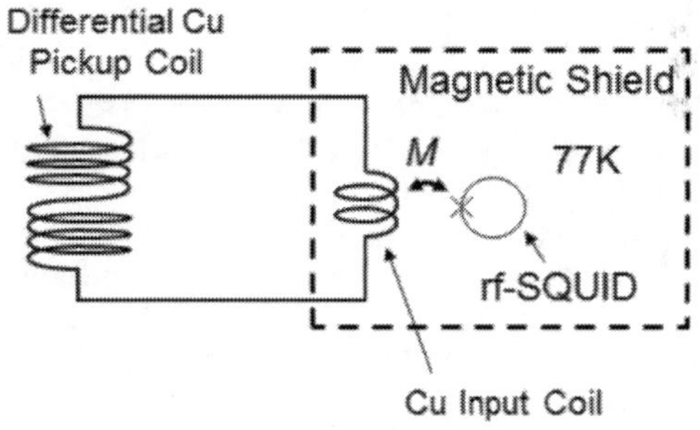

Figure 2. Schematic view of flux transformer

2.2. Optimization of the Flux Transformer

We estimated optimal condition of numbers of turns of input coil and pickup coil by calculation. Figure 3 shows an equivalent circuit of the flux transformer. The voltage V across the pickup coil generated by the motion of flux Φ_{ex} is written by eq. (1), which is known as Faraday's electromagnetic law. The current in the transformer is expressed by eq. (2). As a result, the flux Φ_s detected by the rf-SQUID is obtained as eq. (3). The calculated results are shown in figure 4. The results indicate that there is the best number of turns of an input coil for each pickup coil. The combination of the pickup coil with $N_p = 500$ and the input coil with $N_i = 500$ gives the maximum but it is too sharp and unstable. The combination of $N_p = 2000$ and $N_i = 800$ to 1000 seems to give the stable and better results.

$$V = -j \cdot \Phi_{ex} \cdot \omega \cdot N_p \cdot \frac{1}{2} \tag{1}$$

Here, ω is angler frequency of the external flux Φ_{ex}.

$$|I| = \frac{|V|}{\sqrt{(R_p + R_i)^2 + \omega^2 (L_p + L_i)^2}} \tag{2}$$

$$\Phi_s = |I| \cdot M = \frac{1}{2} \cdot \frac{\Phi_{ex} \cdot \omega \cdot N_p \cdot L_i \cdot k\sqrt{L_i L_s}}{\sqrt{(R_p + R_i)^2 + \omega^2 (L_p + L_i)^2}} \tag{3}$$

Here, *k* is coefficient of coupling between the input coil and rf-SQUID.

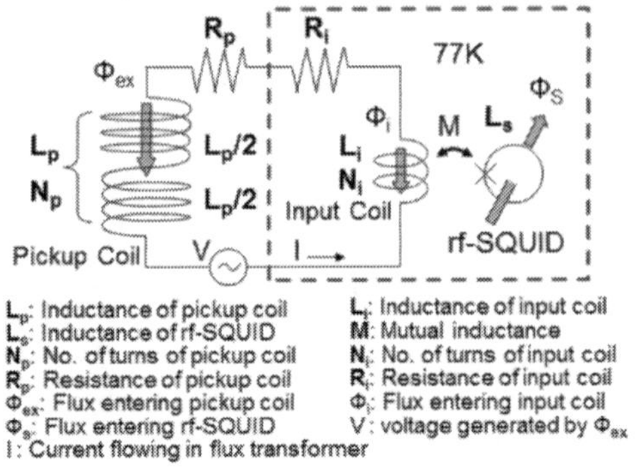

Figure 3. Equivalent circuit of flux transformer. The flux Φ_{ex} is supposed to interlink either of the pickup coil and generate a voltage.

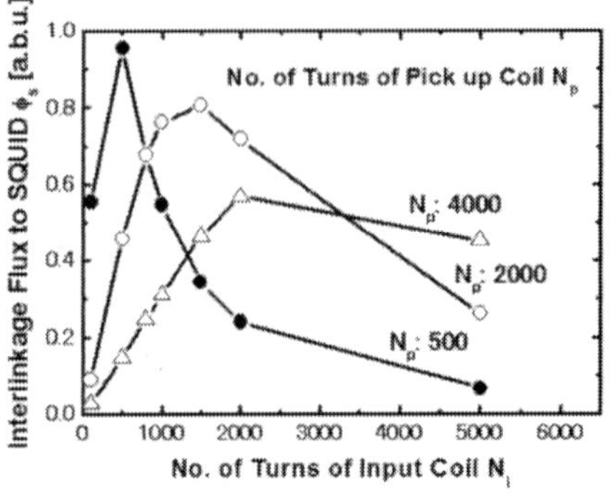

Figure 4. Calculated results. There is a peak for each pickup coil.

3. Experiment and Discussion

3.1. Performance of the Flux Transformer

A flux transformer was made and examined. It consisted of a pickup coil with N_p = 1000 x 2 and an input coil with different N_i. The N_i was 470, 800, 1000 1200 and 1800. The experimental system is shown in figure 2. An ac sinusoidal magnetic field of 390 $nT_{p\text{-}p}$ with frequency of 1270 Hz was applied to one side of differential pickup coil. Then the signal for an input coil with different N_i was detected by the rf-SQUID. The peak-to-peak values were plotted as a function of the N_i. The results were shown in figure 5. There is a maximum value at around N_i of 800-1000, which is almost consistent with the calculation as shown in figure 4. The reason for the moderate reduction over N_i of 1000 is due to the lower real inductance than the expectation in the calculation. Therefore the combination of the pickup coil with N_p = 2000 and the input coil with N_i = 800 was employed in the following experiments. The inductances of the pickup coil and the input coil were 23 mH and 18 mH, respectively. These values are close and almost the same.

3.2. Measurement of NMR Spectra of Water Phantom

First we measured the NMR spectrum of a water phantom, which was placed in the pickup coil; 10 mL of tap water was filled in the phantom. The measurement field coil generated a measurement field B_m in the z-direction, whose field amplitude was about 30 μT including an ambient earth magnetic field in this work. The diameter of the measurement coil was 611 mm. The magnetization M of the sample (tap water in this study, i.e., ^{1}H) in B_m precesses around the z-axis at a Larmor frequency $\nu = \gamma B_m$, where γ is a gyromagnetic ratio (42.6 MHz/T). The permanent magnet (0.8 T) is located at about 2 m away from the pickup coil. In NMR measurements, a water phantom, which is pre-polarized in the permanent magnet for 5 s, is transferred into the one side of the pickup coil, and then exposed in a measurement field B_m from the measurement coil in the z direction. Subsequently, by applying a 90° pulse field B_{AC} in the y direction from the AC pulse coil, which flips the magnetized vector M from the z-axis to the x-axis, ^{1}H precesses about the z-axis with the magnetized vector M at the Larmor frequency of ν while undergoing spin-lattice T_1 and spin-spin T_2 relaxations. During the precession, ^{1}H radiates a NMR signal, which is called a free induction decay (FID) signal, to be measured by the pickup coil. The FID signal is magnetically transferred to the SQUID by the flux transformer and converted to a NMR spectrum by fast Fourier transform (FFT) and recorded by the spectrum analyzer. The gradient coils generate $\partial B_Z / \partial z$ and $\partial B_Z / \partial y$ gradients of the order of 2-10 μT/m to compensate an environmental noise. In the measurements, we compensated for the homogeneity of B_m so as to make the ^{1}H-NMR signal larger and sharper by using the gradient coils as compensation coils

Figure 6 shows the measured ^{1}H-NMR spectra. The harmonics of appliance frequency 60Hz was hidden for clarification. The peak value is 18 pT/Hz$^{1/2}$. Since the Larmor frequency correspond to the applied field B_m of 30 μT is 1278 Hz, the spectra is reasonable. The ^{1}H-NMR spectrum without the field gradient was not shown but it was much smaller and one third of this peak.

3.3. Measurement of T_1 relaxation

Next, we measured the NMR spectra for water, while changing the polarizing time of B_p from 0 to 10 s to estimate the respective T_1 values. In the measurements, we also compensated for the homogeneity of B_m by the gradient coils [6]. The amplitude of the ^{1}H-NMR signal for the water saturated with a polarization was measured. From this result, we estimated that T_1 of water was 2.8 s. This result becomes the base of T_1-weighted contrast imaging in the near future by using this system.

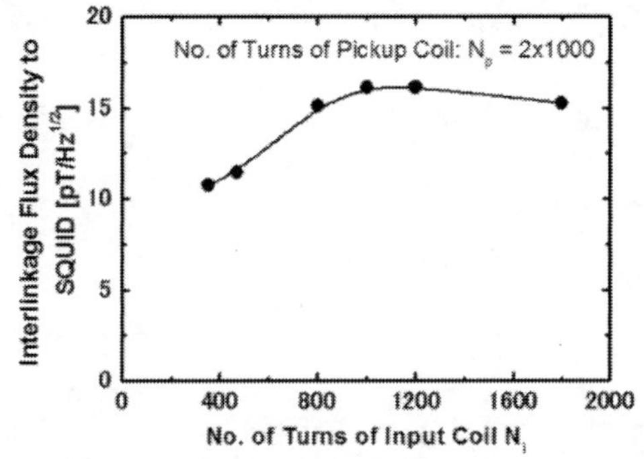

Figure 5. SQUID signal v.s. N_i. There is a maximum value at around N_i of 800-1000, which is consistent with the calculation as shown in figure 4.

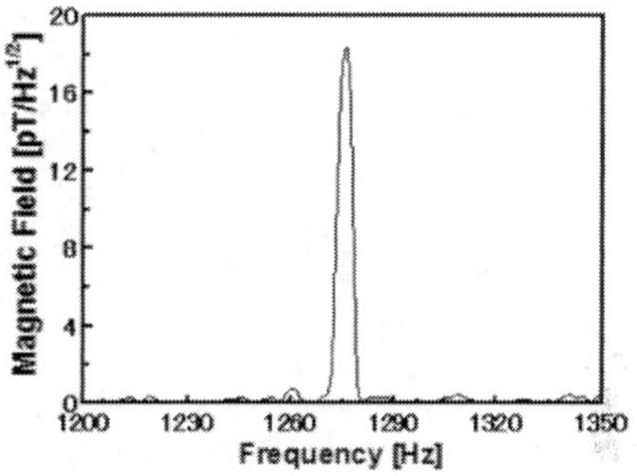

Figure 6. Measured spectra of ^{1}H in water. The peak value is 18 pT/Hz$^{1/2}$. The frequency 1278 Hz is consistency with the Larmor frequency correspond to the applied field B_m of 30 μT.

4. Summary

In this study, we constructed a ULF NMR/MRI system using an HTS-rf-SQUID and a Cu flux transformer. It was found that the combination of the pickup coil with N_p = 2000 and the input coil with N_i = 800 is the best by experiments and calculations. Using this system, the ^{1}H-NMR spectrum was measured under B_m of 30 μT and T_1 value of 2.8 s was experimentally estimated.

References

[1] McDermott R, Trabesinger A H, Mück M, Hahn E L, Pines A and Clarke J 2002 Science **295** 2247

[2] Matlachov A N, Volegov P L, Espy M A, Stolz R, Fritzsch L, Zakosarenko V, Meyer H G, and Kraus J R H 2005 IEEE Trans. Appl. Supercond. **15** 676

[3] Burghoff M, Hartwig S, and Trahms L 2005 Appl. Phys. Lett. **87** 054103

[4] Zhang Y, Qiu L Q, Krause H J, Hartwig S, Burghoff M, and Trahms L 2007 Appl. Phys. Lett., **90** 054103

[5] Yang H C, Liao S H, Horng H E, Kuo S L, Chen H H, and Yang S Y 2006 Appl. Phys. Lett. **88** 252505

[6] Snigirev O, Hayashi M, Fukumoto S, Hatsukade Y, Katsu Y, and Tanaka S 2011 J. of Supercond. & Novel Magn **24** 1033

[7] Hatsukade Y, Abe T, Tsunaki S, Yamamoto M, Murata H, and Tanaka S 2013 IEEE Trans. Appl. Supercond **23** 1602204

Journal of Physics: Conference Series **507** (2014) 042043 doi:10.1088/1742-6596/507/4/042043

Deep sub-micron stud-via technology for superconductor VLSI circuits

Sergey K. Tolpygo, V. Bolkhovsky, T. Weir, L.M. Johnson, W.D. Oliver, and M.A. Gouker

Lincoln Laboratory, Massachusetts Institute of Technology, Lexington, MA 02420, USA

Sergey.Tolpygo@ll.mit.edu

Abstract. A fabrication process has been developed for fully planarized Nb-based superconducting inter-layer connections (vias) with minimum size down to 250 nm for superconductor very large scale integrated (VLSI) circuits with 8 and 10 superconducting layers on 200-mm wafers. Instead of single Nb wiring layers, it utilizes Nb/Al/Nb trilayers for each wiring layer to form Nb pillars (studs) providing vertical connections between the wires etched in the bottom layer of the trilayer and the next wiring layer that is also deposited as a Nb/Al/Nb trilayer. This technology makes possible a dramatic increase in the density of superconducting digital circuits by reducing the area of interconnects with respect to presently utilized etched contact holes between superconducting layers and by enabling the use of stacked vias. Results on the fabrication and size dependence of electric properties of Nb studs with dimensions near the resolution limit of 248-nm photolithography are presented. Superconducting critical current density in the fabricated stud-vias is about 0.3 A/μm^2 and approaches the depairing current density of Nb films.

1. Introduction

In order to realize the tremendous advantages of superconducting digital integrated circuits over semiconductor circuits in speed and reduction of energy dissipation [1], their integration scale must be increased from its current medium level to very large scale integration (VLSI) levels and beyond by increasing the density of Josephson junctions (JJs) from the present density of ~ 10^4 - 10^5 JJs per cm^2 to 10^6 - 10^7 JJs/cm^2, with a corresponding increase in the density of interconnects. This can be achieved by scaling down the dimensions of all circuit elements and increasing the number of superconducting metal layers available for circuit integration. In all reported fabrication processes [2]-[6], superconducting interconnects between wiring layers (vias) are made by etching contact holes in the inter-metal dielectric and depositing the next Nb layer by physical vapor deposition (PVD). This limits the minimum size of contact holes to ~ 0.5 μm, as it becomes difficult to achieve reliable superconducting contacts with sufficient critical currents in smaller holes. In practice, the contact hole also needs to be surrounded by metal overlay to allow for photolithography misalignment and etch bias, making the typical via size around 1 μm and yet larger when many vias need to be stacked on top of each other to make low-inductance connections between the bottom and top layers in a 10-metal layer circuit. As a result, vias became one of the largest components in superconducting digital circuits, occupying up to 30% of the circuit area because the number of vias is much larger than the number of Josephson junctions.

In this paper we propose new processes for deep-submicron vias, which uses etched Nb pillars (studs) between wiring layers instead of etched contact holes with metal filling. The studs are formed by Nb etching with subsequent dielectric deposition followed by dielectric chemical mechanical planarization (CMP). We present results of the practical realization of one of the processes, including electrical characterization of the resultant vias. We have found that this process allowed us to form Nb stud-vias with design sizes down to our photolithography resolution limit (~250 nm) and with superconducting critical current density approaching the depairing critical current density of Nb.

2. Stud-via process description

Instead of etched contact holes with metal filling we propose to use etched Nb pillars (studs) from a deposited Nb film in contact with bottom wiring layer M_i. An interlayer dielectric could be deposited over them and planarized using CMP to make access to the top of the stud. Then, the top wiring layer would be deposited. A detailed description of the dielectric CMP processes can be found in [7].

There are two possible realizations. The first is shown in figure 1 and referred to as a single etch and planarization (SEAP) process. Here, the bottom Nb wiring layer M_i is deposited and patterned, figure 1(a). This layer is then planarized by depositing a thick layer of SiO_2 [figure 1(b)] with subsequent CMP to get the planar structure shown in figure 1(c). After this, Nb deposition for subsequent studs follows, figure 1(d). Since Nb studs need to be etched while stopping on the bottom Nb wiring, it is useful to have a good etch stop (ES) layer – a material with a lower etch rate than Nb. This ES layer must not degrade the superconducting properties of the interface in order to maintain high critical current in the superconducting state. There are very few materials suitable for this, and a thin layer of aluminium (a few nanometers thick) is one of them. After etching the studs, figure 1(e), the process flow repeats planarization steps (b) and (c) to planarize studs and open access to their top surface, figure 1(g). Finally, the top Nb wiring layer is deposited and patterned to complete the via-stud interconnects between a pair of Nb layers M_i and M_{i+1}. The process can further be repeated for the next pair of wiring layers M_{i+1} and M_{i+2} and so on in a straightforward manner. It also allows for stacking studs on top of each other.

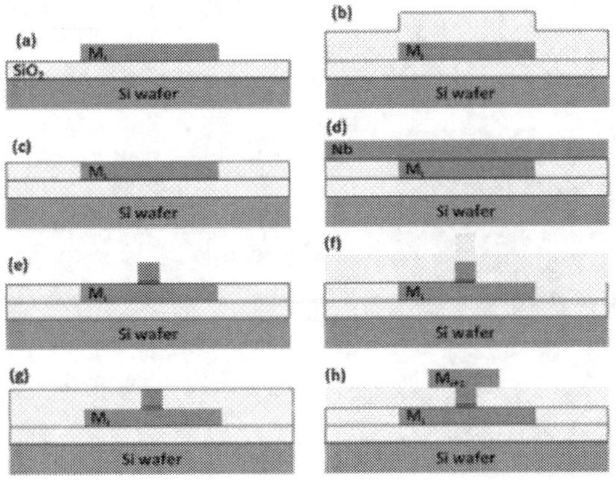

Figure 1. Cross sections of the SEAP version of a stud-via process for superconducting circuits: (a) – deposition and pattering of (Nb) wiring layer M_i over the wafer; (b) – deposition of a blanket SiO_2 layer for CMP planarization; (c) CMP of the SiO_2 layer; (d) deposition of an etch-stop layer (ES) and Nb stud metal bi-layer; (e) patterning of Nb studs and ES layer; (f) blanket SiO_2 deposition for CMP planarization; (g) CMP of SiO_2 to access the top of Nb studs; (h) deposition and pattering of the next Nb wiring layer, M_{i+1}. Prior step (a) the wafer may already contain several patterned and planarized layers. A convenient etch stop layer is Al. The bi-layer deposition should be done in-situ to minimize contamination of the interface between Al and Nb stud.

The described process can be simplified by noticing that the structure formed after stud etching, figure 1(e), resembles the standard trilayer-type SNS Josephson junction where the base electrode is bottom wiring layer M_i, the counter (top) electrode is the stud, and the Al etch stop layer is N-layer or Al-AlO_x barrier layer in tunnel junction trilayers [6] without oxidation. Therefore, the stud-via formation can be done in the same manner as the formation of Josephson tunnel junctions in a whole-wafer trilayer process [6] with the addition of planarization [8]. This second version of the stud-via process, shown in figure 2, is referred to as a dual-etch and planarization (DEAP) process.

Instead of depositing a single wiring layer as in the previous (SEAP) version, in this process each wiring layer is deposited in-situ as a Nb/Al/Nb trilayer, figure 2(a). The top layer of the trilayer is then patterned by photolithography and dry etching to form Nb studs as shown in figure 2(b). Aluminium serves as an excellent etch-stop layer for etching Nb in F-based chemistries. Then, the bottom electrode of the trilayer is patterned in Cl-based chemistry to etch through the Al layer and form the wiring pattern M_i. Etching is followed by blanket SiO_2 deposition and CMP to access the tops of Nb studs, figure 2(e). Then, the next layer M_{i+1} is deposited as a Nb/Al/Nb trilayer and patterned in the same manner to form the next layer of interconnects, and so on. If no more stud-vias are required above, a single metal wiring layer can be deposited. The described processing unit needs to be repeated as many times as there are wiring layers in the full process.

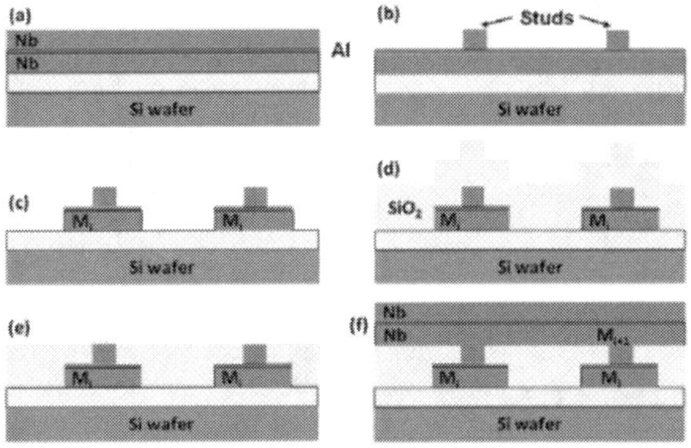

Figure 2. Cross-sections of Nb/Al/Nb trilayer-based (DEAP) stud-via process: (a) Nb/Al/Nb in-situ trilayer deposition; (b) patterning of the top electrode to etch Nb studs; (c) pattering of the bottom electrode to form wiring layer M_i; (d) blanket SiO_2 deposition for planarization by CMP; (e) CMP down to the tops of Nb studs; (f) Nb/Al/Nb wiring layer deposition to form wiring layer M_{i+1} and Nb stud to contact next layer M_{i+2}. If no additional stud-vias are required above, a single Nb wiring layer M_{i+1} can be deposited instead. This process is to be repeated as many times as there are wiring layers in the entire process.

3. Stud-via fabrication

Nb/Al/Nb trilayers were deposited over 200-mm Si wafers with 500 nm of thermal oxide. The target thickness of the top Nb layer (Nb studs) was 250 nm, and of the bottom layer and the wiring layer was 200 nm. The thickness of the Al etch-stop layer was 8 nm. The photolithography was done using a positive deep-UV photoresist, bottom antireflection coating, Canon FPA-3000 EX4 248-nm stepper with 5x reduction and 250-nm nominal resolution. The circular shape for stud definition was implemented with diameters from 200 nm to 15 μm. The typical SEM images are shown in figure 3.

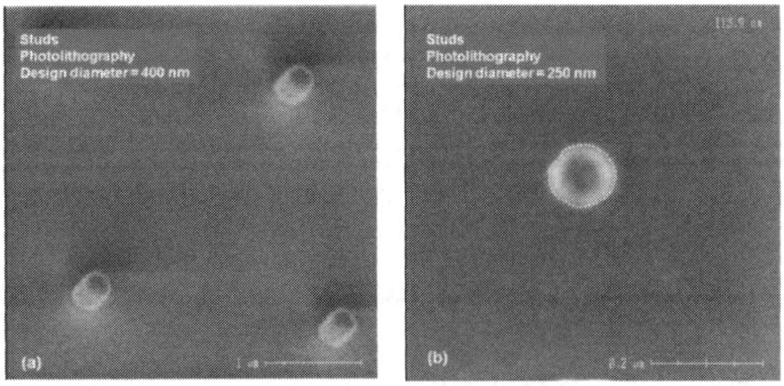

Figure 3. SEM images of the photoresist mask and etched Nb features: (a) tilted view of 400-nm photoresist posts masking Nb/Al/Nb trilayer used for stud-via definition; (b) top view of a photoresist post with design diameter of 250 nm, showing the automated CD measurement of the mean diameter using 48 points around the feature perimeter and giving 113.9 nm (right upper corner).

Nb etching was done using the high density plasma etching chamber of an Applied Materials Centura system with end-point detection. The results are shown in figure 4. Etching of the bottom wire was done in a Cl-based chemistry to break through the Al etch-stop layer first. A SiO_2 film was deposited using plasma-enhanced chemical vapor deposition (PECVD) from $SiH_4/N_2O/Ar$ mixture for the subsequent CMP using Applied Materials Mirra polisher.

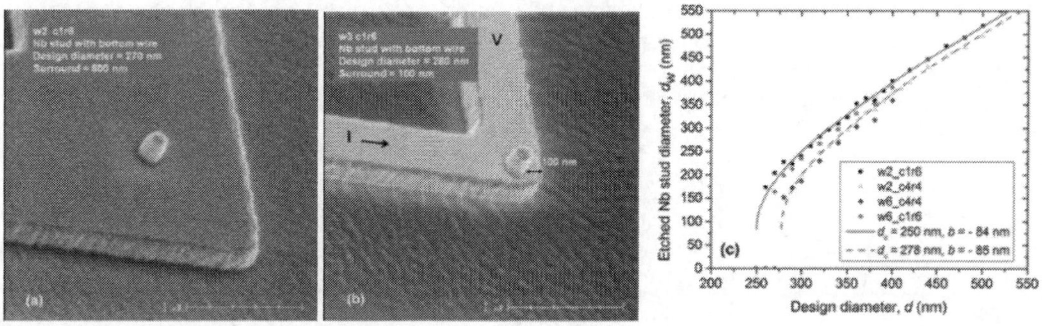

Figure 4. Etched Nb studs with various diameters on top of bottom Nb wire: (a) 270-nm design diameter, 600-nm surround; (b) 280-nm design diameter, 100-nm surround. The bottom wire has an L-shape that with the corresponding L-shape in the top wire will form a cross-bridge Kelvin resistor (CBKR) configuration for electric measurements of studs. (c) The bottom diameter of the etched Nb studs as a function of the design diameter.

We have found that the size of the photoresist features and etched Nb studs can be well described by the relationship $d_w = k(d^2 - d_c^2)^{1/2} - b$ for $d > d_c$, and $d = 0$ for $d \leq d_c$, as shown in figure 4(c). Here d is the design diameter, d_c is a photolighography cut-off size that depends on the photolithography parameters, and b is the process bias. The scaling factor k characterizes the accuracy of the projection optics and SEM magnification. We have found that for the stable and optimized process $d_c \approx 250$ nm, $k = 1\pm0.7\%$, and b is within the range of $\pm$ 50 nm.

The top wiring layer of Nb was deposited and patterned similarly to the bottom wiring and with the same overlap. The SEM picture of the completed stud-via structure is shown in figure 5.

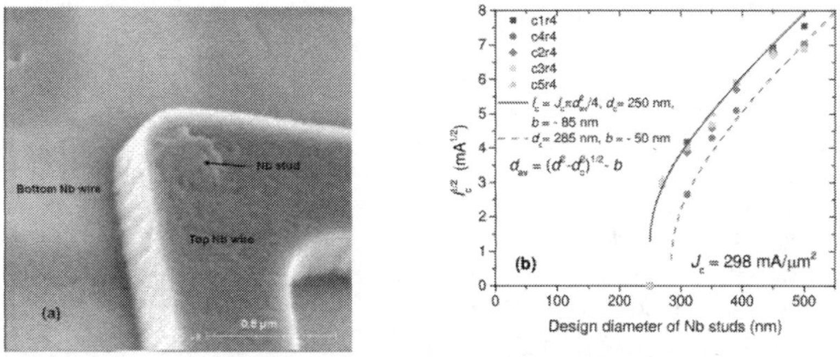

Figure 5. (a) SEM image of the completed stud-via structure with 350-nm stud between two L-shaped wires. Clearly visible is the top Nb wire, lighter blurry image is the bottom wire as seen through the planarized SiO_2 dielectric. The outline of the stud can be also seen due to a difference in niobium grain structure. (b) Critical current, I_c, of the fabricated Nb stud-vias along with the fits to $I_c = J_c\pi d_{av}^2/4$, where $J_c = 0.3$ A/μm^2 is the critical current density and d_{av} is the mean of the top and bottom diameters of the fabricated studs, shown for two values of the lithography cut-off parameter and process bias: $d_0 = 250$ nm $b = -85$ nm, and $d_0 = 290$ nm $b = -50$ nm.

4. Electrical test results and discussion

The critical currents, I_c, of the fabricated stud-vias are shown in figure 5(b). I_c scales with the actual area of the studs. The current density J_c was found to be $3{\cdot}10^7$ A/cm^2 (0.3 A/μm^2), only about a factor of 2.5 lower than the Ginzburg-Landau depairing critical current density $J_c^{GL} = (2/3)^{3/2}B_c/(\mu_0\lambda)$ in our films, where $\mu_0 = 4\pi{\cdot}10^{-7}$ H/m, B_c - the thermodynamic critical magnetic field, λ – the magnetic field penetration depth [9]. Indeed, for Nb at 4.2 K, $B_c \approx 0.145$ T [10] and λ is in the range from 80 nm to 90 nm for our films. This gives J_c^{GL} in the range from $7\text{x}10^7$ A/cm^2 to $8\text{x}10^7$ A/cm^2. In the actual structure, the J_c should be somewhat lower than J_c^{GL} of the film due to the presence of an 8-nm Al etch-stop layer at the interface between the Nb stud and the bottom Nb wire. On the other hand, in the studied range, the diameter of the studs is larger than 4ξ, where ξ is the coherence length that in our films is about 20 nm. Hence, instead of the depairing, the stud critical current can be caused by the entry and motion of Abrikosov vortices when the current-induced magnetic field at the stud surface reaches H_{c1}. This gives an estimate for the critical current density $j_c = H_{c1}/(\mu_0\lambda)$, where $H_{c1} = H_c\ln(\kappa+0.08)/(\sqrt{2}\kappa)$ [11]. Using $\kappa = \lambda/\xi = 4$ for our films, we get $j_c \approx 0.36$ A/μm^2 in a perfect agreement with the J_c observed in the fabricated Nb stud-vias.

Independently of the actual critical current mechanism, the observed critical currents are more than sufficient for the use of stud-vias in superconducting integrated circuits for interlayer connections. At $d = 0.5$ μm the critical currents of stud-vias exceed those observed in etched contact holes of the same diameter filled with deposited Nb metal.

5. Conclusions

We have developed and demonstrated a novel process for making deep sub-μm superconducting multilayer interconnects for use in VLSI and ULSI superconductive digital circuits. These interconnects are formed using Nb/Al/Nb trilayer wiring layers by etching Nb studs in the top layer and Nb wires in the bottom layer of the trilayers with subsequent planarization of the formed interconnects by dielectric CMP. The purpose of this process development is to replace the currently used etched-contact holes filled with sputtered Nb that are too big for VLSI SFQ circuits. Nb stud-vias with diameters as small as 150 nm have been fabricated by the developed process. For the design diameters ≥280 nm, the yield of the fabricated Nb stud-vias was 100% on the test structures available. The critical currents of the obtained Nb stud-vias approach the maximum possible superconducting currents for Nb and are certainly sufficient for interconnects in superconducting VLSI circuits.

Acknowledgement

The authors would like to thank Dr. Marc Manheimer and Dr. Scott Holmes for their interest and support of this work. This work was sponsored by the IARPA under Air Force Contract FA8721-05-C-0002. Opinions, interpretations, conclusions, and recommendations are those of the authors, and not necessarily endorsed by the United States Government.

References

[1] Likharev K and Semenov V 1991 *IEEE Trans. Appl. Supercond.* **1** 3

[2] Nagasawa S, Satoh T, Hinode K, *et al.* 2009 *Physica C* **469** 1578; Hidaka M, Nagasawa S, Hinode K, and Satoh T 2008 *IEICE Trans. Electron.* **E91-C** 318; Nagasawa S, Hinode K, Satoh T, Akaike H, Kitagawa Y, and Hidaka M 2004 *Physica C* **412-414** 1429

[3] Tolpygo SK, Amparo D, Hunt RT, *et al.* 2011 *IEEE Trans. Appl. Supercond.* **21** 119; Tolpygo SK, Yohannes D, Hunt RT, *et al.* 2007 *IEEE Trans. Appl. Supercond.* **17** 946

[4] Kerber GL, Abelson LA, Edwards K, *et al.* 2003 *IEEE Trans. Appl. Supercond.* **13** 82; Abelson LA and Kerber GL 2004 *Proc. IEEE* **92** 1517

[5] Numata H and Tahara S 2001 *IEICE Trans. Electron.* **E84-C** 2

[6] Gurvitch M, Washington MA, and Huggins HA 1983 *Appl. Phys. Lett.* **42** 472

[7] Steigervald JM, Murarka SP, and Gutmann RJ 1997 *Chemical Mechanical Planarization of Microelectronic Materials* (John Wiley & Sons, Inc., New York) 129

[8] Ketchen M, Pearson D, Kleinsasser AW *et al.* 1991 *Appl. Phys. Lett.* **59** 2609; Bao Z, Bhushan M, Han S, and Lukens JE 1995 *IEEE Trans. Appl. Supercond.* **5** 2731

[9] Ginzburg VL and Landau LD 1950 *Zh. Experim. i Teor. Fiz.* **20** 1064; Ginzburg VL 1958 *Doklady Akademii Nauk SSSR*, **118** 467; Tinkham M *1996 Introduction to Superconductivity* (New York: McGraw-Hill) pp 124-125, 2nd ed; Bardeen J 1962 *Rev. Mod. Phys.* **34** 667

[10] French RA 1968 *Cryogenics* **8** 301

[11] Abrikosov AA 1957 *Sov. Phys. JETP* **5** 1174

Development of contaminant detection system based on ultra-low field SQUID-NMR/MRI

S Tsunaki, M Yamamoto, J Hatta, Y Hatsukade, and S Tanaka

Toyohashi University of Technology, 1-1 Hibarigaoka, Tenpaku-cho, Toyohashi, Aichi 441-8580, Japan

E-mail: hatukade@ens.tut.ac.jp

Abstract. We have developed an ultra-low field (ULF) NMR/MRI system using an HTS-rf-SQUID and evaluated performance of the system as a contaminant detection system for foods and drinks. In this work, we measured 1D MRIs from water samples with or without various contaminants, such as aluminum and glass balls using the system. In the 1D MRIs, changes of the MRI spectra were detected, corresponding to positions of the contaminants. We measured 2D MRIs from food samples with and without a hole. In the 2D MRIs, the hole position in the sample was well visualized. These results show that the feasibility of the system to detect and localize contaminants in foods and drinks.

1. Introduction

Nuclear magnetic resonance (NMR) and magnetic resonance imaging (MRI) in ultra-low field (ULF) using SQUID have attracted attention not only in the medical fields [1-2] but also in the other applications. For example, explosive liquid detection in airport [3-4] and disease diagnosis of cancer tissues have been proposed and researched [5]. On the other hand, we have developed an ULF NMR/MRI system using HTS-rf-SQUID for a new application, which aims to detect contaminants in foods and drinks [6]. By using NMR/MRI for the application, various kinds of contaminants may be detected, taking advantages of the matured techniques, such as T_1 and T_2 contrasts. Use of the HTS-SQUID offers lower running cost, easiness to handle, and inexpensive system. In this paper, we showed feasibility of the contaminants detection by our system by measuring one-dimensional (1D) MRIs on water samples with small contaminants. The 1D MRI can determine whether contaminant exists or not, but it is difficult to localize contaminant position. Therefore, we developed two-dimensional (2D) MRI system and evaluated the performance of the system by measuring 2D MRIs of vegetable samples.

2. ULF SQUID-NMR/MRI system

2.1. *System configuration*

The ULF SQUID-NMR/MRI system developed in this study is shown in Figure 1. The system consists of an HTS-rf-SQUID, cryostat, SQUID electronics unit, Helmholtz-type measurement coil, AC pulse coil, three field gradient coils, permanent magnet of 1.1 T, sample transfer mechanism, delay pulse generator, current source, function generator, and computer for data acquisition. All of the coils used in the system are room-temperature (RT) coils. The HTS-SQUID is positioned at the bottom of the cryostat, in which liquid nitrogen is filled. The RT coils and the cryostat are located in a magnetically shielded room (MSR) with opened door. The permanent magnet is located outside the MSR and about 2 m away from the SQUID. The sample transfer mechanism utilizes aluminum frames and a plastic track, on which a sample is transferred. A water sample, which is pre-polarized in the permanent magnet, is transferred to under the SQUID, and then exposed in a measurement field B_m from the measurement coil in the z direction. Subsequently, by applying a 90° pulse and 180 ° pulse field B_{AC} in the y direction from the AC pulse coil, a FID signal or an echo signal from the sample is measured by the SQUID. The signals are converted to 2D MRIs and recorded by the computer. The permanent magnet used for the pre-polarization magnetic field B_p is shown in Figure 1 (b). In order to reduce influence of a leakage field from the magnet on the measurement field, yoke magnets are arranged around the magnet.

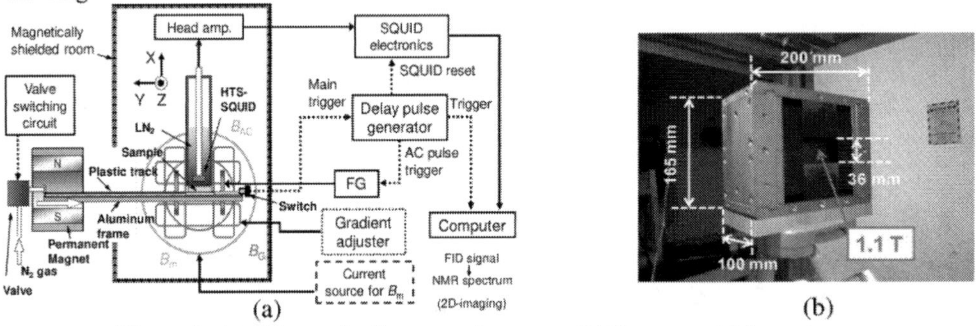

(a) (b)

Figure 1. (a) Schematic diagram of system. (b) Permanent Magnet.

2.2. *Pulse sequence*

In the system, a 1D MRI measurement pulse sequence shown in Figure 2(a) is used to obtain the ^{1}H-NMR signals from water samples. The pre-polarization field B_p pre-polarizes the sample in the magnet for 5 or more seconds. Then the sample is transferred under the SQUID in about 0.5 s. The measurement field B_m is applied continuously to the area around the SQUID in the MSR. In 1D MRI measurements, a gradient field G_y (dB_z/dy) of 52 nT/cm (corresponding to 2.3 Hz/cm) in the y direction is applied to the sample continuously. After the transfer, a 90° pulse field B_{AC} is applied from the AC pulse coil. At approximately 1 ms after applying the B_{AC}, the SQUID is set in the lock mode to measure the FID signal. At 16 ms after locking the SQUID electronics unit, the signal is recorded for 2 s by the computer. After recording the FID signal, the spectrum analyzer converts the FID signal to a NMR spectrum by fast-Fourier-transform (FFT), and records the spectrum.

A 2D MRI measurement pulse sequence shown in Figure 2(b) is used to obtain the ^{1}H-NMR signals from food samples. In 2D MRI measurements, a combinational gradient field G (gradient field G_y (dB_z/dy) and G_z (dB_z/dz)) of 280 nT/cm (corresponding to 11.9 Hz/cm) in the y and z directions is applied to the sample continuously. To obtain spin echo (SE) signals, the 180° pulse field is applied at 0.5 s after the 90° pulse field is applied. The SQUID is set in the unlock mode until the 180° pulse field is applied as well as 1D MRI. Data acquisition time is 0.512 s to measure a symmetric SE signal.

3. Experiments and results

3.1. *1D MRI measurements*

We measured 1D MRIs from a tap water sample of 10 ml with contaminants using the above mentioned system and sequence. A glass bottle of 10 ml was used as a container. Balls of various contaminants were prepared as shown in Figure 3(a). Each contaminant was set in the center of the bottom of the bottle. 1D MRI spectra of only water and water with the aluminum, glass, and ceramic, balls of 1.5 mm in diameter are shown in Figure 3(b). Compared to the MRI spectrum from the water only, the spectra from the water with the contaminant balls changed around the central frequency of about 1914 Hz. This is because distribution of protons of the water was changed with the existence of the contaminant balls. For every kind of the contaminants, the changes of the signal spectra were measured at the central frequency, where corresponded to the positions of the contaminants. From these results, it is shown that existence of contaminants can be distinguished by this technique.

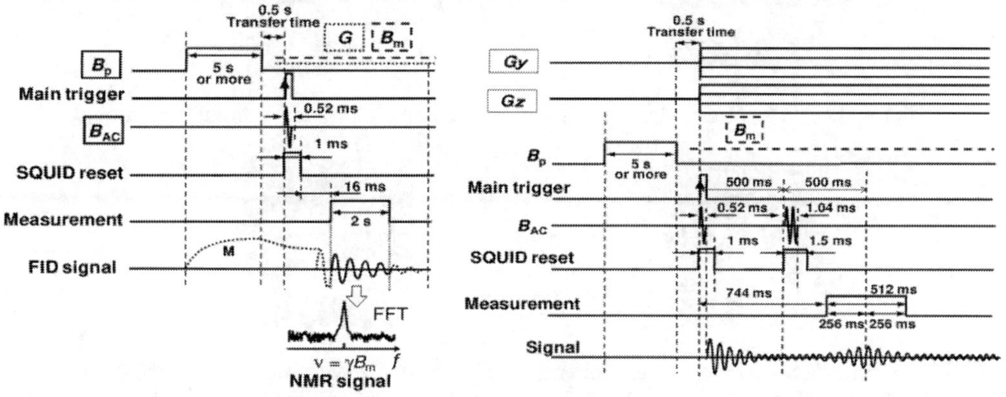

(a) (b)

Figure 2. (a) Typical pulse sequence for 1D MRI data acquisition. (b) Typical pulse sequence for 2D MRI data acquisition.

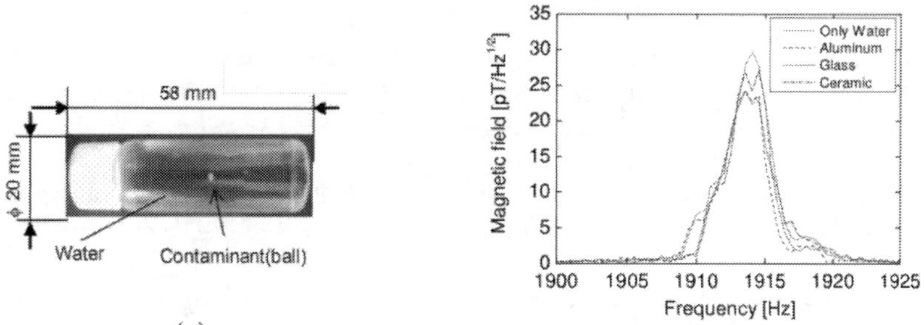

(a) (b)

Figure 3. (a) Water sample containing a contaminant. (b) 1D MRI measurement results of water with and without contaminant balls of 1.5 mm in diameter.

3.2. 2D MRI measurements

We measured 2D MRIs in the *y-z* plane from cucumber samples using the above mentioned system and sequence. The filtered back projection reconstruction was utilized to obtain the 2D MRIs. Projection number was 12, because the gradient field directions were rotated for 15° step by step to cover 180°. Cover angle is 180°, which is half of 360° (full angle) because we utilized the spin echo sequence. For all projections, SE signals were recorded without averaging. The space resolution is 1.953 Hz/pixel (corresponding to 0.16 cm/pixel), while the field of view (FOV) is 51.2 ×51.2 mm in area ,which correspond to 62.5 Hz in bandwidth, or 32 × 32 points in pixel. Measured sample is a cucumber of 33 mm in diameter and 8 mm in thickness, which was cut perpendicular to the longitudinal direction. We measured the cucumber samples with and without a hole using the 2D MRI sequence. An air hole in raw foods like cucumbers can be a quality problem. The photograph of the cucumber samples with and without a hole in the center are shown in Figure 4(a) and (b). The results of 2D MRI measurements are shown in Figure 5(a) and (b). The sizes of the 2D MRIs of the cucumbers (about ϕ22 mm) are smaller than the actual sample sizes (about ϕ33 mm) because the samples sizes are larger than the SQUID's size (about 10 mm square). In the images shown in Figure 5(b), the image intensity at the hole position was clearly lower compared to Figure 5(a). From this result, it is shown that the hole position can be localized by this technique. As the next challenge, we plan to utilize flux transformer or increase SQUID channel, to expand the SQUID's detection area.

(a) (b)

Figure 4.(a) cucumbers sample without a hole. (b) cucumbers sample with a hole.

(a) (b)

Figure 5.(a) 2D MRI measurement results of cucumbers without the hole. (b) 2D MRI measurement results of cucumbers with the hole. Dotted lines indicate the actual sample size.

4. Discussions and Conclusions

In this research, we developed the ULF NMR/MRI system using HTS-SQUID and evaluated the performance of the system as the contaminant detection system for foods and drinks. It was shown that the contaminants such as aluminum, glass and ceramics balls in the water sample could be detected by the 1D MRI measurements. By using the cut cucumber with the air hole, it was shown that the hole position in the sample can be localized by 2D MRI measurements. From these results, the possibility of applying the ULF SQUID-NMR/MRI system to the contaminant detection was successfully demonstrated. We have planned to introduce the ULF-NMR/MRI system into production line in the future. Since an object is transported in our system, our system will suit to contaminant detection on the production line, compared to the other systems which measure a static object. On the other hand, we had developed contaminant detection systems for fine magnetic contaminants using HTS-SQUIDs for production line [7]. In comparison with these systems, the ULF-NMR/MRI system has possibility to detect many materials other than magnetic materials. In addition, localization and material identification of contaminants are needed to trace the contamination on the production line. Low cost and convenience must be considered for practical use of the system. Using HTS-SQUID is a significant advantage for these requirements.

Acknowledgements

This work was supported in part by the "Knowledge Hub Project" managed by Aichi prefecture in Japan.

References

[1] Vesanen P T et al. 2013 *Magnetic Resonance in Medicine* **69** 1795
[2] Magnelind P E, Gomez J J, Matlashov A N, Owens T, Sandin J H, Volegov P L and Espy M A 2011 *IEEE Trans. Appl. Supercond.* **21** 456
[3] Matlashov A N, Schultz L J, Espy M A , Kraus R H, Savukov I M, Volegov P L and Wurden C J 2011 *IEEE Trans. Appl. Supercond.* **21** 465
[4] Espy M A et al. 2010 *Supercond. Sci. Technol.* **23** 034023
[5] Adolphi N L et al. 2012 *Contrast Media & Molecular Imaging* **7** 308
[6] Tsunaki S, Yamamoto M, Abe T, Hatta J, Hatsukade Y and Tanaka S 2013 *Physica C* http://dx.doi.org/10.1016/j.physc.2013.04.004
[7] Tanaka S, Kitamura Y, Hatsukade Y, Ohtani T, Suzuki S 2012 *Journal of Magnetism and Magnetic Material* **324** 3487

Effectiveness of $BaTiO_3$ dielectric patches on $YBa_2Cu_3O_7$ thin films for MEM switches

J Vargas[1], Y Hijazi[1], J Noel[2], Y Vlasov[3] and G Larkins[3]

[1] Department of Electrical Engineering, Universidad del Turabo, PO Box 3030, Gurabo, PR 00778, USA
[2] Department of Electrical Engineering, Universidad de Ingeniería y Tecnología, Lima 33, Perú
[3] Department of Electrical Engineering, Florida International University, Miami, FL 33199, USA

Email: jvargas@suagm.edu

Abstract. A micro-electro-mechanical (MEM) switch built on a superconducting microstrip filter will be utilized to investigate $BaTiO_3$ dielectric patches for functional switching points of contact. Actuation voltage resulting from the MEM switch provokes static friction between the bridge membrane and $BaTiO_3$ insulation layer. The dielectric patch crystal structure and roughness affect the ability of repetitively switching cycles and lifetime. A series of experiments have been performed using different deposition methods and RF magnetron sputtering was found to be the best deposition process for the $BaTiO_3$ layer. The effect examination of surface morphology will be presented using characterization techniques as x-ray diffraction, SEM and AFM for an optimum switching device. The thin film is made of $YBa_2Cu_3O_7$ deposited on $LaAlO_3$ substrate by pulsed laser deposition. For this work, the dielectric material sputtering pressure is set at $9.5x10^{-6}$ Torr. The argon gas is released through a mass-flow controller to purge the system prior to deposition. RF power is 85 W at a distance of 9 cm. The behavior of Au membranes built on ultimate $BaTiO_3$ patches will be shown as part of the results. These novel surface patterns will in turn be used in modelling other RF MEM switch devices such as distributed-satellite communication system operating at cryogenic temperatures.

1. Introduction

MEMS play an important role due to the benefits that MEM switches can offer over PIN-diode of FET counterpart in terms of fabrication, cost operation and functionality [1]. In this work, we will focus on the $BaTiO_3$ dielectric patches for functional switching points of contact and fabrication of $BaTiO_3/YBa_2Cu_3O_7/LaAlO_3$ substrates using laser ablation and RF magnetron sputtering deposition to build optimum MEM switches. Therefore, characterization process is a key to resolve the morphology of the die surface using techniques such as X-ray diffraction, SEM and AFM which will be presented. After the device characterization is done we will be set for the MEMS device layout patterning process. This device is designed to be switched operating at cryogenic temperatures.

2. Design

These switches execute switching by shunting the signal by the $BaTiO_3$ patch acting as capacitive contact to ground. Physically, the suspended membrane collapses when biased thus allowing the $BaTiO_3$ path and transmission line to make electrical short to ground as shown in Figure 1. Force acting in between two metal electrodes has been used to actuate switching. A bias voltage of 36 V was enough to activate the switches in [2], it was only when it reached 80 V that a contact resistance of less than 1 ohm was measured. However, a high voltage severely impacts the switch lifetime by increasing the risk of a mechanical or dielectric failure [3]. After calculations it was determined that the line width of transmission line was 169 μm, the length varies between 14.90 and 18.574 mm depending on the frequency, and ε_{eff} equals to 14.7. Four MEM switches working in pairs were necessary to actuate and deactivate a bandpass hairpin filter [4] which will be presented later on.

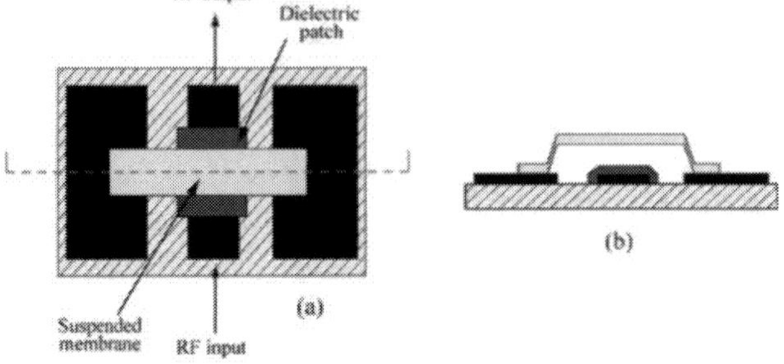

Figure 1. Shunt MEMS switch on coplanar transmission line: (a) top view, (b) cross section.

3. Fabrication

The deposition of $YBa_2Cu_3O_7$ on $LaAlO_3$ using PLD of high-temperature superconductor materials provides high quality superconducting thin films because of the ability to stoichiometrically transfer the composition of the target to the deposited film [5].

3.1. Preparation of $BaTiO_3$ patches

3.1.1. Lift off process. This process involves two layers. The underlaying layer is a polymer (PMGI SF15) to ensure proper undercut for an ideal liftoff. The top layer is a coat of thin photoresist AZ5214.

3.1.2. $BaTiO_3$ and gold bridge liftoff depositions. Planar magnetron sputtering with RF potential on the surface enables the use of non-conducting materials as targets and allows reactive deposition of films as conventional RF sputtering. For this work the pressure is set at $9.5x10^{-6}$ Torr. RF power is 85 W at a distance of 9 cm. A double coated PMGISF15 is required to have an ideal suspended bridge support above the dielectric patches. The thickness of this layer is key for this process because it defines the height at which the gold bridge is suspended, and consequently controls the actuation voltage required to pull down the gold membrane. The gold bridge liftoff deposition process is very similar to the $BaTiO_3$ dielectric mask but the difference is that AZP4620 thick photoresist is utilized. A resistively heated of a tungsten filament vacuum chamber is used. The gold film thickness is between 0.12 and 0.25 μm.

4. Film characterization

The $BaTiO_3$ dielectric patch crystal structure and roughness affect the ability of repetitively switching cycles and lifetime. A series of experiments have been performed using different deposition methods to examine the effects of surface morphology and RF magnetron sputtering was found to be the best deposition process for the $BaTiO_3$ dielectric layer. Figure 2 is an X-ray diffraction pattern for a non-uniform epitaxial $BaTiO_3$ growth using RF magnetron sputtering. This film was sputtered on and off axis at 36 °C in 260 mTorr of argon gas with 80 W RF power. The resulting film was slightly more than 0.2 μm thick. As seen in Figure 3, a scanning electron microscope (SEM) micrograph confirmed a thickness of 0.3 μm. and the roughness of the surface derived from the X-ray diffraction analysis.

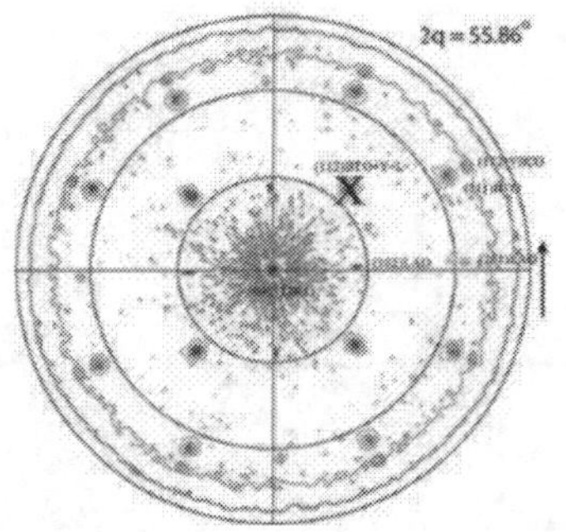

Figure 2. X ray diffraction using RF magnetron sputtering deposition.

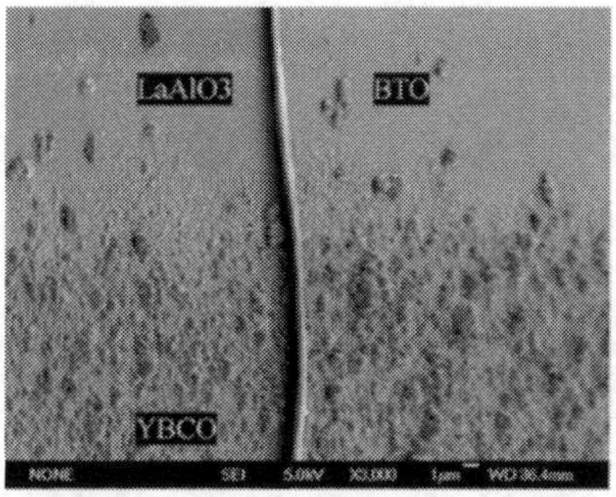

Figure 3. SEM of $YBa_2Cu_3O_7$ on $LaAlO_3$ substrate and sputtered $BaTiO_3$.

In Figure 4, the roughness of the substrate is consistent with the X-ray diffraction and the SEM results. The atomic force microscope (AFM) analysis shows the $BaTiO_3$ film deposited by RF magnetron sputtering of the contact area when the suspended gold membrane is in the down position or "on" state. Figure 5 shows a 3-D analysis of the $BaTiO_3$ film deposited by RF magnetron sputtering in the up position or "off" state. Notice that the surface in Figure 5 is smoother than the contact area analysis in Figure 4 due to the electro-mechanical contact between the gold and the dielectric material.

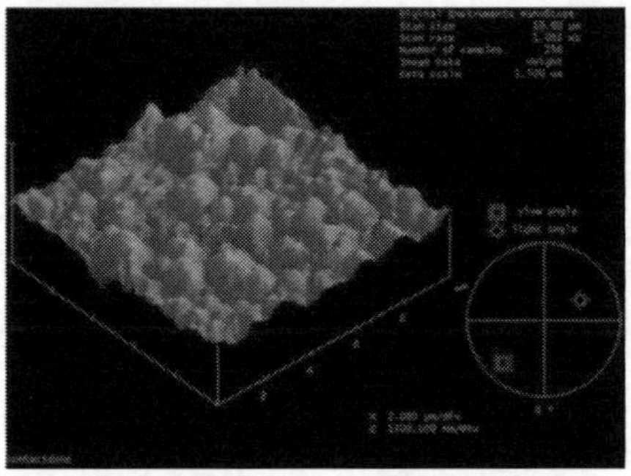

Figure 4. AFM of $BaTiO_3$ film of the contact area in the "on" state.

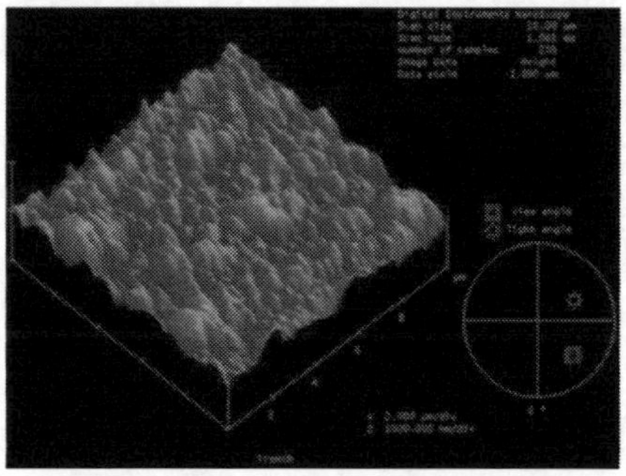

Figure 5. AFM of $BaTiO_3$ film of the contact area in the "off" state.

In Figure 6, the flatted area of gold (Au) membrane in contact with $BaTiO_3$ film has bigger chunks of the dielectric compared to the sputtered deposition in Figure 7. The bigger the chunks present in a film the more irregular and inconsistent is its surface. These chunks or large particle sizes increase stiction forces upon the gold membrane as it collapses on them creating irregularities in its surface. Stiction occurs when drying the substrate after dissolving the sacrificial material or after the fabrication is finished. In some cases, capillary forces of the rinsing liquid pull the released structures together while drying and cause them to join. These forces are predominantly due to the liquid's surface tension [4]. In this work, a second PMGI or sacrificial layer as stated in chapter four was found to smooth out irregularities and improve the gold suspended membrane's profile thus reducing film stress and lowering the required actuation voltage.

5. Conclusions

A chebyshev bandpass filter is presented to illustrate the effectiveness of $BaTiO_3$ patches as shown in Figure 8. The 2.1 GHz bandpass filter (dotted-curve 1) switched at 68 V at 40 K while the 2.6 GHz filter

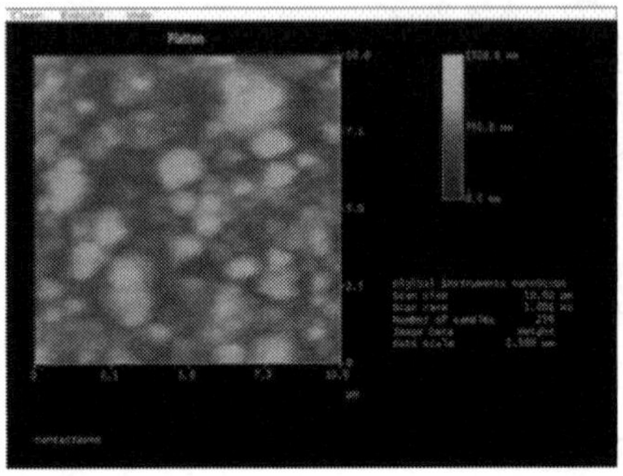

Figure 6. Flatted area of Au membrane in contact with $BaTiO_3$ film.

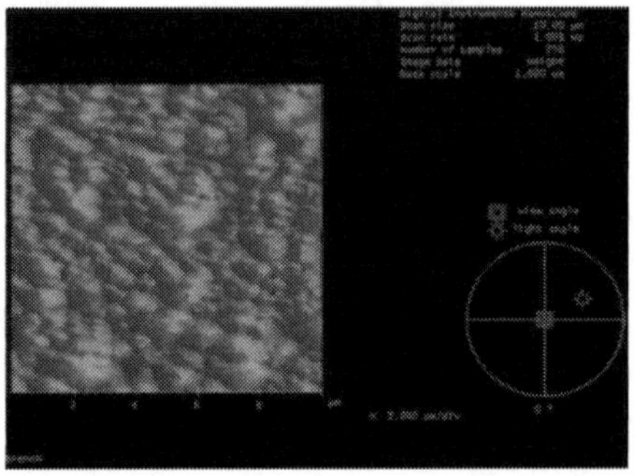

Figure 7. Figure 7: Flatted area of $BaTiO_3$ patch before switching.

(solid-curve 2) switched at 75 V at 30 K and a surface mode resonance (3) caused by cross coupling from the other filter. The use of a $BaTiO_3$ dielectric layer as an insulator under the suspended gold membrane has shown suitable and reliable switching cycles. The sacrificial layer was essential to form a proper undercut for the fabrication of the bridge. $LaAlO_3$ is a suitable microwave substrate for use with high T_c superconductors. Results have demonstrated the feasibility of connecting superconductive MEMS switches with bandpass filters.

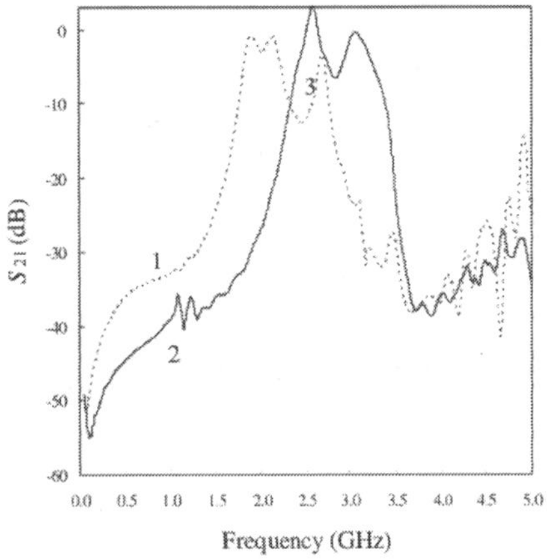

Figure 8. Experimental response (dotted curve 1) at 2.1GHz, T = 40 K and experimental response (solid curve 2) at 2.6 GHz, T = 30 K. Surface mode resonance (3).

References

[1] Nguyen C, Katehi L, and Rebeiz G 1998 *Wireless Comm. IEEE Proc.* **86** 1756-1768
[2] Duffy S, Bozler C, Rabe S, Knecht J, Travis L, Wyatt P, Keast C and Gouker M 2001 *IEEE Microwave and Wireless Components Letters* **11** 106-108
[3] Goldsmith C, Ehmke J, Malczewski A, Pillans B, Eshelman S, Yao Z, Brank J and Eberly M 2001 *IEEE MTT-S Int. Microwave Symp. Dig.* **1** 227-230
[4] Vargas J, Bogozi A, Noel J, Hijazi Y, Vlasov Y and Larkins G 2011 *IEEE Trans. On Appl. Supercond.* **21** 567-570
[5] Nakamiya T, Ikegami T and Ebihara K 1993 *IEEE Trans. On Appl. Supercond.* **3** 1057-1060

Acknowledgments

This work was supported by the Air Force Office of Scientific Research (AFOSR)

Journal of Physics: Conference Series **507** (2014) 042046 doi:10.1088/1742-6596/507/4/042046

Induced change of critical current density profile in Nb/Al-AlOx/Nb Josephson junctions.

C Granata, L Petti, M Rippa, S Rombetto, B Ruggiero, M Russo, R Russo[1], and A Vettoliere[2]
Istituto di Cibernetica "E. Caianiello" del Consiglio Nazionale delle Ricerche, Via Campi Flegrei 34, 80078 Pozzuoli (Napoli), Italy

a.vettoliere@cib.na.cnr.it

Abstract. A technique to induce spatial modulation of critical current density in niobium based Josephson devices by using a selective thermal annealing is reported. By depositing a carbon film onto selected region of the Josephson element it is possible to induce a localized heating, with a spatial resolution less than 1μm, exploiting the much higher absorbance coefficient of carbon than the niobium one. The effectiveness of such technique is demonstrated by experimental measurement of the critical current vs. magnetic field, measured at T = 4.2 K, showing that the change of critical current density occurs only in the region corresponding to the absorber film area. Furthermore, the theoretical behaviour, by modelling a suitable step-like junction barrier shape, has been carried out to fit the experimental data in order to verify the selective modulation of critical current. This technique can be very useful in view of quantum computing experiments, Majorana fermions detection and superconducting magnetic sensors.

1. Introduction

Josephson junctions are widely employed in superconducting circuits for both applications and basic physics research, so the possibility to control the main parameters of a Josephson device like the critical current density can be very useful. As the Josephson critical current density depends mainly on the thickness of insulation barrier, typically obtained by oxidation process, it is possible to modify the critical current density after the sample fabrication by using thermal or selective laser annealing [1-4]. However, these techniques allow us to modify the critical current density of the whole chip or, at most, of the single Josephson element in the case of selective laser annealing [5]. In any case, the critical current decreases as a consequence of tunnelling barrier thickness expansion produced by thermal effect. The increase of AlOx barrier thickness is due to several phenomena, thermally activated, such as diffusion through channels formed in Nb granular structure of oxygen coming from surface niobium oxide, reaction of Al atoms with free oxygen embedded at interstitial places close to the tunnel barrier and dissociation of aluminium hydroxides formed during the oxidation process. Here, a technique to change the critical current density within the single niobium based Josephson junction with a good spatial resolution by a selective annealing is presented. This technique allows us to induce spatial modulation of critical current density in Josephson elements paving the route to the control of critical current density profile and/or to the production of controlled "defects" inside the junction area. It is

[1]Present address: Consiglio Nazionale delle Ricerche, Istituto per la Microelettronica e i Microsistemi, UoS di Napoli, via Pietro Castellino 111, 80131 Napoli, Italy
[2]To whom any correspondence should be addressed

worth to note that a technique based on Ion implantation could be also employed to modify the critical current density profile, but with an unavoidable degradation of the involved superconducting films, leading to a local depression of the critical temperature and energy gap [6,7].

2. Methods and experimental results

The localized heating inside a single circuit element is made possible by depositing a carbon film on the desired region exploiting the larger absorbance coefficient of carbon (close to 1) with respect to niobium (about 0.2). At later stage, the Josephson junction is uniformly exposed to a heating radiation generated by an Argon laser beam aligned by an optical system. The difference in the power absorption by the different surfaces under irradiation allows reaching higher temperatures under the carbon film than in the surrounding areas, leading to a selective annealing with good spatial resolution down to 1 μm. In figure 1, the basic principle and the picture of the sample under investigation are shown.

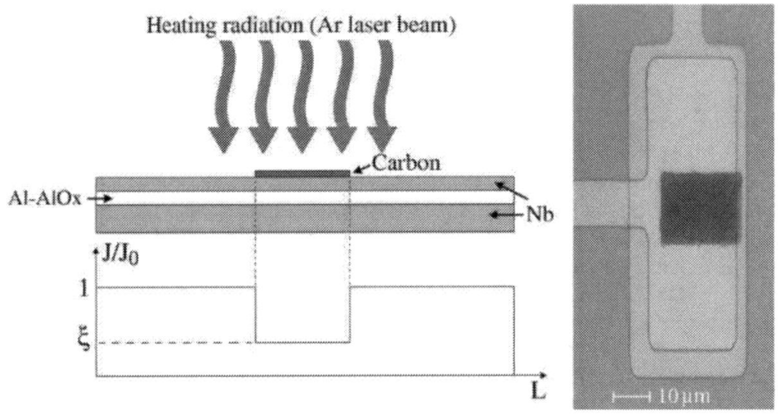

Figure 1. (Left) Scheme of basic procedure for selective annealing in a Josephson device. Thanks to absorber film, the heating radiation produces a selective annealing only under carbon covered area. It is also reported the expected critical current density profile with a step-like behaviour after annealing process. (Right) Picture of the Josephson junction subject to selective annealing process.

The sample fabrication process, based on trilayer (Nb/Al-AlOx/Nb) technology, is well described elsewhere [8, 9]. High quality window type junctions with different sizes have been fabricated having a critical current density of about 50 A/cm^2.
The Josephson junction of presented experiment has a rectangular shape with a side length of 80 μm and 20 μm. The carbon absorber has been deposited by thermal evaporation and patterned by lift off process resulting in a film thickness of 300 nm. It has a length of 30 μm and is centred at middle of the junctions covering about the 37% of its length.
To heat the Josephson element, an argon laser beam (λ=514.5 nm), has been used at room temperature. The laser power and time exposition was 2W and 30 s respectively. The laser spot of 200×80 μm^2 has been focused on the junction by using an optical system [5]. It is worth noting that the laser spot is larger than the junction size being the device annealed region determined by the absorber dimensions, therefore a different radiation heating source could be also employed such as an UV lamp. In these conditions a temperature of about 330 °C was estimated in correspondence of insulating tunnel barrier in the region covered by the absorber.
In figure 2, the current to voltage characteristic of the junction before annealing process is shown, measured in a shielded cryostat containing liquid helium [10,11]. In the same figure, the critical current as a function of magnetic flux is also shown, where the very good agreement with the Fraunhofer pattern (black line) is evident, indicating an uniform current density profile and a good junction quality. The measured critical current is of about 900 μA in the untreated Josephson junction.

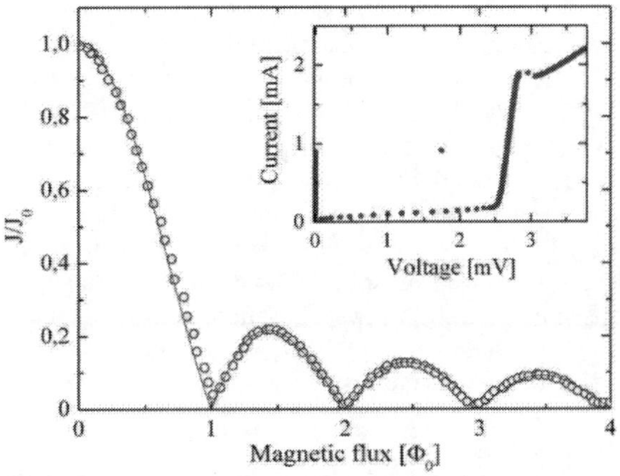

Figure 2. Critical current as a function of magnetic flux before thermal annealing of the Josephson junction measured at T = 4.2 K; the continuous line represents the calculated Fraunhofer pattern. In the inset, current-voltage characteristic before the annealing process is shown.

After the selective annealing process, the measurements of the critical current as a function of magnetic flux treading the junction and current to voltage characteristics of the device have been performed again and the results are displayed in figure 3. From the current to voltage characteristic a reduction of the overall critical current is measured which, after exposition, has a value of 630 μA corresponding to a decrease of about 30%. In figure 3, the critical current vs. magnetic flux is shown in which the increase of secondary slope's height is observed, due to the change of the current density profile and, in particular, to a depression of the critical current in the middle area of the junction. In order to better verify that the modification area corresponds to the absorber one the experimental data have been compared with theoretical predictions obtained by using a suitable junction barrier model based on step-like barrier shape taking into account the possible asymmetries due to imperfect alignment of the absorber with respect to junction centre [12]. The fitting curves (continuous line in figure 3) are obtained fixing all values to the experimental ones and using ξ (see figure 1) as free parameter [13].

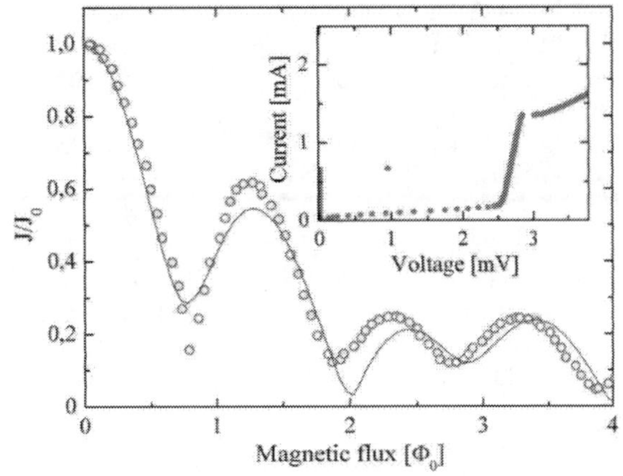

Figure 3. Critical current vs. magnetic flux after thermal annealing of the Josephson junction measured at T = 4.2 K. The continuous line represents the theoretical behaviour of a step-like shape junction model. The inset shows current-voltage characteristic after the annealing process showing a reduction of the critical current of about 30% compared to non annealed junction.

The good agreement obtained for $\xi = 0.2$ demonstrates that the critical current density was changed only in the absorber covered region, leaving unchanged the critical current density outside that region ensuring the capability to locally modify the Josephson critical current density. Finally, by fixing ξ to obtained value and leaving free the asymmetry parameter Δs, the absorber misalignment has been estimated finding that its value is about of 10 μm which is comparable with the misalignment measured on the junction under investigation. It is worth noting that it is possible to vary the annealing parameters, although in a very narrow range to avoid damage to the insulating barrier. In particular, by varying the timing of exposition different values for the parameter ξ are obtained. Such results demonstrate that it is possible to control some fundamental parameters of the Josephson device (junction and SQUID) after the fabrication process, obtaining a very tunable device. This peculiarity could be very useful in view of interesting applications in quantum computing, Majorana fermions detection and superconducting magnetic sensors.

3. Conclusions

In conclusion, the spatial modification of the profile of critical current density within a Josephson device has been obtained by using a selective annealing technique based on the use of a carbon film acting as heat absorber deposited on the device region to modify. The effectiveness of such method has been proved by experimental results and confirmed by theoretical predictions based on an asymmetric step-like barrier profile model. It has been demonstrated also that the change of current profile, with an estimated resolution of about 1 μm, occurs only in the region of the Josephson junction corresponding to the area covered by the absorber.

On the basis of these results, the here reported selective annealing technique provide the possibility to control some fundamental parameters of the devices based on Josephson effect after the fabrication process, obtaining a very tunable device. This peculiarity is very useful in view of recent basic physics research and applications involving the Josephson devices. The results have been obtained making use of the different material absorption coefficients and, with a suitable choice of reflecting/absorbing materials, the reported method could be also applied on other materials and devices whose properties can be modified by thermal annealing [14, 15].

References

[1] Lehnert T, Billon D, Grassi C and Gundlach K H 1992 J. Appl. Phys. **72**, 3165

[2] Shiota T, Imamura T and Hasuo S, 1992 IEEE Trans. Appl. Supercond. **2**, 222

[3] Oliva A and Monaco R 1994 IEEE Trans. Appl. Supercond. **4** 25

[4] Migacz J V and Huber M E 2003 IEEE Trans. Appl. Supercond. **13** 123

[5] Granata C, Vettoliere A, Petti L, Rippa M, Ruggiero B, Mormile P and Russo M 2007 Appl. Phys. Lett. **90** 232503

[6] Camerlingo C, Russo M and Tosello C 1988 J. Appl. Phys. **64** 2759

[7] Camerlingo C, Ruggiero B, Russo M and Sarnelli E 1989 IEEE Trans. Magn. **25** 1251

[8] Granata C, Di Russo C, Monaco A and Russo M, 2001 IEEE Trans Appl. Supercond. **11** 95

[9] Granata C, Vettoliere A and Russo M 2007 Appl. Phys. Lett. **91** 122509

[10] Granata C, Vettoliere A, Russo R, Russo M and Ruggiero B 2011 Phys. Rev. B **83** 092504.

[11] Russo R, Granata C, Esposito E, Peddis D, Cannas C and Vettoliere A 2012 Appl. Phys. Lett. **101** 122601

[12] Barone A and Paternò G 1982 *Physics and Application of the Josephson Effect* (John Wiley and Sons).

[13] Granata C, Petti L, Rippa M, Rombetto R, Ruggiero B, Russo M, Russo R and Vettoliere A 2013 Appl. Phys. Lett. **102** 222603

[14] Seidel P 2001 J. Supercond. **14** 305

[15] Schneidewind H, Wunderlich S, Schmidl F, Dorrer L and Seidel P in Physics and Material Science of High Temperature Superconductors IV, ed Kossowsky R, Jelinek M and Novak J (Springer, Dordrecht, 1997), pp. 311–322.

Journal of Physics: Conference Series **507** (2014) 042047 doi:10.1088/1742-6596/507/4/042047

Improvements of a Beam Current Monitor by using a High Tc Current Sensor and SQUID at the RIBF

T Watanabe[1], N Fukunishi[1], M Kase[1], O Kamigaito[1], S Inamori[2], and K Kon[2]

[1] RIKEN (The Institute of Physical and Chemical Research), Wako-shi, Saitama 351-0198, Japan

[2] TEP Corporation, 2-20-4 Kosuge, Katsushika, Tokyo 124-0001, Japan

E-mail: wtamaki@riken.jp

Abstract. To measure a beam current non-destructively, a conventional DC current transformer (DCCT) has been used at accelerator facilities. However, the current resolution of the DCCT is worse than 1μA. This current resolution is sufficient for electron and proton accelerators in which the beam intensity is high, but it is not sufficient for lower intensity heavy-ion beams. Thus, superconducting technology has been applied to the precise measurement of the beam current. In particular, to measure the DC current of high-energy heavy-ion beams non-destructively at high resolution, a high critical temperature (HTc) superconducting quantum interference device (SQUID) beam current monitor (HTc SQUID monitor) has been developed for use in the radioactive isotope beam factory (RIBF) at RIKEN in Japan. Beginning this year, the magnetic shielding system has been greatly reinforced. The measurement resolution is determined by the signal to noise ratio, that is improved by attenuating the external magnetic noise and is mainly produced by the distribution and transmission lines from the high current power supplies. The new strong magnetic shielding system can attenuate the external magnetic noise to 10^{-10}.

1. Introduction

The radioactive isotope beam factory (RIBF) project to accelerate all elements from hydrogen to uranium up to energies of 440 MeV/u for light ions and 350 MeV/u for heavy ions started in April 1997 [1]. The research activities in the RIBF project make extensive use of the heavy-ion accelerator complex, which consists of two linear accelerators and four ring cyclotrons including a superconducting ring cyclotron (SRC). Energetic heavy-ion beams are converted into intense RI beams via the projectile fragmentation of stable ions or the in-flight fission of uranium ions using a superconducting isotope separator, BigRIPS [2]. The combination of these accelerators and BigRIPS is greatly expanding our knowledge of the nuclear world into the previously inaccessible regions on the nuclear chart. For these facilities, since it is essential to measure the beam current non-destructively, an HTc SQUID monitor has been developed for the RIBF at RIKEN. As recently reported [3, 4], the high-Tc superconducting quantum interference device (SQUID) monitor allows us to measure the DC of high-energy heavy-ion beams nondestructively in such a way that the beams are diagnosed in real time and the beam current extracted from the cyclotron is recorded without interrupting beam user experiments. Unlike at other existing facilities, a low-vibration, pulse-tube refrigerator cools the HTc fabrications that includes the SQUID in

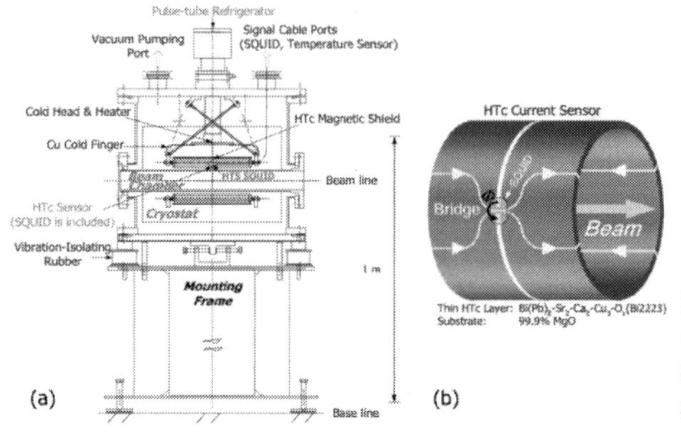

Figure 1. Schematic drawing of (a) the high-Tc SQUID monitoring system and (b) the current sensor [4].

such a way that the size of the system is reduced and the running costs are lowered, which is its largest advantage.

2. Reinforcement of the magnetic shielding

2.1. Fundamental principles of the HTc SQUID monitor

A schematic drawing of the HTc SQUID monitor system is shown in Fig. 1(a). Both the HTc magnetic shielding and the HTc current sensor were fabricated by dip-coating a thin Bi_2-Sr_2-Ca_2-Cu_3-O_x (Bi-2223) layer on a 99.6% MgO ceramic substrate. When a charged particle beam passes along the axis of the HTc tube, a shielding current produced by the Meissner effect flows in the opposite direction along the wall of the HTc tube so as to screen the magnetic field generated by the beam. Since the outer surface is designed to have a bridge circuit (Fig. 1(b)) [4], the current generated by the charged particle beam is concentrated there and forms an azimuthal magnetic field Φ around the bridge circuit. The HTc SQUID is set close to the bridge circuit and can detect the azimuthal magnetic field with a high S/N ratio. In particular, we adopt a SQUID gradiometer [5] to increase the S/N ratio.

2.2. Development of the hybrid magnetic shielding system

The performance of the current monitor is determined by the S/N ratio. To improve the measurement resolution, it is important to attenuate the external magnetic noise, that is mainly produced by the distribution and transmission lines from the high current power supplies in the RIBF. Therefore, we developed a hybrid magnetic shielding method based on the properties of perfect diamagnetic materials and ferromagnetic materials. By applying this method to the HTc SQUID monitor, a high shielding effect could be obtained despite the compact system. This system consists of two shielding parts: one for the HTc current sensor and the other for ferromagnetic shielding materials. The HTc current sensor also works as the superconducting shield via the Meissner effect (perfect diamagnetism). The ferromagnetic shielding materials are made with high permeability alloys (Permalloy, Mu-metal, etc.) and consist of three parts: frame, cap, and band (Fig. 2). The HTc SQUID is installed inside the frame and onto the HTc current sensor, and the frame is covered by the cap. Consequently, the HTc SQUID is almost completely surrounded by the hybrid magnetic shielding system. As mentioned in the previous section, since the dip coated superconducting material of the HTc current sensor is peeled off except for the bridge circuit so as to concentrate the current induced by the beam passage, the external magnetic noise can penetrate through the peeled part. To prevent this penetration of external magnetic noise, the band plays an important role.

To design a hybrid magnetic shielding system, the attenuated magnetic field was calculated by using the electromagnetic field simulation program Opera-3 [6]. This simulation assumes

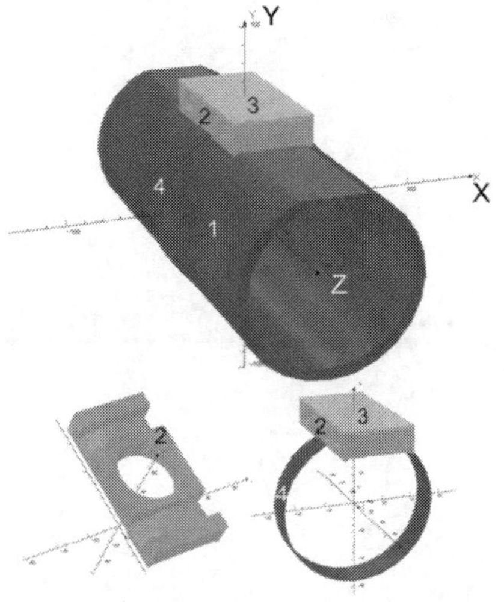

Figure 2. 1: HTc current sensor with ferromagnetic shielding materials, 2: frame, 3: cap, and 4: band.

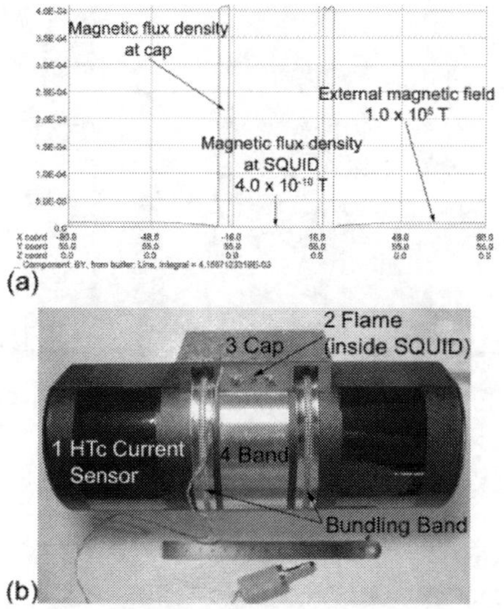

Figure 3. Calculated result by Opera-3d and a photograph of the completed hybrid magnetic system.

there is a uniform magnetic field of 1 $\times 10^{-5}$ T inside the boundary conditions and that the relative permeability of the ferromagnetic material is 1 $\times 10^{6}$. Since Opera-3d cannot deal with superconductivity, the value of 1 $\times 10^{-12}$ was substituted as the relative permeability of the HTc current sensor for the approximate calculation. The calculated result indicates that the magnetic field was attenuated from 1 $\times 10^{-5}$ to 4 $\times 10^{-10}$ T by applying the hybrid magnetic shielding system (Fig. 3(a)). Consequently, this calculation shows that the hybrid magnetic shielding system can achieve a magnetic attenuation factor of 10^{-4}. A photograph of the completed hybrid magnetic system is shown in Fig. 3(b).

2.3. Active magnetic field canceller system

In the acceleration facility, since there exist 50 Hz and higher order AC magnetic noises that are much stronger than terrestrial magnetism, an active magnetic field canceller system [6] was designed and introduced to the HTc SQUID monitor. This system is comprised of a magnetic field control unit, combined AC/DC magnetic field sensors, and compensation coils. The compensation coils consist of three pairs of coils that are arranged perpendicular to each other. Each of these pairs form a so-called Helmholtz-Coil-Pair , able to produce a homogenous magnetic field in between them, and each pair controls one direction (along x-, y-, or z- axis). Specifications of the active magnetic canceller system are listed in Table 1 and a photograph is shown in Fig. 3. Furthermore, by hooking up a terminal via RS-232C serial interface, access to the menu-driven interface of the system is achieved.

Table 1. Specifications of active magnetic field canceller system

Magnetic field attenuation	Max. -40 dB	Max. compensation	6 μT
Bandwidth	DC - 1,000 Hz	AD/DA converter	16 bit
Magnetic sensor(DC)	3 axes flux gate	Magnetic sensor(AC)	Search coil
DC Noise	$\leq$ 50 pT/$\sqrt{Hz}$	AC Noise @1 Hz	$\leq$ 150 pT/$\sqrt{Hz}$

Figure 4. HTc SQUID monitor with active magnetic field canceller system.

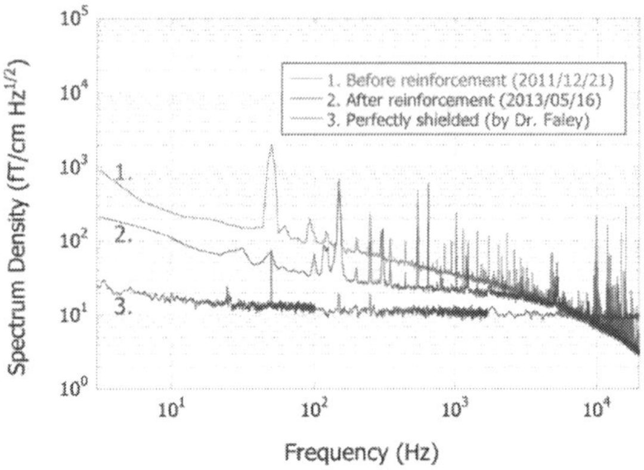

Figure 5. Measured spectra in the frequency domain.

3. Measured spectra in the frequency domain

To evaluate the performance of the hybrid magnetic shielding system and the active magnetic canceller system, the output signals of the HTc SQUID were analyzed in the time- and frequency domains. Measured spectra in the frequency domain are shown in Fig. 5. Each of the three lines represents a measurement under a different condition as follows: 1. red line: before reinforcing the magnetic shielding, 2. blue line: after the reinforcing of the magnetic shielding, and 3. green line: perfectly shielded by a superconducting shield and threefold permalloy shields [8]. The same setting parameters of the HTc SQUID controller were used in these measurements. When the outputs of the HTc SQUID monitor were measured (lines 1 and 2), the RIBF experiment was performed: ^{238}U was being accelerated and the output powers of the high current power supplies was almost at maximum level. The signal was measured in the room next to where the power supplies were located, where the leakage magnetic field of the 50 Hz component was measured by a gauss meter as 4.5×10^{-4} T. In contrast, the output signal of the 50 Hz component of the HTc SQUID monitor was 6×10^{-14} T as shown in Fig. 5. Based on these findings, we consider that the combination of the hybrid magnetic shielding system and the active magnetic canceller system can attenuate the external magnetic noise to 10^{-10}.

The authors are grateful to Dr. M. I. Faley from Forschungszentrum Juelich GmbH for valuable discussions on the HTc SQUID, and to E. Nemoto of JEOL DATUM, Ltd., for cooperation regarding the noise cancellation system. We also thank Mr. H. Fujimori from the High Energy Accelerator Research Organization, KEK for valuable advice on Opera-3d. This work was supported by JSPS KAKENHI (Grants-in-Aid for Scientific Research) Grant Number 23600015.

References

[1] Y. Yano, Nucl. Instrum. Methods B **261** (2007) p. 1009.
[2] T. Kubo et al., IEEE Trans. Appl. Supercond. **17** (2003) p. 97.
[3] T. Watanabe et al., Proc. 2010 Beam Instrumentation Workshop (BIW10), Santa Fe, New Mexico, U.S.A (2010) p. 523.
[4] T. Watanabe et al., J. Supercond. Nov. Magn. **26**, (2013) p. 1297-1300, DOI 10.1007/s10948-012-1943-0.
[5] M. I. Faley et al., J. Phys. **43** (2006) p. 1199.
[6] COBHAM http://www.cobham.com/home.aspx
[7] JEOL Ltd. http://www.jeol.co.jp/en/
[8] M. I. Faley private communication (2012).

Journal of Physics: Conference Series **507** (2014) 042048 doi:10.1088/1742-6596/507/4/042048

YBCO grain boundary Josephson junction coupled with a slot dipole antenna for terahertz wave detectors

H Yamada, T Hayasaka, G Toya, A Saito, S Ohshima and K Nakajima

Yamagata University, 4-3-16 Jonan, Yonezawa-shi, Yamagata 992-8510, Japan

E-mail: hyamada@yz.yamagata-u.ac.jp

Abstract. We examined terahertz wave detectors that used $YBa_2Cu_3O_{7-\delta}$ (YBCO) grain boundary Josephson junctions (GBJJs) coupled with a slot dipole antenna (SDA). The detectors consisted of a 220-GHz full-wavelength SDA patterned on a Au layer and a GBJJ patterned on an YBCO/bicrystal MgO film, which were separated by an insulating benzocyclobutene layer. The microbridge of the fabricated junction was 5-μm wide and was trimmed to 2 μm using an ultraviolet laser cutter to modify the junction parameters. The critical current I_C, normal resistance R_N, and I_CR_N product after the trimming at 30 K were 0.62 mA, 1.42 Ω, and 0.88 mV, respectively. The current–voltage characteristics and radio frequency (RF) wave responses of these detectors for a millimeter wave of 180–240 GHz were measured at 30 K. The coupling efficiency between the GBJJ and the SDA and the system sensitivity were obtained as -19.0 dB and 630 V/W, respectively, at 193 GHz. For the RF wave response of 180–240 GHz, the coupling efficiency was relatively flat.

1. Introduction

Terahertz wave, which is an electromagnetic wave of approximately from 100 GHz to 10 THz, has the transparency of radio wave and straightness of optical wave. It features absorption lines that vary with the materials. Therefore, the terahertz wave can be used in some applications such as in imaging [1] and biological and chemical analyses [2]. Terahertz wave can be detected using semiconductor detectors, e.g., Si bolometer and Schottky barrier diode detector, or low-temperature superconductor (LTS) detectors, e.g., superconductor bolometer [3] and kinetic inductance detector [4]. These detectors have their own advantages and disadvantages. For example, the Si bolometer has a wide detection-frequency range, but its working temperature is only a few Kelvin because of thermal noise. The Schottky barrier diode detector can operate at room temperature, but its detection-frequency range is limited. Finally, the LTS detectors have higher sensitivity than the semiconductor detectors; however, their working temperature is only a few Kelvin owing to superconducting transition. Therefore, more sensitive general-purpose detectors are desired.

We have been examining high-temperature superconductor Josephson junction detectors because they can operate at liquid nitrogen temperature (77 K) and have high sensitivity around the terahertz wave band. We have reported on detectors coupled with two sets of half-wavelength (λ/2) slot dipole antenna (SDA) and a coplanar waveguide (CPW) [5]. The λ/2 SDAs have relatively low feed-point impedance, but as a detector, the CPWs have a limited bandwidth. Therefore, to develop wide band and sensitive terahertz detectors, we propose the grain boundary Josephson junction (GBJJ) detectors with a full-wave length (1λ) SDA without a CPW [6]. In this paper, we report the results of our

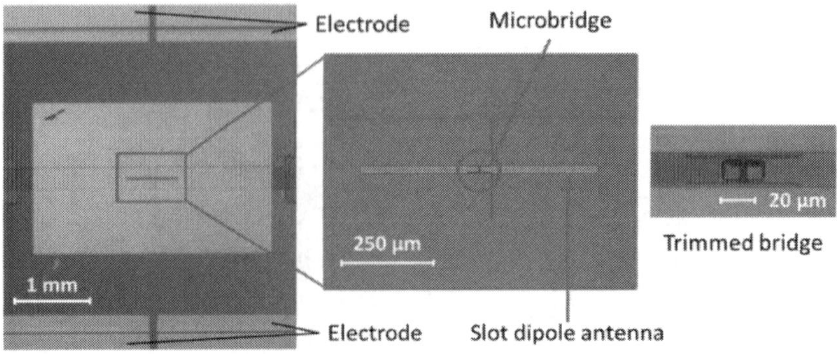

Figure 1. Optical microscope images of the fabricated single 1λ-SDA-coupled GBJJ detector.

Table 1. Characteristics of the junction to the bridge width.

Bridge width w (μm)	Critical current I_C (mA)	Normal resistance R_N (Ω)	$I_C R_N$ (mV)
5	1.30	0.62	0.81
3	0.68	1.06	0.72
2	0.62	1.42	0.88

examination on the coupling efficiency between a 1λ SDA designed for 220 GHz and the GBJJ, as well as the system sensitivity of the detector.

2. Experimental

The 1λ-SDA-coupled GBJJ detectors were made of Au (50 nm)/$YBa_2Cu_3O_{7-\delta}$ (YBCO, 100 nm)/bicrystal MgO (0.5 mm) films using Ar ion etching with photolithography. Initially, a 5-μm-wide YBCO microbridge was fabricated across the grain boundary of the MgO substrate. Next, the Au layer was etched to remove the Au on the junction and to fabricate four electrodes. Then, an insulating benzocyclobutene (BCB) layer was fabricated on the bridge and the bared MgO. Finally, a Au layer was deposited on the BCB layer by sputtering, and a 1λ SDA designed at 220 GHz, whose feed-point impedance Z_{ANT} was simulated as $3 + j20$ Ω, was fabricated.

The microbridges of the fabricated detectors were trimmed using ultraviolet laser etching to improve the junction characteristics. Thereafter, the current–voltage (I–V) characteristics of the detectors were evaluated. The detectors were cooled down to approximately 30 K using a pulse tube cryocooler, and an alternating current was applied to the junction to measure the voltage using an oscillator. In addition, the response was measured for the radio frequency (RF) wave of 180–240 GHz. The RF wave was generated using a Gunn diode signal generator (30–40 GHz) that reached the detector via an amplifier (30 dB), frequency multipliers (a total of six times), horn antenna, polymethilpentene lens, parabolic mirror, and hyperhemispherical Si lens.

3. Results and discussions

Figure 1 shows the optical microscope images of the fabricated single 1λ-SDA-coupled GBJJ detector. The length and width of the antenna slot and the thickness of the BCB layer were 700, 20, and 2.5 μm, respectively. Table 1 lists the characteristics of the junction for the 5-, 3-, and 2-μm-wide microbridges, which were obtained from the I–V characteristics. As the width of the bridge became thinner, the critical current I_C decreased, and the normal resistance R_N increased. However, the value of the product of I_C and R_N was almost constant, which was approximately 0.8 mV.

Figure 2 shows the RF wave responses measured at 30 K and 193 GHz for the 5- and 2-μm-wide bridges. Although the RF power values P_{RF} irradiated to the antenna were almost the same, the

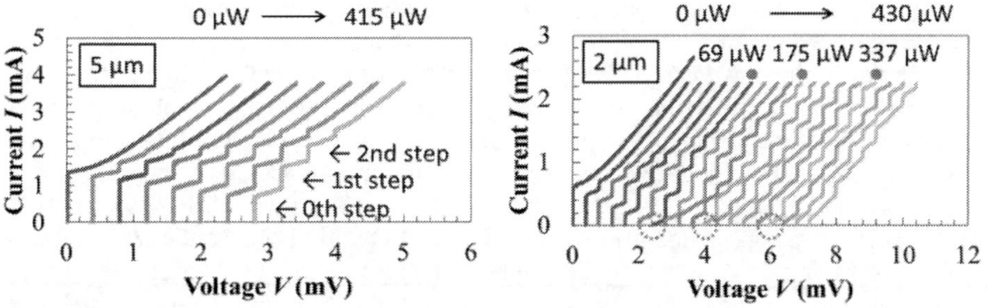

Figure 2. RF wave responses of the detectors for the 5- and 2-μm-wide bridges at 30 K and 193 GHz. Each characteristic is shifted to the right-hand side by 0.4 mV as the power decreased.

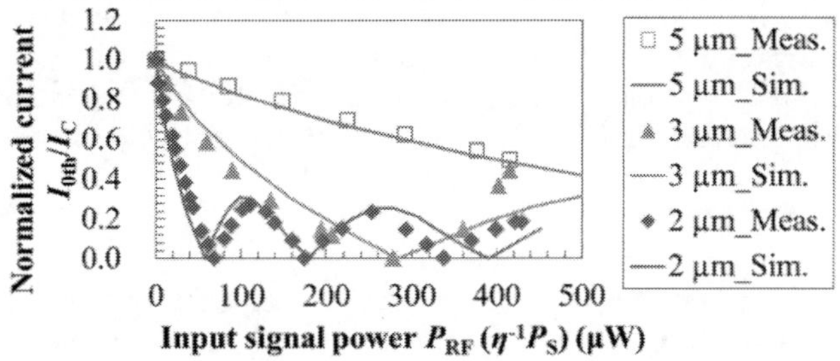

Figure 3. Characteristics of the current at the zeroth step I_{0th} to irradiated RF power P_{RF}. The I_{0th} values are normalized by each critical current I_C measured at $P_{RF} = 0$.

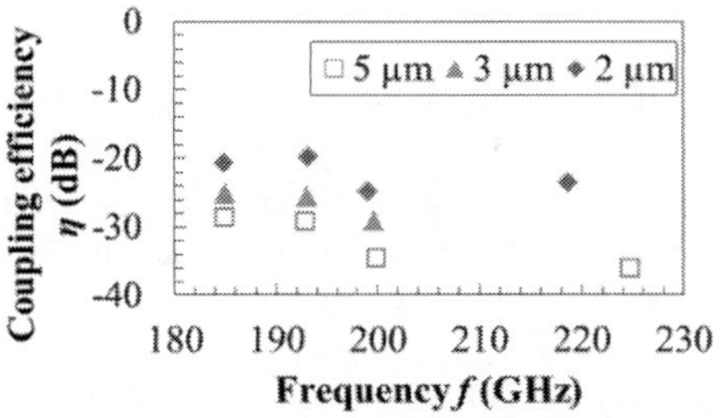

Figure 4. Frequency characteristics of coupling efficiency η for the bridge width at 30 K.

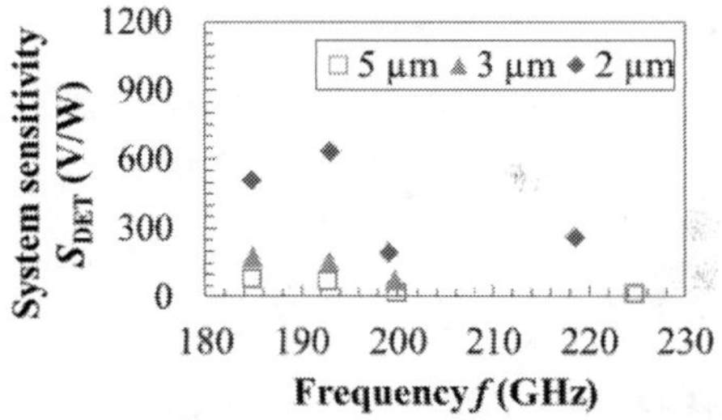

Figure 5. Frequency characteristics of system sensitivity S_{DET} for the bridge width at 30 K.

variations in the Shapiro steps to the power were different. Figure 3 shows the characteristics of the current in the zeroth step I_{0th} to the P_{RF}. For the 5-μm width, the I_{0th} values decreased monotonically with the RF power and did not reach 0 mA. Meanwhile, for the 2-μm width, the I_{0th} values alternately decreased and increased and became 0 mA three times, which means that the coupling between the junction and the antenna was improved by the trimming of the bridge.

To quantitatively evaluate the coupling between the junction and the antenna, the RF response of the Josephson junction was simulated using a resistively shunted junction model, which yielded the following equation:

$$I + I_S \sin 2\pi f_S t = I_C \sin\theta + \frac{V}{R_N}, \quad (1)$$

where I, I_S, f_S, t, θ, and V are the current, input signal current, input signal frequency, time, phase difference, and voltage, respectively. From the simulation, the I–V characteristics and the power introduced to the junction P_S (= $R_N I_S^2/2$) were obtained. The coupling efficiency η was obtained by fitting P_{RF} to P_S and was defined as P_S/P_{RF}. The I_{0th} value of P_S, which was fitted by multiplying η^{-1}, is also shown in Figure 3. Because the variations in the measured I_{0th} values agreed well with those of the fitted I_{0th} values, we believe that the effect of noise on the measured values was low. Figure 4 shows the frequency dependence of the coupling efficiency. The coupling efficiency increased as the bridge became thinner and was approximately -19 dB at 193 GHz for the 2-μm-wide bridge. We believe that this effect was caused by the improvement in the impedance matching because the normal resistance of the junction was enlarged by approximately 2.3 times and approached the impedance of the antenna. The values of the impedance mismatching loss M, which were estimated from the following equation:

$$M = 10\log_{10}\left(1-|\Gamma|^2\right), \quad \Gamma = \frac{Z_{ANT} - R_N}{Z_{ANT} + R_N}, \tag{2}$$

were -17.4 and -13.9 dB for the 5- and 2-μm-wide bridges, respectively. Although the impedance mismatching loss improved by approximately 3.5 dB, each loss value could be high. We believe that the Z_{ANT} reactance enlarged the impedance; thus, we need to examine the structure of the antenna or the whole detector. The frequency characteristics of the coupling efficiency slightly decreased with the frequency, but they were relatively flat. We believe that this is beneficial for wide-band detector.

Figure 5 shows the system sensitivity S_{DET} calculated by the following equations:

$$S_{DET} = \eta S, \quad S = \frac{\Phi_0}{16\pi k_B T}\frac{2f_C^2}{2f_S^2 + 3f_C^2}, \quad f_C = \frac{I_C R_N}{\Phi_0}. \tag{3}$$

In these equations, S, Φ_0, k_B, T, and f_C are the sensitivity, flux quantum, Boltzmann constant, device temperature, and junction characteristic frequency, respectively. The highest S_{DET} value was obtained as approximately 630 V/W at 193 GHz for the 2-μm-wide bridge. This value is yet lower than that of the Schottky barrier diode detectors. We are going to improve the system sensitivity by examining in details the effects of thinning the bridge width on the improvement of the coupling efficiency.

4. Summary

A single 1λ-SDA-coupled YBCO GBJJ detector (220 GHz, 5-μm-wide microbridge) has been fabricated. The bridge was thinned by laser etching, and the highest coupling efficiency of approximately -19 dB was obtained for the 2-μm-wide bridge at 30 K and 193 GHz. Here, the highest system sensitivity was approximately 630 V/W. The coupling efficiency increased as the bridge became thinner. For an RF wave response of 180–240 GHz, the coupling efficiency was relatively flat.

Acknowledgments

This work was supported by JSPS KAKENHI Grant Number 24760313. A part of this work was carried out in the clean room of Yamagata University.

References

[1] Du J, Hellicar A D, Li L, Hanham S M, Macfarlane J C, Leslie K E, Nikolic N, Foley C P and Greene K J 2009 *Supercond. Sci. Technol.* **22** 114001
[2] Kawase K, Ogawa Y, Watanabe Y and Inoue H 2003 *Opt. Express* **11** 2549-2554
[3] Liu L, Xu H, Percy R R, Herald L L, Lichtenberger A W, Hesler J L and Weikle R M II 2009 *IEEE Trans. Appl. Supercond.* **19** 282-286
[4] Day P K, LeDuc H G, Mazin B A, Vayonakis A and Zmuidzinas J 2003 *Nature* **425** 817-821
[5] Nakajima K, Ebisawa N, Sato H, Chen J, Yamashita T and Sawaya Y 2005 *IEEE Trans. Appl. Supercond.* **15** 549-551
[6] Yamada H, Hayasaka T, Saito A, Ohshima S and Nakajima K 2013 *Physics Procedia* **45** 217-220

Development of Robust HTS-SQUID for Non-destructive Inspection System in Unshielded Environment

K Yoshida[1], T Kage[1], T. Suzuki[1], Y Hatsukade[1] and S Tanaka[1]
[1]Toyohashi University of Technology, 1-1 Hibarigaoka, Tempaku-cho, Toyohashi, Aichi 441-8580, Japan

E-mail: hatsukade@ens.tut.ac.jp

Abstract. Cross-shaped $YBa_2Cu_3O_{7-x}$ (YBCO) film was overlapped on directly-coupled multi-pickup-coil HTS-dc-SQUID magnetometer in flip-chip configuration as a superconducting shield, and the characterestics of the SQUID were examined in DC and AC magnetic fields. We created slots in the YBCO film and pickup-coil of HTS-SQUID magnetometer for suppression of flux trapping. In low magnetic field environment, we measured the characteristics of the SQUID without and with HTS film. The most of the parameters were same with and without the HTS films, while $S_\phi^{1/2}$ with the HTS film was a bit smaller than that of the bare SQUID. In the DC field, I_c of the HTS-SQUID covered with the HTS film did not change until a DC field over 140 μT was applied to the SQUID, while the bare SQUID's I_c decreased with the increase of the applied DC field. In the AC fields at 10 MHz, the SQUID with the HTS film showed degradation of the parameters such as V_{pp} and $S_\phi^{1/2}$ at lower field amplitude than the bare SQUID.

1. Introduction

High temperature superconducting (HTS) superconducting quantum interference devices (SQUIDs) have very high magnetic field sensitivity and wide frequency range. Bicrystal grain boundary $SrTiO_3$ (STO) substrates are commonly used for the HTS-SQUIDs [1]. Fabrication process of single-layer HTS-SQUID based on the bicrystal STO substrates is simple, and offers relatively good characteristics [2]. Therefore, HTS-SQUIDs with bicrystal Josephson junctions have been widely utilized in many field of such as non-destructive inspection (NDI), magnetocardiogram (MCG), geophysical exploration and nuclear magnetic resonance (NMR)/magnetic resonance imaging (MRI) [3-6]. Most of the applications using the HTS-SQUIDs are generally conducted in magnetic shield. However, it is desirable that no magnetic shields are used for practical use. When a conventional bicrystal STO based HTS-SQUID is cooled in an unshielded environment, flux trapping occurs to increase SQUID's noise [7]. In the applications such as NDI, when a bicrystal STO based HTS-SQUID is moved in environmental magnetic field, critical current I_c of the SQUID changes due to the Earth's field (about 50 μT), and flux locked loop circuit cannot keep locking [8]. Therefore, it is difficult to operate the bicrystal STO based HTS-SQUIDs stably and with low noise for such advanced NDI application in the Earth's field [9]. In order to solve these problems, we developed bicrystal STO based HTS-SQUID that can be cooled, operated and moved in unshielded environment for the NDI application.

2. Design of HTS-SQUID magnetometer and HTS film

We designed a directly-coupled HTS-SQUID magnetometer based on a bicrystal STO substrate with misorientation angle of 30 degrees and $YBa_2Cu_3O_{7-x}$(YBCO) thin film of about 200 nm in thickness. For suppression of the change of critical current I_c by moving a HTS-SQUID in the Earth's field, we designed another HTS film that covered on the SQUID's Josephson junctions. Figure 1 (a) shows design of the directly-coupled HTS-dc-SQUID magnetometer. This SQUID magnetometer had no grain boundary in the pickup-coil because the SQUID had multi-pickup-coil. The SQUID ring is located in the center of the device, and is coupled with the quadruple pickup-coil. The SQUID ring inductance was estimated to be about 52 pH. The pickup-coil of the SQUID was slotted so that the minimum width in the pickup-coil became 5 μm for suppression of flux trapping in cooling in the Earth's field (see Figure 1(c)) [7]. Figure 1(b) shows a pattern of the HTS film. The shape of the HTS film was a cross-shape and had slots such as Figure 1(c). For comparison, we prepared a normal HTS film without slots. The HTS films were designed to cover the SQUID ring and apertures between the pickup-coil. Using the conventional sputtering method, 200 nm-thick YBCO film, for the SQUID and the HTS film, were deposited on the bicrystal and normal STO substrates, respectively. For suppression of the change of I_c, these HTS films were overlapped on the SQUID in flip-chip configuration in experiments in section 3. The position of the HTS films are indicated in Figure 1(a).

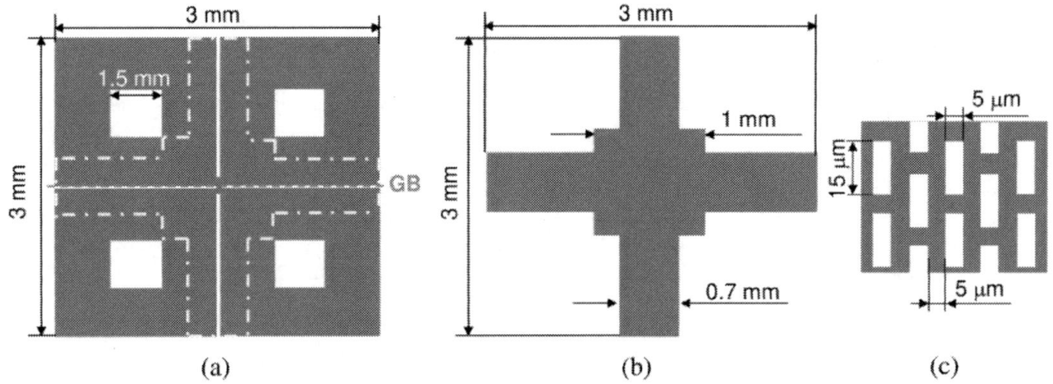

Figure 1. Directly-coupled HTS-dc-SQUID magnetometer with quadruple square pickup-coil. (a) Whole design. Bold chain line indicates the position where the HTS film is overlapped in flip-chip configuration. (b) HTS film. (c) Slots in pickup-coil and HTS film.

3. Experimental Setup

At first, to measure the characteristics of the bare HTS-SQUID magnetometer in a low magnetic field, we set up low field environment in a tri-layer μ-metal shielding case that was set in a moderate magnetically shielded room (MSR). The residual DC field in the case was about 50 nT. We filled liquid nitrogen in a Dewar in the shielding case, and cooled the SQUID in this environment. We measured the characteristics of the bare SQUID such as critical current I_c, normal resistance R_n, modulation voltage V_{pp} and flux noise $S_\phi^{1/2}$. Most of them were measured using a commercial SQUID (Star Cryoelectrnics PCI-100) with dc current bias. As shown in the Figure 2(a), to measure the changes in the characteristics of the SQUID in DC field, we applied an DC field from 1 to 500 μT.

Next, for insulation and adhesion, we painted Apiezon grease between the magnetometer and the film in the flip-chip configuration, and pressed them so that the distance between both films became as close as possible. Figure 2(b) shows that the magnetometer and film were tightly connected using polymer threads for reinforcement. The same characteristics of the SQUID magnetometer with HTS film were measured in the same way as above-mentioned. We measured V_{pp} and $S_\phi^{1/2}$ of the uncovered SQUID and the covered SQUID with the HTS films in an AC field at 10 MHz.

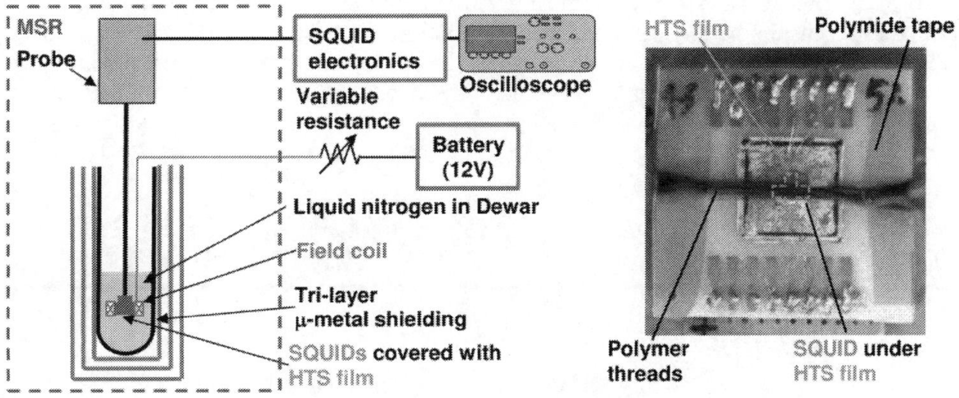

(a) (b)

Figure 2. Experimental setup. (a) Measurement setup for SQUID characteristics in moderate MSR. (b) Photograph of HTS-SQUID and HTS film in flip-chip configuration.

4. Results and Discussions

The *V-I* characteristics, *V-Φ* characteristics, and flux noise spectra of the HTS-SQUID magnetometer without and with the HTS films in a low magnetic field are summarized in Table 1. I_c, R_n, V_{pp} of the SQUID without and with the HTS films were roughly the same. As shown in the Table 1 and Figure 3, $S_\phi^{1/2}$ decreased a bit with the HTS films on the SQUID. Figure 4 shows the change of I_c when a DC field from 1 to 500 μT was applied to the SQUID. As the DC field increased, I_c of the SQUID without the HTS film immediately decreased. On the other hand, in the case of the SQUID covered with the normal HTS film, the SQUID's I_c did not change until a DC field over 140 μT was applied. It is thought that the HTS film protected the Josephson junctions from DC field. I_c of the SQUID with slot HTS film also did not change until DC field of 140 μT was applied. In comparison with I_c of the SQUID covered with the normal HTS film, I_c of the SQUID covered with slot HTS film decreased gently with the increase of the DC field. Figure 5(a) shows change of V_{pp} when an AC field at 10 MHz was applied to the SQUID with and without the films. When AC field more than 200 nT_{rms} was applied to the SQUID without the HTS film, V_{pp} of the SQUID started to decrease. In contrast, when AC field more than 120 nT_{rms} was applied to the SQUID with the HTS films, SQUID's V_{pp} started to decrease. Figure 5(b) shows change of $S_\phi^{1/2}$ when the AC field was applied. The tendency of the change of $S_\phi^{1/2}$ has similar with those of V_{pp} in Figure 5(a). $S_\phi^{1/2}$ started to increase at 200 nT_{rms} without HTS film, while it started to increase at 120 nT_{rms} with the HTS films.

Table 1. Characteristics of HTS-SQUID without and with HTS films.

Parameter	Without HTS film	With normal HTS film	With slot HTS film
I_c	80 μA	81 μA	80 μA
R_n	1.4 Ω	1.2 Ω	1.3 Ω
V_{pp}	10.7 μV	10.5 μV	10.3 μV
$S_\phi^{1/2}$	50 $\mu\phi_0/Hz^{1/2}$	35 $\mu\phi_0/Hz^{1/2}$	45 $\mu\phi_0/Hz^{1/2}$

5. Conclusions

We overlapped the slot or normal HTS film on the multi-pickup-coil HTS-dc-SQUID magnetometer with slots in the flip-chip configuration. The critical current I_c of the SQUID did not change until a DC field over 140 μT was applied to the SQUID. When an AC field at 10 MHz was applied to the SQUID, V_{pp} of the SQUID decreased and $S_\phi^{1/2}$ increased significantly when the amplitude of the AC field was about 120 nT_{rms}. The AC field applied to the SQUID at that time was less than that of the bare SQUID. Finally, the experimental results demonstrate that the bicrystal STO based HTS-SQUID covered with the HTS film can be operated and moved in the Earth's field. Further experiments are currently in progress aiming to better understand the reproducibility of the results.

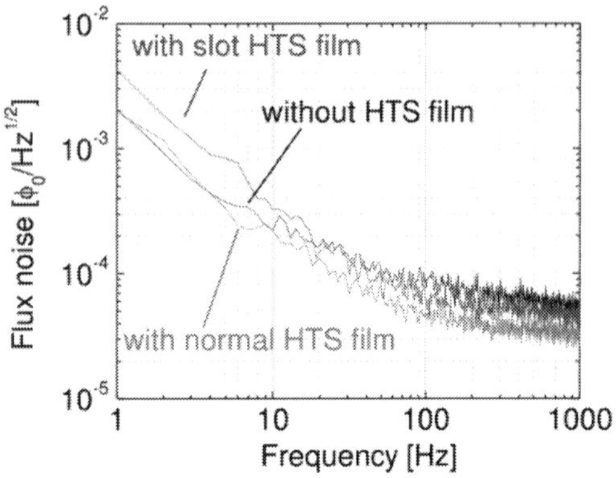

Figure 3. Flux noise spectra of HTS-SQUID without and with HTS films in MSR.

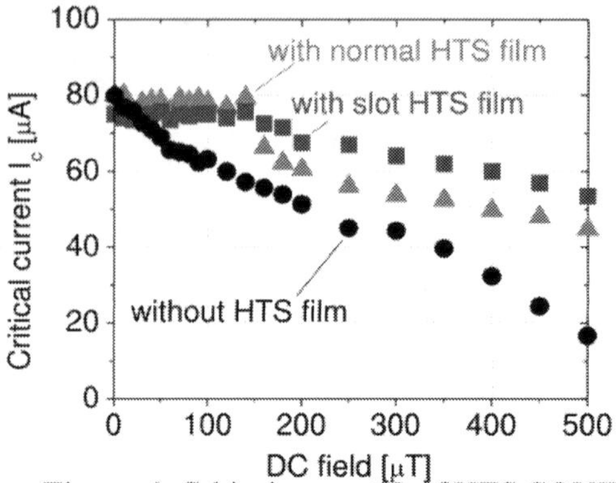

Figure 4. Critical current I_c of HTS-SQUID without and with HTS films in DC field.

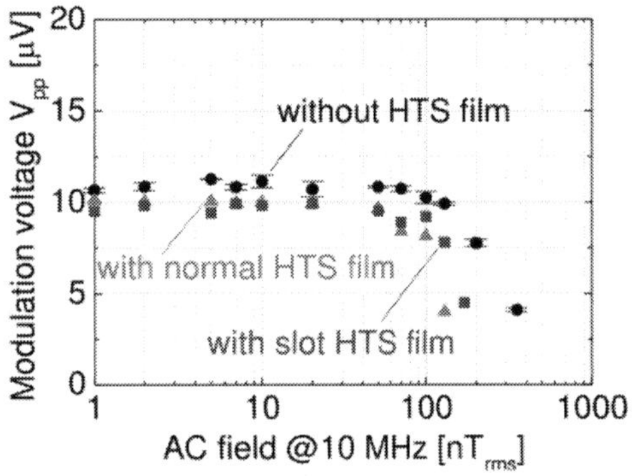

(a)

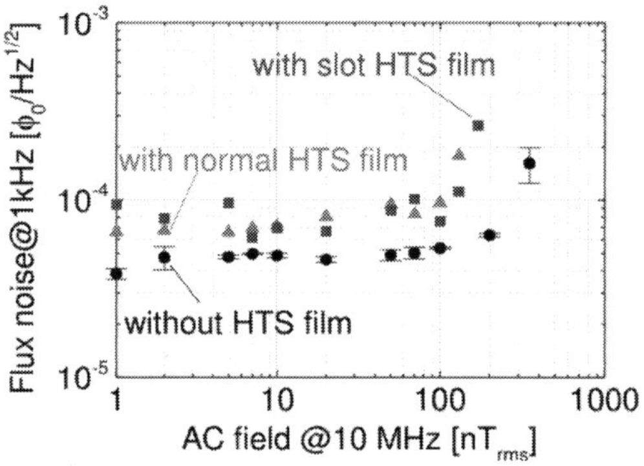

(b)

Figure 5. Characteristics of HTS-SQUID without and with HTS films in AC field at 10 MHz. (a) Modulation voltage V_{pp}. (b) Flux noise $S_{\phi}^{1/2}$ at 1 kHz.

References

[1] Dimos D, Chaudhari P and Mannhart J 1990 *Phys. Rev. B* **41** 4038
[2] Clarke J and Braginski A I (ed.) 2004 *The SQUID Handbook*, (Weinheim: WILEY-VCH)
[3] Barone A (ed.) 1992 *Principles and Application of Superconducting Quantum Interference Device* (Singapore: World Scientific)
[4] Kobayashi T, Hayakawa H and Tonouchi M (ed.) 2003 *Vortex Electronics and SQUIDs* (Berlin: Springer)
[5] Kreutzbruck MV, Baby U, Theiss A, Mueck M and Heiden C 1999 *IEEE Trans. Appl. Supercond.* **8** 3805
[6] Hatsukade Y, Shinyama Y, Yoshida K, Takai Y, Aly-Hassan M S, Nakai A, Hamada H, Adachi S, Tanabe K and Tanaka S 2013 *Physica C* **484** 195
[7] Dantsker E, Tanaka S, Nilsson P-A, Kleiner R and Clarke J 1996 *Appl. Phys. Lett.* **69** 4099
[8] Hatsukade Y, Hayashi K. Takemoto M and Tanaka S 2009 *Supercond. Sci. Technol.* **20** S385
[9] Hatsukade Y, Hayashi K, Shinyama Y, Adachi S, Tanabe K and Tanaka S 2011 *Physica C* **471** 1228

Ultra-low-noise amplifier for ultra-low-field MRI main field and gradients

Koos CJ Zevenhoven and Sarianna Alanko

Department of Biomedical Engineering and Computational Science (BECS), Aalto University, Finland

E-mail: `koos.zevenhoven at aalto.fi`

Abstract. In ultra-low-field (ULF) MRI, applied fields on the order of 100 µT are present during the measurement with one or more SQUID sensors. The current noise in the coils that produce the fields must be extremely small in order not to add noise to the measurement, especially when measured using magnetometer pickup coils. In addition, to allow other than the most basic pulse sequences, the applied fields must often be ramped up and down at millisecond time scales, requiring relatively high voltages and a sufficient bandwidth. Since commercial power amplifiers are far from satisfying these requirements, we designed and implemented a dedicated ULF MRI current amplifier. We demonstrate that, using our amplifier, it is possible to make ULF MRI measurements without increased noise even with magnetometer pickup coils that are oriented to directly measure the magnetic field component along the applied field. Using tailored amplifiers for pulsing gradients and uniform fields opens many opportunities in developing and implementing advanced ULF MRI pulse sequences.

1. Introduction

In contrast to the radio-frequency signals of conventional tesla-range MRI, an ultra-low main magnetic field $\vec{B}_0$ on the order of 100 µT results in nuclear magnetic resonance (NMR) in the kilohertz range. MRI at such low frequencies is made feasible by the use of superconducting quantum interference devices (SQUIDs), since they can be configured as highly sensitive magnetic-field sensors with sensitivities independent of frequency [1]. Over the past few years, interest in ultra-low-field (ULF) MRI has spread rapidly (see e.g. Refs. [2–5]). Motivations for studying ULF MRI include its safety, cost-effectiveness, and its compatibility with other electromagnetically sensitive technologies such as magnetoencephalography (MEG) [2–3].

ULF MRI has further possibilities that are not available in conventional MRI. Generally, in MRI, it is the design of magnetic field pulse sequences that enables the continuing development of new innovative contrast mechanisms in addition to the traditional ones based on T_1 and T_2 relaxation times. Besides the more established diffusion-tensor imaging and functional MRI, examples are contrasts based on magnetic permeability and electric current density, of which the latter may be particularly suitable for ULF MRI [6–7]. A notable advantage of ULF-MRI is to ability to independently change the polarizing and measurement field strengths, even during the sequence. However, ULF-MRI-specific sequence development is hindered by the lack of amplifier electronics capable of producing the sequences.

A central issue in the amplifier electronics is that extremely low-noise currents are required in order not to increase the noise level of the measurement, as at least two applied fields are usually

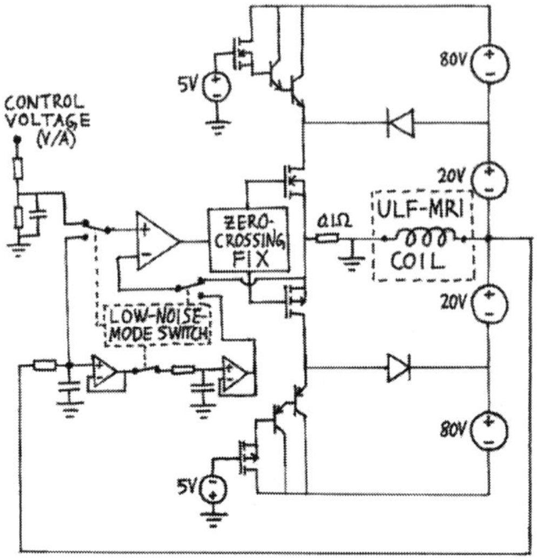

Figure 1. Simplified circuit diagram of the amplifier. The amplifier is currently in its normal mode. To switch into ultra-low-noise mode, the three semiconductor switches are toggled, which changes the feedback scheme of the main operational amplifier.

Figure 2. Photograph of the amplifier enclosure with the top plate removed.

present during signal acquisition. While the problem exists even with gradiometric sensors [8], it is particularly challenging when magnetometer pickup-coils are used. The required dynamic range from fT-level noise to 100 μT fields is as high as 180 dB (10 kHz bandwidth). Additional difficulties arise from ramping the current of multiple amperes up and down in the inductive load at millisecond time scales. The pulse sequences extend into the kilohertz range, overlapping with the signal frequency band, within which the system must have extremely low noise (see Ref. [9] for more discussion). It is difficult, however, to achieve both low noise and precise current feed for an inductive load within the same frequency band. This overlap of frequency bands is virtually nonexistent in conventional MRI, which explains why electronics thereof is not very suitable for ULF MRI.

To solve the current-feed issue, we have developed a specialized current amplifier for the $\vec{B}_0$ and gradient coils in ULF-MRI for use in our combined ULF MRI and MEG system at Aalto University [2]. In this paper, we describe design details of the amplifier and show measurement results of its properties and, in particular, its noise performance.

2. Amplifier circuit

A simplified circuit diagram of the amplifier is depicted by Fig. 1. The circuit was designed and implemented as a tailored linear amplifier with a field-effect-transistor (FET) output stage. Its ULF-MRI-specific features include the ability to freeze the current into an ultra-low-noise state for signal acquisition, effectively providing the required dynamic range. In addition, the amplifier has circuitry that increases the available voltage temporarily to allow millisecond-scale ramping of the current I in the inductive load with currents up to around 10 A. Neglecting the

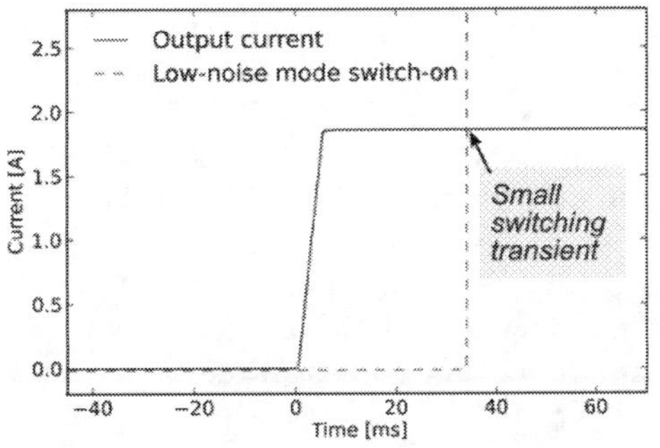

Figure 3. A sample current pulse with switching to ultra-low-noise mode. The mode switching transient is very small and therefore not visible in the figure.

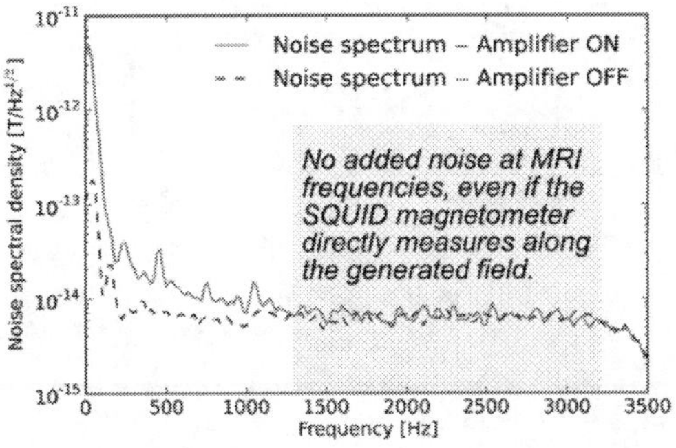

Figure 4. The noise amplitude spectrum seen by a SQUID magnetometer with no B_0 coil current (dashed) and with the magnetometer directly measuring the field produced by the B_0 coil with a 2-A current in ultra-low-noise mode (solid).

parasitic capacitance of the coil, the voltage U over the coil is given by

$$U = RI + L\frac{dI}{dt}\ , \tag{1}$$

where I is the current in the coil, R the coil resistance ($\sim$ 1 Ω), and L the coil inductance ($\sim$ 10 mH). Consequently, a 80-V voltage boost corresponds to a ramp time of 1 ms for a 8-A current.

The circuit is built around a single low-noise precision operational amplifier, which drives the FET output stage through a network that ensures the flow of at least a small quiescent current in the FETs at all times. This 'zero-crossing fix' allows distortion-free switching of the output polarity, and consists of two Zener diodes and resistors. The output stage is powered from separate floating voltage sources of e.g. 20 V. Additional Darlington configurations control the use of higher-voltage sources during ramps when needed. To provide this 'intelligent' behavior, we introduce a relatively simple trick, in which constant ground-referenced voltages are provided at the gates of two FETs that control the bipolar Darlington pairs. As soon as the remaining voltage over an output stage transistor becomes too low, the higher voltage source begins to provide current. Diodes are used to force this current into the output stage.

The ultra-low-noise mode, on the other hand, is implemented by switching into another feedback scheme. A sample-and-hold circuit is set to sample and remember the potential at the non-grounded end of the load coil, and the main operational amplifier is configured to steadily maintain the sampled state. Consequently, the current noise decreases by orders of magnitude, because the inductive reactance of the coil now suppresses the noise of the operational amplifier. In addition, since the amplifier is decoupled from the control voltage signal, any noise in the signal does not affect the current in the coil. At least at low frequencies, the remaining noise is likely to originate dominantly from the low-noise operational amplifiers.

A photograph of the amplifier can be found in Fig. 2.

3. Results

Figure 3 shows the beginning of a measured trapezoidal current pulse fed into our B_0 coil. During the pulse, the amplifier is given a signal to switch into ultra-low-noise mode. The mode switching is fast, and the resulting transient is so small that it cannot be seen in the figure. Also the current offset between the two amplifier modes has been adjusted to nearly zero. A small transient can be seen at the end of the ramp-up, which should be removed by changing some component values.

The noise performance of the amplifier in ultra-low-noise mode is shown in Fig. 4. We used one SQUID magnetometer in our system to measure the field generated by a current in our B_0 coil, which has a field efficiency of 16.7 μT/A at the SQUID sensor. At our typical NMR signal frequencies between 2 and 3 kHz, the current fed by the amplifier does not increase the noise of the measurement, suggesting that the noise is below roughly 1 $\mathrm{fT}/\sqrt{\mathrm{Hz}}$, corresponding to less than 60 $\mathrm{pA}/\sqrt{\mathrm{Hz}}$ in the spectral density of the amplifier output current noise. However, at lower frequencies, where the inductive reactance of the coil is smaller, the measurement noise is somewhat elevated even in ultra-low-noise mode.

4. Conclusions

In the absence of suitable commercially available amplifier candidates, we have designed and constructed a tailored $\vec{B}_0$ and gradient coil amplifier for ULF MRI, providing both fast current ramping and low noise. By switching between normal and ultra-low-noise states, the amplifier effectively achieves a dynamic range of over 180 dB at our typical NMR frequencies, adding no noise to the measurement. At low frequencies the measurement noise is somewhat elevated, but still low enough that, in a multichannel measurement, signal processing techniques or so-called software gradiometers could be used to reject the noise. It may also be possible to decrease the amplifier noise even further.

The capabilities of existing amplifiers have been limiting the progress of ULF MRI. In the future, ULF MRI systems could potentially use a single amplifier model for all coils except perhaps the polarizing coil. This would provide a highly flexible platform for a rich variety of pulse sequences, including sequence types that are not possible in traditional MRI.

Acknowledgment

This work has received funding from the Academy of Finland, from the iBioMEP graduate school, from the Finnish Cultural Foundation, and from the European Community's Seventh Framework Programme (FP7/2007-2013) under grant agreement No. 200859.

References

[1] McDermott R, Lee SK, ten Haken B, Trabesinger AH, Pines A and Clarke J 2004 *Proc. Natl. Acad. Sci.* **101** 7857

[2] Vesanen PT, Nieminen JO, Zevenhoven KCJ, Dabek J, Parkkonen LT, Zhdanov AV, Luomahaara J, Hassel J, Penttilä J, Simola J, Ahonen AI, Mäkelä JP and Ilmoniemi RJ 2013 *Magn. Reson. Med.* **69** 1795

[3] Zotev VS, Matlashov AN, Volegov PL, Savukov IM, Espy MA, Mosher JC, Gomez JJ and Kraus Jr RH 2008 *J. Magn. Reson.* **194** 115

[4] Burghoff M, Hartwig S, Kilian W, Vorwerk A and Trahms L 2007 *IEEE Trans. Appl. Supercond.* **17** 846

[5] Lee SJ, Kim K, Kang CS, Hwang SM and Lee YH 2010 *Supercond. Sci. Technol.* **23** 115008

[6] Nieminen JO, Zevenhoven KCJ, Vesanen PT, Hsu Y-C, Ilmoniemi RJ 2013 Current-density imaging using ultra-low-field MRI with adiabatic pulses *Magn. Reson. Imaging*, http://dx.doi.org/10.1016/j.mri.2013.07.012

[7] Vesanen PT, Nieminen JO, Zevenhoven KCJ, Hsu Y-C, Ilmoniemi RJ 2013 Current-density imaging using ultra-low-field MRI with zero-field encoding *Magn. Reson. Imaging*, in review

[8] Oyama D, Hatta J, Miyamoto M, Adachi Y, Higuchi M, Kawai J, Fujihira JI, Tsuyuguchi N, Uehara G 2013 Investigation of magnetic interference induced via gradient field coils for ultra-low field MRI systems, Poster 2P-EL1-15, European Conference on Applied Superconductivity, Genoa, Italy

[9] Zevenhoven KCJ 2011 Solving transient problems in ultra-low-field MRI, MSc thesis, Aalto University, Finland, http://urn.fi/URN:NBN:fi:aalto-201305163099

11th European Conference on Applied Superconductivity (EUCAS2013) IOP Publishing
Journal of Physics: Conference Series **507** (2014) 042051 doi:10.1088/1742-6596/507/4/042051

Investigation of Helium-Cooled Planar Transformer-Coupled SQUID Magnetometer

Jia Zeng[1,2,3,4], Yi Zhang[2,3] , Michael Mück[5], Chao Liu[1,2,3,4], Hans-Joachim Krause[2,3], Xiangyan Kong[1,3], Xiaoming Xie[1,3], Andreas Offenhäusser[2,3]

[1]State Key Laboratory of Functional Materials for Informatics, Shanghai Institute of Microsystem and InformatioTechnology (SIMIT), Chinese Academy of Sciences (CAS), Shanghai 200050, China

[2]Peter Grünberg Institute (PGI-8), Forschungszentrum Jülich (FZJ), D-52425 Jülich, Germany

[3]Joint Research Laboratory on Superconductivity and Bioelectronics, Collaboration between CAS-Shanghai,Shanghai 200050, People's Republic of China and FZJ, D-52425 Jülich Germany

[4]University of Chinese Academy of Sciences, Beijing 100049, China

[5]ez SQUID, Herborner Strasse 9, D-35764 Sinn, Germany

E-mail: y.zhang@fz-juelich.de

Abstract. We investigated helium-cooled planar transformer-coupled SQUID magnetometers with regard to their field resolution δB by varying the SQUID loop inductance L_s, input coils and the pick-up loop L_p. It was found that the pick-up-loop area A_p is the most important parameter for δB of transformer-coupled magnetometers. δB with $A_p = 10 \times 10$ mm^2 reached about 3 fT/√Hz, even using direct readout scheme without any feedback circuitries.

1. Introduction

Our transformer-coupled SQUID magnetometer consists of three parts: a dual-washer SQUID with two series-opposite integrated input coils L_{in} connected to an on-chip pickup loop L_p.

An important figure of merit for a magnetometer is the magnetic field resolution δB, which is a product of the SQUID system flux noise $\delta\Phi$ and the flux-to-field transfer coefficient $\partial B/\partial\Phi$, i.e., $\delta B = \delta\Phi \times (\partial B/\partial\Phi)$. $\delta\Phi$ consists of two parts, the SQUID intrinsic noise $\delta\Phi_s$ and the preamplifier noise contribution $\delta\Phi_{preamp}$, i.e., $\delta\Phi^2 = \delta\Phi_s^2 + \delta\Phi_{preamp}^2$. Minimizing $\delta\Phi$ of SQUIDs was analyzed in detail in many previous studies, e.g., in reference [1]. For transformer-coupled magnetometers, $\partial B/\partial\Phi$ can be represented as $\partial B/\partial\Phi = (L_p/\sqrt{L_{in,eff}} + \sqrt{L_{in,eff}})/(k\sqrt{L_{s,eff}} \times A_{p,eff})$ [2], whereby L_p and $A_{p,eff}$ denote the inductance and the effective area of the pick-up loop, $L_{in,eff}$ and $L_{s,eff}$ the effective inductances of the input coil and the SQUID loop, and k the corresponding coupling coefficient. The value of $L_{in,eff}$ should be designed to equal L_p in order to achieve a minimum of $\partial B/\partial\Phi$ for a given $L_{s,eff}$. Generally, larger L_s leads to a larger $\delta\Phi$, but a smaller (*i.e.*, better) value of $\partial B/\partial\Phi$. In practice, most magnetometers are designed with $L_s > 100$ pH to meet the high field resolution requirements [3].

We experimentally studied the properties of magnetometers employing SQUIDs with large $\beta_c \approx 3$ to find the combination of inductances L_s, L_{in} and L_p that is optimum for minimizing δB.

2. Experiments and results

2.1. SQUID magnetometer and readout electronics

In our experiments, mostly SQUID magnetometers with Steward-McCumber parameter $\beta_c \approx 3$ were employed. In our earlier work we have shown that large β_c leads to a large flux-to-voltage transfer coefficient $\partial V/\partial\Phi > 300\ \mu V/\Phi_0$ at $L_s = 350$ pH [4], thus reducing $\delta\Phi_{preamp} = V_n/(\partial V/\partial\Phi)$, whereby the preamplifier voltage noise V_n is 0.9 nV/√Hz (AD797). In this case, the SQUIDs can be connected to the preamplifier directly without any feedback circuitries in order to construct a simple SQUID system with an acceptable $\delta\Phi$ [5], measured in flux-locked loop (FLL) inside a niobium shielding tube. The general layout of our SQUID magnetometer is shown in reference [6]. In this layout, we used the dual-washer gradiometric SQUID to increase the coupling between L_s and L_{in}.

2.2. Different input coil types

We designed and fabricated three types of input coils, with the input coil consisting of 4.5 × 2 turns (see figure 1(a)) at $L_s = 350$ pH and $A_p = 5 \times 5$ mm²: (i) whole coil placed on the SQUID washer of "Ketchen-type" [7]; (ii) whole coil located inside the hole of the SQUID washer; (iii) a part of coil is overlaying the SQUID washer and the other part is inside the hole. Subsequently, we measured $\delta\Phi$ and $\partial B/\partial\Phi$ to determine δB. We found that $\delta\Phi \approx 5\ \mu\Phi_0/\sqrt{Hz}$ for all types of magnetometers investigated, while $\partial B/\partial\Phi$ strongly depended on the positions of the input coils. The transfer function $\partial B/\partial\Phi$ increased from 1.5 (for type (i)), and 1.6 (for type (iii)) to 2.7 nT/Φ_0 (for type (ii)). Of course, the "Ketchen-type" magnetometer provides the best coupling method for the transformer-coupled SQUID magnetometer. However, it is surprising that placing all of the coil turns inside the SQUID hole (in the case of type (ii)) reduced the effective area of the magnetometer only to about 50 %, because a much smaller value of the effective area was speculated. In the work described below, only input coils of "Ketchen-type" were utilized

2.3. SQUID effective inductance

In a transformer-coupled magnetometer, the design (geometrical) value of the SQUID inductance L_s is reduced due to the screening effect when the input coils are connected to the pick-up loop. Therefore, we introduce the SQUID effective inductance $L_{s,eff}$, because it (and not the geometrical inductance L_s) is responsible for δB of magnetometers. The value of $L_{s,eff}$ can be determined by the SQUID screening parameter $\beta_L = 2I_0L_{s,eff}/\Phi_0$, where I_0 is the critical current of one junction and Φ_0 is the flux quantum. The value of β_L is determined by the ratio of I_{cmin}/I_{cmax}. The value of I_{cmax} is $2I_0$ in the above expression of β_L, when two junctions are identical. The dependence of β_L on I_{cmin}/I_{cmax} shown in the inset in figure 1(b) is reproduced from [1]. The screening parameter β_L increases monotonously with increasing

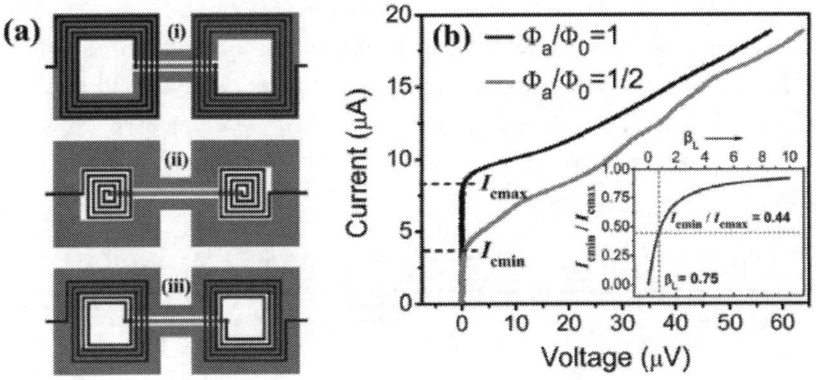

Figure 1. (a) Schematic layout of dual-washer SQUID and input coils (i) fully overlaying the SQUID washer ("Ketchen-type"), (ii) fully inside the SQUID holes, and (iii) partly overlaying the washer and holes. (b) Measured *I* - *V* characteristics of the SQUID magnetometer with L_s = 350 pH and L_{in} of 4.5 × 2 turns of type (i) in (a). I_{cmax} and I_{cmin} denote the critical currents at integer and half integer flux quantum, and Φ_a is the applied flux. The inset shows the dependence of the screening parameter β_L on I_{cmin}/I_{cmax} from [1].

I_{cmin}/I_{cmax}. Its β_L = 0.75 is obtained from the curve of β_L *vs.* (I_{cmin}/I_{cmax}) and the measured I_{cmin}/I_{cmax} = 0.44 (3.65 μA/8.3 μA), leading to $L_{s,eff} \approx$ 181 pH, almost half of L_s [2].

2.4. Field resolution δB

To investigate δB of the magnetometer, we varied the SQUID layout parameter, such as L_s, or the turn number of the input coil. Two SQUID pick-up areas of A_p = 5 × 5 mm^2 (table 1) and 10 × 10 mm^2 (table 2) were employed.

In table 1, three important parameters of the magnetometers, $L_{s,eff}$, $\delta\Phi$ and $\partial B/\partial\Phi$, are listed. The value of $L_{s,eff}$, for example, at L_s = 350 pH, obviously increases with decreasing the turn number of L_{in}, thus increasing $\delta\Phi$, but concurrently improving $\partial B/\partial\Phi$. In fact, for a given L_s of 350pH, $L_{in,eff}$ slightly changes with the turns of L_{in} varying in a certain range shown in table I and II, since $L_{in,eff}$ depends on both the turns of L_{in} and $L_{s,eff}$ [2], which leads to $L_{in,eff} \approx$ 14 nH to match $L_p \approx$ 15nH [6]. Consequently, δB was almost constant for such transformer-coupled magnetometers, even when varying L_s from 350 up to 620 pH.

Table 1. Parameters of SQUID magnetometers with A_p of 5 × 5 mm^2.

L_s	[pH]	350	350	350	480	480	620	620
Turns of L_{in}	—	3.5 × 2	4.5 × 2	5.5 × 2	2.5 × 2	5.5 × 2	2.5 × 2	4.5 × 2
$L_{s,eff}$	[pH]	235	180	140	365	190	420	310
$\partial B/\partial\Phi$	[nT/Φ_0]	1.47	1.50	1.63	1.24	1.53	1.09	1.28
$\delta\Phi$	[μΦ_0/√Hz]	5.4	5	4.5	6.6	5.2	7.4	5.6
δB	[fT/√Hz]	7.9	7.5	7.3	8.2	7.9	8.1	7.2

For A_p of 10 × 10 mm^2 (table 2), we increased the turn number of L_{in}, *e.g.*, to 7.5 × 2 at L_s = 350 pH, to improve the matching between L_{in} and L_p, thus leading to the reduction of $\partial B/\partial\Phi \approx 0.55$ nT/Φ_0, although $L_{s,eff}$ further reduced to 120 pH. Indeed, making A_p larger is very useful for the improvement of δB. For different L_s, $\delta B \approx 3$ fT/√Hz of such magnetometers was achieved. Note that $\delta\Phi$ of the present SQUID magnetometers in table 2 was larger than that listed in table 1, due to $\beta_c > 3$. Larger β_c led to a higher $\delta\Phi_s$, which dominated $\delta\Phi$ [5].

Table 2. Parameters of SQUID magnetometers with A_p of 10 × 10 mm^2

L_s	[pH]	350	480	620
Turns of L_{in}	—	7.5 × 2	6.5 × 2	5.5 × 2
$L_{s,eff}$	[pH]	120	192	265
$\partial B/\partial\Phi$	[nT/Φ_0]	0.55	0.47	0.4
$\delta\Phi$	[$\mu\Phi_0$/√Hz]	5.5	7	9.5
δB	[fT/√Hz]	3	3.3	3.8

3. Conclusion

We studied different layouts of transformer-coupled SQUID magnetometers with varying L_s, L_{in} and L_p. An input coil fully overlaying the SQUID washer ("Ketchen type") provides the best coupling between L_s and L_{in}, whereas placing the input coil fully inside the SQUID hole will reduce the pick-up area by about 50%. The effective SQUID inductance $L_{s,eff}$ obviously increases with decreasing turn number of L_{in}, thus increasing $\delta\Phi$, but at the same time reducing $\partial B/\partial\Phi$. Consequently, δB is almost constant for a certain pick-up-loop area A_p. In other words, L_s, L_{in} and the matching between L_{in} and L_p will not influence greatly δB. Only by increasing A_p will the field resolution improve.

Acknowledgement

This work was Supported by the "Strategic Priority Research Program (B)" of the Chinese Academy of Sciences (Grant No: XDB04010100).

References

[1] Tesche C D and Clarke J 1977 *J. Low Temp. Phys.* **29** pp 301–331

[2] Cantor R and Koelle D 2004 "Practical DC SQUIDs: Configuration and Performance" in *The SQUID Handbook* edited by Clarke J and Braginski A I (Weinheim: Wiley) Vol I pp 172–210

[3] Schmelz M, Stolz R, Zakosarenko V, Schönau T, Anders S, Fritzsch L, Mück M and Meyer H-G 2011 *Supercond. Sci. Technol.* **24** 065009

[4] Liu C, Zhang Y, Mück M, Krause H-J, Braginski A I, Xie X, Offenhäusser A, Jiang M 2012 *Appl. Phys. Lett.* **101** 222602

[5] Zeng J, Zhang Y, Mück M, Krause H-J, Braginski A I, Kong X, Xie X, Offenhäusser A and Jiang M 2013 *Appl. Phys. Lett.* **103** 042601

[6] Zhang Y, Liu C, Schmelz M, Krause H-J, Braginski A I, Stolz R, Xie X, Meyer H-G, Offenhäusser A and Jiang M 2012 *Supercond. Sci. Technol.* **25** 125007

[7] Jaycox J M and Ketchen M B 1981 *IEEE Trans. Magn.* **MAG 17** 400–403

AUTHOR INDEX

AUTHOR INDEX

AUTHOR INDEX

AUTHOR INDEX

AUTHOR INDEX

AUTHOR INDEX

AUTHOR INDEX

AUTHOR INDEX

AUTHOR INDEX

AUTHOR INDEX